Lake Basins through Space and Time

Edited by
E. H. Gierlowski-Kordesch
and
K. R. Kelts

AAPG Studies in Geology #46

Published by
The American Association of Petroleum Geologists
Tulsa, Oklahoma, U.S.A.
Printed in the U.S.A.

Printed in the U.S.A.
Published May 2000

ISBN: 0-89181-052-8

Association Editor: Neil F. Hurley
Science Director: Robert C. Millspaugh
Publications Manager: Kenneth M. Wolgemuth
Managing Editor, Publications: Anne H. Thomas
Production: Custom Editorial Productions, Inc., Cincinnati, Ohio

This and other AAPG publications are available from:

The AAPG Bookstore
P.O. Box 979
Tulsa, OK 74101-0979
Telephone: 1-918- 584-2555 or 1-800-364-AAPG (USA)
Fax: 1-918-560-2652 or 1-800- 898-2274 (USA)
www.aapg.org

Geological Society Publishing House
Unit 7, Brassmill Enterprise Centre
Brassmill Lane, Bath, U.K.
BA1 3JN
Tel +44-1225-445046
Fax +44-1225-442836
www.geolsoc.org.uk

Australian Mineral Foundation
AMF Bookshop
63 Conyngham Street
Glenside, South Australia 5065
Australia
Tel. +61-8-8379-0444
Fax +61-8-8379-4634
www.amf.com.au/amf

Affiliated East-West Press Private Ltd.
G-1/16 Ansari Road Darya Ganj
New Delhi 110 002
India
Tel +91 11 3279113
Fax +91 11 3260538
e-mail: affiliat@nda.vsnl.n

Preface

The geology of lake basins was a popular subject in the last century, fired by interest in the discoveries during the exploration of the American West. G.K. Gilbert (1890) recognized climate, tectonic, and sediment facies implications from the shorelines of vast paleolake Bonneville, 300 m above the modern Great Salt Lake. The Fortieth Parallel Survey under Clarence King discovered the extensive lacustrine deposits of the Eocene Green River Formation that have become the archetypal model of a giant ancient lake basin and oil-shale deposit. W.H. Bradley's lifelong study of a wide diversity of geoscience problems based on the Green River Formation of Wyoming, Utah, and Colorado established a role model that helps define the discipline of limnogeology (McKelvey, 1983). His interests ranged from the basinal tectonic structure and geologic history, to paleoenvironmental reconstructions, stratigraphy, paleogeography, paleoclimate, solar and orbital rhythms, nonmarine paleontology, algae and microbial paleoecology, organic geochemistry, carbonate petrology, hydrothermal activity, diagenesis, the evolution of saline brines and minerals, and, of course, the origin of the kerogen.

This book builds on the experience of an international group of limnogeology enthusiasts. In 1984, an IGCP project 219 (International Geological Correlation Program) began to galvanize a resurgent interest in lake basin geology, partly as a result of the recognition of vast hydrocarbon potential sourced from lacustrine deposits, especially in China, the south Atlantic marginal basins, and southeast Asia (Fleet et al., 1988). The activities of IGCP projects 219 and 324 brought together a loosely-organized, worldwide group of sedimentologists to compile, compare, and understand the global geologic record of lake basins and their facies within a paleogeographic context. A series of meetings and special publications ensued (cf. Gierlowski-Kordesch and Kelts, 1994); more recently a meeting in Antifagasto, Chile, looked at lake facies in foreland basins (Cabrera and Saéz, 1999). The IGCP projects evolved into the International Association of Limnogeology (IAL), which held its first congress in 1995 in Copenhagen (Nøe-Nygaard, 1998) followed by the second in Brest, France, in 1999. The results of these meetings clearly confirm the original hypothesis that lacustrine deposits were much more common in the geologic record than perceived, and that their sediment facies contain a wealth of information on regional environmental dynamics, paleoclimate, and paleohydrology of the continents; endemic fauna and flora evolution; resource evaluation; and on tectonic styles, basin formation, and early rifting processes. The aim has been to discover global commonalities among the processes and patterns of lacustrine deposits through time and space, rather than to emphasize the regional peculiarities. A detailed strategy for global lacustrine comparisons and differentiation from marine features is presented in an earlier IGCP compilation (Gierlowski-Kordesch and Kelts, 1994).

The science of limnogeology is of importance to petroleum geology. Although not every limnic deposit is an exploration target, a comprehensive understanding of diverse lacustrine environments of deposition can help exploration strategies. The multi-dimensional lacustrine matrix is confusing. Simplifying models such as the Over-, Under-, Balance-Fill model of Bohacs et al. (this volume) are based on a broad comparative approach to provide a clearer framework for the interpretation, for example, of lacustrine seismic sequence stratigraphy. Much as the discovery of extensive carbonate reservoirs in the Middle East stimulated basic research on dolomitization, an increasing awareness of hydrocarbons from lacustrine basins poses new questions about the fundamental processes controlling lake facies, showing the need for better actualistic models. A result of compiling the geologic record of lake basins shows a correlation of greater hydrocarbon potential with certain time windows; those with the optimal climate concurrent with major transtensional shear or rifting phases. The distribution of lakes in our present world is dominated by settings related to Pleistocene glaciation. During various geological time slices, climatic, tectonic, and physiographic conditions supported the long-term persistence of lacustrine giants covering thousands of square kilometers in tropical and subtropical regions. We still lack an adequate understanding of these ancient deposits and modern facies. Most lakes sequester organic carbon with high-quality petroleum potential at rates far greater than most marine basins (Kelts, 1988). Even Lake Superior, a dilute, low-productivity, oligotrophic, clastic system, has Holocene muds with 1 percent TOC. Application of the Green River analog is consistent with evidence that many optimal lacustrine source rocks are associated with transitional, closed-basin water chemistries, neither hypersaline nor dilute but mesohaline with alkaline affinities. Anoxia is helpful, but not a requirement as shown by well-mixed, shallow, Lake Victoria sediments with 10–20 percent TOC and hydrogen indices above 800 (Talbot, 1988).

This volume is also complementary to an ongoing IAL community effort to create a database of all lacustrine basins and their deposits organized according to geologic time and paleogeographic position, called the Global Geological Record of Lake Basins (GGLAB). This ambitious project is iterative and based on voluntary participation. As new investigations and summaries are published, these should encourage studies of other sequences with improved standards. We lack standard or traditional conventions and classifications to describe lake deposits. Investigators for GGLAB are given guidelines and suggestions to encourage coherent presentations (see for example, the GGLAB legend, Figure 1). As the GGLAB database grows, more information will become available on the distribution and paleoenvironmental reconstructions from sequences for certain time intervals and continental configurations.

The basic GGLAB questionnaire for the database (cf. Gierlowski-Kordesch and Kelts, 1994) requires for each lake deposit: geographic region (10 choices), country, region, formation or member name, latitude/longitude, maximum and minimum size and area estimate, dating method, age of deposit (Ma), tectonic setting, salinity ranges, typical facies types, and environmental index features, biota, organic matter type and percent, sedimentology, mineralogy, any isotopic or geochemical information, complete references, an index map, and simplified litho-stratigraphic section of the lake deposit. The objective is not to create a quantitative reduction for mapmaking or facies distribution, but rather to spur syntheses and comparisons with a meta-database that searches unique lacustrine sequences by name, age, and location in a simple code system; for example: Ebro Basin/Cz, Mio/EUR, E, for Miocene gypsum and chalk deposits of the Spanish Ebro Basin. Green River Laney Shale is coded simply as Cz, Eo/NA, WY. This helps users zoom in on literature references because typically lacustrine studies are basin-oriented. Participating in the database concept helps guide investigators toward a coherent perspective. Because of the inherent dynamics in lake environments, it is usually not possible to simply assign a lake sequence a designation such as "fresh," "organic-rich," or "evaporite." Such deposits may exist in many sequences, perhaps only centimeters apart. We strive to have consistent conventions. For example, we have suggested lithologic/biotic symbols in the GGLAB legend (Figure 1). Age designations follow the 1986 Geological Society of America DNAG time scale.

The volume presents 60 new basin summaries, essentially volume 2 of the Global Geological Record of Lake Basins, following a similar format and organization for the first volume compilation of 70 other basin summaries (Gierlowski-Kordesch and Kelts, 1994). The introductory chapter by Bohacs et al. (this volume) reviews basic features of lake systems and presents a new unifying model to guide basin sequence and seismic exploration. The intervening years between the two volumes reflect the beginning fruits of exploration in central Asia. Sladen (this volume) synthesizes the time-lacustrine record from Mongolia and several other papers outline sequences from Junggar and Kazakhstan. Investigations by various teams continue well-represented across the Iberian penninsula, while the interests in the South Atlantic marginal basins remain high. Figure 2 shows the spatial distribution of the basin investigations reported by chapter number in the present volume, reflecting current research of contributing limnogeologists.

Precambrian sequences are poorly represented, partly as a result of ongoing controversy over marine versus lacustrine facies, for example, from the Belt series of North America, and partly from a lack of awareness by researchers. Our earlier hypothesis is yet untested that relatively undeformed rift sediments from the Precambrian are likely relicts of intrashield rifting, and thus likely to contain significant lacustrine deposits. Some of the oldest source rocks, for example from the Nonesuch Shale of the Midcontinent Rift, or within the MacArthur River basin of north Australia, may be indeed lacustrine facies (Gierlowski-Kordesch and Kelts, 1994).

The large Permian Pangea and Cretaceous Gondwana provide plate tectonic settings for widespread lacustrine deposition, particularly during phases of continental breakup. New views of the Junggar Basin of western China are highlighted by two papers (Zhao and Tang, Wartes et al., this volume). In recent years, the upper Permian deposits of western China have become recognized as among the true lacustrine giants, as large as the modern Caspian Sea. Similarly, thick sequences on opposing margins would predict the entire basin is underlain by contiguous source rocks. The middle Jurassic is another period with significant lacustrine deposition in basinal facies (Green et al., this volume) corresponding to times with widespread humid lake environments across China (eg., Buatois et al., this volume).

Tanner (this volume) presents a stratigraphic view of the Fundy Basin as a representative of extensive lacustrine strata in numerous Triassic-Jurassic rift basins along the eastern North America and North Africa to southern Europe. These are all linked with the early opening of the North Atlantic. Early Jurassic alkaline lake deposits of Namibia (Stollenhofen et al., this volume) add new evidence for the connection of flood basalts with rifting of Karoo deposits across southern Africa.

The Early Cretaceous is an interval preserving widespread lacustrine deposits, particularly in relation to the break-up of Gondwana. Lacustrine rift facies extend along the marginal basins of the South Atlantic, and along southern Australia, India, and likely Antarctica. In addition, oil exploration has given evidence of widespread intraplate rift lakes across Africa and South America that appear to cease activity around the late Aptian when ocean spreading was in full swing in the South Atlantic. Many of the detailed reports on these lacustrine deposits have yet to be published, nor do we have a comprehensive synthesis evaluating the interconnections among these basins with similar lacustrine sequence stratigraphy in the Cretaceous or Permian of Gondwana. A first such attempt correlates the Permian in Namibia and Brazil (Stollhofen et al., this volume). Summaries for the Lower Cretaceous Brazilian margin range from the NE Araripe-Potiguar basins (Mabesoone et al., this volume), a study of deep water lacustrine facies in the Recôncavo rift (da Silva et al., this volume), and the more southerly Campos Basin stratigraphy and facies (Rangel and Carminatti; de Carvalho et al., this volume) are included.

Early Cretaceous lacustrine facies are also an important part of the Iberian basins, for example Soria et al., Meléndez and Gómez-Fernández, and Soria et al., Liesa et al., (this volume) and the unique lacustrine Lagerstätten fauna of the Hoyas Basin (Fregenal et al., this volume). Late Cretaceous lacustrine deposits tend to be shallow, calcareous facies in Spain and southern France (Cojan, this volume). Several have the potential of a section crossing the Cretaceous-Tertiary boundary in a lacustrine facies, one of the Holy Grails left to test for evidence of K/T impact consequences. New sites are described from near Hong Kong (Owen, this volume), Argentina (Palma, this volume), and Kazakhstan (Lucas et al., this volume).

Cenozoic deposits are, of course, best represented in the geologic record and extensive on all the continents, linked, for example, with the translational megashearing of eastern China, transtension in southeast Asia, Basin and Range tectonics of western North America, and nonmarine basins from active tectonic regions of the circum-Mediterranean across Turkey and through central Asia. These illustrate more diverse environments and include smaller basins or even crater lake deposits.

Quaternary and modern lakes are of particular importance for paleoclimate reconstructions, although they also offer a wide spectrum of facies examples for petroleum exploration, for instance the Lake Bogoria environment of East Africa (Renaut et al., this volume). The study of lacustrine sequences of the Olorgesailie Formation of the southern Kenya rift (Owen and Renaut, this volume) illustrates a frontier direction of linking the evolution of hominids with oscillating environmental pressures.

Basic data-gathering is still progressing around the world and will continue well into the next millennium. The number of deposits awaiting further investigation is awesome. The reader can imagine that the examples in this volume present only a small selection of the deposits worldwide. We estimate conservatively that some 1500 basins are known from literature, but only a limited number have adequate investigations for a coherent overview, including diagenesis studies.

Parallel with the patterns emerging from regional basin studies, numerous basic geoscience questions can be pursued with an awareness and access to appropriate sediment sequences. Many lake sequences have characteristic annual to millennial-scale rhythms and life-cycle patterns. Detailed investigations can critically evaluate evidence and differentiation of solar, orbital, or tectonic rhythms. New tools are needed for age dating, global correlations, and a better understanding of lacustrine flora and faunal evolutionary processes, endemism, and paleoecological distribution. How important have lacustrine giants been as incubators of evolution? Applications of environmental isotopes and organic biomarkers will be increasingly used to reconstruct paleoclimate, paleohydrologic, and paleoenvironmental scenarios, but these must still be carefully calibrated with understanding of modern lake processes and sedimentologic facies context. Ostracodes are becoming the "foraminifera of the continent" providing a mainly benthic microfossil that archives precise environmental information with taxonomy, shell structure, trace element geochemistry, and stable isotopes of carbon and oxygen. They are also excellent archives of strontium isotopic ratios that can be used to trace the hydrologic continuity of lacustrine giants. Each lake can be viewed as a natural experiment with lacustrine lessons (models) of clastic, chemical, and biogenic processes. Much remains to be learned about the microbial processes, components, and diagenesis, for the wide range of chemistry, concentrations, and pH values found in lakes. Many modern lakes present extreme microbial habitats of interest to researchers of Precambrian origins of life, deep biosphere, methanogenesis, and Type I lacustrine kerogen.

Lacustrine sequence comparisons and correlations, both seismic and drilled, provide critical insight into the tectonic mechanisms, timing, and vectors of early stage continental rifting. Developments continue toward better facies models that differentiate tectonic and climatic signals. A major goal continues to better tie nonmarine paleoenvironmental records closer to coeval marine history and paleogeography.

The record of lake deposits will see increasing utility to test global to regional climate models for past plate configurations. Surface paleohydrology patterns on the continents can be complex however. Confusion is bound to arise for situations such as in the modern Caspian Sea, which sits in a rather extreme desert, sustained as a brackish water body only by the long distance river input from the Volga River draining northern latitudes. The modern analogy extends to other examples in the Near East where large freshwater lakes such as Sevran coexist near large, shallow, saline lakes such as Urmia and the largest and deepest soda-lake Van. Their deposits thus archive the signatures of the regional geology and topography.

Viewed with an integrated, holistic, system perspective, the investigation of modern and ancient lake basins and sequences promises lively challenges to geoscientists for the next century. We caution only to view these deposits fully cognizant of differences in processes and principles between marine and lacustrine systems.

MINERALOGY	
Analcime	Ac
Anhydrite	Anh
Ankerite	Ak
Aragonite	Ar
authigenic Feldspar	aF
authigenic Quartz	aQ
Barite	Ba
Bischofite	Bis
Bloedite	Blo
Burkeite	Bur
Carnallite	Crn
Celestite	Ce
Chert	Ch
Copper	Cu
Corrensite	Cor
Dawsonite	Daw
Diatomite	Diat
Dolomite	Dol
Epsomite	Ep
Galena	Gl
Gaylussite	Gay
Glauberite	Gla
Gypsum	Gyp
Halite	Hal
high Mg-calcite	h-cc
Iron	Fe
Lead	Pb
low Mg-calcite	l-cc
Magadiite	Ma
Magnesite	Mag
Marcasite	Mr
Mirabilite	Mir
Nahcolite	Nah
Nickel	Ni
Phosphate	Ph
Pirssonite	Pir
Pyrite	Py
Siderite	Sd
Sphalerite	Zn
Sulfur	S
Sylvite	Syl
Tachyhydrite	Thy
Thenardite	The
Trona	Tr
Uranium	U
Vivianite	Viv
Zeolite	Z

CLAYS	
Bentonite	Ben
Chlorite	Chl
Illite	Ill
Kaolinite	Kao
Smectite	Sm
Sepiolite	Sep
Palygorskite	Pgk

COAL	
Hard Coal	HD
Anthracite	AT
Humic Coal	HU
Sapropelic Coal, Cannel Coal, Boghead	SO

For more precise data-add abbreviation of method plus classification value:
I = International System
F = Fixed Carbon
B =BTU/lb.
C = Kcal/kg

(Example: Hard Coal - Class 7 of International System = HD I.7)

LITHOLOGIC SYMBOLS-SILICICLASTICS

Shales
Oil Shales
Siltstones
Sandstones
Mudstones, claystones
Bioturbated mudstones
Grain-supported conglomerates
Matrix-supported conglomerates
Breccias
Sand
Silt
Mud
Gravel

LITHOLOGIC SYMBOLS-CARBONATES

Limestone (micrites, wackestones)
Limestones (packstones, grainstones)
Limestones (algal bindstones, algal reefs, stromatolitic ls., bioherms)
Bioclastic limestones
Silty/sandy limestones
Limestones breccias
Limestone conglomerates
Dolomite
Tufa, travertine
Chalk
Marlstone
Soft carbonate ooze (75-100% carbonate)
Marl (20-75% carbonate)

LITHOLOGIC SYMBOLS MISCELLANEOUS

Crystalline basement
Volcanics
Tuff, ash
Evaporites Halite
Gypsum
Lignite, brown coal
Coal
Diatomite
Varves
Loess
Chert

Hal Thy Gyp Bis

Hal, Sy, Bis, Tr, Gyp, Gla, Thy, etc.

Dol Ph Py Z

Zeolite, Phosphate, Pyrite, Dolomite, e

Figure 1. GGLAB Legend: suggested legend symbols for lithology, sedimentary structures, fossils, and mineral abbreviations for use in geologic columns.

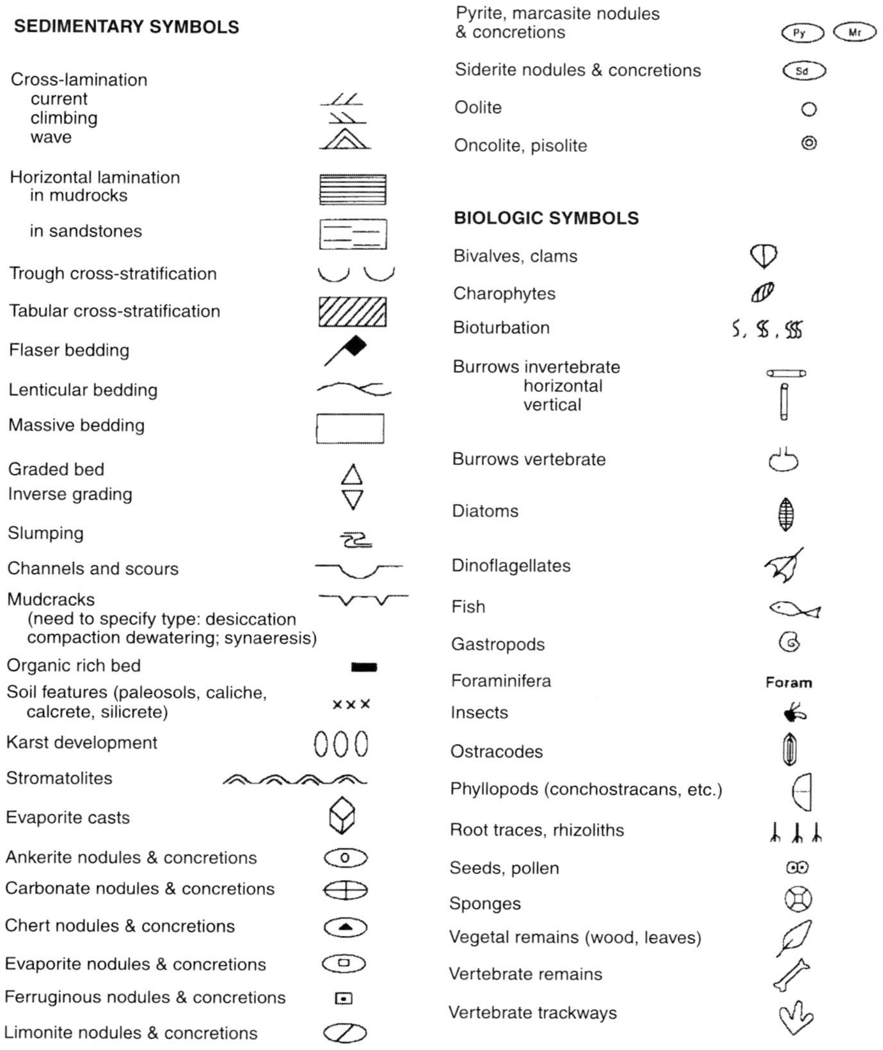

SEDIMENTARY SYMBOLS

Cross-lamination
current
climbing
wave

Horizontal lamination
in mudrocks

in sandstones

Trough cross-stratification

Tabular cross-stratification

Flaser bedding

Lenticular bedding

Massive bedding

Graded bed
Inverse grading

Slumping

Channels and scours

Mudcracks
(need to specify type: desiccation
compaction dewatering; synaeresis)

Organic rich bed

Soil features (paleosols, caliche,
calcrete, silicrete)

Karst development

Stromatolites

Evaporite casts

Ankerite nodules & concretions

Carbonate nodules & concretions

Chert nodules & concretions

Evaporite nodules & concretions

Ferruginous nodules & concretions

Limonite nodules & concretions

Pyrite, marcasite nodules
& concretions

Py Mr

Siderite nodules & concretions

Sd

Oolite

Oncolite, pisolite

BIOLOGIC SYMBOLS

Bivalves, clams

Charophytes

Bioturbation

Burrows invertebrate
horizontal
vertical

Burrows vertebrate

Diatoms

Dinoflagellates

Fish

Gastropods

Foraminifera

Foram

Insects

Ostracodes

Phyllopods (conchostracans, etc.)

Root traces, rhizoliths

Seeds, pollen

Sponges

Vegetal remains (wood, leaves)

Vertebrate remains

Vertebrate trackways

Figure 1. Continued.

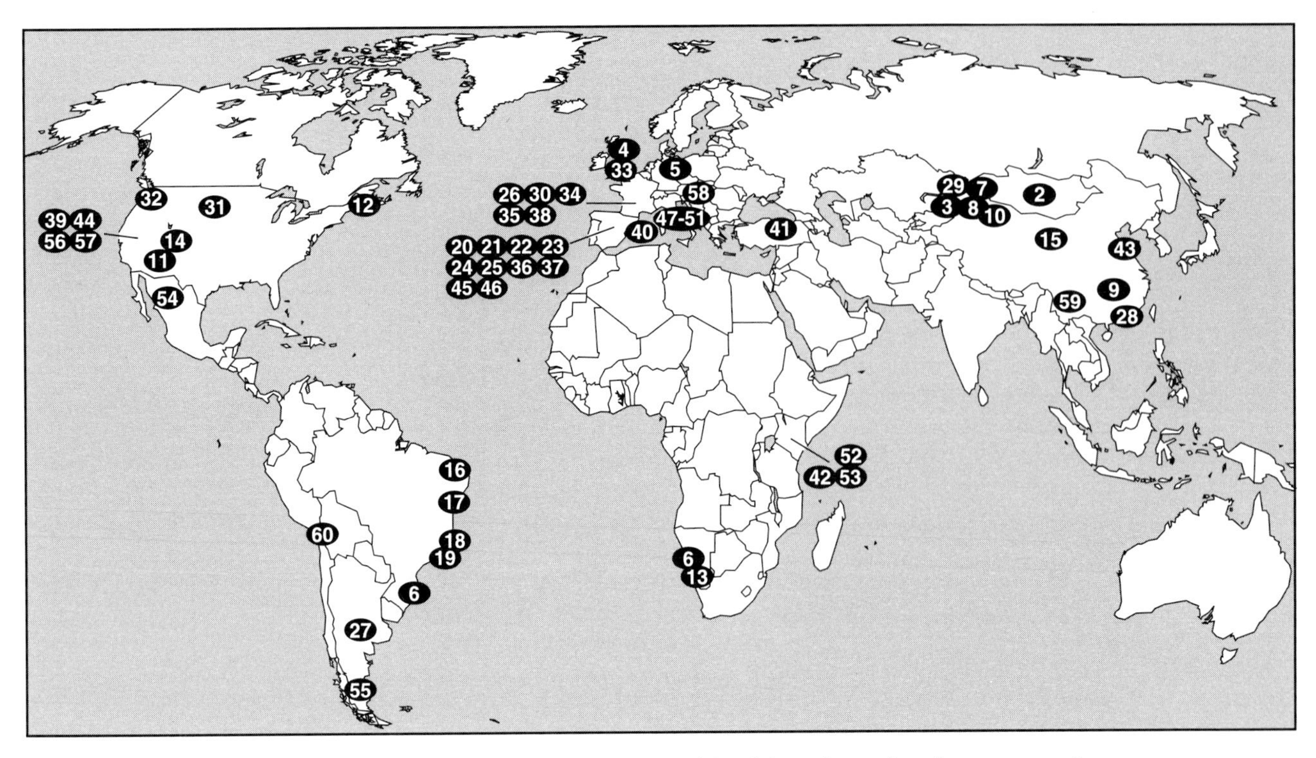

Figure 2. Global distribution of lacustrine basins presented in this volume by chapter numbers.

ACKNOWLEDGMENTS

This volume is a group effort to increase our knowledge of the richness and diversity of lacustrine deposits preserved in the geologic record. The entire compilation, reviewing, and editing depended on voluntary work without benefit of specific grants. We much appreciate efforts of the manuscript reviewers, a long list that includes most authors. Further, we gladly acknowledge the multiple review contributions of Keddy Yemane, Michael Talbot, Lluís Cabrera, Paul Buchheim, Dorothy Sack, Greg Nadon, Bill Last, John Parnell, Lisa E. Park, Martin Gibling, and Dan Textoris. Alan Carroll kindly compiled Figure 2. Thanks to everyone who contributed to the publishing costs. We are particularly grateful to the help and encouragement of AAPG and their editors Ken Wolgemuth and Anne Thomas.

Elizabeth Gierlowski-Kordesch
Kerry Kelts
2000

REFERENCES

Cabrera, Ll. and A. Sáez, eds., 1999, Ancient and recent lacustrine systems in convergent margins, Palaeogeography, Palaeoclimatology, Palaeoecology, v. 151, no. 1-3, 239 p.

Fleet, A., K. Kelts, and M. Talbot, eds., 1988, Lacustrine petroleum source rocks, Geological Society Special Publication, v. 40, 391 p.

Gierlowski-Kordesch, E., and K. Kelts, eds., 1994, Global geological record of lake basins, v. 1, Cambridge University Press, Cambridge, 427 p.

Gilbert, G.K., 1890, Lake Bonneville, USGS Monograph 1.

Kelts, K., 1988, Environments of deposition of lacustrine petroleum source rocks: an introduction, *in* A. Fleet, K. Kelts, and M. Talbot, eds., Lacustrine petroleum source rocks, Geological Society Special Publication, v. 40, p. 3-26.

McKelvey, V.E., 1983, Wilmont Hyde Bradley, April 4, 1899-April 12, 1979, Biographical memoirs, National Academy of Science, v. 54, p. 75-86.

Noe-Nygaard, N., ed., 1998, Limnogeology—research and methods in ancient and modern lacustrine basins, Palaeogeography, Palaeoclimatology, Palaeoecology, v. 140, no. 1-4, 478 p.

Talbot, M., 1988, The origins of lacustrine source rocks: evidence from the lakes of tropical Africa, *in* A. Fleet, K. Kelts, and M. Talbot, (eds.), Lacustrine petroleum source rocks, Geological Society Special Publication, v. 40, p. 20-43.

AAPG

wishes to thank the following

for their generous contributions

to

LAKE BASINS THROUGH SPACE AND TIME

❖

CNR—Centro di Studio per il Quaternario e l'Evoluzione Ambientale
Dipartimento Scienze della Terra, Università La Sapienza

Department of Geology, Bowling Green State University

Dra. Gabriela C. Cusminsky, Centro Regional Universitario, Bariloche, Argentina,
and Prof. Robin C. Whatley, University of Wales, Great Britain

Ecole des Mines de Paris

ExxonMobil Upstream Research Company

Geological Institute of Hungary, *and*
Geodesign B.t. (Hungary)

Institut für Geologie Universität Würzburg

Petrobras

❖

Contributions are applied against the production costs of publication, thus directly reducing the book's purchase price and making the volume available to a greater audience.

About the Editors

Elizabeth H. Gierlowski-Kordesch

Elizabeth H. Gierlowski-Kordesch is an associate professor of geological sciences at Ohio University (1989-present). She received a B.A. honors degree (1978) in geophysical sciences from the University of Chicago and an M.S. degree (1982) and Ph.D. (1985) in geological sciences from Case Western Reserve University in Cleveland, Ohio. She then spent some time in Germany learning the language as well as the geology of Europe and won a post-doctoral position at the Institut für Paläontologie at the Freie Universität of Berlin from the Deutsche Forschungsgemeinschaft (1986-1988).

Dr. Gierlowski-Kordesch's research has always centered on limnogeology, especially the sedimentology of lake basin deposits. Her doctoral research detailed the sedimentology of the East Berlin Formation of the Hartford basin (Newark Supergroup) while her post-doctoral research centered on the non-marine carbonate deposits of the Weald (Lower Cretaceous) of central Spain. Her more recent research involves the lacustrine/palustrine limestones within the Pennsylvanian cyclothems of the northern Appalachian basin and the carbonates of the Newark Supergroup. As one of the founders of the International Association of Limnogeology (IAL), Dr. Gierlowski-Kordesch has been involved in the Global Geological Record of Lake Basins (GGLAB) database and book series project from its inception through ICGP Project 219: Comparative Lacustrine Sedimentology in Space and Time and IGCP Project 324: Global Paleoenvironmental Archives in Lacustrine Systems

Kerry R. Kelts

Kerry R. Kelts is a professor in the Department of Geology and Geophysics at the University of Minnesota and has been director of the Limnological Research Center from 1990 to 2000. He has a B.A. in geophysics from the University of California (1967), and a Diplom (1970) and Ph.D. (1978) in geology from the Swiss Federal Institute of Technology, Zürich (ETH), where he developed a comprehensive program of lake sediment studies inspired by participation in three Deep Sea Drilling Legs (33, 51, 53). Following postdoctoral research on the Gulf of California (Leg 64) with the Deep Sea Drilling Project (1978–1980) and a fellowship at the University of Paris (1982), he continued as lecturer at ETH-Zürich where he initiated IGCP Project 219: Comparative Lacustrine Sedimentology in Space and Time. Dr. Kelts has been influential in recognizing the significance of lacustrine source rocks for hydrocarbon exploration, first with studies of South Atlantic marginal basins, then by organizing a special conference of the Geological Society of London (1985) on lacustrine petroleum source rocks, and finally in spearheading a strategic industry report on exploration of lacustrine basins in Africa, in addition to regular research expeditions in China from 1984–1993. He was the director of the Geology Group at the Swiss Federal Institute of Water Resources (1985–1988) and the first director of the National Climate Program of the Swiss Academy of Sciences (1988–1990). Dr. Kelts is currently the W. H. Bradley fellow and President of the International Association of Limnogeology (IAL) and author of over 85 scientific publications. He received the 1993 outstanding publication award from the Association of Earth Science Editors for the book "Climate—Our Future?"

Table of Contents

Section I

Selected Topics

Bohacs, K. M., A. R. Carroll, J. E. Neal, P. J. Mankiewicz, 2000, Lake-basin type, source potential, and hydrocarbon character: an integrated-sequence-stratigraphic–geochemical framework, *in* E. H. Gierlowski-Kordesch and K. R. Kelts, eds., Lake basins through space and time: AAPG Studies in Geology 46, p. 3–34.

Chapter 1

Lake-Basin Type, Source Potential, and Hydrocarbon Character: an Integrated Sequence-Stratigraphic–Geochemical Framework

Kevin M. Bohacs
Alan R. Carroll[1]
John E. Neal
Paul J. Mankiewicz
Exxon Production Research Company, Houston, Texas, U.S.A.

INTRODUCTION

Rocks associated with lakes probably account for more than 20% of current worldwide hydrocarbon production (Kulke, 1995; Calhoun, 1999), and lacustrine organic-rich rocks are significant sources of these hydrocarbons. Lacustrine sources and reservoirs are important in many areas of current and future exploration opportunities: Africa, South America, southeast Asia, China (Hedberg, 1968; Powell, 1986; Smith, 1990; Katz, 1995).

The years since the last AAPG lake Memoir (Katz, 1990) have seen both an expansion of work on modern and ancient lake systems and a focusing on their hydrocarbon potential. Through the efforts of individual workers and teams in academia and industry, along with collaborative efforts (e.g., IGCP-GLOPALS, International Association of Limnogeology), we have significantly increased our knowledge of lake systems on two fronts: key processes and sedimentary response-record (e.g., Anadon et al., 1991; Gierlowski-Kordesch and Kelts, 1994; Katz, 1995).

Particularly enlightening have been the increase in (1) basin-scale studies of ancient systems that integrate stratigraphy, sedimentology, biofacies, and inorganic and organic geochemistry, (2) the use of reflection seismic data to gain large-scale 3-D perspectives on basin-fill history, (3) studies by petroleum-industry scientists that benefit from this large-scale perspective and integration of physical, chemical, and biological processes and responses, and (4) studies of modern lakes and closely associated Quaternary deposits focused on key elements of sediment delivery and dispersal, organic production and preservation, and temporal evolution of lake hydrology (again aided by seismic-scale perspective, especially in east Africa) (e.g., Johnson et al., 1987; Scholz, 1995). Analytical advances and a broader experience base integrated into geological context have also contributed; geochemists have better tools for difficult nonmarine organic-matter mixtures (e.g., GC-MS/MS, isotope-ratio-monitoring GC-MS), stable isotope records are more widely available, and more sophisticated studies of the interactions among inorganic and organic geochemistry and sedimentation have been completed (e.g., Horsfield et al., 1994; Renaut and Last, 1994).

These studies allow construction of process-based models strongly conditioned by the geological record and provide a solid foundation for extending hypotheses into predictive realms. We can thus better and more naturally group many and disparate observations to reveal genetic relations and construct better predictive models based on our increased understanding of the links and feedback among process, response, and record of lake systems. This paper presents a summary of key observations and an overview of a framework for integrating process, response, and record we find useful for understanding and predicting source character and hydrocarbon distribution in lake basins.

LEADING FACTORS OF LACUSTRINE BASIN-FILL EVOLUTION

Numerous modern studies and ancient observations reveal that lakes are not just small oceans. Lakes

[1]Present address: University of Wisconsin-Madison, 1215 W. Dayton Avenue, Madison, WI 53706

differ from oceans in several significant ways (e.g., discussions in Kelts, 1988; Sladen, 1994). These differences strongly influence the occurrence, distribution, and character of hydrocarbon source, reservoir, and seal play elements. Recognizing these differences is essential to successful exploration and exploitation in lake basins. Major differences between lake and marine depositional systems include the following.

- Lakes contain much smaller volumes of sediment and water; hence, lake systems are much more sensitive to changing accommodation and climate. Lake levels vary more widely and rapidly than sea level—300 m in 15,000 yr is not uncommon (e.g., Currey and Oviatt, 1985; Manspeizer, 1985; Hayberan and Hecky, 1987; Johnson et al., 1987). In systems with very low relief over wide areas, small short-term lake level changes (weeks to months) can move shorelines large distances; the shoreline of Lake Chad recently retreated up to 18 km in 9 months due to a 3-m fall and lake area fluctuated 92% between 1966 and 1985 (Mohler et al., 1995). The result in the ancient record is shoreline strata that are commonly poorly developed and relatively thin (see discussions in Smoot, 1983 and Sladen, 1994). Water chemistry and lake ecology can also vary greatly over short stratigraphic intervals (e.g., Gierlowski-Kordesch and Kelts, 1994) with great impact on source and seal character.
- Lake level and sediment supply are directly linked in lake systems (e.g., Schumm, 1977; Perlmutter and Matthews, 1990). Lake level rises when river discharge is high and falls when discharge drops. The strength of this linkage varies according to lake-basin type, with the strongest in closed-hydrology basins and the weakest in open-hydrology basins. This contrasts greatly with marine systems, where sea level and sediment supply are only weakly linked at most and most models assume no linkage (e.g., Posamentier and Vail, 1988). This linkage is the prime cause of the variety of expression of depositional sequences among lake-basin types and of their contrast with marine sequences. For example, significant thicknesses of strongly progradational lowstand strata are unlikely to form in most lake basins.
- Lake shorelines can move basinward either by progradation or simple withdrawal of water. Progradation deposits a distinct rock package, whereas withdrawal leaves little record other than desiccation features on previously deposited strata.
- The nature and existence of a lake is fundamentally controlled by the relative rates of potential accommodation change and supply of sediment+water. Potential accommodation is the space available for sediment accumulation below the basin's outlet or spillpoint (it is a function of basin subsidence, outlet height, and inherited basin shape) (Carroll and Bohacs, 1995, 1999). The type of lake system is controlled by how much of that space is filled by some combination of sediment+water and over what time span (Gilbert, 1890); hence, climate (∞sediment+water) and tectonics/inherited topography (∞potential accommodation), commonly exert co-equal control on the nature and distribution of lacustrine depositional systems tracts and their source, reservoir, and seal lithofacies.

The contrast among lake and marine systems also makes it inappropriate to directly apply one unmodified marine sequence-stratigraphic model to all lake systems. The sequence-stratigraphic approach, looking at a hierarchy of rock packages bounded by various surfaces, works very well in lake strata; however, the expression of depositional sequences varies as a function of lake depositional system, just as shallow-marine carbonate sequences look different from shallow-marine siliciclastic sequences. Indeed, one lacustrine model is not applicable to all lake-basin types. Failure to appreciate these differences has led to some of the difficulties encountered in exploring in lacustrine basins. It is possible, however, to understand and predict lake-basin type, as well as source and hydrocarbon distribution and character, using first principles of sequence stratigraphy and lake depositional environments.

It also became clear that there were underlying major controls when we examined modern and ancient lake examples with a wide variety of ages, tectonic settings, climates, and latitudes (Carroll and Bohacs, 1995, 1999). The many good studies of modern lakes (e.g., Johnson et al., 1987; Cohen, 1989; Scholz and Rosendahl, 1990) reveal an almost bewildering array of process, interactions, and feedbacks that portend a corresponding large complexity in ancient lake deposits. This complexity, however, is not completely recorded in ancient lake strata from many ages and basins. We, along with several others (e.g., Bradley, 1931; Glenn and Kelts, 1991; Gore, 1989; Olsen, 1990), see three major facies associations in lake strata at the tongue to member scale (depositional sequence to sequence-set scale; meters to hundreds of meters) and a characteristic stacking of these facies associations as a basin fills (Figure 1). (Most workers attributed these three motifs or lake types to endemic, generally climatic, causes, hindering their general applicability and even their recognition.)

Figure 1—Comparison of lake-basin-type facies associations in vertical section with examples from the Eocene Green River Formation, Washakie basin, Wyoming. Each lake-basin type has distinctive associations of lithologies, sedimentary structures, biota, and geochemical nature. These characteristics appear to be strongly controlled by the interaction of potential accommodation and sediment+water supply and record the integrated history of a lake's hydrologic state. These associations are relatively independent of overall thickness, age, and inferred water depth.

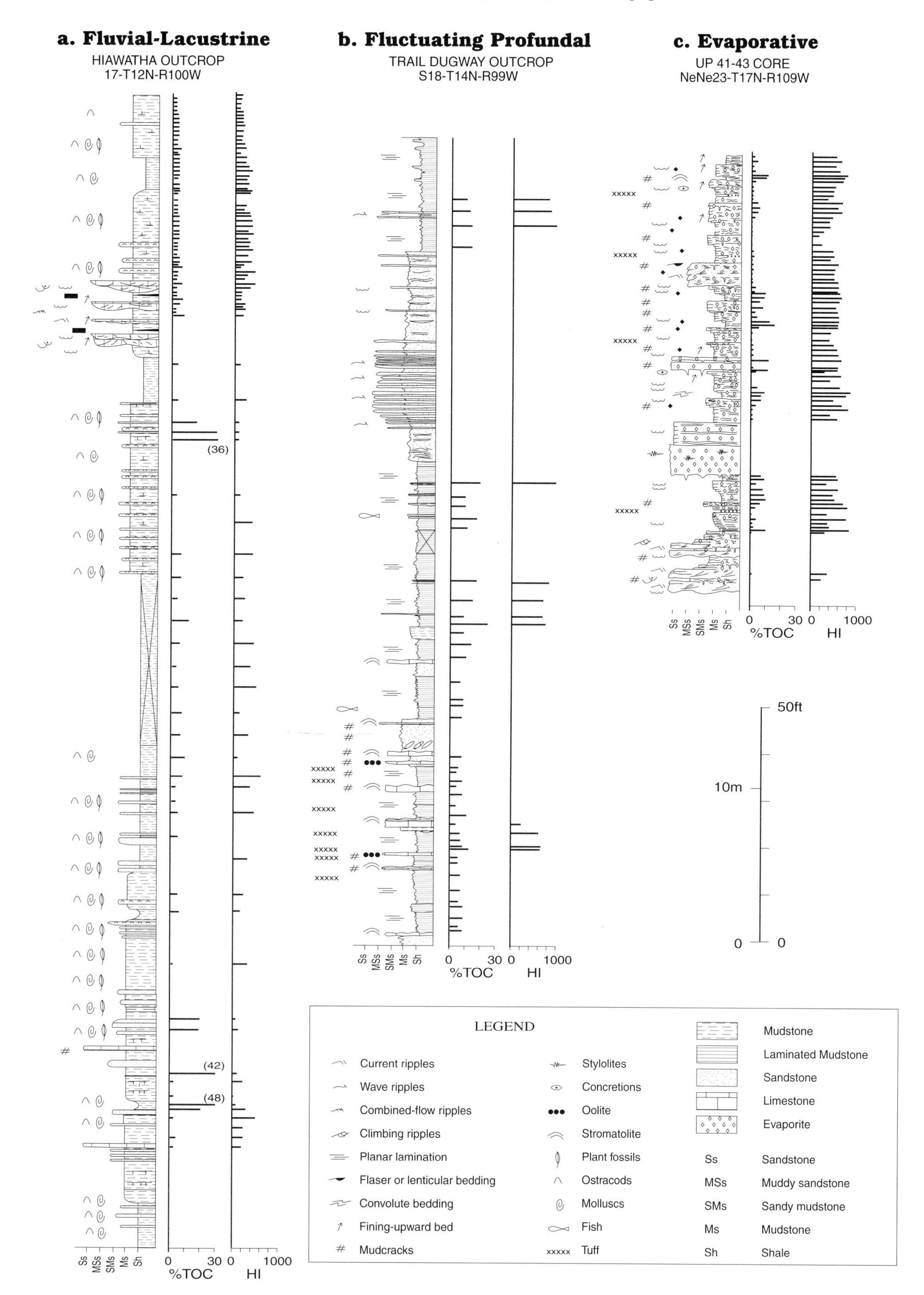
a. Fluvial-Lacustrine
HIAWATHA OUTCROP
17-T12N-R100W
(36)
(42)
(48)
Ss MSs SMs Ms Sh
0 30 0 1000
%TOC HI
b. Fluctuating Profundal
TRAIL DUGWAY OUTCROP
S18-T14N-R99W
Ss MSs SMs Ms Sh
0 30 0 1000
%TOC HI
c. Evaporative
UP 41-43 CORE
NeNe23-T17N-R109W
Ss MSs SMs Ms Sh
0 30 0 1000
%TOC HI
50ft
10m
0 0
LEGEND
Current ripples
Wave ripples
Combined-flow ripples
Climbing ripples
Planar lamination
Flaser or lenticular bedding
Convolute bedding
Fining-upward bed
Mudcracks
Stylolites
Concretions
Oolite
Stromatolite
Plant fossils
Ostracods
Molluscs
Fish
Tuff
Mudstone
Laminated Mudstone
Sandstone
Limestone
Evaporite
Ss Sandstone
MSs Muddy sandstone
SMs Sandy mudstone
Ms Mudstone
Sh Shale

Table 1. Representative Attributes of Three Major Lacustrine Facies Associations .

Lacustrine Facies Association	Strata Stacking Patterns	Sedimentary Structures	Lithologies	Organic Matter
Fluvial-Lacustrine	Dominantly progradation Indistinctly expressed parasequences	Physical Transport: ripples, dunes, flat bed Root casts Burrows (infaunal & epifaunal)	Mudstone, marl Sandstone Coquina Coal, coaly shale	Freshwater biota Land-plant, charophytic and aquatic algal OM* Low to moderate TOC* Terrigenous & algal biomarkers
Fluctuating Profundal	Mixed progradation & aggradation Distinctly expressed parasequences	Physical & Biogenic: flat bed, current, wave, & wind ripples; stromatolites, pisolites, oncolites Mudcracks Burrows (epifaunal)	Marl, mudstone Siltstone, sandstone Carbonate grainstone, wackestone, micrite Kerogenite	Salinity tolerant biota Aquatic algal OM* Minimal land plant Moderate to high TOC Algal biomarkers
Evaporitive	Dominantly aggradation Distinctly to indistinctly expressed parasequences	Physical, Biogenic, & Chemical: climbing current ripples, flat bed, stromatolites, displacive fabrics, cumulate textures	Mudstone, kerogenite Evaporite Siltstone, sandstone Grainstone, boundstone, flat-pebble cgl*	Low-diversity, halophytic biota Algal-bacterial OM* Low to high TOC Hypersaline biomarkers

*OM = organic matter, TOC = total organic carbon, cgl = conglomerate.

These three end-member facies associations are recognized based on objective physical, chemical, and biological criteria (Carroll and Bohacs, 1999; cf. Olsen, 1990, and references therein). Specifically, each end member has characteristic lithologic successions, sedimentary structures, geochemical indicators, fossils, and stratal stacking patterns. Table 1 lists the most common attributes of each of these lake facies associations; more details on stratal stacking patterns are found in later sections of this paper. In summary, the three lacustrine facies associations were named fluvial-lacustrine, fluctuating profundal, and evaporative by Carroll and Bohacs (1999), according to their most generally recognizable characteristics. The fluvial-lacustrine facies association tends to be composed dominantly of marlstone, argillaceous coquina, and bioclastic grainstone, along with sandstone, carbonaceous mudstones, and coals that contain freshwater fauna and mixtures of aquatic and terrigenous organic matter in beds and bedsets dominated by physical sedimentary structures (Figures 1a, 2). The fluctuating-profundal lacustrine facies association typically comprises a complex interbedding of heterogeneous lithologies (carbonate, siliciclastic, argillaceous, organic-rich mudstone) containing fresh- to saline-water biota and dominantly aquatic organic matter in beds and bedsets with both biogenic and physical sedimentary structures (Figures 1b, 3). The evaporative lacustrine facies association actually contains a wide variety of lithofacies (clastic sandstone and grainstone, kerogenite, evaporite) that contains a low-diversity, salinity-tolerant flora in beds and bedsets with sedimentary structures due to physical transport, biogenic precipitation, desiccation, and crystallization (e.g., ripples, stromatolites, mudcracks, displacive fabrics, and cumulate textures) (Figures 1c, 4). These three end-members appear to capture a large portion of the variation in essential attributes of lacustrine strata and are useful for exploration-scale summaries despite the observed wide variation in accidental attributes, such as evaporite or clay mineralogy, thickness, clastic composition, color, or absolute area.

Seeking to explain the common and widespread occurrence of these three lacustrine lithofacies associations, we sought to uncover the fundamental controls on the preserved geological record of lakes. One's intuition might lead one to suppose that wetter climates have lakes that are larger, deeper, and in the ancient thicker deposits; however, observations of both modern and ancient lakes do not corroborate this intuition. Modern lake size, depth, and character or ancient lake-strata thickness, extent, and attributes do not correlate with measured or inferred climatic humidity (precipitation/evaporation = P/E) (Carroll and Bohacs, 1995, 1997, 1999). No correlation exists between P/E and any measure of lake size (e.g., depth, surface area, volume) for large modern lakes, even those of tectonic origin (Figure 5a). Also, predictions

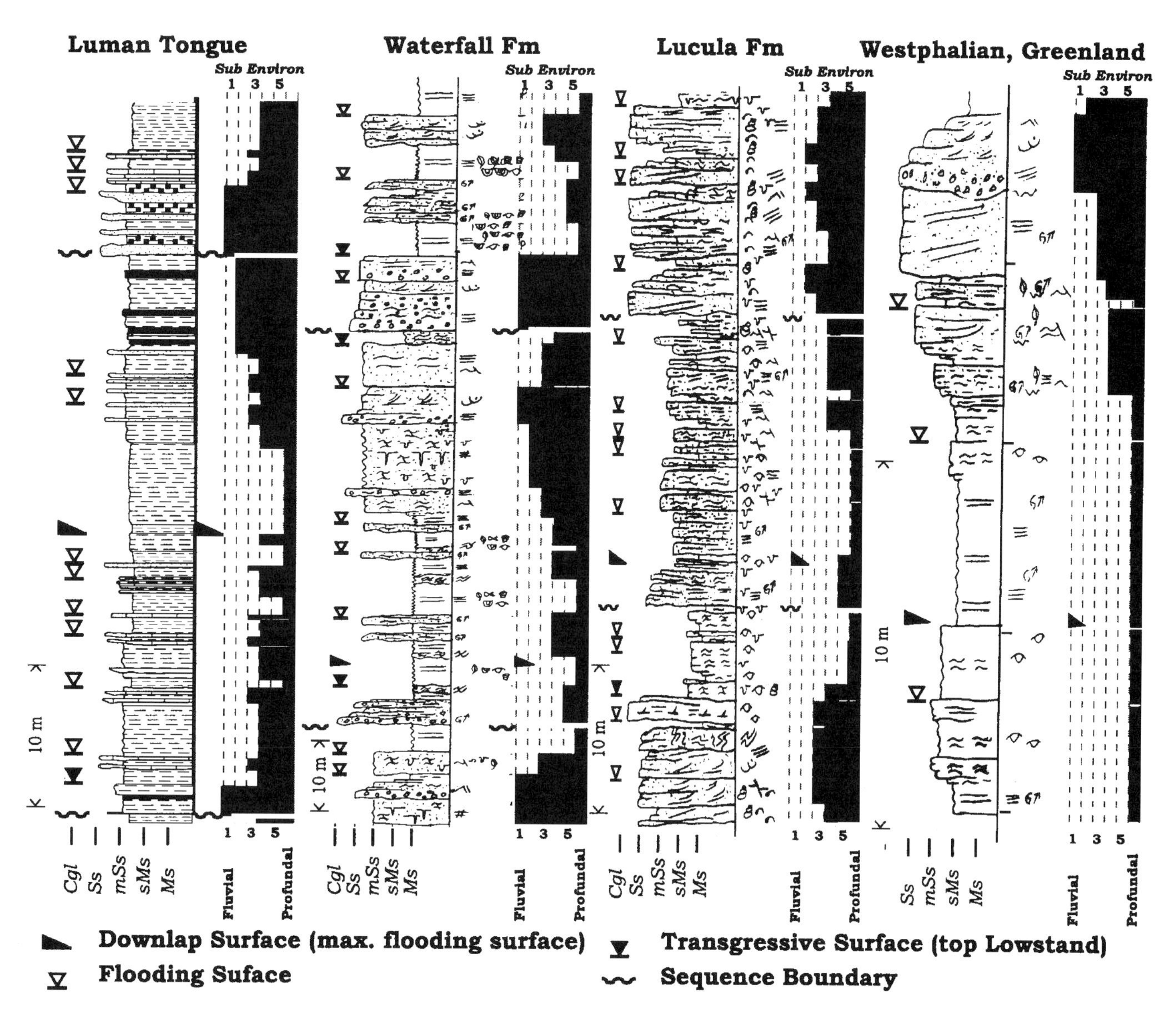

Figure 2—Examples of fluvial-lacustrine facies association, typical lithologies, and stratal packages (depositional sequences). Note dominance of clastic lithologies and physical sedimentary structures and similar stratal stacking patterns despite different ages, paleolatitudes, and overall thickness (note different scales). Subenvironments: 1 = fluvial/floodplain, 2 = lake plain and supralittoral, 3 = littoral, 4 = sublittoral, 5 = lake-floor "fan"/turbidite channel, and 6 = profundal. Subenvironments delineated based on indices of bottom energy, subaerial exposure (pedogenesis), and trace fossils; preservation of vertebrate fossils; and organic and inorganic geochemistry. Data derived from Eocene Luman Tongue, Green River Formation, Hiawatha outcrop (Sec 17-T12N-R100W), Washakie basin, Wyoming (Horsfield et al., 1994), Triassic Waterfall Formation, I-66 outcrop (Thoroughfare Gap, Virginia), Culpeper basin, Virginia (Hentz, 1981; Gore, 1988), Cretaceous Lucula Formation, 73-69 core (Malongo North field), offshore Cabinda, Angola (Bracken, 1994), Westphalian lacustrine shales, Traill Ø outcrop (ca. 72°N, 24°W), East Greenland (Stemmerik et al., 1990).

based on climatic humidity alone fail to explain the wide variety of modern lakes within a single climatic zone, where they can range from saline to freshwater, eutrophic to organically barren, and purely siliciclastic to carbonate forming (Herdendorf, 1984) (Figure 5b). Modern Utah Lake contains fresh water, but drains into the adjacent, hypersaline Great Salt Lake, only 70 km away within the same semi-arid area (Stansbury, 1852). Also note that the drainage basins of the world's largest rivers contain only few, relatively small lakes (e.g., Amazon, Congo, Orinoco, Mississippi).

A similar discordance between climate and lake size is also apparent in ancient systems. For example, detailed mapping of the Green River Formation of Wyoming reveals that strata of the freshest water lake phases (e.g., the Luman Tongue) are actually the thinnest and least areally extensive (Sullivan, 1980; Roehler, 1992) (Figure 6). In contrast, more saline lake members such as the lower LaClede Bed and upper

Tipton member are the most widespread; furthermore, the most evaporitic Wilkins Peak member is also the thickest stratigraphically, although it had the shallowest depth lakes (Smoot, 1983). All these observations strongly indicate that changes among lacustrine facies associations do not result solely from climatically driven changes. Additionally, studies of organically enriched lake deposits show little if any correspondence with inferred paleoclimate or climatic zone (e.g., Smith, 1990; Carroll, 1998).

LAKE-BASIN TYPES

To reconcile these seemingly paradoxical observations, we focused on the fundamental controls on lake strata. We interpret that it is the relative balance of rates of potential accommodation change (mostly tectonic) with sediment+water supply (mostly climatic) that controls lake occurrence, distribution, and character. Our model addresses lake types and hydrocarbon potential in the context of these two fundamental controls, incorporating meso- and macroscale stratal stacking, sediment-supply variations, and subsidence history. We propose that climate and tectonics, through their strong effects on these controls, exert coequal influence on the occurrence, distribution, and character of preserved lake strata at both meso- and macroscales (one to tens of meters and hundreds of meters) (see also Manspeizer, 1985), mainly through their influence on the time-integrated history of lake hydrology.

Based on numerous empirical observations of ancient systems from Cambrian to Holocene, we therefore propose that the three most common lacustrine facies associations correspond to distinctive lake-basin types: overfilled lake basins, balanced-fill lake basins, and underfilled lake basins (Figure 7). Although named for interpreted genetic factors (to provide predictive, as well as descriptive, utility), each type is characterized by readily observable features, such as lithofacies association and stratigraphic packaging, and possesses predictable hydrocarbon generation characteristics (Carroll and Bohacs, 1995, 1999).

A short summary of the major characteristics of each lake-basin type follows (Table 2); subsequent sections of this paper expand on stratal attributes, hydrocarbon source potential, and associated hydrocarbon character. These sections highlight the most commonly occurring attributes of each lake-basin type useful for recognition and mapping in exploration data, and are not intended to be exhaustive. Note that many of the same factors that control source quality also influence the distribution of lacustrine reservoir and seal lithofacies.

Overfilled lake basins (Figure 8) occur when the rate of supply of sediment+water consistently exceeds potential accommodation (usually when P/E is relatively high or rates of tectonic subsidence are relatively low). The resulting lake hydrology is open either permanently or dominantly over the time span of accumulation of depositional sequences. Climatically driven lake-level fluctuations are minimal because water inflows are in equilibrium with outflows. These lakes are very closely related to perennial river systems; their deposits are commonly interbedded with fluvial deposits and coals, which is the fluvial-lacustrine facies association. Parasequence development is driven predominantly by shoreline progradation and delta-channel avulsion.

Balanced-fill lake basins (Figure 9) occur when the rates of sediment+water supply and potential accommodation are roughly in balance over the time span of sequence development. Water inflows are sufficient to periodically fill available accommodation, but are not always matched by outflow. As a result, climatically driven lake level fluctuations are common. Lake hydrology is closed during deposition of basinally restricted lowstand strata and open during highstand deposition of obliquely prograding strata. Depositional sequences record a combination of progradation of clastic sediments and aggradation of mostly chemical sediments due to desiccation, which is the fluctuating profundal facies association. These deposits typically have the highest organic enrichments of any noncoaly source rocks due to an optimal combination of primary production, mean water depth, chemical stratification, and rates of burial.

Underfilled lake basins (Figure 10) occur when rates of accommodation consistently outstrip available water and sediment supply, resulting in a persistently closed basin hydrology with deposits of ephemeral lakes or brine pools and playas interspersed with those of relatively "perennial" lakes. Individual lakes are geologically short-lived, hence parasequences and sequences are commonly very thin (on the scale of decimeters). Parasequence stacking mainly records vertical aggradation of the products of desiccation cycles, which is the evaporative facies association. These deposits are composed of highly contrasting lithologies, commonly associated with evaporites.

Due to changes in climate or tectonic subsidence, nonmarine basins commonly evolve from one lake type to another through a variety of time scales in a predictable pattern (cf. Lambiase, 1990). Predictable successions of lake-basin types can arise from either a climatic or tectonic cycle, following a mostly vertical or horizontal trajectory on the lake-basin-type phase diagram (Figure 7). These changes are commonly recorded within a single formation. For example, the Green River Formation shown in Figure 6 records a primarily tectonic cycle (horizontal trajectory) in its subdivisions: fluvial (Wasatch Formation) to overfilled (Luman and Niland tongues) to balanced-fill (Tipton Member) to underfilled (Wilkins Peak Member) to balanced-fill (lower Laney/LaClede) to overfilled (upper Laney/LaClede) to fluvial (Washakie Formation). Different lake types can also coexist in adjacent basins (e.g., Eocene Green River Formation in Washakie, Uinta, and Piceance Creek basins) (Bradley, 1964); chains of modern lakes especially illustrate that the relation of sediment+water supply to potential accommodation change in each individual lake controls lake type and character.

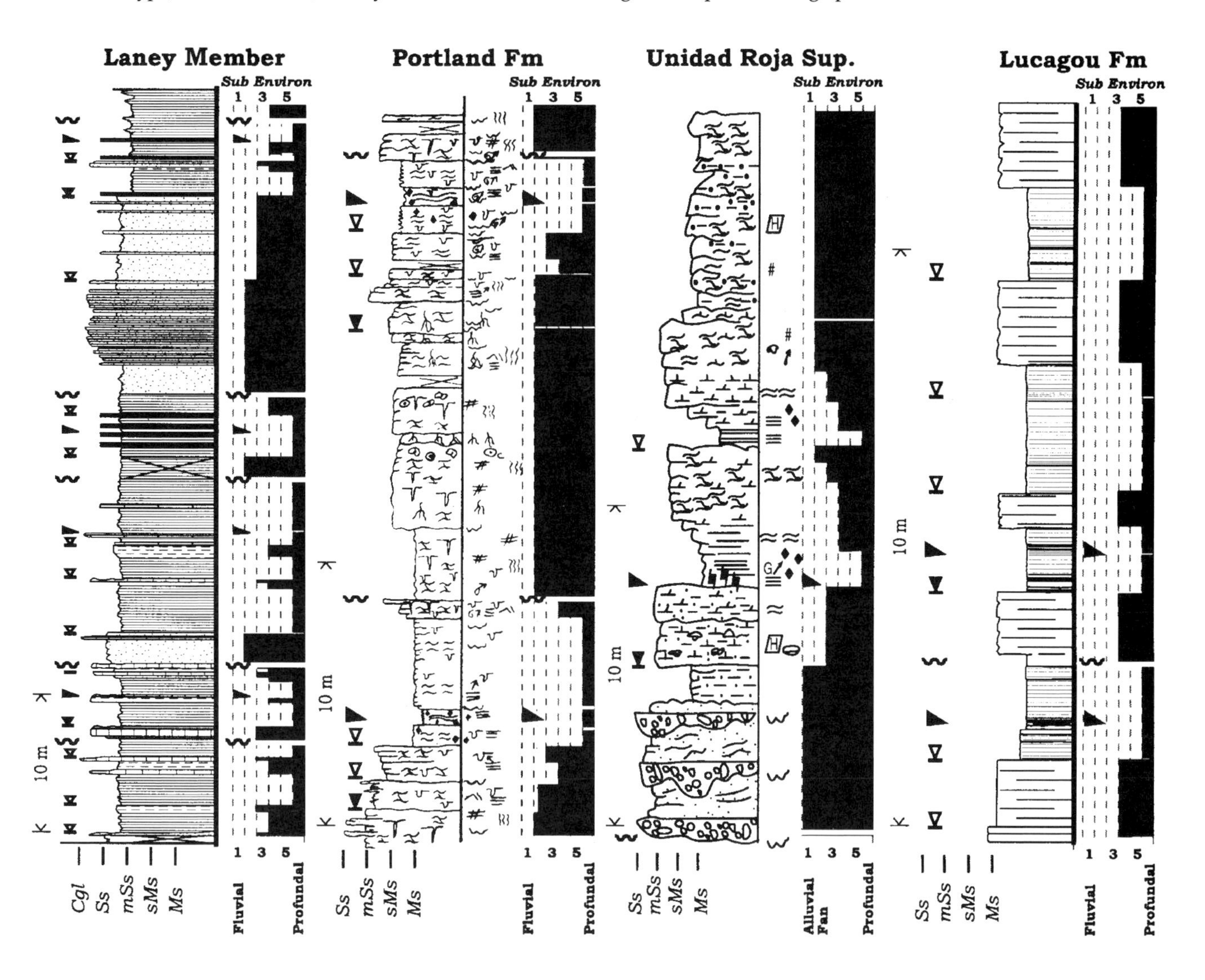

Figure 3—Examples of fluctuating-profundal lacustrine facies association, typical lithologies, and stratal packages (depositional sequences). Note complex interbedding of clastic and carbonate lithologies in both physical and biogenic sedimentary structures and similar stratal stacking patterns despite different ages, paleolatitudes, and overall thickness (note different scales). Subenvironments: 1 = fluvial/floodplain, 2 = lake plain and supralittoral, 3 = littoral, 4 = sublittoral, 5 = lake-floor "fan"/turbidite channel, and 6 = profundal. Subenvironments delineated based on indices of bottom energy, subaerial exposure (pedogenesis), and trace fossils; preservation of vertebrate fossils; and organic and inorganic geochemistry. Data derived from Eocene lower Laney member, Green River Formation, Trail Dugway outcrop (Sec 18-T14N-R99W), Washakie basin, Wyoming (Horsfield et al., 1994), Jurassic lower Portland Formation, Park River Tunnel core (41° 45.4′N, 72° 42′W), Hartford basin, Connecticut (K. M. Bohacs, 1992, unpublished company report based on Park River Tunnel project cores), Permian Lucaogou Formation, Tianchi aqueduct outcrop (43.7°N 84.3°E), Junggar basin, China (Carroll, 1998), and Permian Unidad Roja Superior, Fuentes de Izas outcrop (42.45°N 0.28°W), Aragón-Béarn basin, Spain (Valero Garcés, 1991).

Ancient deposits may not be directly comparable to modern lakes, for which only a synoptic ("snapshot") view is available, because the record of a lake system is mainly controlled by the response of its sedimentary systems to changing hydrology over time (e.g., Kelts, 1988). The shorthand lake-basin-type names are intended to summarize the relation of potential accommodation and supply of sediment+water and the resulting lake hydrology over the time span of accumulation of the common associations of lithofacies and stratal stacking patterns. This time span typically corresponds to the scale of depositional sequence to sequence set (or tongue to member; several meters to hundreds of meters). The relative simplicity of lacustrine facies associations points out the importance of the geological "filter" in determining what modern processes leave preservable records (see also discussion in Shanley and McCabe, 1994). All sedimentary rocks represent a time-averaging of related but diachronous and evolutionary depositional environments. Lakes, in particular, are extremely dynamic, with high rates of change of environmental parameters, and vary widely in their sensitivity to climatic and hydrodynamic changes (Hutchinson, 1957; Cole, 1979; Gierlowski-Kordesch and Kelts, 1994).

Within our framework, water depth is a secondary attribute of lakes. Water depth can be portrayed as a third, independent axis within the lake-basin-type phase diagram (Figure 7), separating the percent of sediment+water supply that is water from the potential accommodation that remains unfilled with sediment. Although terms such as "deep" and "shallow" are relative, it is possible to subdivide each lake-basin type into shallow and deep subtypes based mainly on coarse clastic facies and stratal geometry. Table 3 lists some key examples of each type. In general, compared to their shallow subtype, deep lake-basin types tend to have more deposits associated with high-relief processes, thicker stratal packages, and better expressed stratal geometries. Basinally restricted strata are more distinct and aggradational and basin-margin erosion, downlap, and onlap are better developed; however, basinward shifts of facies tend to be shorter and more subtly expressed. Clearly, "deep" and "shallow" do not control the essential attributes of each lake-basin type, but "deep" and "shallow" can strongly influence many accidental attributes, such as sequence thickness, areal extent, shoreline type, and geographic distribution (especially reservoir facies) (see Neal et al., 1997).

Our use of the terms "overfilled" and "underfilled" is directly analogous to that employed in marine basins (Marzo et al., 1996; Ravnas and Steel, 1998) and evolved within our company at just about the same time (first published in Carroll and Bohacs, 1995). It is, however, distinctly different in application and consequences for lake-basin fill because of the fundamental differences between lake and ocean basins, detailed in the previous section. Workers on marine rift basins based their terms on the interaction of total accommodation and sediment supply alone, and not on potential accommodation and supply of sediment+water. Most important, the existence of a sill controls the very existence of a lake and its absolute maximum amount of accommodation. It necessitates the definition of a balanced-fill lake state and enables the explanation of the fluctuating-profundal lacustrine facies association, reconciling the close interbedding of desiccational and progradational shoreline parasequences.

Other workers have rightly pointed out the variety of controls on lake strata and organic-rich rock accumulation (e.g., Zhou, 1981; Powell, 1986; Kelts, 1988; Watson et al., 1987; summary in Katz, 1990). What we found most helpful was the close examination of the profundal and sublittoral strata associated with the organic-rich rocks. Their nature and distribution record the complex interaction of depositional and preservational controls integrated over space and time. This integrated physical-chemical-biological approach allows robust recognition and prediction of lake-basin types and their associated source and hydrocarbon potential. By concentrating on fundamental controls, one can see how primary interactions affect all aspects of a lake system and its geological record.

SEQUENCE STRATIGRAPHY OF LAKE-BASIN TYPES

Introduction

Lake-basin type strongly affects the physical character and distribution of strata and the expression of the parasequences and sequences they accumulate. Each type has characteristic expression and distribution of hydrocarbon play elements (Figures 8–10). Their expression ranges from similar to shallow-marine sequences in some overfilled lake basins to different in underfilled lake basins. The range of expression is detailed below.

The sequence-stratigraphic approach, looking at packages of rocks bounded by physical surfaces to construct a chronostratigraphic framework, works well in lacustrine and alluvial strata. The expression of depositional sequences in these settings can vary widely from the "normal" marine case, but these variations are readily understood and predicted within a sequence framework. As with all depositional sequences, their character is controlled by the combined influence of base level (lake level and the groundwater table), sediment supply, and tectonics. The influence of climate is recorded in both water supply (lake level and groundwater table) and the sediment supply. The expression of lacustrine depositional sequences varies widely because lake systems themselves are greatly variable. Lakes are much more responsive due to the smaller volumes of water and sediment involved, their frequently closed-basin nature, and their generally closer tie of sediment supply to lake level (Gierlowski-Kordesch and Kelts, 1994; Neal et al., 1997, 1998; Carroll and Bohacs, 1999). Although lacustrine systems can vary more rapidly and widely (Figure 7), this does not affect the utility of the sequence-stratigraphic approach because the rocks are still deposited in layers bounded by physical surfaces that can be used as time lines. Indeed, sequence stratigraphy is especially appropriate for lakes because of its focus on integrating observations of many scales of sedimentary changes and hiatuses in a hierarchy that spans millimeters to kilometers.

Each lake-basin type tends to have characteristic stratal patterns that arise from typical phase relations between sediment+water supply and lake level (accommodation) summarized as follows. Lake types have distinctive facies stacking patterns at the sequence and parasequence scale that reveal the link of sediment+water flux to changing lake level.

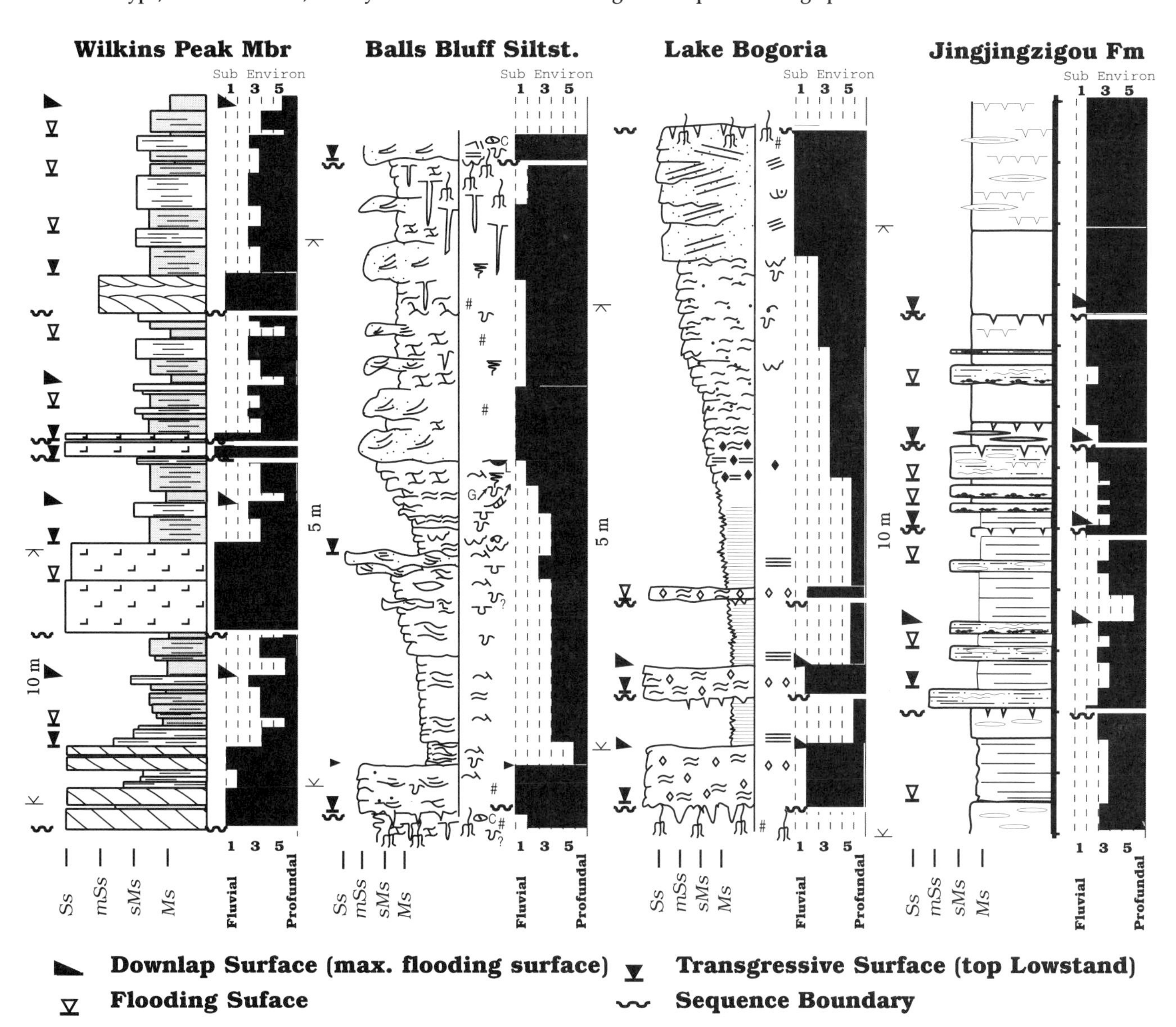

Figure 4—Examples of evaporative lacustrine facies association, typical lithologies, and stratal packages (depositional sequences). Note wide variety of lithofacies in sedimentary structures due to physical transport, desiccation, and crystallization, and similar stratal stacking patterns despite different ages, paleolatitudes, and overall thickness. Subenvironments: 1 = fluvial/floodplain, 2 = lake plain and supralittoral, 3 = littoral, 4 = sublittoral, 5 = lake-floor "fan"/turbidite channel, and 6 = profundal. Subenvironments delineated based on indices of bottom energy, subaerial exposure (pedogenesis), and trace fossils; preservation of vertebrate fossils; and organic and inorganic geochemistry. Data derived from Eocene Wilkins Peak member, Green River Formation, UPRR 41-43 core (NENE Sec 23-T17N-R109W), Bridger basin, Wyoming (Bohacs, 1998), Triassic Balls Bluff siltstone, Culpeper Crushed Stone Quarry (Stevensburg, Virginia), Culpeper basin, Virginia (Gore, 1988, 1989), Permian Jingjingzigou Formation, Tianchi aqueduct outcrop (43.7°N 84.3°E), Junggar basin, China (Carroll, 1998), and Pleistocene–Holocene Lake Bogoria, cores and outcrops (0.15°N 36.06°E), (Renaut and Tiercelin, 1994).

Understanding this link leads to predictability in the sequence stratigraphy of each lake-basin type.

Sequence Expression In Lake Types

Overfilled lake basins tend to accumulate fluvial-lacustrine facies associations in depositional sequences that can appear very similar in geometry and development to both Vail/Exxon-type-1 and Vail/Exxon-type-2 sequences in shallow-marine siliciclastic settings, although they are generally thinner (Figure 8) (Bohacs, 1995, 1998). Observations show that flooding surfaces tend to have relatively minimal lithologic contrast across them. Depositional sequence boundaries tend to show either minimal or very distinctive erosional reworking. Depositional geometries are generally well expressed due to the relatively large sediment supply and are readily recognized on seis-

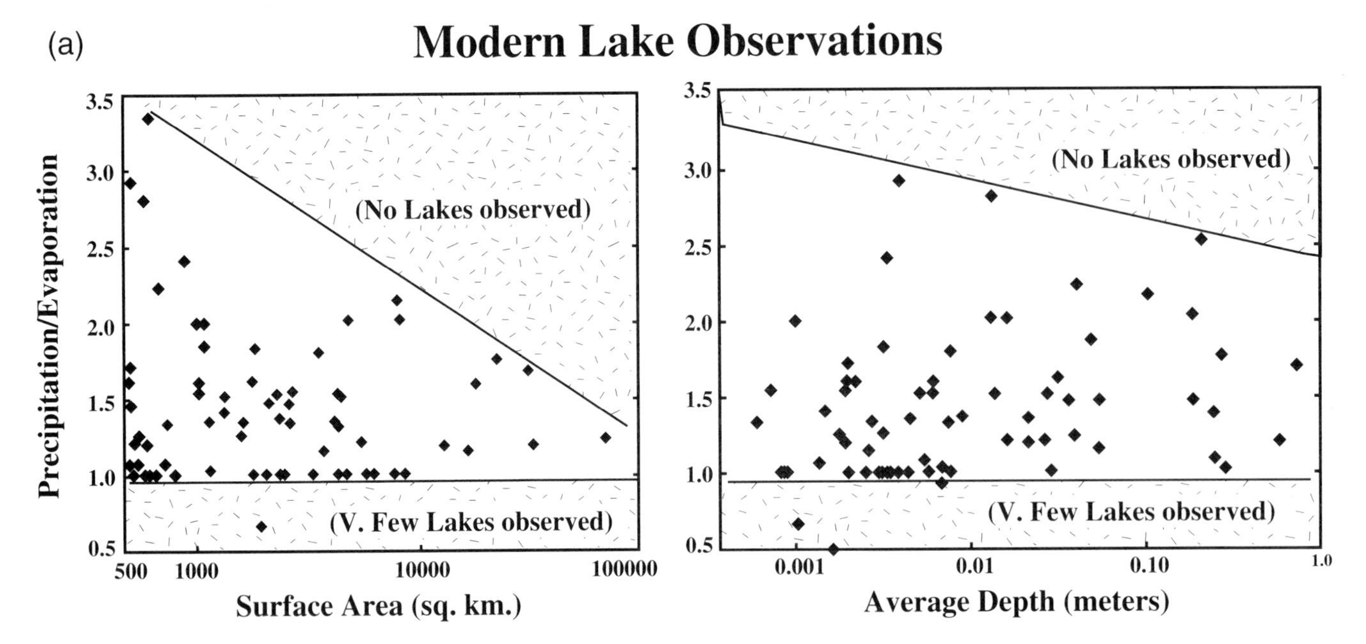

Figure 5—(a) Comparison of climatic humidity (precipitation/evaporation or P/E) with surface area and mean depth of large modern lakes. No correlation exists between P/E and any measure of lake size for large modern lakes, even those of tectonic origin. (b) Comparison of climatic humidity with lake chemistry and mixis, showing little relation. Both graphs indicate that predictions of lake existence or character based on climatic humidity alone fail to explain the bewildering complexity of modern lakes, which can range from freshwater to saline, organically barren to eutrophic, and carbonate-precipitating to purely siliciclastic within a single climatic zone. Data for both figures from Herdendorf (1984).

mic and in well-log cross sections and outcrop. Based on observations of two distinctive stratal-stacking patterns of the fluvial-lacustrine facies association, we interpret two states of overfilled lake basins, based on whether and how far lake level falls below sill height. In both overfilled lake states, observations indicate that shoreline progradation dominates parasequence development, resulting in the stacking of indistinctly expressed parasequences up to 10 m thick. Sequence boundaries and systems tracts are what vary in development between the two lake states. In permanently overfilled lake basins, wherein lake level remains at or very near sill height, stratal patterns are dominated by variations in sediment supply, and erosive sequence boundaries are nonexistent to minimally developed (Neal et al., 1997). This is analogous to a classic type 2 depositional sequence boundary, wherein no significant downward shifts of facies belts or basinally restricted lowstand deposits are developed (Posamentier and Vail, 1988). Flooding surfaces are enhanced in this case because they form by decreased sediment supply. Depositional geometries in this overfilled lake state reflect changing sediment flux at fixed lake level. By contrast, in dominantly overfilled lake basins, lake level occasionally falls significantly below sill height, enabling extensive erosion and incised-valley formation during lake-level fall. These lakes form depositional sequences that look similar to classic type 1 sequences in shallow-marine siliciclastic settings. Relatively low organic richness and an abundance of terrestrial organic matter are typical. Reservoirs generally are best developed in highstand clastic shoreline strata, and occasionally in charophytic algal lithosomes and in lowstand incised valley fills and lakefloor "fans" (basinally restricted turbidite and mass-flow deposits). Seal facies tend to be best and most extensively developed in distal transgressive and highstand prodelta strata. Examples of this lake type are listed in Table 4.

Balanced-fill lake basins generally accumulate fluctuating-profundal lacustrine facies associations in depositional sequences that can be somewhat similar to or distinctly different from shallow-marine siliciclastic sequences (Figure 9) (Bohacs, 1995, 1998). Observations indicate that parasequences record a combination of progradation of clastic sediments and vertical aggradation of chemical sediments due to desiccation cycles. Carbonates are generally abundant in these lakes, hence their parasequences and depositional sequences are more similar to shallow-marine carbonate or mixed carbonate-clastic settings. Flooding surfaces are commonly well expressed in distinct lithologic contrasts. Sequence boundaries tend to be marked by large basinward shifts of depositional environments with minimal erosion and incision. Lowstands (basinally restricted strata) in shallow balanced-fill lake basins accumulate relatively thin aggradational parasequence sets, with common evidence of desiccation, that can be either carbonate or clastic dominated, whereas deep balanced-fill lake basins can accumulate relatively thick aggradational

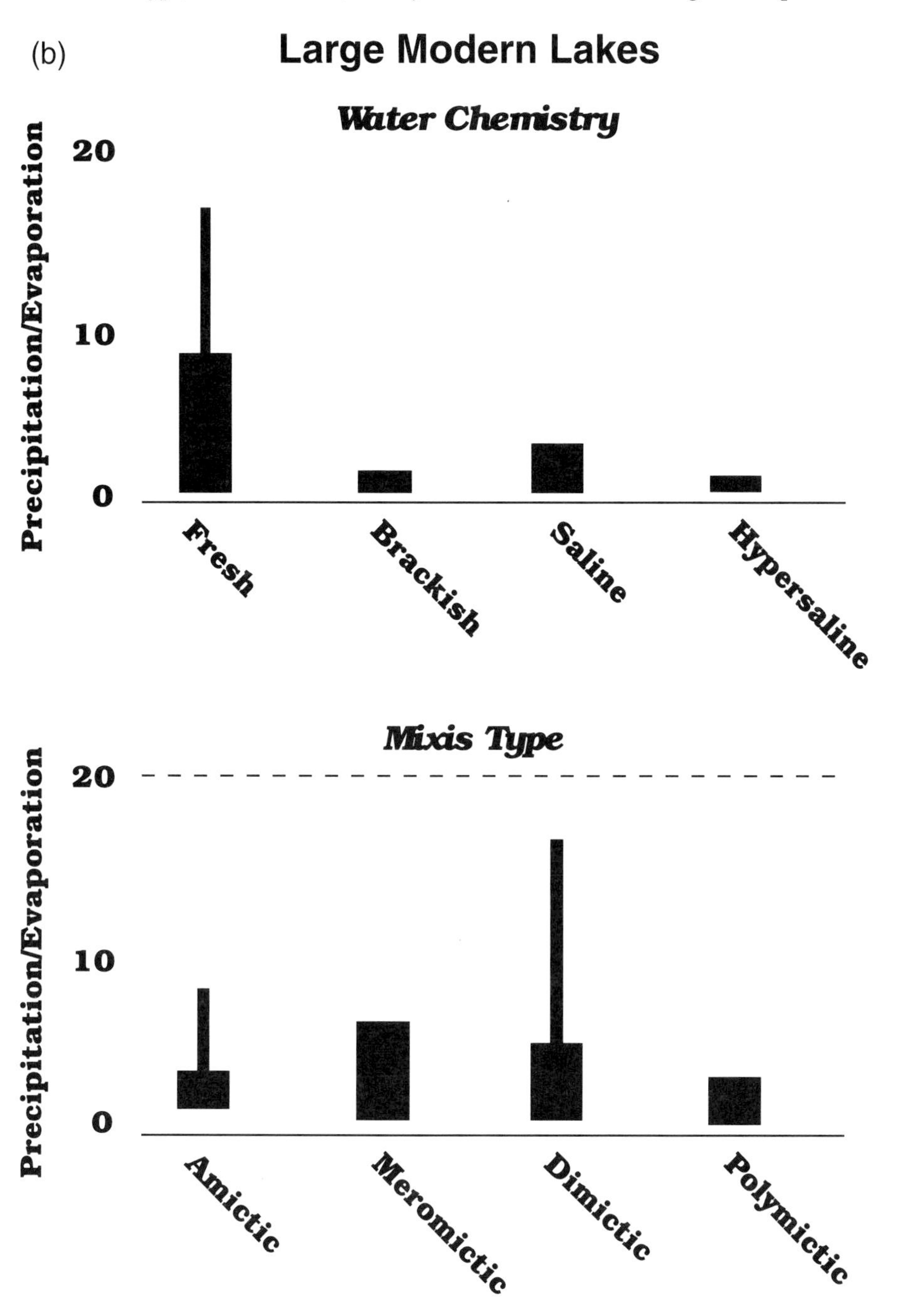

Figure 5—Continued.

lake-floor turbidite "fans." Indeed, the great variability of lowstand expression is a key defining trait of this lake-basin type. Lake hydrology is interpreted to be closed commonly during this stage. Transgressive systems tracts tend to be relatively thick and well developed, marked by significant retrogradational stacking and some basal erosion. Highstands range from relatively thin to moderately thick and from carbonate to clastic dominated, interpreted as typically accumulated under open hydrologic conditions. They can show well-developed aggradation and some progradation, expressed as sigmoidal to oblique progradation on seismic. Strata generally thin basinward by widely spaced downlapping. For this lake-basin type, we interpret that sediment supply is more closely linked to lake level because often the same rivers supply both water and sediment to the lake in an intermittently open hydrologic system. When the rivers dry up, lake level falls due to decreased water input. The erosive power of the river and its sediment supply would also decrease with lake-level fall, but increase during lake-level rise. This contrasts with the explicit assumptions for shallow-marine depositional sequences (Posamentier and Vail, 1988), thus giving rise to the contrasting expression of depositional sequences in this lake type. Organic matter types can be mixed algal and terrigenous, although commonly dominated by type I algal-bacterial kerogen. Reservoir facies can include lake-floor "fans," incised-valley fills, and shoreline clastics or carbonates deposited during transgressions and highstands. Seal-prone facies are widespread and well developed in late transgressive

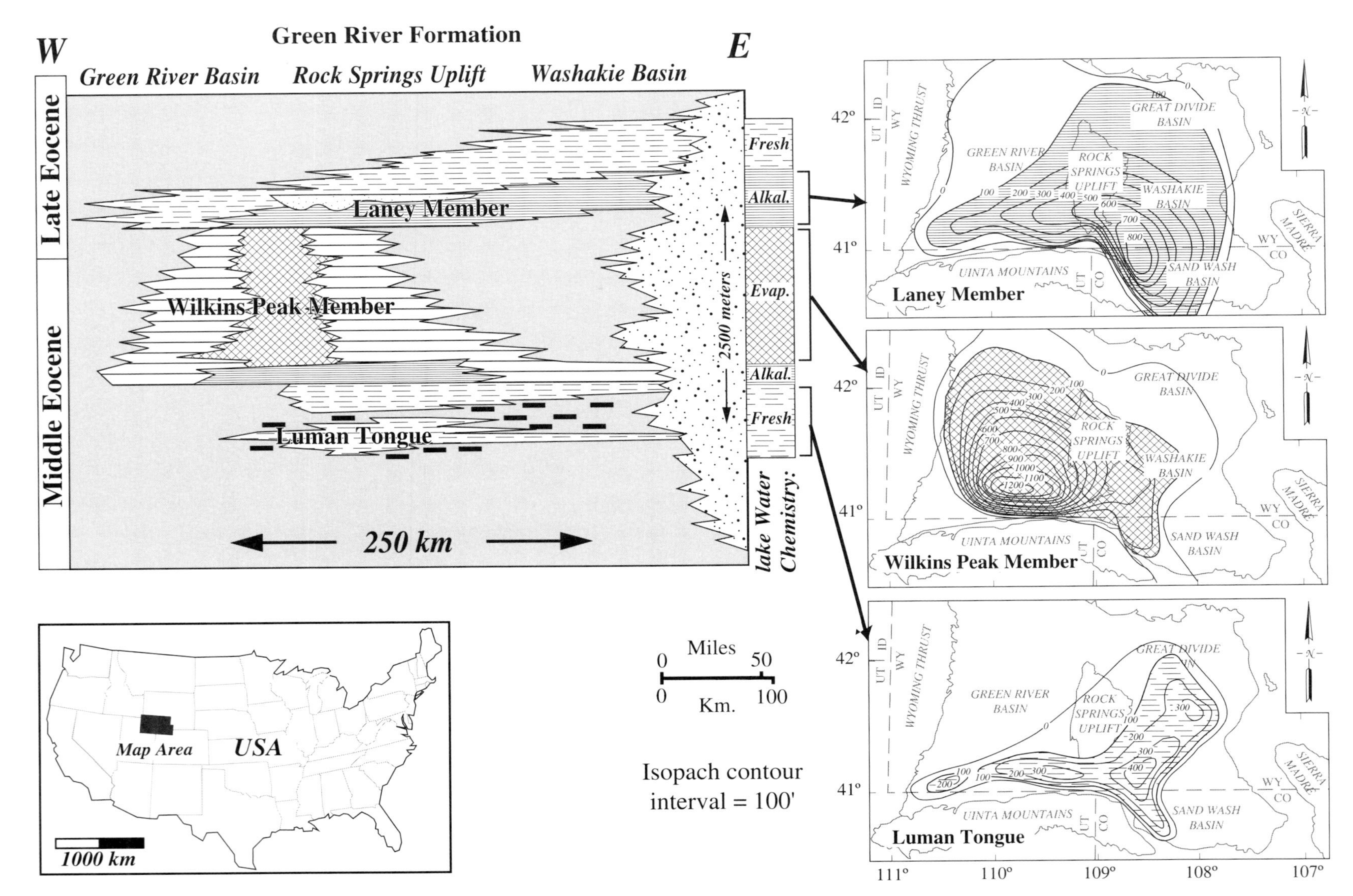

Figure 6—Cross section and maps of three major lake units in the Eocene Green River Formation of Wyoming (after Sullivan, 1980; Roehler, 1992). The hypersaline lake strata, deposited in the most shallow lakes, are stratigraphically thickest, the saline-alkaline lake strata are the most areally extensive, and the freshwater lake strata are relatively thin and of smallest areal extent. All these observations indicate that climate alone cannot explain variations in lake water chemistry and stratal character.

Table 2. Characteristics of Lake-Basin Types: Strata, Source Facies, and Hydrocarbons.

Lake Type & Lacustrine Facies Association	Stratigraphy	Source Potential	Hydrocarbon Characteristics
OVERFILLED *Fluvial-Lacustrine Facies Association*	Maximum progradation: • Parasequences related to lateral progradation (relatively subtle) • Maximum fluvial input	• Low to moderate TOC • Mixed type I-III kerogen • Marked organic facies contrasts • Distinct lateral changes in organic facies	• Generate both oil and gas • Very waxy, low-sulfur oils • Terrigenous biomarker assemblage dominant
BALANCED-FILL *Fluctuating-Profundal Facies Association*	Mixed progradation and desiccation: • Distinct shoaling cycles common • Fluvial input variable	• Moderate to high TOC • Predominantly type I kerogen, with type I-III mixtures near flooding surfaces • Relatively homogeneous and laterally consistent organic facies	• Mostly oil generative • Paraffinic but relatively nonwaxy oils; low sulfur • Algal biomarker assemblage dominant
UNDERFILLED *Evaporative Facies Association*	Maximum desiccation: • High-frequency wet-dry cycles • Minimum fluvial input	• Low overall TOC (w/ some high TOC intervals) • Type I kerogen • Minimum organic facies contrasts • Laterally consistent organic facies	• Mostly oil generative • Paraffinic oils; moderate to high sulfur • Distinctive "hypersaline" biomarker assemblage

and early highstand systems tracts as prodelta mudrocks and sublittoral marls or micrites. Examples of this lake-basin type are listed in Table 5.

Underfilled lake basins accumulate evaporitic lacustrine facies associations in depositional sequences that are distinctly different from shallow-marine siliciclastic or carbonate settings (Figure 10) (Bohacs, 1995, 1998). The nature and distribution of depositional environments within this lake-basin type can change drastically from small evaporitic ponds on broad playa mud/salt flats at lowstand to broad "perennial" lake with lake-plain streams at highstands. Observations indicate that sequence boundaries are subtly expressed in underfilled lake basins, whereas flooding surfaces are marked by distinct lithologic contrasts. Flooding surfaces are commonly coincident with sequence boundaries (FS/SB) across large portions of the basin area. Depositional geometries are generally parallel to subparallel and dominated by aggradational stacking. Strata thin basinward by convergence, although some basinward thickening can occur in strongly evaporitic settings. Lowstand deposition is commonly restricted to evaporites or other chemical/biogenic sediments formed in remnant pools in the highest subsidence area of the basin; the bulk of the lowstand record is only desiccation features modifying underlying highstand or other strata (mudcracks, soil formation, etc.). Transgressive systems tracts are typically recorded by thin, widespread clastic sheet-flow deposits at the base (some local erosion is possible), reflecting the rejuvenation of river input. These deposits are overlain directly by distal lake strata, commonly organically enriched, marking the rapid spread of the lake over a low-relief surface. Highstands are relatively thick, but typically composed of only one or two parasequences. Channel-fill sandstones are found in some highstand lake-plain strata, reflecting the development and integration of drainage systems during this wet phase of

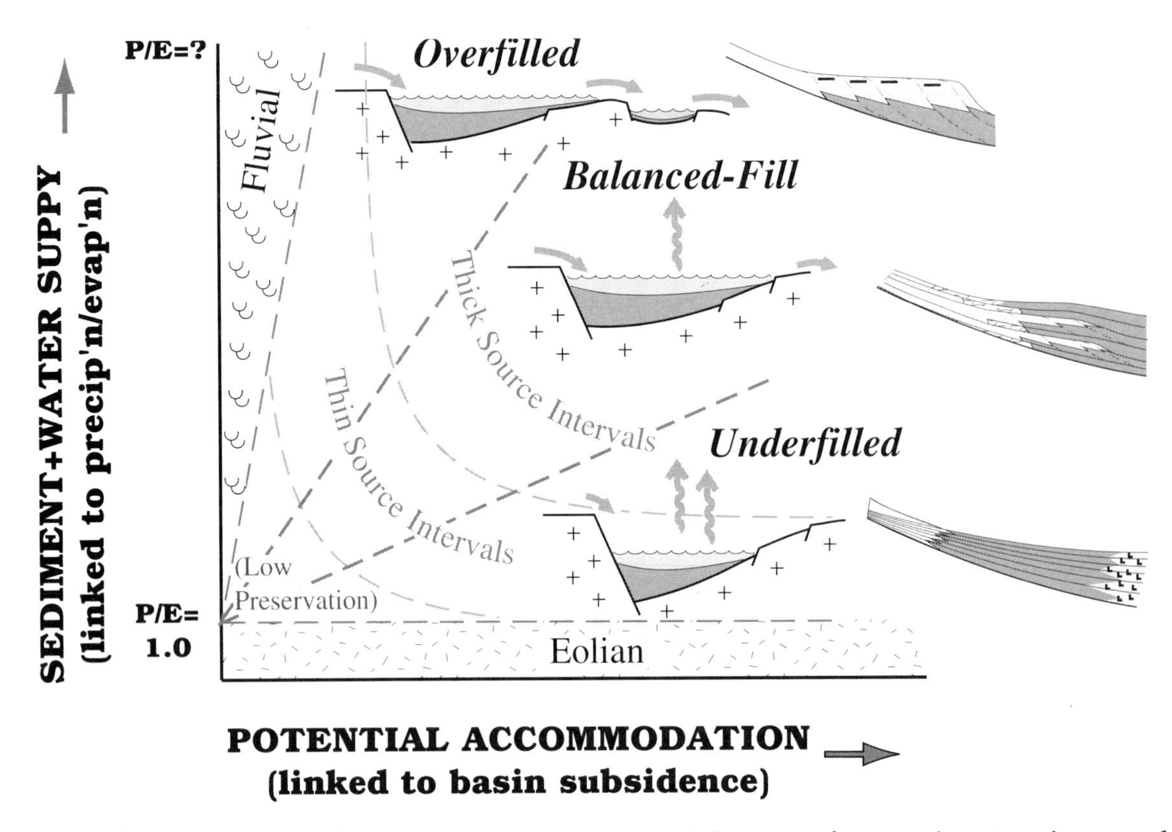

Figure 7—Lake-basin type phase diagram showing existence and character of nonmarine strata in general and lacustrine strata in particular as a function of both sediment+water supply and potential accommodation. Interaction of these two controls is reflected in the lithology, stratal stacking, biota, and geochemistry of lake deposits. Potential accommodation is the space available for sediment accumulation below the basin's outlet or spillpoint (a key difference from marine systems) (Carroll and Bohacs, 1995), and it is mainly influenced by basin tectonics, along with sill uplift and erosion, and inherited topography. Sediment+water supply is primarily a function of climatic humidity, along with seasonality, local relief, and bedrock geology. (Clastic sediment yield is a nonlinear, nonmonotonic function of climatic factors, generally peaking in semi-arid, distinctly seasonal climates) (Einsele, 1992).

lake development. We interpret that parasequence development and stacking in the lake center primarily reflects vertical aggradation of the products of desiccation cycles, with minimal advected clastic input. Sediment supply appears tightly linked to lake level. Lake level falls by the net withdrawal of water by evaporation and percolation, hence sequence boundaries are marked by extensive indicators of subaerial exposure and minimal erosion. The high primary production of these lakes, low fluvial input, and paucity of terrestrial vegetation in the drainage basin result in production of rich type I algal organic matter, but ultimate preservation is relatively low due to frequent desiccation. Reservoir facies are best developed in transgressive sheetflood clastics, early highstand fluvial channels, and late highstand shoreline carbonate grainstones. Seal-prone facies are most widespread in upper transgressive and basal-highstand systems tract strata. Examples of this lake-basin type are listed in Table 6.

Another key element to recognize is that some lake basins become connected to the marine realm at highstand. This allows lake level to vary independently of sediment+water supply to the lake basin, resulting in a variety of stratal patterns that are controlled by the phase relations between sea level and sediment+water supply. (These relations vary by paleolatitude, thereby providing some predictive capability) (e.g., Perlmutter and Matthews, 1990). Intermittent marine connection can be beneficial for the accumulation of organic-rich rocks (Kelts, 1988; Mello and Maxwell, 1990; Higgs, 1991). Examples of marine-connected lake basins are listed in Table 7.

The practical application of recognizing this diversity of lake-basin types boils down to the necessity of using the sequence-stratigraphic approach to understand the lake system and to select appropriate models for predicting the distribution of play elements. A two-phased approach is most helpful: (1) Use the standard sequence-stratigraphic approach to make sense of all the data to recognize and correlate the hierarchy of surfaces (flooding surfaces/parasequence boundaries, channel bases, parasequence-set boundaries, downlap surfaces, and sequence boundaries) and (2) apply the understanding derived from constructing the sequence framework to select appropriate lake-basin type and sequence-stratigraphic models for making predictions away from control.

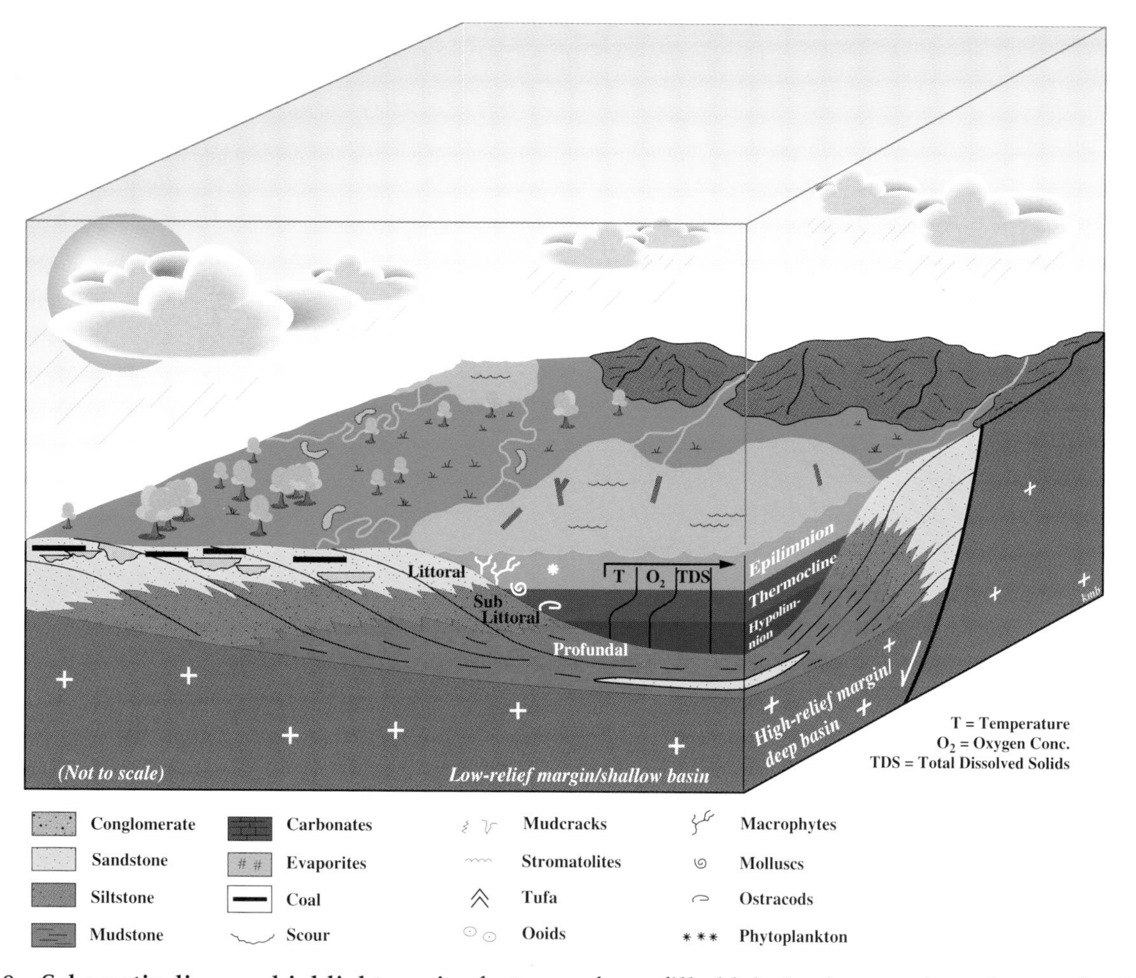

Figure 8—Schematic diagram highlights major features of overfilled lake basins: persistently open hydrology, freshwater lake chemistry, high groundwater table, progradational shoreline architecture, close relation to fluvial systems, and commonly interbedded fluvial deposits and coals. (Not shown are bioclastic and other freshwater carbonate lithosomes.) This lake-basin type occurs when the rate of supply of sediment+water consistently exceeds potential accommodation (usually when P/E is relatively high compared to rates of tectonic subsidence). Climatically driven lake-level fluctuations are minimal because water inflows are in equilibrium with outflows. (Left side of the diagram schematically represents shallow overfilled lake basin or low-relief margin; right side of the diagram schematically represents deep overfilled lake basin or high-relief margin.)

CONTROLS ON ORGANIC-RICH ROCK DEPOSITION

Lake-basin type also strongly influences production and accumulation of organic matter that forms source rocks as detailed in Table 8. Recognizing the lake type allows one to predict geochemical attributes from geologic data.

Organic enrichment in lacustrine rocks is a function of the same basic factors as for other environments, and can be expressed as a simple relation (e.g., Bohacs, 1990, 1998):

$$\text{Organic Enrichment} = \frac{\text{Production} - \text{Destruction}}{\text{Dilution}}$$

Optimum organic enrichment occurs where production is maximized and destruction and dilution are minimized. Any appropriate combination of these factors can produce potential source rocks.

Production refers chiefly to the photosynthetic fixation of CO_2 by organisms and can include both autochthonous organic matter derived from algae and aquatic plants and allochthonous organic matter transported from land (Figure 11). The primary production of organic matter in a lake is a function of many variables, including solar input, wind, precipitation, water chemistry, and temperature (Kelts, 1988). Of these, solar input and water chemistry have the largest effect on overall primary production (Katz, 1990).

Solar input controls the energy available for photosynthetic production. Production decreases with increasing latitude, area/depth ratio, and turbidity due to the decrease in duration and intensity of solar energy available for photosynthesis. In general, primary

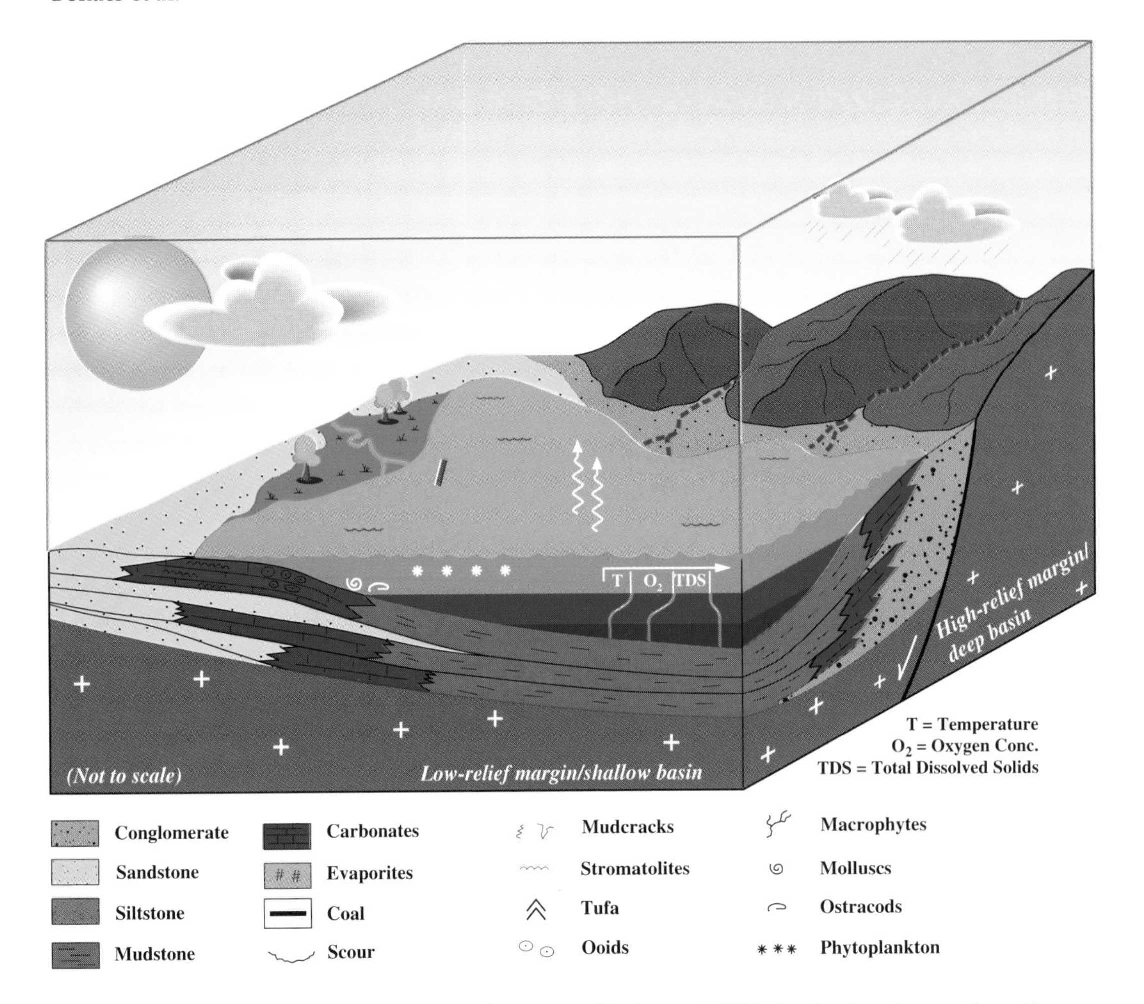

Figure 9—Schematic diagram highlights major features of balanced-fill lake basins: intermittently open hydrology, fluctuating groundwater table and lake water chemistry, interaction of thermal and chemical stratification, mixed progradational and aggradational shoreline architecture, and varied interbedding of clastic and carbonate strata. This lake-basin type occurs when the rates of sediment+water supply and accommodation are roughly in balance over the time span of sequence development. Water inflows are sufficient to periodically fill available accommodation, but are not always balanced by outflow; hence, climatically driven lake level fluctuations are common. (Left side of the diagram schematically represents shallow balanced-fill lake basin or low-relief margin; right side of the diagram schematically represents deep balanced-fill lake basin or high-relief margin.)

production in modern lakes peaks at a mean depth of about 18 m (Cole, 1979).

The water chemistry of a lake controls the availability of nutrients to support primary production of organic matter. Water chemistry is strongly influenced by basin hydrology and climate (Hardie and Eugster, 1970), whose integrated effects may be discerned through lake-basin type. The nutrients available to a lake are brought in by overland flow from the lake catchment area or by eolian transport, so production is also, in part, related to bedrock geology (e.g., Cerling, 1994). Most nutrients are recycled during mixing of lakes; permanently stratified lakes, in fact, may require an external nutrient load to support intense primary production (Katz, 1990). Alkaline waters help support a much higher level of primary production than waters of neutral pH because of the abundance of CO_3 ions available for incorporation by plants in addition to atmospheric CO_2 (Kelts, 1988). For these reasons, the highest natural primary productivities on earth occur in tropical alkaline lakes (Likens, 1975).

The autochthonous organic matter of a lake may be composed of benthic, higher plants, planktic and benthic algae, and bacteria (Figure 11). Size and water depth control the proportion of each of these contributions. Small, shallow overfilled lakes are commonly dominated by benthic higher plants, and the organic sediments tend to be peat rich (Cole, 1979). In contrast, large, deep balanced-fill lakes can have a relatively limited area of littoral benthic plant growth, and so the dominant organic matter deposited in the lake tends to be planktic and benthic algal and bacterial matter.

Allochthonous organic matter originates mostly from higher land plants. Deposition of allochthonous organic matter is greater in the parts of the lake that are proximal to deltas and fluvial input. Input may be seasonal or episodic, related to weather/climatic events in the basin. Allochthonous organic matter may

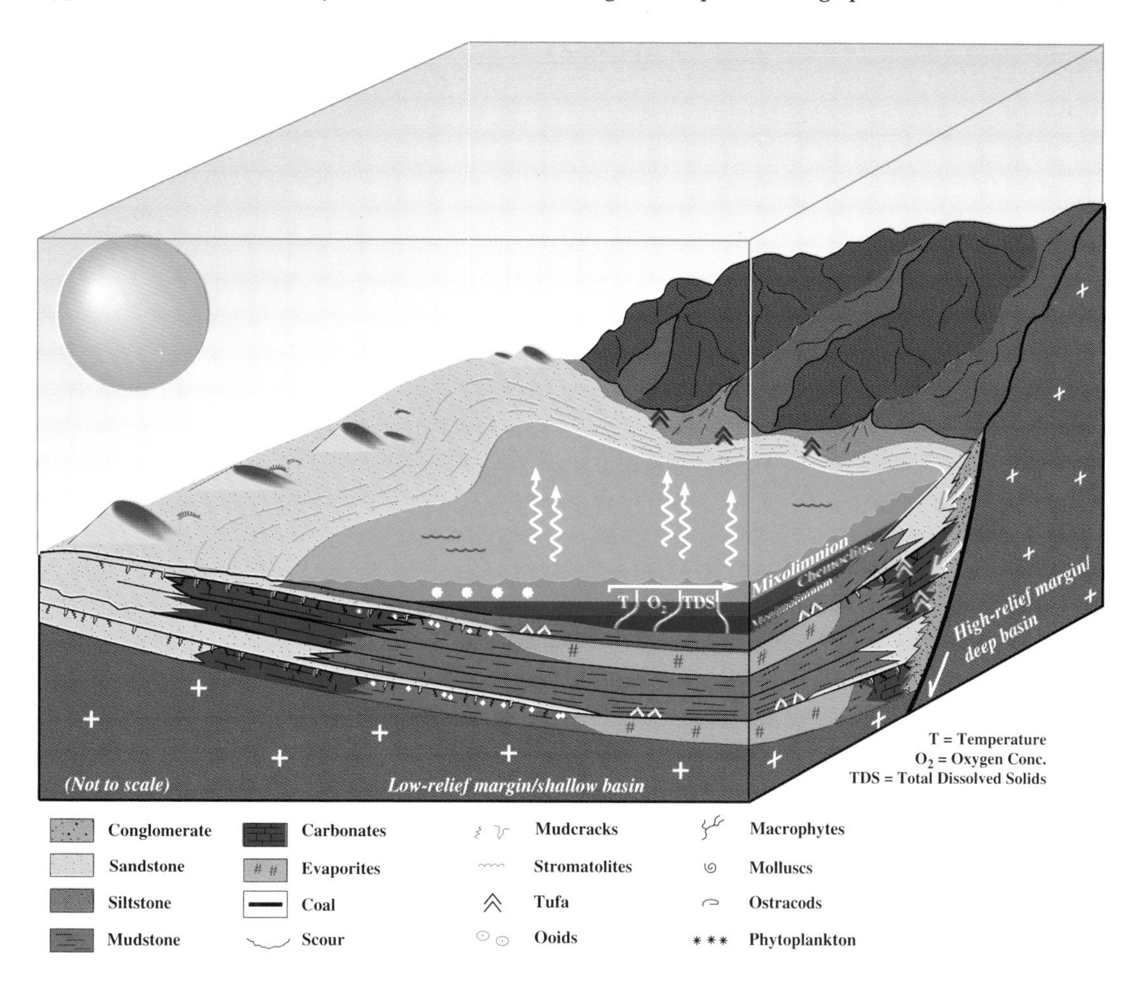

Figure 10—Schematic diagram highlights major features of underfilled lake basins: persistently closed hydrology, characteristic chemical stratification, low groundwater table, high solute content of lake waters, extensive desiccation features, highly contrasting lithologies, common association with evaporite deposits, and dominantly aggradational shoreline architecture. This lake-basin type occurs when rates of accommodation consistently outstrip available water and sediment supply, commonly resulting in a persistently closed basin with ephemeral lakes interspersed with playas or brine pools or both. (Left side of the diagram schematically represents shallow underfilled lake basin or low-relief margin; right side of the diagram schematically represents deep underfilled lake basin or high-relief margin.)

also be blown into the lake by wind, although this is generally a small contribution to the overall deposition of organic sediments in large lakes. Overfilled lake basins are the most likely to receive input of allochthonous organic matter due to the proliferation of forests in humid climates and to the existence of well-integrated fluvial systems (e.g., Kelts, 1988; Sladen, 1994).

For source potential, then, autochthonous aquatic organic matter is most important because its lipid-rich cell membranes usually form oil-prone kerogens. These membranes are sensitive to lake type because they are used to control cellular osmotic pressure in aqueous environments; biomarker distributions therefore reflect the prevailing water chemistry (e.g., Meyers and Ishiwatari, 1993; Peters and Moldowan, 1993). Allochthonous organic matter from land plants is, in general, more likely to create gas-prone kerogens.

Destruction of organic matter is primarily a function of the efficiency of various scavengers, particularly bacteria, which, in turn, is largely controlled by the availability of oxygen. The supply of oxygen to lake waters occurs via exchange with the atmosphere and as a byproduct of photosynthesis. In well-oxygenated water columns, most of the primary organic production is destroyed by microbial respiration, converting the organic matter back into CO_2 (Cole, 1979). Oxygen is depleted by biological and inorganic oxidation, and if the oxygen is not renewed, the waters will become anoxic, favoring preservation of organic matter. Anoxic bottom waters enhance organic preservation by limiting the activity of scavengers and bacterial respiration (e.g., Demaison and Moore, 1980; Kelts, 1988). The extent and duration of oxygen deficiency in a lake are controlled by the intensity and frequency of mixing, as well as by primary

Table 3. Examples of Shallow and Deep Lake-Basin Types.

Lake Basin Type	Formation	Basin, County	Age
Overfilled, Shallow	Modern	Sudd, Sudan	Pleistocene–Holocene
	Luman Tongue, GRF*	Washakie, U.S.A.	Eocene
	Quantou-3	Songliao, China	Albian
Overfilled, Deep	Modern	Lake Baikal	Pleistocene–Holocene
	Kissenda	N'Komi, Gabon	Berriasian
	Mako to Nagyaföld	Pannonian, Hungary	U. Miocene–M. Pliocene
Balanced Fill, Shallow	Modern	Lake Victoria	Pleistocene–Holocene
	Laney Mbr, GRF*	Washakie, U.S.A.	Eocene
	Lucaogou	Junggar, China	Permian
Balanced Fill, Deep	Modern	Lake Malawi	Pleistocene–Holocene
	Bucomazi	Cabinda/Gabon	Neocomian
	Candeias	Reconcavo, Brazil	Neocomian
Underfilled, Shallow	Modern	Dabusun Lake	Pleistocene–Holocene
	Wilkins Peak Mbr, GRF*	Bridger, U.S.A.	Eocene
	Blanca Lila	Pastos Grandes **	Pleistocene
	Lagoa Feia (upper)	Campos, Brazil	Aptian
Underfilled, Deep	Modern	Issyk-Kul Lake	Pleistocene–Holocene
	Lisan	Dead Sea, Israel	Pleistocene–Holocene

* GRF = Green River Formation.
** Argentina.

production of organic matter and water chemistry.

Mixing of lake waters is inhibited when a lake is thermally or chemically stratified, hence oxygen in the bottom waters of the lake cannot be renewed. Chemical stratification is common in balanced-fill and underfilled lake basins. It is especially well developed in deep balanced-fill lake basins and in marine-connected overfilled and balanced-fill lake basins. Intermittent input of denser sea water generates persistent ectogenic meromixis resulting in excellent organic preservation (e.g., Raisz, 1937; Mello and Hessel, 1998). Thermal stratification is practically essential for organic-rich rock accumulation in deep overfilled lake basins (e.g., Dean, 1981; Demaison and Moore, 1980). Very deep lakes, however, can have high rates of organic destruction due to the longer time required for particles to settle through deep, well-oxygenated surface waters (Katz, 1990, 1995).

Some lakes remain stratified because high contents of dissolved organic matter in bottom waters deplete the oxygen content (see discussion in Katz, 1990). These biologically stratified lakes are associated with high levels of organic production and currently occur most commonly in rift basins in tropical climates. They may start out as thermally stratified oligomictic lakes; however, the buildup of dissolved organic carbon in the bottom waters enhances the stability of stratification. Excessive primary production (hypertrophication) can also apparently overwhelm the delivery of oxygen to the lake bottom, with good organic preservation in relatively shallow lakes (e.g., Horsfield et al., 1994; Bohacs et al., 1996). These processes are most important in shallow balanced-fill lake basins.

Anaerobic degradation is strongly influenced by the concentration of sulfate in lake waters (e.g., Powell, 1986; Kelts, 1988). Sulfate content of lake waters is a function of both catchment area input and lake-basin concentration. Sulfate-reducing bacteria dominate anaerobic consumption in the presence of excess sulfate, a condition most common in marine-connected and some underfilled lake basins. Less efficient fermenters and methanogens prevail in lakes with low sulfate concentrations (more likely in overfilled and balanced-fill lake basins).

Dilution by mineral sediments can limit source rock richness by decreasing the proportion of organic matter relative to inorganic matrix. Organic matter in profundal sediments is diluted by mainly detrital sediments from rivers at deltas or from wave reworking of the shorelines and littoral zone. Deltas represent a large addition of sediment as bedload and suspended load to a lake. Organic-rich sediments form farther away from points where rivers enter a lake (Kelts, 1988; Sladen, 1994). In many elongate lakes in rift and wrench settings, drainage is axial to the basin, and rivers and deltas are formed at one or both ends of the lake. In these cases, deposition of organic-rich sediments is displaced away from the clastic-dominated ends of the lake (e.g., Crossley, 1984; Nichols 1987; Cohen, 1989). If organic concentration were low, the quantity of hydrocarbons (particularly oil) expelled during generation would be limited by retention on mineral surfaces (Sandvik et al., 1992). In extreme cases, organic matter can be completely overwhelmed by clastic sediment input.

Table 8 summarizes the effect of different lake types on each of these factors. Each lake type has characteristic ranges of total organic carbon contents (TOC) and hydrogen indices (HI) and associations of organic matter (noted on the right side of the table); a specific example from the Green River Formation is shown on Figure 12. The optimal lake for source rock deposition is a compromise among production, destruction, and dilution: it is shallow enough to have high primary production, deep enough to preserve a significant proportion of the resultant organic matter, and starved of sediment but not of nutrients. These conditions appear from the geological record of ancient lakes to coincide with balanced-fill lake basins, in which deposits commonly attain more than 20% TOC in oil-prone kerogens. For example, in the Washakie basin, the balanced-fill lower Laney Member has significantly richer potential source rocks than the other intervals of the Green River Formation (Luman Tongue, Wilkins Peak Member) (Figure 12).

Overfilled lake basins have good aquatic production and abundant land-plant input. The main challenges to organic enrichment are preservation and dilution. Good preservation requires thermal stratification because solute-controlled density stratification is unlikely to develop under persistently open hydrology (unless marine connection occurs). Dilution is common due to strong fluvial influence. Resultant organic enrichment is therefore most likely above flooding surfaces, especially around maximum flooding surfaces (mid-sequence downlap surfaces), in profundal strata, and within some intervals of lake-plain strata associated with mire and pond environments (Bohacs, 1998).

Balanced-fill lake basins experience optimal combinations of production, preservation, and dilution; production is boosted by intermittent fluvial input of nutrients subsequently concentrated by evaporation. Shallow balanced-fill lake basins have the largest portion of their volume in the photic zone for the longest portion of their history. This lake-basin type records the highest primary production levels (e.g., Horsfield et al., 1994; Sladen, 1994; Carroll, 1998). Preservation is enhanced by commonly developed stable chemical stratification and dilution is minimized because the maximum fluvial input rates occur on transgressions, which traps most clastics near shore. Organic enrichment is most likely above flooding surfaces at the parasequence scale and in the lower portion of highstand systems tracts (Horsfield et al., 1994; Bohacs, 1998).

Underfilled lake basins can have high primary production. Recent shallow saline lakes can have surficial sediments with TOCs over 6% (Burne and Ferguson, 1983; Boon et al., 1983). Clastic dilution tends to be minimal because most fine-grained clastics are trapped in the lower lake plain and nearshore environments at lake highstand. Peak organic enrichment occurs just above the initial transgressive surface (base of transgressive systems tract) (Bohacs, 1998). Long-term preservation can be extremely problematic because frequent and prolonged desiccation and exposure degrades most organic matter. In present-day solar lakes in Sinai, organic matter is degraded to a depth of at least 66 cm (surface TOC = 9.7% decreases to 3.0%) (Boon et al., 1983), which is about the thickness of typical underfilled lake-basin parasequences. For example, low preservation of organic matter prevails in Cambrian lake shales in south Australia (TOC <1.1%), and Jianghan basin in China (TOC <1%) (Powell, 1986; Carroll, 1998). Other underfilled lake basins preserve significant organic-rich rocks, such as the Wilkins Peak Member (Wyoming; TOC <19%, HI <1054 mg HC/g) (Bohacs, 1998) and Jingjingzigou Formation (Junggar basin; TOC < 6.6%, HI <794 mg HC/g) (Carroll, 1998).

INFLUENCE OF LAKE-BASIN TYPE ON HYDROCARBON TYPE

Lake-basin type influences hydrocarbon generation through both the type of organic matter produced and how it is preserved during burial. Overfilled lake deposits are likely to contain, on average, higher quantities of advected terrestrial organic matter, and therefore can be expected to generate relatively more gas than other lake types. In contrast, kerogens from balanced-fill and underfilled lake basins both appear to be principally oil-prone algal-bacterial material. Sulfur content of most lacustrine oils is low due to the low sulfur content of most lake waters and generally high availability of reactive iron associated with clays (Tissot and Welte, 1984). In some underfilled lake basins, however, the combination of extreme evaporative concentration and low influx of clastic sediments can result in the preservation of sulfur-rich kerogens, resulting in oils with up to 12% sulfur (Shi et al., 1982; Fu et al., 1985; Sheng et al., 1987).

Especially useful in tying physical and geochemical character are specific biomarker assemblages of oils and rock extracts that are associated with organic matter input (terrestrial, aquatic) and lacustrine depositional–preservational environments (solute species and concentrations). These associations can help to interpret lake-basin type from oils and to predict biomarker distributions from lake-basin type.

Biomarkers can be thought of as molecular fossils whose basic carbon skeleton is derived from once-living organisms. Biomarkers are found in petroleum and rock extracts, and provide information about depositional environment, thermal maturity, migration pathways, and hydrocarbon alteration (Peters and Moldowan, 1993). Biomarker data can be used to interpret the relative contributions of terrigenous and aquatic organic matter and the environmental conditions under which source rocks were deposited. These, in turn, can be related to lake-basin type (see below). The major diagnostic biomarker compounds used to distinguish lacustrine-sourced oils are tricyclic terpanes, β-carotane, gammacerane, and, to a lesser extent, the 4-methyl steranes.

Please note that most of the following biomarker distributions are from source rocks tied directly to lake-basin type. Oils, however, can show less distinction because they result from maturing an interval of source strata that can include several lake-basin types and mixing of several source facies types. Another

Table 4. Examples of Overfilled Lake Basins.

Formation	Basin, Country	Age	References
Anthrocosia shale	North Sudetic basin, Poland	Permian	Mastalerz, 1994
Hongyanchi Formation	Junggar basin, China	Permian	Carroll, 1998
Chinle Formation, Monitor Butte member	Colorado plateau, U.S.A.	Triassic	Dubiel, 1994
Waterfall Formation	Culpeper basin, Virginia, U.S.A.	Jurassic	Hentz, 1981; Gore, 1989
Mangara shale	Doseo basin, Chad	Cretaceous	Genik, 1993
Kissenda Formation	N'Komi basin, Gabon	Cretaceous	Kou, 1994
Pematang shale	Kutei subbasin, Sumatra, Indonesia	Eocene	Kelley et al., 1995
Luman Tongue & upper LaClede bed, Green River Formation	Green River basin, Wyoming, U.S.A.	Eocene	Horsfield et al., 1994
Terengganu shale	Malay Basin, Malaysia	Miocene	Madon-Mazlan, 1992; Creaney et al., 1994
Upper part of Snake River basin fill	Snake River basin, Idaho, U.S.A.	Holocene	Wood, 1994; Wood and Squires, 1998
Lake Baikal	Russia	Modern	Flower et al., 1995

Table 5. Examples of Balanced-fill Lake Basins.

Formation	Basin, Country	Age	References
Lucagou Formation	Junggar basin, China	Permian	Carroll, 1998
Lockatong Formation, lower members	Newark basin, New Jersey, U.S.A	Jurassic	Olsen et al., 1989
Alternance de Sokor Formation	Termit graben, Niger	Cretaceous	Genik, 1993
Qingshankou 1 Formation	Songliao basin, China	Cretaceous	Schwans et al., 1997
Lagoa Feia Formation, upper Jiquiá portion	Campos basin, Brazil	Cretaceous	Carneio de Castro et al., 1981, Bertani and Carozzi, 1985, Abrahão and Warme, 1990
Brown shale, middle portion	Aman trough, central Sumatra	Oligocene	Kelley et al., 1995
Green River Formation, upper Tipton and lower Laney members	Green River basin, Wyoming, U.S.A.	Eocene	Surdam et al., 1980; Horsfield et al., 1994
Middle part of Snake River basin fill	Snake River basin, Idaho, U.S.A.	Pleistocene	Wood, 1994
Units C and D	Ribesalbes basin, Spain	Miocene	Anadón, 1994
Lakes Malawi and Victoria	East Africa rift system	Pleistocene–Holocene	Scholz and Rosendahl, 1990; Scholz , 1995

Table 6. Examples of Underfilled Lake Basins.

Formation	Basin, Country	Age	References
Jingjingzigou Formation	Junggar basin, China	Permian	Carroll, 1998
Passaic Formation	Newark basin, New Jersey, U.S.A.	Triassic	Olsen et al., 1989
Balls Bluff Siltstone	Culpeper basin, Virginia, U.S.A.	Triassic	Gore, 1989
East Berlin Formation	Hartford basin, Connecticut, U.S.A.	Jurassic	Gierlowski-Kordesch and Rust, 1994
Lagoa Feia Formation, Alagoas portion	Campos basin, Brazil	Cretaceous	Mello and Maxwell, 1990
Argille de Sokor Formation	Termit Graben, Niger	Cretaceous	Genik, 1993
Shahejie 4	Bohai basin, China	Oligocene	Zhou, 1981; Hu et al., 1989; Chang, 1991; Remy et al., 1995
Green River Formation, Wilkins Peak member	Green River basin, Wyoming, U.S.A.	Eocene	Smoot, 1983
Lake Bogoria	Kenyan rift	Pleistocene–Holocene	Renaut and Tiercelin, 1994
Dabusun Lake	Qaidam basin, China	Modern	Yang et al., 1995

Table 7. Examples of Marine-Connected Lake Basins.

Formation	Basin, Country	Age	References
Bude Formation	Westphalian basin, England	U. Carboniferous	Higgs, 1991
Lagoa Feia Formation, Alagoas portion	Campos basin, Brazil	Cretaceous	Mello and Hessel, 1998
Coquiero Seco Formation, Moro do Chaves member	Sergipe-Alagoas basin, Brazil	Cretaceous	deAzambuja Fihlo et al., 1997
Oligocene lake strata of Daban Basin	Daban Basin, Sudan	Oligocene	Sagri, et al., 1994
Coorong Lakes	Younghusband peninsula, South Australia	Holocene	Warren, 1994
Lake Maracaibo	Maracaibo basin, Venezuela	Modern	Redfield, 1958

complicating factor is that hydrocarbons in reservoirs are subject to postemplacement alteration. The numerical values in this section are an attempt to summarize representative ranges of oil properties as originally emplaced with data culled from published literature.

Much of the biomarker signature is due to the two general categories of primary producers (autotrophs) of organic matter in lakes: prokaryotes (blue-green algae and bacteria) and eukaryotes (higher plants, algae). The majority of triterpanes are associated with prokaryotic sources, whereas steranes are produced by eukaryotic organisms. Thus, the triterpane/sterane ratio is a rough measure of the prokaryote/eukaryote contribution to the organic material. As salinity increases in a lake one expects that more sensitive eukaryotic organisms (e.g., sterane-producing green algae) would give way to the more tolerant bacteria or cyanobacteria (tricyclic and hopane producing) with a corresponding increase in the triterpane/sterane ratio. Possible exceptions could occur with an

Table 8. Controls on Organic-rich Rock Development in Lakes.*

Lake Type	Production	Destruction	Dilution	Source Potential
Overfilled	+ Nutrient input increased – Fresh water input dilutes nutrients – Overall production decreases with increasing lake volume	– Increased oxygen supply to bottom – Homogeneous water mass makes wind mixing more effective – Cold underflow – Increased turbulence	– Abundant clastic detritus ± Abundant advected terrigenous clastics	• Moderate to poor oil/gas • Mixed gas/oil • Marked lateral variability TOC: <1–7% (muds) < 80% (coals) OMT: mixed algal/terrigenous (I/II) HI: 50–600 mg HC/g Relatively thick (< tens of meters)
Balanced fill	+ Appreciable nutrient input + Nutrients concentrated by episodic drying + Larger percent of lake volume in photic zone	+ Closed basin and episodic drying promotes density stratification + Large amount of production consumes oxygen at bottom	+ Varying, but relatively minor clastic detritus + Minor component of advected terrigenous organic matter – Episodic floods or flashy discharge may deliver significant clastic debris	• Moderate to excellent oil • Mostly oil, some gas? • Little lateral variation TOC: 1–30% OMT: mostly algal (I), some terrigenous (II) HI: 500–700 mg HC/g Relatively thin (1–10m)
Underfilled	± Variable nutrient input + Nutrients concentrated by episodic drying – Extreme concentration of solutes kills organisms – Water available for production only part of time	- Episodic drying oxidizes organic matter - Episodic freshening introduces oxygen, consumers	– Semi-arid climates yield highest clastic input + Minimal input of terrigenous organic matter + Significant amount of fill due to precipitated minerals	• Poor to excellent oil • Mostly oil • Minimal lateral variation TOC: <0.5–20% OMT: Algal (Type I), HI: 650–1150 mg HC/g Relatively thin (meters)

* + Positive for organic enrichment, – Negative for organic enrichment, ± Variable influence on organic enrichment, TOC = total organic carbon, OMT = organic matter type, HI = hydrogen index.

increased input of a limiting nutrient such as nitrogen, or in a shallow balanced-fill to underfilled lake that is highly stratified, with a fresh surface layer and a saline bottom layer. In the latter case, both freshwater organisms and salinity-tolerant organisms could coexist in the photic zone with little change in the triterpane/sterane ratio. The other major component of the biomarker fraction comes from reworking of primary organic matter (photosynthetically generated) by heterotrophs, bacteria whose signature are recorded mostly in hopanes.

Stressed environments (high alkalinity/salinity, typical of underfilled and balanced-fill lake basins) are characterized by tricyclics (C_{20}–C_{24}; m/z 191), β-carotane (a C_{40} compound; m/z 125), and gammacerane (a C_{30} triterpane; m/z 191), all with prokaryote sources. Gammacerane and β-carotane are specifically associated with nonmarine, highly saline environments (Peters and Moldowan, 1993). Gammacerane is thought to be derived from protozoa (ciliates) or bacteria or both. Gammacerane is derived from tetrahymanol, which is produced by protozoa in an anoxic environment; however, if free sterols are available in the anoxic environment, protozoa will assimilate them and not synthesize the tetrahymanol precursor of gammacerane (Tissot and Welte, 1984). In addition, if the environment is oxic (to possibly dysoxic), then the protozoa can generate their own sterols to rigidify their cell membranes. The implications are that if a source rock has a low triterpane/sterane ratio (i.e., possibility of free sterols), then gammacerane may not be abundant even though the environment could be both saline and anoxic. In this case one needs to look for other indicators of salinity (β-carotane) and anoxia (C_{35}/C_{34} triterpane ratio). As ever, one needs to place all geochemical data into geological context for most robust interpretations.

Most dominantly overfilled lake basins contain a mix of terrigenous and aquatic organic matter deposited in freshwater, suboxic to oxic conditions (Table 8) (Carroll and Bohacs, 1995, 1999). The resulting hydrocarbons are typically oil plus some gas and gas condensate (e.g., Pannonian, Eromanga basins) (Clayton et al., 1994; Powell, 1986). The oils are paraffinic with large contents of high-molecular-weight n-alkanes (waxes >20%) mainly generated from cuticular tissues of vascular plants and membrane lipids from some freshwater algae (Tissot and Welte, 1984; Goth et al., 1988; Tegelaar et al., 1989). API gravities range widely (24–57°API), as do pour points (–5 to >20°C) mainly as a function of gas/oil ratio (GOR) (Powell, 1986; Clayton et al., 1994; Telnaes et al., 1992; Kulke, 1995). Sulfur contents are low and NSO contents are low. The molecular character is dominated by terrigenous organic matter input and relatively oxic preservational conditions: pristane/phytane (Pr/Ph) and hopane/sterane ratios are high and C_{29} desmethyl steranes are relatively abundant (Powell, 1986; Isaksen, 1991; Carroll, 1998). Some overfilled lake systems contain a predominance of 4-methyl steranes (e.g., Brassel et al., 1986). Gammacerane contents are low and β-carotane is usually not detected. Tricylic indices range from 12 to 100 (Carroll, 1998). If an overfilled lake is not in an appropriate climate and deep enough to maintain thermal stratification and anoxic bottom waters, the resulting organic matter generally will have a low hydrogen content and be completely swamped by terrigenous organic matter. In the extreme case, the organic matter is largely terrigenous with characteristics similar to coals or coaly shales

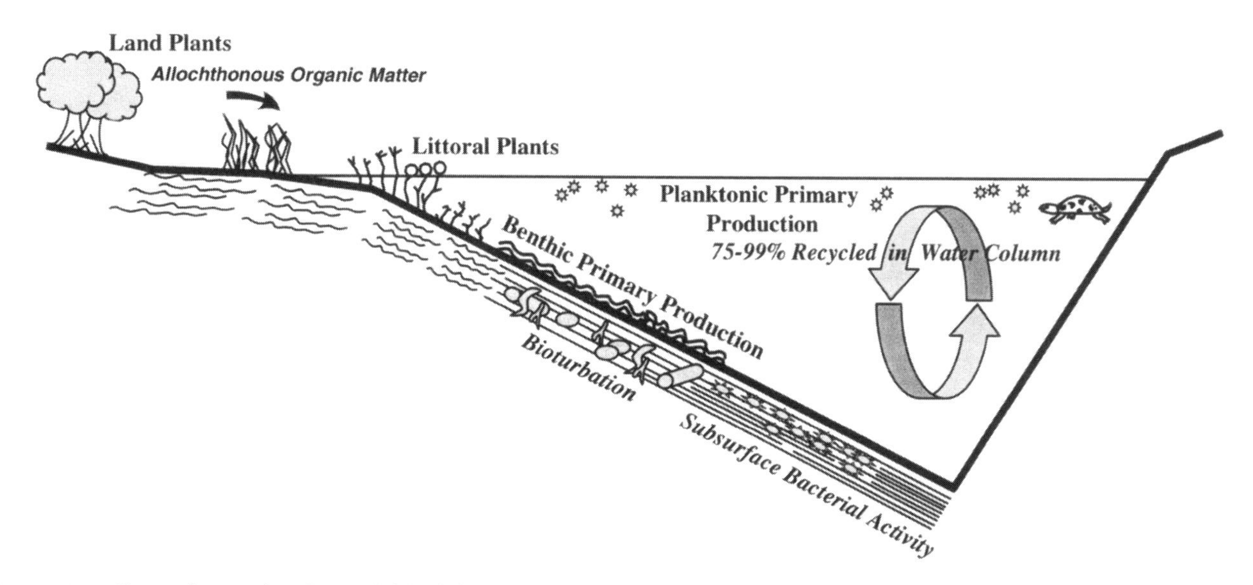

Figure 11—Organic production within lakes is chiefly due to photosynthetic fixation of CO_2 by organisms and includes both autochthonous organic matter derived from algae and aquatic plants and allochthonous organic matter transported from land. The primary production of organic matter in a lake is a function of many variables, including climate, solar input, wind, precipitation, water chemistry, and temperature. Of these, solar input and water chemistry have the largest effect on overall primary production.

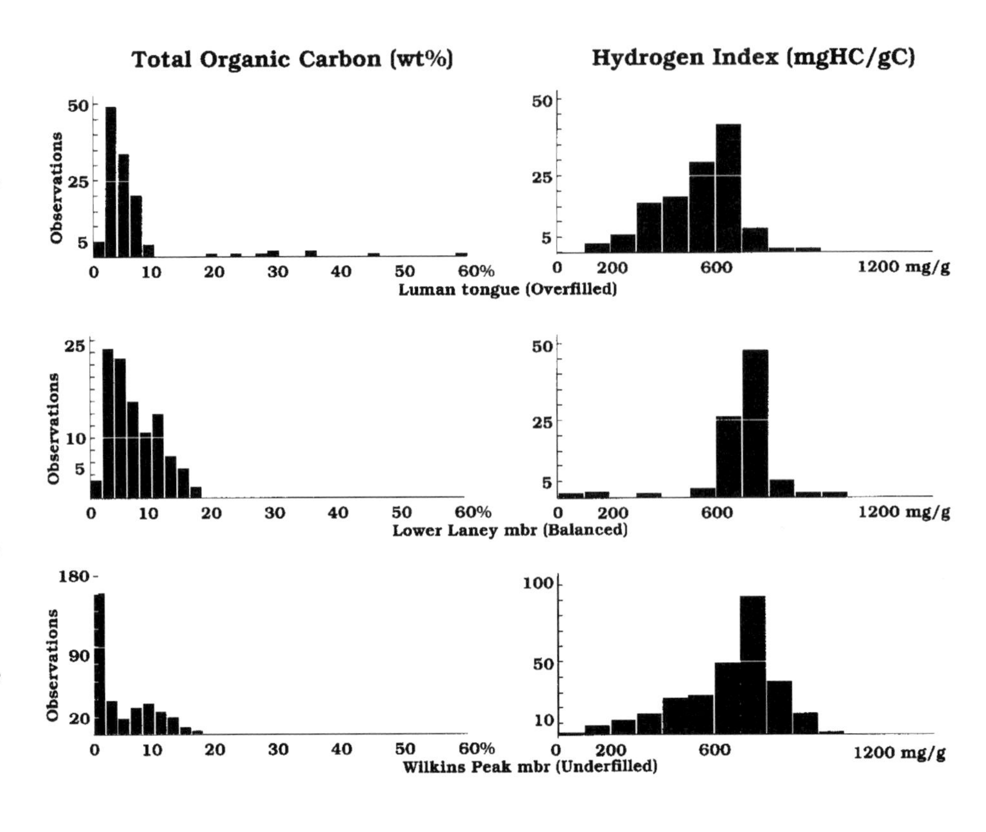

Figure 12—Each lake-basin type has characteristic ranges of total organic carbon and hydrogen indices and associations of organic matter, illustrated here with examples from the Eocene Green River Formation of Wyoming. The optimal lake for source rock deposition is a compromise between production, destruction, and dilution; it is shallow enough to have high primary production, deep enough to preserve a significant proportion of the resultant organic matter, and starved of sediment, but not of nutrients. These conditions are most common in balanced-fill lake basins, here shown that the balanced-fill lower Laney Member has significantly richer potential source rocks than the other intervals of the Green River Formation.

(e.g., Eromanga basin). In these cases, typical of permanently overfilled lake basins, Pr/Ph is even higher (>3), the proportion of isoprenoids to n-alkanes is high, and diterpenoids (from higher plants) are common (Powell, 1984).

Both deep and shallow balanced-fill lake basins can have significant hydrocarbon generation potential. Organic matter in both is dominated by aquatic organisms, with some land-plant material. TOC and HI can be high (≤27%) (Horsfield et al., 1994). Hydrocarbons are dominantly oil with low GORs (e.g., 40.4–48.5 m^3/t, Songliao basin) (Yang, 1985). Oils are typically paraffinic and rich in n-alkanes from membrane lipids of aquatic organisms. Wax contents range from 5 to >25% (e.g., Daqing field, Songliao basin), API gravities from 18° to 45°API, and pour points from 25° to 59°C (mostly >35°C), with surface-condition viscosities from 6 to 24 cp (Tissot et al., 1978; Yang, 1985; Powell, 1986; Schull, 1988; Kulke, 1995). Sulfur contents are uniformly low, although NSO+aromatic contents can range from 5 to 40% (Tissot and Welte, 1984; Yang et al., 1985). Their molecular character is dominated by compounds derived from aquatic organisms and their microbial degradation, commonly under suboxic to oxic depositional conditions (e.g., Powell, 1986). Pr/Ph ratios range from <1 to 2.0, reflecting persistent anoxia and moderate to low terrigenous input. Also in contrast with freshwater overfilled lake systems, CPI (odd-carbon number compound preference) is lower and n-C_{30}+ compounds are less abundant. Enhanced algal procaryote input is recorded in moderate amounts of β-carotane and increased amounts of tricyclic terpanes. Gammacerane is typically elevated due either to overall higher salinities (e.g., Mello et al., 1988) or well-developed chemical stratification (e.g., Schoell et al., 1994; Sinninghe Damsté et al., 1995). C_{29} steranes are most common, tricyclic indices range from 80 to 200, and hopane/sterane ratios range from 0.5 to 15. Dinosteranes, along with other 4-methyl steranes, dominate biomarker distributions of several shallow balanced-fill lake systems, consistent with preponderant aquatic organic matter input in salinity stratified lakes with anoxic bottom waters (Horsfield et al., 1994; Carroll, 1998).

Underfilled lake basins have variable overall hydrocarbon generation potential, although primary production can be very high, preservation is problematic. For example, the potential of the underfilled Wilkins Peak Member (mean HI ~ 730 mg HC/g) is much greater than that observed in the Jingjingzigou Formation of the Junggar basin (mean HI ~ 351 mg HC/g) (Grabowski and Bohacs, 1996; Carroll, 1998), but their detailed petroleum character is generally similar. Hydrocarbons are typically oils with minimal associated gas (Tissot and Welte, 1984; Powell, 1986). Oils can have significant asphaltic and aromatic contents and wax contents are relatively low (<5–10%) (Yang et al., 1985; Powell, 1986). API gravities range from 12 to 37°API, pour points from –5 to 23°C, and GORs from 10.6 to 149 m^3/t (Chen-yong Chang, 1991; Kulke, 1995). Sulfur contents are higher than in other lake-basin types, typically around 1%, but ranging as high as 12% (Shi et al., 1982; Fu et al., 1985; Sheng et al., 1987). The molecular character of

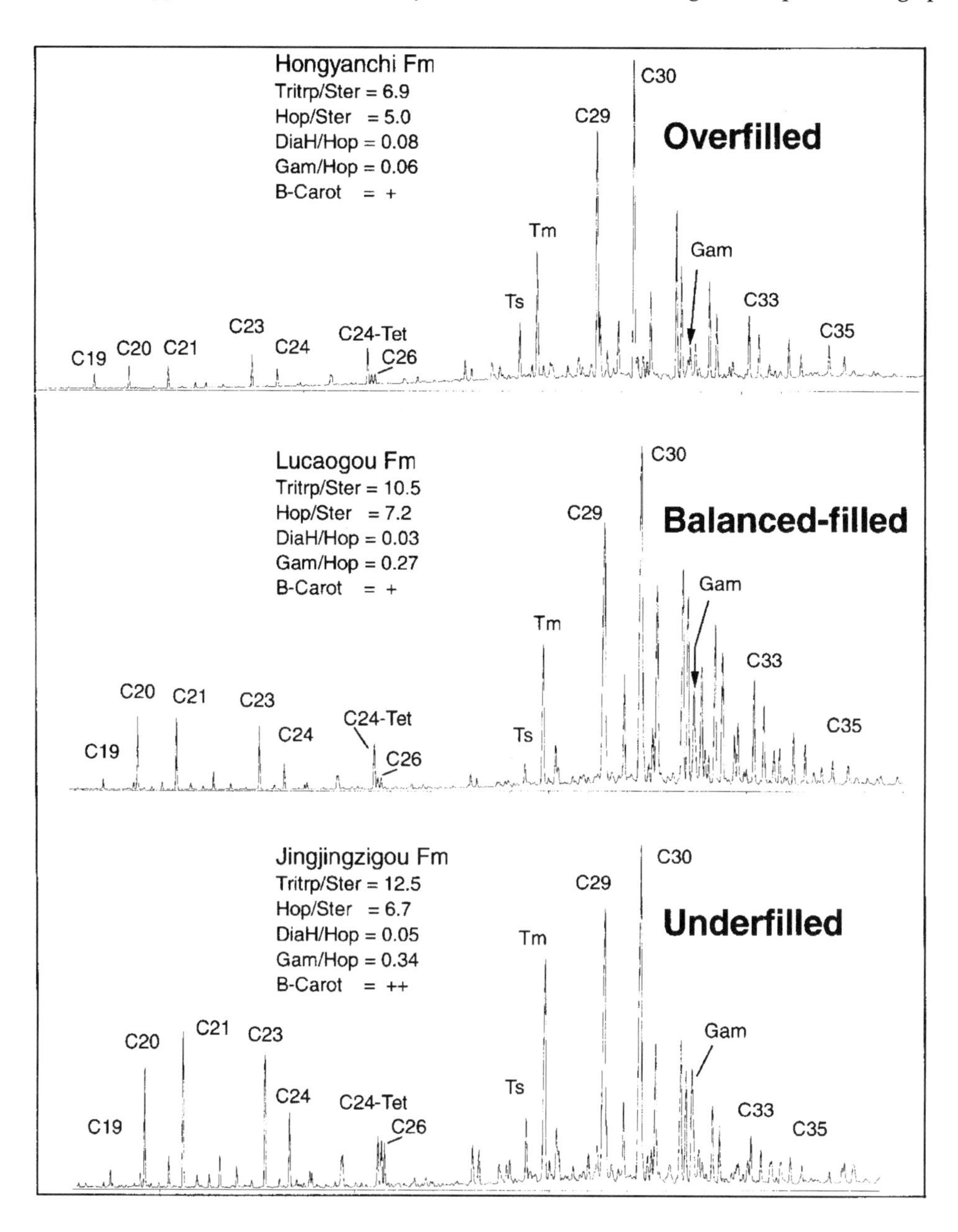

Figure 13—Examples of variations in biomarker distribution among lake-basin types from a single basin. In the Junggar basin (Upper Permian) the three lake types are represented by the Hongyanchi Formation (overfilled), the Lucaogou Formation (balanced-fill), and the Jingjingzigou Formation (underfilled). The triterpane chromatograms (m/z 191) show an increase in the tricyclics (C_{20}–C_{25}) and gammacerane from the overfilled lake to the underfilled lake. Also note that the C_{24} tetracyclic triterpane is more abundant relative to the adjacent C_{26} tricyclics in the overfilled and balanced-fill lakes in comparison to the underfilled lake.

the oils is derived from low-diversity assemblages of organisms highly specialized for rapidly changing and often hypersaline conditions. Pr/Ph ratios are typically <1, probably due to distinctive bacterial contributions, as well as persistent anoxic depositional conditions. Commonly β-carotane is dominant, and concentrations of gammacerane and tricyclic terpanes are elevated (Powell, 1986). Hopane/sterane ratios are low, consistent with a restricted diversity of producers and excellent syndepositional preservational conditions.

Figure 13 demonstrates these variations in organic-matter character. In the Upper Permian of the Junggar basin the three lake types are represented by the Hongyanchi Formation (overfilled), the Lucaogou Formation (balance-fill), and the Jingjingzigou Formation (underfilled). The triterpane chromatograms (m/z 191) show an increase in the tricyclics (C_{20}–C_{25}) and gammacerane from the overfilled lake to the underfilled lake. Also note that the C_{24} tetracyclic triterpane is more abundant relative to the adjacent C_{26} tricyclics in the overfilled and balanced-fill lakes in comparison to the underfilled lake. Similarly, the GC (gas chromatogram) traces of the saturate fraction show the underfilled lake basin extracts to contain elevated β-carotane and a Pr/Ph ratio of <1, in contrast to the overfilled lake basin extracts, which show little or no β-carotane and Pr/Ph ratios of >3. The best source potential for this series is the balanced-fill (mean HI = 693 mg HC/g) followed by the underfilled (mean HI = 351 mg HC/g). The overfilled Hongyanchi Formation is essentially a nonsource for liquids. Similar trends between lake types are observed in the Washakie basin of the United States, except that the overall oil potential of the underfilled lake basin is better. (Remember, however, oils may demonstrate less distinction because they result from maturing an interval of source strata that can include several lake types and consequent mixing of several source facies types.)

SUMMARY AND CONCLUSIONS

So, where are we in our understanding of lacustrine systems? Studies over the last century have pretty well boxed the compass of variations of lake behavior and stratal records, explored the wide range of variation, and seen enough examples to begin to see essential common elements. We stand on the shoulders of people who have done a giant amount of work (e.g., Lyell, 1830, 1847; Livingston, 1865; Gilbert, 1890; Bradley, 1929, 1964; Picard and High, 1972; Manspeizer and Olsen, 1981). We can start synthesizing all these observations now because we also have the large-scale 3-D perspective of reflection seismic data and the framework of sequence stratigraphy with which to integrate all the disparate data on physical, biological, and chemical attributes from both modern and ancient examples. We found it essential to look through the geologic/stratigraphic filter of preserved strata to sort out and reconcile the essentials of process, response, and record aspects of lake systems. This approach was necessary because modern process studies, so helpful in marine systems, have yet to yield broadly successful models for ancient strata. This makes sense, because lakes are not just small ocean systems. Their much smaller volumes make them more dynamic and sensitive to changing boundary conditions, and potentially subject to closely linked processes of accommodation creation and fill. These many and complex interactions and feedbacks should not be reduced to consideration of just one variable such as climate.

By starting with the rock record we are able to focus on what processes at what time scales are most important to generating preserved strata. We herein have presented a short synthesis of the range of lake systems and strata that addresses the essential controls on the occurrence, distribution, and character of lake strata and hydrocarbon play elements—source, reservoir, and seal. We proposed a lake-basin type classification based on observations of lake strata of many ages worldwide that focuses on the major controls on preserved strata. The interaction of the rates of sediment+water supply (mostly climatic) with potential accommodation (mostly tectonic) that mainly controls lake occurrence, the distribution of their strata, and the character of their inorganic and organic components. We presented these lake-basin types within a sequence-stratigraphic framework that facilitates the integration of all aspects of the depositional system: physical, chemical, and biological. This approach allows one to understand the nature of lake strata and their hydrocarbon potential from data at all scales, i.e., from basin architecture, through seismic, well log, outcrop/core, to molecules and isotopes. As ever, this framework works best when informed by regional data and an appreciation for local and areal variations.

ACKNOWLEDGMENTS

Our understanding of lakes has benefited from discussions and cooperative projects with L. Cabrera, Y. Y. Chen, K. S. Glaser, G. J. Grabowski, N. B. Harris, N. C. de Azambujo Filho, L. Magnavito, M. Marzo, K. Miskell-Gerhardt, P. E. Olsen, D. J. Reynolds, C. A. Scholz, and K. O. Stanley. We thank R. P. Steinen (University of Connecticut) for introducing KMB to the wonderful world of lake deposits, insightful interactions, and for access to the Park River Tunnel Project cores. We are grateful to Exxon Production Research Company for permission to publish. We also thank our two anonymous reviewers for their very thorough and helpful comments and E. Gierlowski-Kordesch for her editorial encouragement and patience. Alan Carroll thanks the donors of the Petroleum Research Fund, American Chemical Society for financial support.

REFERENCES CITED

Abrahão, D., and J. E. Warme, 1990, Lacustrine and associated deposits in a rifted continental margin—Lower Cretaceous Lagoa Feia Formation, Campos basin, offshore Brazil, *in* B. J. Katz, ed., Lacustrine Basin Exploration—Case Studies and Modern Analogues: American Association of Petroleum Geologists Memoir 50, p. 287-306.

Anadón, P., 1994, The Miocene Ribesalbes basin (eastern Spain) *in* E. Gierlowski-Kordesch and K. Kelts, eds., Global Geological Record of Lake Basins, Vol. 1: Cambridge, Cambridge University Press, p. 315-318.

Anadon, P., L. Cabrera, and K. Kelts, eds., 1991, Lacustrine Facies Analysis: International Association of Sedimentologists Special Publication No. 13: Oxford, Blackwell, 327 p.

de Azambujo Filho, Nilo C., F. E. G. da Cruz, and S. C. Hook, 1997, Guidebook to the rift-drift Sergipe-Alagoas passive margin basin, Brazil. AAPG-ABGP Hedberg Research Symposium on Petroleum Systems of the South Atlantic Margins, Rio de Janeiro, 16-19 November 1997, 93 pages.

Bertani, R. T., and A. V. Carozzi, 1985, Lagoa Feia Formation (Lower Cretaceous), Campos basin, offshore Brazil: rift valley stage lacustrine carbonate reservoirs—I: Journal of Petroleum Geology, v. 8, p. 37-58.

Bohacs, K. M., 1995, Mudrocks from marine to non marine; contrasting expressions of depositional sequences: Abstracts with Programs, Geological Society of America, v. 27, p. 399.

Bohacs, K. M., B. Horsfield, D. J. Curry, A. R. Carroll, 1996, Control of lake type on physical and geochemical nature of depositional sequence development; examples from the Green River Formation, Wyoming: Annual Meeting Abstracts, AAPG and SEPM, v. 5, p. 17.

Bohacs, K. M., 1990, Sequence stratigraphy of the Monterey Formation, Santa Barbara County: Integration of physical, chemical, and biofacies data from outcrop and subsurface: SEPM Core Workshop No. 14, San Francisco, California, p. 139–201.

Bohacs, K. M., 1998, Contrasting expressions of depositional sequences in mudrocks from marine to non marine environs, *in* J. Schieber, W. Zimmerle, and P. Sethi, eds., Mudstones and Shales, vol. 1, Characteristics at the basin scale: Stuttgart, Schweizerbart'sche Verlagsbuchhandlung, p. 32-77.

Boon, J. J., H. Hines, A. L. Burlingame, J. Klok, W. I. C. Rijpstra, J. W. DeLeeuw, K. E. Edmunds, and G. Eglington, 1983, Organic geochemical studies of Solar Lake cyanobacterial mats, *in* M. Bjorøy et al., eds., Advances in Organic Geochemistry, 1981: London, Wiley, p. 207-227.

Bracken, B. R., 1994, Syn-rift lacustrine beach and deltaic sandstone reservoirs- pre-salt (Lower Cretaceous) of Cabinda, Angola, west Africa, *in* A.J. Lomando, B. C. Schrieber, and P. M. Harris, eds., Lacustrine Reservoirs and Depositional Systems: SEPM Core Workshop No. 19, p. 173-200.

Bradley, W. H., 1929a, The varves and climate of the Green River epoch: U.S. Geological Survey Professional Paper 158-E, p. 87-110.

Bradley, W. H., 1931, Origin and microfossils of the oil shale of the Green River Formation of Colorado and Utah: U.S. Geological Survey Professional Paper 1931, 58 p.

Bradley, W. H., 1964, Geology of Green River formation and associated Eocene rocks in southwestern Wyoming and adjacent parts of Colorado and Utah: U.S. Geological Survey Professional Paper 1964, p. A1-A86.

Brassell, S. C., G. Eglinton, and J. Fu, 1986, Biological marker compounds as indicators of the depositional history of the Maomin oil shale: Organic Geochemistry, v. 10, p. 927-941.

Burne, R. V., and J. Ferguson, 1983, Contrasting marginal sediments of a seasonally flooded saline lake, Lake Eliza, South Australia; significance for oil shale genesis: BMR Journal of Australian Geology and Geophysics, v. 8, p. 99-108.

Calhoun, D. R., ed., 1999, 1999 Britannica book of the year, Chicago, Illinois, Encyclopedia Britannica, 920 p.

Carneio de Castro, J., N. C. de Azambuja Filho, and A. A. P. G. Xavier, 1981, Facies e analise estratigrafica da formacao Lagoa Feia, Cretaceo Inferior da bacia de Campos, Brazil: VIII Congreso geológico Argentino, San Luis (20-26 Setiembre, 1981) Actas II, p. 567-576.

Carroll, A. R. and K. M. Bohacs, 1997, Lacustrine source quality and distribution; lake type controls on hydrocarbon generation: Annual Meeting Abstracts, AAPG and SEPM, v. 6, p. 18.

Carroll, A. R., 1998, Upper Permian lacustrine organic facies evolution, southern Junggar basin, NW China: Organic Geochemistry v. 28, p. 649-667.

Carroll, A. R., and K. M. Bohacs, 1995, A Stratigraphic Classification of Lake Types and Hydrocarbon Source Potential: Balancing Climatic and Tectonic Controls: First International Limno-geological Congress, Geological Institute, University of Copenhagen, Denmark, August 21-25th, 1995, p. 18-19.

Carroll, A. R., and K. M. Bohacs 1999, Stratigraphic classification of ancient lakes: balancing tectonic and climatic controls: Geology v. 27, p. 99–102.

Cerling, T. E., 1994, Lake Turkana and its precursors in the Turkana Basin, East Africa (Kenya and Ethiopia), *in* E. Gierlowski-Kordesch and K. Kelts, eds., Global Geological Record of Lake Basins, Vol. 1: Cambridge University Press, p. 341-343.

Cheng-yong Chang, 1991, Geological characteristics and distribution patterns of hydrocarbon deposits in the Bohai Bay basin, east China: Marine and Petroleum Geology v. 8, p.98-106.

Clayton, J. L., C. W. Spencer, and I. Koncz, 1994, Tótkomlós-Szolnok (.) Petroleum System of southeastern Hungary, *in* L. B. Magoon and W. G. Dow, eds., The petroleum system—from source to trap: AAPG Memoir 60, p. 587-598.

Cohen, A. S., 1989, Facies relationships and sedimentation in large rift lakes and implications for hydrocarbon exploration: examples from Lake Turkana and Tanganyika: Palaeogeography, Palaeoclimatology, Palaeoecology, v. 70, p. 65-80.

Cole, G. A., 1979, Textbook of Limnology (2nd ed.): St. Louis, C.V. Mosby, 426 pp.

Creaney, S., A. H. Hussein, D. J. Curry, K. M. Bohacs, and R. Hassan, 1994, Source facies and oil families of the Malay Basin, Malaysia: AAPG Bulletin, v. 78, p. 1139.

Crossley, R., 1984, Controls of sedimentation in the Malawi rift valley, central Africa: Sedimentary Geology, v. 40, p. 33-50.

Currey, D. R., and C. G. Oviatt, 1985, Durations, average rates, and probable causes of Lake Bonneville expansion, still-stands, and contractions during the last deep-lake cycle, 32,000 to 10,000 years ago, *in* P. A. Kay and H. F. Diaz, Problems of and Prospects for Predicting Great Salt Lake Levels—Proceedings of a NOAA Conference held March 26-28, 1985: Salt Lake City, Center for Public Affairs and Administration, University of Utah, p. 9-24.

Dean, W. E., 1981, Carbonate minerals and organic mater in sediments of modern north temperate hard water lakes, *in* E. G. Ethridge and R. M. Flores, eds., Recent and Ancient nonmarine depositional environments: models for exploration: SEPM Special Publication 31, p. 213-232.

Demaison, G. J., and G. T. Moore, 1980, Anoxic environments and oil source bed genesis: AAPG Bulletin, vol. 64, p. 1179-1209.

Dubiel, R. F., 1994, Lacustrine deposits of the Upper Triassic Chinle Formation, Colorado Plateau, USA, *in* E. Gierlowski-Kordesch and K. Kelts, eds., Global Geological Record of Lake Basins, v. 1: Cambridge University Press, p. 151-154.

Einsele, G., 1992, Sedimentary Basins: Evolutions, Facies, and Sediment Budget: Berlin, Springer-Verlag, 628 p.

Flower, R. J., A. W. Mackay, N. L. Rose, J. L. Boyle, J. A. Dearing, P. G. Appleby, A. E. Kuzmina, L. Z. Granina, 1995, Sedimentary records of Recent environmental change in Lake Baikal, Siberia: The Holocene, v. 5, p. 323-327.

Fu, J., G. Sheng, and J. Jiang, 1985, Immature oil from a saline deposit-bearing basin, Shiyou Yu Tianranqi

Dizhi: Oil and Gas Geology, v. 6, p. 150-158.
Genik, G. J., 1993, Petroleum geology of Cretaceous-Tertiary rift basins in Niger, Chad, and Central African Republic: AAPG Bulletin v. 77, p. 1405-1434.
Gierlowski-Kordesch, E., and B. R. Rust, 1994, The Jurassic East Berlin Formation, Hartford basin, Newark Supergroup (Connecticut and Massachusetts): a saline lake-playa-alluvial plain system *in* R. W. Renaut and W. M. Last, Sedimentology and geochemistry of modern and ancient saline lakes: SEPM Special Publication 50, p. 249-265.
Gierlowski-Kordesch, E., and K. Kelts, 1994, Global Geological Record of Lake Basins, v. 1: Cambridge, Cambridge University Press.
Gilbert, G. K., 1890, Lake Bonneville: U.S. Geological Survey Monograph 1, 427 p.
Glenn, C., and K. Kelts, 1991, Sedimentary rhythms in lake deposits, *in* G. Einsele, W. Ricken, and A. Seilacher, eds., Cycles and Events in Stratigraphy: Berlin, Springer-Verlag, p. 188-221.
Gore, P. J. W., 1988, Lacustrine sequences in an early Mesozoic rift basin: Culpeper Basin, Virginia, USA, *in* A. J. Fleet, K. Kelts, and M. R. Talbot, eds., Lacustrine Petroleum Source Rocks: Oxford, Geological Society of London Special Publication 40, p. 247-278.
Gore, P. J. W., 1989, Toward a model for open- and closed-basin deposition in ancient lacustrine sequences: the Newark Supergroup (Triassic–Jurassic), eastern North America: Palaeogeography, Palaeoclimatology, Palaeoecology, v. 70, p. 29-51.
Goth, K., J. W. deLeeuw, W. Puettmann, and E. W. Tegelaar, 1988, Origin of Messel oil shale kerogen: Nature, v. 336, p. 759-761.
Grabowski, G. J., and K. M. Bohacs, 1996, Controls on composition and distribution of lacustrine organic-rich rocks of the Green River Formation, Wyoming. Annual Meeting Abstracts, AAPG and SEPM, v. 5, p. 55.
Hardie, L. A., and H. P. Eugster, 1970, The evolution of closed-basin brines: Fiftieth Anniversary Symposia, Mineralogy and Geochemistry of Non-Marine Evaporites, Special Paper, Mineralogical Society of America, v. 3, p. 273-290.
Hayberan, K. A., and R. E. Hecky, 1987, The late Pleistocene and Holocene stratigraphy and paleolimnology of lakes Kivu and Tanganyika: Palaeogeography, Palaeoclimatology, Palaeoecology, v. 61, p. 169-197.
Hedberg, H. D., 1968, Significance of high-wax oils with respect to genesis of petroleum: AAPG Bulletin, v. 52, p. 736-750.
Herdendorf, C. E., 1984, Inventory of the morphometric and limnologic characteristics of the large lakes of the world: The Ohio State University Sea Grant Program, Technical Bulletin OHSU-TB-17, 78 p.
Higgs, R., 1991, The Bude Formation (lower Westphalian), SW England; siliciclastic shelf sedimentation in a large equatorial lake: Sedimentology, v. 38, p. 445-469.
Hentz, T. F., 1981, Sedimentology and structure of Culpeper Group lake beds (Lower Jurassic) at Thoroughfare Gap, Virginia: M.Sc. Thesis, University of Kansas, Lawrence, Kansas, 166 p.
Horsfield, B., D. J. Curry, K. M. Bohacs, A. R. Carroll, R. Littke, U. Mann, M. Radke, R. G. Schaefer, G. H. Isaksen, H. J. Schenk, E. G. Witte, and J. Rullkötter, 1994, Organic geochemistry of freshwater and alkaline lacustrine environments, Green River Formation, Wyoming: Organic Geochemistry, v. 22, p. 415-450.
Hu, J., S. Xu, X. Tong, and H. Wu, 1989, The Bohai Bay basin, *in* X. Zhu, ed., Chinese Sedimentary Basins: Amsterdam, Elsevier, p. 89-106.
Hutchinson, G. E., 1957, A Treatise on Limnology, Vol. 1—Geography, Physics and Chemistry: New York, John Wiley and Sons, 1015 p.
Isaksen, G. H., 1991, Molecular geochemistry assists exploration: Oil and Gas Journal, p. 127-131.
Johnson, T. C., J. D. Halfman, B. R. Rosendahl, and G. S. Lister, 1987, Climatic and tectonic effects on sedimentation in a rift-valley lake; evidence from high-resolution seismic profiles, Lake Turkana, Kenya: Geological Society of America Bulletin, v. 98, p. 439-447.
Katz, B. J., 1990, Lacustrine Basin Exploration—Case Studies and Modern Analogues: American Association of Petroleum Geologists Memoir 50, 340 p.
Katz, B.J., 1995, Lacustrine source rock systems—is the Green River Formation an appropriate analog?: Geologiya i Geofizika v. 36, p. 26-41.
Kelley, P.A., B. Mertani, and H.H. Williams, 1995, Brown Shale Formation: Paleogene lacustrine source rocks of central Sumatra, *in* B.J. Katz, ed., Petroleum Source Rocks: Berlin, Springer-Verlag, p. 283-308.
Kelts, K., 1988, Environments of deposition of lacustrine petroleum source rocks: an introduction, *in* A.J. Fleet, K. Kelts, and M. R. Talbot, eds., Lacustrine Petroleum Source Rocks: Oxford, Geological Society Special Paper 40, p. 3-26.
Kulke, H., 1995, Regionale Erdöl-und Erdgasgeologie der Erde: Berlin, Gerbrüder Borntrager, 2 volumes, 942 p.
Kuo, L. C., 1994, Lower Cretaceous lacustrine source rocks in northern Gabon: effect of organic facies and thermal maturity on crude oil quality: Organic Geochemistry, v. 22, p. 257-273.
Lambiase, J. J., 1990, A model for tectonic control of lacustrine stratigraphic sequences in continental rift basins, *in* B.J. Katz, ed., Lacustrine Basin Exploration—Case Studies and Modern Analogues: AAPG Memoir 50, p. 265-276.
Likens, G. E., 1975, Primary production of inland aquatic ecosystems, *in* H. Lieth and R. H. Whittaker, eds., Primary Productivity of the Biosphere: New York, Springer-Verlag, p. 186-215.
Livingston, D., 1865, Narrative of an Expedition to the Zambezi and its tributaries: London, 227 p.
Lyell, C., 1830, Principles of Geology, vol. 1: London, J. Murray, 511 pp.
Lyell, C., 1847, On the structure and probable age of the coal field of the James River, near Richmond, Virginia: Quarterly Journal of the Geological Society of London, v. 3, p. 261-280.
Madon-Mazlan, B. H., 1992, Depositional setting and origin of berthierine oolitic ironstones in the lower

Miocene Terengganu shale, Tenggol Arch, offshore peninsular Malaysia: Journal of Sedimentary Petrology, v. 62, p. 899-916.

Marzo, M., J. Verges, M. Lopez-Blanco, J. A. Munoz, T. Santaeularia, and D.W. Burbank, 1996, Third-order underfilling-overfilling foreland basin sequences: American Association of Petroleum Geologists 1996 Annual Convention, San Diego, California, Annual Meeting Abstracts, v. 5, p. 92.

Manspeizer, W., and P. E. Olsen, 1981, Rift basins of the passive margin: tectonics, organic-rich lacustrine sediments, basin analyses, *in* G. W. Hobbs, III, ed., Field Guide to the Geology of the Paleozoic, Mesozoic, and Tertiary Rocks of New Jersey and the Central Hudson Valley: New York, Petroleum Exploration Society of New York, p. 25-105.

Manspeizer, W., 1985, The Dead Sea rift: impact of climate and tectonism on Pleistocene and Holocene sedimentation, *in* K. T. Biddle and N. Christie-Blick, eds., Strike-slip deformation, basin formation, and sedimentation: SEPM Special Publication 37, p. 143-158.

Mastalerz, K., 1994, Anthracosia shale (Lower Permian), North Sudetic basin, Poland, *in* E. Gierlowski-Kordesch and K. Kelts, eds., Global Geological Record of Lake Basins, Vol. 1: Cambridge, Cambridge University Press, p. 97-100.

Mello, M. R., and M. H. Hessel, 1998, Biological mark and paleozoological characterization of the early marine incursion in the lacustrine sequences of the Campos basin, Brazil: Extended Abstracts, 1998 American Association of Petroleum Geologists Annual Convention, Salt Lake City, Utah, v. 2, p. A455.

Mello, M. R., and J. R. Maxwell, 1990, Organic geochemical and biological marker characterization of source rocks and oils derived from lacustrine environments in the Brazilian continental margin, *in* B. J. Katz, ed., Lacustrine Basin Exploration—Case Studies and Modern Analogues: AAPG Memoir 50, p. 77-97.

Mello, M. R., N. Telnaes, P. C. Gaglianone, M. I. Chicarelli, S. C. Brassel, and J.R. Maxwell, 1988, Organic geochemical characterisation of depositional paleoenvironments of source rocks and oils in Brazilian margin basins: Organic Geochemistry, v. 13, p. 31-45.

Meyers, P. A., and R. Ishiwatari, 1993, Lacustrine organic geochemistry— an overview of indicators of organic matter sources and diagenesis in lake sediments: Organic Geochemistry, v. 20, p. 867-900.

Mohler, R. R., M. J. Wilkinson, and J. R. Giardino, 1995, The extreme reduction of Lake Chad surface area: input to paleoclimatic reconstructions: Geological Society of America Abstracts with Program, 1995 Annual Meeting, New Orleans, p. A-265.

Neal, J. E., K. M. Bohacs, D. J. Reynolds, and C. A. Scholz, 1997, Sequence stratigraphy of lacustrine rift basins: critical linkage of changing lake level, sediment supply, and tectonics: Abstracts with Programs, Geological Society of America, v. 29, p. 73.

Neal, J. E., K. M. Bohacs, D. J. Reynolds, and C. A. Scholz, 1998, Tectonic and sequence-stratigraphic controls on facies distribution in a deep balanced-fill lake: seismic and drop-core evidence form Lake Malawi, East Africa: AAPG Hedberg Conference, Cairo, Egypt, p. 40.

Nichols, G. J., 1987, Structural controls on fluvial distributary systems — the Luna system, northern Spain: SEPM Special Publication 39, p. 269-277.

Olsen, P. E., 1990, Tectonic, climatic, and biotic modulation of lacustrine ecosystems—examples from Newark supergroup of eastern North America, *in* B. J. Katz, ed., Lacustrine Basin Exploration, AAPG Memoir 50, p. 209-224.

Olsen, P. E., R. W. Schlische, and P. J. W. Gore, 1989, Tectonic, depositional, and paleoecological history of Early Mesozoic rift basins, eastern North America: Twenty-eighth International Geological Congress Field Trip Guidebook T351, Washington, D.C., American Geophysical Union, 158 p.

Perlmutter, M. A., and M. D. Matthews, 1990, Global cyclostratigraphy—a model, *in* T. A. Cross,ed., Quantitative Dynamic Stratigraphy: Englewood Cliffs, New Jersey, Prentice Hall, p. 233-260.

Peters K. E., and J. M. Moldowan, 1993, The Biomarker Guide: Interpreting Molecular Fossils in Petroleum and Ancient Sediments: Englewood Cliffs, New Jersey, Prentice Hall, 363 p.

Picard, M. D., and L. R. High, 1972, Criteria for recognizing lacustrine rocks, in Rigby, J. K., and Hamblin, W. K., eds., Recognition of Ancient Sedimentary Environments: SEPM Special Publication 16, p. 108-145.

Posamentier, H. W. and P. R. Vail, 1988, Eustatic controls on clastic deposition II— sequence and systems tracts models, *in* C. K. Wilgus, et al., eds., Sea-level changes: an integrated approach: SEPM Special Publication 42, p. 125-154.

Powell, T. G., 1984, Developments in concepts of hydrocarbon generation from terrestrial organic matter: Beijing Petroleum Symposium 20-24 September 1984, Beijing, China, 40 pp.

Powell, T. G., 1986, Petroleum geochemistry and depositional setting of lacustrine rocks: Marine and Petroleum Geology v. 3, p. 200-219.

Raisz, E. R., 1937, Rounded lakes and lagoons of the coastal plain of Massachusetts: Journal of Geology v. 42, p. 839-848.

Ravnas, R. and R. J. Steel, 1998, Architecture of marine rift-basin successions: AAPG Bulletin v. 82, p. 110-146.

Redfield, A. C., 1958, Preludes to the entrapment of organic matter in the sediments of Lake Maracaibo *in* L. G. Weeks, ed., Habitat of oil—a symposium: AAPG, p. 968-981.

Remy, R. R., Ming Pang, D. Nummedal, and Fuping Zhu, 1995, Lithologic, well log, and seismic facies of lacustrine rift-fill sequences, Bohai basin, China: AAPG Hedberg Research Symposium on Lacustrine Basin Exploration in China, southeast Asia, and Indonesia, October 15-19, 1995, Dongying City, China.

Renaut, R. W. and W. M. Last, 1994, Sedimentology and geochemistry of modern and ancient saline lakes: SEPM Special Publication 50, Tulsa.

Renaut, R. W., and J. J. Tiercelin, 1994, Lake Bogoria,

Kenya Rift valley, a sedimentological overview, *in* R.W. Renaut and W.M. Last, eds., Sedimentology and geochemistry of modern and ancient saline lakes: SEPM (Society for Sedimentary Geology) Special Publication 50, p. 101-123.
Roehler, H. W., 1992, Correlation, composition, areal distribution, and thickness of Eocene stratigraphic units, greater Green River basin, Wyoming, Utah, and Colorado: U.S. Geological Survey Professional Paper 1506-E, 49 p.
Sagri, M., E. Abbate, P. Bruni, 1989, Deposits of Ephemeral and Perennial Lakes in the Tertiary Daban basin (northern Somalia): Palaeogeography, Palaeoclimatology, Palaeoecology, v. 70, p. 225-233.
Sandvik, E. I., W. A. Young, and D. J. Curry, 1992, Expulsion from hydrocarbon sources: the role of organic absorption, *in* M.C. Eckardt, et al., eds., Advances in Organic Geochemistry, 1991, part 1, Advances and applications in energy and the natural environment: Organic Geochemistry, v. 19, p. 77-87.
Schoell, M., B. R. T. Simoneit, and T. G. Wang, 1994, Organic geochemistry and coal petrology of Tertiary brown coal in the Zhoujing Mine, Baise Basin, South China; 4, Biomarker sources inferred from stable carbon isotope compositions of individual compounds: Organic Geochemistry, v. 21, p. 713-719.
Scholz, C. A., 1995, Deltas of the Lake Malawi rift, east Africa: seismic expression and exploration implications: AAPG Bulletin, v. 79, p. 1679-1697.
Scholz, C. A., and B. R. Rosendahl, 1990, Coarse-clastic facies and stratigraphic sequence models from Lakes Malawi and Tanganyika, east Africa, *in* B. J. Katz, ed., Lacustrine Basin Exploration: AAPG Memoir 50, p. 151-168.
Schull, T. J., 1988, Rift basins of interior Sudan; petroleum exploration and discovery: AAPG Bulletin, v. 72, p. 1128-1142.
Schumm, S. A., 1977, Applied fluvial geomorphology, *in* J. R. Hails, ed., Applied geomorphology; a perspective of the contribution of geomorphology to interdisciplinary studies and environmental management: Amsterdam, Elsevier Scientific Publishing Company, p. 119-156.
Schwans, P., K. M Bohacs, V. K. Hohensee, F. C. Lin, R. Q. Gao, J. Y. Qiu, S. H. Sun, and F. J. Li, 1997, Nonmarine sequence stratigraphy; impact on source, reservoir, and seal distribution in a rift setting; Cretaceous Songliao Basin, China: AAPG Bulletin, v. 81, p. 1411.
Shanley, K. W., and P. J. McCabe, 1994, Perspectives on the sequence stratigraphy of continental strata: AAPG Bulletin, v. 78, p. 544-568.
Sheng, G., J. Fu, S. C. Brassel, A. P. Gowar, G. Eglington, J. S. Sinninghe Damsté, J. W. De Leeuw, and P. A. Schenk, 1987, Sulphur-containing compounds in sulphur-rich crude oils from hypersaline lake sediments and their geological implications: Geochemistry (English Language Edition), v. 6, p. 115-126.
Shi, J. Y., A. S. Mackenzie, R. Alexander, G. Eglinton, A. P. Gowar, G. Wolff, and J. R. Maxwell, 1982, A biological marker investigation of petroleums and shales from the Shengli oilfield, the People's Republic of China, Chemical Geology, v. 35, p. 1-31.
Sinninghe Damsté, J.S., F. Kenig, M.P. Koopmans, J. Köster, S. Schouten, J.M. Hayes, and J.W. de Leeuw, 1995, Evidence for gammacerane as an indicator of water column stratification: Geochimica et Cosmochimica Acta, v. 59, p. 1895-1900.
Sladen, C.P., 1994, Key elements during the search for hydrocarbons in lake systems, *in* E. Gierlowski-Kordesch and K. Kelts, eds., Global Geological Record of Lake Basins, Vol. 1: Cambridge, Cambridge University Press, p. 3-17.
Smith, M., 1990, Lacustrine oil shale in the geological record, *in* B.J. Katz, ed., Lacustrine Basin Exploration—Case Studies and Modern Analogues: AAPG Memoir 50, p. 43-60.
Smoot, J. P., 1983, Depositional subenvironments in an arid closed basin; the Wilkins Peak Member of the Green River Formation (Eocene), Wyoming, USA: Sedimentology, v. 30, p. 801-827.
Stansbury, H., 1852, Exploration and survey of the valley of the Great Salt Lake of Utah, including a reconnaissance of a new route through the Rocky Mountains: Senate Executive Document, No. 3, p. 487, maps.
Stemmerik, L., F. G. Christiansen, and S. Piasecki, 1990, Carboniferous lacustrine shale in East Greenland—additional source rock in northern North Atlantic?, *in* B. J. Katz, ed., Lacustrine Basin Exploration—Case Studies and Modern Analogues: AAPG Memoir 50, p. 277-286.
Sullivan, R., 1980, A stratigraphic evaluation of the Eocene rocks of southwestern Wyoming: Geological Survey of Wyoming, Report of Investigations 20, 50 p.
Surdam, R. C., K. O. Stanley, and H. P. Bucheim, 1980, Depositional environments of the Laney Member of the Green River Formation, southwestern Wyoming: SEPM (Society for Sedimentary Geology) Field Trip No. 5 Guidebook, Denver, Colorado, 72 p.
Tegelaar, E. W., R. M. Matthezing, J. B. H. Jansen, B. Horsfield, and J. W. de Leeuw, 1989, Possible origin of *n*-alkanes in high-wax crude oils: Nature, v. 342, p. 529-531.
Telnæs, N., G. H. Isaksen, and P. Farrimond, 1992, Unusual triterpane distributions in lacustrine oils: Organic Geochemistry, v. 18, p. 785-789.
Tissot, B. P., and D. H. Welte, 1984, Petroleum Formation and Occurrence (2d ed.): Berlin, Springer-Verlag, 699 p.
Tissot, B. P., G. Deroo, and A. Hood, 1978, Geochemical study of the Uinta basin: formation of petroleum from the Green River Formation: Geochimica et Cosmochimica Acta, v. 42, p. 1469-1486.
Valero Garcés, B. L., 1991, Los sistemos lacustres carbonatodos del Stephaniense y Pérmico en el Pirineo Central y Occidental: Ph.D. dissertation, University of Zaragoza, Zaragoza, Spain, 292 p.
Warren, J. K., 1994, Holocene Coorong Lakes, South Australia, *in* E. Gierlowski-Kordesch and K. Kelts, eds., Global Geological Record of Lake Basins, Vol. 1: Cambridge, Cambridge University Press, p. 387-394.

Watson, M. P., A. B. Hayward, D. N. Parkinson, and Z. M. Zhang, 1987, Plate tectonic history, basin development and petroleum source rock deposition onshore China: Marine and Petroleum Geology, v. 4, p. 205-225.

Wood, S. H., 1994, Seismic expression and geological significance of a lacustrine delta in Neogene deposits of the western Snake River Plain, Idaho: AAPG Bulletin, v. 78, p. 102-121.

Wood, S. H., and E. Squires, 1998, History of Neogene Lake Idaho form geological mapping of the NE margin and subsurface of the western Snake River Plain: Extended Abstracts, 1998 AAPG Annual Convention, Salt Lake City, Utah, v. 2, p. A709.

Yang W., 1985, Daqing oil field, People's Republic of China: a giant field with oil of non-marine origin: AAPG Bulletin, v. 69, p. 1101-1111.

Yang W., Y. Li, and R. Gao, 1985, Formation and evolution of non-marine petroleum in Songliao basin, China: AAPG Bulletin, v. 69 p. 1112-1122.

Yang, W., R. J. Spencer, H. R. Krouse, T. K. Lowenstein, and E. Casas, 1995, Stable isotopes of lake and fluid inclusion brines, Qaidam Basin, western China., hydrology and paleoclimatology in arid environments: Palaeogeography, Palaeoclimatology, Palaeoecology, v. 11, p. 279-290.

Zhou Guangia, 1981, Character of organic matter in source rocks of continental origin and its maturation and evolution, *in* J.F. Mason, ed., Petroleum Geology in China, Tulsa, PennWell Books (for the United Nations), p. 26-47.

Sladen, C., and J. J. Traynor, Lakes during the evolution of Mongolia, 2000, *in* E. H. Gierlowski-Kordesch and K. R. Kelts, eds., Lake basins through space and time: AAPG Studies in Geology 46, p. 35-57.

Chapter 2

Lakes During the Evolution of Mongolia

Chris Sladen
BP Amoco
Irkutsk, Russia

J. J. Traynor
Deutsche Bank
Edinburgh, United Kingdom

INTRODUCTION

The Mongolian People's Republic in central Asia (frequently called Mongolia, but often referred to in the past as Outer Mongolia in English language literature) encompasses some 2 million km^2 of steppe, desert, and mountains. Knowledge and understanding of Mongolian geology is limited, and in particular very little is known outside of Mongolia and Russia. The purpose of this paper is to highlight the development of lake basins, lakes, and lacustrine sequences during the geologic evolution of Mongolia and thereby stimulate further interest and research. Many characteristic features and apparent key controls on lake basin and lacustrine sequence development are described. Lake sediments can be identified in all geologic periods from the Carboniferous to Quaternary. Collectively, they include examples of all the principal types of tectonically formed lake basins and lacustrine depositional environments.

The geology of Mongolia is not known in any detail in western literature, partly due to the country's remoteness and inaccessibility. Pioneering work reported by Berkey and Morris (1927) remains the single most important source of documentation in the English language. Following the changes in Mongolia and the former Soviet Union during the late 1980s, geoscientists can now access the remotest areas of the country. Mongolian and Soviet surface geology maps, made from the 1950s onward, are a key source of data. Most of the stratigraphic units and nomenclature in this paper are based on these maps. The maps, most of which are not officially published, do notably contain some inconsistencies in the use of terms such as "early" vs. "lower" and "late" vs. "upper."

LAKE BASINS IN A TECTONOSTRATIGRAPHIC FRAMEWORK AND PLATE TECTONIC CONTEXT

Geologic maps of Mongolia immediately reveal a complexity of structures and an abundance of rock types with representatives from Precambrian to Tertiary. Mongolia is an important geologic link between the Siberian craton in the north, which is essentially an amalgamation of lower Paleozoic terranes, and northern China to the south, which is an area of complex middle Paleozoic–Tertiary tectonics and sutures (Figure 1). Basin formation has thus repeatedly been controlled by preexisting crustal-scale weaknesses.

During the Paleozoic, terrane fragments were successively welded to the much larger Siberian craton. By the late Paleozoic, lakes were developing in compressional basins formed during terrane accretion (Figure 2). The structures and sutures that formed during these terrane-docking episodes were repeatedly reactivated during the Mesozoic and Tertiary in response to further suturing south of Mongolia and in western and central China. These episodes of reactivation typically involved extension allowing lakes to develop in rift and strike-slip basins (Figures 2 and 3).

Deformation from the Tertiary onward includes both compression and strike-slip extension and has been focused along the belts of Mesozoic basins in southern and eastern Mongolia. Present lakes exist mostly in successor basins that can be related to strike-slip tectonics exploiting the inherited crustal weaknesses.

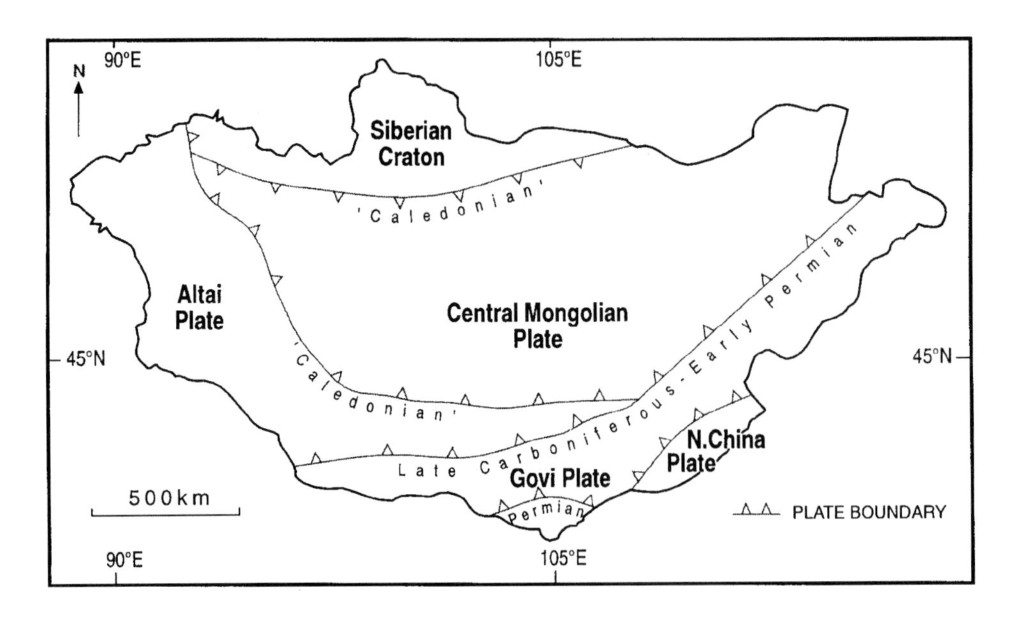

Figure 1—Principal tectonic plates in Mongolia (modified from Traynor and Sladen, 1995).

CARBONIFEROUS AND PERMIAN LAKES

Tectonic Overview

Devonian–Permian facies reflect progressive shoaling and emergence on the late Paleozoic shelf margins. Basinal areas that existed across southern Mongolia became more confined and nonmarine in character in response to compression and uplift, especially during the Permian. A basal Late Permian unconformity in many areas marks the termination of any extensive marine deposition in Mongolia, and the subsequent dominance of nonmarine sedimentation that has continued to the present day.

By the Late Permian, a major fold belt was established by the collision of the amalgamated North China, Govi, and Central Mongolian plates (Traynor and Sladen, 1995). Locally, nonmarine sedimentation included alluvial anastomosing(?) systems with floodplain lakes, coal swamps, and shallow swamp lakes occurring in small compressional basins. In the Altai Mountains, small remnants of Late Carboniferous–Early Permian coal-rich basins are preserved in thrust slices. Some lakes may have been in piggyback basins on top of thrust sheets, but lateral thrust sheet movements and basin configuration are poorly constrained due to subsequent deformation.

Unlike the large Junggar lake basin in nearby northwest China, a large lake basin failed to develop in Mongolia in the Late Permian. A large lake occupying many tens of thousands of square kilometers developed in the Junggar basin in a flexural foreland setting between growing fold belts to the northeast (part of which was in Mongolia) and to the south and northwest. This lake contained deep, oxygen-deficient stratified waters and accumulated thick mudstones that are excellent hydrocarbon source rocks (Carroll et al., 1992). Drainage systems in the small basins in southwest Mongolia may have flowed toward this lake.

Facies Details

In Mongolia, the Carboniferous and Permian nonmarine sequences are dominated by fluvial siliciclastics. Alluvial fan and fluvial channel sequences predominate with conglomerates that commonly are reddened and locally contain large pieces of silicified wood. They have only limited porosity and permeability because deep burial, in most cases, has caused considerable compaction and quartz cementation. Subordinate coals and dark-colored carbonaceous mudstones developed in interchannel temporary lakes and ponds. Thick, high-quality (bituminous and anthracitic) coals and interbedded mudstones and siltstones of floodplain and swamp lake origin are found locally, e.g., at Ulan Gom, Hoshoort, Maan Yt, and Tavantolgoi (Figures 4 and 5).

Biostratigraphic Information

In the late Carboniferous–Early Permian coals and carbonaceous mudstones, relatively rich, but generally poorly preserved, assemblages of pollen and miospores are present. Spores identified include *Camptotriletes warchianus*, *Densoisporites playfordii*, *Horriditriletes ramosus*, *Apiculatisporites cornutus*, *Acanthotriletes superbus*, *Lophotriletes novicus*, *Raistrickia* cf. *crenata*, and *Lundbladisporites* spp. A common monosaccate pollen is *Cordaitina* spp. and a disaccate pollen includes *Striatoabietes multistriatus*.

Coal beds contain large fragments of tree trunks, branches, and twigs that include *Paracalamites* cf. *frigidus*, *Rufloria* cf. *tasmyrica*, *Cordaites* cf. *latifolus*, *Sphenopteris* ex gr. *tunguskama*, *Cordaites* ex gr. *singularis*, *Pecopteris* aff. *anthriscifola*, *Cordaites* cf. *zalessky*, and *Claudostrobus lutuginii*. Permian palynoflora show that Mongolia probably constituted part of the Angaran palynofloral province during the Permian. This indicates latitudes of 30–60°N with a warm temperate climate (Broutin et al., 1990).

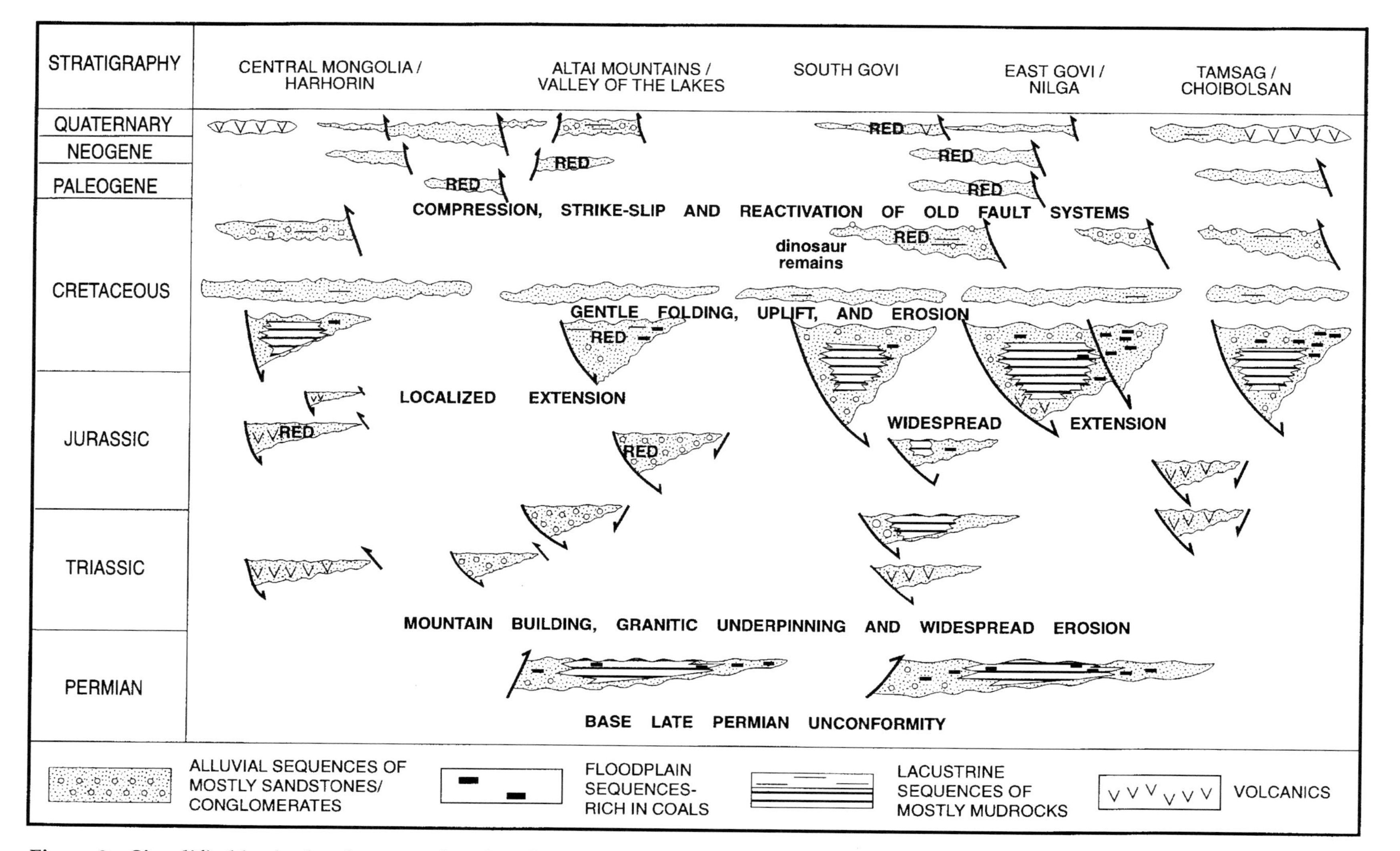

Figure 2—Simplified basin development showing the presence of lacustrine sequences (modified from Traynor and Sladen, 1995).

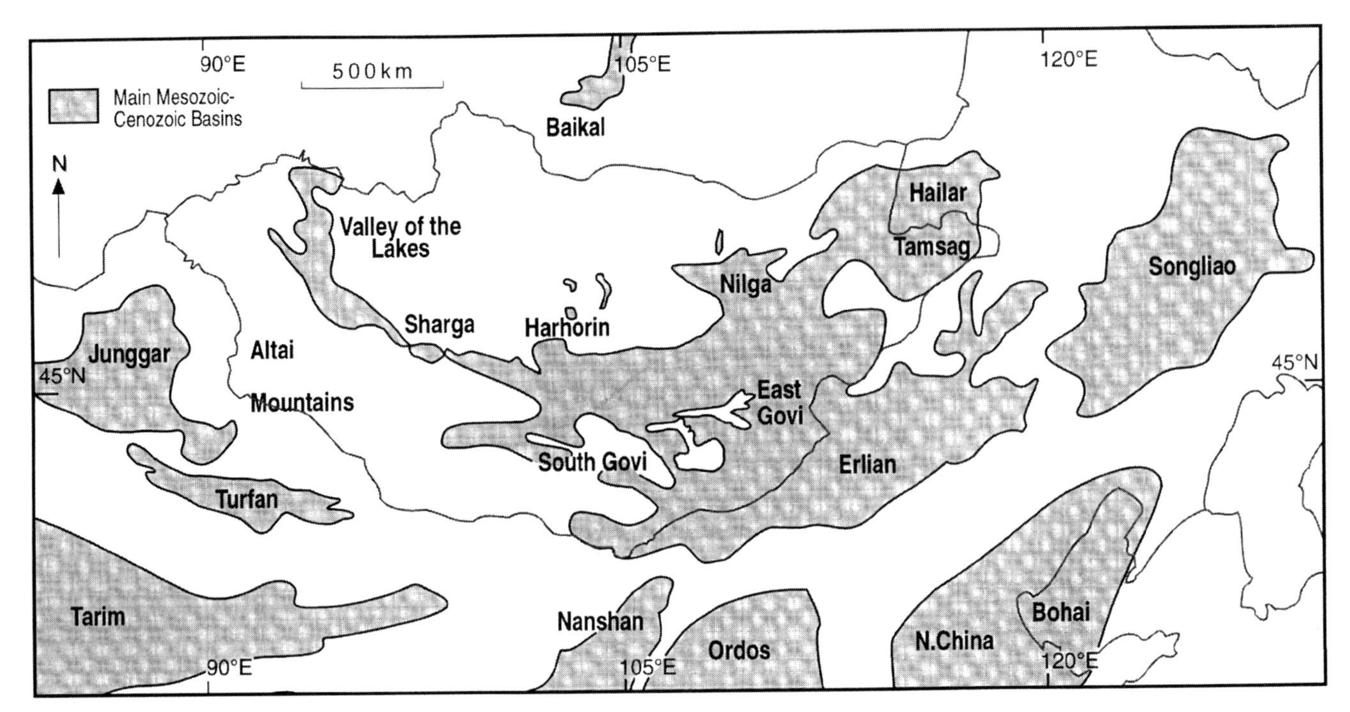

Figure 3—Distribution of the main Mesozoic–Cenozoic basins in Mongolia and northern China. Note that many small basins are not shown (modified from Traynor and Sladen, 1995).

Geochemical Characterization

Geochemical data indicate that the coal and carbonaceous mudstone sequences are both gas prone and have in the past frequently reached the gas window during deep burial. Pyrolysis gas chromatogram (PGC) analysis indicates a high gas-to-oil generation index (Figure 6a). Analyses indicate high total organic carbon (TOC) content up to 83.3% and excellent source potential with the generation capability (S2) of coals up to 193.6 kg/t, and S2 of mudstones up to 54.3 kg/t (Tables 1 and 2, Figures 4 and 7).

Unraveling the migration history of the hydrocarbon charge is difficult because of the complex tectonic deformation. Much or all of the charge may have been lost by breaching of traps and seepage to surface; however, opportunities for producing coal-bed methane have yet to be investigated.

TRIASSIC, EARLY JURASSIC, AND MIDDLE JURASSIC LAKES

Tectonic Overview

Most of Mongolia was experiencing widespread erosion during the Triassic and Early Jurassic due to continued compression, folding, and uplift. The Early and Middle Jurassic probably represented the final stages of the progressive shoaling, uplift, and emergence of the late Paleozoic shelf margin and its conversion into a large landmass with mountain ranges and fold belts. This transition changed the region from one of marine sedimentation and net deposition to one of emergence and an eroding landmass. This created more possibilities for the development of nonmarine environments and, in particular, lakes.

A major mountain chain existed across Mongolia by the Triassic, extending through the present Altai Mountains and across the Govi desert. Extensive Late Triassic–Early Jurassic granitic intrusions underpinned the mountains, and rhyolitic and andesitic volcanics were extruded. Erosion of the mountains proceeded rapidly. It is common to observe Upper Jurassic–Lower Cretaceous volcanic or sedimentary rocks resting directly on Upper Triassic–Lower Jurassic granites. This implies erosion and granite unroofing on a massive scale, with removal of perhaps 5 km or more of section during the Late Triassic and Early Jurassic, exposing plutons at the surface. A deeply dissected and rapidly eroding mountain range probably extended across Mongolia, forming a major Asian watershed. Eroded sediments probably found their way to distant areas to the north and south of Mongolia; for example, a contribution to the Triassic and Lower Jurassic sediments in the Ordos basin and Junggar basin in China could have come from large rivers draining from these mountain ranges.

Triassic–Middle Jurassic sedimentation in Mongolia was limited to foreland basin settings in front of this major mountain chain and small intermontane basins (Figure 2). Basins capable of maintaining large long-lived lakes appear to have been absent, and remnants of basins usually cover only a few tens of square kilometers (Marinov et al., 1973). Early and Middle Jurassic basin remnants are scattered across most regions of Mongolia, but these also are rarely more

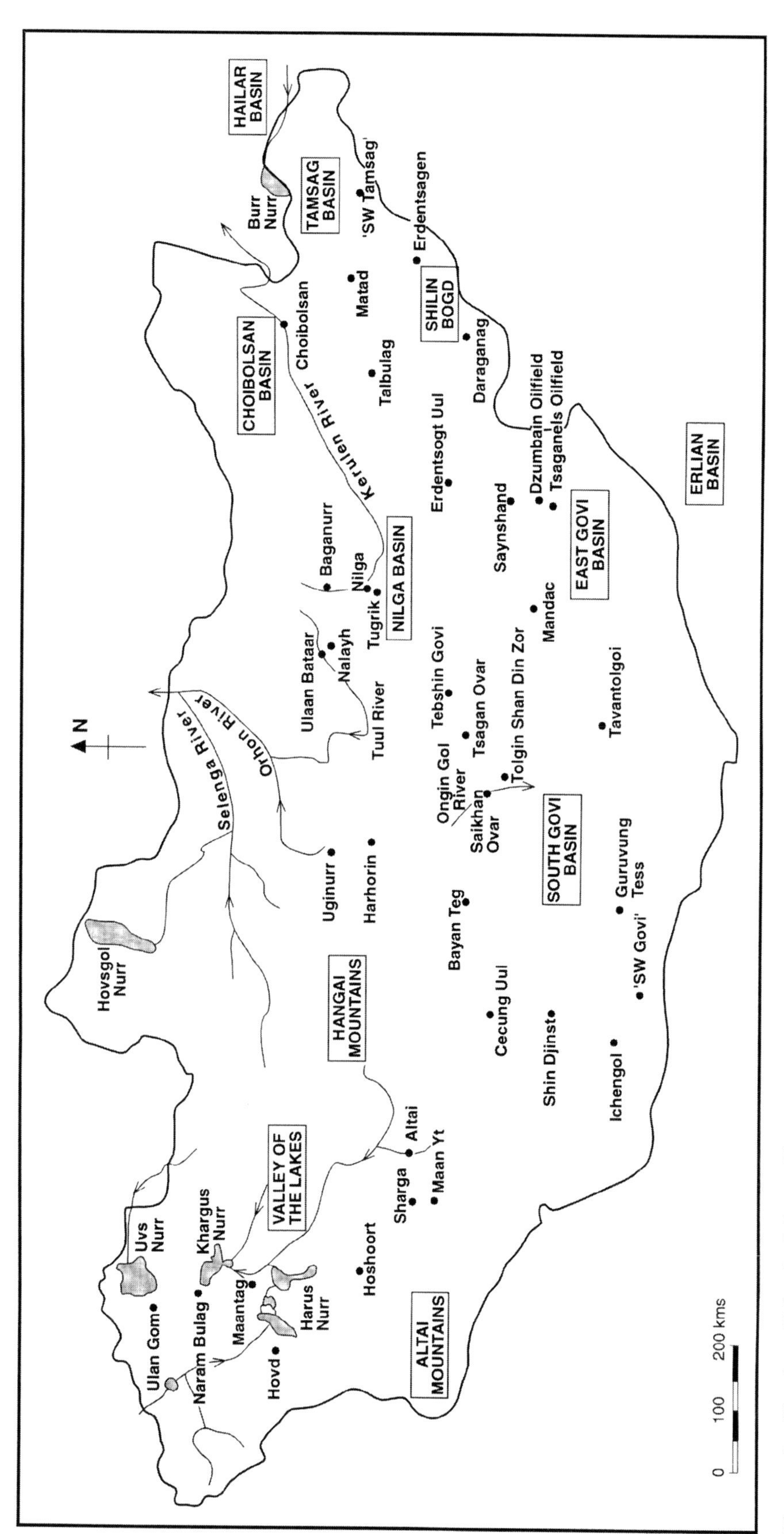

Figure 4—Key localities in Mongolia.

Figure 5—Upper Permian coals, interbedded mudstones, and siltstones of floodplain and swamp lake origin at the Tavantolgoi opencast coal mine.

than a few tens of square kilometers in size and are probably fragments of intermontane basins. Sequences could also be preserved in basins beneath younger Mesozoic and Tertiary cover, but there is insufficient detailed seismic and gravity data.

Facies Details

Many sedimentary sequences are thought to be Triassic in age based solely on field relationships because they fail to yield any fauna or flora. Some may be Late Permian or even Early Jurassic in age, but are probably best referred to as Triassic(?). Preserved remnants of these Triassic(?) sequences are overwhelmingly parts of alluvial systems usually comprising fining-upward, polymictic cross-bedded conglomerates and sandstones. Sandstones are multicolored including gray, red, purple, gray-green, and pink varieties. Compositions are immature and porosity is very low; most are arkoses with pervasive, poikilotopic carbonate cement or locally a quartz cement. Fine-grained Triassic(?) rocks are particularly scarce and there is rarely evidence for more than a few short-lived Triassic(?) floodplain lakes. These fine-grained rocks are also multicolored and include dark gray, purple, green, and red mudstones and siltstones up to 30 cm thick, typically with near-conchoidal fracture.

An interesting exception to the dominantly coarse alluvial sediments of the Triassic(?) occurs in the southeast Govi, just east of Ichen Gol. At this locality, there are more than 100 m of hard, light gray and gray, calcareous mudstones, siltstones, and sandstones arranged in coarsening-upward sequences 3–10 m thick (Figures 4 and 8). The mudstones contain 20–45% micritic limestone and millimeter-scale laminations. The thickest mudstone reaches 30 m with lateral continuity of more

Table 1. Data on Carboniferous source rock quality from selected sites shown on Figure 4.

Location	Sample	S1 (kg/t)	S2 (kg/t)	HI	TOC (%)	Pr/Ph
Hoshoort						
AT/6/10B	Coal	0.1	36.9	48	77.4	–
AT/6/10G	Coal	0.0	49.2	59	88.3	–
Ulan Gom						
AT/13/3A	Coal	1.4	17.2	27	16.0	–

S1 = free hydrocarbons, S2 = cracked hydrocarbons, HI = hydrogen index, TOC = total organic carbon, and Pr/Ph = pristane/phytane ratio.

Table 2. Data on Permian source rock quality from selected sites shown on Figure 4.

Location	Sample	S1 (kg/t)	S2 (kg/t)	HI	TOC (%)	Pr/Ph
Maan Yt						
AT/4/6D	Coal	3.4	51.5	314	16.4	2.79
AT/4/6E	Coal	5.0	193.6	302	64.1	6.65
Tavantolgoi						
GV/6/7D	Coal	5.2	102.6	175	58.6	–
GV/6/11A	Coal	1.1	99.8	132	75.6	–
GV/6/6L	Mudstone	1.7	19.4	87	22.3	–
GV/6/8B	Mudstone	4.0	54.3	156	34.9	–
Nilga						
TS/19/3C	Coal	2.0	0.6	3	20.1	–
TS/19/3A	Mudstone	0.2	0.1	10	1.0	–

S1 = free hydrocarbons, S2 = cracked hydrocarbons, HI = hydrogen index, TOC = total organic carbon, and Pr/Ph = pristane/phytane ratio.

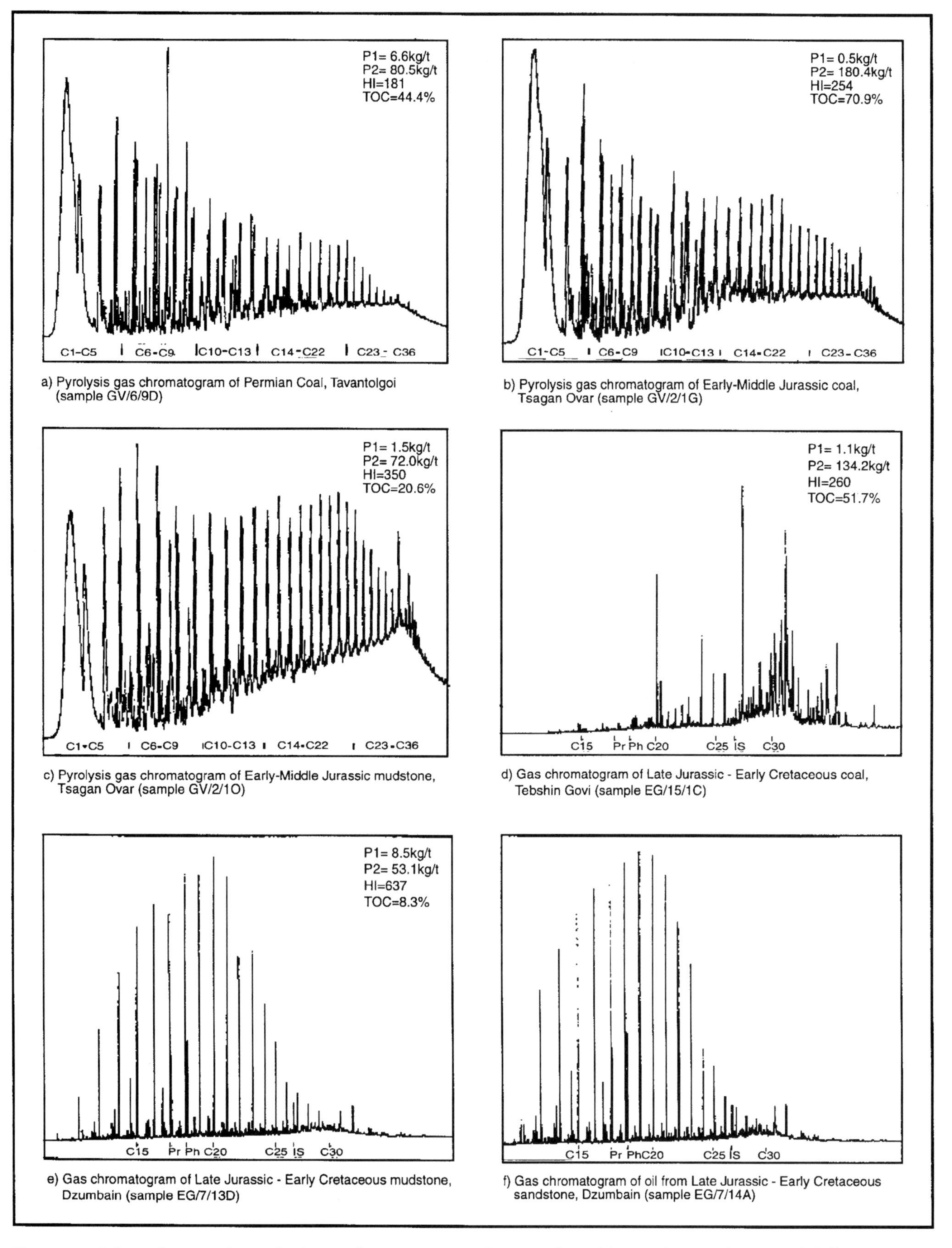

Figure 6—Selected gas and pyrolysis gas chromatograms from coals and lacustrine mudstones of different ages in Mongolia. Sample localities shown in Figure 4. P1 (normally S1) is the amount of free hydrocarbons, P2 (normally S2) is the amount of cracked hydrocarbons, HI = hydrogen index, and TOC = total organic carbon.

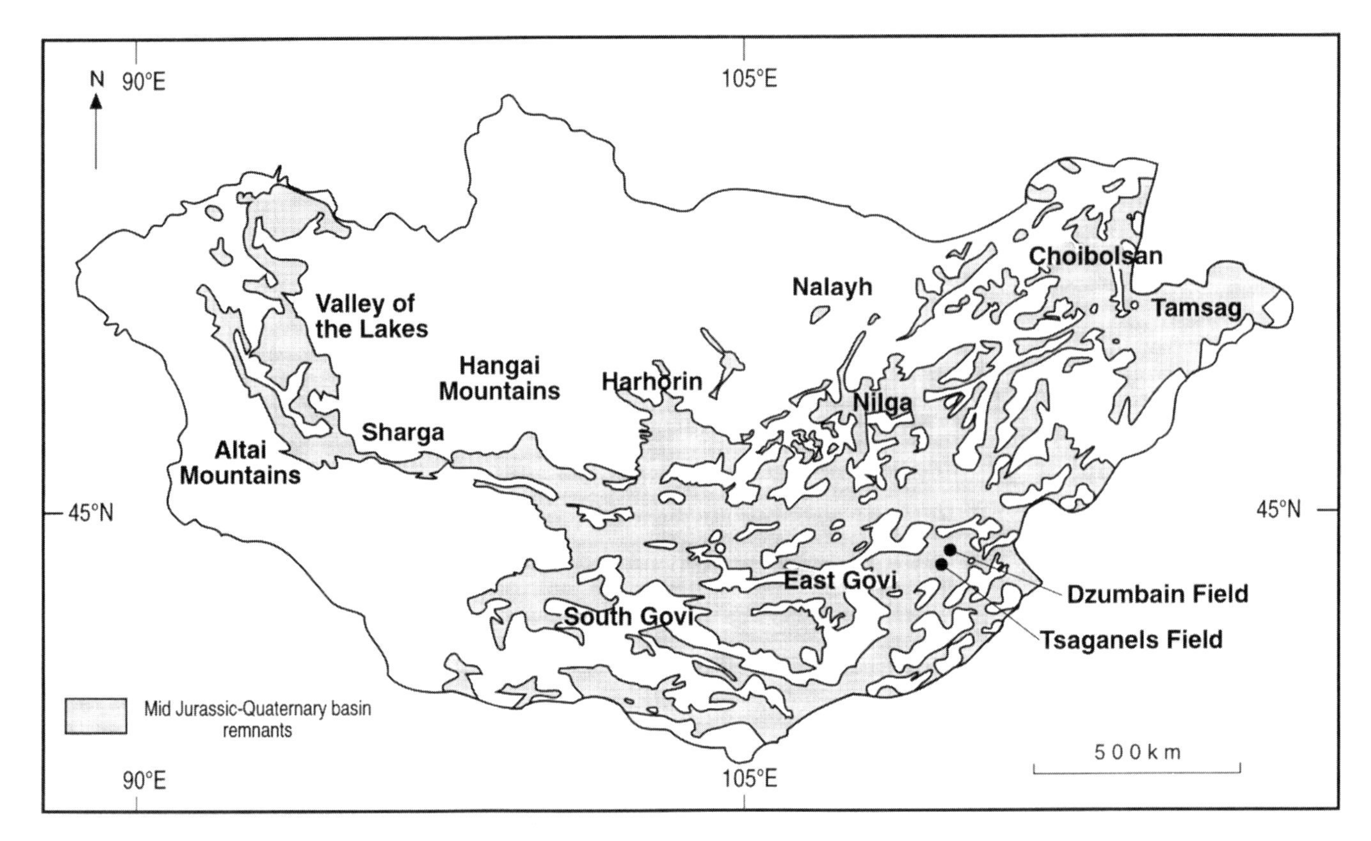

Figure 7—Distribution of major Middle Jurassic–Quaternary basins with preserved sequences (modified from Traynor and Sladen, 1995).

than 5 km. Sandstones contain symmetric and asymmetric ripples, migrating ripple sets, rare flecks of plant debris, and bioturbation that exploits bedding planes. The preferred interpretation is wave-agitated and bioturbated, alkaline shallow-lake environments.

The Lower and Middle Jurassic successions are dominated by continental red beds and acid-intermediate volcanics; however, in southern Mongolia, Lower Jurassic sequences are commonly more than 500 m thick and contain bituminous coals and dark floodplain lake mudstones interbedded with lacustrine mudstones. The sequences share many similarities to the coal-rich Lower and Middle Jurassic sequences nearby in the Tarim, Junggar, and Turpan basins of northwestern China (Hendrix et al., 1995). In these basins, there are more than 1500 m of strata that pass basinward from braidplain and meander to anastomosing fluvial facies into lake margin deltas and then into permanent lake facies.

Coal beds up to 5 m thick can be found in coal mines at Tsagan Ovar, Saikhan Ovar, and Bayan Teg (Figure 4). The sequence at Tsagan Ovar includes about 50 m of lacustrine mudstone suggesting an interplay between lake margin deltas and mud-rich lake facies. The lake mudstones possess varvelike laminations, presumably due to seasonal climatic variations, and sideritic horizons, and various brown-colored mudstones have oil shale potential (Figure 9). At Tsagan Ovar "clinker rock" has formed by recent in situ burning of bituminous coals and oil shales, most likely resulting from spontaneous combustion of outgassed methane from the coals. Actively smoldering fissures are sometimes lined with a crust of yellow sulfurous mineralization.

Figure 8—Triassic(?) fine-grained lacustrine sequences of light gray and gray calcareous mudstones, silty mudstones, and siltstones in coarsening-upward sequences capped by very fine and fine-grained sandstones, east of Ichen Gol.

Figure 9—Lower–Middle Jurassic lacustrine mudstones at the Tsagan Ovar opencast coal mine include gray, dark gray, and brown mudstones, and siderite-rich mudstones.

Biostratigraphic Information

As mentioned, many sequences thought to be Triassic in age fail to yield any significant fauna or flora; sparse plant fragments include *Neocalamites* sp., *Equisetites* sp., and *Phlebopteris* sp. Early–Middle Jurassic red bed sequences also yield little fauna or flora. In contrast, the coal and mudstone sequences of southern Mongolia contain rich palynomorph assemblages. Taxa include *Neoraistrickia gristhorpensis*, *Chasmatosporites* spp., *Quadraeculina anellaeformis*, and *Jiaohepollis* pollen. A very distinctive spore assemblage may be present. This includes numerous ornamented spores such as *Klukisporites variegatus*, *Densoisporites perinatus*, *Leptolepidites* spp., *Duplexisporites* cf. *gyratus*, and *Osmundacidites* spp., together with various verrucose spores of indeterminate affinity and disaccate pollen.

Accurate dating is complicated because there are few age-diagnostic palynomorphs. Worldwide ranges of *Quadraeculina anellaeformis* and *Chasmatosporites hians* confirm an age of Early–Middle Jurassic for some outcrops. These palynomorph assemblages support other facies evidence for nonmarine depositional environments. The algal taxon *Botryococcus* spp. occurs locally and is consistent with the development of at least temporary lakes. The dominance of disaccate pollen probably reflects aridity in the paleoclimate.

Geochemical Characterization

With S2 <1.5 kg/t and TOC <1%, mudstones of Triassic(?) age appear to have little or no hydrocarbon source potential (Table 3). Most organic matter was probably oxidized prior to deep burial. Maturity data indicate that the sequences have reached high maturity levels, and some samples with moderate residual organic richness may originally have had higher source potential for hydrocarbons. Both the hydrogen indices and visual analysis of kerogen indicate potential only for gas.

Lower and Middle Jurassic bituminous coals and mudstones have high TOCs and very good source potential with S2 of coals up to 221.3 kg/t and S2 of mudstones up to 72.0 kg/t (Table 4). The sequences bear several geochemical similarities to the coal-rich sequences of the Jurassic coal measures found in basins in northwestern China (Hendrix et al., 1995). PGC analyses of coals indicate potential for predominantly gas and potential exists for coal-bed methane. Many mudstones show potential for both oil and gas (Figure 6b). At Tsagan Ovar and Bayan Teg richness approaches that of oil shales with TOCs around 20% and S2 to 72 kg/t (Table 4). PGC data indicate a waxy oil-prone source, most likely deposited in a freshwater lake (Figure 6c). Vitrinite data indicate that many of these sequences have generated oil during deep burial.

LATE JURASSIC AND EARLY CRETACEOUS LAKES

Tectonic Overview

Lake basin development was widespread during the Late Jurassic and Early Cretaceous. Rifting and synrift deposition produced an extensive belt of Mesozoic basins across southern and eastern Mongolia (Figure 7). Extension also affected large areas of neighboring China; the Tamsag extends eastward into China as the Hailar basin, and basins constituting the south and east Govi extend southward into China as the Erlian basin (Figure 3). Immediately east of Mongolia, in northeastern

Table 3. Data on Triassic source rock quality from selected sites shown on Figure 4.

Location	Sample	S1 (kg/t)	S2 (kg/t)	HI	TOC (%)	Pr/Ph
SW Govi						
GV/15/1F	Mudstone	0.4	0.1	91	0.11	–
EAST Ichen Gol						
GV/16/12B	Mudstone	0.2	0.5	56	0.89	–
GV/16/13C	Mudstone	0.4	0.9	150	0.60	–

S1 = free hydrocarbons, S2 = cracked hydrocarbons, HI = hydrogen index, TOC = total organic carbon, and Pr/Ph = pristane/phytane ratio.

Table 4. Data on Lower–Middle Jurassic source rock quality from selected sites shown on Figure 4.

Location	Sample	S1 (kg/t)	S2 (kg/t)	HI	TOC (%)	Pr/Ph
Shin Djinst						
GV/18/2C	Coal	2.0	57.2	89	64.0	–
GV/18/3B	Mudstone	1.1	39.3	235	16.7	–
Cceceng Uul						
GV/19/1D	Coal	2.5	221.3	396	55.8	–
Bayan Teg						
GV/20/1B	Mudstone	1.1	31.5	406	7.8	–
GV/20/1I	Coal	2.0	218.9	340	64.4	–
Tsagan Ovar						
GV/2/1G	Coal	0.5	180.4	254	70.9	–
GV/2/1O	Mudstone	1.5	72.0	350	20.6	–
Saikhan Ovar						
GV/2/10G	Mudstone	0.5	1.3	159	0.8	–
GV/2/11A	Coal	5.0	165.4	287	57.7	–

S1 = free hydrocarbons, S2 = cracked hydrocarbons, HI = hydrogen index, TOC = total organic carbon, and Pr/Ph = pristane/phytane ratio.

China, numerous other small lake basins formed in rifts, and slightly farther east in China, the Songliao basin developed a giant lake in a rift setting (Li Sitian et al., 1984; Yang Wanli, 1985). The rifts in Mongolia, however, were not especially large, deep, or well developed in comparison to famous rifts elsewhere in the world. The amount of extension associated with these basins is generally only around 20–40%, and a β factor of 1.2 has been calculated for the Hailar basin (Wang Bing, 1990).

Origins of the regional scale extension that led to the widespread lake basin development is currently poorly understood due to limited structural and chronostratigraphic control. Postorogenic collapse is possible, over the preexisting Triassic mountain belt. Back-arc extension related to subduction of the Pacific plate beneath the North China plate is thought to be a more likely cause. Watson et al. (1987) postulated a convergent Andean-type subduction margin to the east at this time. The presence of significant early synrift volcanics in the Mongolian basins supports this, but the area would have been a very long way (~2000 km) from the subduction zone.

The full cycle of rifting commonly seen in extensional lake basins appears to be incomplete in Mongolia. There appears to have been limited thermal subsidence and failure to develop any significant postrift section (similar to nearby Erlian, but unlike the more distant and famous lacustrine Songliao basin) (see Wang Tonghe, 1986; Yang Wanli et al., 1985). This might be attributable to a coincident jump in the focus of rifting activity due to changes in plate motions, or more likely the effects of subduction rollback of the Pacific plate (cf. Watson et al., 1987; Tian Zaiyi, 1990). Arguably, subduction rollback has been causing extension in northeast China from the Late Jurassic to present-day and accounts for the large number of rift-related lacustrine sections progressively younging eastward toward the Pacific plate.

Alternatively, much of the extension may have resulted from strike-slip motion along the many major strike-slip faults that bounded the plate fragments in China and Mongolia (e.g., Liu Hefu, 1986; Watson et al., 1987). Strike-slip extension conceivably used these old deep-seated structural weaknesses, generating a host of small rhomb-shaped basins. These basins, where subsidence typically rapidly outpaces sedimentation, are well known as ideal sites for lacustrine basins (see Sladen, 1994). As is often the case, a combination of causes may have been responsible for basin development, i.e., both subduction-related and strike-slip–related extension led to conditions ideal for lake basins.

Although there are no seismic data in many areas to assist interpretation, the structural evidence is consistent with inversion of a number of grabens and half-grabens, and the sequences and facies arrangements are consistent with deposition in alluvial and lacustrine synrift environments. Possible analogs to the East Govi and Tamsag basins that are present in nearby China are, in contrast, well studied and explored. These clearly show synrift sections developing during the Late Jurassic and Early Cretaceous in Erlian to the south and Hailar to the east (Figure 10). It thus seems quite likely that the Upper Jurassic and Lower Cretaceous lacustrine sequences in Mongolia represent true synrift deposits.

In many cases, the edges of the extensional lake basins are poorly constrained at the surface because they are masked at outcrop by a thin fluvial blanket of younger Cretaceous and Tertiary siliciclastics. More detailed seismic and gravity data, as well as deep wells, are needed to define the true extent of these synrift basins, although the currently available seismic and gravity data do demonstrate the presence of deep basins with typical characteristics of synrift lacustrine sequences.

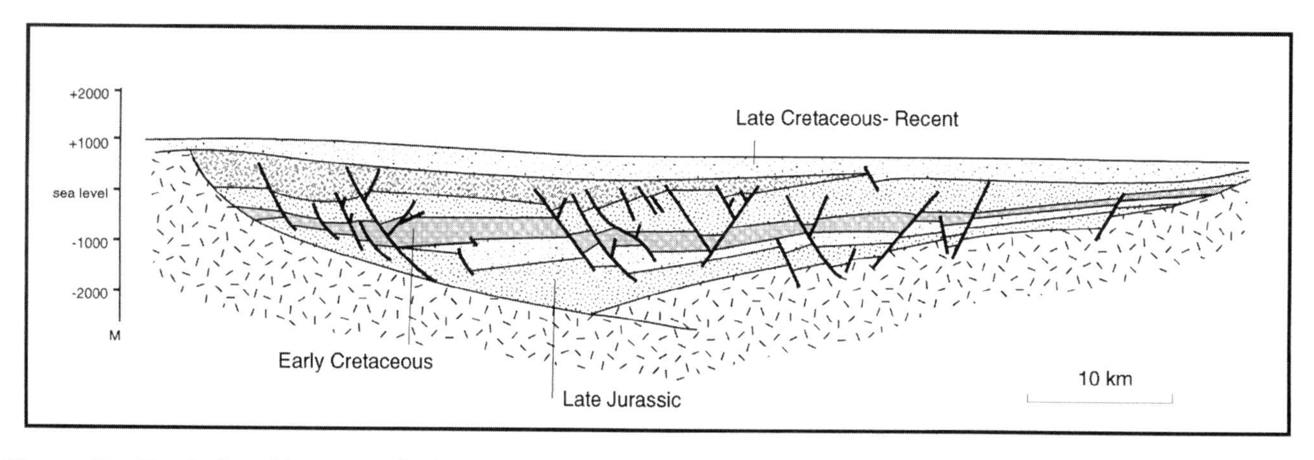

Figure 10—Typical architecture of a late Mesozoic extensional lake basin. Extent of strike-slip and inversion tectonics varies from basin to basin (modified from Traynor and Sladen, 1995).

Basin-Scale Facies Arrangements

On a basin scale, there are typically early synrift alkaline volcanics (basalts, ignimbrites, and tuffs) and coarse alluvial siliciclastics that interdigitate with, and are overlain by, synrift lacustrine mudstones in basin centers and coal-rich alluvium toward the basin margins. Classic synrift facies arrangements are exposed in inverted half-grabens at the basin margins, and nearby well data confirm the rapid basinward facies variations that are so characteristic of synrift lake basins. The youngest synrift sequences indicate a return to dominantly alluvial sedimentation as the lakes were infilled and choked with sediments (Figure 11). Similar sequences are described in lake basins to the east in adjacent parts of China (Li Sitian et al., 1984).

Transpressive inversion destroyed many lake basins during the middle Cretaceous and brought any thermal subsidence to an abrupt end. This tectonism created open folds, thrusts, and flower structures. Many of these structures were partially eroded either before or during Late Cretaceous–Tertiary deposition (Figure 11).

Lithostratigraphy of individual rift sequences in Mongolia varies considerably, and there are few obvious correlations between remnant parts of the rift system. Note analogies to lacustrine basins in rift systems elsewhere, for example, the rift systems of Central Africa, Niger, and Chad (Genik, 1993). Variation in alluvial and lacustrine facies among individual rift basins appears to be a similar characteristic to that seen in the synchronous Erlian and Hailar rifts in neighboring China (see Du Yonglin and Sheng Zhiwei, 1984; Long Yongwen and Zhang Jiguang, 1986). This variety was no doubt a function of different patterns and timings of rift inception together with differences in basin architecture, sedimentation rate, climate, hinterland geology, volcanic activity, and subsidence.

Analysis of the systems tracts and facies mosaics in the rifts indicates that a number of relatively small lakes (<10,000 km^2) developed in discrete rift basins (Figure 12). Most lake basins were <40,000 km^2 in area, and these are in contrast to the giant Songliao lake basin that was ~260,000 km^2 in area and contained a lake of ~100,000 km^2. Some rift basins in Mongolia contain thick lacustrine mudstones and others appear to be filled by coarse siliciclastics or thick coals. These differences presumably result principally from subtle differences in the type of sediment input and in subsidence vs. sedimentation rates among the basins. Late synrift fluvial facies, commonly interbedded with thick lignitic coals and floodplain lake mudrocks, onlap the basement around many rift margins. Outcrop patterns show that these sediments began to link formerly discrete rift basins, probably because thermal subsidence began to replace rifting as the dominant basin-forming mechanism.

Facies Details

Late Jurassic and Early Cretaceous stratigraphy in southern and eastern Mongolia contains two sequences identified as an older Tsagansab sequence (J3-K1"cc") and a younger Dzumbain sequence (K1"dz"). Locally other sequences have been mapped, for example, the Shinhudag (K1"sh"). These are probably lateral facies equivalents of the Tsagansab and Dzumbain sequences or they may reflect the effect of localized rifting and rift infill occurring at different times. It is clear from the facies patterns and sequence distribution that the region comprised a collection of small, partially linked graben, half-graben, and strike-slip basins each with their own alluvial and lacustrine depocenters, as opposed to one large lacustrine basin covering the entire region.

The Tsagansab sequence is a diverse collection of volcanics, alluvial and lacustrine siliciclastics, lacustrine carbonates, and coals that in total are normally less than 1000 m thick. Volcanics often represent the main rock type, and sometimes there are no sediments at all. Basalts are most common, followed by andesites. There are commonly also ashes, tuffs, ignimbrites, and tuffaceous sandstones related to the lavas.

Alluvial siliciclastics are mostly sandstones with subsidiary siltstones and mudstones. Large pieces of

Figure 11—Typical evolution of nonmarine basins from Late Jurassic to the present in Mongolia.

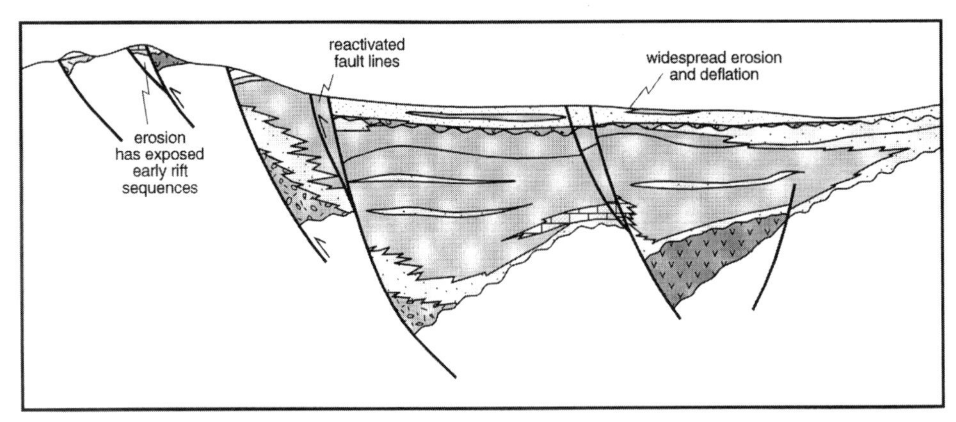

3. LATE TERTIARY - PRESENT
Reactivation of structural lineaments causing local compression has been accompanied by regional uplift and gradual erosion

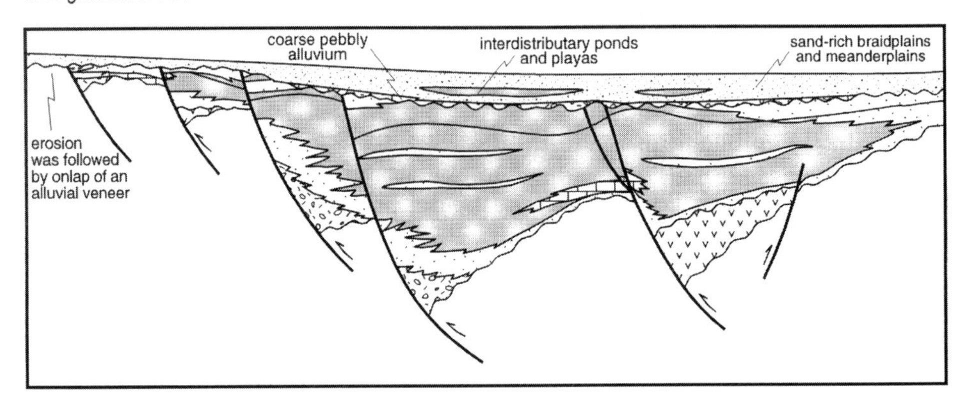

2. LATE CRETACEOUS
Mild mid-Cretaceous inversion and erosion was followed by gentle regional subsidence and widespread deposition of a thin Late Cretaceous alluvial veneer

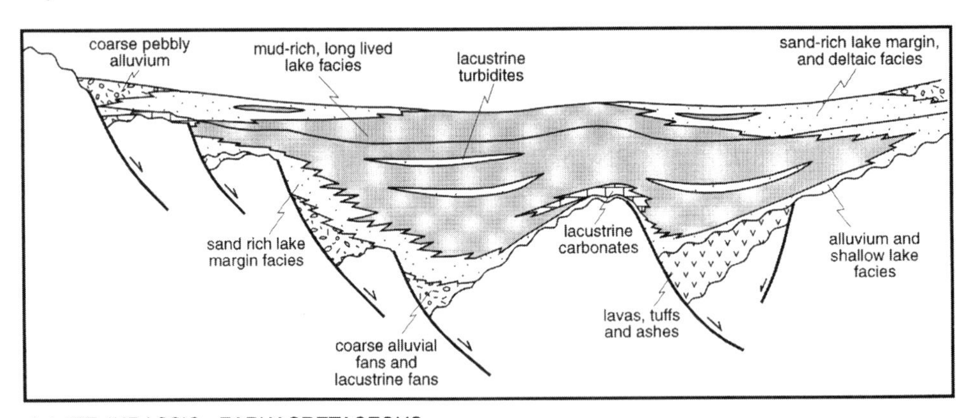

1. LATE JURASSIC - EARLY CRETACEOUS
Local extension along old plate sutures and fault lines formed basins filled with a variety of alluvial and lacustrine syn-rift sequences and in some places volcanics

wood are present in many sandstones representing pieces of tree trunk, branches, and twigs; many have been silicified. Coals are usually poorly developed, comprising only a few meters of brown coal. At Maantag, however, there is up to 40 m of coal (Figure 4). Lacustrine siliciclastics are normally strongly interbedded sandstones, siltstones, and mudstones; there are no extremely thick and extensive lacustrine mudstones in outcrop, although these might exist in some deeper parts of the basins. Locally, in the Nilga basin for example, there are also outcrops of shallow lacustrine carbonates comingling with volcanics. Facies include algal micrites, oncolitic packstones, encrustations with abundant gastropod and bivalve fragments, and carbonate conglomerates and breccias together with opaline silica and chert (Figure 13). These appear to have been deposited in shallow and lake margin environments situated on structural highs

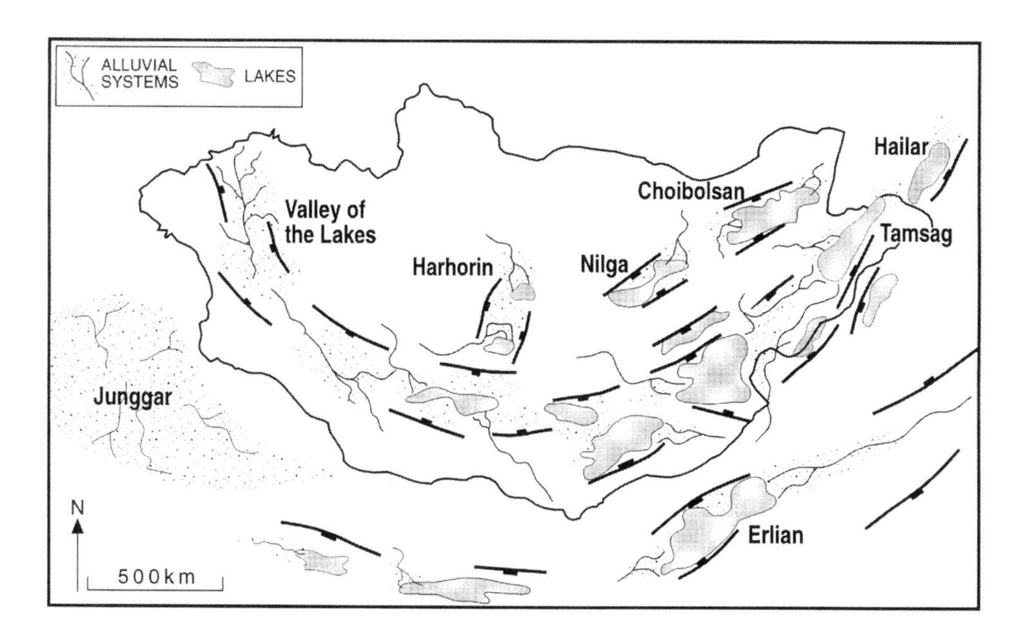

Figure 12—Late Jurassic–Early Cretaceous paleogeographic reconstruction (modified from Traynor and Sladen, 1995).

in the rift basins, such as over the crests of fault blocks. Carbonate diagenesis is complex with extensive secondary porosity. Facies are reminiscent of Lower Cretaceous lacustrine carbonates from Brazil and modern-day lakes in east Africa (see Bertani and Carozzi, 1985; Cohen, 1989).

The Dzumbain sequence is characterized by abundant alluvial siliciclastics, coals, and lacustrine siliciclastics, particularly lacustrine mudstones. In outcrop, the sequence reaches 1000 m thick in places, and gravity and seismic data indicate that it could be up to 2000 m or more in thickness in some areas of the Govi and Tamsag basins.

Alluvial siliciclastics in the Dzumbain sequence are predominantly sand-grade braidplain sequences and mixed sand-mud meanderplain/anastomosing(?) sequences. Most of these alluvial systems probably drained into, and toward, lake sites. Dzumbain sequence coal deposits fringe the main basins and appear to have developed around the edges of the main lake sites. They typically comprise up to 40 m or sometimes more, of brown and brownish black, plant-rich beds often rich in charcoal (Figure 14). In exposures, the coals are usually found to comprise almost entirely wood and plant debris, with large fragments of tree trunks, branches, twigs, and leaves identifiable (Figure 15). Coal-rich sequences are predominantly laminated coals, but in places there are lenses and thin layers of algal coal. Interbedded rocks include mudstones, siltstones, and sandstones, pebbly channeled sandstones, and a variety of seatearth layers, siderite nodules, and concretions. Many exposures contain evidence for synsedimentary faulting. Locally, particularly in small basins away from the main rifts, there may be up to 200 m of coal, for example at Tebhsin Govi, and other commercially significant coals are present at Baganurr, Nalayh, Talbulag, Tugrik, and Erdentsogt Uul (Figure 4).

Facies mosaics indicate that the coals are lateral facies equivalents to alluvial and lacustrine sediments. There are many close similarities to the sequences and facies described from basins immediately east of Mongolia inside China (see Li Sitian et al., 1984). No significant coal deposits are known in the central parts of the larger basins where thicker Dzumbain sequences occur; consequently, where lacustrine mudstones are well developed in basin depocenters, coals are likely to be poorly developed, and in contrast, coals are better developed in alluvial and marginal lacustrine facies belts lying around the main lacustrine depocenters.

Dzumbain sequence lacustrine rocks are best known for their lacustrine mudstones, some of which

Figure 13—Upper Jurassic–Lower Cretaceous lacustrine carbonates on the northern edge of the Nilga basin include conglomerates and breccias with oncolites and algal encrustations, opalline silica, and chert.

Figure 14—Upper Jurassic–Lower Cretaceous coal deposits at the Talbulag opencast coal mine.

Figure 15—Upper Jurassic–Lower Cretaceous coal deposits often include large pieces of tree trunks, branches, and wood, as seen in this example from the Tebshin Govi opencast coal mine.

are excellent oil-prone hydrocarbon source-rocks. In depocenters, mudstones form up to 50% of the Dzumbain sequence, and individual mudstone intervals are up to 150 m thick. Mudstones include laminated gray, black, and brown varieties, with rare thin carbonates, incipient siderite cementation, and siltstone or very fine sandstone stringers. Many remnants of the rifts preserve sequences of highly fissile lacustrine mudstones, sometimes with well-preserved fossil fish, e.g., in the Nilga basin. Laminae have a waxy feel and are slightly plastic and flexible. They are known locally as "paper shales" (Figure 16). Elsewhere, for example at Harhorin, there are "combustible shales" of oil-shale quality that are finely laminated, brownish gray in color, and contain small gastropods, siderite concretions, and rare plant fragments.

The lacustrine mudrocks confirm the existence of significant lacustrine depocenters; however, although lacustrine mudstones did accumulate, it appears that subsidence did not outpace sedimentation for significant periods. Lengthy phases of stable water stratification were rare, and the lakes did not experience long periods of freedom from coarse siliciclastic influxes; consequently, thick, monotonous sequences of mudstones (often >500 m) that are a common feature of many long-lived lake sites appear to be absent.

In the Altai basins in western Mongolia, sequences are localized, but some contain thick continental red beds deposited in alluvial fan, braidplain, meanderbelt, and playa environments. Outcrop geometry, facies continuity, and systems tract analysis indicate that laterally derived coarse alluvial fans in early synrift settings were succeeded by axially flowing braided rivers and more mature meanderbelt systems, linking some of the half-graben basins across transfer fault zones during the late synrift. Meanderbelt sequences include overbank deposits with well-developed caliche layers and sometimes a few minor coals. Lacustrine sequences comprise rare thin reddened playa mudstones. Overall the facies suggest a more arid climate and higher evaporation than in basins to the east in the Govi and Tamsag.

Biostratigraphic Information

Various fossils that can be found in the mudstones include *Arguinella compacta*, *A. elongata*, *A. quadrata*, *Bairdestheria jeholensis*, *Viviparus onogiensis*, *Cypridea polita*, and *Mongolianella* cf. *gigantea*. Palynomorph assemblages in the lacustrine mudrocks and coaly swamp/temporary lake mudrocks can be subdivided into three distinctive facies associations:

In the *Botryococcus* dominated assemblage, *Botryococcus* algae and disaccate pollen are dominant. The East Govi region has also yielded a rich facies-controlled assemblage, dominated by the green algae *Schizosporis reticulatus* and *Schizosporis parvus*. Pollen include *Jiaohepollis* spp., *Rotundipollis punctatus*, and associated forms. Spores that are generally few in numbers include *Densoisporites microrugulatus* and *Kuylisporites lunaris*. Mudstones with abundant *Botryococcus* yield slightly more diverse miospore assemblages, including small numbers of *Pilosisporites trichopapillosus*, *Impardecispora minor*, *Cicatricosisporites* spp., and *Leptolepidites* spp.

Palynofacies work indicates the presence of lakes in which one or two algal species dominated. This situation compares closely with the distribution of algae in many present-day lakes; dominance of freshwater algae indicate high algal productivity. *Botryococcus* develops as colonial planktonic lacustrine green algae, is resistant to a wide variety of environmental conditions, and can withstand a pH range of 4.5 to 8. *Botryococcus* may form such a dense growth over the

Figure 16—Upper Jurassic–Lower Cretaceous finely laminated lacustrine mudstones ("paper shales") from the northern edge of the Nilga basin have a brownish gray color, commonly have laminae with a waxy feel, contain fish remains, and are combustible. They weather rapidly when exposed at the surface, as seen in this geological survey trench.

surface of lakes that it exhausts the oxygen supply and thick oily gelatinous and highly decay-resistant algal deposits may accumulate. *Schizosporis* spp. is also recorded in present-day lacustrine environments.

Algal occurrences most likely reflect a succession of blooms over a number of years. As in modern environments, blooms could have been initiated in a variety of ways depending on ecologic and local climatic circumstances. For instance, by stimulation from nutrient enrichment of the lakes due to fluvial input after a recent rain, alterations in water temperature, and seasonal changes in weather can cause seed populations to be activated.

The diverse spore assemblage contains a high number of disaccate pollen grains, but differs from the *Botryococcus* dominated assemblage in that spores are greater than or equal to disaccate pollen in number and occur in sufficient diversity to indicate different environmental conditions. Taxa include *Pilosisporites trichopapillosus, P. verus, Cicatricosisporites* spp., including *C. australiensis, Rouseisporites reticulatus, Crybelosporites striatus, Scortea granulatus, Impardecispora marylandensis, Concavissimisporites variverrucatus, Aequitriradites spinulosus,* and *A. verrucosus.*

Greater amount and diversity of spores indicate conditions favorable for the development of swamp environments supporting ferns. The presence of coniferous forest pollen indicates a regional montane forest hinterland, whereas the fern spores indicate more local conditions with the development of coals. The presence of *Crybelosporites striatus* is especially significant because this taxon represents the spores of the aquatic water fern *Arcellites.* A lacustrine/swamp basin environment with run-off from mountain streams contributing organic debris, bisaccate pollen, and local fern spores appears likely.

In the *Rotundipollis* pollen assemblage, pollen grains (other than disaccates) are more dominant and more important than spores, and disaccate pollen and algae are scarce or absent. Taxa include *Rotundipollis punctatus, Rotundipollis* spp., *Corollina/Classopollis* spp., *Quadraeculina* spp., and *Jiaohepollis* spp. Spores include *Cicatricosisporites* spp. Taxa unique to China and Mongolia occur in distinctive numbers within this assemblage, including *Rotundipollis punctatus* and related *Rotundipollis* spp., and a variety of taxa assigned to the genus *Jiaohepollis.* Overall the assemblage is thought to represent alluvial and floodplain environments fed by streams and rivers draining a mountainous hinterland.

Geochemical Characterization

Analyses of coals and mudstones show that there are excellent potential hydrocarbon source rocks (Table 5). There is a clear distinction between the coals that possess potential to generate gas and the lacustrine mudstones that primarily have potential to generate oil.

Mudstones with potential to generate oil were deposited in lacustrine environments in the Dzumbain sequence. They possess S2 richnesses sometimes up to 98.4 kg/t, high TOCs, and low pristane/phytane ratios. Gas chromatogram analyses and hydrogen indices indicate clear potential to generate oil (Table 5 and Figure 6e). Lean mudstones typically show only potential for gas. Of the various basins analyzed, Dzumbain, Nilga, and Harhorin clearly contain high-quality oil-prone lacustrine source rocks (Table 5, Figures 4 and 7). Most, if not all, of the bitumen and oil seeps reported in Mongolia (Nemec et al., 1991; Penttila, 1994) probably derive from these lacustrine source rocks. In the Dzumbain area, the thickest and most widespread mudrocks appear to be developed, with lacustrine mudrock sections in excess of 100 m in some wells and in exposures; however, it is common within the lacustrine mudstones to find thin, rich intervals interbedded in thicker intervals of mediocre source potential. Much of the 100-m-thick sections of mudrocks may, therefore, have only limited source potential.

Lacustrine mudstones are sometimes so rich that they are combustible and potential exists for extraction and processing oil from oil shales. For example at Harhorin, where there are S2 yields of ~95–100 kg/ton and TOCs of 14–19% (Table 5). Pristane/phytane ratios are very low (Table 5 and Figure 6e). The mudstones contain a significant algal input, and geochemical fingerprinting shows that the lakes were sometimes fresh and sometimes slightly saline/alkaline. Similar lake waters appear to have existed in the Songliao basin (see Yang Wanli, 1985; Yang Wanli et al., 1985). Data from oils from the Dzumbain oil field can be geochemically typed to local lacustrine mudstone source rocks (compare Figure 6e and f), as can a bitumen deposit on the edge of the Nilga basin and oil

Table 5. Data on Upper Jurassic–Lower Cretaceous source rock quality from selected sites shown on Figure 4.

Location	Sample	S1 (kg/t)	S2 (kg/t)	HI	TOC (%)	Pr/Ph
Nilga						
UB/1/3D	Mudstone	1.1	40.3	541	7.4	0.04
UB/1/3H	Mudstone	2.9	66.9	589	11.4	0.12
UB/1/3AA	Mudstone	0.3	4.2	452	0.9	–
Baganurr						
UB/3/6A	Coal	0.7	51.5	108	47.6	4.50
Nalayh						
TS/19/4A	Coal	0.9	68.5	127	53.8	2.78
TS/19/4C	Mudstone	0.0	47.0	132	35.6	–
Uginurr						
UB/7/6A	Coal	1.1	61.4	103	59.5	–
Harhorin						
UB/8/1B	Mudstone	1.2	34.3	433	7.9	0.21
UB/8/2A	Mudstone	5.0	98.4	539	18.3	–
UB/8/2B	Mudstone	4.6	95.1	508	18.7	0.11
Dzumbain						
EG/7/13D	Mudstone	8.5	53.1	637	8.3	0.80
EG/8/4C	Mudstone	0.5	11.0	431	2.5	0.45
EG/8/4D	Mudstone	0.1	2.1	350	0.6	–
Mandac						
EG/14/5A	Mudstone	0.4	1.3	138	0.9	–
Tebshin Govi						
EG/15/1A	Coal	1.5	35.9	99	36.2	–
EG/15/1C	Coal	1.1	134.2	260	51.7	–
EG/15/1J	Mudstone	0.5	1.4	28	5.0	–
Choibolsan						
TS/2/3C	Coal	0.9	53.2	94	56.8	0.83
Matad						
TS/7/2A	Coal	6.0	57.4	115	49.9	0.99
TS/7/2B	Coal	2.7	32.4	83	39.2	–
Tal Bulag						
TS/17/10D	Coal	0.8	65.9	122	53.9	1.95
TS/17/10E	Mudstone	0.3	5.3	67	7.9	–
SW Tamsag						
TS/11/5B	Coal	2.4	268.1	407	65.9	–
TS/11/6A	Coal	7.3	224.6	454	49.4	4.33
Maantag						
AT/9/10J	Mudstone	1.4	91.1	283	32.2	0.48
AT/9/10Q	Coal	2.4	44.1	98	45.0	–
AT/9/10R	Coal	1.0	101.2	158	64.2	–

S1 = free hydrocarbons, S2 = cracked hydrocarbons, HI = hydrogen index, TOC = total organic carbon, and Pr/Ph = pristane/phytane ratio.

seeping in sandstones at Harhorin. These oils show the typical characteristics of oils expelled from source rocks that have reached moderately low maturity levels and temperatures of ~110°C. Sterane and triterapane distributions are characterized by a full suite of C_{27}–C_{29} steranes, a moderate hopane to hopane+sterane ratio, low pristane/phytane ratios, and the presence of gammacerane and β-carotene (Table 5 and Figure 6e).

The oils show typical lacustrine characteristics and have medium to high gravities (25°+API), low gas/oil ratios, low viscosity, low sulfur, high wax contents, high temperature pour points, high nitrogen, and high Ni/V ratios (see Traynor and Sladen, 1995). There are many similarities to the classic Early Cretaceous lacustrine oils and source rocks of the Songliao basin in China (Yang Wanli, 1985; Yang Wanli et al., 1985). A reducing lacustrine environment (which would aid preservation of organic matter) is indicated by pristane/phytane ratios. Data on sterane and triterapane distributions, a scarcity of C_{25+} molecules, and the presence of gammacerane and the scarcity of high-molecular-weight hopanes together suggest algal lacustrine source rocks deposited in salinities that ranged from freshwater to slightly saline and alkaline. The gas/oil ratio is extremely low at both the Dzumbain and Tsaganels oil fields. Neither field has recorded any significant gas during past production, suggesting the ratio is <10 ft^3/bbl. This too is consistent with generation from oil-prone algal lacustrine source rocks that have reached very limited maturity.

Coals with potential to generate gas are common in the Dzumbain sequence around basin margins and have richnesses ranging up to S2 268.1 kg/t and TOCs to 68.7% (Table 5). N-alkanes are back-end biased and have high odd-over-even preference (Figure 6d). Compared to the lacustrine mudstones, the pristane/phytane ratios are high (Table 5). Mudstones that are interbedded with the coals normally have poor source potential. Many coals are brown and lignitic at the surface and have limited maturity. Thermal values range from 3000 to 6500 Kcal/t. Low vitrinite reflectance values suggest that the coals on the basin margins have had little burial, generally only 1–2 km; consequently, they may never have generated and expelled any significant volumes of gas. This limited depth of burial on the basin margins is consistent with subsequent basin evolution and overlying stratigraphy (Figure 11).

Reservoir and Cap Rock Properties

Proven sandstone reservoirs are present in the Dzumbain and Tsaganels oil field (see Traynor and Sladen, 1995, for details). Porosities are variable (ranging from 5 to 20%), reflecting facies variations and variable, commonly complex diagenesis. Similar sandstones are proven reservoirs in the Erlian and Hailar basins in nearby China (e.g., Zhang Jinliang and Shen Feng, 1990; Yu Jiaren, 1989). There are a variety of other potential reservoirs, including alluvial and lacustrine conglomerates, lacustrine carbonates, and volcanic rocks. These are all characterized by heterogeneity including variable porosity and permeability, limited lateral extent, and fluctuating net/gross ratios.

In the Tsagansab sequence, fractured and vesicular volcanics are also a potential reservoir target. Similarly aged volcanics form an important reservoir target in the Erlian basin in China, and similar possibilities exist in Mongolia. In Erlian, vesicular and fractured andesites, basalts, and tuffs possess up to 10% porosity and 1000 md permeability; well flow rates are typically higher than from the siliciclastic reservoirs (Du Yonglin et al., 1984; Yu Jiaren, 1989).

Lacustrine and fluvial overbank mudstones are proven effective cap rocks for hydrocarbon accumulations in the Dzumbain and Tsaganels accumulations. Again, there are similar proven cap rocks in the nearby Erlian and Hailar basins in China. Lacustrine mudstones appear to be the thickest and most laterally extensive cap rock facies, whereas mudstones in the alluvial sequences simply contribute layering and heterogeneity to reservoirs.

In the basins of the Govi and Tamsag that extend across the Mongolian border into China, oil is produced from various fault traps including paleostructures (e.g., Wang Bing, 1990; Du Yonglin and Sheng Zhiwei, 1984; Long Yongwen and Zhang Jiguang, 1986; Zhang Jiguang, 1990). The Chinese parts of this Mesozoic basin trend are relatively well explored, with detailed seismic coverage (often 1 × 1 km grids) and considerable drilling. This gives an insight into petroleum potential in the Govi and Tamsag (Figure 17). In Erlian, the largest accumulation, Aershan, is a cluster of five small fields and approaches 100 million bbl of reserves; current total production is around 20,000 bbl of oil per day from a total of over 100 wells. Multiple thin hydrocarbon pay zones are typical.

LATE CRETACEOUS, TERTIARY, AND PRESENT-DAY LAKES

Tectonic Overview

An unconformity separates the Lower and Upper Cretaceous sequences, and associated with this there is evidence for inversion of the Early Cretaceous basins, folding, and erosion (Figure 11). Following elimination of the extensional lake basins of the Early Cretaceous, Upper Cretaceous continental siliciclastics were spread over large areas of southern and eastern Mongolia. These sediments were deposited during widespread gentle subsidence. The sequences commonly give the appearance of having developed in a postrift setting. They markedly onlap and overstep the former rift basin margins, passively infill older topography, and create a thin sedimentary veneer (Figure 11). Over wide areas, dips are low, typically <5°, and the total Upper Cretaceous–Tertiary thickness is rarely more than 500 m. During both the Late Cretaceous and the Tertiary, most of Mongolia was one of limited, intermittent deposition and frequent nondeposition or erosion. Similar thin sequences extend across the border into the Erlian and Hailar basins in China (see Du Yonglin and Sheng Zhiwei, 1984).

In Mongolia, Upper Cretaceous rocks are most widespread in the East Govi, occupying large areas north, south, and southwest of Saynshand. Around the basin margins, sequences passively onlap and overstep onto older rocks that range back to Precambrian in age. There is no evidence to support a continuation of extension and synrift deposition apparent in the Lower Cretaceous sequence below. It is likely that the Late Cretaceous represents a form of poorly developed postrift deposition. Although classic "steers head" geometries, characteristic of the full rifting process, exist in the Songliao basin to the west, they are absent in Mongolia (cf. Liu Hefu, 1986; Yang Wanli et al., 1985). Their absence may be explained by a rapid jump in the focus of rifting, and perhaps also due to similarities between crustal and lithospheric stretching. Late Cretaceous sediments were derived by erosion from remnants of the Early Cretaceous

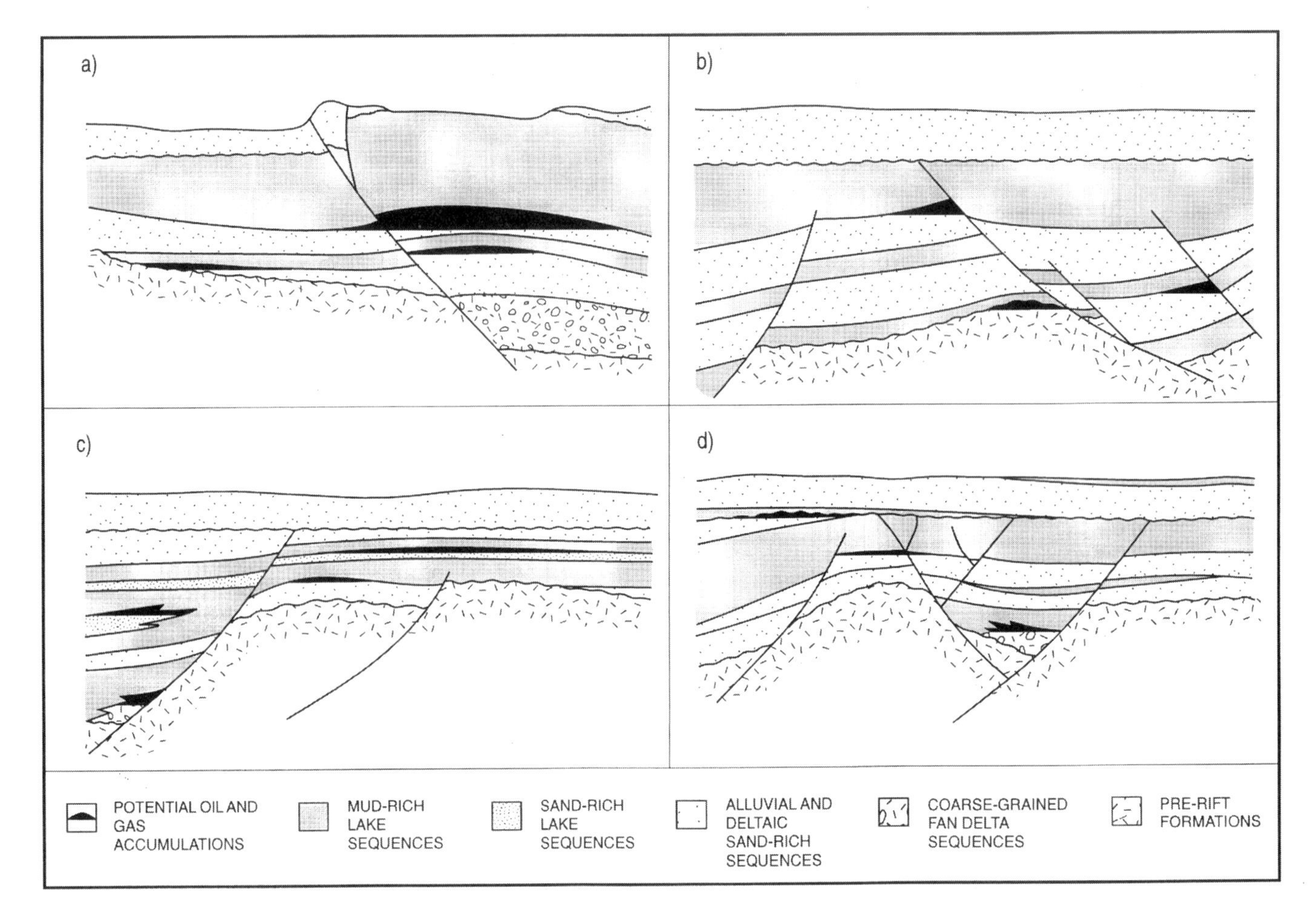

Figure 17—Common hydrocarbon trapping styles in late Mesozoic and Cenozoic lake basins of Mongolia.

topography and locally from newly formed inversion structures.

Upper Cretaceous and Lower Tertiary siliciclastics are rarely preserved in the Altai Mountains. This area seems to have undergone substantial uplift and was dominated largely by active erosion. There are, however, more widespread Quaternary–Holocene sediments and a number of present-day lakes (described in a following section).

Facies Details

Late Cretaceous sequences in southern and eastern Mongolia are often mapped as the Saynshand sequence (K2 "ss") and Bayunshirere sequence (K2 "bs"). In western parts of southern Mongolia, an additional sequence known as the Barungoi (K2 "br") is recognized. Mapping indicates that the sequences young in the order Saynshand, Bayunshirere, Barungoi. Dating of the component sequences is difficult because they are dominated by poorly fossiliferous, coarse-grained nonmarine siliciclastics. Many sequences are reddened, further hampering biostratigraphy. The Upper Cretaceous in southern Mongolia has yielded world-renowned dinosaur bones and eggs but few other fossils (see Berkey and Morris,1927; Gradzinski and Jerzykiewicz, 1974).

With limited subsidence, Late Cretaceous and Tertiary paleoconditions were conducive neither to long-lived "permanent" lake sites, nor the formation of thick lacustrine sequences. Intermittently active braided rivers drained through the topography depositing pebble-beds and channel bar sandstones (the most common depositional feature), or less common fine-grained extrachannel and playa sediments. Wind action created eolian dunes in some areas.

The predominant braidplain facies are fine-, medium-, and coarse-grained porous sandstones that are fawn-colored, reddened, or light gray in color, and commonly form >75% of sequences. Continuous layers of pebble beds are also common, with polymictic conglomerates containing clasts up to 2 cm in diameter. Sedimentary structures and lithologies clearly indicate dominantly alluvial deposition from braided streams and sheetfloods. Rarely there are laterally discontinuous thin siltstones or red bed mudstones, but these are typically only 1 or 2 m thick. These probably represent fine-grained sheetflood deposition or short-lived playa lake mudflats.

Many sequences show evidence of soil processes with a variety of caliche profiles, often with irregular nodules and root forms, gypcretes, and a variety of brown, yellow, and red color mottling. No doubt the alluvial plains were intermittently active sites of deposition and there were frequent opportunities for soil formation. Many of the mature soil profiles probably indicate many thousands or hundreds of thousands of years of nondeposition.

There do not appear to have been any substantial thickness of Tertiary sedimentary sequences in Mongolia. Color-mottled alluvium and lacustrine veneers characterize some of the larger basinal areas, but these sediments rarely attain a few hundred meters in total thickness. A combination of shallow borehole information and field observations indicate that sequences are usually only 20–100 m thick. There are typically no fossils present; exact age dating is difficult, and the sequences are variably assigned a Paleogene, Neogene, or Quaternary age based on outcrop relationships. Sequences are characterized by brown and red siliciclastics lying uncomformably on Cretaceous or older sequences. There are also locally developed basaltic volcanics.

Tertiary siliciclastics include coarse alluvial breccias, conglomerates, and fanglomerates deposited close to mountains and fault scarps, to alluvial and playa lake sandstones, siltstones, and mudstones deposited in basinal areas usually overlying Upper Cretaceous sequences. The sediments are normally poorly consolidated, although locally there is calcrete cementation in sandstones, and calcrete and gypcrete profiles in red bed muds (Figure 18). Tertiary sequences rarely contain thin (<20 m) reddened mudstones that may represent short-lived playa lake deposits forming on distal parts of alluvial plains and at terminal drainage sites; however, none of these deposits are areally extensive, and in terms of hydrocarbon geology, there are no significant cap rocks nor any notable hydrocarbon source rock potential.

Figure 18—Upper Cretaceous and Tertiary sediments are predominantly reddened alluvium and commonly contain caliche profiles, as seen in this example from the Neogene south of Tsogt Tsigsee in the South Govi.

Present-Day Lakes

Strike-slip motion, basement-involved thrusting, and regional uplift continue to affect large areas of Mongolia. Many of the structures that lie around the current depocenters appear to be complex flower structures and involve reactivation of older deep-seated faults. Overall there is very limited deposition and widespread nondeposition or erosion. Regional base level elevation is about 1000 m above sea level, with some areas still rising. Presently, significant sediment exits Mongolia via rivers draining to the north into Lake Baikal (Selenga and Orhon rivers) and to the east into Hulun Nurr (Kerulen River), while in the south, deflation removes considerable sediment from the Govi desert contributing to the loess of north-central China (Figure 4).

In the East Govi, there are neotectonic strike-slip fault strands up to 30 km long. These are clearly visible on Landsat images and in the field, where they form prominent hill lines often bringing much older rocks to the surface. Field evidence indicates that the most recent motion is sinistral strike-slip. Temporary lakes have formed in depressions developed adjacent to these hills in areas where the faulting has created subsidence and downwarping. When traced farther to the northeast, the fault strands pass into the Tamsag basin, and some of these control the position of the modern lake Burr Nurr on the eastern Mongolian border with China.

There are several small Late Tertiary compressional basins in the Altai Mountains, including piggyback basins riding on reactivated Paleozoic structures. The much larger Valley of the Lakes basin today lies between southerly-dipping thrust slices in the Altai to the south and northerly-dipping thrust slices on the margins of the Central Mongolian plate to the north (Figure 19).

Many faults are still active; the Altai Mountains are presently rising and are an active earthquake belt. Recorded earthquakes are dominantly shallow and of intermediate depths and appear to reflect a shuffling of the complex mosaic of small, angular plate fragments and tectonic slivers. Evidence of compression can often be found in gentle folds in Neogene rocks, and more dramatically north of Ichen Gol where lower Paleozoic ophiolite is thrust over Upper Cretaceous red bed sediments (Figure 4).

In the Altai Mountains, high-energy streams flow down the mountains into the Valley of the Lakes and then flow longitudinally following the structural grain toward closed lows that contain areally extensive lakes (Figures 20 and 21). Salinity of these lakes is very variable. Lakes that are full-to-spill and have open drainage are fresh, such as Harus Nurr near Hovd. In contrast, Khargus Nurr, which is a terminal lake site lying at the end of the same drainage system, is quite brackish (Figure 21).

In the desert areas of southern Mongolia, evaporation is so intense that many small rivers simply

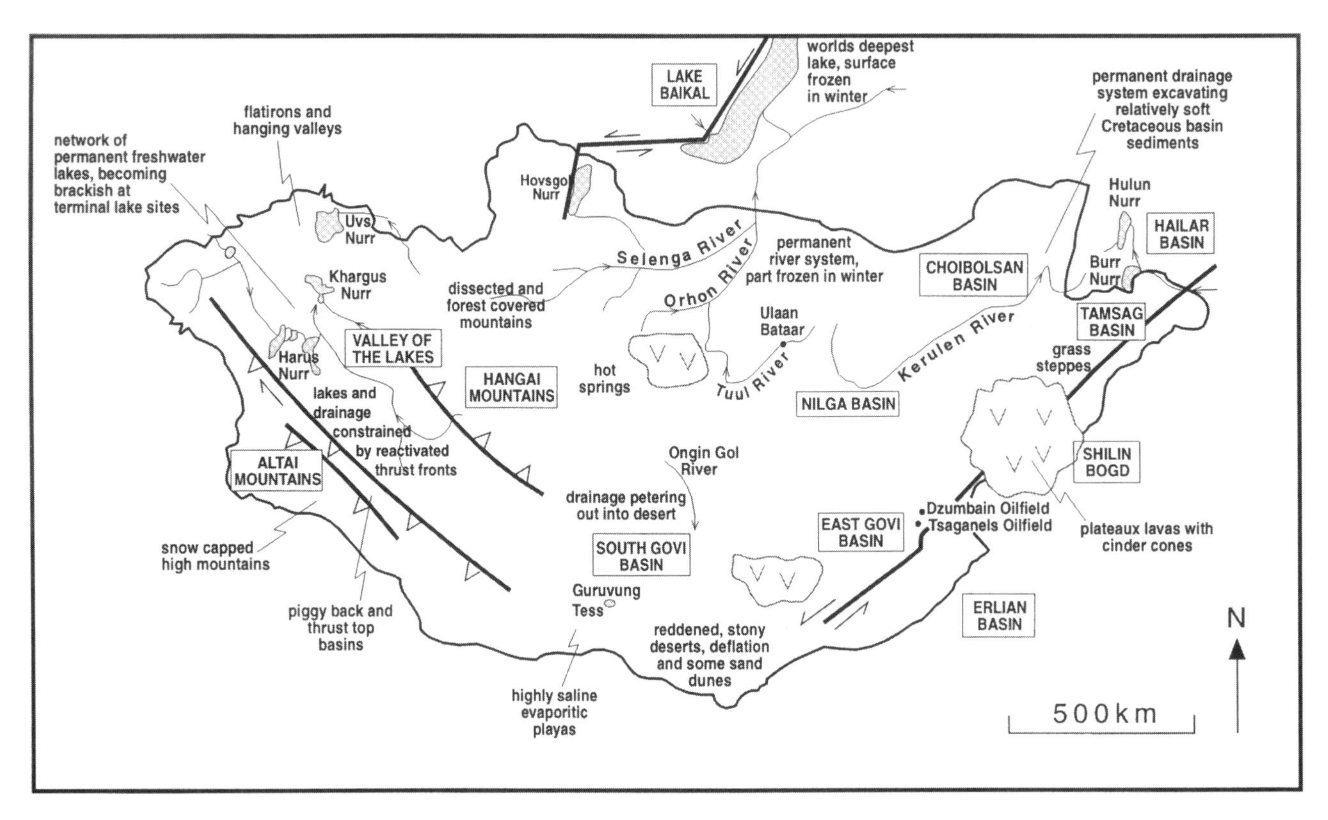

Figure 19—Present day geography/geology of Mongolia showing alluvial systems draining into a variety of lakes in strike-slip and compressional basins, together with the main Quaternary volcanic centers (modified from Traynor and Sladen, 1995).

Figure 20—Uvs Nurr is a freshwater lake in the Valley of the Lakes in northwest Mongolia and occupies a structurally controlled closed topographic low.

Figure 21—Khargus Nurr is a brackish to saline lake in the Valley of the Lakes that occupies a topographically closed low at the end point of the drainage system.

evaporate away in the desert plains, for example the Ongin Gol river south of Saikhan Ovar. Elsewhere, there are highly saline playa lakes depositing salts. At Guruvung Tess, in the south Govi, the end point of the alluvial system tracts is a small playa lake 2–3 km^2 in size surrounded by sand dunes (Figures 4 and 22). Intense evaporation, particularly during the hottest, driest months of April to October, precipitates mirabalite ($Na_2SO_4.10H_2O$). These deposits are sufficiently pure and extensive to be mined in open pits and exported (Figure 23).

Lake Baikal, immediately north of Mongolia in east Siberia, is a Holocene sinistral pull-apart basin, which

Figure 22—At Guruvung Tess in the South Govi basin, the topographic low at the end point of the alluvial systems tracts is a small playa lake 2–3 km^2 in area and surrounded by sand dunes.

Figure 23—The playa lake at Guruvung Tess precipitates monoclinic mirabalite crystals ($Na_2SO_4 \cdot 10H_2O$) that are mined in open pits.

contains the world's deepest lake (Figure 19). This lake has formed along a major crustal scale weakness that has been periodically reactivated since the Precambrian. Vertical fault throws in places reach 6 km. In the pre-Tertiary, strike-slip motion was predominantly dextral, but in the Tertiary, motion has been mostly sinistral (Sherman, 1978). The fault system that controls the position of Lake Baikal extends southwestward into Mongolia and various transfer faults relay motion. Earthquakes are a common phenomenon (Stepanov and Volkhonin, 1969). In northern Mongolia, the fault system controls the position of lakes such as Hovsgol Nurr and its related drainage system. Both Hovsgol Nurr and Lake Baikal, as well as many of the river systems in northern Mongolia, have their surfaces permanently frozen in the winter, often for 3 months or more.

Large plateaus of Neogene and Quaternary flood basalts exist in eastern and central Mongolia, as well as smaller accumulations in north Mongolia. Most flood basalts are 50–250 m thick. The largest volcanic plateaus are southwest of the Tamsag basin around the Shilin Bogd mountains and the town of Daraganag (Figures 4 and 19). The plateaus cover an area of some 10,000 km^2 and comprise mostly vesicular ropy basaltic lava flows and pumice surmounted by scattered volcanic cones, up to 300 m high; however, there is no current volcanic activity. These basalts were probably extruded in response to extension on steep crustal-scale faults along the old plate margins in a generally transtensional tectonic regime. Volcanic flows are not conducive to forming volcanic lakes, and even where there are volcanic cones, no volcanic crater lakes appear to have developed. Although lakes in many countries are frequently found associated with recent volcanic rocks, in this instance none are known.

Effects of glacial activity in Mongolia appear to have been minor. Overall there is little evidence for ice sheets or glaciers, probably because low humidity associated with its continental interior location limited their development. Ice sheets may have covered the higher parts of the Altai Mountains. The large size of some present-day alluvial fans in this region appears inconsistent with current arid climate, size of their sediment source area, and sediment transport mechanisms. The growth of these fans was perhaps much greater while the region was experiencing glacial outwash. In the far northwest, north of Ulan Gom, drainage systems include some examples of hanging valleys and flatirons (Figure 19). Cryoturbation is a common feature throughout Mongolia and can be seen in many soil profiles exposed around the rims of quarries; however, evidence for the existence of large glacial lakes is absent.

SUMMARY

Mongolia represents a case study in lacustrine diversity with various kinds of lake sediments identified in all geologic periods from the Carboniferous to Quaternary.

By the late Paleozoic, following shoaling and emergence of a broad continental shelf, lakes began to develop in compressional basins. Emergence culminated in a large mountain range developing across Mongolia in the early Mesozoic. These mountains created a major Asian watershed. Within the eroding mountain range, lakes formed in intermontane basins. Later on, during the late Mesozoic and Tertiary, reactivation of inherited crustal weaknesses, largely by strike-slip tectonics, created the opportunity for lakes to develop in rift and strike-slip extensional basins. Lake basins were typically <40,000 km^2 in area and the lakes within them were usually <10,000 km^2 in size. Subsidence rarely outpaced sedimentation for long periods, and so conditions were not suitable for long-lived lake basins containing extremely thick lake sequences. Many lake basins were rapidly infilled with sediment. Volcanic rocks were also partly responsible for infilling basins.

Late Jurassic and Early Cretaceous lacustrine sequences that developed in relatively long-lived lakes include algal-rich, oil-prone mudstones, algal micrites, and carbonate encrustations and oncolites, together with chert, bivalves, and gastropods. There are various other subaqueous lacustrine facies of sandstones, siltstones, and mudstones. Many lakes were fringed with swamps and marshes that accumulated coals and elsewhere some basins developed reddened mudstones in playas.

The Late Jurassic and Early Cretaceous lake basins appear most significant for generating and trapping hydrocarbons. Many of these basins accumulated oil-prone mudstones and oil shales that have locally generated lacustrine oils. Oil has been trapped in interbedded lacustrine and alluvial reservoirs, and the potential exists for a number of structural and stratigraphic accumulations.

In addition to lacustrine sequences, there are alluvial fan sequences rich in polymict conglomerates, braidplain sequences dominated by sandstones, and meanderplain/anastomosing(?) sequences of interbedded sandstones, siltstones, mudstones, and coals. Sequences of carbonaceous mudstones and coals developed in temporary floodplain lakes on floodplains and the tops of deltas. Soil processes on the alluvial plains formed caliche, gypcretes, and extensive color mottling. Some eolian sequences are also present; these comprise dune sandstones that locally contain world-famous dinosaur skeletons and eggs.

Volcanic rocks commingle with many of the alluvial and lacustrine sequences. These include basalts (sometimes forming plateaus), andesites, rhyolites, felsic lavas, cinder cones, ignimbrites, chloritized and flow-banded tuffs, and tuffaceous sandstones; however, no volcanic lakes have been identified.

The position of modern lake basins in Mongolia is largely controlled by reactivation of crustal weaknesses under the influence of compression and strike-slip tectonics. These "successor basins" contain lakes that range from fresh to highly saline, developing in a region of arid to semiarid continental interior climate. In the winter, many of these lakes are frozen for more than 3 months; however, no evidence for the previous existence of large glacial lakes has been found.

Mongolia contains a diverse assemblage of lake basins, lacustrine facies, and depositional environments spanning more than 250 m.y. Understanding of the lacustrine record is still at an early stage, and Mongolia's own particular set of geologic, logistical, and environmental challenges will largely dictate the pace at which our understanding improves. The detailed tectonic and stratigraphic history of Mongolia is no doubt more complicated than the relatively simple concepts and models presented here. Key questions remain concerning basin development, and more detailed data and interpretation will be required to resolve the paleogeography and stratigraphic correlations. Detailed controls on sediment thicknesses and facies can only become better understood with improved seismic and gravity data coverage combined with subsurface control from deep wells and further detailed studies.

ACKNOWLEDGMENTS

We would particularly like to acknowledge the guidance and foresight of Ross Stobie and Guy Stubbs throughout our studies, which also involved a large number of BP Exploration staff. The Mongolian state oil company, Mongol Gazarin Toss, provided unpublished data, exhaustive field guidance, and general support; this project could not have been undertaken without their assistance. In particular, the efforts of Doobatyn Sengee are much acknowledged and appreciated. This paper is published with the permission and assistance of BP Exploration.

REFERENCES CITED

Berkey, C. P., and F. K. Morris, 1927, Geology of Mongolia. Natural History of Central Asia, vol 2. American Museum of Natural History, New York, 475 p.

Bertani, R. T., and A. V. Carozzi, 1985, Lagoa Feia Formation (Lower Cretaceous), Campos Basin offshore Brazil: Rift Valley stage lacustrine carbonate reservoirs—I. Journal of Petroleum Geology, v. 8, p. 37–58.

Broutin, J., J. Doubinger, M. O. El Hamet, and J. Lang, 1990, Comparitive palynology in the Permian of Niger (West Africa) and the Peritethysian realm: stratigraphic and phytogeographic implications. Review of Palaeobotany and Palynology, v. 66, p. 243–261.

Carroll, A. R., S. C. Brassell, and S. A. Graham, 1992, Late Permian lacustrine oil shales, southern Junggar Basin, Northwest China. AAPG Bulletin, v. 76, p. 1874–1902.

Cohen, A. S., 1989, Facies relationships and sedimentation in large rift lakes and implications for hydrocarbon exploration: examples from Lakes Turkana and Tanganyika. Palaeogeography, Palaeoclimatology, Palaeoecology, v. 70, p. 65–80.

Du Yonglin and Sheng Zhiwei, 1984, The stratigraphic characteristics in the Erlian Basin. Petroleum Exploration and Development, v. 11, no. 1, p. 1–5 (in Chinese).

Du Yonglin, Sheng Zhinwei, Chen Liuhua and Xiong Xiaogin, 1984, Type and characters of reservoir rock in Erlian Basin. Petroleum Exploration and Development, v. 11, no. 3, p. 1–7 (in Chinese).

Genik, G. J., 1993, Petroleum geology of Cretaceous–Tertiary rift basins in Niger, Chad, and Central African Republic—1993. AAPG Bulletin, v. 77, p. 1405–1434.

Gradzinski, R., and T. Jerzykiewicz, 1974, Sedimentation of the Barun Goyot Formation. Palaeontologia Polonica, v. 30, p. 111–146.

Hendrix, M. S., S. C. Brassell, A. R. Carroll, and S. A. Graham, 1995, Sedimentology, organic geochemistry and petroleum potential of Jurassic Coal Measures:

Tarim, Junggar, and Turpan Basins, Northwest China. AAPG Bulletin, v. 79, p. 929–959.
Li Sitian, Li Baofang, Yang Shigong, Huang Jiafu, and Li Zhen, 1984, Sedimentation and tectonic evolution of Late Mesozoic faulted coal basins in north-eastern China. Special Publication, International Association of Sedimentologists, no. 7, p. 387–406.
Liu Hefu, 1986, Geodynamic scenario and structural styles of Mesozoic and Cenozoic basins in China. AAPG Bulletin, v. 70, p. 377–395.
Long Yongwen and Zhang Jiguang, 1986, Characteristics of petroleum geology and potential of Hailar Basin. Oil and Gas Geology, v. 7, p. 59–67 (in Chinese).
Marinov, N. A., R. A. Hasin, Y. A. Borzakovsky, and L. P. Zonnenshain, 1973, Geology of the Mongolian Peoples Republic. I. (Stratigraphy). Nedra, Moscow, 583 p. (in Russian).
Nemec, M. C., M. M. Page, and W. C. Penttilla, 1991, Sedimentary basins and petroleum occurrences, Mongolia. AAPG Bulletin, v. 75, p. 646.
Penttila, W. C., 1994, The recoverable oil and gas resources of Mongolia. Journal of Petroleum Geology, v. 17, p. 89–98.
Sherman, S. I., 1978, Faults in the Baikal Rift Zone. Tectonophysics, v. 45, p. 31–39.
Sladen, C. P., 1994, Key elements during the search for hydrocarbons in lake systems, *in* E. Gierlowski-Kordesch and K. Kelts, eds., Global Geological Record of Lake Basins, Cambridge University Press, Cambridge, p. 3–17.
Stepanov, P. P., and V. S. Volkhonin, 1969, Present-day structure and deep structure of the Earth's crust of Mongolia according to petrophysical data. Soviet Geology, v. 5, p. 47–73 (in Russian).
Tian Zaiyi, 1990, The formation and distribution of Mesozoic–Cenozoic sedimentary basins in China. Journal of Petroleum Geology, v. 13, p. 19–34.
Traynor, J. J., and C. P. Sladen, 1995, Tectonic and stratigraphic evolution of the Mongolian People's Republic and its influence on hydrocarbon geology and potential. Marine and Petroleum Geology, v. 12, p. 35–52.
Wang Bing, 1990, Characteristics of tectonic geology and hydrocarbons for Mesozoic Erlian Basins. Experimental Petroleum Geology, v. 12, p. 8–20 (in Chinese).
Wang Tonghe, 1986, A preliminary study on the Erlian Basin characteristics for structural petroleum geology. Experimental Petroleum Geology, v. 8, p. 313–324 (in Chinese).
Watson, M. P., A. B. Hayward, D. N. Parkinson, and M. Zhang, 1987, Plate tectonic history, basin development and petroleum source rock deposition onshore China. Marine and Petroleum Geology, v. 4, p. 205–225.
Yan Wenbo, 1984, Preliminary study of oil generation in Baiyan-Hua group of Erlian Basin. Acta Petrolei Sinica, v. 5, p. 29–35 (in Chinese).
Yang Wanli, 1985, Daqing oil field, Peoples Republic of China: a giant field with oil of nonmarine origin. AAPG Bulletin, v. 69, p. 1101–1111.
Yang Wanli, Li Yongkang, and Gao Ruiqi, 1985, Formation and evolution of non-marine petroleum in Songliao Basin, China. AAPG Bulletin, v. 69, p. 1112–1222.
Yu Jiaren, 1989, Water saturation of the pays in Menggulin Breceia of Erlian Basin. Experimental Petroleum Geology, v. 11, p. 273–276 (in Chinese).
Zhang Jiguang, 1990, A discussion on the formation of Urxun downfaulted structural belt and its relation to hydrocarbons. Petroleum Geology and Oil Field Development in Daqing, v. 9, p. 1–7 (in Chinese).
Zhang Jinliang and Shen Fang, 1990, Diagenenis and reservoir properties of Damoguaihe Fm., Late Jurassic, Wurxun depression, Hailar Basin. Petroleum Exploration and Development, v. 17, no. 5, p. 75–86 (in Chinese).

Lucas, S. G., B. Z. Aubekerov, A. K. Dzhamangaraeva, B. U. Bayshashov, L. A. Tyutkova, 2000, Cenozoic lacustrine deposits of the Ili Basin, Southeastern Kazakhstan, *in* E. H. Gierlowski-Kordesch and K. R. Kelts, eds., Lake basins through space and time: AAPG Studies in Geology 46, p. 59-64.

Chapter 3

Cenozoic Lacustrine Deposits of the Ili Basin, Southeastern Kazakhstan

Spencer G. Lucas
New Mexico Museum of Natural History
Albuquerque, New Mexico, U.S.A.

Bolat Zh. Aubekerov
Institute of Geology
Almaty, Kazakhstan

Ayzhan K. Dzhamangaraeva
Institute of Botany
Almaty, Kazakhstan

Bolat U. Bayshashov
Lyubov A. Tyutkova
Institute of Zoology, Akademgorodok
Almaty, Kazakhstan

INTRODUCTION

The Ili basin (Figures 1 and 2) is one of the smaller Cenozoic successor basins associated with the Tien Shan ranges of southeastern Kazakhstan and western China (Allen et al., 1991; Graham et al., 1993). Nonmarine Cenozoic strata deposited in the Ili basin are best exposed at and around Aktau Mountain in the southern foothills of the Dzhungarian Alatau of southeastern Kazakhstan (Figures 1 and 2). The Cenozoic section exposed at Aktau Mountain is about 2.5 km thick and mostly of Neogene age (e.g., Lavrov and Rayushkina, 1983); however, the lower third of this section has yielded fossil mammals of late Eocene, late Oligocene, and early Miocene age (e.g., Russell and Zhai, 1987; Abdrakhmanova et al., 1989; Tleuberdina et al., 1993).

The Aktau Mountain section preserves sedimentary rocks deposited in an array of lacustrine and fluvial environments. Here, we present a preliminary overview of these strata, including their lithostratigraphy, paleontology, and depositional systems.

LITHOSTRATIGRAPHY

Seven lithostratigraphic units of formational rank (i.e., mappable at the 1:25,000 scale) of Cenozoic age can be recognized at Aktau Mountain (ascending order): Kyzylbulak, Aktau, Chuladyr, Aygyrzhol, Santash, Ili, and Khorgos formations (Figure 3) (Bazhanov and Kostenko, 1961; Dmitryeva and Nesmeyanov, 1982; Lavrov and Rayushkina, 1983; Aubekerov, 1995).

The oldest exposed stratigraphic unit, the Kyzylbulak Formation (Figure 3), is at least 130 m thick (its base is not exposed). About half of the formation is sandstone and conglomeratic sandstone; the other half is mostly mudstone and shale. Conglomerate is a minor rock type, and one bed of gypsum is present near the top of the formation. The Kyzylbulak Formation is mostly red beds of moderate reddish brown and pinkish gray color, although some beds are shades of gray or dark yellowish orange. Kyzylbulak Formation sandstones are mostly litharenites. Conglomerate clasts are mostly gneissic metamorphic rocks, and a few clasts are rhyolitic volcanic rocks, quartzite, and chert. Trough crossbedding is the dominant bedform. There is a lithologic basis for dividing the Kyzylbulak Formation into three members, but no such subdivision is attempted here.

The Aktau Formation disconformably overlies the Kyzylbulak Formation at Aktau Mountain (Figure 3). The Aktau Formation is about 120–134 m thick and almost totally conglomeratic sandstone and sandstone. Only a few thin beds of mudstone and shale are

Figure 1—Map of Kazakhstan showing location of the Ili basin.

present. Dominant colors of Aktau Formation sandstone and conglomeratic sandstone are dark yellowish orange, grayish orange, and pale yellowish gray. All are trough crossbedded. Most sandstones are hematitic litharenites. Conglomerate clasts are mostly gneissic metamorphics. A laterally extensive, 2-m-thick interval of sandy mudstone near the middle of the Aktau Formation is a prominent marker bed. Extensive color mottling and root casts suggest it has undergone pedogenic modification. A 0.9-m-thick bed of gray shale high in the formation contains fossil leaves (Lavrov and Rayushkina, 1983; Rayushkina, 1987, 1993). Uppermost sandstones of the Aktau Formation grade upward into muddy, red-bed sandstones of the overlying Chuladyr Formation.

The Chuladyr Formation is 150–200 m of muddy, red-bed sandstone, sandy mudstone, and siltstone with thin interbeds of gypsum. Greenish gray and bluish clays are common, as are epigenetic veins of selenite from 1–2 to 50–70 cm thick that create a complex lattice-like structure in some outcrops. Bedding is laterally persistent and repetitive. The Chuladyr Formation occupies the central part of the Aktau Mountain anticline and forms the areas of highest topographic

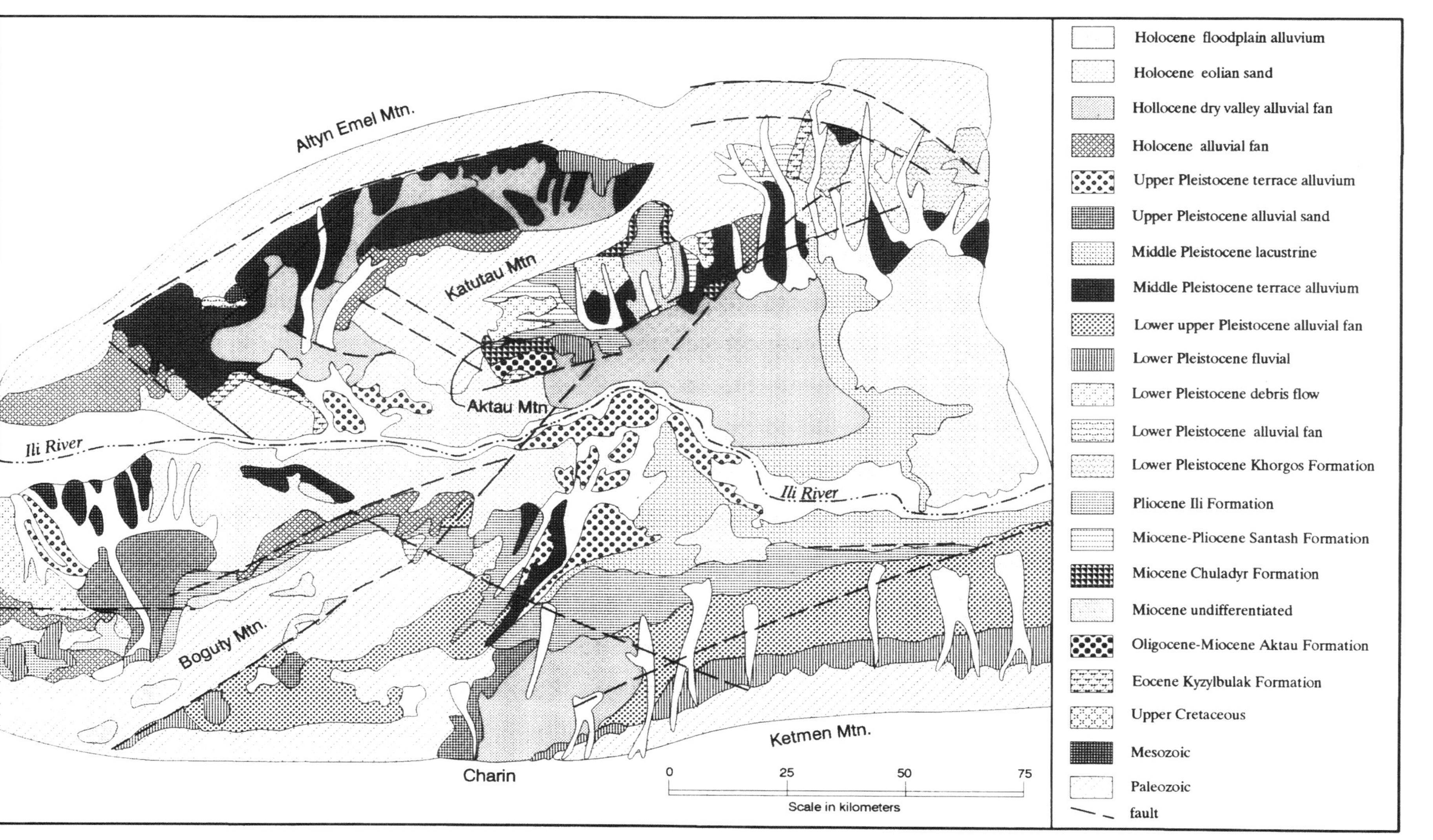

Figure 2—Geological map of the Ili basin (compiled by B. Aubekerov).

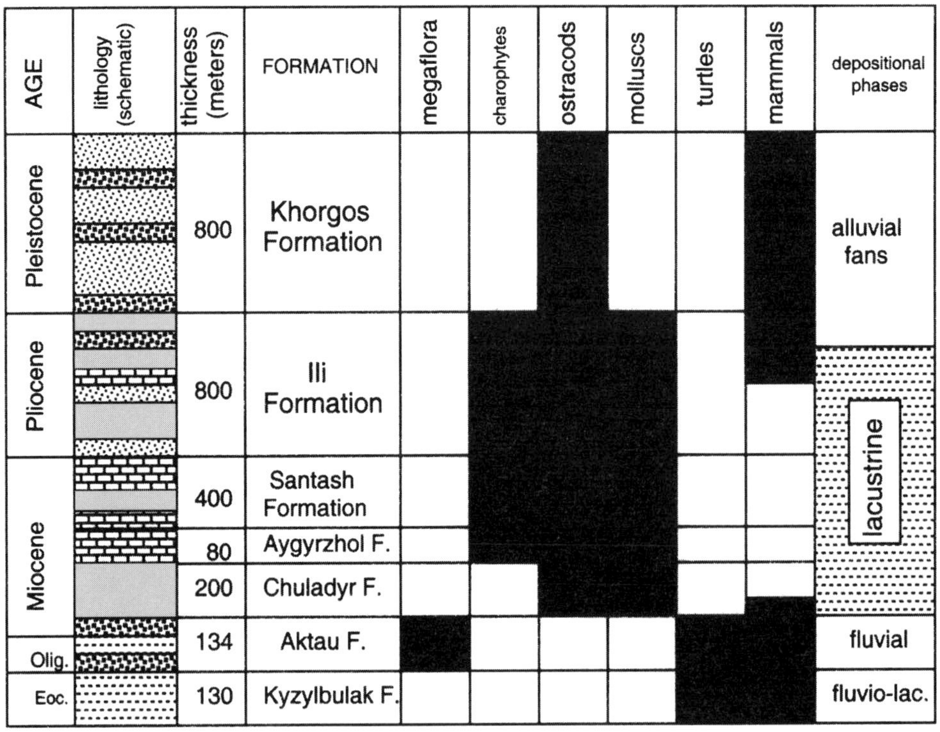

Figure 3—Summary of lithostratigraphy, paleontology, and depositional phases at Aktau Mountain.

AGE	lithology (schematic)	thickness (meters)	FORMATION	megaflora	charophytes	ostracods	molluscs	turtles	mammals	depositional phases
Pleistocene		800	Khorgos Formation							alluvial fans
Pliocene		800	Ili Formation							lacustrine
Miocene		400	Santash Formation							
		80	Aygyrzhol F.							
		200	Chuladyr F.							
Olig.		134	Aktau F.							fluvial
Eoc.		130	Kyzylbulak F.							fluvio-lac.

relief; however, at some other outcrops, the Chuladyr Formation has layers of limestones, marls, and salts (mirabilite) and/or is mostly clayey with gypsum and carbonate concretions.

The Aygyrzhol Formation is as much as 80 m of thinly bedded limestone, dolomite, and gypsum. Green and gray are the dominant colors. It is repetitively bedded, bedforms are laterally persistent, and the unit forms castellate cliffs.

The Santash Formation is about 400 m thick and represents a transition from typical Aygyrzhol Formation to typical Ili Formation lithotypes. The Santash consists of interbedded reddish brown and green claystones, limestones, and dolomites with minor sandstone interbeds. Beds of blocky claystone, massive limestone/dolomite, and trough crossbedded sandstone are laterally persistent.

The lower part of the Ili Formation consists of thick, Gilbert-delta sandstones, green and gray siltstones and claystones, and thin charophyte- and ostracod-limestones. The upper part is beds of green and blue siltstone and claystone interbedded with sandstones and sedimentary breccias of Khorgos lithology. The Ili Formation is up to 800 m thick.

The Khorgos Formation is up to 800 m of tan and terra cotta sandstones, sedimentary breccias, and conglomerates. These coarse-grained units typically are trough crossbedded. Minor, interbedded siltstones contain extensive pedogenic calcretes.

PALEONTOLOGY

Fossil bones of trionychid turtles, crocodilians, and mammals of late Eocene age are the only fossils known from the Kyzylbulak Formation (Lucas et al., 1997). Two fossil-mammal-producing layers are known from the Aktau Formation, one of late Oligocene and the other of early Miocene age. The upper part of the Aktau Formation produced an extensive leaf flora of early Miocene age (Rayushkina, 1987, 1990, 1993). The Chuladyr Formation produces ostracods and molluscs (Bodina, 1961; Dmitriyeva and Nesmeyanov, 1982).

At Aktau Mountain, the Aygyrzhol, Santash, and Ili formations contain extensive fossil assemblages of charophytes, ostracods, and gastropods (Bodina, 1961; Nikolskaya, 1990; Dzhamangaraeva, 1996). At other localities in the Ili basin, the Ili Formation produces charophytes and fossil mammals of Pliocene age (e.g., Tyutkova and Nikolskaya, 1990). The Khorgos Formation produces ostracods and fossil mammals of Pleistocene age (Dmitriyeva and Nesmeyanov, 1982).

DEPOSITIONAL SYSTEMS

At present, no detailed sedimentological analysis has been done of the Cenozoic strata in the Ili basin; nevertheless, we can identify four distinct phases of deposition in the Ili basin section (Figure 3).

Strata of the Kyzylbulak Formation are mostly siliciclastic red beds dominated by laterally extensive, texturally and mineralogically immature trough-crossbedded sandstones and conglomeratic sandstones. We interpret these strata as deposits of a high-energy fluvial system of low sinuosity. They grade upward to lacustrine shales culminated by evaporites.

Aktau Formation conglomerates and sandstones also represent a high-energy, low-sinuosity fluvial system. Imbricated pebbles and channel axes suggest paleoflow to the southwest. Presumably the source was a proto-Dzhungarian Alatau to the north–northeast. The lower part of the Aktau Formation is one tectonosequence capped by an extensive paleosol. A major unconformityunderlies upper Aktau deposits similar to those of the lower part of the formation. Chuladyr Formation red beds (fine sandstones, sandy mudstones, and, higher in the section, anhydrite) represent the initiation of lacustrine deposition.

An extensive, long-lived, and complex lake occupied the Ili basin from the early Miocene through the Pliocene. Details of the evolution of this lake are little understood at present.

By the Pleistocene, the lake basin disappeared and deposition was on alluvial fans and by high-energy streams (Khorgos Formation). This change occurred in response to Quaternary orogeny in the northern Tien Shan ranges, which formed the current Dzhungarian, Zaili, Kungei, Terskei Alatau and Ketmen ranges (Aubekerov, 1995).

ACKNOWLEDGMENTS

The National Geographic Society (grant 5412-95) supported this research. A. Slovar and his staff provided logistical support in the field.

REFERENCES CITED

Abdrakhmanova, L. T., B. U. Bayshashov, and N. N. Kostenko, 1989, Novyye dannyye po palyeontologii Dzhungarskovo Aktau (Vostochnyi Kazakhstan) [New data on the paleontology of Dzhungarian Aktau (eastern Kazakhstan)]. Vestnik Akademiya Nauk Kazakhskoy SSR, 3: 76–78.

Allen, M. B., Windley, B. F., Zhang, C., Zhao, Z, and Wang, G., 1991, Basin evolution within and adjacent to the Tien Shan range, NW China: Journal of the Geological Society, London, 148: 369–378.

Aubekerov, B. Zh, 1995, The Quaternary deposits of Kazakhstan. Occasional Publication ESRI, new series 12B, University of South Carolina, 16 p.

Bazhanov, V. S., and N. N. Kostenko, 1961, Geologichyeskii razryez Dzhungarskovo Aktau i yevo palyeozoologichyeskoye obosnovaniye [Geological section of the Dzhungarian Aktau and its paleozoological basis]. Akademiya Nauk Kazakskoi SSR, Material po Istorii Fauny i Flory Kazakhstan, 3: 47–51.

Bodina, L. E., 1961, Ostrakody tretichnykh otlozhenii Zaisanskoi i Iliiskoi depressi. Trudy VNIGRI, 170: 43–153.

Dmitriyeva, E. L., and S. A. Nesmeyanov, 1982, Mlyekopitayushchiye i stratigrafiya kontinyentalnykh tryetichnykh otlozhyenii yugo–vostoka Srednii Azii [Mammals and stratigraphy of the continental Tertiary outcrops of southeastern Middle Asia]. Akademiya Nauk SSSR Trudy Paleontologicheskovo Instituta, 193: 1–138.

Dzhamangaraeva, A. K., 1996, Pliocene charophytes from Aktau Mountain, eastern Kazakhstan: Geobios, in press.

Graham, S. A., Hendrix, M. S., Wang, L. B., and Carroll, A. R., 1993, Collisional successor basins of western China: impact of tectonic inheritance on sand composition: Geological Society of America Bulletin, 105: 323–344.

Lavrov, V. V., and G. S. Rayushkina, 1983, Oligotsyenmiotsyenoviye floronosnii gorizont v razryezye Aktau (Iliinskaya vpadina, Yuzhnii Kazakhstan) [Oligocene-Miocene floral horizon at the Aktau section (Ili depression, southern Kazakhstan)]. Doklady Akadyemiya Nauk SSSR, 1983(2): 397–399.

Lucas, S. G., Bayshashov, B. U., Tyutkova, L. A., Zhamangara, A. K., and Aubekerov, B. Z., 1997, Mammalian biochronology of the Paleogene–Neogene boundary at Aktau Mountain, eastern Kazakhstan. Paläontologische Zeitschrift, 71: 305–314.

Nikolskaya, V. D., 1990, Pozdneneogenovye kharovye vodorosli gor Aktau i Koibyn (Ilinskaya vpadina) [Late Neogene charophytes of Aktau Maountain and Koibyn (Ili depression)]. Fauna Pozvonochnykh i Flora Mezozoya i Kainozoya Kazakhstana, 2: 80–104.

Rayushkina, G. S., 1987, Perviye matyeriali k rannyemiotsyenovoi florye Ilinskoi vpadini [First material of the early Miocene flora of the Ili depression]. Akademiya Nauk Kazakskoi SSR, Matyerial po Istorii Fauny i Flory Kazakhstan, 9: 140–152.

_____, 1990, K istorii paleobotanichyeskikh isslyedovanii v Kazakhstanye (1946–1989 gg.) [Toward the history of studies of fossil flora in Kazakhstan (years 1946–1989)]. Akademiya Nauk Kazakskoi SSR, Matyerial po Istorii Fauny i Flory Kazakhstan, 11: 164–186.

_____, 1993, Miotsyenovaya flora Dzhungarsovo Aktau (Iliiskaya vpadina). [Miocene flora from Dzhungarian Aktau (Ili depression)]. Akademiya Nauk Kazakskoi SSR, Matyerial po Istorii Fauny i Flory Kazakhstan, 116–131.

Russell, D. E., and R. Zhai, 1987, The Paleogene of Asia: mammals and stratigraphy. Mémoires du Muséum National d'Histoire Naturelle Série C Sciences de la Terre, 52: 1–487.

Tleuberdina, P. A., L. T. Abdrakhmanova, and B. U. Bayshashov, 1993, Rannemiotsyenovaya fauna mlyekopitayushikh Dzhungarskovo Alatau (G. Aktau) [Early Miocene mammalian fauna of the Dzhungarian Alatau (Aktau Mountain)]. Akadyemiya Nauk Kazakskoi SSR, Matyerial po Istorii Fauny i Flory Kazakhstan, 12: 92–115.

Tyutkova, L. A., and Nikolskalaya, V. D., 1990, Pozdnepliotsenovyye kharofity i pishchuki Charyna [Late Pliocene charophytes and pikas of Charyn]: Matyerial po Istorii Fauny i Flory Kazakhstan, 11: 17–26.

Section II

Carboniferous to Permian

Brookfield, M. E., 2000, Temporary desert lake deposits lower permian (Rotliegendes) southern Scotland, U. K., *in* E. H. Gierlowski-Kordesch and K. R. Kelts, eds., Lake basins through space and time: AAPG Studies in Geology 46, p. 67–74.

Chapter 4

Temporary Desert Lake Deposits, Lower Permian (Rotliegendes) Southern Scotland, U.K.

M. E. Brookfield
Land Resource Science, Guelph University
Guelph, Ontario, Canada

INTRODUCTION

During the Late Carboniferous to Early Permian, a large lowland desert developed over the whole of southern Scotland.

Several desert basins were eroded into softer Carboniferous sediments preserved in postdepositional grabens within the lower Paleozoic Southern Uplands massif (Figure 1) (Brookfield, 1978, 1980; Glennie, 1982). Early Permian eruptions of basaltic lavas occurred in the Thornhill basin where they were accompanied by pediment formation and the deposition of thin pediment and desert floor stream and lake sediments. These interfinger with, and are overlain by, thick eolian sands.

Northwesterly directed faulting then formed the isolated grabens of the Moffat, Lochmaben, and Dumfries basins to the south and east. In these grabens, marginal alluvial fan sequences are dominated by immature streamflood and sheetflood breccias and sandstones with interbedded eolian sandstones that pass basinward into massive dune sandstones. Depositional facies are those of very arid intermontane basins summarized in Figure 2 (Brookfield, 1980; Nilsen 1982). The fan deposits have angular, poorly sorted clasts and often contain abundant well-rounded, reworked, coarse eolian sand and reworked ventifacts derived from the fan surfaces. Silt and clay are rare and probably were mostly removed by the wind; nevertheless, rare silt and clay beds are occasionally interbedded with the pediment, alluvial fan, and eolian deposits.

These fine-grained sediments were deposits in ephemeral ponds and lakes and provide additional data on paleoenvironments. Such deposits are rarely described from sections of ancient arid desert deposits because the most impressive units are the alluvial and eolian deposits (cf. Brookfield, 1984).

The purpose of this paper is to record the facies and paleoenvironments of the rare lake and pond deposits of an ancient arid intermontane desert, compare them with modern examples, and note their significance for paleoenvironmental interpretation. Detailed descriptions of the associated facies and justification of the assigned processes and environments are in Brookfield (1978, 1979, 1980, 1989).

ALLUVIAL FAN AND PEDIMENT PONDS

After floods, silts and clays can be seen settling out of suspension in small ponds on pediments and alluvial fans. Due to the slope of pediments and the slope and porosity of fans in very arid environments, such pond deposits are exceedingly small, short-lived, and likely to be removed during later floods; nevertheless, such pond deposits can occasionally be preserved beneath overlying deposits.

In the Thornhill basin, thin pond deposits occur between basalt eruptions on the incised pediments. Figure 3A shows fine-grained sediments deposited in a depression between two successive basalt flows. The sediments consist of two fining-upward cycles, each of which starts with an erosion surface on which rests cross-bedded, graded pebbly sandstone. This is overlain by alternating thin beds of graded sandstones passing up into planar laminated micaceous siltstones overlain by mud cracked silty mudstones. Each fining-upward sandstone-siltstone bed is between 20 and 50 cm thick and probably marks the waning stages of successive sheetfloods across the pediment due to overflows from incised pediment channels. Successive mudcracked surfaces within the overlying mudstone unit indicate sporadic rainfall and redistribution of fine sediment before the next major flood. The end of this particular pond is marked by a thin eolian lag, covered by the next basalt eruption.

On the western side of the Dumfries basin, fan lobes built out spasmodically over eolian sands occupying the desert floor (Figures 1, 2). Each lobe represents a catastrophic avulsion of fan feeder channels during rare storms separated by wind reworking of the fan

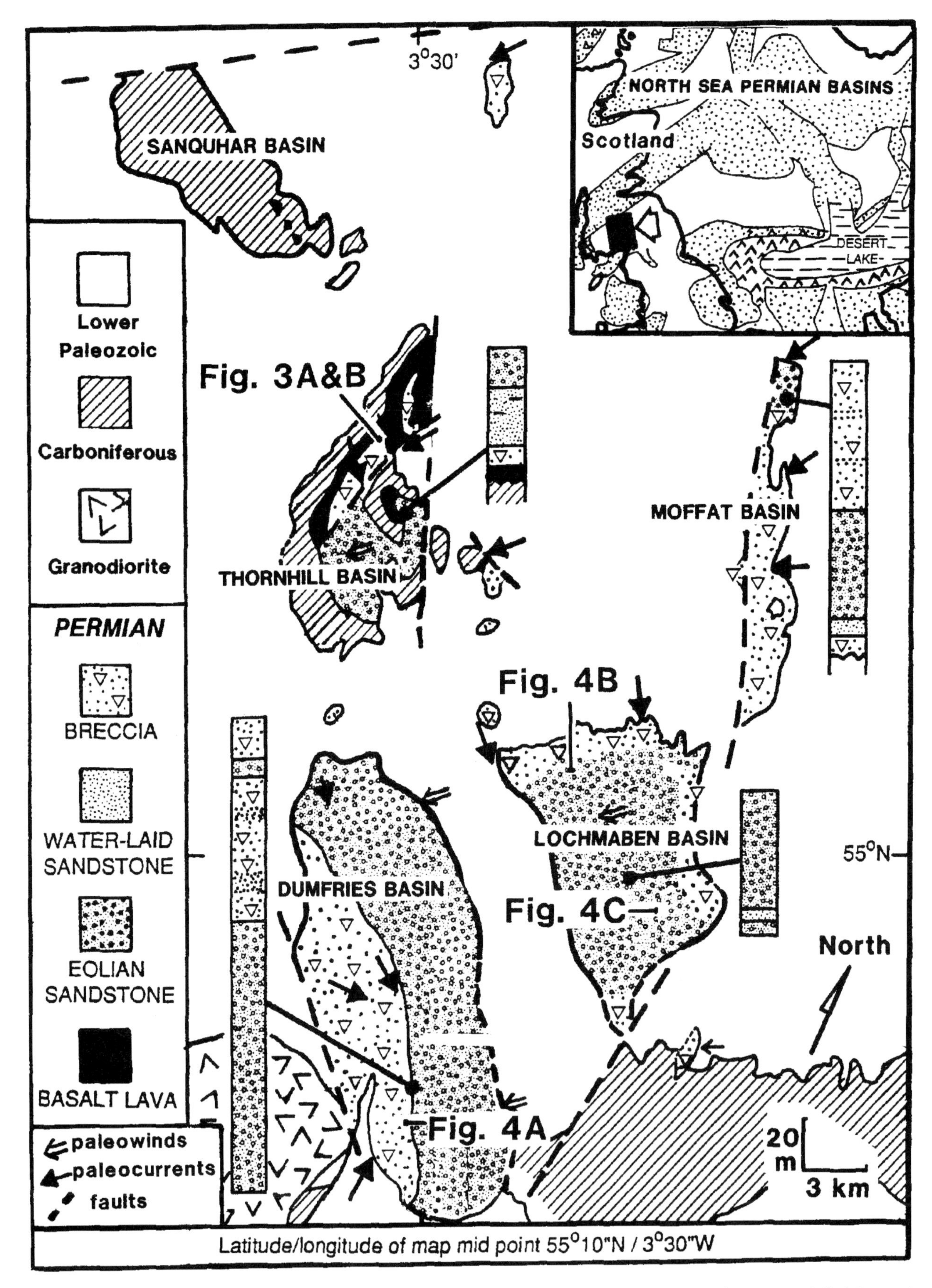

Figure 1—Southern Scotland basins with representative stratigraphic sections (from Brookfield, 1978). Inset shows location map with Lower Permian facies map of the North Sea area. Dotted areas are dune and fan sands; chevrons are lake margin gypsum deposits.

DIAGRAMATIC RECONSTRUCTION	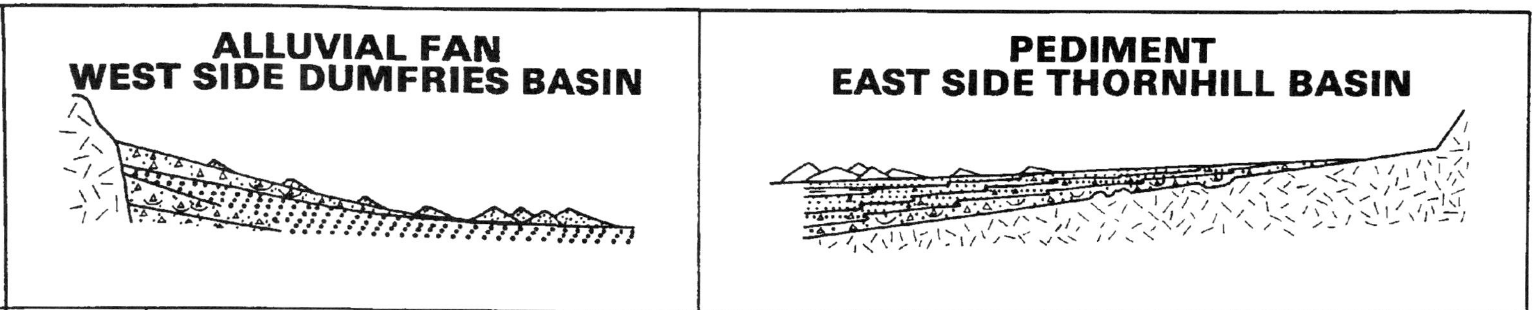 ALLUVIAL FAN WEST SIDE DUMFRIES BASIN						PEDIMENT EAST SIDE THORNHILL BASIN				
		FAN AND PEDIMENT						DESERT FLOOR			
FACIES	SCREE	DEBRIS FLOW TYPE I	DEBRIS FLOW TYPE II	STREAM FLOOD	BRAIDED STREAM TYPE I	BRAIDED STREAM TYPE II	SHEET FLOOD	EPHEMERAL STREAM	TEMPORARY LAKE	AEOLIAN ACCRETION	AEOLIAN 'DUNE'
SYMBOL											
LITHOLOGY	BRECCIA	BRECCIA THIN LITHIC SANDSTONE	BRECCIA THIN LITHIC SANDSTONE	BRECCIA	SANDY BRECCIA LITHIC PEBBLY SANDSTONE		PEBBLY SANDSTONE LITHIC SANDSTONE	LITHIC SANDSTONE MINOR PEBBLY SANDSTONE	SILTSTONE CLAYSTONE MINOR LITHIC SANDSTONE	QUARTZ SANDSTONE	QUARTZ SANDSTONE
TEXTURE AND FABRIC	BOULDERS—CLAY UNSORTED UNROUNDED CLASTS NO ORIENTATION	COBBLES—CLAY POORLY SORTED PEBBLES SUBANGULAR SANDY SILTY MATRIX FLOATING CLASTS	COBBLES—SILT POORLY SORTED PEBBLES ANGULAR SANDY MATRIX FLOATING CLASTS	COBBLES—SAND MODERATELY SORTED SANDY MATRIX CLASTS IN CONTACT	COBBLES—SAND MODERATELY WELL SORTED PEBBLES SUBANGULAR <5%W.R. COARSE SAND IN MATRIX	SMALLER ANGULAR PEBBLETS >5% W.R. COARSE SAND IN MATRIX	PEBBLES—SILT PEBBLES ANGULAR SUBANGULAR MODERATELY WELL SORTED ALTERNATING PEBBLY GRANULAR AND SANDY LAMINAE	SAND—SILT WELL SORTED SUBANGULAR SAND	SILT—CLAY WELL SORTED LAYERS OF WELL ROUNDED COARSE (EOLIAN) SAND	COARSE QUARTZ SAND—SILT WELL SORTED WELL ROUNDED COARSE SAND	MEDIUM—FINE QUARTZ SAND WELL SORTED SUB ROUNDED TO ROUNDED
BEDDING AND SEDIMENTARY STRUCTURES / THICKNESS		PLANER—GRADED OFTEN LOW ANGLE CROSS-STRATIFIED SANDSTONE AT TOP WEAK IMBRICATION BEDS 1 - 2 METERS THICK		PLANAR—TROUGH CROSS-STRATIFIED BEDS >1 METER THICK STRONGLY LENSITIC	PLANAR — TROUGH CROSS-STRATIFIED TROUGH CROSS-STRATIFIED GRAVELS IMBRICATION BEDS <1 METER THICK		THIN PLANAR LOW ANGLE CROSS-LAMINATION SMALL SCALE SCOURS IMBRICATION BEDS 5 - 100cm THICK	PLANAR LOW ANGLE CROSS-LAMINATION FINING UPWARD BEDS 30 - 100cm THICK	THIN PLANAR MUDCRACKS SMALL SAND DIKES DISRUPTED BEDDING BEDS 1 - 50cm THICK	THIN PLANAR	HIGH - LOW ANGLE PLANAR - TROUGH CROSS-STRATIFIED BEDS 0.3 - 1 METER THICK
DISTRIBUTION OF SAND FRACTION (Weight % 99.5, 50, 10, 1, 0.1; ø Scale -2, 0, 2, 4)											
CLAST ROUNDNESS (2-4cm)	0.2	0.5-0.7	0.4	0.4	0.6	0.4	0.4	0.4			
MAXIMUM CLAST SIZE (cm)	3×10^3	30 - 100		10 - 50 (200)	6 - 30		2 - 10	1 - 5	COARSE SAND		
SEQUENCES	NONE	SANDSTONE TOP PEBBLE CONCENTRATION GRADED		LARGE CHANNELS	CHANNELING FINING UPWARD SEQUENCES		planar lam.	FINING UPWARD			

Figure 2—Facies descriptions and interpretations of Permian alluvial fan, pediment, and desert floor facies.

Figure 3—(A) Pediment and (B) fan toe pond and lake deposits of the Thornhill basin.

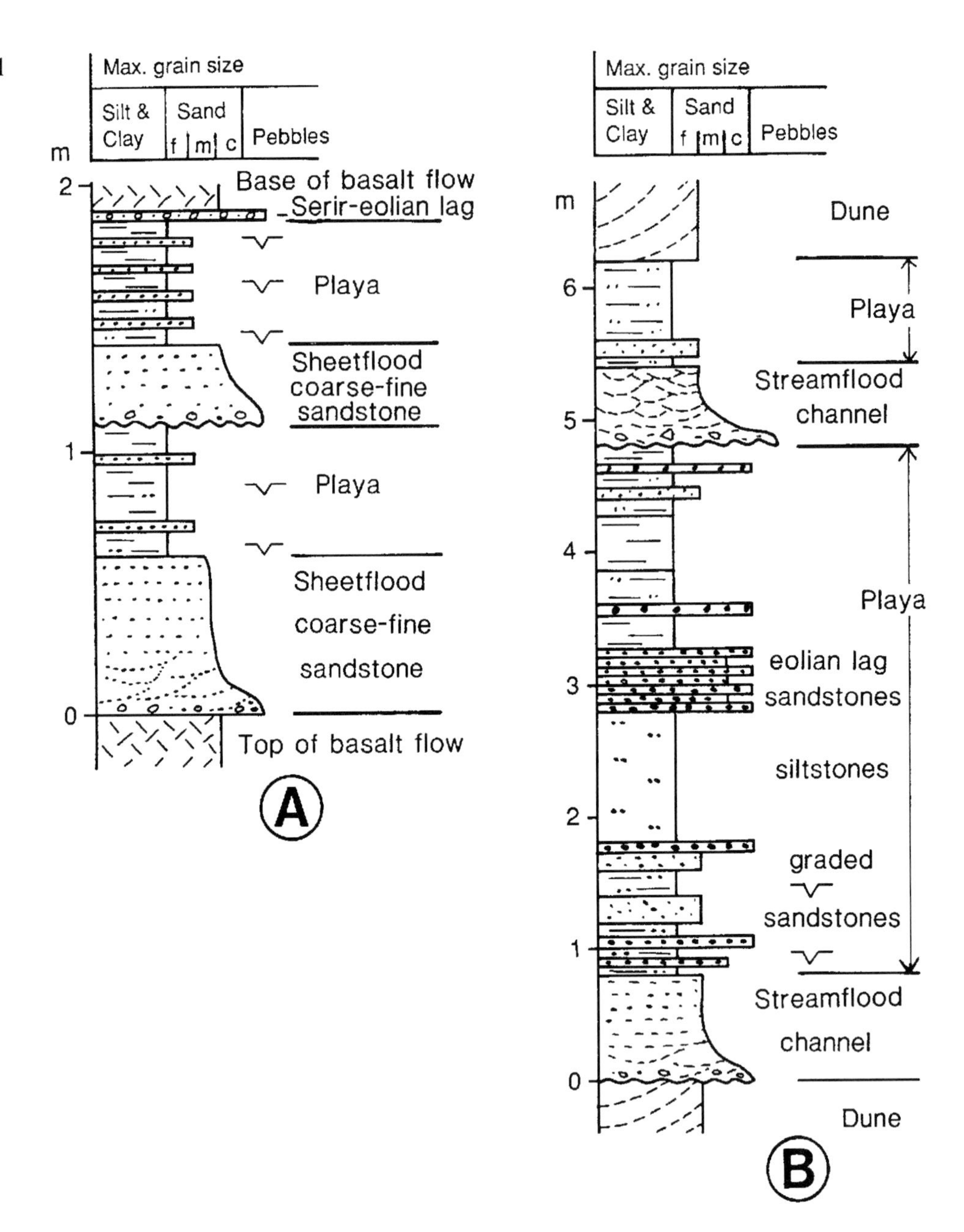

surface (Brookfield, 1980). Many of the breccias and sandstones contain reworked ventifacts formed during long periods of wind ablation on the fan surface. Figure 4A shows a section through a thin alluvial fan lobe that prograded over eolian sands in the Dumfries basin. The section starts with coarse streamflood channel and sheetflood deposits, marking the advance of the fan toe over the desert floor. These are overlain and cut into by the deposits of a major streamflood channel at the top of the section. Overlying eolian sands were probably deposited after upstream avulsion and abandonment of this channel. Multiple stream and sheetflood beds can be deposited by the same flood. Thus, the entire deposit could have been deposited in a very short time during one or several storm floods, or over a long period of time by slow aggradation of thin beds. How can this be determined?

In several places, small, thin, lenticular graded sand, silt, and clay lenses lie between successive sheetflood deposits. In one place the top clay is mudcracked and filled with eolian sand. These lenses are interpreted as residual pond deposits formed after waning of each flood. They thus allow distinct flood events to be separated and the detailed development of the lobe to be inferred. On Figure 4A, only one pond deposit is preserved, suggesting two streamflood-sheetflood episodes between the initial cutting of the channel and incision of the final large streamflood channel.

PLAYA LAKES

Floods that reach the desert floor form temporary playa lakes between the pediment and alluvial fan toes

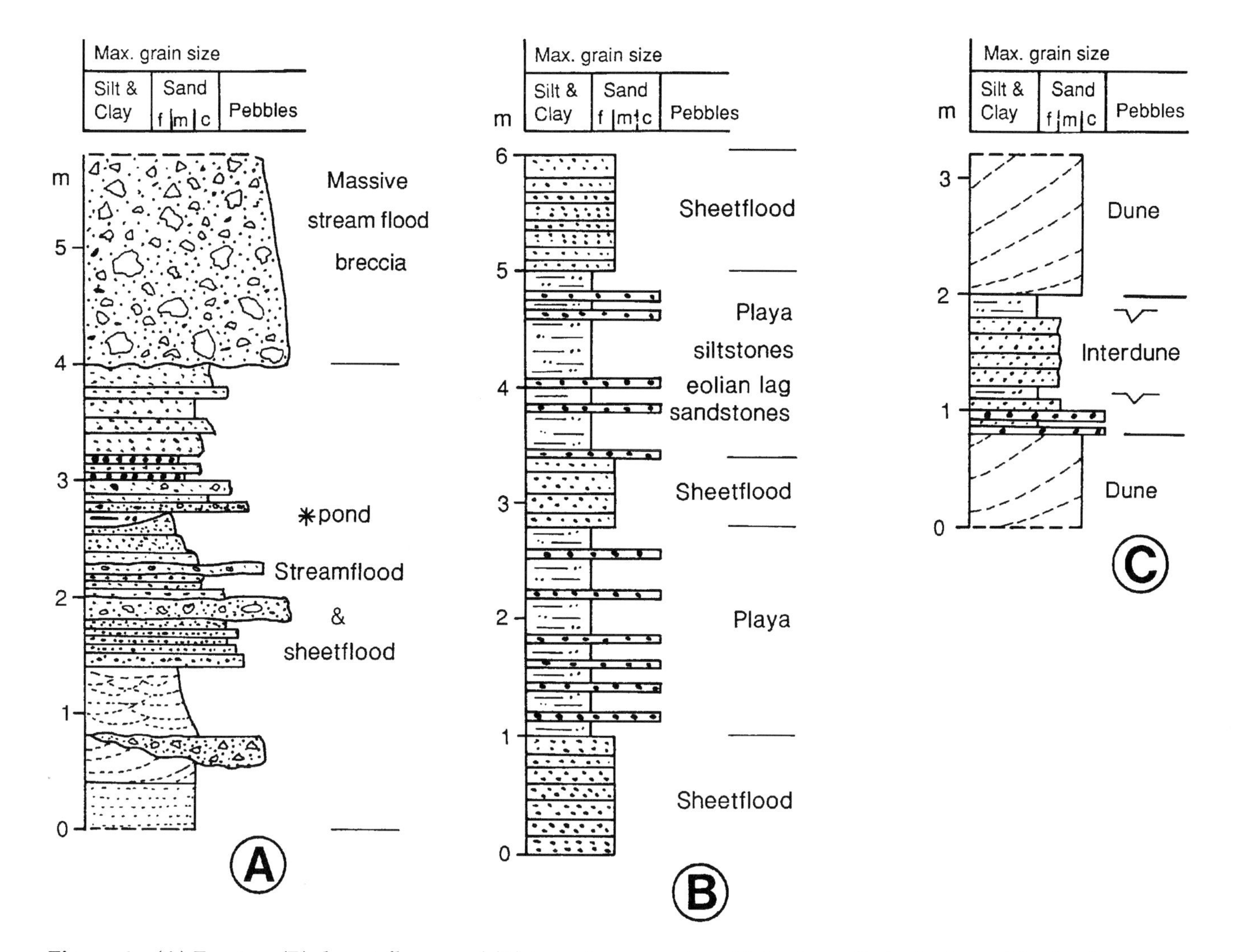

Figure 4—(A) Fan toe, (B) desert floor, and (C) interdune pond and lake deposits in the Dumfries basin.

and the desert floor dunes. Recent playa surfaces vary from hard, dry and smooth to soft, friable, flaky, and puffy depending on the hydrology of the playa (Cooke and Warren, 1973; Langer and Kerr, 1966).

Soft, friable, flaky, and puffy surfaces occur where groundwaters rise seasonally through capillary action or direct discharge to the surface. Precipitation of chemical salts within and on the surface of the playa, causes disruption of both the surface and internal structure. Playa deposits of this type are relatively common in many ancient rift basins (e.g., Smooth and Olsen, 1985).

Hard, dry, and smooth surfaces commonly occur on arid playas dominated by surface run-off and with little groundwater recharge. They consist of relatively impermeable silt and clay with very little chemical precipitates. Playa deposits of this type have rarely been described, possibly because they are confounded with stream overbank deposits (but see Hubert and Hyde, 1982).

A variety of temporary playa lake deposits occur in the Permian valley floor environments of the Thornhill and Dumfries basins. All show features of hard, smooth surfaces and show no evidence for evaporite minerals or caliche soil development. They are thus compatible with the extreme aridity indicated by the other facies (Brookfield, 1980).

In the Thornhill basin, relatively thick playa and streamflood deposits lie sandwiched between dune sandstones (Figure 3B). These deposits lie at the ends of incised pediment channels where they debouched onto the desert floor. The playa deposits consist of graded thin-bedded, very fine-grained sandstones, siltstones, and mudstones alternating with thin, coarse very well-rounded and well-sorted eolian sandstone laminae. The siltstones and mudstones are frequently mudcracked. The coarse eolian sand laminae form very thin, occasionally multiple laminae, sometimes infilling the mudcracks in underlying mudstones. To form such extensive thin laminae consisting of well-sorted, well-rounded coarse sand, such coarse sands have to be spread across the playa surface and concentrated as a residual lag by deflation during the passage of ripples or dunes across the playa surface. Ripples are more likely than dunes because all the laminae are very thin. Simply blowing sand across

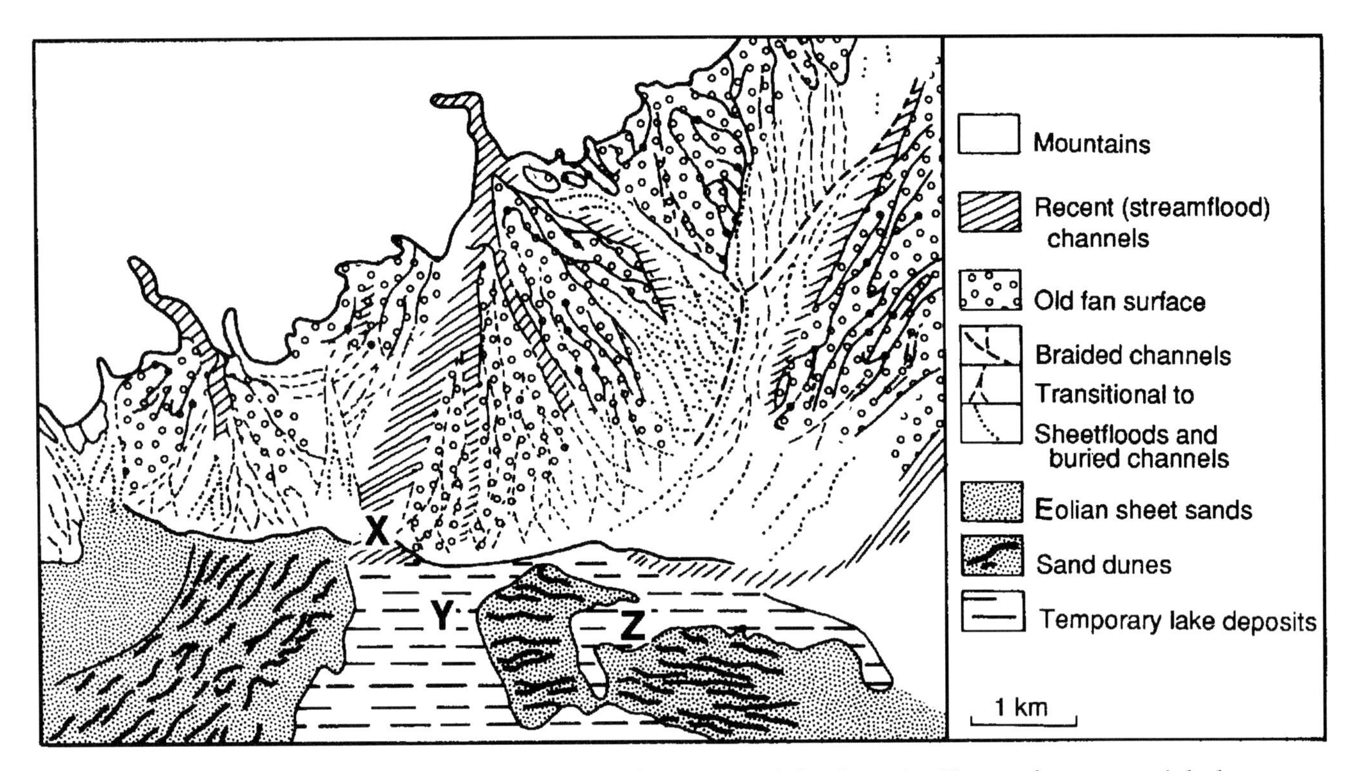

Figure 5—Recent association of wadi fan, dune, and temporary lake deposits (drawn from an aerial photograph of Death Valley, California, U.S.A.). Note how streamflood channels occasionally reach the desert floor. Modern association with ancient Dumfries basin equivalents are marked as follows: X=fan toe association, equivalent to Figure 4A; Y=desert floor association equivalent to Figure 4B; Z=interdune association equivalent to Figure 4C.

and onto the playa surface to form adhesion ripples will not work because the result of that is relatively poorly sorted layers usually showing some sort of internal structure (Hubert and Hyde, 1982). There are a large number of such coarse laminae in the Thornhill section, indicating that the playa must have alternatively flooded and had ripples migrating across its surface for many years, despite the thinness of the deposit. There is a tendency for a consistent fining-upward cycle from coarse cross-bedded streamflood pebbly sandstones through medium- to fine-grained parallel laminated sandstones and siltstones, into laminated siltstones and mudcracked mudstones alternating with coarse eolian sandstone laminae. This suggests that the two complete cycles shown are due to unusually large flooding events extending far into the desert basin; nevertheless, the playa must have flooded many times to deposit silt and mudstone between the eolian and laminae. This thick deposit lies between thick dune sandstones of large draas (Brookfield, 1979) and probably marks a large interdraa depression.

In the Dumfries basin, thicker units of regularly alternating sheetflood and playa deposits are fairly common and are interbedded with eolian sandstones at the edge of fan toes. Figure 4B shows a part of such a unit. The fine-grained sheetflood sandstones have nonerosive bases and are overlain by thinly laminated siltstones alternating with coarse well-sorted eolian sand laminae identical to those at Thornhill. The large number of alterations of eolian lag and lake siltstones suggests many periods of wetting and drying of the playa. It is unlikely that large dunes migrated over a wet sabkha surface, as is found in places in Arabia, because there are no adhesion ripples or thicker eolian sands preserved.

Farther into the center of the Dumfries basin, similar but thin deposits mark the interdune depressions between migrating simple barchanoid dunes (Figure 4C). Silt and clay laminae are scarce and probably represent fines washed out of the adjacent dunes and atmosphere during rare rainstorms.

MODERN ANALOGIES

Three main ephemeral lake and pond associations can be recognized in the recent alluvial fan–desert floor environments of Death Valley, California (Figure 5). These environments can be used, almost unmodified, to interpret the ancient Permian equivalents above.

CONCLUSION

Ephemeral lake and pond deposits greatly aid in interpreting depositional facies and paleohydrology of the dominant wind, stream, and debris flow deposits of arid environments. Such lake and pond deposits give further environmental details unobtainable in any other way.

ACKNOWLEDGMENTS

Field work was funded by NSERC (Canada). I thank E. H. Gierlowski-Kordesch for criticism of earlier drafts of the manuscript.

REFERENCES CITED

Brookfield, M. E., 1978. Stratigraphy of the Permian and supposed Permian rocks of southern Scotland. Geologische Rundschau, 67, 110–149.

Brookfield, M. E., 1979. Anatomy of a Lower Permian aeolian sandstone complex. Scottish Journal of Geology, 15, 81–96.

Brookfield, M. E., 1980. Permian intermontane basin sedimentation in southern Scotland. Sedimentary Geology, 27, 167–194.

Brookfield, M. E., 1984. Eolian facies. In R.G. Walker (ed.), Facies Models, Geological Association of Canada, St. Johns, Newfoundland, 91–118.

Brookfield, M. E., 1989. Rotliegendes desert sedimentation in Scotland. Zeitschrift der geologische Wissenschaft Berlin, 17, 204–241.

Cooke, R. U., and Warren, A., 1973. Geomorphology in Deserts. Bastford, London, 374p.

Glennie, K. W., 1982. Early Permian (Rotliegendes) paleowinds of the North Sea. Sedimentary Geology, 34, 245–265.

Hubert, J. F., and Hyde, M. G., 1982. Sheet-flow deposits of graded beds and mudstones on an alluvial sandflat-playa system: Upper Triassic Blomidon redbeds, St. Mary's Bay, Nova Scotia. Sedimentology, 29, 457–474.

Langer, A. M., and Kerr, P. F., 1966. Mojave playa crusts: physical properties and mineral content. Journal of Sedimentary Petrology, 36, 377–396.

Lorenz, J. C., 1988. Triassic-Jurassic rift-basin sedimentology: history and methods. Van Nostrand Reinhold Company, New York, 315p.

Nilsen, T. H., 1982. Alluvial fan deposits. American Association of Petroleum Geologists Memoir, 31, 49–86.

Smooth, J. P., and Olsen, P. E., 1985. Massive mudstones in basin analysis and paleoclimatic interpretation of the Newark supergroup. United States Geological Survey Circular, 946, 29–33.

Gaupp, R., R. Gast, C. Forster, 2000, Late Permian playa lake deposits of the Southern Permian Basin (Central Europe) *in* E. H. Gierlowski-Kordesch and K. R. Kelts, eds., Lake basins through space and time: AAPG Studies in Geology 46, p. 75–86.

Chapter 5

Late Permian Playa Lake Deposits of the Southern Permian Basin (Central Europe)

R. Gaupp
Institute of Earth Sciences, University of Jena
Germany
e-mail: gaupp@geo.uni-jena.de

R. Gast
BEB Erdöl Erdgas GmbH
Hannover, Germany

C. Forster
Hamburg, Germany

GEOLOGIC SETTING OF THE SOUTHERN PERMIAN BASIN

The Southern Permian basin of central Europe (Figure 1) was superimposed on the Late Carboniferous Variscan foredeep. This elongated depocenter comprises several interconnected en echelon sub-basins (Silverpit/Dutch, North German, and Polish basins), where maximum thicknesses of >2000 m of German Rotliegende red beds and halites coincide with areas of Late Carboniferous–Early Permian volcanism (Ziegler, 1982, 1988). Superimposed on regional thermal subsidence, a post-Variscan dextral shear system is widely recognized, inferred from basin subsidence patterns (Ziegler, 1990). Early Permian extrusive magmatism, located mainly at intersections of fault systems, resulted from deep crustal fracturing (Plein, 1978). Related to this magmatic event, Early Permian thermal uplift apparently caused rifting and a regional stratigraphic hiatus generally referred to as the Saalian unconformity. After magmatic activity had terminated, the central portions of this future Southern Permian basin began to subside thermally and Rotliegende deposition was initiated within graben systems (Drong et al., 1982; Gast, 1988). Accelerated thermal subsidence during the Late Permian was accompanied with an increasing depositional area, sedimentation onlapping, and overstepping of the southern and western margins of the large basin. With a basin floor that was below global sea level, rapid flooding of this large intracontinental closed basin occurred during the Zechstein marine transgression (Glennie and Buller, 1983).

Both biostratigraphy (Schneider, 1988; Hoffmann and Kamps, 1989; Gebhardt et al., 1991; Gebhardt, 1988, 1994; Gebhardt and Plein, 1995; Schneider et al., 1995) and magnetostratigraphy (Lützner and Menning, 1980; Menning, 1986, 1994; Menning et al., 1988) indicate a Tatarian age for the upper Rotliegende (Figure 2) of the North German basin (NGB), designated as the central part of the Southern Permian basin. Exact biostratigraphic dating is hampered by the scarcity of fossils in the generally barren red beds. Conchostracans, ostracods, hydromedusae, insect and fish remains, macroflora, and tetrapod imprints provide rare traces of life in an arid saline environment. The significant marker horizon of the Illawara magnetic reversal has been identified (Menning, 1986) within synrift clastic deposits of the Havel Subgroup (Figure 2). Chronostratigraphic data are limited to isotope dating of late Tatarian (Zechstein) potash salt (Lippolt et al., 1993) and Rotliegende volcanic rocks (Lippolt et al., 1982; Lippolt and Hess, 1989). Lithostratigraphic investigations in Rotliegende successions of the NGB, based mainly on subsurface information, were reported by Drong et al. (1982), Gralla (1988), Gast (1991, 1993), Gralla et al. (1991), Plein (1978, 1990, 1993, 1995), and Forster (1996). Sedimentologic interpretations of Rotliegende sedimentary rocks of central and western Europe were published by Glennie (1972, 1983), Glennie et al. (1978), Nagtegaal (1979), Gralla (1988), Gast (1991, 1993, 1995), and Gast and Gaupp (1991). A new paleogeographic and depositional model for the upper Rotliegende, offshore from the Netherlands, has been presented by George and Berry (1994). Dudley et al. (1994) reported on sequence stratigraphy in the Rotliegende of the southern North Sea.

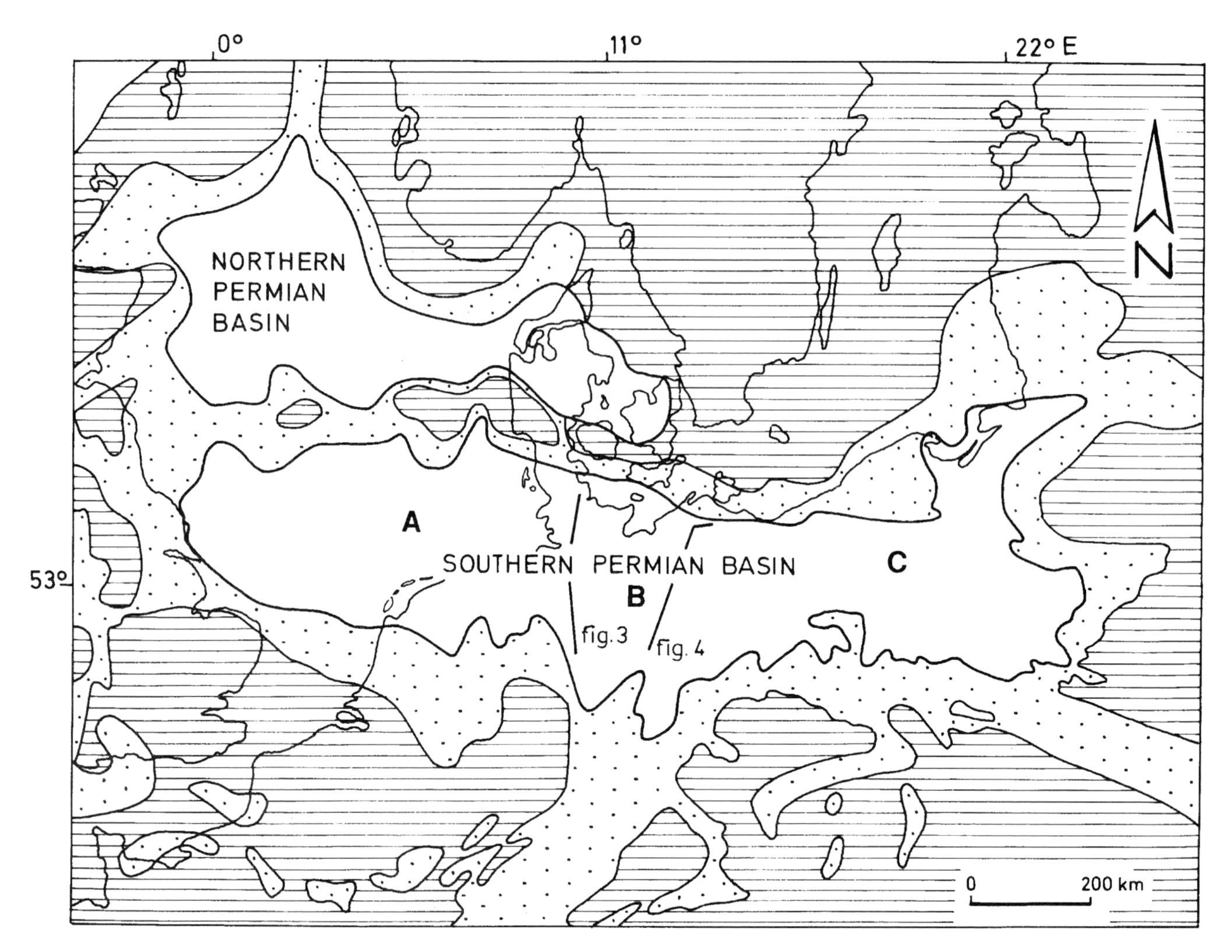

Figure 1—Paleogeographic reconstruction of the Southern Permian basin (SPB) of central Europe (after Ziegler, 1980), hatched = nondepositional areas, stippled = marginal siliciclastic facies, white (in the SPB) = predominantly fine siliciclastic and evaporitic lacustrine deposits; position of cross-sections in Figures 3 and 4 is indicated; Sub-basins of the Southern Permian basin: A = Dutch/Silverpit basin, B = North German basin, C = Polish basin.

This paper focuses on Late Permian upper Rotliegende sediments of the North German basin (NGB) (B, in Figure 1), where the most extensive lacustrine deposits are encountered.

SEDIMENTOLOGY AND FACIES ASSOCIATIONS

Lacustrine and playa mudflat deposits of the Southern Permian basin have their maximum lateral and vertical distribution in northern Germany (NGB) and adjacent offshore areas of the North Sea (Figure 1). Rotliegende sandstones, which are developed in a broad belt along the southern margin of the basin, form the reservoir of major gas accumulations in the southern North Sea and the Dutch and north German onshore areas (Glennie, 1986).

Fine siliciclastic and evaporitic sediments (Elbe subgroup, Figure 2) reflect the transition from rift-related continental deposition (Havel subgroup and older lower Rotliegende rocks, Figure 2) to the marine transgression of the late Tatarian (Zechstein). Flat marginal areas of the basin were dominated by terminal fan (outwash plain), ephemeral stream, and eolian sedimentation with rare inundations during periods of maximum highstands of the lake level (lower division of the Dethlingen Formation, Figure 2). Playa and lacustrine deposition possibly prevailed relatively early in central parts of the basin (Havel Subgroup, Figure 2), but it covers large parts of the entire basin during late stages of Rotliegende sedimentation (upper division of Dethlingen and Hannover Formation, Figure 2).

A brief description of lacustrine and related lithofacies is summarized in Table 1 with inferred depositional paleoenvironments. The following subenvironments of lacustrine and related deposition can be tentatively differentiated within the upper Rotliegende succession (Elbe subgroup):

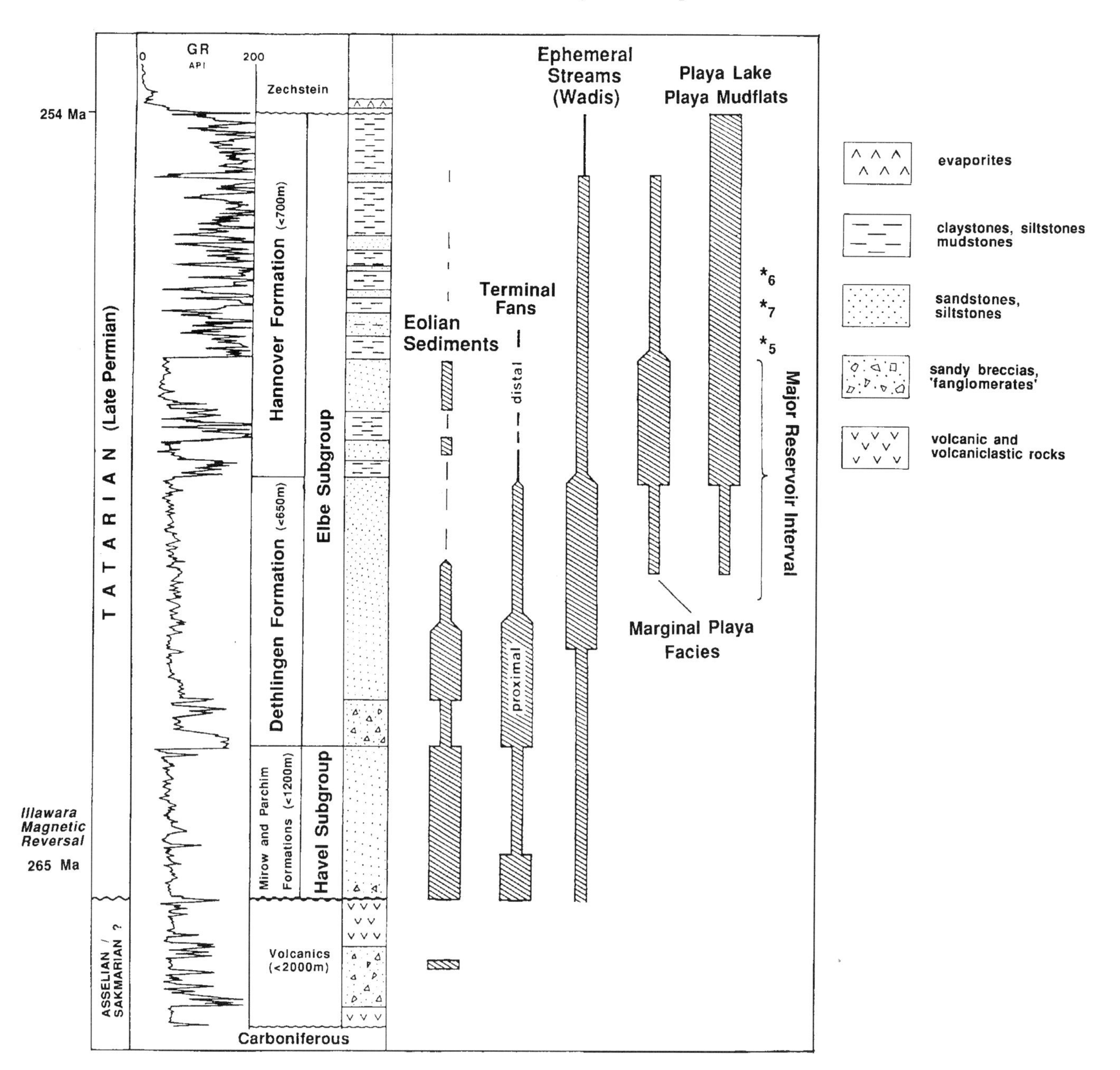

Figure 2—Summary stratigraphic column of the Late Permian (Rotliegende) succession with distribution of major depositional facies, modified after Gaupp et al. (1993); stratigraphic subdivision and thicknesses according to Plein (1993; 1995); position of Illawara magnetic reversal after Menning et al. (1988). The column is representative for basin marginal to medial areas at the central southern part of the basin (equivalent to southern ends of the cross-sections of Figures 3 and 4 (see Figure 1). Lacustrine deposition is restricted there to the Elbe subgroup and top parts of the Dethlingen Subgroup. Low gradient terminal fans/outwash plains and ephemeral streams are limited to marginal fringes of the basin. The upper Rotliegende comprises the Havel and Elbe subgroups. Older formations with graben-related local distribution are not shown in this schematic column. GR = gamma ray well log, API = units of spectral measurements (K, Th, U). *5, *6, *7 = approximate position of lithologic columns of Figures 5, 6, and 7, respectively.

(1) Distal sandflat (dry, damp, wet eolian, and rare terminal fan deposition)
(2) Lacustrine shoreline sand accumulation, mainly eolian with subordinate wave reworking (major reservoir lithology)
(3) Ephemeral lake/playa, evaporitic mudflat, and evaporitic sandy mudflat
(4) Perennial lake (basin floor below wave base), salina (hypersaline lake)

Amalgamated halite deposits of the salina reach individual thicknesses of 60 m and constitute up to 65% of the succession in the center of the basin (Figure 3). In the eastern part of the North German basin

Table 1. Summary of Sedimentary Structures, Bed Thicknesses, and Paleoenvironmental Interpretation of Rotliegende Lithofacies, Southern Permian Basin, Europe.

Lithofacies	Sedimentary Structures		Bed Thickness	Environmental Interpretation
	Primary	Secondary		
Horizontally laminated mudstone, claystone	Laminae normally a few mm thick Four subtypes: Clay in clay laminations Clay and finest sand Thin anhydritic laminae Calcitic and anhydritic aminae Graded laminae Erosive contacts between clay laminae, transitions to small-scale ripples	Rare convolution Rare cracks Microfractures Haloturbation	< 1 m	Perennial lake with basin floor below wave base
Brecciated mudstone with halite	Halite between clay laminae	Brecciation Disruption		Ephemeral hypersaline lake
Massive mudstone	No primary or secondary structures preserved		< 2.20 m	Ephemeral lake
Brecciated mudstone	No primary structures preserved	Brecciated, clasts consist of laminated or massive mudstone Cracks sometimes filled with anhydrite Evaporitic dissolution structures	< 3 m	Ephemeral hypersaline lake
Diffusely structured mudstone	Gradual variations of grain size (laterally and vertically)	Convolution Haloturbation	< 1.70 m	Ephemeral lake/ evaporitic mudflat, efflorescent salt crust clastic accumulation
Sandy mudrock	Irregularly wavy, patchy lamination Evaporitic adhesion structures: irregularly shaped lenses of silt or fine sand, rarely with very small mudclasts at their base Mud flasers Rare current ripples or thin sand layers	Convolution Haloturbation	< 4.20 m	Ephemeral lake/ sandy evaporitic mudflat, efflorescent salt crust clastic accumulation
Muddy sandstone	Irregularly wavy, flaser, or lenticular lamination Salt ridges, evaporitic adhesion structures Mud flasers	Haloturbation	< 3 m	Distal sandflat
Massive halite	Crystal size variable (3 mm to 5 cm) Mud between crystals and mudchip inclusions Nearly no primary texture		Several meters	Hypersaline lake, salina

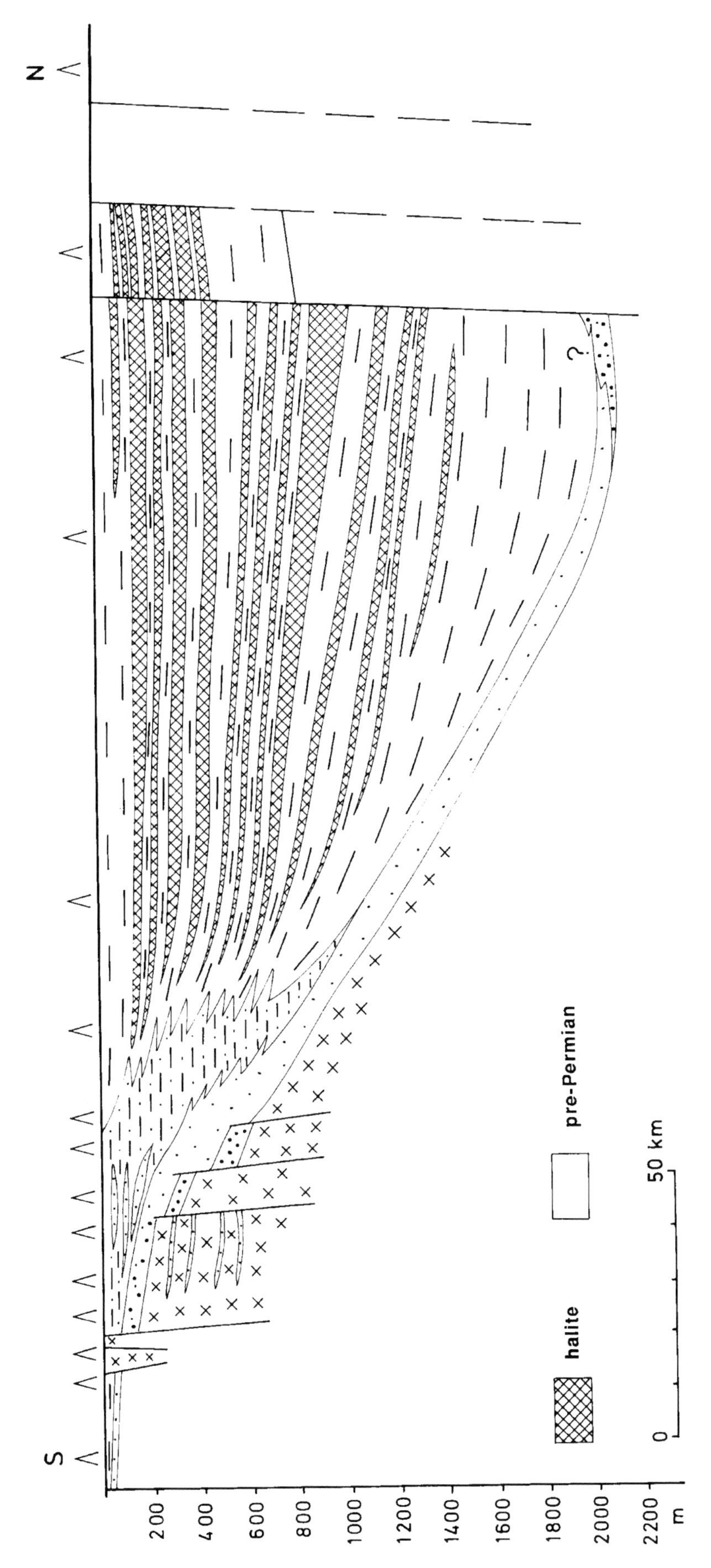

Figure 3—Cross-section through the the sedimentary succession of the Rotliegende in the central part of the Southern Permian basin (west). Location of the cross-section is indicated in Figure 1. For legend, see Figure 4.

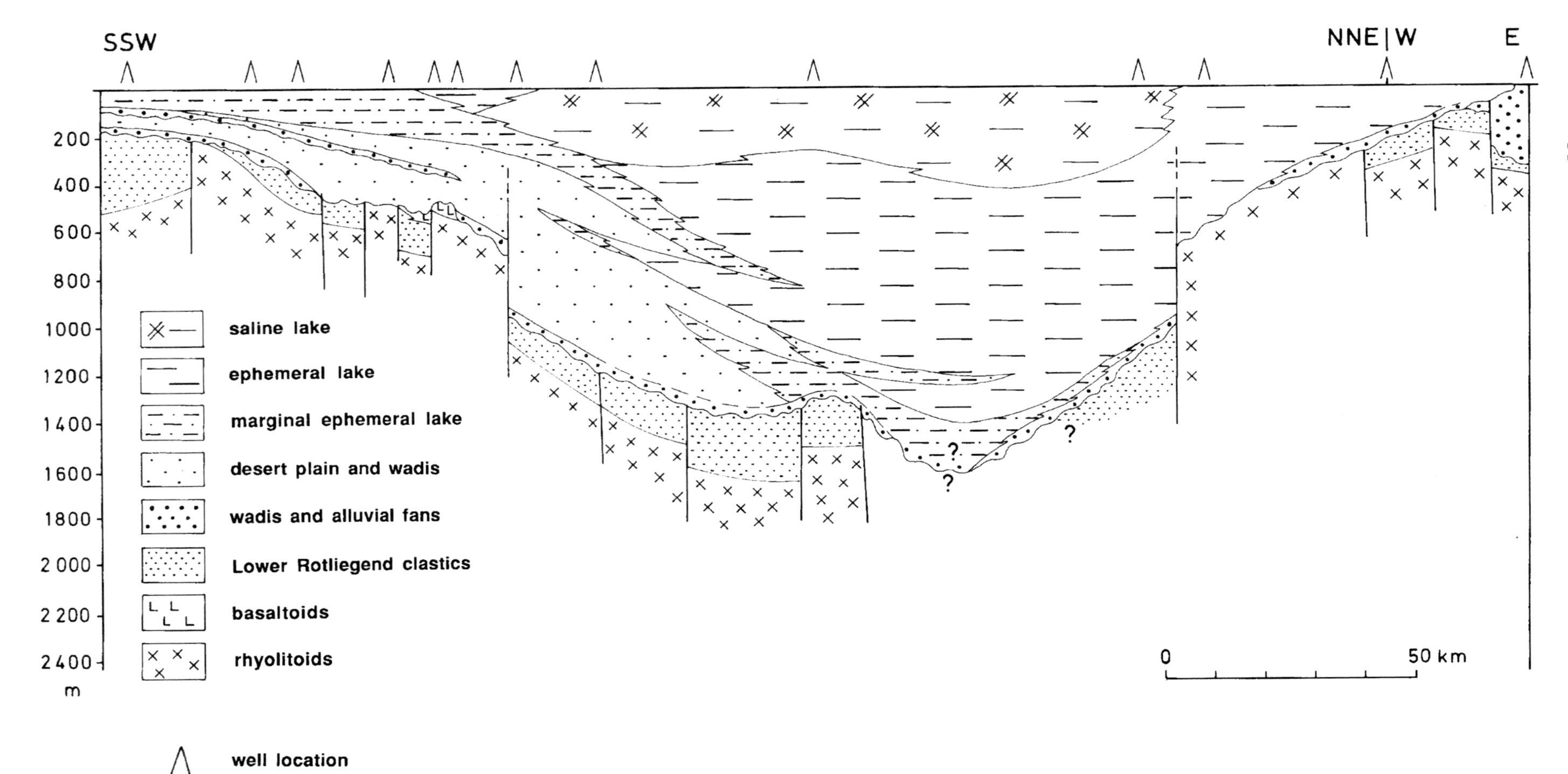

Figure 4—Cross-section through the sedimentary succession of the Rotliegende in the central part of the Southern Permian basin (east); after Lindert et al. (1990). Location of the cross-section is indicated in Figure 1.

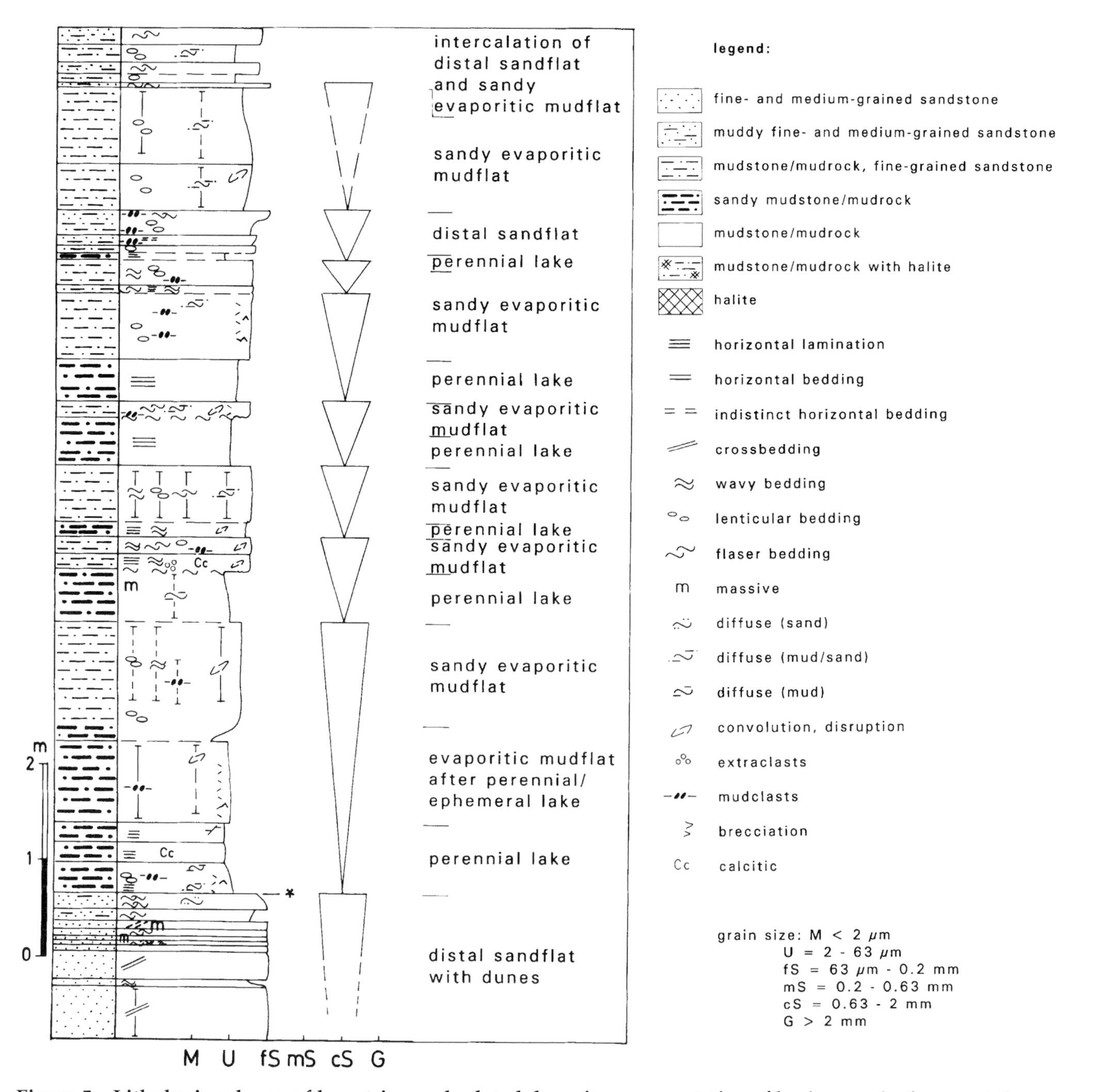

Figure 5—Lithologic column of lacustrine and related deposits representative of basin marginal areas at the southern part of the North German basin. The section is dominated by sandy mudflat and silty laminated perennial lake deposits. The basal 1.5 m of the profile comprise the top of a major regressive sandbody and the transition to a period of frequent (perennial) lake highstands. Triangles (vertex down) indicate drying-upward minor cycles. The bounding surface of the major cycles is situated at the base of the first perennial lake interval (*).

(Figure 4), halite deposition is far less pronounced. A possible marine influence on the deposition of these continental evaporites and associated lacustrine deposits, especially in the upper part of the Elbe Subgroup, is discussed in Plumhoff (1966), Trusheim (1971), and Holser (1979). A continental origin of the Rotliegende halite deposits with increasing marine influence up-section is inferred by Gebhardt (1994).

Fourteen major sedimentary cycles with predominantly lacustrine and related deposition and a total thickness of 500–1200m can be distinguished and laterally correlated across the central parts of the basin within the Elbe subgroup (Gast, 1991). Each major cycle (<10–80 m) documents the basinward progradation of marginal siliciclastic facies during major regressive episodes (times of frequent lowstands of the lake level).

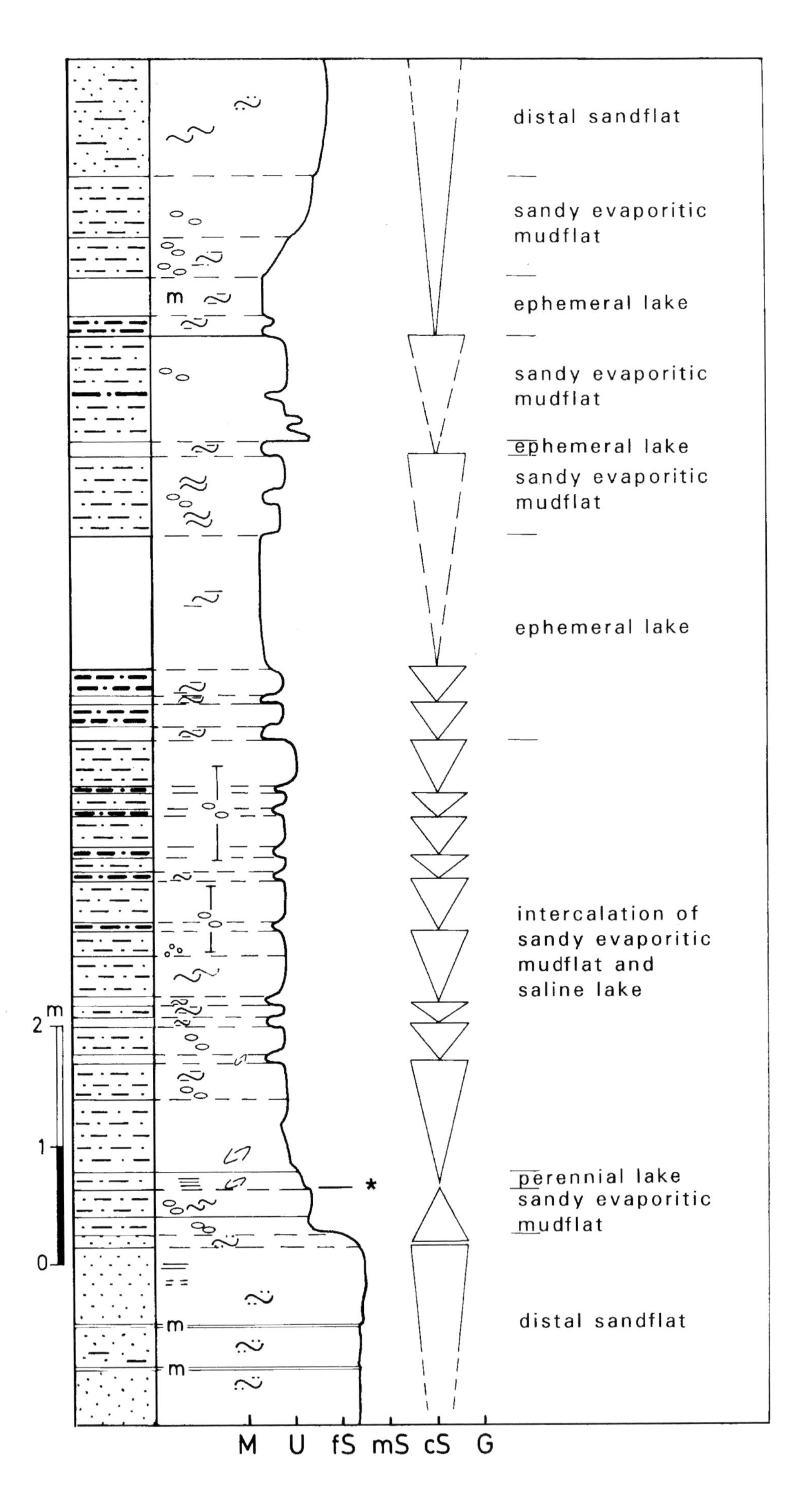

Figure 6—Representative lithologic column of areas transitional between marginal and central saline lake facies. The section comprises the top of a major cycle and the basal part of a successive major cycle that shows the general regressive or "drying-upward" trend. Only rare indications of perennial lake conditions are preserved. The triangles (vertex down) indicate drying-upward minor cycles, vertex up indicates "wetting-upward" minor cycles. The bounding surface of the major cycles is situated at the base of the perennial lake interval (*). For legend, see Figure 5.

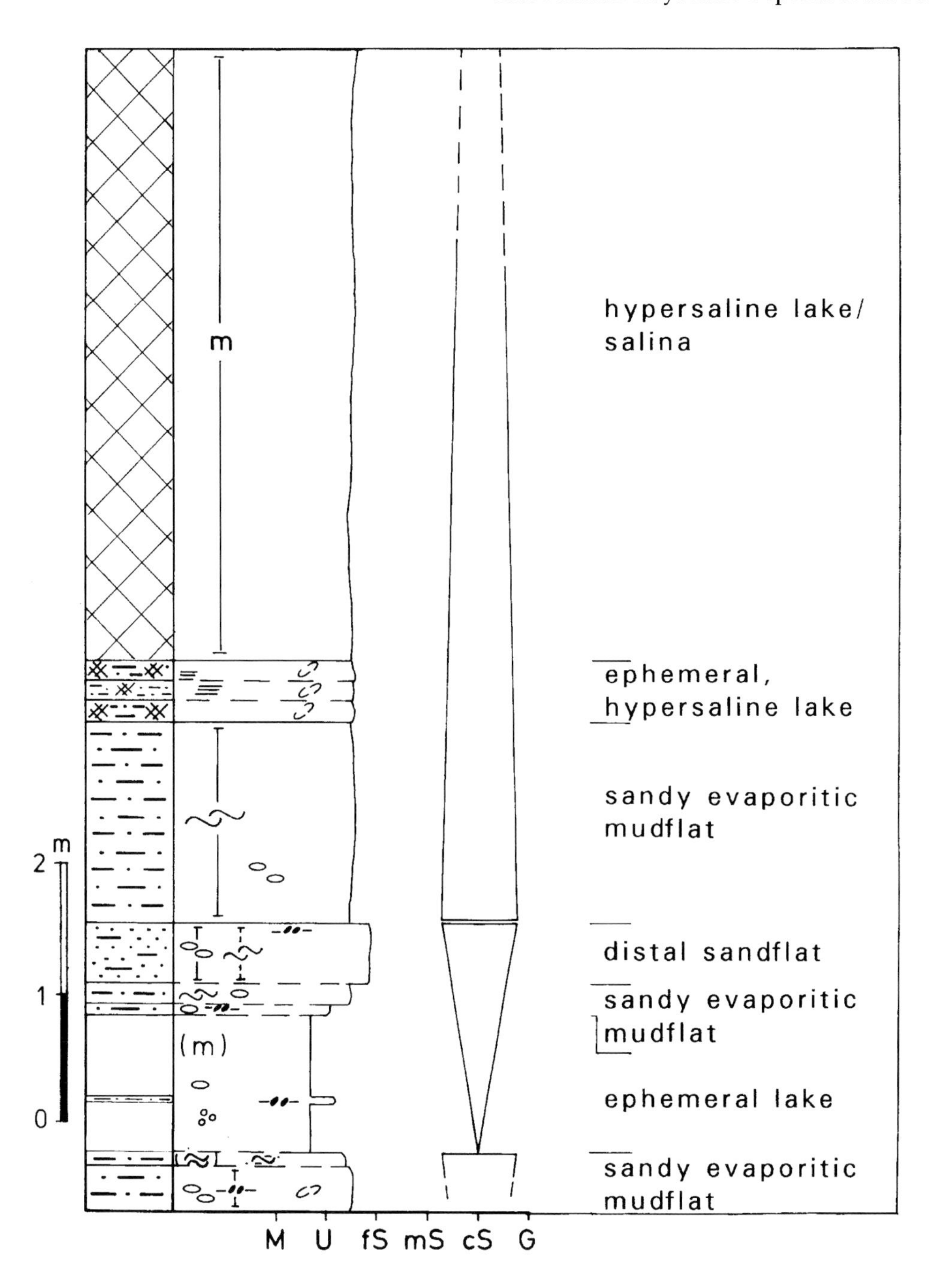

Figure 7—The section shows the top part of a major regressive cycle (distal sandflat underlain by shoreline sands), and the initiation of transgressive (frequently wet) conditions. Ephemeral lake sedimentation changes through hypersaline ephemeral lake to perennial salina halite deposition (characteristic of central basinal areas). Triangles show minor cycle trends as explained in Figures 5 and 6. For legend, see Figure 5.

Minor cycles (sand-shale intervals of some meters) represent subordinate fluctuations in lake level and groundwater table. Subaqueous evaporite deposition probably filled depressions in the sediment-starved central parts of the basin where the highest subsidence rates occurred (compare Figures 3 and 4).

Climatic change due to orbital forcing is the most likely parameter controlling lake and groundwater levels and, therefore, the apparent cyclic stratigraphic evolution (Gast and Gaupp, 1991). This assumption is substantiated by spectral analysis of upper Rotliegende gamma-ray logs from the Dutch offshore (Yang and Baumfalk, 1994) and the NGB (Bitzer and Gaupp, 1996) that indicate Milankovitch periods. Tectonic differentiation of the subsidence pattern during Elbe subgroup deposition appears to be minor.

Cyclostratigraphic subdivisions in major cycles rely on thick (>2 m) claystone/mudstone intervals (Gast, 1991, 1995), which can be recognized in cores and well logs (Figures 5 and 6). These intervals of fine siliciclastics define the correlatable bounding surfaces of major cycles and are considered to represent perennial lake conditions (most frequent lake-level highstands, relatively wet climatic conditions). Major cycles are comparable to the third-order sequences as defined by Yang and Baumfalk (1994) in the upper Rotliegende succession of offshore Netherlands; however, these authors define bounding surfaces of major cycles not in conjunction with the definition above.

Recognition of the precise nature and primary features of subaqueous lacustrine deposits is hampered by frequent changes in paleolake levels. Flash flooding of the large and flat basin with pelitic deposition from suspension was generally succeeded by intense disruption of sedimentary structures during partial or total desiccation, haloturbation, or re-deposition. Laminated subaqueous pelitic and carbonate/anhydritic sediments were preserved only under exceptional conditions. Disrupted fabrics indicate the ubiquitous presence of these laminated shallow lake sediments that were possibly deposited below wave base. Disrupted laminated pelitic sediments from the southern margin of the NGB rarely have anhydrite or halite cements and can be correlated by gamma-log configurations with these halite deposits of much greater thickness in the basin center (Forster, 1996).

The lack or scarcity of preserved mudcracks possibly reflects the generally high salinity of the mudflats or the pervasive distortion of primary and secondary sedimentary structures. Patches of eolian sand grains in massive and disrupted mudstones testify to dry intervals. Mudflat deposits consist of mudrocks with irregular patchy fabric indicating a sand, silt, and mud pellet accumulation on efflorescent salt crusts (cf. Smoot and Castens-Seidell, 1994). Wind-blown siliciclastic and evapoclastic material of a wide range of grain sizes adhered to salt efflorescences or was trapped in depressions of the irregular surfaces on evaporitic crusts. Secondary distortion of primary aggradational structures occurred after evaporite dissolution during flooding.

Intervals of intense and prolonged aridity are recorded in the basinward progradation of sandflat facies. Adhesion and accumulation on salt ridges were a major factor of sand accumulation in sandflats. Precipitation and preservation of halite layers appear to be related to the initial stages of transgressive periods (increasingly wet climatic conditions), when interstitial halite cements within marginal siliciclastic deposits were likely to be dissolved and transported toward the basin center. Halite precipitation occurred in central basinal depressions from a basinward flux of brines (Figure 7).

Subsurface brine flux from marginal basinal areas was reduced during increasingly frequent rises in paleolake level, which also resulted in an increase in dissolution of precipitated halite layers, accompanied by increasing siliciclastic deposition. Perennial lake conditions were established during times of a relatively wet climate in the source (catchment) areas.

ACKNOWLEDGMENTS

Special thanks go to Ken Glennie and Blas L. Valero-Garcés for critical reviews of the manuscript and constructive comments, and to Elizabeth Gierlowski-Kordesch for editorial handling.

REFERENCES CITED

Bitzer, F., and R. Gaupp, 1996, Stoffumlagerung in der spätorogenen Phase am Beispiel des Norddeutschen Beckens: Sedimentvolumetrie und Isostasie in Verbindung mit Backstripping-Verfahren und Methoden der Beckenmodellierung, Terra Nostra, v. 96 (2), p. 24–25.

Drong, H. J., E. Plein, D. Sannemann, M. A. Schüpbach, and J. Zimdars, 1982, Der Schneverdingen-Sandstein des Rotliegenden—eine äolische Sedimentfüllung alter Grabenstrukturen, Zeitschrift der deutschen geologischen Gesellschaft, v. 133, p. 699–725.

Dudley, G., E. J. Jolley, and M. L. Sweet, 1994, Sequence stratigraphy in mixed aeolian/lacustrine/fluvial depositional systems—two examples from the Rotliegendes Group (Permian), southern North Sea, Liverpool Sequence Stratigraphy Conference1994, abstracts, p. 122–125.

Forster, C., 1996, Faziesentwicklung und Zyklostratigraphie in der oberen Hannover-Formation (Oberrotliegend) im Raum Bremen bis Wustrow, Doctoral Dissertation, University Mainz, 107 p.

Gast, R. E., 1988, Rifting im Rotliegenden Niedersachsens, Die Geowissenschaften, v. 6, p. 115–122.

Gast, R. E., 1991, The perennial Rotliegend saline lake in NW Germany, Geologisches Jahrbuch, v. 119A, p. 25–59.

Gast, R. E., 1993, Sequenzanalyse von äolischen Abfolgen im Rotliegenden und deren Verzahnung mit Küstensedimenten, Geologisches Jahrbuch, v. 131A, p. 117–139.

Gast, R. E., 1995, Sequenzstratigraphie als Grundlage für die Neugliederung der Elbe-Subgruppe im Rotliegenden, *in* E. Plein, ed., Norddeutsches Becken, Rotliegend Monographie Teil II, Frankfurt am Main, Courier Forschungs-Institut Senckenberg, p. 47–54.

Gast, R. E., and R.H. Gaupp, 1991, The sedimentary record of the Late Permian saline Lake in N.W. Germany. Symposium on Sedimentary and Paleoclimatological Records of Saline Lakes, Saskatoon 1991, abstracts, p. 17.

Gaupp, R. H., A. Matter, J. Platt, K. Ramseyer, and J. Walzebuck, 1993, Diagenesis and fluid evolution of deeply buried Permian (Rotliegend) gas reservoirs, northwest Germany, AAPG Bulletin, v. 77, p. 1111–1128.

Gebhardt, U., 1988, Mikrofaziesanalyse und stratigraphisch-regionalgeologische Interpretation terrestrischer Karbonate der varistischen Molasse (Mitteleuropa, Permokarbon), Freiberger Forschungshefte, v. 427C, p. 30–59, Leipzig.

Gebhardt, U., 1994, Zur Genese der Rotliegend-Salinare in der Norddeutschen Senke (Oberrotliegend II). Freiberger Forschungshefte, v. 452C, p. 3–22, Leipzig.

Gebhardt, U., and E. Plein, 1995, Neugliederung des Rotliegenden im Norddeutschen Becken, *in* E. Plein, ed., Norddeutsches Becken, Rotliegend Monographie Teil II, Frankfurt am Main, Courier Forschungs-Institut Senckenberg, p. 18–23.

Gebhardt, U., J. Schneider, and N. Hoffmann, 1991, Modelle zur Stratigraphie und Beckenentwicklung

im Rotliegenden der Norddeutschen Senke, Geologisches Jahrbuch, v. 127A, p. 405–427.
George, G. T., and J. K. Berry, 1994, A new palaeogeographic and depositional model for the Upper Rotliegend, offshore The Netherlands, First Break, v. 12 (3), p. 147–58.
Glennie, K. W., 1972. Permian Rotliegendes of Northwest-Europe interpreted in light of modern desert sedimentation studies, AAPG Bulletin, v. 56 (8), p. 932–943.
Glennie, K. W., 1983, Lower Permian Rotliegend desert sedimentation in the North Sea area, *in* M. E. Brookfield and T. S. Ahlbrandt, eds., Eolian Sediments and Processes, Developments in Sedimentology 30B, Blackwell Scientific, Oxford, p. 521–541.
Glennie, K. W., 1986, Development of NW-Europe's Southern Permian Gas Basin, *in* J. Brooks, J. Goff, and B. Van Hoorn, eds., Habitat of Palaeozoic Gas in N.W. Europe, Geological Society (London) Special Publication 23, Blackwell Scientific, Oxford, p. 3–22.
Glennie, K. W., and A. T. Buller, 1983, The Permian Weissliegend of NW-Europe, the partial deformation of eolian sands caused by the Zechstein transgression, Sedimentary Geology, v. 35, p. 43–81.
Glennie, K. W., G. C. Mudd, and P. J. C. Nagtegaal, 1978, Depositional environment and diagenesis of Permian Rotliegendes sandstones in Leman Bank and Sole Pit areas of the UK, southern North Sea, Journal of the Geological Society (London), v. 135, p. 25–34.
Gralla, P., 1988, Das Oberrotliegende in NW-Deutschland, Lithostratigraphie und Faziesanalyse, Geologisches Jahrbuch, v. 106A, p. 3–59.
Gralla, P., F. Nieberding, and R. Sobott, 1991, Der Wustrow-Sedimentationszyklus des Oberrotliegenden im Bereich einer nordwestdeutschen Erdgaslagerstätte, Zeitschrift der deutschen geologischen Gesellschaft, v. 142, p. 13–21.
Hoffmann, N., and H.-J. Kamps, 1989, Neuerkenntnisse zur Biostratigraphie und Paläodynamik des Perms in der Norddeutschen Senke—ein Diskussionsbeitrag, Zeitschrift für Angewandte Geologie, v. 35 (7), p. 198–207.
Holser, W. T., 1979, Rotliegend evaporites, Lower Permian of northwest Europe, geochemical confirmation of the nonmarine origin, Erdöl und Kohle, v. 32, p. 159–162.
Lindert, W., D. Warncke, and M. Stumm, 1990, Probleme der lithostratigraphischen Gliederung des Oberrotliegenden (Saxon) im Norden der DDR, Zeitschrift für Angewandte Geologie, v. 36, p. 368–375.
Lippolt, H. J., and Hess, J. C., 1989, Isotopic evidence for the stratigraphic position of the Saar-Nahe Rotliegende Volcanism III. Synthesis of results and geological implications, Neues Jahrbuch für Geologie und Paläontologie, Monatshefte 1989, v. 9, p. 553–559.
Lippolt, H. J., J. Raczek, and H. Schneider, 1982, Isotopenalter eines Unteren Rotliegend-Biotits aus der Bohrung Wrzesnia/Polen, Aufschluß 33, p. 13–25.
Lippolt, H. J., S. Hartmann, and J. Pilot, 1993, $^{40}Ar/^{39}Ar$-dating of Zechstein potash salts: new constraints on the numerical age of the Latest Permian and the P-Tr boundary, TERRA abstracts, v. 1, p. 591.
Lützner, H., and M. Menning, 1980, Erste Ergebnisse zur Magnetostratigraphie des Rotliegenden der Saale-Senke, Permian of the West Carpathians, Symposium 1979, Bratislava, abstracts, p. 41–51.
Menning, M., 1986, Zur Dauer des Zechsteins aus magnetostratigraphischer Sicht. Zeitschrift für geologische Wissenschaften, Berlin, v. 14, p. 395–404.
Menning, M., Katzung, G., and Lützner, H., 1988, Magnetostratigraphic Investigations in the Rotliegendes (300–252 Ma) of Central Europe. Zeitschrift für geologische Wissenschaften, Berlin, v. 16, p. 1045–1063.
Menning, M., 1994, A numerical time scale for the Permian and Triassic periods: an integrated time analysis, *in* P. A. Scholle, T. M. Peryt, and D. S. Ulmer-Scholle, eds., The Permian of Northern Pangea, Paleogeography, Paleoclimates and Stratigraphy, Vol. 1, Springer-Verlag, Heidelberg, p. 77–97.
Nagtegaal, P. J. C., 1979, Relationship of facies and reservoir quality in Rotliegendes desert sandstones, southern North Sea region, Journal of Petroleum Geology, v. 2, p. 145–158.
Plein, E., 1978, Rotliegend-Ablagerungen im Norddeutschen Becken. Zeitschrift der deutschen geologischen Gesellschaft, v. 129, p. 71–97.
Plein, E., 1990. The Southern Permian basin and its paleogeography, *in* D. Heling, P. Rothe, U. Förstner, and P. Stoffers, eds., Sediments and Environmental Geochemistry, Springer-Verlag, Heidelberg, p. 124–133.
Plein, E., 1993, Bemerkungen zum Ablauf der paläogeographischen Entwicklung im Stefan und Rotliegenden des Norddeutschen Beckens, Geologisches Jahrbuch, v. 131A, p. 99–116.
Plein, E. (ed.), 1995, Norddeutsches Becken—Rotliegend Monographie Teil II.-Courier Forschungs-Institut Senckenberg, v. 183, 1–193.
Plumhoff, F., 1966, Marines Oberrotliegendes (Perm) im Zentrum des NW-deutschen Rotliegend-Beckens. Neue Beweise und Folgerungen, Erdöl, Kohle, Erdgas und Petrochemie, v. 19, p. 713–720.
Schneider, J., 1988, Grundlagen der Morphogenie, Taxonomie und Biostratigraphie isolierter Xenacanthodier-Zähne (Elasmobranchii). Freiberger Forschungs-Hefte, v. 419C, p. 71–80.
Schneider, J., U. Gebhardt, and B. G. Gaitzsch, 1995, Fossilführung und Biostratigraphie. In: Norddeutsches Becken, *in* E. Plein, ed., Norddeutsches Becken, Rotliegend Monographie Teil II, Frankfurt am Main, Courier Forschungs-Institut Senckenberg, p. 25–39.
Smoot, J. P., and Castens-Seidell, B., 1994, Sedimentary features produced by efflorescent salt crusts, Saline Valley and Death Valley, California, -SEPM Special Publication 50, p. 73–90.
Trusheim, F., 1971, Zur Bildung der Salzlager im Rotliegenden und Mesozoikum Mitteleuropas,

Beihefte Geologisches Jahrbuch, v. 112, p. 1–51.
Yang, C. S., and Y. A. Baumfalk, 1994, Milankovitch cyclicity in the Upper Rotliegend Group of the Netherlands offshore, Special Publication of the International Association of Sedimentologists (IAS), v. 19, p. 47–61.
Ziegler, P. A., 1980, Northwestern European basin: geology and hydrocarbon provinces, *in* A. D. Miall, ed., Facts and Principles of World Petroleum Occurrence, Canadian Society of Petroleum Geologists Memoir (Calgary), v. 6, p. 653–706.
Ziegler, P. A., 1982, Geological Atlas of Western and Central Europe. SIPM , The Hague, 2nd edition 1990, 130p.
Ziegler, P. A., 1988, Evolution of the Arctic-North Atlantic and the Western Tethys, AAPG Memoir, v. 43, p. 1–198.
Ziegler, P. A., 1990, Tectonic and paleogeographic development of the North Sea rift system, *in* D. J. Blundel and A. Gibbs, eds., Tectonic Evolution of the North Sea Rifts, Clarendon Press, Oxford, p. 1–36.

Stollhofen, H., I. G. Stanistreet, R. Rohn, 2000, The Gai-As Lake system, Northern Namibia and Brazil, *in* E. H. Gierlowski-Kordesch and K. R. Kelts, eds., Lake basins through space and time: AAPG Studies in Geology 46, p. 87-108.

Chapter 6

The Gai-As Lake System, Northern Namibia and Brazil

Harald Stollhofen
Institut für Geologie, Universität Würzburg
Würzburg, Germany

Ian G. Stanistreet
Department of Earth Sciences, University of Liverpool
Liverpool, England, United Kingdom

Rosemarie Rohn
Departamento de Geologia Sedimentar/IGCE, Universidade Estadual Paulista Rio Claro
São Paulo, Brazil

Frank Holzförster
Ansgar Wanke
Institut für Geologie, Universität Würzburg
Würzburg, Germany

INTRODUCTION

The separation of South America from Africa during the Early Cretaceous isolated equivalent stratigraphic sequences on both continents. This is well established for rock sequences, including flood basalts, which were deposited prior to oceanic onset; however, earlier extensional events are also recorded by the resulting intracontinental basins. Of these, the depositional area containing the Late Permian–earliest Triassic Gai-As Lake is a prime example.

The aims of this paper are (1) to record facies generated within and outside the lake body and (2) to compare them with correlative bodies on the other side of the present-day South Atlantic Ocean, and (3) to record the controls of fault structures on facies architecture and lake margins. The advantages of good exposures produced by river dissection of the continental margin in northern Namibia allows good access for identifying synsedimentary fault controls on the Gai-As Lake. We suggest that these can be extrapolated to correlative sequences at the conjugate South American side where exposure and thus the potential for recognition of synsedimentary structural activity is limited; consequently, the Paraná "basin," commonly dealt with as an intracratonic sag basin, may be underlain by a complex of stacked rift and thermal subsidence-controlled depositional centers.

GEOLOGIC CONTEXT

In northern Namibia, sections of the Gai-As and overlying Doros formations are well exposed in the Huab area (Figure 1B, C), located next to the present Atlantic coastline some 400 km northwest of Namibia's capital city, Windhoek. The structural setup of this area in central Gondwana is characterized by northerly trending sets of westerly dipping extensional faults, interpreted as synthetic to a more major westerly dipping detachment system (Stollhofen, 1999). Synsedimentary extensional activity associated with these faults expresses one of several successive periods of intracontinental rifting during the early extensional history of the pre-southern South Atlantic rift zone (Figure 2A).

The Gai-As and Doros lake sediments are correlatives of the Beaufort Group (Karoo supergroup), a stratigraphic unit widespread in whose equivalents are various Late Carboniferous–Early Jurassic depositional areas over a large portion of the Gondwana supercontinent (Figure 1A). Of these depositional centers, two major basin types are represented: (1) those such as the main Karoo basin of South Africa, which developed as a flexurally deformed foredeep in front of the northerly overthrusting Cape fold belt (Veevers et al., 1994a) and (2) the more elongate graben and half-graben basins, known from Namibia (Stollhofen et al., 1998, 1999),

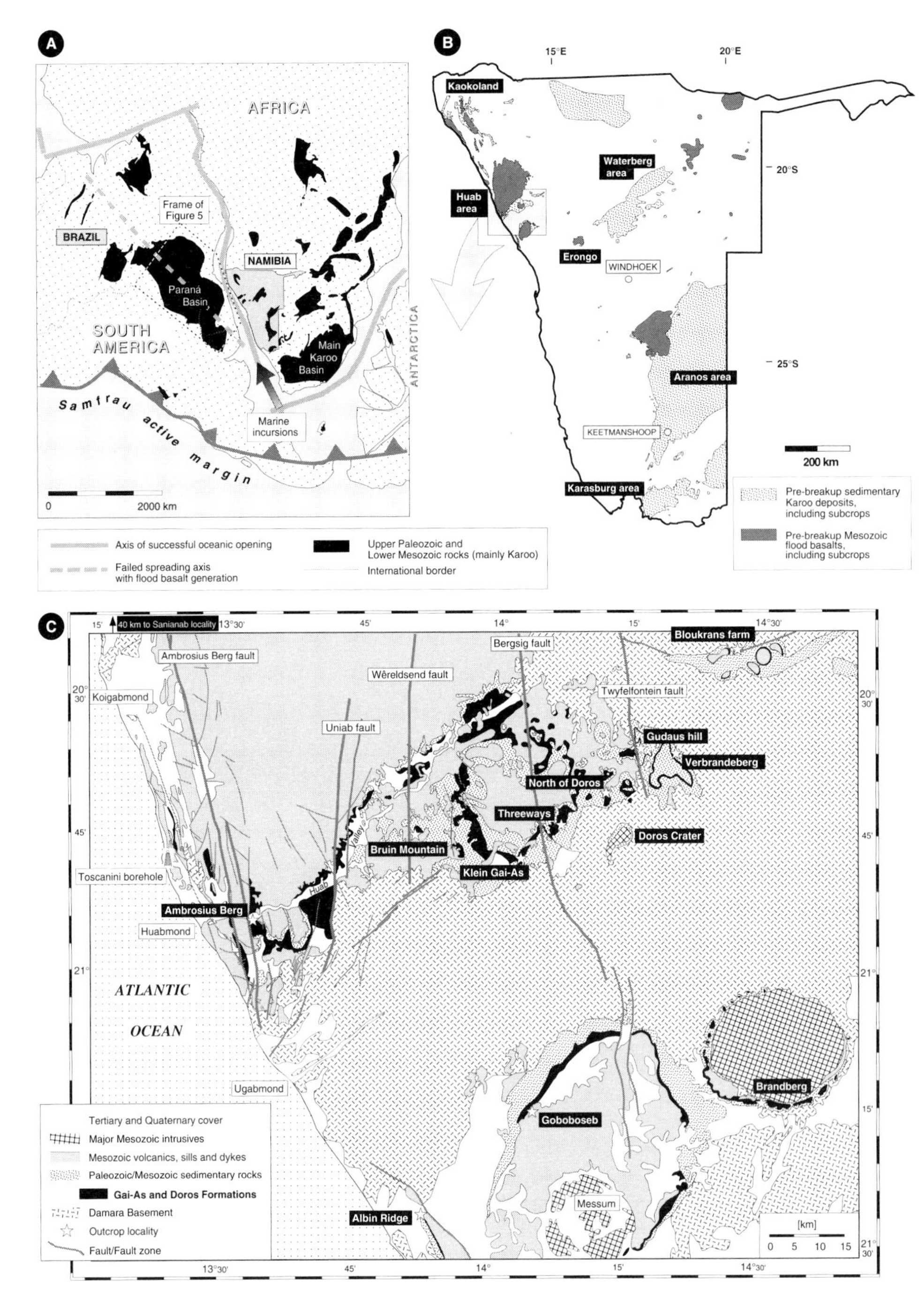

Figure 1—Maps showing (A) distribution of upper Paleozoic–Lower Mesozoic (Karoo) strata in south Gondwana (based on de Wit et al., 1988) and (B) in Namibia (based on Miller and Schalk, 1980). (C) The simplified geology (based on Miller and Schalk, 1980; Miller, 1988), the structural setting, and locations of measured sections in the Huab area of northern Namibia, including outcrops of the Gai-As and Doros Formations.

Botswana (Smith, 1984), Zimbabwe, Zambia, Malawi (Ring, 1995), Tanzania (Wopfner, 1991), and Madagascar (Hankel, 1994), which define intracontinental rift systems affected by repeated phases of extensional faulting (cf. Lambiase, 1989; Tankard et al., 1995).

GAI-AS AND DOROS FORMATIONS

The Gai-As Formation is up to 170 m thick and breaks naturally into two subunits that are informally termed the lower and upper Gai-As Formation (Figure 2B). This lacustrine succession is separated by an unpronounced hiatal unconformity (Figure 2B) from the underlying, mesosaurid-bearing marine deposits of the Huab Formation, an equivalent of the Lower Permian (Artinskian) Whitehill Formation of South Africa and the Irati Formation of Brazil (Anderson, 1981; Oelofsen, 1987). In the Huab area, the top of the Gai-As Formation is most commonly defined by a widespread erosional unconformity below the overlying eolian dune deposits of the Lower Cretaceous Twyfelfontein Formation (Stanistreet and Stollhofen, 1999). Adjacent to the pronounced northerly trending faults, however, a more continuous deposition is recorded in local hanging-wall traps by semi-conformable contacts of

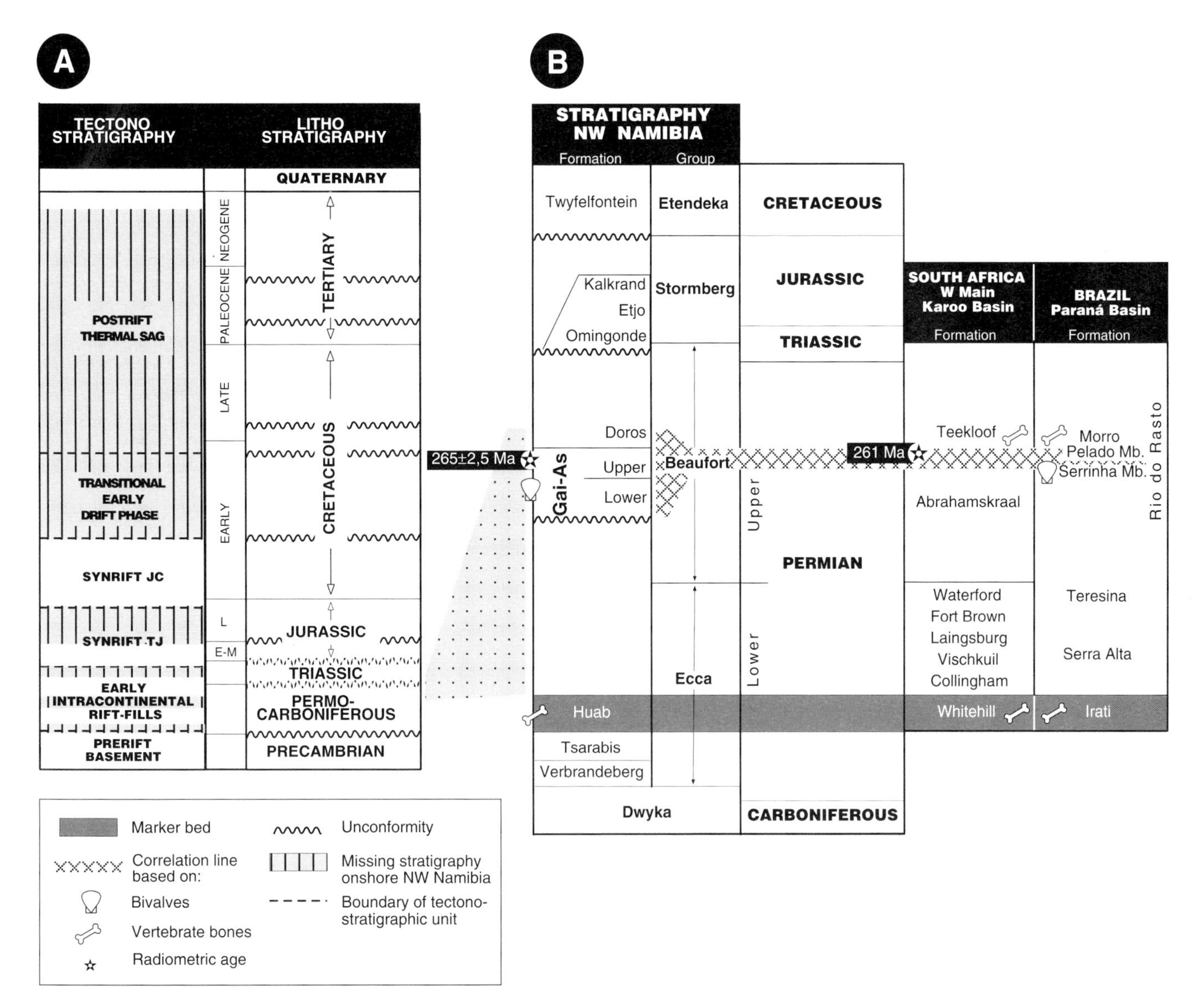

Figure 2—The Gai-As and Doros Formations of northwestern Namibia (A) placed within the regional tectonostratigraphic framework, which is modified from Light et al. (1993), originally based on the stratigraphies inferred from seismic sections in the Namibian offshore area. (B) Correlation of Gai-As and Doros Formations with equivalent strata in South Africa and Brazil by using radiometric and biostratigraphic age constraints.

the mudstone-dominated lacustrine Gai-As to the overlying sandstone and gravel-dominated (<30 m) Doros Formation (Stanistreet and Stollhofen, 1999). A time-stratigraphic gap above the Gai-As and Doros units is caused by a Late Jurassic to Early Cretaceous unconformity. The time gap is reduced toward the east to south–southeast of the central Huab area, where an intervening Triassic to Lower Jurassic sequence is preserved, involving the Omingonde and Etjo Formations (Figure 2B) in the Goboboseb, Brandberg, Erongo-Karibib, and Waterberg areas (Figure 1B). The Jurassic/Cretaceous unconformity is conspicuously offset by faults and further accentuated by the development of paleosols and subvertical eolian sandstone-filled fissures up to 20 cm wide and reaching up to 15 m deep into the uppermost Gai-As, Doros, and Omingonde successions, respectively.

EXTENT AND AGE OF THE GAI-AS LAKE

The geometry and extent of the Gai-As and Doros Formations are controlled by earlier basin geometries, in particular those that hosted Lower Permian sequences. In this regard, equivalents of the Lower Permian Whitehill Formation are widespread in southern Africa and South America and form a prominent marker horizon (Oelofsen and Araujo, 1983), which records the latest major marine transgression into southern west Gondwana during the Permian. We use this marker horizon to set up a stratigraphic framework for the comparison of the overlying lacustrine sequence in various parts of the pre-southern South Atlantic rift zone and the neighboring foreland basin of South Africa. Because outcrop areas are in two areas on either side of the Atlantic, great care was taken in comparing lake sediments in Namibia with lake sediments in Brazil by using radiometric and biostratigraphic constraints.

Based on its conspicious red coloring, the discovery of an isolated cynodont tooth (von Huene, 1925) and supposed dinosaur vertebra (Keyser, 1973), the Namibian Gai-As Formation had originally been assigned a Late Triassic age. Later, Horsthemke (1992) and Ledendecker (1992) correlated the Gai-As Formation both with the Brazilian Serra Alta, Teresina, and Lower Rio do Rasto Formations and the South African Collingham to Waterford Formations (cf. Figure 2B) by using endemic bivalve associations, which suggest a Late Permian age. Only recently, the Late Triassic age constraints of von Huene (1925) and Keyser (1973) were shown to be based on misinterpretations of the fossil fauna (B. S. Rubidge, 1997, personal communication).

New radiometric dating of zircon separates from tuff beds interlayered with the top part of the Gai-As succession (cf. Figure 2B) revealed U/Pb ages of 265±2.5 Ma. This would principally confirm the stratigraphic positioning of Horsthemke (1992) and Ledendecker (1992) and place this part of the Gai-As Formation as uppermost Lower Permian according to the time scale of Harland et al. (1990). This datum is close to U/Pb zircon ages of 261 M (de Wit, 1998, personal communication) derived from tuffs of the *Cistecephalus* assemblage zone sensu Rubidge et al. (1995) of the Upper Permian Teekloof Formation (Beaufort Group) in the western Cape region of South Africa. In contrast with the Gai-As Formation in Namibia, this stratigraphic unit is placed more than 2000 m above the Huab-equivalent Whitehill Formation.

As the entire Namibian Gai-As Formation only contains endemic bivalves of the *Terraia altissima* biozone, characterized by *Leinzia similis* (cf. Figure 2B), it can be confidently correlated with the Upper Permian Serrinha Member of the Rio do Rasto Formation in the Brazilian Paraná basin. Comparable to the stratigraphic relationships in South Africa, this unit is located about 500 m above the Whitehill/Huab equivalent Irati Formation. Remains of the dicynodont *Endothiodon* occur at the base of the Upper Permian Morro Pelado Member of the Rio do Rasto Formation, immediately above the Serrinha Member (Barberena et al., 1991). This suggests a faunal affinity to the biostratigraphically well-constrained *Pristerognathus, Tropidostoma,* and *Cistecephalus* assemblage zones sensu Rubidge et al. (1995), located in the Upper Permian Teekloof Formation (Beaufort Group) of the main Karoo basin in South Africa. Such a biostratigraphic relationship coincides perfectly with the previously discussed correlation of the tuff beds between South Africa and Namibia; however, age assumptions based on amphibian remains (including vertebrae) and wood fragments, both from the top of the upper part of the Namibian Gai-As and Doros Formations still remain problematic. The vertebrae were determined to derive from Triassic mastodonsauroid or matoposaurid amphibians, but teeth occurring at the same stratigraphic level are more rhinesuchid in shape, suggesting an Upper Permian age (B. S. Rubidge, 1998, personal communication). These stratigraphic relationships might imply a discrepancy between the traditional biostratigraphy and radiometric ages, which is also obvious from log fragments of silicified *Podocarpoxylon* wood occuring in the Doros Formation. In South Africa and elsewhere, the latter is only known from Early Triassic and younger strata (Bamford, 1998); consequently, it appears that the Gai-As/Doros succession contains some faunal and floral elements that are apparently biostratigraphically younger than their numerical correlatives in South Africa.

In summary, radiometric and biostratigraphic age constraints suggest a Late Permian–earliest Triassic age for the Namibian Gai-As and Doros Formations. Correlative beds are the upper Abrahamskraal and lower Teekloof Formations in South Africa and the Serrinha Member of the Rio do Rasto Formation in Brazil. Additional equivalents of these units might be represented in Argentina, Paraguay, and Uruguay, and in the Kaokoland area of northwest Namibia (cf. Figure 1A, B). This implies that equivalent beds of the entire post-Whitehill Ecca Group (Collingham to Waterford Formations) in South Africa and of the Serra Alta and Teresina Formations in Brazil are poorly or not preserved in northern Namibia; however, in southern Namibia, they may be partly represented by the up to 740-m-thick Aussenkjer and Amibberg formations of the Karasburg area (cf. Miller, 1992).

LAKE DEVELOPMENT

The trans-Atlantic correlation between the Gai-As and Rio do Rasto formations would imply the existence of an elongated inland water body extending over more than 1.5 million km^2 along an overall northwest–southeast trend (Porada et al., 1996) during the Late Permian–Early Triassic. Other extensive lakes existed at the same time in eastern central Africa (Yemane and Kelts, 1990), and some were probably strung along comparable rift networks with intervening fluvial deposits or basement highs. Considering the global paleogeographic reconstructions of Scotese and McKerrow (1990), Scotese and Barrett (1990), and Lottes and Rowley (1990), the Gai-As lake area was situated at a latitude of about 40° south during the Permian. Such a paleogeographic and paleolatitudinal setting would be susceptible to intense mid-latitude winter cyclones as reviewed by Duke (1985). Climates in southern Gondwana have been inferred as cool-temperate and seasonal on the basis of isotope geochemistry, clay mineral assemblages, and palynoflora of Late Permian sequences in northern Malawi (Yemane et al., 1996; Yemane and Kelts, 1996). With the paleolatitudinal setting of Malawi about 15° south of Namibia, warm-temperate to subhumid/semiarid climates can be predicted for the area occupied by the Gai-As paleolake.

The Gai-As and Doros Formations together record an overall shallowing- and coarsening-upward trend

(Figure 3): (1) the lower Gai-As Formation characterized by laminated claystones and mudstones, (2) the upper Gai-As Formation comprising dominantly mudstones with interbedded limestones, sandstones, and fallout tuffs, and (3) the Doros Formation dominated by sandstones and gravels with minor interbeds of mudstones and limestones. The base of the Doros is normally unconformable with the underlying Gai-As Formation, but shows an angular unconformity due to the activity of synsedimentary faults; however, the Doros Formation is integral to Gai-As Lake development and therefore will be included in our study.

Facies architecture of the lithologic units outlined varies laterally and was affected by contemporaneous activity of extensional fault systems. The spectrum of environments is generally best represented in the central part of the Huab area, whereas toward the east (Verbrandeberg region), a condensed sequence development is recorded, and toward the west (Atlantic coastline and Paraná basin), enhanced sequence developments are encountered; therefore, the lithologic sequence of the Gai-As Formation is described on the basis of the central Huab area type sections (Figure 3), well exposed in cliffs at localities Klein Gai-As, Threeways, and north of Doros (cf. Figure 1C). We will examine their lateral variation eastward toward the lake margin and then westward toward the basin center. Finally, we relate lake development to the contemporaneous geodynamic evolution of the pre-southern South Atlantic rift zone.

Lower Gai-As Formation: Offshore Lacustrine Setting

Early lake history is recorded by a 35-m-thick succession (Figure 3) of reddish to violet, mostly laminated mudrocks containing thin (1–3 cm) tabular interbeds of normally graded, fine- to medium-grained sandstone layers and a few laminated limestone beds, 10–50 cm thick. At least two widespread shale horizons (Figure 3A) are associated with conspicious concentrations of articulated and disarticulated shells of the *Terraia* molluscan fauna. Concentrations occur in several layers, up to 4 cm thick, with a dominance of concave-down shells reflecting the influence of traction.

Thick shaly units were deposited dominantly from suspension fallout in a quiet offshore lacustrine hemipelagic setting. We interpret the limestone beds as deposited by storm-initiated flows transporting clastic carbonate grains into the offshore region. Sand was only sparsely delivered to the offshore to form tempestites and turbidites, the majority of which contain abraded fish bones, teeth, scales, and fin spines. Such thin, sandy units acted as nuclei for early diagenetic formation of abundant disc-shape dolomitic concretions measuring up to 3 m in diameter (Horsthemke, 1992). Wave activity was counteracted by the stabilizing effects of microbial mats at the lake margin. Major storms seem to have rarely affected this part of the sequence. Concentrations of bivalve shells reveal a low diversity of population, which could happen in an ecologically stressed setting such as a saline lake where lake overturns due to thermal density instabilities or storms may cause mass mortality of bivalves, later to be reworked by gravity flows.

Upper Gai-As Formation: Upward-Shallowing Lake with Tempestites

This succession is about 20 m thick and comprises red violet mudstones containing minor tabular sandstone interbeds, laminated or small-scale cross-bedded limestone, and laminated rhyolitic to dacitic fallout tuff (Figure 3B). The fine- to medium-grained sandstones, 1–10 cm thick, are normally graded and show rare basal load casts and wave-rippled tops. Commonly, such beds contain concentrations of fish bones, scales, and teeth in a single layer at their top surface, which reflects density grading of biogenic particles. They are interpreted as tempestite or turbidite beds arranged in a thickening- and coarsening-upward architecture, the latter suggesting a gradual overall shallowing of the lake toward the top of the middle unit.

Lenticular to tabular limestone beds up to 35 cm thick are found particularly in the lower half of the unit. Both laminated stromatolitic limestone beds and micritic limestone/sandstone interlaminated beds are developed, the latter showing rhythmic wavy and lenticular bedding with isolated current and wave ripples made of sand. Such bedding is well known from intertidal and shallow subtidal areas (Reineck and Wunderlich, 1968; Reineck and Singh, 1980), but has also been described from non-tidal deltaic settings (e.g., Coleman, 1966). We agree with McCave (1970) that processes associated with storm and calm conditions are generally the primary cause of wavy and lenticular bedding. One of the interlaminated limestone/sandstone beds is characterized by conspicuous concentrations of articulated and disarticulated shells of the *Leinzia similis* molluscan fauna in a single layer 2–5 cm thick. A biological picture of what was happening in the lake is provided by disarticulated skeletal fragments of large mastodonsauroid or matoposaurid amphibians and abundant fragmentary fish remains of *Namaichthys* and *Atherstonia*. Such large organisms required the lake to be permanent to develop well-established food chains. Also required were low and stable salinities and temperatures as well as an oxygen-rich environment.

The unit further comprises up to five tuff beds, each 2–5 cm thick, showing a poorly developed normal grading. Tabular geometry of the tuff beds and their conspiciously bright color makes them important chronostratigraphic markers with the lowermost and uppermost tuff beds coinciding broadly with base and top of the upper Gai-As Formation. According to XRD (x-ray diffraction) analyses, the tuff beds usually consist of quartz, albite, orthoclase, sanidine, clinopyroxene, and illite-montmorillonite mixed-layer minerals with some tuff beds containing analcime, calcite, dolomite, and ankerite. Thin sections reveal a microcrystalline groundmass of recrystallized quartz and

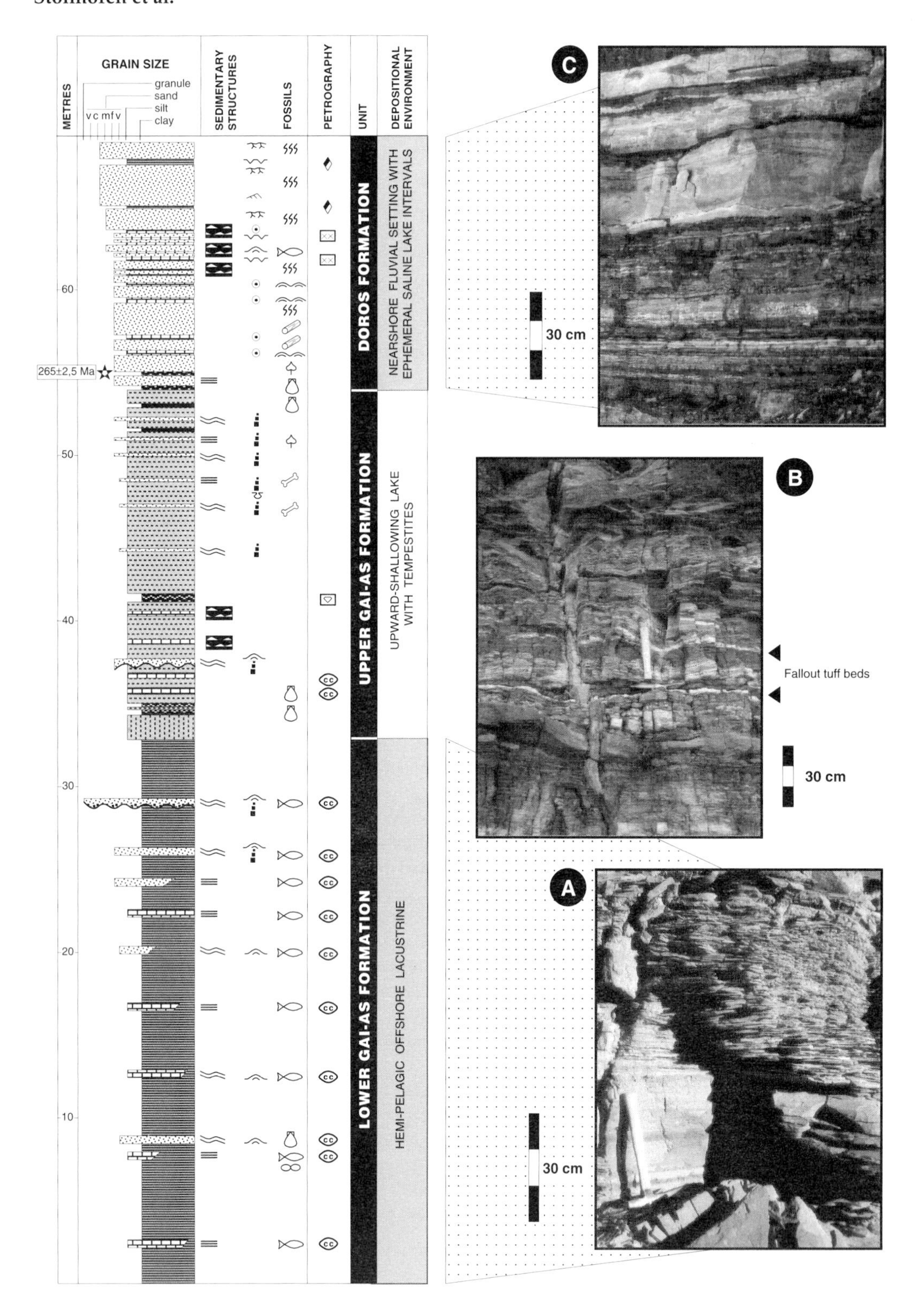

Figure 3—Sedimentologic log of the Gai-As and Doros Formations compiled from sections measured at the Klein Gai-As and the north of Doros type localities (see Figure 1C for locations) in the central Huab area. Outcrop photographs of the three subunits illustrate the typical field appearances of (A) laminated mudstones of the LowerGai-As Formation, (B) mudstones interbedded with thin, normally graded fine-grained sandstones (tempestites or turbidites) and thin fallout tuff beds (arrows) of the upper Gai-As Formation, and (C) mudstones overlain by stromatolitic limestones interbedded with the otherwise fluvially dominated Doros Formation. See Figure 4 for legend of symbols.

analcime with a few juvenile crystals of biotite and sanidine. Heavy mineral separates of the tuffs comprise hornblende, monazite, titanite, zircon, and apatite that fit well with the rhyolitic to dacitic composition determined by XRF (x-ray fluorescence). Overall grain size and sorting characteristics of the tuff beds show that they comprise multiple pyroclastic fallout units in areas distal from the volcanic source region.

The upper Gai-As Formation records a pronounced shallowing of a paleolake, with deposition dominantly at water depths below fair weather but above storm wave base. This setting resulted in an increasing number of siliciclastic and mixed carbonate/siliciclastic tempestite/turbidite interbeds bearing shallow-water features, such as wavy and lenticular bedding and wave-rippled surfaces. Consistent thicknesses of the thin limestone/sandstone laminae suggest seasonally controlled deposition of siliciclastics and carbonate, including lenses of small-scale cross-bedded clastic carbonate. Longer term intervals of warm climate and low precipitation suppressing siliciclastic input, but favoring microbial activity, are then indicated by tabular, stromatolite-textured limestone beds, up to several decimeters thick. Considering the tectonic framework of the Gai-As Lake, carbonate productivity could have been further influenced by groundwater-controlled input of Ca-rich waters along fault and fracture zones (cf. Gierlowski-Kordesch, 1998).

Doros Formation: Nearshore Fluvial Setting with Ephemeral Saline Lake Intervals

This formation is up to 30 m thick with an abundance of 60–80% sheet-like sandstone beds, up to 2 m thick, dominating the succession. The medium- to coarse-grained sandstones show either planar or low-angle scoured basal contacts and may contain low-angle accretion surfaces, small-scale hummocky cross-stratification, or plane-bedding. Hummocky beds are 5–25 cm thick with wave lengths ranging between 20 and 130 cm. Several of the upper surface contacts are wave-rippled with the majority of ripple crests striking 50–60°. In places extensive bioturbation is preserved with burrows of *Planolites* ichnosp., *Skolithos* ichnosp., *Beaconites* ichnosp., *Palaeophycus tubularis*, and *Rosselia* ichnosp., in addition to larval and nematode traces. Also observed are plant debris, imprints of rootlets, and rare layers with concentrations of disarticulated bivalve shells. A thin (max. 4 cm) but widespread marker bed about 8 m above the dated tuff bed contains extraordinarily abundant fish remains, including scales and teeth.

A few channelized, massive or faintly plane-bedded sandstone units preserve imprints of rootlets and small fragments and trunks (max. 1.5 m length and 10 cm diameter) of silicified wood, identified as *Araucarioxylon* sp.1 and *Podocarpoxylon* i spp. (Bamford, 1998). The latter suggests an age comparable with that of the upper Beaufort (Bamford, 1998), which would place the Doros Formation in the Lower Triassic. Mudstones are only rarely interlayered in this part of the succession and contain abundant mudcracks and carbonate concretions.

Limestone beds at this stratigraphic level (Figure 3C) are characterized by domal stromatolites (up to 25 cm high) with interdomal areas infiltrated by ooliths and oncolites. Horsthemke (1992) measured in bulk samples ^{18}O values of 3.01–5.62 in the stromatolitic beds. This suggests that considerable salinities developed through evaporation, which would be particularly favored in a hydrologically closed lake system during periods of low rainfall (cf. Cerling et al., 1977). Locally, the stromatolitic limestones contain v-shaped grooves and erosional surfaces that are mantled by thin, normally graded calcareous sandstone layers of laterally uniform thickness. These features are interpreted as storm deposits, interrupting biogenic growth and mobilizing material in the shoreline area of the lake. Particularly above the stromatolitic limestone units, intercalations of laminated micritic dolomite beds are found, averaging between 0.4 and 1.5 cm thick, but can be up to 18 cm thick. Their sharp planar bases define either vertical amalgamation or a rhythmic alternation with erosionally based, small-scale, cross-bedded fine-grained sandstone layers. Silicification is common in the micritic beds, and halite and gypsum pseudomorphs are rarely preserved in the top parts of the layers. We agree with Horsthemke (1992) in interpreting the micritic limestones as evaporitic deposits that formed in alkaline-saline lake environments.

Sandstone units are interpreted as regressive sheet-sands (cf. Heward, 1981) and poorly channelized fluvial sheetflood deposits. These were emplaced in a shallow, wave-dominated lake margin/shoreline environment that favored minor lag concentrates of fossil material with interlayered mudstones, preserving abundant subaerial exposure features. Pseudomorphs of evaporite minerals and the $\delta^{18}O$ values of stromatolitic limestones record periods of enhanced salinity of the lake with more alkaline conditions dissolving silica. Freshwater flushing led to chert precipitation comparing with processes described by Knauth (1994). Development of subaerial exposure features and erosional surfaces suggests that lake level was, in general, significantly lowered during deposition, and the regularity of limestone/sandstone alternations may account for a lower order climatic variation controlling lake level fluctuations, involving evaporation and influx.

CONTRASTING MARGIN AND BASINAL SETTINGS OF THE GAI-AS LAKE IN NAMIBIA

Lateral variability of Gai-As and Doros Formation thicknesses occurs across the study area. Maximum thicknesses attain 180 m in the western Huab area, close to the present Atlantic coastline, 70 m in the central Huab type section area (Klein Gai-As), and only 15 m down to total pinch-out around Verbrandeberg in the eastern Huab area. Condensed sequence development in the eastern Huab area particularly contrasts to thicknesses of more than 700 m attained for the Gai-As/Doros equivalent Rio do Rasto Formation in the southeastern Paraná basin of Brazil, but is mirrored by a complete pinch-out in the northeastern part of the Paraná basin. Such thickness variations coincide with pronounced facies changes along southwest–northeast trends both in the Huab and Paraná area, as well as in areas close to contemporaneously active extensional fault systems.

Transitional Offshore Lake Facies of the Western Huab Area

To the west of the Gai-As type section, the Gai-As sequences rapidly increase in thickness with considerable facies changes at a distance of 20 km east of the present coastline. Figure 4 (section Ambrosius Berg) shows the threefold subdivision of the Gai-As and Doros Formations can still be identified at this locality in the thick, shale-dominated succession, but limestone, sandstone, and tuff interbeds are much less abundant compared to the sequences containing the marginal facies farther east. Thin, normally graded interbeds of massive, fine-grained sandstones occur exclusively in the middle unit and have been interpreted as turbidite deposits (Horsthemke, 1992). Carbonates occur as calcareous concretionary horizons and thin lenticular unfossiliferous beds. The topmost 30 m of the section is dominated by plane-bedded or hummocky cross-bedded fine-grained sandstones, some of which contain mud clasts, gutter casts, and elongate groove marks at their bases. Single vertical and subvertical burrows, millimeters to centimeters in diameter, are sparsely developed and concentrate in the plane-bedded sandstones. No subaerial exposure features are present.

Hummocky cross-stratification is commonly viewed as forming under storm conditions, where waves interact with bottom currents (Harms et al., 1982; Cheel and Leckie, 1993); however, contrasting views exist whether hummocky cross-stratification may be best preserved below fair-weather base (Walker, 1984), in very shallow lacustrine environments (Duke, 1985), or the surf zone of lakes (Greenwood and Sherman, 1986). Compared to the type section, the facies association measured at the Ambrosius Berg locality (Figure 1C) nevertheless records deposition in an offshore basinal setting. These deeper water units are succeeded by a weakly bioturbated shoreface facies comprising sedimentary structures that indicate the influence of considerable wave activity associated with low-frequency storms.

Marginal Facies of the Verbrandeberg Area

The section measured at Gudaus hill (Figure 4), about 9 km southwest of the more easily accessible Verbrandeberg locality (Figure 1C), is taken as representative of facies patterns in the eastern Huab area with a total formation thickness attaining 15 m on average. The Gai-As Formation comprises interbedded reddish massive mudstones, fine-grained sandstones, and abundant nodular carbonate with the whole succession heavily affected by a network of desiccation and shrinkage cracks along with voids, many of which are filled by chert. The Doros Formation is represented by an erosionally based, trough cross-bedded, pebbly medium- to coarse-grained sandstone unit that contains *Planolites* ichnosp. and *Phycodes curvipalmatum* burrows at its base and larval and nematode traces at the top. Trough cross-beds reveal a low to moderate variation of paleocurrent directions toward the southwest (210–280°), and flat lens-shape channel geometries indicate rivers characterized by relatively high width/depth ratios. Considering the scarcity of mudstone interbeds, the restricted variation of paleocurrent data, and the channel morphologies, this unit is interpreted as an amalgamated braided fluvial channel sandstone complex that formed at the subaerially exposed lake margin in a setting relatively proximal to the clastic source.

In situ breccia formation observed in the lower part of the section is a process particularly abundant in palustrine lacustrine carbonates (Freytet and Plaziat, 1982). There, lake level fluctuations may have caused alternating periods of subaerial exposure and flooding of lake-fringing marshes. Carbonate nodules within the section comprise fibrous, micritic, and sparitic types, as well as ferruginous calcitic nodules (Horsthemke, 1992), with the latter preferentially forming in the vadose zone under the influence of groundwater level fluctuations (cf. Sehgal and Stoops, 1972). Replacement chert nodules and layers up to 60 cm thick are concentrated in the carbonate-poor host-lithologies. The silica source required for their formation is believed to result not solely from dissolved detrital quartz grains, but also from surface weathering of highly reactive dacitic volcanic glass shards. Enhanced silica concentrations would have been formed by the high pH of saline, alkaline lake waters coinciding with the fallout of volcanic ash. Dissolved silica then would precipitate after lowering of pH due to enhanced freshwater discharge or in sub-lacustrine sediment from groundwaters rich in dissolved sodium carbonate/bicarbonate (cf. Parnell, 1988).

The lower part of the Gudaus hill section is interpreted as hydromorphic calcareous soils that developed contemporaneously with the formation of stromatolitic limestones in the more basinward lake areas. These soils can be viewed as pedogenically modified mudflats of a moderately steep lake margin that is influenced by considerable fluctuations in lake and groundwater levels. Abundant root casts and rhizocretions (cf. Klappa, 1980) indicate that the mudflats of these marginal lake areas were locally vegetated. This fits well with the observation of upright in situ embedded tree trunks and logs, 0.3–0.6 m in diameter, at Bloukrans farm area, about 30 km northeast of Gudaus hill locality (Figure 1C) (Bruhn and Jäger, 1991).

In fact, the facies architecture of the Verbrandeberg area is rather complex because it appears to be controlled not only by contrasting basinal-marginal aspects, but also to a considerable degree by the complex structural framework of the eastern Huab area. Due to the

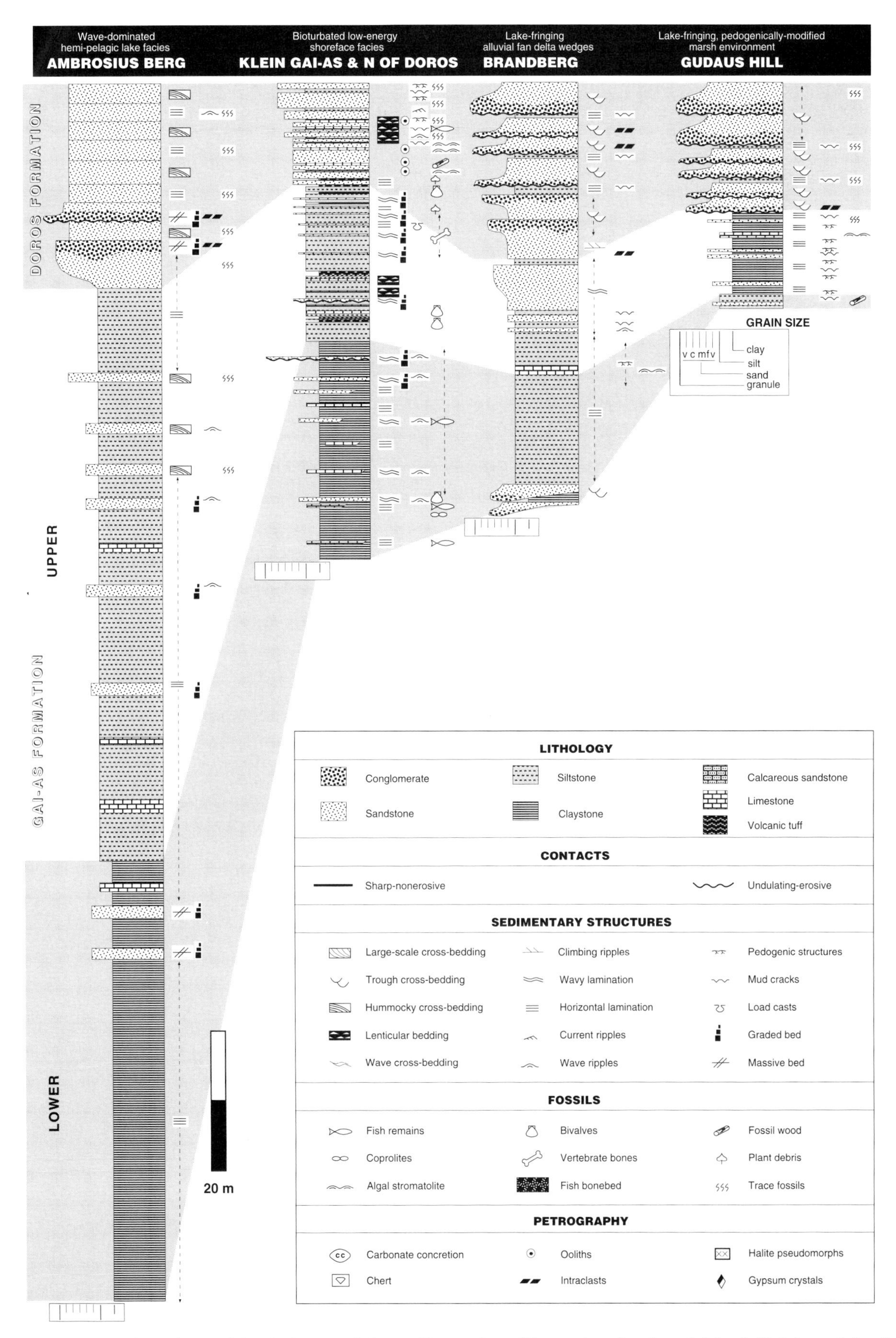

Figure 4—Measured sections of the Gai-As and Doros Formations illustrating the threefold subdivision of the fluvio-lacustrine succession and lateral and vertical changes in facies architecture. See Figure 1C for location of sections.

overlap of major north-northwest–trending structures, such as the Twyfelfontein fault zone and synchronously active east-northeast–trending faults from reactivated neo-Proterozoic Damaran basement structures, a mosaic framework of footwall and hanging-wall blocks developed contemporaneous with Gai-As deposition. This is reflected by rapid changes in facies and thickness, with maximum thicknesses on the order of 70 m in hanging-wall block positions and total formation pinch-out associated with footwall blocks. Development of condensed sequences in association with footwall block positions, however, was not all restricted to the Verbrandeberg area but was identified at various other localities, namely the Albin Ridge, Sanianab, and Brandberg localities (Figure 1C).

Fault-Controlled Marginal Fan-Deltas

The section (Figure 4) measured at Brandberg locality (cf. Figure 1C) illustrates another important aspect of Gai-As Lake deposition. The lower part of the Brandberg section correlates with the lower Gai-As Formation at its Klein Gai-As and north of Doros type locations. A thickness of 15 m and a pronounced pedogenic overprinting of the upper part of the Brandberg section compare to the condensed sequences of the lake margin facies described from the Verbrandeberg area. The overlying Doros Formation comprises a 25-m-thick coarsening-upward succession of trough cross-bedded, pebbly, medium- to coarse-grained sandstones (Figure 4). These are characterized by enhanced feldspar and tourmaline contents and by abundant lithoclasts eroded from both the underlying Gai-As units and the Huab Formation. The sandstones represent deposits of proximal mouth bars and of braided fluvial channels with paleocurrents to the southwest (180–296°) from measured trough cross-beds. A comparable coarse-grained siliciclastic facies, 2–15 m thick, has been observed at the Doros Crater, Bloukrans farm, Sanianab, Albin Ridge, and Gudaus hill localities (Figure 1C), with the latter two showing a similar association to an underlying pedogenically modified marginal lake facies. On a regional scale, this coarse-grained siliciclastic facies grades basinward into the previously described sheetlike, plane-bedded, or hummocky cross-bedded sandstone units that are interpreted as regressive sheetsands and fluvial sheetflood deposits of a lake-margin/shoreline setting at the Gai-As type localities.

Considering the structural setting of the localities described, it is significant that all of them are situated on the hanging-wall sides of large-scale northerly trending faults in association with their east-northeast–trending conjugates. The faults were identifiably active from Lower Permian Tsarabis Formation deposition onward (Stollhofen, 1999). The influence of faulting on facies architecture is particularly well recorded at the north of Doros locality (Figure 1C), 8 km north of Doros Crater where a roll-over anticline developed adjacent to the hanging-wall west side of the Twyfelfontein fault zone (Figure 7). The Doros Formation, forming the coarse-grained top unit, is separated from the underlying sequence by an angular unconformity of up to 4° and contains strata that wedge out conspicuously against the developing roll-over. By analogy, the coarse-grained facies at Sanianab locality in the western Huab area (Figure 1C) is associated with the major Ambrosius Berg fault zone subparalleling the present coastline. In contrast, the Doros Formation and in places the entire Huab, Gai-As, and Doros successions are eroded at the footwall immediately east of the Bergsig fault.

A picture emerges of a wave-dominated Gai-As lake margin that is fringed by coarse siliciclastic alluvial braided fan-deltas of the Doros Formation prograding off emerging fault scarps. Progradation of such fan-delta wedges would be particularly favored during lake level lowstands, with a proximal fan facies triggered by newly generated fault scarps. The fact that the Huab and Gai-As Formations are partly eroded in footwall positions and the Doros Formation is preferentially preserved in hanging-wall blocks emphasizes that active tectonism occurred contemporaneously with Gai-As lake deposition.

THE CONTINUATION OF THE GAI-AS LAKE INTO BRAZIL

Sections measured in the Brazilian portion of the Paraná basin (Figure 5C) include the entire Passa Dois Group (Figure 6), which attains thicknesses of more than 1300 m (Mühlmann, 1983) and subdivides into the Irati, Serra Alta, Teresina (or Estrada Nova), and Rio do Rasto Formations. Only the Serrinha Member of the basal Rio do Rasto Formation can be correlated to the Namibian Gai-As and Doros Formations on the basis of the *Terraia altissima* biozone (Rohn, 1994), characterized by *Leinzia similis*; however, for the understanding of Gai-As Lake initiation and development as a whole, it is important to study the underlying sequence development as well because this records the paleoecologically well-constrained transition from the marine-influenced mesosaurid-bearing Irati Formation (Whitehill/Huab Formation equivalents) to the lacustrine Gai-As Formation equivalents.

Transition from Marine Irati to Lacustrine Rio Do Rasto

The Irati Formation is conformably overlain by the 10–150-m-thick Serra Alta Formation (Figures 2, 6), which is dominated by dark shales and siltstones containing abundant fish scales. Endemic bivalves of the *Barbosaia angulata* assemblage especially occur in the northeastern part of the basin, near the base of the formation. Several authors interpret this unit as marine (e.g., Gama, 1979; França et al., 1995), despite the fact that no unequivocal marine fossils are known and paleogeographic data point to a basin isolated from the world ocean; therefore, we attribute the depositional environment to a relatively deep, large lake with occurrences of wavy lamination, wave ripples, and bioturbation features in the upper part of

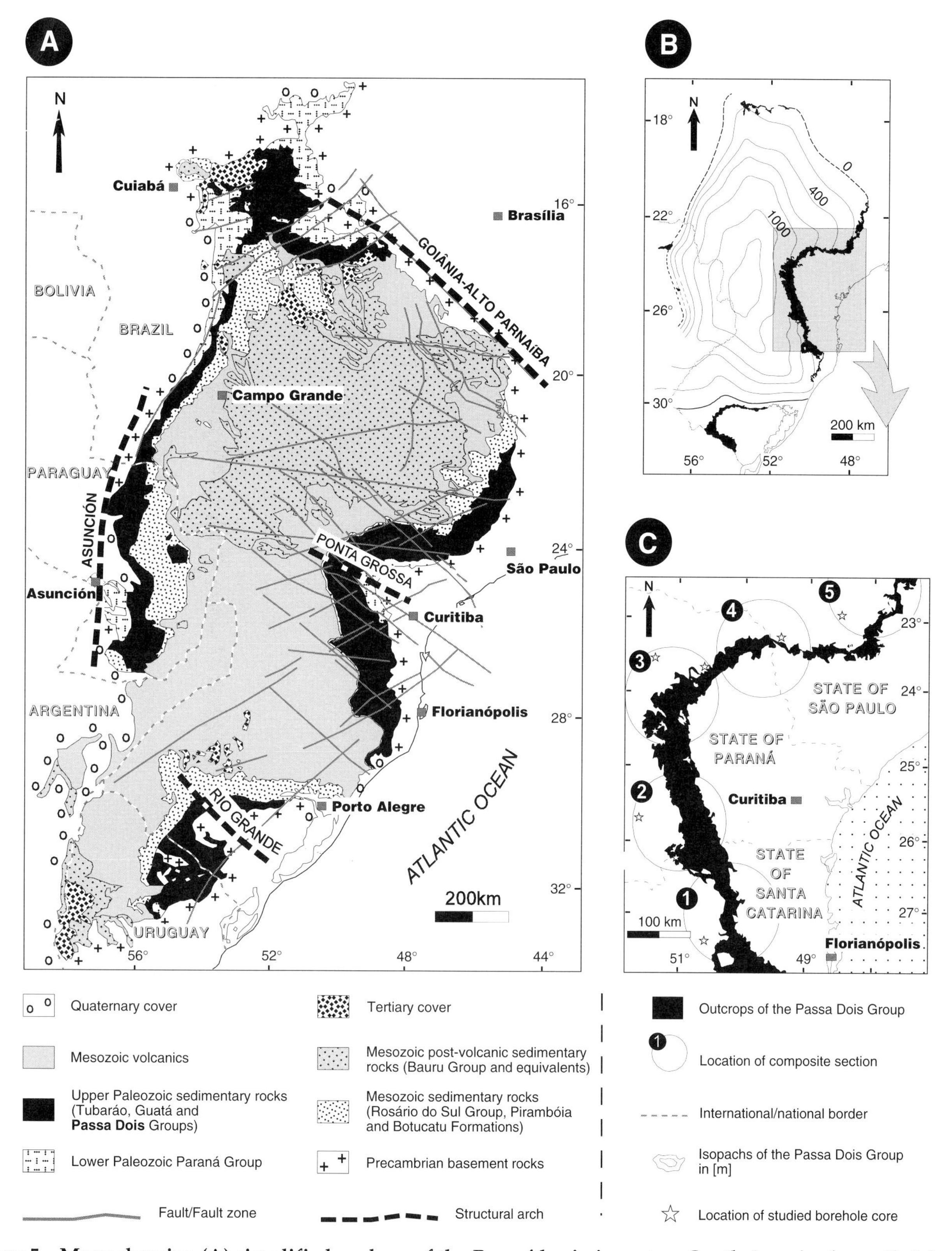

Figure 5—Maps showing (A) simplified geology of the Paraná basin in eastern South America (compiled from Bossi et al., 1975; Schobbenhaus et al., 1984; Zalán et al., 1990, and Fulfaro et al., 1995), (B) the isopach map of the Passa Dois Group (redrawn from Mühlmann, 1983), which also includes stratigraphic units below the lake beds of the Rio do Rasto Formation, with (C) outcrops of the Passa Dois Group in detail. For comparison of structural trends with Namibia, a counter-clockwise rotation of 58.2° (Powell and Li, 1994) has to be performed. Outcrop and isopach maps of the Passa Dois Group include the Irati Formation (Whitehill/Huab Formation equivalent), which is only a few tens of meters thick and therefore does not modify an interpretation based on the illustration.

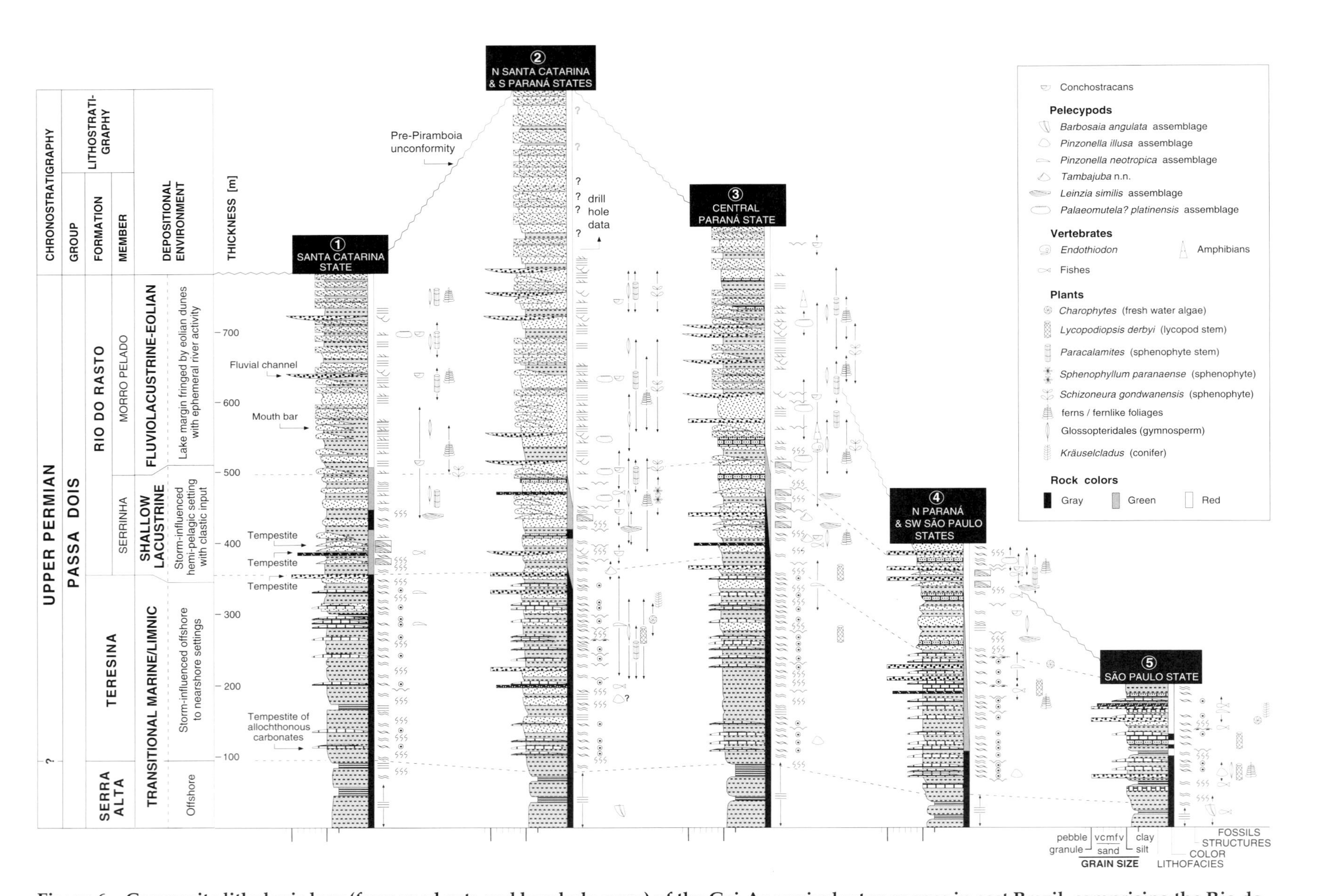

Figure 6—Composite lithologic logs (from road cuts and borehole cores) of the Gai-As equivalent sequence in east Brazil, comprising the Rio do Rasto Formation (subdivided into the Serrinha and Morro Pelado Members) based on Rohn (1994). Also included are logs of the Serra Alta and Teresina Formations, which are not well represented in Namibia. Only fossil occurrences that are important for correlation purposes are included. See Figure 4 for explanation of additional lithologic symbols and Figure 5C for location of sections. Examples of the signatures used to distinguish the sedimentologic interpretation of individual depositional units are given in the left column. Distances between logs indicated on Figure 5C.

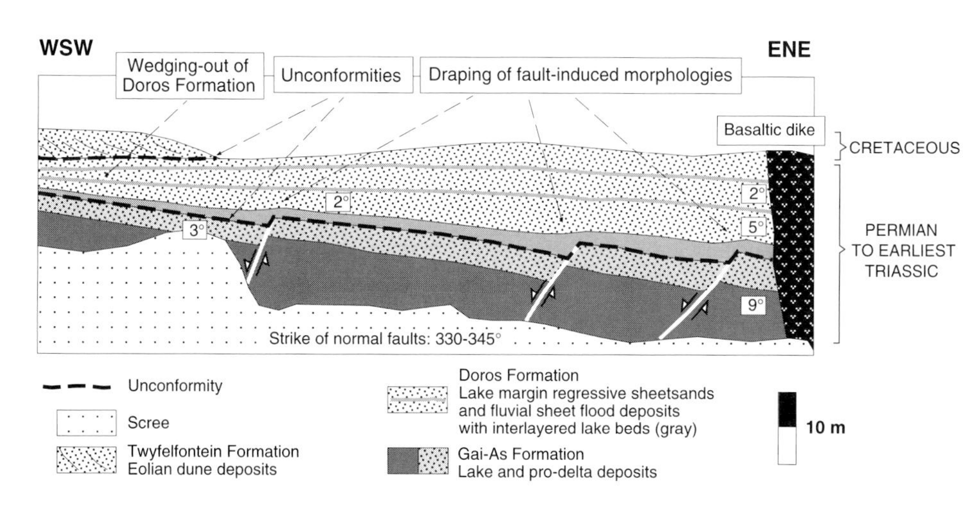

Figure 7—Detail of a synsedimentary, faulted roll-over anticline exposed north of Doros (Figure 1C), 8 km north of Doros Crater in the hanging-wall of the Twyfelfontein fault. The Gai-As and Doros Formations are separated by an angular unconformity, and both formations wedge out sequentially against the developing roll-over structure.

the succession recording a gradual shallowing trend and deposition above wave base.

The overlying Teresina Formation, 150–350 m thick, is almost entirely composed of interbedded gray mudstones and minor, very fine-grained sandstones showing wavy, lenticular, and flaser bedding. In places, the nearshore mudstones contain mudcracks and shallow burrows and trails of a yet unnamed ichnofacies, including *Planolites* ichnosp. Thicker sandstone units (up to 0.4 m) concentrate in the northeast Paraná basin and are characterized by wave-ripple cross-bedding or hummocky cross-bedding with the latter attributed to a storm wave origin. Few massive, gray siltstones bear freshwater shark remains and fossil plants of the *Glossopteris* flora, including lycopod stems and leaves as typical representatives. Intercalations (<3 m) of bivalve-bearing, oolitic grainstones exist at many levels throughout the more silty parts of the succession. Several intervals that are particularly rich in limestone intercalations are well marked in geophysical borehole logs and can be correlated across the basin. The oolitic grainstones commonly show wavy bedding and are interpreted as tempestites. We suggest that the ooids formed along the high-energy lake shoreline and then became redistributed toward the basin center during storms. Conditions for carbonate precipitation, such as lake water saturation, temperature, pH, and salinity, were probably favored during more arid climatic intervals or periods of enhanced calcium-rich groundwater input. The bivalves, endemic to the basin, belong to the *Pinzonella illusa* and *Pinzonella neotropica* assemblages; the latter of which can be traced as far as eastern Paraguay. This indicates a closed basin cut off from the sea with a relatively uniform lateral facies development (cf. Rohn et al., 1995); however, in the southernmost exposures of the Paraná basin, the Teresina Formation and equivalent Yaguari Formation in Uruguay consist almost entirely of shales (França et al., 1995). This contrasts with the facies development toward the northern parts of the basin (São Paulo state), where the Teresina Formation is gradually replaced by the more marginal facies association of the Corumbataí Formation. This trend is indicated by dominantly red siltstones containing some interbeds of domal stromatolites and ostracod-rich calcilutites, both up to a few decimeters thick. Geochemical analyses of the limestones suggest high salinities, but charophytes indicating freshwater conditions were also found at several levels and may register alternating humid and arid periods with higher salinities resulting from periodically enhanced evaporation. Previously, the Teresina Formation had been interpreted to record a coastal marine environment (e.g., França et al., 1995), a tidal flat, lagoon, and prodelta setting (Gama, 1979), or a storm-dominated sea-lake (e.g., Rohn, 1994); however, as with the Serra Alta Formation, no definitive marine fossils have yet been found, and sedimentary structures, such as hummocky cross-bedding, are not restricted to marine environments (cf. Greenwood and Sherman, 1986).

The Gai-As Lake in Brazil

The Rio do Rasto Formation is subdivided into the Serrinha and Morro Pelado Members (Figure 6). The Serrinha Member, 150–250 m thick, displays a wide facies variability. Greenish, mudcracked muddy siltstones dominate the section, with (<1 m) tabular interbeds of fine-grained sandstone increasing in density into the upper part of the succession. Some are arranged within fining- and thinning-upward cycles, each 4–8 m thick, consisting of massive to hummocky cross-bedded fine-grained sandstones that grade upward into wavy-bedded, interlaminated sandstones and mudcracked shales (Castro, 1994). Such successions are attributed to proximal to distal tempestites, but sandstones deposited by turbidites, lobate mouth bars, and rare small eolian dunes are also present in the sections. Fluvial facies are almost absent and only a few, thin massive limestones occur. The biogenic record of the Serrinha Member is quite different from that of the underlying Teresina Formation (cf. Figure 6). Such a pronounced environmental change perhaps relates to a paraconformity between the two stratigraphic units as discussed by Rohn (1994). The bivalves of the *Leinzia similis* assemblage are endemic and associated with conchostracans, indicating freshwater conditions. Abundant fossil plants include

Glossopteris. Lycopods disappear with the onset of the Serrinha Member, and the sphenophyte *Sphenophyllum* becomes one of the typical fossil plants. As inferred from the abundance of plant remains, the paleoclimate in general appears to have been more humid than during deposition of the Teresina Formation, with interludes of aridity evidenced by development of eolian dune sandstones toward the top of the unit (Rohn, 1994).

The 250–600(?) m thick Morro Pelado Member (Figure 6) is characterized by red mudstones, local rhythmites, and a higher density of 1–4-m-thick lenticular fine-grained sandstones. The latter are interpreted as eolian dune deposits that are intercalated with shallow lacustrine and floodplain deposits. Typical fluvial sediment structures are only rarely preserved, probably due to wind abrasion. Fluvial deposits, however, are particularly well documented in the area around Ponta Grossa arch (cf. Figure 5A), which was probably affected by uplift at that time (cf. Gama et al., 1982; Williams, 1995). Fossil bivalves and macrophytic remains are rare (*Palaeomutela? platinensis* assemblage), whereas conchostracans are relatively abundant and diverse (including the genus *Leaia*). The sphenophyte *Schizoneura gondwanensis* is an important widespread element in many other contemporaneous Gondwanan deposits. These plants are representatives of a marginal hygrophilous vegetation (Rohn, 1994), with aquatic fauna becoming increasingly localized upsection to floodplain lakes and interdune ponds; however, basal parts of the Morro Pelado Member contain bones of the endemic amphibians *Rastosuchus* and *Australerpeton*, which are associated with occurrences of the first known terrestrial tetrapods of the basin. The reptile *Endothiodon* from the base of the Morro Pelado Member indicates an affinity to the Beaufort Group *Pristerognathus, Tropidostoma*, and *Cistecephalus* biozones sensu Rubidge et al. (1995) of the main Karoo basin. Fossil plants and the leaiid conchostracans also support a Late Permian age of the Rio do Rasto Formation.

Gama (1979) suggested a large deltaic system sourced from the west and prograding toward the southeast as a sedimentologic model for the Passa Dois Group. Compared to the paleogeographic setting of the Paraná basin (Figure 1A), this delta progradation would be subparallel to the northwest-southeast to north-northwest–south-southeast–trending axis of the pre-southern South Atlantic rift that initially trended predominantly through the Paraná basin area (Figure 1A). The overall lateral continuity of facies, development of considerable thicknesses, and higher preservation potential of strata compared to their Namibian correlatives suggest a more basinal setting for the Paraná area associated with enhanced subsidence within the central parts of the rift zone. Because of this enhanced subsidence, signatures of contemporaneous fault activity are generally less pronounced in the Paraná basin when compared to the more marginal Namibian settings; however, the influence of several important structural elements is still registered by the basin fill. Activity of the Asunción Arch in Paraguay (Figure 5A), for instance, is evidenced by both the asymmetric isopach pattern of the Passa Dois Group (Figure 5B) and a wedge of coarse sediments paralleling the western basin margin (cf. Gama, 1979). The Ponta Grossa arch (Figure 5A) became uplifted during deposition of the upper parts of the Passa Dois Group, when fluvial intraformational boulders became deposited in the Reserva-Cândido de Abreu region, southwest of the arch (Rohn, 1994).

TECTONIC EVOLUTION OF THE LAKE SYSTEM

Stollhofen (1999) postulates that the area of rifting prior to the opening of the South Atlantic experienced a series of extensional rift events since the Late Carboniferous, interspersed by relatively quiescent periods. The ensuing extrusion of voluminous Jurassic and Cretaceous flood basalts in southern Africa and South America (Baksi and Archibald, 1997) thus represent only the peaks of a long-term extensional history culminating in oceanic opening during the Early Cretaceous. The area that was to become the present Namibian continental margin was controlled by a long-lived north-northwest–trending tectonic zonation comprising from west to east: (1) a rift valley depression later bisected to initiate the continental shelf, (2) an adjacent rift shoulder that underwent maximum thermal uplift, (3) an extensionally faulted zone floored by a breakaway detachment, and, farther inland, (4) the relatively stable cratonic continental interior. During the Permian, this tectonic zonation had not achieved its maximum expression; however, the north-northwest–trending rift shoulder and the adjacent rift basin were already defined. In addition, the width of this linear intracontinental extensional zone was not constant throughout its long-lived evolution, but focused toward the center of the rift during its more advanced stages of evolution.

Prior to the establishment of the Gai-As Lake, the intracontinental rift valley was repeatedly affected by northwardly directed marine incursions following the Carboniferous–Permian Dwyka interglacial sea level highstands (Martin, 1975; Horsthemke et al., 1990; Visser, 1997; Stollhofen et al., in press). Comparable intermittent episodes of marine flooding are well known along other intracontinental rift zones such as the Oligocene Upper Rhine graben in Germany (Schreiner, 1977). Causes of marine incursions may relate to phases of enhanced subsidence or global sea-level rise. Low or no subsidence or renewed activity of tectonic barriers are potential causes for isolation from the marine, in turn resulting in a closed inland lake and establishment of freshwater conditions. During the Early Permian, the rift valley depression finally extended from the western part of South Africa as far north as northern Brazil to accommodate the Irati-Whitehill sea (Oelofsen, 1987; Williams, 1995). Following this marine incursion and deposition of thick bituminous shales of the Irati/Whitehill Formation and its equivalents, the sedimentary environment

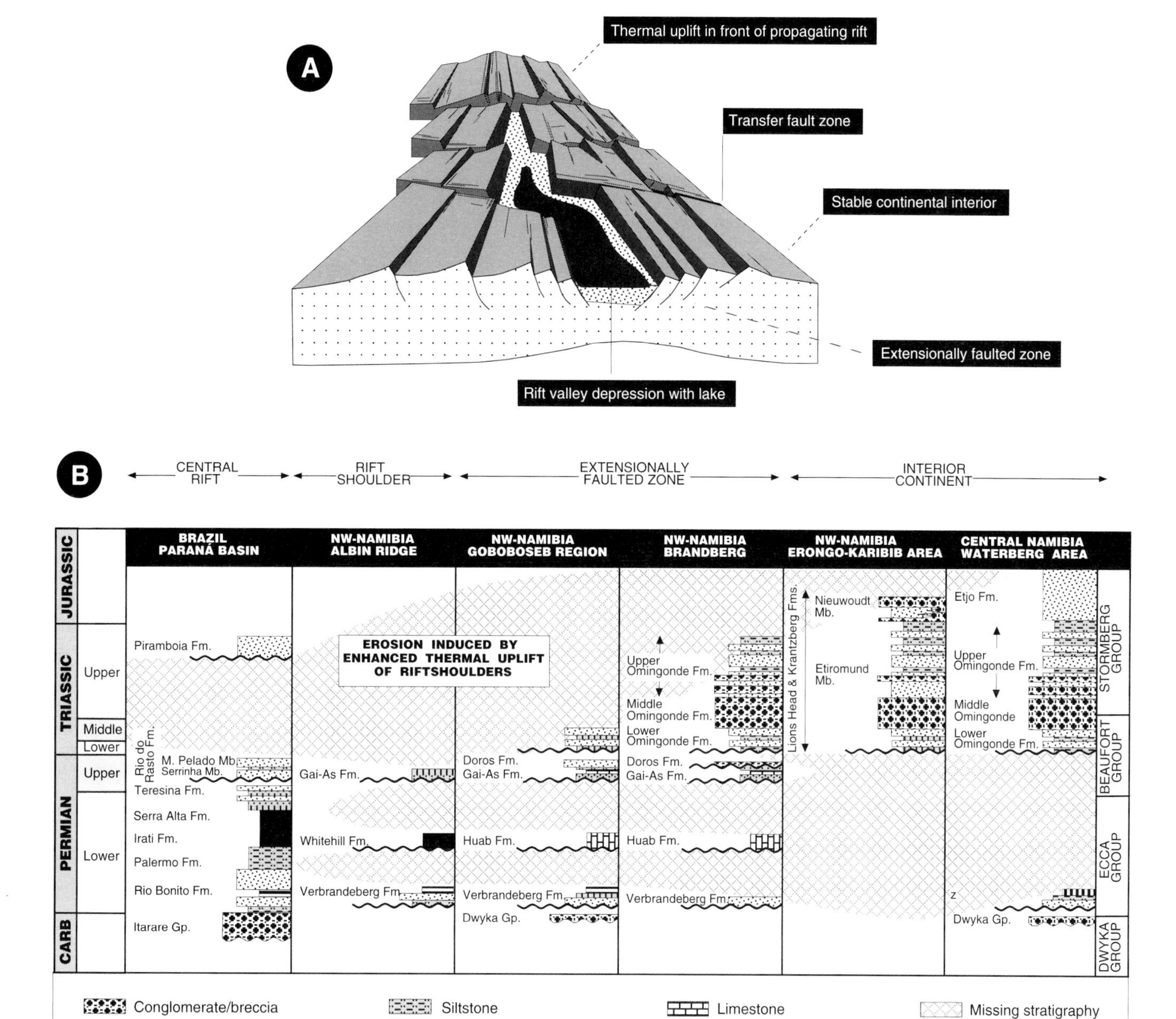

Figure 8—(A) Generalized architectural model of the rift zone, as visualized in Namibia, compartmentalized by orthogonal transfer faults. (B) The distribution of the Gai-As unit illustrates confinement of the lake largely to the rift valley itself, whereas above the rift shoulder (Albin Ridge) the Gai-As and Doros Formations are poorly preserved, with the sequence only starting later in the Late Permian–earliest Triassic.

within the rift valley depression changed gradually into a freshwater lake. This cut-off from the marine realm was perhaps a consequence of large-scale uplift and structural inversion of the Argentinian Puna highlands associated with the Cape-Ventana and San Rafael orogenies (cf. Veevers et al., 1994b; Porada et al., 1996). The depositional sequence recording the marine-nonmarine transition is only fully preserved in the Karoo foreland basin and the Paraná basin, the latter including the more central parts of the rift valley. Thinner and sand-dominated deltaic units of the Corumbatai Formation in the northeastern Paraná basin document the more marginal facies of the rift, which prograded southward, along the rift axis. In northwestern Namibia, however, a hiatus, probably caused by early thermal uplift of the rift shoulder along with the entire rift zone, separates the marine Whitehill equivalents (Huab Formation) from the overlying nonmarine Gai-As Lake deposits.

Comparison of the lateral extent of the Gai-As equivalent deposits to the framework of the outlined structural zonation (Figures 8B, 9) shows the elongation of the extensive lake basin coincides with the axis of the early, north-northwest–striking branch of

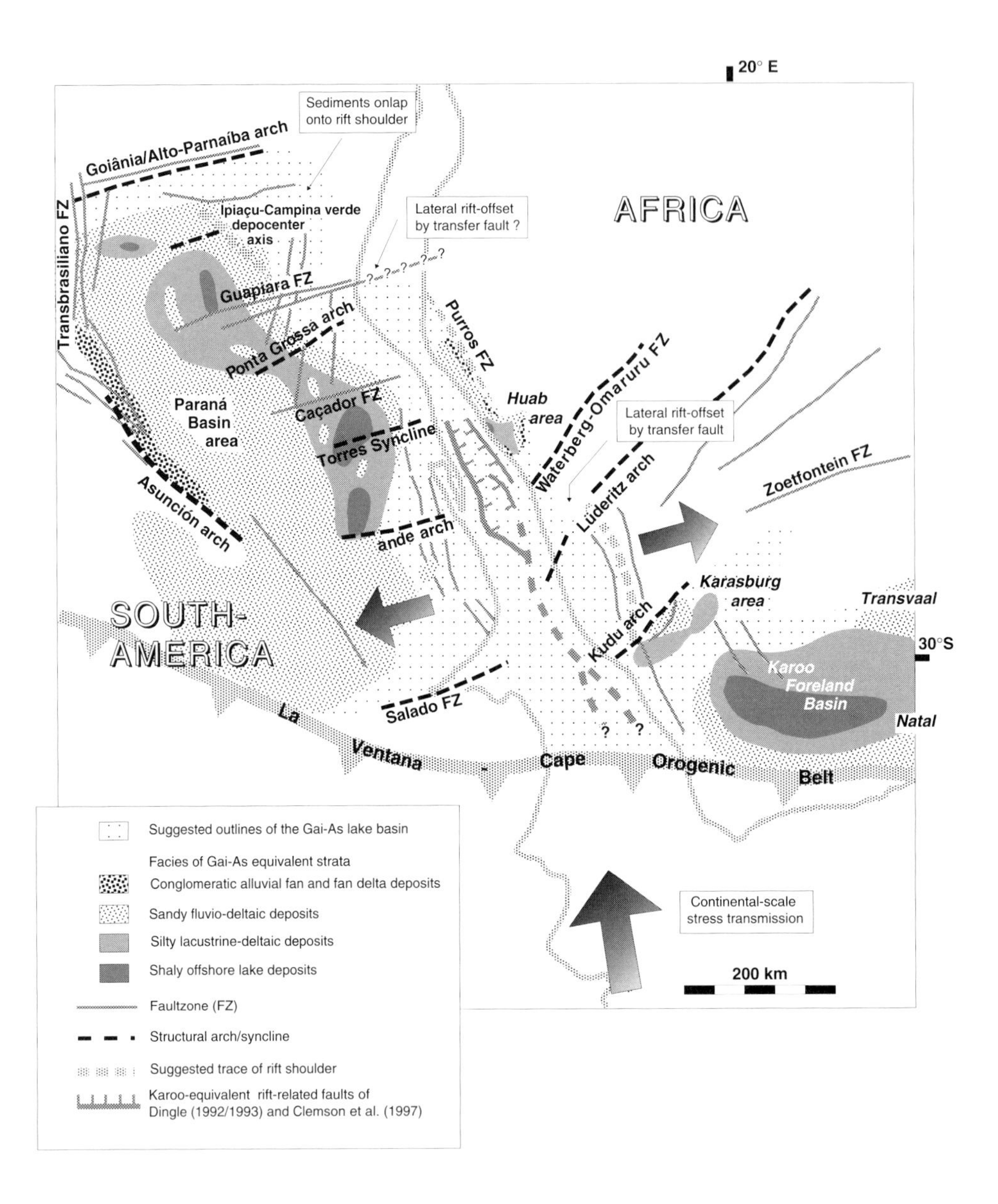

Figure 9—Latest Paleozoic paleogeographic map of the pre-southern South Atlantic rift within west Gondwana (compiled from de Wit et al., 1988; Zalán et al., 1990; Milani, 1992; Cobbold et al., 1992; Dingle, 1992/1993; Clemson et al., 1997) illustrating how the distribution and facies of Gai-As equivalent strata (compiled from Ledendecker, 1992; Visser, 1992; França et al., 1995) coincide with the axis of the early pre-southern South Atlantic rift system and associated structural elements. Karoo-equivalent rift-related faults between Africa and South America were particularly active during the Early Jurassic (de Wit and Ransome, 1992; Dingle, 1992/1993), but were also active during the late Paleozoic (Stollhofen et al., in press), probably due to impact tectonics along the southern convergent margin of Gondwana (cf. de Wit and Ransome, 1992; Cobbold et al., 1992).

the pre-southern South Atlantic rift zone (Figure 1A). Timing and the considerable amount of Permian–Carboniferous tectonic subsidence, followed by thermal cooling, is clearly displayed by the backstripped subsidence curve from a well in the Paraná basin area (cf. Zalán et al., 1990). An important aspect during the period of mechanical subsidence is the tectonic reactivation of basement structures both along the Namibian faulted margin (Clemson et al., 1997; Stollhofen, 1999) and along the conjugate South American side (Brito Neves et al., 1984); however, Permian faulting in the Paraná area appears to have developed more diffusely, spread over a relatively wide area when compared to the Namibian counterparts. Linear gravity lows paralleling the Asunción arch along with the indications of crustal thinning suggest that remnants of one rift basin system lie along the west-central Paraná province (Vidotti et al., 1995). Following Early Cretaceous flood basalt extrusion and oceanic onset along a north-south trend, the rift branch penetrating the Paraná area was finally aborted and resulted in a failed rift (cf. Sibuet et al., 1984). In comparison with equivalent successions in Africa (Stanistreet and Stollhofen, 1999), the fill of the Paraná basin area is subdivided into (1) a Permian-Carboniferous, (2) a Triassic–Jurassic, and (3) a Cretaceous megasequence, each separated by time-stratigraphic gaps. Zalán et al. (1990) viewed the Paraná area not as a single basin in the strict sense, but as a combination of tectonically different types of basins with varying outlines stacked on top of one another developing a cumulative geometry, which is commonly referred to as the Paraná "basin." Some of the evolutionary steps of the Paraná (cf. Klein, 1995) compare well with the development of African Karoo rifts involving sequential stages of extensional faulting, heating, and mechanical fault-controlled subsidence succeeded by subsidence related to periodic thermal cooling and contraction. A comparative example is the North Sea basin, recognized by Sclater and Christie (1980) from backstripped subsidence curves to involve two phases of rifting and thermal cooling superimposed on top of one another. The present North Sea "basin" geometry is superficially that of the last phase of thermal cooling and subsidence.

Lateral extent and the coincidence of thickness and facies variations of the Gai-As and Doros Formations with regional structural elements in Namibia (Figures 1C, 7, 9) illustrate the structural control on Gai-As Lake deposition. By considering the pronounced northwest-southeast to north-northwest–south-southeast elongation and facies architectural pattern of the lake deposits (Figure 9), the huge Gai-As Lake was essentially confined to a rift valley depression, but frequently onlapped onto its shoulders. The Gai-As Lake hosted pelecypods, fishes, and amphibians, the latter growing to considerable size, and would compare well to modern permanent rift valley lakes such as Lakes Tanganyika and Malawi in East Africa (cf. Crossley, 1984; Tiercelin, 1991; Baltzer, 1991). The Gai-As and Doros Formations are not preserved from sections above the rift shoulder where much of the Triassic–Jurassic successions are also missing due to enhanced thermal uplift. Alluvial fans of the Doros Formation prograded into the lake off marginal faults and the rift shoulder, which emphasizes that at least locally considerable relief existed in the lake margin areas. With the establishment of less humid climatic conditions and associated lake retreat, the lake shorelines increasingly fell dry and were modified by pedogenic processes as recorded by the Gai-As Formation at the Gudaus Hill locality (Figure 4). In Brazil, the more sandy parts of the shoreline and fluvial tributaries became significantly affected by wind reworking, resulting in the eolian dune deposits of the upper Serrinha Member and, in particular, the Morro Pelado Member (Figure 6). The development of eolian processes toward the northern end of Lake Gai-As compares well to modern Lake Malawi where southerly winds generate eolian sands from beach ridges, particularly behind southerly facing beaches (Crossley, 1984). As Lake Gai-As most probably experienced predominantly southerly winds (cf. Kutzbach and Ziegler, 1994), the northern end of the lake should be the one potentially most affected by such winds.

In Brazil, stratigraphic sections combined with an isopach map (Figures 5B, 6) of the Passa Dois Group suggest the influence of rifting on depositional patterns there, as well. The area of maximum subsidence, indicated by contours of maximum thickness development (Figure 5B) and offshore facies associations (Figure 6, sections 1, 2, and 3) are localized in the states of Santa Catarina and Paraná. Toward the northeastern Paraná area (Figure 6, sections 4 and 5), the whole Passa Dois Group developed a more marginal facies with a reduced thickness. In addition, its upper part became partially eroded prior to the onset of deposition of the Upper Triassic Pirambóia Formation. Pinch-out of composite sections seems equivalent to the variation on the Namibian basin edge from Klein Gai-As and north of Doros locations eastward to Twyfelfontein and from Brandberg eastward to the continental interior. Facies and thickness development in the northeastern Paraná area is mirrored across the depositional area but, according to the slightly asymmetric isopach pattern (Figure 5B), more pronounced towards its southwestern and western margin. The north-south–trending Asunción arch forms the predominant structural element. Although the arch is not exposed in Argentina, its southward extent can be inferred from the almost complete pinch-out of the Teresina Formation and the Serrinha Member in southern Brazil and adjacent Uruguay. Within the Brazilian portion of the pre-southern South Atlantic rift zone, the Asunción structural arch reflects a western rift shoulder, whereas a less pronounced eastern rift shoulder can be expected in eastern Brazil to be associated with the Ipiaçu-Campina Verde depocenter axis (cf. Zalán et al., 1990).

Facies development within the elongate Gai-As lake basin was neither uniform nor was it consistently parallel with the rift axis, either in the main basinal area or in the area transitional with the main Karoo foreland basin. As described by Stollhofen (1999) and Holzförster et al. (in press), transfer fault zones sensu Gibbs (1990) compartmentalized the rift zone into a chain of individual depocenters (Figure 8A), which were linked only during times of highstands. Fulfaro et al. (1982), Zalán et al. (1990), and Light et al. (1993) illustrate how "arches" influence the sediment thickness distribution in the Paraná basin area and the rift valley fill buried in the present Namibian offshore area. Examples of such structures are represented by the Goiânia/Alto Parnaíba, the Ponta Grossa, the Rio Grande, and Lüderitz and Kudu arches (cf. Figures 5A, 9), all of which are arranged perpendicular to the main rift trend. They record dominantly strike-slip and oblique-slip deformation (Zalán et al., 1990). The arches became particularly active during the Triassic and developed along strike with both transpressional effects causing linear zones of uplift and erosion and transtensional effects initiating rapidly subsiding half-graben and pull-apart basin geometries.

The Gai-As lake area was probably cut off from the more southerly parts of the rift zone because of the intervening Kudu and Lüderitz arches. The latter, recognized by Dingle (1992/1993), was operative during the Jurassic, but seems to have already initiated in preceding tectonic events (cf. Light et al., 1993). For instance, on the basis of paleogeographic and facies reconstructions (Figures 8, 9), the Lüderitz arch may divide post-Whitehill/Irati facies associations in the north (Gai-As Formation and Passa Dois Group), from those in the south (Collingham to Teekloof formations) (Figure 9). The area north of the Lüderitz and Kudu arches then became a separate rift valley lake center extending northward into the Paraná area.

Transfer zones in rifts are known to contain complex structural geometries that make them suitable locations for structural hydrocarbon traps (cf. Morley et al., 1990). In terms of the Gai-As Lake, zones of uplift associated with the transfer faults had considerable influence on the development of thickness and facies patterns within the basin. Bituminous black shales of the Whitehill/Irati Formation equivalents, just below the Gai-As Lake deposits, form the major hydrocarbon source rock in the South Atlantic region, and oil-impregnated sandstones of the Pirambóia Formation occur immediately above the Gai-As Lake correlative beds in Brazil (cf. França et al., 1995). Lopatin-type calculations for the Paraná basin indicate that the Permian and older rocks entered the oil window during the Late Jurassic–Early Cretaceous phase of subsidence (Zalán et al., 1990), which was ultimately followed by the onset of drift and continental break-up (Figure 2A). Thus, knowledge of structural elements that were active during the Permian–Triassic is fundamental for understanding hydrocarbon wandering paths and future exploration strategies.

DISCUSSION AND CONCLUSIONS

The Gai-As Lake formed during an early stage of the long-term extensional history of the pre-southern South Atlantic rift zone and recorded one of a series of early extensional episodes (Figure 2A) that ultimately led to an Early Cretaceous ocean basin (Dingle, 1992/1993). Permian–Carboniferous rifting was succeeded by a widespread phase of non-deposition and erosion. Along the central parts of the rift zone, early thermal uplift phases caused a time-stratigraphic gap comprising the Upper Permian–Upper Triassic, prior to extensive Early Jurassic flood basalt extrusions (Stanistreet and Stollhofen, 1999; Stollhofen et al., this volume).

Such an early stage of rifting compares for example to Paleocene–Eocene rift basin developments in the Bohai region of northeast China (Chen Changming et al., 1981) with red beds and evaporitic limestones being deposited within elongate lakes flanked by alluvial fan deltas. Because the marginal rift shoulders in Namibia were tilted away from the rift axis during initial Gai-As Lake deposition, the majority of drainage systems were in turn directed away from the central rift depression. Ingress of coarse detritus into the early rift valley therefore is relatively restricted. Discharge and preservation of clastic detritus is limited to areas neighboring fault scarps and elevated footwall blocks within the rift. Crossley (1984) showed that faults transecting the margins of the modern, central African Malawi rift valley lake were crucial in causing fans to prograde almost directly into the lake. In terms of ancient Lake Gai-As, such a scenario is particularly well recorded in the Brandberg section where the abundance of lithoclasts derived from lithologies stratigraphically below the Gai-As reveals partial erosion of nearby footwall blocks and hanging-wall–confined braided river channels sourcing alluvial fan deltas. In contrast, the Gudaus hill, Sanianab, and Albin Ridge localities (Figure 1C) record condensed sequences associated with areas of reduced tectonic subsidence, characterized by paleosol development. It is remarkable that increases in clastic sediment supply at the base of both the upper Gai-As and Doros Formations coincided with the deposition of pyroclastic fallout material (Figure 3).

A good analogy of the sedimentologic relationships between the Doros and Gai-As Formations is provided by the example of Pliocene–Pleistocene fans prograding into a rift valley setting recorded by Vondra and Burggraf (1978) from Lake Turkana, Kenya. There, widespread fluvial deposition began shortly after deposition of pyroclastic material and localized subsidence led to the relative elevation of basin margins with subsequent initiation of erosion in that area. A concurrent climatic change toward less humid conditions was inferred from oxygen-isotope studies of carbonates, and an apparent increase in alkalinity of paleo–Lake Turkana was recorded by molluscan extinction and changes in authigenic mineral contents of both pyroclastic and siliciclastic sediments. Adjacent to the basin margins, broad alluvial fans accumulated wedge-shaped gravel units. Shifts from low- to high-energy fluvial systems resulted in the erosion of underlying lithologies, downcutting of up to 11 m.

Much of the sediment within Lake Gai-As was probably redistributed by longshore currents, which is common in modern lakes (Cohen et al., 1986; Renaut and Owen, 1991) and has been inferred in a variety of ancient lakes (e.g., Allen, 1981; Martel and Gibling, 1991). The longshore currents themseves may have been induced by strong southerly winds (cf. Kutzbach and Ziegler, 1994) that became locally funnelled along the north-northwest– to northwest–trending depression traced by the lake axis. In addition, bottom currents comparable to those recorded by Crossley (1984) from modern Lake Malawi may have caused mixing within a thermally stratified lake resulting in mass mortalities of fish and bivalves, both preserved as significant accumulation horizons in the Gai-As sedimentologic record.

Due to its paleogeographic setting at about 40°S, Lake Gai-As would have been influenced by intense winter storms (cf. Duke, 1985) and high rates of evaporation during the summer. Such a paleolatitudinal setting would compare well to the modern Caspian Sea. The low diversity of fish and bivalve populations indicates ecologic stress exacerbated by temporarily saline lake waters established during periods of low precipitation and enhanced evaporation. The presence of amphibians and lake-fringing mudflats vegetated by trees suggest, however, that low-salinity lake waters prevailed and were only interrupted by such high-salinity interludes.

ACKNOWLEDGMENTS

Research has been undertaken under the auspices of the postgraduate research program Interdisciplinary Geoscience Research in Africa of the Julius-Maximilians University of Würzburg, Germany, in cooperation with the Geological Survey of Namibia. The study forms part of the Karoo tectonics project and was prepared and edited by Harald Stollhofen and Ian G. Stanistreet with sedimentologic and paleontologic sections provided by Rosemarie Rohn, Frank Holzförster, and Ansgar Wanke.

We thank all those who have contributed to this paper through their discussions, in particular Bruce Rubidge, Volker Lorenz, Marion Bamford, Richard Armstrong, Alfred Uchmann, Wulf Hegenberger, and

Dougal Jerram. Funding by the German Research Foundation (DFG) and logistic support by the Geological Survey of Namibia; Nature Conservation, Namibia; and the Ministry of Tourism, Namibia; is gratefully acknowledged. We thank Elizabeth Gierlowski-Kordesch, Keddy Yemane, and Greg Nadon for thoughtful and constructive reviews of the manuscript.

REFERENCES CITED

Allen, J. R. L., 1981, Devonian lake margin environments and processes, SE Shetland, Scotland: Journal of the Geological Society (London), v. 138, p. 1–14.

Anderson, J. M., 1981, World Permo-Triassic correlations: Their biostratigraphic basis, *in* M. M. Cresswell and P. Vella, eds., 5th International Gondwana Symposium, Proceedings, p. 3–10.

Baksi, A. K., and D. A. Archibald, 1997, Mesozoic igneous activity in the Maranhão province, northern Brazil: $^{40}Ar/^{39}Ar$ evidence for separate episodes of basaltic magmatism: Earth and Planetary Science Letters, v. 151, p. 139–153.

Baltzer, F., 1991, Late Pleistocene and recent detrital sedimentation in the deep parts of northern Lake Tanganyika (East African rift), *in* P. Anadón, L. Cabrera, and K. Kelts, eds., Lacustrine Facies Analysis: International Association of Sedimentologists, Special Publication 13, p. 147–173.

Bamford, M. K., 1998, Fossil woods of Karoo age deposits in South Africa and Namibia as an aid to biostratigraphical correlation (abs.): Journal of African Earth Sciences, v. 27, p. 16.

Barberena, M. C., D. C. Araujo, E. L. Lavina, and U.F. Faccini, 1991, The evidence for close paleofaunistic affinity between South America and Africa, as indicated by Late Permian and Triassic tetrapods: 7th International Gondwana Symposium, São Paulo, 1988, Proceedings, p. 455–467.

Bossi, J., L. A. Ferrando, A. Fernandez, G. Elizalde, H. Morales, J. Ledesma, E. Carballo, E. Medina, I. Ford, and J. Montana, 1975, Carta Geologica del Uruguay a escala 1:1.000.000, Montevideo.

Brito Neves, B. B. de, R. A. Fuck, U.G. Cordani, and F. A. Thoma, 1984, Influence of basement structures on the evolution of the major sedimentary basins of Brazil: a case of tectonic heritage: Journal of Geodynamics, v. 1, p. 495–510.

Bruhn, D., and R. Jäger, 1991, Geologie, Sedimentologie und Fazies der Karoo-Sequenz westlich "Petrified Forest", Damaraland, Namibia: Diploma-thesis, Institut für Geologie und Dynamik der Lithosphäre, University of Göttingen, 128 p.

Castro, J. C., 1994, Field trip guide to Paraná Basin: 14th International Sedimentological Congress, Recife, Brazil, 42 p.

Cerling, T. E., R. L. Hay, and J. R. O'Neil, 1977, Isotopic evidence for dramatic climatic changes in East Africa during the Pleistocene: Nature, v. 267, p. 137–138.

Cheel, R. J. and D. A. Leckie, 1993, Hummocky cross-stratification, *in* V. P. Wright, ed., Sedimentology Review 1, p. 103–122.

Chen Changming, Huang Jiakuan, Chen Jingshan, Tian Xingyou, Chen Ruijun, and Li Li, 1981, Evolution of sedimentary tectonics of Bohai rift system and its bearing on hydrocarbon accumulation: Scientia Sinica, v. 24, p. 521–529.

Clemson, J., J. Cartwright, and J. Booth, 1997, Structural segmentation and the influence of basement structure on the Namibian passive margin: Journal of the Geological Society (London), v. 154, p. 477–482.

Cohen, A. S., D. S. Ferguson, P. M. Gram, S. L. Hubler, and K. W. Sims, 1986, The distribution of coarse-grained sediments in modern Lake Turkana, Kenya: implications for clastic sedimentation models of rift lakes, *in* L. E. Frostick, R. W. Renaut, I. Reid, and J.-J. Tiercelin, eds., Sedimentation in the African Rifts: Geological Society (London) Special Publication 25, p. 127–139.

Coleman, J. M., 1966, Ecological changes in a massive freshwater clay sequence: Transactions of the Gulf Coast Association, v. 16, p. 159–174.

Cobbold, P. R., D. Gapais, E. R. Rossello, E. R. Rossello, E. J. Milani, and P. Szatmari, 1992, Permo-Triassic intracontinental deformation in SW Gondwana, *in* M. J. De Wit and I. G. D. Ransome, eds., Inversion Tectonics of the Cape Fold Belt, Karoo and Cretaceous Basins of Southern Africa: Balkema, Rotterdam, p. 23–26.

Crossley, R., 1984, Controls of sedimentation in the Malawi Rift Valley, Central Africa: Sedimentary Geology, v. 40, p. 33–50.

De Wit, M. J., M. Jeffery, H. Bergh, and L. Nicolaysen, 1988, Geological map of sectors of Gondwana: 2 sheets, 1:10.000.000: Tulsa, American Association of Petroleum Geologists.

De Wit, M. J., and I. G. D. Ransome, 1992, Regional inversion tectonics along the southern margin of Gondwana, *in* M. J. De Wit and I. G. D. Ransome, eds., Inversion Tectonics of the Cape Fold Belt, Karoo and Cretaceous Basins of Southern Africa: Balkema, Rotterdam, p. 15–21.

Dingle, R. V., 1992/1993, Structural and sedimentary development of the continental margin off south-western Africa. Communications of the Geological Survey of South West Africa/Namibia, v. 8, p. 35–43.

Duke, W. L., 1985, Hummocky cross-stratification, tropical hurricanes, and intense winter storms: Sedimentology, v. 32, p. 167–194.

França, A. B., E. J. Milani, R. L. Schneider, M. Lopez-Paulsen, R. Suárez-Soruco, H. Santa-Ana, F. Wiens, O. Ferreiro, E. A. Rossello, H. H. Bianucci, R. F. A. Flores, M. C. Vistalli, F. Fernandez-Seveso, R. P. Fuenzalida, and N. Munoz, 1995, Phanerozoic correlation in southern South America, *in* A. J. Tankard, R. Suárez-Soruco, and H. J. Welsink, eds., Petroleum Basins of South America: AAPG Memoir 62, p. 129–161.

Freytet, P., and J.-C. Plaziat, 1982, Continental carbonate sedimentation and pedogenesis–Late Cretaceous and early Tertiary of Southern France: Contributions to Sedimentology, v. 12, p. 1–213.

Fulfaro, V. J., J. A. J. Perinotto, A. R. Saad, and P. A. Souza, 1995, The Upper Paleozoic sedimentary sequence of Eastern Paraguay (abs.): 6 Simposio Sul-Brasileiro de Geologia, 1. Encontro de Geologia do Cone Sul, Porto Alegre, Brasil, 1995, Boletim de Resumos Expandidos, p. 262–263.

Fulfaro, V. J., A. R. Saad, M. V. Santos, and R. B. Vianna, 1982, Compartimentação e evolução tectonica da Bacia do Paraná: Revista Brasileira de Geociências, v. 12, p. 590–610.

Gama, E., Jr., 1979, A sedimentação do Grupo Passa Dois (exclusive Formação Irati): um modelo geomórfico: Revista Brasileira de Geociências, v. 9 (1), p. 1–16.

Gama, E., Jr., A. N. Bandeira, Jr., and A.B. Franca, 1982, Distribuição especial e temporal das unidades litoestratigráficas Paleozoícas na parte central da bacia do Paraná, *in* R. Yoshida and E. Gamma, Jr., eds., Geologia da bacia do Paraná, Reavaliação da potencialidade e prospectividade em hidrocarbonetos: Pauli Petro Consorcio, São Paulo, p. 19–40.

Gibbs, A. D., 1990, Linked fault families in basin formation: Journal of Structural Geology, v. 12, p. 795–803.

Gierlowski-Kordesch, E. H., 1998, Carbonate deposition in an ephemeral siliciclastic alluvial system: Jurassic Shuttle Meadow Formation, Newark Supergroup, Hartford Basin, USA: Palaeogeography, Palaeoclimatology, Palaeoecology, v. 140, p. 161–184.

Greenwood, B., and D. J. Sherman, 1986, Hummocky cross-stratification in the surf zone: flow parameters and bedding genesis: Sedimentology, v. 33, p. 33–45.

Hankel, O., 1994, Early Permian to Middle Jurassic rifting and sedimentation in East Africa and Madagascar: Geologische Rundschau, v. 83, p. 703–710.

Harland, W. B., R. L. Armstrong, A. V. Cox, L. E. Craig, A. G. Smith, and D. G. Smith, 1990, A geologic time scale: Cambridge, Cambridge University Press, 263 p.

Harms, J. C., J. B. Southard, and R. G. Walker, 1982, Structures and Sequences in Clastic Rocks: SEPM, Short course 9, 249 p.

Heward, A. P., 1981, A review of wave-dominated clastic shoreline deposits: Earth Science Reviews, v. 17, p. 223–276.

Holzförster, F., H. Stollhofen, and I. G. Stanistreet, in press, Waterberg-Erongo area lithostratigraphy and depositional environments, Central Namibia and correlation with the main Karoo Basin, South Africa: Journal of African Earth Sciences, v. 29.

Horsthemke, E., 1992, Fazies der Karoosedimente in der Huab-Region, Damaraland, NW-Namibia: Göttinger Arbeiten zur Geologie und Paläontologie, v. 55, p. 1–102.

Horsthemke, E., S. Ledendecker, and H. Porada, 1990, Depositional environments and stratigraphic correlation of the Karoo Sequence in northwestern Damaraland: Communications of the Geological Survey of South West Africa/Namibia, v. 6, p. 63–73.

Keyser, A. W., 1973, A new Triassic vertebrate fauna from South West Africa: Palaeontologia Africana, v. 16, p. 1–15.

Klappa, C. F., 1980, Rhizolithes in terrestrial carbonates: classification, recognition, genesis and significance: Sedimentology, v. 27, p. 613–629.

Klein, G. D., 1995, Intracratonic basins, *in* C. J. Busby and R. V. Ingersoll, eds., Tectonics of sedimentary basins: Cambridge, Blackwell, p. 459–478.

Knauth, L. P., 1994, Petrogenesis of chert, *in* P. J. Heaney, C. T. Prewitt, and G. V. Gibbs, eds., Silica; physical behavior, geochemistry and materials applications: Reviews in Mineralogy, v. 29, p. 233–258.

Kutzbach, J. E., and A. M. Ziegler, 1994, Simulation of Late Permian climate and biomes with an atmosphere-ocean model: comparisons with observations, *in* J. R. L. Allen, B. J. Hoskins, B. W. Sellwood, R. A. Spicer, and P. J. Valdes, eds., Palaeoclimates and their modelling. With special reference to the Mesozoic era: London, Chapman & Hall, p. 119–132.

Lambiase, J. J., 1989, The framework of African rifting during the Phanerozoic: Journal of African Earth Sciences, v. 8, p. 183–190.

Ledendecker, S., 1992, Stratigraphie der Karoosedimente der Huabregion (NW-Namibia) und deren Korrelation mit zeitäquivalenten Sedimenten des Paranàbeckens (Südamerika) und des Großen Karoobeckens (Südafrika) unter besonderer Berücksichtigung der überregionalen geodynamischen und klimatischen Entwicklung Westgondwanas. Göttinger Arbeiten zur Geologie und Paläontologie, v. 54, p. 1–87.

Light, M. P. R., M. P. Maslanyj, R. J. Greenwood, and N. L. Banks, 1993, Seismic sequence stratigraphy and tectonics offshore Namibia, *in* G. D. Williams and A. Dobb, eds., Tectonics and Seismic Sequence Stratigraphy: Geological Society (London), Special Publication 71, p. 163–191.

Lottes, A. L., and D. B. Rowley, 1990, Early and Late Permian reconstructions of Pangea, *in* W. S. McKerrow, and C.R. Scotese, eds., Palaeozoic Palaeogeography and Biogeography: Geological Society London, Memoir 12, p. 383–395.

Martel, A. T., and M. R. Gibling, 1991, Wave-dominated lacustrine facies and tectonically controlled cyclicity in the Lower Carboniferous Horton Bluff Formation, Nova Scotia, Canada, *in* P. Anadón, L. I. Cabrera, and K. Kelts, eds., Lacustrine Facies analysis: International Association of Sedimentologists, Special Publication 13, p. 223–243.

Martin, H., 1975, Structural and palaeogeographical evidence for an Upper Palaeozoic sea between southern Africa and South America, *in* K. S. W. Campbell, ed., Gondwana Geology: papers from the 3rd Gondwana symposium: Canberra, Australian National University Press, p. 37–51.

McCave, I. N., 1970, Deposition of fine-grained suspended sediments from tidal currents: Journal of Geophysical Research, v. 75, p. 4151–4159.

Milani, E.J., 1992, Intraplate tectonics and the evolution of the Paraná basin, SE Brazil, *in* M. J. and I. G. D. Ransome, eds., Inversion Tectonics of the Cape Fold Belt, Karoo and Cretaceous Basins of Southern

Africa: Balkema, Rotterdam, p. 101–108.
Miller, R. McG., 1988, Geological map area 2013 Cape Cross: scale 1:250.000: Windhoek, Geological Survey of Namibia.
Miller, R. McG., 1992, Pre-breakup evolution of the western margin of southern Africa (abs.): Mini-conference Geological Survey of Namibia "Southwestern African continental margin: evolution and physical characteristics," Windhoek 1992, Abstracts, p. 37–42.
Miller, R. McG., and K. E. L. Schalk, 1980, South West Africa/Namibia Geological Map: scale 1:1.000.000: Windhoek, Geological Survey of Namibia.
Morley, C. K., R. A. Nelson, T. L. Patton, and S. G. Munn, 1990, Transfer zones in the East African Rift System and their relevance to Hydrocarbon Exploration in Rifts: AAPG Bulletin, v. 74, p. 1234–1253.
Mühlmann, H. (coord.), 1983, Integração dos dados geológicos e geofisicos da Bacia do Paraná (mapas e seções estratigráficas). Relatorio Interno, Petrobras (unpublished report of Petrobras).
Oelofsen, B. W., 1987, The biostratigraphy and fossils of the Whitehill and Irati shale formations of the Karoo and Paraná Basins, *in* G. D. McKenzie, ed., Gondwana six: stratigraphy, sedimentology, and paleontology: American Geophysical Union, Geophysical Monograph 41, p. 131–138.
Oelofsen, B. W., and D. L. Araujo, 1983, Palaeoecological implications of the distribution of mesosaurid reptiles in the Permian Irati sea (Parana basin), South America: Revista Brasileira de Geociências, v. 13, p. 1–6.
Parnell, J., 1988, Significance of lacustrine cherts for the environment of source-rock deposition in the Orcadian Basin, Scotland, *in* A. J. Fleet, K. Kelts, and M. R. Talbot, eds., Lacustrine Petroleum Source Rocks: Geological Society of London, Special Publication 40, p. 205–217.
Porada, H., T. Löffler, E. Horsthemke, S. Ledendecker, and H. Martin, 1996, Facies and Palaeo-Environmental trends of northern Namibian Karoo sediments in relation to West Gondwanaland palaeogeography: Ninth International Gondwana Symposium, Proceedings, p. 1101–1114.
Powell, C. McA., and Z. X. Li, 1994, Reconstruction of the Panthalassan Margin of Gondwanaland, *in* J. J. Veevers, and C. McA Powell, eds., Permian–Triassic Pangean Basins and Foldbelts along the Panthalassan Margin of Gondwanaland: Geological Society of America Memoir 184, p. 5–9.
Reineck, H.-E. and I. B. Singh, 1980, Depositional sedimentary environments, with reference to terrigenous clastics, 2nd edition: Berlin, Heidelberg, Springer, 549 p.
Reineck, H.-E., and F. Wunderlich, 1968, Classification and origin of flaser and lenticular bedding: Sedimentology, v. 11, p. 99–104.
Renaut, R. W, and R. B. Owen, 1991, Shore-zone sedimentation and facies in a closed rift lake: the Holocene beach deposits of Lake Bogoria, Kenya, *in* P. Anadón, L. Cabrera, and K. Kelts, eds., Lacustrine Facies Analysis: International Association of Sedimentologists, Special Publication 13, p. 175–195.
Ring, U., 1995, Tectonic and lithological constraints on the evolution of the Karoo graben of northern Malawi (East Africa): Geologische Rundschau, v. 84, p. 607–625.
Rohn, R., 1994, Evolução ambiental da Bacia do Paraná durante o Neopermiano no leste de Santa Catarina e do Paraná. Ph.D. dissertation, Universidade de São Paulo, Instituto de Geociências, São Paulo, Brazil, 386 p.
Rohn, R., J. A. J. Perinotto, V. J. Fulfaro, A. R. Saad, and M. G. Simões, 1995, On the significance of the *Pinzonella neotropica* assemblage (Upper Permian) for the Paraná Basin—Brazil and Paraguay: VI Simpósio Sul-Brasileiro de Geologia, I Encontro de Geologia do Cone Sul, Porto Alegre, Brazil, Sociedade Brasileira de Geologia, Boletim de Resumos Expandidos, p. 260–261.
Rubidge, B. S., M. R. Johnson, J. W. Kitching, R. M. H. Smith, A. W. Keyser, and G. H. Groenewald, 1995, An introduction to the biozonation of the Beaufort Group, *in* B. S. Rubidge, ed., Biostratigraphy of the Beaufort Group (Karoo Supergroup): South African Committee for Stratigraphy. Biostratigraphic Series 1, p. 1–45.
Schobbenhaus, C., D. A. Campos, G. R. Derze, and H. E. Asmus (coords.), 1984, Geologia do Brasil. MME/DNPM, 501 p., mapa (1:2.500.000), Brasília.
Schreiner, A., 1977, Tertiär, *in* R. Groschopf, G. Kessler, J. Leiber, H. Maus, W. Ohmert, A. Schreiner, and W. Wimmenauer, eds., Geologische Karte von Baden-Württemberg 1:50.000, Erläuterungen zu Blatt Freiburg i. Br. und Umgebung: Stuttgart, Landesvermessungsamt Baden-Württemberg, p. 133–153.
Sclater, J. G., and P. A. F. Christie, 1980, Continental stretching: an explanation of the post-mid Cretaceous subsidence of the central North Sea: Journal of Geophysical Research, v. 85, p. 3711–3739.
Scotese, C. R., and S. F. Barrett, 1990, Gondwana's movement over the South Pole during the Palaeozoic: evidence from lithological indicators of climate, *in* W. S. McKerrow, and C.R. Scotese, eds., Palaeozoic Palaeogeography and Biogeography: Geological Society (London), Memoir 12, p. 75–85.
Scotese, C. R, and W. S. McKerrow, 1990, Revised world maps and introduction, *in* W. S. McKerrow, and C.R. Scotese, eds., Palaeozoic Palaeogeography and Biogeography: Geological Society (London), Memoir 12, p. 1–21.
Sehgal, J. L. and G. Stoops, 1972, Pedogenic calcite accumulation in arid and semi-arid regions of the Indo-Gangetic alluvial plain of erstwhile Punjap (India)—their morphology and origin: Geoderma, v. 8, p. 59–72.
Sibuet, J.-C., W. W. Hay, A. Prunier, L. Montadert, K. Hinz, and J. Fritsch, 1984, Early evolution of the South Atlantic: Role of the rifting phase: Initial Reports of the Deep-Sea Drilling Project, v. 75, p. 469–481.
Smith, R. A., 1984, The lithostratigraphy of the Karoo Supergroup in Botswana: Bulletin of the Geological Survey of Botswana, v. 26, p. 1–239.

Stanistreet, I. G., and H. Stollhofen, 1999, Onshore equivalents of the main Kudu gas reservoir in Namibia, *in* N. Cameron, R. Bate, and V. Clure, eds., Oil and Gas Habitats of the South Atlantic: Geological Society (London), Special Publication 153, p. 345–365.

Stollhofen, H., S. Gerschütz, I. G. Stanistreet, and V. Lorenz, 1998, Tectonic and volcanic controls on early Jurassic rift-valley lake deposition during emplacement of Karoo flood basalts, southern Namibia: Palaeogeography, Palaeoclimatology, Palaeoecology, v. 140, p. 185–215.

Stollhofen, H., I. G. Stanistreet, and S. Gerschütz, this volume, Early Jurassic rift valley-related lake deposits interbedded with Karoo flood basalts, Southern Namibia, *in* E. H. Gierlowski-Kordesch and K. R. Kelts, eds., Lake Basins Through Space and Time: AAPG Studies in Geology 46, p. 87–108.

Stollhofen, H., 1999, Karoo Synrift-Sedimentation und ihre tektonische Kontrolle am entstehenden Kontinentairand Namibias: Zeitschrift der Deutschen Geologischen Gesellschaft, v. 149, p. 519–632.

Stollhofen, H., I. G. Stanistreet, B. Bangert, and H. Grill, in press, Tuffs, tectonism and glacially related sea-level changes, Carboniferous–Permian, Southern Namibia: Palaeogeography, Palaeoclimatology, Palaeoecology.

Tankard, A. J., M. A. Uliana, H. J. Welsink, V. A. Ramos, M. Turic, A. B. França, E. J. Milani, B. B. de Brito Neves, N. Eyles, J. Skarmeta, H. Santa Ana, F. Wiens, M. Cirbián, O. López Paulsen, G. J. B. Germs, M. J. De Wit, T. Machacha, and R.McG. Miller, 1995, Structural and tectonic controls of basin evolution in southwestern Gondwana during the Phanerozoic, *in* A. J. Tankard, R. Suárez-Soruco, and H. J. Welsink, eds., Petroleum Basins of South America: AAPG Memoir 62, p. 5–52.

Tiercelin, J.-J., 1991, Natural resources in the lacustrine facies of the Cenozoic rift basins of East Africa, *in* P. Anadón, L. Cabrera, and K. Kelts, eds., Lacustrine Facies Analysis: International Association of Sedimentologists, Special Publication 13, p. 3–37.

Veevers, J. J., D. J. Cole, and E. J. Cowan, 1994a, Southern Africa: Karoo Basin and Cape Fold Belt, *in* J. J. Veevers, and C. McA. Powell, eds., Permian–Triassic Pangean basins and foldbelts along the Panthalassan margin of Gondwanaland: Geological Society of America, Memoir 184, p. 223–279.

Veevers, J. J., C. McA. Powell, J. W. Collinson and O. R. López-Gamundi, 1994b, Synthesis, *in* J. J. Veevers, and C. McA. Powell, eds., Permian–Triassic Pangean basins and foldbelts along the Panthalassan margin of Gondwanaland: Geological Society of America, Memoir 184, p. 331–353.

Vidotti, R. M., C. J. Ebinger, and J. D. Fairhead, 1995, Lithospheric structure beneath the Paraná Province from gravity studies; is there a buried rift system?: EOS, American Geophysical Union, Transactions, v. 76, p. 608.

Visser, J. N. J., 1992, Basin tectonics in southwestern Gondwana during the Carboniferous and Permian, *in* M. J. De Wit and I. G. D. Ransome, eds., Inversion Tectonics of the Cape Fold Belt, Karoo and Cretaceous Basins of Southern Africa: Balkema, Rotterdam, p. 109–115.

Visser, J. N. J., 1997, Deglaciation sequences in the Permo-Carboniferous Karoo and Kalahari basins of southern Africa: a tool in the analysis of cyclic glaciomarine basin fills: Sedimentology, v. 44, p. 507–521.

Vondra, C. F., and D. R. Burggraf, 1978, Fluvial facies of the Plio-Pleistocene Koobi Fora Formation, Karari Ridge, East Lake Turkana, Kenya, *in* A. D. Miall, ed., Fluvial Sedimentology: Canadian Society of Petroleum Geologists, Memoir 5, p. 511–529.

Von Huene, F., 1925, Triassischer Säugetierzahn aus Südwestafrika: Centralblatt Mineralogie, Geologie und Paläontologie, Abt. B, Jahrg. 1925, p. 174–181.

Walker, R. G., 1984, Shelf and shallow marine sands, *in* R. G. Walker, ed., Facies Models: Geoscience Canada, Reprint Series 1, p. 141–170.

Williams, K. E., 1995, Tectonic subsidence analysis and Paleozoic paleogeography of Gondwana, *in* A.J. Tankard, R. Suárez Soruco, and H. J. Welsink, eds., Petroleum Basins of South America: AAPG Memoir 62, p. 79–100.

Wopfner, H., 1991, Structural development of Tanzanian Karoo basins and the break-up of Gondwana: Gondwana eight. Assembly, Evolution and Dispersal, Proceedings, p. 531–539.

Yemane, K., and K. Kelts, 1990, A short review of palaeoenvironments for lower Beaufort (Upper Permian) Karoo sequences from southern to central Africa: A major Gondwana lacustrine period: Journal of African Earth Sciences, v. 10, p. 169–185.

Yemane, K., and K. Kelts, 1996, Isotope geochemistry of Upper Permian early diagenetic calcite concretions; implications for Late Permian waters and surface temperatures in continental Gondwana: Palaeogeography, Palaeoclimatology, Palaeoecology, v. 125, p. 1–4.

Yemane, K., G. Kahr, and K. Kelts, 1996, Imprints of post-glacial climates and palaeogeography in the detrital clay mineral assemblages of an Upper Permian fluviolacustrine Gondwana deposit from northern Malawi: Palaeogeography, Palaeoclimatology, Palaeoecology, v. 125, p. 27–49.

Zalán, P. V., S. Wolff, M. A. M. Astolfi, I. S. Vieira, J. C. J. Conceição, V. T. Appi, E. V. S. Neto, J. R. Cerqueira, and A. Marques, 1990, The Paraná basin, Brazil, *in* M. W. Leighton, D. R. Kolata, D. F. Oltz, and J. J. Eidel, eds., Interior Cratonic Basins: AAPG Memoir 51, p. 681–708.

Section III

Permian-Triassic to Jurassic

Zhao Xiafei and Tang Zhonghua, 2000, Lacustrine deposits of the Upper Permian Pingiquan Formation in the Kelameili Area of the Junggar Basin, Xinjiang, China, in E. H. Gierlowski-Kordesch and K. R. Kelts, eds., Lake basins through space and time: AAPG Studies in Geology 46, p. 111–122.

Chapter 7

Lacustrine Deposits of the Upper Permian Pingdiquan Formation in the Kelameili Area of the Junggar Basin, Xinjiang, China

Zhao Xiafei
State Key Laboratory of Oil and Gas Reservoir and Exploitation
Department of Petroleum, Chengdu Institute of Technology
Sichuan, People's Republic of China

Tang Zhonghua
Geological Survey of Xinjiang Administration Bureau of Petroleum Industry
Xinjiang, People's Republic of China

INTRODUCTION

The Junggar basin is one of the most important oil-producing areas in China. It is located in the northern part of Xinjiang Province and occupies an area of about 140,000 km^2. Commercial production comes mainly from the giant Karamay and associated oil fields at the northwest margin of the basin (Figure 1). Extensive exploration since 1980 has resulted in a series of new discoveries in other regions of the basin. The purpose of this paper is to document the lacustrine deposits of the Upper Permian Pingdiquan Formation in the eastern Junggar basin, which serves as both source and reservoir for the Huoshaoshan oil field (Figure 1).

In the eastern part of Junggar basin, the thrust-faulted Kelameili mountains are composed of Devonian and Carboniferous volcanic rocks plus derivative volcaniclastic sedimentary rocks (Carroll et al., 1990). To the southwest, Permian and Mesozoic strata are exposed, forming several broad, gentle anticlines and synclines. Quaternary sands and gravels compose the dry river channels derived from the foothills of Kelameili. Zou (1984) first reported on the fan-delta deposits of Pingdiquan Formation. In 1986, Zhao Xiafei, together with his students, described and measured two lithologic sections in outcrop (Figure 2). At the same time, we did seismic stratigraphic studies to reveal the subsurface features of the Pingdiquan Formation. Liu and Zhao (1992) discussed the gravelly fan delta of Figure 2a and Wang et al. (1992) published the microfacies of the six producing horizons of the Pingdiquan Formation in the Huoshaoshan oil field. This paper documents the sedimentary paleoenvironment of the formation and examines the delta and fan delta occurrences in the foreland depression of Kelameili.

GEOLOGIC SETTING

There has been widespread disagreement concerning the origin, nature, and evolution of the Junggar basin (e.g., Wu, 1986; Peng and Zhang, 1989; Hsü, 1988; Coleman, 1989; Carroll et al., 1990, 1992; Hendrix et al., 1992; Graham et al., 1993; Allen et al., 1995). Many Chinese geologists thought that the central and western part of the basement of the Junggar basin are composed of stable and rigid pre-Sinian crystalline rocks (e.g., Zhang, 1982; Tian, 1989; You et al., 1992), constituting a microcontinent in the paleo-Asian ocean, and is covered by early Paleozoic platform-type deposits. The rest of the Junggar basement rocks and the western and eastern Junggar folded-thrusted mountains are thought to be Devonian and Carboniferous island arcs and accretionary complexes. The amalgamation of these microcontinents, island arcs, and accretionary complexes resulted in the consolidation of the south margin of paleo-Asia (e.g., Coleman, 1989; Tang et al., 1997); however, we postulate that the Junggar area was underlain by oceanic crustal material behaving as a back-arc or relict oceanic basin by the Early Permian. So, during the Late Permian, owing to the initial uplift of the northern Tianshan ridge, a flexurally subsiding foreland basin was formed (Watson, 1987; Carroll et al., 1990; Hendrix et al., 1992; Tang et al., 1997). Recently, You et al. (1992) and Wang (1997) even imagined a Permian–Early Triassic rifted Junggar basin. Unfortunately, whether the Junggar basin's connections with the Tethys were severed by the Late Permian (e.g., Peng and Zhang, 1989) or by the end of the Triassic (e.g., Hsü, 1988) is not yet clear.

As shown in Table 1, in the Kelameili Mountains, the Devonian and Carboniferous are exposed with a

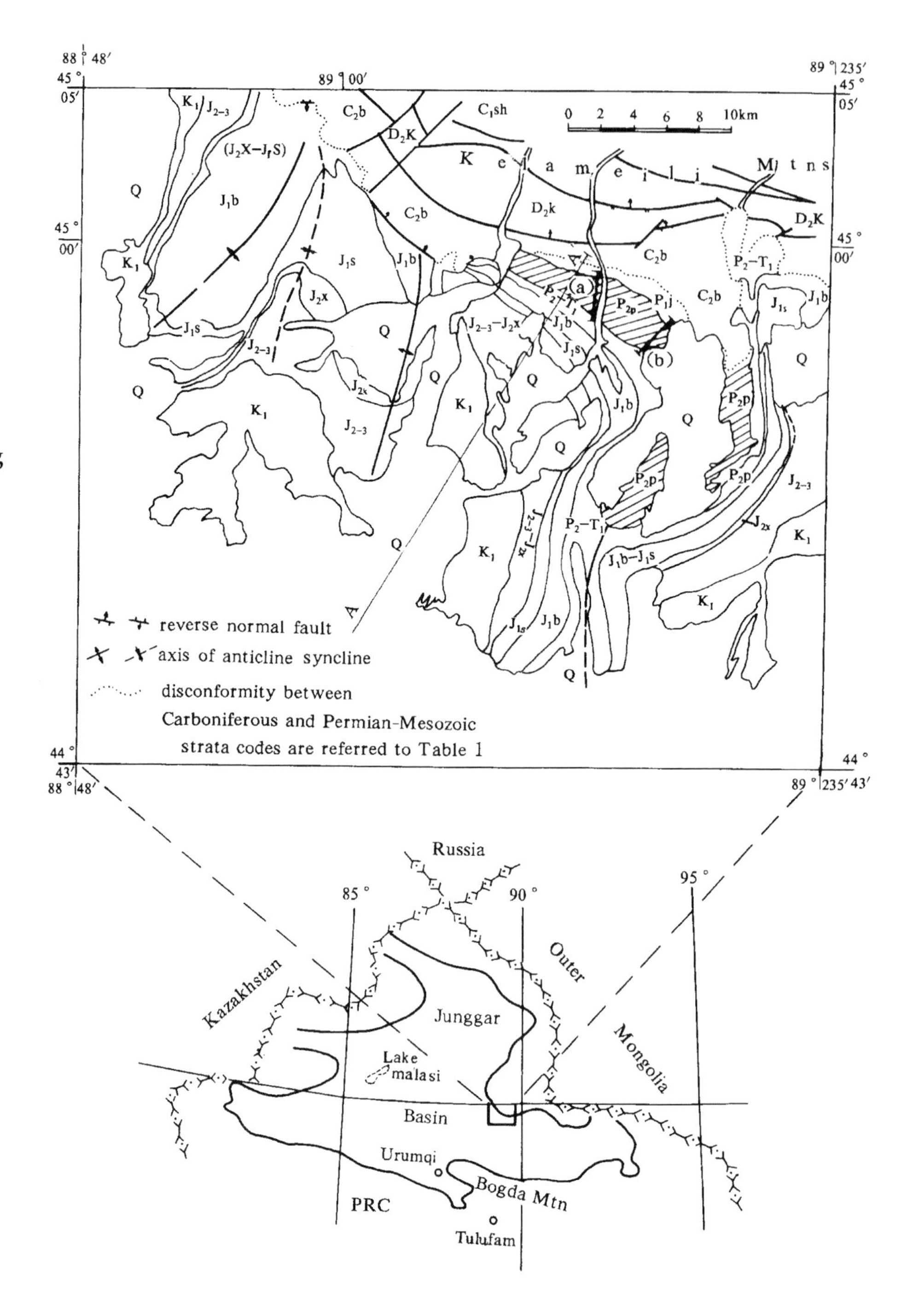

Figure 1—Maps showing the position of the Junggar basin in Asia and geology of the Kelameili area in the eastern portion of the basin. Q denotes Quaternary deposits. Lines (a) and (b) show locations of measured lithologic sections. Cross-section A-A′ is shown in Figure 5. Shaded areas highlight the exposed Pingdinquan Formation. Formation codes are listed in Table 1. Courtesy of the Geological Survey of Xinjiang Adminstrative Bureau of Petroleum Industry.

thickness greater than 5000 m, including thick strata of andesite and tuffite. The Middle–Upper Carboniferous consists of shallow marine deposits and the Permian of continental alluvial and lacustrine sediments. This regressive succession can be correlated to the Upper Carboniferous–Permian in the southern Junggar basin where Upper Carboniferous marine turbidites grade upward into Lower Permian shallow marine sandstones and Upper Permian fluvial-lacustrine deposits, suggesting the final southern and eastern retreat of the paleo-Tethys during consolidation of paleo-Asia (Carroll et al., 1990; Coleman, 1989).

The isopach map of the Upper Permian (Figure 3) shows there are three foreland depressions in the Junggar basin: those along the Tienshan-Bogda, western Junggar Mountains, and the Kelameili. Besides some alluvial-delta systems in the latter two depressions, Upper Permian deposits over most of these areas are represented by profundal lacustrine facies consisting of organic-rich mudstone, shale, and oil shale with intercalations of siltstone and sandstone. These rocks rank among the richest and thickest petroleum source rock intervals in the world (Carroll et al., 1992). During the Late Permian, the Junggar basin might have undergone an extensional stress regime. You et al. (1992) suggested that during Permian–Early Triassic, a large rifting trough developed from Urumqi to Malasi

Table 1. Generalized Stratigraphy of Kelameili Area of the Eastern Junggar Basin.*

Age	Formation		Thickness (m)	Lithology
NEOGENE			330	Brown mudstone & sandstone
PALEOGENE			340	Mudstone, sandstone, & basal conglomerate
UPPER CRETACEOUS	Donggou		486	Red mudstone, siltstone, & sandstone
LOWER CRETACEOUS	Tugulu	(K_1t)	1300	Red mudstone, siltstone, & sandstone
MIDDLE–UPPER JURASSIC	Shishugou	($J_{2-3}sh$)	1500	Red mudstone with sandstone & conglomerate lenses & layers
MIDDLE JURASSIC	Xishanyao	(J_2x)	1120	Coal measures
LOWER JURASSIC	Sangonhe	(J_1s)	809	Gray mudstone, yellowish-gray sandstone, & rare coal seams coal measures
	Badaowan	(J_1b)	1080	
MIDDLE–UPPER TRIASSIC	Xiaoquangou	($T_{2-3}X$)	1246	Gray mudstone & siltstone with rare coal seams
LOWER TRIASSIC–UPPER PERMIAN	Cangfanggou Group	(P_2-T_1)	1140	Purple to gray mudstone dominates (lacustrine)
UPPER PERMIAN	Pingdiquan	(P_2p)	600–2690	Dark gray mudstone, siltstone, & conglomerate (lacustrine)
	Jiangjunmiao	(P_2j)	982	red muddy conglomerates
LOWER PERMIAN	Xaijijichao Group	(P_1x)	762	Yellowish-green sandy mudstone, sandstone (continental)
UPPER CARBONIFEROUS	Liukeshu		1700	Motley tuff
MIDDLE–UPPER CARBONIFEROUS	Shiqiantan		840	Biogenic limestone, carbonaceous, mudstone, siltstone, & sandstone (marine)
MIDDLE CARBONIFEROUS	Batamainei Mountain	(C_2b)	1010	Andesite & volcanic breccia
LOWER CARBONIFEROUS	Shanliang-lishi	(C_1sh)	887	Yellowish-brown conglomerate, sandstone, & siltstone with rare coal seams
	Tamogang		877	Dark siltstone & sandstone with carbonaceous shale & coal seams
UPPER DEVONIAN	Sujiquan		230	Gray-green tuff & fine sandstone (marine)
MIDDLE DEVONIAN	Kelimeili	(D_1k)	210	Motley tuff & tuffaceous siltstone (marine)

*Formation codes are used in Figure 1. Reliability index: B.

Lake in response to hot spot activity. Allen et al. (1995) proposed that during the Late Permian–Early Triassic (?), the Junggar, Turfan, and Alakol (in Kazakhstan) basins were subjected to roughly south to north extension in response to the sinistral movement of the two strike-slip faults in the Altaid orogenic collage. This event is the only important phase of extension in a region otherwise dominated by compressional tectonics throughout the Phanerozoic.

The Late Permian deep lake basin of the Junggar basin might have originated from this unique crustal extension period in central Asia. At that time, the basin is postulated to have been a relict back-arc basin behind the Tienshan-Bogda volcanic island arc. This is similar to the conditions that formed the Black Sea and southern Caspian Sea basins, as remnants of back-arc basins originating from Late Cretaceous–Paleogene volcanic island arc systems (Zonenshain

Figure 2—Measured lithologic sections (a) and (b) through portions of the Pingdinquan Formation from the eastern fan-delta (data from Zhao et al., 1987). Sedimentary cycles are illustrated in Roman characters (from Liu and Zhao, 1992). See Figure 1 for exact locations.

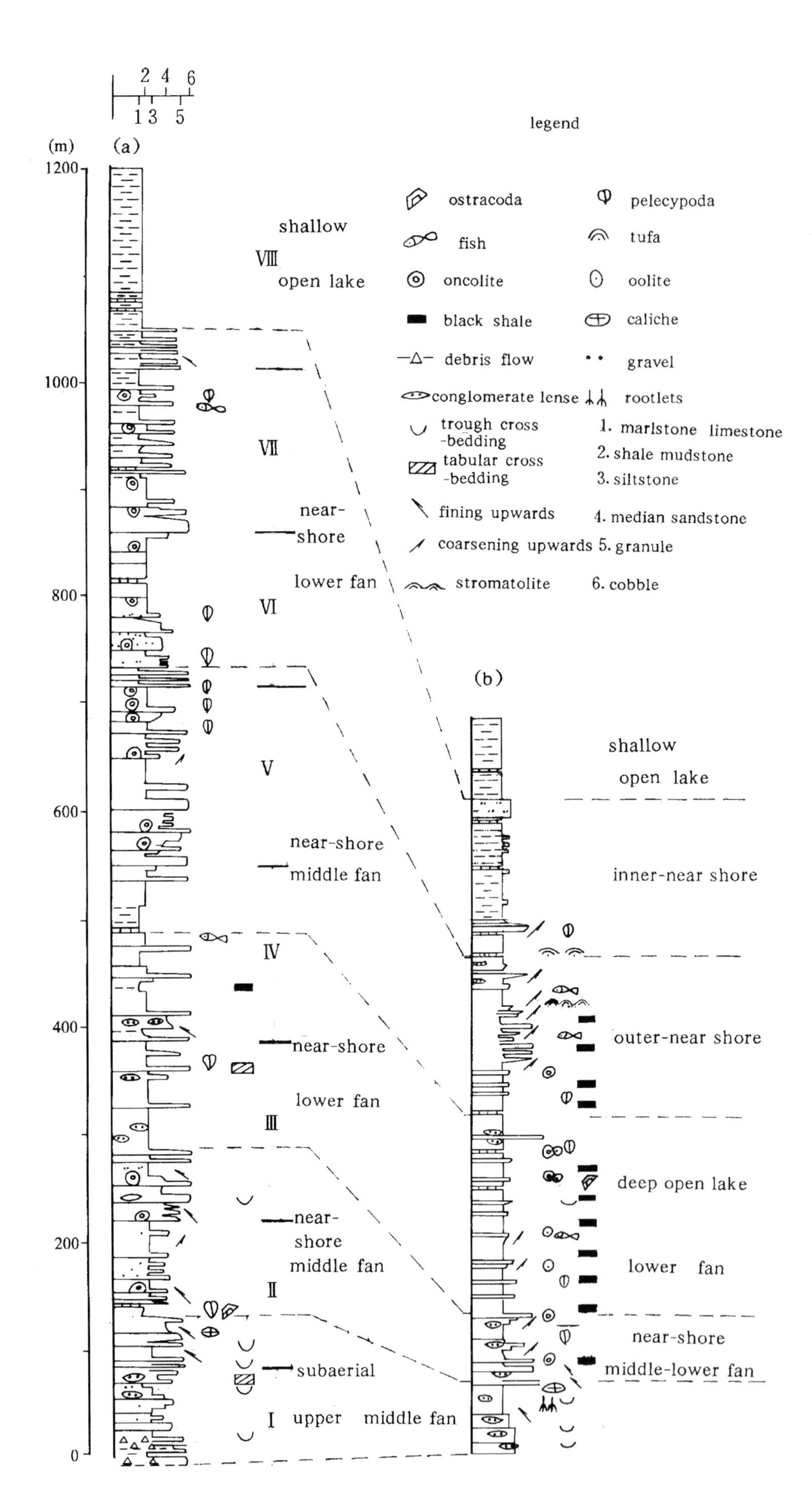

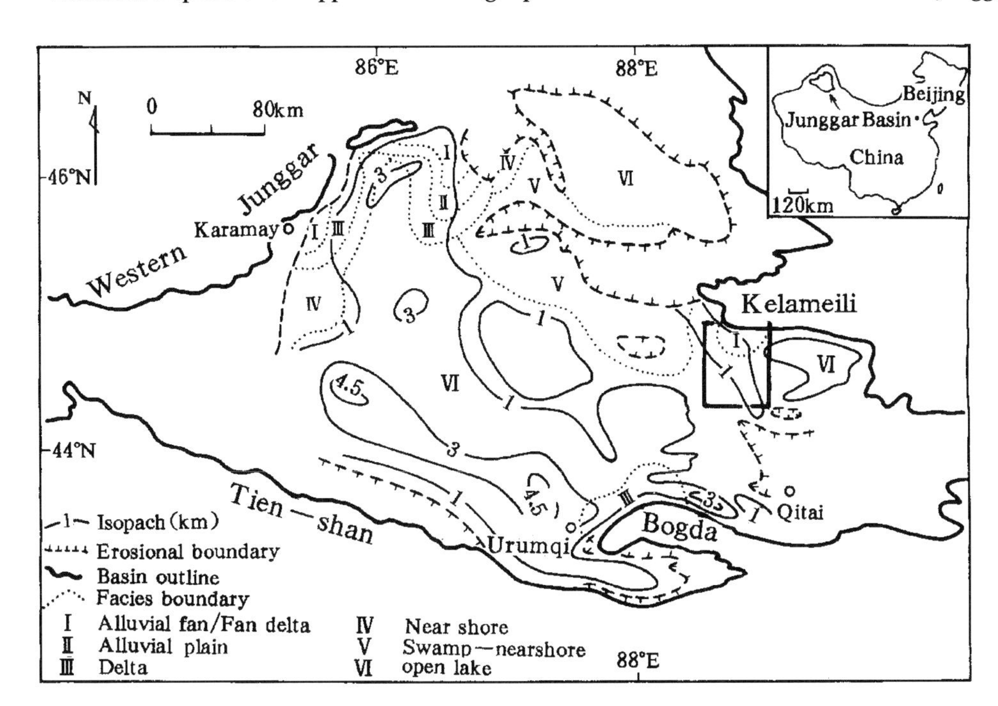

Figure 3—Lithofacies-paleogeography map of the Late Permian of the Junggar basin (from Tian, 1989). Boxed area is shown in Figure 4.

and LePichon, 1986). The Permian Junggar lake may have had a connection to the paleo-Tethys south of Tienshan-Bogda, although geologic data are too limited to confirm this theory.

Based on the bivalve assemblages and sedimentary features of the Late Permian, Xinjiang Province is presumed to be in the transitional climate zone between subtropical and temperate zones with a more humid climate in its earlier stage (Wei, 1982). Early Upper Permian sediments in Junggar and its adjacent area are characterized by organic-rich deposits (dark carbonaceous mudstone and oil shale), correlatable coal measures into western Siberia (e.g., Kuznetsk basin), and the relatively abundant and highly variable bivalve fauna *Microdontella*, suggesting a warm and humid climate (Wei, 1982); however, the latest Permian sediments containing redbed intercalations imply a shift to arid and semi-arid climate, which is supported by the occurrences of the bivalve assemblage *Microdontella-Palaeomutela-Anthraconai-Oligodon*, alternating with those of bivalves adapted to fresher environments (Wei, 1982).

The Cangfanggou Group (Table 1), overlying the Pingdiquan Formation, is also lacustrine in origin, representing continuous sedimentation from the Late Permian to the Early Triassic.

SEDIMENTOLOGY AND PALEOENVIRONMENT

Lithology and Facies

The Upper Permian Pingdiquan Formation is composed of conglomerates, sandstones, siltstones, mudstones, and shales. The shales contain fine microlamination of alternating claystone and siltstone, and are rich in organic matter with total organic carbon (TOC) ranging from 1.25 to 1.47% (Zhou et al., 1992). The underlying Jiangjunmiao Formation consists of brownish-red muddy breccia, poorly bedded with normal and reverse size grading. At its top, large calcareous concretions are present. The Pingdiquan Formation has been subdivided downward into three lithostratigraphic members: the first, second, and third members. The third member is characterized by the presence of a great deal of dark mudstone, shales/oil shales, and oolitic limestones. The second member contains alternating deposits of dark mudstone, oil shale, and marlstone, and the first member is marked by purple and brown-red mudstone beds.

Seventeen facies types have been differentiated (Table 2). They are grouped into upper fan, middle fan, and lower fan facies associations of the sublacustrine fan-delta system and lake beach, nearshore lacustrine zone, and open lake facies associations of the paleolake system; however, these two systems cannot be clearly differentiated from each other; the transition is gradual. McPherson et al. (1988) pointed out that beach, bars, and spits are normal constituents of a fan-delta as well as a beach. So facies 5, 6, and 8 in Table 2 can be regarded as facies members of the fan-delta system, as well.

Figure 2 shows lithology and facies of the Pingdiquan Formation for the two measured sections. The beds cannot be traced between these sections, so stratigraphic correlation was based on lithology and sequence boundaries. At both sections, the brownish-gray, relatively thin beds of the basal Pingdiquan Formation are differentiated from the underlying red massive thick beds of the Jiangjunmiao Formation. The first member of the Pingdiquan Formation at both sections is featured by light gray-green mndstones, contrasting with dark gray shales and oil shales in the

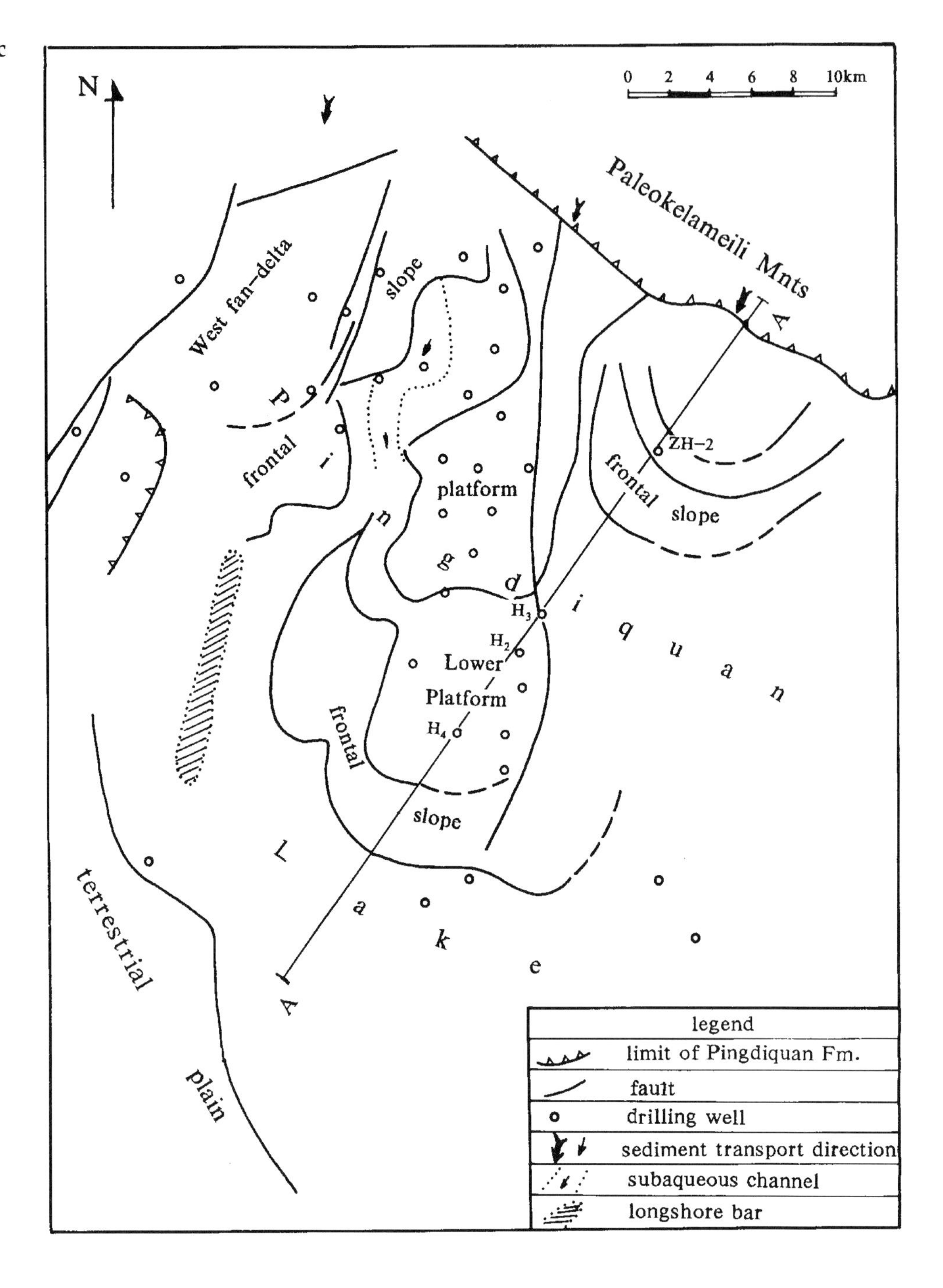

Figure 4—Paleogeographic map of subenvironments within the paleolake Pingdinquan (from Zhao et al., 1987). The approximate position of cross-section A–A′ is indicated. Lines (a) and (b) denote the two measured lithologic sections of Figure 1.

second and third members; however, exact correlation between the two sections was based on the principles of sequence stratigraphy. Shanley and McCabe (1994) wrote that sequence boundaries reflecting an allocyclic forcing function might be identified from an understanding of those stratal geometries that signified an abrupt change in the rate at which accommodation space was created. Criteria included regional alluvial incision, marked changes in size and composition of grains, and stacking pattern of fluvial channel sandstones. Multistory, multilateral sandbodies are thought to generally overlie sequence boundary unconformities associated with allocyclic phenomena (Shanley and McCabe, 1994). So, the lower boundaries of first and second members were set at the base of the most notable alluvial sandstones of both sections.

In Figure 2, a contrast is seen between measured sections in Figure 2a and 2b from the Pingdiquan Formation. Section (a) (Figure 2a) contains abundant coarse siliciclastic sediments, i.e., cobble to granule sandstones with basal scours and lenticular bed morphology. These coarse beds grade upward into siltstone and mudstone that lack black shale. Section (b) (Figure 2b) is located 4 km east of section (a) and characterized by upward-coarsening sequences, an abundance of laminated organic-rich black shale, and the scarcity of coarse deposits. Farther east of section (b), the Pingdiquan Formation is almost completely composed of black shales. Such a pattern of lithologic

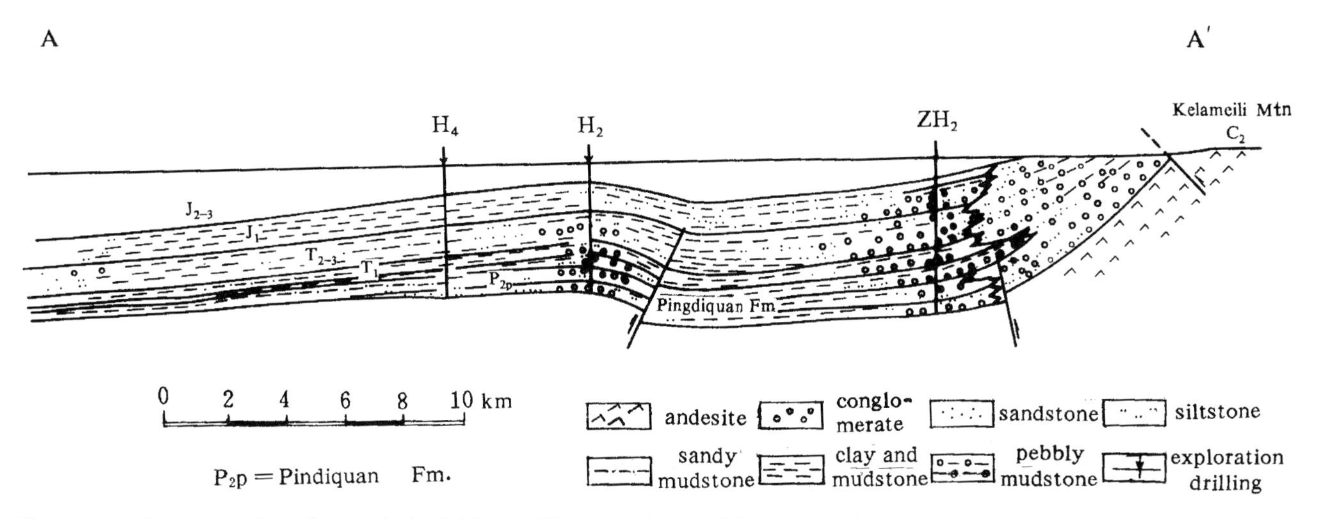

Figure 5—Cross-section through A–A′ from Figure 4 derived from geophysical data in the Junggar basin. See Figure 1 for exact location of the cross-section in the Kelameili area.

trends can be interpreted as a progression from proximal sublacustrine delta deposits in section (a) to distal deeper water lake deposits in section (b).

Based on the maximum particle size (MPS), section (a) (Figure 2a) was subdivided into eight upward-fining units (Liu and Zhao, 1992). Unit I contains a great deal of debris flow deposits. Units II, III, and IV are characterized by the poor sorting, trough cross-bedding, and basal scours that might be attributed to subaqueous channel deposits; however, the coarse members of Units V, VI, VII, and VIII are largely horizontally bedded and well sorted with rounded grains containing calcareous cements. These sequences are interpreted to be lake beach, shoal, or more distal deposits. Section (b) (Figure 2b) is lithologically distinct from section (a), but both show similar vertical trends in variability. We assume that such alternation of units can be related to channel switching and to tectonic changes in the source area.

It is not difficult to discriminate the subaqueous from the terrestrial sediments in Pingdiquan Formation. Terrestrial siliciclastic deposits do not contain allochemical elements; they are often cemented by calcite and can contain abundant calcareous concretions. These deposits are interpreted as soil-zone carbonates formed under a semi-arid climate (Collinson, 1978); however, the subaqueous deposits commonly contain allochemical grains and calcite cement.

Carbonate Rocks

Carbonate rocks amount to less than 3% by volume of the Pingdiquan Formation. They occur mainly in the third and second members of the formation and are represented by grainstones and tufa. In addition, three thin marlstone beds with cone-in-cone structures are present in the first member of the formation.

Grainstones are composed of micrite and allochems of bioclasts, oolites, oncoids, and cryptalgal grains. Bioclasts normally include whole ostracode shells, single valves or fragments of bivalve shells, fish vertebrae fragments, and rare plant or algal debris. Oolites commonly have a nucleus surrounded by a single lamina, and in rare cases they have three concentric laminae. Oncoids have incomplete laminae, occurring in grainstones as well as in siltstones and sandstones. In some cases, cryptalgal grains and peloids are also constituents of the grainstone.

Stromatolitic limestone occurs rarely in the Pingdiquan Formation. Small hemispheroids are laterally linked with lamellae whose domes are closely packed, so the stromatolites can be assigned to the type LLH-C of Anstey and Chase (1974, after Flügel, 1978).

One tufa bed occurs in the section shown in Figure 2b, as a white porous calcareous layer 0.4 m in thickness. It is covered by a stromatolite (0.2 m thick) and underlain successively by thin beds of calcareous siltstone, brown mudstone, and a well-sorted, bimodal gravelly sandstone. The sequence containing the tufa bed and affiliated carbonate layers is interpreted as beach sediments associated with springs exposed during a contraction of the paleolake. Riding (1979) pointed out that a common feature of many lacustrine algal bioherms is the veneer formed on the algal tufa and adjacent sediments by laminated carbonate during emergence phases due to a temporary fall in lake level and the probable influx from springs or seeps.

From these lithology and facies associations, the Pingdiquan Formation can be interpreted as deposits of lacustrine fan-deltas adjacent to active faults. The paleolake is inferred to be hydrologically open. This type of lake has a stable shoreline (Eugster and Kelts, 1983), which is evidenced by the occurrences of lake-beach gravels and shoreface coarse sandstones (see Table 2). Based on the ecology of the nonmarine bivalves in the Pingdiquan Formation, Zhang et al. (1992) pointed out the lake may have been alkaline at times. Carbonate input from springs and seeps appear to be the source for the biochemical rocks. Considering that oolites are not known from modern

Table 2. Facies Description and Facies Associations with Interpretations of the Upper Permian Pingdiquan Formation, Junggar Basin.

No. Facies	Description	Interpretation	Facies Association: Fan-delta	Facies Association: Lake
1. breccia	brownish red, argillaceous, very poorly sorted, upward-fining (0.1–0.4 m thick)	debris flow	upper fan	
2. breccia-siltstone	argillaceous, normal or reverse grading, trough cross-strata rare (0.5–1.2 m thick)	sheet flow, may be channelized	upper fan	
3. trough cross-bedded conglomerate and gravelly sandstone	(a) with calcareous concretions, upward-fining, rootlets in fines, with interbeds of breccia and breccia-siltstone facies (0.3–7 m thick)	subaerial braided channel deposits	middle fan	
	(b) with oncolites, well-sorted, basal scours, upward-fining (1–2 m thick)	subaqueous distributary channel	middle fan	
4. large-scale foresets of conglomerate and sandstone	set thickness up to 3 m	Gilbert Delta	lower fan	
5. well-sorted conglomerate	horizontally bedded, well sorted with rounded clasts, no matrix, 1–2 m thick, clasts may be coated by calcite	shoreline gravel		lake beach zone
6. well-sorted coarse sandstone	with calcareous cement (1–2 m thick)	shoreface, shoal		lake beach zone, nearshore zone
7. oncolite-bearing sandstone and siltstone	oncolites in well-sorted medium sandstone and siltstone, pelecypoda-bearing mudstone, pisolites very rare, up to 3 m thick	shoreface, shoal		nearshore zone
8. shelly siltstone and sandstone	shells with minor oolites and granules, millimeter-scale bed thickness	shoal		nearshore zone
9. shaly siltstone	gray-green, plant debris, and abundant shells of pelecypoda, with interbeds of conglomerate, 3–30 m thick	low-energy shoal		nearshore zone
10. light-colored mudstone	(a) gray-green mudstone bearing plant debris, ostracoda, and pelecypoda, 2–4 m thick (b) thick successions of gray-green mudstone with interbeds of brownish red, dark gray mudstone and siltstone	interdistributary bay	middle fan	shallow open lake
	(c) brownish-red silty mudstone, fossil fish bearing		nearshore zone	
11. dark gray mudstone	Massive, thick successions, with rare cone-in-cone marlstone interbeds, uppermost part of section (b) is >175 m thick	lake bottom sediments		open lake
12. black shale and oil shale	fine millimeter-scale laminae, rich in organic matter, intercalated with yellowish-gray mudstone (units 1–5 m thick)	deposits on stratified lake bottom		deep open lake
13. tufa, stromatolites	decimeter-thick units, rare	spring deposits		lake beach zone
14. marlstone	(a) intercalated with siliciclastic mudstone, decimeter-thick beds (b) terrigenous clasts up to 30–40%, associated with siltstones, decimeter-thick beds (3 m maximum thickness)	palustrine deposits		nearshore zone
15. oolitic limestone	may bear plant debris and fish scales, decimeter-thick units	clear water shoal		nearshore zone
16. oncolitic limestone	decimeter-thick units, may bear plant debris and granules	shoreface, shoal		nearshore zone
17. shelly limestone	thin beds intercalated with black shale (decimeter-thick units)	shell beach		beach zone

dilute lakes (Eugster and Kelts, 1983), Pingdiquan lake water might have been saline to some extent.

Fossils and Paleoecology

In the underlying Jiangjumiao Formation, only rare ostracodes, fish scales, and plant debris are present; however, the Pingdiquan Formation contains abundant fossils of bivalves, ostracoda, fish, and *Estheria*. The ostracode fauna includes *Palaeocypridopsis sp.*1, *Tomiella, Permiana, Kelameilia, Hongyanchiella, Qitaina,* and *Darwinula*. Except for *Palaeocypridopsis sp.*1 (brackish?), all the ostracodes are interpreted as living mainly in shallow transitional waters and *Darwinula* in fresh to brackish waters (Zhang et al., 1992). The bivalve fauna includes *Palaeomutela, Palaeanodonta, Senderzoniella, Abiella, Anthraconauta, Pseudomodiolus, Procopievskia,* and *Mrassiella*. Among them, *Palaeomutela* and *Palaeanodonta* are interpreted as living in turbulent freshwater and the others in more quiet semipelagic brackish water (Zhang et al., 1992). The fish fauna includes Palaeoniscidae and palaeoniscoid scales (Chen, 1988). Based on the relative abundance of these families and genera (species) and their living habits and characteristics in the Pingdiquan Formation, Zhang et al. (1992) inferred that the salinity of Pingdiquan lake changed through time. They suggested that during deposition of the lower third member, the lake water had a salinity between that of freshwater and sea water. During deposition of the middle and upper third member and the second member, the lake basin was enlarged and deepened and waters decreased in salinity to brackish. With deposition of the first member, freshwater conditions were established.

Trace Elements

Boron (B) is the most reliable element to be used in differentiating between continental freshwater and ocean environments. It is rare in modern freshwater; however, its concentration in sea water is proportional to the salinity (Couch, 1971). In general, freshwater lake deposits have a low boron content, whereas the marine sediments concentrations are greater than 100 ppm, with clays in salt-forming lagoons having boron concentrations more than 1000 ppm. Gallium (Ga) concentrations in modern marine and freshwater ooze (Keith and Degens, 1959) are similar, so the B/Ga ratio can be used as an index to differentiate freshwater from saline environments. Table 3 lists data on B and Ga contents of Pingdiquan core sections in the studied area (from Zhang et al., 1992). These data show apparent fluctuations between maximum and minimum values for the three members of the formation; however, from the third through the first member, both the mean value of the B content and B/Ga ratio tend to decrease, indicating a reduction in salinity through time.

Glauconite Occurrences

The mineral glauconite has been revealed in the Pingdiquan core sections of several drillings in eastern Junggar (Zhang et al., 1992). Glauconite is thought to be formed in a suboxic environment with slow sedimentation, and is thought to be connected with the sea (Liu and Ceng, 1985); however, in the modern Fuxian lake in Yunnan, which is far from the sea, several glauconite phases have been reported. Fuxian Lake is a freshwater deep lake with a perennial water temperature of about 13°C. With stable temperature and pressure, slightly oxic conditions, and favorable pH and Eh values in the substrate, conditions are thought to be advantageous for the formation of glauconite (Zhu et al., 1989). Similar paleoenvironmental conditions are interpreted for the Pingdiquan glauconite, a deep lake with stable bottom conditions and perhaps a connection to some extent for some period of time with the open sea, especially during the deposition of the third member of the formation.

FACIES ZONATION AND RESERVOIR PROPERTIES

Extensive seismic stratigraphic studies have been conducted in combination with well-log data to reveal the subsurface structure and facies of the Pingdiquan

Table 3. Boron (B) and Ballium (Ga) Concentrations and B/Ga Ratios Analyzed from Cores from Upper Permian Pingdiquan Formation, Eastern Junggar Basin.*

Strata	Amount of Sample Analyzed	B (ppm)			Ga (ppm)			
		Max	Min	Average	Max	Min	Average	B/Ga
1st member	6	300	150	210	40	30	37	5.7
2nd member	26	1000	50	290	100	30	40	7.3
3rd member	12	5000	50	590	50	7	27	21.9
Pingdiquan Fm. in total	44	5000	50	360	100	7	38	9.5

*From Zhang, 1992.

Formation (Zhao et al., 1987); the results are presented in Figures 4 and 5. The northern boundary of Pingdiquan Lake is defined by a huge thrust fault along the foothills of the paleo-Kelameili Mountains. The southern boundary is unknown. The lake basin is inferred to be open to the west and east, based on structural depressions in these directions (Li and Jiang, 1987). Paleocurrent directions measured on outcrops and estimated from seismic foreset configurations indicate that the sediments were transported roughly from north to south with a decrease in grain size southward. Particular horizons, of course, show considerable variations in the transport direction.

Three sublacustrine deltas (Liu, 1988) have been differentiated. The eastern delta is a fan-delta, small in size with its northern portion exposed. Lithostratigraphic sections shown in Figure 2 were measured here. The western delta is defined by the sinuous edge of its frontal slope. The central delta is a relatively large delta where the Huoshaoshan oil field is located. Three structural sedimentary units have been identified in the central delta based on seismic data, slope angles, and minor faults: platform, lower platform, and front slope. The platform is a structural high and characterized by chaotic seismic reflection configurations. Its proximal end adjacent to the boundary thrust consists of coarse siliciclastics and debris flow deposits that fine rapidly southward to form a relatively sandstone-rich platform. The frontal slope constitutes the distal part of the sublacustrine delta and is characterized by mounded and shingled seismic reflection configurations with few sandstone beds. The lower platform is a transitional zone between the former two units and characterized by weak seismic reflections with more common sandstone beds. All these units are less well defined in the other two deltas.

Based on detailed studies of the core sections and well logs, Wang et al. (1992) depicted the subfacies of six horizons of the Pingdiquan Formation in the area roughly including the platform of the middle delta and its adjacent zone. Oil-producing sandstones are feldspathic litharenite with a porosity of 7–27% and permeability of $0.5–40 \times 10^{-3}$ d. Production capacity of Pingdiquan horizons are mainly controlled by sedimentary subfacies distribution and to a lesser extent by fractures (Liu et al., 1992). The well-sorted fine sandstones of channel deposits are the main producing reservoirs. Stream mouthbar sandstones are fairly good as well, with wells on bar ridges having higher and relatively stable output. Besides the oil accumulations being exploited now at the margins of the Junggar basin, reservoirs consisting of sandy middle to lower sublacustrine delta beds and lowstand deposits can be expected in the inner parts of the basin. The relatively sand-rich fan-deltas tend to form structural highs, owing to post-depositional compaction, and are producing some oil.

ACKNOWLEDGMENTS

This paper is sponsored by the National Natural Science Foundation of China (No. 49070109). Special thanks to Dr. A. Carroll and the editors for their dedication and help in improving this manuscript for publication.

REFERENCES CITED

Allen, M. B., A. M. C. Sengor, and B. A. Natal'in, 1995, Junggar, Turfan and Alakol basins as Late Permian to? Early Triassic extensional structures in a sinistral shear zone in the Altaid orogenic collage, central Asia: Journal of the Geological Society, London, v. 152, p. 327–338.

Anstey, R. L., and T. L. Chase, 1974, Geographic diversity of late Ordovician corals and bryozoans in North America: Journal of Paleontology, v. 48, p. 1141–1148.

Carroll, A. R., Y. Liang, S. A. Graham, X. Xiao, M. S. Hendrix, J. Chu, and C. L. McKnight, 1990, Junggar basin, northwest China: trapped late Paleozoic ocean: Tectonophysics, v. 181, p. 1–14.

Carroll, A. R., Brassell, S. C., and Graham, S. A., 1992, Upper Permian lacustrine oil shales, southern Junggar basin, northwest China: AAPG Bulletin, v. 76, p. 1874–1902.

Chen, L., 1988, Petrographic and sedimentologic studies of the Upper Permian Pingdiquan Formation at Laoshangou section in the southern foothills of Kelameili, Xinjiang: M.S. Thesis, Chengdu College of Geology, Chengdu, 133 p. (in Chinese).

Coleman, R. G., 1989, Continental growth of northwest China: Tectonics, v. 8, p. 621–636.

Collinson, J. D., 1978, Alluvial sediments, in H. G. Reading, ed., Sedimentary Environments and Facies: London, Blackwell Science, 49 p.

Couch, E. L., 1971, Calculation of paleosalinities from boron and clay mineral data: AAPG Bulletin, v. 55, p. 1829–1837.

Eugster, H. P., and K. Kelts, 1983, Lacustrine chemical sediments, in A. S. Goudie and K. Pye, eds., Chemical Sediments and Geomorphology: London, Academic Press, p. 321–368.

Flügel E., 1978, Microfacies of Limestones (translated by K. Christenson, 1982): Berlin, Springer-Verlag, 633 p.

Graham, S. A., M. S. Hendrix, L. B. Wang, and A. R. Carroll, 1993, Collisional successor basins of western China: impact of tectonic inheritance on sand composition: Geological Society of America Bulletin, v. 105, p. 323–344.

Hendrix, M. S., S. A. Graham, A. R. Carroll, E. K. Sobel, C. L. McKnight, B. J. Schulein, and Z. Wang, 1992, Sedimentary record and climatic implications of recurrent deformation in the Tian Shan: evidence from Mesozoic strata of north Tarim, south Junggar, and Turpan basins, northwest China: Geological Society of America Bulletin, v. 104, p. 53–79.

Hsü, K. J., 1988, Relict back-arc basins: principles of recognition and possible new examples from China, in K. L. Kleinpell and C. Paola, eds., New Perspectives in Basin Analysis: New York, Springer-Verlag, p. 245–263.

Keith, M. L., and E. T. Degens, 1959, Geochemical

indicators of marine and fresh-water sediments: Researches in Geochemistry, p. 38–61.
Li, X., and J. Jiang, 1987, Petroleum geology and oil-gas distribution controlls of eastern Junggar basin (in Chinese with English abstract): Oil and Gas Geology, v. 8, no. 1, p. 99–107.
Liu, B., and Y. Ceng, 1985, Basis and applications of lithofacies-paleogeography (in Chinese): Beijing, Geology Press, 220 p.
Liu, L., 1988, Fan-delta deposits of Permian Pingdinquan Formation at Xidagou Section in southern foothills of Kelameili, Xinjiang (in Chinese): M.S. Thesis, Chengdu College of Geology, Chengdu, 94 p.
Liu, L., and X. Zhao, 1992, Conglomerate fan-delta sediments of the Pingdiquan Formation (Upper Permian) in Xidagou area of southern piedmont of Kelameili Mountains (in Chinese): Xinjiang Petroleum Geology, v. 13, p. 149–158.
Liu, Z., S. Yang, X. Han, and S. Shan, 1992, Reservoir properties and sensitivity evalution of eastern Junggar basin, in B. Luo et al., eds., Comprehensive Studies of the Petroleum Geology of Junggar Basin (in Chinese): Gansu Science and Technology Press, 322 p.
Nemec, W., and R. J. Steel, 1988, What is a fan-delta and how to recognize it, in W. Nemec and R. J. Steel, eds., Fan Deltas: Glasgow, Blackie and Son, p. 3–13.
Peng, Z., and G. Zhang, 1989, Tectonic features of the Junggar basin and their relationship with oil and gas distribution, in X. Zhu., ed., Chinese sedimentary basins: Amsterdam, Elsevier, p. 17–31.
Riding, R., 1979, Origin and diagenesis of lacustrine algal bioherms at the margin of the Ries crater, Upper Miocene, southern Germany: Sedimentology, v. 26, p. 645–680.
Shanley, K. W., and P. McCabe, 1994, Perspectives on the sequence stratigraphy of continental strata, AAPG Bulletin, v. 78, no. 4, p. 544–568
Tang, Z., J. Parnell, and F. J. Longstaffe, 1997, Diagenesis and reservoir potential of Permian-Triassic fluvial/lacustrine sandstones in the Southern Junggar basin, northwestern China: AAPG Bulletin, v. 81, p. 1843–1865.
Tian, Z., 1989, The analysis of oil and gas prospects in the Junggar basin from geological development history (in Chinese with English abstract): Xinjiang Petroleum Geology, v. 10, no. 3, p. 3–14.
Wang, Y. F., M. Feng, S. Wang, H. Ni, L. Zhang, and X. Li, 1992, Detailed studies of the sedimentary facies of the reservoir in Haoshaoshan oil field, in B. Luo et al., eds., Comprehensive Studies of the Petroleum Geology of Junggar basin (in Chinese): Gansu Science and Technology Press, 322 p.
Wang, Y. C., 1997. The aeromagnetic anomalies and rift system of northern Xinjiang (in Chinese): Xinjiang Petroleum Geology, v. 18, no. 4, p. 295–301.
Watson, M. P., A. B. Hayward, D. N. Parkinson, and Z. M. Zhang, 1987, Plate tectonic history, basin development and petroleum source rock deposition onshore China: Marine and Petroleum Geology, v. 4, p. 205–225.
Wei, J., 1982, Fossil assemblages of the Late Permian through Cenozoic bivalves of Xinjiang and their implications to stratigraphical division, correlation and paleoclimate (in Chinese): Xinjiang Petroleum Geology, v. 3, no. 1, p. 1–58.
Wu, Q., 1986, On evolution stages, tectonic elements, and origin of local structures of the Junggar Basin(in Chinese): Xinjiang Petroleum Geology, v. 7, no. 1, p. 29–37
You, Q., X. He, and J. Liu, 1992, The tectonic unit subdivision and perspective for each tectonic phase of Junggar basin, in B. Luo et al., eds., Comprehensive Studies of the Petroleum Geology of Junggar Basin (in Chinese): Gansu Scientific and Technological Press, 322 p.
Zhang, Y. J., X. Qi, and X. Cheng, 1992, The sedimentary environment and correlation of the Pingdiquan Formation of Zhangpenggou area, eastern Junggar (in Chinese): Xinjiang Petroleum Geology, v. 13, no. 3, p. 217–225.
Zhang, Y. R., 1982, The gravity-magnetic field and basement texture of Junggar basin (in Chinese): Xinjiang Petroleum Geology, v. 3, p. 1–5.
Zhao, X., Z. You, and Z. Cheng, 1987, Lithofacies and oil perspective of the Permian Pingdiquan Formation in the Shaqiuhe-Zhangpenggou area of the Junggar Basin (in Chinese): Report, Chengdu College of Geology & Xinjiang Administrative Bureau of Petroleum Industry, 52 p.
Zhou Zhongyi, C. Pan, L. Han, and S. Fang, 1992, Evalution methoddogy for the heat history, productivity and out-put of the gas-source beds in Junggar basin, in B. Luo et al., eds., Comprehensive Studies of the Petroleum Geology of Junggar basin (in Chinese): Gansu Science and Technology Press, 322 p.
Zhu, H., Y. Chen, and P. Pu, eds., 1989, Environments and sedimentation of fault lakes, Yunnan province (in Chinese): Beijing, Science Press, 513 p.
Zonenshain, L. P., and X. Le Pichon, 1986, Deep basins of the Black Sea and Caspian Sea as remnants of Mesozoic back-arc basins: Tectonophysics, v. 123, p. 181–211.
Zou, Y., 1984, Sedimentology and petroleum geology of the Upper Permian Pingdiquan Formation in the southern foothills of the Kelameili, Junggar Basin (in Chinese): Symposium, Xinjiang Petroleum College.

Wartes, M. A., A. R. Carroll, T. J. Greene, Keming Cheng, and Hu Ting, 2000, Permian lacustrine deposits of northwest China, *in* E. H. Gierlowski-Kordesch and K. R. Kelts, Eds., Lake basins through space and time: AAPG Studies in Geology 46, p. 123–132.

Chapter 8

Permian Lacustrine Deposits Of Northwest China

Marwan A. Wartes
Alan R. Carroll
Department of Geology and Geophysics, University of Wisconsin-Madison Madison, Wisconsin, U.S.A.

Todd J. Greene
Department of Geological and Environmental Sciences, Stanford University Stanford, California, U.S.A.

Keming Cheng
Research Institute of Petroleum Exploration and Development, China National Petroleum Corporation Beijing, People's Republic of China

Hu Ting
Research Institute of the Tu-Ha Petroleum Exploration and Development Bureau, China National Petroleum Corporation Hami, People's Republic of China

INTRODUCTION

Permian deposits of the Junggar and Turpan-Hami basins of the Xinjiang Uygur Autonomous Region of northwest China preserve some of the thickest and most areally extensive lake strata on Earth. In the south Junggar depocenter, these nonmarine deposits are up to 5 km thick and organic-rich facies rank among the thickest and richest petroleum source rocks in the world (Graham et al., 1990; Lawrence, 1990; Demaison and Huizinga, 1991; Carroll et al., 1992). In addition, Permian lacustrine deposits are estimated to span 1000 km along strike, indicating that widespread lakes represented a major paleogeographic feature of central Asia (Figure 1). Unfortunately, the remote location of these deposits has hindered detailed studies, and the western literature contains only sparse reference to this important record of continental sedimentation. The purpose of this paper is to briefly review the Permian nonmarine stratigraphy and report on recent field-based studies documenting the Permian lacustrine stratigraphy exposed along the north and south flanks of the Bogda Shan (Figure 2).

SETTING

Paleocurrents and other data indicate that the development of Permian lakes pre-date the uplift of the Bogda Shan, which now partitions the Junggar and Turpan-Hami basins (Figures 1, 2) (Hendrix et al., 1992; Greene et al., 1997, in press; Shao et al., 1999); therefore, the Bogda Shan provide an oblique cross-section of Permian lake deposits of the unified Junggar-Turpan-Hami basin. In addition to lacustrine sedimentation, the southern margins of this basin periodically received coarse clastic sediments shed from the ancestral Tian Shan (Figure 1) (Carroll et al., 1990, 1995; Greene et al., 1997). Coarse clastics were also shed southward into the basin from the Kelameili and west Junggar Shan, uplifts bounding the northeastern and northwestern edges of the basin respectively (Figure 1) (Lee, 1985; Carroll et al., 1990; Tang et al., 1997a; and Zhao and Tang, this volume).

Up to 5 km of total Permian subsidence is recorded over a 40–50 m.y. span, yet the tectonic significance of this basin remains controversial. Hypotheses provided to explain Permian subsidence are numerous, often mutually incompatible, and reflect the paucity of first-hand geologic observations available to constrain the Permian tectonic setting. Proposed hypotheses range from foreland basin flexure (Watson et al., 1987; Carroll et al., 1995), to extension (Bally et al., 1986; Allen et al., 1991), to regional transtension (Allen et al., 1995).

STRATIGRAPHY

The Junggar and Turpan-Hami basins portray a complicated association of facies, particularly between

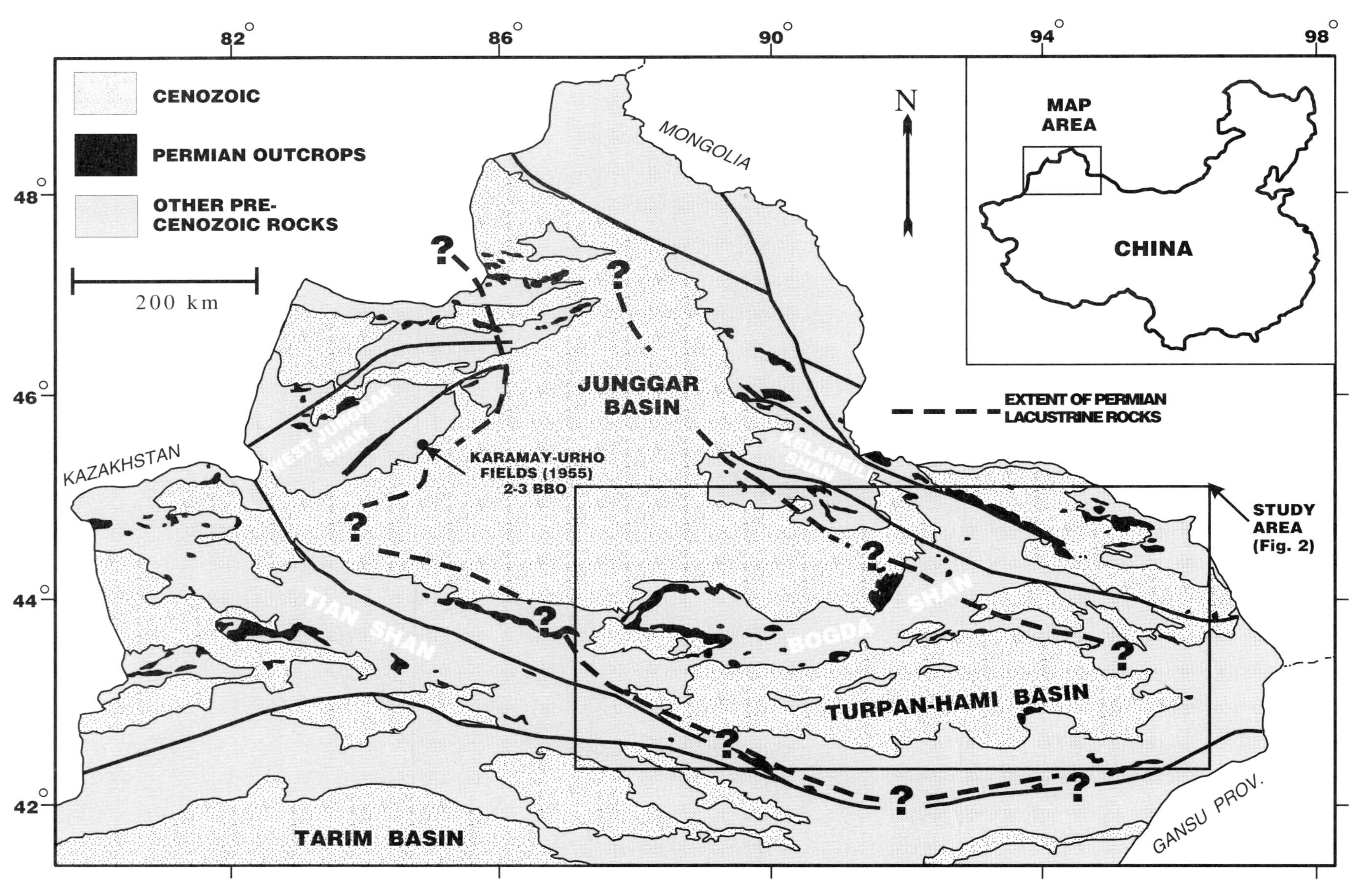

Figure 1—Simplified geologic map of the Junggar and Turpan-Hami basins illustrating the distribution of Permian outcrops (modified from Chen et al., 1985) and the known extent of Permian lacustrine facies. Box indicates the location of Figure 2.

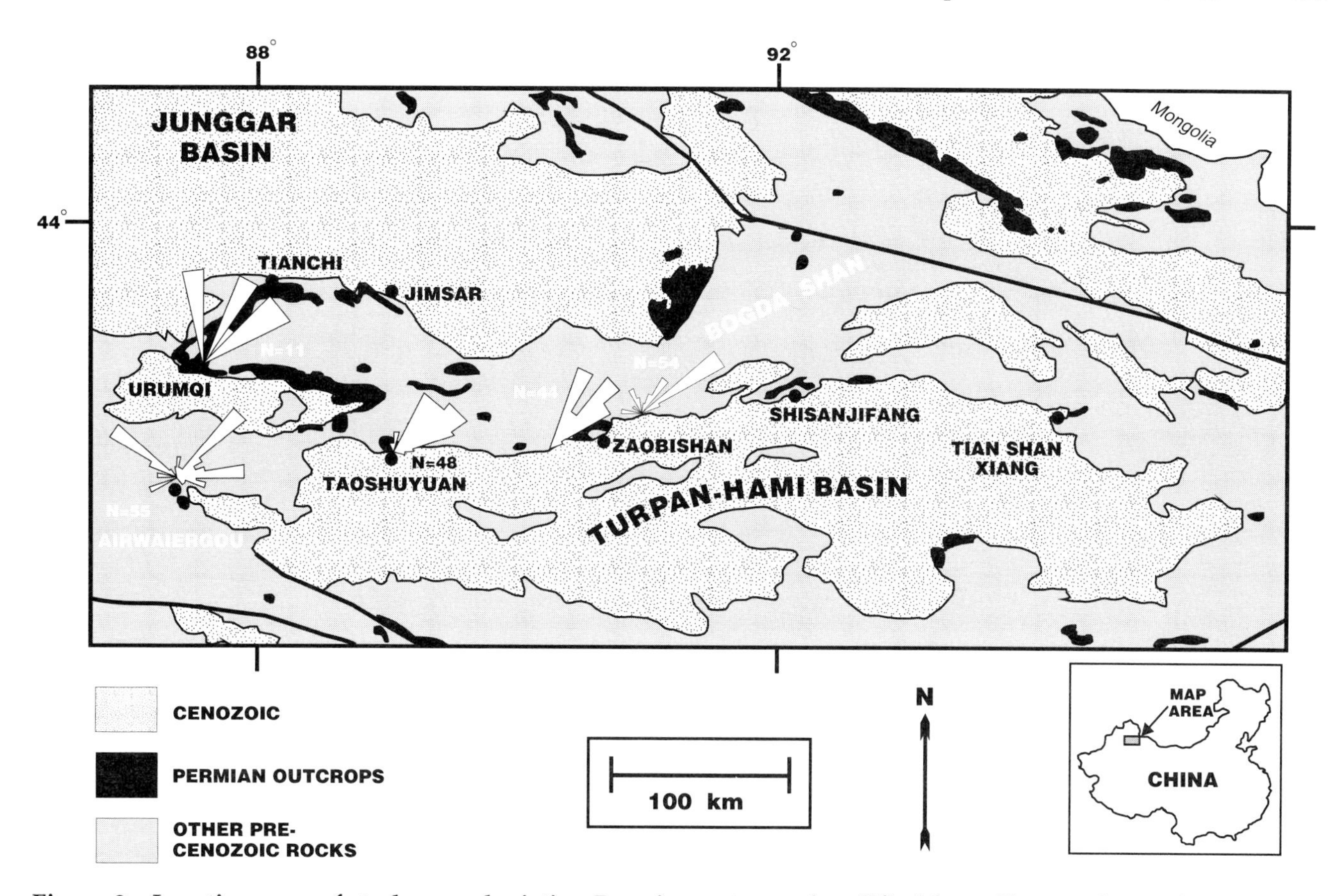

Figure 2—Location map of study area depicting Permian outcrops (modified from Chen et al., 1985). Stratigraphic details for named localities are shown in the regional correlation (Figure 3). Note paleocurrent data taken from clast imbrications and trough cross-beds indicating a north-northeasterly paleoflow (Carroll, 1991; Carroll et al., 1995; Greene et al., in press). These data are inconsistent with a Permian physiographic expression of the Bogda Shan, supporting a hypothesis for a united Permian lake basin spanning both the Junggar and Turpan-Hami basins.

lake marginal and basinal environments. In addition, precise understanding of the temporal framework for these deposits is limited and regional correlations, such as that shown in Figures 3 and 4, await further detailed work. Nonetheless, a synthesis of existing and new data are summarized in a basin-oblique west to east chronostratigraphic cross-section of the Permian (Figure 3). The stratigraphy and physical characteristics illustrated by the correlation allow for a simplified tripartite division of these deposits into phases of lacustrine development (Figure 3).

LOWER PERMIAN

The first phase of lacustrine deposition spans the Lower Permian and is the least well documented, having been noted in just two localities along the Bogda Shan (phase I on Figure 3). These deposits follow an overall marine regression marked by a diachronous retreat of marine waters from the region. Marine deposition continued in deeper portions of the basin, particularly near the south Junggar depocenter, well into the Early Permian (Carroll et al., 1990) (Figure 3).

An 80 m thick succession of lacustrine rocks of the Lower Permian Yierxitu Formation of the Aqibulake Group are exposed at Zaobishan (Greene et al., 1997; Wartes et al., 1999) (Figures 2, 3). The section consists of fine-grained sandstones, limestones, and dark gray calcareous mudstones organized in a series of parasequences 1–3 m thick. These cycles generally grade upward from stromatolitic limestone, to laminated mudstones, and are capped by a coarsening-upward succession of siltstone to fine-grained sandstone. This cyclicity is interpreted to reflect transgressive-regressive fluctuations in lake-level (Wartes et al., 1998), perhaps in response to varying climatic aridity. Several 20–30 m intervals of Lower Permian lacustrine facies have also been documented in the eastern end of the Turpan-Hami basin at Tian Shan Xiang (Figures 2, 3, 5). They are expressed as laminated, dark gray mudstone containing fish scales, algal lamination, and localized soft-sediment deformation (Figure 5).

The Early Permian age designation for these deposits is based largely on stratigraphic position; however, interbedded volcanic rocks occur at both localities (Figure 3). Future radiogenic age determinations of these igneous rocks will provide a critical control on the chronostratigraphy of Lower Permian lacustrine deposits.

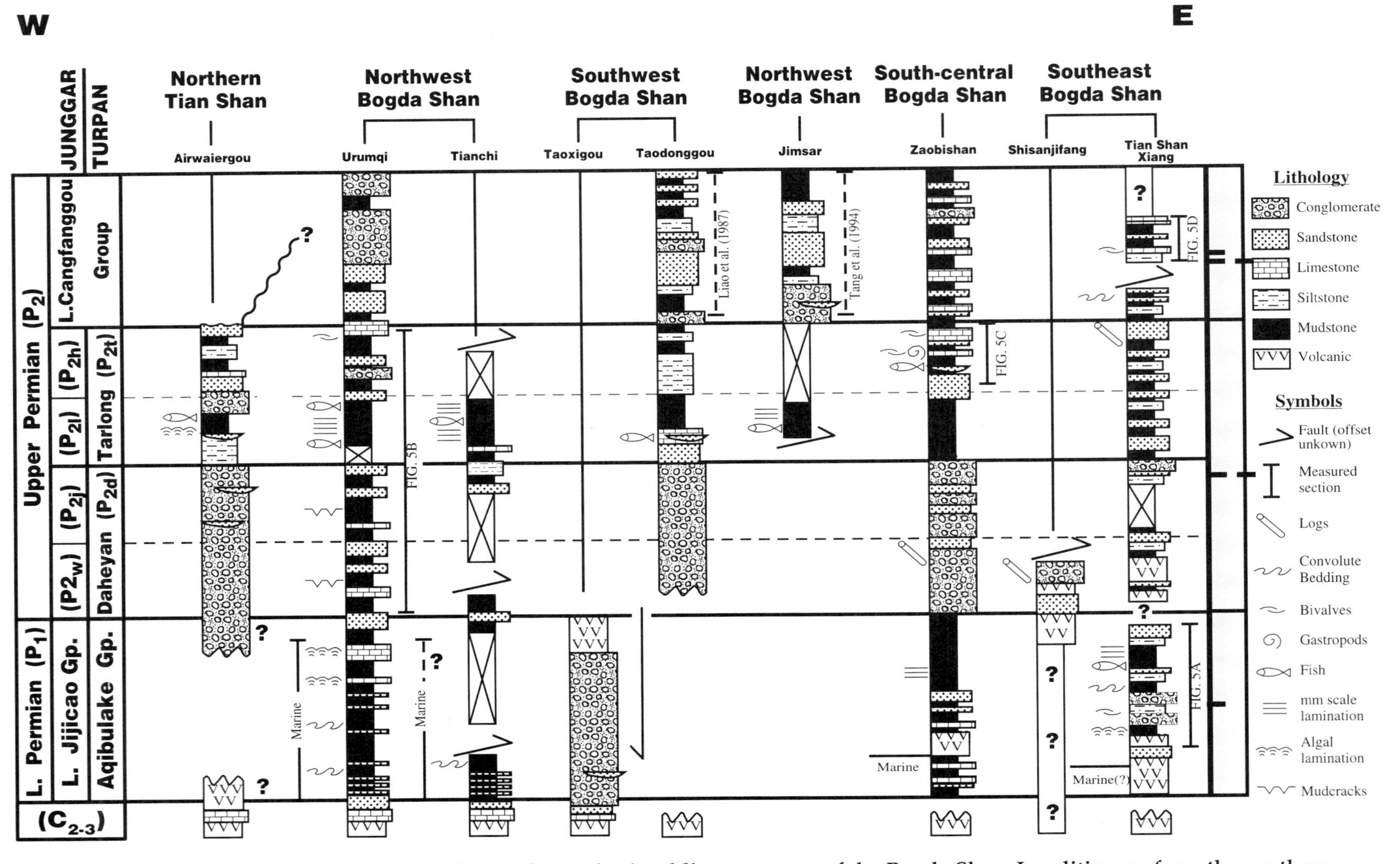

Figure 3—Permian chronostratigraphic correlation along a basin-oblique transect of the Bogda Shan. Localities are from the southern Junggar (northern Bogda Shan), western Turpan-Hami (northern Tian Shan), and northern Turpan–Hami (southern Bogda Shan) areas. Lacustrine development can be divided into three phases, which are shown at right as I, II, and III, and correspond to the organization of the text description. The phases are based on age relationships, sedimentary facies, lake type, and other diagnostic characteristics. Note the indicated stratigraphic location of detailed sections displayed in Figure 5. Distances between localities are not shown; refer to Figure 2 for true lateral separation. Formation subdivisions and regional nomenclature are shown in Figure 4.

System	Series	Turpan-Hami Basin		Southern Junggar Basin		Northeastern Junggar Basin	Northwestern Junggar Basin
PERMIAN	UPPER	Guodikeng	L. Cangfanggou Group	Guodikeng	L. Cangfanggou Group	Lower Cangfanggou Group	Lower Wuerhe
		Wutonggou		Wutonggou			
		Quanzijie		Quanzijie			Xiazhijie
		Tarlong	Taodonggou Group	Hongyanchi	Upper Jijicao Group	Pingdiquan	Fengcheng
				Lucaogou			
		Dayehen		Jingjingzigou		Jiangjunmiao	Jiamuhe
				Wulapo			
	LOWER	Yierxitu	Aqubulake Group	Tashikula	Lower Jijicao Group	Jingou	
		Shirenzigou (?)		Shirenzigou			

Figure 4—Subdivision and correlation of Permian units in the Junggar and Turpan-Hami basins (modified from Zhao, 1982; Chen et al., 1985; and Liao et al., 1987).

LOWER-UPPER PERMIAN

The next phase of lacustrine development (labeled II on Figure 3) is early-Late Permian in age and has received the most attention due to the extremely rich source rocks occurring in this interval along the northwest flank of the Bogda Shan (Graham et al., 1990; Carroll et al., 1992, 1998). Recent studies have begun to correlate these organic-rich facies with widespread, coeval lacustrine mudstones preserved south of the Bogda Shan in the Turpan-Hami basin (Greene et al., 1997; Wartes et al., 1998). The following description of this phase of lacustrine deposition will begin with rocks of the Upper Jijicao Group in the southern Junggar basin, followed by equivalent strata of the Taodonggou Group in the Turpan-Hami basin. This separation is warranted due to the differing formation names and, to a lesser degree, lithologic character (Figures 3, 4).

Upper Jijicao Group

The lower Upper Permian section of the southern Junggar begins in the Wulapo Formation with variable amounts of sandstone, rippled siltstone, and plane-laminated mudstone interpreted to represent shallow oxic lacustrine deposition interbedded with occasional fluvial deposits (Carroll et al., 1995). Fossils reported from the Wulapo are rare but include the bivalve *Palaeonodonta pseudolongissima* and flora such as *Walchia sp., Dadoxylon teilhardii,* and *Paracalamites sp.* (Liao et al., 1987; XPGEG, 1991).

The Wulapo Formation grades up into siltstones, mudstones, and fine-grained sandstones of the Jingjingzigou Formation. The Jingjingzigou Formation is characterized by highly visible outcrop cyclicity punctuated by periodic desiccation surfaces (Figure 5). Biomarker geochemistry of mudstone extracts record a specialized, salinity-tolerant biota, which, when considered in concert with other physical characteristics such as mudcracks and diagenetic dolomite, may suggest evaporative conditions (Carroll et al., 1992; Carroll, 1998). Hypersaline facies similar to the Jingjingzigou Formation appear to have generated the oils produced from the giant Karamay and related fields along the northwestern margin of the basin (Jiang and Fowler, 1986; Carroll et al., 1992; Clayton et al., 1997; Carroll, 1998). Another potential source for these oils is the slightly older Pingdiquan Formation of the northeastern Junggar basin (Tang et al., 1997a) (Figure 4). The Jingjingzigou Formation is slightly more fossiliferous than the Wulapo Formation with reports of numerous ostracode genera including *Tomiella, Darwinula,* and *Darwinuloides,* as well as palynomorphs such as *Cordaites sp.* (Zhang, 1981; XPGEG, 1991).

The Jingjingzigou Formation grades upward into the thick oil shales characteristic of the Lucaogou Formation (Figure 5). The Lucaogou Formation consists of more than 800 m of finely laminated, organic-rich mudstone. Dolomite also occurs as occasional nodular beds and as cement in very fine to fine-grained sandstones. Total organic carbon (TOC) commonly exceeds 20% in the richest facies and averages 4% over one 800 m interval (Carroll et al., 1992). Coeval organic-rich lacustrine facies of the Pingdiquan Formation (Figure 4) are the dominant source rocks in the subsurface of the northeastern Junggar basin (Peng and Zhang, 1989; Tang et al., 1997a), implying a widespread distribution of lacustrine strata correlative to the Lucaogou Formation. The organic richness, biomarker characteristics, and lack of bioturbation all suggest deposition in a deep, stratified lake with anoxic bottom waters. Deposits of the Lucaogou Formation differ from many other organic-rich lacustrine facies, however, in that they were deposited at a relatively high paleolatitude (39–43°N) (Sharps et al., 1992; Nie et al., 1993) and in low to moderately productive lakes (Carroll, 1998). The Lucaogou Formation is rich in fossils including fish, such as *Cichia sp., Tienchaniscus longipterus,* and *Turfania taoshuyuanensis*; bivalves *Anthraconauta pseudophilipsii* and *A. iljinskiensis*; ostracods *Darwinula parallela, D. monitoria, Darwinulides ornata, and Sinusuella polita*; and flora including *Rufloria (Neoggerathiopsis) derzavinii* and *Paracalamites sp.,* (Zhang, 1981; Liao et al., 1987; Carroll et al., 1992).

Overlying the Lucaogou is the Hongyanchi Formation, which includes nonlaminated mudstone, limestones, and conglomerates. Pebble imbrications from within the conglomerates exposed near Urumqi are the source of one of the paleocurrent insets in Figure 2. The Hongyanchi Formation is generally interpreted as being deposited in freshwater lakes that were associated with

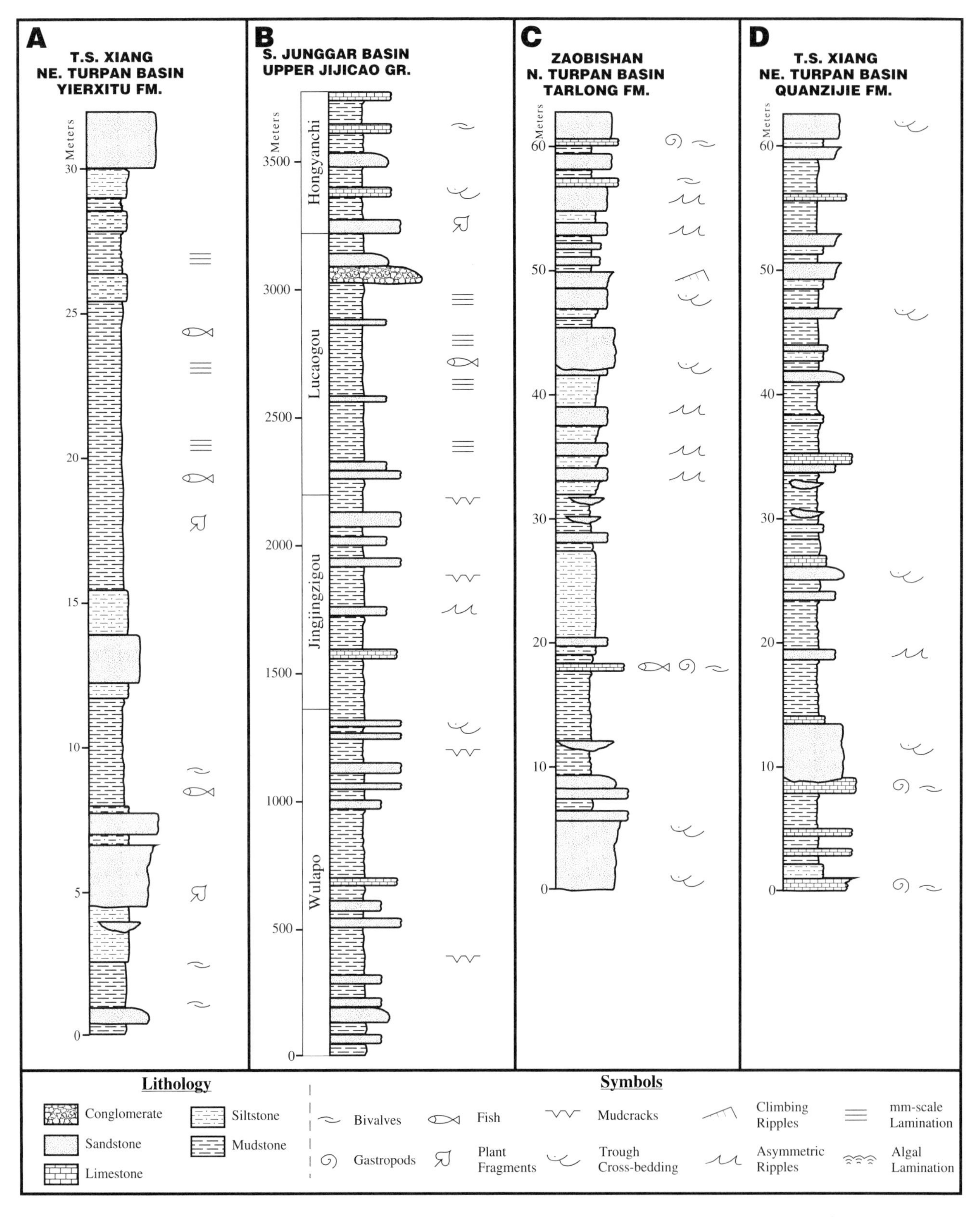

Figure 5—Representative sections of Permian lacustrine deposits in northwest China (see Figure 2 for locations and Figure 3 for stratigraphic position). All measured section thicknesses are expressed in meters. (A) Detailed measured section of the Lower Permian exposures of the Yierxitu Formation exposed in the easternmost Turpan-Hami basin at Tian Shan Xiang. (B) Schematic stratigraphy of the lower-Upper Permian upper Jijicao Group of the southern Junggar basin (modified from Liao et al., 1987; Carroll et al., 1992, 1995). (C) Detailed measured section of the upper portion of the lower Upper Permian Tarlong Formation of the Turpan-Hami basin exposed at Zaobishan along the south-central Bogda Shan. (D) Detailed measured section of the lower portion of the Upper Permian Quanzijie Formation of the lower Cangfanggou Group exposed in the southeastern Bogda Shan at Tian Shan Xiang.

fluvial systems (Carroll et al., 1992; Carroll, 1998). The Hongyanchi Formation is also fossiliferous containing the palynomorphs *Cordaitina* and *Hamiapollenites*; bivalves *Anthraconauta ilijinskiensis* and *Microdontella elliptica*; ostracodes such as *Darwinuleides ornata*; and the plant *Pecopteris anthriscifolia* (Yang et al., 1986; XPGEG, 1991).

Taodonggou Group

The Daheyen Formation, preserved farther south in the Turpan-Hami basin, is recorded by thick sections of conglomerates and minor sandstones (Figure 3). Fine-grained facies are very rare but have been reported from one well in the southern Turpan (Greene et al., 1997) and in the easternmost locality at Tian Shan Xiang (Figure 3). We interpret the Daheyen Formation to be approximately time-equivalent to the Wulapo and Jingjingzigou formations deposited farther north, although biostratigraphic verification of this correlation is still pending (Figure 3). The Daheyen Formation is often well organized with pebble imbrications and trough cross-stratification, both of which form the basis for paleocurrent measurements along the southern margin of the Bogda Shan (Figure 2). Sedimentologic features and stratigraphic style suggest deposition within braided fluvial systems derived from exposed Carboniferous rocks of the ancestral Tian Shan.

The Daheyen Formation grades up into the Tarlong Formation, which is well exposed in several localities along the southern flanks of the Bogda Shan (Figures 2, 3). The Tarlong is characterized by numerous coarsening upward cycles that begin with mudstones and siltstones and grade up through coarse sandstone or conglomerate (Figures 3, 5). This unit also contains several limestone intervals, many of which are fossiliferous, preserving bivalves, gastropods, and fish such as *Turfania taoshuyuanensis*. The Tarlong is interpreted as varying from profundal lacustrine facies to cyclic lake-marginal facies, which periodically shallow up into fluvial deposits (Figure 5) (Wartes et al., 1998). Several coarsening-upward packages are suggestive of prograding deltaic sedimentation. Biostratigraphic control in Tarlong Formation is also poor; however, the upper age is limited by overlying rocks of the Lower Cangfanggou Group, which contain well-described, age diagnostic fauna and flora (Yang et al., 1986; Liao et al., 1987).

Regional lower Upper Permian lake deposits, particularly those exposed on the northwest flank of the Bogda Shan, portray an overall decrease in salinity through time. This trend is indicated by early, more hypersaline deposits of the Jingjingzigou Formation which grade up into the deeper basinal laminated facies of the Lucaogou Formation, and finally up into more freshwater conditions recorded by the preservation of several intervals of fossiliferous freshwater limestones in the Hongyanchi and upper Tarlong Formations (Figure 3).

UPPER-UPPER PERMIAN

The final phase of lacustrine deposition occurs in the uppermost Permian Quanzijie, Wutonggou, and Guodikeng formations of the lower Cangfanggou Group (phase III on Figure 3). As noted by Tang et al. (1994), the lower Cangfanggou Group is a highly variable lithostratigraphic unit. Conglomerate intervals range from organized to weakly organized deposits and are often associated with pebbly sandstones. Well-bedded sandstones, rippled siltstones, and thinly bedded carbonates are common (Figure 5). A variety of fine-grained intervals are also present throughout the lower Cangfanggou Group. The heterogeneous lithologies reflect a variety of depositional environments and flow regimes, including alluvial, braided and meandering fluvial, and shallow lacustrine (Yang et al., 1986; Zhao et al., 1991; Tang et al., 1994, 1997b). This phase of lake development was dominated by freshwater lacustrine systems, effectively continuing the trend of decreasing salinity noted in the previous interval. Correlation of these strata from both sides of the Bogda Shan is facilitated by the best developed biostratigraphy of all three phases. In particular, many Chinese researchers have focused on characterizing conformable Permian–Triassic boundary sections that are well-exposed along the northern flank of the Bogda Shan (e.g., Liu, 1993; Cheng, 1993), compiling impressive fossil lists consisting of dozens of flora and fauna. Typical fossils include a diverse assemblage of flora, such as *Callipteris zeilleri, Comia dentata, Iniopteris sibirica,* and *Pecopteris anthriscifolia*; bivalves *Palaeonodonta psuedolongissima, P. parallela, P. fischeri,* and *Microdontella plotnikovskiensis*; ostracodes such as *Darwinula implana, D. kuznetskiensis,* and *Darwinuloides dobrinkaensis*; and several Late Permian vertebrates including *Dicynodon tienshanensis, Jimusaria sinkiangensis,* and *Kunpania scopulusa* (see Yang et al., 1986, for complete listing).

DISCUSSION

Rapid Permian subsidence allowed for the preservation of areally extensive lacustrine deposits covering approximately 200,000 km^2 of central Asia. These lake deposits reach thicknesses approaching 2000 m and thus represent one of the most impressive, yet least well understood, accumulations of lacustrine strata on Earth. The association of Lower Permian lacustrine facies with volcanic rocks (see Figure 3) is most consistent with rift-related basin subsidence during this time; however, the thickest nonmarine deposits, presumably responding to the maximum rate of subsidence, were deposited during the Late Permian. Preliminary thermal history modeling based on stratigraphic thicknesses and vitrinite reflectances in the northwest Bogda Shan does not support the existence of elevated Permian heat flows anticipated in a rift environment (Carroll et al., 1992; see also King et al., 1994). Finally, folds exposed beneath a sub-Middle Triassic angular unconformity in the western Turpan basin appear to have been compressional in origin (Wartes et al., 1998). These apparent contradictions suggest a complex basin history, possibly related to regional strike-slip tectonics as suggested by Allen et al. (1995).

ACKNOWLEDGMENTS

This research was supported by a grant from the Donors of The Petroleum Research Fund, administered by the American Chemical Society. Additional support was provided by the Stanford-China Industrial Affiliates, including Agip, Arco, Chevron, Exxon, JNOC, Mobil, Phillips, Shell, Statoil, Texaco, and Triton. We also acknowledge funding from the Graduate School at the University of Wisconsin-Madison. We are grateful for the insightful advice and assistance of S. A. Graham. We thank A. Hessler and Yongjun Yue for their valuable field assistance. This paper benefited from reviews by K. Kelts and J. Parnell.

REFERENCES CITED

Allen, M. B., B. F. Windley, C. Zhang, Z. Y. Zhao, and G. R. Wang, 1991, Basin evolution within and adjacent to the Tien Shan range, NW China: Journal of the Geological Society, London, v. 148, p. 369–378.

Allen, M. B., A. M. C. Sengor, and B. A. Natal'in, 1995, Junggar, Turfan and Alakol basins as Late Permian to ?Early Triassic extensional structures in a sinistral shear zone in the Altaid orogenic collage, central Asia: Journal of the Geological Society, London, v. 152, p. 327–338.

Bally, A. W., I. M. Chou, R. Clayton, H. P. Eugster, S. Kidwell, L. D. Meckel, R. T. Ryder, A. B. Watts, and A. A. Wilson, 1986, Notes on sedimentary basins in China-Report of the American Sedimentary Basins Delegation to the People's Republic of China: United States Geological Survey Open-File Report 86–327, 108 p.

Carroll, A. R., Y. Liang, S. A. Graham, X. Xiao, M. S. Hendrix, J. Chu, and C. L. McKnight, 1990, Junggar basin, northwest China: trapped late Paleozoic ocean: Tectonophysics, v. 181, p. 1–14.

Carroll, A. R., 1991, Late Paleozoic tectonics, sedimentation, and petroleum potential of the Junggar and Tarim basins, northwest China: Ph.D. dissertation, Stanford University, Stanford, California, 405 p.

Carroll, A. R., S. C. Brassell, and S. A. Graham, 1992, Upper Permian lacustrine oil shales, southern Junggar basin, northwest China: AAPG Bulletin, v. 76, p. 1874–1902.

Carroll, A. R., S. A. Graham, M. S. Hendrix, D. Ying, and D. Zhou, 1995, Late Paleozoic tectonic amalgamation of northwestern China: sedimentary record of the northern Tarim, northwestern Turpan, and southern Junggar basins: GSA Bulletin, v. 107, p. 571–594.

Carroll, A. R., 1998, Upper Permian organic facies evolution, southern Junggar Basin, NW China: Organic Geochemistry, v. 28, p. 649–667.

Clayton, J. L., J. Yang, J. D. King, P. G. Lillis, and A. Warden, 1997, Geochemistry of oils from the Junggar basin, northwest China: AAPG Bulletin, v. 81, p. 1926–1944.

Chen, Z., N. Wu, D. Zhang, J. Hu, H. Huang, G. Shen, G. Wu, H. Tang, and Y. Hu, 1985, Geologic map of Xinjiang Uygur Autonomous Region, Scale 1:2,000,000: Beijing, Geologic Publishing House.

Cheng, Z., 1993, On the discovery and significance of the nonmarine Permo–Triassic transition zone at Dalongkou in Jimusar, Xinjiang, China, *in* S. G. Lucas and M. Morales, eds., The Nonmarine Triassic: New Mexico Museum of Natural History & Science Bulletin No. 3, p. 65–67.

Demaison, G., and B. J. Huizinga, 1991, Genetic classification of petroleum systems: AAPG Bulletin, v. 75, p. 1626–1643.

Graham, S. A., S. Brassell, A. R. Carroll, X. Xiao, G. Demaison, C. L. McKnight, Y. Liang, J. Chu, and M. S. Hendrix, 1990, Characteristics of selected petroleum source rocks, Xinjiang Uygur Autonomous Region, northwest China: AAPG Bulletin, v. 74, p. 493–512.

Greene, T. J., A. R. Carroll, and M. S. Hendrix, 1997, Permian–Triassic basin evolution and petroleum system of the Turpan-Hami basin, Xinjiang Province, Northwest China: AAPG Annual Meeting with Abstracts, v. 6, p. 42.

Greene, T. J., M. S. Hendrix, A. R. Carroll, and S. A. Graham, in press, Middle to Late Triassic uplift of the Bogda Shan: constraints on initial partitioning of the Turpan-Hami and southern Junggar Basins, Xinjiang Uygur Autonomous Region, NW China, *in* M. S. Hendrix and G. A. Davis, eds., Paleozoic and Mesozoic tectonic evolution of central Asia—from continental assembly to intracontinental deformation: GSA Special Paper.

Hendrix, M. S., S. A. Graham, A. R. Carroll, E. R. Sobel, C. L. McKnight, B. J. Schulein, and Z. Wang, 1992, Sedimentary record and climatic implications of recurrent deformation in the Tian Shan: evidence from Mesozoic strata of the north Tarim, south Junggar, and Turpan basins, northwest China: GSA Bulletin, v. 194, p. 53–79.

Jiang, Z., and M. G. Fowler, 1986, Carotenoid-derived alkanes in oils from northwestern China: Organic Geochemistry, v. 10, p. 831–839.

King, D. J., J. Yang, and F. Pu, 1994, Thermal history of the periphery of the Junggar basin, northwestern China: Organic Geochemistry, v. 21, p. 393–405.

Lawrence, S. R., 1990, Aspects of the petroleum geology of the Junggar basin, northwest China, *in* J. Brooks, ed., Classic petroleum provinces: GSA Publication no. 50, p. 545–557.

Lee, K. Y., 1985. Geology of the petroleum and coal deposits in the Junggar basin, Xinjiang Uygur Zizhiqu, northwest China: United States Geological Survey Open-File Report 85-230, 53p.

Liao, Z., L. Lu, N. Jiang, F. Xia, F. Sung, Y. Zhou, S. Li, and Z. Zhang, 1987, Carboniferous and Permian in the western part of the east Tianshan mountains: Eleventh Congress of Carboniferous Stratigraphy and Geology, Guide Book Excursion 4, 50 p.

Liu, S., 1993, Some Permian–Triassic Conchostracans from the mid-Tianshan mts. of China and the significance of their geological dating, *in* S. G. Lucas and M. Morales, eds., The Nonmarine Triassic: New Mexico Museum of Natural History & Science Bulletin No. 3, p. 277–278.

Nie, S., D. B. Rowley, R. Van der Voo, and M. Li, 1993, Paleomagnetism of late Paleozoic rocks in the Tianshan, northwestern China: Tectonics, v. 12, p. 568–579.
Peng, X., and G. Zhang, 1989, Tectonic features of the Junggar basin and their relationship with oil and gas distribution, *in* K. J. Hsu, and X. Zhu, eds., Chinese sedimentary basins: Amsterdam, Elsevier, p. 17–31.
Shao, L., K. Stattegger, L. Wenhou, and B. J. Haupt, 1999, Depositional style and subsidence history of the Turpan Basin (NW China): Sedimentary Geology, v. 128, p. 155–169.
Sharps, R., Y. Li, M. McWilliams, and Y. A. Li, 1992, Paleomagnetic investigations of Upper Permian sediments in the south Junggar basin, China: Journal of Geophysical Research, v. 97, p. 1753–1765.
Tang, Z., J. Parnell, and A. H. Ruffell, 1994, Deposition and diagenesis of the lacustrine-fluvial Cangfanggou Group (uppermost Permian to Lower Triassic), southern Junggar basin, NW China: a contribution from sequence stratigraphy: Journal of Paleolimnology, v. 11, p. 67–90.
Tang, Z., J. Parnell, and F. J. Longstaffe, 1997a, Diagenesis of analcime-bearing reservoir sandstones: the Upper Permian Pingdiquan Formation, Junggar basin, northwest China: Journal of Sedimentary Research, v. 67, p. 486–498.
Tang, Z., J. Parnell, and F. J. Longstaffe, 1997b, Diagenesis and reservoir potential of Permian–Triassic fluvial/lacustrine sandstones in the southern Junggar basin, northwestern China: AAPG, v. 81, p. 1843–1865.
Wartes, M. A., T. J. Greene, and A. R. Carroll, 1998, Permian lacustrine paleogeography of the Junggar and Turpan–Hami basins, northwest China: AAPG Annual convention, Extended Abstracts, v. 2, p. A682.
Wartes, M. A., A. R. Carroll, and T. J. Greene, 1999, Permian sedimentary evolution of the Junggar and Turpan-Hami basins, northwest China [abs.]: GSA Abstracts with Programs.
Watson, M. P., A. B. Hayward, D. N. Parkinson, and Z. M. Zhang, 1987, Plate tectonic history, basin development and petroleum source rock deposition onshore China: Marine and Petroleum Geology, v. 4, p. 205–225.
XPGEG (Xinjiang Petroleum Geology Editing Group), 1991, Petroleum geology of China volume 15, part I, Junggar basin: Beijing Petroleum Industry Press, 390 p.
Yang, J., L. Qu, H. Zhou, Z. Cheng, T. Zhou, J. Hou, P. Li, S. Sun, Y. Li, Y. Zhang, X. Wu, Z. Zhang, and Z. Wang, 1986, Permian and Triassic strata and fossil assemblages in the Dalongkou area of Jimsar, Xinjiang, Ministry of Geology and Mineral Resources Geologic Memoir, Series 2, No. 3: Beijing, Geological Publishing House, 262 p. (in Chinese with English summary).
Zhang, X., 1981, Regional stratigraphic chart of northwestern China, branch of Xinjiang Uygur Autonomous Region: Beijing, Geological Publishing House, 496 p. (in Chinese).
Zhao, B., 1982, The prospects of petroleum exploration of Permo–Carboniferous in the Junggar Basin. Oil and Gas Geology, v. 3, p. 1–14 (in Chinese with English abstract).
Zhao, X., Z. Lou, and Z. Chen, 1991, Response of alluvial-lacustrine deposits of the Permian–Triassic Cangfanggou Group to climate and tectonic regime in Dalonggou, Junggar basin, Xinjiang, *in* K. R. Kelts and C. Yang, eds., Comparative lacustrine sedimentation in China: Part 2, China Earth Sciences, v. 1, p. 343–354.

Mángano, M. G., L. A. Buatois, Wu Xiantao, Sun Junmin, and Zhang Guocheng, 2000, Triassic lacustrine sedimentation from the Tanzhuang Formation, Jiyuan-Yima basin, southeastern China, *in* E. H. Gierlowski-Kordesch and K. R. Kelts, eds., Lake basins through space and time: AAPG Studies in Geology 46, p. 133–140.

Chapter 9

Triassic Lacustrine Sedimentation from the Tanzhuang Formation, Jiyuan-Yima Basin, Southeastern China

María Gabriela Mángano
Luis Alberto Buatois
CONICET Instituto Superior de Correlación Geológica
Facultad de Ciencias Naturales, Universidad Nacional de Tucumán
Tucumán, Argentina

Wu Xiantao
Sun Junmin
Zhang Guocheng
Jiaozuo Institute of Technology
Jiaozuo City, Henan Province, People's Republic of China

INTRODUCTION AND REGIONAL SETTING

The Tanzhuang Formation represents the uppermost Triassic unit in the Jiyuan-Yima basin. It is well exposed southwest of Jiyuan City, western Henan Province, central China (Figure 1). The Jiyuan-Yima basin, which is part of the Ordos megabasin, contains about 1500 m of Middle Triassic–Middle Jurassic continental sediments. This basin is divided into two subbasins, Jiyuan and Yima, located north and south of the Yellow River, respectively. Stratigraphy and paleontologic content of the deposits have been summarized by Zhou and Li (1980), Kang et al. (1984, 1985), and Hu (1991) (Figure 2). Sedimentologic and stratigraphic evidence suggest that Jurassic sedimentation took place in a pull-apart basin (Buatois et al., 1994a, this volume); however, the Triassic tectonic framework is still poorly understood. Triassic deposits of the northern Jiyuan area and southern Yima area show significant changes in facies associations. These facies variations may have resulted from asymmetry of the basin. Triassic lacustrine sediments were most likely developed within a post-collisional intracratonic downwarp basin that formed as a result of early Indo-Sinian tectonic movements during the Middle Triassic. Strike-slip tectonism occurred in the study area coincident with the Triassic–Jurassic boundary (cf. Wang, 1985; Mángano et al., 1994; Buatois et al., 1994a).

The Tanzhuang Formation consists of mudstones, carbonaceous shales, thin-bedded carbonates, siltstones, and sandstones that host a *Danaeopsis-Bernoulia* flora considered to be of Late Triassic age (Hu, 1991); however, palynologic analysis of the Jiyuan section suggests a late Middle–early Late Triassic age for the upper part of the Tanzhuang Formation (G. Ottone, personal communication, 1993). The upper part of this formation (about 180 m) has been interpreted as a shallow permanent lake according to facies associations and paleontologic data (Mángano et al., 1994). The lower part of the Tanzhuang Formation has been interpreted as comprising alluvial plain deposits by previous Chinese geologists. The Tanzhuang Formation (Figure 2) is underlain by the Chuenshuyao Formation (fluvial and alluvial plain materials with interbedded lacustrine intervals) and is paraconformably overlain by the Jurassic Anyao Formation, which consists of deep lacustrine turbidites (Wu, 1985; Hu, 1991; Buatois et al., 1994b, this volume).

SEDIMENTARY FACIES AND FACIES ASSOCIATIONS

Mángano et al. (1994) recognized six sedimentary facies in the upper part of the Tanzhuang Formation: (A) laminated siltstones, (B) current-ripple-laminated sandstones, (C) interbedded fine-grained carbonates and siltstones, (D) deformed and wave-reworked sandstones, (E) organic-rich shales, and (F) clayey mudstones (Figure 3).

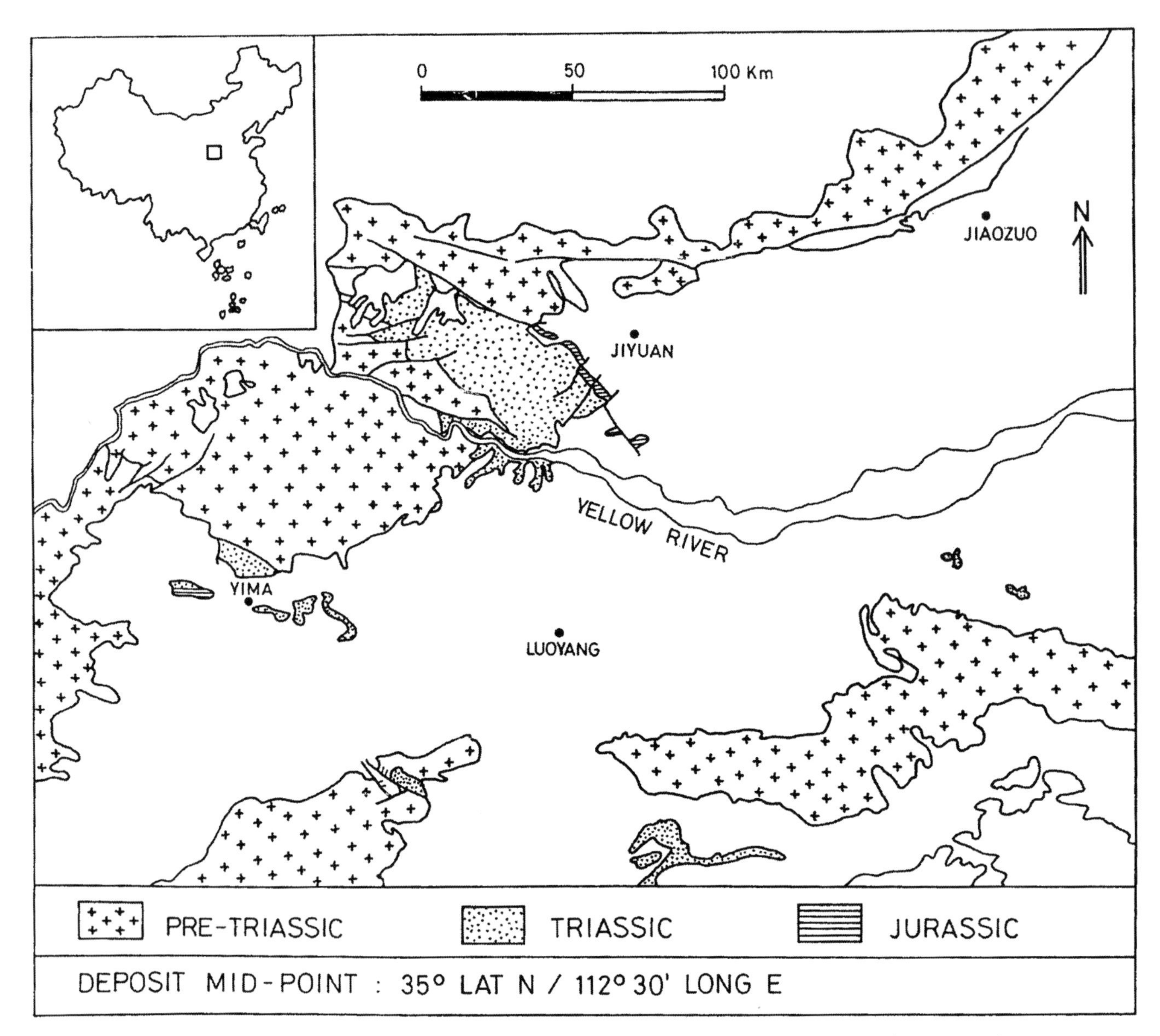

Figure 1—Geologic map of the Triassic lacustrine deposits of the Jiyuan-Yima basin of eastern China.

Facies A: Laminated Siltstones

Facies A is formed by laterally continuous, horizontally laminated, greenish-gray siltstones, forming units up to 5 m thick, which host a well-preserved ostracode fauna dominated by *Darwinula* (Wu Linyouling, personal communication, 1992). Delicate rootlets, plant debris, and coaly horizons are present locally. Absence of current and wave-induced structures and the fine grade of the material involved indicate suspension fallout in a low-energy setting. The presence of rooted coaly horizons suggests that part of these deposits record sedimentation in interdistributary bay areas.

Facies B: Current-Ripple-Laminated Sandstones

Sandstone units, 0.6–8 m thick, lenticular at the scale of several tens of meters form facies B. Facies A commonly grades upward into facies B sandstones, which usually coarsens upward from very fine- to medium- or, exceptionally, coarse-grained sandstones. An ideal sequence of sedimentary structures within this facies shows current-ripple or climbing-ripple lamination in the finer lower part. This passes up into planar lamination and culminates with planar cross-bedded or trough cross-bedded sandstones. Trace fossils (*Arenicolites* isp.), fish scales, and plant-rich horizons are commonly found at the top of the bedset. In some cases, mudcracks were detected at the silty top of the sandstone unit. Facies B is interpreted as comprising lacustrine deltaic mouthbar deposits. Trough cross-bedded sandstones at the top of the coarsening-upward package most likely record small distributary channels.

Facies C: Interbedded Fine-Grained Carbonates and Siltstones

Facies C occurs as interbedded yellowish-white carbonate laminae (0.05–0.4 cm thick) and grayish-green

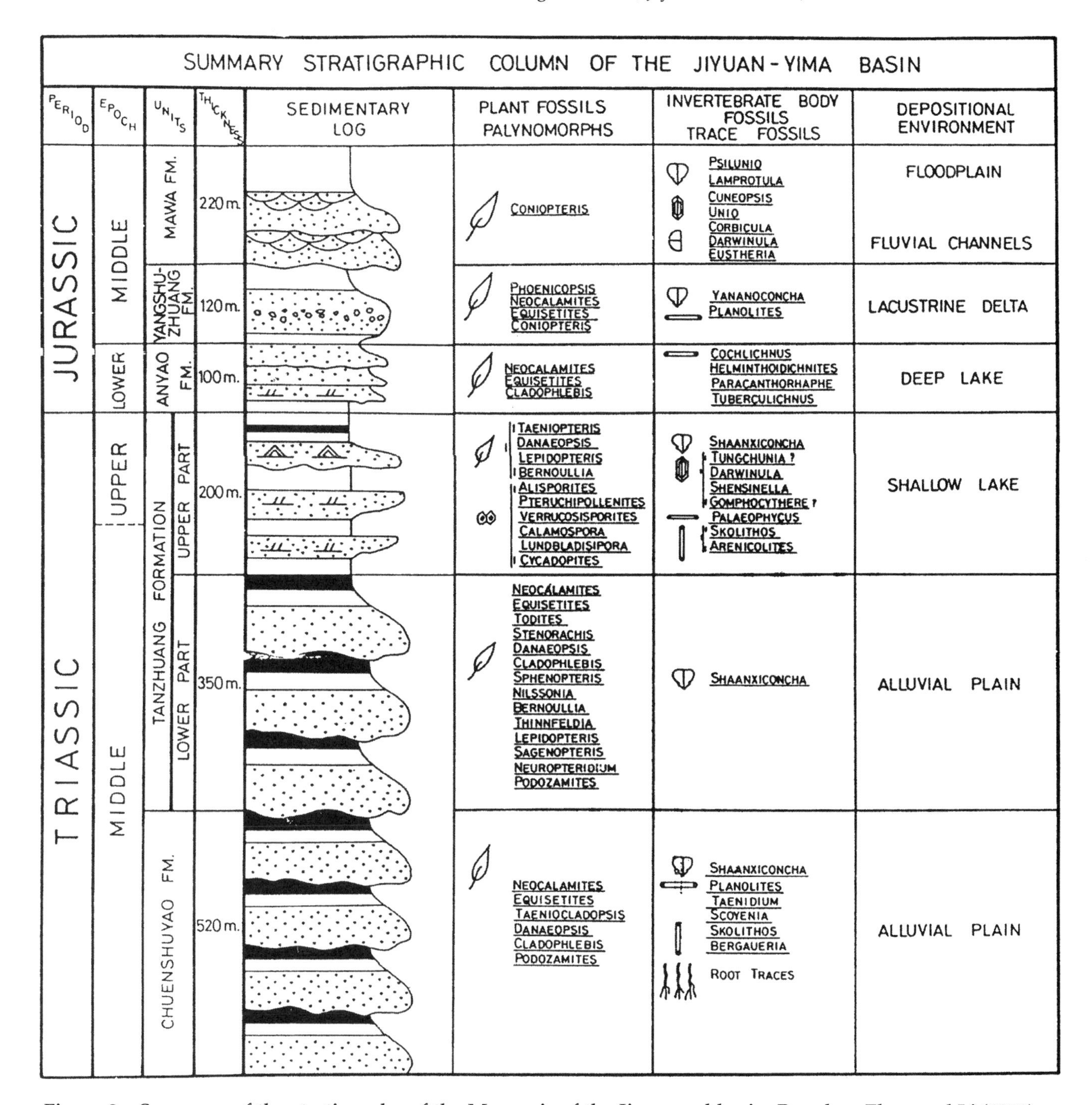

Figure 2—Summary of the stratigraphy of the Mesozoic of the Jiyuan subbasin. Based on Zhou and Li (1980), Kang et al. (1984, 1985), Wu (1985), and Hu (1991). See Figure 4 for legend for sedimentary log.

siltstones (0.5–1.5 cm thick), forming intervals up to 2.6 m thick. Presence of these carbonate and interbedded siliciclastic sediments perhaps reflects dramatic seasonal changes in carbonate concentration or siliciclastic to carbonate detritus influx, with periodic carbonate precipitation in low-energy settings. Persistent lamination and paucity of bioturbation suggest sedimentation under oxygen-depleted conditions. Fine-grained carbonate-siltstone siliciclastic couplets may reflect strong seasonality, with periods of extremely low detritus influx (cf. Platt and Wright, 1991). Facies C is associated with facies D (sediment gravity flow deposits).

Facies D: Deformed and Wave-Reworked Sandstones

Facies D consists of laterally persistent, syndepositionally deformed, load-based sandstone units, up to 1.8 m thick. Individual beds range in thickness from a few centimeters to 0.7 m. An ideal facies D sandstone bed is formed by three divisions: a parallel-laminated

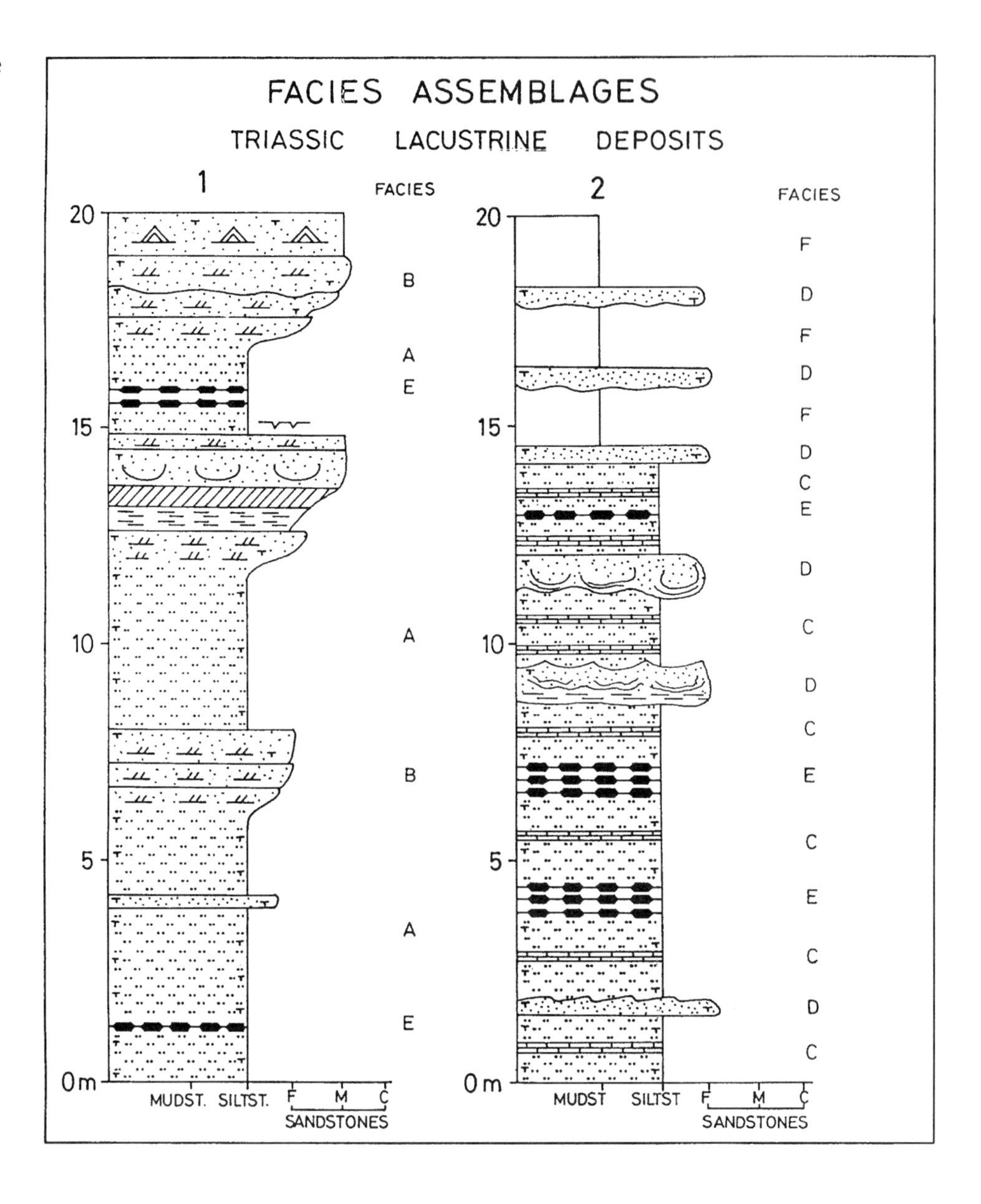

Figure 3—Facies assemblage 1 and 2 of the Triassic lacustrine deposits of the Tanzhuang Formation of the Jiyuan-Yima basin of China. For composition of the facies assemblage and the facies constituents, see text. See Figure 4 for legend for sedimentary log.

basal division that passes upward into a deformed interval and culminates with a wave-ripple-laminated division capped by oscillatory ripples. Incomplete variants are the rule rather than the exception and, in many cases, this facies is only represented by a load-based, deformed sandstone bed. Post-depositional trace fossils (*Skolithos* isp., *Palaeophycus* isp.) occur at the top of some sandstones. Facies D was probably deposited from episodic sheet floods entering into a shallow lacustrine setting, where they transformed into turbidity currents (after Smoot, 1985).

Facies E: Organic-Rich Shales

Facies E is formed by dark gray to black, thinly laminated organic-rich shales, which occur in packages from 2 to 50 cm thick. According to palynologic content and deposit thickness, two distinct types of organic-rich shales (ORS) can be distinguished. Type I-ORS consists of thin packages of shales (2–20 cm thick), which contain a palynomorph assemblage dominated by bisaccate pollen grains of pteridosperms and coniferophytes (75–85%). Spores of pteridophytes, sphenophytes, and lycophites; monosulcate pollen grains; and *Botryococcus* sp. are minor components. Wood remains and cuticles are also common (G. Ottone, written communication, 1993). Total organic carbon is 1.08 (TOC; wt % of dry sediment) and thermal maturity, estimated from vitrinite reflectance, is 0.93% (H. Villar, written communication, 1993). Type II-ORS is formed by packages of shales (15–50 cm thick) that host a monotonous palynomorph assemblage composed of the green algae *Botryococcus* sp. (>90%). *Tasmanites* sp., spores, cuticles, and wood remains are minor constituents (E. G. Ottone, written communication, 1993). Total organic carbon is 1.87 (TOC; wt % of dry sediment) and thermal maturity, estimated from vitrinite reflectance, is 0.84%. Sedimentologic features of the carbonaceous shale facies suggest fallout deposition in a relatively quiet basin under anoxic or near-anoxic bottom conditions.

Facies F: Clayey Mudstones

Facies F consists of packages of structureless light gray clayey mudstones, 0.6–2.0 m thick, and occurs in the uppermost part of the Tanzhuang Formation. It is characterized by the presence of plant fragments. Fine grain size indicates suspension deposition. Poorly exposed outcrops and the absence of primary features preclude a reliable interpretation, though a very shallow water environment, such as a mudflat, is the most likely scenario.

FACIES ASSOCIATIONS AND VERTICAL DISTRIBUTION

The facies described can be clustered into two distinct facies assemblages (Figure 3). Detailed vertical sections of the Tanzhuang Formation, showing the relative proportions of the various facies, were presented by Mángano et al. (1994). Facies assemblage 1 is characterized by 2–8 m thick coarsening-upward sequences that consist, from base to top, of type I organic-rich shales, laminated siltstones (facies A), and fine- to medium-, exceptionally coarse-grained sandstones (facies B). Vertical variability of assemblage 1 packages is shown in Figure 4. Coarsening-upward successions are commonly stacked, forming two stratigraphic intervals (at 0–35 m and 71–95 m) (Mángano et al., 1994) within the lacustrine succession in the type section of the Tanzhuang Formation. Each succession represents a parasequence that records the progradation of mouthbar deltas in a hydrologically open lake basin. The base and top of each succession are sharp and record flooding episodes.

Facies assemblage 2 consists of thick packages of type II organic-rich shales (facies E) interbedded with fine-grained carbonates and cohesive siltstones (facies C) punctuated by event sandstone beds (facies D). Toward the top of the succession, an interval dominated by clayey mudstones (facies F) occurs. In the type section, this assemblage forms two distinctive stratigraphic intervals (at 35–71 m and 95–175 m) (Mángano et al., 1994), but lacks a clear vertical facies pattern. The bulk of the assemblage represents sedimentation in a stratified lake below wave base level; however, the presence of wave-ripple sandstones shows that part of the event sediments were deposited above wave base level.

DISCUSSION

The existence of deep lacustrine turbidites in the Jurassic Anyao Formation overlying shallow perennial lacustrine facies of the Triassic Tanzhuang Formation suggests a rapidly subsiding basin, most likely in a transtensile regime linked to tectonism linked to the Triassic–Jurassic boundary. The Jurassic lake was finally later filled with deltaic and fluvial siliciclastics under a subsequent transpressive regime (Buatois et al., 1994a).

The upper part of the Tanzhuang Formation is interpreted as having formed in a shallow perennial open lake. Alternation of both types of facies assemblages is thought to reflect climatic modulation because no clear evidence of tectonic activity has been identified in the Triassic. Stratigraphic recurrence of assemblage 1 suggests humid temperate climatic phases, whereas assemblage 2 probably records drier climatic phases, characterized by strong periodicity marked by rainfall and runoff into the lake. This paleoclimatic pattern suggests monsoonal influence within the paleogeographic framework of Late Triassic Pangaea. Mángano et al. (1994) noted that several tectonic events could have promoted a monsoonal influence in the Ordos megabasin during the Middle–Late Triassic, including connection of the south and north China block to form a single continental mass, closure of the northern branch of the Tethys, and elevation of orographic chains. Strong climatic periodicity evidenced toward the middle and upper part of the Tanzhuang Formation most likely records the influence of Pangaean megamonsoons.

Estimations based on the thickness of the asymmetric cycles of assemblage 1 suggest water depths of about 15 m previous to mouthbar progradation. Mouthbar deltas suggest hyperpicnal or friction-dominated flows. The underflow would have carried sediment away from the point of discharge, preventing the development of steep foresets (Gilbert type deltas), and resulted in bed-load deposition away from the shoreline. The palynologic content of type I-ORS indicates a significant extrabasinal supply, which is consistent with the presence of river-derived detritus in the associated facies.

Existence of deformed sandstone bodies within facies assemblage 2 suggests the action of episodic floods associated with catastrophic rainfalls (inundites). Absence of evaporite deposits and well-developed short-term transgressive-regressive successions argues against a hydrologically closed lake interpretation, however (Smoot, 1985). Presence of relatively thick organic-rich shale deposits suggests the existence of a permanent oxygen-depleted layer. This probably records incomplete turnovers of a stratified, meromictic lake. Palynologic content of type II-ORS may reflect high endogenic algal production (*Botryococcus* sp.), probably enhanced by a low fluvial detritus influx during periods between floods. This fact is also supported by the absence of terrigenous prograding successions in facies association 2.

ACKNOWLEDGMENTS

We would like to thank the Chinese Academy of Sciences and the Third World Academy of Sciences (TWAS), who supported the research visit to China of Mángano and Buatois. We also would like to express our gratitude to Guillermo Ottone for palynologic information, Wu Linyouling for ostracode identifications, and Hector Villar for organic geochemistry analysis. The paper has benefited from comments by Elizabeth Gierlowski-Kordesch and two anonymous reviewers. Illustrations were kindly made by

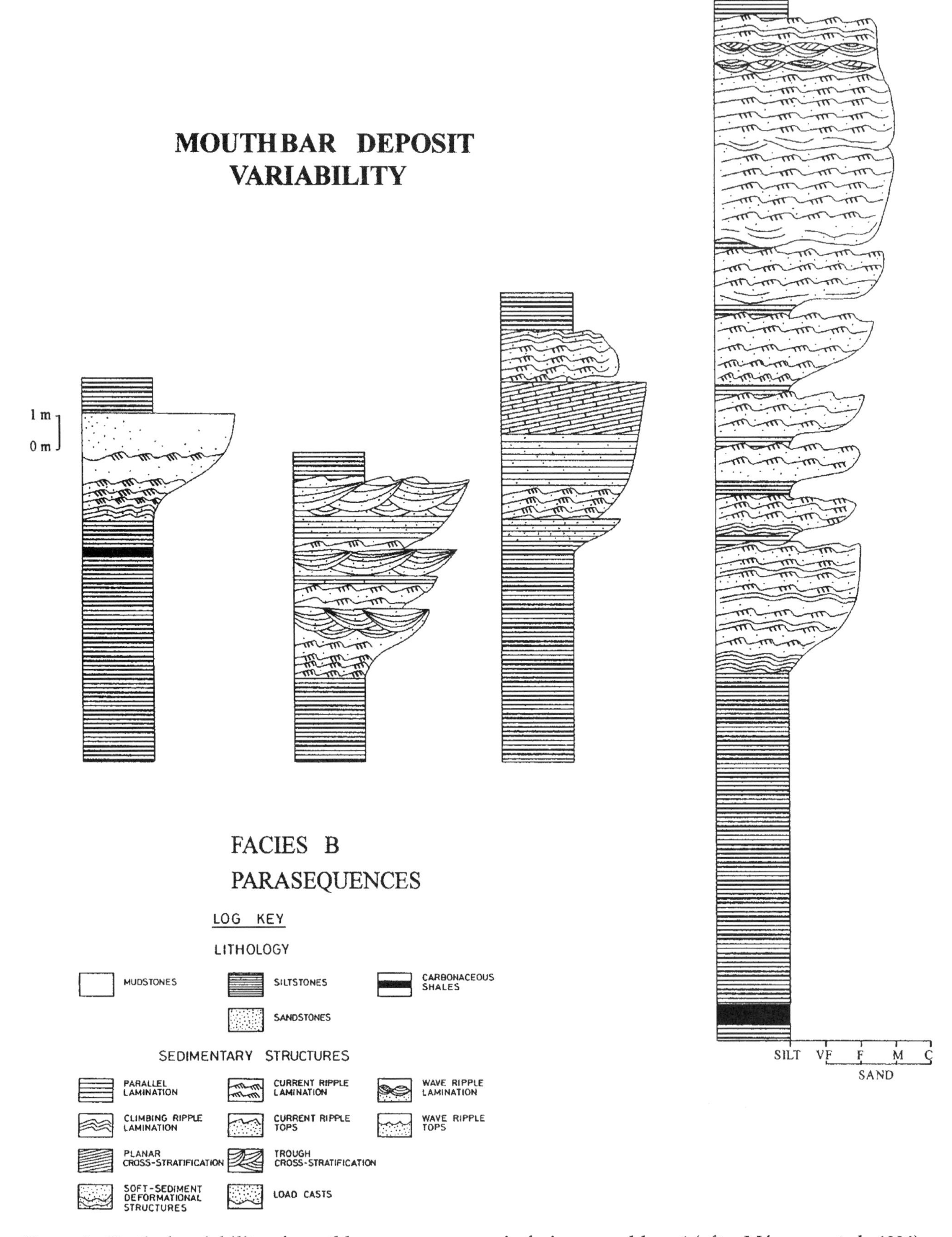

Figure 4—Vertical variability of mouthbar parasequences in facies assemblage 1 (after Mángano et al., 1994) with kind permission from Kluwer Academic Publishers.

Guillermo Guirado and Daniel Ruiz Holgado, whose help is greatly acknowledged.

REFERENCES CITED

Buatois, L. A., M. G. Mángano, X. Wu, and G. Zhang, 1994a, Evolution of lacustrine depositional systems from the early Mesozoic Jiyuan-Yima Basin, Central China, Actas V Reuniun Argentina de Sedimentología, San Miguel de Tucumán, p. 241–246.

Buatois, L. A., M. G. Mángano, X. Wu, and G. Zhang, 1994b, Turbidite lacustrine sedimentation from the Jurassic Anyao Formation, Henan Province, central China, 14th International Sedimentological Congress, Abstracts of Papers, Recife, p. S6–1.

Hu, B., 1991, On the Late Triassic–Middle Jurassic continental stratigraphy from Jiyuan, western Henan Province, Journal of Stratigraphy, v. 15, p. 48–52. (In Chinese with English summary).

Kang, M., F. Meng, B. Ren, and B. Hu, 1984, The age of Yima Formation and the establishment of the Yangshuzhuang Formation, Journal of Stratigraphy, v. 8, p. 194–198. (In Chinese with English summary.)

Kang, M., F. Meng, B. Ren, B. Hu, and Z. Ching, 1985, Division and correlation of the Mesozoic coal-bearing stratigraphy from Henan Province, Journal of Jiaozuo Mining Institute, v. 5, p. 1–146. (In Chinese with English summary).

Mángano, M. G., L. A. Buatois, X. Wu, J. Sun, and G. Zhang, 1994, Sedimentary facies, depositional processes and climatic controls in a Triassic lake, Tanzhuang Formation, western Henan Province, China, Journal of Paleolimnology, v. 11, p. 41–65.

Platt, N. H., and V. P. Wright, 1991. Lacustrine carbonates: facies models, facies distributions, and hydrocarbon aspects, *in* P. Anadón, Ll. Cabrera, and K. Kelts, eds., Lacustrine Facies Analysis, International Association of Sedimentologists Special Publication No. 13, p. 57–74.

Smoot, J. P., 1985. The closed-basin hypothesis and its use in facies analysis of the Newark Supergroup, United States Geological Survey Bulletin 1176, p. 4–10.

Wang, H., 1985, Atlas of the Palaeogeography of China, Cartographic Publishing House, Beijing, 283 p.

Wu, X., 1985, Trace fossils and their significance in non-marine turbidite deposits of Mesozoic coal and oil bearing sequences from Yima-Jiyuan basin, western Henan, China, Acta Sedimentologica Sinica, v. 3, p. 23–31. (In Chinese with English summary.)

Zhou, Z., and P. Li, 1980, Discussion on the division, correlation and age of the Mesozoic continental stratigraphy of China based on plant fossils, Geological Symposium for International Exchange, v. 4, Stratigraphy and Paleontology, Geological Publishing House, Beijing.

Greene, T. J., A. R. Carroll, M. S. Hendrix, Cheng Keming, Zeng Xiao Ming, 2000, Sedimentology and paleogeography of the Middle Jurassic Qiketai formation, Turpan-Hami Basin, Northwest China, *in* E. H. Gierlowski-Kordesch and K. R. Kelts, eds., Lake Basins through space and time: AAPG Studies in Geology 46, p. 141–152.

Chapter 10

Sedimentology and Paleogeography of the Middle Jurassic Qiketai Formation, Turpan-Hami Basin, Northwest China

Todd J. Greene
Department of Geological and Environmental Sciences, Stanford University
Stanford, California, U.S.A.

Alan R. Carroll
Department of Geology and Geophysics, University of Wisconsin
Madison, Wisconsin, U.S.A.

Marc S. Hendrix
Department of Geology, The University of Montana
Missoula, Montana, U.S.A.

Cheng Keming
Zeng Xiao Ming
China National Petroleum Corporation
Department of Geology, Research Institute of Petroleum Exploration and Development
Beijing, People's Republic of China

INTRODUCTION

The occurrence and distribution of Early and Middle Jurassic lake systems in northwest China (Figure 1) are poorly represented in the lacustrine literature. This is surprising because lacustrine environments would appear to have flourished in this geologic setting: intracontinental foreland-style, internally drained basins, with rapid, periodic basin floor subsidence, and a thriving humid climate supplying abundant fresh water into the many available basins. Although the Chinese literature reports mostly coaly, swamp/marsh, fluvial/alluvial, fan delta, deltaic, and marginal lacustrine environments for Lower and Middle Jurassic strata, the areal extent and descriptive sedimentology of the lacustrine systems are virtually unknown (Wang et al., 1994; Qiu et al., 1997; Wu and Zhao, 1997; Wu et al., 1997).

This study will briefly describe one of the many Middle Jurassic lacustrine systems contained within northwest China: the Qiketai Formation (J_2q) of the Turpan-Hami basin (Figures 1–3). Based on measured outcrop stratigraphic sections, detailed borehole core descriptions, and well log data, Qiketai Formation deposits covered an area of nearly 300 km × 40 km along depositional strike and dip, respectively (Figure 4). Facies represent littoral to profundal freshwater to saline lacustrine and lake-plain environments and reach a maximum thickness of 250 m in the basin center (Figure 4). The finer grained facies are organic-rich, ranging from 2 to 21% total organic carbon (TOC), with hydrogen indices (HI) reaching 811 mg hydrocarbons/gram rock (Table 1). The data presented here represent the first documentation of this large lake system outside the Chinese literature, and emphasize the importance of future detailed studies of Jurassic lake deposystems throughout northwest China.

GEOLOGIC SETTING

Knowledge of the geology and petroleum resource potential of the Turpan-Hami basin is somewhat limited due to its geographic isolation and harsh environmental conditions; nevertheless, previous collaborative reconnaissance studies by Stanford University and Chinese workers are beginning to build a good geologic understanding of the regional tectonic setting for

Jurassic deposition (e.g., Qiu et al., 1997; Wu and Zhao, 1997; Cheng et al., 1996; Carroll et al., 1995; Hendrix et al., 1992, 1995; Huang et al., 1991; Carroll et al., 1990; Graham et al., 1990).

The Turpan-Hami basin is underlain by accreted volcanic and volcaniclastic rocks deposited within and adjacent to late Paleozoic volcanic arcs (Wu and Zhao, 1997; Carroll et al., 1995). Marine waters retreated in the Permian, coincident with deposition of thick Permian lacustrine facies in the combined Junggar-Turpan-Hami basin (Greene et al., 1997; Wu and Zhao, 1997; also see Wartes et al., this volume). Subsequent uplift of the Bogda Shan during the Middle/Late Triassic partitioned the Junggar and Turpan-Hami basins, thereby placing the Turpan-Hami basin in its current intermontane position nestled within the Tian Shan (Greene et al., 1997; Carroll et al., 1995; Allen and Windley, 1993; Graham et al., 1993; Hendrix et al., 1992).

Lower–Middle Jurassic sedimentary fill of the Turpan-Hami basin is entirely nonmarine, ranging from fluvial/alluvial through profundal lake facies (Qiu et al., 1997; Wu et al., 1997; Wu and Zhao, 1997; Cheng et al., 1996; Hendrix et al., 1992). Cenozoic thrusting has provided excellent outcrop exposures of this full range of facies. Stratigraphic age control is based on nonmarine faunal and floral assemblages including pollen, lacustrine bivalves, ostracods, and vertebrates (XBGMR, 1993).

STRATIGRAPHY AND SEDIMENTOLOGY

Outcrop

Flaming Mountain

The 172 m measured interval at Flaming Mountain (latitude 42° 49′, longitude E89° 51′) represents a complete Qiketai Formation section, and is divided into three parts (Figure 5, see Figure 2 for location): (1) 34 m of uninterrupted profundal lacustrine facies overlain by (2) a partially covered 72 m section of indistinct gray/red, massive siltstone capped with (3) 64 m of continuous section containing more shallow and lake-plain facies.

The first section begins with a 6 m series of amalgamated sand packages that generally coarsen upward and contain abundant organic debris; several mudstone interbeds disrupt the upper half of the sequence (Figure 5). This is followed by a 4 m series of thin, tabular, rippled sandstones, which, in turn, are overlain by 22 m of mostly green/gray mudstone with weak millimeter-scale planar laminations. A 2–3 m thick coarsening-upward sand package containing trough cross-stratification caps the first section.

Following a 50 m thick section of partially to fully covered gray/red siltstone, the third and final section represents the most shallow facies in the formation (Figure 5). The lowermost 10 m contain alternating packages of 2 m thick red/gray siltstone with mudcracks, and 1–2 m of fine to medium sandstones that fine upwardly. The subsequent 20 m contain up to 8 m of red/gray silty mudstone packages with millimeter-scale laminations, interrupted by 0.5–1 m thick fine sandstones that have symmetrical ripples, and several 10 cm scale normally graded sequences. An abrupt change occurs at the 138 m mark, to 1 m of fine sandstone with mud-draped ripples and interbedded mudstone containing desiccation cracks. This is followed by an 18 m thick package of gray siltstone containing two 1–2 m thick fine sands with symmetrical ripples. The 2 m thick fine sandstone at the 144 m mark contains some soft-sediment deformation features and 50 cm scale, low-angle cross-sets that are loaded at their base. The uppermost 15 m begin with 5 m of gray, silty mudstone with mudcracks and small fine sand beds, followed by 10 m of red/gray mudstones containing symmetrical rippled 1 m thick sandstone beds, along with syndepositional wavy laminations.

Qiketai

At Qiketai (N42° 57′, E90° 34′), another complete Qiketai Formation section exposes 62 m of lacustrine deposits that also can be separated into three sections (see Figure 2 for location): (1) a 30 m thick high-energy sand and mud/siltstone sequence capped by a sand unit with cobbles, (2) 14 m of mostly covered siltstone, and (3) 18 m of silty mudstone with paleosols, mudcracks, and root traces (Figure 6).

The lowermost section contains 20 m of tan/gray siltstone with four tabular fine sandstone beds that grade both normally and inversely, containing trough-stratification and ripples. Above this lies 5 m of medium to fine sandstones with troughs, upper plane bed laminations, and several 10 cm scale normally graded sequences. Broken bivalve shells and granules are scattered throughout. Directly overlying are 4 m of red mudstone, capped by 1 m of a sandy conglomerate bed with clasts up to 6 cm in size (Figure 6).

Following a 14 m thickness of covered section, a final 18 m thick package contains mostly paleosol layers. Purple/gray mudstones with common red/gray mottling are overlain by purple/red mudstones with $CaCO_3$ stringers, root traces, up to 50 cm of fine sandstone lenses, scattered shelly material, and possible mudcracks (Figure 6).

Subsurface

Based on well log character and lithofacies of selected core intervals, Wu and Zhao (1997) and Qiu et al. (1997) used marine sequence-stratigraphic nomenclature and seismic facies analysis to place the Qiketai Formation in the final Middle Jurassic lake transgressive system tract. Unfortunately, limited access to well log and other borehole data for this study do not allow for a rigorous facies analysis for the lacustrine system in the subsurface; however, detailed sedimentologic description of approximately 8 m thick intervals of Qiketai Formation rocks in three wells, in addition to limited biomarker analysis on a few core samples, yields useful depositional environment information.

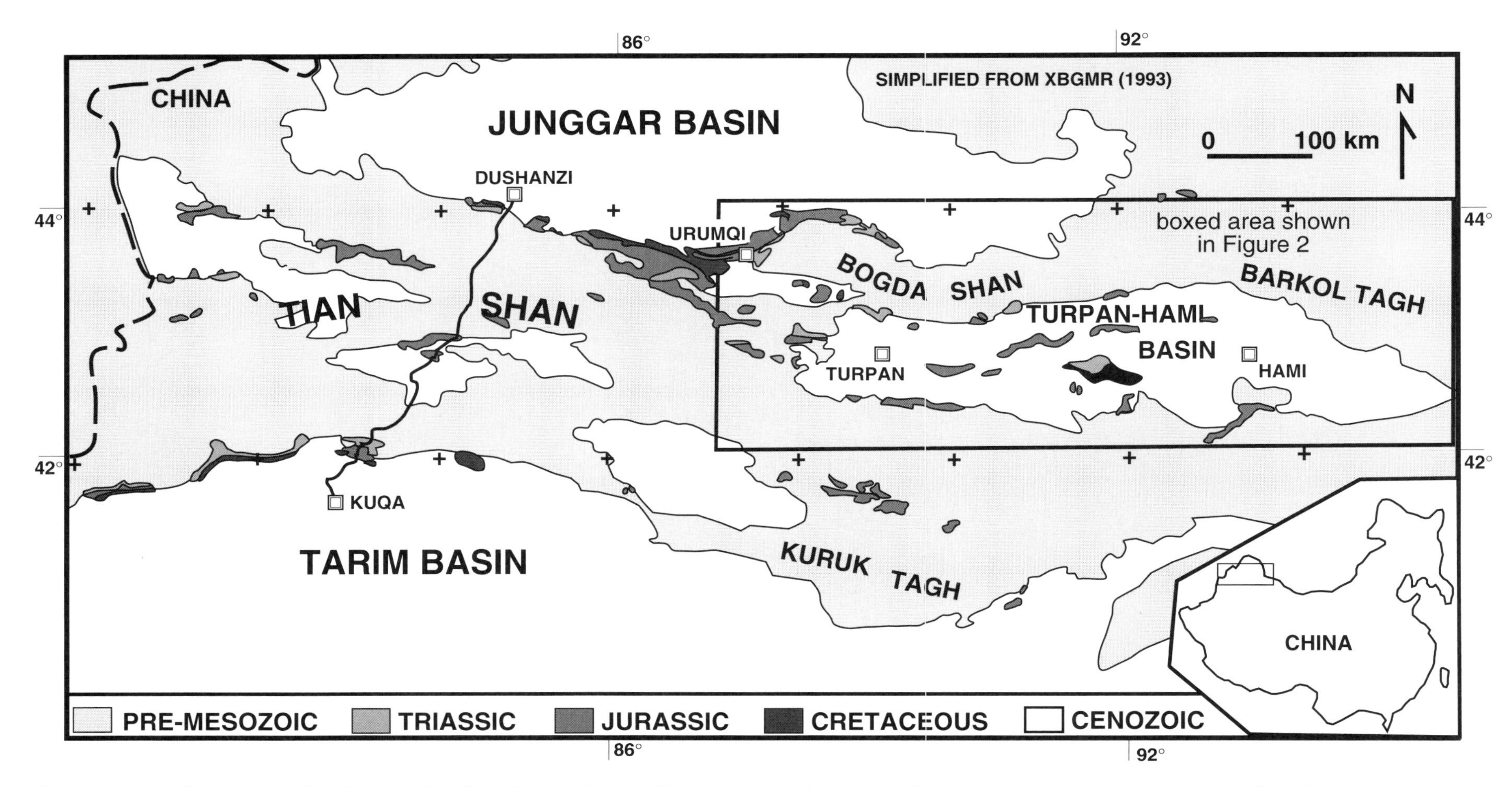

Figure 1—Location map of the Turpan-Hami basin, northwest China showing its current intermontane position, nestled within the Tian Shan.

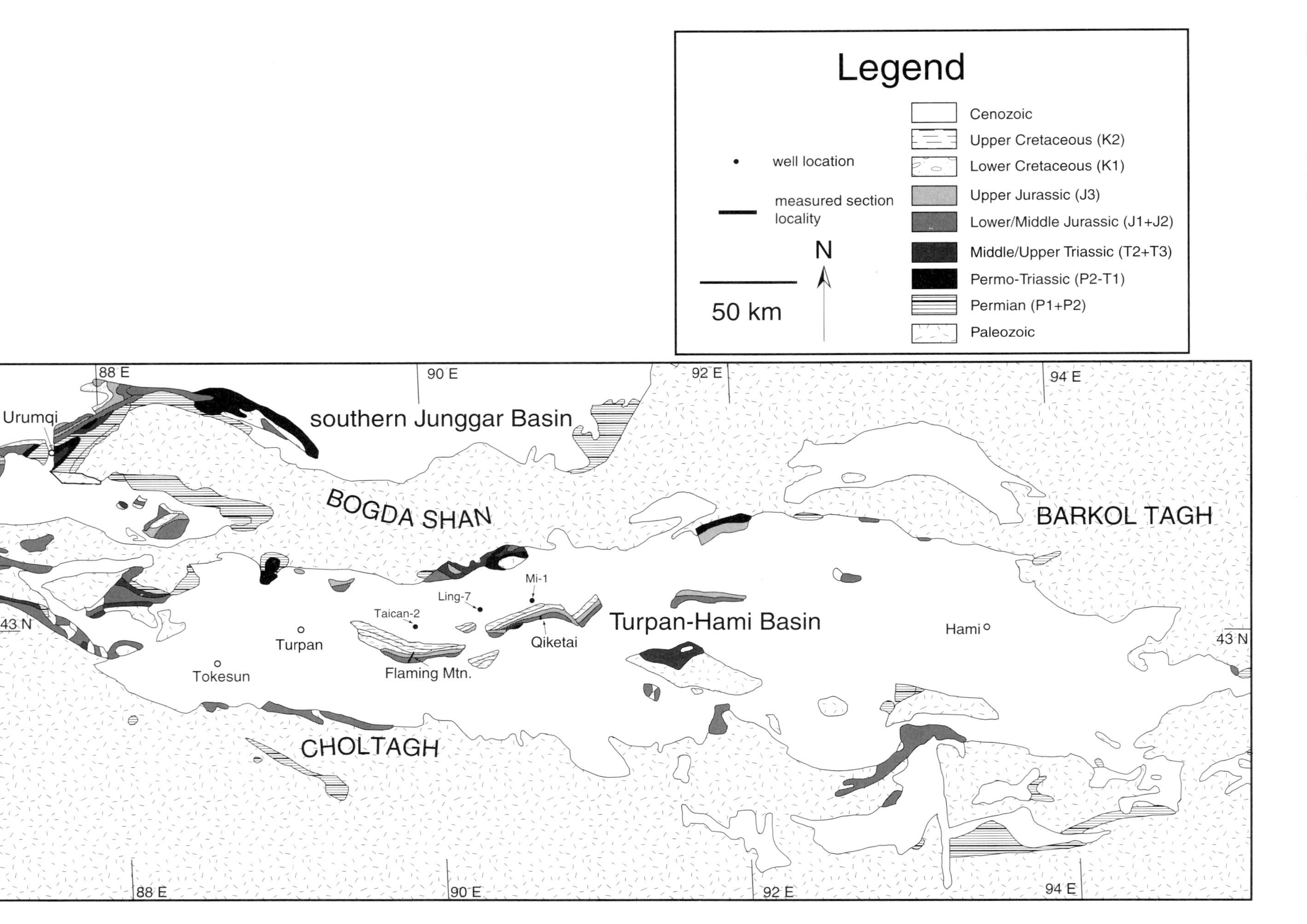

Figure 2—Time-slice geologic map of the Turpan-Hami basin, northwest China, with synthesis of acquired data (see legend; adapted from XBGMR; Xinjiang Bureau of Geology and Mineral Resources, 1993).

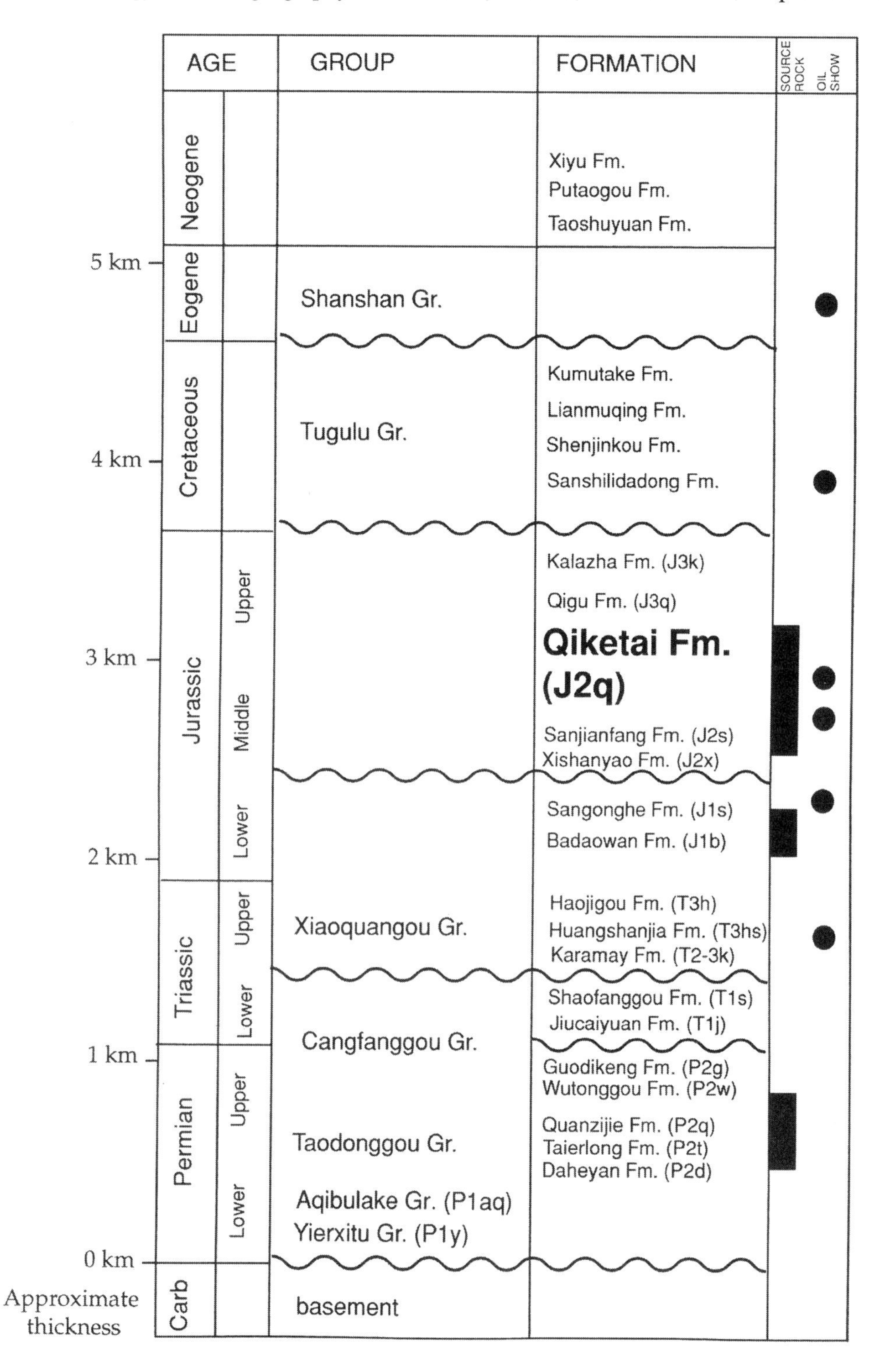

Figure 3—Stratigraphic chart of the Turpan-Hami basin, northwest China, with basinwide unconformities. Important source intervals include Upper Permian and Lower/Middle Jurassic nonmarine strata. The Middle Jurassic lacustrine deposits of the Qiketai Formation (J2q) are in boldface.

Taican-2 Well

Biomarker analysis was conducted on sample 96-TC2-4, a laminated carbonaceous mudstone from the upper Qiketai Formation, cored in the Taican-2 well (Figure 7, Table 1). After solvent extraction of bitumen, saturate hydrocarbons were separated and concentrated using standard liquid chromatography and silicaliting techniques (cf. Peters and Moldowan, 1993). Gas chromatography-mass spectrometry (GC-MS) and metastable reaction monitoring (MRM) were used to identify specific biomarker compounds indicating depositional environment (using the m/z 125 and 191 mass fragmentograms and the m/z 414-231 transition).

Biomarkers are complex molecular fossils derived from once-living organisms (Peters and Moldowan, 1993). For example, beta-carotane (Figure 7) derives from organisms that favor shallow, well-lit, evaporative environments, and frequently is associated with hypersaline lakes (e.g., Jiang and Fowler, 1986). Gammacerane (Figure 7) commonly derives from organisms living near the chemocline in saline, stratified water columns

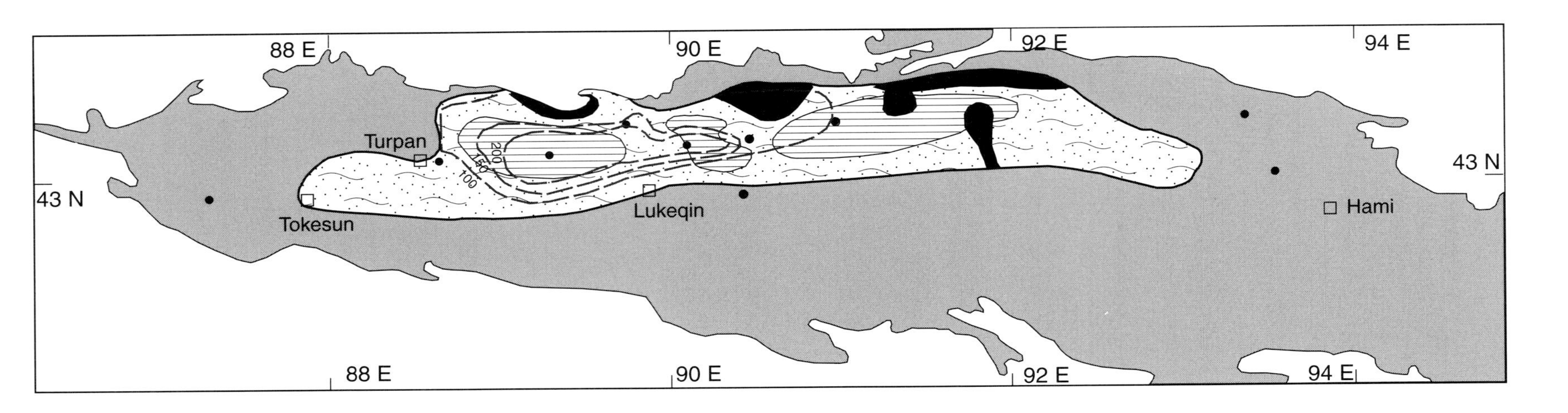

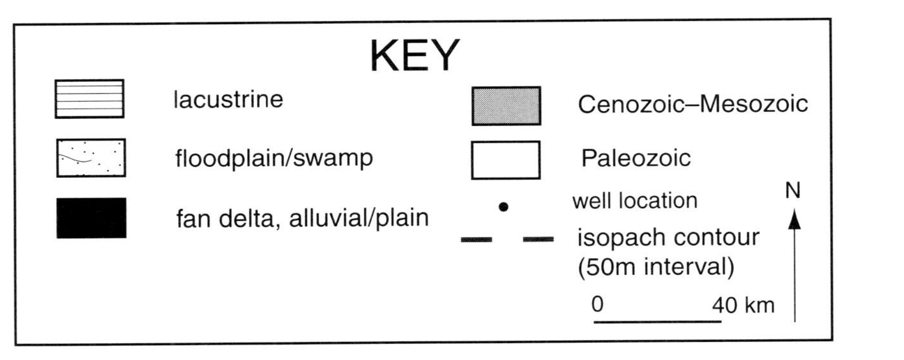

Figure 4—Depositional facies and isopach map for the Middle Jurassic Qiketai Formation, Turpan-Hami basin, northwest China (adapted from Cheng et al., 1996)

Table 1. Rock-Eval and thermal maturation data for Middle Jurassic Qiketai Formation Rocks.

Sample #	Well Name	Latitude (N)	Longitude (E)	Depth (m)	Fm.	Lithology	%TOC	S1	S2	HI	T_{max}	$\%R_o$	T.A.I.
96-TC2-4	Taican-2	42.59.11	89.51.34	3865.5	J2q	mudstone	3.85	0.47	31.22	811	438	IR	2.3-2.8
96-TC2-5	Taican-2	42.59.11	89.51.34	3887.7	J2q	mudstone	1.91	0.35	12.4	649	437	NA	NA
96-TC2-6	Taican-2	42.59.11	89.51.34	3997.4	J2q	mudstone	18.27	2.99	88.62	485	432	0.63	2.4-2.6
96-TC2-7	Taican-2	42.59.11	89.51.34	4000.5	J2q	mudstone	21.74	4.04	78.36	360	431	NA	NA
96-TC2-8	Taican-2	42.59.11	89.51.34	4002	J2q	mudstone	4.85	0.62	11.17	230	433	NA	NA
96-TC2-9	Taican-2	42.59.11	89.51.34	4006	J2q	coal	49.63	10.27	183.5	370	433	0.55	IR

S1 = free hydrocarbons, S2 = cracked hydrocarbons,
HI = hydrogen indes. HI = hydrogen index
IR = insufficient recovery T.A.I. = thermal alteration index
$\%R_o$ = vitrinite reflectance NA = not available

(Sinninghe Damsté et al., 1995). Also abundant are 4-methyl steranes and dinosteranes in sample 96-TC2-4, and they are commonly associated with algal species (e.g., dinoflagellates) derived from freshwater and brackish water lacustrine environments (Fu et al., 1990). Together, the presence of these compounds is consistent with physical evidence for saline or hypersaline environmental conditions during deposition of the upper Qiketai Formation.

Mi-1 and Ling-7 Wells

In addition, facies descriptions from Qiketai Formation core rocks obtained from the Mi-1 and Ling-7 wells represent terrestrial, lake-plain facies (Figure 8; see Figure 2 for well locations). Abundant plant material and pedoturbation, and wavy laminations of lenses of silt, mud, and fine sands on a centimeter to millimeter scale all argue for a terrestrial, low-energy depositional environment.

FACIES ANALYSIS AND PALEOGEOGRAPHY

Periodic sandstones and subaerial exposure surfaces in littoral lacustrine facies, even at the basin center (e.g., Flaming Mountain), suggest that the freshwater input into the lake was sporadic; if the Flaming Mountain rocks represent some of the most distal facies of the Qiketai Formation (i.e., basin center), then lake waters must have been quite shallow at times (saline) and at other times completely disconnected into smaller bodies of water. This scenario is supported by the biomarker analysis and documented sedimentology at both localities (Figures 5–7).

Two main sequences within the Qiketai Formation occur at both Flaming Mountain and Qiketai; in each sequence, the Flaming Mountain locality consists of more distal, finer grained facies and Qiketai has more lake-plain and foreshore facies consisting of coarser, more proximal deposits (Figures 5, 6).

The lowermost 34–44 m of Sequence 1a and 1b (see Figures 5, 6 for explanation of sequence labels) represents primarily freshwater lake deposits with stratigraphic packaging reflecting deltaic channel avulsion and progradation. This is consistent with the terrestrial organic matter input, cross-bedding, and lack of mudcracks and cyclicity (Figures 5, 6).

The second interval (Section 2) is mostly covered and contains massive, gray/red silt/mudstones. The third and final sequence (overlying the covered Section 2) appears to record a generally more littoral, saline lake environment (Figures 5, 6). The Flaming Mountain section contains at least six packages reflecting periodic flooding and desiccation suggesting a more cyclic character (Sequences 3c–f; Figure 5). Contemporaneous deposits at the Qiketai locality contain paleosols and littoral and lake-plain facies (Sequence 3), also suggesting shallow water depths (Figure 6).

The Qiketai Formation marks the end of the Middle Jurassic and is overlain by hundreds of meters of Upper Jurassic massive red beds of the

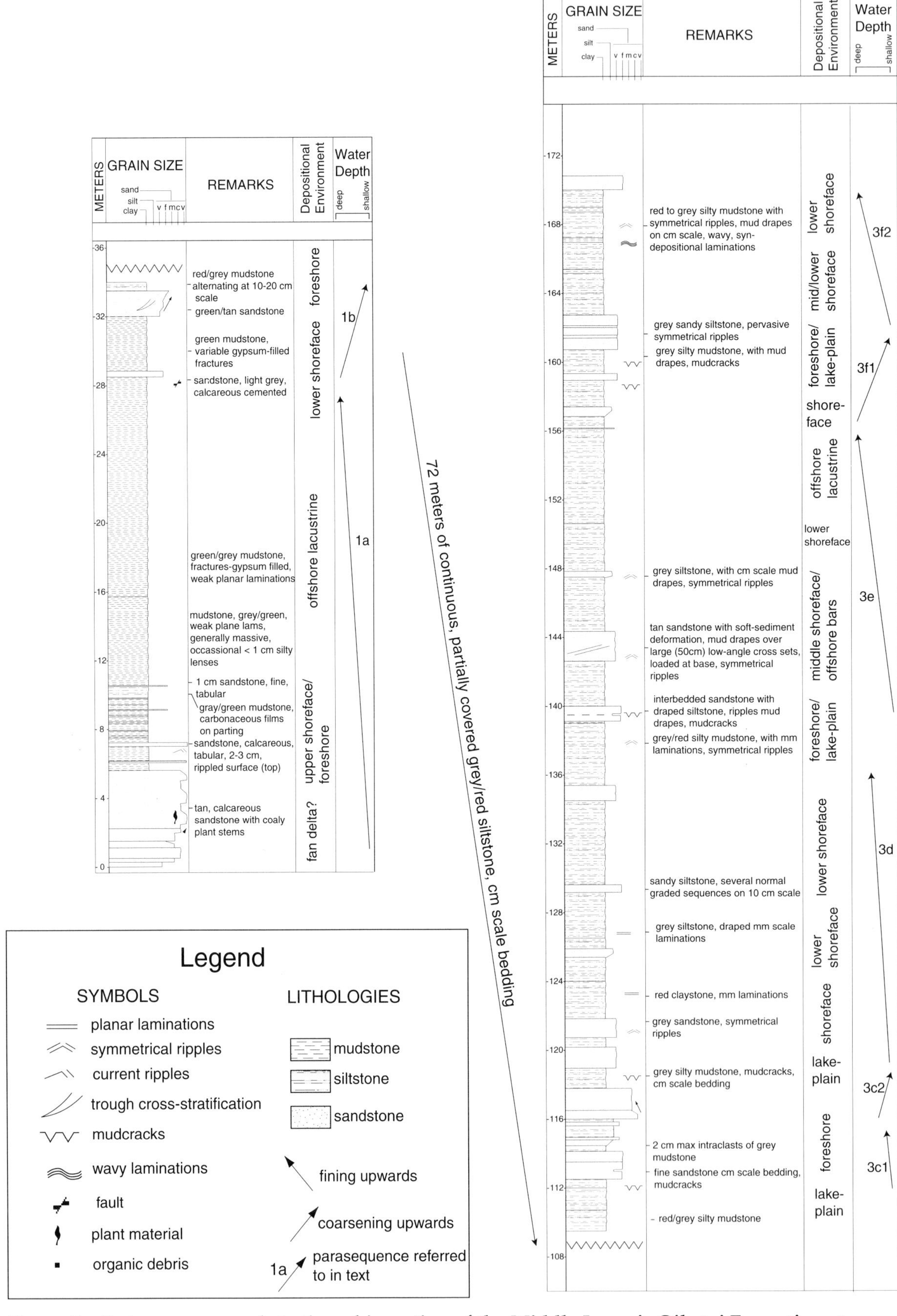

Figure 5—Outcrop measured stratigraphic section of the Middle Jurassic Qiketai Formation at Flaming Mountain (see Figure 2 for location).

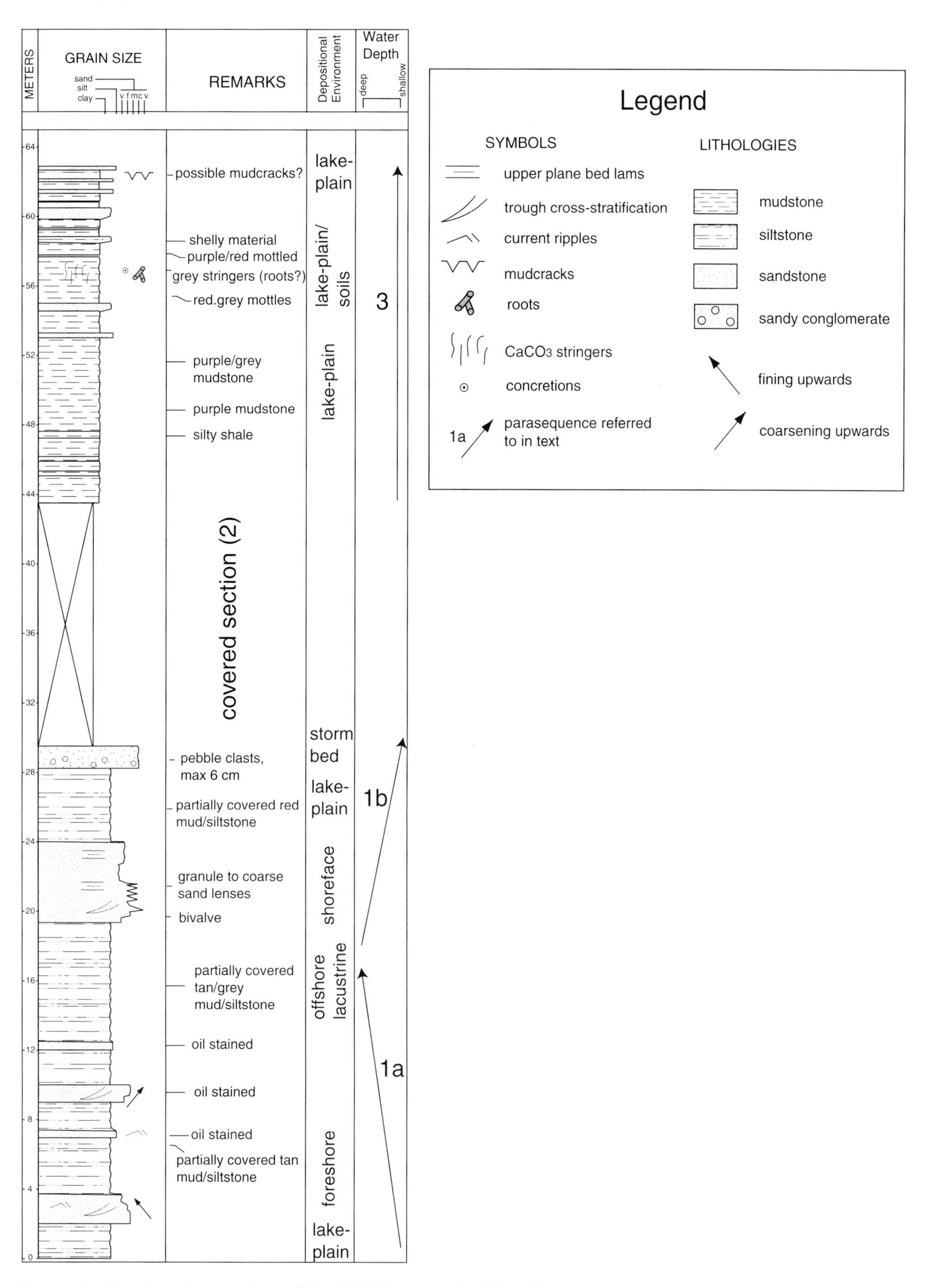

Figure 6—Stratigraphic section of the Middle Jurassic Qiketai Formation at the Qiketai locality, east of Flaming Mountain. See Figure 2 for location.

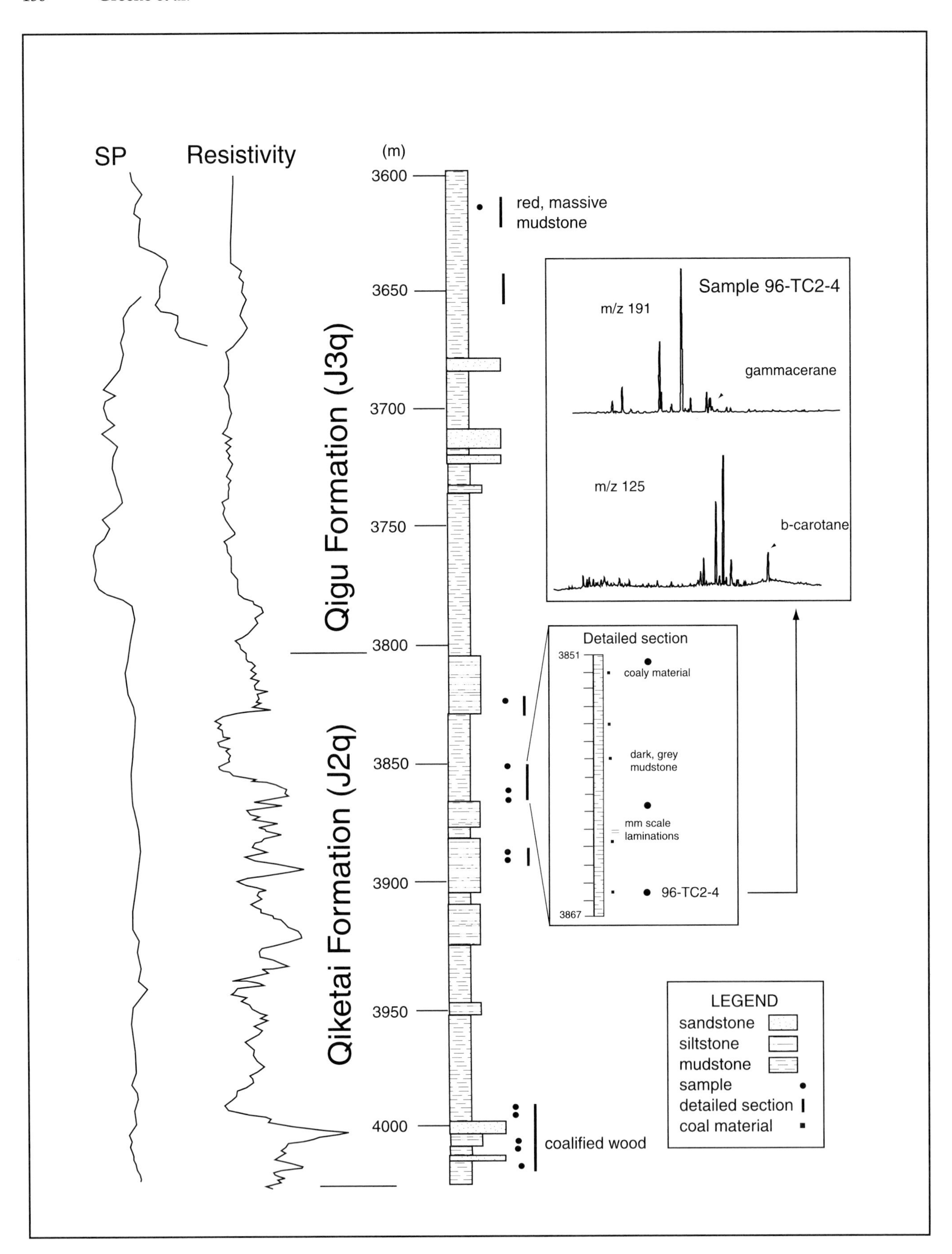

Figure 7—Log of the Taican-2 well (see Figure 2 for location). Biomarker data is from sample 96-TC2-4 (see Table 1).

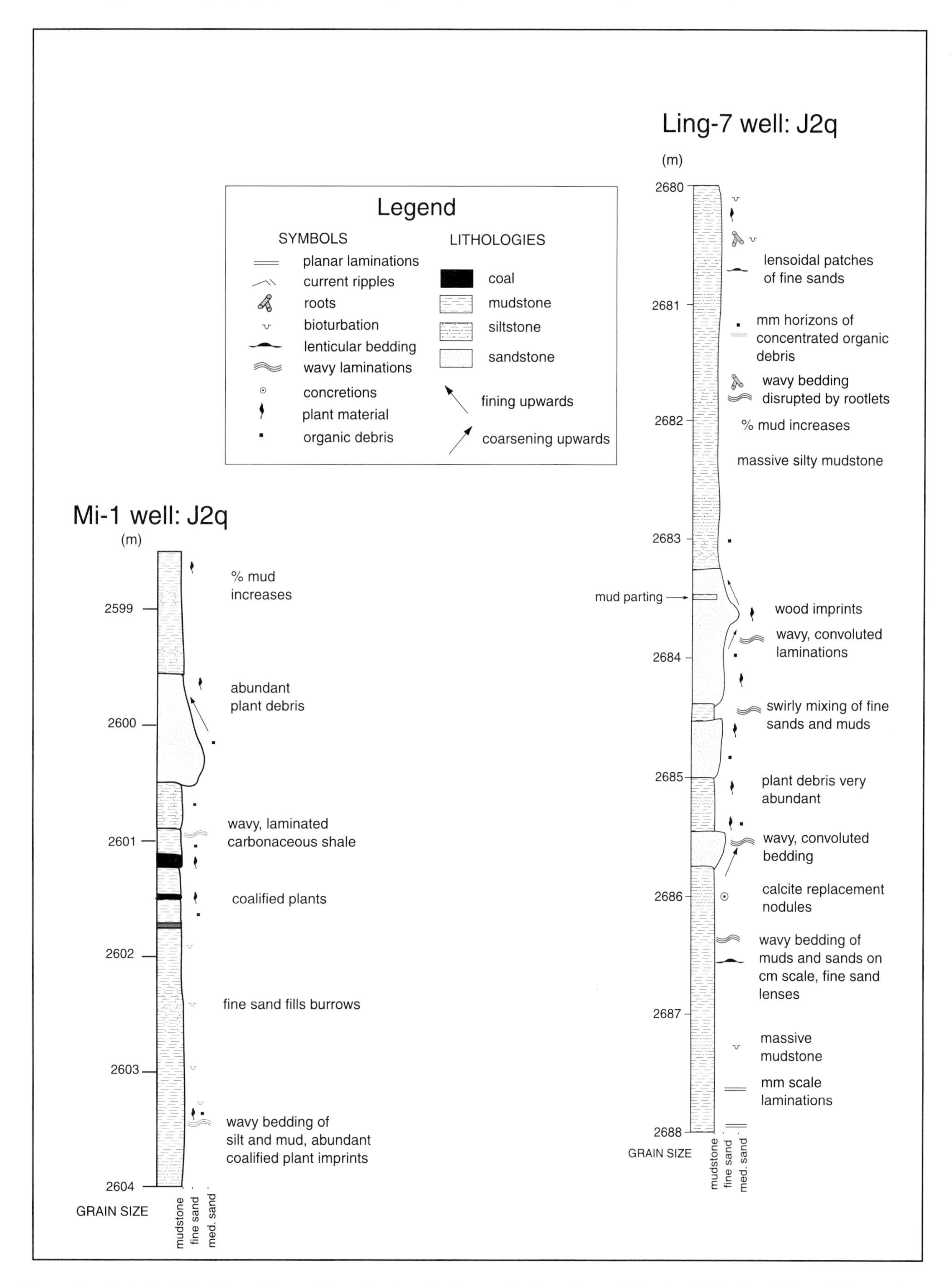

Figure 8—Detailed stratigraphic sections for borehole core intervals contained within the Qiketai Formation (J2q). See Figure 2 for well locations.

Qigu Formation (J_3q; see Figure 3). This is a common boundary marker throughout north-west Chinese basins and has been inferred to reflect a regional shift from Middle Jurassic wet climates to Late Jurassic arid climates (Hendrix et al., 1992; Zhao et al., 1992). The lake shallowing trend of Sequence 3 could represent the precursor to this major climatic change.

REFERENCES CITED

Allen, M. B., and B. F. Windley, 1993, Evolution of the Turfan basin, Chinese central Asia: Tectonics, v. 12, p. 889–896.

Carroll, A. R., Y. Liang, S. A. Graham, X. Xiao, M. S. Hendrix, J. Chu, and C. L. McNight, 1990, Junggar basin, northwest China: trapped late Paleozoic ocean: Tectonophysics, v. 181, p. 1–14.

Carroll, A. R., S. A. Graham, and M. S. Hendrix, 1995, Late Paleozoic amalgamation of northwest China: sedimentary record of the northern Tarim, northwestern Turpan, and southern Junggar basins: Geological Society of America Bulletin, v. 107, p. 571–594.

Cheng Keming, Su Aiguo, Zhao Changyi, and He Zhonghua, 1996, Study of coal-generated oil in the Tuha basin: 30th International Geological Congress, Progress in Geology of China, p. 796–799.

Fu Jiamo, Sheng Guoying , Xu Jiayou, G. Eglinton, A. P. Gowar, Jia Rongfen, Fan Shanfa, and Peng Pingan, 1990, Application of biological markers in the assessment of paleoenvironments of Chinese non-marine sediments: Organic Geochemistry, v. 16, p. 769–779.

Graham, S. A., S. C. Brassell, A. R. Carroll, X. Xiao, G. Demaison, C. L. McKnight, Y. Liang, J. Chu, and M. S. Hendrix, 1990, Characteristics of selected petroleum source rocks, Xinjiang Uygur Autonomous Region, northwest China: American Association of Petroleum Geologists Bulletin, v. 74, p. 493–512.

Graham, S. A., M. S. Hendrix, L. B. Wang, and A. R. Carroll, 1993, Collisional successor basins of western China: impact of tectonic inheritance on sand composition: Geological Society of America Bulletin, v. 105, p. 323–344.

Greene, T. J., A. R. Carroll, M. S. Hendrix, and Jing Li, 1997, Permian-Triassic basin evolution and petroleum system of the Turpan-Hami Basin, Xinjiang Province, northwest China: American Association of Petroleum Geologists 1997 Annual Meeting Abstracts, v. 6, p. 42.

Hendrix, M. S., S. A. Graham, A. R. Carroll, E. R. Sobel, C. L. McKnight, B. J. Schulein, and Z. Wang, 1992, Sedimentary record and climatic implications of recurrent deformation in the Tian Shan: evidence from Mesozoic strata of north Tarim, south Junggar, and Turpan basins, northwest China: Geological Society of America Bulletin, v. 105, p. 53–79.

Hendrix, M. S., S. C. Brassell, A. R. Carroll, and S. A. Graham, 1995, Sedimentology, organic geochemistry, and petroleum potential of Jurassic coal measures: Tarim, Junggar, and Turpan basins, northwest China: American Association of Petroleum Geologists Bulletin, v. 79, p. 929–959.

Huang Difan, Zhang Dajiang, Li Jinchao, and Huang Xiaoming, 1991, Hydrocarbon genesis of Jurassic coal measures in the Turpan Basin, China: Organic Geochemistry, v. 17, p. 827–837.

Jiang, Z. S., and M. G. Fowler, 1986, Carotenoid-derived alkanes in oils from northwestern China: Organic Geochemistry, v. 10, p. 831–839.

Moldowan, J. M., W. K. Seifert, and E. J. Gallegos, 1985, Relationship between petroleum composition and depositional environment of petroleum source rocks: American Association of Petroleum Geologists Bulletin, v. 69, p. 1255–1268.

Peters, K. E., and J. M. Moldowan, 1993, The Biomarker Guide: Interpreting Molecular Fossils in Petroleum and Ancient Sediments: Englewood Cliffs, New Jersey, Prentice Hall, 363 p.

Qiu Yinan, Xue Shuhao, and Ying Fengxiang, 1997, Continental hydrocarbon reservoirs of China: Beijing, Petroleum Industry Press, p. 87–104.

Sinninghe Damsté, J. S., F. Kenig, M. P. Koopmans, J. Koster, S. Schouten, J. M. Hayes, and J. W. de Leeuw, 1995, Evidence for gammacerane as an indicator of water column stratification: Geochimica et Cosmochimica Acta, v. 59, p. 1895–1900.

Wang Shangwen, Zhang Wanxuan, Zhang Houfu, and Tan Shidian, 1994, Petroleum Geology of China: Beijing, Petroleum Industry Press, p. 411–414.

Wu Tao and Zhao Wenzhi, 1997, Formation and Distribution of Coal Measure Oil-gas Fields in Turpan-Hami Basin: Beijing, Petroleum Industry Press, 262 p.

Wu Chongyun, Xue Shuhao, et al., 1997, Sedimentology of Petroliferous Basins in China: Beijing, Petroleum Industry Press, p. 384–400.

XBGMR, 1993, Bureau of Geology and Mineral Resources of Xinjiang Uygur Autonomous Region, in Regional Geology of Xinjiang Uygur Autonomous Region: Geological Memoirs, Series 1, no. 32, Beijing, Geological Publishing House, 841 p.

Zhao Xiwen, et al., 1992, The Paleoclimate of China: Beijing, Geological Printing House, 132 p.

Lucas, S. G., and O. J. Anderson, 2000, The Todilto salina Basin, Middle Jurassic of the U.S. Southwest, *in* E. H. Gierlowski-Kordesch and K. R. Kelts, eds., Lake basins through space and time: AAPG Studies in Geology 46, p. 153–158.

Chapter 11

The Todilto Salina Basin, Middle Jurassic of the U.S. Southwest

Spencer G. Lucas
New Mexico Museum of Natural History
Albuquerque, New Mexico, U.S.A

Orin J. Anderson
New Mexico Bureau of Mines and Mineral Resources
Socorro, New Mexico, U.S.A.

INTRODUCTION

One of the most distinctive Jurassic lithostratigraphic units in the American Southwest is the Todilto Formation of northern New Mexico and southwestern Colorado (Figure 1). This relatively thin (>75 m) unit is mostly carbonates and evaporites in a thick section otherwise dominated by siliciclastic eolianites (Figures 2, 3). The Todilto Formation is extremely significant economically as a source rock for petroleum (Vincelette and Chittum, 1981) and uranium (Chenoweth, 1985); it also provides all the gypsum mined in New Mexico (Weber and Kottlowski, 1959).

Some earlier workers regarded the Todilto as having been deposited in a marine embayment of the Middle Jurassic Curtis seaway (e.g., Harshbarger et al., 1957; Ridgley and Goldhaber, 1983), but more recent studies of stratigraphy, paleontology, and geochemistry indicate that any marine connection to the Todilto Basin was short-lived or intermittent (Lucas et al., 1985; Kirkland et al., 1995). Todilto deposition took place in a paralic salina culminated by a gypsiferous evaporitic lake.

STRATIGRAPHY

The Todilto Formation crops out and is present in the subsurface across most of northern New Mexico and southwestern Colorado (Figure 1), an area of about 100,000 km^2 (Lucas et al., 1985, 1995; Lucas and Kietzke, 1986). Throughout its extent, the Todilto rests on the Entrada Sandstone with minor disconformity and is overlain disconformably to conformably by the Summerville Formation (Figures 2, 3).

Two members of the Todilto Formation are the Luciano Mesa (limestone) and Tonque Arroyo (gypsum) (Figure 2). The Luciano Mesa Member has a maximum thickness of 13.3 m and is mostly microlaminated, kerogenic limestone. It crops out over an area of 88,000 km^2 in northern New Mexico and southwestern Colorado. The unit can be divided at most outcrops into a lower, thinly laminated limestone and an upper, massive limestone. The lower limestone has fine (millimeter-scale) laminations that are locally microfolded and contorted. It typically overlies Entrada eolianite sandstone with erosional unconformity marked by a thin (20-cm or less) interval of limy sandstone, which is often pebbly.

The upper, massive limestone is a ledge-forming, poorly laminated carbonate containing numerous vugs filled with secondary carbonate. The vugs are selectively located in rounded, "algal"-like structures, and some contain thin needles of secondary gypsum. The structures, however, lack the finely laminated texture typical of most algal stromatolites. Anderson and Kirkland (1960) suggested the microlaminae of the Luciano Mesa Member form varved couplets and counted these couplets to estimate a duration of about 14,000 years for their deposition.

The Tonque Arroyo Member is as much as 61 m thick and mostly massive and brecciated white gypsum. In its lower part the Tonque Arroyo Member contains some 1–2-mm-thick carbonate layers. The Luciano Mesa Member has a much broader distribution than the Tonque Arroyo Member and is a continuous unit across the Todilto depositional basin (Figure 1). The Tonque Arroyo Member has a more limited outcrop belt, widely varying thickness, and numerous local pinchouts.

TECTONICS AND SEDIMENTATION

The Todilto Basin developed in a tectonically passive downwarp between two east-west oriented

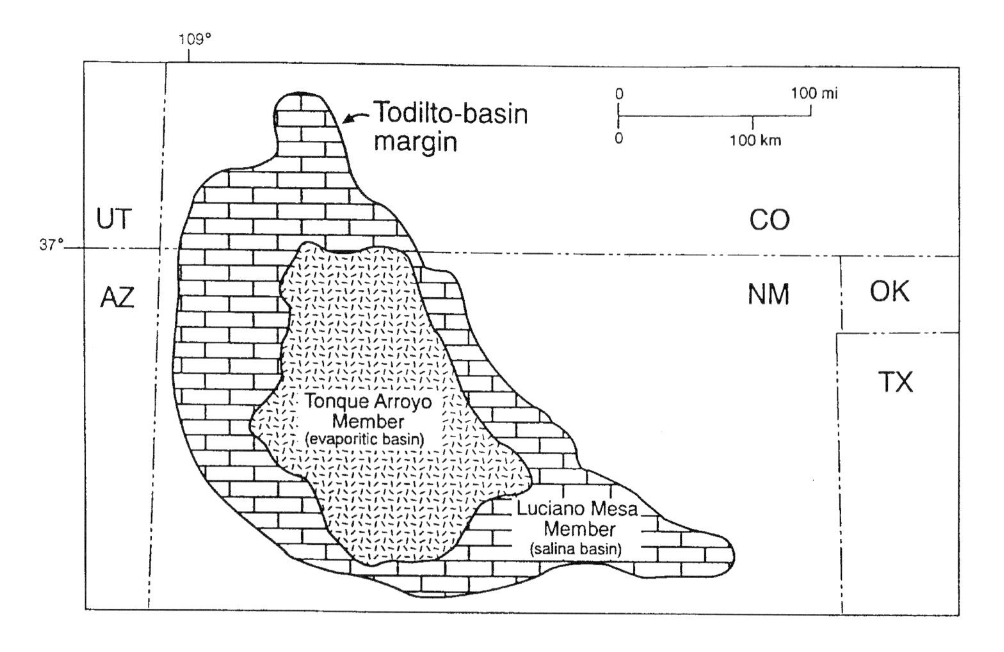

Figure 1—Location and distribution of the two members of the Todilto Formation in northern New Mexico and southwestern Colorado. The distribuition of the Luciano Mesa Member essentially defines the extent of the Todilto depositional basin.

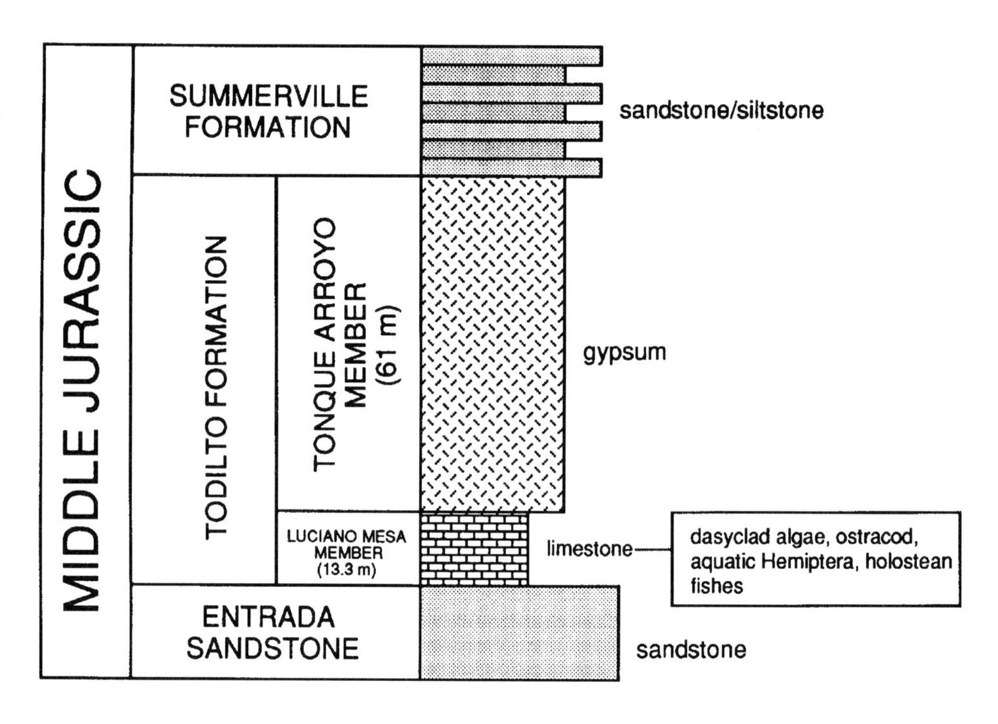

Figure 2—Summary of the stratigraphy and paleontology of the Todilto Formation.

positives, the Mogollon highland of east-central Arizona and west-central New Mexico and the Uncompaghre highland of north-central New Mexico and south-central Colorado (Kirkland et al., 1995). Laterally persistent laminae of calcite and organic matter in the Luciano Mesa Member of the Todilto Formation suggest deposition in quiet water with no bioturbating benthos. Kirkland et al. (1995) estimate the Todilto waterbody to have been stratified and less than 91.5 m deep.

A regional drop in base level cut the Todilto depositional basin off from waters in the Curtis seaway to the northwest and intitiated evaporite deposition (Lucas et al., 1985). The sub-Todilto paleotopography of Entrada dunes controlled evaporite deposition, confining it largely to brine pools that developed in inter-dunal lows (Vincelette and Chittum, 1981).

PALEONTOLOGY, AGE, AND CORRELATION

Fossils are not abundant in the Todilto Formation. No megafossil plants are known, and attempts to extract identifiable palynomorphs have failed (Anderson and Kirkland, 1960). Algal structures are present, as are dascycladacaean algae at one locality in west-central New Mexico (Armstrong, 1995).

Invertebrate fossils are limited to the ostracod *Cytheridella* and aquatic Hempitera, including *Xiphenax jurassicus* (Cockerell, 1931; Kietzke, 1992; Kirkland et al., 1995). Fossil vertebrates are the holostean fishes *Hulettia americana, Todiltia schowei,* and *Caturus dartoni* (Koerner, 1930; Dunkle, 1942; Schaeffer and Patterson, 1984; Lucas et al., 1985).

Only the fossil fishes provide a possible basis for correlation of the Todilto Formation because they are also known from the Hulett Member of the Sundance Formation in Wyoming. The Hullett is in the zone of *Kepplerites macleami* of early Callovian age (Imlay, 1980).

Regional stratigraphic relationships indicate that the Todilto is homotaxial with the marine Curtis Formation in Utah (Anderson and Lucas, 1992, 1994) (Figure 3). Both units are between the Entrada and Summerville formations, and both have pebbly zones (transgressive lag deposits) at their bases. The regional rise in base level that led to transgression of the Curtis seaway and the resultant highstand produced a paralic salina in northern New Mexico-southwestern Colorado, immediately southeast of the seaway (Imlay, 1980; Anderson and Lucas, 1994) (Figure 4). Both Todilto members were deposited in this short-lived salina.

SALINA DEPOSITION

Several lines of evidence suggest that the Todilto depositional basin had little connection to the Jurassic seaway and instead was a vast, paralic salina.

(1) No direct stratigraphic continuity of Todilto strata and marine Jurassic strata exists, either on outcrop or in the subsurface. The Todilto pinches out around its basin periphery into eolianites (Lucas et al., 1985; Anderson and Lucas, 1992).

(2) No clearly marine fauna or flora are known from the Todilto Formation. Instead, a very low diversity fish fauna, as is characteristic of saline lakes, is known from the Todilto (Barbour and Brown, 1974; Lucas et al., 1985). Indeed, the low diversity fish and invertebrate fauna of the Todilto are strikingly similar to that of Quaternary salinas in Australia (e.g., Warren, 1982; Warren and Kendall, 1985). Armstrong (1995) claimed that the dascyclad algae found in the Todilto near Grants, New Mexico, indicate a marine environment, but today these blue green algae tolerate a wide range of salinity from fresh to hypersaline waters.

(3) Carbon and sulfur isotope ratios calculated for Todilto limestones have a wide range of values compatible with a nonmarine, marine or mixed waterbody (Kirkland et al., 1995); however, strontium isotope ($^{87}Sr/^{86}Sr$) ratios for the Todilto do not match those of sediments deposited from normal marine Callovian seawater (Kirkland et al., 1995).

Reconstructing the Todilto paleoenvironment as a salina is consistent with all data. Todilto deposition began with initial flooding of marine waters during transgression of the Curtis-Sundance seaway. After the initial flooding, the Todilto Basin was separated from the seaway by coastal ergs (Figure 4). Freshwater stream runoff, influx of seawater by seepage through the erg, and possible short-lived overtopping of the erg maintained the Todilto salina. Increased aridity promoted evaporation, which eventually produced a smaller, evaporitic basin in which gypsum precipitated.

REFERENCES CITED

Anderson, O. J., and S. G. Lucas, 1992, The Middle Jurassic Summerville Formation, northern New Mexico. New Mexico Geology, v. 14, p. 79–92.

Anderson, O. J., and S. G. Lucas, 1994, Middle Jurassic stratigraphy, sedimentation and paleogeography in the southern Colorado Plateau and southern High Plains, *in* M. V. Caputo, J. A. Peterson, and K. J. Franczyk, eds., Mesozoic Systems of the Rocky Mountain Region, USA: Denver, Rocky Mountain Section, Society for Sedimentary Geology, 391 pp.

Anderson, R. Y., and D. W. Kirkland, 1960, Origin, varves and cycles of Jurassic Todilto Formation. AAPG Bulletin, v. 44, p. 37–52.

Armstrong, A. K., 1995, Facies, diagenesis and mineralogy of the Jurassic Todilto Limestone Member, Grants uranium district, New Mexico. New Mexico Bureau of Mines and Mineral Resources Bulletin, v. 153, p. 1–41.

Barbour, C. D., and J. H. Brown, 1974, Fish species diversity in lakes. The American Midland Naturalist, v. 108, p. 423–489.

Chenoweth, W. L., 1985. Historical review of uranium production from the Todilto Limestone, Cibola and McKinley counties, New Mexico. New Mexico Geology, v. 7, p. 80–83.

Cockerell, T. D. A., 1931, A supposed insect larva from the Jurassic. Bulletin of the Brooklyn Entomological Society, v. 26, p. 96–97.

Dunkle, D. H., 1942, A new fossil fish of the family *Leptolepidae*. Cleveland Museum of Natural History, Science Publications, v. 8, p. 61–64.

Harshbarger, J. W., C. A. Repenning, and J. H. Irwin, 1957, Stratigraphy of the uppermost Triassic and Jurassic rocks of the Navajo Country. U. S. Geological Survey Professional Paper, no. 291, p. 1–74.

Imlay, R. W., 1980, Jurassic paleobiogeography of the conterminous United States in its continental setting. U. S. Geological Survey Professional Paper, no. 1062, p. 1–134.

Kietzke, K. K., 1992, Reassignment of the Jurassic Todilto Limestone ostracode *Metacypris todiltoensis* Swain, 1946, to *Cytheridella* with notes on the phylogeny and environmental implications of this ostracode. New Mexico Geological Society Guidebook, v. 43, p. 173–183.

Kirkland, D. W., R. E. Denison, and R. Evans, 1995,

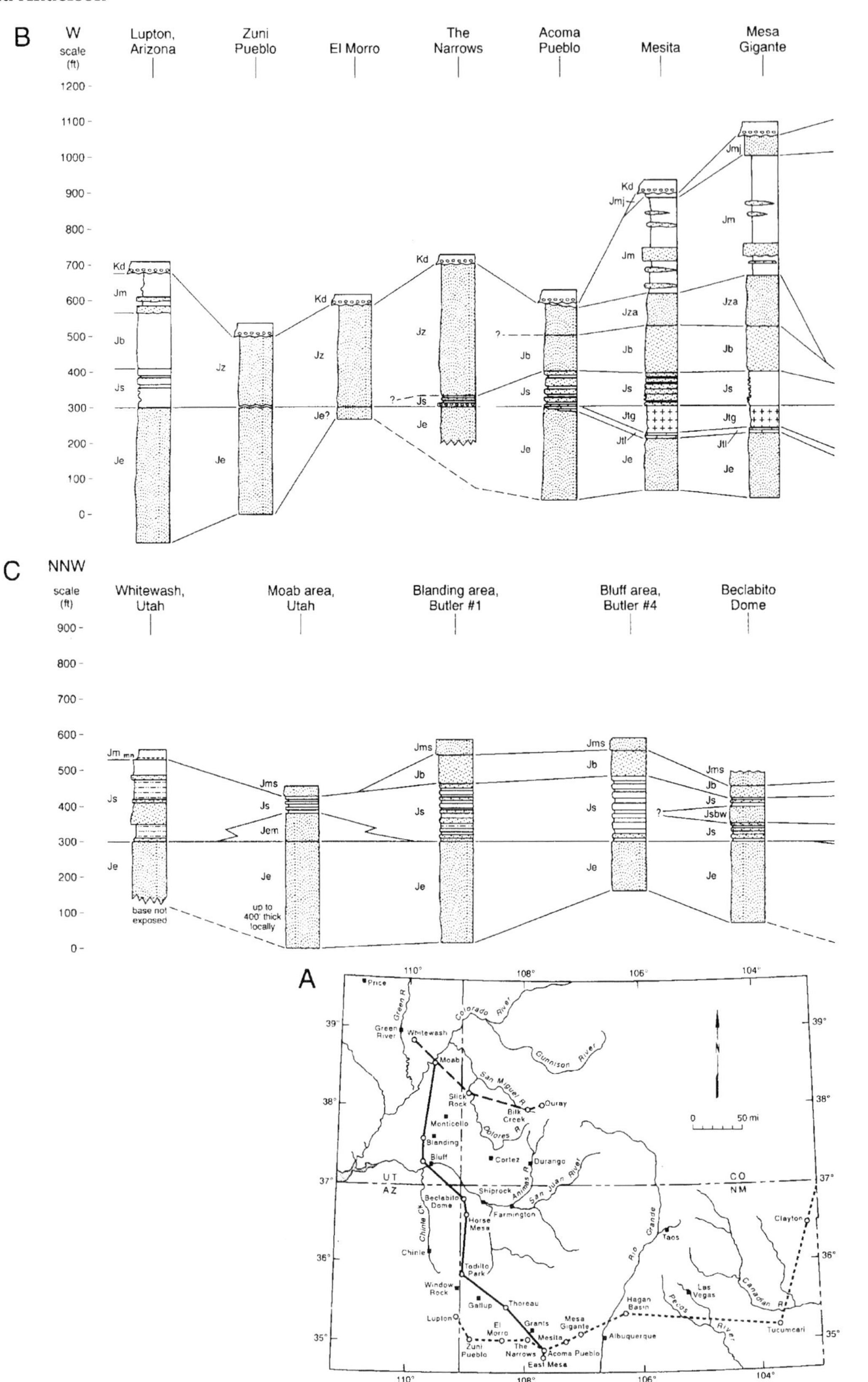

Figure 3—Regional correlation of the Todilto Formation and other San Rafael Group strata. (A) Lines of stratigraphic cross sections shown in B–D. (B) Lupton, Arizona, to northeastern New Mexico. (C) Whitewash, Utah, to Acoma Pueblo, New Mexico. (D) Whitewash, Utah, to Ouray, Colorado.

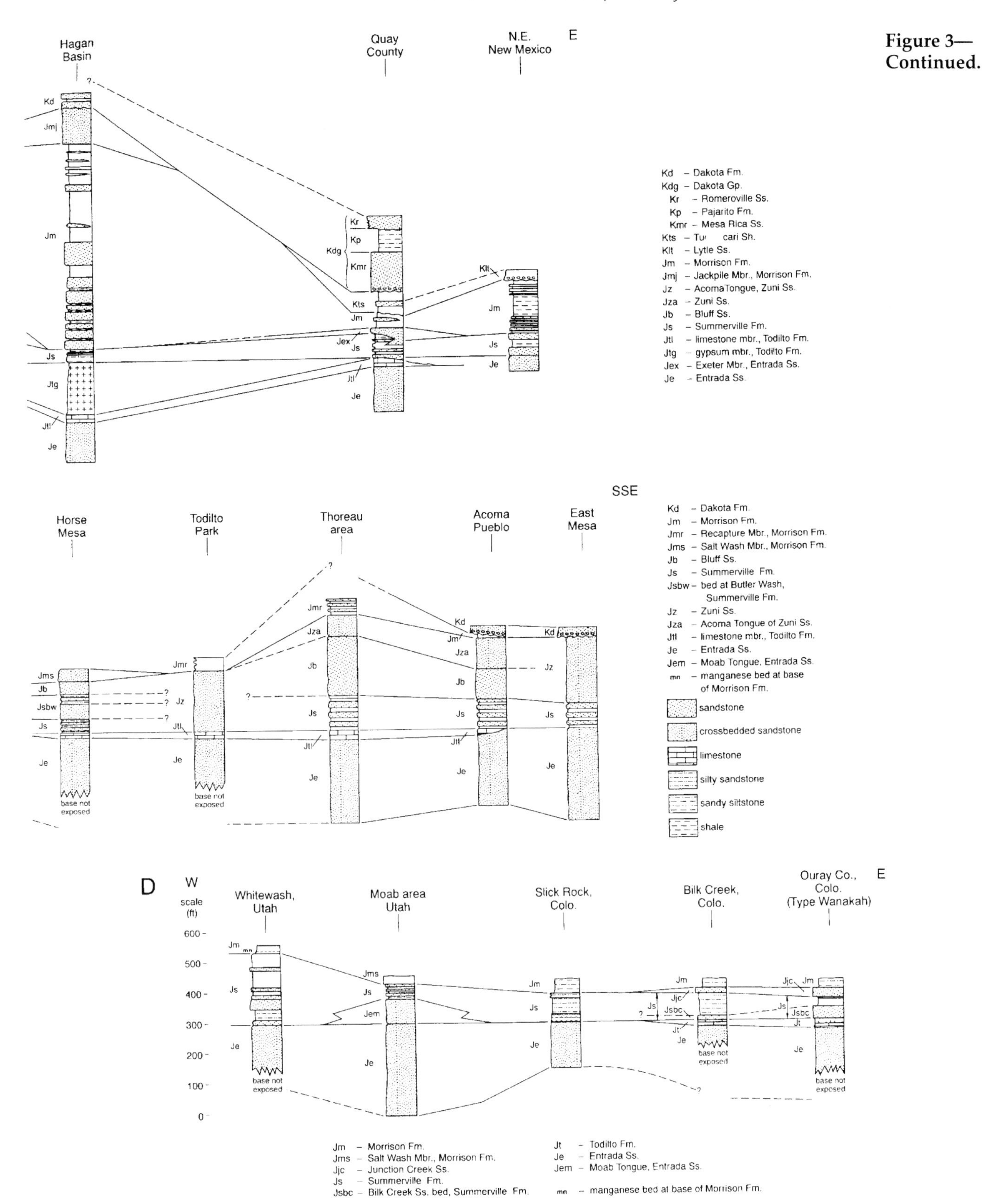

Figure 3— Continued.

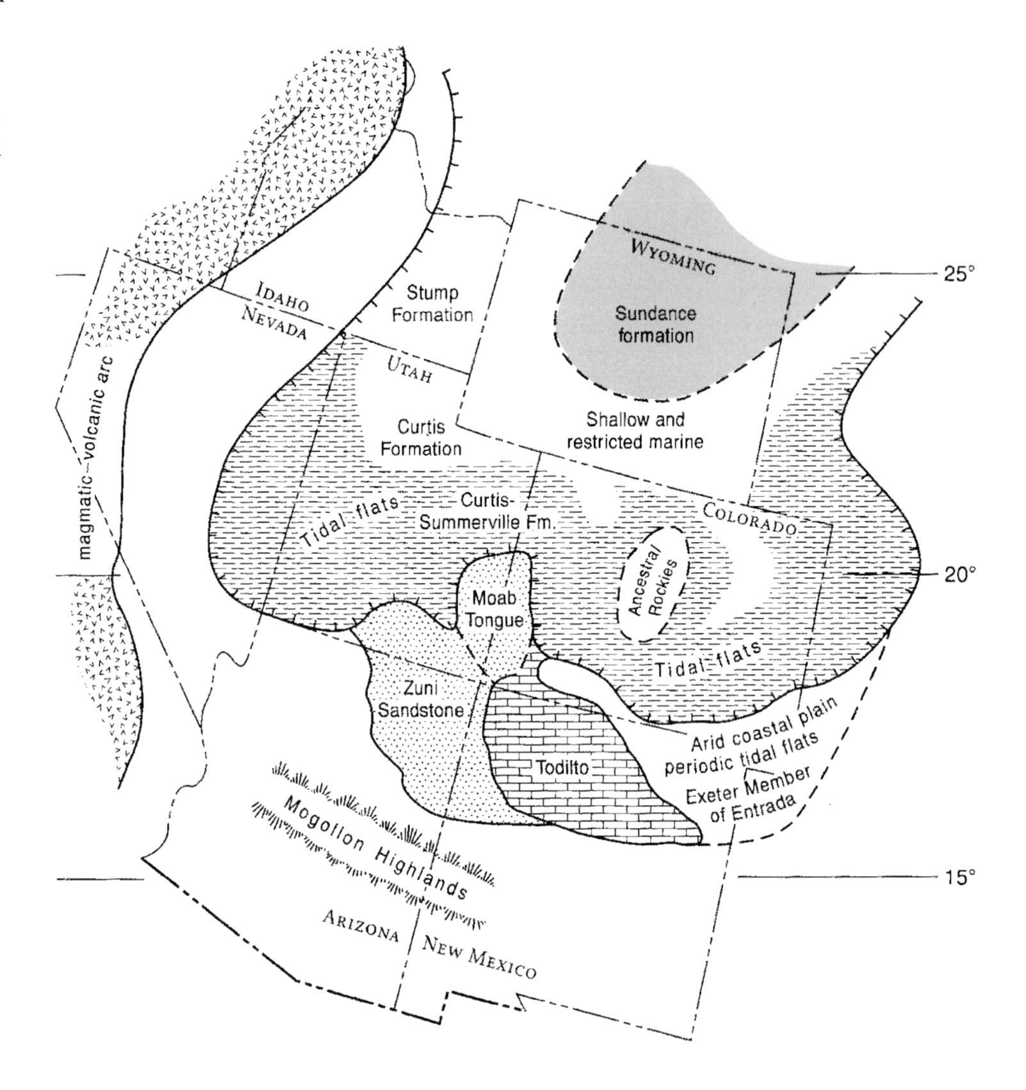

Figure 4—Late Middle Jurassic (Callovian) paleogeography of the American Southwest. After Anderson and Lucas (1994).

Middle Jurassic Todilto Formation of northern New Mexico and southwestern Colorado: marine or nonmarine? New Mexico Bureau of Mines and Mineral Resources Bulletin, v. 147, p. 1–39.

Koerner, H. E., 1930, Jurassic fishes from New Mexico. American Journal of Science, v. 19, p. 463.

Lucas, S. G., and K. K. Kietzke, 1986, Stratigraphy and petroleum potential of the Jurassic Todilto Formation in northeastern New Mexico, *in* J. L. Ahlen, M. E. Hanson, and J. Zidek, eds., Southwest Section of AAPG Transactions and Guidebook of 1986 Convention, Ruidoso, New Mexico: Socorro, New Mexico Bureau of Mines and Mineral Resources, 98 pp.

Lucas, S. G., 0. J. Anderson, and C. Pigman, 1995, Jurassic stratigraphy in the Hagan basin, north-central New Mexico. New Mexico Geological Society Guidebook, v. 46, p. 247–255.

Lucas, S. G., K. K. Kietzke, and A. P. Hunt, 1985, The Jurassic System in east-central New Mexico. New Mexico Geological Society Guidebook, v. 36, p. 213–242.

Ridgley, J. L., and M. Goldhaber, 1983, Isotopic evidence for a marine origin of the Todilto Limestone, north-central New Mexico. Geological Society of America Abstracts with Programs, v. 15, p. 414.

Schaeffer, B., and C. Patterson, 1984, Jurassic fishes from the western United States, with comments on Jurassic fish distribution. American Museum Novitates, no. 2796, p. 1–86.

Vincelette, R. R., and W. E. Chittum, 1981, Exploration for oil accumulations in Entrada Sandstone, San Juan Basin, New Mexico. AAPG Bulletin, v. 65, p. 2546–2570.

Warren, J. K., 1982, The hydrological setting, occurrence sand significance of gypsum in late Quaternary salt lakes in South Australia. Sedimentology, v. 29, p. 609–637.

Warren, J. K., and C. S. G. Kendall, 1985, Comparison of sequences formed in marine sabkha (subaerial) and salina (subaqueous) settings—modern andf ancient. AAPG Bulletin, v. 69, p. 1013–1023.

Weber, R. H., and F. E. Kottlowski, 1959, Gypsum resources of New Mexico. New Mexico Bureau of Mines and Mineral Resources Bulletin,v. 64, p. 1 68.

Tanner, L. H., Triassic-Jurassic lacustrine deposition in the Fundy Rift Basin, Eastern Canada, 2000, *in* E. H. Gierlowski-Kordesch and K. R. Kelts, eds., Lake basins through space and time: AAPG Studies in Geology 46, p. 159–166.

Chapter 12

Triassic-Jurassic Lacustrine Deposition in the Fundy Rift Basin, Eastern Canada

Lawrence H. Tanner
Department of Geography and Earth Science, Bloomsburg University
Bloomsburg, Pennsylvania, U.S.A.

OVERVIEW

The Fundy rift basin, comprising the contiguous Minas, Fundy, and Chignecto structural subbasins (Figure 1), is filled by terrestrial redbed siliciclastics, minor carbonates, and tholeiitic basalts of the Fundy Group, Newark Supergroup. The Minas subbasin is a shallow transtensional basin formed by left-oblique slip (Olsen and Schlische, 1990) on the reactivated Minas fault zone, a transform along which the Meguma and Avalon terranes were superimposed during the late Paleozoic. Outcrops of Fundy Group strata occur almost exclusively as seacliffs along the shores of the Bay of Fundy. In the Minas subbasin, a maximum of approximately 1 km of Fundy Group section is exposed along the northern and southern shores of the geographic Minas Basin and the eastern side of the Blomidon Peninsula (Figure 1).

The Fundy and Chignecto subbasins are simple half-grabens formed when regional extension caused reactivation of Paleozoic thrusts as southeast dipping normal faults with displacement locally exceeding 10 km (Withjack et al., 1995). Fundy Group strata are exposed along the Nova Scotia shore of the Bay of Fundy, forming the Fundy subbasin margin. A rider block along the faulted northwestern margin of the Fundy subbasin at Point Lepreau in New Brunswick exposes nearly 2.5 km of strata. Over 3.5 km of Mesozoic strata were penetrated in the Chinampas N-37 well (Figure 1) drilled offshore in the Fundy subbasin, and interpretation of seismic data suggests that the thickness of the Mesozoic section may exceed 8 km in the Fundy subbasin depocenter (Brown and Grantham, 1992; Wade et al., 1996). The Chignecto subbasin is bounded to the south by the Minas fault zone, separating it from the Fundy subbasin to the south. Outcrops in the Chignecto subbasin consist of rider blocks along the faulted northwestern margin in New Brunswick, exposing a maximum of 2.7 km of strata at St. Martins, and sections of 100–200 m at Martin Head and Waterside (Figure 1).

FUNDY GROUP STRATIGRAPHY

The stratigraphy of the Fundy Group is not unified among the subbasins (Figure 2), owing in large part to the discontinuous nature of the outcrop exposures in the Fundy and Chignecto subbasins. Olsen (1997) defines four tectonostratigraphic sequences (TS) for the strata of the central Atlantic margin basins, including the Fundy rift basin. Strata of TS I and II in the Fundy rift basin are mainly fluvial strata of Middle–Late Triassic age, with lacustrine and marginal lacustrine deposition limited to Late Triassic–Early Jurassic TS III and IV.

Wolfville Formation and Equivalents

The basal unit in the Fundy Group is the Wolfville Formation of the Minas subbasin and its equivalents in the Fundy and Chignecto subbasins. One fault-bounded outcrop on the north shore of the Minas subbasin, comprising cyclically interbedded cross-bedded sandstones and mudstones, has been dated as Anisian (Baird, 1986; Baird and Olsen, 1983, 1986) and is interpreted as the deposits of fluvial channels and floodplains (Skilliter, 1996). Most of the formation, which unconformably overlies pre-Triassic basement, is assigned an age of Carnian to Norian on the basis of osseus remains and ichnotaxa (Olsen, 1988).

These strata comprise basal conglomerates overlain by cross-bedded sandstones and minor interbedded mudstones, and are interpreted as the deposits of alluvial fans, alluvial braid-plains, and eolian dunes forming a general fining-upward sequence (Hubert and Forlenza, 1988; Olsen et al., 1989). Lithologically similar strata of roughly equivalent age are exposed in the Chignecto subbasin at St. Martins, Martin Head, and

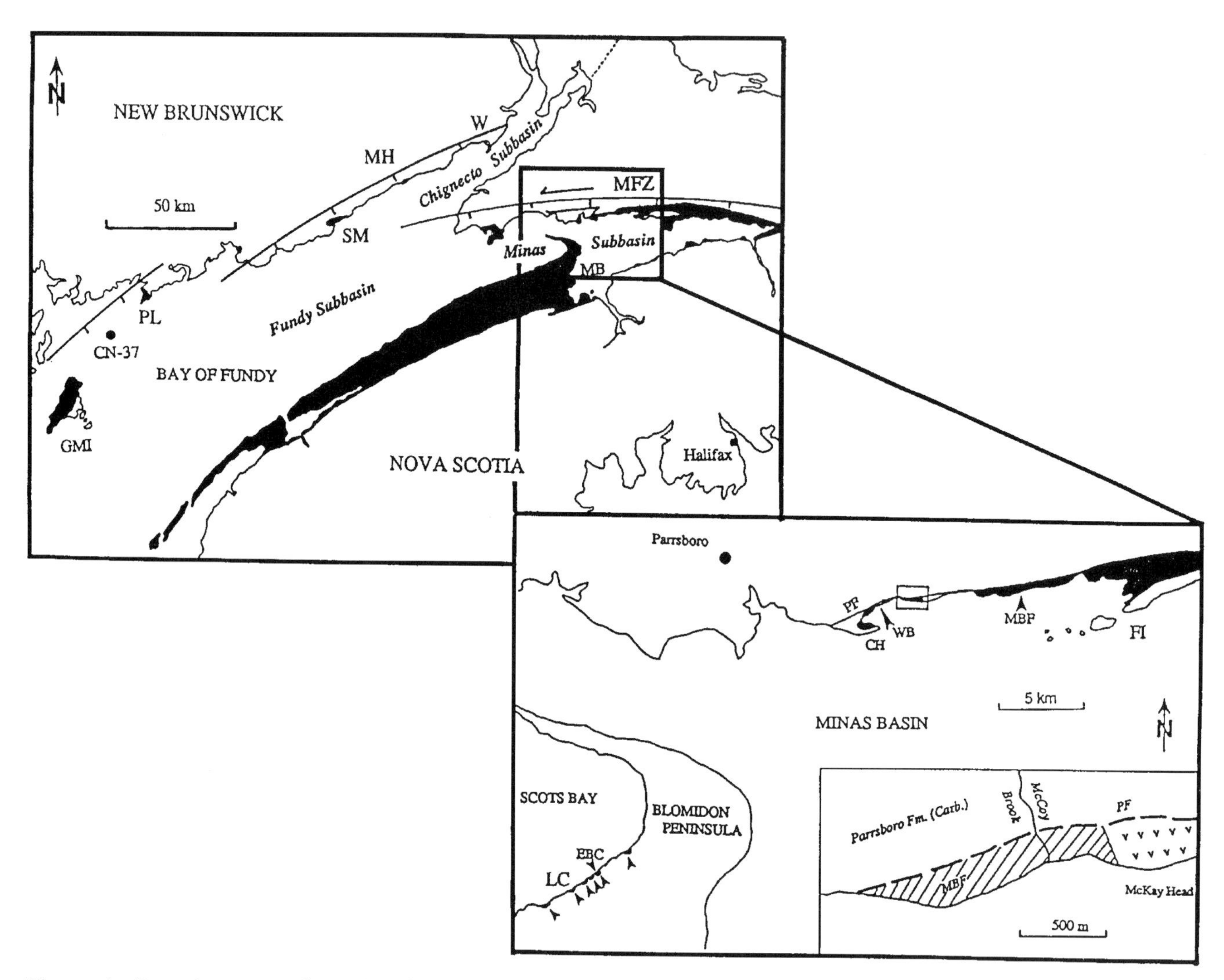

Figure 1—Location map of Fundy rift basin showing locations of the Fundy, Minas, and Chignecto subbasins, main structural elements, and key locations discussed in text. Black-shaded areas are outcrop exposures of Fundy Group strata. MFZ = Minas fault zone, MB = Minas Basin, CN-37 = Chinampas well, PL = Point Lepreau, SM = Saint Martins, MH = Martin Head, W = Waterside, GMI = Grand Manan Island. Inset maps: location map of outcrops (shaded and cross-hatched areas) of the Jurassic McCoy Brook formation (MBF) in the Minas subbasin. PF = Portapique fault, CH = Clark Head, WB = Wasson Bluff, EBC = East Broad Cove, LC = Lime Cove. McKay Head is the basalt headland to the east in the inset map. The type section of the McCoy Brook formation is to the west. Adapted from Tanner (1996).

Waterside, New Brunswick; and at Point Lepreau in the Fundy subbasin (Figure 2). These strata comprise coarse or interbedded coarse and fine redbed facies of alluvial and eolian origin (Nadon, 1981; Nadon and Middleton, 1984, 1985).

LACUSTRINE STRATA OF THE FUNDY RIFT BASIN

Blomidon Formation

Strata of the Blomidon Formation conformably overlie the Wolfville Formation except for unconformably overlying basement on the north shore of the Minas basin (Olsen and Schlische, 1990). A Norian age for most of the formation is assigned on the basis of freshwater fauna, ichnotaxa, and osseus remains (Olsen, 1988; Olsen et al., 1989).

A Hettangian palynoflora dominated by *Corollina meyeriana* at the top of the Blomidon Formation indicates the position of the Triassic–Jurassic boundary within the upper 10 m (Cornet, 1977). The formation comprises interbedded meter-scale beds of reddish-brown to gray, massive to cross-bedded sandstone, shale, and massive mudstone containing gypsum, relict evaporite crusts, sandpatch fabrics (Smoot and Olsen, 1988), and interpreted solution-collapse breccias (Mertz and Hubert, 1990; Ackerman et al., 1995). These strata are interpreted as the deposits of sediments in playa, sandflat, minor lacustrine, eolian, and fluvial

Period	Epoch	Age	Minas subbasin	Fundy subbasin	Chignecto subbasin	TS (Olsen, 1997)
JURASSIC	Early	Pliensbachian	*erosional top* ~~~?~~~			
		Sinemurian				
		Hettangian	*McCoy Brook Fm.* * / *North Mountain Basalt*			IV
TRIASSIC	Late	Norian	*Blomidon Formation* *	?	?	III
		Carnian	*Wolfville Formation*	*Lepreau Formation* ?	*Echo Cove Formation* *Quaco Formation* ?	II
	Middle	Ladinian			*Honeycomb Point Fm.*	
		Anisian	*"lower Wolfville"*		?	I

Figure 2—Stratigraphy of the Fundy Group formations. The question marks indicate the uncertainty in the ages of Fundy and Chignecto subbasin strata. TS = tectonostratigraphic sequences of Olsen (1997). * contains lacustrine strata.

environments under semiarid to arid conditions (Hubert and Hyde, 1982; Mertz and Hubert, 1990). Cyclic alternations of coarse and fine facies have been interpreted as representing tectonic autocycles controlled by differential basin subsidence (Mertz and Hubert, 1990) or Milankovitch-frequency climatic cycles (Olsen et al., 1989).

Interpreted lacustrine deposits consist of red shale and mudstone up to 1 m in thickness, locally containing a fragmentary freshwater fauna consisting of crustacean (darwinulid and candonid ostracodes, and conchostracans) and fish (redfieldiid) remains (Olsen, 1988; Olsen et al., 1989), interbedded with mudstones and sandstones of playa/sandflat origin as parts of asymmetric coarsening–fining-upward cycles (Figures 3, 4). The laminated fine-grained beds represent deposition in shallow lakes that periodically occupied the valley floor, perhaps during intervals of moister climate (Olsen et al., 1989; Ackerman et al., 1995). An increase in the abundance and organic content of lacustrine claystone at the top of the formation suggests an increasingly humid climate at the end of the Triassic (Mertz and Hubert, 1990; Fowell and Traverse, 1995). Dark gray, palyniferous mudstones at the top of the formation are the most organic-rich facies cropping out in the basin. The Blomidon Formation is overlain by up to 300 m of tholeiitic lavas of the North Mountain Basalt.

McCoy Brook Formation

Jurassic sedimentary rocks overlie the North Mountain Basalt along the north and south shores of the Minas Basin and beneath the Bay of Fundy (Figure 1). An Early Jurassic age is supported by the stratigraphic relation with the underlying North Mountain Basalt of Hettangian age; ichnotaxa, including therapod and fabrosaurid dinosaurs (Olsen, 1981, 1988), fish remains; and the osseus remains of synapsid, saurian, and therapod reptiles (Olsen et al., 1987; Olsen, 1988). Palynology suggests an age range of Hettangian to Pliensbachian for these sediments (Traverse, 1987). McCoy Brook strata comprise sandstones, mudstones, conglomerates, and basalt breccias of interpreted fluvial, lacustrine, playa, sandflat, alluvial-fan, eolian, debris-flow, and talus origin (Hubert and Mertz, 1984; Olsen et al., 1989; Tanner and Hubert, 1991, 1992) exposed in fault-bounded sections with a maximum thickness of approximately 230 m. The type section of the formation near McKay Head on the north shore of the Minas Basin (Figure 1) comprises sandstones in which the sandpatch fabric is common. Massive mudstones and minor claystones (Figure 5) represent deposition on laterally adjacent and shifting sandflats and playas, and intermittent shallow lakes. This interpretation of depositional environments is similar to that of the Blomidon Formation.

Lacustrine sedimentation is most fully developed in the basal Scots Bay Member of the McCoy Brook Formation (sensu Tanner, 1996). Outcrops of the Scots Bay Member occur on the north side of the basin, particularly at Wasson Bluff (Figure 6) (Olsen et al., 1989), and along the shore of Scots Bay on the west side of the Blomidon Peninsula in isolated synclinal basins (Figures 1, 7). These basins have been

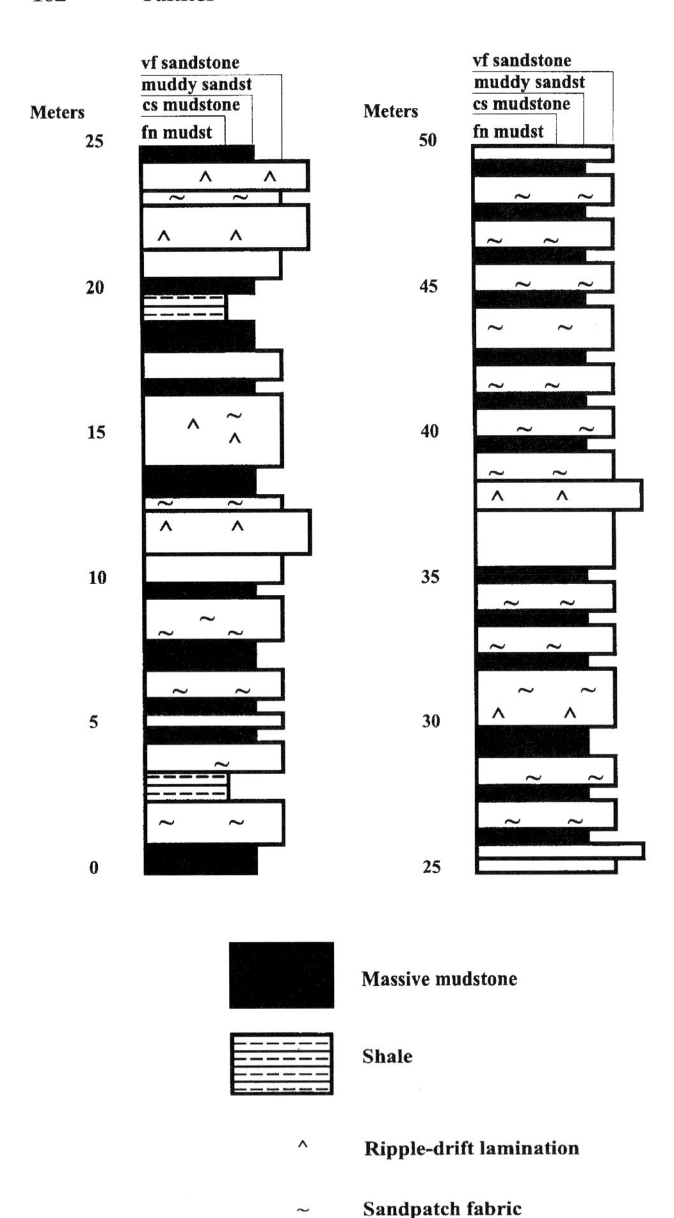

Figure 3—Portion of a section of the Blomidon Formation measured on the east side of the Blomidon Peninsula. Section comprises mostly alternating beds of coarse mudstone and muddy sandstone with ripple lamination and the sandpatch fabric.

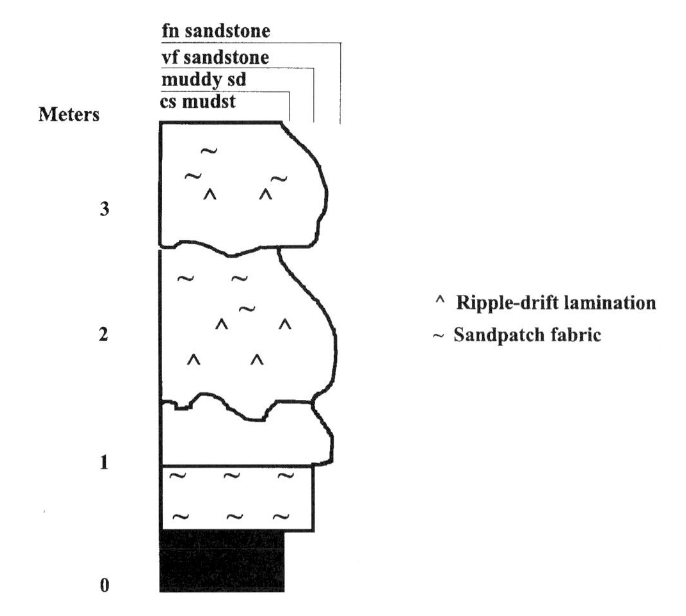

Figure 4—Detail of a representative section of the Blomidon Formation displaying sandy mudstones and massive to ripple-laminated to sandpatch sandstones arranged in asymmetric coarsening–fining-upward cycles.

interpreted as circular collapse structures in the North Mountain Basalt (Stevens, 1987). Outcrops of the Scots Bay Member lie directly on the surface of the North Mountain Basalt and range in thickness from 2 to over 9 m. At East Broad Cove along the shore of Scots Bay, the type section of the member comprises crudely bedded, red and green calcareous siltstone, overlain by partially silicified limestone, thick-bedded chert, stromatolitic limestone, and topped by brown, cross-bedded sandstone (Figure 8) (De Wet and Hubert, 1989; Olsen et al., 1989). At various outcrops, strata contain domal stromatolites, silicified logs, charophyte debris, fish bones, ostracodes, gastropods, and conchostracans. These features suggest deposition of shoreline to offshore sediments in a shallow, oxygenated, oligotrophic to eutrophic lake (Birney, 1985). This lake may or may not have been widespread across the rift valley (Good et al., 1994). The cherts are attributed to hydrothermal vents on the lake floor (De Wet and Hubert, 1989).

HYDROCARBON POTENTIAL

In contrast to the organic-rich mudstones that are abundant in lacustrine facies of the Newark Supergroup in basins to the south, lacustrine facies cropping out in the Fundy rift basin are primarily oxidized. Seismic profiles from the Chignecto subbasin, however, exhibit laterally continuous high-amplitude reflections, equivalent to the Wolfville Formation, that may indicate well-developed deep-water lacustrine strata that are potentially organic-rich (Withjack et al., 1995; Olsen, 1997). Wade et al. (1996) model deep lacustrine sedimentation centered along the basin axis during deposition of the Blomidon and McCoy Brook Formation and speculate that potential hydrocarbon source rocks exist in the depocenter of the basin in equivalents of these formations as well. If present, these source rocks would exist at a depth of over 3 km over most of the basin and would be thermally mature, based on geothermal gradients of 2.24–2.44°C/100 m measured in exploratory wells drilled in the Bay of Fundy (Wade et al., 1996). To date, no wells have been drilled in the basin depocenter to test this hypothesis.

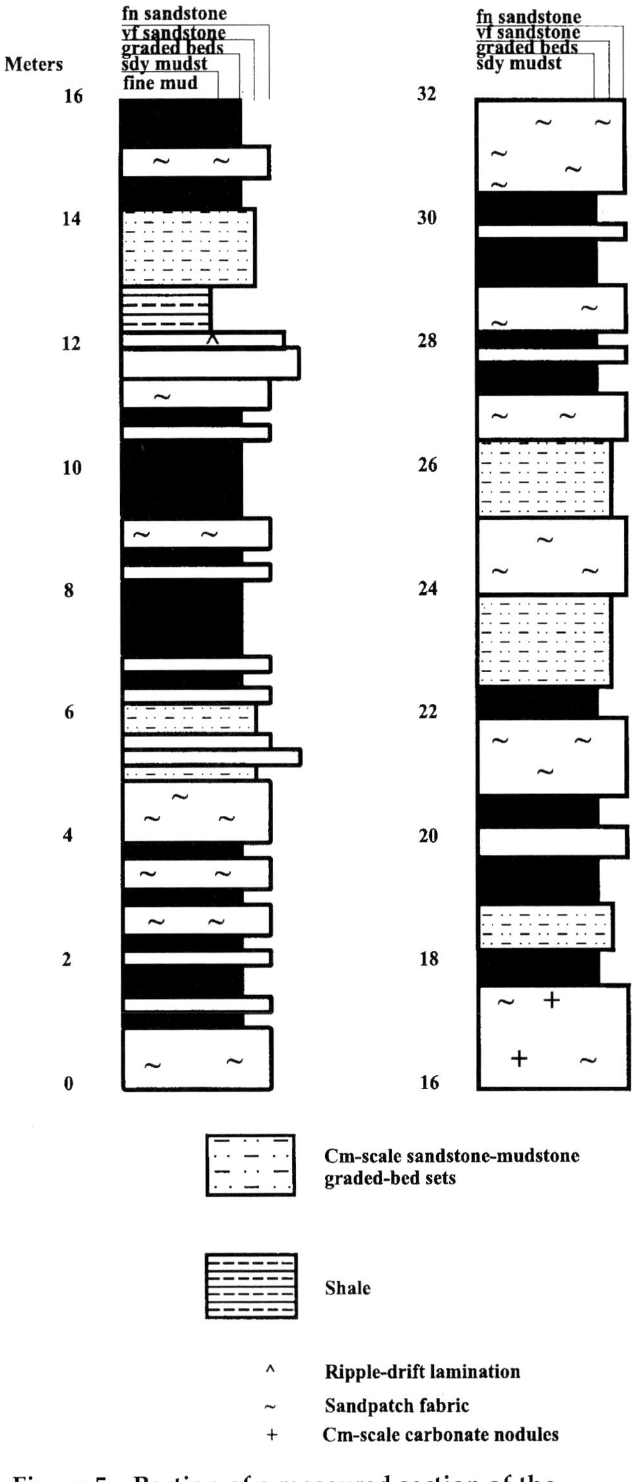

Figure 5—Portion of a measured section of the McCoy Brook Formation at the type location near McKay Head. Section consists of coarse mudstones and sandstones with the sandpatch fabric interbedded with sequences of cm-scale sandstone-mudstone graded beds.

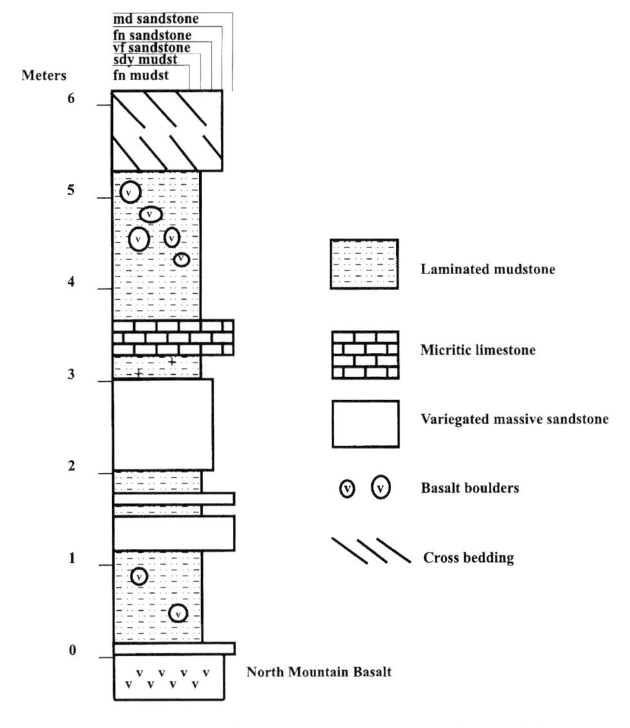

Figure 6—Section of the Scots Bay Member of the McCoy Brook Formation measured near Wasson Bluff on the north shore of the Minas Basin. Section consists of a basal volcaniclastic sand overlain by locally calcareous, laminated mudstone, sandstone, and micritic limestone, with locally interbedded basalt boulders.

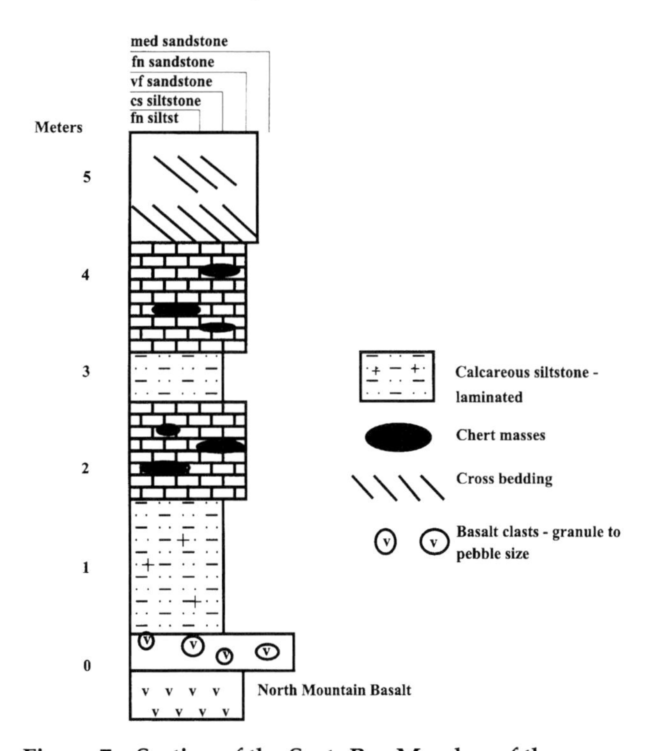

Figure 7—Section of the Scots Bay Member of the McCoy Brook Formation measured at the type location at East Broad Cove on the Blomidon Peninsula (Figure 1). Section consists of basal volcaniclastic sand, siltstone, calcareous siltstone, cherty limestone, and cross-bedded sandstone. Adapted from De Wet and Hubert (1989).

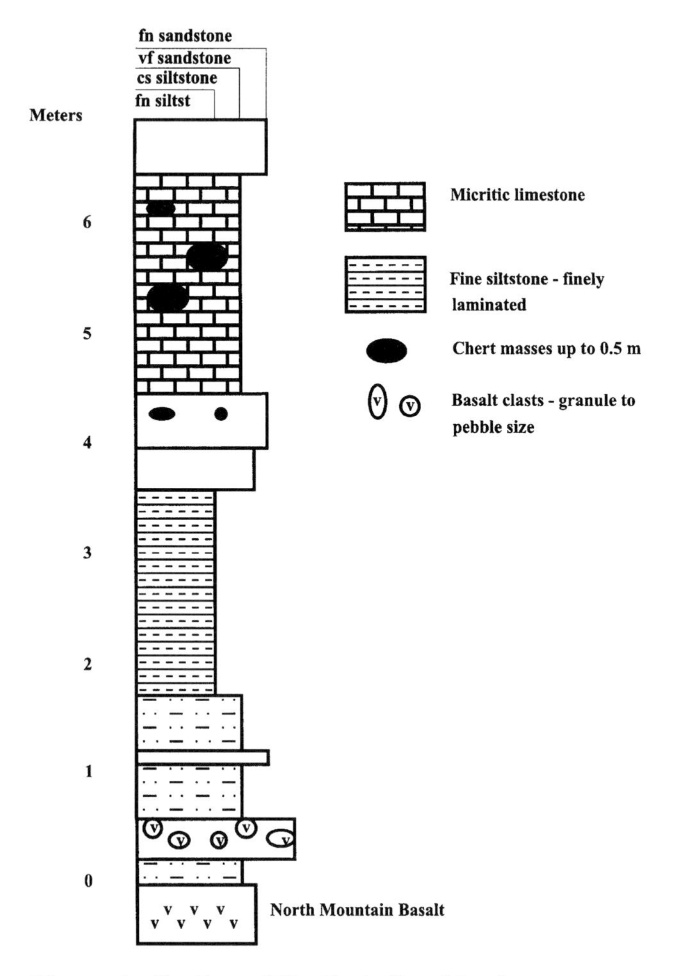

Figure 8—Section of the Scots Bay Member measured at Lime Cove (Figure 1). Facies are identical to those observed at East Broad Cove (Figure 7).

REFERENCES CITED

Ackerman, R. V., R. W. Schlische, and P. E. Olsen, 1995, Synsedimentary collapse of portions of the lower Blomidon Formation (Late Triassic), Fundy rift basin, Nova Scotia: Canadian Journal of Earth Science, v. 32, p. 1965–1976.

Baird, D. M., 1986, Middle Triassic herpetofauna in Nova Scotia: Friends of the Newark Newsletter, v. 10, p. 10.

Baird, D. M., and P. E. Olsen, 1983, Late Triassic herpetofauna from the Wolfville Formation of the Minas basin (Fundy basin) Nova Scotia, Canada: Geological Society of America (Abstracts with Programs) v. 15, no. 3, p. 122.

Birney, C. C., 1985, Sedimentology and petrology of the Scots Bay Formation (Lower Jurassic), Nova Scotia, Canada: MS thesis, University of Massachusetts, Amherst, 325 p.

Brown, D. E., and R. G. Grantham, 1992, Fundy basin rift tectonics and sedimentation, field excursion A-3: Guidebook, Geological Association of Canada, Mineralogical Association of Canada, Atlantic Geoscience Society, Wolfville, Nova Scotia, 264 p.

Cornet, B., 1977, The stratigraphy and age of the Newark Supergroup: PhD dissertation, Pennsylvania State University, University Park, Pennsylvania, 505 p.

De Wet, C. C. B., and J. F. Hubert, 1989, The Scots Bay Formation, Nova Scotia, Canada, a Jurassic carbonate lake with silica-rich hydrothermal springs: Sedimentology, v. 36, p. 857–873.

Fowell, S. J., and A. Traverse, 1995, Palynology and age of the upper Blomidon Formation, Fundy basin, Nova Scotia: Reviews in Paleobotany and Palynolology, v. 86, p. 211–213.

Good, S. C., L. A. Yenik, P. E. Olsen, and N. G. McDonald, 1994, Non-marine molluscs from the Scots Bay Formation, Newark Supergroup (Early Jurassic), Nova Scotia: taxonomic assessment and paleoecologic significance: Geological Society of America (Abstracts with Programs), v. 26, no. 3, p. 20.

Hubert, J. F., and M. F. Forlenza, 1988, Sedimentology of braided-river deposits in Upper Triassic Wolfville redbeds, southern shore of Cobequid Bay, Nova Scotia, Canada, *in* W. Manspeizer, ed., Triassic–Jurassic rifting: continental breakup and the origin of the Atlantic passive margins, part A: Amsterdam, Elsevier, p. 231–248.

Hubert, J. F., and M. G. Hyde, 1982, Sheet-flow deposits of graded beds and mudstones on an alluvial sandflat-playa system; Upper Triassic Blomidon redbeds, St. Mary's Bay, Nova Scotia: Sedimentology, v. 29, p. 457–474.

Hubert, J. F., and K. A. Mertz, 1984, Eolian sandstones in Upper Triassic–Lower Jurassic red beds of the Fundy basin, Nova Scotia: Journal of Sedimentary Petrology, v. 54, p. 798–810.

Mertz, K. A., and J. F. Hubert, 1990, Cycles of sand-flat sandstone and playa-lacustrine mudstone in the Triassic–Jurassic Blomidon redbeds, Fundy rift basin, Nova Scotia: implications for tectonic and climatic controls: Canadian Journal of Earth Science, v. 27, p. 442–451.

Nadon, G. C., 1981, The stratigraphy and sedimentology of the Triassic at St. Martins and Lepreau, New Brunswick: MSc thesis, McMaster University, Hamilton, Ontario, 279 p.

Nadon, G. C., and G. V. Middleton, 1984, Tectonic control of Triassic sedimentation in southern New Brunswick: local and regional implications: Geology, v. 12, p. 619–622.

Nadon, G. C., and G. V. Middleton, 1985, The stratigraphy and sedimentology of the Fundy Group (Triassic) of the St. Martins area, New Brunswick: Canadian Journal of Earth Science, v. 22, p. 1183–1203.

Olsen, P. E., 1981, Comment and reply on "eolian dune of field of Late Triassic age, Fundy basin, Nova Scotia": Geology, v. 9, p. 557–559.

Olsen, P. E., 1988, Paleontology and paleoecology of the Newark supergroup (early Mesozoic, eastern North America), *in* W. Manspeizer, ed., Triassic–Jurassic rifting: continental breakup and the origin of the Atlantic passive margins, part A: Amsterdam, Elsevier, p. 185–230.

Olsen, P. E., 1997, Stratigraphic record of the early Mesozoic breakup of Pangea in the Laurasia–Gondwana rift system: Annual Review of Earth and Planetary Sciences, v. 25, p. 337– 401.

Olsen, P. E., and D. M. Baird, 1986, The ichnogenus *Atreipus* and its significance for Triassic biostratigraphy, *in* K. Padian, ed., The beginning of the age of the dinosaurs: New York, Cambridge University Press, p. 66–87.

Olsen, P. E., and R. W. Schlische, 1990, Transtensional arm of the early Mesozoic Fundy rift basin: penecontemporaneous faulting and sedimentation: Geology, v. 18, p. 695–698.

Olsen, P. E., N. H. Shubin, and M. H. Anders, 1987, New Early Jurassic tetrapod assemblages constrain Triassic–Jurassic tetrapod extinction event: Science, v. 237, p. 1025–1029.

Olsen, P. E., R. W. Schlische, and P. J. W Gore, 1989, Tectonic, depositional, and paleoecological history of early Mesozoic rift basins, eastern North America: Guidebook to International Geological Congress Field Trip T–351, 174 p.

Plint, A. G., and H. W. van de Poll, 1984, Structural and sedimentary history of the Quaco Head area, southern New Brunswick: Canadian Journal of Earth Science, v. 21, p. 753–761.

Skilliter, C. C., 1996, The sedimentology of Triassic fluvial and aeolian deposits at Carrs Brook, Colchester County, Nova Scotia: BSc thesis, Saint Mary's University, Halifax, Nova Scotia, 61 p.

Smoot, J. P., and P. E. Olsen, 1988, Massive mudstones in basin analysis and paleoclimatic interpretation of the Newark Supergroup, *in* W. Manspeizer, ed., Triassic–Jurassic rifting: continental breakup and the origin of the Atlantic passive margins, part A: Amsterdam, Elsevier, p. 249–274.

Stevens, G. R., 1987, Jurassic basalts of northern Bay of Fundy region, Nova Scotia, *in* D. C. Roy, ed., Centennial Field Guide Volume 5: Northeastern Section of the Geological Society of America, p. 415–420.

Tanner, L. H., 1996, Formal definition of the Lower Jurassic McCoy Brook Formation, Fundy rift basin, eastern Canada: Atlantic Geology, v. 32, p. 127–135.

Tanner, L. H., and J. F. Hubert, 1991, Basalt breccias and conglomerates of the Lower Jurassic McCoy Brook Formation, Fundy basin, Nova Scotia: differentiation of talus and debris–flow deposits: Journal of Sedimentary Petrology, v. 61, p. 15–27.

Tanner, L. H., and J. F. Hubert, 1992, Depositional facies, palaeogeography and palaeoclimatology of the Lower Jurassic McCoy Brook Formation, Fundy rift basin, Nova Scotia: Palaeogeography, Palaeoclimatology, Palaeoecology, v. 96, p. 261–280.

Thompson, J. P., 1974, Stratigraphy and geochemistry of the Scots Bay Formation, Nova Scotia: MS thesis, Acadia University, Wolfville, Nova Scotia, 358 p.

Traverse, A., 1987, Pollen and spores date origin of rift basins from Texas to Nova Scotia as early Late Triassic: Science, v. 236, p. 1469–1472.

Wade, J. A., D. E. Brown, A. Traverse, and R.A. Fensome, 1996, The Triassic–Jurassic Fundy basin, eastern Canada: regional setting, stratigraphy and hydrocarbon potential: Atlantic Geology, v. 32, p. 189–231.

Withjack, M. O., P. E. Olsen, and R. W. Schlische, 1995, Tectonic evolution of the Fundy rift basin, Canada: evidence of extension and shortening during passive margin development: Tectonics, v. 14, p. 390–406.

Stollhofen, H., I. G. Stanistreet, S. Gerschütz, 2000, Early Jurassic rift-valley–related alkaline lake deposits interbedded with Karoo flood basalts, southern Namibia, *in* E. H. Gierlowski-Kordesch and K. R. Kelts, eds., Lake basins through space and time: AAPG Studies in Geology 46, p. 167–180.

Chapter 13

Early Jurassic Rift-Valley-Related Alkaline Lake Deposits Interbedded with Karoo Flood Basalts, Southern Namibia

Harald Stollhofen
Institut für Geologie, Universität Würzburg
Würzburg, Germany

Ian G. Stanistreet
Department of Earth Sciences, University of Liverpool
Liverpool, United Kingdom

Stephan Gerschütz
Institut für Geologie, Universität Würzburg
Würzburg, Germany

GEOLOGICAL CONTEXT

The lake sediments are preserved as two separate layers interleaved with Early Jurassic Kalkrand flood basalts (Duncan et al., 1984; Gerschütz, 1996) and are best exposed around Hardap reservoir, 15 km north of Mariental town in southwestern Namibia (Figure 1). The 55–300 m thick volcano-sedimentary sequence unconformably oversteps underlying Triassic (Stormberg) and latest Carboniferous–Early Permian (Dwyka and Ecca) sediments in the east onto latest Proterozoic–Early Cambrian (Nama) basement toward the west (Heath, 1972; Schalk and Germs, 1980). The top of the sequence is defined by the erosive Cretaceous land surface with continental sediments of the Cenozoic Kalahari thermal sag basin draped thinly over the top. The 183.0 ± 0.6 Ma to 186.0 ± 0.8 Ma dated olivine-tholeiitic basalts (Duncan et al., 1997) of the Kalkrand Formation were extruded during extensional rifting between South America and Africa (Miller, 1992; Dingle, 1993). This was one of a series of early rifting episodes from the Permian onward (Figure 2) prior to ultimate extensional rift phases during the Early Cretaceous. The latter caused extrusion of the Etendeka-Paraná flood basalts and onset of South Atlantic oceanic opening (Hawkesworth et al., 1992; Renne et al., 1996; Gladczenko et al., 1997).

Extensional tectonism was experienced in the study area along a northerly trending set of extensional faults dipping toward the east. This is inferred (Stollhofen et al., 1998) to be conjugate to a westerly dipping break-away detachment system that connected with the more major detachment toward the west, along which oceanic opening was eventually effected (Maslanyj et al., 1992; Light et al., 1992, 1993). The Hardap area is therefore characterized by local depocenters that involved half-graben geometries. The extensional fault system is demonstrably synvolcanic and synsedimentary (Stollhofen et al., 1998), and we infer that faults acted as conduits passing magma through fissure-type feeders to the surface and near-surface. The lava sequence has been important in protecting and preserving the alkaline lake sequences from subsequent erosion. In addition to the build-up of composite flood basalt flows, magma was transported considerable distances laterally through subvolcanic lava feeder ridges (Stollhofen et al., 1998). These were generated when magma inflated an existing, partly cooled pahoehoe flow-unit (cf. Hon et al., 1994) to form a linear ridge feature along the axis of magma transport. Quite commonly such ridge structures followed regional slopes, parallel to the axis of the half-graben. Ultimately, the lava feeder ridges would have fed secondary, subaerial breakout flows in areas both lateral to and distal from the elevated ridge structure. Where subaerial flows developed laterally, they onlap onto the ridge structure, lessening its topographic relief.

Figure 1—Maps showing locality of Hardap area and offshore Kudu wells in southern Namibia, and geology, structure, and type sections (localities 1–4), together with the distribution of volcanically related depressions ("pool structures") of the Jurassic Kalkrand Formation northwest of Mariental.

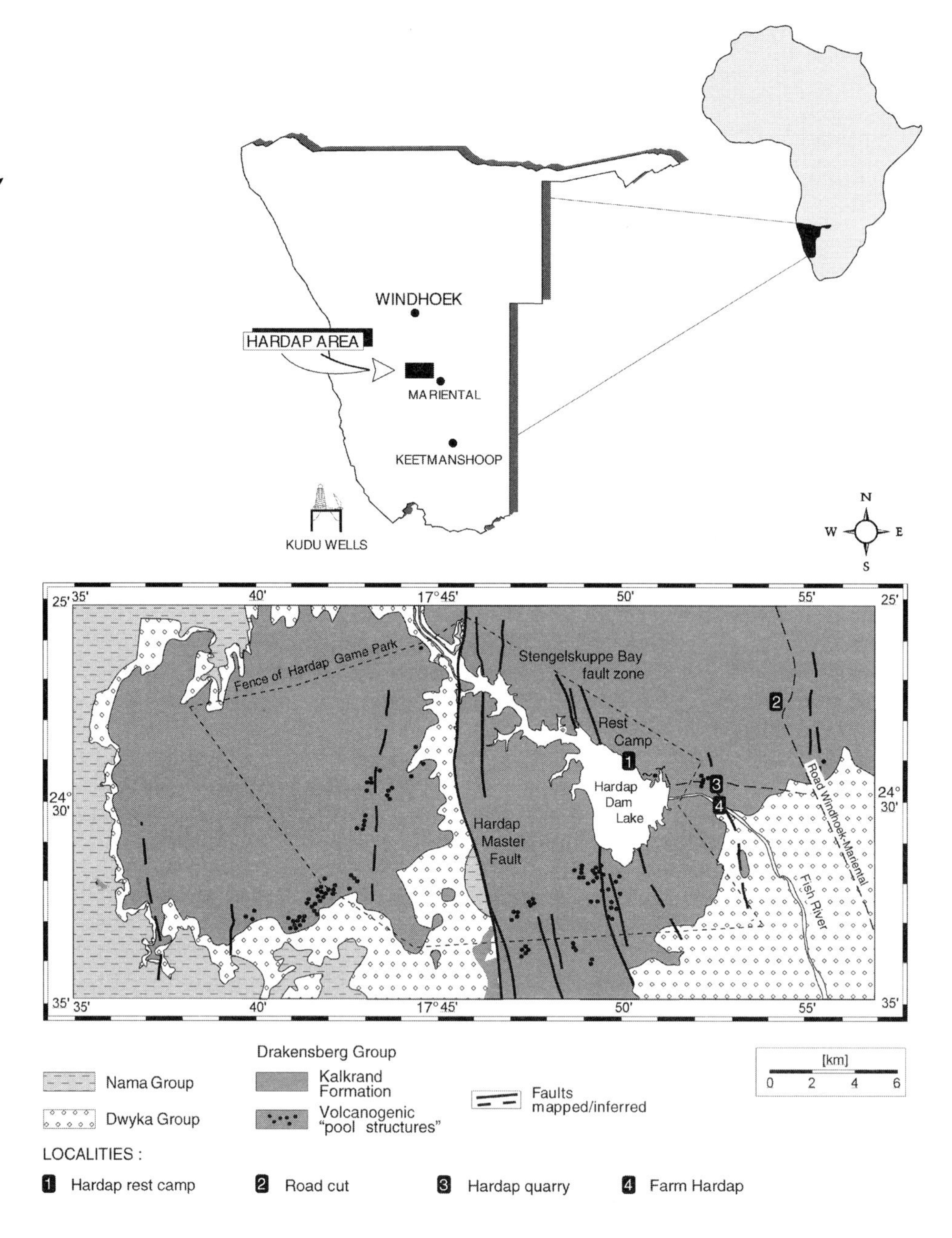

INTERLEAVED LACUSTRINE SEQUENCES

The volcano-sedimentary sequence was heralded by the deposition of coarse conglomerates derived from the newly uplifted Nama basement footwall of the extensional fault system along the Hardap master fault and confined to valley fills incised into the underlying Dwyka sequence. Figure 2 illustrates this relationship and the arrangement of the three overlying lava sequences (8–25 m thick) with the two interleaved sediment layers, ranging in thickness from 12 m to areas of total pinch out, with the latter often developed against lava feeder ridges that have not collapsed subsequently.

Type sections (Figure 3) for sediment interlayers I and II were measured where they are both well exposed in the same section, in the cliffs below the resort area of Hardap Nature Park (locality 1 in Figure 1). After initial weathering of the lava surface, both are characterized by (phase 1) early eolian/fluvial layers, when the basin was still connected with the regional drainage system. Faulting and volcanic damming subsequently cut the basin progressively off from the drainage, resulting in (phase 2) more local derivation of sediment and ultimately a closed lake system in which evaporites were precipitated. Eventually, decreased tectonic subsidence rates led to the reestablishment of a regional drainage and (phase 3) siliciclastic sedimentation resumed, leading to the termination of lake development.

Sediment Interlayer I

The early basin history (phase 1) is characterized by quartz-dominated, poorly sorted sandstones containing

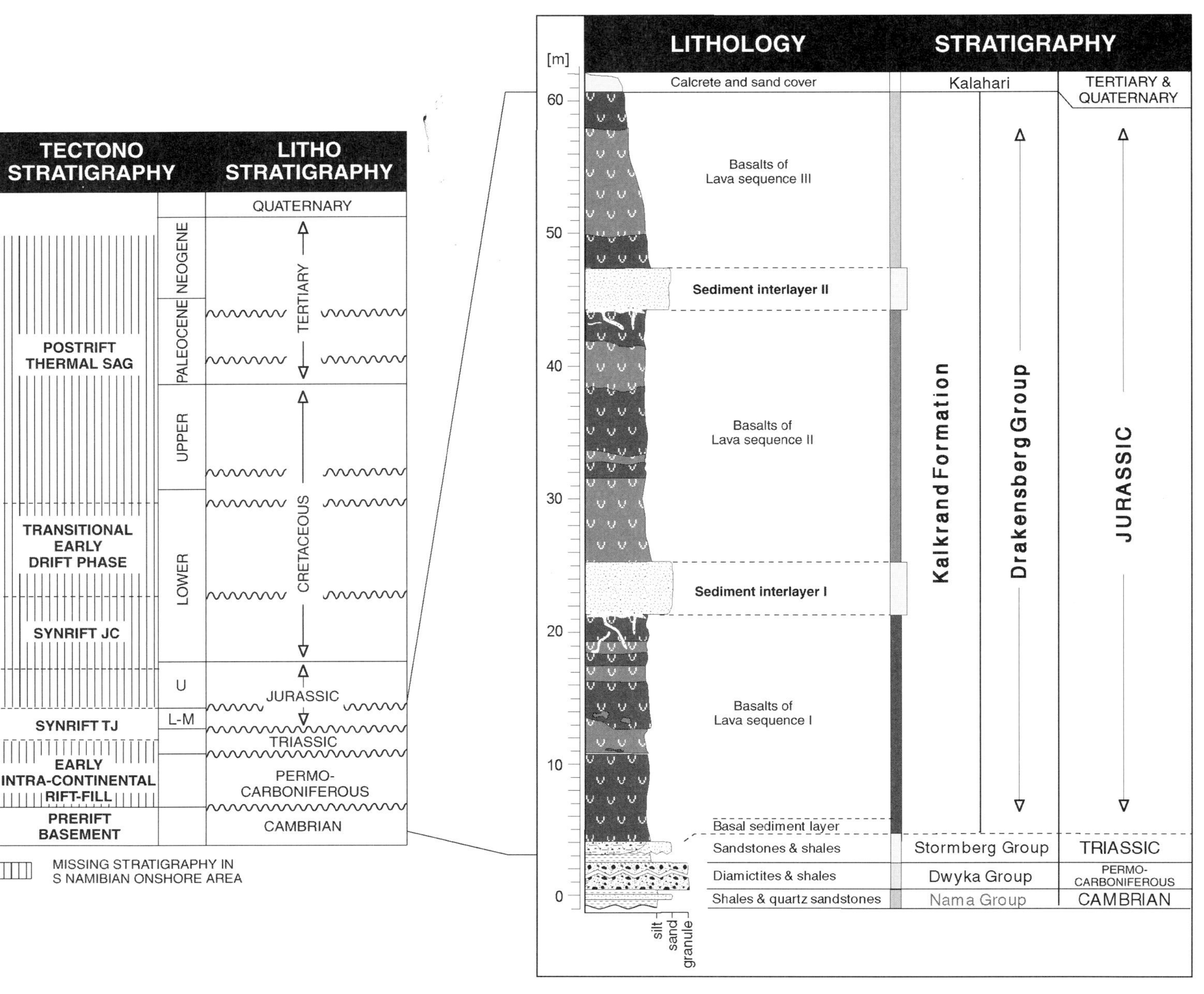

Figure 2—Generalized stratigraphy of the Kalkrand Formation and associated strata (pre-Jurassic strata not to scale) within the regional tectonostratigraphic framework, which is modified from Light et al. (1993) and largely based on the stratigraphy observed in the Namibian offshore area.

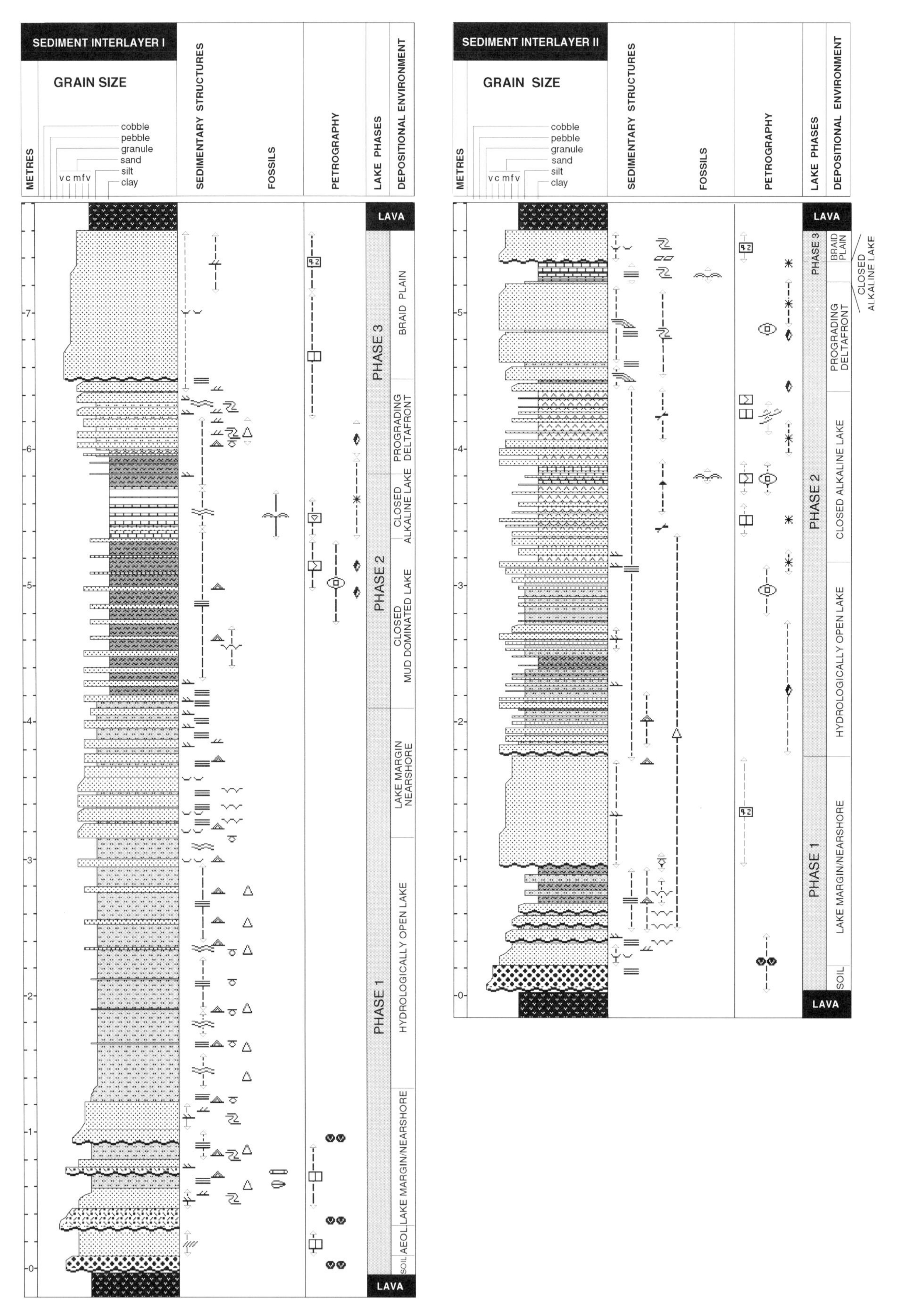

Figure 3—Detailed lithologic logs of sediment interlayer I and II at the cliffs below the resort area of Hardap Nature Park (locality 1 in Figure 1).

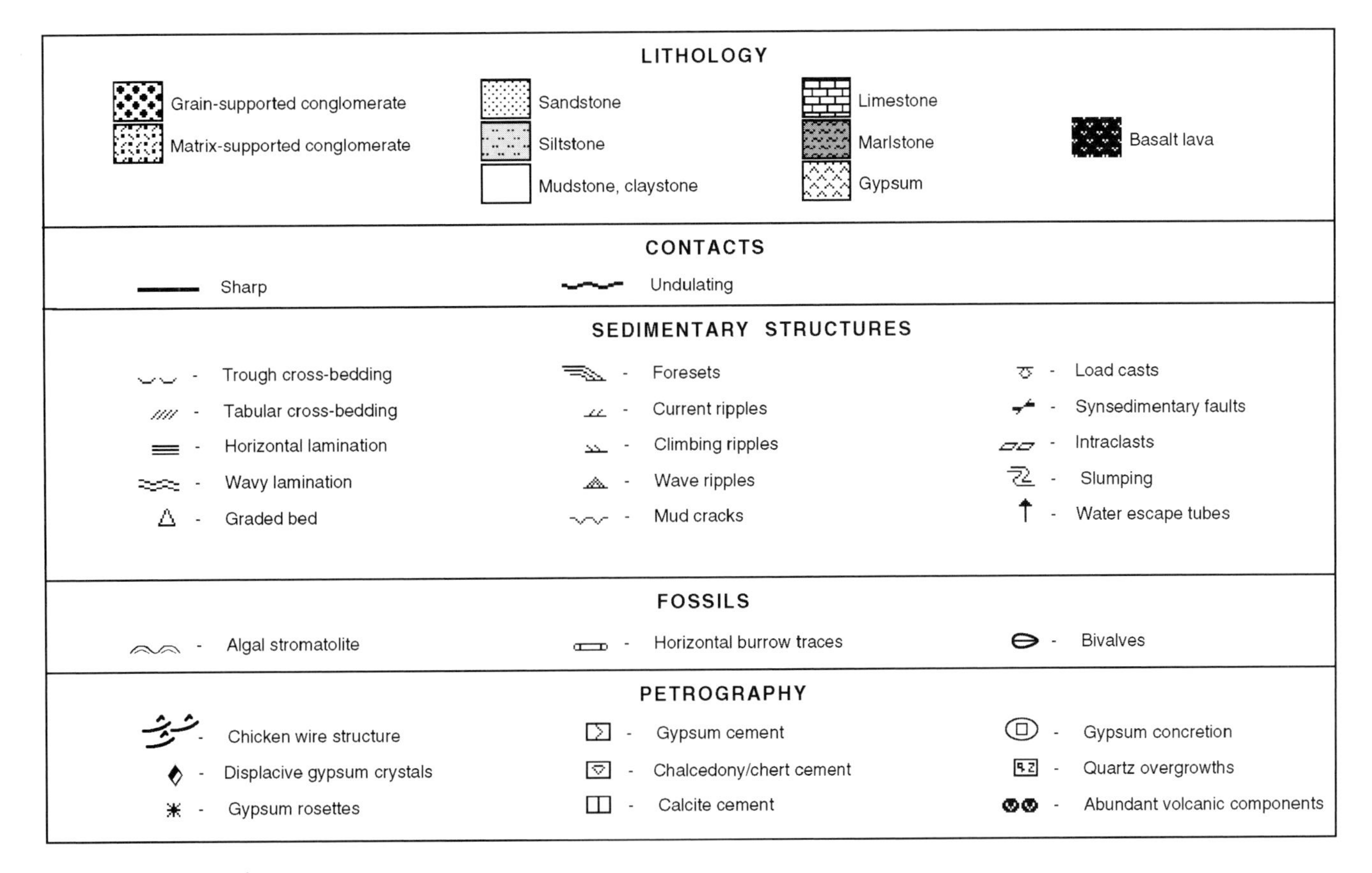

Figure 3—Legend.

basaltic lithic grains and gravel fragments interpreted as sheet flood deposits. More quartz-rich planar cross-bedded sandstone lenses are interpreted to be eolian or eolian-derived. With increasing subsidence, muds were deposited in a hydrologically open lake system, subsequently filled by tempestites and/or turbidites associated with the progradation of fluvio-deltaic complexes. The latter comprise trough cross- and planar-bedded, reddish, calcareous, medium sandstones, rarely yielding unidentified, strongly flattened shells of bivalved organisms on bedding planes at the base of sediment interlayer I. The only trace fossils found in either of the sediment interlayers were meandering burrow systems preserved in this part of interlayer I. Sedimentary structures indicate rather uniform paleocurrents toward the west-northwest to north (280–360°), and wave-rippled bedding surfaces show ripple crests aligned northwest-southeast (150–170°).

The hydrologically closed lake that followed (phase 2) is heralded by the deposition of greenish finer-grained mudstones in which an increasing amount of gypsum is observed of both displacive subsurface and free-surface growth origins. As clay input decreased, algally laminated limestones and evaporite layers were deposited, recording the maximum development of closed lake conditions.

The reestablishment of the regional drainage (phase 3) brought mud back into the lake, forming the base of a coarsening-upward sequence that represents deltaic advance. This was fed by a northward-flowing (340–355°) braided fluvial system, transporting and depositing quartz-rich, trough cross-bedded, medium to coarse sand that caps and terminates the sequence.

Sediment Interlayer II

Early deposition (phase 1) exhibits alluvially deposited units at the base in which well-rounded grains appear to have eolian source characteristics. Subsidence led to the development of the lake that was hydrologically open, depositing reddish, desiccation-cracked muds and thin turbidites/tempestites. The subsequent hydrological response to the subsidence was the progradation of a climbing-ripple cross-laminated yellowish sandstone that loaded into the underlying muds, interpreted as mouth bar progression into the shallow lake.

Subsequently (phase 2), the setting up of the major lake system is recorded by interbedded sandstones and reddish mudstones in which cyclic changes in sandstone bed thickness can be recognized. Such sandstone beds are normally graded, showing load casts at their basal contact and well-developed wave ripples crests (strike 120–150°) associated with the upper bedding surfaces. We suggest deposition dominantly by tempestites and/or turbidites. This implies that the lake was still hydrologically open, receiving large quantities of quartz-rich sediments from marginal fluvio-deltaic systems. As subsidence progressed, the influence of this external source decreased, and a hydrologically closed system established. At a well-defined set of maximum flooding surfaces, algally laminated cherty limestones indicate hemipelagic basinal conditions. Following the flooding a pronounced mouth bar sequence filled large areas of the lake, testified by classic toeset, foreset, and topset relationships. A second, more major flooding period rapidly restricted this siliciclastic input and a carbonate-evaporite unit was deposited. Stromatolitic cherty limestones, in which algal domes are identified, were interbedded with gypsum layers exhibiting fibrous textures together with minor, laminated, dark gray claystone layers.

The reestablishment of the regional drainage system (phase 3) is heralded by a major erosional surface overlain by braided, fluvial, trough cross-bedded quartzarenites in which grain shapes and textures are indicative of an ultimate aeolian source. A silcrete cap and a major erosion surface indicate a considerable time gap prior to the extrusion of lava sequence III.

VARIABILITY OF SEDIMENT INTERLAYERS

Although the sections presented above can be viewed as type sections, a pronounced variability is evident in both of the sediment interlayers. This can be illustrated with specific reference to sediment interlayer II. Two sets of cross-sectional profiles (Figures 4, 5) are reproduced of areas in which profound differences are apparent. The first is of a road cutting 7 km east of the type section just described on the Mariental-Kalkrand highway B1 15 km north of Mariental (locality 2 in Figure 1). The second comprises two profiles across pool structures (localities 3 and 4 in Figure 1) respectively 1.8 km and 2 km east of Hardap dam, both adjacent to the main access road to Hardap resort.

The Road Section

Figure 4 shows a totally different aspect of lake sedimentation from that displayed by the type section, although the same sequence-stratigraphic patterns can be established. The sequence is dominated by debris flow deposition, but the profile also explains why such a difference should be apparent. On the southern end of the section, the sequence (4 m maximum) pinches out almost entirely (<1 m) against an adjacent lava feeder ridge. Two subsidiary lava flows initially lapped against the ridge, but an appreciable topography still remained to provide a local source for debris flows. The sequence is most completely preserved directly adjacent to the lava feeder ridge.

Phase 1 is initiated with weathered in situ and transported soils preserved particularly well in the depression between the lava ridge and the topmost onlapping subsidiary flow. In contrast with the fluviodeltaic deposits of the type section, phase 1 is completed at the road section by a series of debris flow deposits dominated by lava clasts up to 25 cm in diameter. That many of the debris flows were subaqueous is shown by interbedded muds and thin turbidites without any sign of emergence.

Phase 2 is represented only by a thin, mainly eroded wedge of algally laminated carbonate with replacement chert and some gypsum layers.

Phase 3 is initiated by erosional down-cutting that correlates with the erosional surface beneath phase 3 at the type section. At the road section, however, debris flows are deposited on top of this surface prior to the deposition of a trough cross-bedded quartzarenite that correlates with the topmost braided fluvial unit at the type section. Although the quartz-rich sand was still unlithified, it loaded chaotically into the underlying diamictites; however, it is preserved with a tabular geometry on top of the stable lava feeder ridge to the south and where the underlying sequence thins substantially to the north. A thin amygdaloidal lava flow caps the sedimentary sequence and is eroded prior to the overlying volcanic sequence. We relate this to the local flows that still originated from lava feeder ridges.

The overlying lava sequence III was initially punctuated by pronounced phases of erosion, producing surfaces on top of which were deposited thick lava breccias. By the time the next major lava flow was extruded, erosion had cut down to break blocks of quartzite away from the top of the sediment interlayer 2 sequence, as shown at the northern end of the profile (Figure 4). This indicates that the quartzite had lithified by this time and that a considerable time gap is represented by the boundary between sediment layer II and the overlying volcanics. The whole profile proves that a volcanically induced topography was still preserved throughout and that the damming and source effects on lake sedimentation styles were pronounced.

The Hardap Quarry "Pool Structure" Profile

Within the second set of profiles, Figure 5A shows the opposite extreme aspect of lake evolution in which little record is preserved of siliciclastic deposition. The sequence was perched on top of a lava feeder ridge that locally subsided due to magma withdrawal and decreasing magmatic pressure, to generate a linear series of pool structures. The pool-filling sequence is a 6 m preserved remnant of the overall lake deposits that subsided progressively into the resulting depression.

Phase 1 is represented only by the weathered top of the underlying lava sequence, involving in situ weathering and transported soils. The remainder of the sequence is made up of phase 2 (bio)chemical sediments that must have onlapped onto the feeder ridge, thus excluding phase 1 siliciclastics from the locality. The (bio)chemical sedimentary unit comprises cherty algal limestones showing plane and domal stromatolitic lamination interbedded with gypsum-dominated evaporitic layers. The lower part of the sequence displays well-developed slump structures directed toward the interior of the pool. In addition, down-tilted unconformity-bounded subunits subsiding into the pool structure are onlapped by subsequent units. The younger parts of the sequence have been removed by erosion and thus provide no information about potential phase 3 deposits.

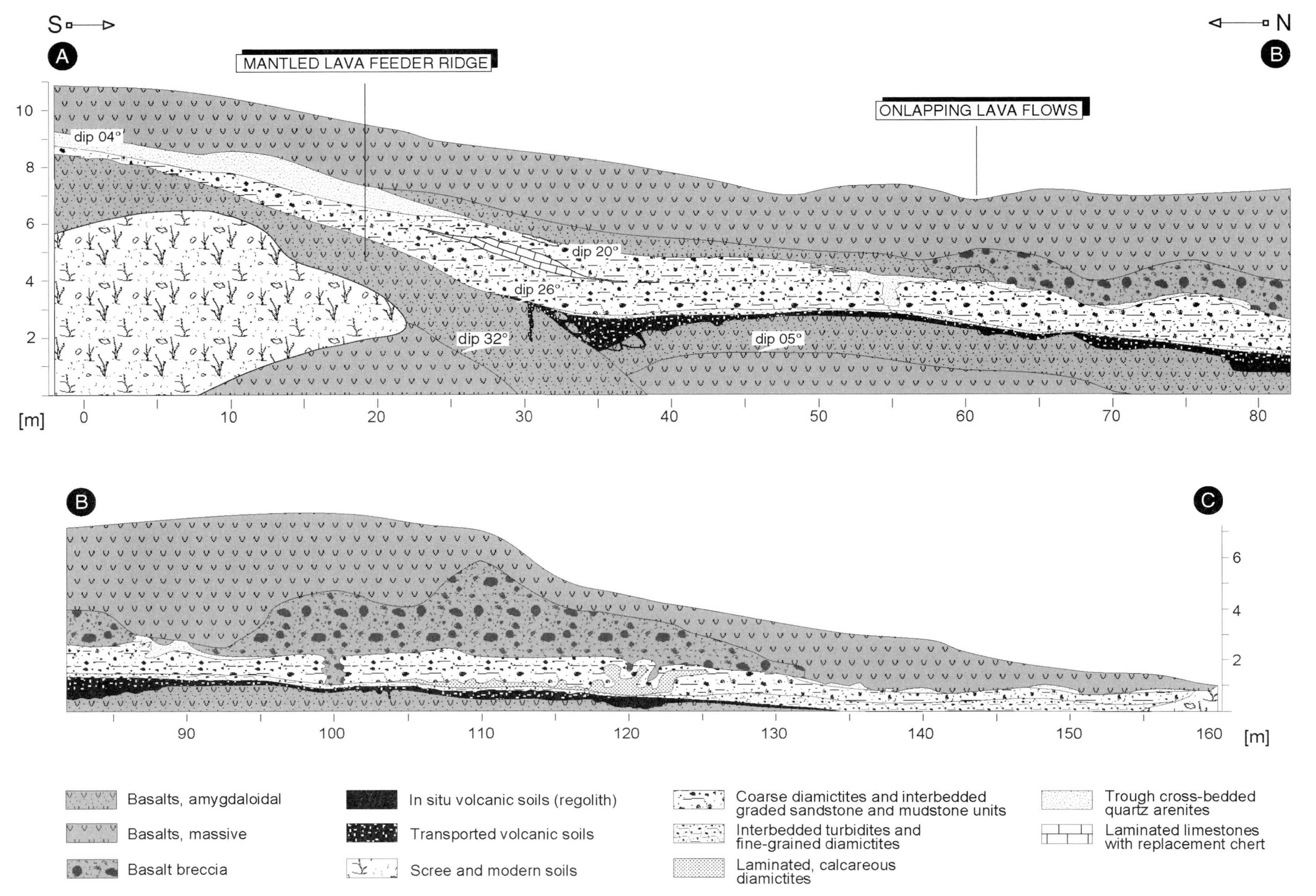

Figure 4—Cross-section of sediment interlayer II in the road cutting at Mariental-Kalkrand highway B1 (locality 2 in Figure 1), 7 km east of the previously described type at Hardap test camp (Figure 3) sections, showing changes in facies architecture against an adjacent lava feeder ridge.

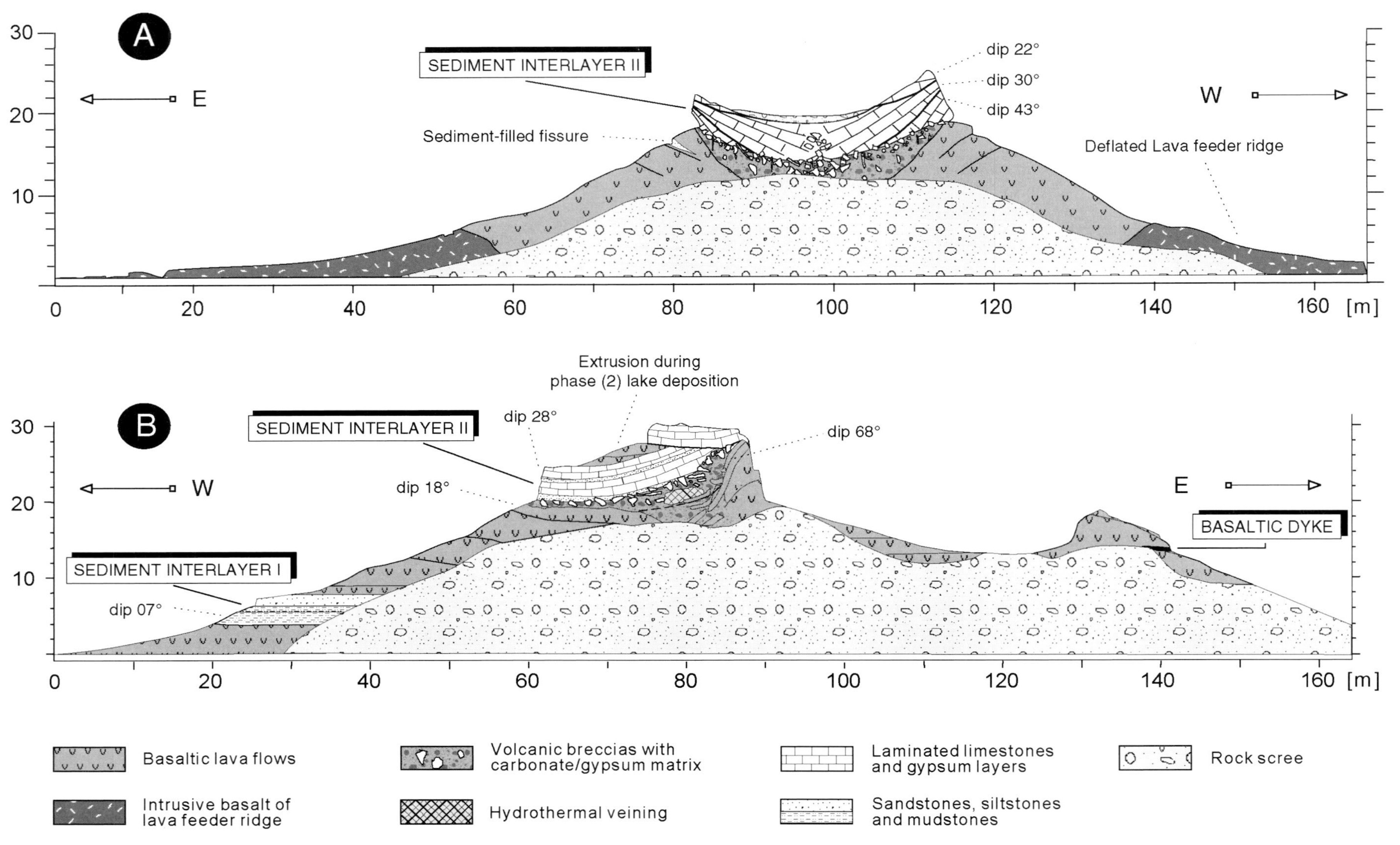

Figure 5—Cross-sections of "pool structures" at (A) Hardap quarry (locality 3 in Figure 1), and (B) at Farm Hardap (locality 4 in Figure 1), illustrating lateral variability of sediment interlayer II.

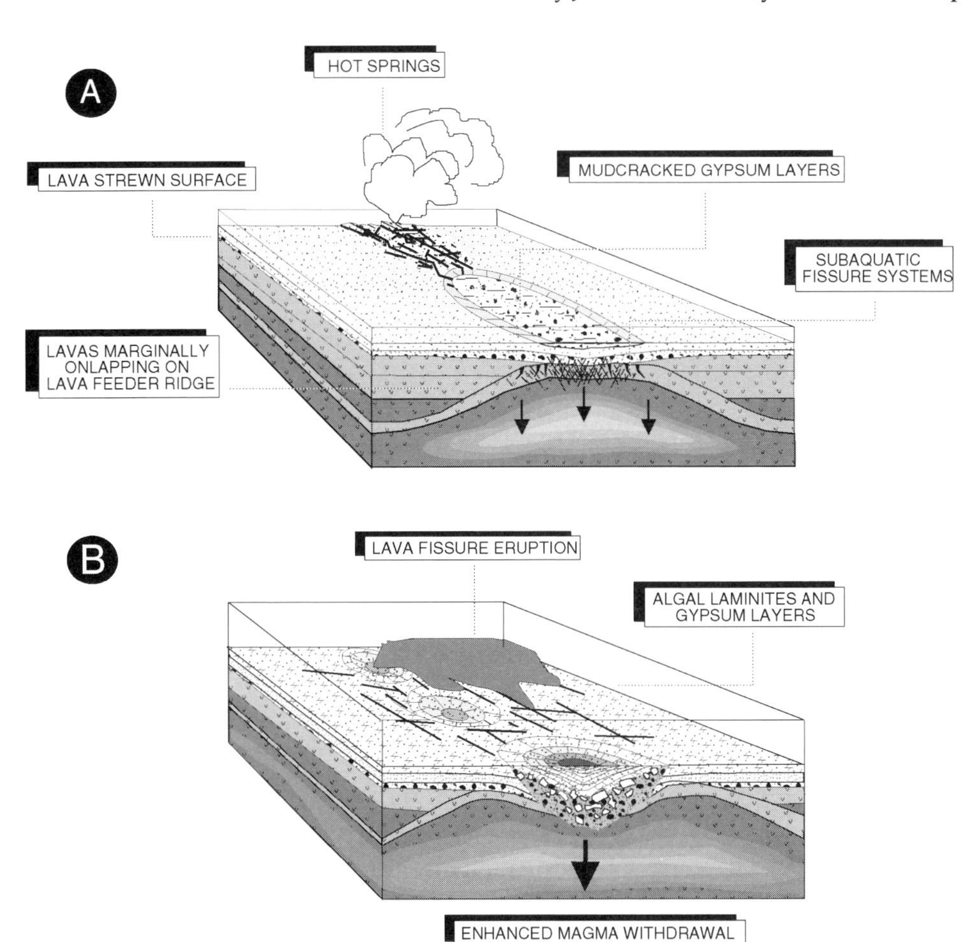

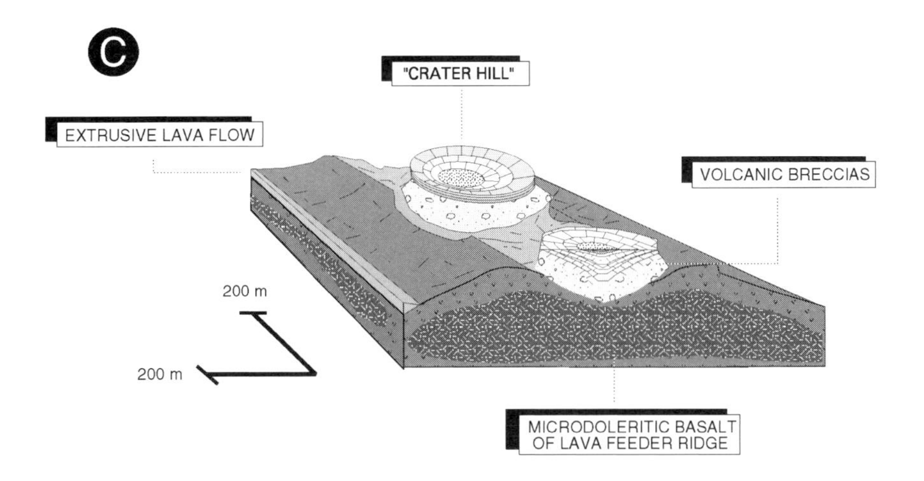

Figure 6—Sketch, showing "pool structure" development. At stage (A) a decrease in magma pressure led to initial deflation of the roof of the lava feeder ridge. Carbonates and evaporites were still deposited in the entire lake. Enhanced magma withdrawal from the ridge caused localized areas of enhanced subsidence during stage (B). Evaporites up to 10 m thick interbedded with cherty algal laminites and lacustrine muds preferentially accumulated in pool structures. Erosion created at a later stage (C) so-called "crater hills," which are remnants of the previously described pool structures.

The Farm Hardap "Pool Structure" Profile

The second profile through a pool structure (Figure 5B) shows similar collapse phenomena. The fill sequence, however, provides a tie with the previously described styles of sediment interlayer II. Thin siliciclastic interlayers reflect phases of sediment input into the lake system established at the type section. An intercalated lava flow within the pool sequence filled the depression, locally coordinating with, and probably responsible for, a phase of lava withdrawal subsidence. This synsedimentary volcanism compares with the amygdaloidal lava flow just above the quartzarenite at the top of sediment layer II in the road cut. This emphasizes that localized lava flows were extruded into both (bio)chemical and siliciclastically dominated areas of lake deposition even at quite a late stage. The most probable source of the lavas was the still actively flowing magma within the feeder ridges. Again, phase 3 deposits are not preserved.

Volcanic and Associated Hydrothermal Interaction with the Lake

The pool structure sequences described emphasize the way in which volcanic and lake evolution are

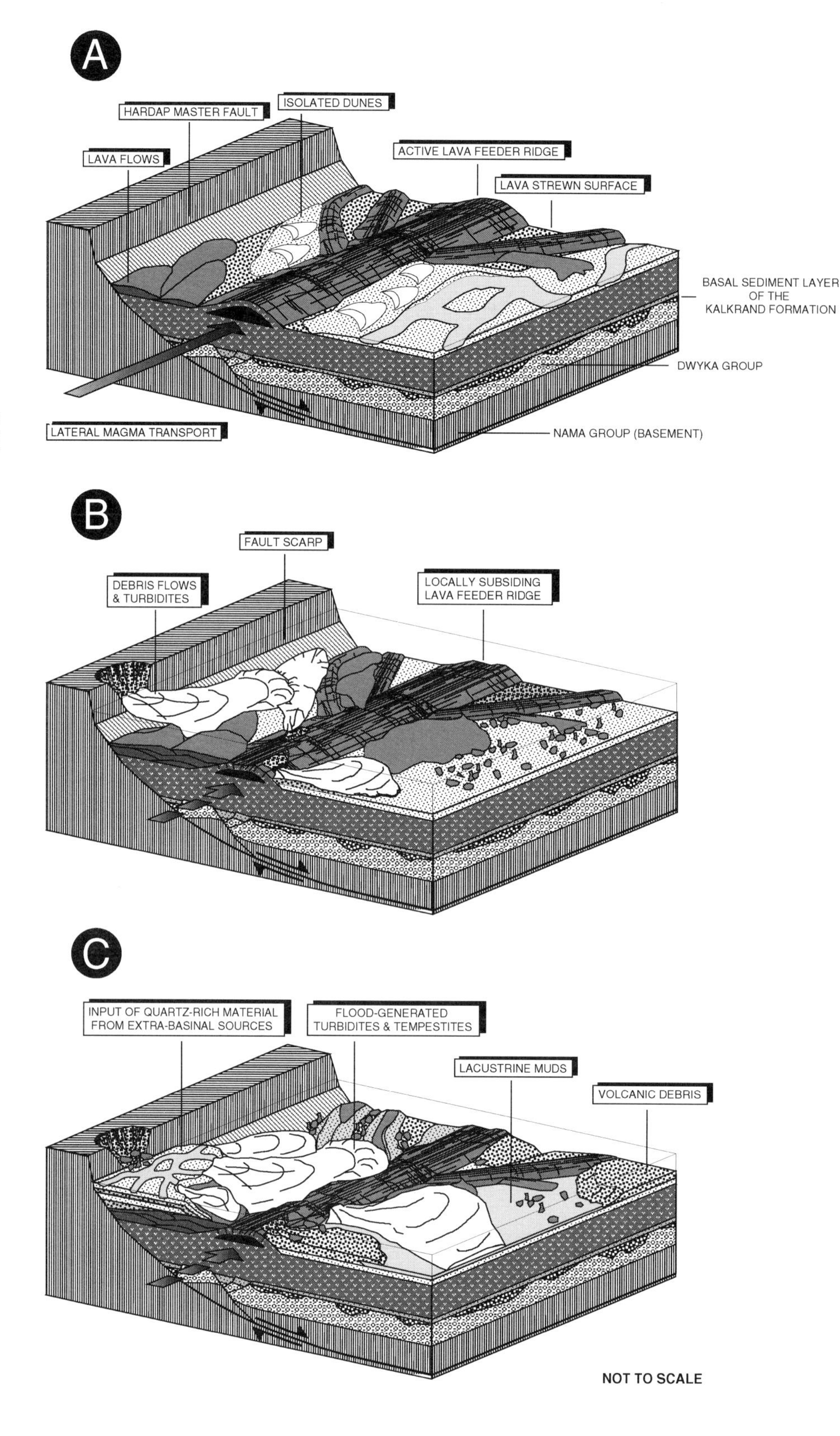

Figure 7—Sedimentological model for the Kalkrand Formation illustrating lake evolution and associated processes. (A) The regional drainage system was still attached to extra-basinal sources, and fluvial and eolian processes dominated the depositional environment. (B) Regional tectonic subsidence and local damming by lava topography accommodated establishment of a lake, and activity of lava feeder ridges and tectonic faulting created topography providing gravity flows. (C) Later fluvio-deltaic systems introduced siliciclastic detritus from extra-basinal sources.

intertwined. Localized collapse of lava ridges (Figure 6) would have allowed the lateral extrusion of thin lavas and also provided hydrothermal pathways. This is evidenced by the concentration of hydrothermally, lava-, and sediment-filled fissure systems that occur beneath them within the contemporaneously active lava feeder ridges. Although they do not occur in such abundance, fissure systems are also evident beneath the sedimentary interlayers throughout the study area. Algal blooms to some extent can explain the alkaline conditions that developed within the lake; however, it is additionally probable that a major control on the

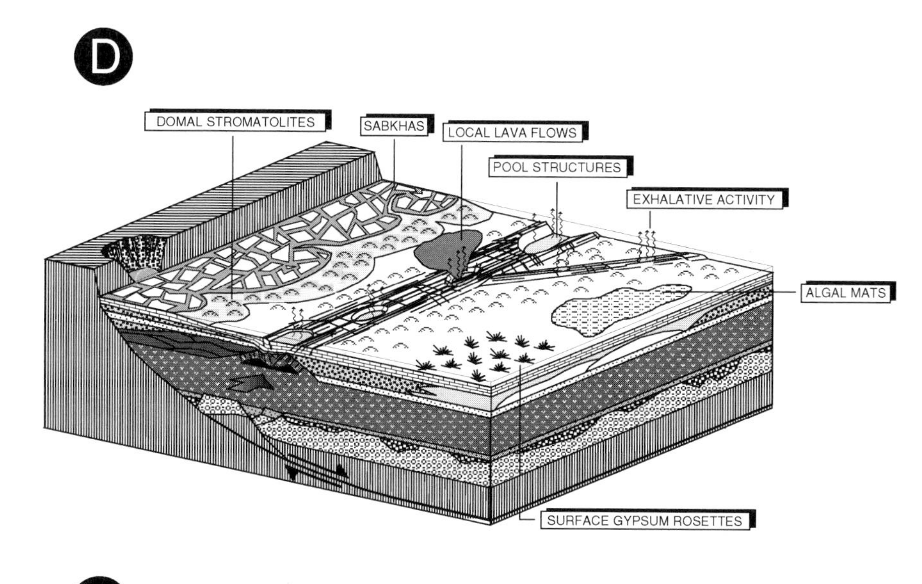

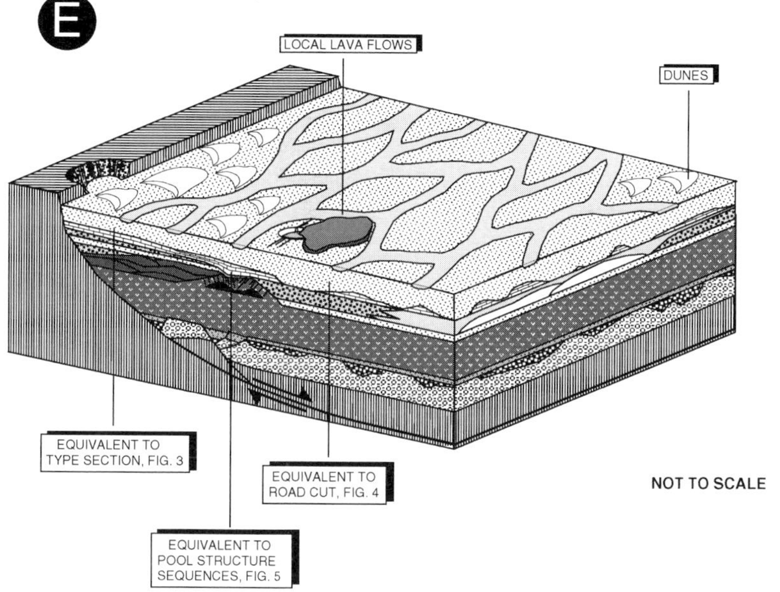

Figure 7—Continued. (D) With increasing subsidence and lake cut-off from the regional drainage, these sources were almost terminated and (bio)chemical processes dominated the shallow lake environment. Pressure release in the lava feeder ridges caused subsidence of circular pool structures, and associated exhalative activity favored alkaline lake water conditions. (E) Reattachment to the regional drainage finally caused the filling-up of the depository and the reestablishment of braided fluvial combined with eolian conditions.

alkalinity of the lake was provided by hydrothermal recharge through vein network systems developed within the underlying lava sequence. That the lava feeder ridges still released heat during lake deposition is also indicated by the formation of authigenic, well-crystallized montmorillonite together with palygorskite within the sediment interlayers only at locations proximal to lava ridges (Himmerkus, 1996, personal communication).

TECTONIC AND VOLCANIC EVOLUTION OF THE LAKE SYSTEM

The broad-scale evolution of the generalized lake system can be represented by a series of schematic block diagrams (Figure 7). The tectonic provision of accommodation space by low-angle extensional fault systems has been simplified in the diagrams to be represented by a single fault surface. The occurrence of footwall-derived Nama clasts within the basal conglomerates of the volcano-sedimentary sequence is indicated as fills of valleys incised into the underlying Dwyka Group.

Prior to lake sedimentation (Figure 7A), a volcanic topography segmented the tectonically created accommodation space. Features such as lava feeder ridges and fanned lava flow lobes provided local dams. The continuation of the regional drainage system in some of these areas deposited initial alluvial units from which eolian deposited material was deflated. This sets the scene for the subsequent considerable variation in facies during phases 1 and 2 of basin development. The setting up of the lake within the available accommodation space (Figure 7B) promoted gravity flows from the various topographic elements, including lava feeder ridges and fault scarps. Induced debris flows and turbidites were interleaved with the lacustrine muds. Subsequently (Figure 7C), in parts of the lake, the still active lava ridge topography continued to provide gravity flow deposits elsewhere; however, fluviodeltaic systems introduced a new source of quartz-rich detrital material from extra-basinal regions. Flood-generated turbidites and tempestites provided prodelta and offshore deposits. With continued faulting activity and regional lake cut-off from the regional drainage, siliciclastic input was almost terminated. Closed basin conditions evolved into (bio) chemically dominated environmental settings (Figure 7D). Laminated and domal stratified algal mats dominated sedimentation. Evaporation and exhalative activity resulted in alkaline lake water chemistry, causing a lack of algal grazers and other macroinvertebrates, under which conditions algal mats can thrive. Surface and subsurface gypsum precipitation was associated with dolomitization under evaporative reflux conditions (cf. McKenzie et al., 1980) associated with sabkha settings. At this time, pressure variation in the lava feeder ridges caused progressively enhanced subsidence and collapse of pool structures within the lake. Locally, small basalt flows were extruded from the ridges themselves. Finally, reconnection of the basin with a reestablished regional drainage (Figure 7E) allowed capping of the sequence by sheet-like, braided fluvial quartzarenites. Although feeder ridge activity was at a minimum, some small basalt flows were still extruded on top of this fluvial blanket. The regional drainage subsequently eroded parts of the sedimentary sequence prior to the extrusion of the next major lava sequence.

POTENTIAL GAS RESERVOIR EQUIVALENTS OFFSHORE

Similar to the sedimentary interlayers described, the main reservoir rocks in the giant Kudu offshore gas field (Figure 1) are interbedded with flood basalts. Although these altered lavas are so far undatable, they were traditionally assumed to correlate with the Lower Cretaceous Etendeka flood basalts of northern Namibia, which are underlain and interbedded at multiple levels with almost exclusively eolian interbeds. By contrast, highly variable facies and fossils described by Wickens and McLachlan (1990) from paleoenvironmentally continental interlayers containing the main Kudu gas reservoir suggest that these are not exclusively eolian, but comprise a considerable proportion of lacustrine elements. Bivalves and calcareous algae occur in shale units, and volcaniclastic sandstones are present, preserved in fining-upward mass flow units. Additionally, anhydrite is ubiquitously preserved in the reservoir sandstone, which might indicate high groundwater levels, tying in with the lacustrine paleoenvironmental associates described.

None of the elements described have been reported from sedimentary sequences associated with the Etendeka flood basalts (Mountney et al., 1998), which seem to have been deposited in generally more arid desert conditions during their extrusion. By contrast all of the elements and the generally more "wet" desert milieu is characteristic of the facies described in this paper from the Early Jurassic interbeds of the Kalkrand flood basalts. Mass flow units of volcaniclastic sandstones have been described, similar fauna have been discovered in a similar stratigraphic position, and evaporitic gypsum is ubiquitous. Additionally, the Kalkrand flood basalts are the closest correlative units to the Kudu field and both units are located 400–450 km outside the Huab area into which the Etendeka Basalts were later extruded (Ledendecker, 1992; Milner et al., 1994). When this line of reasoning is taken together with the results of regional sedimentological and biostratigraphic studies of the extents of both Lower Jurassic and Lower Cretaceous eolian sandstones in northern, central, and southern Namibia, the conclusion can be drawn that the Kudu main gas reservoir represents most probably an offshore correlative equivalent of the Kalkrand Formation interlayers. This correlation is considered in detail in a subsequent manuscript by Stanistreet and Stollhofen (1999) who contrast the "wet" desert nature of the Lower Jurassic eolian related sandstones, including Kudu and Kalkrand units, with the enhanced aridity of the Lower Cretaceous units. They further speculate that the Kalkrand Formation interlayers would have had surrounding more eolian-dominated source regions that would have acted as a source of eolian sand grains throughout their depositional history. This speculation now meets a likely candidate source region in the existence of the eolian cross-bedded sandstones of the Kudu gas field.

ACKNOWLEDGMENTS

This work has been supported by the Geowissenschaftliches Graduiertenkolleg Würzburg and the Jubiläumsstiftung of the University of Würzburg. We owe particular thanks to the Geological Survey of Namibia, Nature Conservation, the Ministry of Tourism, and the Department of Fishery Affairs for their support and logistic assistance with fieldwork. We also have benefited from many discussions in the field with our colleagues Volker Lorenz and Andy Duncan, which are gratefully acknowledged.

REFERENCES CITED

Dingle, R. V., 1993, Structural and sedimentary development of the continental margin off southwestern Africa: Communications of the Geological Survey of South West Africa/Namibia, v. 8, p. 35–43.

Duncan, A. R., A. J. Erlank, and J. S. Marsh, 1984, Regional geochemistry of the Karoo Igneous Province, in A. J. Erlank, ed., Petrogenesis of the Volcanic Rocks of the Karoo Province: Geological Society of South Africa, Special Publication 13, p. 355–388.

Duncan, R. A., P. R. Hooper, J. Rehacek, J. S. Marsh, and A. R. Duncan, 1997, The timing and duration of the Karoo igneous event, southern Gondwana: Journal of Geophysical Research, v. 102, p. 18127–18138.

Gerschütz, S., 1996, Geology, volcanology, and petrogenesis of the Kalkrand Basalt Formation and the Keetmanshoop dolerite complex, southern Namibia: Dr.rer.nat.-thesis, Faculty of Earth Sciences, University of Würzburg, 186 p.

Gladczenko, T. P., K. Hinz, O. Eldholm, H. Meyer, S. Neben, and J. Skogseid, 1997, South Atlantic volcanic margins: Journal of the Geological Society of London, v. 154, p. 465-470.

Hawkesworth, C. J., K. Gallagher, S. Kelley, M. Mantovani, D. W. Peate, M. Regelous, and N. W. Rogers, 1992, Paraná magmatism and the opening of the South Atlantic, *in* B. C. Storey, T. Alabaster, and R. J. Pankhurst, eds., Magmatism and the causes of continental break-up: Geological Society of London, Special Publication 68, p. 221–240.

Heath, D. C., 1972, Die Geologie van die Sisteem Karoo in die Gebied Mariental-Asab, Suidwes Afrika: Geological Survey of South Africa, Memoir 61, p. 1–44.

Hon, K., J. Kauahikaua, R. Denlinger, and K. Mackay, 1994, Emplacement and inflation of pahoehoe sheet flows: Observations and measurements of active lava flows on Kilauea Volcano, Hawaii: GSA Bulletin, v. 106, p. 351–370.

Ledendecker, S., 1992, Stratigraphie der Karoosedimente der Huabregion (NW-Namibia) und deren Korrelation mit zeitäquivalenten Sedimenten des Paranàbeckens (Südamerika) und des Großen Karoobeckens (Südafrika) unter besonderer Berücksichtigung der überregionalen geodynamischen und klimatischen Entwicklung Westgondwanas: Göttinger Arbeiten zur Geologie und Paläontologie, 54, p. 1–87.

Light, M. P. R., M. P. Maslanyj, and N. L. Banks, 1992, New geophysical evidence for extensional tectonics on the divergent margin offshore Namibia, *in* B. C. Storey, T. Alabaster, and R. J. Pankhurst, eds., Magmatism and the causes of continental break-up: Geological Society of London, Special Publication 68, p. 257–270.

Light, M. P. R., M. P. Maslanyj, R. J. Greenwood, and N. L. Banks, 1993, Seismic sequence stratigraphy and tectonics offshore Namibia, *in* G. D. Williams and A. Dobb, eds., Tectonics and Seismic sequence stratigraphy: Geological Society of London, Special Publication 71, p. 163–191.

Maslanyj, M. P., M. P. R. Light, R. J. Greenwood, and N. L. Banks, 1992, Extension tectonics offshore Namibia and evidence for passive rifting in the South Atlantic: Marine and Petroleum Geology, v. 9, p. 590–601.

McKenzie, J. A., K. J. Hsü, and J. F. Schneider, 1980, Movement of subsurface waters under the sabkha, Abu Dhabi, UAE, and its relation to evaporative dolomite genesis, in D. H. Zenger, J. B. Dunham, and R. L. Ethington, eds., Concepts and models of dolomitization, SEPM Special Publication 28, p. 11–30.

Miller, R. McG., 1992, Pre-breakup evolution of the western margin of southern Africa (abs.): Mini-Conference, Southwestern African continental margin: evolution and physical characteristics, Abstracts, p. 37–43.

Milner, S. C., A. R. Duncan, A. Ewart, and J. S. Marsh, 1994, Promotion of the Etendeka Formation to Group status: A new integrated stratigraphy: Communications of the Geological Survey of South West Africa/Namibia, v. 9, p. 5–11.

Mountney, N., J. Howell, S. Flint, and D. A. Jerram, 1998, Aeolian and alluvial deposition within the mesozoic Etjo Sandstone Formation, northwest Namibia: Journal of African Earth Sciences, v. 27, p. 175–192.

Renne, P. R., J. M. Glen, S. C. Milner, and A. R. Duncan, 1996, Age of Etendeka flood volcanism and associated intrusions in southwestern Africa: Geology, v. 24, p. 659-662.

Schalk, K. E. L. and G. J. B. Germs, 1980, The geology of the Mariental area. Explanation of Sheet 2416 (scale 1:250 000): Geological Survey of Namibia, Windhoek.

Stanistreet, I. G. and H. Stollhofen, 1998, Onshore equivalents of the main Kudu gas reservoir in Namibia, *in* N. Cameron, R. Bate, and V. Clure, eds., Oil and gas habitats of the South Atlantic: Geological Society of London, Special Publication 53, p. 345–365.

Stollhofen, H., S. Gerschütz, I. G. Stanistreet, and V. Lorenz, 1998, Tectonic and volcanic controls on early Jurassic rift-valley lake deposition during emplacement of Karoo flood basalts, southern Namibia: Palaeogeography, Palaeoclimatology, Palaeoecology, v. 140, p. 185–215.

Wickens, H. de V., and I. R. McLachlan, 1990, The stratigraphy and sedimentology of the reservoir interval of the Kudu 9A-2 and 9A-3 boreholes: Communications of the Geological Survey of Namibia, v. 6, p. 9–22.

Dunagan, S. P., 2000, Lacustrine carbonates of the Morrison Formation (Upper Jurassic, western interior), East-Central Colorado, U.S.A., in E. H. Gierlowski-Kordesch adn K. R. Kelts, eds., Lake basins through space and time: AAPG Studies in Geology 46, p. 181-188.

Chapter 14

Lacustrine Carbonates of the Morrison Formation (Upper Jurassic, Western Interior), East-Central Colorado, U.S.A.

Stan P. Dunagan
Department of Geology and Geography, Austin Peay State University
Clarksville, Tennessee, U.S.A.

INTRODUCTION

Siliciclastic- and carbonate-dominated lacustrine systems are present within the Morrison Formation (Upper Jurassic), Western Interior basin, U.S.A. The siliciclastic-dominated systems, common on the Colorado Plateau, have been studied intensely in association with the extensive Lake T'oo'dichi' complex, an alkaline-saline lake in the Brushy Basin Member of the Morrison Formation (Bell, 1983, 1986; Peterson and Turner-Peterson, 1987; Turner and Fishman, 1991). Conversely, the carbonate-dominated lacustrine systems prevalent in east-central Colorado, New Mexico, Kansas, and Wyoming have received limited attention (Frazier et al., 1983; Sweet 1984; Lockley et al., 1986; Sweet and Donovan, 1988; Johnson, 1991), with few regional-scale and even fewer detailed sedimentologic and stratigraphic investigations (exceptions include West, 1978; Jackson, 1979; Dunagan et al., 1996, 1997; Dunagan, 1997, 1998). This study describes the initial results of a systematic, regional-scale investigation of lacustrine carbonate deposits in the Morrison Formation focusing on the well-developed lacustrine complex in east-central Colorado, which will ultimately provide important insights into one of the most poorly understood depositional environments in the Morrison paleoecosystem. Detailed results are forthcoming in other publications.

REGIONAL GEOLOGIC SETTING

During the Late Jurassic, the Morrison Formation was deposited as part of the thick Middle Jurassic–early Eocene succession in the back-bulge depozone of the Cordilleran foreland basin system (DeCelles and Currie, 1996). The remnants of this broad, shallow sedimentary basin are present across a wide expanse of the Western Interior basin from Arizona to Montana and Kansas to Utah (Figure 1), which occupied a paleolatitude of approximately 31–35°N (Peterson, 1988). Within this Late Jurassic depozone, Morrison Formation consists of alluvial-plain, eolian, fluvial, lacustrine, and marine lithofacies (Peterson and Turner-Peterson, 1987; Peterson, 1994), bound at the base and top of the formation by the J-5 and K-1 unconformities, respectively (Pipiringos and O'Sullivan, 1978).

In east-central Colorado, the Morrison Formation consists of a complex sequence of interbedded floodplain, fluvial, and lacustrine deposits with minor evaporitic and marine deposits (Figure 2). In the study area, the Morrison contains a particularly well-developed lacustrine carbonate complex. This complex is different from the large alkaline-saline lake complex typically associated with Morrison Formation. In east-central Colorado, the Morrison Formation is predominantly undifferentiated with the exception of nearshore marine deposits of the basal Windy Hill Member, which is present only in the northernmost portions of the study area (Figure 2). The undifferentiated Morrison Formation consists of (1) an extensive succession of overbank, floodplain, and lacustrine mudstone interbedded with lacustrine carbonate deposits, (2) fluvial and lacustrine sandstone and siltstone units, and (3) red bed paleosols and siltstones (Figure 2) (Dunagan et al., 1996).

AGE OF THE MORRISON FORMATION

The Morrison Formation has been dated by isotopic and biostratigraphic methods. Bentonite beds are present near the base and top of the formation in the Colorado Plateau (Peterson 1994) and single-crystal $^{40}Ar/^{39}Ar$ isotopic dates from multiple bentonite beds indicate an age for the Morrison Formation ranging from 155 to 148 Ma (Kowallis et al., 1998). Fossil charophytes, ostracodes, and palynomorphs indicate that most of the Morrison Formation is Kimmeridgian in age and that the upper one-quarter of the formation is Tithonian (Schudack et al., 1998; Litwin et al., 1998). The lack of microfossils or isotopically datable bentonite beds in the lower 3 m of the formation means that the basal Morrison Formation could, conceivably, be latest

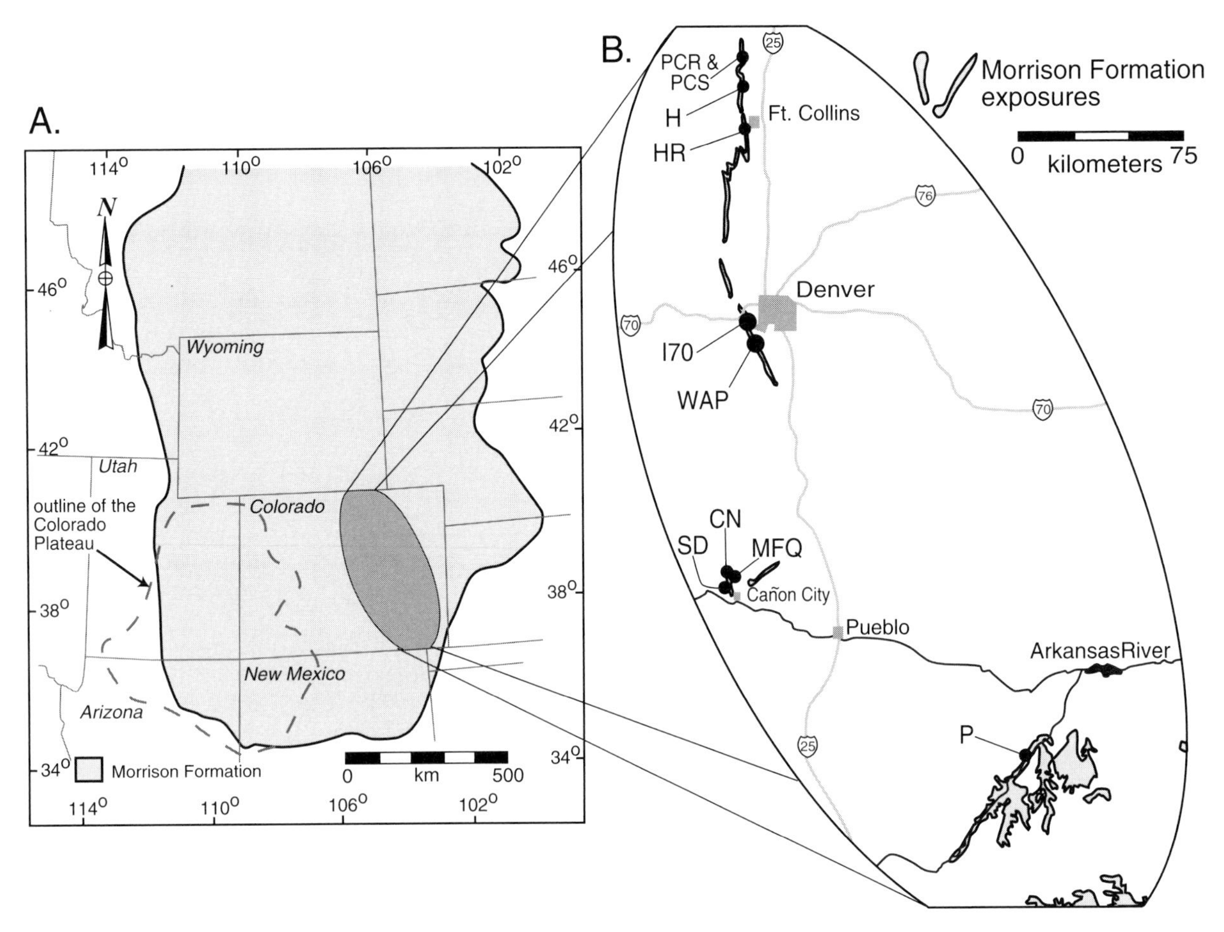

Figure 1—(A) Distribution of the Morrison Formation (lightly shaded area) in the Western Interior of North America (based on Peterson, 1972). (B) Enlargement of shaded area in (A) shows location of measured stratigraphic sections in east-central Colorado. PCR = Park Creek Reservoir, 6.5 km northeast of Livermore, H = Highway 287, 1.5 km north-northeast of Ted's Place, HR = Horsetooth Reservoir, near Spring Canyon Dam, west of Ft. Collins, I70 = Interstate 70 roadcut, west of Denver, WAP = West Alameda Parkway and Dinosaur Ridge National Historic Landmark, 3 km north of Morrison, P = Purgatoire River area, 44 km south-southwest of La Junta, MFQ = Marsh-Felch quarry, Garden Park Paleontological Area, 14 km north of Cañon City, SD = Skyline Drive roadcut, Cañon City, CN = Cope's Nipple, Garden Park Paleontological Area, 14 km north of Cañon City.

Oxfordian in age (F. Peterson, personal communication, 1997). Thus, the Morrison Formation ranges from Oxfordian–Kimmeridgian to Tithonian in age (scale after Harland et al., 1990).

LACUSTRINE SEDIMENTOLOGY

The Morrison Formation contains a well-developed lacustrine carbonate succession in east-central Colorado (Figure 3) (Dunagan et al., 1996; Dunagan, in press). Siliciclastic lacustrine deposits include mudstone, siltstone, and sandstone lithofacies. The mudstone deposits are greenish gray and calcareous with a restricted and a partially allochthonous biota composed of gastropods, charophyte gyrogonites, ostracodes, and conchostracans. The mudstones are massively bedded, probably due to intense bioturbation or pedoturbation. The lacustrine siltstone and sandstone deposits are commonly calcareous, bioturbated, thinly bedded, and locally are characterized by ripple marks, thinly interbedded evaporites, or more rarely teepee structures.

Petrographic observations indicate that lacustrine carbonates in the Morrison Formation consist of eight microfacies (Table 1). These microfacies contain widely varied sedimentologic features and a diverse paleontologic component (discussed in following paragraphs). Pseudomicrokarst, microkarst, and evaporites are commonly associated with the carbonate microfacies. Pseudomicrokarst includes features such as (1) complex voids filled with vadose and internal sediment, micritic clasts, and calcite cement, (2) root traces (millimeter to decimeter scale), columnar and stacked rhizoconcretions, (3) circumgranular, desiccation, horizontal, and septarian cracks, and (4) brecciation and grainification (Dunagan et al., 1996). Evaporites are present locally as pore-filling anhydrite

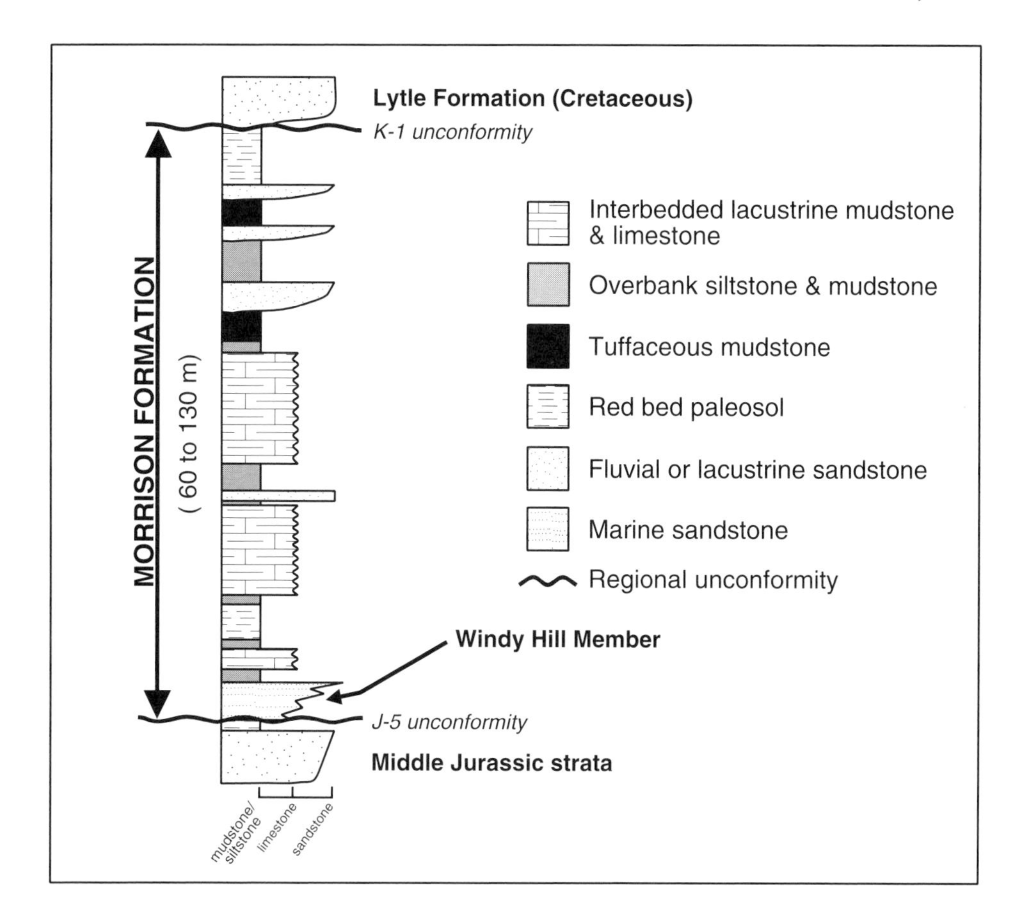

Figure 2—Schematic stratigraphic section of the Morrison Formation in east-central Colorado.

(locality PCR; Figure 1), small (1–2 cm, diameter) gypsum nodules or rosettes (localities H, I70, and MFQ), and calcite and silica replacement of evaporite pseudomorphs after gypsum, halite, and trona? (localities H, HR, I70, and MFQ).

Magadi-type cherts have also been reported (localities HR, P, and SD) in the lacustrine carbonates from Colorado (Dunagan et al., 1997; Dunagan, in press). The cherts typically occur as thin (1–8 mm) to thick (0.1–5.0 cm), red to black, discontinuous layers and nodules in skeletal mudstone-wackestone and microbialite deposits associated with open lake settings. The Morrison cherts display characteristic Magadi-type chert textural features, including buckled chert layers, small lobate protrusions, soft-sediment deformation, and evaporite pseudomorphs, as well as petrographic criteria (Dunagan et al., 1997; see discussion in Schubel and Simonson, 1990).

Open-lacustrine carbonate and siliciclastic lithofacies characterized by laminated deposits are rare. The majority of open lake deposits are faintly laminated to bioturbated carbonate lithofacies, commonly associated with laminae of disarticulated ostracode valves and quartz silt. At locality MFQ, a dark siliciclastic mudstone is present that contains laminations and soft-sediment deformation structures. The definition of these deposits as open lacustrine does not imply deposition in a deep-water body with a permanently stratified water column even though this facies occupied the most distal portion of the lacustrine system. On the contrary, open lacustrine carbonates locally display pseudomicrokarst and desiccation cracks; this suggests deposition in a lake that was extremely shallow even in its open areas.

PALEONTOLOGY

A diverse biota was present in Morrison carbonate lakes and ponds. Body fossils include charophytes (calcified stems and gyrogonites), ostracodes, spongillids, unionid bivalves, gastropods (prosobranch and pulmonate), microbialites, and conchostracans (Dunagan, 1997, 1999a, 1999b); bone fragments indicate crocodiles, dinosaurs, and fish were present. Trace fossil evidence also suggests the presence of insects, crustaceans, plants, and dinosaurs (Hasiotis and Demko, 1996).

Based on paleontologic evidence, east-central Colorado contained numerous perennial to ephemeral lakes and ponds that were predominantly freshwater. Perennial lacustrine conditions are suggested by the presence of prosobranch gastropods, unionid bivalves, and fish remains, where restricted faunas, such as conchostracans and pulmonate gastropods, dominate, ephemeral conditions are indicated. Mild currents, probably driven by winds over the shallow lakes, are also suggested due to the presence of unionid bivalves and spongillids in Morrison lakes.

The freshwater nature of Morrison lakes and ponds is demonstrated by the abundance of freshwater ostracodes

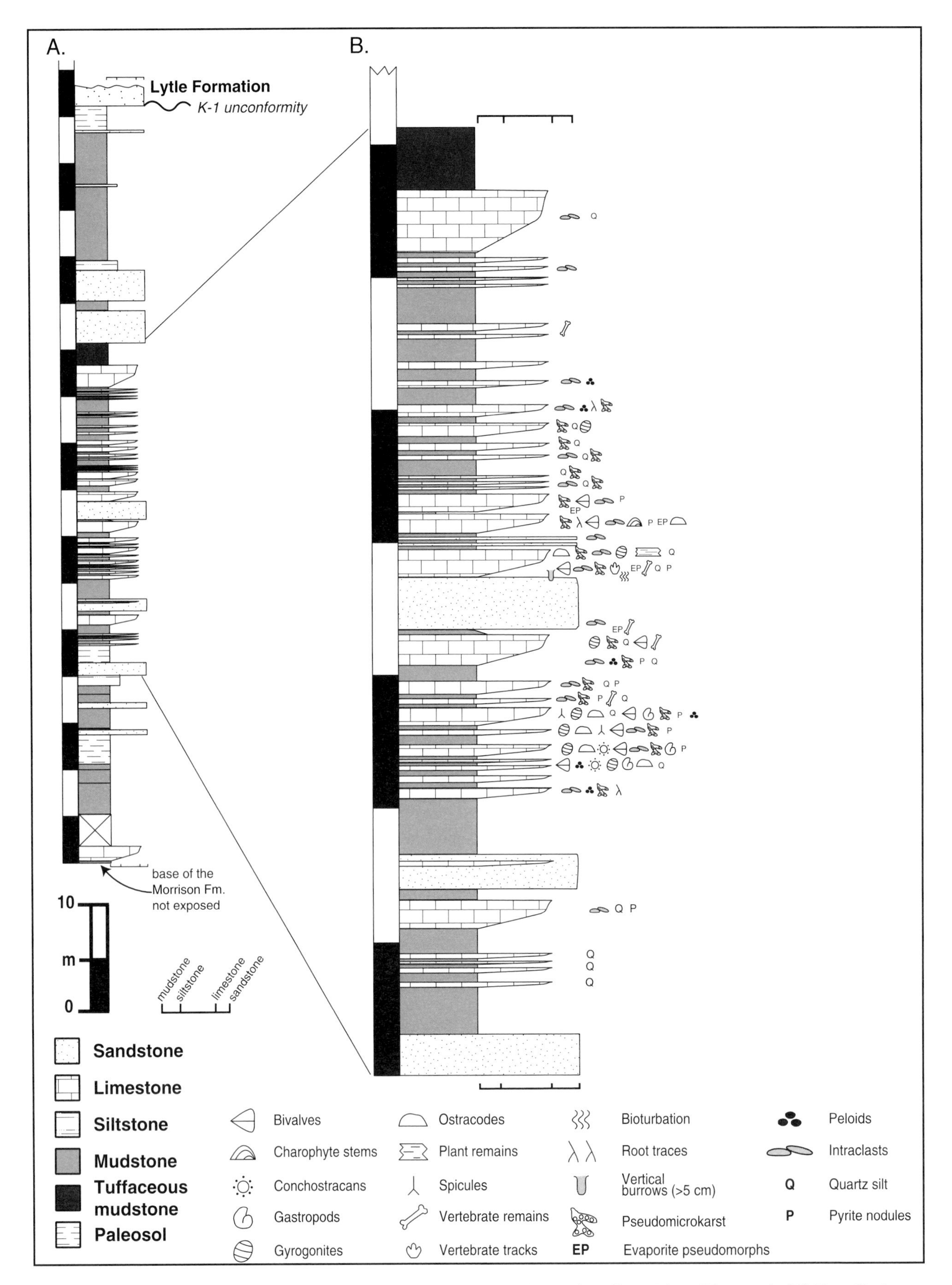

Figure 3—(A) Stratigraphic section of the Morrison Formation at locality H (see Figure 1). (B) Detailed portion of major lacustrine interval.

Table 1. Carbonate Microfacies from Marginal and Open Lake Settings in the Morrison Formation

Microfacies	Biota	Other Depositional Features	Lacustrine Setting
Carbonaceous packstone	Plant fragments (xylem and cortical vascular tissue), minor gastropods	Diffuse laminations, pyrite	Marginal
Intraclast wackestone-grainstone	Gyrogonites, ostracodes	Peloids and green mudstone rip-up clasts; rare dark intraclasts; minor ooids and volcanic shards; chert; detrital quartz; pyrite	Marginal
Micritic mudstone	Rare unionid bivalves, ostracodes, gyrogonites	Starved ripples of detrital quartz and peloids; pseudomorphs after halite, gypsum, and trona, plant fragments, detrital quartz, volcanic shards, pyrite	Marginal
Ooid packstone-grainstone	Bones (fish and reptile), unionid bivalves, ostracodes	Minor ripples marks; dinoturbation and trackways; nuclei composed of quartz, bone, ooids, and bioclasts	Marginal
Peloid skeletal mudstone-wackestone	Charophytes (stems and gyrogonites), ostracodes, unionid bivalves, sponges, gastropods	Rare plant fragments, oncoids, and bone debris (fish and reptile); rounded to elliptical peloids and rare dark intraclasts; minor green mudstone rip-up clasts; Magadi-type cherts; pseudomorphs after gypsum; pyrite; detrital quartz	Marginal
Peloid skeletal packstone-grainstone	Charophyte stems and gyrogonites, unionid bivalves, ostracodes	Minor cross-laminations; laminae of alternating grainstone and packstone; pseudomorphs after gypsum; Magadi-type cherts; rounded to elliptical peloids; minor intraclasts; pyrite; detrital quartz	Marginal
Skeletal mudstone-wackestone	Gyrogonites, ostracodes, gastropods (prosobranch and pulmonate), fish bone, algal and cyano-bacterial filaments	Minor intraclasts and detrital quartz; Magadi-type cherts; diffuse to bioturbated laminations	Open lake
Microbialites	Stromatolites, algal, and cyanobacterial filaments, oncoids, gyrogonites, ostracodes, fish bone fragments	Minor intraclasts and detrital quartz; Magadi-type cherts; diffuse to bioturbated laminations; composite microbial biostromes (planar, wrinkled, laterally-linked, and columnar stromatolites	Open lake

and charophytes, with salinity ranges typically <9‰ (Schudack et al., 1988). Magadi-type cherts indicated that the lakes experienced periods of high alkalinity (high pH > 9.0–9.5) (Hay, 1968; Eugster, 1969), which was probably due, in part, to high productivity associated with charophyte photosynthesis. These cherts, however, are restricted to the lower Morrison Formation and their association with microbialites and freshwater ostracodes and charophytes indicates highly alkaline, but not necessarily saline, conditions. The presence of unionid bivalves, prosobranch gastropods, and fish remains suggest that Morrison lake deposits experienced at least temporary periods where the lakes were hydrologically open.

DEPOSITIONAL SYNTHESIS

Relatively shallow paleolakes were common features in the Morrison paleoecosystem of east-central Colorado during the Late Jurassic. Lateral and vertical stratigraphic relationships of the Morrison Formation suggest an idealized Morrison lake was characterized by a relatively shallow, open-lake setting surrounded by a wide marginal lacustrine zone that laterally passed into adjacent alluvial environments. The lacustrine carbonate deposits were bordered by a wide lacustrine mudflat characterized by abundant mudstone and more rarely by siltstone and sandstone lithofacies (Figure 4). Yuretich et al. (1984) interpreted

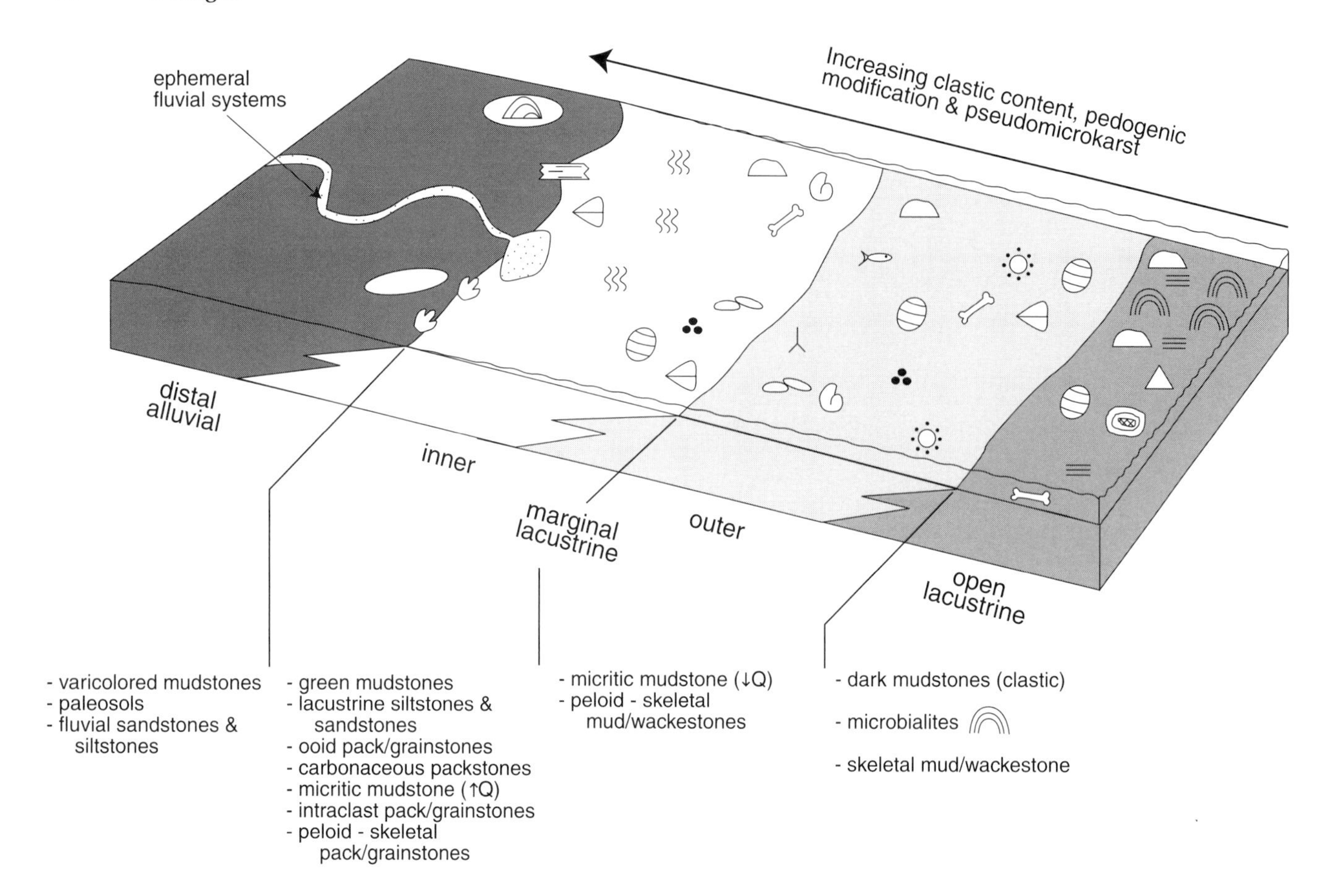

Figure 4—Schematic block diagram illustrating the lacustrine facies relationships for the Morrison Formation. The hypothesized vertical relationships for the open lacustrine, marginal lacustrine, and distal alluvial facies are also shown. Symbols from Figure 3. Not to scale.

similar lacustrine mudstones from the Fort Union Formation (Paleocene) as marginal lacustrine deposits.

Lacustrine carbonate sedimentation occurred in a marginal to dominantly shallow-water, open-lacustrine setting that probably experienced periods of increased alkalinity (pH >9.0–9.5) and salinity. This interpretation is based on (1) the abundance of pseudomicrokarst and microkarst within the carbonate microfacies, (2) the abundance of shallow-water biota, (3) the occurrence of Magadi-type cherts and minor evaporite deposits, and (4) the lack of "oil shale" and paucity of laminated lithofacies.

The abundance of shallow-water carbonate deposits with low detrital contents and extensive lake margin facies is consistent with lakes dominated by low-energy, low-gradient ramp-type margin settings (sensu Platt and Wright, 1991); ooid grainstone and cross-stratified peloid-skeletal packstone to grainstone facies were locally present along wave-influenced margins. Oscillations in lake level resulted in the exposure of large areas of the lake margin. Pedogenic activity, vegetative reworking, and meteoric diagenesis modified lacustrine fabrics through processes, such as rooting, mottling, pseudomicrokarst, desiccation and pedogenic crack formation, and microkarst on carbonate substrates in both the inner and outer marginal lacustrine zones. The paucity of well-defined laminations with significant bioturbation suggests oxygenated bottom waters and indicates that the lakes were probably polymictic to holomictic. The general absence of well-preserved fish remains is also a good indicator of holomictic lakes (Freytet and Plaziat, 1982). The mudstone and wackestone textures suggest predominantly low-energy sedimentation.

ACKNOWLEDGMENTS

Lacustrine carbonate research was supported by student research grants to SPD from the Colorado Scientific Society (Ogden Tweto Memorial Fund) and Geological Society of America and by discretionary funds from the Morrison Extinct Ecosystems Project (U.S. Geological Survey and National Parks Service), the National Science Foundation (EAR-9418183, S.G. Driese and C.I. Mora), and the University of Tennessee Mobil Carbonate Research Fund.

S.G. Driese and K.R. Walker provided helpful comments on an earlier version of this manuscript. The author also benefited from numerous discussions on Morrison depositional systems with T.M. Demko, F. Peterson, and C.E. Turner.

REFERENCES CITED

Bell, T. E., 1983, Deposition and diagenesis of the Brushy Basin and upper Westwater Canyon Member of the Morrison Formation in northwest New Mexico and its relationship to uranium mineralization [unpublished Ph.D. dissertation]: Berkley, California, University of California, 102 p.

Bell, T. E., 1986, Deposition and diagenesis in the Brushy Basin and upper Westwater Canyon Member of the Morrison Formation, San Juan basin, New Mexico, *in* Turner-Peterson, C. E., Santos, E. S., and Fishman, N. S., eds., A Basin Analysis Case Study—The Morrison Formation, Grants Uranium Region, New Mexico: AAPG Studies in Geology 22, p. 77–91.

DeCelles, P. G., and Currie, B. S., 1996, Long-term sediment accumulation in the Middle Jurassic–early Eocene Cordilleran retroarc foreland-basin system: Geology, v. 24, p. 591–594.

Dunagan, S. P., 1997, Jurassic freshwater sponges from the Morrison Formation, U.S.A.: Geological Society of America, Abstracts with Programs, v. 29, no. 3, p. 14–15.

Dunagan, S. P., 1998, Lacustrine and palustrine carbonates from the Morrison Formation (Upper Jurassic), east-central Colorado, U.S.A.: implications for Late Jurassic depositional patterns, paleoecology, paleohydrology, and paleoclimatology [unpublished Ph.D. dissertation]: The University of Tennessee, Knoxville, 302 p.

Dunagan, S. P., 1999a, A North American freshwater sponge (*Eospongilla morrisonensis* new genus and species) from the Morrison Formation (Upper Jurassic), Colorado: Journal of Paleontology, v. 73, no. 3, p. 389–393.

Dunagan, S. P., 1999b, Paleosynecology and taphonomy of freshwater carbonate lakes and ponds from the Upper Jurassic Morrison Formation (Western Interior, U.S.A.): Geological Society of America, Abstracts with Program, v. 31, p. 365.

Dunagan, S. P., in press, Constraining Late Jurassic paleoclimate within the Morrison paleoecosystem: insights from the terrestrial carbonate record of the Morrison Formation (Colorado, U.S.A.): Proceedings of the Fifth International Symposium on the Jurassic System.

Dunagan, S. P., Driese, S. G., and Walker, K. R., 1997, Paleolimnological implications of Magadi-type cherts from the lacustrine carbonates in the Morrison Formation (Upper Jurassic), Colorado, U.S.A: Geological Society of America, Abstracts with Programs, v. 29, no. 6, p. 270.

Dunagan, S. P., Demko, T. M., Driese, S. G., and Walker, K. R., 1996, Lacustrine and palustrine carbonate facies of the Morrison Formation (Upper Jurassic): implications for paleoenvironmental reconstructions: Geological Society of America, Abstracts with Programs, v. 28, no. 7, p. 336.

Eugster, H. P., 1969, Inorganic bedded cherts from the Magadi area, Kenya: Contributions to Mineralogy and Petrology, v. 22, p. 1–31.

Frazier, F., Houck, K., Lockley, M., Prince, N., Vest, W., and Coringrato, V., 1983, Interpretations of some depositional environments and paleoecology in the Morrison Formation of southeastern Colorado: Geological Society of America, Abstracts with Programs, v. 15, no. 5, p. 333.

Freytet, P., and Plaziat, J. C., 1982, Continental Carbonate Sedimentation and Pedogenesis—Late Cretaceous and Early Tertiary of Southern France: Contributions to Sedimentology, no. 12, 213 p.

Harland, B. W., Armstrong, R. L., Cox, A. V., Craig, L. E., Smith, A. G., and Smith, D. G., 1990, A Geologic Time Scale: Cambridge, Cambridge University Press, 263 p.

Hasiotis, S. T., and Demko, T. M., 1996, Terrestrial and freshwater trace fossils, Upper Jurassic Morrison Formation, Colorado Plateau, *in* Morales, M., ed., The Continental Jurassic: Museum of Northern Arizona Bulletin 60, p. 355–370.

Hay, R. L., 1968, Chert and its sodium-silicate precursors in sodium-carbonate lakes of East Africa: Contributions to Mineralogy and Petrology, v. 17, p. 255–274.

Jackson, T. J., 1979, Part B. Lacustrine deltaic deposition of the Jurassic Morrison Formation of north-central Colorado, *in* Ethridge, F. G., ed., Guidebook for Field Trips 1979: Colorado State University, Rocky Mountain Section of the GSA, p. 31–54.

Johnson, J. S., 1991, Stratigraphy, sedimentology, and depositional environments of the Upper Jurassic Morrison Formation, Colorado Front Range [unpublished Ph.D. dissertation]: The University of Nebraska, Lincoln, 180 p.

Kowallis, B. J., Christiansen, E. H., Deino, A. L., Peterson, F., Turner, C. E., Kunk, M. J., and Obradovich, J. D., 1998, The age of the Morrison Formation, *in* Carpenter, K., Chure, D. J., and Kirkland, J. I., eds., The Upper Jurassic Morrison Formation: An Interdisciplinary Study: Modern Geology, v. 22, no. 1–4, p. 235–260.

Litwin, R. J., Turner, C. E., and Peterson, F, 1998, Palynological evidence on the age of the Morrison Formation, Western Interior U.S.: a preliminary report, *in* Carpenter, K., Chure, D. J., and Kirkland, J. I., eds., The Upper Jurassic Morrison Formation: An Interdisciplinary Study: Modern Geology, v. 22, no. 1–4, p. 297–320.

Lockley, M. G., Houck, K. J., and Prince, N. K., 1986, North America's largest dinosaur trackway site: implications for Morrison Formation paleoecology: Geological Society of America Bulletin, v. 97, p. 1163–1176.

Peterson, F., and Turner-Peterson, C. E., 1987, The Morrison Formation of the Colorado Plateau: recent advances in sedimentology, stratigraphy, and paleotectonics: Hunteria, v. 2, no. 1, p. 1–18.

Peterson, F., 1988, Pennsylvanian to Jurassic eolian transportation systems in the western United States: Sedimentary Geology, v. 56, p. 207–260.

Peterson, F., 1994, Sand dunes, sabkhas, streams, and shallow seas: Jurassic paleogeography in the southern part of the Western Interior basin, *in* Caputo, M. V., Peterson, J. A., and Franczyk, K. J., eds.,

Mesozoic Systems of the Rocky Mountain Region, U.S.A.: Rocky Mountain Section, SEPM, p. 233–272.

Peterson, J. A., 1972, Jurassic system, *in* Mallory, W. W., ed., Geologic Atlas of the Rocky Mountain Region: Rocky Mountain Association of Geologists, Denver, p. 177–189.

Pipiringos, G. N., and O'Sullivan, R. B., 1978, Principal unconformities in Triassic and Jurassic rocks, Western Interior United States—A preliminary report: U.S. Geological Survey Professional Paper 1035A, p. A1–A29.

Platt, N. H., and Wright, V. P., 1991, Lacustrine carbonates: facies models, facies distribution, and hydrocarbon aspects, *in* Anadon, P., Cabrera , Ll., and Kelts, K., eds., Lacustrine Facies Analysis: International Association of Sedimentologists, Special Publication No. 13, p. 57–74.

Schubel, K. A., and Simonson, B. M., 1990, Petrography and diagenesis of cherts from Lake Magadi, Kenya: Journal of Sedimentary Petrology, v. 60, p. 761–776.

Schudack, M. E., Turner, C. E., and Peterson, F., 1998, Biostratigraphy, paleoecology, and biogeography of charophytes and ostracodes from the Upper Jurassic Morrison Formation, Western Interior, U.S.A., *in* Carpenter, K., Chure, D. J., and Kirkland, J. I., eds., The Upper Jurassic Morrison Formation: An Interdisciplinary Study: Modern Geology, v. 22, no. 1–4, p. 379–414..

Sweet, R. G., 1984, Evidence for playa sedimentation in the Morrison Formation, Canon City, Colorado: Geological Society of America, Abstracts with Programs, 16(6): 672.

Sweet, R. G., and Donovan, R. N., 1988, Lacustrine environments in the Jurassic Morrison Formation a carbonate record in the Cañon City area, Colorado: Geological Society of America, Abstracts with Programs, 20:A52.

Turner, C. E., and Fishman, N. S., 1991, Jurassic Lake T'oo'dichi': a large alkaline, saline lake, Morrison Formation, eastern Colorado Plateau: Geological Society of America Bulletin, v. 103, p. 538–558.

West, E. S., 1978, Biostratigraphy and paleoecology of the lower Morrison Formation of Cimarron County, Oklahoma [unpublished M.S. thesis]: Wichita, Wichita State University, 83 p.

Yuretich, R. F., Hickey, L. J., Gregory, B. P., and Hsia, Y. L., 1984, Lacustrine deposits in the Paleocene Fort Union Formation, northern Bighorn basin, Montana: Journal of Sedimentary Petrology, v. 54, p. 836-852.

Buatois, L. A., M. G. Mángano, Wu Xiantao, and Zhang Guocheng, 2000, Jurassic lake deposits from the Anyao Formation, Jiyuan-Yima Basin, central China, in E. H. Gierlowski-Kordesch and K. R. Kelts, eds., Lake basins through space and time: AAPG Studies in Geology 46, p. 189–194.

Chapter 15

Jurassic Lake Deposits from the Anyao Formation, Jiyuan-Yima Basin, Central China

Luis Alberto Buatois
María Gabriela Mángano
CONICET
Facultad de Ciencias Naturales, Universidad Nacional de Tucumán
Casilla de correo 1 (CC), Argentina

Wu Xiantao
Zhang Guocheng
Jiaozuo Mining Institute
Jiaozuo City, Henan Province, People's Republic of China

INTRODUCTION AND REGIONAL SETTING

The Lower Jurassic Anyao Formation crops out southwest of Jiyuan city, western Henan Province, central China (Figure 1). It represents part of the infill of the Late Triassic–Middle Jurassic Jiyuan-Yima Basin, which hosts more than 2000 m of nonmarine sedimentary rocks. The east-west Yellow River fault separates the northern Jiyuan sub-basin from the southern Yima sub-basin. Jurassic sedimentation is thought to represent deposition in a pull-apart basin. This type of sedimentary basin is common in many different areas of China (Hsü, 1989; Lin et al., 1991). The Jiyuan-Yima Basin is characterized by an asymmetric fill, a limited lateral extent, and extreme lateral facies changes, as well as coal seams and oil reservoirs. Data for the Jiyuan-Yima Basin come from both outcrops and boreholes. Stratigraphy of the Jurassic units of the Jiyuan-Yima Basin was summarized by Zhou and Li (1980), Kang et al. (1984, 1985), and Hu (1991). In particular, the sedimentology and ichnology of the Anyao Formation were previously discussed by Wu (1985).

The Anyao Formation is about 100 m thick and consists of light greenish-grey, medium to very fine-grained sandstones and mudstones (Figure 2). The lower part of the formation is well-exposed southwest of Jiyuan city, but the upper part is covered in most places. The Anyao Formation overlies the Triassic Tanzhuang Formation, which consists of fluvial and shallow lacustrine deposits (Mángano et al., 1994) and is overlain by the Middle Jurassic Yangshuzhuang Formation of deltaic origin. The Jurassic succession culminates with alluvial deposits of the Mawa Formation. The age of the Anyao Formation is considered to be Early Jurassic by means of indirect evidence. The underlying Tanzhuang Formation contains the *Danaeopsis-Bernoullia* flora of Late Triassic age and the overlying Yangshuzhuang Formation hosts the *Coniopteris-Phoenicopsis* flora of Middle Jurassic age (Zhou and Li, 1980; Wu, 1985; Hu, 1991).

SEDIMENTARY FACIES

Three sedimentary facies were recognized in the Jurassic Anyao Formation: facies A (thin-bedded turbidites), facies B (classic turbidites), and facies C (massive sandstones).

Facies A consists of very fine-grained sandstones and mudstones, displaying Tc-e or Td-e Bouma sequences. A typical bed is formed, from base to top, by current-rippled, very fine-grained sandstones; parallel-laminated, very fine-grained sandstones and mudstones; and massive mudstones. The basal ripple-laminated interval is commonly absent. Beds have sharp bases and are laterally persistent. In places, small and elongated flute marks, transitional with load casts, occur on the soles of sandstones. Deformational structures are extremely rare. Individual bed thickness ranges from 3 to 26 cm. Beds are commonly stacked, forming up to 2-m-thick sets. In most cases, no trends in grain size and bed thickness have been detected within the sets. Rarely, sets display thinning- and fining-upward trends, with division C well-developed toward the base of the package.

Facies B is composed of fine to very fine-grained sandstones and mudstones containing Bouma sequences beginning with division A (normally graded or massive) or B (parallel-laminated). In most cases, division C is absent. Beds display sharp bases

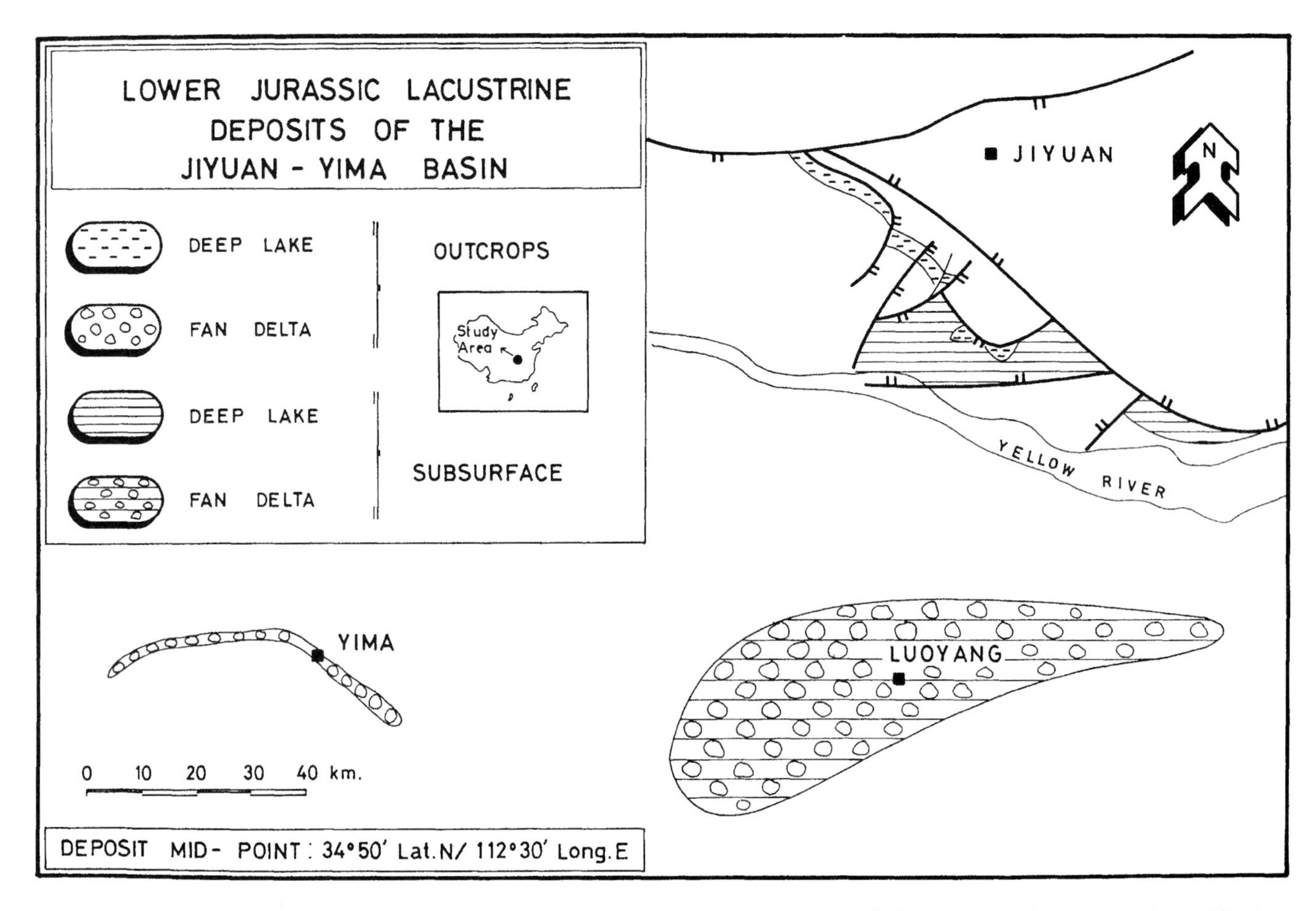

Figure 1—Geologic map of the Jurassic lacustrine and fan delta deposits of the Jiyuan-Yima Basin (modified from Wu, 1985). Strike-slip faults are indicated by bold lines with tic marks.

and tabular geometry. Flute marks are common on the soles of sandstone beds. Individual bed thickness ranges from 3 to 100 cm. Invariably, the thinnest beds begin with division B and lack division A. Conversely, the thickest beds are mainly massive sandstones with poor development of finer grained intervals. Beds are rarely stacked, forming sets of up to 226 cm thick. No trends in grain size and bed thickness have been detected within the sets.

Facies C consists of massive, medium- or fine-grained sandstones. Well-developed, large-scale load structures (up to tens of centimeters in width) are commonly developed at the base of beds. Beds display loaded or erosive bases. Mudstone intraclasts are locally common. Although beds are laterally persistent, lateral variations in thickness are relatively common. Individual bed thickness ranges from 20 to 150 cm. Beds are rarely stacked, forming up to 200-cm-thick sets.

DISCUSSION

The Anyao Formation records siliciclastic sedimentation in a sublacustrine turbidite system (Wu, 1985; Buatois et al., 1994a). Facies A is thought to represent sedimentation from low-density turbidity currents in turbidite lobe fringes. Facies B was interpreted as deposited from "classic" turbidity currents in turbidite lobes. Facies C was deposited in the inner part of lobes from rapid high-density turbidity currents later affected by liquefaction/fluidization. A typical turbidite lobe sequence consists, from base to top, of facies A, B, and C. The general stacking pattern shows coarsening-upward trends, but no clear trends in bed thickness have been detected (Figure 3). The absence of thickening-upward trends may indicate that the lobes were formed by vertical aggradation, rather than by basinward progradation, and that they were fed from multiple sources, rather than a single point-source (Buatois et al., 1994a). Conversely, at the basin scale, the whole system is considered to be progradational, from deep lake to deltaic and fluvial deposits. Wu (1985) reported the presence of wave ripples toward the top of the Anyao Formation, which suggests that shallower lacustrine environments were represented immediately below the deltaic facies of the Yangshuzhuang Formation. The absence of thick mudstone packets indicates that background sedimentation was obscured by high rates of event deposition.

The overlying Yangshuzhuang Formation consists of light gray and yellow coarse to fine-grained sandstones and mudstones. Sandstones display a wide

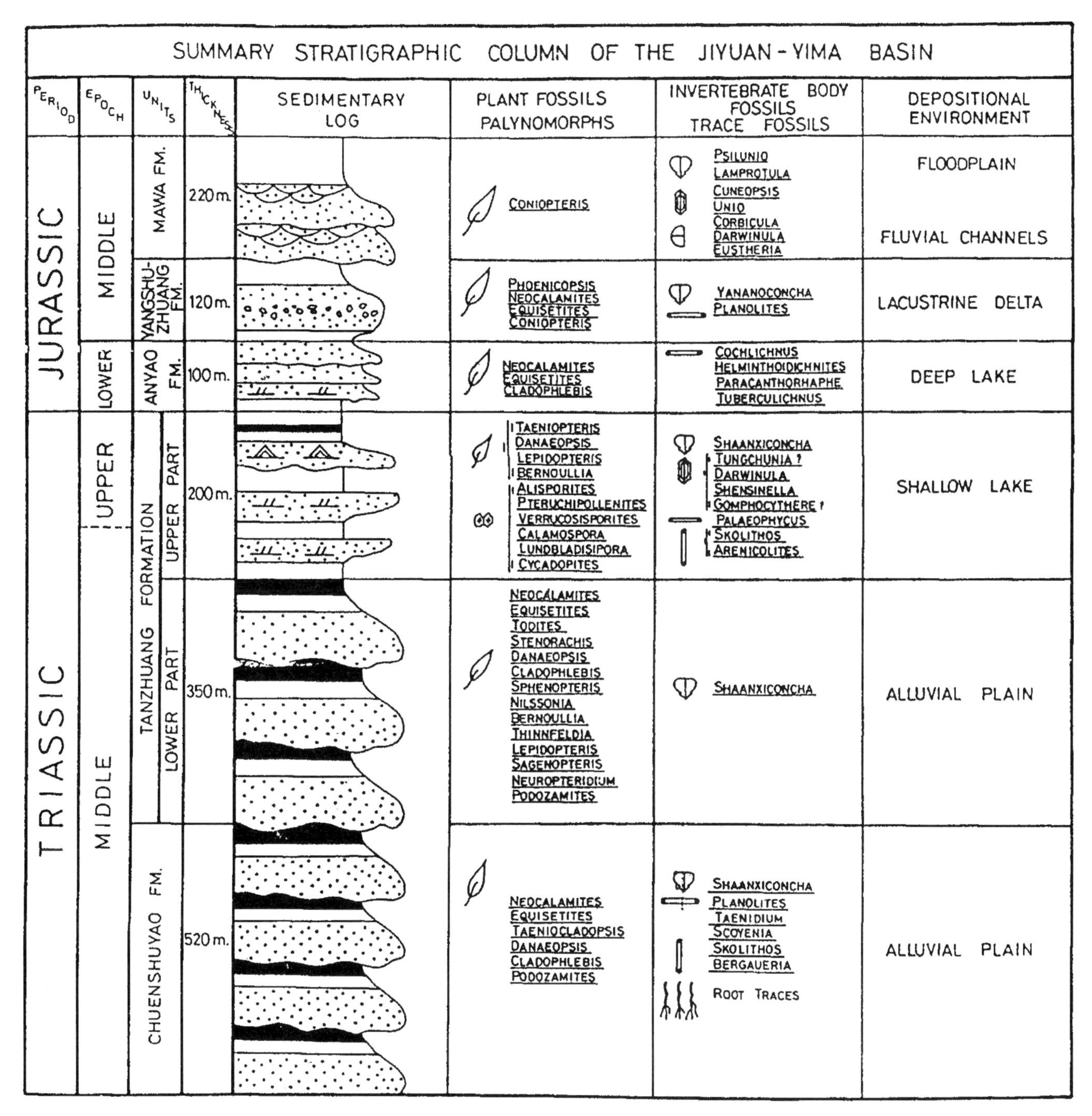

Figure 2—Summary of the stratigraphy of the Mesozoic of the Jiyuan sub-basin. Based on Zhou and Li (1980), Kang et al. (1984, 1985), Wu (1985), and Hu (1991). GGLAB legend for sedimentary log, except shales are black.

variety of primary sedimentary structures, such as planar cross-bedding, parallel bedding, wavy bedding, wave-dominated, combined-flow ripples, and wave-influenced ripples. Plant remains, fish scales, and bivalves occur in the mudstones. This unit represents the progradation of a deltaic system that led to the final infill of the lake basin.

Although no body fossils were found, the lacustrine deposits of the Anyao Formation host a moderately diverse ichnofaun,a which was originally described by Wu (1985) and reviewed by Buatois et al., (1994b, 1995, 1996). The Anayo ichnofauna includes *Cochlichnus anguineus*, *Helminthoidichnites tenuis*, *Helminthopsis abeli*, *H. hieroglyphica*, *Monomorphichnus lineatus*, *Paracanthorhaphe togwunia*, *Tuberculichnus vagans*, *Vagorichnus anyao*, tiny grazing trails, and irregularly branching burrows (Buatois et al., 1996). Trace fossils are mostly preserved as hypichnial casts on the soles of turbidite sandstones. The assemblage is dominated by feeding and grazing traces (Figure 2). Although some ichnofossils show overall similarities with deep sea graphoglyptids, they reflect less specialized feeding strategies and, in some cases, they are of postdepositional origin. Presence of less complex forms may be

Figure 3—Schematic reconstruction of facies sequences from turbidite lacustrine deposits of the Anyao Formation. GGLAB legend for sedimentary log, except shales are black. From Buatois et al., 1996, with permission from Gordon and Breach Science Publishers.

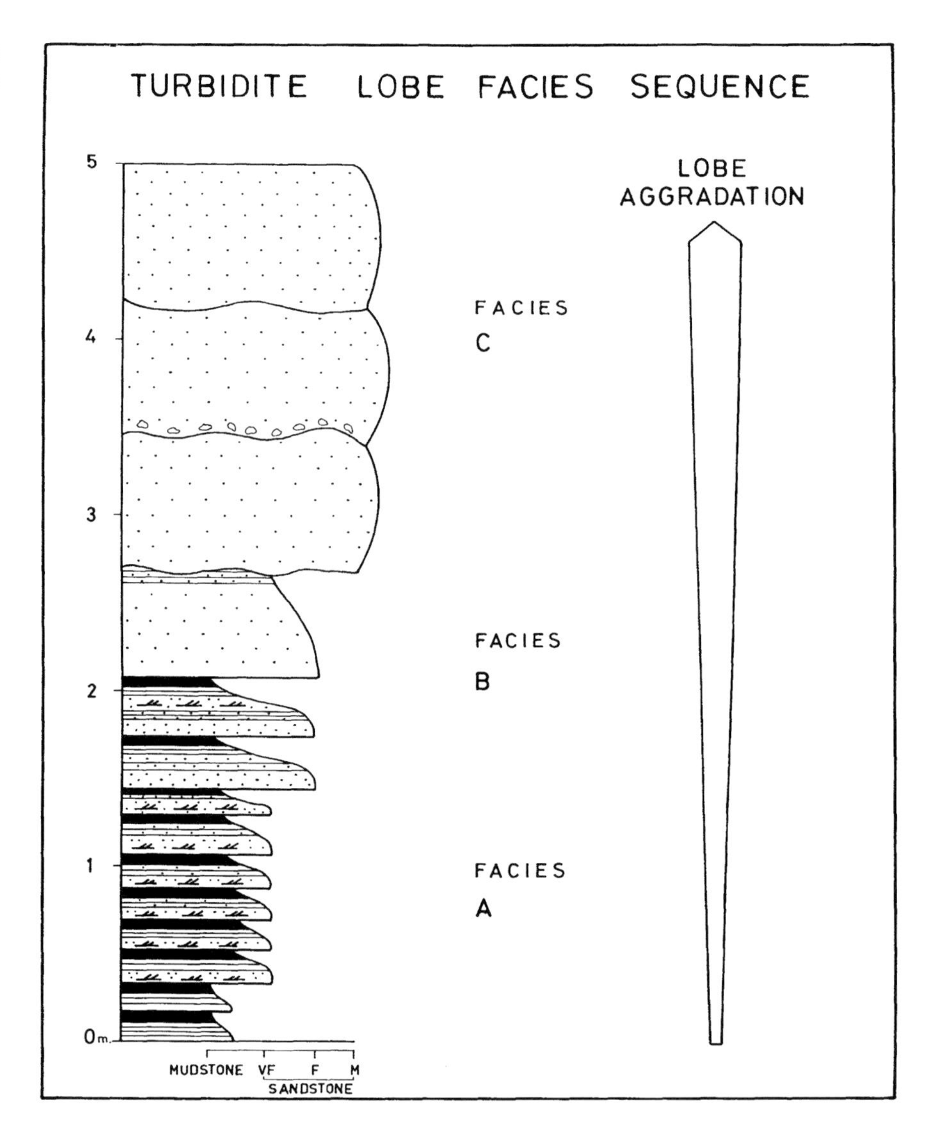

related to less stable conditions, typical of lake settings. Occurrence of this relatively diverse ichnofauna indicates oxygenated conditions in the lake bottom, probably linked with the action of turbidity currents.

Paleocurrents from flute marks in turbidites of the Anyao Formation indicate alternating provenances from the southwest and southeast (Wu, 1985). Coeval strata locally known as the Basal conglomerate bed crop out farther southwest, south of the Yellow River (Figure 1). They consist mainly of conglomerates, coarse-grained sandstones, and carbonaceous shales deposited in fan deltas and alluvial fans (Wu, 1985). Coarse-grained sediments are thought to record deposition from debris flows close to the faulted margin of the basin.

During the Triassic–Jurassic transition a slight uplift was recorded in the area, related to Indian-Chinese plate movements. The southern basin margin was bounded by a strike-slip fault, typified by a high-relief fault scarp, and the northern side probably had a less-pronounced topography. Early fan delta and lacustrine sediments represent the onset of tectonic subsidence along the basin-bounding faults, probably during a transtensional regime (Buatois et al., 1994a). In the southern areas, close to the margin faults, coarse-grained alluvial fan and fan delta systems were developed; however, deposition in deep lakes, fed from the southern fan deltas, was predominant in the northern Jiyuan area. Sedimentation in this early stage is recorded by outcrops of the Anyao Formation in the Jiyuan area and the basal conglomerate bed south of the Yellow River, as well as from several borehole localities. During the Early–Middle Jurassic transition, the Yanshanian movements took place and the Taihang Mountains were formed. These mountains are located in the northeast China highlands, northeast of the study area. At this stage, a modified drainage of fluvial systems funneled large amounts of siliciclastics into the basin, overcoming the effects of subsidence. This later stage is probably related to a transpressive regime and is documented by the deltaic deposits of the Yangshuzhuang Formation and the overlying fluvial sediments of the Mawa Formation in the Jiyuan sub-basin and by the comparable Yima and Dongmengchun Formations in the Yima sub-basin. New movements during the Jurassic–Cretaceous transition were probably responsible for basin closure.

ACKNOWLEDGMENTS

The research visit to China of L. A. Buatois and M. G. Mángano was supported by the Chinese Academy of Sciences and the Third World Academy of Sciences, whose continuous help is gratefully acknowledged. The authors would like to thank Sun Junmin for his assistance in the field and Wang Guanzhong for his comments on the Anyao ichnofauna. Reviews by Elizabeth Gierlowski-Kordesch and John Parnell are gratefully acknowledged. L. A. Buatois and M. G. Mángano particularly wish to compliment the politeness and efficiency of the Foreign Affairs Office Staff of the Jiaozuo University and the hospitality of the people of Jiaozuo. Daniel Ruiz Holgado kindly made the illustrations.

REFERENCES

Buatois, L. A., M. G. Mángano, X. Wu, and G. Zhang, 1994a, Turbidite lacustrine sedimentation from the Jurassic Anyao Formation, Henan Province, central China. 14th International Sedimentological Congress, Abstracts of Papers, Recife, p. S6-1.

Buatois, L. A., M. G. Mángano, X. Wu, and G. Zhang, 1994b, Ichnology of a lacustrine turbidite system: The Anyao Formation, Lower Jurassic of Henan Province, central China. 14th International Sedimentological Congress, Abstracts of Papers, Recife, p. S5-1.

Buatois, L. A., M. G. Mángano, X. Wu, and G. Zhang, 1995, *Vagorichnus*, a new ichnogenus for feeding burrow systems and its occurrence as discrete and compound ichnotaxa in Jurassic lacustrine turbidites of Central China. Ichnos, v. 3, p. 265-272.

Buatois, L. A., M .G. Mángano, X. Wu, and G. Zhang, 1996, Trace fossils from Jurassic lacustrine turbidites of the Anyao Formation (central China) and their environmental and evolutionary significance. Ichnos, v. 4, p. 287-303.

Hsü, K., 1989, Origin of sedimentary basins of China, *in* X. Zhu, ed., Chinese Sedimentary Basins, Amsterdam, Elsevier, p. 207-227.

Hu, B., 1991, On the Late Triassic-Middle Jurassic continental stratigraphy from Jiyuan, western Henan Province. Journal of Stratigraphy, v. 15, p. 48-52. (In Chinese with English summary.)

Kang, M., F. Meng, B. Ren, and B. Hu, 1984, The age of Yima Formation and the establishment of the Yangshuzhuang Formation. Journal of Stratigraphy, v. 8, p. 194-198. (In Chinese with English summary.)

Kang, M., F. Meng, B. Ren, B. Hu, and Z. Ching, 1985, Divission and correlation of the Mesozoic coal-bearing stratigraphy from Henan Province. Journal of Jiaozuo Mining Institute, v. 5, p. 1-146. (In Chinese with English summary.)

Mángano, M. G., L. A. Buatois, X. Wu, J. Sun, and G. Zhang, 1994, Sedimentary facies, depositional processes and climatic controls in a Triassic lake, Tanzhuang Formation, western Henan Province, China. Journal of Paleolimnology, v. 11, p. 41-65.

Wu, X., 1985, Trace fossils and their significance in non-marine turbidite deposits of Mesozoic coal and oil bearing sequences from Yima-Jiyuan basin, western Henan, China. Acta Sedimentologica Sinica, v. 3, p. 23-31. (In Chinese with English summary.)

Zhou, Z., and P. Li , 1980, Discussion on the division, correlation and age of the Mesozoic continental stratigraphy of China based on plant fossils, Geological Symposium for International Exchange 4, Stratigraphy and Paleontology, Beijing, Geological Publishing House.

Section IV

Early to Middle Cretaceous

Mabesoone, J. M., M. S. S. Viana, and V. H. Neumann, 2000, Late Jurassic to mid-Cretaceous lacustrine sequences in the Araripe-Potiguar Depression of Northeastern Brazil, *in* E. H. Gierlowski-Kordesch and K. R. Kelts, eds., Lake basins through space and time: AAPG Studies in Geology 46, p. 197–208.

Chapter 16

Late Jurassic to Mid-Cretaceous Lacustrine Sequences in the Araripe-Potiguar Depression of Northeastern Brazil

Jannes M. Mabesoone
Maria Somália S. Viana
Virginio H. Neumann
Department of Geology, Federal University of Pernambuco
Recife, Brazil

INTRODUCTION

During the Late Jurassic and Early Cretaceous, a series of aborted extensional rift basins developed in northeastern Brazil as a consequence of the separation of South America and Africa. This basin series occurs in a southwest-northeast–trending depression (Figure 1), called the Araripe-Potiguar Depression (Mabesoone, 1994). From southwest to northeast, these basins are the Araripe, composed of two fault-limited depressions; the Rio do Peixe with three subbasins; Iguatu-Icó with four subbasins; and the onshore Potiguar or Pendência graben composed of three depressions, as well as a number of intermediate smaller outcrops. Table 1 presents the stratigraphy of the basin fill for this series of basins; two extensive lacustrine phases are recorded: one in the latest Jurassic–Barremian and the other in the late Aptian–early Albian. During the first phase (latest Jurassic–Barremian), a chiefly fine-grained section accumulated during rather humid climates in a slowly subsiding area. Record of the second phase (late Aptian–early Albian) is only found in the Araripe Basin, represented by laminated limestones and shales deposited under fairly dry climatic conditions. This latter sequence is extremely rich in fossils (Viana et al., 1994).

TECTONIC SETTING

The basins of the Araripe-Potiguar Depression are situated on the western flank of the so-called Borborema dome, extending in a southwest-northeast trend, which probably continues into Africa (Berthou, 1990). Differential fault movements, as a result of a crustal extension due to rifting, were mainly controlled by the lithologic and structural differences in the Precambrian crystalline basement. The resulting Late Jurassic–Early Cretaceous sedimentary basins, subsiding as half-grabens, are postulated to outline original fault depressions, although an initial single basin, now eroded, was also possible. During evolution of these basins, subsidence rates varied with time (see Table 1). During slower subsidence or quiet phases, lacustrine paleoconditions developed.

LACUSTRINE PHASES

Two phases of lacustrine sedimentation can be distinguished. The first phase lasted from about the Tithonian (equals Portlandian) until the beginning of the Barremian, which is between the late Dom João and Aratu stages (Table 1) (Arai et al., 1989). This lake phase is represented in all larger basins by fine-grained siliciclastics, with ubiquitous coverage at the beginning of the phase leading to more localized occurrences toward the end of the phase (see Table 1). The second lacustrine phase is only represented in the Araripe Basin, deposited during the late Aptian–early Albian, with chiefly calcareous sediments. During this second lacustrine phase, the sea had already transgressed from the east into the Potiguar Basin.

Portlandian-Barremian Phase

Araripe Basin

In this basin (Figure 2), the first lake section corresponds to the Brejo Santo Formation (Table 1, Figure 3), represented by red- to brown-colored mudstones with intercalations of medium to fine clayey sandstones, siltstones, and greenish marls to limestones, all well-bedded. Fossils include ostracodes of the *Bisulcocypris pricei* biozone of the Dom João and conchostracans (Ponte and Appi, 1990). Thickness of the unit as known from subsurface data may reach up to 400 m. Because outcrops are poorly preserved due to intense tropical weathering and well data are limited, no

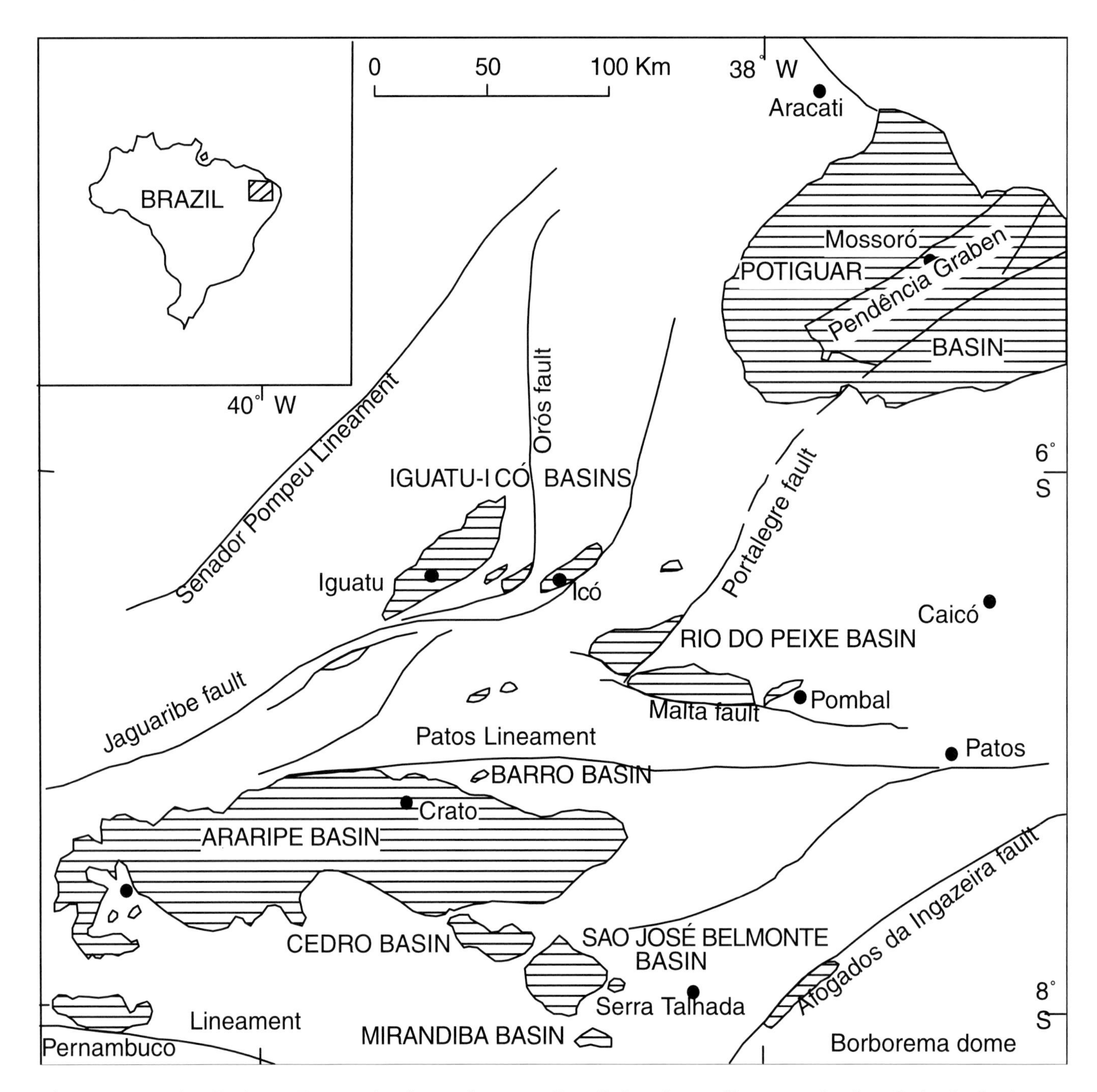

Figure 1—Araripe-Potiguar Depression in northeastern Brazil showing sedimentary basins. Only the basins with lake deposits are highlighted.

detailed sedimentologic log of this lake section yet exists.

The upper portion of the first lacustrine phase is found in the Missão Velha Formation (Table 1, Figure 3). With an increase in tectonic activity, character of the lacustrine deposition changed. Proximal deposits included fine to silty sandstones, which are well sorted, lenticular to sigmoidally bedded with planar cross-bedding topped by climbing ripple lamination amalgamated with mudstone drapes between sets. These sandstones are interpreted as deposits of prograding lake deltas and fluvial floodplains. Distal deposits record intercalations of mudstones and shales with coarser siliciclastics similar to those in the Brejo Santo Formation; the fine-grained intercalations are interpreted as lacustrine and referred to as the Olho d'Água Comprido Member (Cavalcanti and Viana, 1992a) (Figure 3). In the subsurface, lacustrine shales are black to grey in color, with some microsparite, rarely dolomitized; common are recrystallized ostracodes, pisolites, and thin microbial mats around ostracodes and pollen (Mabesoone, 1986). The lacustrine sequence of the Olho d'Água Comprido Member is presented in more detail in Figure 4; thickness of this lake section is about 30 m. Age has been determined as

Table 1. Stratigraphic Correlation of the Sedimentary Sequences of the Basins of the Araripe-Potiguar Depression*

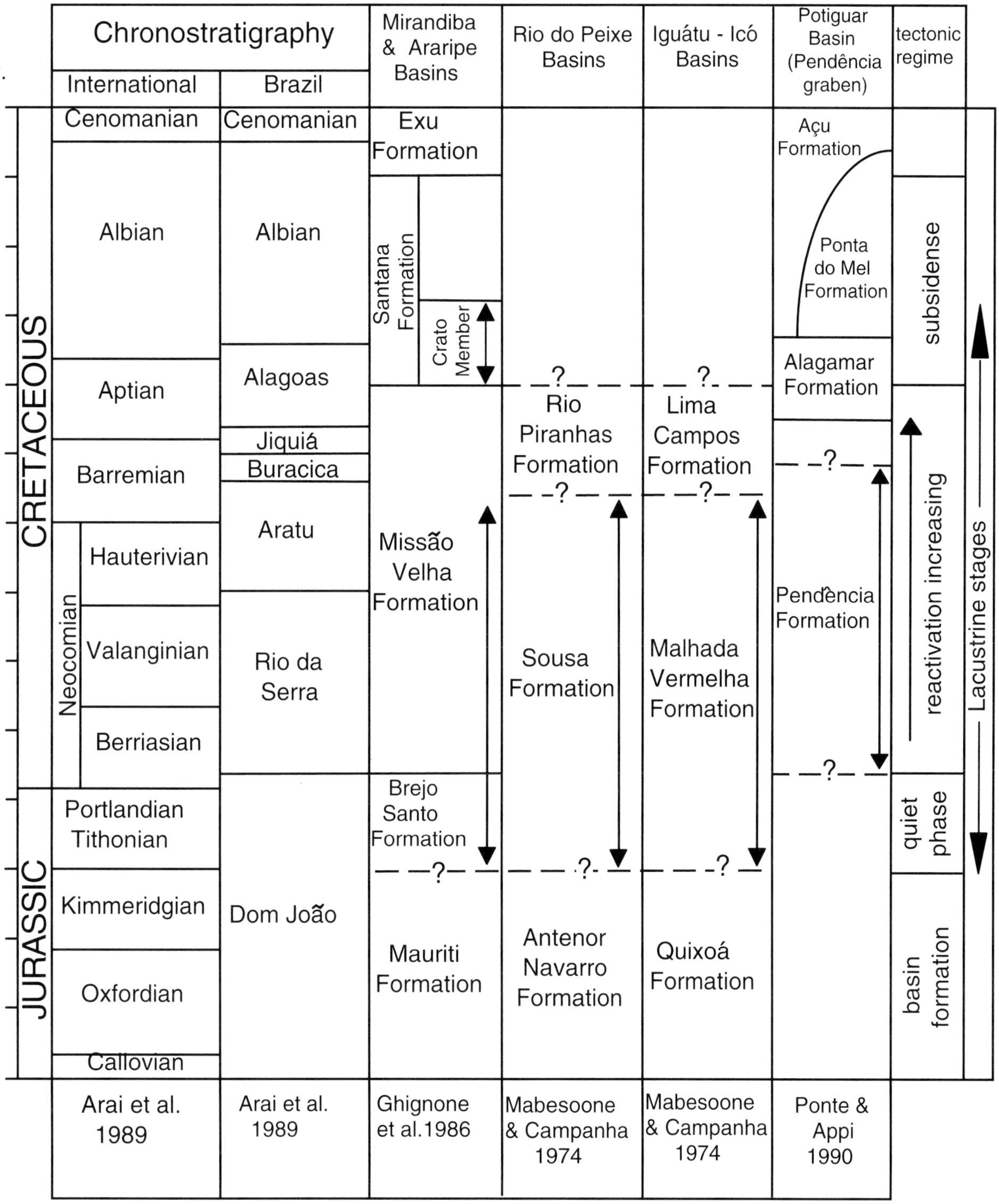

m.y.	Period	Chronostratigraphy: International	Chronostratigraphy: Brazil	Mirandiba & Araripe Basins	Rio do Peixe Basins	Iguátu - Icó Basins	Potiguar Basin (Pendência graben)	tectonic regime	
95	CRETACEOUS	Cenomanian	Cenomanian	Exu Formation			Açu Formation		
100–110	CRETACEOUS	Albian	Albian	Santana Formation; Crato Member			Ponta do Mel Formation	subsidense	Lacustrine stages
115	CRETACEOUS	Aptian	Alagoas	Crato Member	?	?	Alagamar Formation	subsidense	Lacustrine stages
120	CRETACEOUS	Barremian	Jiquiá / Buracica	Missão Velha Formation	Rio Piranhas Formation	Lima Campos Formation	?	reactivation increasing	Lacustrine stages
125	CRETACEOUS (Neocomian)	Hauterivian	Aratu	Missão Velha Formation	? Sousa Formation	? Malhada Vermelha Formation	Pendência Formation	reactivation increasing	Lacustrine stages
130–135	CRETACEOUS (Neocomian)	Valanginian	Rio da Serra	Missão Velha Formation	Sousa Formation	Malhada Vermelha Formation	Pendência Formation	reactivation increasing	Lacustrine stages
140	CRETACEOUS (Neocomian)	Berriasian	Rio da Serra	Missão Velha Formation	Sousa Formation	Malhada Vermelha Formation	?	reactivation increasing	Lacustrine stages
145	JURASSIC	Portlandian Tithonian	Dom João	Brejo Santo Formation	Sousa Formation	Malhada Vermelha Formation		quiet phase	Lacustrine stages
150–155	JURASSIC	Kimmeridgian	Dom João	? Mauriti Formation	? Antenor Navarro Formation	? Quixoá Formation		basin formation	
160	JURASSIC	Oxfordian	Dom João	Mauriti Formation	Antenor Navarro Formation	Quixoá Formation		basin formation	
165	JURASSIC	Callovian	Dom João	Mauriti Formation	Antenor Navarro Formation	Quixoá Formation		basin formation	
Source		Arai et al. 1989	Arai et al. 1989	Ghignone et al. 1986	Mabesoone & Campanha 1974	Mabesoone & Campanha 1974	Ponte & Appi 1990		

*Slightly modified from Mabesoone and Viana (1994). Intervals marked with arrows indicate times of lake deposition.

Rio da Serra–Aratu (Neocomian–Barremian), based on ostracodes. Brito et al. (1994) recorded a rich, but poorly preserved, fauna in the Missão Velha Formation, including actinopterygians, sarcopterygians, pelomedusid turtles, mesosuchian crocodiles, sauropods, theropods, lizards, and perhaps amphibians. Bivalves, conchostracans, and ostracodes, as well as many types of trace fossils, were also found.

Iguatu-Icó Basins

In the Iguatu-Icó Basin series (Ceará State, Figure 5), the lake sequence (Table 1) is represented by the

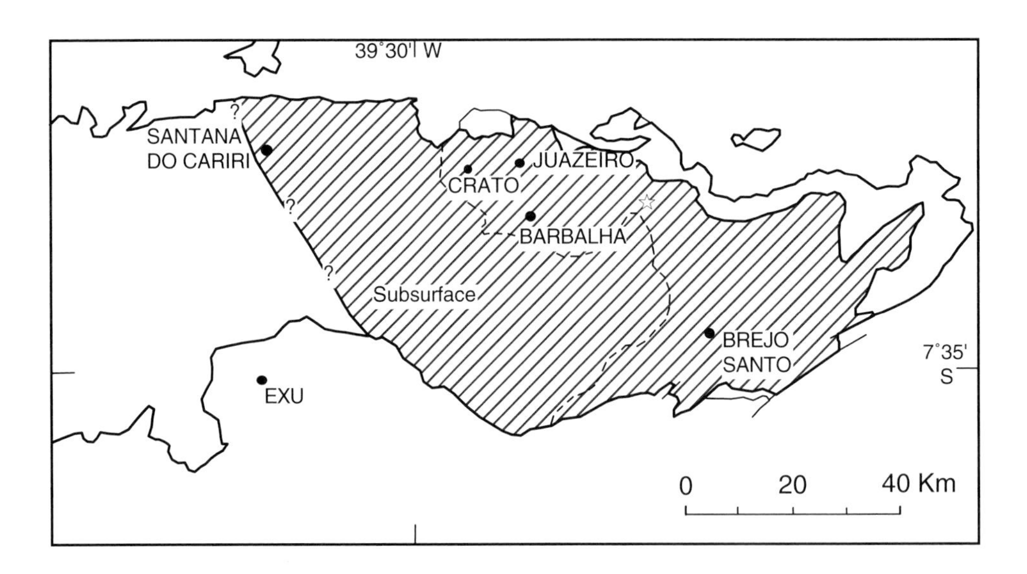

Figure 2—Geographic extent of the first phase lake sedimentation in the Araripe Basin, from outcrop and subsurface data. The star shows the location of the detailed section of Figure 4.

Malhada Vermelha Formation (Mabesoone and Campanha, 1974). Many studies have been published on stratigraphy, sedimentation, and tectonic setting, with recent summaries by Srivastava (1990) and Cavalcanti and Viana (1992b). All descriptions are based on outcrop studies. The inferred total thickness is about 1000 m, with a sand-shale ratio of 1:4.

In the Iguatu Basin, the following lithofacies and depositional paleosystems have been determined (Figure 6). (1) Sandy-pelitic tabular heterolitic facies with fining-upward sequences, mudcracks, low-angle cross-stratification, ripple marks, and small bioturbation features that represent vertical accretion deposits of river floodplains. (2) Coarse to medium sandstones, with lenticular geometry, medium-size trough cross-bedding with lag gravels interpreted as lateral accretion bars. (3) Calcareous greenish to reddish mudstone having tabular bedding and intercalations of calcilutites and lenses of white claystone. Fossils here include casts of conchostracans and bivalves, vertebrate fragments, and ostracodes. Rare coquina-beds of ostracodes (Iguatu Basin) and gastropods/conchostracans (Lima Campos and Malhada Vermelha basins) are present. The depositional realm was lacustrine. (4) Micaceous, well-sorted, fine to very fine sandstones that have lenticular to sigmoidal geometry and muddy intercalations between sets. Toward the top, this facies sequence turns coarser due to a decrease in mudstone layers. These deposits have been interpreted as lake deltas. (5) Grain-supported conglomerates containing pebble imbrication, normal grading, and gravel fragments of quartz, sandstone, and basement rocks up to 10 cm in length. They occur unconformably on top of facies 1 as localized alluvial fan deposits. Sandstones in all these facies are chiefly quartzose with rare plagioclase feldspars, corroded by calcite. Mineralogy of the mudstones includes smectites dominating over illite and irregularly inter-stratified minerals.

In the Malhada Vermelha and Lima Campos (Figures 5, 6) Basins, the Malhada Vermelha Formation exhibits the same lithology, but in different proportions. In the Icó Basin (Figure 5), the section is absent. Srivastava (1990) postulated that the continental complex represented by the Malhada Vermelha Formation represents only one lake; however, it seems more probable that the lake of the Iguatu Basin was separated from that of the Malhada Vermelha and Lima Campos Basins.

Rio do Peixe Basin

The Rio do Peixe Basin in Paraíba State is one of the best studied of the smaller intracontinental basins of the Araripe-Potiguar Depression. This basin is composed of three subbasins: Pombal, Sousa, and Uiraúna-Brejo das Freiras (or Triunfo) subbasins. The lake sections occur in the two latter subbasins (Figure 7) and are well-studied chiefly because of a great number of fossilized dinosaur footprints. The subbasins contain three stratigraphic units: lower sandstone and conglomerate, the Antenor Navarro Formation; middle fine-grained siliciclastics, the Sousa Formation; and upper sandstone, the Rio Piranhas Formation (Table 1). For detailed descriptions of the entire sequence, see Mabesoone and Campanha (1974) and Carvalho and Leonardi (1992). The Sousa Formation is interpreted as a lake sequence and has been studied in detail by Vasconcelos (1980).

The Sousa Formation crops out along the Peixe River (Figure 7) and has been studied in subsurface. A well was drilled in the "deepest" part of the Sousa subbasin where it is about 1200 m thick (800 m present in the well). The well sequence (Figure 8) is essentially fine-grained, characterized by a succession of flat-bedded, thin layers of mudrock, with minor shale, limestone, siltstone, and fine sandstone layers. The mudrocks are variegated with red, green, and grey colors. Sedimentary structures include wave and current ripple marks, mudcracks, convolute lamination, and trough cross-bedding in the sandstone.

A detailed petrographic and microfacies study by Vasconcelos (1980) distinguished the following four microfacies of the Sousa Formation (Figure 8): silty shales (dominant), clayey siltstones, sandy siltstones, and fine and medium quartzose sandstone (rare). Fossils

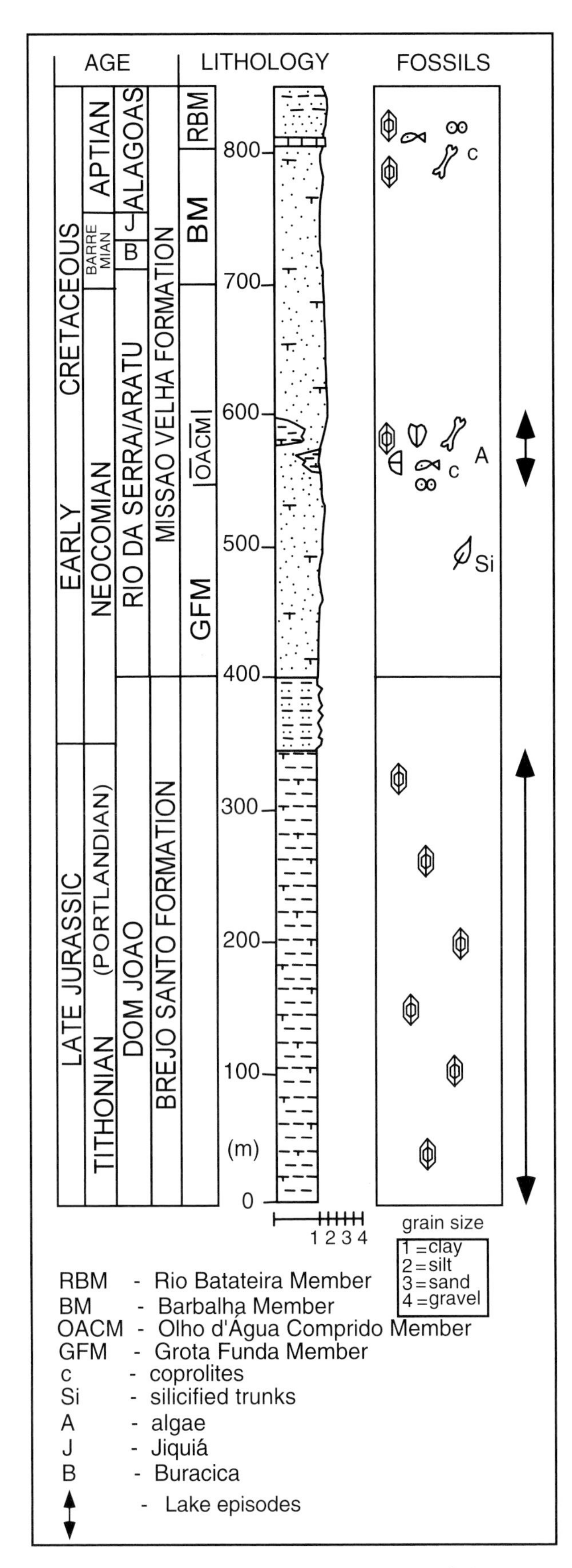

Figure 3—Schematic lithologic column of the Brejo Santo and Missão Velha Formations in the Araripe Basin.

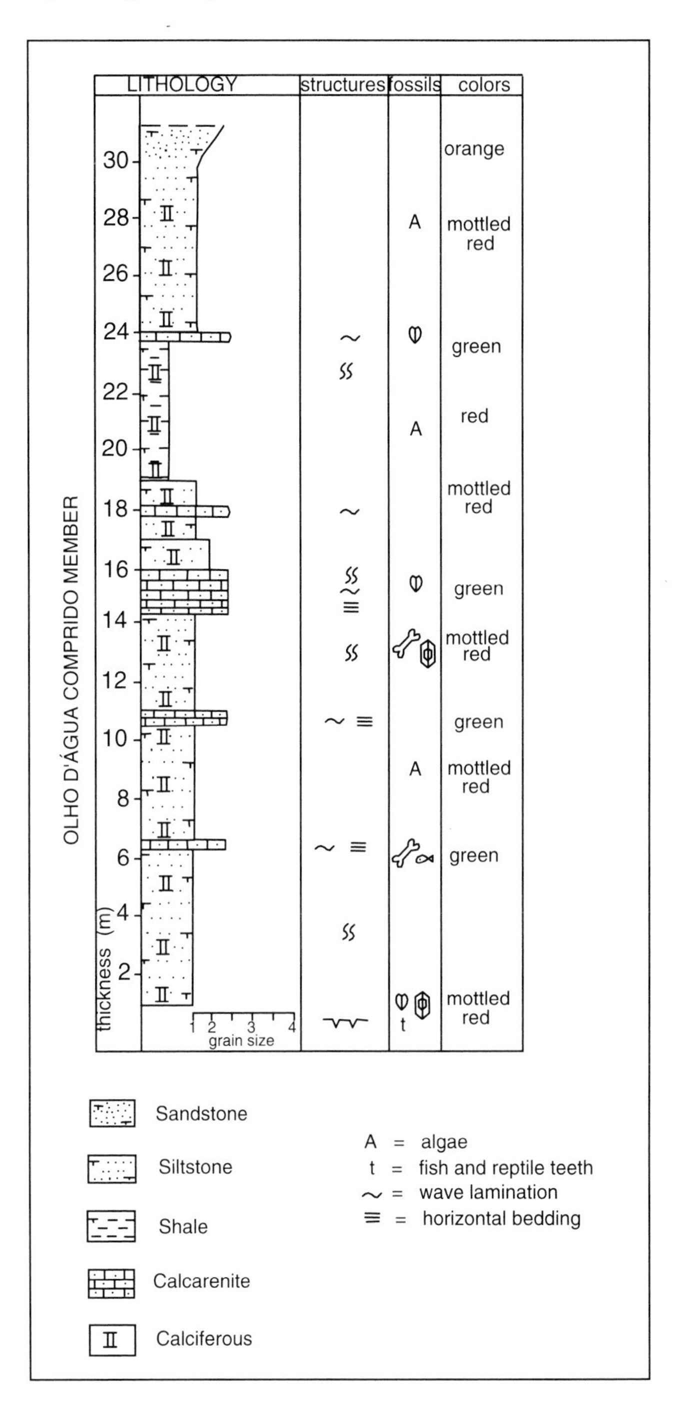

Figure 4—Detailed lithologic column for the lake sequence of the Olho d'Agua Comprido Member, after Cavalcanti and Viana (1992a). See location in Figure 2.

occur at certain levels, including ostracodes, conchostracans, bone fragments, common carbonized macrophyte remains (conifers and cicadoids), spores, pollen, algal filaments, sclerondonts, and vertebrate tracks (for a complete list see Carvalho and Leonardi, 1992). Numerous dinosaur footprints are found at 21 localities around the Sousa subbasin, with 77 tracks and 350 isolated prints of Carnosauria (dominant), Theropoda, Iguanodontidae, Coelurosauria, Sauropoda, and

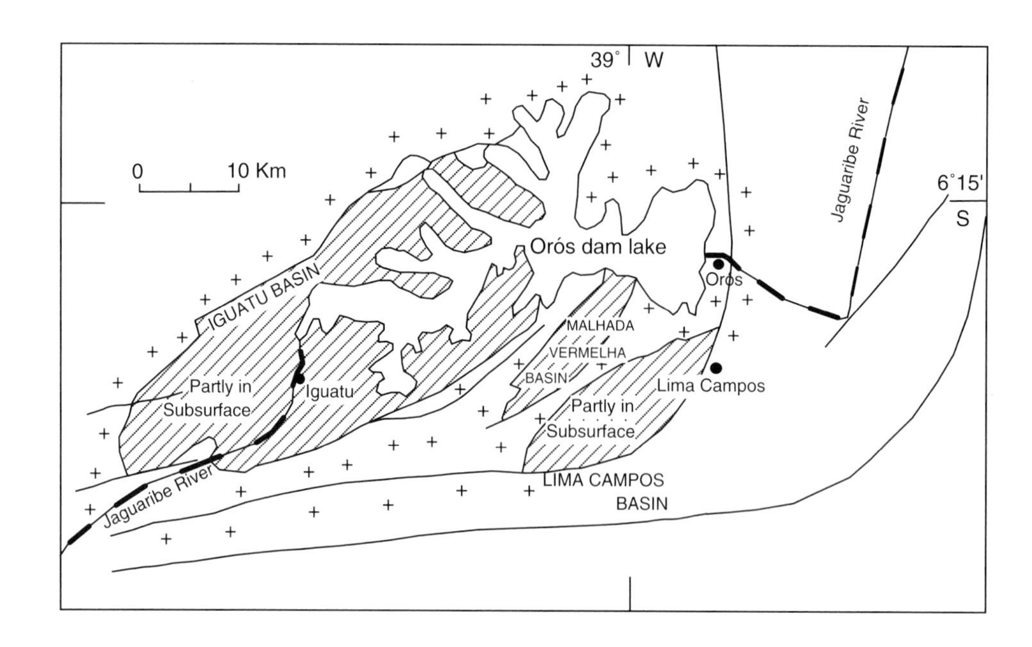

Figure 5—Geographic extent of first phase lacustrine sedimentation in the Iguatu, Malhada Vermelha, and Lima Campos Basins, from outcrop and subsurface data.

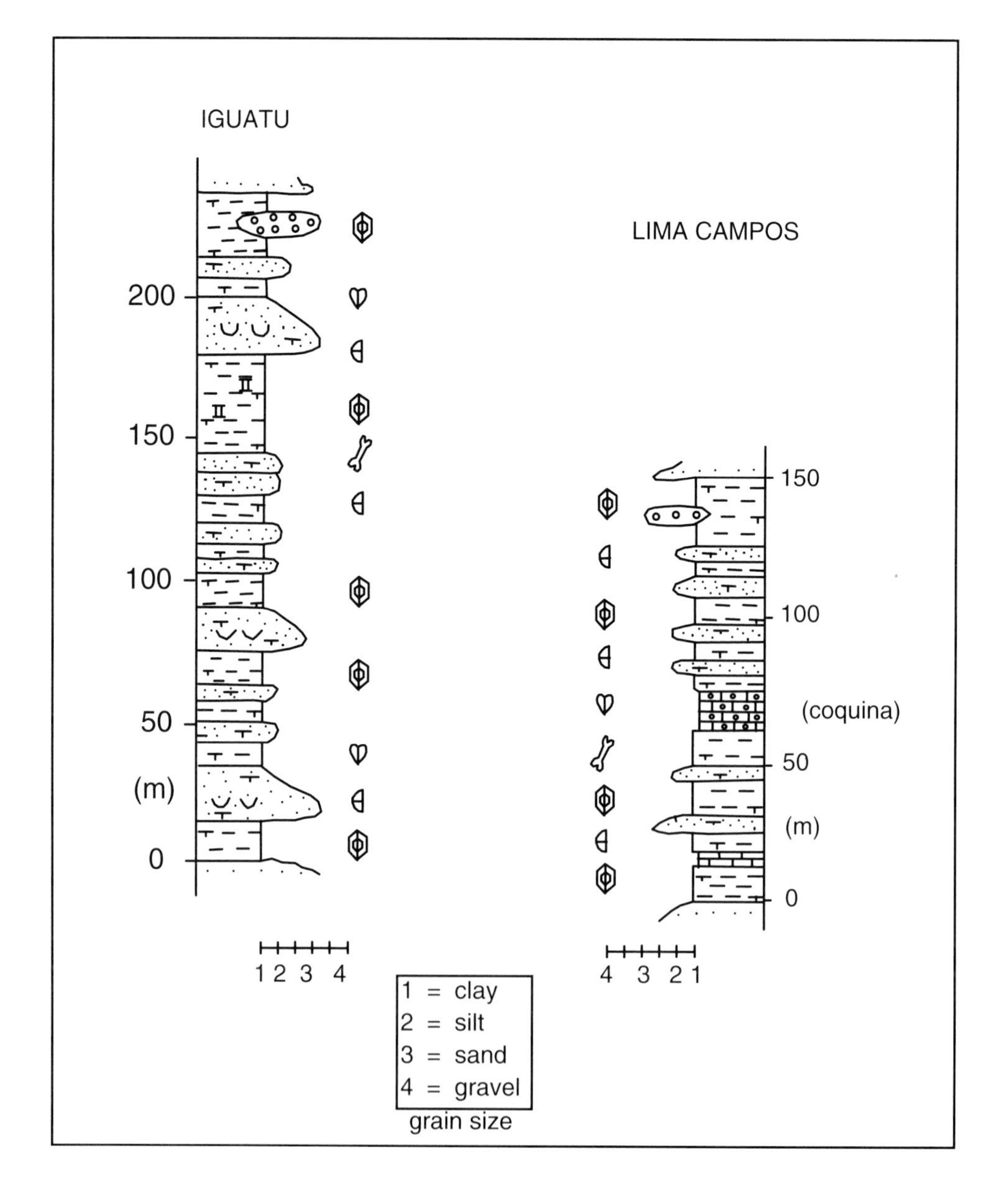

Figure 6—Schematic lithologic columns of the Malhada Vermelha Formation in the Iguatu and the Malhada Vermelha-Lima Campos Basins.

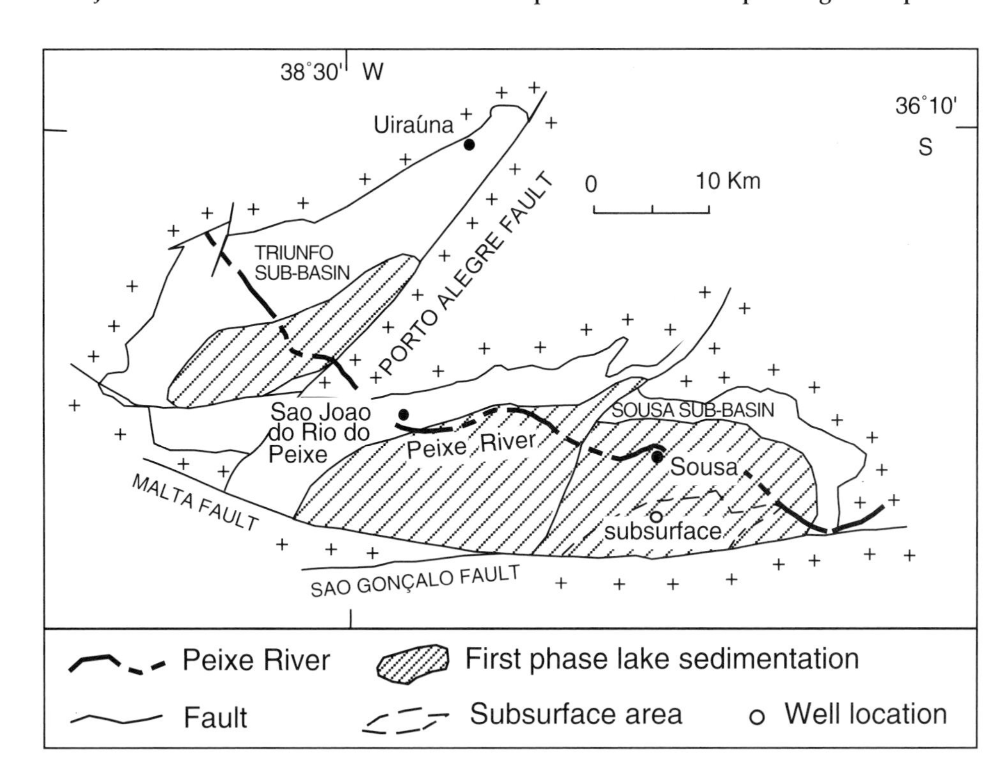

Figure 7—Geographic extent of the first phase lake sedimentation in the Rio do Peixe Basin based on outcrop and subsurface data. The open circle shows the well location of the core sequence shown in Figure 8.

others (Carvalho, 1989). Age of the section was determined chiefly by ostracodes, representing three biostratigraphic zones ranging from uppermost Jurassic to Hauterivian (in local chronostratigraphic terms, uppermost Dom João, Rio da Serra, and Aratu) (cf. Arai et al., 1989) (Table 1).

Based on sedimentary structures, microfacies, fossils, and biostratigraphic zones data, four phases in the depositional history of the Sousa Formation were distinguished: (1) lacustrine to palustrine in dry climate; (2) lacustrine and fluvial in less dry climate; (3) lacustrine with alluvial plains bordering the subbasin; and (4) low- to high-energy fluvial in the uppermost part transitional to renewed fluvial accumulation of the Antenor Navarro Formation above.

Pendência Graben (Potiguar Basin)

The lake sequence in this graben is found only in subsurface. The Pendência graben, although belonging to the Araripe-Potiguar Depression, opens toward the equatorial Atlantic Ocean where its presence has been confirmed (Figure 9). The Pendência Formation (Souza, 1982), which lies unconformably on the Precambrian crystalline basement, represents a lake sequence (Table 1), which is characterized by alternations of greenish-grey shales, grey siltstones, and whitish-grey, very fine to medium sandstones with a small carbonate component. Lima Neto et al. (1986) in Neves (1989), based on a litho-, bio-, and seismostratigraphy, proposed five chronostratigraphic units (I–V) for the Pendência Formation. These units are separated by surfaces that represent variations in the sedimentation regime of the Pendência lake, interpreted as caused by climatic changes (Figure 10). Along the borders of the Pendência rift, polymictic conglomerates interbedded with some sandstones occur. In the southern part of the graben (more inland), the measured thickness of the formation is 1100 m, although total thickness has been estimated at 3000 m, increasing oceanward to about 3500 m. Lake deposits have also been actually identified in the offshore basin cores. The eastern extent of the graben has not yet been studied in detail.

The depositional paleoenvironment of the Pendência Formation is interpreted as fluvial at the base and progressively more lacustrine-dominated up-section. Along the borders of the graben, fan delta deposits were deposited, passing into turbidites toward the deeper parts of the lake (Silva, 1993), with sliding and grain flows as dominant depositional processes (Figure 10). Rodrigues and Takaki (1992) determined ancient salinity and organic matter production through isotopic analyses of carbonate and organic fractions, as well as esterane and terpane isoprenids of the soluble fraction of the organic matter, in parts of the lacustrine sequence. They concluded that the lake was initially fresh water (units I, II, III), became larger, and then turned slowly more saline (units IV, V) due to excess evaporation in a more arid climate. Organic production did not appear to change very much during lake history. Based on pollen and spores, age of the section has been determined as Rio da Serra–Aratu (Table 1), maybe somewhat younger (until Jiquiá) in the direction of the equatorial ocean (Regali and Gonzaga, 1982, in Souza, 1982).

Albian Phase

Araripe Basin

The second lacustrine phase became established due to increased tectonic reactivation during the late Barremian, followed by a period of slow subsidence in

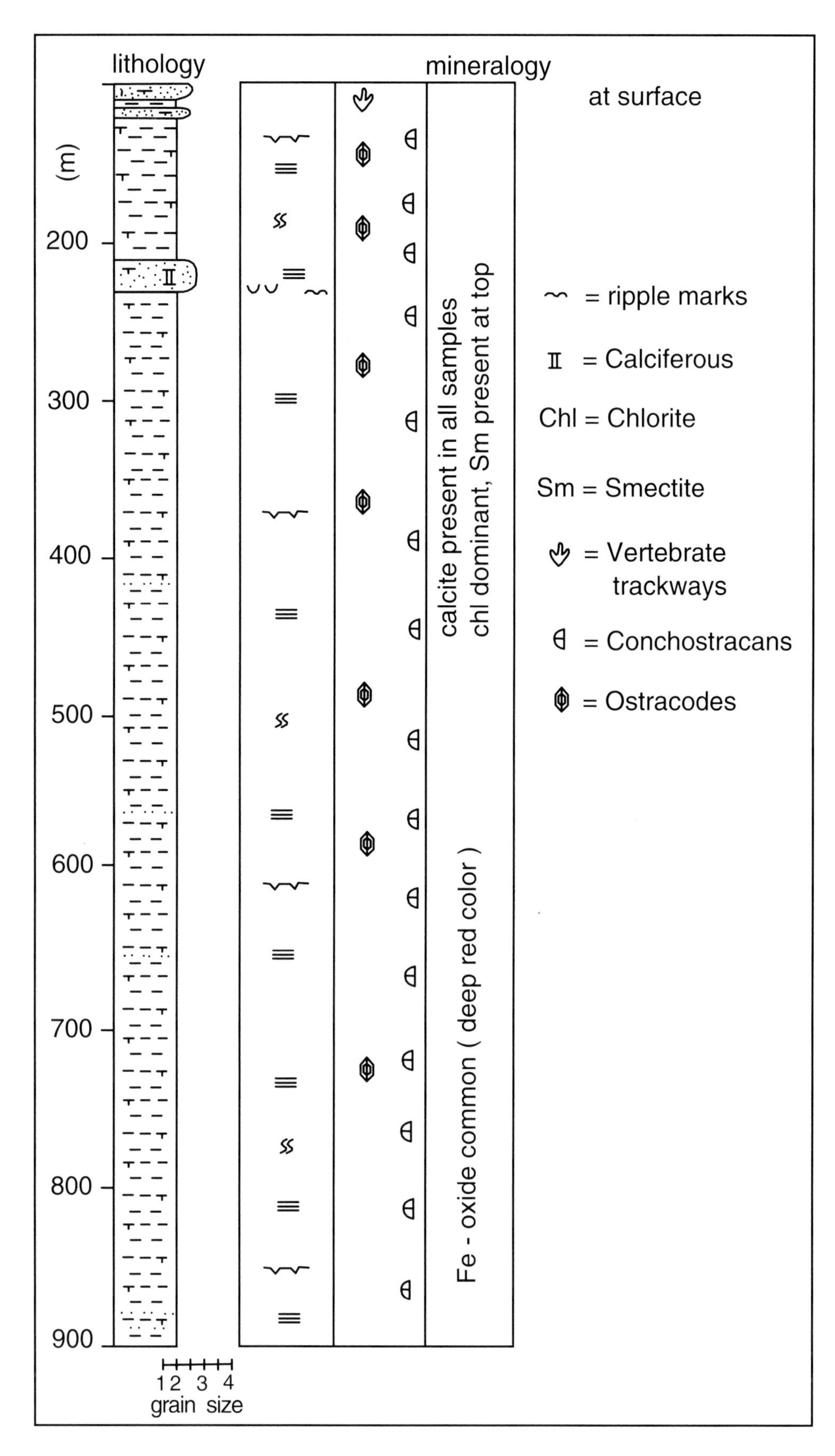

Figure 8—Schematic lithologic column of a well core from the Sousa Formation in the "deepest" portion of the Rio do Peixe Basin, after Vasconcelos (1980). Location of well shown in Figure 7.

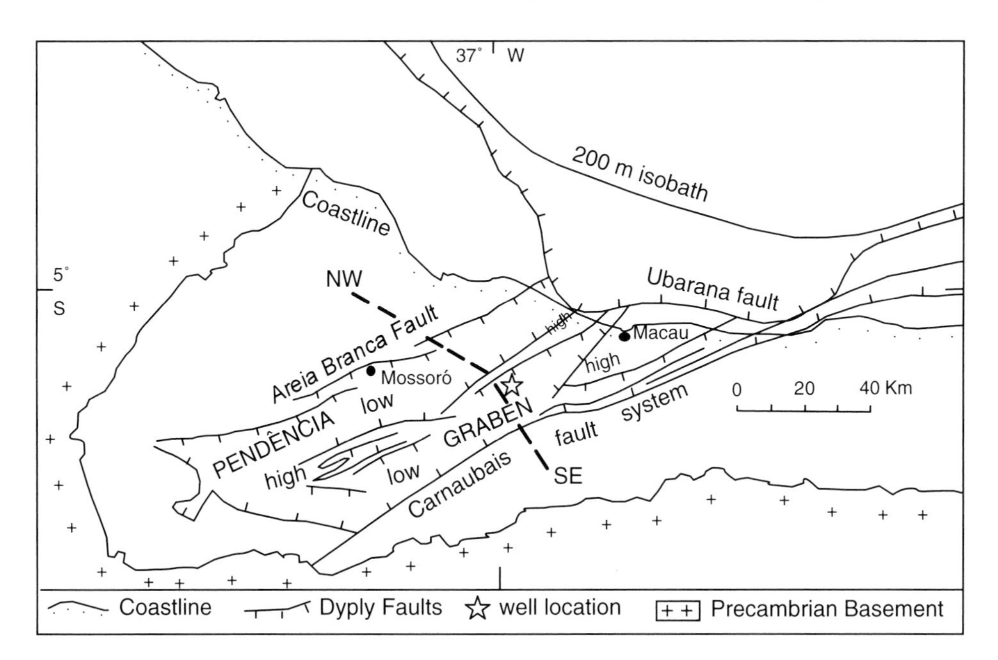

Figure 9—Tectonic details of the Pendência graben of the Potiguar Basin; lacustrine sedimentation extended across the entire graben. Northwest-southeast cross section shown in Figure 10.

the late Aptian, which continued into the Albian (Table 1). The only lacustrine sequence of this time is represented by the Crato Member of the Santana Formation, with a thickness of about 30 m. It is well exposed at various places, for example, in quarries near the town of Barbalha (Figure 11). Figure 11 presents the possible extension of the lake based on surface and subsurface data. Figure 12 gives a schematic lithologic log of the lake sequence of the Crato Member. It is composed of the following from base to top (Viana and Cavalcanti, 1991). (1) Cream-colored micritic calcilutite containing parallel lamination, salt pseudomorphs, wavy laminations, and algal filaments, with thin intercalations of ostracod claystones and algal laminae. (2) Grey calcarenite having parallel lamination and fining-upward sequences containing bivalves. (3) Brown-greenish micaceous shales with parallel lamination and ostracodes alternates with whitish laminated calcilutites. In thin section the clay laminae contain organic matter, traces of macrophytic fossils with algal remains, and poorly preserved ostracodes. Centimeter-sized mudcracks are present. Soft sediment deformation and slumps evidence sliding of lake sediments after deposition. Mineralization of lead and zinc sulfides in the organic-rich laminae (Cassedanne, 1965) are present at Taboca, north-northwest of Crato (Figure 11). In the carbonate laminae, fine- to medium-grained microsparite is strongly recrystallized and commonly dolomitized. Rarely observed are phosphatized fossil remains of algal peloids, intraformational microconglomerates,

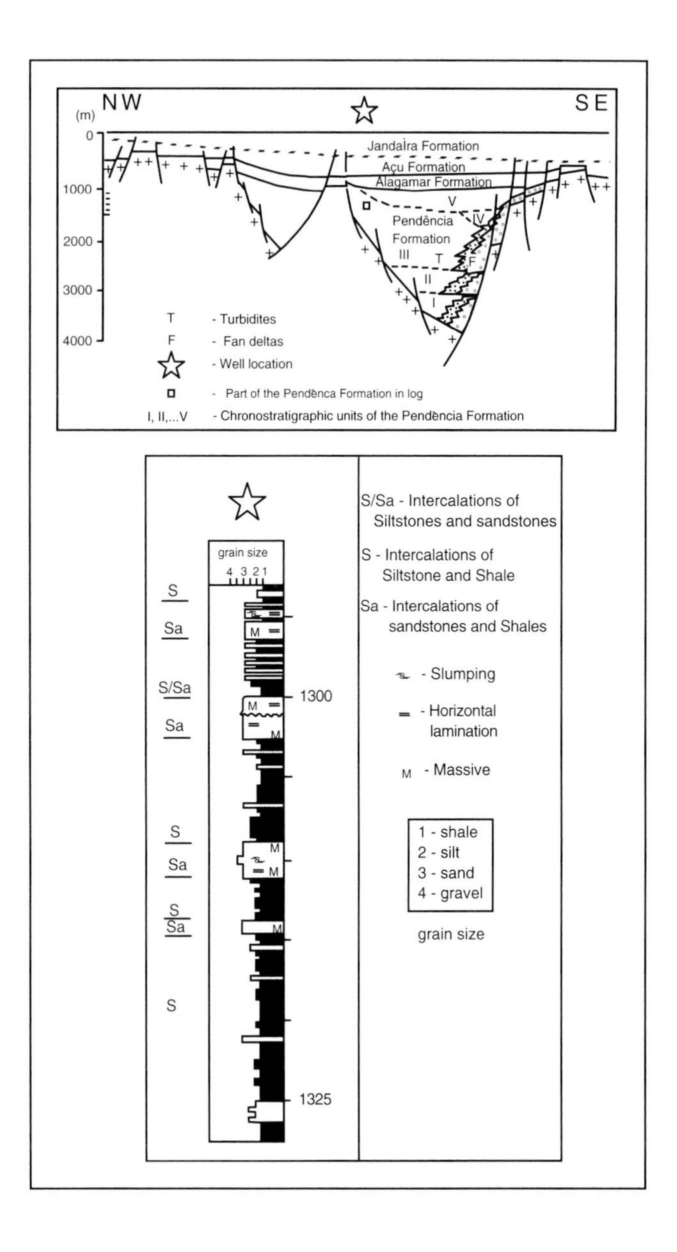

Figure 10—Schematic cross-section diagram (top) northwest-southeast across the Potiguar Basin. See Figure 9 for exact location in basin. Detailed sedimentologic log (bottom) of the shallow lacustrine system. Sh = shale, S = siltstone, Sa = sandstone.

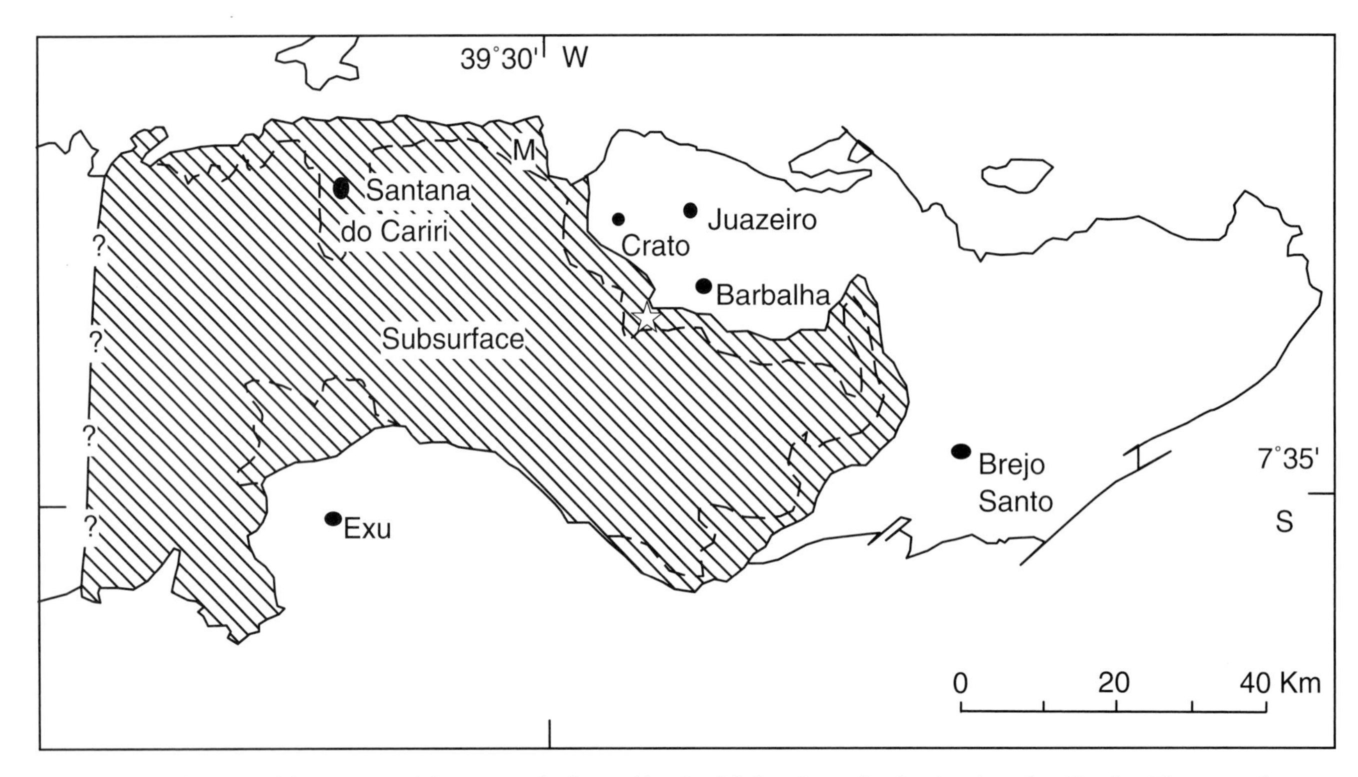

Figure 11—Geographic extent of the second phase (Aptian) lake deposits in the Araripe Basin. The star shows the location of the detailed section of Figure 12, and the letter M shows the location of the mineralization of lead and zinc sulfides. After Cassedanne (1965).

probable compactional cracks, and some bioturbation features. The upper part of the lacustrine sequence in some localities is dominated by fossil-rich calcilutites.

Age of the lake sequence was determined by pollen analysis and ranges from the Aptian to early Albian. An extremely rich fauna, chiefly of insects, is found in the lower laminated limestones. Fauna and flora are currently being described. The following fossil groups have been found to date (for a detailed list, see Martill, 1993): Crustacea (decapods, ostracods, conchostracans), Insecta (Coleoptera = beetles, Heteroptera = bugs, Ephemeroptera = mayflies, Odonata = damselflies and dragonflies, Dermoptera = earwigs, Isoptera = termites, Homoptera = leafhoppers and allies, Blattodea = cockroaches, Hymenoptera = ants and wasps, Raphidioptera = snakeflies, Diptera = flies, Orthoptera = grasshoppers, crickets, etc., Neuroptera = lacewings, etc., Lepdoptera = moths, Megaloptera = alder flies, Mecoptera = scorpions flies, Psocoptera = booklice, Trichoptera = caddis flies, Phasmatodea = stick insects), Arachnida (Araneae = spiders, Scorpionida = scorpions), Vertebrata (Osteichthyes, among which *Dastilbe elongatus*, amphibians, reptiles as crocodiles and pterosaurians, bird feathers), flora (pollen, spores, and macroplants = gymnosperms and angiosperms, and possible gnetaleans).

Environmental interpretation of the Crato Member points to a fresh shallow lake in which two energy levels alternated in the lowest part of the sequence: (1) a low-energy period, characterized by high productivity, and (2) a high-energy period, characterized by wave agitation and tempestite layers. The top of the sequence may have accumulated under long periods of quiet sedimentation, as well.

REFERENCES CITED

Arai, M., A. T. Hashimoto, and N. Uesugui, 1989, Significado cronoestratigráfico da associação microflorística do Cretáceo Inferior do Brasil. Boletin Geociências Petrobrás, v. 3, p. 87-103.

Berthou, P. Y., 1990, Le bassin d'Araripe et les petits bassins intracontinentaux voisins (NE du Brásil): formation et évolution dans le cadre de l'ouverture de l'Atlantique Equatorial. Comparison avec les bassins ouest-africains dans le même contexte. Atas I Simpósio Bacia Araripe e Bacias Interiores Nordeste, Crato (CE), p. 113-134.

Brito, P. M., R. J. Bertini, D. Martill, and L. O. Salles, 1994, Vertebrate fauna from the Missão Velha Formation (Lower Cretaceous, NE Brazil). 3º Simpósio sobre o Cretáceo do Brasil, Rio Claro (SP), Boletin Resumos Expandidos, p. 139-140.

Carvalho, I. S., 1989, Icnocenoses continentais: bacias de Sousa, Uiraúna-Brejo das Freiras e Mangabeira. Universidade Federal do Rio de Janeiro, Rio de Janeiro, Master's Thesis, 167 p.

Carvalho, I. S., and G. Leonardi, 1992, Geologia das bacias de Pombal, Souza, Uiraúna-Brejo das Freiras e vertentes (Nordeste do Brasil). Anais Academia Brasileira de Ciências, v. 64, p. 231-252.

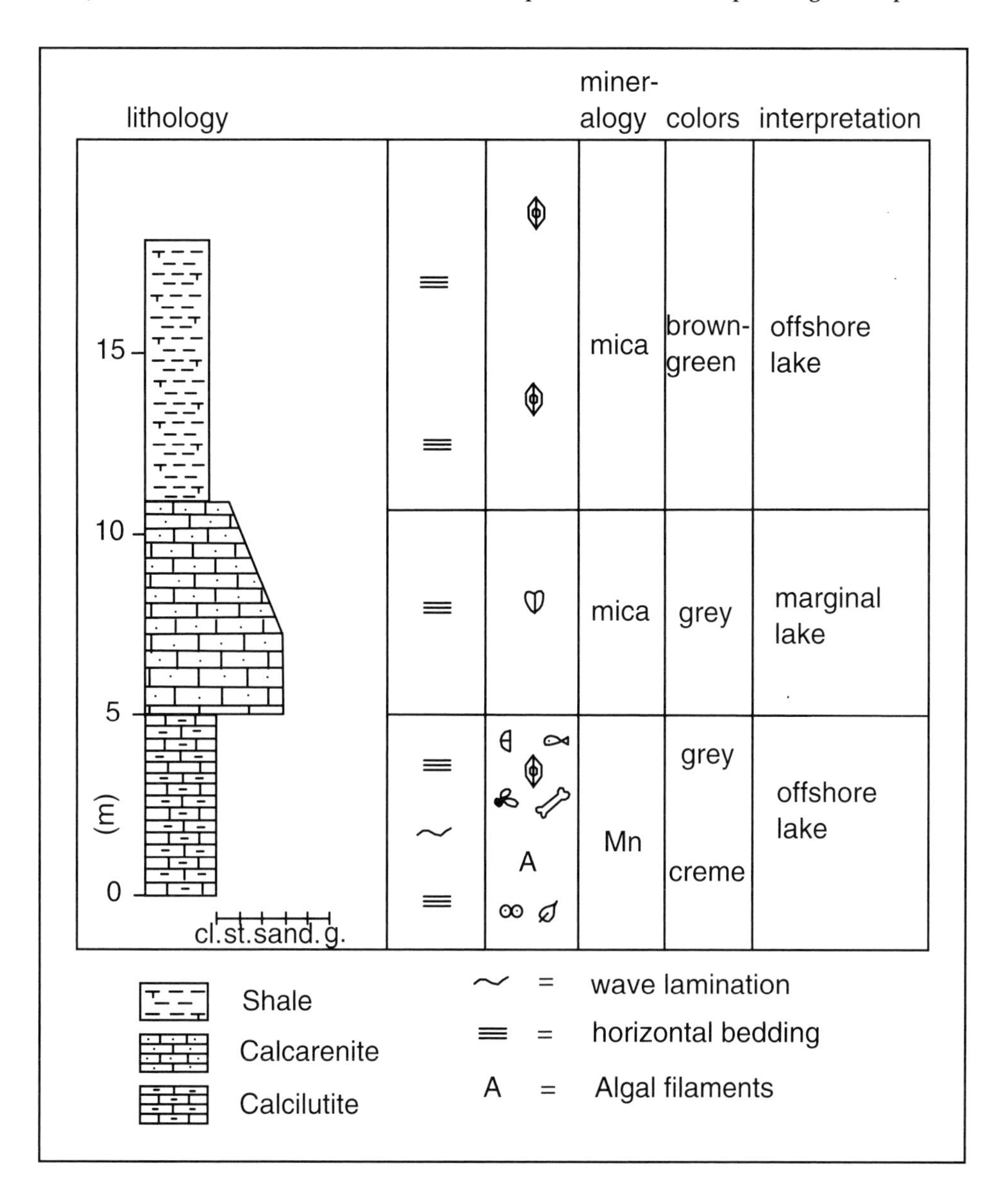

Figure 12—Detailed schematic lithologic column for the lake sequence of the Crato Member, after Viana and Cavalcanti (1991). See location in Figure 11.

Cassedanne, J. P., 1965, Indice de sulfures sédimentaires de Taboca (Municipe de Crato, Etat du Ceará, Brásil). Bulletin Société Geologie France (7th série), v. 7, p. 177-186.

Cavalcanti, V. M. M., and M. S. S. Viana, 1992a, Revisão estratigráfica da Formação Missão Velha, Bacia do Araripe, Nordeste do Brasil. Anais Academia Brasileira de Ciências, v. 64, p.155-168.

Cavalcanti, V. M. M., and M. S. S. Viana, 1992b, Considerações sobre o Cretáceo da Bacia de Iguatu, Nordeste do Brasil. Anais Academia Brasileira de Ciências, v. 64, p. 63-70.

Ghignone, J. I., E. A. Couto, and M. L. Assine, 1986, Estratigrafia e estruturas das bacias do Araripe, Iguatu e Rio do Peixe. Anais 34 Congresso Brasileiro de Geologia, Goiânia (Go), v. 1, p. 271-285.

Lima Neto, F. F., R. M. D. Matos, U. M. Praca, and A. T. Hashimoto, 1986, Análise éstratigráfica da Formação Pendêcia. Report of the Stratigraphy Working Group of Petrobas, Rio de Janeiro.

Mabesoone, J. M., 1986, Petrografia de alguns sedimentos calcários e microclásticos do Grupo Araripe (Nordeste do Brasil). Anais 34 Congresso Brasileiro de Geologia, Goiânia (Go), v. 1, p. 262-270.

Mabesoone, J. M., 1994, Sedimentary basins of northeast Brazil. Federal University Pernambuco, Recife, Geology Department, Special Publication, no. 2, 308 p.

Mabesoone, J. M., and V. A. Campanha, 1974, Caracterização estratigráfica dos Grupos Rio do Peixe e Iguatu. Estudos Sedimentológicos, no. 3/4, p. 21-41.

Mabesoone, J. M., and M. S. S. Viana, 1994. Intracratonic Mesozoic basins of northeastern Brazil. 14th International Sedimentological Congress, Recife (PE), Brazil; Field Trip Guide B-12, 25 p.

Martill, D.M., 1993, Fossils of the Santana and Crato Formations, Brazil. The Paleontological Association; (London) Field Guides to Fossils, no. 5, 159 p.

Neves, C. A. O., 1989, Análise regional do trinômio geração-migração-acumulação de hidrocarbonetos na sequência continental eocretácica da bacia Potiguar. Boletin Geociências Petrobrás, Rio de Janeiro, v. 3 (3), p.131-145.

Ponte, F.C., and C.J. Appi, 1990, Proposta de revisão da coluna litoestratigráfica da Bacia Araripe. Anais 36 Congresso Brasileiro de Geologia, Natal (RN), v. 1, p. 211-236.

Regali, M. S. P., and S. M. Gonzaga, 1982, Palinocronoestratigrafia da bacia Potiguar. Report of the Paleontologic Working Group of Petrobas, Rio de Janeiro.

Rodrigues, R., and T. Takaki, 1992, Estratigrafia química de sequências lacustres: Cretáceo Inferior da Bacia Potiguar. II Simpósio sobre as Bacias Cretáceas Brasileiras, Rio Claro (SP), Resumos Expandidos, p. 26.

Silva, L. G., 1993, Modelo de sedimentação turbidítica em bacia lacustre tipo rifte—um exemplo no graben de Umbuzeiro. Boletin Núcleo Nordeste, Sociedade Brasileira de Geologia, Atas XV Simpósio Geol. Nordeste, Natal (RN), no. 13, p. 37-38.

Souza, S. M., 1982, Atualização da litoestratigrafia da Bacia Potiguar. Anais 32 Congresso Brasileiro de Geologia, Salvador (BA), v. 5, p. 2392-2406.

Srivastava, N. K., 1990, Aspectos geológicos e sedimentológicos das bacias de Iguatu, Lima Campos e Malhada Vermelha (Ceará). Atas I Simpósio Bacia Araripe e Bacias Interiores Nordeste, Crato (CE), p. 209-222.

Vasconcelos, E. C., 1980, Estudo faciológico da Formação Souza. Universidade Federal de Pernambuco, Recife, Masters Thesis, 143 p.

Viana, M. S. S., and V. M. M. Cavalcanti, 1991, Sobre a estratigrafia da Formação Santana, Bacia do Araripe. Revista Geologia, Universidade Federal do Ceará, Fortaleza, no. 4, p. 51-60.

Viana, M. S. S., J. M. Mabesoone, and V. H. Neumann, 1994, Lacustrine phases in the Araripe Basin (NE Brazil). 14th International Sedimentological Congress, Recife (PE), Brazil, Abstracts, p. S6-11–S6-12.

da Silva, H. T. F., J. M. Caixeta, L. P. Magnavita, and C. P. Sanches, 2000, Syn-rift lacustrine deep-water deposits: Examples from the Berriasian sandy strata of the Recôncavo Basin, northeastern Brazil, *in* E. H. Gierlowski-Kordesch and K. R. Kelts, eds., Lake basins through space and time: AAPG Studies in Geology 46, p. 209–224.

Chapter 17

Syn-Rift Lacustrine Deep-Water Deposits: Examples from the Berriasian Sandy Strata of the Recôncavo Basin, Northeastern Brazil

Hercules T. F. da Silva
Petrobras E&P
Salvador, Bahia, Brazil

José M. Caixeta
Luciano P. Magnavita
Christovam P. Sanches
Petrobras E&P
Salvador, Bahia, Brazil

INTRODUCTION

Hydrocarbon exploration in rift basins increased during the 1980s mainly because of (1) the recognition of lacustrine shales as good source rocks and (2) notable exploration successes (Begawan and Lambiase, 1995). Petroleum production from rift basins is particularly important for Brazil (Bruhn et al., 1988), Indonesia (Williams and Eubank, 1995), and China (Desheng, 1995) and for African margins. In addition to good source rocks, rift basins commonly contain sandy reservoirs including turbidites, deltaic, and fluvial types.

A typical vertical succession for nonmarine rift basins includes basal syn-rift lacustrine strata overlain by deltaic deposits, then fluvial sediments. Basin fill follows from the interplay between tectonics and short-term climatic fluctuations (Cohen, 1990; da Silva, 1993; Lambiase and Bosworth, 1995; and others). These two mechanisms control important parameters regulating basin evolution like basin-floor subsidence rates, main sediment entry points, and lake-level fluctuations.

Cohen (1990) recognized three typical associations of margins and drainage zones for rift basins: (1) flexural or shoaling margin, related to more stable, low subsidence areas, (2) axial margin, running parallel to the main elongation of the rift basin (usually) draining axial depocenters, and (3) escarpment margin, related to the asymmetry of the main fault system. Generally, shoaling and axial margins deliver a higher sediment load.

During early rifting, subsidence rates may overcome sediment input and starve the rift basin of coarse-grained sediments. If climate is adequately wet, a perennial lake may develop. During early phases, main rivers may be impounded behind structural barriers with coarse-grained sediments derived from the uplifted footwall region and/or from elevated blocks on the flexural border. Such conditions are favorable for gravity-induced flows because of both tectonic activity and high topographic gradients. Along the escarpment margin, fan-deltas and sublacustrine fans form. On the shoaling margin, major faults may control sedimentation, acting as hinge-lines, and promote the development of near-fault sublacustrine fans (da Silva et al., 1989).

Sublacustrine fans have been reported for Early Cretaceous basins of eastern Brazil (Bruhn et al., 1988; da Silva, 1993; Magnavita and da Silva, 1995; among others), for Early Cretaceous–Tertiary basins of eastern and western Africa (Scholz and Rosendahl, 1990; Scholz, 1995; Smith, 1995), and for Permian–Triassic basins in central-southern Africa (Banks et al., 1995).

The Recôncavo Basin is the most prolific rift basin of Brazil. Sandy sublacustrine reservoirs occur encased in organic-rich, lacustrine shales that provide both seal and source. A basinwide fresh to brackish water lake spanned the Berriasian and Early Valanginian. Most of the coeval coarse-grained deposits were formed by gravity-induced transport. In this paper we focus only on the sandy strata of the Berriasian Recôncavo Basin.

REGIONAL SETTING

The Recôncavo Basin is located in northeastern Brazil (Figure 1) in the southern portion of the Recôncavo-Tucano-Jatobá rift (Milani and Davison, 1988;

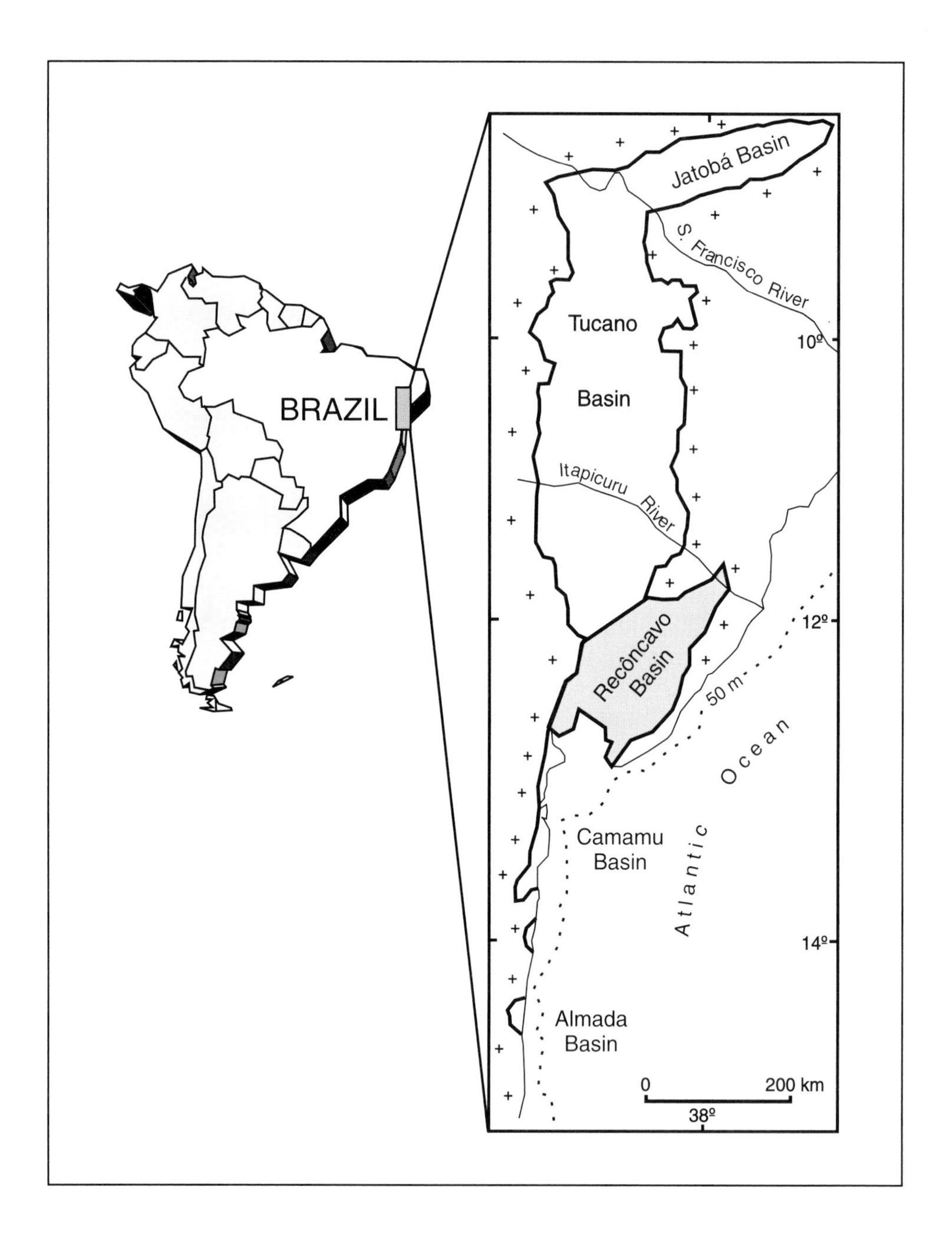

Figure 1—Location map of the Recôncavo Basin, Brazil.

Magnavita et al., 1994). The basin developed prior to the opening of the South Atlantic Ocean, evolving as an aborted branch of the eastern Brazilian continental margin. The Recôncavo Basin covers an area of 11,500 km^2 and consists of a southeastward-dipping continental half-graben. Its structural framework is characterized by normal faults trending N30°E, parallel to basin elongation, with a transversal N30–40°W transfer fault, the Mata-Catu fault (Figure 2).

Basin-fill of the Recôncavo Basin contains strata deposited during pre-, syn- and post-rift phases (Figure 3). Late Paleozoic–Jurassic pre-rift deposits are mainly red beds and coarse-grained fluvial deposits. Neocomian syn-rift strata are the volumetrically most important fill of the basin and encompass lacustrine (main source rocks) to deltaic to fluvial deposits. Late Cretaceous–Cenozoic post-rift deposits are relatively thin (about 100 m thick) and constituted mostly of medium- to coarse-grained alluvial and fluvial sandstones. Total sedimentary thickness along basinal depocenters exceeds 5 km. Age control on the Recôncavo Basin strata is based mainly on nonmarine ostracod biostratigraphy (Krömmelbein, 1966; Viana et al., 1971). Correlation of the nonmarine strata of the Recôncavo Basin with the geologic time scale (Harland et al., 1982) is somewhat uncertain due to the absence of good index fossils (Arai et al., 1989; Regali and Viana, 1989).

The Recôncavo Basin, the oldest petroleum-producing basin in Brazil, is at a mature stage of exploration. Commercial oil production dates back to the early 1940s, and massive investments led to the discovery of approximately 85 oil and gas fields. Petroleum accumulation in the Recôncavo basin can be grouped into

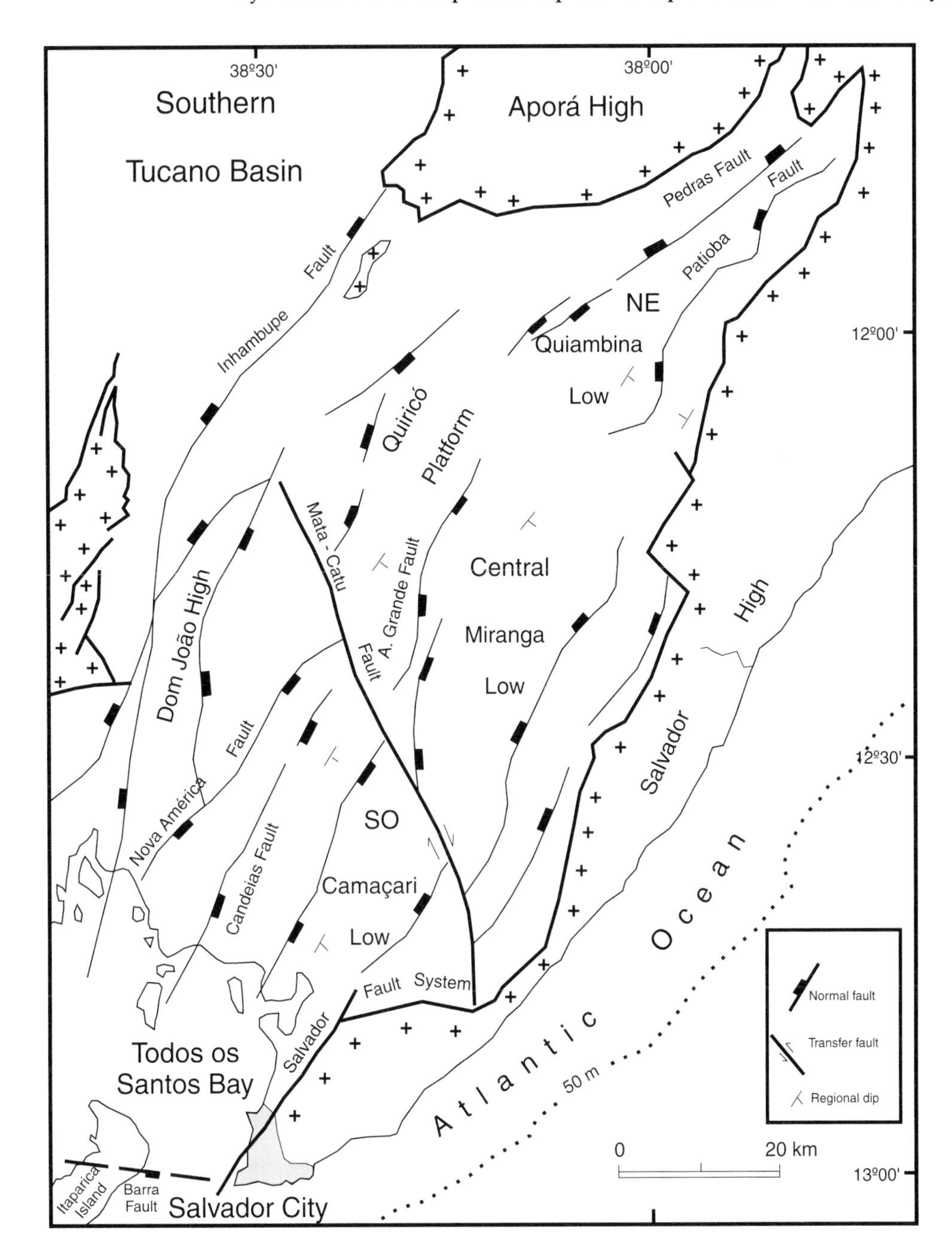

Figure 2—Structural architecture of the Recôncavo Basin (modified from Netto et al., 1984). The most important fault system in the basin is represented by normal faults trending N30°E. The prominent Mata–Catu fault is interpreted as a transfer zone.

three basic systems (Figure 4): (1) the pre-rift Sergi-Água Grande system, (2) the syn-rift Candeias system, and (3) the syn-rift Ilhas system (Figueiredo et al., 1994). Principal reservoirs of the pre-rift system are the fluvial Sergi and Água Grande Formations (see Figure 3). Trapping is mainly structural (horsts and tilted blocks). This system accounts for 57% of the proven oil volume of the basin (Figueiredo et al., 1994). The second most important petroleum system is the syn-rift Ilhas system. Main reservoirs are deltaic sandstones in structural (anticlinal), combined, and stratigraphic traps. Growth faulting is abundant. Oil volume of this system corresponds to 27% of the total proven oil volume of the basin (Figueiredo et al., 1994). Finally, the syn-rift Candeias system is characterized by stratigraphic and combined trapping. Reservoirs are mainly sublacustrine turbidites in northeastern and southern Recôncavo. Subordinate oil production comes from fractured shales. This petroleum system comprises 16% of the proven oil volume of the basin (Figueiredo et al., 1994). Sublacustrine sands discussed here are part of this petroleum system.

Da Silva (1993), studying the syn-rift interval of the Recôncavo Basin, coined the term "Lower Cretaceous tectonosequence," meaning the spatial and temporal assemblage of genetically related strata deposited during the syn-rift phase of the Recôncavo Basin. The tectonosequence can be broken down into six tectono-depositional intervals (tdi I through VI; see Figure 3). A tdi designates a stratigraphic package that comprises genetically related deposits created during a distinct tectonic period within the syn-rift phase of the basin.

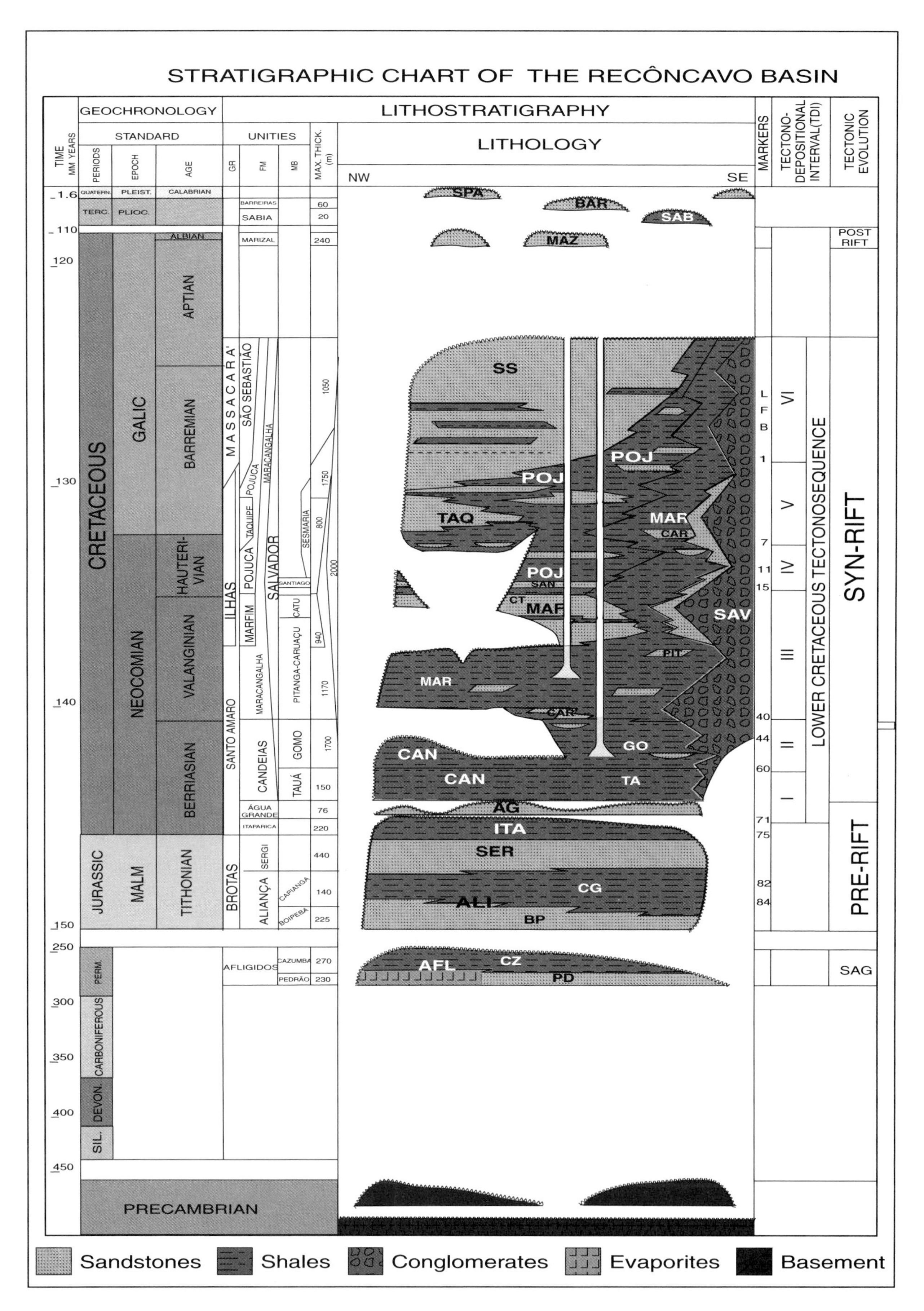

Figure 3—Stratigraphic chart of the Recôncavo Basin (adapted from da Silva, 1993; Caixeta el al., 1994).

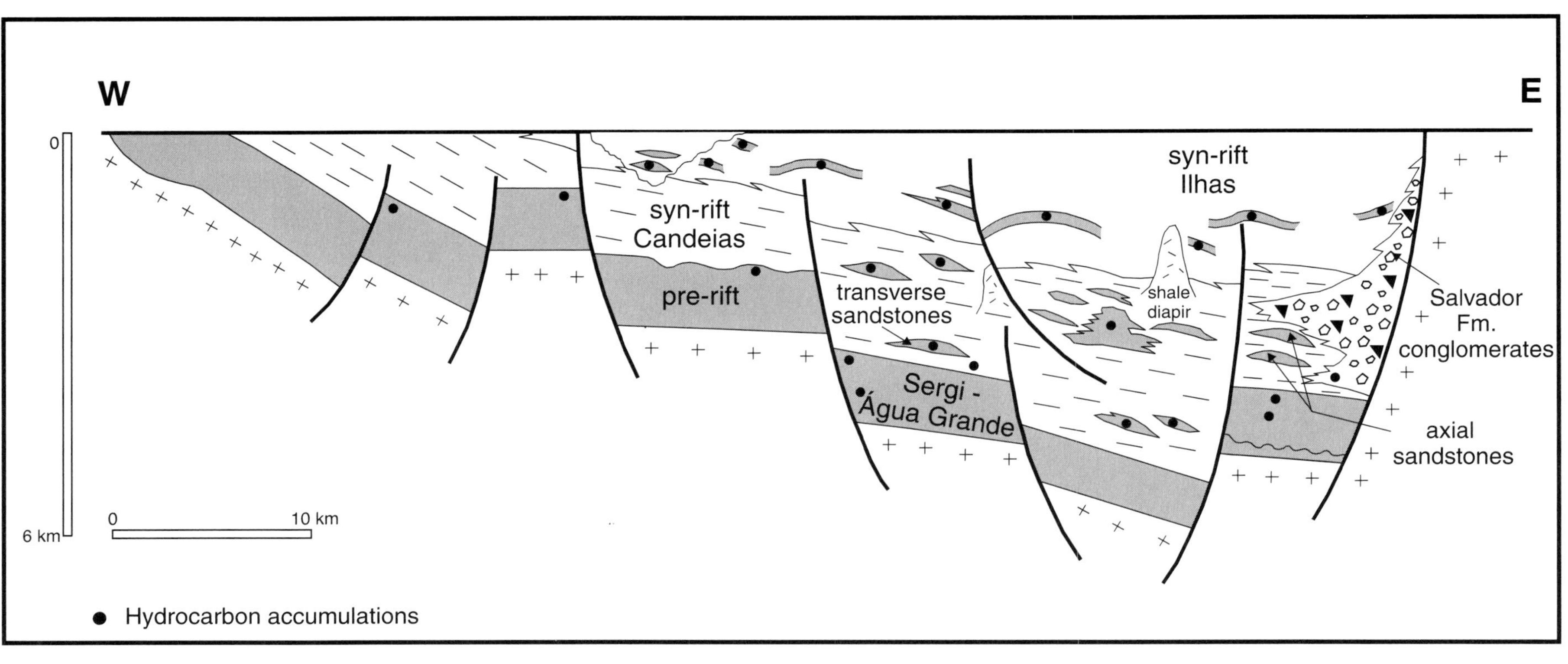

Figure 4—Main petroleum systems of the Recôncavo Basin (modified after Figueiredo et al., 1994). This idealized dip cross-section shows different existing plays in the basin.

SYN-RIFT BERRIASIAN TURBIDITES OF THE RECÔNCAVO BASIN

Depositional Characteristics

Syn-rift Berriasian sublacustrine turbidites of the Recôncavo Basin occur within tectono-depositional interval II (tdi II, Figure 3); (da Silva, 1993). The lithostratigraphic unit that makes up this tdi is the Gomo Member of the Candeias Formation (Figure 3), which consists mainly of shales with subordinate sandstone and limestone beds. Sandstones within organic-rich shales constitute reservoirs in the Recôncavo Basin. Limestone and shale are present at the base of tdi II; sandstone occurs preferentially in the middle part. Limestone commonly occurs as mudstone with a high carbonate content and conspicuous levels of calcitic concretions. Mainly muddy strata constitute the top portion of tdi II. Tectonic activity coeval with tdi II was strong and reflects the onset of major faulting and block tilting in the basin, characterizing a starved-basin phase.

Isopach values for tdi II delineate two distinct areas in the Recôncavo Basin (Figure 5). One with thicknesses up to 200 m occurs at the northwestern portion of the basin. The other area is characterized by a narrow, northeast–southwest oriented region with maximum thicknesses over 1600 m. It is located close to the eastern margin of the basin, within axial depocenters. Basinwide syn-rift conglomerates (Salvador Formation; see Figure 3) are contemporaneous with tdi II.

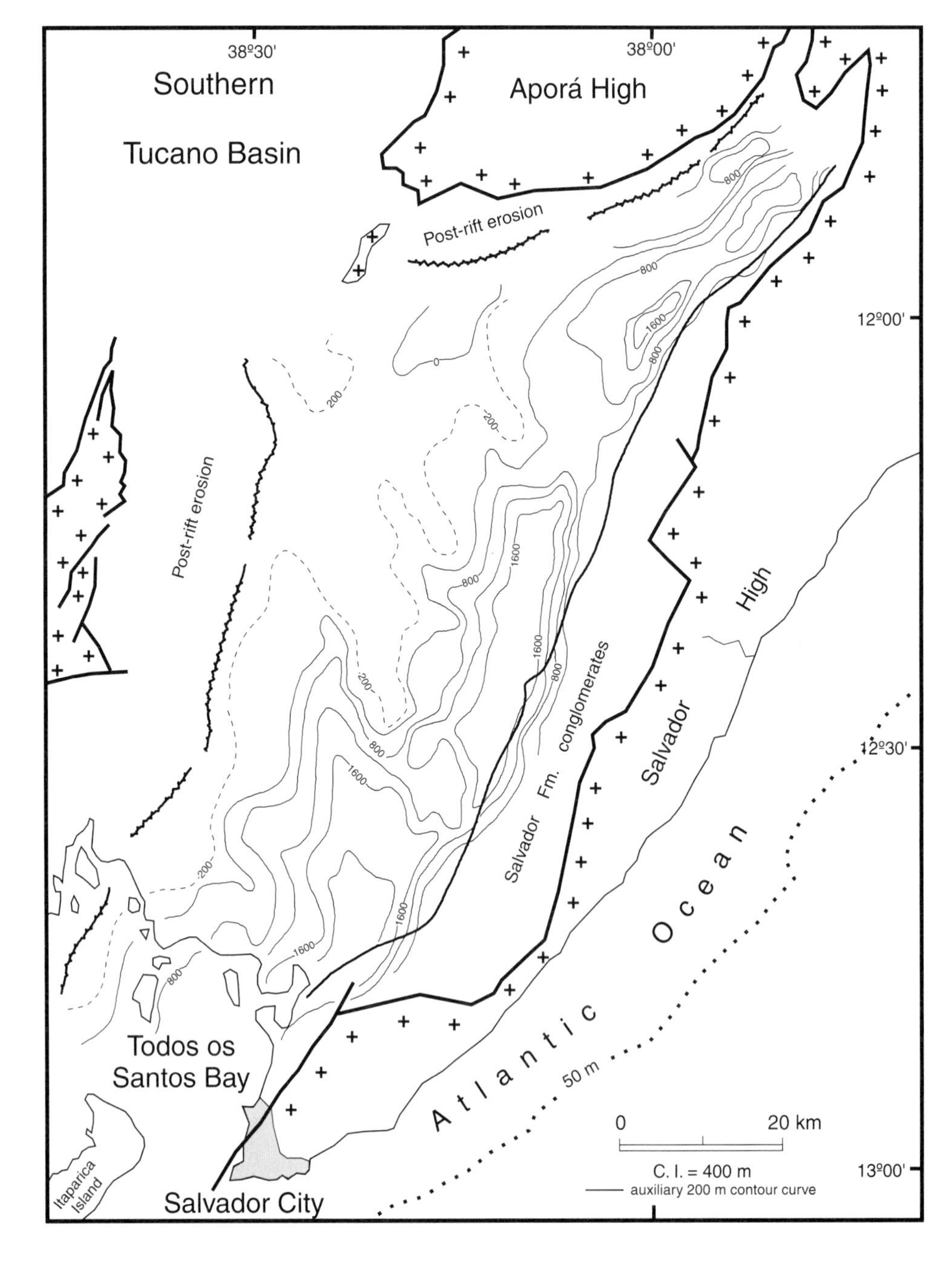

Figure 5—Isopach map of tdi (tectonic-depositional interval) II, Berriasian. Observe highest isopach values present along the Quiambina, Miranga, and Camaçari lows (see Figure 2 for comparison).

Analysis of sand occurrence for tdi II (Figure 6) reveals two major sandy trends. One northeast–southwest (axial), parallel to the principal elongation of the basin, and the other oriented northwest–southeast (transverse). Axial sandstones are mainly present in northeast and central Recôncavo (Figure 2), whereas transverse sandstones occur mostly in southern Recôncavo. In this case, the Mata-Catu transfer fault (see Figure 2) seems to have acted as the main controlling element on the distribution of these two sandy systems. Sand occurrence to west–northwest of the basin was strongly influenced by faulting (Figure 6). As a consequence, those sandstones rapidly pinch out toward the footwall of northeast–southwest trending faults.

Compositionally, axial sandstones range from subarkose to sublitharenite (basal deposits) to lithic arkose and feldspathic litharenite (top deposits). This compositional change is accomplished by enrichment in lithic fragments. Sandstone is moderately to well sorted, medium-grained, and may be intercalated with conglomeratic beds of the Salvador Formation. Sedimentary structures recognized within these sandstones fit those described for turbidite beds (Bouma, 1962): massive bedding, T_{AB}, T_{ABE}, T_{BC}, T_{BE}, and T_{BCE} (Bruhn et al., 1985; Bruhn, 1985; Souza et al., 1990) (Figure 7). Axial sandstone bodies are up to 40 m thick, and sand body associations have variable width of between 2 and 4 km (Figure 8).

Transverse sandstone composition includes quartzarenite, subarkose, and sublitharenite (Carozzi et al., 1976; Carozzi and Fonseca, 1989; Menezes, 1991). Sandstone is moderately sorted and fine- to medium-grained.

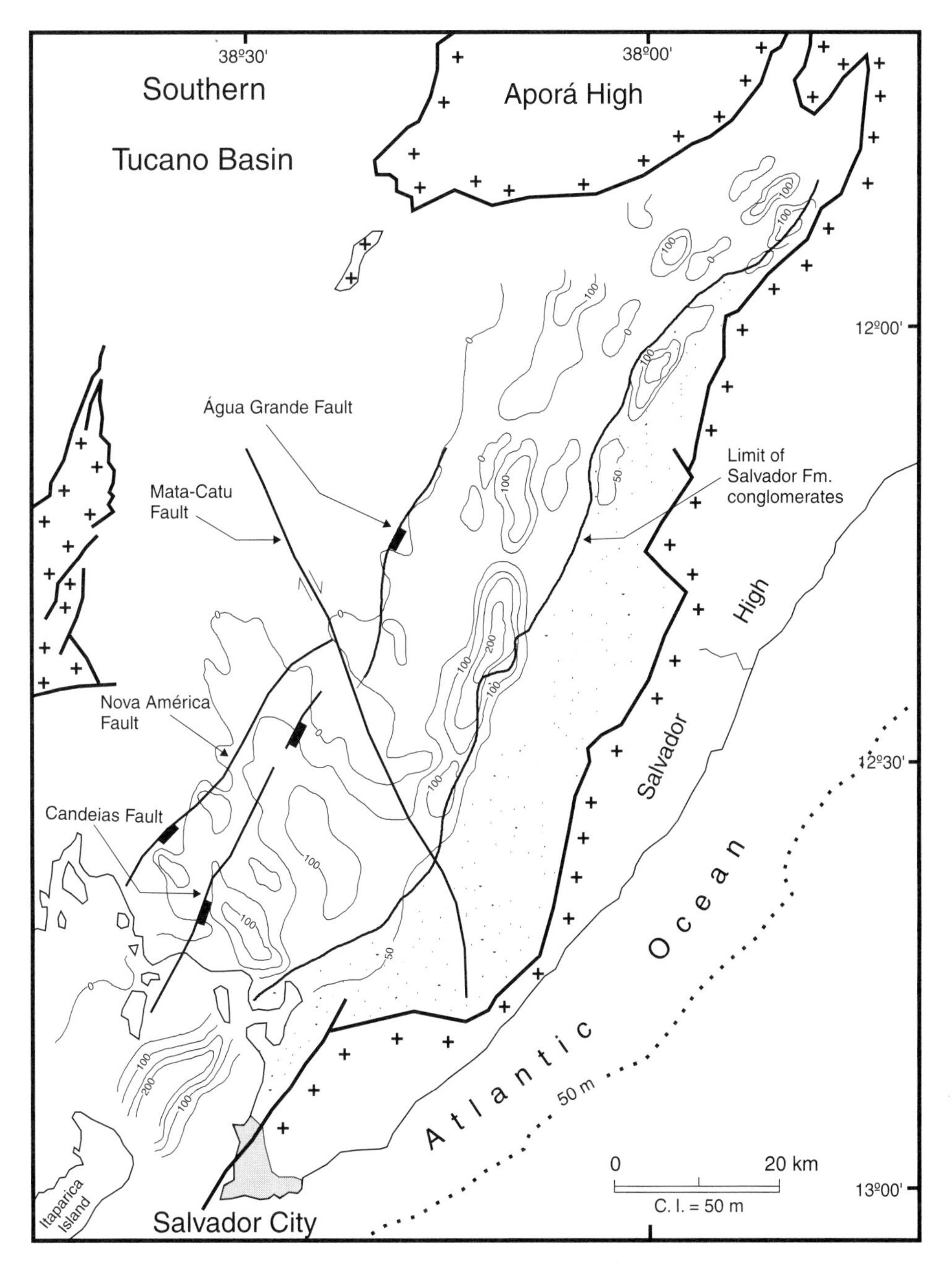

Figure 6—Sandstone isolith map of tdi II. Sand occurrence in western Recôncavo Basin suggests that both Água Grande and Nova América faults controlled sand distribution during the Berriasian.

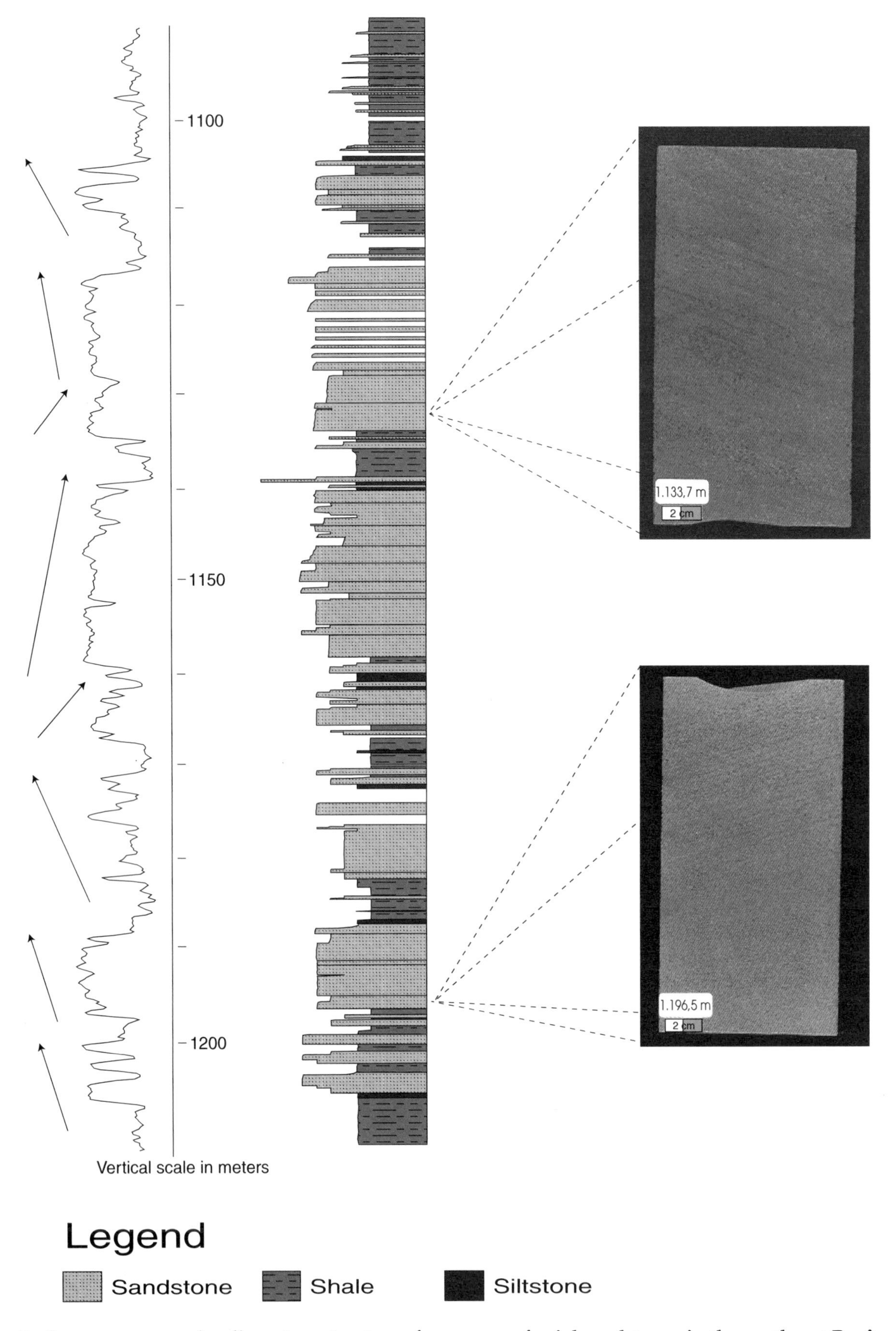

Figure 7—Log response and sedimentary structures from cores of axial sandstones in the northeast Recôncavo Basin. In detail, core photographs showing turbidite beds with T_{AB} and T_{ABE} Bouma sequences. Both core slabs display traction carpet levels (adapted from Bruhn et al., 1985; Sanches, 1993).

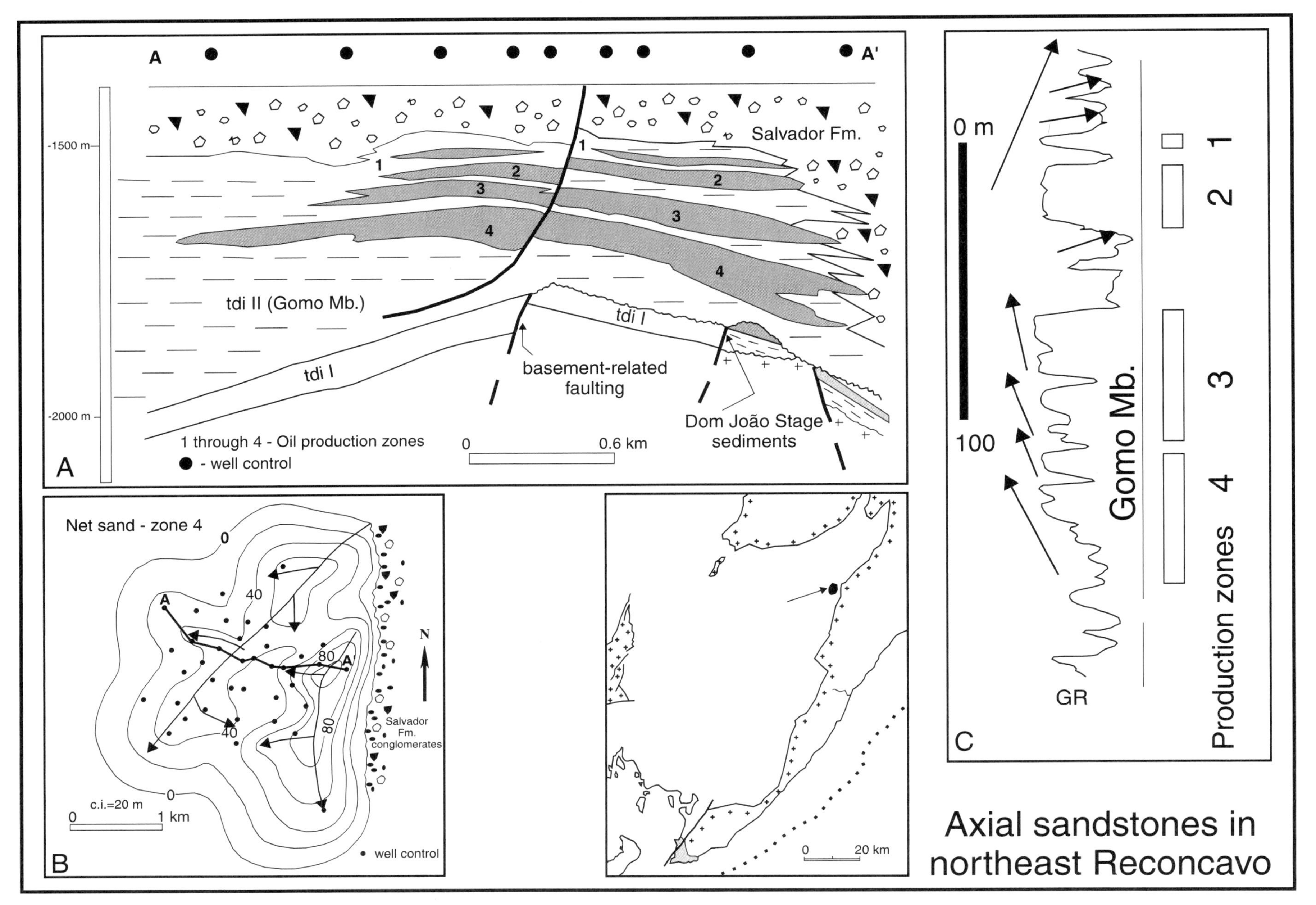

Figure 8—Characteristics of the axial sandstones in the northeast Recôncavo Basin. (A) Intercalation of four production zones of sandstones with conglomerates of the Salvador Formation. (B) Lobate geometry of the production zone 4. (C) Geologic logs of the sandstones (adapted from Bruhn et al., 1985).

Sedimentary structures include Bouma sequences, sole marks, load casts, and rip-up clasts. Reworked carbonate clasts (micrite, pisolites, and oncolites) are associated with these sandstones (Camões and Rigueira, 1987; Carozzi and Fonseca, 1989) (Figure 9). Micrite clasts contain oxidation rims, desiccation cracks, laminated crusts, and pseudo-brecciated structures. Transverse sandstone bodies exhibit individual thickness up to 30 m. The sandy facies association for these sandstones may have a down-dip extent of over 4 km (Figure 10).

Depositional System Synthesis

Facies associations for both axial and transverse sandstones (Figure 6) can be generalized for other portions of the basin. One exception are the axial sandstones that occur along the downthrown blocks of faults located away from the eastern border of the basin, such as the Patioba fault in northeast Recôncavo (see Figures 2 and 6). In this case, axial sandstones are devoid of intercalation with syntectonic conglomerates (Figure 6).

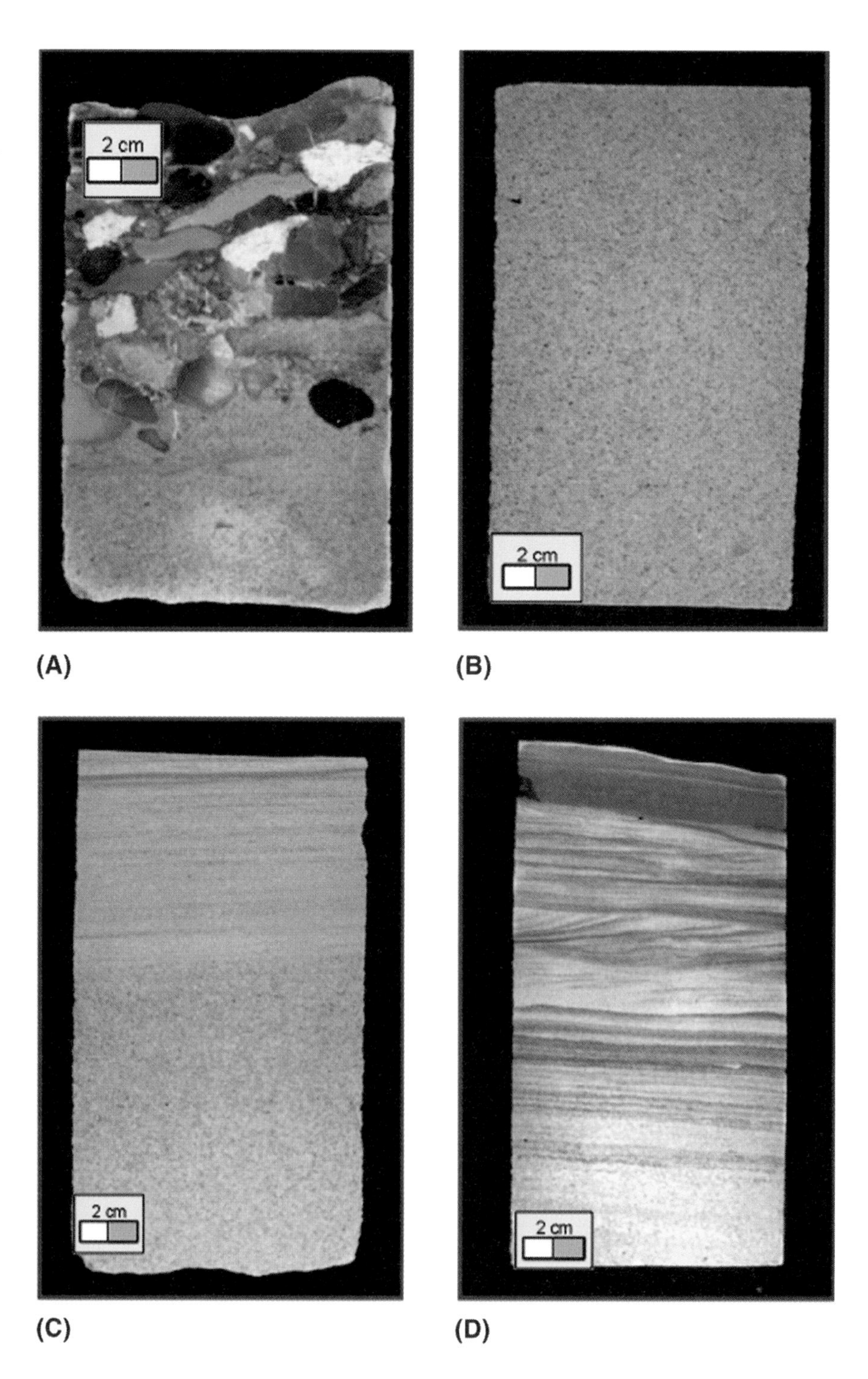

Figure 9—Sedimentary characteristics of transverse sandstones in the southern Recôncavo Basin. (A) Intrabasinal conglomerate with carbonate and clay clasts in fine-grained, massive sandstone. (B) Well-sorted, massive sandstone. (C) T_{AB} Bouma sequence. (D) Complete Bouma sequence in very fine to fine sandstone (adapted from Camões, 1988).

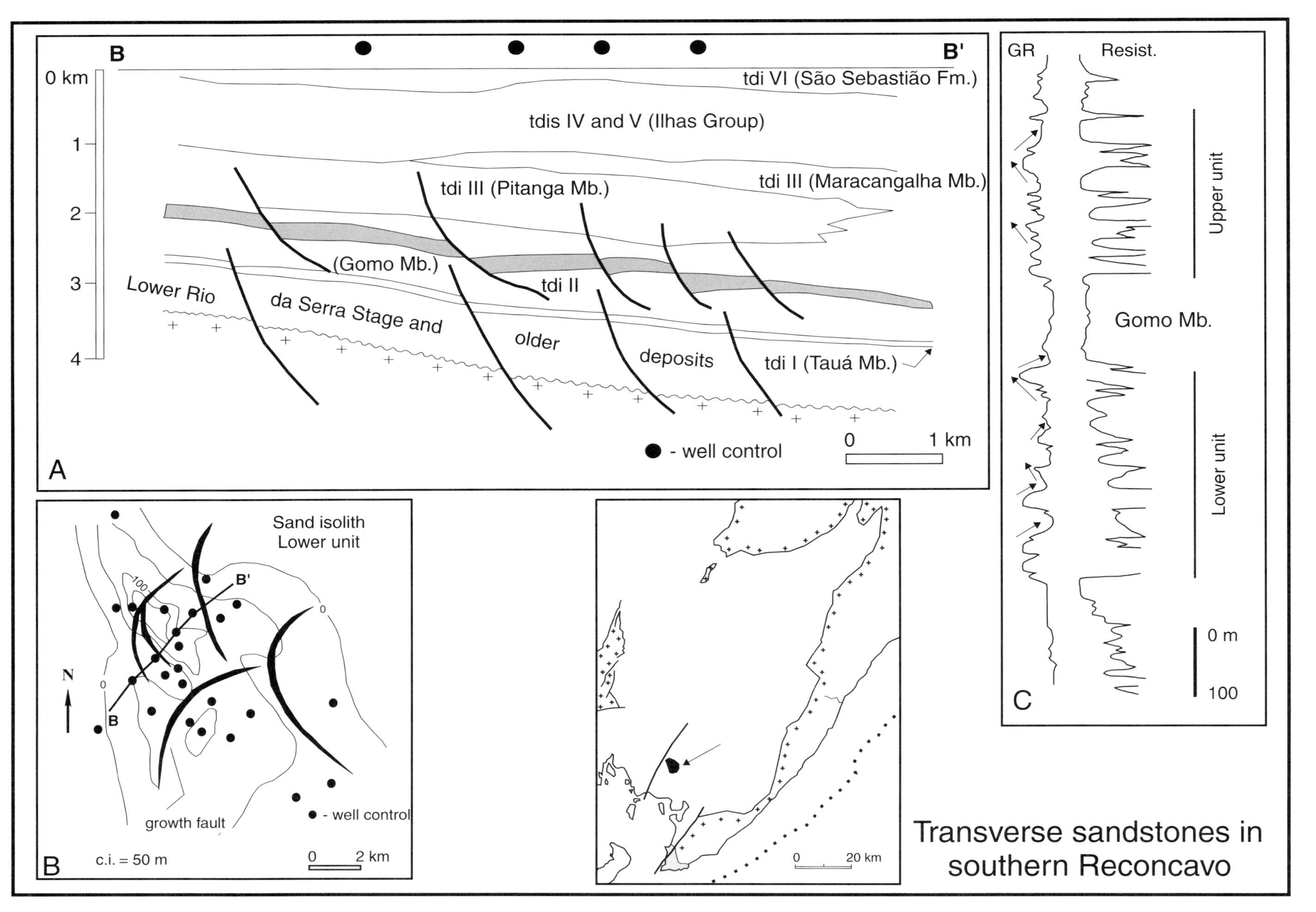

Figure 10—Characteristics of the transverse sandstones in the southern Recôncavo Basin. (A) Dip cross-section showing growth faults controlling sand accumulation. (B) External geometry of the lower unit. (C) Geologic logs associated with transverse sandstones (adapted from Camões, 1988).

Axial Sandstones

Axial sandstones are present at the basal portion of tdi II (see Figure 3). In northeast Recôncavo, axial sandstones have been interpreted as turbidite deposits (Bruhn, 1985; Bruhn et al., 1985; Souza et al., 1990; Sanches, 1993; among others). These sandstones are genetically related to the conglomerates of the Salvador Formation. Evidence of this association includes (1) lateral and vertical relationships (Figure 8), (2) paleocurrent indication (based on processed dipmeter logs), and (3) sand enrichment in lithic fragments derived from the uplifted blocks of the eastern border of the basin. Their spatial distribution is related to the retreat of the border through back-faulting of the boundary fault, a mechanism described by Magnavita and Da Silva (1995). Other characteristics of these axial sandstones include absence of typically deltaic or shallow-water sedimentary structures or emergence features (coals, paleosols, plant roots, and mudcracks). Axial sandstones of northeast Recôncavo were associated with a very active fault system, i.e., the Salvador fault system (see Figure 2). Secondarily, axial sandstones present in central Recôncavo may have had some contribution from the flexural margin on the west (see Figures 2 and 6).

Geologic logs of the axial sandstones display two intervals (Figure 8): (1) basal deposits showing a coarsening- and thickening-upward pattern (compacted thickness approximately 140 m) and (2) upper deposits depicting a fining- and thinning-upward pattern (compacted thickness approximately 80 m). The source for basal deposits would have been pre-rift sandy strata. For the top deposits, sandy material was derived from Paleozoic deposits and Precambrian metasediments and granulitic rocks (see Figure 3).

Gravity-driven flows are common in coarse-grained, high-gradient systems of tectonically active basins, like rift basins. Facies associations of these sandstones with conglomerates suggest a genetic link to two types of depositional models: fan-deltas and gravity-induced, deep-water sublacustrine fans. The sandstone-conglomerate association found in northeast Recôncavo lacks evidence of shelfal components. Thicknesses of the two sandy intervals (Figure 8) point to deep-water sedimentation (>100 m).

Various terms have been applied to deep-water fan-deltas: slope-apron fan or fan-delta front slope (Porebski, 1984), fan-deltoid slope apron (Busby-Spera, 1988), and deep-water fan-delta (Higgs, 1990). Sedimentary structures reported for axial sandstones have been similarly described in modern fjord fan-deltas (Prior and Bornhold, 1988). Gravity-driven sedimentary flows (turbidity currents and debris flows) and various types of sliding and avalanching constitute dominant processes in the development of coarse-grained deltas in fjord areas (Prior and Bornhold, 1988).

An alternative depositional model for axial sandstones involves the development of sublacustrine fans. Characteristics of the basal deposits (Figure 8) allow the interpretation of prograding lobes (mid-fan lobes) with top strata constituting the record of channelized facies (Bruhn et al., 1985) (Figure 8) similar to fan deposits in marine basins (Walker, 1978). Turbidite facies have been reported for lacustrine rift basins, notably in the East African rift (Cohen, 1990; Scholz and Rosendahl, 1990; Prosser, 1993). Facies associations, sedimentary structures, and geologic setting could support an interpretation of either a coalesced fan-delta apron or an axial turbidite channel/lobe complex for the axial sandstones of northeast Recôncavo (Reading, 1991; Stow et al., 1982). A regional synthesis beyond oil field-scale studies is necessary to evaluate these depositional models. Axial sandstones occurring along the downthrown block of the Patioba fault (see Figures 2 and 6) are interpreted as turbidite deposits, sourced from both the eastern and northern portions of the basin.

Transverse Sandstones

Transverse sandstones do not intercalate with and are not overlain by syntectonic conglomerates. They occur in the top part of tdi II (see Figure 3), which implies a change in basin physiography. These sandstones are characterized by the presence of resedimented limestone clasts with subaerial exposure features (Carozzi et al., 1976; Carozzi and Fonseca, 1989) and are associated with growth faulting (Figure 10). The deposits are better developed on the downthrown block of the Candeias fault, which behaved as a hinge-line for deep-water sand sedimentation during tdi II. Studies recognized a sublacustrine turbidite complex (Camões, 1988) with the best reservoirs associated with mid-fan channeled lobes (Carozzi and Fonseca, 1989). Sand for these deposits would have been derived from erosion of pre-rift sandy strata on the flexural margin (uplifted blocks within and bordering the basin).

Transverse sandstones constitute a record of narrow point-sourced, sand-dominated sublacustrine fan deposits not directly associated with deltaic systems. In the vicinity of the downthrown block of the Candeias fault, growth faulting acted as an efficient sand trap (see Figure 10), suggesting steep subaqueous slopes and talus lobes.

VERTICAL STRATIGRAPHIC PROGRESSION

The tdi II (Berriasian) strata was coeval with the onset of major tectonic activity in the Recôncavo Basin (faulting and block tilting). As a consequence, the paleoslope of the basin was reoriented to the east-southeast (see Figure 2). Tectonic activity related to the syn-rift phase redefined the structural framework of the basin and led to development of platform areas and high subsidence depocenters between fault blocks (axial depocenters; Figure 11). The hinge-line defined by the 200 m-line of tdi II (see Figure 5) approximates the Água Grande fault in central Recôncavo and the Nova América fault in southern Recôncavo (see Figures 5 and 6), both synthetic (down-to-the-basin) faults. This

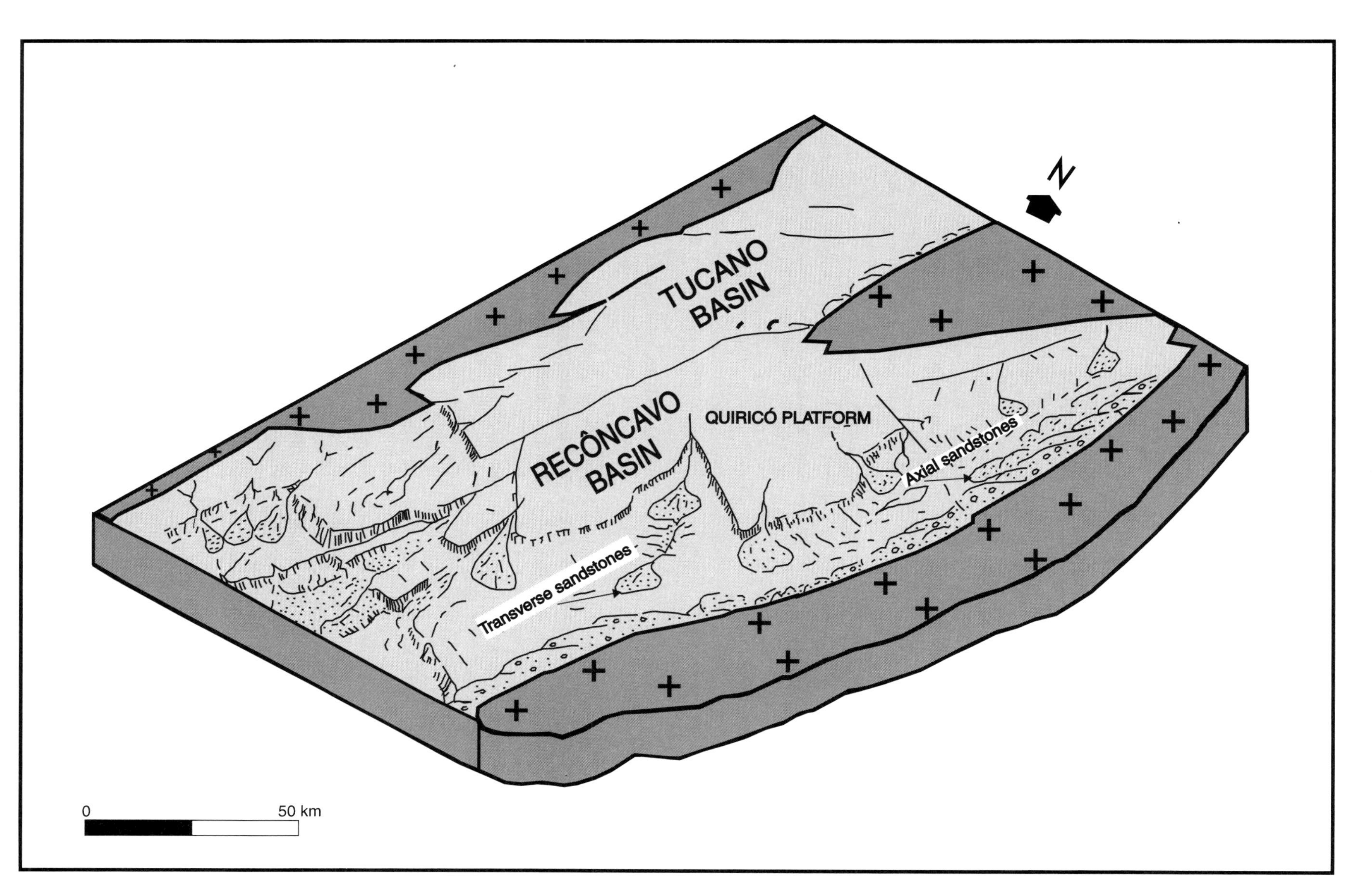

Figure 11—Paleogeography of the Recôncavo Basin of Brazil in the Berriasian (modified from Barroso, 1985).

hinge-line controlled the deposition of the sandy strata sourced from west-northwest into the basin across the Quiricó platform (see Figure 2). The geologic situation during deposition of both axial and transverse sandy bodies in the Berriasian stage in the Recôncavo Basin is illustrated in Figure 11.

Low subsidence areas were characterized by flexural faulting and limited accommodation and subject to intermittent erosion and bypass. Axial depocenters (see Figure 2) provided the ultimate storage sites of the basin because of their high accommodation. The physiographic subdivision between platform and depocenter regions in the Recôncavo Basin occurred within a few tens of kilometers (see Figures 2, 5, and 6), which facilitated downslope transport of sand-size and even coarser sediments. Sand layers account for a minor portion of the bulk of the Berriasian deposits in the basin. This may be a result of an immature fluvial system or structural barriers. Axial sands were mainly derived from the footwall of the Salvador fault (along the eastern border of the basin) (Figure 2). Source for the transverse sands were uplifted blocks located on the flexural margin of the basin.

Tectonics coupled with humid climatic conditions allowed the development of a basinwide, deep (hundreds of meters deep), stratified, fresh to brackish-water lake. Tölderer-Farmer et al. (1989) concluded this based on ostracod and palynomorph analyses. The depositional record of tdi II shows similarity with portions of the Triassic Locktong Formation of the Newark Basin of northeastern United States (Olsen, 1990) and the Cretaceous Abu Gabra Formation of southern Sudan (Schull, 1988).

CONCLUSIONS

Tectonics plus climatic fluctuations controlled the evolution of the syn-rift fill of the Recôncavo Basin. During the Berriasian, tectonic activity was strong and climate was wet, leading to a large, deep, perennial fresh to brackish-water lake. Coeval coarse-grained sediments can be grouped into two main systems: (1) axial, mainly related to the eastern border of the basin and closely linked to the conglomeratic wedge of the Salvador Formation and (2) transverse, sourced from the flexural margin and associated with growth faulting. Axial sandstones are generally older than their transverse counterparts.

Most axial sandstones of the Recôncavo Basin are interpreted as sublacustrine turbidites. An alternative model implies coalescing steep fan-lobe sands from the eastern border of the basin. For transverse sandstones, evidence suggests sublacustrine turbidite facies. The paucity of coarse-grained input during the Berriasian reflects an immature fluvial system and presence of structural barriers along the margins of the basin; consequently, the source for both axial and transverse sandstones was mostly from regions located along the eastern escarpment and from the western flexural margin, respectively.

ACKNOWLEDGMENTS

We are grateful to Kerry Kelts, Clóvis F. Santos, and an anonymous reviewer who critically read the manuscript and to Petrobras for permission to publish this paper. H. T. F. da Silva thanks Wagner Lima for his help in editing some of the illustrations in this paper.

REFERENCES CITED

Arai, M., T. Hashimoto, and N. Uesugui, 1989, Significado cronoestratigráfico da associação microflorística do Cretáceo Inferior do Brasil: Boletim Técnico da Petrobras, v. 9, no. 1, p. 17–45.

Banks, N. L., K. A. Bardwell, and S. Musiwa, 1995, Karoo rift basins of the Luangwa Valley, Zambia, *in* J. J. Lambiase, ed., Hydrocarbon habitat in rift basins: The Geological Society of London Special Publication no. 80, p. 285–295.

Barroso, A. S., 1985, Sedimentologia, diagênese e potencialidades petrolíferas dos arenitos Morro do Barro—Ilha de Itaparica e adjacências: Petrobras Internal Report, 37 p.

Begawan, S. R., and J. J. Lambiase, 1995, Preface, *in* J. J. Lambiase, ed., Hydrocarbon habitat of rift basins: The Geological Society of London Special Publication no. 80, p. vii–viii.

Bouma, A. H., 1962, Sedimentology of some flysch deposits: A graphic approach to facies interpretation: Amsterdam, Elsevier, 168 p.

Bruhn, C. H. L., J. M. Caixeta, and J. C. Scarton, 1985, Sublacustrine reservoirs of the Riacho da Barra Field, Recôncavo rift-basin, Brazil: Petrobras Internal Report, 22 p.

Bruhn, C. H. L., 1985, Sedimentação e evolução diagenética dos turbiditos eocretácicos do Membro Gomo, Formação Candeias, no Compartimento Nordeste da Bacia do Recôncavo, Bahia: M. S. thesis, Ouro Preto Federal University, 210 p.

Bruhn, C. H. L., C. Cainelli, and R. Darros de Matos, 1988, Habitat do petróleo e fronteiras exploratórias nos rifts brasileiros: Boletim de Geociências da Petrobras, v. 2, p. 217–253.

Busby–Spera, C. J., 1988, Development of fan-deltoid slope aprons in a convergent-margin tectonic setting: Mesozoic Baja California, Mexico, *in* W. Nemec and R. J. Steel, eds., Fan-Deltas: Sedimentology and Tectonic Settings: Blackie & Son, Glasgow, p. 419–429.

Caixeta, J. M., G. V. Bueno, L. P. Magnavita, and F J. Feijó, 1994, Bacias do Recôncavo, Tucano and Jatobá: Boletim de Geociências da Petrobras, v. 8, p. 163–172.

Camões, A. M. C., and R. C. Rigueira, 1987, Campo de Cexis-relatório geológico: Petrobras, Internal Report, 73 p.

Camões, A. M., 1988, Modelo tectono-sedimentar do Campo de Cexis, Bacia do Recôncavo: Boletim de Geociências da Petrobras, v. 2, p. 267–275.

Carozzi, A. V., J. D. R. Fonseca, M. B. Araújo, D. N. N. Vasconcelos, and J. R. L. Sandoval, 1976, New genetic interpretation of oil-producing coarse clastics of the Candeias Field, Recôncavo Basin: Petrobras Internal Report, 39 p.

Carozzi, A. V., and J. D. R. Fonseca, 1989, A new technique of locating turbidite fans: Candeias Formation (Lower Cretaceous), Recôncavo Basin, Brazil: Journal of South American Earth Sciences, v. 2, p. 277–293.

Cohen, A. S., 1990, Tectono-stratigraphic model for sedimentation in Lake Tanganyika, Africa, *in* B. J. Katz, ed., Lacustrine Basin Exploration—Case Studies and Modern Analogs: AAPG Memoir No. 50, p. 137–150.

da Silva, H. T. F., A. T. Picarelli, J. M. Caixeta, N. R. Campos, O. B. Silva, and R. C. Rigueira, 1989, Aspectos evolutivos do Andar Rio da Serra, fase rifte, na Bacia do Recôncavo e a Formação Jacuípe: Petrobras Internal Report, 139 p.

da Silva, H. T. F., 1993, Flooding surfaces, depositional elements and accumulation rates: Characteristics of the Lower Cretaceous tectonosequence in the Recôncavo Basin, northeast Brazil, Ph. D. dissertation, The University of Texas at Austin, 313 p.

Desheng, L., 1995, Hydrocarbon habitat in the Songliao rift basin, China, *in* J. J. Lambiase, ed., Hydrocarbon Habitat in Rift Basins: The Geological Society of London Special Publication No. 80, p. 317–329.

Figueiredo, A. M. F., J. A. E. Braga, H. M. C. Zabalaga, J. J. Oliveira, G. A. Aguiar, O. B. Silva, L. F. Mato, L. M. F. Daniel, L. P. Magnavita, and C. H. L. Bruhn, 1994, Recôncavo Basin: A prolific intracontinental rift basin, *in* S. M. Landon, ed., Interior Rift Basins: AAPG Memoir No. 59, p. 157–203.

Harland, W. B., A. V. Cox, P. G. Llewellyn, C. A. G. Pickton, A. G. Smith, and E. Walters, 1982, A Geologic Time Scale: Cambridge University Press, Cambridge, 131 p.

Higgs, R., 1990, Sedimentology and tectonic implications of Cretaceous fan-delta conglomerates, Queen Charlotte Islands, Canada: Sedimentology, v. 37, p. 83–103.

Krömmelbein, K., 1966, On "Gondwana Wealden" ostracoda from NE Brazil and West Africa: Proceedings of the 2nd West African Micropaleontological Colloquium, Ibadan, Nigeria, p. 112–119.

Lambiase, J. J., and W. Bosworth, 1995, Structural controls on sedimentation in continental rifts, *in* J. J. Lambiase, ed., Hydrocarbon Habitat in Rift Basins: The Geological Society of London Special Publication No. 80, p. 117–144.

Magnavita, L. P., I. Davidson, and N. J. Kusznir, 1994, Rifting, erosion, and uplift history of the Recôncavo-Tucano-Jatobá Rift, northeast Brazil: Tectonics, v. 13, no. 2, p. 367–388.

Magnavita, L. P.,and H. T. F. da Silva, 1995, Rift border system: the interplay between tectonics and sedimentation in the Recôncavo Basin, northeastern Brazil: AAPG Bulletin, v. 79, p. 1590–1607.

Menezes, G. M. N., 1991, Análises de lâminas delgadas do poço 7-CX-56-BA: Petrobras Internal Report, 9 p.

Milani, E. J., and I. Davison, 1988, Basement control and transfer tectonics in the Recôncavo-Tucano-Jatobá Rift, northeast Brazil: Tectonophysics, v. 154, p. 41–70.

Netto, A. S. T., J. A. E. Braga, C. H. L. Bruhn, L. P. Magnavita, J. J. Oliveira, H. M. Agle, and J. C. Ribeiro, 1984, Prospectos estratigráficos do Recôncavo: Petrobras Internal Report, 83 p.

Olsen, P. E., 1990, Tectonic, climatic, and biotic modulation of ecosystems-Examples from the Newark Supergroup of Eastern North America, *in* B. J. Katz, ed., Lacustrine Basin Exploration—Case Studies and Modern Analogs: AAPG Memoir No. 50, p. 209–224.

Porebski, S. J., 1984, Clast size and bed thickness trends in resedimented conglomerates: Example from a Devonian fan-delta succession, southwest Poland, *in* E. H. Koster and R. J. Steel, eds., Sedimentology of Gravels and Conglomerates: Memoir Canadian Society of Petroleum Geologists, v. 10, p. 399–411.

Prior, D. B., and B. D. Bornhold, 1988, Submarine morphology and processes of fjord fan-deltas and related high-gradient systems: Modern examples from British Columbia, *in* W. Nemec and R. J. Steel, eds., Fan-Deltas: Sedimentology and Tectonic Settings: Blackie & Son, Glasgow, p. 125–143.

Prosser, S., 1993, Rift-related linked depositional systems and their seismic expression, *in* G. D. Williams and A. Dobb, eds., Tectonics and Seismic Stratigraphy: The Geological Society of London Special Publication No. 71, p. 35–66.

Reading, H. G., 1991, The classification of deep-sea depositional systems by sediment caliber and feeder system: Journal of the Geological Society of London, v. 148, p. 427–430.

Regali, M.S.P., and C. F. Viana, 1989; Sedimentos do Neojurássico—Eocretáceo do Brasil: idade e correlação com a Escala Internacional: Petrobras Internal Report, 95 p.

Sanches, C. P., 1993, Paraffin crystallization in reservoirs: A case history from Fazenda Bálsamo Field, Recôncavo Basin, Bahia, Brazil: International Conference on Fluid Evolution, Migration and Interaction in Rocks: Torquay, England, p. 183–188.

Scholz, C. A., and B. R. Rosendahl, 1990, Development of coarse-grained facies in rift basins-Examples from East Africa: Geology, v. 18, p. 140–144.

Scholz, C. A., 1995, Seismic stratigraphy of an accommodation-zone margin rift-lake delta, Lake Malawi, Africa, *in* J. J. Lambiase, ed., Hydrocarbon Habitat in Rift Basins: The Geological Society of London Special Publication No. 80, p. 183–195.

Schull, T. J., 1988, Rift basins of interior Sudan-Petroleum exploration and discovery: American Association of Petroleum Geologists Bulletin, v. 72, p. 1128–1142.

Smith, R. D. A., 1995, Reservoir architecture of syn-rift lacustrine turbidite systems, early Cretaceous, offshore South Gabon, *in* J. J. Lambiase, ed., Habitat of Hydrocarbon in Rift Basins: The Geological Society of London Special Publication No. 80, p. 197–210.

Souza, E. M., L. F. Mato, and C. F. Conde, 1990, Desenvolvimento e estratégia de produção do Campo de Rio do Bu e análise da zona de produção *II* (Formação Candeias), Bacia do Recôncavo: Petrobras Internal Report, 32 p.

Stow, D. A. V., C. D. Bishop, and S. J. Mills, 1982, Sedimentology of the Brae oilfield, North Sea: Fan models and controls: Journal of Petroleum Geology, v. 5, p. 129–148.

Tölderer–Farmer, M., J. C. Coimbra, J. A. Moura, and H. N. Gilson, 1989, Reconstrução paleoambiental da Bacia do Recôncavo com base em ostracodes-Um estudo preliminar: Petrobras Internal Report, 111 p.

Viana, C. F., E. G. Gama, Jr., I. A. Simões, J. A. Moura, J. R. Fonseca, and R. J. Alves, 1971, Revisão estratigráfica da bacia Recôncavo/Tucano: Boletim Técnico da Petrobras, Rio de Janeiro, v. 14, no. 3–4, p. 157–192.

Walker, R. G., 1978, Deep water sandstone facies and ancient submarine fans: American Association of Petroleum Bulletin, v. 62, no. 5, p. 932–966.

Williams, H. H., and R. T. Eubank, 1995, Hydrocarbon habitat in the rift graben of the Central Sumatra Basin, Indonesia, *in* J. J. Lambiase, ed., Hydrocarbon Habitat in Rift Basins: Geological Society of London Special Publication No. 80, p. 331–371.

Rangel, H. D., and M. Carminatti, 2000, Rift lake stratigraphy of the Lagoa Feia Formation, Campos Basin, Brazil, *in* E. H. Gierlowski-Kordesch adn K. R. Kelts, eds., Lake basins through space and time: AAPG Studies in Geology 46, p. 225–244.

Chapter 18

Rift Lake Stratigraphy of the Lagoa Feia Formation, Campos Basin, Brazil

Hamilton Duncan Rangel
Mario Carminatti
Petrobras E&P
Rio de Janeiro, Brazil

INTRODUCTION

Lower Cretaceous lacustrine basins occur rather continuously along about 3500 km (from Pelotas to Sergipe-Alagoas Basins) of the eastern Brazilian margin (Figure 1). Because these basins accumulated large volumes of organic-carbon-rich sediments, they form source rocks for much of the Brazilian oil discovered to date. Economic and scientific interest in these lacustrine basins has resulted in numerous publications (e.g., Ponte and Asmus, 1978; Ojeda, 1982; Estrella et al., 1984; Bertani and Carozzi, 1985; Baumgarten et al., 1988; Guardado et al., 1989; Van der Ven et al., 1989; Aquino and Lana, 1990; Cupertino, 1990; Mello and Maxwell, 1990; Santos and Braga, 1990; Castro, 1992; Horschutz and Scuta, 1992; Mato et al., 1992; Rangel et al., 1993a, b; Caixeta et al., 1994; Dias et al., 1994; Feijó, 1994a, b; Pereira and Feijó, 1994; Rangel et al., 1994a, b; Santos et al., 1994; Teixeira Netto et al., 1994; Vieira et al., 1994; da Silva et al., this volume).

Several of these studies have attempted comparative syntheses. They emphasize differences in each basin system dependent on the tectonic evolution. For example, continental breakup is accompanied by basaltic extrusions during initial stages of rifting, such as in the Campos Basin (Rangel et al., 1994a, b), or basins may lack basaltic extrusions, such as in the Recôncavo Basin. Based on tectonic evolution, the eastern Brazilian Atlantic margin can be organized into basically two types of rift lake environments: (1) deep-water lakes, in which tectonism outpaced sedimentation, with the Recôncavo Basin, northeastern Brazil, for example, and (2) shallow water lakes in which tectonism and sedimentation maintain a balanced pace, such as in the Campos Basin. Evolution of both rift lake types allowed accumulation of thousands of meters of lacustrine sediments.

Our study deals with the stratigraphy of lacustrine sections in the Campos Basin that extends over an area of 100,000 km^2 (250 km × 400 km), down to the modern bathymetric contour at 3000 m ocean depth. Vertical sequence thickness in the basin measures up to about 2500 m, or almost 4000 m, taking into account lateral tectonic regimes with thick sequences in different structural lows (Rangel et al., 1994a, b). These sedimentary sequences form the Lagoa Feia Formation.

STRATIGRAPHY

The stratigraphic framework of lacustrine and evaporitic deposits of the Lagoa Feia Formation is based on depositional markers and unconformities identified on well and seismic logs. A depositional-stratigraphic marker as used here is a rock unit formed by one or more lithofacies with physical attributes and a depositional nature that are laterally extensive, as recognized by subsurface log patterns.

Following this stratigraphic approach of mainly physical surfaces (depositional-markers and unconformities), it was possible to differentiate seven stratigraphic units above the basalts within the sequence interpreted as deposited during the rifting stage, and three units during a transitional stage. The uppermost transitional stage unit is related to precipitation of evaporites (halite) in conjunction with spreading of the early South Atlantic Ocean (Figure 2).

All units are bounded by stratigraphic discontinuities of erosive origin and constitute, in practice, depositional sequences, such as those proposed by Mitchum et al. (1977). They define a depositional sequence as a stratigraphic unit composed of a relatively conformable succession of genetically related strata and bounded at its top and base by unconformities or their correlative conformities. Such units in the Campos Basin represent a sedimentary response to the expanding and retreating cycles of a lake. These cycles, controlled by tectonics or climate, determine sedimentary dynamics in each compartment or depositional depocenter of the rift, defining genetic associations inherent to each depositional sequence. Compartmentalization is not complete because there are connections with adjacent depocenters. Such cycles impose problems for the direct application of systems-tract

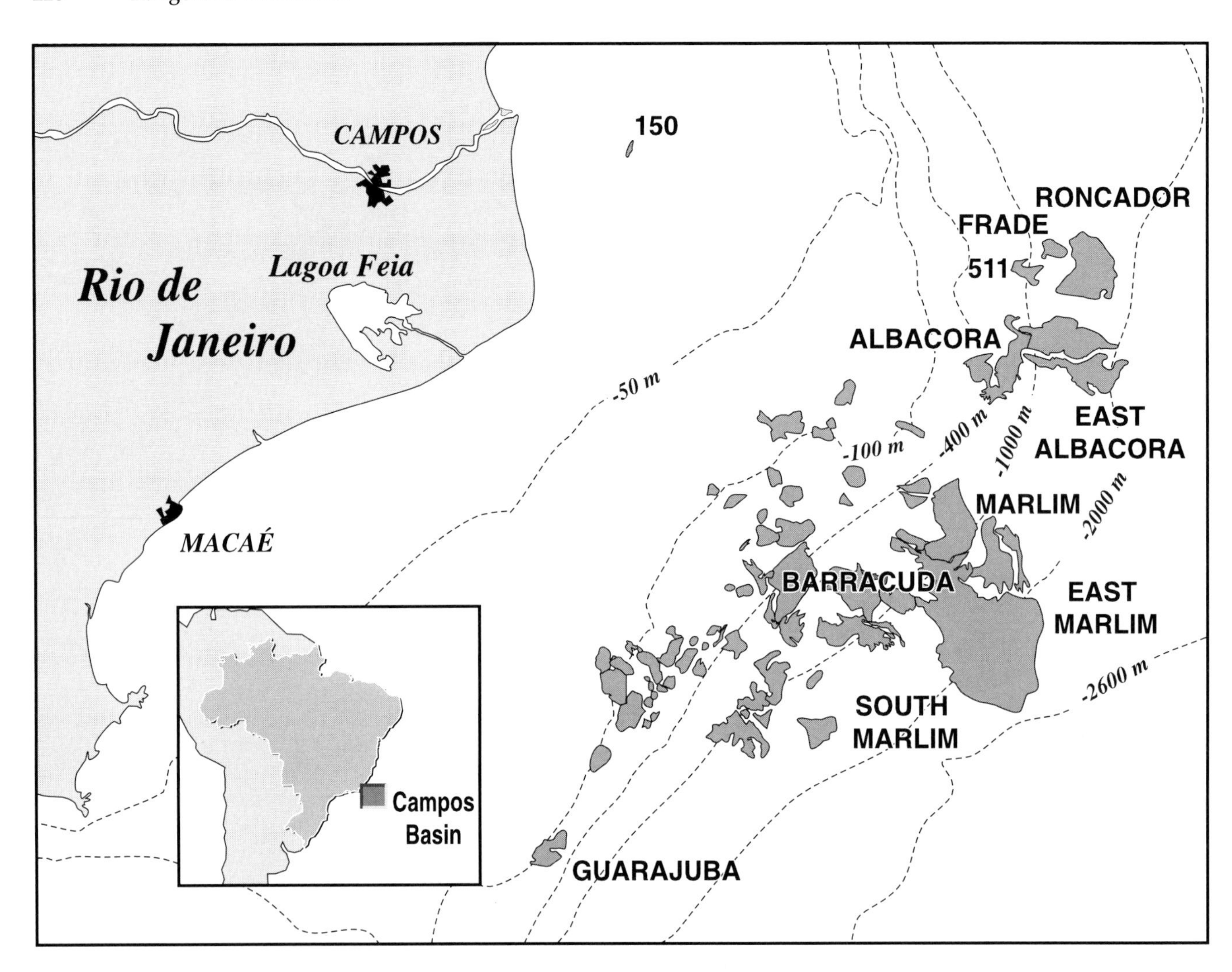

Figure 1—Index map for the Campos Basin, Brazil.

models of sequence stratigraphic concepts (Van Wagoner et al., 1988) in the sedimentary studies of the Campos Basin lacustrine deposits.

Erosion unconformities provide depositional markers for a more detailed stratigraphic framework and represent the fundamental criteria for assigning units. Biostratigraphic correlation was used to establish synchroneity of events. Biostratigraphic assignments and correlations remain necessarily approximate because of the endemism of the fauna in these rift lake deposits and in the regional pollen. Zones are loosely correlated to international standards. In this study the local Brazilian chronostratigraphic units are based on ostracode (OS) assemblages (Dias, 1994). These define lacustrine rift stages in a phase of continental separation of Brazil from Africa while the North Atlantic region already had coastal marine basins.

During the early stages of rifting, the regional climate was humid along the eastern margin of Brazil, as suggested by lithofacies and geochemical parameters. During later stages of rifting and transitional to the marine salt basin phase, the climate evolved toward more predominantly arid conditions.

DESCRIPTION OF ACOUSTIC–LITHOLOGIC UNITS

Unit A

Unit A is related to NRT-005 to NRT-008.1 biozones, which corresponds to the Aratu and lower Buracica local stages (Barremian) and is associated with the Itabapoana Member (proximal facies) and Atafona Member (distal facies) of the Lagoa Feia Formation (Figures 2–5). This unit is bounded at the top and the base by regional unconformities. Its lower limit is defined by a disconformity in which the first lacustrine sediments overlie igneous rocks of the Cabiúnas Formation (Figure 2). The entire unit displays higher sonic velocities than the overlying stratigraphic unit. It is more regionally restricted compared with other rift stratigraphic units. Unit A represents the initial phase of tectonic subsidence during the rifting phase. Several half-graben sub-basins were established during this initial rifting phase which control the distribution of sedimentation and lithofacies.

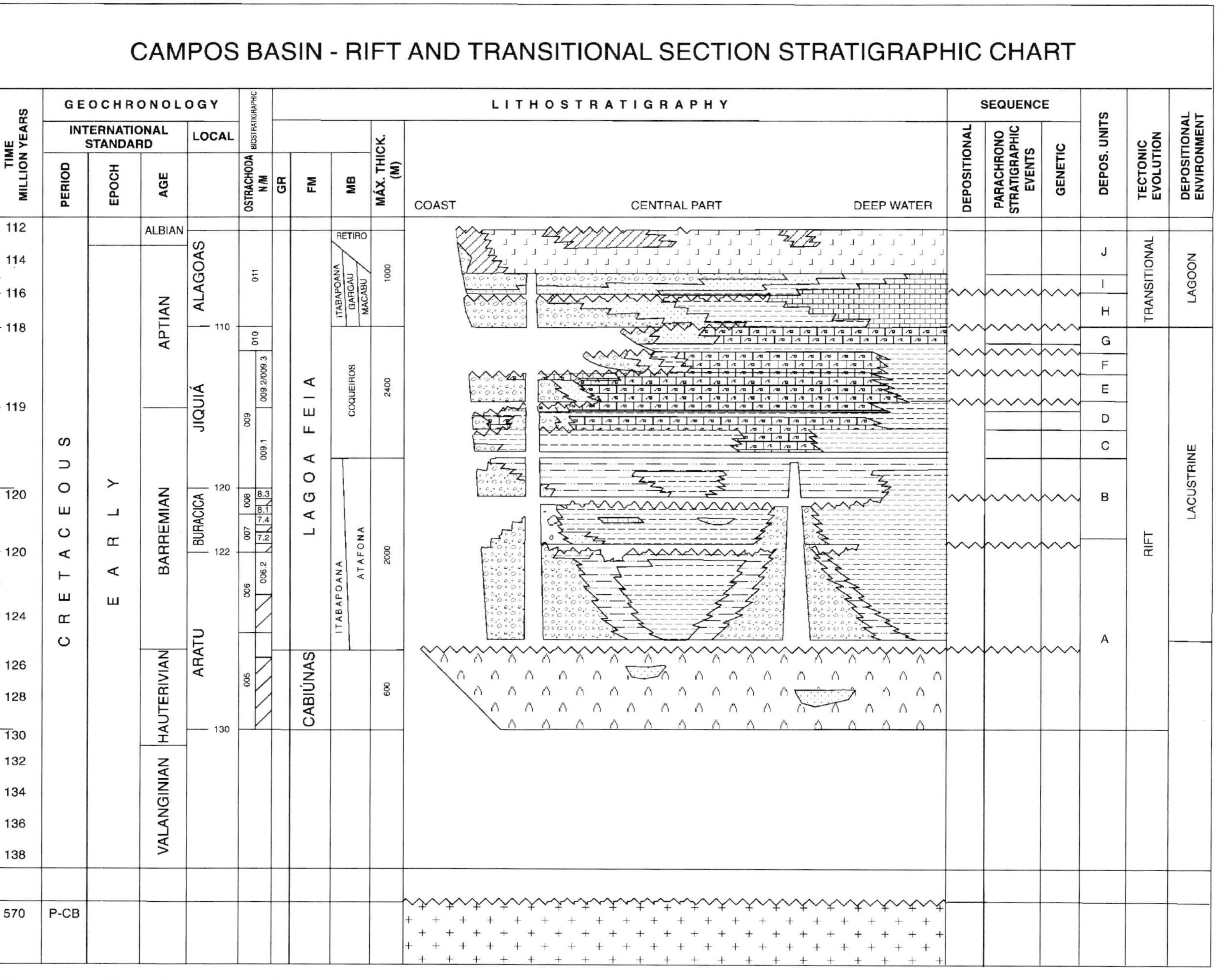

Figure 2—Rift and transitional phase stratigraphy of the Campos Basin, Brazil.

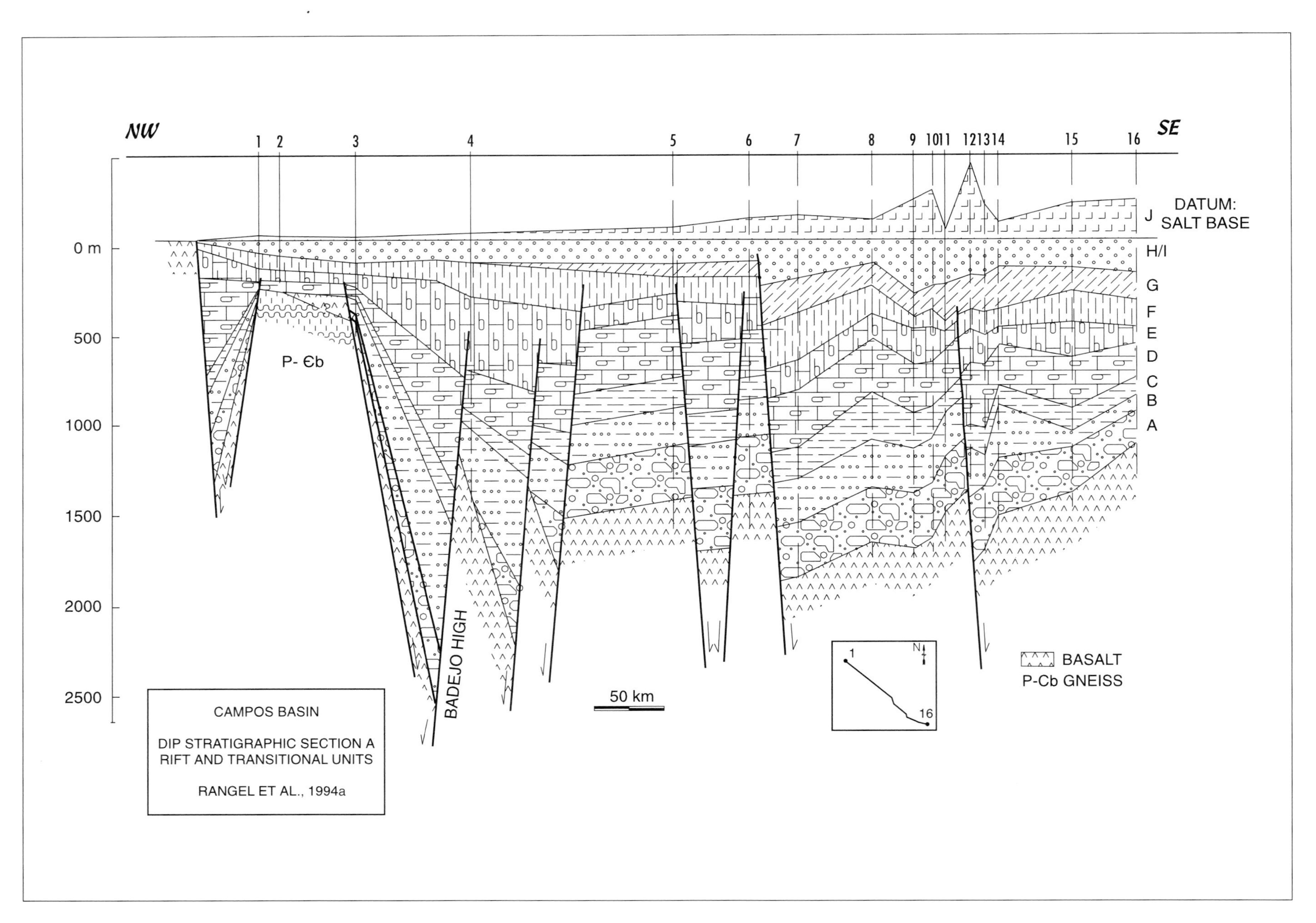

Figure 3—North-dipping structural section of the rift and transitional units (location on Figure 5) of the Campos Basin, Brazil.

The Badejo Structural High (see location in Figures 3, 4) is a very prominent feature that trends northeast-southwest across the western part of the basin. It appears to have been only mildly active during the deposition of unit A. The main depocenters of this interval were located in the northwestern, eastern, and southern areas, with up to 500 m sediment thickness (Figure 5).

Basaltic terrain of the Aratu local stage (Hauterivian) acted as provenance for most of the sedimentary components deposited as unit A, except in the northwestern region where prominent Precambrian gneiss exposures also supplied detrital sediment grains (Rangel et al., 1987). Despite the relatively modest tectonic movements, these features still exercised the main controls on stratigraphic architecture (Dias et al., 1987; Rangel et al., 1993a).

Unit A is made up of predominantly terrigenous sediments. Facies associations include conglomerates, sandstones, and shales genetically related to sediment gravity flows. Tectonic movements resulted in fan-delta depositional systems in areas adjacent to structural highs or in narrow structural lows (Figure 5). In the upper part of unit A, strata are dominated by deep, freshwater mudstones and carbonates (Soldan et al., 1995). Tectonically induced turbidite deposits also occur in this upper section.

Unit B

Unit B is related to NRT-008.3 to NRT-009.2 biozones, which corresponds to Buracica and lower Jiquiá local stages (Barremian) and is associated with the Itabapoana and Atafona Members of the Lagoa Feia Formation (Figures 2–4, 6). The lower contact is defined by an unconformity that is characterized by coarse siliciclastics grading directly to siltstones. At the São João da Barra Low (Figure 4), unit B generally rests directly on basalts of the Cabiúnas Formation. The entire unit exhibits lower sonic velocity values than its overlying unit. Siltstones and shales are the main lithofacies and more widespread than in unit A.

Exposed basalts adjacent to the basin margin or along prominent structural highs within the basin provided the main sediment source. Gneiss, exposed in the northwest, was the secondary source. Alluvial fan and fan-delta deposits occur predominantly along the western margin of the basin, although they are also present adjacent to other structural highs in the basin.

Unit B is absent over prominent structural highs such as the Badejo High, but very thick (over 400 m) and widespread in structural lows, such as the São João da Barra Low, adjacent to the Badejo High (Figure 6). Based on seismic data, unit B attains a thickness of 800 m north of this depression. This marks an important paleogeographic shift in the lake pattern, characterized by deepening and expansion.

Unit B corresponds to special geochemical conditions where stevensite peloids were formed by chemical precipitation in a magnesium-rich siliceous alkaline lake phase. Unit B has higher sulfur content than unit A that is reflected in the molecular organic parameters of the rock extracts. Still, these appear not to be related to major changes in biomass (Rehim et al., 1986, Soldan et al., 1995).

Unit C

Unit C is related to OS 009.3 biozone (Telles, 1992; Carvalho et al., 1993), which corresponds to the upper Jiquiá local stage (Aptian) and is associated with the Coqueiros Member of the Lagoa Feia Formation (Figures 2–4, 7). The basal contact, as defined by an unconformity, is overlain by a stratigraphic marker chacterized by high gamma-ray values. This unit appears related to a general deepening of the lake system. Where encountered, unit C is characterized by uniformly high gamma-ray values for well logs. From the time of deposition of unit A, the lake depositional zone seems to have gradually expanded. Although unit C is widespread throughout the basin, it displays a surprisingly uniform thickness, reaching a maximum of 200 m in structural lows. It is absent over the Badejo High due to non-deposition (Figures 4, 7).

The facies patterns of unit C are uniform, comprising mainly lacustrine shales that are characterized by the high gamma-ray values and low sonic velocities on well logs. Locally, coquinas of mainly pelecypod debris occur as deposits covering and adjacent to structural highs. Conglomerates and sandstone facies occur as more restricted deposits in western areas and near some structural highs. They were related to fan-delta depositional systems, mainly derived from sediment gravity flows. Geochemical parameters and lacustrine lithofacies criteria suggest a more humid climate for the environment of deposition (Soldan et al., 1995).

Unit D

Unit D is related to OS 1010 biozone, which corresponds to the upper Jiquiá local stage (Aptian) and is associated with the Coqueiros Member of the Lagoa Feia Formation (Figures 2–4, 8). This unit is defined at the base by a regional unconformity. Paleoenvironments are interpreted as a part of a regressive regime during the rift phase, in contrast with unit C, which was linked with the culmination of a generally expanding lake sedimentation phase.

The upper contact of unit D is characterized by a strong seismic reflection that represents an abrupt lithologic change from carbonate-rich to terrigenous sediments. On logs, unit D is recognized by high sonic velocities, high resistivity, and low gamma-ray values (coquinas) interbedded with layers of low sonic velocities, low resistivity, and high gamma-ray values (shales). These are interpreted as a high-frequency cyclicity of calcareous lacustrine shelf deposits. Prior seismic studies placed the end of the synrift sequence at this unconformity in some areas of the basin.

Unit D represents the culmination of rifting activity. Not only was major basinal subsidence active, but numerous subbasins formed. Depocenters were widespread. Unit D is 400 m thick in the northwest,

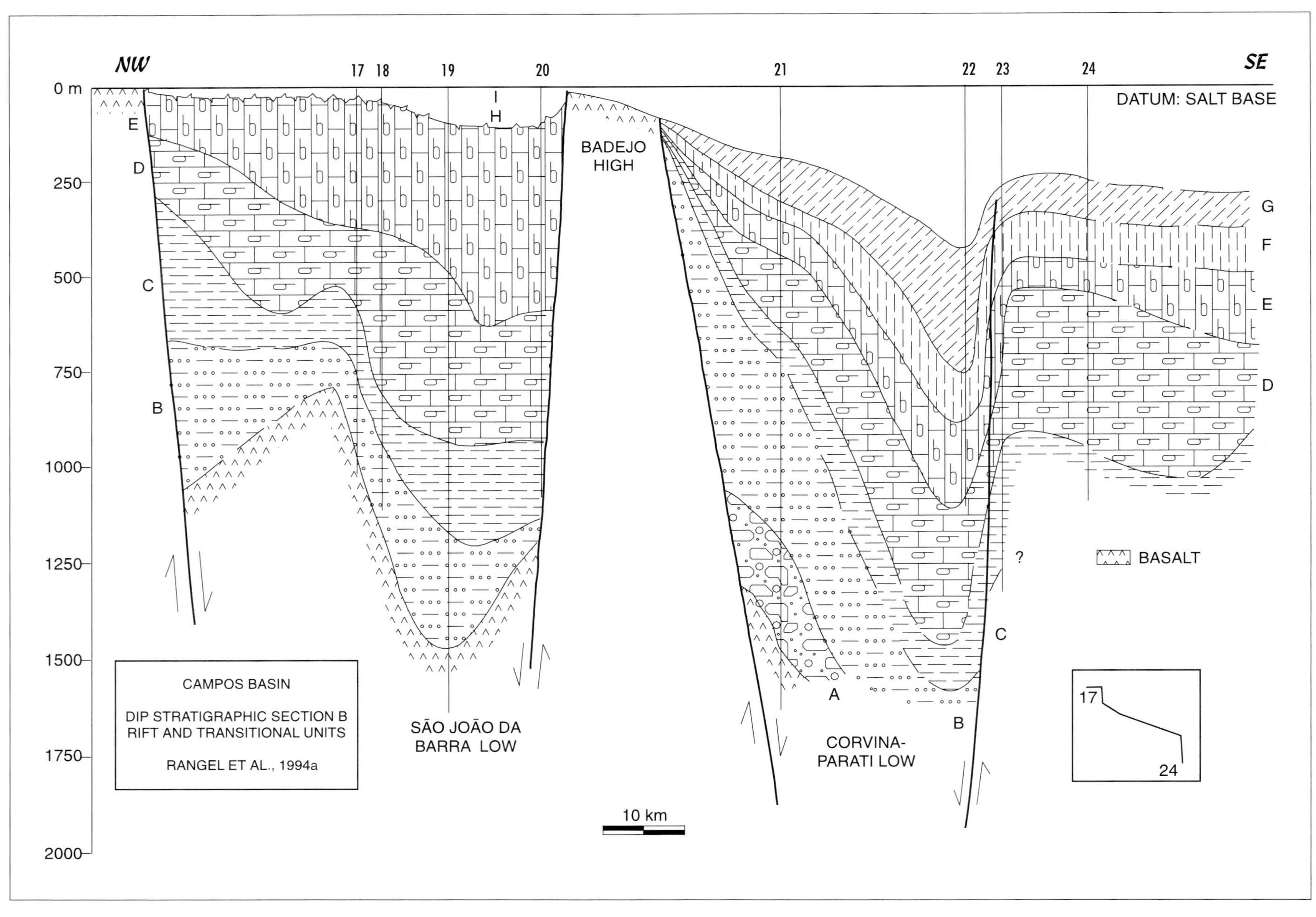

Figure 4—South-dipping structural section of the rift and transitional units (location on Figure 5) of the Campos Basin, Brazil.

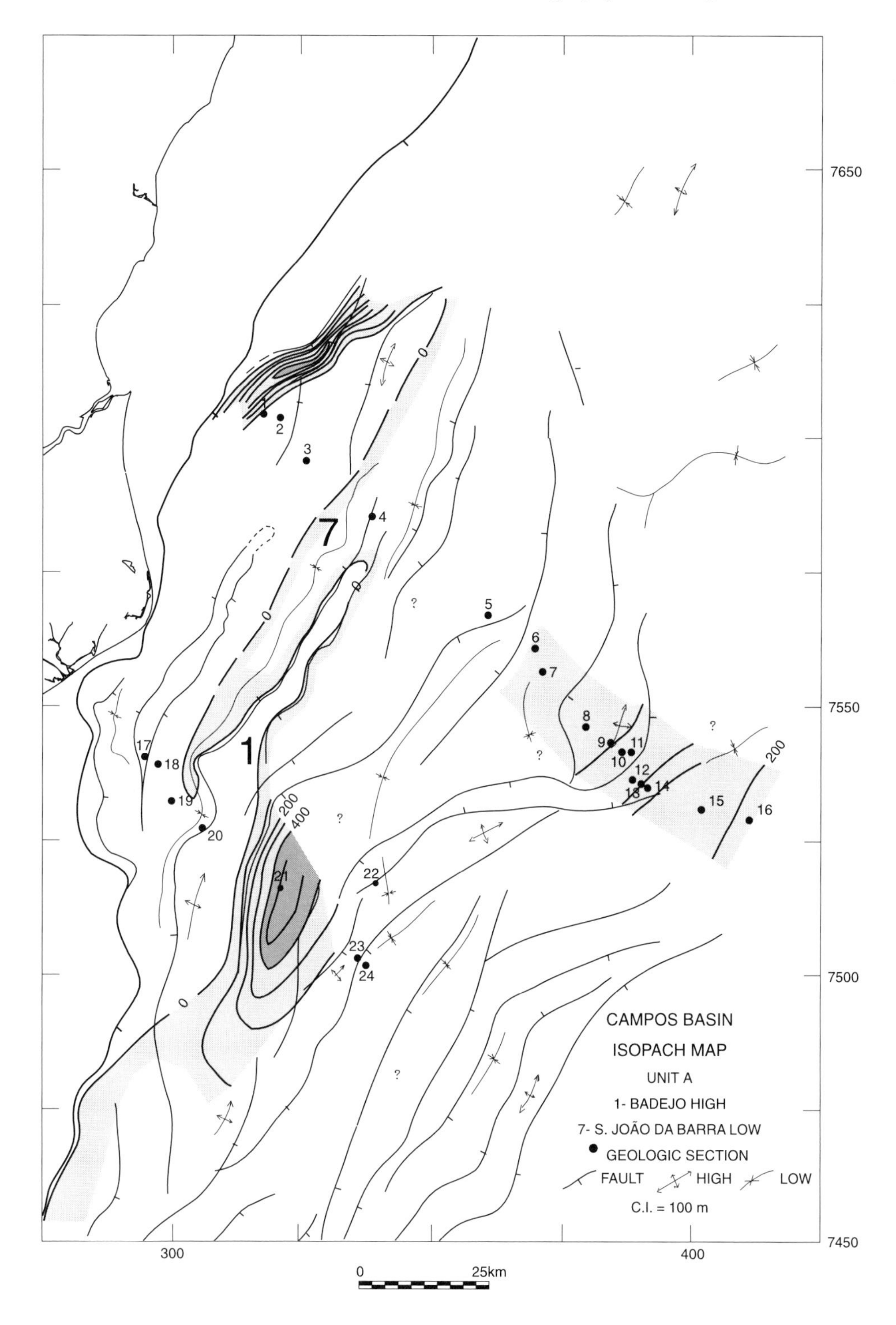

Figure 5—Isopach map of rift stage unit A of the Lagoa Feia Formation, Campos Basin, Brazil.

São João da Barra depression, and over 500 m in other areas of the basin. It pinches out against the Badejo High. Main sources of sediment were basalts outcropping on structural highs within the basin, such as the Badejo High, along with gneiss exposed in the northwest.

Conglomerate and sandstone facies related to fan-delta systems occur on the western side of the lake basin corresponding to cycles of lacustrine regression. Interbedded coquinas and shales form the main lithofacies to the east. Enrichment in calcium carbonate in the lake waters, as sourced from Ca-rich basalts or groundwater through springs and seeps, produced large volumes of shelly biota, leading to pelecypod calcarenites and calcirudites covering some structural highs and flanks.

Deepening of the depositional site allowed dark shale facies to extend well into former proximal areas and onlap over higher energy deposits. Both the lacustrine Buracica and Jiquiá shales of the Lagoa Feia Formation are rich hydrocarbon source rocks (Guardado et

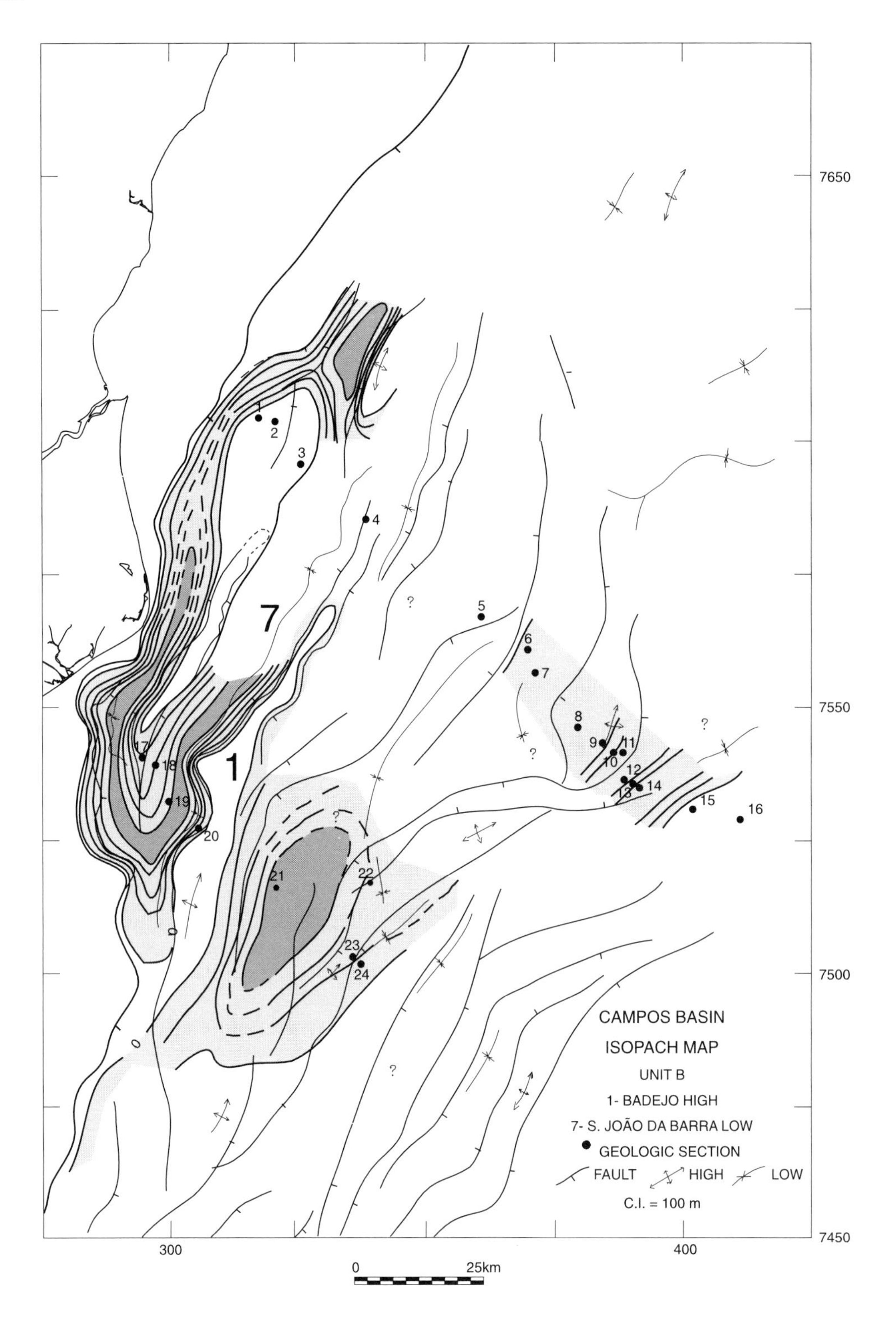

Figure 6—Isopach map of rift stage unit B of the Lagoa Feia Formation, Campos Basin, Brazil.

al., 1989). Sulfur content and other chemical analyses indicate freshwater conditions persisted in the lake, although the interbedding of coquinas and shales may delimit cyclic variations in climate or tectonics.

Unit E

Unit E is related to OS 1020 biozone, which corresponds to the upper Jiquiá local stage (Aptian) and is associated with the Coqueiros Member of the Lagoa Feia Formation (Figures 2–4, 9). The lower part of unit E is characterized by a shale marker-layer with lower sonic velocity, low resistivity, and high gamma-ray values. It is capped by a regional unconformity, difficult to detect on seismic lines. On logs, the upper portion of unit E is generally characterized by high resistivity values, lower gamma-ray values, and higher sonic velocity related to the presence of coquina layers.

The lithofacies association of unit E suggests a general trend toward regression, but depositional extent of the unit seems to have expanded,

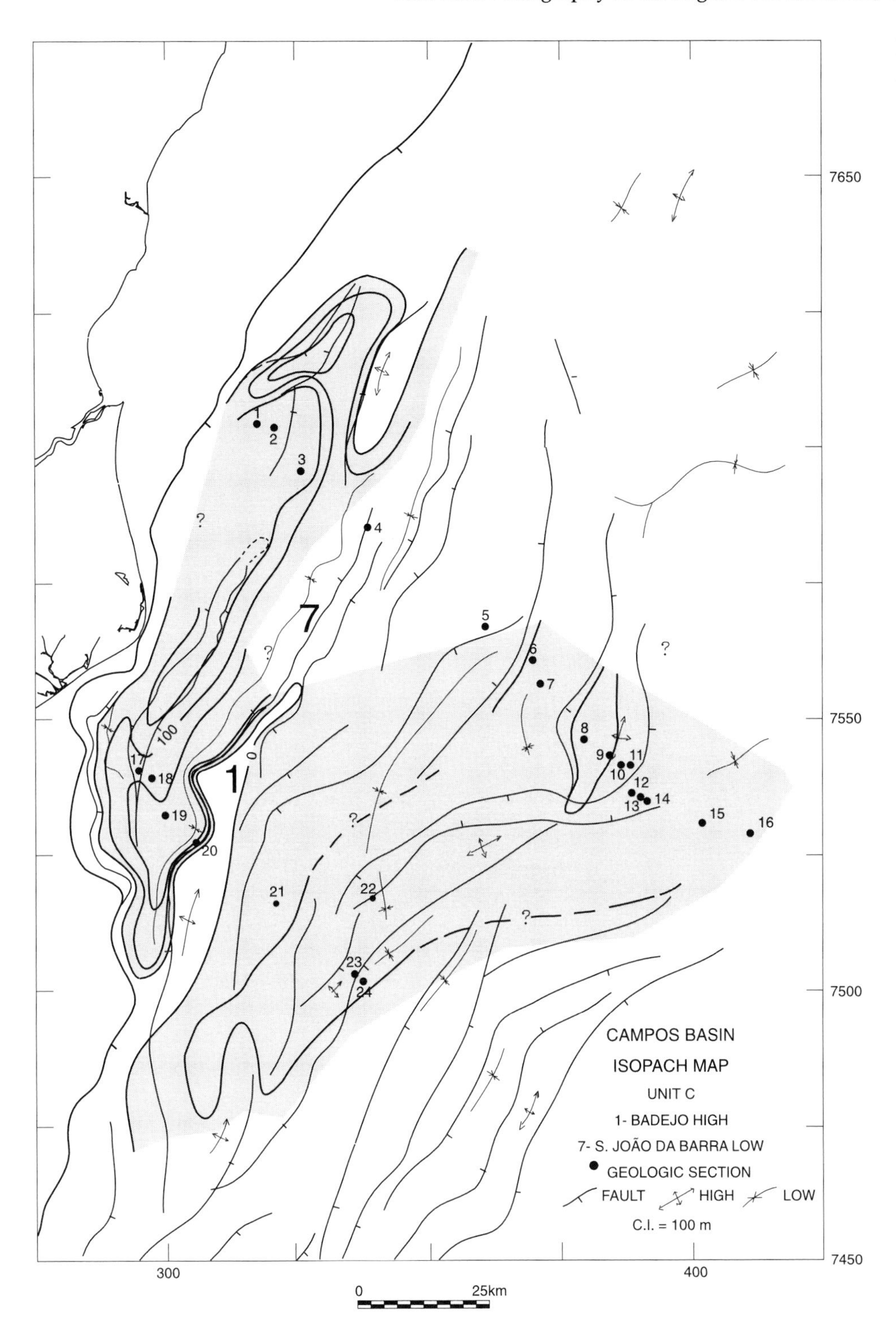

Figure 7—Isopach map of rift stage unit C of the Lagoa Feia Formation, Campos Basin, Brazil.

controlled by a westernmost boundary fault, the Campos fault.

The rift architecture resembles that of the unit D, but subbasins were filled by unit D, leading to more widespread continuity of layering of unit E. The isopach patterns of both units are similar, suggesting that similar tectonic subsidence persisted. The São João da Barra Low became the main depocenter with over 500 m sediment thickness. Provenance derives from the Badejo and adjacent ridges (Figure 9). The western part of the Badejo High is a different depositional environment with coarser grained sediments instead of the finer grained and biogenic sediments of unit D. This suggests a regression with prograding clastics. In both the eastern and southern areas of the basin, the lithofacies persist as interbedded coquinas and shales for both units D and E. Near the end of unit E deposition, the western region of the Badejo High shows evidence of intense erosion, suggesting further regression. Campos fault movements ceased to control sedimentation in the westernmost part of the basin.

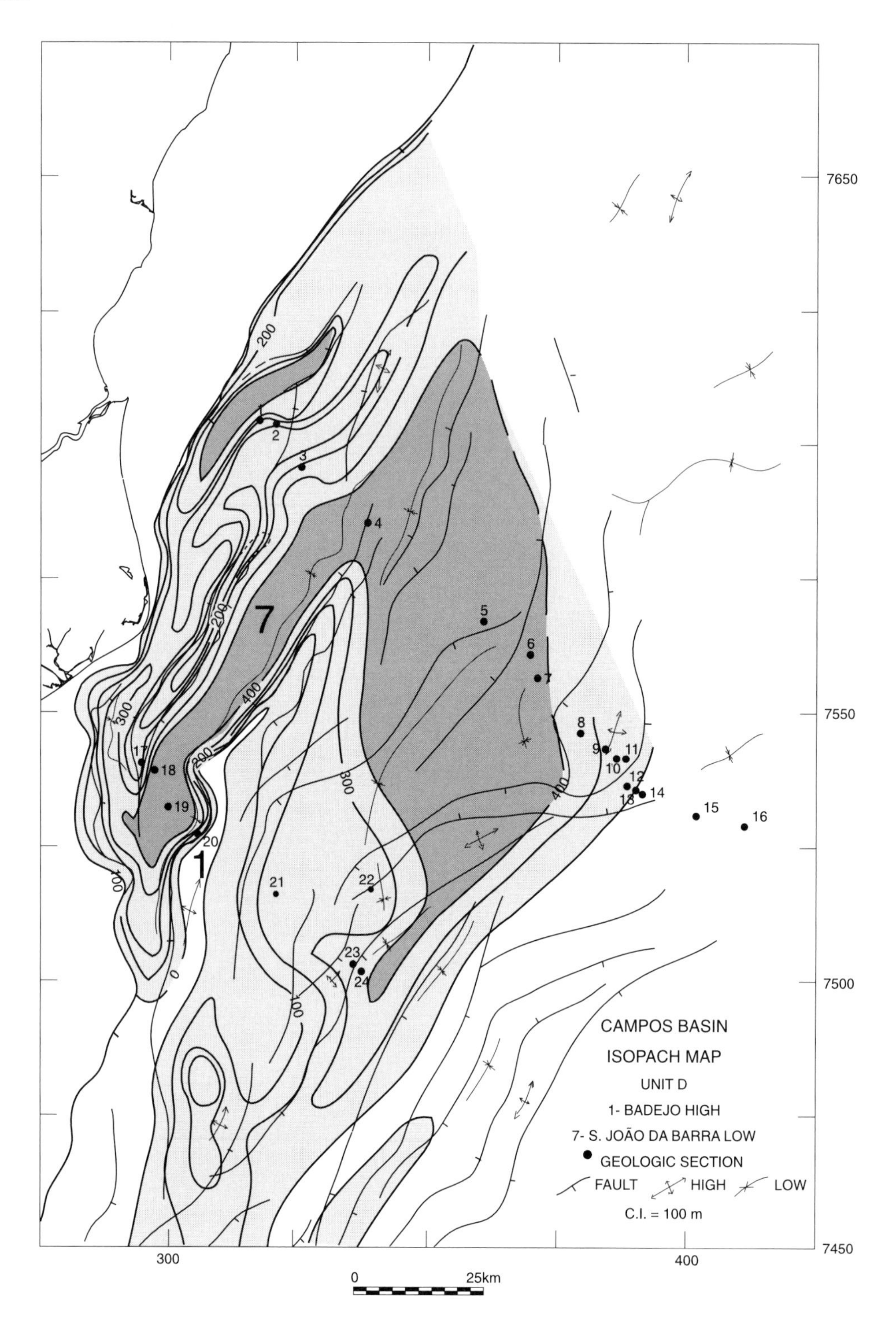

Figure 8—Isopach map of rift stage unit D of the Lagoa Feia Formation, Campos Basin, Brazil.

Higher sulfur content was detected in the upper portion of unit E, interpreted as evidence of more saline conditions, perhaps related to a more arid climate. This is consistent with the presence of *Gymnosperma microflora* palynomorphs (Soldan et al., 1995).

Unit F

Unit F is related to OS 1100 biozone, which corresponds to the upper Jiquiá local stage (Aptian) and is associated with the Coqueiros Member of the Lagoa Feia Formation (Figures 2–4, 10). This unit defines a change in the rift evolution corresponding to a reduction in the extent of rift sedimentation and lake area. In general, structural movements were less active and sedimentation had a more uniform distribution, with about 300 m maximum thickness. A small area in the east, adjacent to the Badejo High, forms an exception with tectonic subsidence leading to 500 m of sediment (Figure 10).

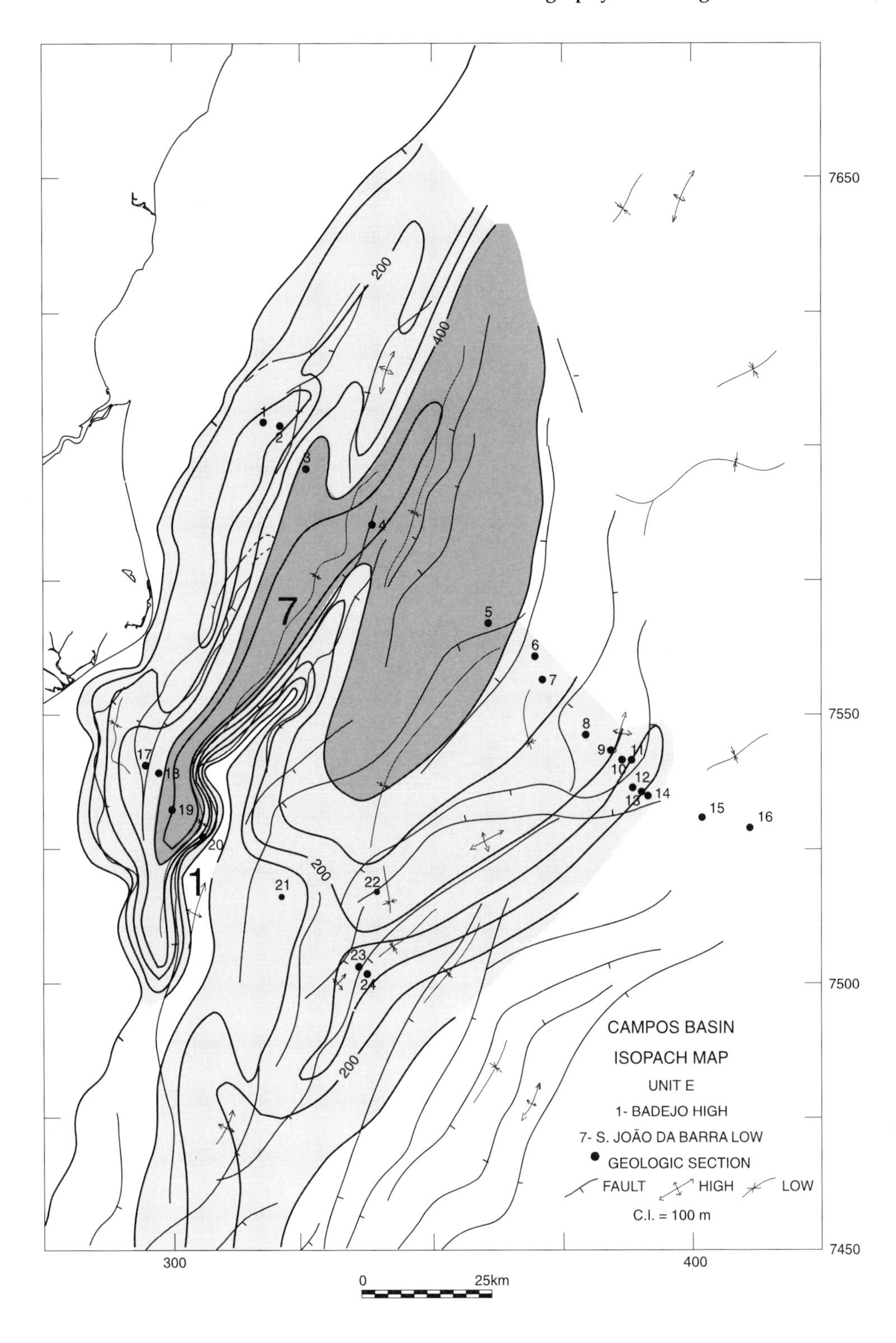

Figure 9—Isopach map of rift stage unit E of the Lagoa Feia Formation, Campos Basin, Brazil.

The upper contact of unit F is an unconformity in the western area, but otherwise stratigraphic continuity persists throughout the basin. The upper contact is generally recognized by a stratigraphic marker on well logs having lower gamma-ray and higher resistivity as well as sonic values. The Badejo High continued as a prominent feature delivering most clastic sediment and limiting sedimentation of unit F to the west. Unit E was exposed and reworked to serve as provenance for some cyclic deposits.

Unit F is characterized by abundant coquinas with thin shaly interbeds, more common in distal basinal areas. The western margin has fan-delta conglomerate and sandstone facies with tectonically induced sediment gravity flows. At the end of unit F deposition, the association of a higher kinetic parameter, low sulfur content, and molecular signatures in the organic matter may reflect evidence of short marine incursions (Soldan et al., 1995).

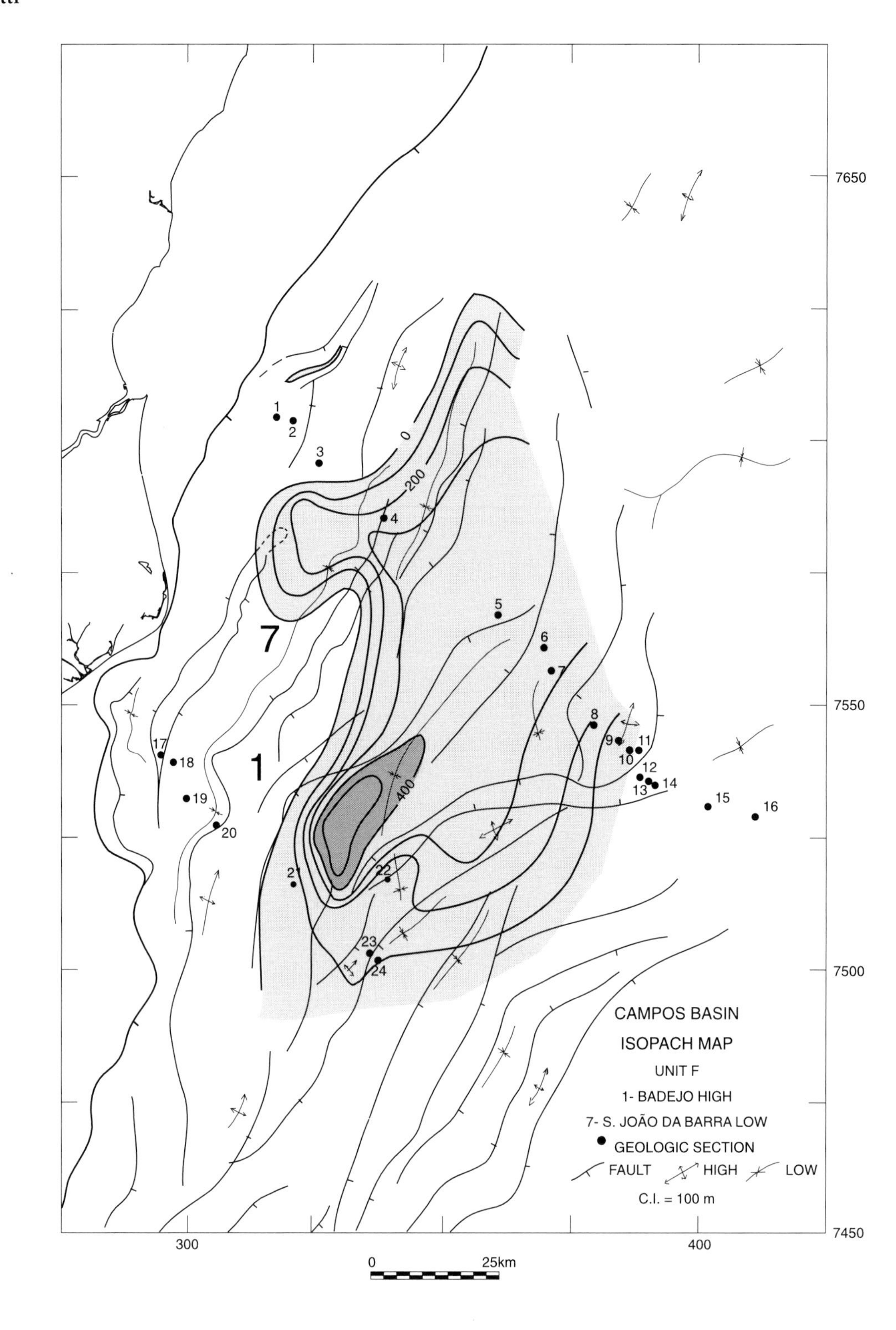

Figure 10—Isopach map of rift stage unit F of the Lagoa Feia Formation, Campos Basin, Brazil.

Unit G

Unit G is related to NRT-010 biozone, which corresponds to the upper Jiquiá local stage (Aptian) and is associated with the Coqueiros Member of the Lagoa Feia Formation (Figures 2–4, 11). This unit has the most restricted extent as a result of reduced tectonic activity, except in the small eastern region near Badejo High with movements similar to unit F. The upper contact in the western region is marked by an unconformity, and the upper part of unit G has lower gamma-ray values than overlying unit H.

Apparently, the lake suffered a major regression in concert with more arid climate. The Badejo High continued to be an exposed source of clastic sediment with coquinas forming the dominant sediment type with interbedded shales. These are more

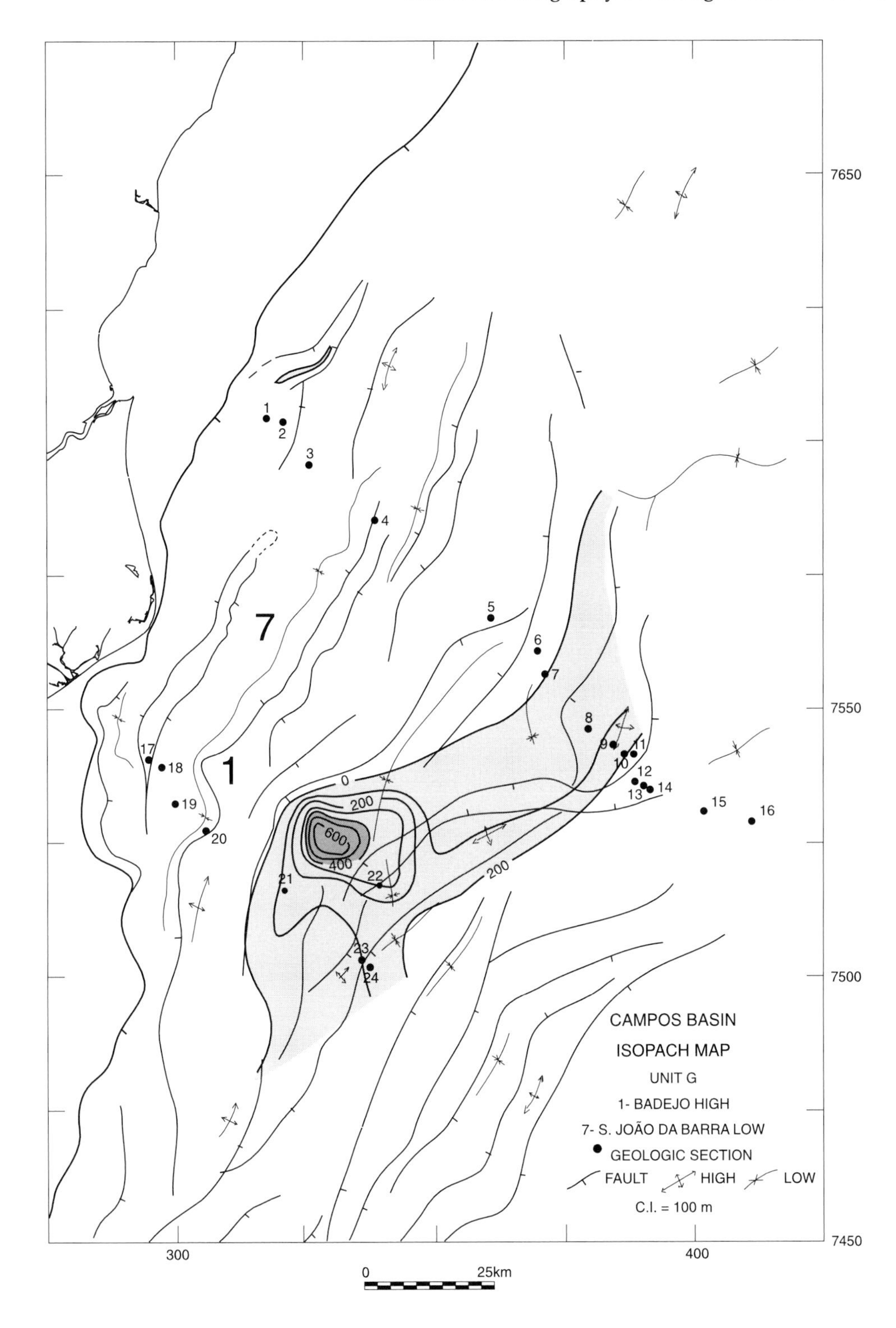

Figure 11—Isopach map of rift stage unit G of the Lagoa Feia Formation, Campos Basin, Brazil.

common in the lower parts of the section. Conglomerates and sandstones deposited in a fan-delta depositional system are more restricted to proximal areas.

Unit G represents the last phase of nonmarine rift deposition. After the deposition of unit G, deposition encompassed broader areas, although the Badejo High continued to shed clastics from its central-southern segment. Local tectonic movements appear to have decreased as the basin broadened.

Unit H

Unit H is related to NRT-011 biozone, which corresponds to the lower Alagoas local stage (Aptian) and is associated with the Itabapoana (conglomerates and sandstones), Gargaú (shales and marls), and Macabú (calcilutites) Members of the Lagoa Feia Formation (Figures 2–4, 12). Unit H is separated by an unconformity from the overlying unit I;

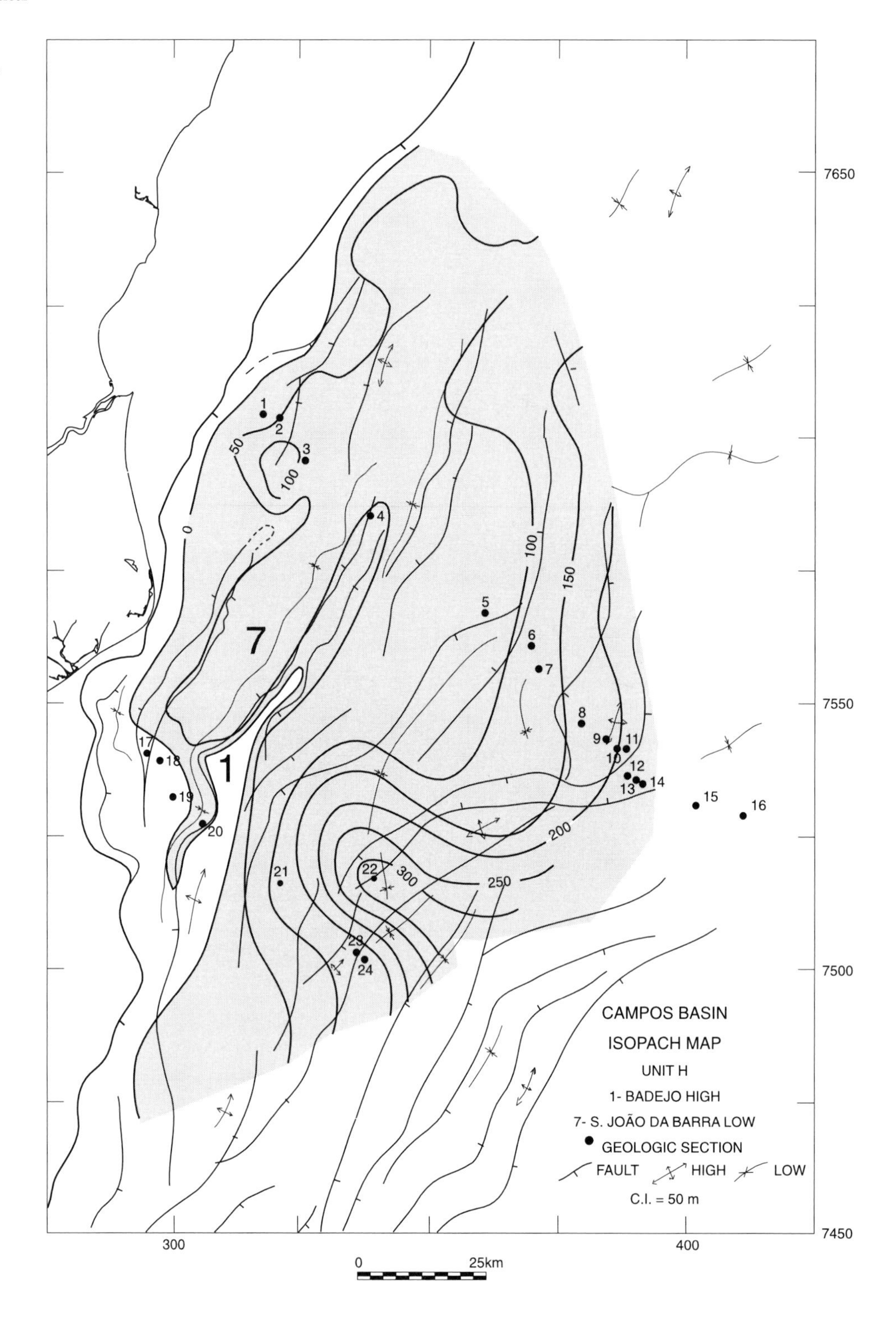

Figure 12—Isopach map of transitional stage unit H of the Lagoa Feia Formation, Campos Basin, Brazil.

together, they display characteristically high gamma-ray values. Detection of this log pattern was important in areas where seismic and biostratigraphic data lack resolution. The unit H gamma-ray values are generally slightly lower than those of unit I.

Differential subsidence on the eastern side of the Campos fault limited the extent of unit H to the west, but tectonism appeared to be uniformly distributed across the basin. The Campos Basin began tilting toward the east, as a precursor to the South Atlantic opening; the Badejo High remained a geomorphologic feature. Thickness of sedimentary sequences increases gradually toward the east, and for the first time, areas to the west of the Campos fault become important sources of sediment.

Sediments are coarse-grained in the western area, characterized by conglomerates and sandstones grading to shales and calcilutites in the central and eastern areas of the basin. Unit H lacks the coquinas

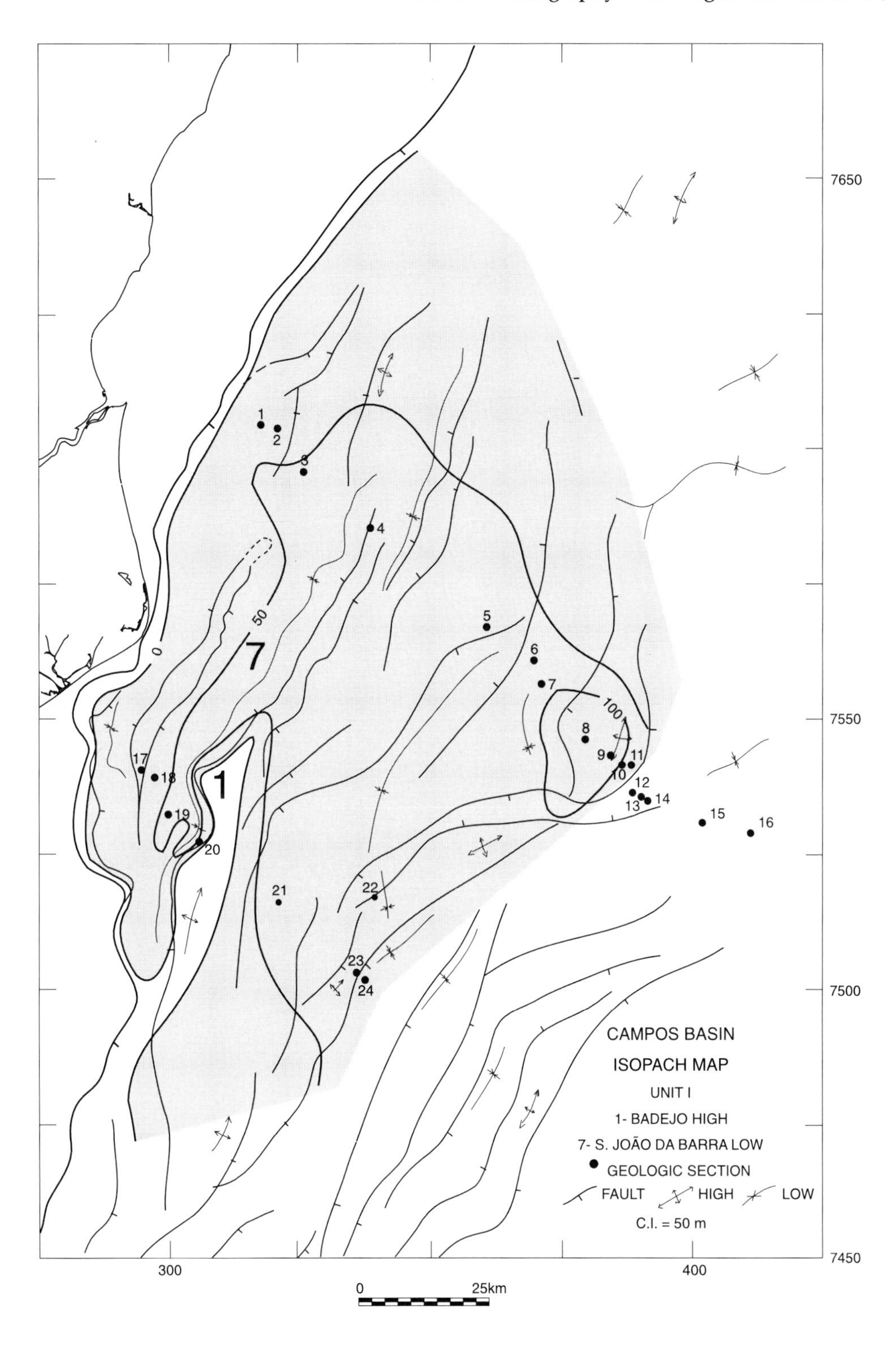

Figure 13—Isopach map of transitional stage unit I of the Lagoa Feia Formation, Campos Basin, Brazil.

common to older units. Climate was likely arid and the transition toward marine conditions began.

Unit I

Unit I is related to NRT-011 biozone, which corresponds to the lower Alagoas local stage (Aptian) and is associated with the Itabapoana, Gargaú, and Macabu Members of the Lagoa Feia Formation (Figures 2–4, 13). Unit I is stratigraphically continuous with the overlying unit J, but facies change from coarse siliciclastics (high gamma-ray log values) into evaporites (low gamma-ray log values).

Unit I and unit H display similar tectonic and distribution patterns. Clastic sediment derived from external areas of the basin to the west of Campos fault. Some small areas along the Badejo High remained exposed. The lithofacies is characterized by coarse siliciclastics in the proximal and medial areas of the basin, passing into finer siliciclastics in deeper regions. Climate became even drier, as a prelude to evaporite deposition.

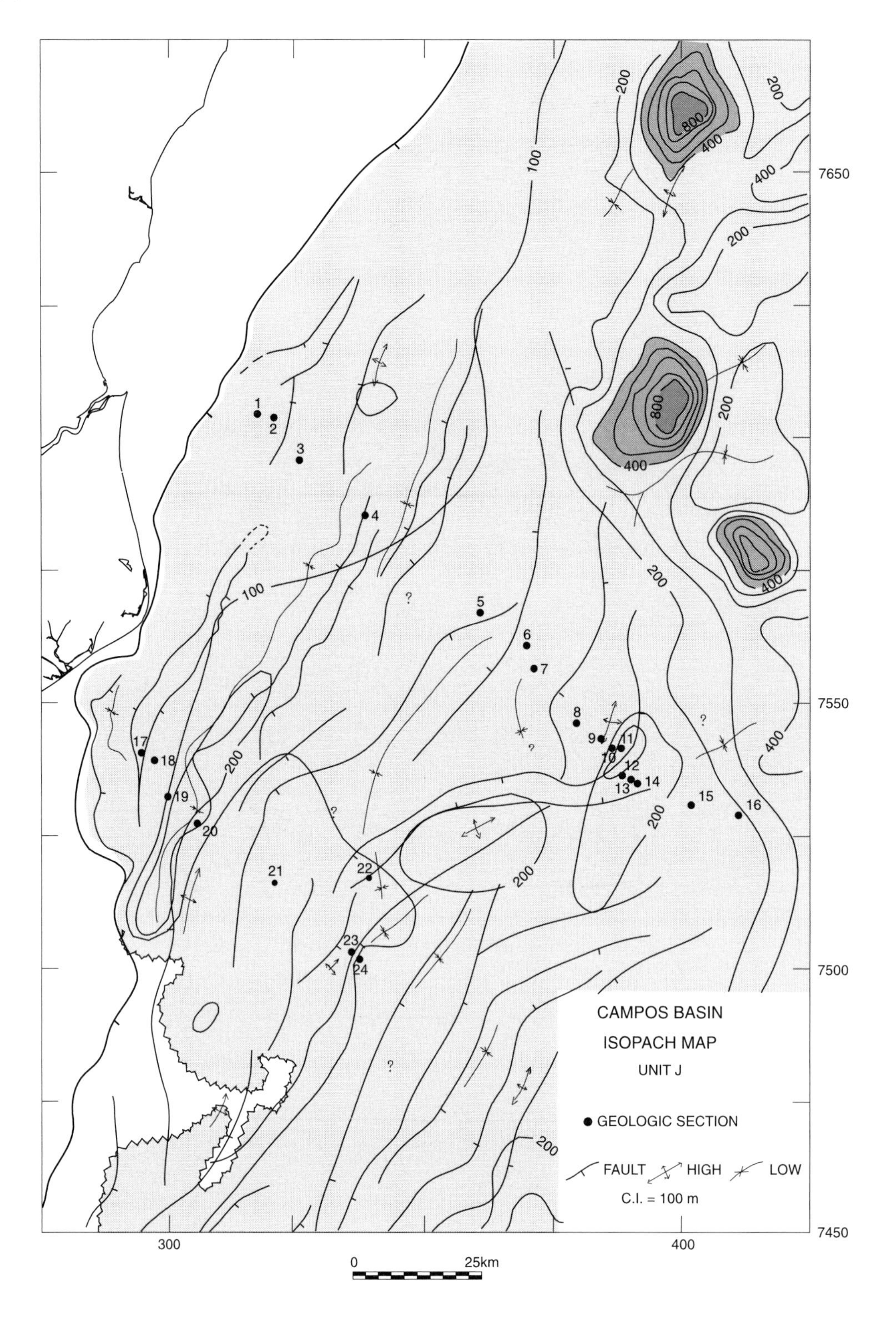

Figure 14—Isopach map of transitional stage unit J of the Lagoa Feia Formation, Campos Basin, Brazil.

Unit J

Unit J is related to NRT-011 biozone, which corresponds to the upper Alagoas local stage (Aptian and lower Albian) and is associated with the Retiro Member of the Lagoa Feia Formation (Figures 2–4, 14).

Unit J represents the marine transgression in concert with the initial stage of a marine South Atlantic Basin. Shallow marine waters flooded the Campos Basin area, restricted by the Campos fault along the western margin. Unit J thickens towards the east due to a higher subsidence with overprints from later salt flowage toward the east as tilting continued. For the first time, the Badejo High was covered by sediments.

A likely unconformity along the lower contact is not distinguished, although an abrupt change occurs from conglomerates to evaporite facies, mainly

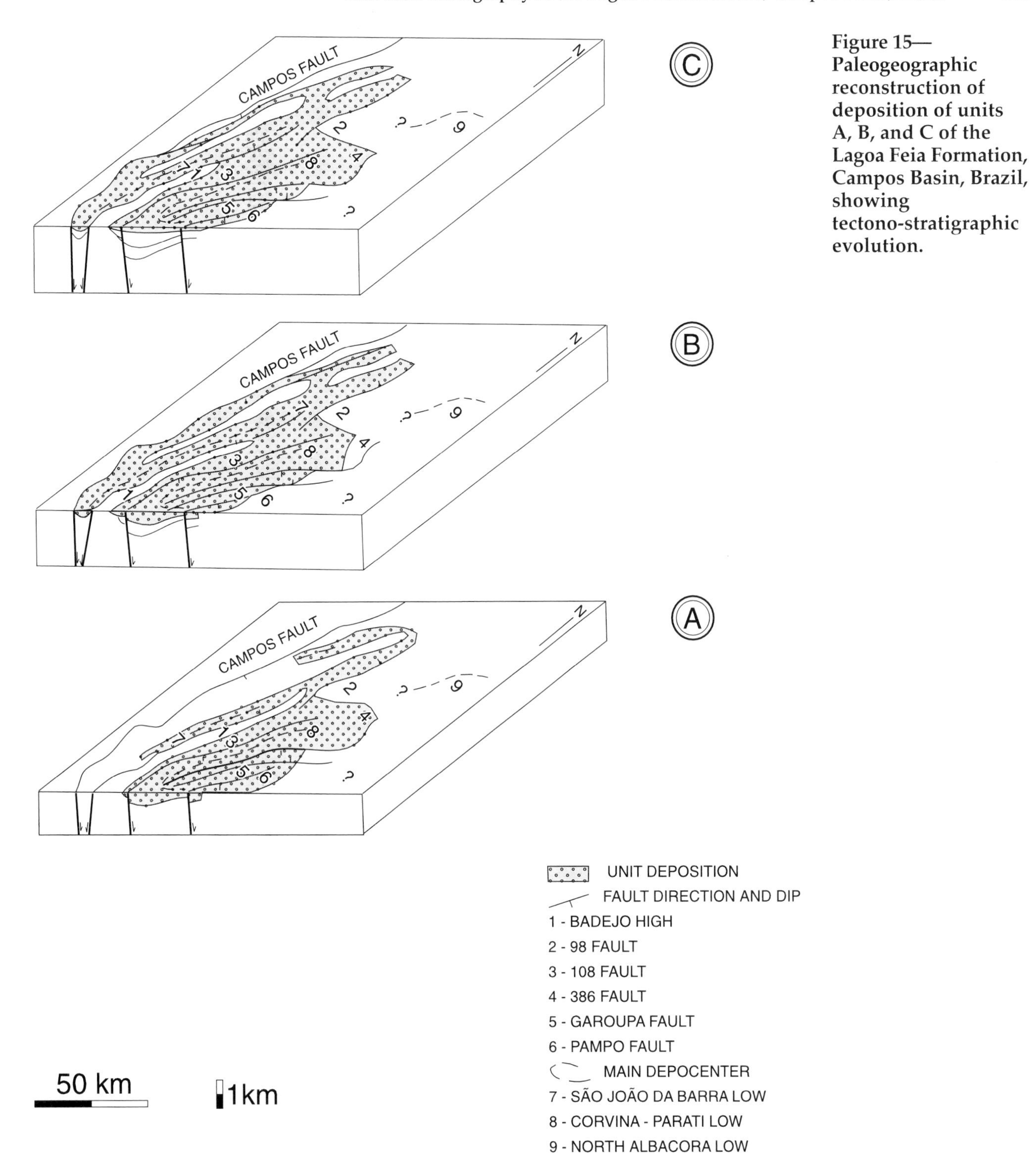

Figure 15—Paleogeographic reconstruction of deposition of units A, B, and C of the Lagoa Feia Formation, Campos Basin, Brazil, showing tectono-stratigraphic evolution.

halite. Halite forms the dominant facies of unit J. Anhydrite layers, if present at the base of the evaporites, are usually very thin. More unstable salts (sylvite and carnalite) have been reported as rare occurrences within the Campos Basin. A gradational facies change is observed for the upper transition with anhydrite, becoming replaced by low-energy limestones, but almost entirely eroded along a lower Albian unconformity.

DISCUSSION AND SUMMARY

This paper provides a brief and systematic summary of the main features and interpretations for each of the major lithostratigraphic and acoustic units recognized on a relatively extensive grid of logging results from the Campos Basin, Brazil. Chronologic correlations are derived from paleontologic and organic geochemical data. Space limitations do not

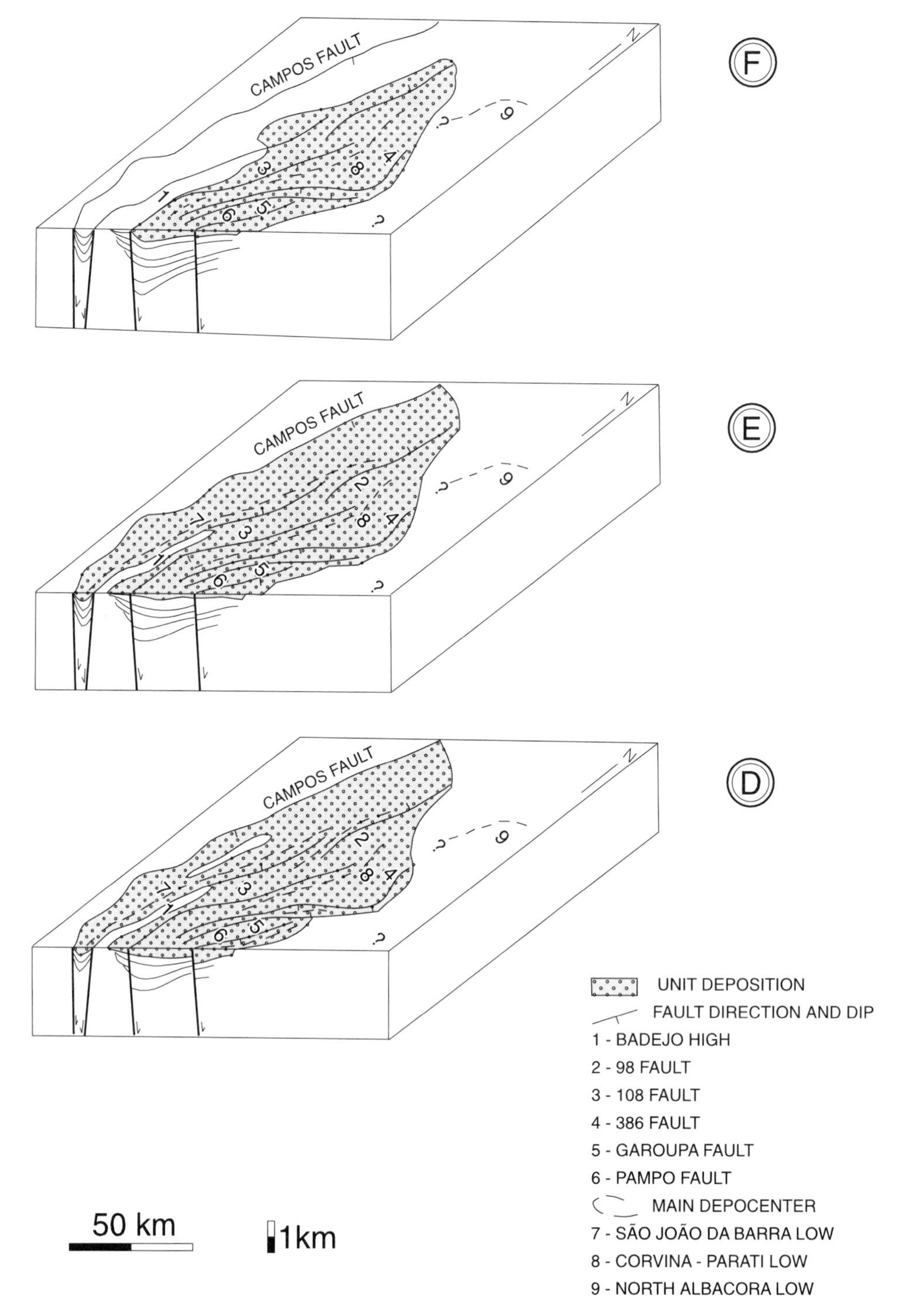

Figure 16—Paleogeographic reconstruction of deposition of units D, E, and F of the Lagoa Feia Formation, Campos Basin, Brazil, showing tectono-stratigraphic evolution.

allow adequate referencing to all studies and internal reports.

The unit boundaries correspond mainly to unconformities with marker horizons. The application of sequence stratigraphic concepts to these rift-lacustrine units is difficult for a number of reasons. Lake basin water levels are subject to rapid fluctuations with concomitant effects on prograding clastic and basinal sediment tracts. Seismic resolution is limited because the synrift and transitional sediments were covered by a thick marine sequence from the Albian to the Pleistocene.

Most depocenters/compartments were tectonically controlled. Fan-delta systems operated well during higher lake levels. Tectonics, however, was overprinted by higher frequency changes in hydrology that resulted in very rapid facies changes and the formation and interfingering of lacustrine coquina limestones and thick interbedded coarse-grained siliciclastic sequences.

Finally, Figures 15, 16, and 17 illustrate a scenario for rift evolution in the basin, representing expanding and

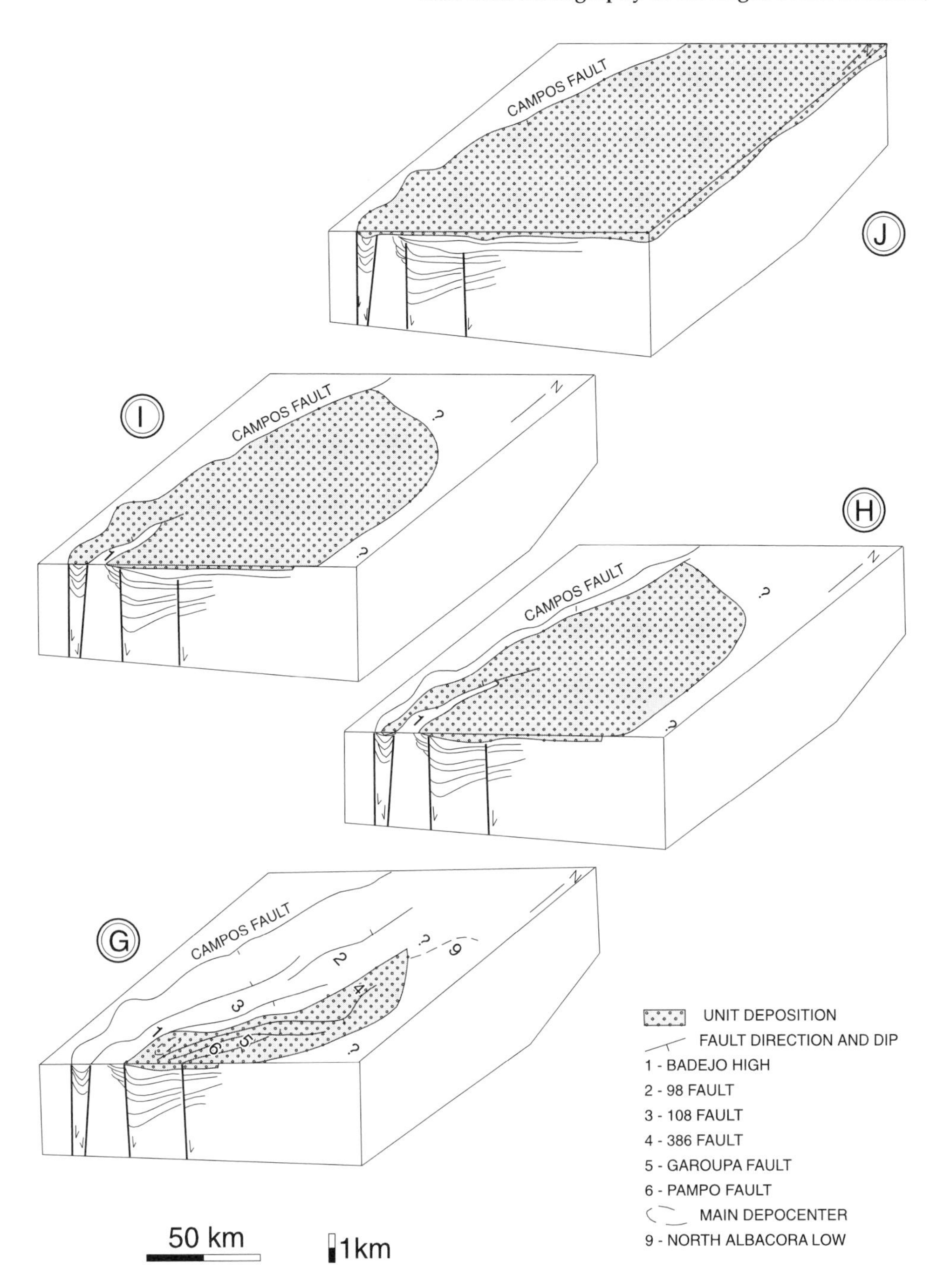

Figure 17—Paleogeographic reconstruction of deposition of units G, H, I, and J of the Lagoa Feia Formation, Campos Basin, Brazil, showing tectono-stratigraphic evolution.

regressive cyclic patterns of the Campos lacustrine sediment systems. These deposits formed the bulk of the Campos Basin fill during rift and transitional stages.

REFERENCES CITED

Aquino, G. S. de, and M. da C. Lana, 1990, Exploração na bacia de Sergipe-Alagoas: O Estado da Arte, Boletim Geociências Petrobras, v. 4 (1), p. 75–84.

Baumgarten, C. S., A. J. C. Dultra, M. da S. Scuta, M. V. L. de Figueiredo, and M. de F. P. Sequeira, 1988, Coquinas da Formação Lagoa Feia, bacia de Campos: Evolução da Geologia de Desenvolvimento. Boletim Geociências Petrobras, v. 2(1), p. 13–26.

Bertani, R. T., and A. V. Carozzi, 1985, Lagoa Feia Formation (Lower Cretaceous), Campos Basin, offshore Brazil—rift valley stage lacustrine carbonate reservoirs, I: Journal of Petroleum Geology, v. 8 (1), p. 37–58.

Caixeta, J. M., G. V. Bueno, L. P. Magnavita, and F. J. Feijó, 1994, Bacias do Recôncavo, Tucano e Jatobá, Boletim Geociências Petrobras, v. 2 (1), p. 163–174.

Carvalho, M. D., J. L. Dias, U. M. Praça, A. C. da S. Jr. Telles, P. Horschutz, M. Hanashiro, M. da S. Scuta, A. D. Sayd, L. C. S. de Freitas, A. S. C. Barbosa, and M. H. Hessel, 1993, Coquina da Formação Lagoa Feia da bacia de Campos, Estudos Sedimentológicos e Diagenéticos na Caracterização da Qualidade do Reservatório, Petrobras/Cenpes/Depex, Internal Report, 151 p.

Castro, J. C. de, 1992, Turbiditos lacustres rasos e profundos na fase rift de bacias marginais Brasileiras, Boletim Geociências Petrobras, v. 6 (1/2), p. 89–96.
Cupertino, J. A., 1990, Estágio Exploratório das Bacias do Tucano Central, Norte e Jatobá, Boletim Geociências Petrobras, v. 4 (1), p. 45–54.
Dias, J. L., J. C. Vieira, A. J. Catto, A. Q. de Oliveira, W. Guazelli, L. A. F. Trindade, R. O. Kowsmann, C. T. Kiang, U. T. Mello, A. M. P. Mizusaki, and J. A. Moura, 1987, Estudo regional da Formação Lagoa Feia, Petrobras/Depex/Cenpes, Internal Report, 152 p.
Dias, J. L., 1994, Stratigraphy, sedimentology and volcanism of the Lower Cretaceous phase along eastern Brazilian continental margin, 14th International Congress of the IAS, expanded Abstracts, p. 1–2 (theme 5). Recife, Brazil.
Dias, J. L., A. R. E. Sad, R. L. Fontana, and F. J. Feijó, 1994, Bacia de Pelotas, Boletim Geociências Petrobras, v. 8 (1), p. 235–246.
Estrella, G. O., M. R. Mello, P. C. Gaglianone, R. L. M. Azevedo, K. Tsubone, E. Rossetti, J. Concha, and I. M. R. A. Bruning, 1984, The Espírito Santo Basin (Brazil) source rock characterization and petroleum habitat, AAPG Memoir 35, p. 252–271.
Feijó, F. J., 1994a, Bacia de Pernambuco-Paraíba, Boletim Geociências Petrobras, v. 8 (1), p. 143–148.
Feijó, F. J., 1994b, Bacias de Sergipe e Alagoas, Boletim Geociências Petrobras, v. 8 (1), p. 149–162.
Guardado, L. R., L. A. P. Gamboa, and C. F. Lucchesi, 1989, Petroleum geology of the Campos Basin, a model for a producing Atlantic type basin, AAPG Memoir, v. 48, p. 3–80.
Horschutz, P. M. C., and M. S. Scuta, 1992, Facies-perfis e mapeamento de qualidade do reservatório de coquinas da Formação Lagoa Feia do Campo de Pampo, Boletim Geociências Petrobras, v. 6 (1/2), p. 45–58.
Mato, L. F., J. M. Caixeta, and M. R. da C. Magalhães, 1992, Padrões de sedimentação na passagem da Formação Marfim para a Formação Pojuca (Andar Rio da Serra/Andar Aratu) e significado estratigráfico do Marco 15, Cretáceo Inferior, Bacia do Recôncavo, Bahia, Boletim Geociências Petrobras, v. 6 (1/2), p. 59–72.
Mello, M. R., and J. R. Maxwell, 1990, Organic, geochemical, and biological marker characterization of source-rocks and oils derived from lacustrine environments in the Brazilian continental margin, *in* B. J. Katz, ed., Lacustrine Basin Exploration—Case Studies and Modern Analogs, AAPG Memoir v. 50, p. 77–98.
Mitchum, R. M., Jr., P. R. Vail, and S. Thompson III, 1977, Seismic stratigraphy and global changes of sea level, part 2: the depositional sequence as a basic unit for stratigraphic analysis, *in* C. E. Payton, ed., Seismic Stratigraphy—Applications to Hydrocarbon Exploration, AAPG Memoir, v. 26, p. 53–62.
Ojeda, H. A. O., 1982, Structural framework, stratigraphy and evolution of Brazilian marginal basins, AAPG Bulletin, v. 66, p. 732–749.
Pereira, M. J., and F. J. Feijó, 1994, Bacia de Santos, Boletim Geociências Petrobrás, v. 8 (1), p. 219–234.
Ponte, E. C., and H. E. Asmus, 1978, Geological framework of the Brazilian continental margin, Geologische Rundschau, v. 68, p. 201–235.
Rangel, H. D., G. A. Corréa, and D. L. Bisol, 1987, Evolução geológica e aspectos exploratórios das áreas norte e central da bacia de Campos, Petrobras/Depex, Internal Report, 57 p.
Rangel, H. D., A. L. Soldan, C. E. da S. Pontes, R. S. de Souza, L. M. Arienti, R. P. Bedregal, A. Bender, N. C. Azambuja, J. C. Ferreira, and R. Jahnert, 1993a, Habitat do petróleo da Porção de água Rasa na Região Central da Bacia de Campos, Petrobras/Depex/Cenpes, Internal Report, 58 p.
Rangel, H. D., and A. Z. N. de Barros, 1993b, Estratigrafia e evolução estrutural da área Sul (adjacente ao Alto de Cabo Frio) da bacia de Campos, Annals of the 3rd Southeastern Geological Symposium, Rio de Janeiro, p. 57–63.
Rangel, H. D., C. E. da S. Pontes, A. L. Soldan, and A. C. da Silva Telles Jr., 1994a, Tectonic control on sedimentation during rifting stage: example from Campos Basin, southeastern Brazil, Expanded Abstracts, 14th International Congress of the IAS, p. 14–15 (theme 5), Recife, Brazil.
Rangel, H. D., F. A. L. Martins, F. R. Esteves, and F. J. Feijó, 1994b, Bacia de Campos, Boletim Geociências Petrobras v. 8 (1), p. 203–218.
Rehim, H. A. A. A., A. M. Pimentel, M. D. Carvalho, and M. Monteiro, 1986, Talco e estevensita na Formação Lagoa Feia da bacia de Campos—possíveis implicações no ambiente deposicional, Annals XXXIV Brazilian Geological Congress, v. 1, p. 416–422.
Santos, C. F., and J. A. E. Braga, 1990, O estado da arte da bacia do Recôncavo, Boletim Geociências Petrobras, v. 4 (1), p. 35–44.
Santos, C. F., R. C. Gontijo, M. B. Araújo, and F. J. Feijó, 1994, Bacias de Cumuruxatiba e Jequitinhonha, Boletim Geociências Petrobras, v. 8 (1), p. 185–190.
Soldan, A. L., H. D. Rangel, A. C. da Silva Telles Jr., S. M. B. de Grande, and G. I. Henz, 1995, Kinetic variability within the Lagoa Feia Formation, palaeoenvironmental influences, 17th International Meeting of Organic Geochemistry, *in* J. O. Grimalt and C. Dorronsoro, eds., Donostia-San Sebastián, the Basque Country, Spain, p. 128–130.
Teixeira Netto, A. S., J. R. Wanderley Filho, and F. J. Feijó, 1994, Bacias de Jacuípe, Camamu e Almada, Boletim Geociências Petrobras, v. 8 (1), p. 173–184.
Telles, A. C., da Silva Jr., 1992, Novo zoneamento da sequência das coquinas da Fm. Lagoa Feia (Neojiquiá da bacia de Campos) com base em ostracodes—aspectos evolutivos, Annals 37 Brazilian Geology Congress, São Paulo, v. 2, p. 489,
Van der Ven, P. H., C. Cainelli, and G. J. F. Fernandes, 1989, Bacia de Sergipe-Alagoas: geologia e exploração, Boletim Geociências Petrobras, v. 3 (4), p. 307–320.
Van Wagoner, J. C., H. W. Posamentier, R. M. Mitchum, P. R. Vail, J. F. Sarg, T. S. Loutit, and J. Hardenbol, 1988, An overview of the fundamentals of sequence stratigraphy and key definitions, SEPM Special Publication, v.42, p. 39–45.
Vieira, R. A. B., M. P. Mendes, P. E. Vieira, L. A. R. Costa, C. V. Tagliari, L. A. P. Bacelar, and F. J. Feijó, 1994, Bacias do Espírito Santo e Mucuri, v. 8 (1), p. 191–202.

Carvalho, M. D., U. M. Praça, A. C. da Silva-Telles, R. J. Jahnert, and J. L. Dias, 2000, Bioclastic carbonate lacustrine facies molds in the Campos Basin (Lower Cretaceous), Brazil, *in* E. H. Gierlowski-Kordesch and K. R. Kelts, eds., Lake basins through space and time: AAPG Studies in Geology 46, p. 245–256.

Chapter 19

Bioclastic Carbonate Lacustrine Facies Models in the Campos Basin (Lower Cretaceous), Brazil

Maria Dolores de Carvalho
Uyara Mundim Praça
Augusto Carlos da Silva-Telles Jr.
Ricardo Jorge Jahnert
Jeferson Luiz Dias
Petrobras/Cenpes/Diger, Cidade Universitária
Rio de Janeiro, Brazil

INTRODUCTION

The Campos Basin is one of the sedimentary basins on the Brazilian continental margin, located offshore of the State of Rio de Janeiro (Figure 1). The basin has an area of 100,000 km^2, with its eastern limit in the ocean at 3000 m water depth with only 500 km^2 onshore. Its structural limits are the Vitória structural high to the north, Cabo Frio structural high to the south, and Campos fault (hinge line) to the west (bounding only on Cretaceous rocks). To the east, the basin theoretically extends to great salt diapirs (of Permian age) presently located between 3500 and 4000 m of water depth.

The Campos Basin originated in the Lower Cretaceous break-up of Gondwana, as part of a rift valley system (Guardado et al., 1988). Stratigraphy of the basin is subdivided, from bottom to top, into the following formations: Cabiúnas, Lagoa Feia, Macaé, Campos and Emboré. This lithostratigraphic sequence reflects its tectonic evolution and is more than 4 km thick. The Lagoa Feia Formation lies unconformably over the volcanic rocks of the Cabiúnas Formation (Figure 2), dated by K-Ar methods as between 125 and 130 m.y. The Lagoa Feia Formation is the oldest sedimentary sequence of the basin and was deposited as a rift-related alluvial-lacustrine complex. At the upper part of the formation, there are thick molluscan shell deposits, known as the Coquinas sequence, which are important oil reservoirs in the Campos Basin, and are the object of the present study.

This study was based on the analysis of core and cuttings of over 100 wells drilled in the proximal portion of the basin. Available data also included geophysical logs and composite logs of all wells. Analysis resulted in the characterization of different paleoenvironments and construction of a depositional evolutionary history for the Coquinas sequence.

GEOLOGIC SETTING AND LOWER CRETACEOUS STRATIGRAPHY

The alluvial-lacustrine sediments of the Lagoa Feia Formation were deposited during the Early Cretaceous (Eobarremian/Eoaptian, Aratu to Alagoas local stages) in a complex system of saline rift lakes. The Lagoa Feia Formation was subdivided by Dias et al. (1988), from bottom to top, into four depositional sequences bounded by stratigraphic markers or unconformities: (1) Clastic basal sequence, (2) Talc-Stevensitic sequence, (3) Coquinas sequence, and (4) Clastic-Evaporitic sequence (Figures 2 and 3).

The Coquinas sequence, object of this study, represents the most important sequence of the Lagoa Feia Formation, not only because it hosts four important oil field reservoirs, but also as the unique source rocks for oil in the Campos Basin. It represents a complex lake sequence, with substantial limestone deposits (coquinas), which interfinger with fluvial-alluvial deposits. Paleoclimate is interpreted as arid, even during times of high lake level. The Coquinas sequence has as its lower limit, log marker LF-35, and as its upper limit, the pre-Alagoas unconformity. Thickness of the sequence ranges from zero at some paleostructural highs to 2400 m in some paleostructural lows of the rift basin.

Main structural features of the rift phase in the Campos Basin are the Campos fault (hinge line), which is a structural northeast alignment and limited Cretaceous sedimentation to the west, and a series of

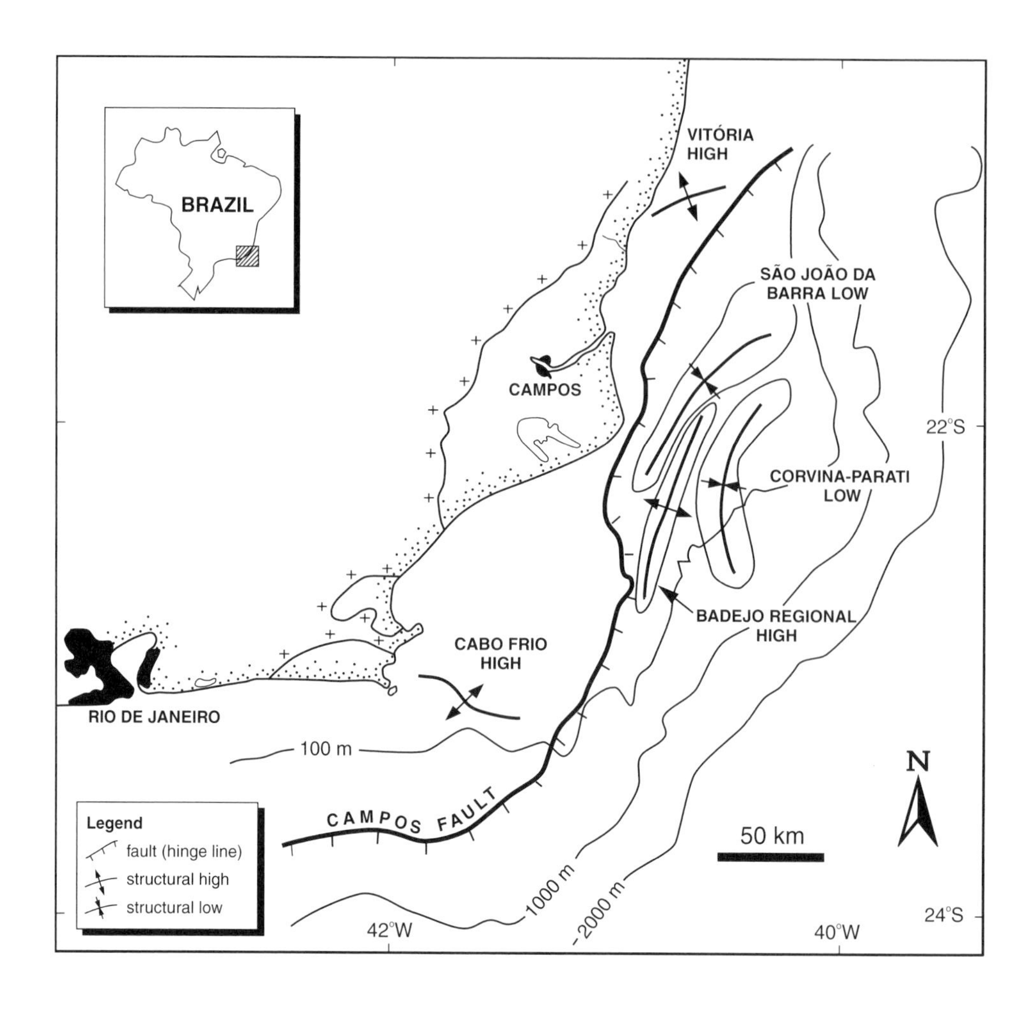

Figure 1—Index map showing the main structural elements of the Campos Basin, Brazil.

half-grabens, common in the northern and central regions of the basin and limited by synthetic and antithetic faults. The Badejo regional high (BRH) is the most important high developed during the rift phase of the basin. It shows a northeasternly direction and structurally dips to the northeast. Basinward, there are other highs parallel to the BRH. Between these highs, the structural lows represent the main depositional sites of the Lagoa Feia, the most important of which are the São João da Barra low, located to the west of the BRH, and the Corvina-Parati low, to the east of the BRH (Figure 1).

Based on the study of the ostracode paleofauna of the Coquinas sequence (middle to upper Jiquiá local stage), five biostratigraphic intervals were characterized (Silva-Telles Jr., 1992): (1) at the base, the *Salvadoriella ? pusilla* subzone (NRT-009.3), defined by the extinction of *Salvadoriella ? pusilla* Krömmelbein and Weber (1971) and characterized by *Petrobrasia diversicostata* Krömmelbein (1965); (2) *Reconcavona ? bateke* Grosdidier (1967) zone (OS-1000), which can be subdivided into two subzones with one at the base, *Reconcavona ? retrosculpturata* n.n. (OS-1010) and one at the top, *Hourcqia africana africana* Krömmelbein (1965) (OS-1020); (3) *Limnocypridea ? subquadrata* n.n. zone (OS-1100), characterized by a monospecific fauna containing noded and smoothed morphotypes; and (4) at the top of the sequence, the *Limnocythere ? troelseni* Krömmelbein and Weber (1971) zone (NRT-010), also characterized by a monospecific fauna.

As a result of the correlation of the upper limits of the five biozones with gamma-ray and resistivity log markers, five chronozones were defined (Figure 3). These are, from bottom to top:

- C009.3—chronozone of the NRT-009.3 biozone
- C1010—chronozone of the OS-1010 biozone
- C1020—chronozone of the OS-1020 biozone
- C1100—chronozone of the OS-1100 biozone
- C010—chronozone of the NRT-010 biozone

FACIES ASSOCIATIONS

Two main series of depositional facies associations of the Coquinas sequence, which were regarded as basic architectural elements of the depositional paleoenvironments, were assigned to the alluvial-lacustrine complex. One was the association dominated by siliciclastic facies, consisting of polymictic conglomerates, lithic to feldspathic sandstones, siltstones, and red shales. These facies represent mainly alluvial fan to sandy and muddy alluvial plain deposits. The other facie was the lacustrine facies association, consisting dominantly of coquinas of bivalves and coquinas of gastropods or ostracodes, locally interfingering with siliciclastic facies. Those coquina facies were deposited between the eulittoral and littoral zones of

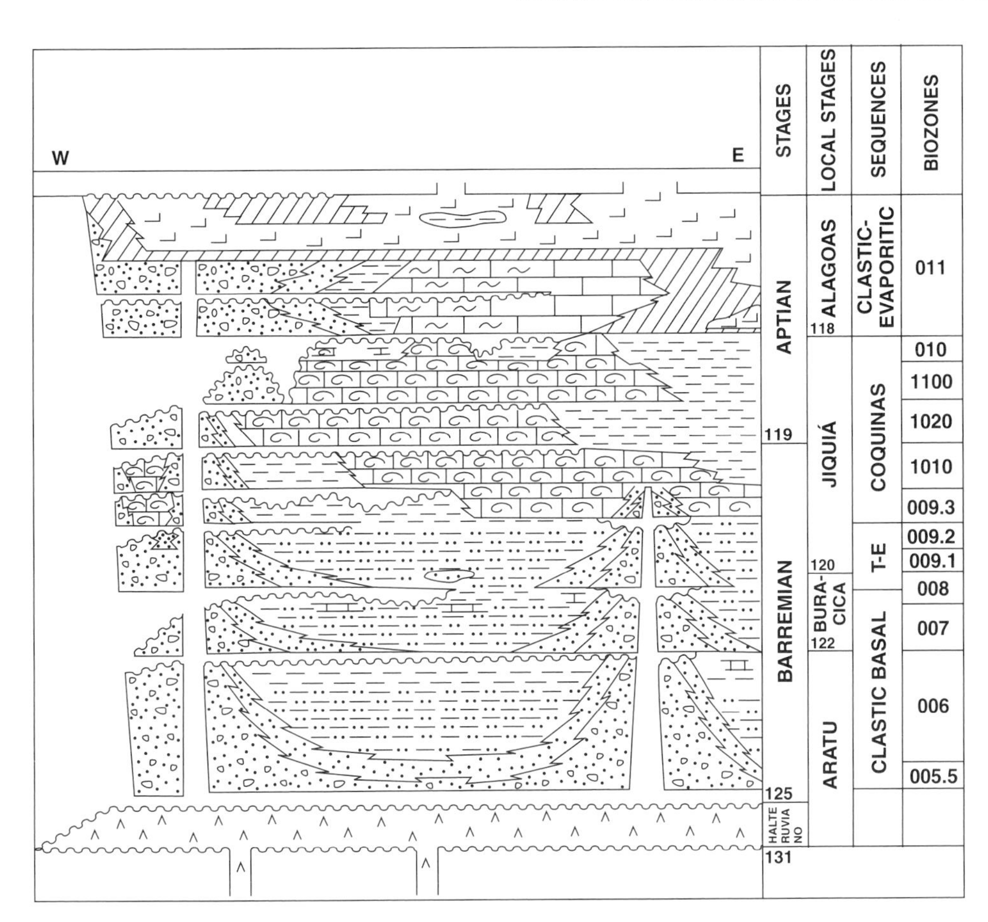

Figure 2—Stratigraphic column of Lagoa Feia Formation of the Campos Basin. Numbers of stages indicate age in millions of years. T-E indicates Talc-Stevensite sequence.

LEGEND
- siltstones
- sandstones
- shales
- conglomerates
- bioclastic limestones
- limestones (stromatolitic ls)
- limestones (micrites)
- halite
- anhydrite
- volcanics

the rift-basin paleolake, rarely at the sub-littoral zone (after Tucker and Wright, 1990).

Lacustrine facies patterns of the Coquinas sequence are vertically and laterally complex, especially at the marginal lacustrine portions. Facies associations of the marginal portions suggest a high frequency of relative lake level fluctuations. As a result of such fluctuations, sedimentary sequences show signs of modifications caused by subaerial exposure and pedogenesis. These changing lake levels suggest that climatic variations might have played an important role in the evolution of the depositional paleoenvironments of the Coquinas sequence.

Seven main facies associations have been defined within the Coquinas sequence and interpreted as the following depositional paleoenvironments: (1) bioclastic sandy beaches, (2) bioclastic calcarenite beaches, (3) marginal, (4) bioclastic bars, (5) bioclastic sheets/bar fringes, (6) bioaccumulation banks, and (7) deep lacustrine. The characteristics of each paleoenvironment are described as follows.

Bioclastic sandy beaches (Figure 4) are characterized mainly by the occurrence of units (<10 m thick) of sandy calcirudites with loosely packed bivalves (grain-supported but with 30–50% shells), that exhibit low-angle cross-bedding (sets 30–70 cm thick). These sandy calcirudite units occur associated with layers (<10 m thick) of massive to graded siliciclastic conglomerates, massive to cross-stratified sandstones, and shales. Successions of this facies association form thick deposits (40–500 m). This facies association was deposited at lake margins with siliciclastic sediments that were mixed with shells carried by storm-induced currents.

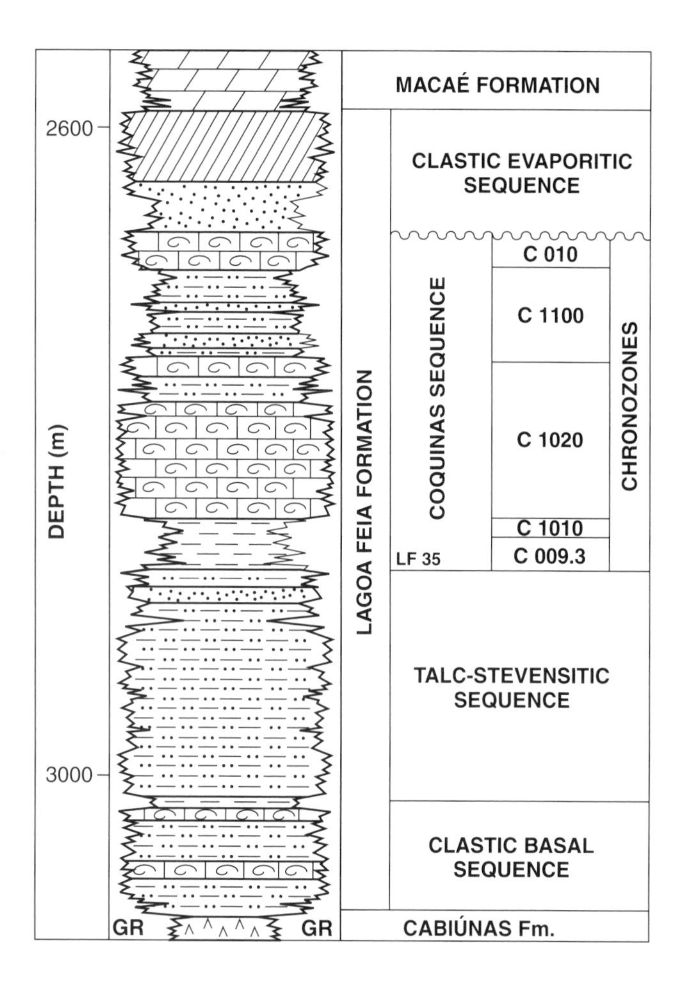

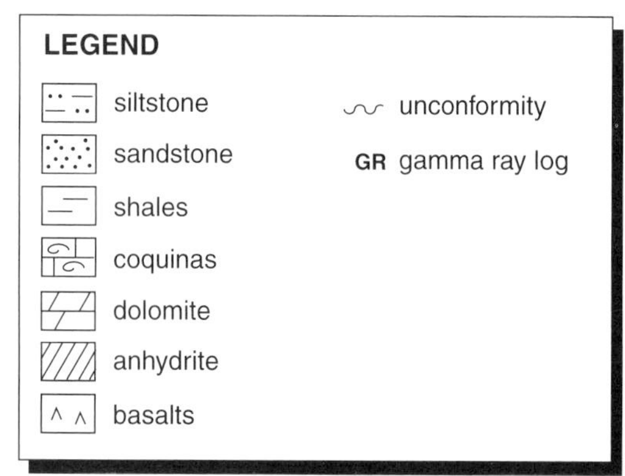

Figure 3—Typical log of the Lagoa Feia Formation showing the four depositional sequences (Clastic Basal, Talc-Stevensitic, Coquinas, and Clastic-Evaporitic) and the five chronozones (C009.3, C1010, C1020, C1100 and C010) of the Coquinas sequence. LF 35 is a stratigraphic marker.

Bioclastic calcarenite beaches (Figure 5) consist of units of bioclastic calcarenites (<10 m thick) associated with siltstone, shale, and calcilutite layers (<5 m thick). The calcarenites, with low-angle cross-bedding (sets of 20–50 cm thick), consist of shell fragments of bivalves or, rarely, gastropods, which commonly exhibit a high degree of abrasion and contain micritic envelopes. Successions of this facies association are 50–150 m thick.

Marginal lake environment (Figure 6) consist of fine siliciclastic or carbonate sediments, which exhibit birds'eyes, mudcracks, and root traces, combined with fine sediments (siliciclastics or ostracodal limestones) containing microripples and lamination. This association (40–100 m thick) is interpreted as deposits of low-gradient to low-energy shallow lacustrine areas.

Bioclastic bars (Figure 7) are characterized by units over 10 m thick of normally packed (grain-supported, with 50–70% shells) to densely packed (grain-supported, with more than 70% shells) calcirudites, associated with thin layers (<5 m thick) of bioclastic calcarenite or calcilutite. No siliciclastic intercalations are present. Successions of this facies association form thick deposits (20–550 m). These units are interpreted to be the results of amalgamation of relatively thin layers (20–100 cm thick) of calcirudites, containing bivalves and rarely gastropods and classified as grainstones, that locally show cross-bedding. These features suggests high-energy and shallow-water paleoconditions. They are associated with paleohighs with steep slopes, and their origin is attributed to storms.

Bioclastic sheets/bar fringes (Figure 8) are characterized by normal to densely packed units of calcirudite less than 10 m thick intercalated with fine siliciclastics or carbonates. Commonly, successions of this facies appear as deposits 100–200 m thick. They represent shell debris of bivalves and rarely gastropods that were spread over the flanks of bars or in areas with gentle slopes and are inferred to be related to storm deposits.

Bioaccumulated banks or bioherms (Figure 9) are comprised of in situ shells from the local community embedded in sandy or muddy deposits. Low-energy and shallow to deep paleoconditions are interpreted. Units are 5 cm–5 m thick.

The deep lacustrine (Figure 10) paleoenvironment consists of intercalations of fine siliciclastic sediments and ostracodal mudstones that are laminated to massive and locally burrowed. Successions of this facies association form thick deposits (100–1000 m thick). These rocks are interpreted as being deposited under low hydraulic energy conditions. The absence of features suggesting subaerial exposure is an important criterion for deep water deposition.

The bivalve limestones, the unique reservoir rocks of the Lagoa Feia Formation, are mainly composed of inarticulate, broken, and reworked shells (Carvalho and Praça, 1994), classified as *Agelasina* cf. *A. plenodonta* Riedel, 1932; *Arcopagella longa* n.sp.; *Camposella rosea* n.gen. et n.sp.; *Desertella acarenata* n.sp.; *Kobayashites brasiliensis* n.sp.; *Remondia (Mediraon) magna* n.sp.; *Sphaerium* cf. *S. ativum* White, 1887; and *Trigonodus camposensis* n.sp. They were found to form 19 calcirudite microfacies, two calcarenite microfacies, and two calcilutite microfacies. The gastropod coquinas, composed of shells of pulmonate gastropods of the *Limneidae* family, were found to form three calcirudite microfacies and one calcarenite microfacies. The

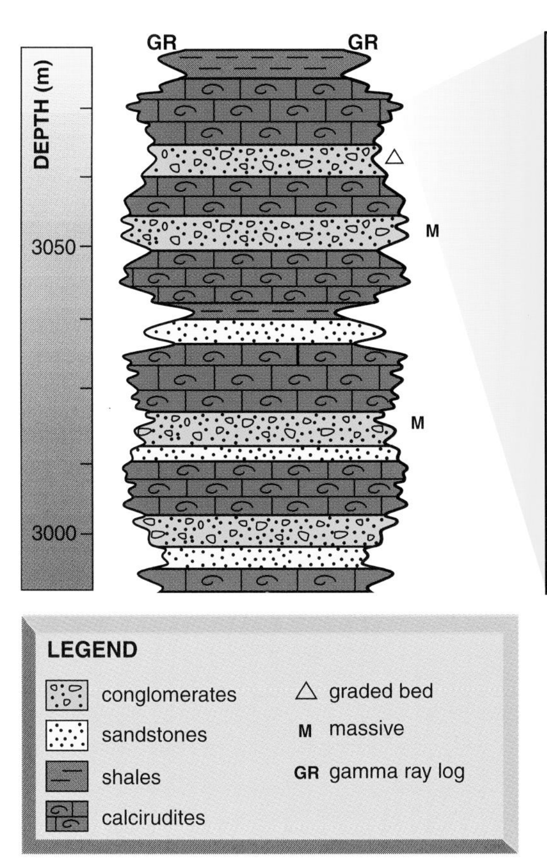

Figure 4—Facies association of the Coquinas sequence of the Lagoa Feia Formation interpreted as bioclastic sandy beach deposits.

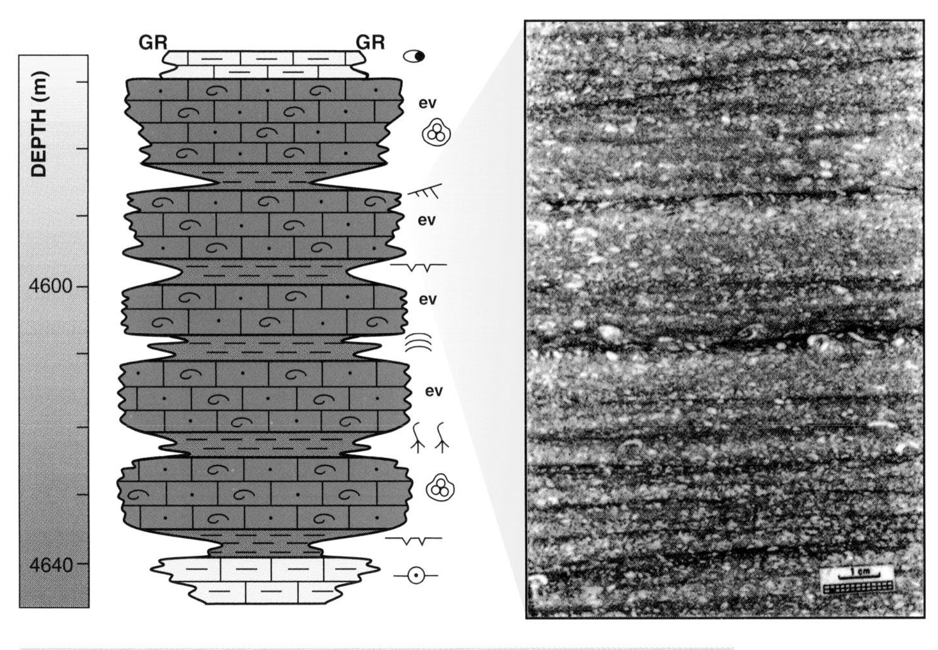

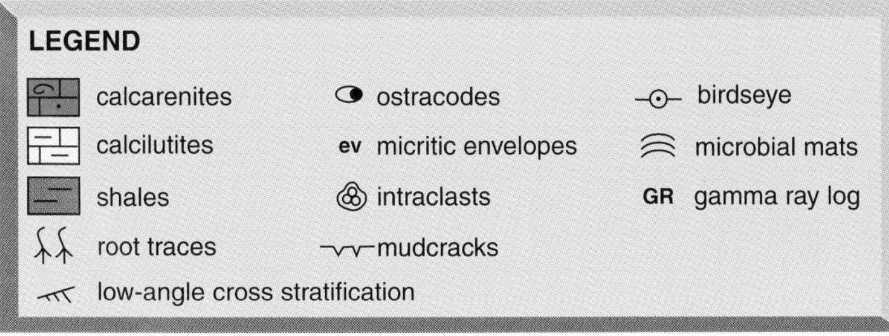

Figure 5—Facies association of the Coquinas sequence of the Lagoa Feia Formation interpreted as bioclastic calcarenite beach deposits.

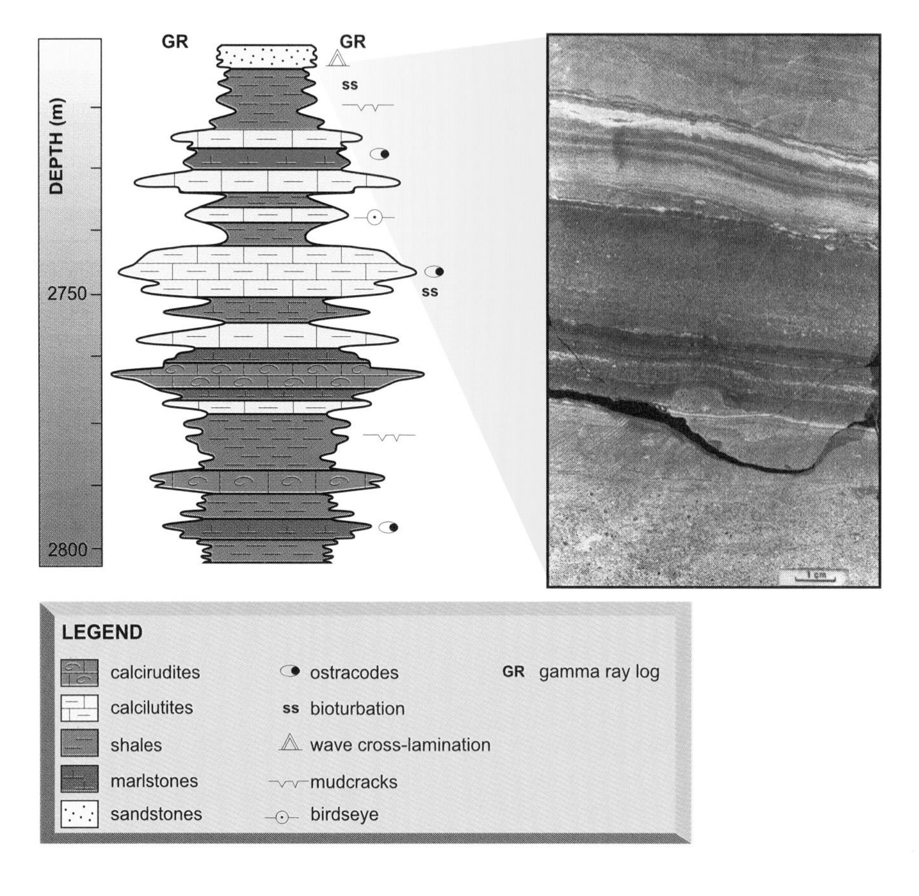

Figure 6—Facies association of the Coquinas sequence of the Lagoa Feia Formation interpreted as a marginal lake environment.

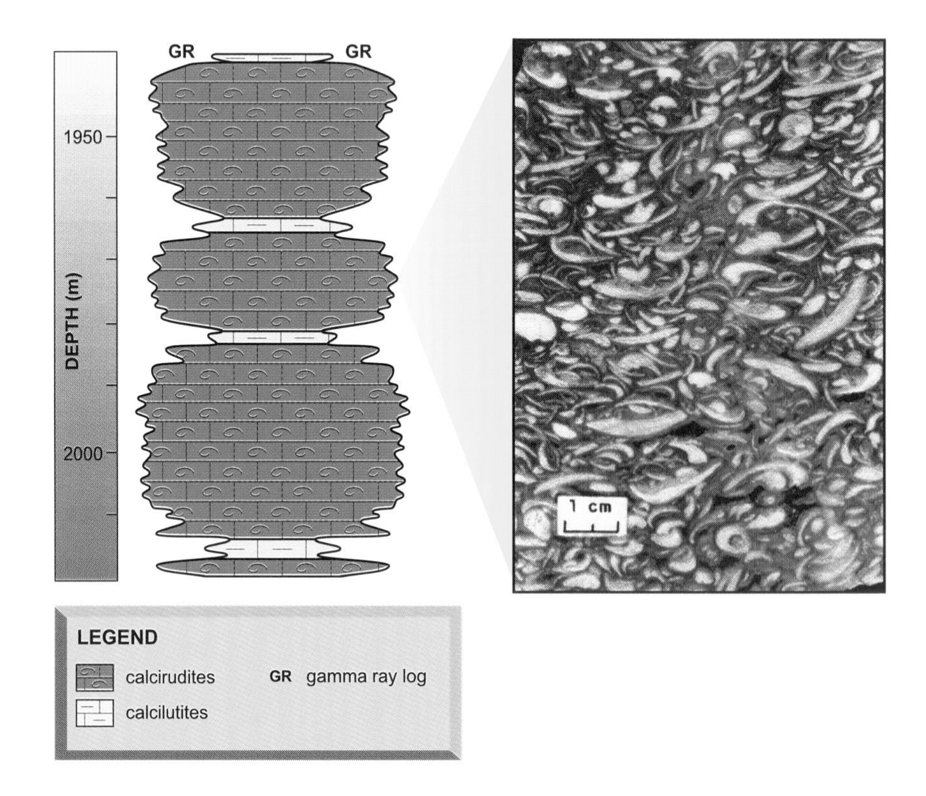

Figure 7—Facies association of the Coquinas sequence of the Lagoa Feia Formation interpreted as bioclastic bar deposits.

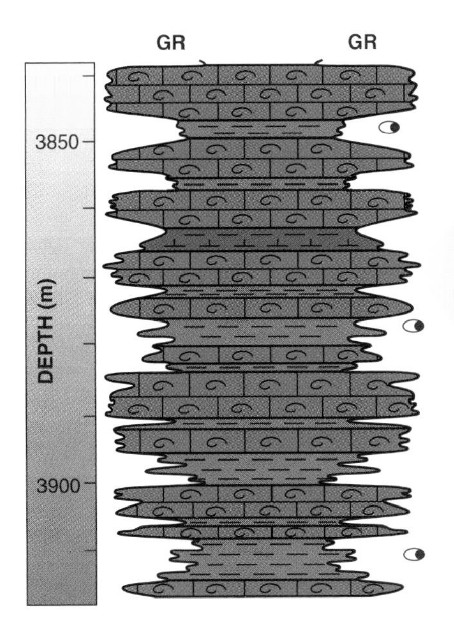

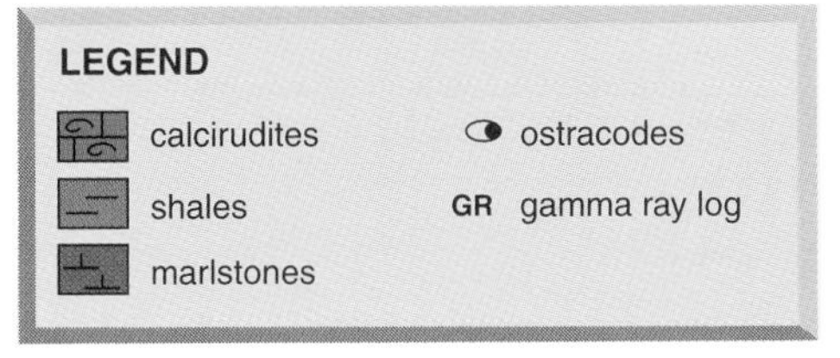

Figure 8—Facies association of the Coquinas sequence of the Lagoa Feia Formation interpreted as bioclastic sheet/bar fringe deposits.

ostracode coquinas form one calcarenite microfacies and two calcilutite microfacies.

DEPOSITIONAL EVOLUTION OF THE COQUINAS SEQUENCE

Depositional evolution of the Coquinas sequence is here described according to chronozones, based on ostracode biozones, from the oldest to the youngest: C009.3, C1010, C1020, C1100, and C010 (see Figures 3, 11).

Chronozone C009.3 encompasses the most basal sediments of the Coquinas sequence. It contains thick deposits in the São João da Barra low (>300 m), with pinch-outs or no sediments occuring over structural highs. This chronozone consists of narrow alluvial fans present at the slopes of the Badejo regional high and spread over a large muddy alluvial plain gradational into a marginal to deep lacustrine environment. Carbonate sedimentation was not important in this period in this area, and coquinas composed of bivalves formed bioclastic bars and bioclastic sandy beaches only in the São João da Barra low (Figure 11). This was the time of the greatest areal extent of lacustrine deposition, as a shallow lake with only one depocenter. Additionally, the absence of significant conglomerate accumulations and clear dominance of fine sediments (siltstone, shale, and marlstone) suggest that during this time, tectonic activity was not intense.

At the time of the chronozone C1010, with tectonic activity increasing, two lacustrine depocenters formed, separated by the Badejo regional high (BRH). Thick packages of conglomerates and sandstones adjacent to the main faults indicate that tectonic activity increased during this time. This chronozone is characterized by thick sedimentary sections in two main structural lows, São João da Barra (100–584 m) and Corvina-Parati (50–22 m), and by pinch-outs or the complete absence of sediments over structural highs. At the São João da Barra low, there is a well-defined depositional system in chronozone C1010, consisting of (1) alluvial facies derived from the Campos fault (hinge line) and the fault that limits the western boundary of the BRH; and (2) lacustrine facies (Figure 11). Bioclastic sandy beaches and bioclastic bars were localized at the central portion of this depression. The paleolake was extremely shallow and received a significant amount of coarse sediments. It also underwent frequent and long periods of subaerial exposure as suggested by pedogenic to weathering features, such as desiccation breccias, skew planes, root traces, and patches of marmorization.

In the eastern portion of the BRH a very large alluvial complex became prominent, spreading from the BRH and fringing a lake with great dimensions. The ancient lake, which was sampled only on its proximal portions, is characterized by a shelf on which accumulated significant bivalve shell deposits. This carbonate shelf exhibits a ramp configuration that slopes to the east, where bioclastic sandy beaches to the south and

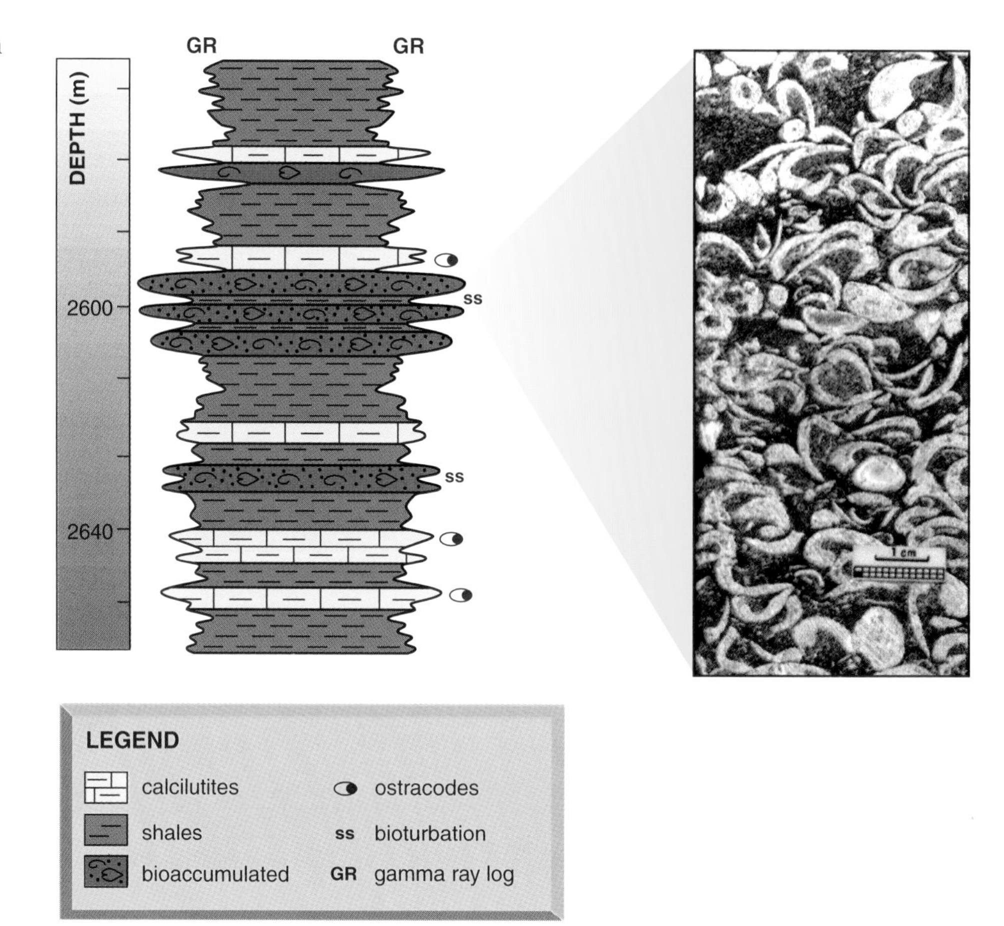

Figure 9—Facies association of the Coquinas sequence of the Lagoa Feia Formation interpreted as bioaccumulated banks or bioherms.

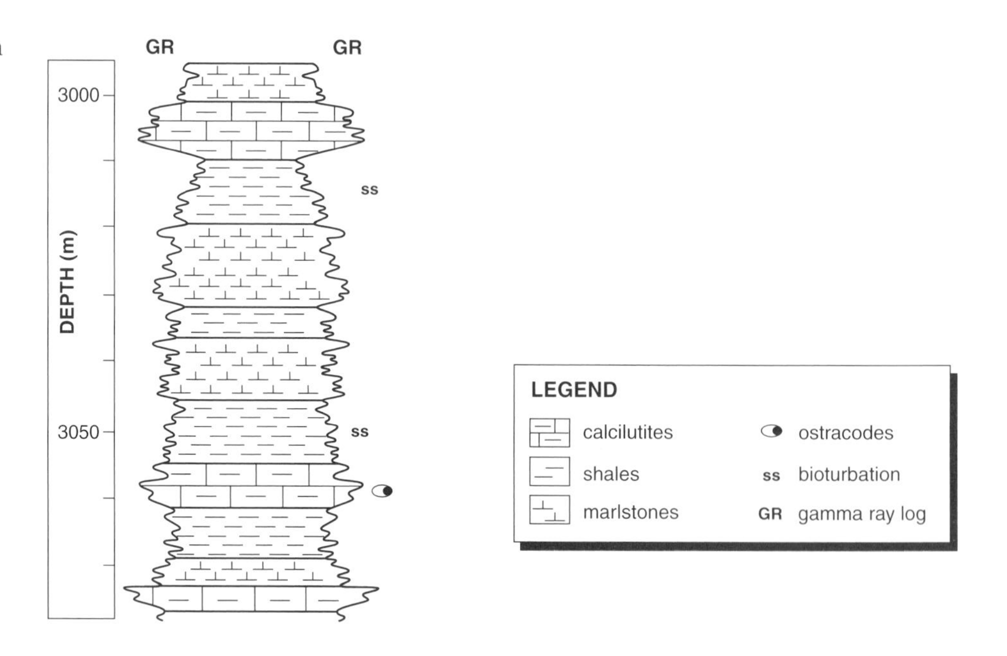

Figure 10—Facies association of the Coquinas sequence of the Lagoa Feia Formation interpreted as deep lacustrine deposits.

bioclastic calcarenite beaches to the north underlie bioclastic sheet deposits. Deep lacustrine sediments are identified only in the southern portion of the basin.

Intensity of tectonic activity remained the same during the time of the next chronozone, C1020, generating a thick sedimentary section consisting of coquinas and associated conglomerates, sandstones, and shales. The greatest sedimentary thicknesses, similar to chronozone C1010, are observed in this chronozone at the São João da Barra (>500 m) and Corvina-Parati (>200 m)

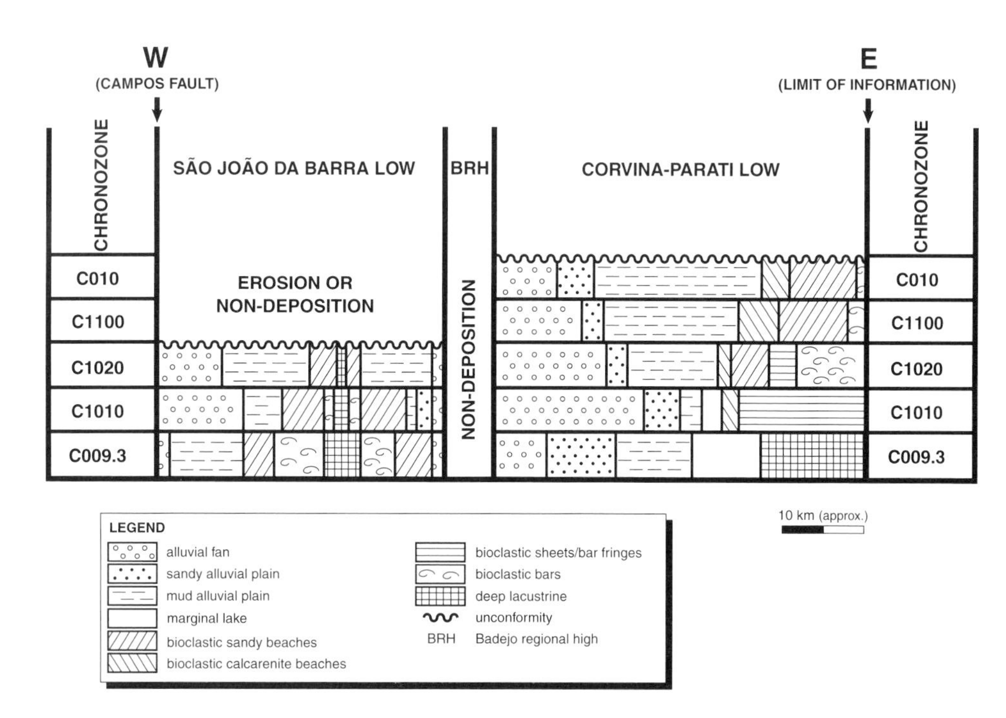

Figure 11—Schematic cross-section west-east paleoenvironment distribution for each chronozone of the Coquinas sequence of the Lagoa Feia Formation through time (vertical thicknesses are not to scale).

lows. Two lacustrine depocenters remained active from chronozone C1010 through chronozone C1020, but underwent significant changes in their sedimentary patterns.

The paleolake of the São João da Barra low covered only a reduced area of active lacustrine sedimentation, dominated by siliciclastics (Figure 11). The system to the east of the BRH in the Corvina-Parati low, containing alluvial and lacustrine deposits, basically maintained the same configuration as that in chronozone C1010; however, the carbonate deposits became more complex. At the shore zone, calcirudite and calcarenite sandy beaches existed. Basinward, bioclastic bars developed over syndepositional highs with steep slopes; bioclastic sheets/bar fringes occurred over areas with gentle slopes. Banks of biological accumulations (bioherms) were situated in marginal areas and even deep lacustrine areas. During the time represented by this chronozone, a strong tectonic tilting to the southeast occurred, exposing sediments of the western margin of the basin to erosion and weathering.

During the time of chronozone C1100, deposition occurred under a new tectonic regime. The areas to the west of the BRH were tectonically stable and did not receive any sedimentary input, which reduced the area of active sedimentation of the rift basin (Figure 11). Alluvial sedimentation occurred only eastward of the BRH in the Corvina-Parati low, grading into lacustrine carbonate sedimentation forming a parallel pattern from west to east (25–400 m thick) along the shoreline. Coquinas formed over a small geographic area, in the northeastern portion of the Corvina-Parati low, as beaches and bioclastic bars with a greater fragmentation of shells in comparison to those in the bioclastic bars of C1020.

Chronozone C010 encompasses the top zone of the Coquinas sequence. It is the chronozone that shows the most restricted areal distribution through nondeposition or erosion. Sedimentation at the Corvina-Parati low (470 m thick) is dominated by alluvial fans, sandy and muddy alluvial plains, calcirudite and calcarenite sandy beaches, and bioclastic bars. Gastropod bioclastic bars occur at the partially eroded southern portion of the basin. At the Corvina-Parati low, where sediments of this age remained preserved, it is plausible that the boundary with the upper unit (clastic-evaporitic sequence) has a continuous, unconformable character. The upper boundary of the Coquinas sequence, the pre-Alagoas unconformity, could have also been the result of superposition of many erosive events during the evolutionary history of the Coquinas sequence (Figure 12).

PALEOENVIRONMENTAL SYNTHESIS

The Lagoa Feia paleolake exhibits characteristics of a hydrologically closed and perennial to open saline lake that underwent marked lake level fluctuations related to climatic and tectonic variations. The sedimentary characteristics of the formation presents complex vertical and horizontal facies patterns linked with common features of subaerial exposure that characterize an evolutionary history as a succession of predominantly closed lakes with frequently changing lake levels and shorelines. Possible changes in the hydrologic balance associated with climatic fluctuations and changes in the drainage patterns due to tectonism caused periods of hydrologically open lakes.

Evidence for saline conditions during deposition of the Lagoa Feia Formation include the occurrence of gypsum and anhydrite molds, absence of charophyte oogonia, high strontium values in carbonates (>1000 ppm), high boron values in illites (100–150 ppm), and the presence of trioctahedral smectites (stevensite) and sepiolite. Ostracode shells are very thick with irregular external morphology and bladed rim calcite, interpreted as orginally Mg-calcite and common in high-energy bivalve coquinas. Oxygen and carbon isotopic values from carbonate rocks and shells range from $\delta^{13}C$ = 0.4 to 3.7‰ and $\delta^{18}O$ = 3.3 to –4.8‰. Diagnostic biomarker features present in the main source rocks, the Coquinas sequence and associated shales, include low concentrations of steranes, presence of ß-carotane, gammacerane, 28,30-bisnorhopane, and 25,28,30-trisnorhopane, high concentrations of hopanes and methylsteranes, and high relative abundance of extended tricyclic terpanes up to C_{45} (Bertani and Carozzi, 1984; Rehim et al., 1986; Trindade et al., 1994).

The tectonics and climatic control of a rift basin give rise to influences in the geometry and sedimentary patterns of the lake, or lakes. Tectonism also determines the subsidence intensity and is one of the major controlling factors of sedimentation rate. In the Coquinas sequence section, thick sedimentary packages are observed that are associated with rapid subsidence rates and a substantial sedimentary supply derived from the rift system borders. During the Coquinas sequence deposition, different tectonic subsidence rates occurred across the Campos Basin. The São João da Barra and Corvina-Parati lows represented areas of higher subsidence and sedimentation rates.

The Lagoa Feia paleolake morphology is defined by patterns of distribution of sedimentary facies. The true lacustrine rocks were deposited at marginal areas under shallow-water conditions. The extensive paleogeographic distribution of the marginal facies indicates that the Lagoa Feia was a low-slope ramp carbonate lake (Platt and Wright, 1991). Presence of calcirudites, associated with high hydraulic energy and deposited mainly along syndepositional highs, characterizes a highly wave-influenced paleolake. Low-energy carbonate facies formed in more protected areas.

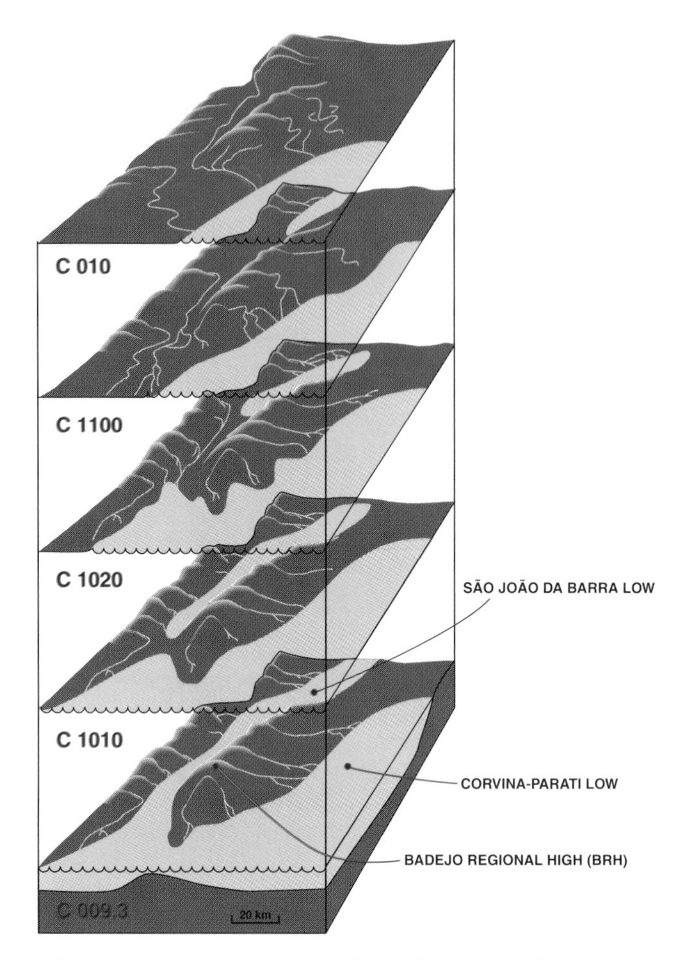

Figure 12—Paleogeographic evolution of the Coquinas sequence of the Lagoa Feia Formation for each chronozone (see Figures 3 and 11 for stratigraphic information). Refer to Figure 1 for exact geographic locations. More details in text.

CONCLUSIONS

- The Coquinas sequence is subdivided into five chronozones, based on ostracode biozones, allowing the construction of a detailed evolutionary history. Chronozones were defined, from bottom to top, as C009.3, C1010, C1020, C1100, and C010.
- For each chronozone, analyses of facies associations allowed recognition of the following depositional paleoenvironments: alluvial fans, sandy fluvial plains, muddy fluvial plains, bioclastic sandy beaches, bioclastic calcarenite beaches, marginal lake, bioclastic bars, bioclastic sheets/bar fringes, bioaccumulation banks (bioherms), and deep lake.
- Paleogeographic mapping between chronozones C009.3 and C1020 showed that the paleolake depositional basin was gradually filled up on the western side of the Campos Basin. The pronounced eastward tilting of the basin during deposition during chronozone C1020 exposed the São João da Barra low area to weathering and erosion, which continued until chronozone C010. Later, tilting of the southern portion of the basin partially exposed the sediments of chronozone C010 to erosion.
- The Coquinas sequence paleolake shows characteristics of a hydrologically closed but perennial to open saline lake. Owing to paleoclimatic and/or tectonic variations, the paleolake water level must have been subjected to marked fluctuations. Expansive distribution of shallow-water sediments indicates that the paleolake contained a low-slope carbonate ramp and was highly influenced by wave action.
- Thick bivalve coquinas deposits (mainly calcirudites and calcarenites) characterize the Coquinas sequence. These coquinas and gastropod coquinas may have accumulated as a result of waves and currents generated by storms, reworking, transporting, and redepositing shells as layers with and without a matrix.

ACKNOWLEDGMENTS

The authors acknowledge Petrobras for the support and permission given for publication. The bivalve classification was performed by Maria Helena Hessel.

REFERENCES CITED

Bertani, R. T., and A.V. Carozzi, 1984, Microfacies, depositional models, and diagenesis of the Lagoa Feia Formation (Lower Cretaceous), Campos Basin, offshore Brazil. Ciênias-Técnica-Petróleo (Petrobras/Cenpes), v. 14, p. 1–104.

Carvalho, M. D., and U. M. Praça, 1994, Criteria for macroscopic descriptions of bioclastic-rich limestones from the Coquinas Sequence, Lagoa Feia Formation, Campos Basin, Brazil. 14th International Sedimentological Congress, Abstracts, p. B2–3.

Dias, J. L., J. Q. Oliveira, and J. C. Vieira, 1988, Sedimentological and stratigraphic analysis of the Lagoa Feia Formation, rift phase of Campos Basin, offshore Brazil. Revista Brasileira de Geociências, v. 18 (3), p. 252–260.

Grosdidier, E., 1967. Quelques ostracodes nouveaux de la série antésalifère ("Wealdienne") des bassins côtiers du Gabon et du Congo. Revue de Micropaléontologie, v. 10 (2), p. 107–118.

Guardado, L. R., L. A. P. Gamboa, and C. F. Lucchesi, 1988, Petroleum geology of the Campos Basin, Brazil, a model for a producing Atlantic type basin *in* D. J. Edwards and P. A. Santogrossi, eds., Divergent/Passive Margin Basins. AAPG Memoir, v. 48, p. 3–79.

Krömmelbein, K., 1965, Zur Taxonomie und Biochronologie stratigraphische wichtige Ostracoden-Arten aus der Oberjurassich?-Unterkretazischen Bahia-Serie (Wealden-Fazies) NE-Brasiliens. Senckenbergiana Lethaea, v. 43 (6), p. 437–528.

Krommelbein, K., and R. Weber, 1971, Ostracoden des "Nordost-Brasilianischen Wealden". Beihefte zum Geologischen Jahrbuch Heft, v. 115, p. 145–162.

Platt, N. H., and V. P. Wright, 1991, Lacustrine carbonates: facies models, facies distributions and hydrocarbon aspects, *in* P. Anadón, Ll. Cabrera, and K. Kelts, eds., Lacustrine Facies Analysis, International Association of Sedimentologists Special Publication, v. 13, p. 57–74.

Rehim, H. A. A., A. M. Pimentel, M. D. Carvalho, and M. Monteiro, 1986, Talco e estevensita na Formação Lago Feia de Bacia de Campos–Possíveis implicações no ambiente depositional. 34 Congresso Brasileiro de Geologia (Goiânia), v. 1, p. 416–422.

Riedel, L., 1932, Die Oberkreide vom Mungofluss in Kamerun und ihre Fauna. Beitrage für Geologische Erforschung. Deutsche Schutzgebirge, v. 16, p. 1–154.

Silva-Telles, A. C., Jr., 1992, Novo zoneamento das coquinas da Formação Lagoa Feia (Neojiquiá da Bacia de Campos) com base em ostracodes - aspectos evolutivos. 37 Congresso Brasileiro de Geologia (Goiânia), v. 2, p. 489.

Trindade, L. A. F., J. L. Dias, and M. R. Mello, 1994, Sedimentological and geochemical characterization of the Lagoa Feia Formation, rift phase of the Campos Basin, Brazil, *in* B. J. Katz, ed., Petroleum Source Rocks, Berlin, Springer-Verlag, p. 149–165.

Tucker, M. E., and Wright, V. P., 1990, Carbonate Sedimentology. Oxford, Blackwell Scientific Publications, 482 p.

White, C. A., 1887, Contribuições à paleontologia do Brazil. Archivos do Museu Nacional do Rio de Janeiro, v. 7, p. 1–124.

Soria, A. R., A. Meléndez, M. Aurell, C. L. Liesa, M. N. Meléndez, and J. C. Gómez-Fernández, 2000, The Early Cretaceous of the Iberian Basin (northeastern Spain), *in* E. H. Gierlowski-Kordesch and K. R. Kelts, eds., Lake basins through space and time: AAPG Studies in Geology 46, p. 257–262.

Chapter 20

The Early Cretaceous of the Iberian Basin (Northeastern Spain)

A. R. Soria
A. Meléndez
M. Aurell
Dpto. Ciencias de la Tierra (Estratigrafía), Universidad de Zaragoza
Zaragoza, Spain

C. L. Liesa
Dpto. Ciencias de la Tierra (Geodinámica), Universidad de Zaragoza
Zaragoza, Spain

M. N. Meléndez
J. C. Gómez-Fernández
Dpto. de Estratigrafía (Fac. de Geología), Universidad Complutense de Madrid, Ciudad Universitaria
Madrid, Spain

INTRODUCTION

The Iberian Basin is an extensional intracratonic basin located in the northeastern part of the Iberian Peninsula (Figure 1). Sequence stratigraphy, subsidence analysis, and integration of basin fill data allow recognition of four successive evolutionary stages in the Iberian Basin during Mesozoic extension (Salas and Casas, 1993): (1) Triassic rift (Late Permian–Hettangian); (2) Early and Middle Jurassic post-rift (Sinemurian–Oxfordian); (3) Late Jurassic–Early Cretaceous rift (Kimmeridgian–middle Albian); and (4) Late Cretaceous post-rift (late Albian–Maastrichtian). Evolution and formation of the Iberian Basin during the Mesozoic correlates well with the evolution of the North Atlantic marginal basins, the Aquitaine Basin, and the central south Pyrennean Basin (Tankard and Balkwill, 1989; Salas and Casas, 1993). The first two stages are related to the opening and spreading of the Neotethys toward the west. The third stage is related to the opening of the central Atlantic, and the fourth stage results from the opening of the Bay of Biscay accompanied by the counterclockwise rotation of the Iberian plate and the consequent opening of the North Atlantic (Salas and Casas, 1993).

During the Early Cretaceous, several isolated and well-differentiated subbasins formed in the Iberian Basin. The origin of these subbasins is related to a rifting phase that took place during the Early Cretaceous concomitant with an increase in tectonic activity (Salas and Casas, 1993; Salas et al., 1995), creating horsts, grabens, and half-grabens controlled by listric faults (Salas, 1983); moreover, reactivation of strike-slip basement faults, inherited from the late Hercynian, seems to have controlled Cretaceous sedimentation (Salas and Casas, 1993). Infill of these small Early Cretaceous subbasins began with continental deposits that were alluvial and lacustrine in origin. Sedimentary records and tectonic evolution of each subbasin have been studied by numerous authors (see references cited). In this paper we sketch a general overview of the Early Cretaceous of the Iberian Basin.

PALEOGEOGRAPHIC DOMAINS

Four important paleogeographic domains are recognized in the Iberian Basin of the Early Cretaceous (see Figure 1). Each domain is divided into several depositional areas (subbasins), each controlled by specific tectonic structures.

(1) The Cameros domain (traditionally known as the Cameros Basin) evolved from the Tithonian until the Aptian. It is located in the northwestern part of the Iberian Basin and is divided into two main areas: west and east Cameros subbasins. Sedimentary evolution and tectonic framework of these subbasins has been studied, in recent years, by Salomon (1982), Guiraud (1983), Guiraud and Seguret (1985), Platt (1986, 1989, 1990, 1994, and 1995), Gómez-Fernández (1992),

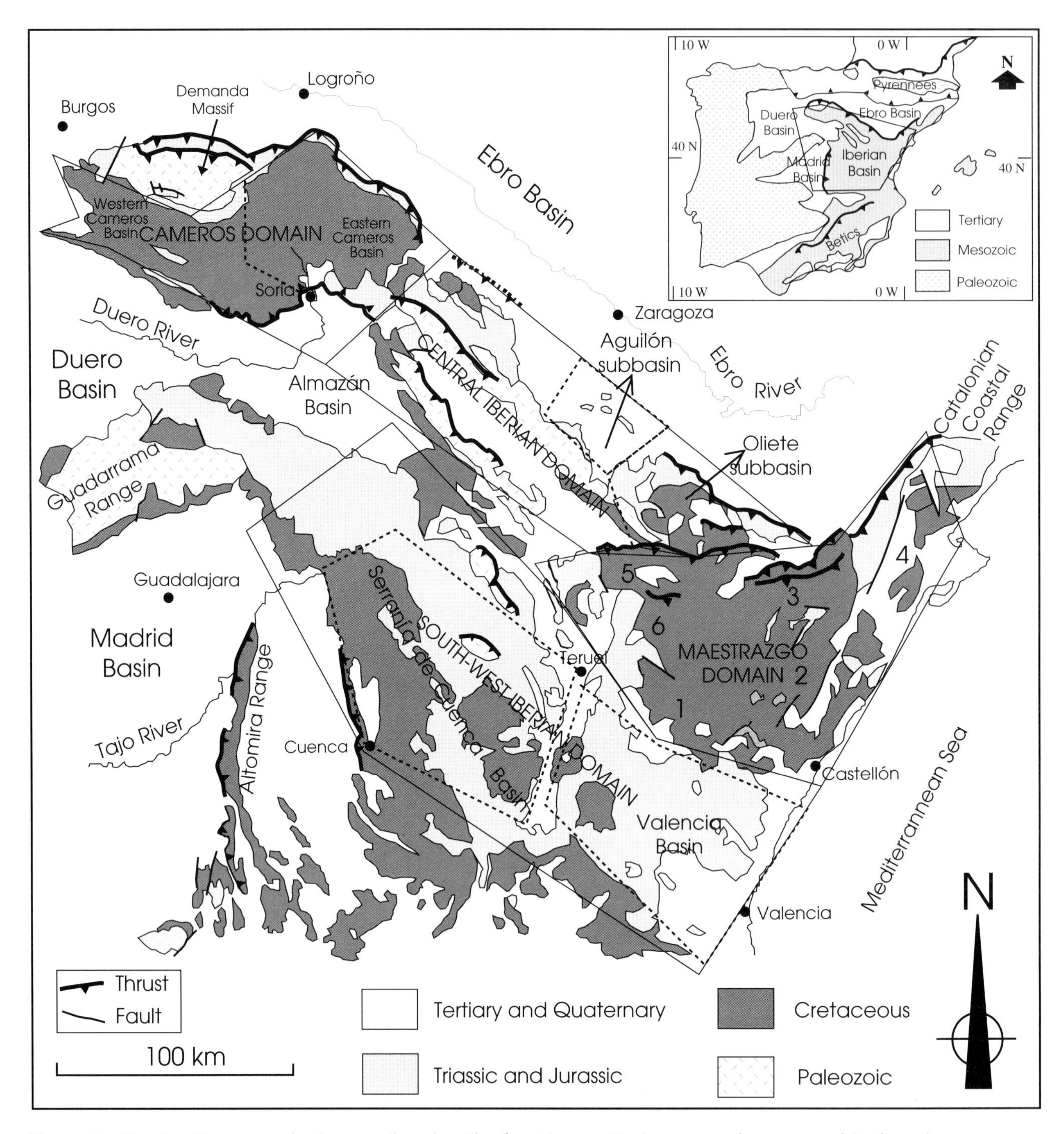

Figure 1—Iberian Range geologic map showing the four Lower Cretaceous paleogeographic domains (Cameros, Central Iberian, Southwest Iberian, and Maestrazgo) and their different subbasins. CW = West Cameros subbasins, CE = East Cameros subbasins, 1 = Peñagolosa subbasin, 2 = La Salzadella subbasin, 3 = Morella subbasin, 4 = El Perello subbasin, 5 = Las Parras subbasin, 6 = Galve subbasin.

Alonso and Mas (1993), Mas et al. (1993), Gómez-Fernández and Meléndez (1994 a, b), and Normati (1994).

(2) The Central Iberian domain, situated in the northeastern part of the Iberian Range, is characterized by two areas of Lower Cretaceous sedimentation: the Aguilón subbasin to the north and the Oliete subbasin to the south. Sedimentary evolution and synsedimentary-tectonic framework have been analyzed by Canerot (1970, 1974, 1980), Canerot et al. (1982), Murat (1983), Mayoral and Sequeiros (1983), Salas (1987), Meléndez and Aurell (1989), Aurell et al. (1990), Soria (1991), Vennin et al. (1993), and Soria et al. (1994, 1995, 1997). Recent studies of Early Cretaceous tectonic controls are found in Soria (1997), Casas et al. (1997), and Soria et al. (1997).

(3) The Southwestern Iberian domain, located in the southwestern part of the Iberian Range, consists of two

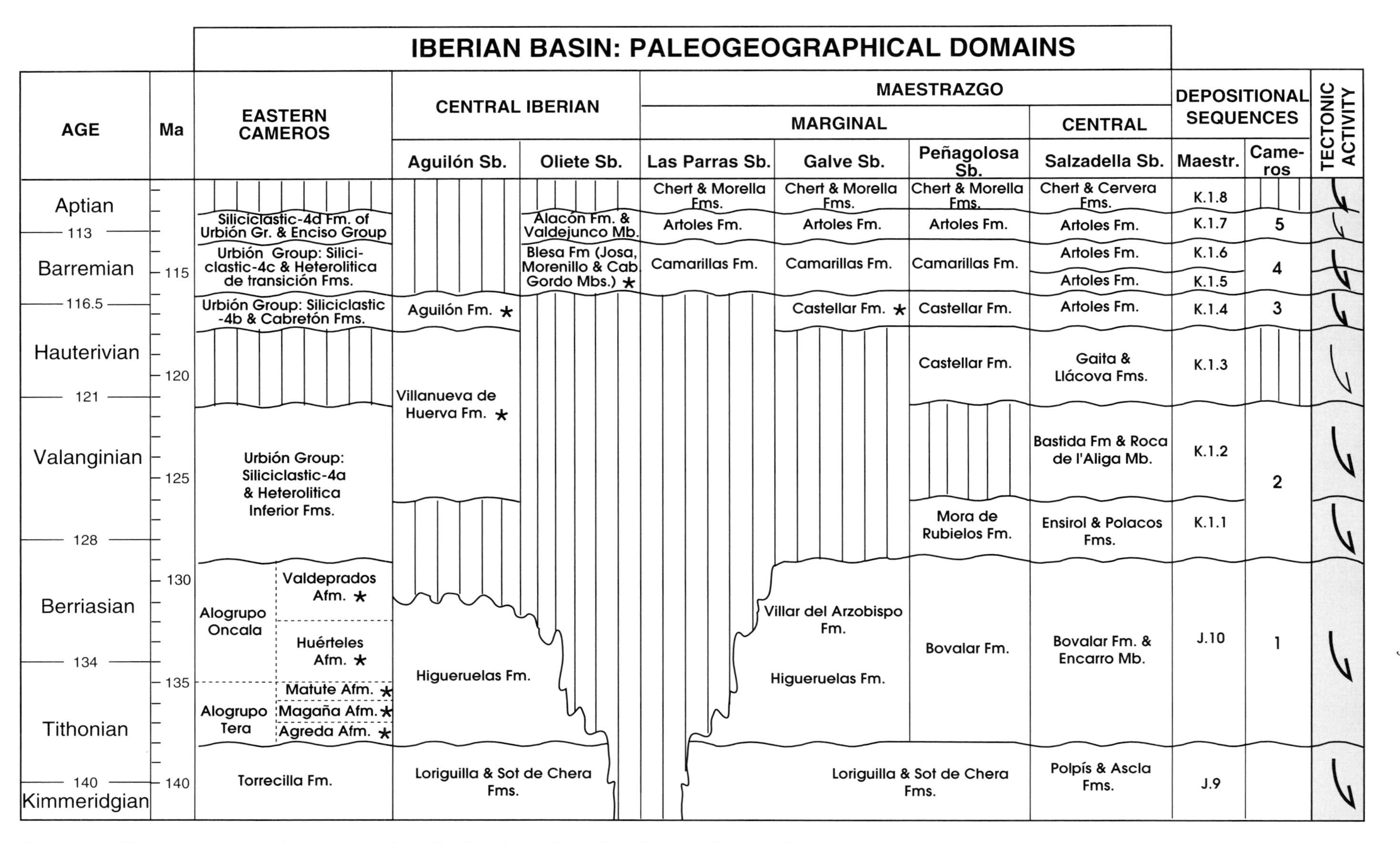

Figure 2—Chronostratigraphic framework of the Iberian subbasins discussed in this book along a northwest-southeast section across the Iberian Basin (Eastern Cameros, Central Iberian, and Maestrazgo domains), compiled from Mas et al. (1984), Salas (1987), Aurell (1990), Mas et al. (1993), Gómez-Fernández and Meléndez (1994a), Salas et al. (1995), and Soria et al. (1995, 1997). * = Units studied in papers in this volume.

areas of sedimentation: the Serranía de Cuenca Basin (Meléndez et al., 1994), located in the northwestern part of this domain, and the Valencia Basin, situated in the southeast. The Southwestern Iberian domain has been studied by Mas (1981), Vilas et al. (1982, 1983), Mas and Alonso (1983), Sanz et al. (1988, 1990), Gómez-Fernández (1988), Meléndez et al. (1989), Gierlowski-Kordesch et al. (1991), Fregenal-Martínez (1991), Gómez-Fernández and Meléndez (1991), Fregenal-Martínez and Meléndez (1993, 1994), and Meléndez et al. (1994).

(4) The Maestrazgo domain (usually known as the Maestrazgo Basin), located in the eastern part of the Iberian Basin, is divided into six areas of Lower Cretaceous sedimentation or subbasins: (1) Peñagolosa, (2) La Salzadella, (3) Morella, (4) El Perello, (5) Las Parras, and (6) Galve. Research on sedimentary evolution and tectonic framework of this sedimentary domain includes Canerot et al. (1982), Salas (1983, 1986, 1987), Salas and Casas (1993), Salas et al. (1995), Salas and Guimerà (1996), Guimerà and Salas (1996), and Soria (1997).

The specifics of our own field studies are summarized in the following contributions on the Eastern Cameros, Central Iberian, and Maestrazgo domains. In these three paleogeographic domains, Cretaceous sedimentation begins with facies traditionally referred to as Weald (continental sedimentation); however, origin and evolution of the different subbasins appear diachronous, but rather congruent with structural development along the Early Cretaceous rifts. This implies locally important variations in age, thickness, and lithology of successive sedimentary units within different subbasins (Figure 2).

Figure 2 outlines a chronostratigraphic framework of a northwest-southeast section across three paleogeographic domains of the Iberian Basin: Eastern Cameros, Central Iberian, and Maestrazgo domains. The Eastern Cameros Basin (to the northwest) and Central Maestrazgo Basin (Salzadella subbasin, to the southeast) show the more complete Early Cretaceous stratigraphic record. Between these two paleogeographic realms, the Central Iberian domain (Aguilón and Oliete subbasins) and the marginal subbasins of the Maestrazgo domain (Las Parras, Galve, and Peñagolosa subbasins) document a discontinuous sequence of Early Cretaceous units.

Early Cretaceous sedimentation was not coeval across the Iberian Basin. Structures evolved ranging from the uppermost Jurassic to Lower Cretaceous rifting. Peñagolosa, Galve, Las Parras, and Oliete subbasins represent progressive expansion of the sedimentary area of the central Maestrazgo Basin toward the north and northwest (Soria, 1997). Maximum extent of sedimentation occurred during the late Barremian–Aptian (Figure 2).

REFERENCES CITED

Alonso, A., and R. Mas, 1993, Control tectónico e influencia del eustatismo en la sedimentación del Cretácico inferior de la Cuenca de los Cameros. España: Cuadernos de Geología Ibérica, v. 17, p. 285–310.

Aurell, M., 1990, El Jurásico superior de la Cordillera Ibérica Central (provincias de Zaragoza y Teruel). Análisis de Cuenca: Ph.D. dissertation, Universidad de Zaragoza, 509 p.

Aurell, M., N. Meléndez, and A. Meléndez, 1990, Evolution of a lacustrine system during the Hauterivian (Early Cretaceous) on the North of Oliete Basin. Central Spain: 13th International Sedimentological Congress, Nottingham. Abstract-Posters, p. 13–14.

Canerot, J., 1970, Stratigraphie et paléogéographie du Crétacé inférieur de la region d'Oliete (Prov. de Teruel, Espagne): Comptes Rendu Sommaire des Scéances de la Societé Géologique de France, v. 3, p. 119.

Canerot, J., 1974, Recherches géologiques aux confins des Chaines ibériques et catalane (Espagne): Ph.D. dissertation, Toulouse, ENADIMSA, 517 p.

Canerot, J., 1980, La cubeta eocretácica de Oliete: un modelo sedimentológico y paleogeológico de una cuenca de borde: Guía de excursión. XIV Curso de Geología Práctica, Teruel, 6 p.

Canerot, J., P. Cugny, G. Pardo, R. Salas, and J. Villena, 1982, Ibérica Central y Maestrazgo, *in* A. García, L. Sánchez de la Torre, V. Pujalte, J. García-Mondejar, J. Rosell, S. Robles, A. Alonso, J. Canerot, L. Vilas, J. A. Vera, and J. Ramírez del Pozo, eds., El Cretácico de España: Universidad Complutense de Madrid, p. 273–344.

Casas, A. M., A. L. Cortés, C. L. Liesa, A. Meléndez, and A. R. Soria, 1997, The structure of the northern margin of Iberian Range between the Sierra de Arcos and the Montalbán anticline: Cuadernos de Geología Ibérica, v. 23, p. 243–268.

Fregenal-Martínez, M. A., 1991, El sistema lacustre de Las Hoyas (Cretácico Inferior, Serranía de Cuenca): Estratigrafía y Sedimentología: Masters Thesis, Universidad Complutense de Madrid, 226 p.

Fregenal-Martínez, M. A., and N. Meléndez, 1993, Sedimentología y evolución paleogeográfica de la cubeta de Las Hoyas (Cretácico Inferior, Serranía de Cuenca): Cuadernos de Geología Ibérica, v. 17, p. 231–256.

Fregenal-Martinez, M. A., and N. Meléndez, 1994, Sedimentological analysis of the Lower Cretaceous lithographic limestones of the Las Hoyas fossil site (Serranía de Cuenca, Iberian Range, Spain): Mémoire Spécial Geobios, v. 16, p. 185–193.

Gierlowski-Kordesch, E., J. C. Gómez-Fernández, and N. Meléndez, 1991, Carbonate and coal deposition in an alluvial-lacustrine setting: Lower Cretaceous (weald) in the Iberian Range (east-Central Spain), *in* P. Anadón, Ll. Cabrera, and K. Kelts, eds., Lacustrine facies analysis: Special Publications International Association of Sedimentologists, v. 13, p. 11–127.

Gómez-Fernández, J. C., 1988, Estratigrafía y Sedimentología del Cretácico Inferior en "Facies Weald" de la Serranía de Cuenca: Masters thesis, Universidad Complutense de Madrid, 228 p.

Gómez-Fernández, J. C., 1992, Análisis de la Cuenca sedimentaria de los Cameros durante sus etapas iniciales de relleno en relación con su evolución paleogeográfica: Ph.D. dissertation, Universidad Complutense de Madrid, 343 p.

Gómez-Fernández, J. C., and N. Meléndez, 1991, Rhythmically laminated lacustrine carbonates in the Lower Cretaceous of La Serranía de Cuenca Basin (Iberian Ranges, Spain), *in* P. Anadón, Ll. Cabrera, and K. Kelts, eds, Lacustrine facies analysis: Special Publications International Association of Sedimentologists, v. 13, p. 247–258.

Gómez-Fernández, J. C., and N. Meléndez, 1994a, Estratigrafía de la "Cuenca de Cameros" (Cordillera Ibérica Noroccidental, N de España) durante el tránsito Jurásico–Cretácico: Revista de la Sociedad Geológica de España, v. 7 (1–2), p. 121–139.

Gómez-Fernández, J. C., and N. Meléndez, 1994b, Climatic control on Lower Cretaceous sedimentation in a playa-lake system of a tectonically active basin (Huérteles Alloformation, Eastern Cameros Basin, North Central Spain): Journal of Paleolimnology, v. 11, p. 91–107.

Guimerà, J., and R. Salas, 1996, Inversión terciaria de la falla normal mesozoica que limitaba la subcuenca de Galve: Geogaceta, v. 20 (7), p. 1701–1703.

Guiraud, M., 1983, Evolution tectono-sedimentaire du bassin Wealdien (Crétacé inferieur) en relais de decrochements de Logroño-Soria (NW Espagne): Ph.D. dissertation, Languedoc, 184 p.

Guiraud, M., and M. Seguret, 1985, A releasing solitary overstep model for the Late Jurassic–Early Cretaceous (Wealdian) Soria strike-slip Basin (Northern Spain), *in* K. T. Biddle and N. Christie-Blick, eds., Strike-slip deformation, basin formation and sedimentation: SEPM Special Publication 37, p. 159–175.

Mas, J. R., 1981, El Cretácico Inferior de la región noroccidental de Valencia: Ph.D. dissertation, Universidad Complutense de Madrid, Seminarios de Estratigrafía (Serie Monografías), v. 8, 408 p.

Mas, J. R., and A. Alonso, 1983, Jurásico terminal y Cretácico, *in* Instituto Geológico y Minero de España (IGME), ed., Memoria de la Hoja Escala 1:200.000: Liria (núm. 55). Madrid.

Mas, J. R., A. Alonso, and N. Meléndez, 1984, La Formación Villar del Arzobispo: un ejemplo de llanuras de Marea siliciclásticas asociadas a plataformas carbonatadas, Jurásico terminal (NW de Valencia y E de Cuenca): Publicaciones de Geología, Universidad Autónoma de Barcelona, v. 20, p. 175–188.

Mas, J. R., A. Alonso, and J. Guimerà, 1993, Evolución tectonosedimentaria de una cuenca extensional intraplaca: La cuenca finijurásica-eocretácica de los Cameros (La Rioja-Soria): Revista de la Sociedad Geológica de España, v. 6 (3–4), p. 129–144.

Mayoral, E., and L. Sequeiros, 1983, El Cretácico inferior en la región de Plou-Cortes de Aragón-Josa (Teruel): Boletín de la Real Sociedad Española de Historia Natural (Geol.), v. 81 (1–2), p. 111–123.

Meléndez, A., and M. Aurell, 1989, Controles en la sedimentación del Cretácico inferior de Aguilón (Zaragoza, Cordillera Ibérica Septentrional): Geogaceta, v. 6, p. 55–58.

Meléndez, N., A. Meléndez, and J. C. Gómez-Fernández, 1994, La Serranía de Cuenca Basin (Lower Cretaceous), Iberian Ranges, Central Spain, *in* E. Gierlowski-Kordesch and K. Kelts, eds., Global Geological Record of Lake Basins: Cambridge University Press, v. I, p. 215–219.

Meléndez, N., A. Meléndez, and J. C. Gómez-Fernández, 1989, Los sistemas lacustres del Cretácico Inferior de la Serranía de Cuenca: Universidad Complutense de Madrid, 70 p.

Murat, B., 1983, Contribution à l'étude stratigraphique, sédimentologique et tectonique du bassin éocrétacé d'Oliete (Prov. de Teruel, Espagne): Masters thesis, Travaux du Laboratoire de Géologie Sedimentaire et Paleontologie, Université Paul Sabatier, Toulouse, 247 p.

Normati, M., 1994, The eastern Cameros Basin (Kimmeridgian to Valanginian)—northern Iberian Ranges (Spain), *in* E. Gierlowski-Kordesch and K. Kelts, eds., Global Geological Record of Lake Basins: Cambridge University Press, v. I, p. 209–213.

Platt, N. H., 1986, Sedimentation and tectonics of the western Cameros Basin, Province of Burgos, northern Spain: Unpublished Ph.D. thesis, University of Oxford, 250 p.

Platt, N. H., 1989, Continental sedimentation in an evolving rift basin: the Lower Cretaceous of the Western Cameros Basin (northern Spain): Sedimentary Geology, v. 64, p. 91–109.

Platt, N. H., 1990, Basin evolution and fault reactivation in the western Cameros Basin, northern Spain: Journal of the Geological Society of London, v. 147, p. 165–175.

Platt, N. H., 1994, The western Cameros Basin, northern Spain: Rupelo Formation (Berriasian), *in* E. Gierlowski-Kordesch and K. Kelts, eds., Global Geological Record of Lake Basins: Cambridge University Press, v. I, p. 195–202.

Platt, N. H., 1995, Sedimentation and tectonics of a synrift succesion: Upper Jurassic alluvial fans and palaeokarst at the late Cimmerian unconformity, western Cameros Basin, northern Spain: Special Publications International Association of Sedimentologists, v. 22, p. 217–234.

Salas, R., 1983, Las secuencias deposicionales del tránsito Jurásico–Cretácico en la zona de enlace Catalánides-Ibérica: Comunicaciones X Congreso Nacional de Sedimentología, Menorca, p. 3.34–3.38

Salas, R., 1986, El cicle Cretaci inferior al marge oriental d'Ibèria. Historia Natural dels Països Catalans. T.1 Geología (I): Fundació Enciclopedia Catalana, p. 368–375.

Salas, R., 1987, El Malm i el Cretaci inferior entre el Massis de Garraf i la Serra d'Espadà: Ph.D. dissertation, Univ. de Barcelona, 345 p.

Salas, R., and A. Casas, 1993, Mesozoic extensional tectonics, stratigraphy and crustal evolution during the Alpine cycle of the eastern Iberian Basin: Tectonophysics, v. 228, p. 33–55.

Salas, R. and J. Guimerà, 1996, Rasgos estructurales principales de la cuenca cretácica inferior del Maestrazgo (Cordillera Ibérica oriental). Geogaceta, v. 20 (7), p. 1704–1706.

Salas, R., C. Martín-Closas, X. Querol, J. Guimerá, and E. Roca, 1995, Evolución tectonosedimentaria de las cuencas del Maestrazgo y Aliaga-Peñagolosa durante el Cretácico inferior, *in* R. Salas and C. Martín-Closas, eds., El Cretácico Inferior del Nordeste de Iberia. Guia de campo de la excursiones científicas del III Coloquio del Cretácico de España (Morella, 1991): Publications Universitat de Barcelona, p. 13–94.

Salomon, J., 1982, Les formationes continentales du Jurassic supérieur-Crétacé inferieur en Espagne du Nord (Chaînes Cantabriques et Nord-Ibérique): Mémoires Géol. de l'Université de Dijon, v. 6, 228 p.

Sanz, J. L., S. Wenz, A. Yébenes, R. Estes, X. Martínez-Delclós, E. Jiménez-Fuentes, C. Diéguez, A. D. Buscalioni, L. J. Barbadillo, and L. Via, 1988, An Early Cretaceous faunal and flora assemblage: Las Hoyas fossil site (Cuenca, Spain): Geobios, v. 21 (5), p. 611–635.

Sanz, J. L., C. Diéguez, M. A. Fregenal-Martínez, X. Martínez-Delclós, N. Meléndez, and F. J. Poyato-Ariza, 1990, El yacimiento de fósiles del Cretácico Inferior de Las Hoyas, provincia de Cuenca (España): Comunicaciones de la 1st Reunión de Tafonomía y Fosilización, p. 337–355.

Soria, A. R., 1991, El Cretácico inferior marino de la Cubeta de Oliete. Análisis de Cuenca: Masters thesis. Universidad de Zaragoza, 120 p.

Soria, A. R., 1997, La sedimentación en las cuencas marginales del surco Ibérico durante el Cretácico Inferior y su control tectónico: PhD dissertation, Universidad de Zaragoza. Servicio de Publicaciones Univ. Zaragoza, 363 p.

Soria, A. R., E. Vennin, and A. Meléndez, 1994, Control tectónico en la evolución de las rampas carbonatadas del Cretácico inferior de la·Cubeta de Oliete (Prov. Teruel): Revista de la Sociedad Geológica de España, v. 7 (1–2), p. 47–62.

Soria, A. R., C. Martín-Closas, A. Meléndez, N. Meléndez, and M. Aurell, 1995, Estratigrafía del Cretácico inferior continental de la Cordillera Ibérica Central: Estudios Geológicos, v. 51 (3–4), p. 141–152.

Soria, A. R., A. Meléndez, N. Meléndez, and C. L. Liesa, 1997, Evolución de dos sistemas continentales en la Cubeta de Aguilón (Cretácico Inferior). Interrelación sedimentaria entre depósitos aluviales y lacustres, y su control tectónico: Cuadernos de Geología Ibérica, v. 22, p. 473–507.

Tankard, A. J., and H. R. Balkwill, 1989, Extensional tectonics and stratigraphy of the North Atlantic margins: Introduction, *in* A. J. Tankard and H. R. Balkwill, eds., Extensional Tectonics and Stratigraphy of the North Atlantic Margins: AAPG Memoir 46, p. 1–6.

Vennin, E., A. R. Soria, A. Preat, and A. Meléndez, 1993, Los sedimentos marinos del Cretácico inferior de la Cubeta de Oliete. Análisis de Cuenca: Cuadernos de Geología Ibérica, v. 17, p. 257–283.

Vilas, L., R. Mas, A. García, C. Arias, A. Alonso, N. Meléndez, and R. Rincón, 1982, Ibérica Suroccidental, *in* A. García, L. Sánchez de la Torre, V. Pujalte, J. García-Mondejar, J. Rosell, S. Robles, A. Alonso, J. Canerot, L. Vilas, J. A. Vera, and J. Ramírez del Pozo, eds., El Cretácico de España: Universidad Complutense de Madrid, p. 457–513.

Vilas, L., A. Alonso, C. Arias, R. Mas, R. Rincón, and N. Meléndez, 1983, The Cretaceous of Southwestern Iberian Ranges (Spain): Zitteliana, v. 10, p. 245–254.

Meléndez, N., and J. C. Gómez-Fernández, 2000, Continental deposits of the eastern Cameros Basin (northern Spain) during Tithonian–Berriasian time, *in* E. H. Gierlowski-Kordesch and K. R. Kelts, eds., Lake basins through space and time:AAPG Studies in Geology 46, p. 263–278.

Chapter 21

Continental Deposits of the Eastern Cameros Basin (Northern Spain) During Tithonian–Berriasian Time

Nieves Meléndez
Juan Carlos Gómez-Fernández
Dpto. Estratigrafía, Facultad de Ciencias Geológicas, Universidad Complutense
Madrid, Spain

INTRODUCTION

The Cameros Basin is located in north-central Spain at the northwestern end of the Iberian Ranges (Figure 1A). See also Figures 1 and 2 in Soria et al. (this volume) concerning an overview of the geology and tectonics of the Iberian Basin. The evolution of the Cameros Basin as a continental rift basin started in the Tithonian. The basin is principally filled by uppermost Jurassic–Lower Cretaceous continental sediments that overlie marine Jurassic deposits (Salomón, 1982a, b; Guiraud and Seguret, 1985; Platt, 1989, 1994a, b; Gómez-Fernández, 1992; Gómez-Fernández and Meléndez, 1994a, b; Normati, 1994; Normati and Salomon, 1989; Mas et al., 1993).

The Cameros Basin is bounded by two important fault systems: one set of extensional faults trending northwest-southeast and the other system with northeast-southwest–trending strike-slip faults. Due to these paleotectonic systems, the basin is divided into two paleogeographic domains: the western and eastern Cameros Basin (Figure 1). Activity of the two fault systems produced a strong subsidence in the Cameros Basin, which led to the accumulation of a thick (up to 800 m) sedimentary record during the Tithonian–Aptian. Metamorphism during the middle Cretaceous, after the cessation of rifting, affected the central part of the Cameros Basin and reached the chloithoid zone (Casquet et al., 1992).

STRATIGRAPHY

This contribution deals with the sedimentary fill of the eastern Cameros Basin during the Tithonian–Berriasian time span. Strong subsidence during this time span allowed accumulation of 4000 m of continental deposits. Synsedimentary tectonic activity and subsequent restructuring of the basin are clearly reflected in the sedimentary record. In this work we use allostratigraphic methodology for subdivision of the sedimentary rocks. This is an alternative approach contained in the North American Stratigraphic Code: "An allostratigraphic unit is a mappable stratiform body of sedimentary rocks that is defined and identified on the basis of its bounding discontinuities" (North American Commission on Stratigraphic Nomenclature, 1983, p. 865). A hierarchy of units, including the allogroup, alloformation, and allomember, was proposed, and rules were established for defining and naming these various types of units. "Allostratigraphic methods enable the erection of a sequence framework that avoids the cumbersome nature of lithostratigraphy, whereby lateral changes in facies within a unit of comparable age require a change in name" (Miall, 1997, p. 12). This methodology is especially relevant for continental deposits of the eastern Cameros Basin because it permits the avoidance of complexity in stratigraphic terminology that continuous lateral changes in facies generate; however, the allostratigraphic units studied here are mappable stratiform bodies as shown by Gómez-Fernández and Meléndez (1994a).

The Tera and Oncala Allogroups (Figure 2), separated by an important unconformity, constitute the major sedimentary units in the early stages of basin infilling. In turn, each of these allogroups can be subdivided into three and two Alloformations respectively, all of them separated by minor rank unconformities that, from base to top, are the Ágreda Alloformation, Magaña Alloformation, Matute Alloformation, Huérteles Alloformation, and Valdeprado Alloformation (Figure 2B). Comparison and equivalence of these allostratigraphic units from Gómez-Fernández (1992) and lithostratigraphy established by Platt (1994a, b; 1995) for the western Cameros and by Normati (1994) for the eastern Cameros Basin are shown in Figure 2A.

The Tera and Oncala Allogroups are differentiated based on a change in the basin configuration as shown in isopach maps (Figure 3). During sedimentation of the Tera Allogroup, the Cameros Basin was less well-defined and two main synsedimentary tectonic directions

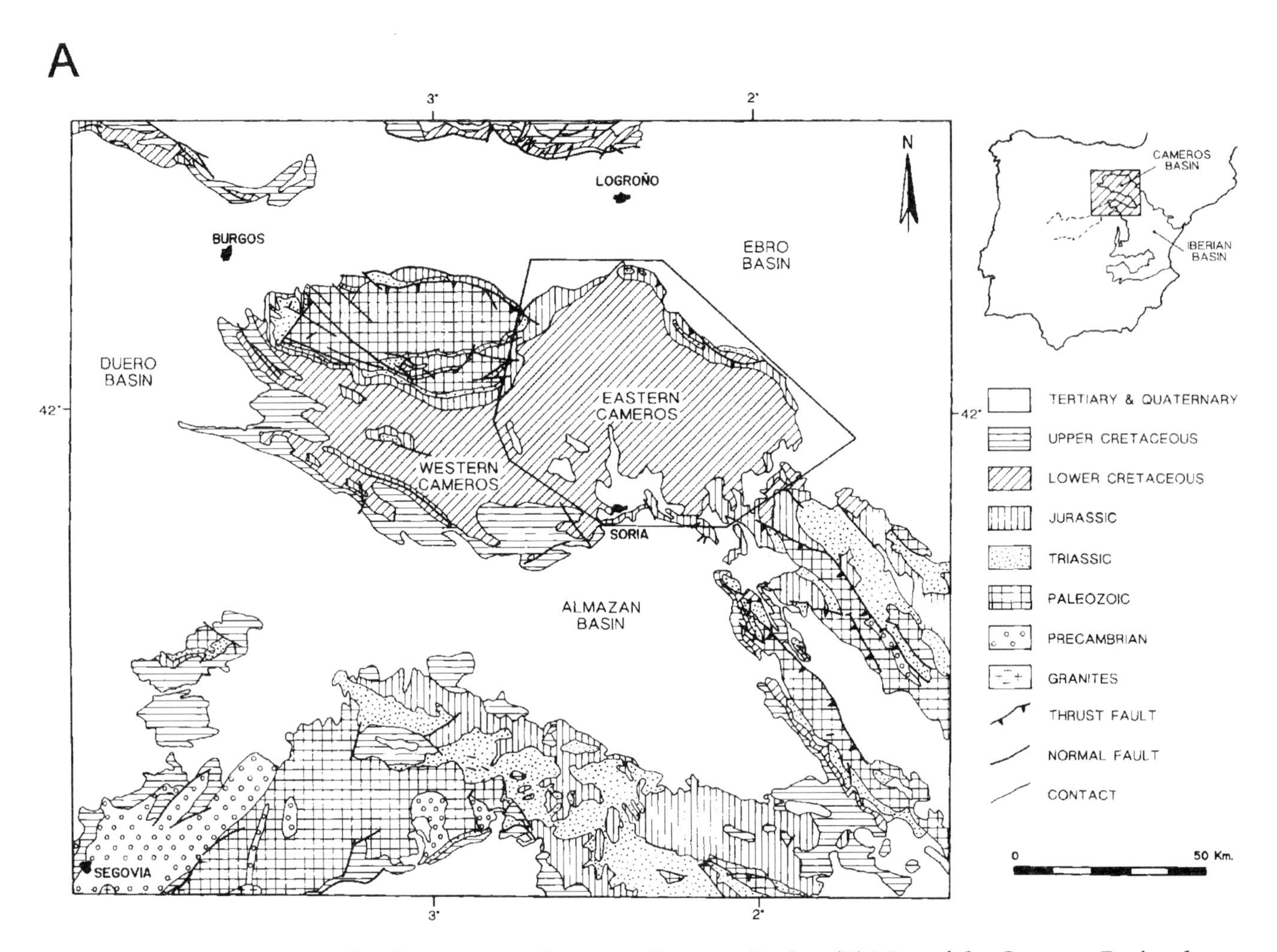

Figure 1—(A) Location map for the eastern and western Cameros Basins. (B) Map of the Cameros Basin, showing the three paleogeographic source areas: Demanda, Castellano, and Moncayo Massifs. Rectangle indicates the area mapped in Figures 3 and 6.

northwest-southeast and northeast-southwest were present. The Oncala Allogroup, however, was deposited in a strongly subsiding trough controlled by a northwest-southeast synsedimentary tectonic direction (see Figure 3). Another important change in the basin evolution that permits differentiation of the two allogroups is a significant change in the paleocurrent directions (Figure 2B). During sedimentation of the Tera Allogroup, source areas show a markedly continuous and progressive variation from southeast to northwest; however, from the Tera Allogroup to Oncala Allogroup, the change in the source area is abrupt, shifting progressively toward the southwest during deposition of the Oncala Allogroup. This change also involves large-scale erosion of the last deposits of the Tera Allogroup (Figure 2B).

Evidence for age correlation includes the discovery of *Dyctioclavator* aff. *fieri* in the Ágreda Alloformation in marly deposits from the Torrecilla outcrop (Figure 3A) and permits assignment of this unit to the Tithonian (Martín i Closas, personal communication, 1992). Rare fossil finds in the Magaña Alloformation and the poor chronostratigraphic value of *Mesochara harrisii*, *Porochara* sp., and *Clavatorites* in the Matute Alloformation make a precise age difficult to assess. Salomón (1982a, b), Martín i Closas (1989), Gómez-Fernández and Meléndez (1994a), and Normati (1994) place equivalent materials between the Tithonian and lower Berriasian strata. A Tithonian–early Berriasian age is acceptable for the Tera Allogroup (Figure 2).

The sparse fossil content of the Huérteles Alloformation and Valdeprado Alloformation makes them difficult to date. As for underlying units, the age is interpolated through lithostratigraphic correlation. Salomón (1982a, b) suggested a Late Berriasian age based on ostracods and Martín i Closas (1989) proposed an early Berriasian–middle Berriasian age based on charophytes. Normati (1994) also dated these materials as Late Berriasian. The age of the Oncala Allogroup therefore ranges from the early Berriasian to the middle–late Berriasian (Figure 2).

SEDIMENTARY ENVIRONMENTS

This section outlines the representative facies present in outcrop for each alloformation. For a more complete

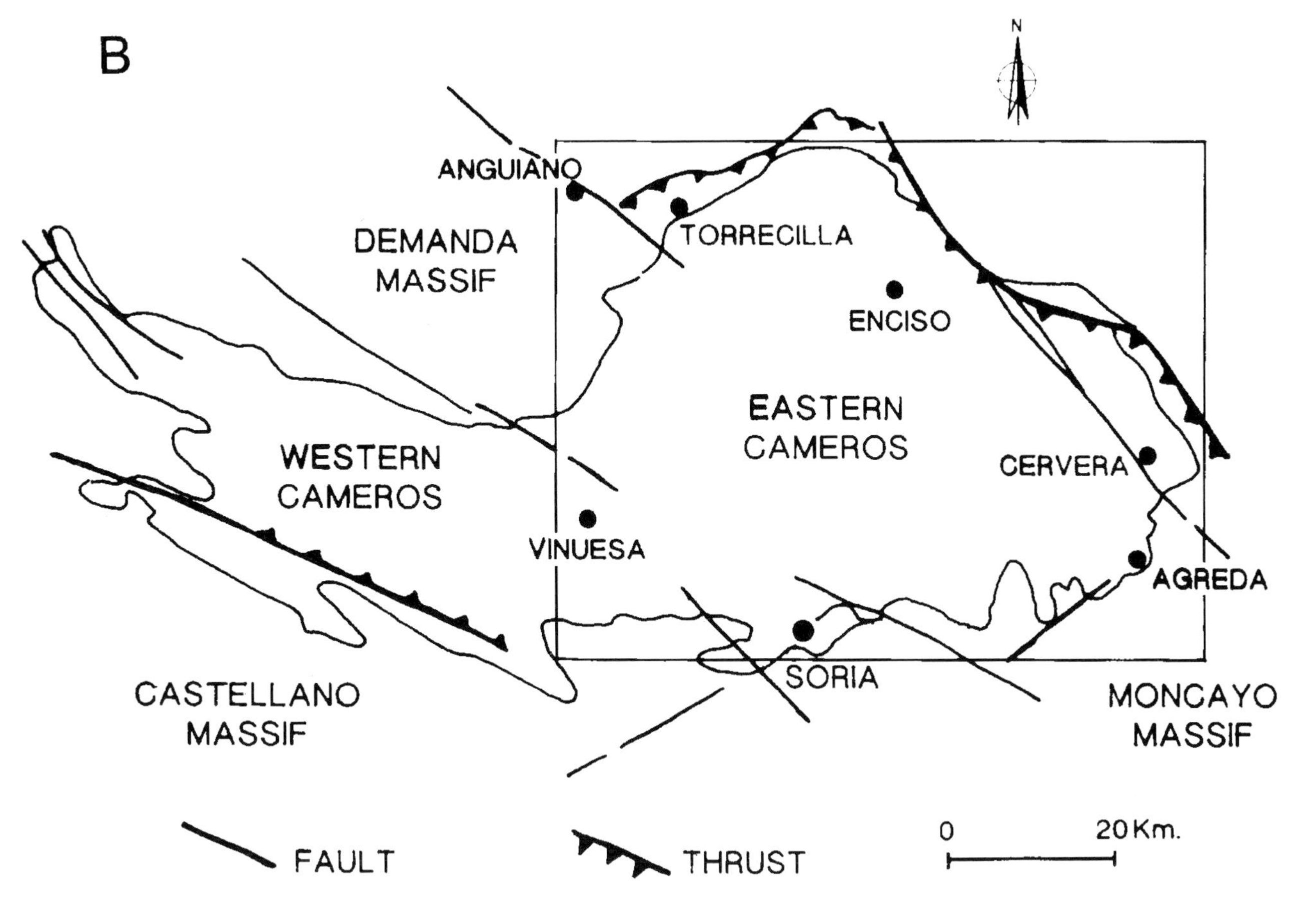

Figure 1—Continued.

description of facies, see Gómez-Fernández (1992) and Gómez-Fernández and Meléndez (1990, 1994b). Facies associations are defined according to classifications from Allen (1983), Miall (1988), and Friend et al. (1979) for siliciclastic materials. For carbonate facies associations, descriptions follow Platt and Wright (1991).

The Tera Allogroup

The Ágreda Alloformation crops out in two isolated lithosomes. The first lithosome is located to the northwest, in the Torrecilla area, where it reaches a thickness of 73 m; the main lithosome of this unit is located toward the southeast in the Ágreda area and reaches a maximum thickness of 255 m (Figure 3A). Predominant lithology of this unit is siliciclastic, but in localized areas carbonates become predominant (Figure 4, section 1). This unit consists mainly of red lutites with interbedded edaphic levels and tabular sandy bodies containing lateral accretion features. Some marlstones with limnic fauna also occur (facies associations C8, I1, L1, Figure 5).

The second lithosome is located toward the southeast and is composed of massive and cross-bedded quartzite conglomerates and sandstones as channel bodies (facies associations C1, C2, C4, C5, C6, Figure 5) separated by red lutites and siltstones containing root traces (facies association I1, Figure 5). Toward the west, these facies assemblages laterally change into carbonate rudstones, sandstones, and red lutites with well-developed caliche horizons (facies associations C7, C11, I1, Figure 5). In small isolated areas northwest of Ágreda, oncolitic limestones; micritic limestones with charophytes, ostracods, gastropods, bivalves, bone fragments, and algae (*Rivularia*); micritic limestones with mud-mound morphologies; and calcareous edaphic profiles are also present (facies associations I2, L1, P3, Figure 5).

Distribution of the Ágreda Alloformation and its facies associations suggests deposition into two isolated subsiding areas (Figure 6A). Alluvial fans and meandering to anastomosed rivers, whose source area was located southeast in the Moncayo massif (Figure 1B), existed in the southeastern area. Drainage was toward the northwest. Laterally related to these alluvial fans, shallow lakes with a low-energy ramp and high organic productivity developed. Carbonate ponds also existed. In the northwestern area, meandering fluvial deposits, coming from the south, were predominant. Fossil content of the Ágreda Alloformation is poor; biota includes scarce molluscs, some ostracods, and charophytes. Also, well-developed paleosol levels and plant remains are present.

The Magaña Alloformation lies unconformably on the Ágreda Alloformation or on the marine Jurassic and covers a greater part of the basin than does the Ágreda Alloformation (Figure 3B). The

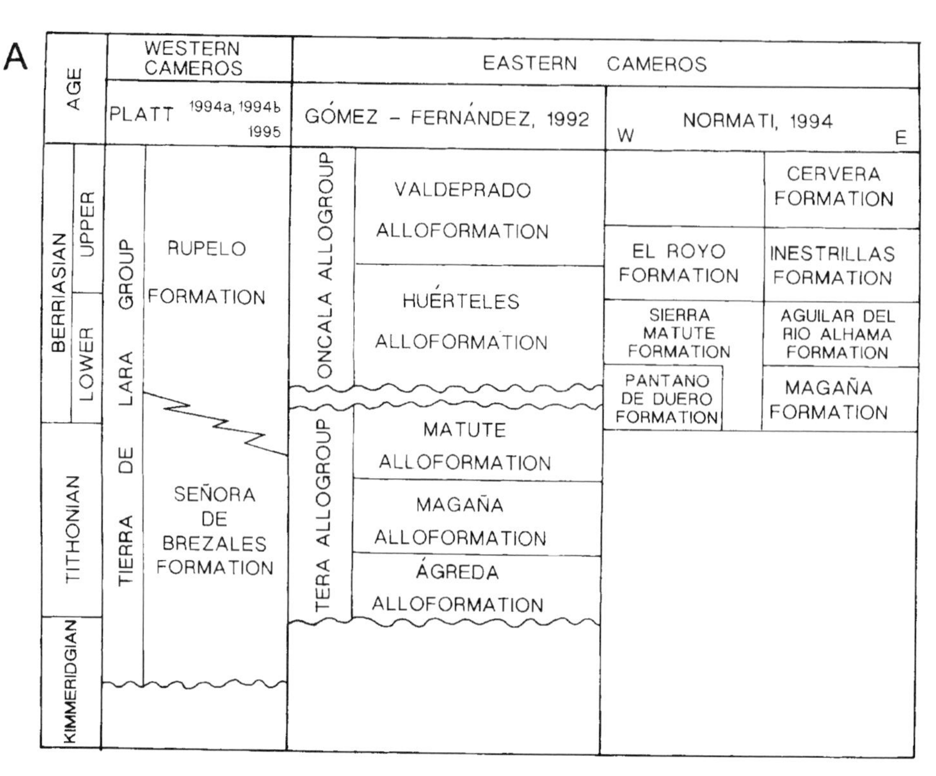

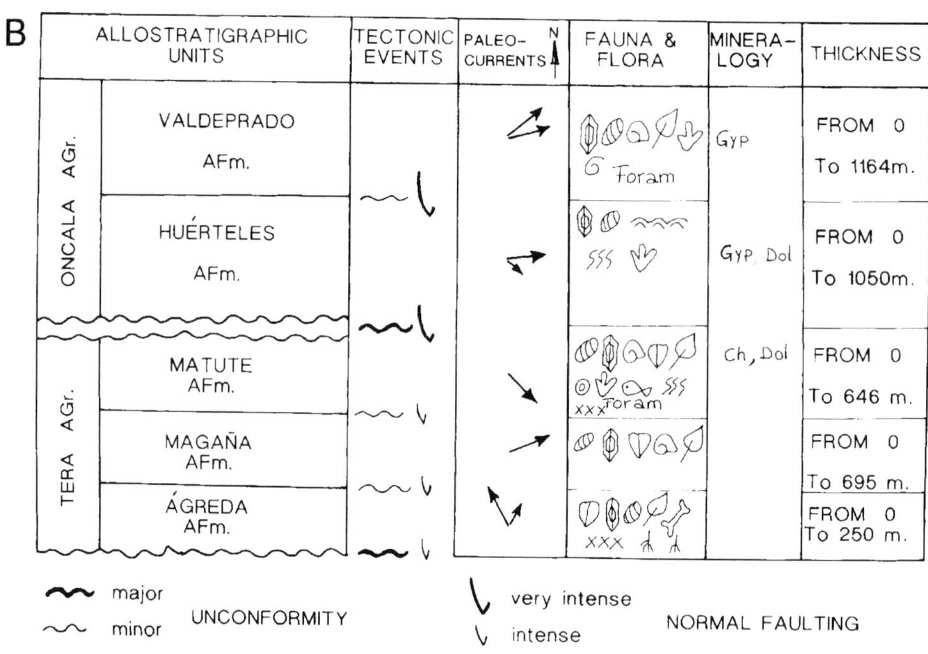

Figure 2—(A) Comparison of stratigraphic units established for the two Cameros subbasins. Allostratigraphic units used in this paper (from Gómez-Fernández, 1992) are compared with lithostratigraphy from Platt (1994a, b; 1995) for the western Cameros Basin and lithostratigraphy from Normatti (1994) for the eastern Cameros Basin. (B) Detailed stratigraphy of the Tera and Oncala Allogroups, eastern Cameros Basin. See Figure 1 for location of the western and eastern Cameros Basins.

Magaña Alloformation reaches a maximum thickness of around 700 m near Almarza, but two depocenters containing more than 500 m of fill are also found near Ágreda in the east and near Vinuesa in the west. The Magaña is a predominantly siliciclastic unit (Figure 4, section 2) that consists of quartzite conglomerate bodies with erosive bases and lenticular sandstone bodies; lutites become more predominant eastward. Facies associations C1, C2, C3, C4, C5, C9, and I9 (Figure 5) characterize these deposits. In the eastern part of the basin, quartzites and sandstones with lateral accretion surfaces and erosive bases (facies associations C8, C10, Figure 5) are present. Red and green siltstones containing interbedded tabular sandstones, carbonate accumulation levels, and black micritic limestones also occur (facies associations I1 and I3).

Distribution of facies associations is shown in the paleogeographic map of Figure 6B. Most proximal facies occur toward the western part of the basin, where gravel to sandy braided rivers developed; toward the east, meandering to anastomosed rivers transporting fine sediments incised a low-gradient alluvial plain. In the central part of the

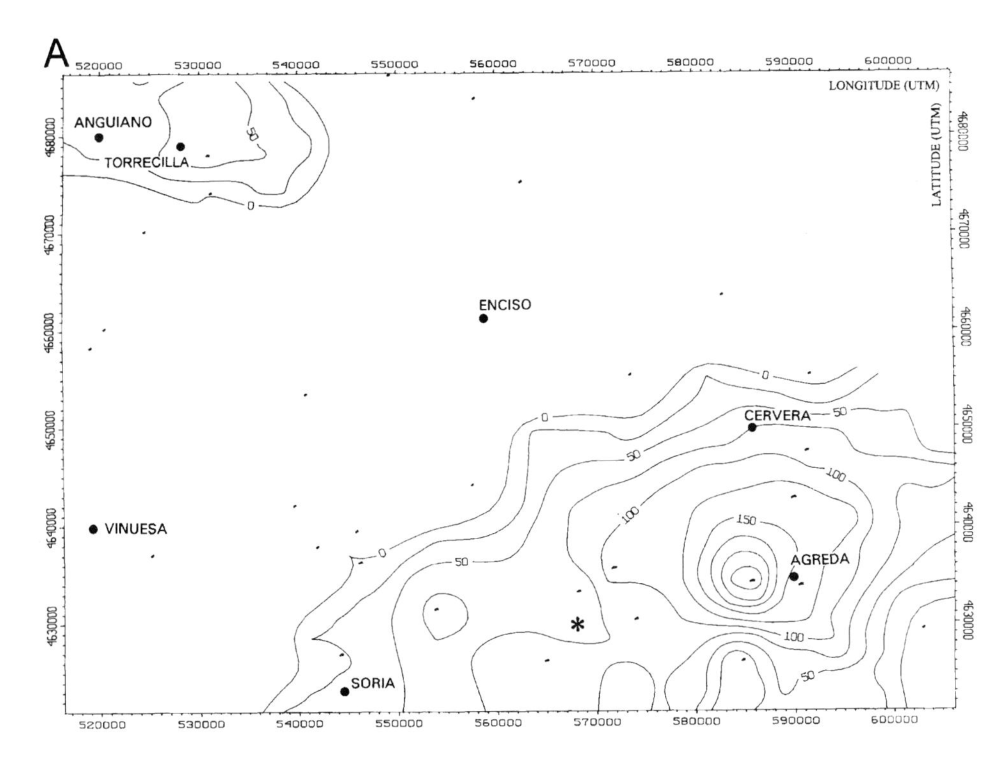

Figure 3—Isopach maps for five alloformations of the eastern Cameros Basin studied here. Points represent the location of logged sections. Asterisk in each map indicates the location of selected section logged in Figure 4. (A) Ágreda Alloformation, * Valdegeña section, (B) Magaña Alloformation, * Embalse de la Cuerda del Pozo section.

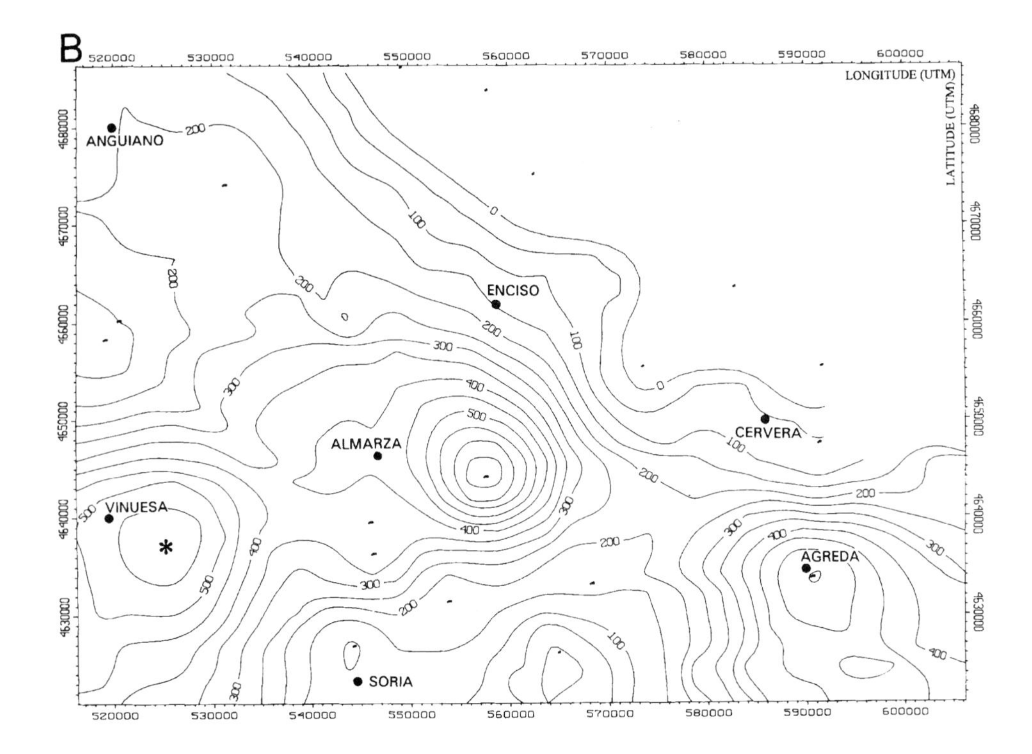

basin, a transition from braided to meandering channels occurred. Source areas of these fluvial systems were located toward the southwest, probably in the Castellano massif (Figure 1B). Fossils are rare in this alloformation because of the siliciclastic nature of its facies. Charophytes, ostracods, bivalves, and gastropods are poorly preserved in the limestone facies; some ferruginous plant remains are also present in sandy bodies.

The Matute Alloformation consists of mixed siliciclastic and calcareous rocks (Figure 4, section 3). It extends across nearly the entire basin and has a maximum thickness of 650 m at Sierra Matute (Figure 3C); however, in this area, other depocenters arranged in an east-west direction contain deposits exceeding 400 m in thickness. In the northern portion of the basin, another depocenter reaches more than 350 m (Figure 3C).

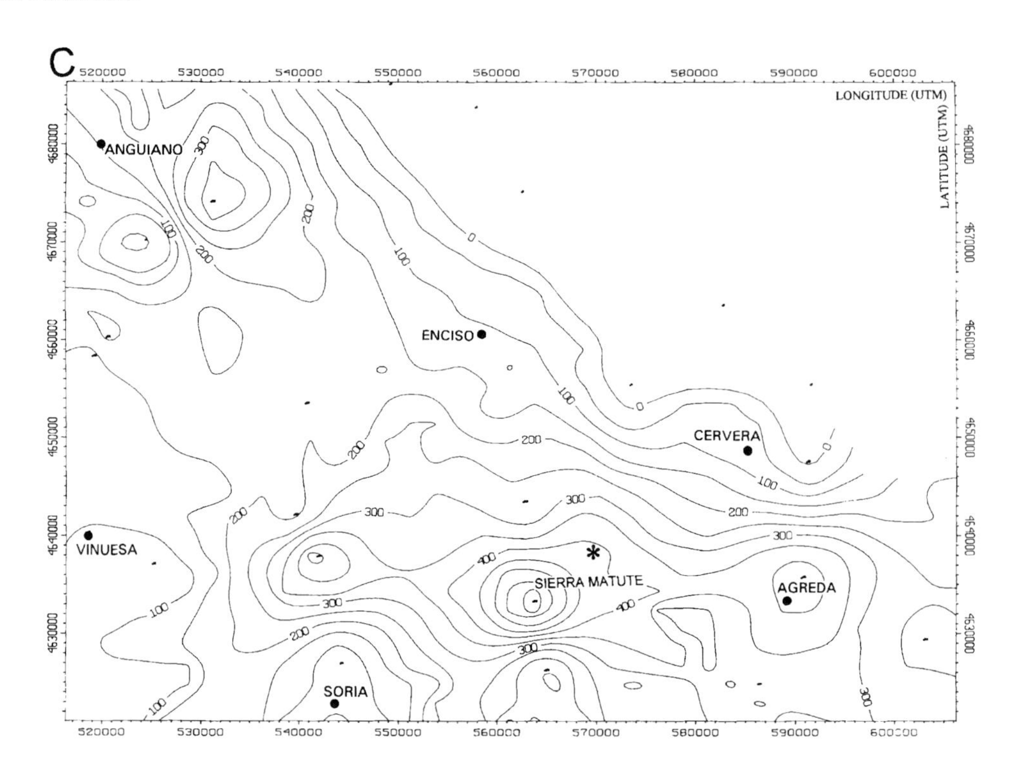

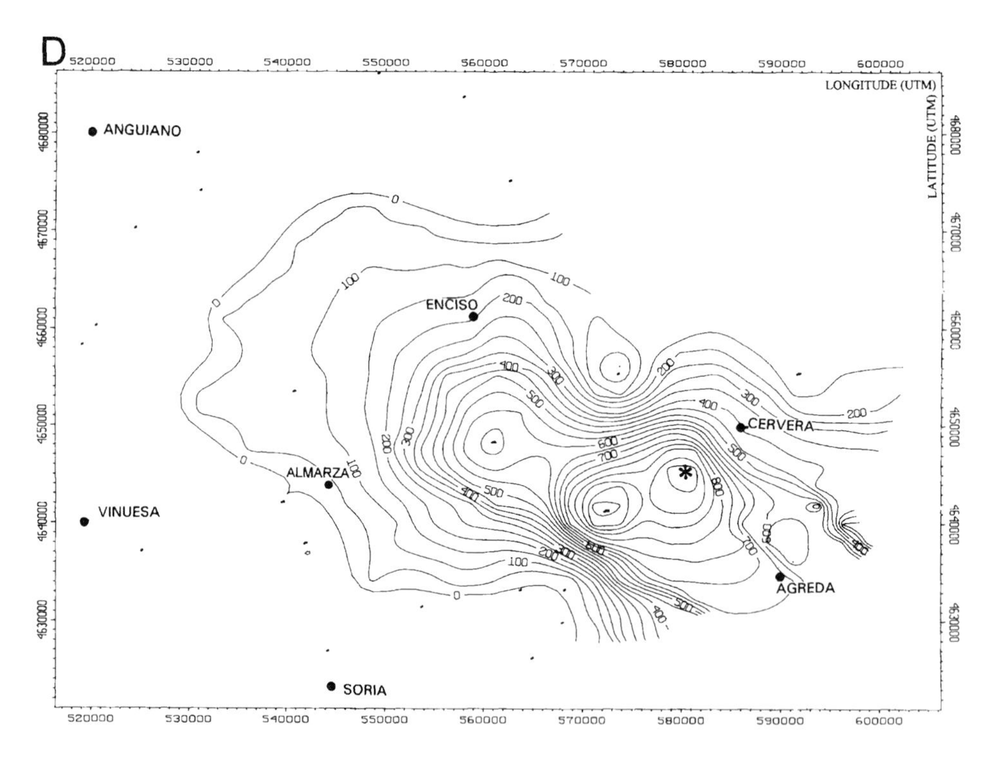

Figure 3—(C) Matute Alloformation, * Arroyo del Reajo section, (D) Huérteles Alloformation,* Aguilar section. From Gomeź-Fernandez and Menendez, 1994, with kind permission from Kluwer Academic Publishers.

In the north, siliciclastic deposits are predominant and consist of quartzite conglomerates outcropping in lenticular bodies with erosive bases, channeled sandstones locally showing lateral accretion features, and laterally related red and gray claystones (facies association C1, C4, C5, C8, I1, Figure 5). Micritic, black bioclastic limestones crop out in thin tabular beds, and massive marlstones occur in locally subsiding areas (facies association L1, Figure 5). This facies assemblage represents a fluvial environment grading to the southeast from braided to an anastomosed system. Locally, in permanently inundated extrachannel areas (wetlands), small shallow lakes with high organic productivity occurred.

In the southwestern zones of the basin, characteristic sediments are micritic bioclastic limestones and oncolitic limestone that show extensive desiccation and diagenetic features (brecciation, nodulization,

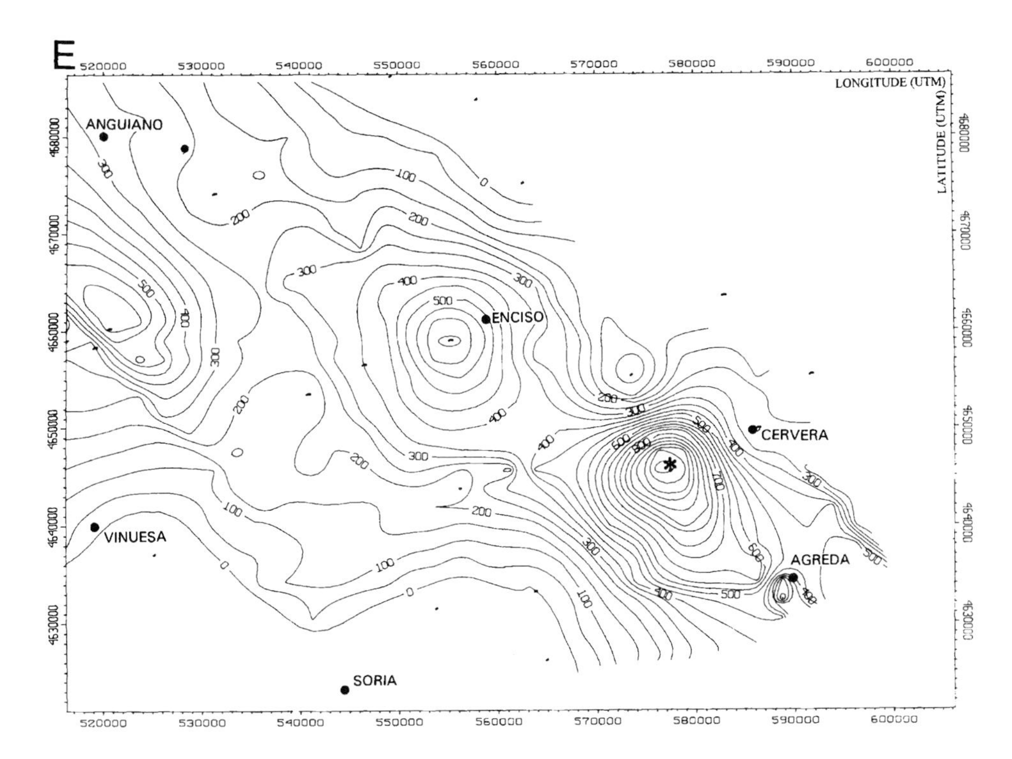

Figure 3—(E) Valdeprado Alloformation,* Navajún section. Isopachs are in meters. Minor intervals are 1 km in scale (i.e., between 520000 and 530000 are 20 km). For location of mapped area see Figure 1B.

pseudomicrokarst and calichification) from facies associations L1 and L2. These facies record deposition in shallow hardwater lakes developing in palustrine areas.

In the central and southern parts of the basin, a vertical variation in composition of the Matute Alloformation can be recognized. At the base, massive lutites with interbedded lenticular sandstones and some caliche levels are present (facies association I1, Figure 5). Sandstone bodies with erosive bases and lateral accretion surfaces, as well as channeled oncolitic limestones, also occur (facies association C8, C9, C10, C11, Figure 5). Rarely, ochre dolomite tabular beds interbedded with lutitic rocks (facies association P3) indicate deposition in small evaporitic ponds on an alluvial plain. Toward the top of the unit, the presence of marly lutites and bioclastic limestones (facies association I2) reflects the progressive flooding of the basin. Shallow carbonate lakes developed with a low-energy ramp; desiccation features are prevalent toward the top of these carbonate lake beds (facies association L1, L2). Bioclastic tabular beds containing freshwater fossils (charophytes, ostracods, gastropods, and bivalves), bioclastic laminated limestones, and micritic limestones (facies association L4) are characteristic of a lacustrine ramp of variable energy. Oncolites and chert nodules also occur. Root traces and brecciation features at the top of micritic limestones, as well as marmorization processes, are common. Scarce foraminifera have been observed in the lacustrine deposits of the southern areas of the basin, indicating the marine shoreline proximity in that direction. Lacustrine deposits are arranged in stacked shallowing-upward sequences at the western lake margin. In the eastern part of the basin, ochre dolomite tabular beds and siltstones of facies association P3 were deposited in muddy plains with evaporitic ponds.

Distribution of sedimentary paleoenvironments is shown in the paleogeographic map of Figure 6C; paleocurrents dominate from the northwest from the Demanda massif (Figure 1B). Likewise, an occasional distal marine influence from the southeast is inferred.

The Oncala Allogroup

The Huérteles Alloformation crops out only in the central and eastern part of the basin. It was deposited in a northwest-southeast–oriented trough with strong subsidence and is divided into several subbasins by tectonic features of northeast-southwest orientation. In one of these subbasins, the sedimentary record exceeds 1050 m in thickness (Figure 3D). Over most of the area, the Huérteles Alloformation lies unconformably on the Matute Alloformation. Locally, in the eastern part of the basin, the Huérteles Alloformation lies directly on marine Jurassic limestones (Kimmeridgian), which reveals onlap geometry to the northeast.

Lithology of the Huérteles Alloformation ranges from siliciclastics in the western area to carbonates and evaporites in the central and eastern part of the basin. Figure 4 shows the Aguilar del Río Alhama section (section 4) selected to represent the Huérteles Alloformation. To the west, the Huérteles is composed of siliciclastics, mainly gray mudstones and tabular fine-grained sandstones, and less commonly black micritic limestones. Some sandstone beds have a channel morphology and lateral accretion features. Rare bodies of quartzite conglomerates occur in the western margin of the basin (facies associations C2, C8, P1, P2, Figure 5). These facies record deposition produced mainly by unconfined sheetfloods in a sandy-muddy plain. Occasionally, and probably in humid periods, this plain was incised by sheetflood ephemeral channels (Figure 6D).

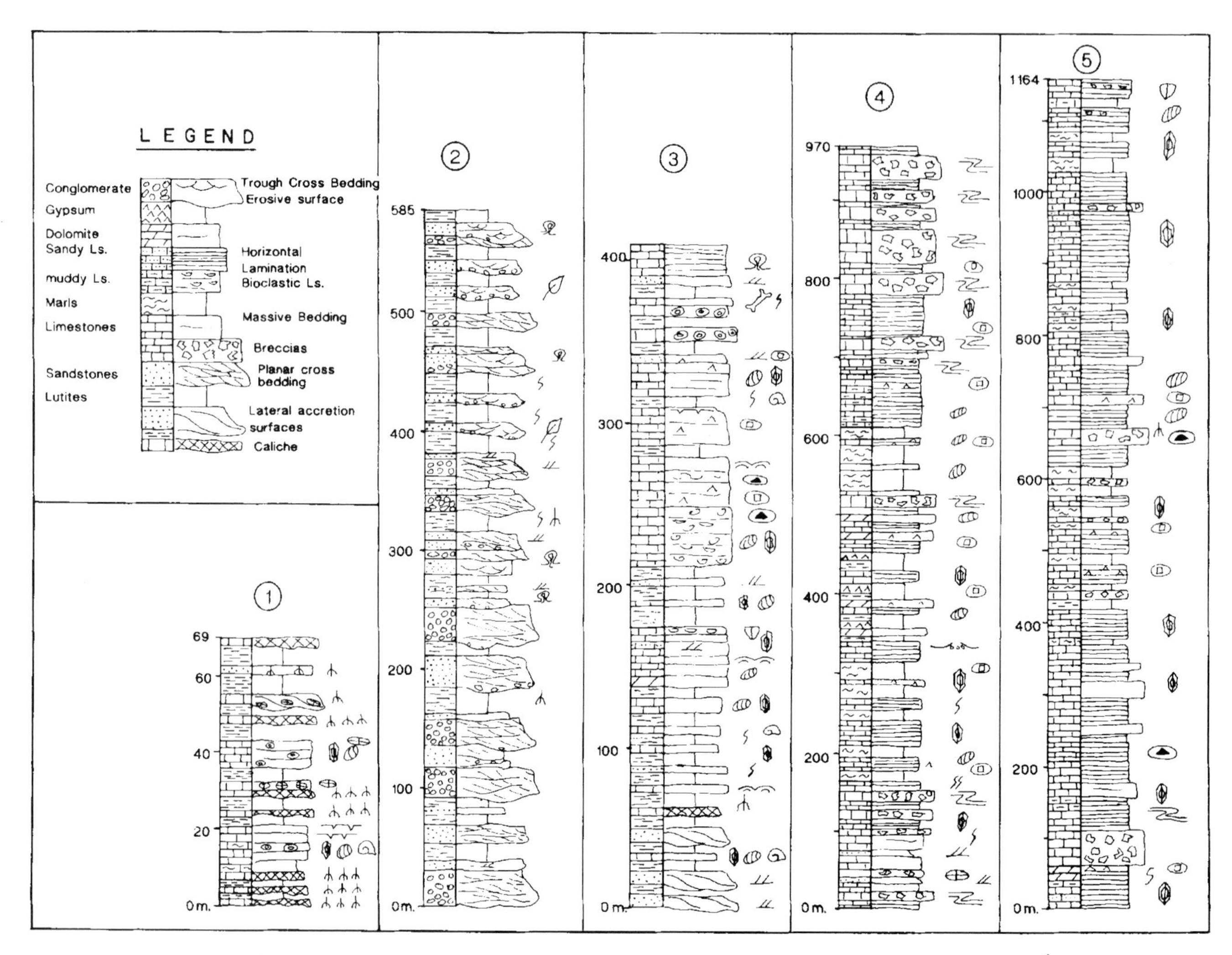

Figure 4—Detailed selected sections for the five alloformations. (1) Valdegeña section from the Ágreda Alloformation, (2) Embalse de la Cuerda del Pozo section from the Magaña Alloformation, (3) Arroyo del Reajo section from the Matute Alloformation, (4) Aguilar section from the Huérteles Alloformation, (5) Navajún section from the Valdeprado Alloformation. For location of each section, see corresponding isopach maps in Figure 3.

Toward the east in the upper part of this unit, a progressive loss of channeled bodies (facies association C8, Figure 5) occurs. The most characteristic sequence is composed of an increase in the number of micritic limestone beds with common gypsum pseudomorphs, decimeter-thick layers of algal-laminated limestones, and mudstones with gypsum that contain tabular decimeter-thick layers of dolostones. On the top of these sequences, relatively abundant reptile footprints can be found (facies associations P2, P3, P4, Figure 5). This facies association mostly reflects deposition in a mudflat paleoenvironment with ephemeral carbonate ponds, some of them saline, recording the transition from an ephemeral lake to a dry mudflat (Figure 6D).

In the central areas of the basin, laminated micritic limestones represent deposition in relatively flat and deep areas of a lake. Upward, micritic and peloidal limestones, with some interbedded algal-laminated limestones, reflect progressive infilling of a deep saline lake. At the top, dolostones, marlstones, and mudstones are present with common gypsum pseudomorphs and rare ostracods characterizing a low-energy lacustrine ramp in a saline lake that rarely was completely desiccated. This saline lake could have received siliciclastics from the northeast episodically. Calcareous breccias, gypsiferous marlstones, and mudstones are relatively abundant. This central area is represented by facies associations P4, L5, and L6 (Figures 5, 6D).

In the eastern margin of the basin, three consecutive detached lithosomes are present; each lithosome is more extensive than the former lithosome to the west. These lithosomes are mainly composed of calcareous dissolution breccias with slump structures and common olistoliths. These slump structures are postulated to originate from several tectonic pulses, related to the activity of synsedimentary normal faults whose movements caused an increase in the gradient of the basin floor and slumping of unconsolidated sediments (containing abundant salts).

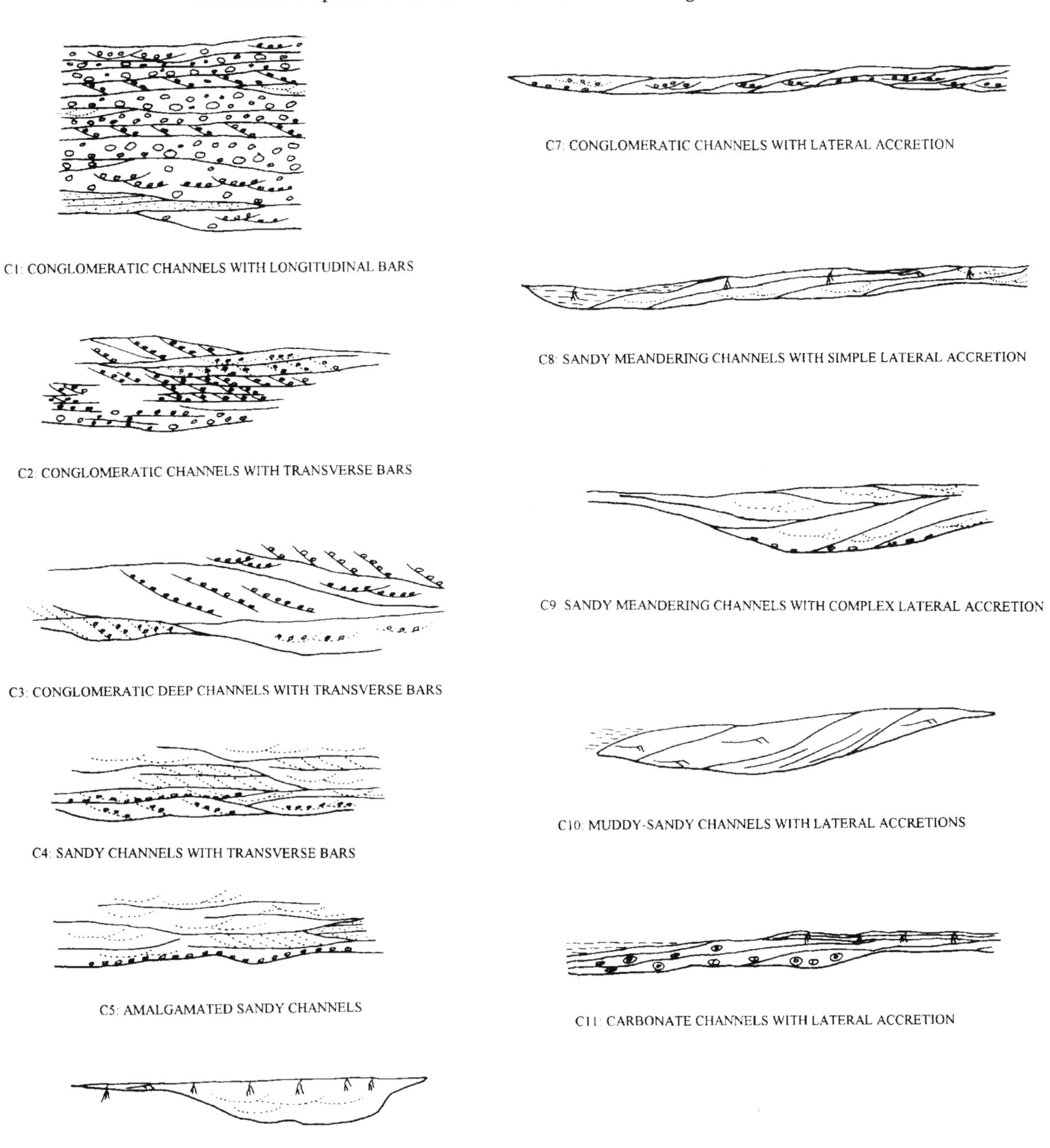

Figure 5—Facies associations and average thicknesses that have been recognized in the five alloformations. For scale, vertical bars in all representations are 1 m in length. Continued on next two pages. From Gomeź-Fernandez and Menendez, 1994, with kind permission from Kluwer Academic Publishers.

In the northeastern part of the basin, the Huérteles Alloformation comprises mudstones and conglomeratic sandstones with a coarsening- and thickening-upward trend. Dolostones and gypsum are intercalated with the mudstones. Dolostones are more abundant in the upper part of the sequence (facies associations C5, P4, I1, Figure 5). This facies assemblage records the progradation of an alluvial fan from the northeast. Intercalated with these alluvial materials, dolostones, gypsum, and mudstones facies are interpreted as saline lake sediments (Figure 6D).

Sedimentation of the Huérteles Alloformation occurred in a playa-lake system with a strongly asymmetric environmental distribution. Unchannelled proximal paleoenvironments, interpreted as a bajada, drained a low-relief provenance area located toward the west, which probably was inside the basin so that sediments were mostly derived from

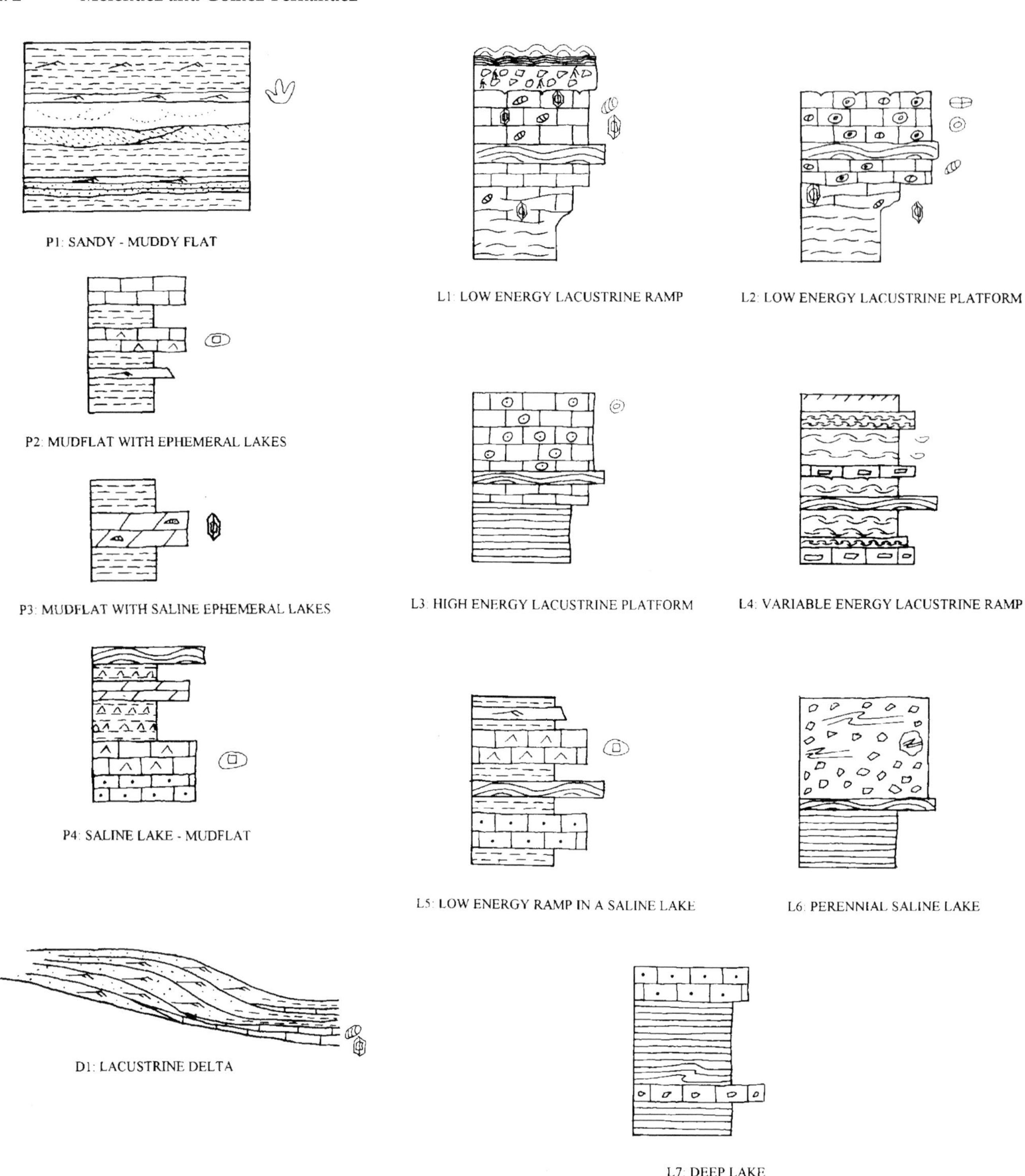

Figure 5—Continued.

erosion of previously sedimented alloformations. Toward the east, a sandflat was progressively transformed into a mudflat with ephemeral ponds, which increased in salinity, until this change culminated in the east into the evolution of a perennial saline lake. This lake received siliciclastics from fan-deltas prograding from high relief located to the northeast.

Major allocyclic controls on sedimentation of the Huérteles Alloformation were sedimentary processes, tectonics, and climate. Internal arrangement of this unit shows a marked cyclicity with primary sequences that represent the filling of a lake with relatively dilute waters that gradually became more saline. This change reflects a transition from a humid to an arid climate. Additionally, another cycle of higher rank is indicated by sequences about 300 m thick formed by minor primary sequences mainly composed of laminated limestones in the lower part of the major

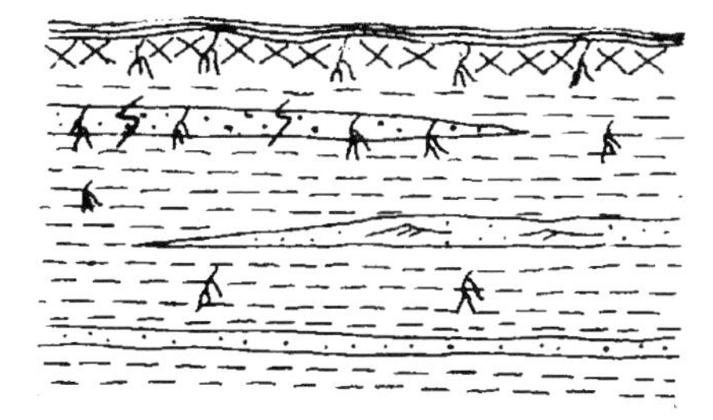

I1: MUDDY FLOOD PLAIN

I2: MUDDY- CARBONATE FLOOD PLAIN

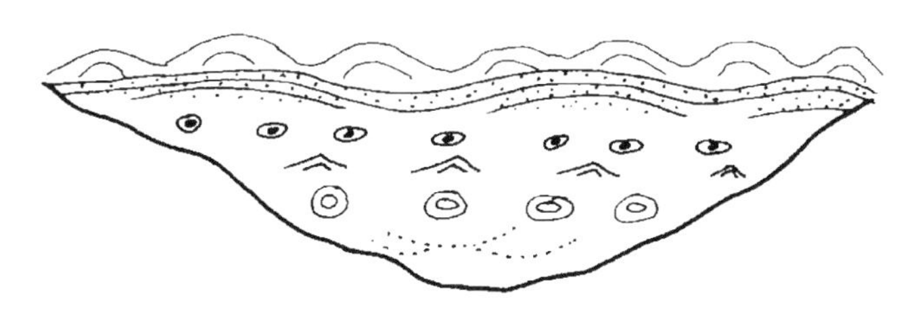

I3: ABANDONED CHANNELS

Figure 5—Continued.

sequences and carbonate breccias in the upper part. These major sequences may indicate longer periods of climatic variation. This sequential arrangement was accentuated by the strong tectonic activity during sedimentation.

The Valdeprado Alloformation extends over most of the basin (see isopach map in Figure 3E). It lies unconformably on different units in different regions. In the central areas, it lies on the Huérteles Alloformation, toward the north on the Matute Alloformation, and in the northwest on the Magaña Alloformation Near Torrecilla, in the northern part of the basin, the Valdeprado onlaps onto marine Jurassic rocks. The Valdeprado Alloformation reaches a maximum thickness of 1164 m in a trough trending northwest-southeast. A depocenter is located to the northwest with a maximum thickness of sediments at 667 m (Figure 3E).

The Valdeprado Alloformation consists of mixed siliciclastic and calcareous sedimentary rocks (Figure 4, section 5). Siliciclastics dominate in the western and northern parts of the basin. Here, transport directions are from west to east, where the Castellano massif relief was reactivated (Figure 1B). These deposits consist of rare red lutites with interbedded tabular quartzose sandstone beds that in the more proximal areas contain lenticular conglomerate bodies with erosive bases (facies associations C1, C4, C5, I1, Figures 5, 6E). Toward the east, sandstones and lutites are predominant. Sandstone bodies show lateral accretion, mainly toward the top of the sections. Some edaphic features, such as nodulization, vertical glaebules, rhizoliths in claystones and macrophytic debris, occur (facies associations C4, C5, C8, I1, L1, Figure 5). These facies associations record deposition in braided fluvial systems that progressively changed into anastomosing rivers. Black micritic limestones with ostracods are also intercalated with the lutites and are interpreted as rare carbonate ponds on the floodplain. The fluvial system (facies associations C8, I1) flowed into a lake as indicated by the presence of facies association D1, composed of fine siliciclastic sediments prograding over micritic and peloidal limestones, containing ostracods and charophytes. This facies association is interpreted as a lacustrine delta. Other facies that characterize the margin of the lake under evaporative conditions consist of limestones with gypsum pseudomorphs and interbedded peloidal limestones and mudstones (facies associations L1, L5).

Toward the east and central areas of the basin, the predominant facies consists of laminated micritic limestones; some slump structures and intraclastic or brecciated levels are common. Continuous beds of peloidal limestones are locally common. Clear predominance of laminated sediments and monotony of this facies mostly reflect the persistence of sedimentation conditions in basinal areas of a relatively deep lake. In these central and eastern areas at the base of the Valdeprado Alloformation, laminated limestones are generally associated with calcareous breccias, in similar sequences to those described in the Huérteles Alloformation. These breccias are interpreted as forming during the onset of sedimentation of this alloformation in a perennial saline lake that evolved through time to a deep lake of more dilute waters. In these central areas of the basin at the upper part of the Valdeprado Alloformation, laminated limestones are less common and the alloformation is mainly composed of siltstones, marlstones, and bioclastic and peloidal micrites. This vertical change represents the shallowing upward of the lake to a low-energy ramp shallow lake. Locally, peloidal micrites with algal lamination and gypsum pseudomorphs also occur. They record deposition in a low-energy ramp of a saline lake, indicating that the lake experienced desiccation due to evaporation. Facies associations L1, L5, L6, and L7 (see Figures 5, 6E) are represented in this alloformation.

Toward Cervera, the Valdeprado Alloformation comprises laminated limestones with minor calcareous breccias and peloidal limestones, indicating the existence of a perennial lake as in the central areas, but oolitic limestones containing miliolid foraminifera, dasycladacean

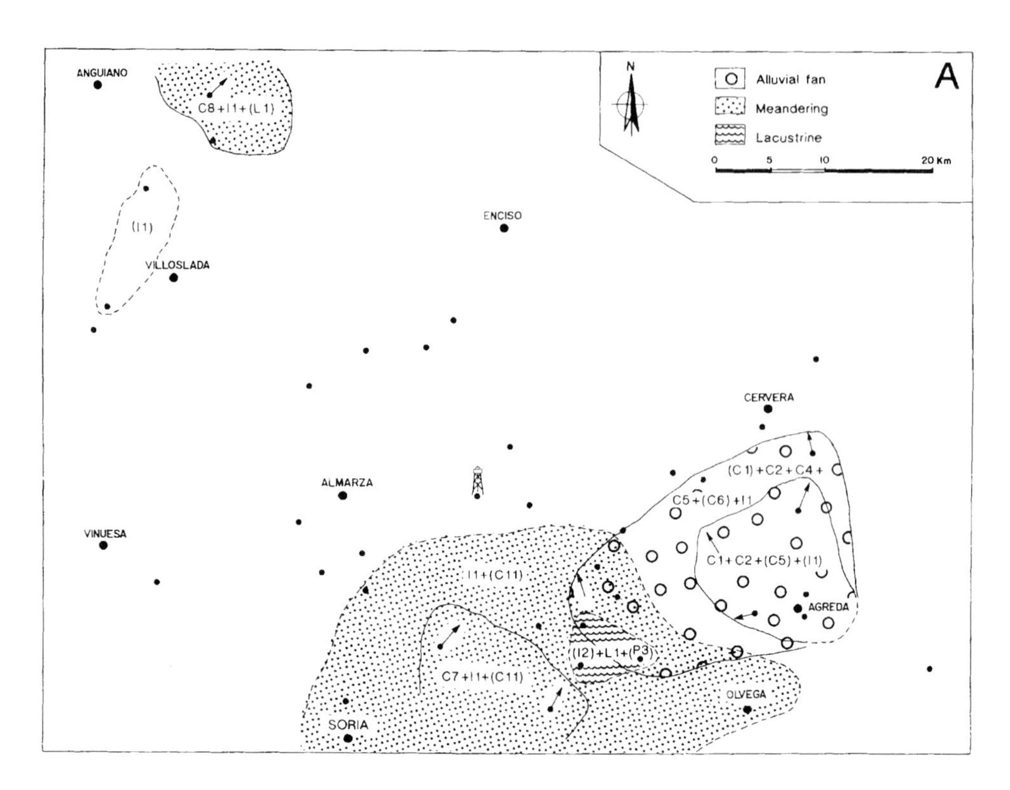

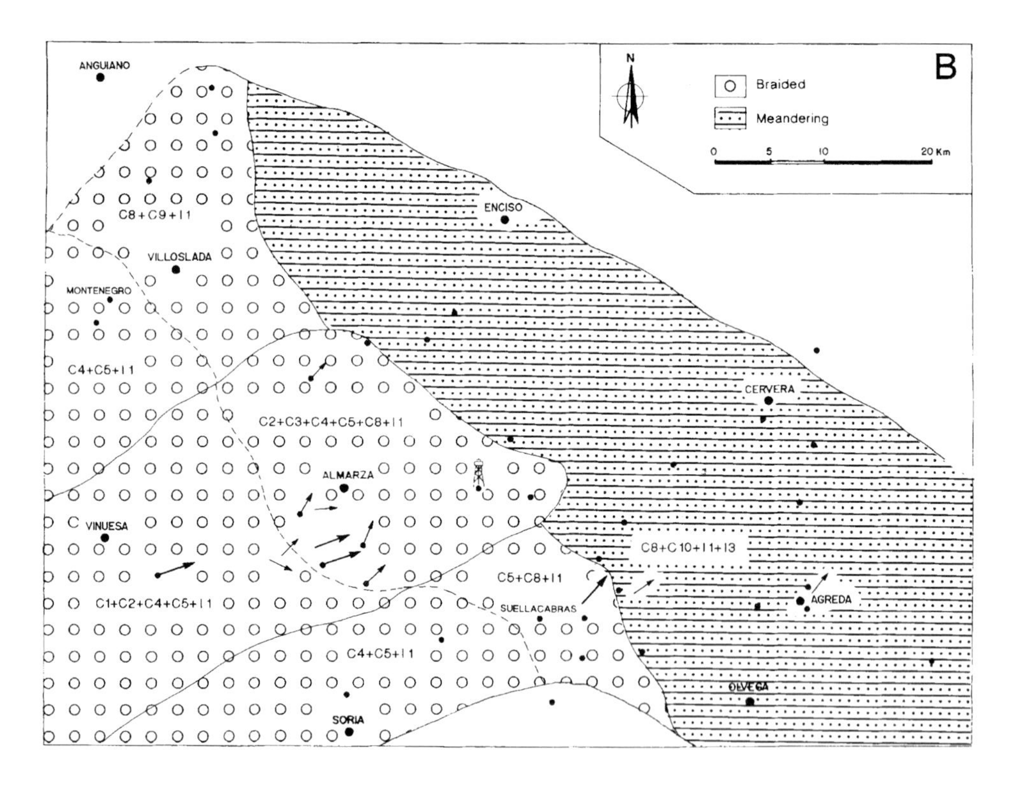

Figure 6—Paleogeographic maps for the five alloformations described. (A) Ágreda Alloformation, (B) Magaña Alloformation, (C) Matute Alloformation, (D) Huérteles Alloformation, from Gomeź-Fernandez and Menendez, 1994, with kind permission from Kluwer Academic Publishers, and (E) Valdeprado Alloformation. Arrows indicate paleocurrents. Biggest arrows in (C) and (E) indicate the direction of marine incursions coming from the Tethys located southeast of the Iberian Basin. Letters correspond to facies associations and are in brackets when not

algae, and algal lamination are present. This facies association records deposition on a high-energy lacustrine platform. The rare presence of marine organisms at the eastern part of the basin indicates occasional marine invasions from the east. Fossil content of the Valdeprado Alloformation is generally poor. Freshwater fossils include ostracods, charophytes, and gastropods. Plant debris and reptile footprints are common in the north.

SUMMARY

The main factor controlling continental infilling of the eastern Cameros Basin during the Tithonian–Berriassian was the tectonic activity, which generated an asymmetric half-graben trending northwest-southeast. This tectonic activity is related to the distensive period of rifting that affected the Iberian plate during

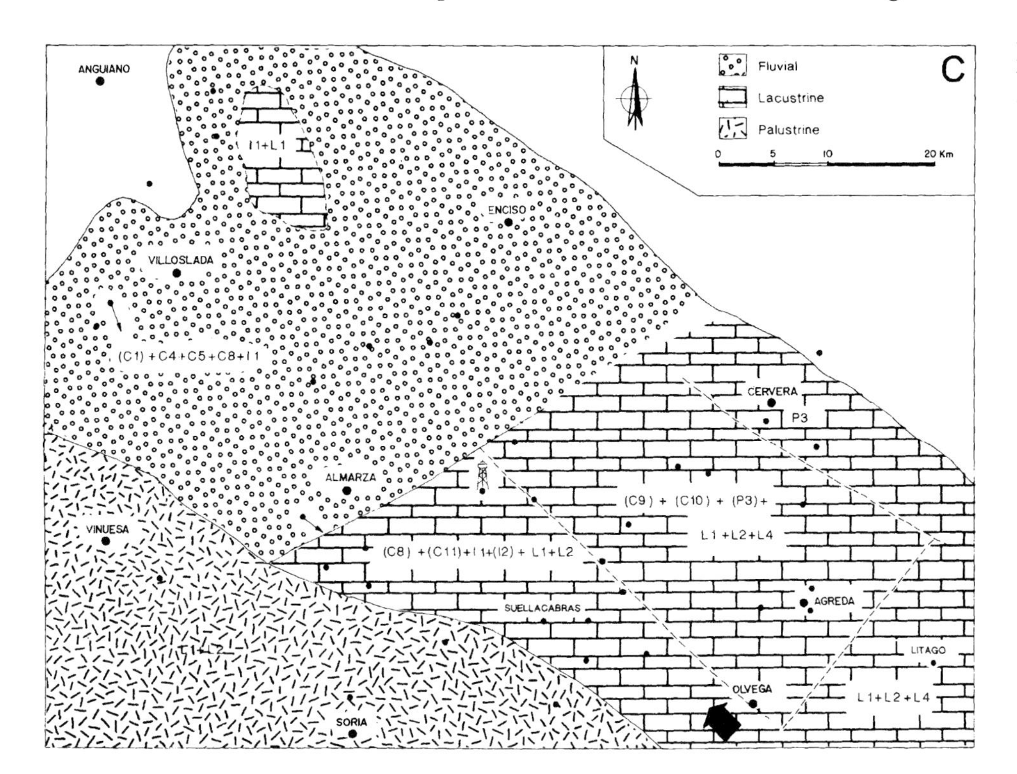

Figure 6— Part (E), Valdeprado Alloformation, located on next page.

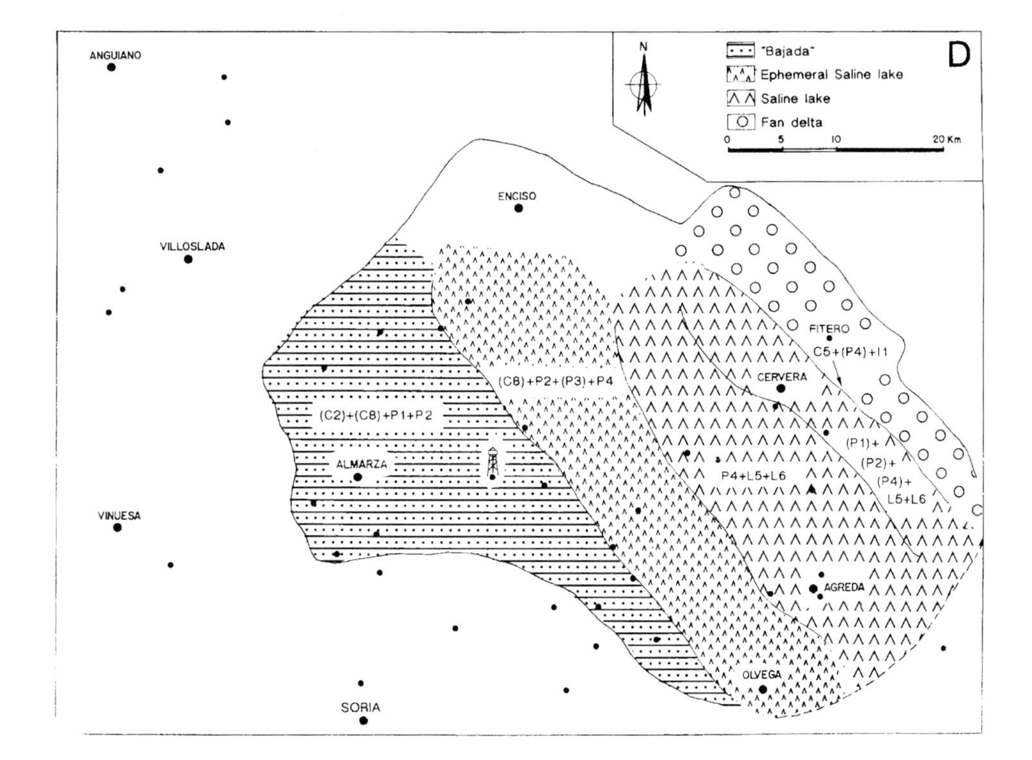

the Late Jurassic–Early Cretaceous. Unconformities resulting from successive basin reorganization allowed the division of 4000 m of sedimentary record into five allostratigraphic mappable units.

Successive isopach maps for each allostratigraphic unit show clear reorganization of the basin in basin shape, orientation, and fill geometry. Also, these isopach maps particularly delineate syndepositional intrabasinal upwarps or swells over which sections are thinned, indicating a successively subdivided basin.

The eastern Cameros Basin contains evidence for lake development throughout the Tithonian–Berriassian; these lacustrine deposits were related to subsiding depocenters as the isopach and paleogeographic maps for each sedimentary unit show. The lacustrine sedimentary records from four of the five alloformations

Figure 6—Continued.

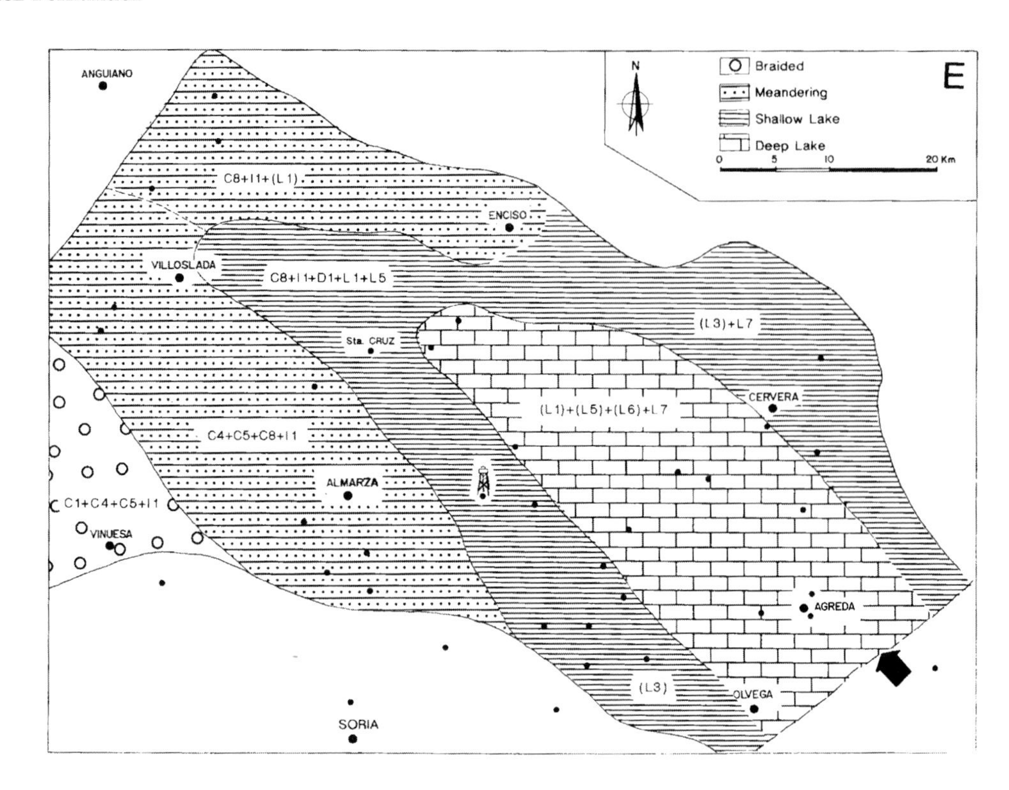

described exhibit a variety of facies that represent different types of lake environments in various stages of lake evolution. During deposition of the Ágreda and Matute Alloformations, shallow carbonate lakes and ponds developed; repeated shallowing-upward sequences of low-energy lacustrine ramp and platform deposits characterize sedimentation in these lakes. The Matute Alloformation also contains deposits from variable energy lacustrine ramps. In both the Ágreda and Matute Alloformations, lakes are laterally related to palustrine and alluvial deposits.

During deposition of the Huérteles and Valdeprado Alloformations, perennial and deep lakes developed, probably due to a higher subsidence rate of the basin. Evidence for synsedimentary activity within the basin is supported by slumps affecting lacustrine laminites and breccias. The Huérteles Alloformation contains deposits from a playa-lake system with ephemeral saline lakes and carbonate ponds in a mudflat environment. These shallow environments pass laterally into a deep saline lake. The Valdeprado Alloformation comprises micritic-laminated limestones deposited in a relatively deep lake. Laminated sediments associated with calcareous breccias have been interpreted as a perennial saline lake.

REFERENCES CITED

Allen, J. R., 1983, Studies in fluviatile sedimentation: bars, bar-complexes and sandstone sheets (low sinuosity braided streams) in the Brown Stones (Lower Devonian), Welsh Borders, Sedimentary Geology, v. 33 p. 237–293.

Casquet, C., C. Galindo, J. M. González-Casado, A. Alonso, J. R. Mas, M. Rodas, E. García-Romero, and J. F. Barrenechea, 1992, El metamorfismo en la Cuenca de Cameros, Geocronología e implicaciones tectónicas, Geogaceta, v. 11, p. 22–25.

Friend, P., M. Slater, and R. Williams, 1979, Vertical and lateral building of river sandstone bodies, Ebro Basin, Spain, Journal of Geological Society of London, v. 136, p. 133–146.

Gómez-Fernández, J. C., 1992, Análisis de la cuenca sedimentaria de los Cameros durante sus etapas iniciales de relleno en relación con su evolución paleogeográfica, Ph.D. Dissertation, Universidad Complutense de Madrid, Spain, 343 p.

Gómez-Fernández, J. C., and N. Meléndez, 1990, Shallow carbonate lakes related to alluvial systems from the Upper Jurassic Cameros Basin (N Spain), 13th Inter. Congr. Sediment. IAS, Nottingham, Abstracts, p. 194–195.

Gómez-Fernández, J. C., and N. Meléndez, 1994a, Estratigrafía de la Cuenca de los Cameros (Cordillera Ibérica Noroccidental, N de España) durante el tránsito Jurásico–Cretácico, Revista de la Sociedad Geológica de España, v. 7 (1–2), p. 121–139.

Gómez-Fernández, J. C., and N. Meléndez, 1994b, Climatic control on Lower Cretaceous sedimentation in a "playa-lake" system of a tectonically active basin "Huérteles alloformation, Cameros Basin, (north-central Spain), Journal of Paleolimnology, v. 11, p. 91–107.

Guiraud, M., and M. Seguret, 1985, A releasing solitary overstep model for the Late Jurassic–Early Cretaceous (Wealdian) Soria strike-slip basin (northern Spain), *in* K. T. Biddle and N. Christie-Blick, eds., Strike Slip Deformation, Basin Formation and Sedimentation, SEPM, Special Publication, v. 37, p. 159–175.

Martín i Closas, C., 1989, Els carofits del Cretaci inferior de las conques periferiques del bloc de l'Ebre, Tesis Doctoral Universidad de Barcelona, Spain, 581 p.

Mas, J. R., A. Alonso, and J. Guimerà, 1993, Evolución tectonosedimentaria de una cuenca extensional intraplaca: la cuenca finijurásica–eocretácica de Los Cameros (La Rioja–Soria), Revista de la Sociedad Geológica de España, v. 6 (3–4), p. 129–144.

Miall, A. D., 1988, Facies architecture in clastic sedimentary basins, *in* K. Kleinspehn and C. Paola, eds., New Perspectives in Basin Analysis, Springer-Verlag, Berlin, p. 67–81.

Miall, A. D. 1997, The Geology of the Stratigraphic Sequences, Springer-Verlag, Berlin, 433 p.

North American Commission on Stratigraphic Nomenclature, 1983, North American Stratigraphic Code (N.A.S.C.), AAPG Bulletin, v. 67, 841–875.

Normati, M., 1994, The eastern Cameros Basin (Kimmeridgian to Valanginian)—northern Iberian Ranges (Spain), *in* E. Gierlowski-Kordesch and K. Kelts, eds., Global Geological Record of Lake Basins, vol. 1, Cambridge University Press, p. 209–213.

Normati, M., and J. Salomon, 1989, Reconstruction of a Berriasian lacustrine paleoenvironment in the Cameros Basin (Spain), Palaeogeography, Palaeoclimatology, Palaeoecology, v. 70, p. 215–223.

Platt, N. H., 1989, Continental sedimentation in an evolving rift basin: the Lower Cretaceous of the Western Cameros Basin (northern Spain), Sedimentary Geology, v. 64, p. 91–109.

Platt, N. H., 1994a, The western Cameros Basin, northern Spain: Rupelo Formation (Berriasian), *in* E. Gierlowski-Kordesch and K. Kelts, eds., Global Geological Record of Lake Basins, vol. 1, Cambridge University Press, p. 195–202.

Platt, N. H., 1994b, The western Cameros Basin, northern Spain: Hortigüela Formation (?Valanginian-Barremian), *in* E. Gierlowski-Kordesch and K. Kelts, eds., Global Geological Record of Lake Basins, vol. 1, Cambridge University Press, p. 203–208.

Platt, N. H., 1995, Sedimentation and tectonics of a synrift succession: Upper Jurassic alluvial fans and paleokarst at the late Cimmerian unconformity, western Cameros Basin, northern Spain, *in* A. G. Plint, ed., Facies Analysis, Special Publication of the International Association of Sedimentologists, v. 22, p. 217–234.

Platt, N. H., and V. P. Wright, 1991, Lacustrine carbonates: facies models, facies distributions and hydrocarbon aspects, *in* P. Anadón, Ll. Cabrera, and K. Kelts, eds., Lacustrine Facies Analysis, Special Publication of the International Association of Sedimentologists, v. 13, p. 57–74.

Salomon, J., 1982a, Les formations continentales du Jurassique supérieur–Crétacé inférieur (Espagne du Nord–Chaînes Cantabrique et NW Ibérique), Memoires Geologiques de l'Université de Dijon, v. 6, p. 1–227.

Salomon, J., 1982b, Cameros–Castilla: El Cretácico inferior, *in* A. Alonso, C. Arias, A. García, R. Mas, R. Rincón, and L. Vilas, eds., El Cretácico de España, Universidad Complutense de Madrid. Madrid, p. 345–387.

Meléndez, A., A. R. Soria, and N. Meléndez, 2000, A coastal lacustrine system in the lower Barremian from the Oliete subbasin, central Iberian Range, northeastern Spain, *in* E. H. Gierlowski-Kordesch and K. R. Kelts, eds., Lake basins through space and time: AAPG Studies in Geology 46, p. 279–284.

Chapter 22

A Coastal Lacustrine System in the Lower Barremian from the Oliete Subbasin, Central Iberian Range, Northeastern Spain

A. Meléndez
A. R. Soria
Dpto. Ciencias de la Tierra, Universidad de Zaragoza
Zaragoza, Spain

N. Meléndez
Dpto. Estratigrafía, Facultad de Ciencias Geológicas, Universidad Complutense de Madrid
Madrid, Spain

INTRODUCTION

The Oliete subbasin is located in the central Iberian Range, northern Teruel Province (see Figure 1 in Soria et al., this volume, concerning an overview of the geology and tectonics of the Iberian Basin). The Oliete subbasin extends over an area of 1500 km^2 and is bounded toward the north by the Sierra de Arcos thrust and toward the south by the Montalbán-Portalrubio thrust (Figure 1). This subbasin is physically separated from two other Cretaceous subbasins: the Aguilón and Las Parras subbasins (see Figure 1 of Soria et al., this volume).

In the Oliete subbasin, during the early Barremian, sedimentation of predominantly continental deposits interbedded with shallow marine layers occurred after a long period of erosion and karstification of the Jurassic basement. These deposits constitute a lithostratigraphic unit called the Blesa Formation (Canérot et al., 1982; Murat, 1983; Salas, 1987; Meléndez, 1991; Soria, 1991, 1997; Soria et al., 1995, 1997).

THE BLESA FORMATION

Four members are recognized in the Blesa Formation (Figure 2), based on data gathered from 11 outcrop sections (Figure 3). These members are the Josa Member, the Cabezo Gordo Member, the Morenillo Member, and the Valdejunco Member.

The Josa Member is represented at the bottom by up to 40 m of red, rarely grey, massive lutites and at the top by up to 73 m of marls and limestones (Table 1). Deposits are arranged in stacked shallowing-upward sequences 1–2 m thick. At the top of these sequences, mudcracks and desiccation breccias are common. The green-black marls are laminated but rarely massive, contain large quantities of organic matter, and are poor in sulfur. The limestones are bioclastic wackestones to packstones, bearing gastropods, bivalves, oysters, and fish teeth.

The Cabezo Gordo Member, with a thickness of up to 25 m, is represented by pisolitic iron lutites (Table 1) grading upward into brown micaceous claystones interbedded with thin (centimeter-scale), bioturbated layers of carbonate mudstones. These carbonate mudstones rarely contain charophytes and ostracods.

The Morenillo Member consists of up to 75 m of successive sequences composed of massive marls and limestones (Table 1). General thicknesses of these sequences range from 3 to 15 m. Sequences comprise from base to top poorly bioturbated marls containing ostracods and charophytes, and bioturbated mudstones and bioclastic wackestones to packstones containing ostracods, charophytes, gastropods, and bivalves. Brecciation layers, ferruginous surfaces, and mudcracks can also be found at the top of these sequences. The marl-unit thicknesses in each sequence decrease upward. Centimeter-scale graded layers of oysters are rarely interbedded with the limestones.

The Valdejunco Member can be up to 30 m in thickness and is represented by both lutites and sandstones. Red lutites show lenticular bedding and parallel lamination. Sandstones (decimeter-scale) show cross-stratification, symmetric ripple lamination, and flaser bedding. Thin (centimeter-scale) graded layers of oysters rarely alternate with claystones at the bottom of this member. Between the Morenillo and Valdejunco members is a well-developed ferruginous surface that can be traced across all of the subbasin.

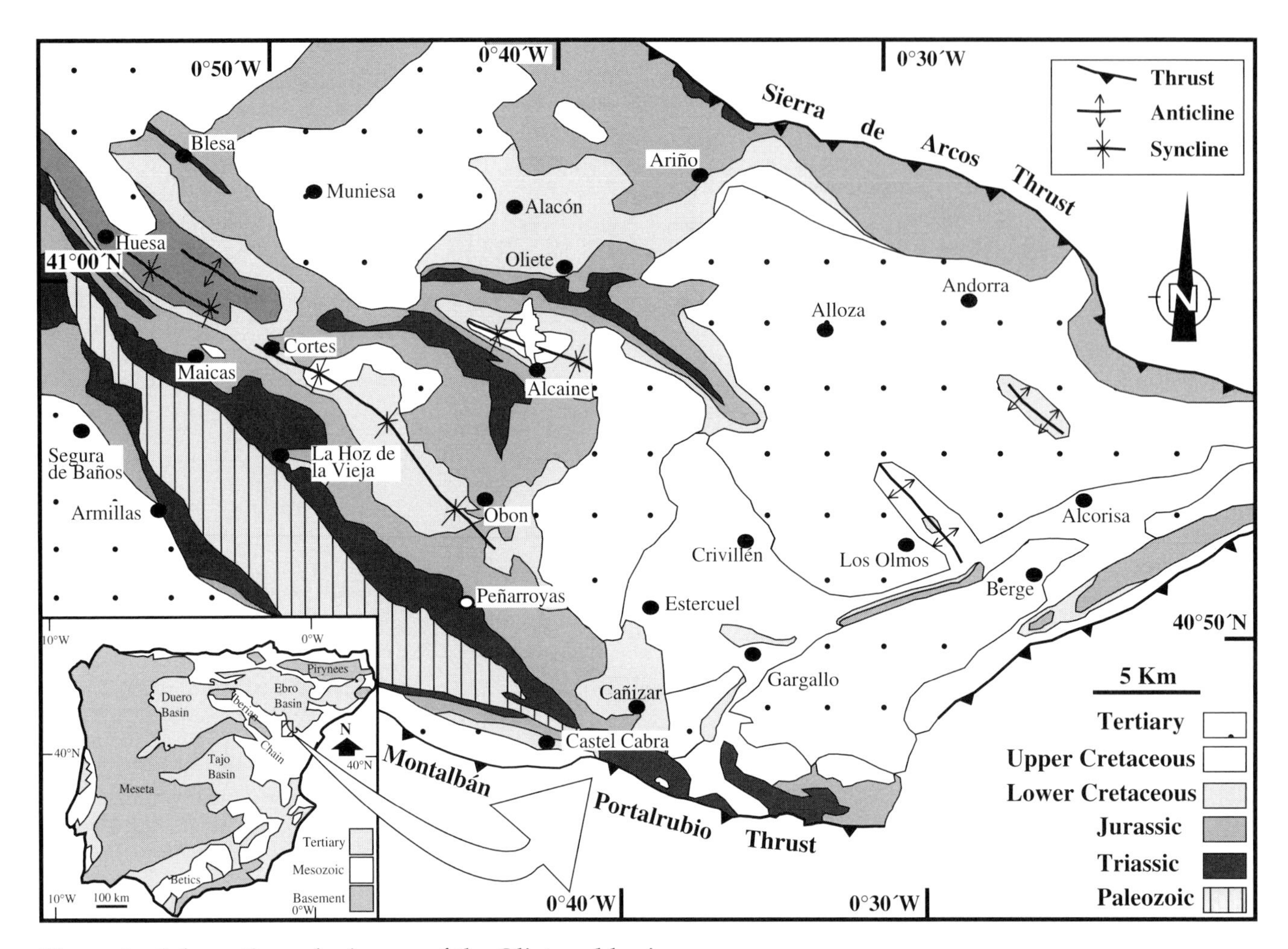

Figure 1—Schematic geologic map of the Oliete subbasin.

SEDIMENTARY ENVIRONMENTS

The Blesa Formation records sedimentation in a transitional paleoenvironment that evolved from coastal conditions toward a lacustrine and palustrine environment (Figure 2). At the base, the Josa Member begins with red-grey massive lutites deposited in a distal muddy alluvial plain. Above, the black laminated marls alternating with bioclastic limestones are interpreted as deposits of a restricted shallow marine lagoon with interbedded storm deposits occurring toward the top of the member. The Cabezo Gordo Member consists of bioturbated, iron lutitic deposits interpreted as a distal alluvial plain with palustrine areas where paleosols developed. Bioturbated claystone facies containing charophytes and ostracods represent shallow lacustrine lutitic deposits. These sediments grade upward into shallow carbonate lacustrine deposits corresponding to the overlying Morenillo Member. The lacustrine deposits of the Morenillo Member are arranged in shallowing-upward sequences documenting the repeated infilling of the carbonate lakes. The interbedded bioclastic limestone graded layers (oyster packstones) have been interpreted as storm events (tempestites), with reworked material being derived from neighboring marine environments toward the southeast. The Valdejunco Member represents a return to coastal marine paleoenvironments. This member is characterized by fining-upward sequences resulting from lateral accretion of subtidal flats over subtidal channels.

DEPOSITIONAL SEQUENCES

The lacustrine system developed in the Oliete subbasin during the Early Cretaceous belongs to the Blesa depositional sequence (Figure 4) of the lower Barremian (Soria et al., 1995). Facies correlation and age assignments are based on charophytes (Martín-Closas, 1989). This depositional sequence includes only the Josa, Cabezo Gordo, and Morenillo Members from the Blesa Formation (Figures 2, 3). Even though the Valdejunco Member was described as part of the same formation (Canérot et al., 1982), it is not related from a genetic point of view to the other members of the Blesa Formation, but to the next lithostratigraphic unit (the Alacón Formation) (Canerot et al., 1982). Thus the Valdejunco Member corresponds to the Alacón depositional sequence (Figure 4), as shown by its sedimentologic

Table 1. Description of the Main Facies Recognized in the Blesa Depositional Sequence in the Oliete Subbasin (Lower Barremian).

Lithology	Facies	Description
Claystones	Massive claystones	Massive but rarely bioturbated claystones with tabular geometry. Unit thickness: m-scale. Ostracods and charophytes present.
	Bioturbated claystones	Massive to bioturbated red claystones with tabular geometry. Unit thickness: m-scale. Paleosols, carbonate nodules, and pisolites.
	Laminated claystones	Claystones with tabular geometry. Unit thickness: m-scale. Lenticular stratification and parallel lamination. Thin sandstones beds intercalated, containing wave ripples on top surfaces.
	Sandy claystones	Sandy claystones with tabular geometry. Unit thickness: m-scale. Lenticular stratification, parallel to ripple cross-lamination (wave).
	Massive marlstones	Massive but rarely bioturbated marlstones with tabular geometry. Unit thickness: dm to m-scale. Ostracods, charophytes, and bivalves.
	Laminated marlstones	Massive to laminated gray to black marlstones with tabular geometry. Unit thickness: dm-scale. Ostracods, charophytes, and fish debris.
Sandstones	Channeled sandstones	Medium- and coarse-grained sandstones in fining-upward sequences. Tabular or channeled geometry. Unit thickness: dm-scale. Tabular and trough cross-stratification, cross-lamination, flaser bedding, and ripples with wave-lamination.
	Tabular sandstones	Medium- and fine-grained sandstones with tabular geometry. Unit thickness: dm to cm-scale. Cross-bedding and ripple lamination.

Lithology	Facies	Description	Microfacies	Components/Characteristics
Limestones	Massive limestones	Bioclastic and rarely micritic limestone with tabular and channeled geometry (m-scale). They show mudcracks and algal laminations.	Mudstone	Charophytes and ostracods
			Bioclastic Wackestone Packstone	Charophytes, ostracods, gastropods, and algae. Rarely bivalves and plant remains.
			Bioclastic Wackestone	Charophytes, ostracods, bivalves, and algae. Rare oncolites.
				Bivalves, gastropods, ostracods, bentonic foraminifera, fish debris, algae, and oolites.
				Bivalves.
			Intraclastic and Bioclastic Packstone	Micritic, bioclastic, peloidal, and oolitic intraclasts. Charophytes, ostracods, bivalves, and rarely oncolites.
	Bioturbated limestones	Micritic limestone with tabular and lenticular geometry. Units: dm- to m-scale thickness. Root traces, vertical prismatic structures, mud mounds, algal laminations, and mudcracks.	Bioturbated Mudstones	Charophytes, ostracods, and gastropods. Fenestral porosity and root traces.
			Bioturbated Mudstone-Wackestone	Charophytes, ostracods, and gastropods. Rarely, algae, plant remains, and intraclasts. Bioturbation.
			Bioclastic Packstone	Charophytes and ostracods. Bioturbation and root traces.
	Nodular limestones	Micritic limestone with tabular geometry and nodular stratification. Units: cm-scale thickness.	Bioturbated Mudstone	Charophytes. Bioturbation and root traces.
	Laminated limestones	Micritic and laminated limestone with tabular geometry. Units: dm-scale thickness. Parallel lamination.	Laminated Mudstone-Wackestone	Alternation of micritic layers (bioclastic mudstone-wackestone with bivalves, ostracods, fish, and oolites) and bioclastic layers (bioclastic packstone with bivalves and ostracods).
	Sandy limestones	Sandy limestone with tabular, channeled, and lenticular geometry. Units: dm- to m-scale thickness. Fining-upward sequences with intraclastic and bioclastic lags at the bottom of channels. Commonly, bioclastic packstone at the bottom (bivalves and intraclasts) and sandy-bioclastic wackestone at the top (bivalves, gastropods, and ostracods) of sequences.		
	Rudstones	Limestone "conglomerates" with tabular or channeled geometry. Units: dm- and cm-scale thickness.		

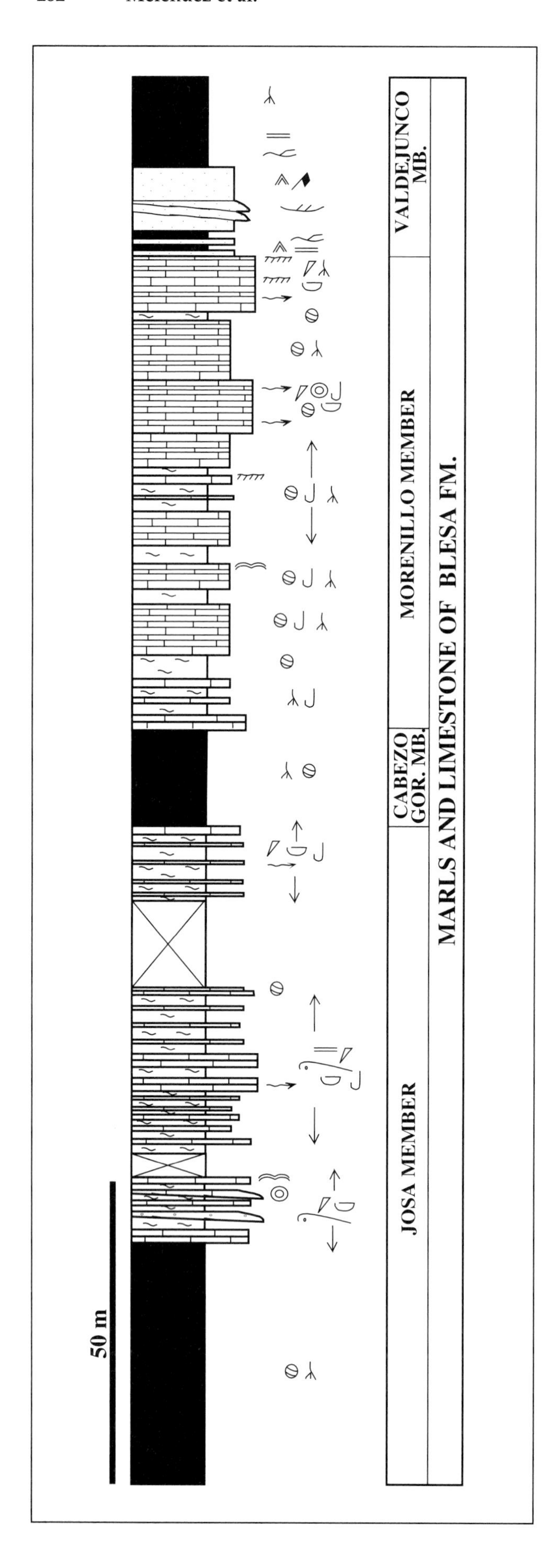

Figure 2—Lithostratigraphic synthesis of the Blesa Formation (Josa, Cabezo Gordo, Morenillo, and Valdejunco Members) in the Oliete subbasin. Global Geological Record of Lake Basins (GGLAB) legend.

evolution and the ferruginous layer between the Morenillo and Valdejunco members (Figure 2) (Soria et al., 1995; Soria, 1997).

The Blesa depositional sequence (lower Barremian) is bounded by two regional unconformities. Its lower boundary is recorded across the entire Oliete subbasin as a regional unconformity or disconformity associated with erosion and karst surface development. The upper boundary occurs either as a paraconformity associated with lithologically and sedimentologically significant changes, or as a disconformity (Vennin et al., 1993; Soria et al., 1994). In the central areas of the basin (Maestrazgo basin; see Figure 1 in Soria et al., this volume, concerning an overview of the geology and tectonics of the Iberian Basin), the lower and upper boundaries of this sequence are characterized by karstification and lateritic soils (Salas et al., 1995; Salas and Casas, 1993).

The Blesa depositional sequence shows a retrogradational-progradational pattern (Figures 3, 4). Synsedimentary tectonics and variations in sediment supply were the principal factors controlling basin fill patterns. Interactions between these controls satisfactorily explain the distribution and thickness of the continental units and nature of the unconformity located over the Jurassic basement. Presence of interbedded marine levels within this depositional sequence indicates that the lacustrine systems developed near a marine paleoshoreline setting. Given such a paleoenvironment, sea level fluctuations may also have been important in controlling the sedimentary regime.

ACKNOWLEDGMENTS

The authors are grateful to B. Owens and the editors for their suggestions and detailed revision of the paper. This work is a contribution to project PB95-1142-C02-02 (Dirección General de Investigación Científica y Técnica).

REFERENCES CITED

Canérot, J., P. Cugny, G. Pardo, R. Salas, and J. Villena, 1982, Ibérica Central y Maestrazgo, *in* A. García, L. Sánchez de la Torre, V. Pujalte, J. García-Mondejar, J. Rosell, S. Robles, A. Alonso, J. Canérot, L. Vilas, J. A. Vera, and J. Ramírez del Pozo, eds., El Cretácico de España: Universidad Complutese de Madrid, p. 273-344.

Martín-Closas, C., 1989, Els caròfits del Cretaci inferior de les conques perifèriques del Bloc de l´Ebre, Ph.D. dissertation, Universidad de Barcelona, 581 p.

Meléndez, A., 1991, Sedimentología del Muschelkalk y Cretácico, *in* E. Ferreiro, coord., Instituto

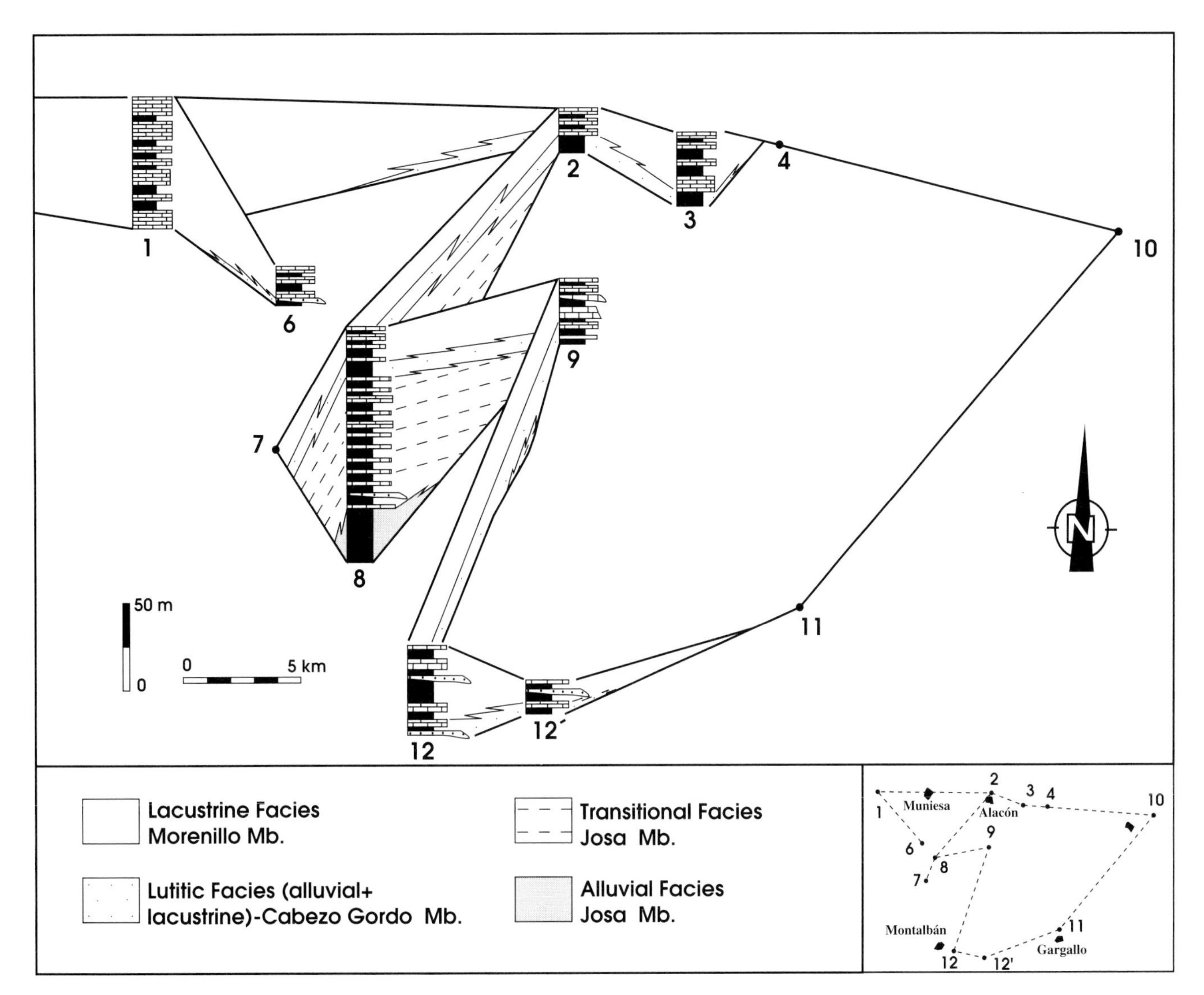

Figure 3—Vertical and lateral distribution of the Blesa depositional sequence (Josa, Cabezo Gordo, and Morenillo Members) in the Oliete subbasin.

Tecnológico y Geominero de España (ITGE), ed., Mapa Geológico de España, escala 1:200.000 (primera edición), Madrid, no. 40 (Daroca), p. 86-101.

Murat, B., 1983, Contribution à l´étude stratigraphique, sédimentologique et tectonique du bassin éocrétacé d´Oliete (Prov. de Teruel, Espagne), Masters thesis, Travaux du Laboratoire de Géologie Sedimentaire et Paleontologie, Université Paul Sabatier, Toulouse, 247 p.

Salas, R., 1987, El Malm i el Cretaci inferior entre el Massis de Garraf i la Serra d´Espadà, Ph.D. dissertation, Universidad de Barcelona, 345 p.

Salas, R., and A. Casas, 1993, Mesozoic extensional tectonics, stratigraphy and crustal evolution during the Alpine cycle of the eastern Iberian Basin, Tectonophysics, v. 228, p. 33-55.

Salas, R., C. Martín-Closas, X. Querol, J. Guimerá, and E. Roca, 1995, Evolución tectonosedimentaria de las cuencas del Maestrazgo y Aliaga-Peñagolosa durante el Cretácico inferior, *in* R. Salas and C. Martín-Closas, eds., El Cretácico Inferior del Nordeste de Iberia. Guia de campo de la excursiones científicas del III Coloquio del Cretácico de España (Morella, 1991), Publications Universitat de Barcelona, p. 13-94.

Soria, A. R., 1991, El Cretácico inferior marino de la Cubeta de Oliete. Análisis de Cuenca, Masters thesis, Universidad de Zaragoza, 120 p.

Soria, A. R., 1997, La sedimentación en las cuencas marginales del surco Ibérico durante el Cretácico Inferior y su control tectónico, Ph.D. dissertation, Universidad de Zaragoza, Servicio de Publicaciones, Universidad de Zaragoza, 363 p.

Soria, A. R., E. Vennin, and A. Meléndez, 1994, Control tectónico en la evolución de las rampas carbonatadas del Cretácico inferior de la Cubeta de Oliete (Prov. Teruel), Revista de la Sociedad Geológica de España, v. 7 (1-2), p. 47-62.

Soria, A. R., C. Martín-Closas, A. Meléndez, M. N. Meléndez, and M. Aurell, 1995, Estratigrafía del

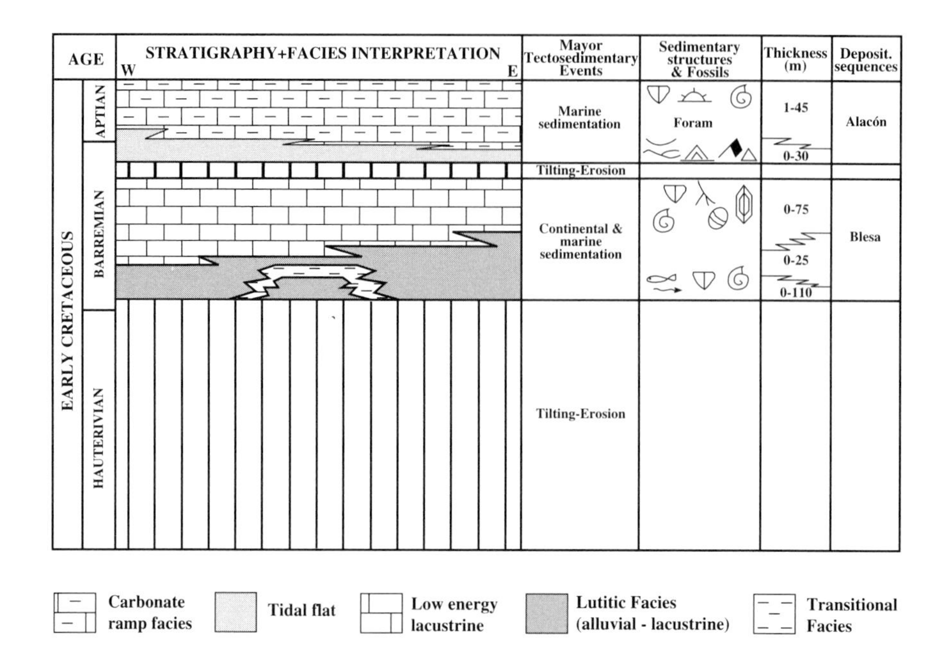

Figure 4—Facies distribution and stratigraphy of the Blesa and Alacón depositional sequences across the Oliete subbasin in a west-east direction. Global Geological Record of Lake Basins (GGLAB) legend for sedimentary structures and fossils.

Cretácico inferior continental de la Cordillera Ibérica Central, Estudios Geológicos, v. 51 (3-4), p. 141-152.
Vennin, E., A. R. Soria, A. Preat, and A. Meléndez, 1993, Los sedimentos marinos del Cretácico inferior de la Cubeta de Oliete. Análisis de Cuenca, Cuadernos de Geología Ibérica, v. 17, p. 257-283.

Meléndez, N., A. R. Soria, A. Meléndez, M. Aurell, and C. L. Liesa, 2000, Early Cretaceaous lacustrine systems of the Aguilón subbasin, central Iberian Range, northeastern Spain, *in* E. H., Gierlowski-Kordesch and K. R. Kelts, eds., Lake basins through space and time: AAPG Studies in Geology 46, p. 285–294.

Chapter 23

Early Cretaceous Lacustrine Systems of the Aguilón Subbasin Central Iberian Range, Northeastern Spain

N. Meléndez
Dpto. de Estratigrafía (Fac. de Geología) Universidad Complutense, Ciudad Universitaria Madrid, Spain

A. R. Soria
A. Meléndez
M. Aurell
Dpto. Ciencias de la Tierra (Estratigrafía), Universidad Zaragoza Zaragoza, Spain

C. L. Liesa
Dpto. Ciencias de la Tierra, (Geodinámica), Universidad Zaragoza Zaragoza, Spain

GEOLOGIC SETTING OF AGUILÓN SUBBASIN

The Aguilón subbasin is one of the Cretaceous sedimentary subbasins of the Iberian basin situated in the Central Iberian paleogeographic domain (Figure 1) (see also Figure 1 in Soria et al., this volume, concerning an overview of the geology and tectonics of the Iberian basin). Extensive facies analysis is based on the study of 13 sections and 10 boreholes (Figures 2, 3). Figure 4 shows the main facies and stratigraphic unit distribution in the basin from correlation of the two most representative and complete sections. Facies correlation and age assignments are based on charophytes. Two lithostratigraphic units are distinguished (Soria et al., 1995a, b, 1997; Soria, 1997): the Villanueva de Huerva Formation (Valanginian–middle Hauterivian) and the Aguilón Formation (late Hauterivian–early Barremian) (Figures 2–4). Both units are separated by a regional unconformity, which has been related to local extensional tectonics (i.e., tilting of blocks along normal faults).

Formation of the Aguilón subbasin was controlled by extensional tectonic events that were related to the rifting that took place during the Early Cretaceous in the Iberian Basin (Salas et al., 1995; Salas and Casas, 1993). This tectonic control allowed the creation of horsts, grabens, and half-grabens (Salas, 1987), which were controlled by listric faults (probably originated in the late Hercynian) and defined isolated intracratonic basins. Two fault systems controlling thickness and facies distribution are identified in the Aguilón subbasin (Figures 2, 3). The main fault system has a northwest-southeast trend and the secondary fault system has a northeast-southwest trend.

Two lake systems are recognized in the Aguilón subbasin: the Villanueva de Huerva and Aguilón formations. These systems developed independently (see Figure 4) of each other during the Valanginian–middle Hauterivian and the latest Hauterivian–earliest Barremian.

VILLANUEVA DE HUERVA FORMATION: FACIES DISTRIBUTION

The Villanueva de Huerva Formation displays a retrogradational pattern in which the facies become progressively more distal at the top of succession (Milton and Emery, 1996) (Figure 4). Table 1 shows the principal facies recognized in this unit. Figure 5 shows the vertical distribution of facies associations of this unit in the stratotype section, as well as some significant lateral variations across the subbasin.

Distal alluvial facies are located at the base of the stratoype section; they consist of claystones with rare pedogenic features, such as root traces, with interbedded tabular, cross-bedded, and channeled sandstones.

Table 1. Description of the Principal Facies in the Villanueva de Huerva Formation of the Aguilón Subbasin, Spain.

Lithology	Facies	Description
Claystones: C1	Red claystones: Clb	Massive and bioturbated red claystones with tabular geometry. Thickness: dm- to m-scale.
	Grey claystones: Clm	Massive grey claystones with tabular geometry (m-scale thickness).
	Marlstones: M	Massive grey marlstones with tabular geometry and dm- and m-scale thickness. Common bioturbation (root traces).
Sandstones: S	Channel sandstones: St	Middle- to coarse-grained channel-geometry sandstones, in fining-upward sequences (dm- to m-scale thickness). Trough cross-bedding as well as cross- and parallel-lamination.
	Tabular and lenticular sandstones: Sm	Middle- to fine-grained sandstones with tabular and lenticular geometry (m-scale thickness) with parallel- and cross-lamination. Bioturbation common.
	Trough cross-bedded sandstones: Ss	Middle- to coarse-grained sandstones with coarsening-upward sequences, tabular or lenticular geometry (m-scale thickness), and trough cross-bedding.

Lithology	Facies	Description	Microfacies	Components/Characteristics
Limestones: L	Tabular limestones: Lt	Bioclastic limestones with tabular geometry (dm- to m-scale thickness). Mud-cracks at tops of limestone layers.	Bioturbated Mudstone-Wackestone	Charophytes and gastropods, quartz, and micritic intraclasts. Fenestral porosity and root traces. Chalcedony and lenticular morphology of primary gypsum.
			Peloidal Wackestone-	Peloids, oolites (types 3, 4, and 6; Strasser, 1986), micritic and bioclastic intraclasts, charophytes, ostracods, and algae. Fenestral porosity.
			Bioclastic Wackestone-Packstone	Charophytes, ostracods, bivalves, gastropods, and algae. Rare oolitic and micritic intraclasts, ooids, and oncoids.
			Intraclastic Packstone-Grainstone	Peloids, bioclasts, oolitic, and micritic intraclasts. Oolites, oncolites, peloids, and quartz grains. Fenestral porosity.
	Channel limestones: Lc	Limestones with channel geometry (dm- to m-scale thickness). Rare calcareous rudstones with tabular and channel geometry. Lag deposits at the bottom of channels.	Peloidal Wackestone	Peloids, oolites (types 3, 4, and 6), micritic intraclasts, charophytes, ostracods, and algae. Fenestral porosity.
			Intraclastic Packstone-Grainstone	Peloids, bioclasts, oolitic, and micritic intraclasts. Oolites, oncolites, peloids, and quartz grains.
	Intraclastic limestones: Li	Intraclastic limestones with tabular geometry (dm- to m-scale thickness).	Intraclastic and Bioturbated Wackestones	Micritic intraclasts, charophytes, gastropods, and quartz grains. Fenestral porosity and root traces.
			Bioclastic Packstone	Charophytes, ostracods, bivalves, gastropods, and algae. Rare oolitic and micritic intraclasts, oolites, and oncolites.
	Bioturbated limestones: Lb	Tabular limestones (dm-scale thickness). Bioturbation with root traces and vertical prismatic structures.	Bioturbated Mudstone-Wackestone	Charophytes, gastropods, quartz grains, and micritic intraclasts. Fenestral porosity and root traces.
	Laminated limestones: Ll	Micritic and laminated limestone with tabular geometry (dm-scale thickness).	Laminated Mudstone	Alternation of micritic layers and thin sparitic layers. Bioturbation and brecciation.
	Sandy limestones: Ls	Sandy and bioclastic limestones with tabular geometry (dm-scale thickness).	Sandy and Bioclastic Packstone	Gastropods, bivalves, ostracods, serpulids, benthic foraminifera, charophytes, and echinoderms. Quartz, feldspar, and mica grains. Glauconite.

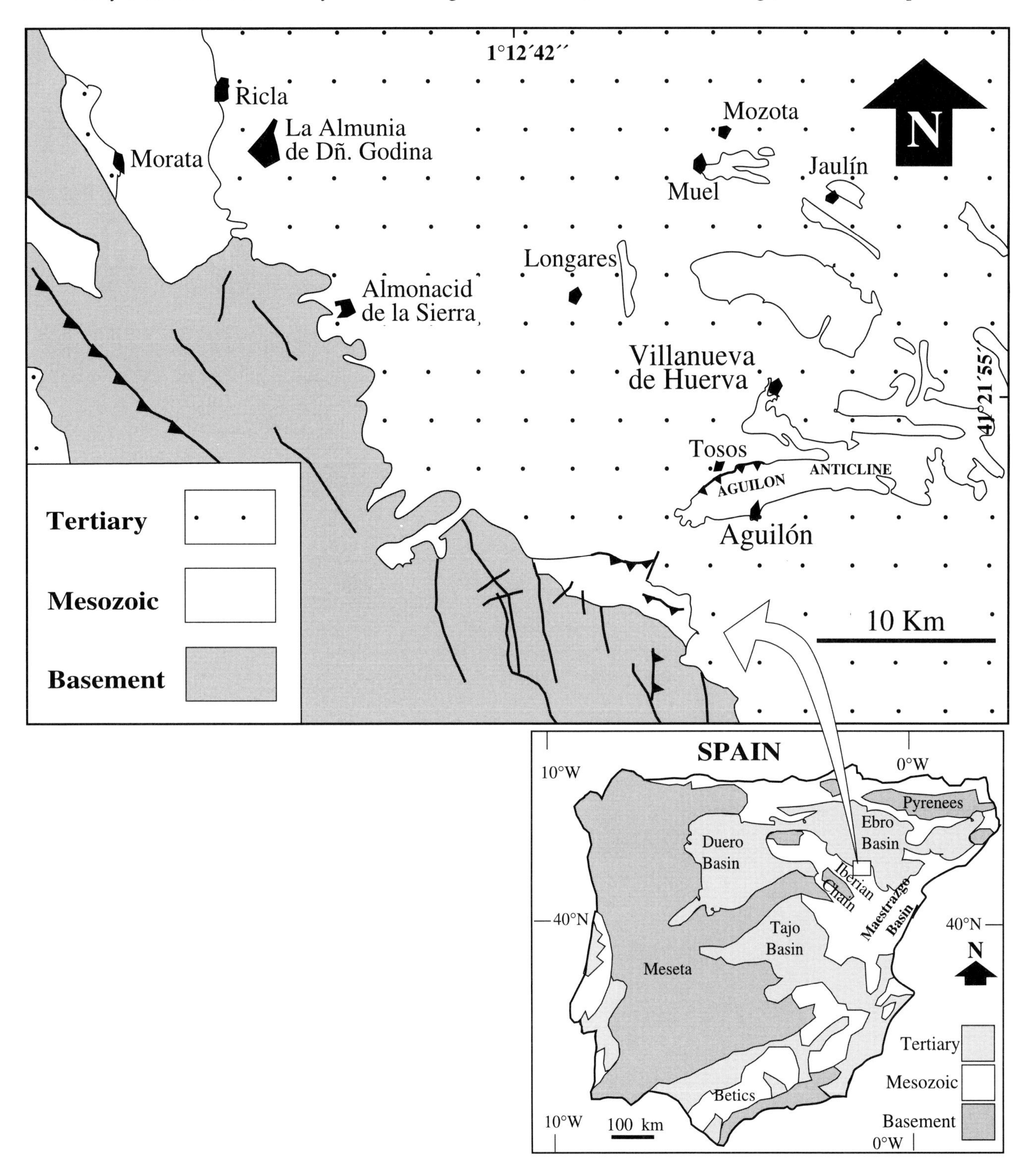

Figure 1—Generalized geologic map (bottom) of the Iberian Peninsula showing the location of the Iberian chain in Spain. Location of outcrops (top) in the Aguilón subbasin of the Iberian Basin.

These alluvial facies grade up into a lacustrine-deltaic progradational system. This system is represented by two different stages (Aurell et al., 1990). The lower stage contains delta deposits with bar progradational sequences at the bottom and delta plain facies at the top. Evidence for marine influence in the lower part of these facies is locally found in the Aguilón section, where a skeletal tempestite level is recognized (sandstone and bioclastic grainstone with serpulids, echinoderms, benthic foraminifera, and oyster shells). The upper stage of the deltaic system is represented by carbonate facies deposited on a lacustrine margin, with some interbedded siliciclastic bars, corresponding to distal

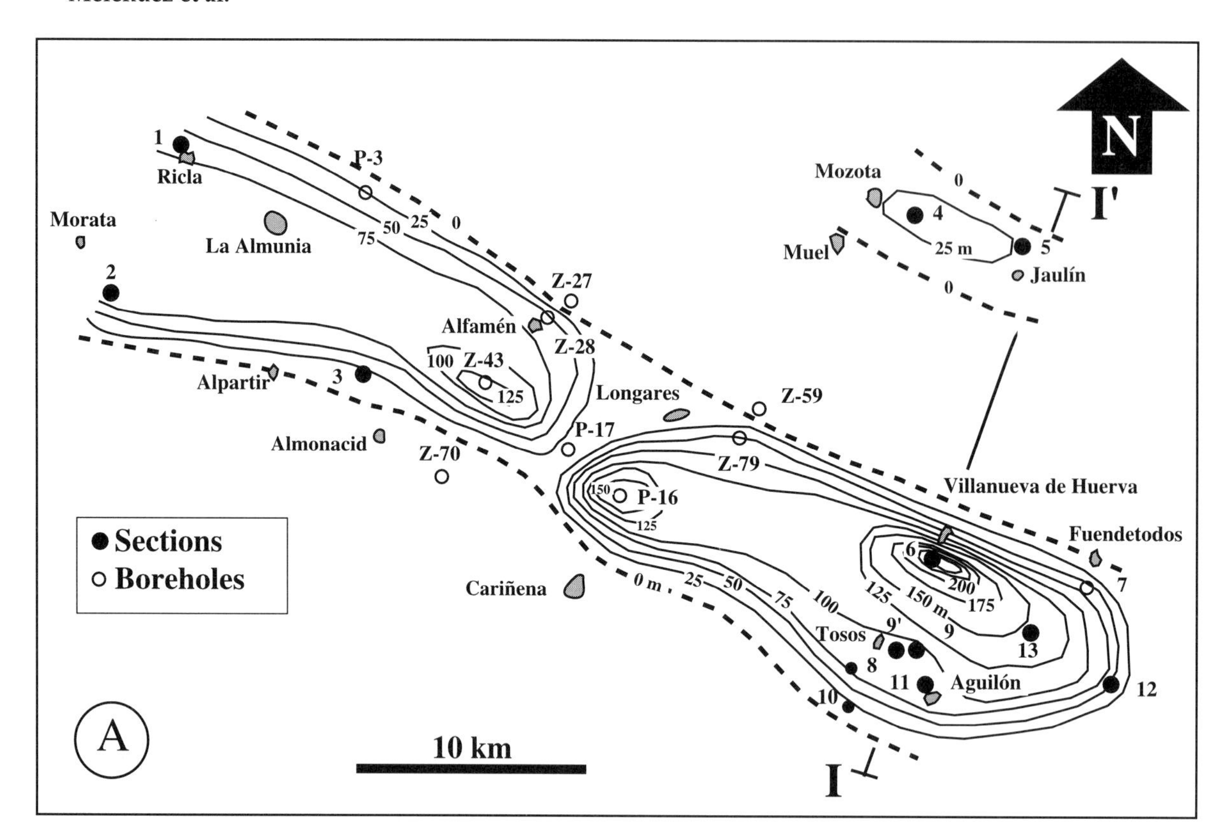

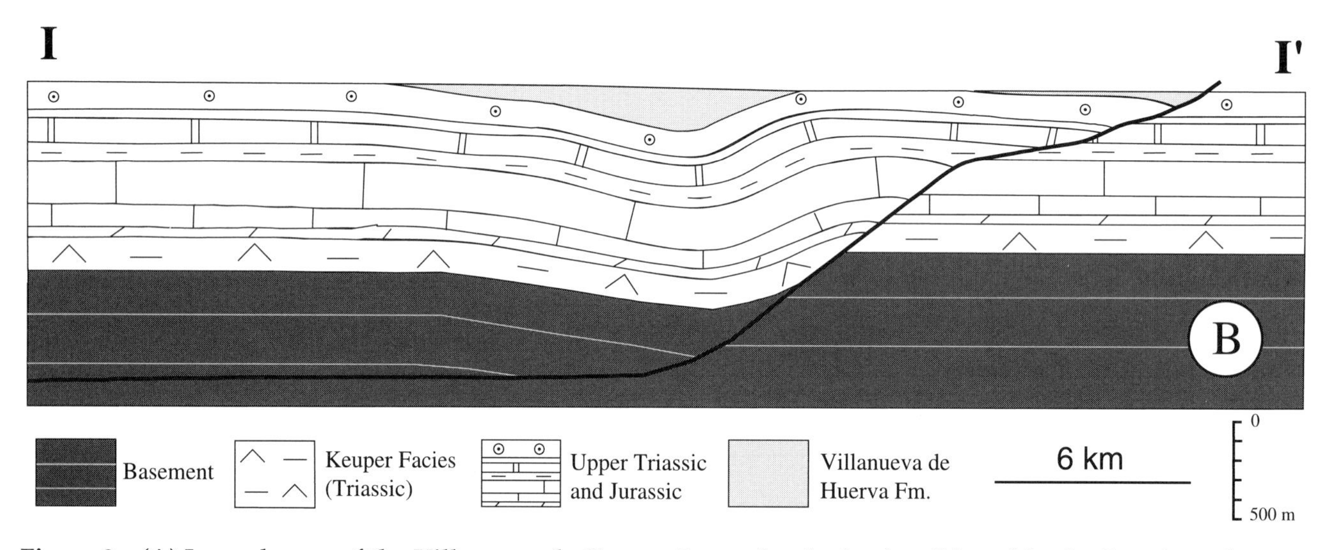

Figure 2—(A) Isopach map of the Villanueva de Huerva Formation in the Aguilón subbasin (location of exposed sections and boreholes are indicated). (B) Cross-section (location in A) showing the main listric fault that controlled basin development.

deltaic deposits. Edaphic processes on repeated shallowing-upward sequences placed over the distal bars indicate expansion of a shallow lake and reduction of the delta system.

The upper part of the Villanueva de Huerva Formation consists of a wide spectrum of carbonate facies that were deposited in different subenvironments of a well-developed lacustrine system (Figure 5, Table 1). These lacustrine deposits are arranged in stacked shallowing-upward sequences resulting from repeated filling of a shallow carbonate lake. The base of these sequences is represented by marlstones (poorly bioturbated) with ostracods and charophytes, and the top consists of bioclastic limestones (wackestone-packstone) with charophytes, ostracods, gastropods, oncoids, bivalves, and gypsum. Algal-laminated limestones (mudstone) with rare vertebrate trackways, brecciation horizons, and mudcracks are present at the top of the uppermost sequences. Palustrine deposits are represented by marlstones that contain ostracods and charophytes. These facies contain vertical prismatic structures formed from the enlargement

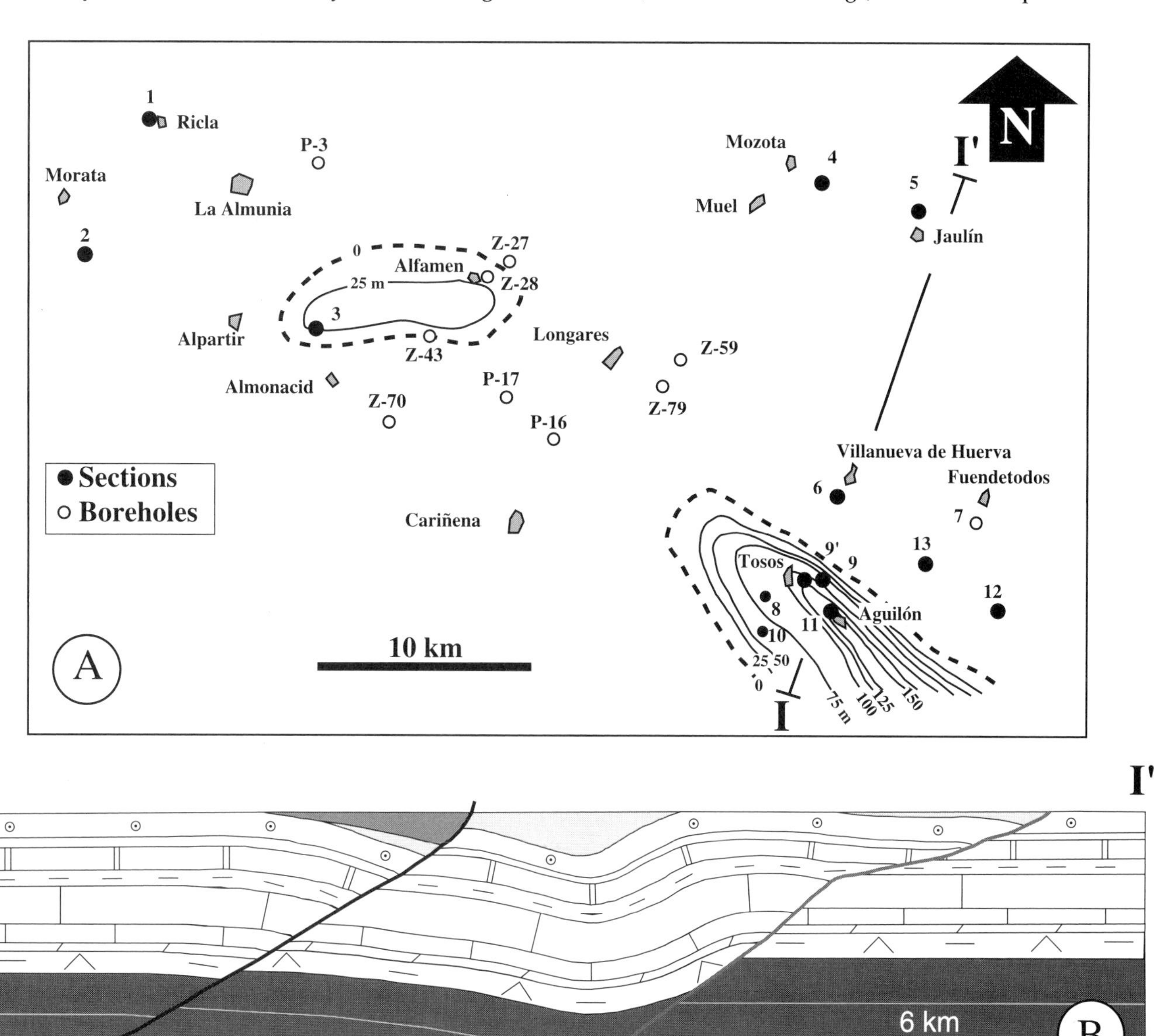

Figure 3—(A) Map showing thickness distribution of the Aguilón Formation in the Aguilón subbasin (location of exposed section and boreholes are indicated). (B) Cross-section (location in A) showing the main fault that controlled basin development.

of the voids created by root decay and their sedimentary infilling (Freytet and Plaziat, 1982), nodulization, and brecciation related to pedogenesis. These marlstones are rarely interbedded with limestone layers of oncolitic and intraclastic packstones related to small streams draining the palustrine areas.

AGUILON FORMATION: FACIES DISTRIBUTION

The Aguilón Formation is late Hauterivian–earliest Barremian in age. Table 2 shows the principal facies recognized in this unit. Figure 6 exhibits the most representative facies associations found in the Aguilón section. The Aguilón sequence displays an alluvial system at its base, with marly and sandy facies containing evidence for sheetflood deposition and edaphic processes, such as root traces and brecciation. This alluvial system grades upward into a well-developed lacustrine system, represented by cross-bedded bioclastic grainstones and packstones, generated by storm-induced waves (Meléndez and Aurell, 1989; Aurell et al., 1990; Soria et al., 1995b, 1997).

In the Aguilón lacustrine system, two types of deposits are distinguished: palustrine and lacustrine

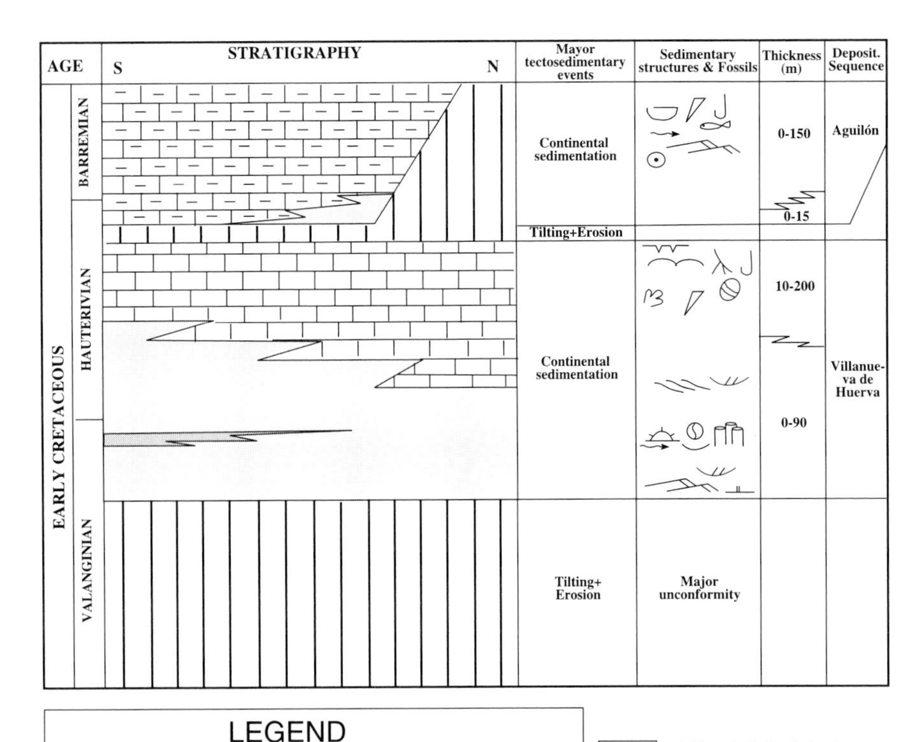

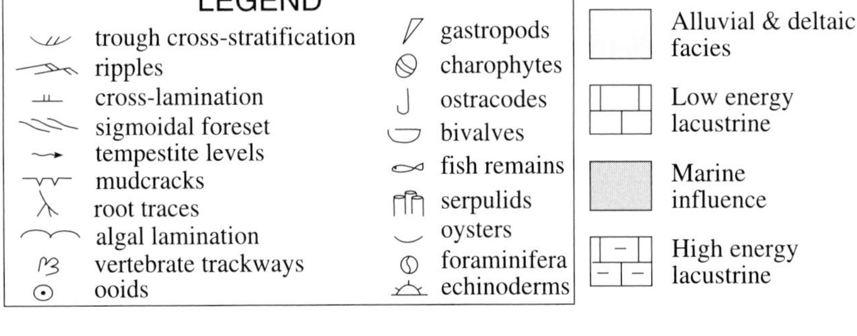

Figure 4—Facies distribution and stratigraphy of the Aguilón subbasin from correlation of the two most representative sections (Aguilón and Villanueva de Huerva) (see Figure 1 for location).

deposits. The palustrine rocks are represented by marlstones with ostracods and charophytes, as well as limestone (intraclastic packstone) with micritic intraclasts and oolites. Early subaerial diagenesis of these deposits is evidenced by root traces and vertical prismatic structures related to edaphic processes. The lacustrine deposits are represented by stacked shallowing-upward sequences, which exhibit a sharp base followed by graded ostracod to oolitic grainstone with parallel and low-angle cross-lamination. These lacustrine sequences correspond to proximal tempestites, deposited in a high-energy littoral lacustrine paleoenvironment. Type 3 and type 4 oolites (Strasser, 1986) preferentially occur in this unit. Proximal areas of the lacustrine system are represented by marlstones containing ostracods and charophytes, as well as limestones (bioturbated mudstone) rarely capped by mudcracks. Mudcracks can be preserved and filled by a thin layer (up to 5 cm) of ostracod and oolite packstones, i.e., storm deposits accumulated in marginal lacustrine areas. These sequences are interpreted as generated in a low-energy littoral and eulittoral lacustrine environments (after Gierlowski-Kordesch and Kelts, 1994) with some sporadic storm episodes.

In the upper part of the Aguilón lake unit, deep lake and eulittoral paleoenvironments are recognized. Deep lake facies are represented by mudstones and laminated marlstones containing fish debris (*Coelodus* sp.). Some interbedded graded layers (centimeter-scale thickness) of ostracod packstones-grainstones are interpreted as distal tempestites (after Aigner, 1985). Bioturbated mudstones interbedded with bioclastic limestone layers (up to 10 cm) with bivalve or gastropod grainstones (fauna low in diversity) indicate a eulittoral paleoenvironment.

SUMMARY

Formation of the Aguilón subbasin was related to the rifting stage that took place during the Early Cretaceous in the Iberian Basin (Salas et al., 1995; Salas and Casas, 1993); activity of two main fault systems with northwest-southeast and northeast-southwest trends defined the boundaries of the subbasin. Extensional tectonics controlled the configuration of the basin, variations in the thickness and facies distribution of the sediments, and governed the location and evolution of Cretaceous lakes (Soria, 1997; Soria et al., 1997).

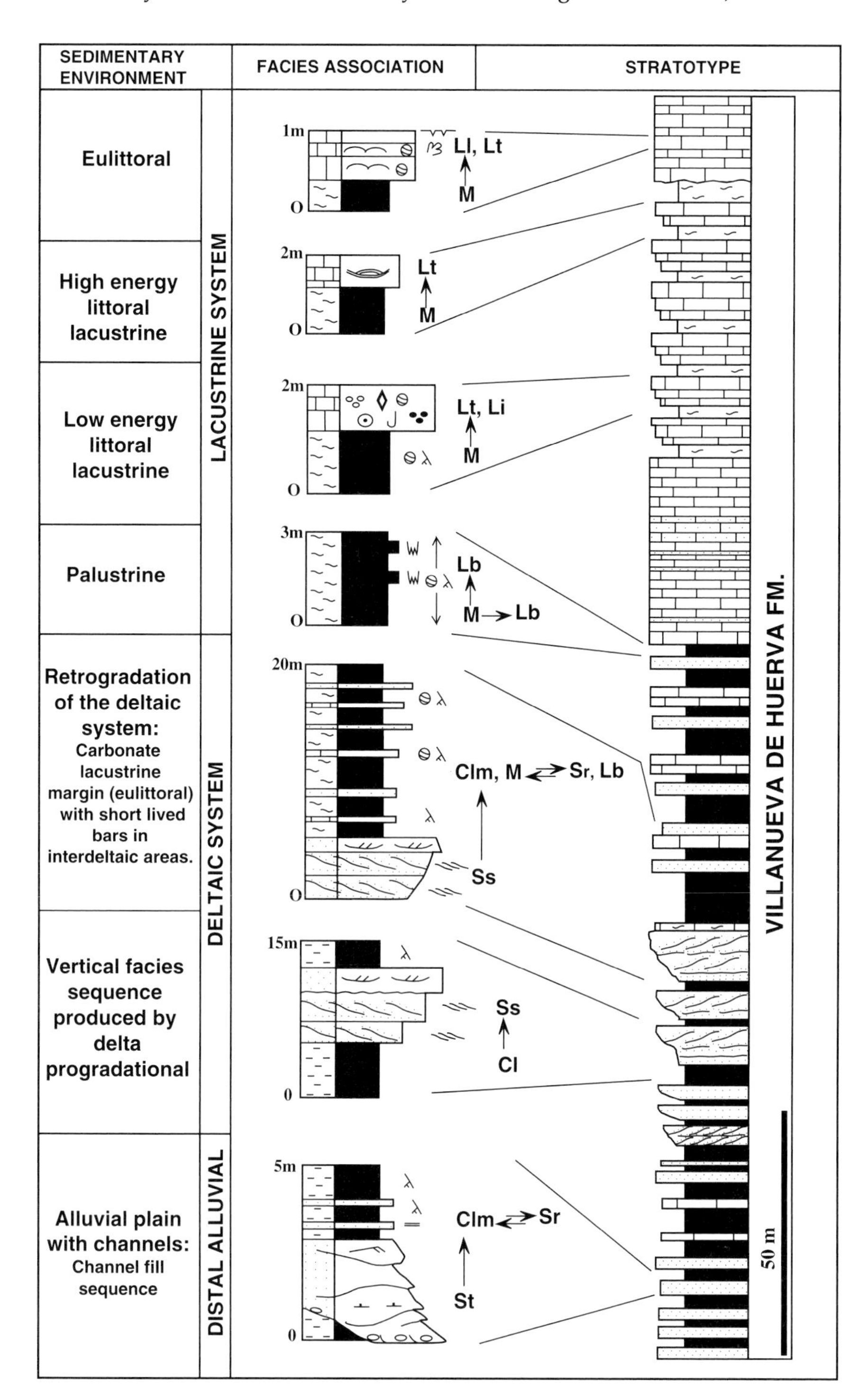

Figure 5—Vertical distribution of the main facies associations and their sedimentary interpretation of the Villanueva de Huerva Formation stratotype in the Aguilón subbasin, Spain (see Table 1 for facies abbreviations).

During the Early Cretaceous, two different lacustrine systems were created in the Aguilón subbasin. The Villanueva de Huerva system consists of stacked shallowing-upward sequences interpreted as a series of low-energy and shallow carbonate lakes. The Aguilón system is represented by cross-bedded bioclastic grainstones and packstones that were generated by storm-induced waves in a high-energy carbonate lake. Carbonate deposition was the main sedimentary process in these lakes because of the carbonate-dominated source area surrounding the basin (hardwater lakes). Climate also may have influenced depositional patterns and significant variations in water-table levels.

ACKNOWLEDGMENTS

The authors are grateful to Brian Daley and the editors for their suggestions and detailed revision of the paper. This work is a contribution to project PB95-1142-C02-02 (Dirección General de Investigación Científica y Técnica).

Table 2. Description of the Principal Facies in the Aguilón Formation of the Aguilón Subbasin, Spain.

Lithology	Facies	Description
Claystones	Claystones: Cl	Massive red and beige claystones (rare gray) with tabular geometry (m-scale thickness). Common pedogenic features.
	Massive and laminated marlstones: Ml	Massive and laminated marlstones with tabular geometry (m-scale thickness) containing charophytes, ostracods, and fish debris.
	Bioturbated marlstones: Mb	Bioturbated and massive marlstones with tabular geometry (dm- and m-scale thickness). Root traces and vertical prismatic structures.
Sandstones: S	Sandstones: S	Fine-grained red sandstones. Lenticular and tabular geometry (dm-scale thickness). Channel basal lags with muddy soft pebbles.

Lithology	Facies	Description	Microfacies	Components/Characteristics
Limestones	Skeletal limestones: Ls	Beige limestones with skeletal layers of bivalves or gastropods. Tabular geometry (dm-scale thickness).	Bioclastic Packstone-Grainstone: Ls	Bivalves and gastropods with low specific diversity.
	Bioclastic and intraclastic limestones: Lb	Bioclastic limestones with tabular and channel geometry (cm- to dm-scale thickness). Carbonate rudstones present at bottom of channels.	Oolitic and intraclastic Packstone-Grainstone: Lbo	Oolites and micritic intraclasts. Oolites correspond to types 3, 4, and rarely, 2 (Strasser, 1986). Nucleus of oolites are micritic or bioclastic (ostracods).
			Ostracod Packstone-Grainstone: Lbb	Ostracods; rare micritic intraclasts and oolites (types 3 and 4).
			Laminated Grainstone: Lbl	Characterized by: (a) oolite and/or ostracod grainstone, with rare intraclasts of laminated micrites or (b) alternation of layers of oolites and/or ostracod grainstone and mudstone-wackestone. Oolites correspond to types 3 and 4 of Strasser (1986).
			Peloidal Wackestone: Lbp	Peloids, ostracods, and oolites (types 3, 4; Strasser, 1986). Rare lamination consisting of alternations of micritic (peloidal wackestone–packstone) and sparitic layers.
	Massive limestones: Lm	Massive limestones with lenticular and tabular geometry (cm- to dm-scale thickness).	Mudstone-Wackestone: Lmm	Gastropods, ostracods, and peloids. Common fenestral porosity. Intense micritization (cyanobacteria).
			Oolitic and intraclastic Packstone: Lmo	Micritic, intraclastic, and oolitic intraclasts. Oolites and quartz grains. Oolites correspond to type 3 of Strasser (1986). Types 1 and 6 are rarely identified.
	Bioturbated limestones: LB	Micritic limestones with tabular geometry (dm- and m-scale thickness). Root traces and vertical prismatic structures.	Bioturbated Mudstone: LB	Micritic microfacies characterized by charophytes and ostracods. Common fenestral porosity.

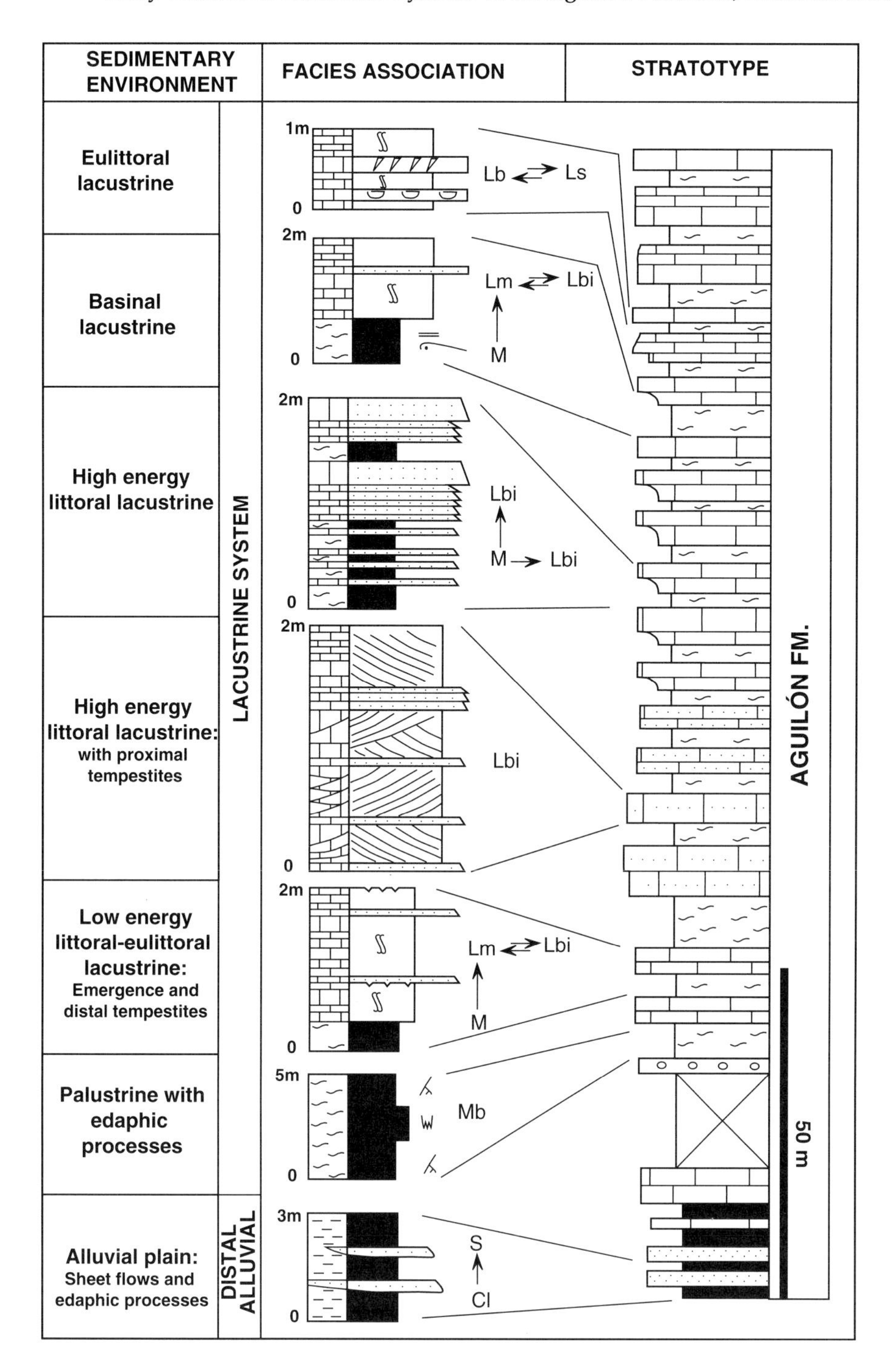

Figure 6—Vertical distribution of the main observed facies associations and their sedimentary interpretation of the Aguilón Formation stratotype in the Aguilón subbasin, Spain (see Table 2 for facies abbreviations).

REFERENCES CITED

Aigner, T., 1985, Storm Depositional Systems: Lecture Notes in Earth Sciences, Springer-Verlag, v. 3, 174 p.

Aurell, M., N. M. Meléndez, and A. Meléndez, 1990, Evolution of a lacustrine system during the Hauterivian (Early Cretaceous) on the North of Oliete Basin, Central Spain: 13th International Sedimentological Congress, Nottingham, Abstract-Posters, p. 13–14.

Freytet, P., and J. Plaziat, 1982, Continental carbonate sedimentation and pedogenesis—Late Cretaceous and Early Tertiary of Southern France: Contributions to Sedimentology, v. 12, 216 p.

Gierlowski-Kordesch, E., and K. Kelts, 1994, Introduction, *in* E. Gierlowski-Kordesch and K. Kelts, eds., Global Geological Record of Lake Basins: Cambrige University Press, v. I, p. xvii–xxxiii.

Meléndez, A., and M. Aurell, 1989, Controles en la sedimentación del Cretácico inferior de Aguilón (Zaragoza, Cordillera Ibérica Septentrional): Geogaceta, v. 6, p. 55–58.

Milton, N. J., and D. Emery, 1996, Outcrop and well data, *in* D. Emery and K. J. Myers, eds., Sequence Stratigraphy: Blackwell Science, p. 61–79.

Salas, R., 1987, El Malm i el Cretaci inferior entre el

Massis de Garraf i la Serra d'Espadà: Ph.D. dissertation, Universidad de Barcelona, 345 p.
Salas, R., and A. Casas, 1993, Mesozoic extensional tectonics, stratigraphy and crustal evolution during the Alpine cycle of the eastern Iberian Basin: Tectonophysics, v. 228, p. 33–55.
Salas, R., C. Martín-Closas, X. Querol, J. Guimerá, and E. Roca, 1995, Evolución tectonosedimentaria de las cuencas del Maestrazgo y Aliaga-Peñagolosa durante el Cretácico inferior, *in* R. Salas and C. Martín-Closas, eds., El Cretácico Inferior del Nordeste de Iberia. Guia de campo de la excursiones científicas del III Coloquio del Cretácico de España (Morella, 1991): Publications Universitat de Barcelona, p. 13–94.
Soria, A. R., 1997, La sedimentación en las cuencas marginales del surco Ibérico durante el Cretácico Inferior y su control tectónico: Ph.D. dissertation, Universidad Zaragoza, Servicio de Publicaciones, Universidad Zaragoza, 363 p.
Soria, A. R., C. Martín-Closas, A. Meléndez, M. N. Meléndez, and M. Aurell, 1995a, Estratigrafía del Cretácico inferior continental de la Cordillera Ibérica Central: Estudios Geológicos, v. 51 (3–4), p. 141–152.
Soria, A. R., M. N. Meléndez, and A. Meléndez, 1995b, Evolución sedimentaria de las unidades genéticas del Cretácico inferior de la Cubeta de Aguilón (Provincia de Zaragoza): 13 Congreso Español de Sedimentología, Libro de resúmenes, p. 169–70.
Soria, A. R., A. Meléndez, M. N. Meléndez, and C. L. Liesa, 1997, Evolución de dos sistemas continentales en la Cubeta de Aguilón (Cretácico Inferior). Interrelación sedimentaria entre depósitos aluviales y lacustres, y su control tectónico: Cuadernos de Geología Ibérica, v. 22, p. 473–507.
Strasser, A., 1986, Ooides in Purbeck limestones (lowermost Cretaceous) of the Swiss and French Jura: Sedimentology, v. 33, p. 711–727.

Liesa, C. L., A. R. Soria, and A. Meléndez, 2000, Lacustrine evolution in a basin controlled by extensional faults: the Galve subbasin, Teruel, Spain, *in* E. H. Gierlowski-Kordesch and K. R. Kelts, eds., Lake basins through space and time: AAPG Studies in Geology 46, p. 295–302.

Chapter 24

Lacustrine Evolution in a Basin Controlled by Extensional Faults: the Galve Subbasin, Teruel, Spain

C. L. Liesa
Dpto. Ciencias de la Tierra (Geodinámica), Universidad de Zaragoza
Zaragoza, Spain

A. R. Soria
A. Meléndez
Dpto. Ciencias de la Tierra (Estratigrafía), Universidad de Zaragoza
Zaragoza, Spain

GEOLOGIC SETTING

The Galve subbasin, which lies in the central part of the Teruel province in east-central Spain, was a region of continental sedimentation during the Early Cretaceous (Figure 1; see also Figure 1 in Soria et al., this volume, concerning an overview of the geology and tectonics of the Iberian basin). The Galve subbasin lies on the margins of the Maestrazgo tectonic domain (Salas and Guimerà, 1996; Soria, 1997; Soria et al., this volume). Outcrops with a northwest-southeast trend are exposed between the villages of Aliaga and Galve north and Alcalá de la Selva and Gúdar in the south (Figure 1).

The Galve subbasin is separated from the Peñagolosa subbasin, which lies to the southwest, by the Sierra of Moratilla uplands. The main structural feature of this range is the Alcalá de la Selva anticline, which has a core of Jurassic and Triassic rocks and was formed by compressional reactivation of the Alcalá de la Selva basement fault. The latter acted as a normal fault during Cretaceous sedimentation (Salas and Guimerà, 1996). The Ejulve-Fortanete high, which represents the eastern edge of the Galve subbasin, is structurally a north–northwest-south–southeast trending syncline with a core of Upper Cretaceous and Tertiary rocks (Soria et al., 1995). During sedimentation, the northern boundary of the subbasin was defined by a series of east–northeast-west–southwest to east-west trending normal listric faults that separated this subbasin from the Las Parras subbasin (Figure 1). These normal faults became inverted during phases of Tertiary compression (Guimerà and Salas, 1996).

SEDIMENTARY RECORD

Where exposed, the Lower Cretaceous deposits of the Galve subbasin lie unconformably upon limestones of the Villar del Arzobispo Formation (Tithonian–Berriasian) (Soria, 1997; Soria et al., 1998), which were deposited in transitional marine to nonmarine environments. The Lower Cretaceous succession (Weald facies) comprises three lithostratigraphic units: the Castellar, Camarillas, and Artoles Formations, which are late Hauterivian to basal Aptian in age (Figure 2). Above these continental and marginal marine units lies the Urgon facies, which includes sediments deposited in marginal marine and full marine environments.

The Castellar Formation (late Hauterivian–basal Barremian) is characterized by fine-grained alluvial plain and low-sinuosity fluvial channel deposits. These facies grade upward into a lacustrine carbonate environment. The Camarillas Formation (early Barremian) represents sedimentation in a fluvial system with low-sinuosity sandy channels (sandstones) and broad muddy floodplains (claystones). The Artoles Formation (late Barremian–basal Aptian) formed in several different depositional environments. In the north of the subbasin, shallow lacustrine and transitional environments are represented, whereas toward the south, marine sedimentation took place. This paper focuses on the Castellar Formation in the Galve subbasin, which was deposited mainly in a lacustrine setting.

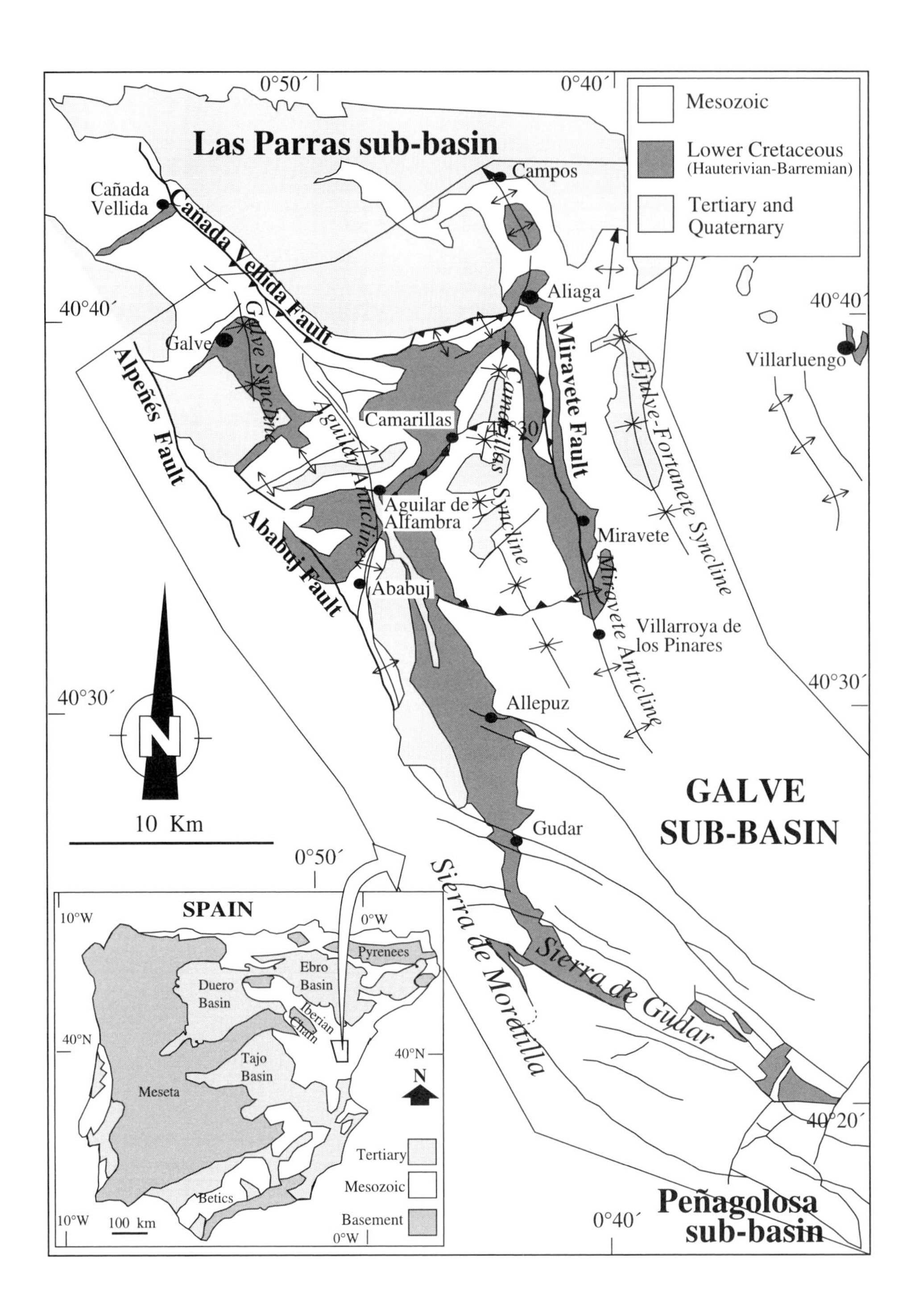

Figure 1—Schematic geologic map of the Galve subbasin located in northeastern Spain.

TECTONIC CONTROL OF SEDIMENTATION

The Galve subbasin was formed by extensional tectonics during an Early Cretaceous phase of rifting (Salas, 1987; Salas and Casas, 1993). The Lower Cretaceous succession displays many local and regional variations in thickness, with several local unconformities (Figure 3). Evidence for tilting and local erosion of the prerift sequence is also present. These features were controlled by tectonic subsidence that resulted from movements on two normal-fault systems that trend north–northwest-south–southeast and east–northeast-west–southwest, respectively (Soria et al., 1998).

The east–northeast-west–southwest growth faults (e.g., the Campos, Santa Bárbara, and Aliaga faults) (Figure 3) have a listric geometry that gave rise to extensional half-grabens. The maximum vertical throw on these faults is about 500 m. This system of normal faults is imbricated at depth toward the south–southeast. The north–northwest-south–southeast trending faults (Alpeñés, Ababuj, Cañada Vellida, and Miravete faults) have higher dips (near vertical) than the east–northeast-west–southwest faults and produce grabenlike structures (Figure 3). The north–northwest-south–southeast trending faults are transfer faults (Gibbs, 1984) that have a significant normal component related to the imbricate system of

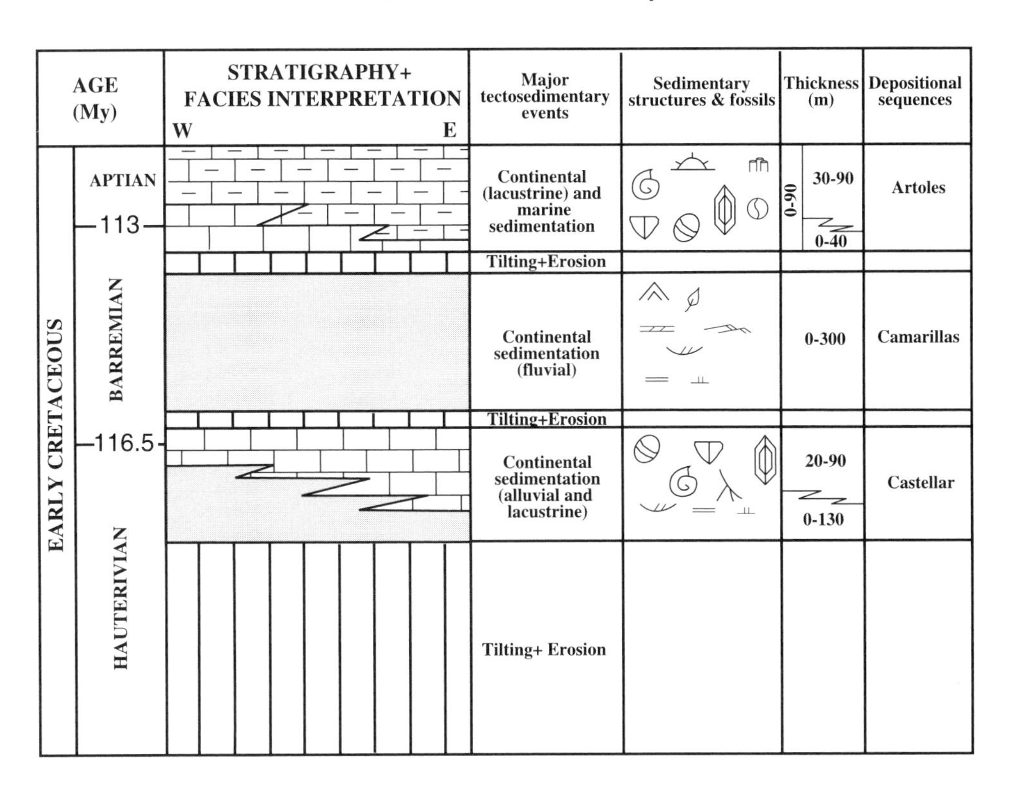

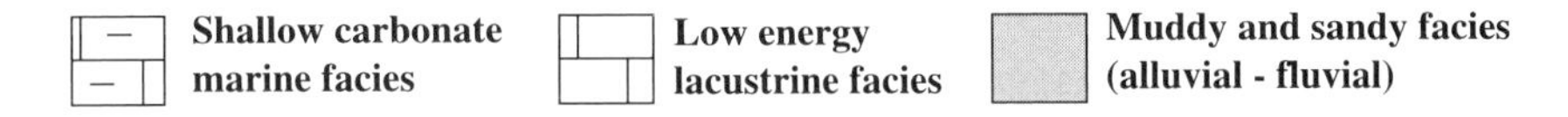

Figure 2—Facies distribution and stratigraphy of the Castellar Formation across the Galve subbasin in a west-east direction. Global Geological Record of Lake Basins (GGLAB) legend for fossils and sedimentary structures.

east–northeast-west–southwest normal listric faults. These faults are probably inherited from the post-Variscan structures of the Iberian chain described by Arthaud and Matte (1975). The vertical Miravete fault (Figure 3) represents the eastern boundary fault of the basin containing the Castellar Formation, whereas the east–northeast-trending Campos fault marks its northern border.

SEDIMENTARY EVOLUTION OF THE CASTELLAR FORMATION

The Castellar Formation displays an overall lacustrine transgressive sequence. Distal alluvial facies, represented by claystones with interbedded sandy channel deposits, are present at the base. These grade upward into a well-developed shallow lacustrine system characterized by alternating marlstones and limestones. Table 1 summarizes the principal facies recognized in this unit.

Stratigraphic sections near the Miravete fault (Miravete and Santa Bárbara, Figure 4), however, show a more complex pattern. In the Santa Bárbara section, the distal alluvial facies, consisting of interbedded red claystones and fine- to medium-grained sandstones, are present at the base. Sandstones display channel-fill geometries and fining-upward sequences with trough cross-stratification and parallel to ripple lamination.

These alluvial facies grade up into a series of lacustrine deposits arranged in stacked shallowing-upward sequences (2–5 m) that are characteristic of sedimentation in a shallow carbonate lake (after Platt and Wright, 1991) (Figures 4, 5). The base of each sequence consists of bioturbated marlstones with charophytes, gastropods, fish, and vertebrate remains, and, in places, benthic foraminifera and oysters. The upper part is characterized by micritic limestone (mudstone and wackestone) with charophytes, bivalves, and gastropods. These sequences represent the sedimentation in a littoral lacustrine environment probably developed near a marine shoreline in a paralic setting.

A massive and bioturbated red claystone facies, with tabular geometry on a meter scale, progrades across the lacustrine deposits and represents sedimentation in a distal muddy alluvial plain. Red gypsum nodules, a few centimeters in diameter, are also present (Figure 4), but only near the Miravete fault. Their restriction to this location can be explained by the supply of sulfate-bearing groundwaters along the permeable fault. The Jurassic prerift sequence is very thin in this zone (200–300 m). Thus the shallow groundwaters could easily reach the Late Triassic clay and gypsum source rock (Keuper facies). Precipitation of interstitial gypsum in the sediment was probably induced by evaporitic pumping of shallow groundwaters.

The upper part of the Castellar Formation is characterized by marlstone-limestone sequences that mark the establishment of a new shallow marginal-lacustrine

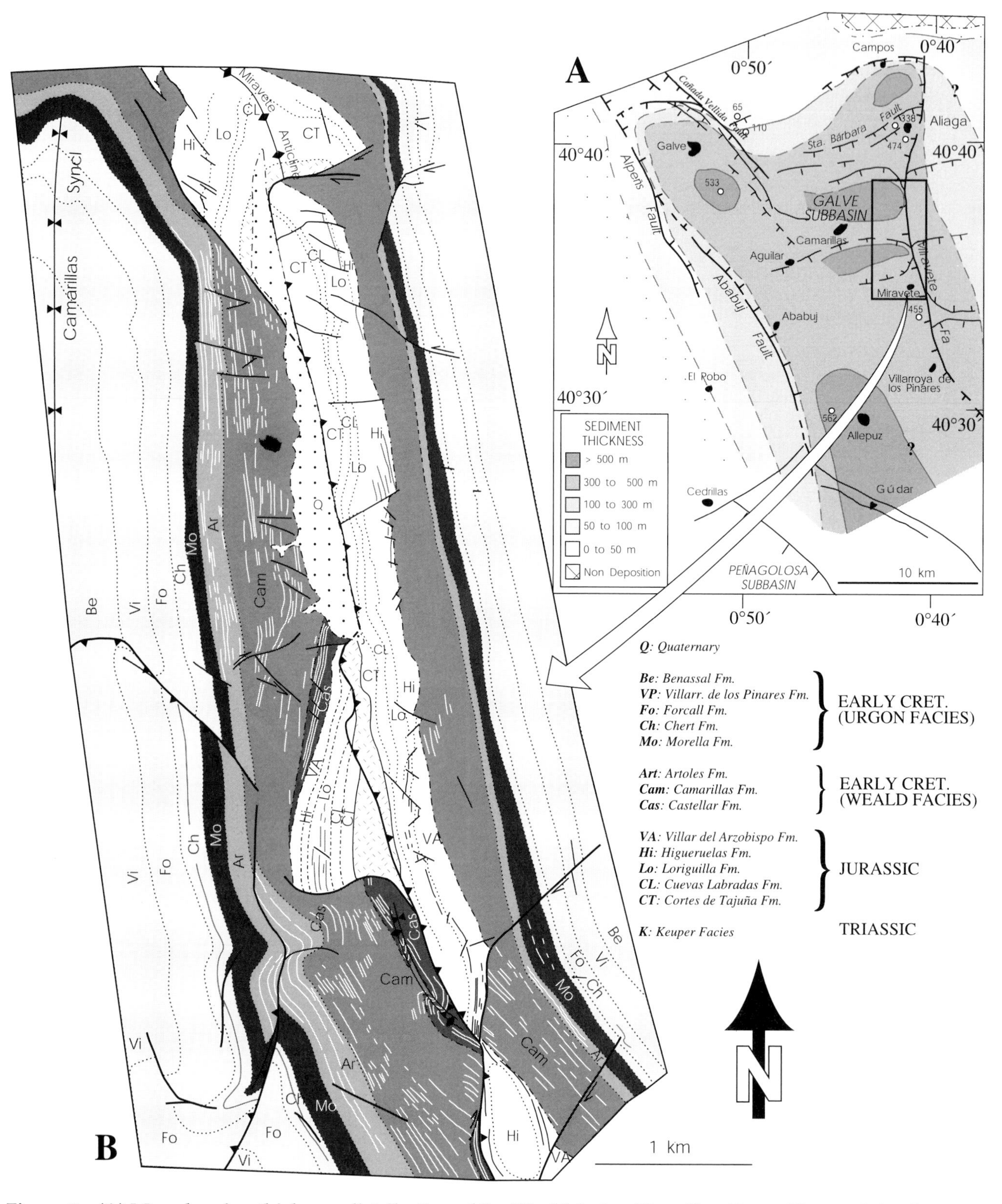

Figure 3—(A) Map showing thickness distribution of the Weald facies (Castellar, Camarillas, and Artoles Formations) across the Galve subbasin. (B) Photogeologic sketch of the Miravete anticline between Aliaga and Miravete showing thickness variation of the Lower Cretaceous sequences between two limbs of the anticline and main normal faults that controlled the basin infilling (north–northwest-south–southeast vertical Miravete fault and east–northeast-west–southwest south-dipping listric faults).

Table 1. Description of the Main Facies Recognized in the Castellar Formation (Upper Hauterivian-Basal Barremian) of the Galve Subbasin, Spain.

Lithology	Facies	Description
Claystones	Massive claystones	Massive red claystones with tabular geometry (m-scale). They contain vertebrate remains and vertical bioturbation.
	Claystones with gypsum	Massive red claystones with tabular geometry (m-scale). They contain gypsum nodules of cm-scale.
	Massive marlstones	Massive and bioturbated marlstones with tabular geometry (dm-scale). Fish, gastropods, charophytes, vertebrate remains, and micrite nodules.
	Bioturbated marlstones	Bioturbated marlstones with tabular geometry (dm-scale). Vertical bioturbation and paleosols.
Sandstones	Channel sandstones	Medium- and coarse-grained channel sandstone with fining-upward sequences (m-scale). Microconglomerate lags at the bottom of channels. Sandstones exhibit trough cross-stratification, parallel and ripple cross-lamination.
	Tabular sandstones	Medium-grained sandstones with tabular geometry (dm to m-scale). Parallel and ripple cross-laminae.

Lithology	Facies	Description	Microfacies	Components/Characteristics
	Tabular micritic limestones	Micritic limestones with tabular geometry (dm-scale). Root traces, vertical prismatic structures, and mudcracks.	Bioturbated Mudstone	Charophytes and bivalves.
			Bioclastic Wackestone	Bivalves and gastropods.
	Tabular bioclastic limestones	Bioclastic limestones with tabular geometry (dm- and cm-scale). Bioturbation, breccias, and ferruginous surfaces.	Bioclastic Wackestone	Charophytes, ostracods, plant remains, bivalves, and algae. Rarely, foraminifera and oysters.
			Bioclastic Wackestone-Packstone	Charophytes, ostracods, gastropods, bivalves, and algae. Rarely, plant remains, micritic intraclasts, foraminifera, and oysters.
	Lenticular limestones	Limestones with channel geometry (3 m thick).	Oncolitic Packstone	Oncolites. Rare charophytes and ostracods.
			Bioclastic Packstone	Charophytes, ostracods, oncolites, micritic intraclasts, and peloids. Bioturbation.

system (Figure 4). These sequences record recurrent episodes of expansion and contraction of the lake. The base of each sequence consists of massive marlstones with root traces. Limestones are mainly bioclastic (packstone and wackestone), containing charophytes, bivalves, and ostracods. Upper surfaces of the limestones are commonly iron-stained and brecciated or exhibit desiccation cracks. These sedimentary sequences are interpreted to have formed in a eulittoral lacustrine environment. Furthermore, oysters, foraminifera, and detrital glauconite are present in the limestone facies, which indicates that these paleolakes probably formed near a marine shoreline in a paralic setting.

ACKNOWLEDGMENTS

The authors are grateful to Robin Renaut and the editors for their suggestions and detailed revision of the paper, which greatly improved the original manuscript. This work was carried out with the financial support of projects PB92-0862-C02-02 and PB95-1142-C02-02 (Dirección General de Investigación Científica y Técnica), Ministry of Education and Research, Spain.

REFERENCES CITED

Arthaud, F., and P. Matte, 1975, Les décrochements tardihercíniens du soud-ouest de l'Europe: geometrie

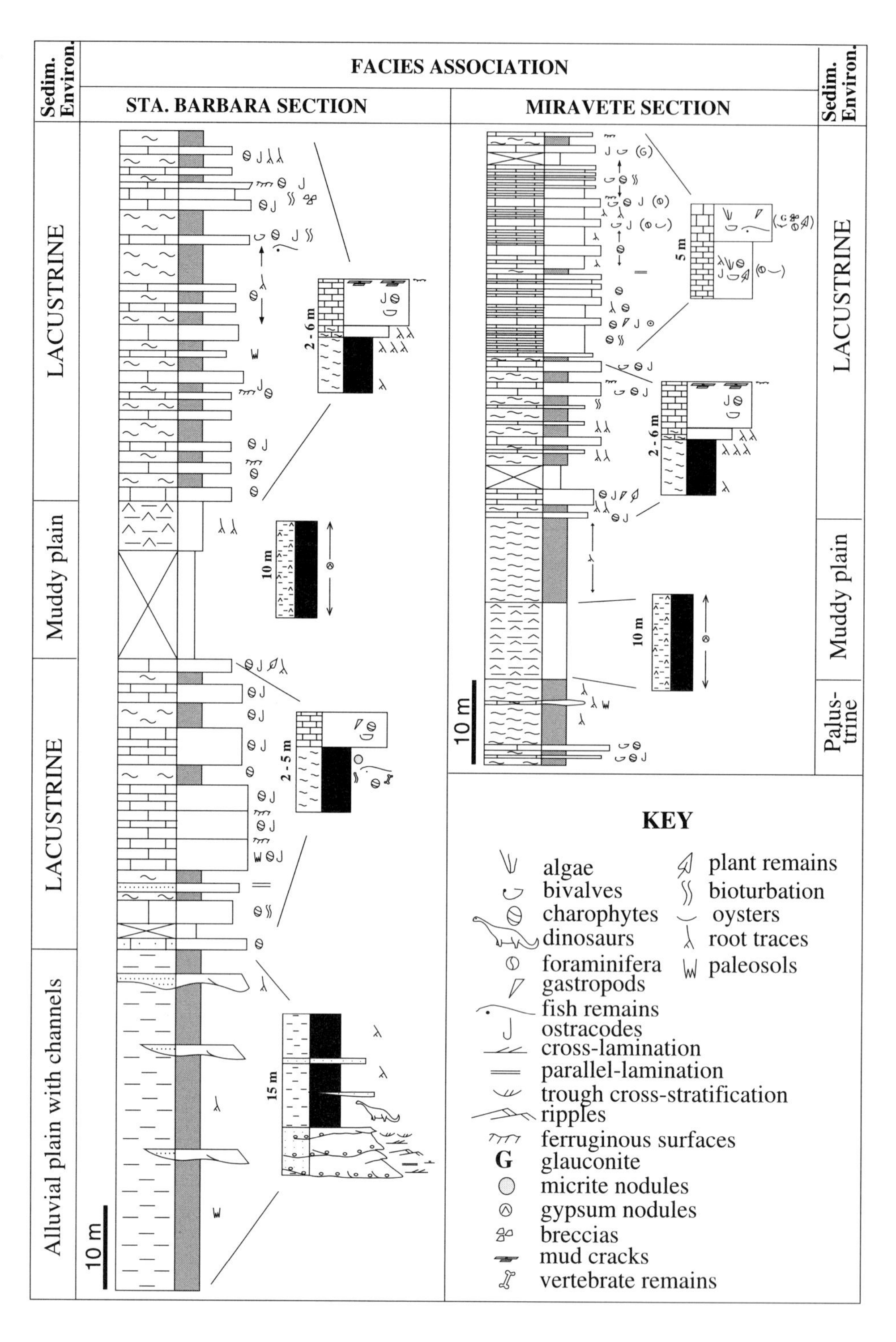

Figure 4—Vertical distribution of the facies associations of the Castellar Formation in the Santa Bárbara and Miravete sections of the Galve subbasin.

et essai de reconstructions des conditions de la deformation: Tectonophysics, v. 25, p. 139–171.

Gibbs, A. D., 1984, Structural evolution of extensional basin margins: Journal Geological Society of London, v. 141, p. 609–620.

Guimerà, J., and R. Salas, 1996, Inversión terciaria de la falla normal mesozoica que limitaba la subcuenca de Galve: Geogaceta, v. 20 (7), p. 1701–1703.

Platt, N. H., and V. P. Wright, 1991, Lacustrine carbonates: facies models, facies distributions and hydrocarbon aspects: Special Publications International Association of Sedimentologists, v. 13, p. 57–74.

Salas, R., 1987, El Malm i el Cretaci inferior entre el Massis de Garraf i la Serra d'Espadá, Ph.D. dissertation, Universitat de Barcelona, 345 p.

Salas, R., and A. Casas, 1993, Mesozoic extensional tectonics, stratigraphy and crustal evolution during the Alpine cycle of the eastern Iberian basin: Tectonophysics, v. 228, p. 33–55.

Salas, R., and J. Guimèra, 1996, Rasgos estructurales principales de la cuenca cretácica inferior del Maestrazgo (Cordillera Ibérica oriental): Geogaceta, v. 20

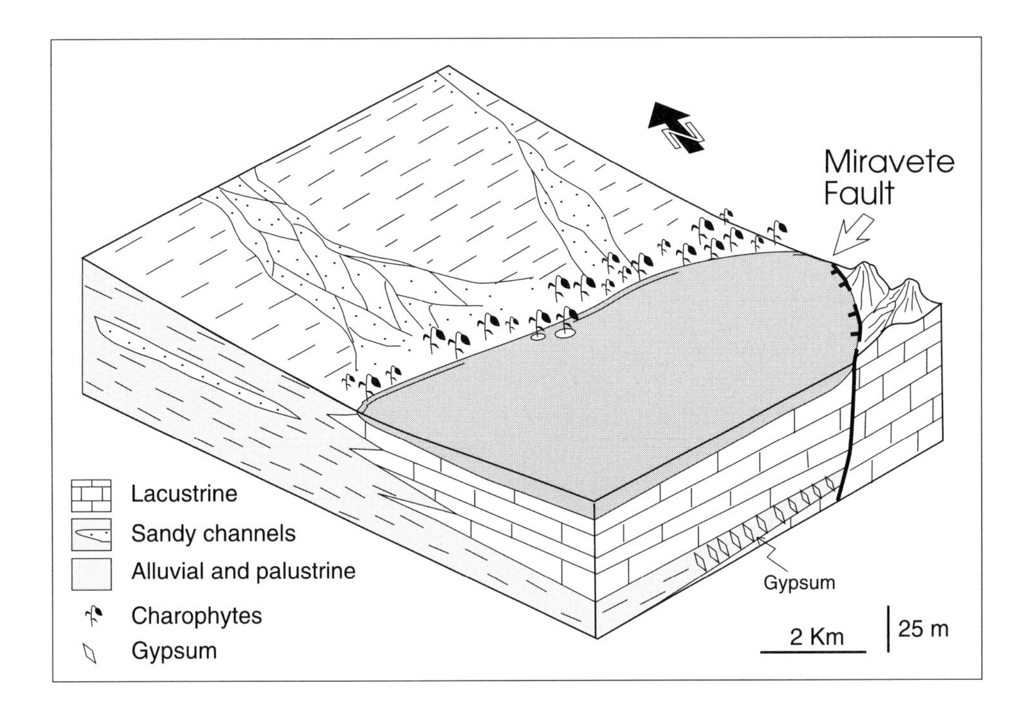

Figure 5—Sedimentologic model for the Castellar Formation. See text for details.

(7), p. 1704–1706.

Soria, A. R., A. Meléndez, G. Cuenca-Bescós, I. Canudo, and C. L. Liesa, 1995, Los sistemas lacustres del Cretácico Inferior de la Cordillera Ibérica Central: La Cubeta de Aliaga: Guía de excursiones XIII Congreso Español de Sedimentología, Teruel, p. 91–141.

Soria, A. R., 1997, La sedimentación en las cuencas marginales del surco Ibérico durante el Cretácico Inferior y su control tectónico: Ph.D. dissertation, Universidad de Zaragoza, Servicio de Publicaciones, Universidad de Zaragoza, 363 p.

Soria, A. R., C. L. Liesa, and A. Meléndez, 1998, Tectonic model of the Galve subbasin (Early Cretaceous; Iberian chain): Half graben controlled by transfer faults: 15th International Sedimentological Congress, Abstracts, Alicante (Spain), p. 731–732.

Fregenal Martínez, M. A., and N. Meléndez, 2000, The lacustrine fossiliferous deposits of the Las Hoyas subbasin (Lower Cretaceous, Serranía de Cuenca, Iberian ranges, Spain), *in* E. H. Gierlowski-Kordesch and K. R. Kelts, eds., Lake basins through space and time: AAPG Studies in Geology 46, p. 303–314.

Chapter 25

The Lacustrine Fossiliferous Deposits of the Las Hoyas Subbasin (Lower Cretaceous, Serranía de Cuenca, Iberian Ranges, Spain)

M.A. Fregenal Martínez
N. Meléndez
Dpto. de Estratigrafía, Facultad de Ciencias Geológicas, Universidad Complutense
Madrid, Spain

INTRODUCTION

The Las Hoyas subbasin is located in the Serranía de Cuenca (southwestern Iberian ranges, east-central Spain) (Figure 1). Its sedimentary record is entirely composed of upper Barremian continental deposits that belong to the La Huérguina Formation (Vilas et al., 1982). It is a half-graben basin (Figures 1–3) that resulted from a period of rifting during the Early Cretaceous that divided the southwestern Iberian Basin into many small subbasins with differential subsidence rates (Vilas et al., 1983; Salas and Casas, 1993; Soria et al., this volume). The climate during the Barremian in this area was subtropical with alternating rainy and dry seasons (Ziegler et al., 1987; Gierlowski-Kordesch et al., 1991).

The Las Hoyas subbasin contains the thickest record of continental deposits of the La Huérguina Formation in the Serranía de Cuenca basin (Meléndez et al., 1994). Three different lacustrine fossiliferous lithosomes are identified throughout the stratigraphic succession of the Las Hoyas subbasin (Figure 3). Each one of those lithosomes is considered to be a *Konservat-Lagerstätte* and has yielded a rich and exceptionally well-preserved assemblage of terrestrial and freshwater flora and fauna (Table 1). The main preservational features of the association are the high density of remains, preservation of soft tissues, high degree of articulation, and skeletal integrity.

STRATIGRAPHY

The sedimentary record of the Las Hoyas subbasin is up to 400 m thick and is divided into four different depositional sequences separated by local unconformities or paraconformities (Figures 2–4). Sedimentation took place in continental paleoenvironments: distal alluvial and palustrine plains to hardwater ponds and lakes. The paleogeographic evolution of the basin was strongly controlled by tectonic activity.

RAMBLA DE LAS CRUCES I SEQUENCE

The first sequence, Rambla de las Cruces I sequence, unconformably overlies Jurassic paleokarstic marine limestones (Figures 2 and 4). This erosive unconformity comprises the Bathonian–Lower Barremian. Maximum thickness of the sequence is 150 m in the central areas of the basin, thinning toward the basin edge where it onlaps Jurassic substrate. Lithologically the sequence is composed of mixed siliciclastic and carbonate deposits. The dominant facies are red to grey claystones and marly claystones that commonly show hydromorphic edaphic features. Interbedded with these facies, rudstones composed of oncolites and carbonate pebbles (mostly intraclasts and Jurassic clasts) occur. Sandstones are medium- to coarse-grained, low to moderately sorted, and mainly composed of quartz, feldspar, and carbonate clasts (comprising both intraclasts and clasts from the Jurassic basement). The sandstones occur as channelized bodies up to 1.5 m thick and 25–100 m long. Lateral accretion surfaces are present in the rudstone and sandstone bodies, especially those exclusively composed of clastic carbonates. Tabular or slightly lenticular beds of biomicritic limestones up to 3 m thick and 200 m in length are also present, containing charophytes, ostracodes, gastropods, and bivalves. Brecciation and abundant vertical voids (up to 50 cm long and 2 cm wide), interpreted as root decay cavities, occur along the upper surfaces of these beds. Channelized bodies dominate the base of the sequence, whereas pond deposits progressively dominate toward the upper part of the sequence. This facies association represents sedimentation in distal alluvial and palustrine plains where high sinuosity streams flowed seasonally. Shallow hardwater ponds commonly developed on the alluvial plain in the extrachannel areas.

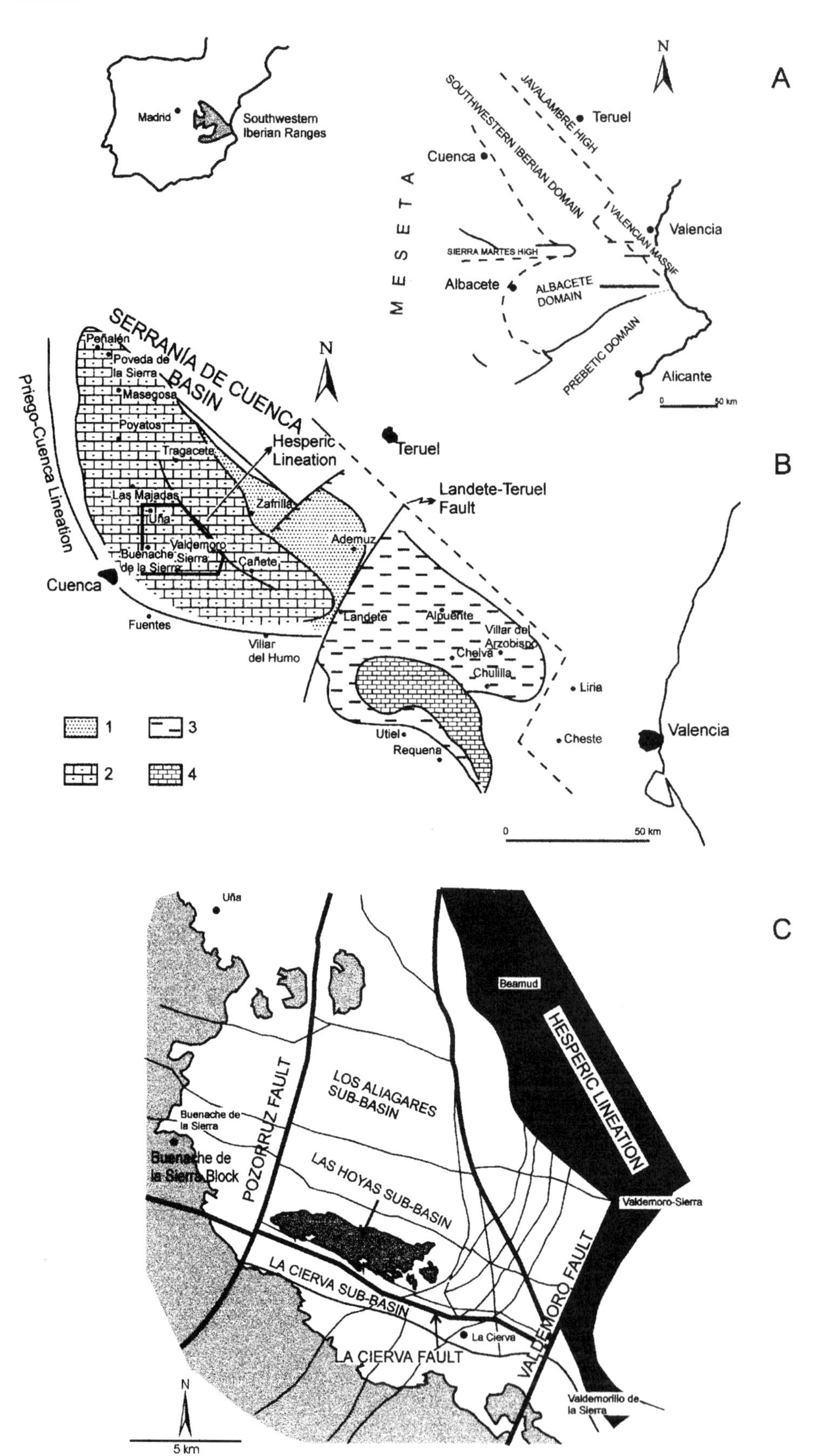

Figure 1—(A) Location of the southwestern Iberian ranges on the Iberian Peninsula and a sketch of the main paleogeographic domains that composed the southern part of the Iberian Basin during the Early Cretaceous (modified from Vilas et al., 1982). (B) Detailed paleogeographic reconstruction of the southwestern Iberian domain during deposition of the La Huérguina Formation. Distribution of the sedimentary environments and main tectonic structures is shown (modified from Mas et al., 1982; Meléndez et al., 1989; Meléndez et al., 1994). Area surrounded by a thick line in the Serranía de Cuenca basin comprises the Las Hoyas and adjacent subbasins. 1= siliciclastic alluvial plains, 2=distal alluvial and palustrine plains to lakes, 3=coastal siliciclastic alluvial plains, tidally influenced, and 4=coastal alluvial and palustrine plains and coastal lakes. (C) General sketch of the Las Hoyas area showing the main fault systems that are interpreted to have played an important role during infilling of the Las Hoyas and adjacent subbasins; see (A) and (B) for general location. The most important fractures have been marked with thick black lines. Lower and Upper Cretaceous stratigraphic units overlying the La Huérguina Formation have been mapped in light shading. Triassic outcrops in the east and northeast of the map are shown in dark shading. This area coincides with the Hesperic Lineation and was the primary source area for the Las Hoyas subbasin during sedimentation. The central area of the map is exclusively composed of Jurassic deposits and outcrops of the La Huérguina Formation, which is not shown in this sketch. Only the deposits of the Las Hoyas syncline are differentiated.

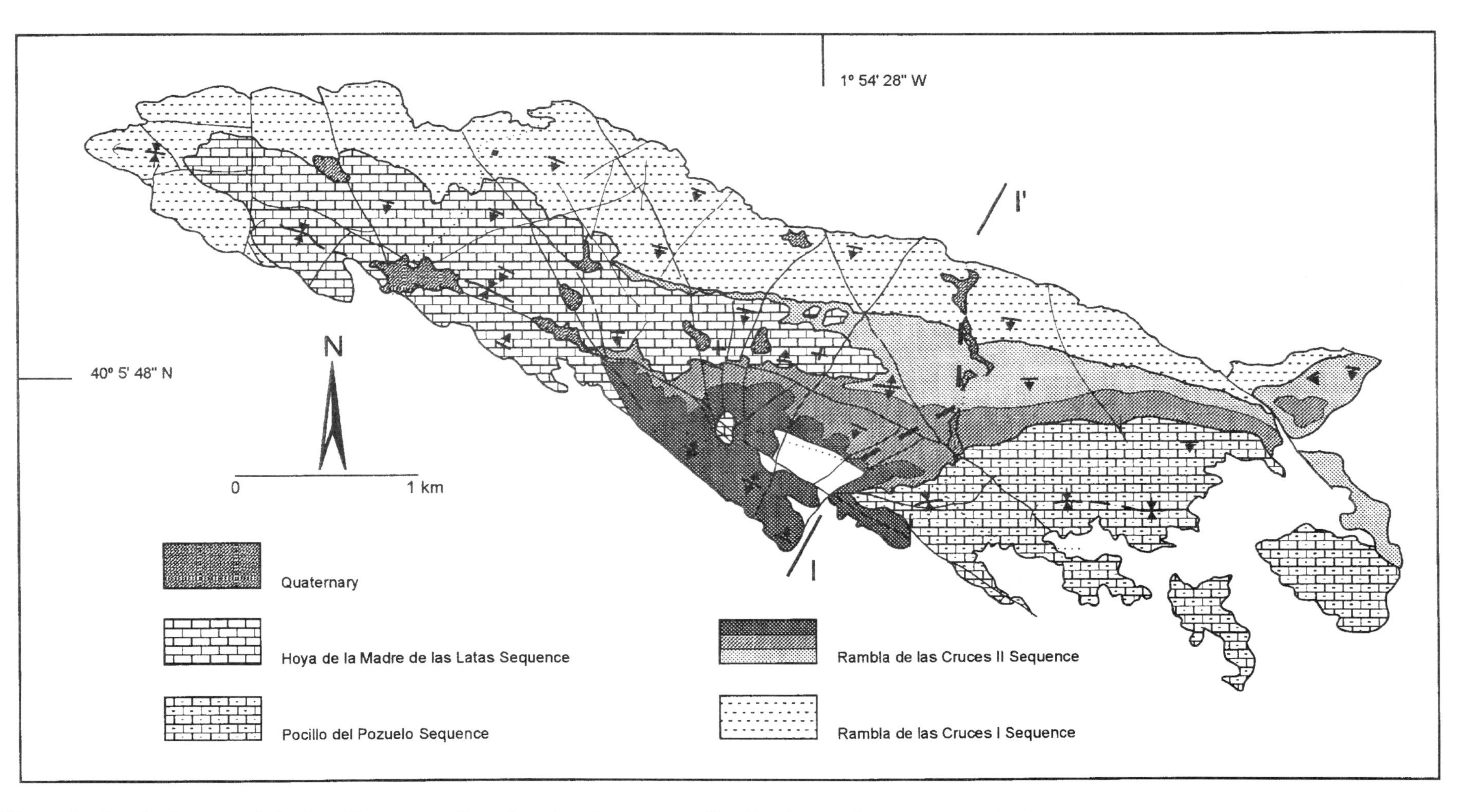

Figure 2—Geologic map of the Las Hoyas syncline showing the current distribution and geometry of the four stratigraphic units that compose the sedimentary record of the Las Hoyas subbasin. Surrounding terrains are completely composed of Jurassic limestones. Line I-I′ corresponds to the position of the cross-section in Figure 3. Thick dashed line shows the location of the logged section in Figure 5.

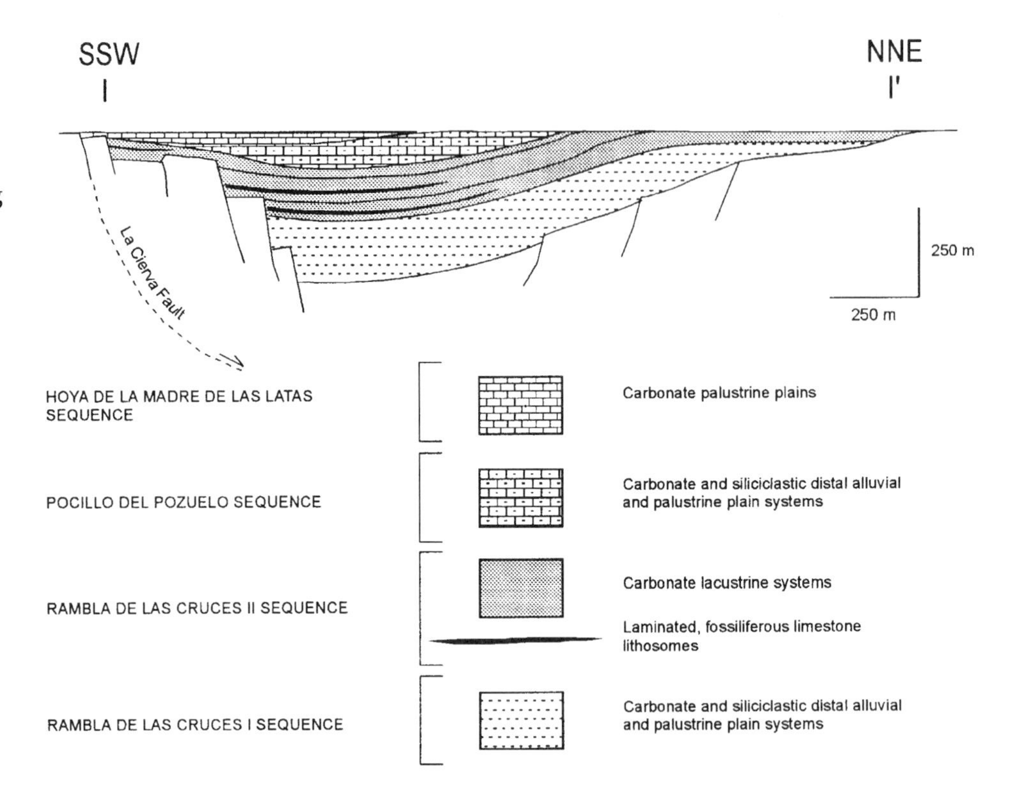

Figure 3—Cross-section showing an idealized reconstruction of the sedimentary infilling of the Las Hoyas subbasin and its geometry, preceding erosion. The section is perpendicular to the main tectonic structures that controlled sedimentation. See Figure 2 for location.

RAMBLA DE LAS CRUCES II SEQUENCE

This second sequence overlies the Rambla de las Cruces I sequence and onlaps Jurassic substrate toward the southern and eastern limits of the basin (Figures 2 and 4). Its maximum thickness is 170 m in the central part of the basin, thinning toward the edge of the basin. This second sequence is entirely composed of limestones and rare marlstones and arranged into three minor sequences (Figures 2–4), each composed of three facies belts. Rhythmically laminated limestones are observed at the base with lamination consisting of varve-like millimeter-scale couplets of dark and light laminae. Synsedimentary deformation structures are abundant, especially small fractures and decimeter- to centimeter-scale slumps. These laminated limestones contain the exceptionally well-preserved assemblage of flora and fauna of the Las Hoyas *Konservat-Lagerstätte*. This laminated facies is overlain by irregular slabby and thin-bedded limestones. The slabby limestones are grainstones to packstones composed of highly reworked bioclastic fragments, peloids, and plant debris, and show centimeter-scale cross-bedding interpreted as ripple migration. The thin-bedded limestones (centimeter-scale beds) are characterized by ostracodal and peloidal mudstones.

Overlying these facies, cross-bedded and massive charophyte limestones can be observed. Cross-bedded limestones are composed of reworked bioclastic fragments, mainly charophytes and ostracodes. Massive limestones are wackestones containing charophytes, ostracodes, gastropods, and bivalves. Both facies may show brecciation and vertical voids, interpreted as root activity, at the top of beds. This facies succession represents sedimentation in a shallow and permanent hardwater lake. Following nomenclature of Gierlowski-Kordesch and Kelts (1994), the first facies belt (rhythmically laminated limestones) is interpreted to correspond to sedimentation at the basinal areas of the lake, the second (irregular slabby and thin-bedded limestones) to the intralittoral and sublittoral realm, and the third facies belt (cross-bedded and massive charophyte limestones) to the supralittoral and eulittoral realm (Figure 5).

POCILLO DEL POZUELO SEQUENCE

This third depositional sequence crops out in the eastern part of the basin (Figure 2) and unconformably overlies the Rambla de las Cruces II sequence and Jurassic basement. It reaches a maximum thickness of 100 m, thinning toward the basin edge, and is composed of carbonate and siliciclastic deposits. Facies include claystones and marlstones with hydromorphic edaphic features, interbedded sandstones, sandy calcarenites, oncolite conglomerates, carbonate rudstones, and quartzite pebble conglomerates arranged in channelized bodies up to 3 m thick showing cross-bedding. Tabular and lenticular beds of biomicritic massive limestones containing charophytes, ostracodes, gastropods, and bivalves are also common. These beds usually show vertical voids and brecciation at the top, interpreted as root activity. Sandy calcarenites with abundant intraclasts and sigmoidal stratification are present toward the top of the sequence. This facies association is interpreted as the result of sedimentation in distal alluvial and palustrine plains with carbonate

Table 1. Composition and Diversity of the Las Hoyas *Konservat-Lagerstätten*.

FLORA	
Charophyta:	Nitellaceae indet.; Characae indet.; Clavatoraceae indet. (A and B)
Bryophyta:	Hepatophyta: *Hepaticites* sp.
Pteridophyta:	Filicales: Schizeaceae (*Ruffordia* sp.), Dicksoniaceae (Onychiopsis sp.), Osmundaceae (*Cladophlebis* sp.) Matoniaceae (Weichselia reticulata).
Spermatophyta:	Bennettitales: *Zanites* sp.
Coniferales:	Cheirolepidiaceae (*Cupressinocladus, Frenelopsis, Pagiophyllum, Brachyphyllum*), Taxodiaceae (*Sphenolepis*), Incertae sedis (*Podozamites*).
Gnetales:	*Drewria potomacensis*
Angiospermae:	Angiospermae indet.
Incertae sedis:	*Montsechia vidalia*

FAUNA	
Mollusca:	Bivalvia indet. (Unionidae?), Gastropoda indet. (A and B).
Arthropoda:	Crustacea: Ostracoda indet.
Peracarida:	Isopoda indet., Spelaeogrphaceae indet.
Decapoda:	Astacidae (*Austrapotanobious llopisi*), Atyidae (*Delclosia martinelli*).
Hexopoda:	Ephemeroptera: Leptophlebiidae: *Huergoneta ciervansis; Hispanoneta hoyaensis.*
Odonata:	Aeschniidae: *Gigantoaeschnidium ibericus, Nannoaeschnidium pumilio; Iberoaeschnidium conquensis*, Aeschniidae indet.
Gomphidae:	*Ilerdaegomphus torcae*
Aeshnidae:	*Hoyaschna cretacica*
Blattodea:	Mesoblattinidae: *Hispanoblatta sumptuosa*, Mesoblattinidae indet.
Isoptera:	*Meiatermes bertrani*
Ortoptera:	Gryllidae: *Torcagryllus apexreditus, Hoyagryllus huecarensis.*
Orthopteroidea:	Chresmodidae.
Heteroptera:	Belostomatidae: *Hispanepa conquensis, Iberonepa romerali, Torcanepa magnapes.*
Homoptera:	Cixiidae.
Raphidioptera:	Mesoraphiidae
Coleoptera:	Cupedidae, ?Ademosynidae, Escarabaeoidea
Hymenoptera:	Apocrita indet.
Diptera:	Nemestrinidae, Stratiomyiidae
Mecoptera:	Panorpidae
Neuroptera:	Chrysopidae (six different forms), Kalligammatidae
Chordata:	Sarcopterygii: Coelacantiformes: Coelacantidae: *"Holophagus"?.*
Actinopterygii:	Amiiformes: Caturidae (*Caturus* sp), Amiidae (*Amiopsis woodwardi, Vidalamia catalunica*), Macrosemiidae (*Notagogus ferreri, Notagogus* aff. *N. ferreri, Propterus* aff. *P. vidali, Propterus* sp.)
Pycnodontiformes:	Pycnodontidae (*Macromesodon* aff. *M. bernissartensis*, cf. *Eomesodon* sp.).
Semionotiformes:	Semionotidae (three forms of *Lepidotes* sp.)
Teleostei:	Pholidophoriformes: Pleuropholidae (*Pleuropholis* sp.).
Gonorynchiformes:	Chanidae: Rubiesichthyinae (*Rubiesichthys gregalils, Gordichthys conquensis*).
Teleostei *incertae sedis*:	four forms.
Amphibia:	Albanerpetontidae: *Celtedens ibericus*
Anura:	Discoglossidae, Anura indet.
Caudata:	*Hylaeobatrachus, Valdotriton*; Caudata indet.
Amniota:	Chelonia: Centrocryptodira indet.
Squamata:	*Meyasaurus*, squamata indet (A and B).
Crocodylomorpha:	Crocodyliformes indet.
Neosuchia:	Atoposauridae (*Montesechosuchus*); *Unasuchus, Goniopholis*, Neosuchia indet.
Dinosauria:	Theropoda: Ornithomimosauria (*Pelecanimimus poliodon*), Theropoda indet.
Aves:	*Iberomesornis romerali, Concornis lacustris, Eoalulavis hoyasi.*

Compiled by Ortega et al., 1998.

pond development. The largest limestone bodies deposited in pond paleoenvironments are recognized toward the top of the sequence, where they are represented by a facies succession composed of massive charophyte limestones and bioclastic limestones that represent sedimentation in lacustrine areas with agitated littoral conditions. Crevasse splay deposits composed of intraclastic sandy calcarenites interfingering

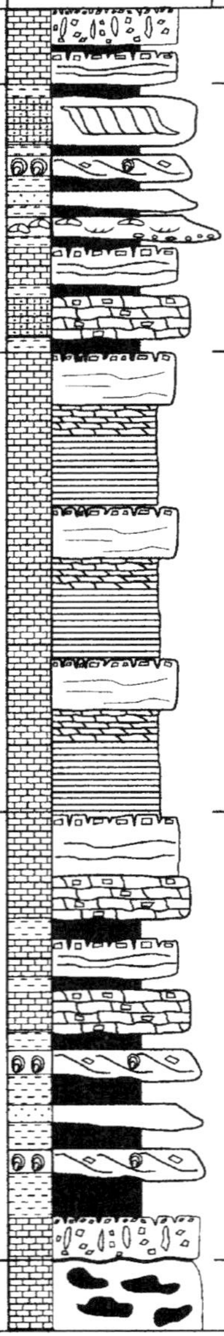

AGE	LTH.	L.ST.	STRATIGRAPHIC COLUMN	FOSSIL CONTENT	THICKNESS	SEDIMENTARY ENVIRONMENTS	EXPOSURE FEATURES	SEQUENCE BOUNDARIES	MAIN PALEOGEOGRAPHIC EVENTS AND CONTROLS
UPPER BARREMIAN	LA HUÉRGUINA FORMATION	H.M.L. SEQU.		CHAROPHYTES OSTRACODES	20 m	PALUSTRINE PLAINS	ROOT VOIDS BRECCIATION NODULIZATION MARMORIZATION	LOCAL UNCONFORMITY	TILTING OF THE BASIN TOWARDS THE NE MINIMUM SUBSIDENCE ABSENCE OF SILICICLASTIC INPUT
		POCILLO DEL POZUELO SEQUENCE		CHAROPHYTES OSTRACODES GASTROPODS BIVALVES	20 - 100 m	DISTAL ALLUVIAL AND PALUSTRINE PLAINS CARBONATE PONDS	VERTICAL ROOT VOIDS BRECCIATION RARE MARMORIZATION	LOCAL UNCONFORMITY	TILTING OF THE BASIN TOWARDS THE EAST-SOUTHEAST UPLIFTING OF THE SOURCE AREAS LOCATED TO THE EAST-NORTHEAST SUBSIDENCE RATE SLOWS DOWN TOWARDS THE TOP OF THE SEQUENCE AXIAL DRAINAGE PATTERN
		RAMBLA DE LAS CRUCES II SEQUENCE		FOR FOSSIL FLORA AND FAUNA SEE TABLE 1 ICHNOFOSSILS: -*Cruziana* isp. -*Helminthoidichnites tenuis* -*Lockeia* isp. -*Palaeophycus* isp. -*Treptichnus pollardi*	30 - 170 m	PERMANENT, SHALLOW, HARDWATER LAKES	VERTICAL ROOT VOIDS BRECCIATION AT THE TOP OF THE SHALLOWING UPWARDS SEQUENCES	PARACONFORMITY	EXPANSION OF THE AREA OF SEDIMENTATION STAGE OF MAXIMUM SUBSIDENCE RATE AND MAINTENANCE OF THE RATIO SUBSIDENCE RATE *VERSUS* SEDIMENTATION RATE DEPOCENTER LOCATED AT THE CENTRAL AREA OF THE BASIN. PROGRESSIVE SOUTHWARDS MIGRATION ABSENCE OF SILICICLASTIC INPUT FORMATION OF THE *KONSERVAT-LAGERSTÄTTEN*
		RAMBLA DE LAS CRUCES I SEQUENCE		CHAROPHYTES OSTRACODES GASTROPODS BIVALVES	25 - 150 m	DISTAL ALLUVIAL AND PALUSTRINE PLAINS CARBONATE PONDS	VERTICAL ROOT VOIDS BRECCIATION NODULIZATION MARMORIZATION	REGIONAL UNCONFORMITY	HIGH SUBSIDENCE RATE: TENDENCY TOWARDS VERTICAL AGGRADATION OF THE ALLUVIAL PLAINS DEPOCENTER LOCATED AT THE CENTRAL-WESTERN AREA OF THE BASIN AXIAL DRAINAGE PATTERN
DOGGER (JURASSIC)				STRONGLY KARSTIFIED MARINE LIMESTONES					

Figure 4—General stratigraphy and main characteristics of the Las Hoyas subbasin fill. LTH. = Lithostratigraphy, L.ST. = Local stratigraphy, H.M.L. SEQU. = Hoya de la Madre de las Latas sequence. Legend in Figure 5.

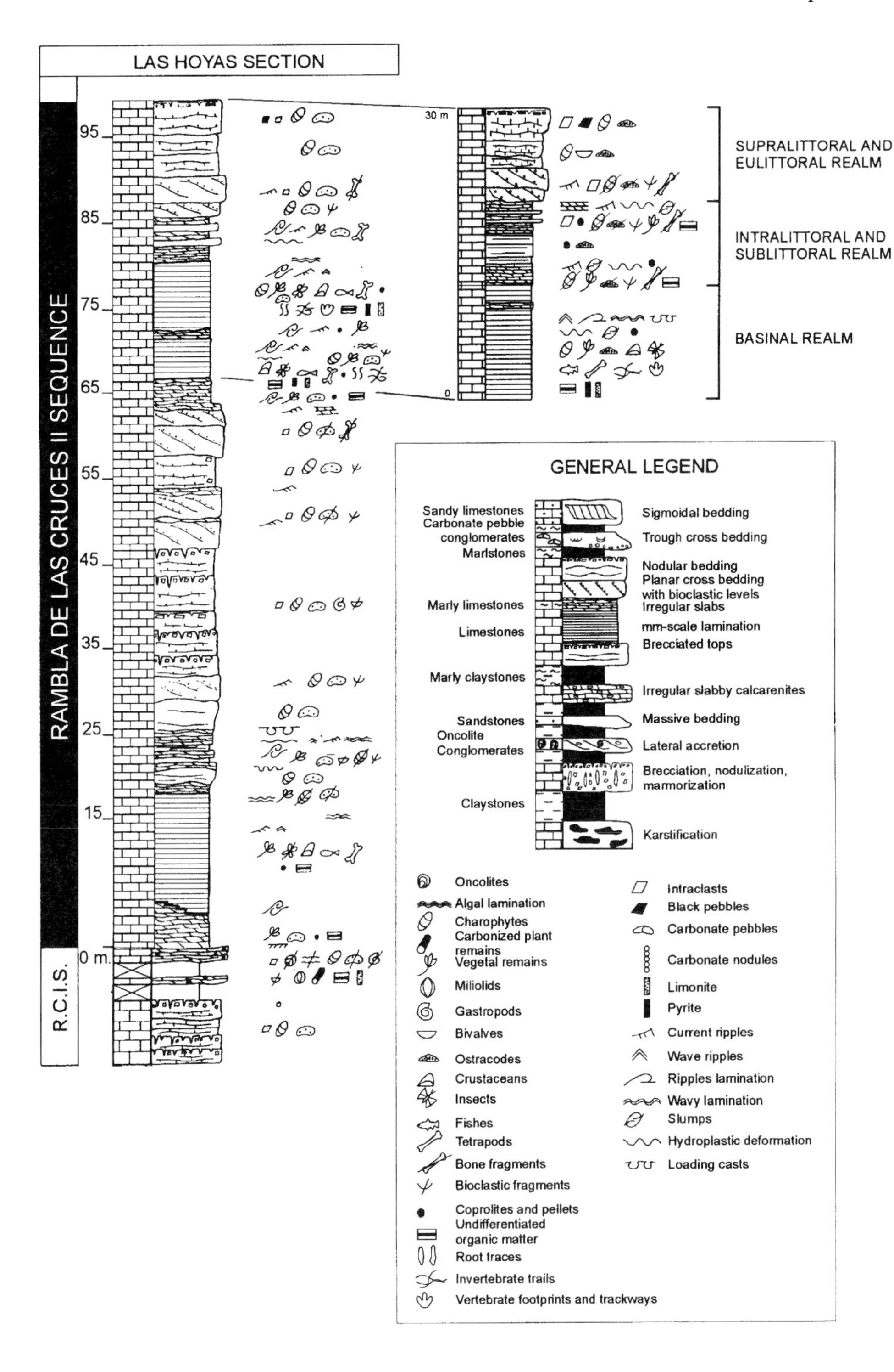

Figure 5—Detailed logged section of the Rambla de Las Cruces II sequence. This sequence is entirely composed of lacustrine carbonate deposits that contain *Konservat-Lagerstätten*. At upper right, a complete shallowing-upward sequence of the lacustrine systems is shown. RCIS. = Rambla de las Cruces I sequence. See Figure 2 for location.

limestone pond deposits are also common toward the top of the sequence.

HOYA DE LA MADRE DE LAS LATAS SEQUENCE

This fourth sequence unconformably overlies all the previous sequences, as well as Jurassic substrate in the southern part of the area. The Hoya de la Madre de las Latas sequence crops out in the western half of the basin (Figure 2) and its thickness rarely exceeds 20 m, probably due to subsequent erosion. Lithologically, the sequence is completely composed of limestones; massive biomicritic limestones with charophyte and ostracodes, in tabular or slightly lenticular discontinuous beds up to 1 m thick, dominate the lower part of this sequence. Subaerial exposure features and possible root activity are common at the top of beds, mainly represented by features such as vertical voids and brecciation. These massive biomicritic limestones are interpreted as being deposited in hardwater pond paleoenvironments. Toward the top of this sequence, nodular and marmorized limestones dominate, interpreted as palustrine plain deposits.

PALEOENVIRONMENTAL MODEL FOR THE LAS HOYAS PALEOLAKE AND ITS ASSOCIATED *KONSERVAT-LAGERSTÄTTEN*

The presence of three laminated lithosomes containing rich and well-preserved assemblages of flora and fauna in the Rambla de Las Cruces II sequence (Figure 5) has led to a detailed sedimentologic analysis of these facies to develop a paleoenvironmental model to explain preservation. Fossils are preserved in rhythmically laminated limestones interpreted as deposits of the basinal areas of permanent and hardwater lakes. Each basinal laminated lithosome is homogeneous in time and space; therefore, general paleoenvironmental conditions may not have changed significantly throughout the deposition of each laminated lithosome. Microscopic analysis of the rhythmically laminated deposits, however, has allowed the discernment of at least three different microfacies that record subtle changes in paleoenvironmental paleoconditions linked to cyclic oscillations in lake level (Fregenal Martínez and Meléndez, 1998).

The first microfacies consists of dark-light couplets composed of small bioclastic fragments, peloidal intraclasts, and micritic mud interpreted as density flows. This represents sedimentation under high lake level conditions and after flooding probably linked to rainy seasons. The carbonate components came from littoral areas, as well as ponds and palustrine plains that surrounded the lake. The second microfacies is entirely composed of micrite and exhibits current structures and small wave ripples, interpreted as low-density or low-concentrated turbidites (Stow and Bowen, 1992; Stow et al., 1996) representing sedimentation under low lake level conditions. The third microfacies is composed of alternating laminae with features typical of the first and second microfacies, probably representing sedimentation during periods of fluctuations in lake level or simply intermediate lake level conditions.

Ichnologic data also support this interpretation of lake level oscillations. Besides some invertebrate trace fossils (see Figure 4) (Fregenal Martínez et al., 1995a), two significant vertebrate trails occur associated with laminated basinal deposits: fish trails and crocodile trackways. Fish need a minimum water layer to produce such trails, and these trails need a viscous substrate to be preserved (De Gibert et al., 1999); however, the crocodile trackways of Las Hoyas are interpreted to have been produced by a crocodile walking on a quite firm substrate of laminated sediments (Moratalla et al., 1995). Because no indications of subaerial exposure have ever been found in the fossiliferous laminated sediments, not even associated with levels bearing crocodile trackways, complete and long-lasting desiccation of the entire lake does not appear to have been possible. Signs of subaerial exposure are also absent in the sublittoral and intralittoral deposits (irregular slabby and thin-bedded limestones). Voids and cavities created by root decay and brecciation are exclusively associated with cross-bedded and massive charophyte limestones of the eulittoral realm. This type of possible short-term, but otherwise seemingly extreme oscillations in water level, affecting even basinal areas can be explained in the context of a shallow lake with a very gentle slope and a high ratio of surface area versus depth, where a small vertical change in the water level may induce exposure of extensive areas around the lake margin for short periods. Vertical arrangement of the facies that compose the shallowing-upward sequence of the Las Hoyas lake confirm this hypothesis of a shallow yet extensive lake. Infilling sequences of deep lakes with steep margins and well-developed talus tend to show abrupt changes of facies in the transition between the basinal deposits and sublittoral or bench deposits. The sequences of Las Hoyas, however, show a very gradual transition from the basinal facies (rhythmically laminated limestones) to the sublittoral and intralittoral facies (irregular slabby and thin-bedded limestones).

Preservation of laminated facies, as well as body and trace fossils, requires a special set of environmental conditions to avoid destruction of lamination, decay, disarticulation, and dispersion of organic remains. Processes usually invoked to explain the formation of this type of fossil preservation (Seilacher et al., 1985) include (1) anoxic bottom conditions preventing decay of organic remains and limiting bioturbation and its associated disarticulation of carcasses; (2) coating by bacterial mats creating biogeochemical microenvironments necessary for mineralization and preservation of organic tissues; and (3) event sedimentation providing rapid burial. The *Konservat-Lagerstätten* at Las Hoyas were formed through similar processes (Fregenal Martínez et al., 1995b).

Development of anoxia at the bottom usually involves water stratification. Because thermal water stratification processes are unlikely to occur in shallow freshwater lakes, a high rate of organic productivity is the most plausible mechanism to account for permanent or semipermanent anoxia in the bottom sediments and dysaerobic to oxic conditions in the lower layers of the water column. Short events of oxygenation of the bottom sediments are represented by some layers bearing a characteristic invertebrate ichnologic association (see Figure 4), which is characterized by dwarfism, dominance of horizontal traces of deposit feeders, and low ichnodiversity but high ethologic diversity; this ichnoassociation indicates a low-energy, permanent subaqueous and shallow lacustrine environment with stressful ecologic conditions linked to oxygen deficiency (Fregenal Martínez et al., 1995a).

Figure 6 shows a model from the sedimentologic analysis that explains most of the geologic evidence while outlining the main characteristics, processes, and controls in the paleoenvironmental dynamics of the lacustrine system (Fregenal Martínez, 1998). The Las Hoyas lake system can be characterized as an extensive, perennial, shallow, hardwater lake with a gentle slope profile. Organic productivity was high and dysaerobic conditions prevailed at the lower levels of the water column, whereas bottom sediments

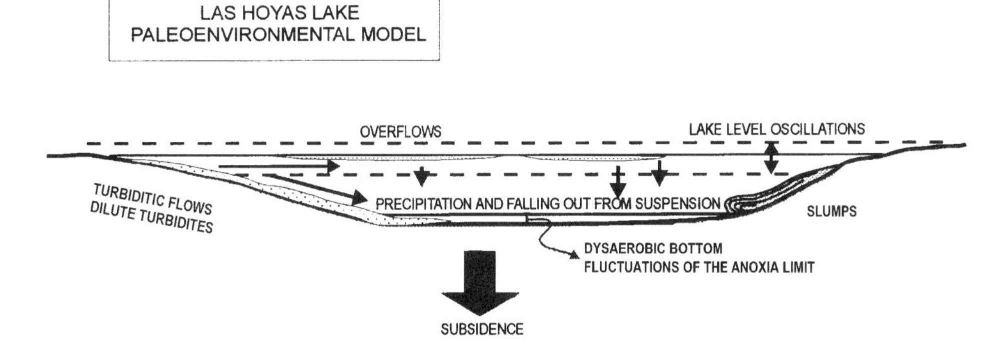

CHARACTERISTICS

- SHALLOW LAKE
- GENTLE SLOPE
- POSITIVE HYDROLOGICAL BALANCE
- HARDWATERS
- HIGH ORGANIC PRODUCTIVITY
- DYSAEROBIC BOTTOM

PROCESSES

- CARBONATE PRECIPITATION
- CLASTIC CARBONATE SEDIMENTATION
- DILUTE TURBIDITIC FLOWS
- REWORKING AND REMOVAL OF LITTORAL DEPOSITS BY STORMS

CONTROLS

- TECTONIC SUBSIDENCE
- CLIMATIC SEASONALITY
- CARBONATE SOURCE AREAS
- KARSTIC DRAINAGE
- LATERAL SWITCHING OF STREAMS ENTERING THE LAKE

Figure 6—General paleoenvironmental model proposed for the Las Hoyas lake (after Fregenal Martínez, 1998).

probably were anoxic most of the time (Fregenal Martínez and Meléndez, 1995b). Hydrologic input was from surface and groundwaters and the lake was surrounded by palustrine areas. The main sources of carbonate were bio-induced precipitation, allochthonous influxes of carbonate leached from the Jurassic source areas, and reworked bioclastic and intraclastic carbonate fragments. Carbonate facies deposited by traction processes are abundant in all the subenvironments of the lake because the system was periodically fed by ephemeral streams and seasonal storms; therefore, tectonics, climatic seasonality, and the composition of the source area were the main controls on the sedimentologic evolution of the Las Hoyas lake.

Distensive basins are commonly associated with lacustrine *Konservat-Lagerstätten*. The Lower Cretaceous fossil sites of the Santana Formation located at the Araripe basin (northeastern Brazil) (Maisey, 1991), the Lower Cretaceous localities of El Montsec (south-central Pyrenees, northern Spain) (Fregenal Martínez and Meléndez, 1995c), and the Tertiary fossil site of the Rubielos de Mora basin (central Spain) (Anadón et al., 1988) are *Konservat-Lagerstätten* linked to extensional tectonic settings that share some preservational features with the Las Hoyas fossil site.

REFERENCES CITED

Anadón, P., Ll. Cabrera, and R. Julià, 1988, Anoxic-oxic cyclical lacustrine sedimentation in the Miocene Rubielos de Mora Basin, Spain, *in* A. J. Fleet, K. Kelts, and M. R. Talbot, eds., Lacustrine Petroleum Source Rocks, Geological Society (London) Special Publication No. 40, p. 353-367.

De Gibert, J. M., L. A. Buatois, G. Mangano, M. A. Fregenal Martínez, F. Ortega, F. J. Poyato Ariza, and S. Wenz, 1999, The fish trace fossil *Undichna* from the Cretaceous of Spain: taphonomic and paleoenvironmental implication for the ichnogenus, Paleontology, v. 42, part 3, p. 409–427.

Fregenal Martínez, M. A., 1998, Análisis de la cubeta sedimentaria de Las Hoyas y su entorno paleogeográfico (Cretácico Inferior, Serranía de Cuenca). Sedimentología y aspectos tafonómicos del yacimiento de Las Hoyas, Ph.D. Dissertation (Unpublished), Universidad Complutense de Madrid, 354 p.

Fregenal Martínez, M. A., and N. Meléndez, 1995b, Autocyclic rhythms in the carbonate lacustrine laminites of the Las Hoyas Sub-basin (Lower Cretaceous, Serranía de Cuenca, Central Spain), Abstracts, First International Limnogeological Congress, Copenhagen, p. 39.

Fregenal Martínez, M. A., and N. Meléndez, 1995c, Geological Setting, *in* X. Martínez Delclòs, ed., Montsec and Montral-Alcover, two *Konservat-Lagerstätten*, Catalonia, Spain, Institut d'Estudis Ilerdencs, Lleida, p. 12-24.

Fregenal Martínez, M. A., and N. Meléndez, 1998, Signals of lake level oscillations recorded in basinal laminated deposits, Las Hoyas lacustrine system (Lower Cretaceous, Serranía de Cuenca, Spain), Abstracts 15th International Sedimentological Congress, Publicaciones de la Universidad de Alicante, p. 347-348.

Fregenal Martínez, M. A., L. A. Buatois, and G. Mángano, 1995a, Invertebrate trace fossils from Las Hoyas fossil site (Serranía de Cuenca, Spain). Paleoenvironmental interpretations, Extended Abstracts II International Symposium on Lithographic Limestones, Cuenca, Ediciones de la Universidad Autónoma de Madrid, p. 67-70.

Fregenal Martínez, M. A., A. D. Buscalioni, C. Diéguez, S. Evans, G. MacGowan, X. Martínez Delclòs, N. Meléndez, F. Ortega, B. P. Perez Moreno, F. J. Poyato Ariza, D. Rabadà, J. L. Sanz, and S. Wenz, 1995b,

Taphonomy, *in* N. Meléndez, ed., Las Hoyas. A Lacustrine Konservat-Lagerstätte, Cuenca, Spain, Editorial Universidad Complutense, Madrid, p. 21-27.

Gierlowski-Kordesch, E., J. C. Gómez Fernández, and N. Meléndez, 1991. Carbonate and coal deposition in an alluvial-lacustrine setting: Lower Cretaceous (Weald) in the Iberian Range (east-central Spain), *in* P. Anadón, Ll. Cabrera, and K. Kelts, eds., Lacustrine Facies Analysis, International Association of Sedimentologists Special Publication No. 13, p. 109-125.

Gierlowski-Kordesch, E., and K. Kelts, 1994. Introduction, *in* E. Gierlowski-Kordesch and K. Kelts, eds., Global Geological Record of Lake Basins, Volume 1, Cambridge University Press, Cambridge, p. xvii-xxxiii.

Maisey, J. G., ed., 1991, Santana fossils: An illustrated atlas, T.F.H. Publications, 459 p.

Mas, J. R., A. Alonso, and N. Meléndez, 1982, El Cretácico basal "Weald" de la Cordillera Ibérica Suroccidental (NW de la provincia de Valencia y E de la de Cuenca), Cuadernos de Geología Ibérica, v. 8, p. 309-335.

Meléndez, N., A. Meléndez, and J. C. Gómez Fernández, 1989, Los sistemas lacustres del Cretácico inferior de la Serranía de Cuenca, Cordillera Ibérica, Guía de campo de la IV reunión del grupo español de trabajo del IGCP-219, Editorial Universidad Complutense, Madrid, 70 p.

Meléndez, N., A. Meléndez, and J. C. Gómez Fernández, 1994, La Serranía de Cuenca Basin (Lower Cretaceous), Iberian Ranges, central Spain, *in* E. Gierlowski-Kordesch, and K. Kelts, eds., Global Geological Record of Lake Basins, Volume 1, Cambridge University Press, Cambridge, p. 215-219.

Moratalla, J. J., M. Lockley, A. D. Buscalioni, M. A. Fregenal Martínez, N. Meléndez, F. Ortega, B. P. Pérez Moreno, E. Pérez Asensio, J. L. Sanz, and R. J. Schultz, 1995, A preliminary note on the first tetrapod trackways from the lithographic limestones of Las Hoyas (Lower Cretaceous, Spain), Geobios, v. 28 (6), p. 777-782.

Ortega, F., J. L. Sanz, L. J. Barbadillo, A. D. Buscalioni, C. Diéguez, S. E. Evans, M. A. Fregenal Martínez, M. de la Fuente, J. Madero, C. Martín Closas, X. Martínez Delcl˜s, N. Meléndez, J. J. Moratalla, B. P. Pérez Moreno, E. Pinardo Moya, F. J. Poyato Ariza, J. Rodríguez Lázaro, B. Sanchiz, and S. Wenz, 1998, El yacimiento de Las Hoyas (La Cierva, Cuenca). Un *Konservat-Lagerstätte* del Cretácico Inferior, *in* E. Aguirre and I. Rábano, eds., La Huella del Pasado. Fosiles de Castilla-La Mancha, Colección Patrimonio Histórico, Arqueología, Castilla-La Mancha, p. 195–215.

Salas, R., and A. Casas, 1993, Mesozoic extensional tectonics, stratigraphy and crustal evolution during the Alpine cycle of the eastern Iberian Basin, Tectonophysics, v. 228, p. 33-55.

Seilacher, A., W. E. Reif, and F. Westphal, 1985, Sedimentological, ecological and temporal patterns of fossil lagerstätten, Philosophical Transactions of the Royal Society of London, v. B311, p. 5-23.

Stow, D. A. V., and A. J. Bowen, 1992, A physical model for the transport and sorting of fine-grained sediment by turbidity currents, *in* D. A. V. Stow, ed., Deep-Water Turbidite Systems, International Association of Sedimentologists, Blackwell Scientific Publications, Reprint Series, v. 3, p. 25-40.

Stow, D. A. V., H. G. Reading, and J. D. Collinson, 1996, Deep Seas, *in* H. G. Reading, ed., Sedimentary Environments: Processes, Facies and Stratigraphy, 3rd edition, Blackwell Science, London, p. 395-453.

Vilas, L., R. Mas, A. García, C. Arias, A. Alonso, N. Meléndez, and R. Rincón, with the collaboration of E. Elizaga, C. Fernández Calvo, C. Gutiérrez, and F. Meléndez, 1982, Ibérica Suroccidental, *in* El Cretácico de España, Editorial de la Universidad Complutense de Madrid, p. 457-513.

Vilas, L., A. Alonso, C. Arias, R. Mas, R. Rincón, and N. Meléndez, 1983, The Cretaceous of the Southwestern Iberian Ranges (Spain), Zitteliana, v. 10, p. 245-254.

Ziegler, A. M., A. L. Raymond, T. C. Gierlowski, M. A. Horrell, D. B. Rowley and A.L. Lottes, 1987, Coal, climate and terrestrial productivity: the present and Early Cretaceous compared, *in* A. C. Scott, ed., Coal and Coal-bearing Strata: Recent Advances. Geological Society (London) Special Publication No. 32, p. 25-49.

ADDITIONAL REFERENCES

Buscalioni, A. D., F. Ortega, B. P. Pérez-Moreno, and S. E. Evans, 1996, The Upper Jurassic Maniraptoran Therapod *Lisbosaurus estesi* (Guimarota, Portugal) reinterpreted as a Crocodylomorph, Journal of Vertebrate Paleontology, v. 16 (2), p. 358-362.

Buscalioni, A. D., F. Ortega, D. Rasskin-Gutman, and B. P. Pérez-Moreno, 1997, Loss of carpal elements in crocodilian limb evolution: morphogenetic model corroborated by paleobiological data, Biological Journal of the Linnean Society, v. 62, p. 133-144.

Evans, S. E., and A. R. Milner, 1996, A metamorphosed salamander from the Early Cretaceous of Las Hoyas, Spain, Philosophical Transactions of the Royal Society of London, v. B351, p. 627-646.

Evans, S. E., and L. J. Barbadillo, 1997, Early Cretaceous lizards from Las Hoyas, Spain, Zoological Journal of the Linnean Society, v. 119, p. 23-49.

Fregenal Martínez, M. A., 1996, El sistema lacustre de Las Hoyas (Cretácico inferior, Serranía de Cuenca): Estratigrafía y Sedimentología, Serie Paleontología, v. 1, Publicaciones de la Excelentísima Diputación provincial de Cuenca, 181 p.

Fregenal Martínez, M. A., and N. Meléndez, 1993, Sedimentología y evolución paleogeográfica de la cubeta de Las Hoyas, Cuadernos de Geología Ibérica, v. 17, p. 231-256.

Fregenal Martínez, M. A., and N. Meléndez, 1994, Sedimentological analysis of the Lower Cretaceous lithographic limestones of the "Las Hoyas" fossil site (Serranía de Cuenca, Iberian Range, Spain), Geobios Mémoire Special, v. 16, p. 185-193.

Fregenal Martínez, M. A., and N. Meléndez, 1995a, Paleotectonic controls of the origin of the Las Hoyas fossil site (Serranía de Cuenca, Spain), Extended

Abstracts II International Symposium on Lithographic Limestones, Cuenca, Ediciones de la Universidad Autónoma de Madrid, p. 71-74.

Garassino, A., 1997, The macruran decapod crustaceans of the Lower Cretaceous (Lower Barremian) of Las Hoyas (Cuenca, Spain), Atti Società Italiana Science Naturale Museo Civico Storia Naturale Milano, v. 137 (I-II), p. 101-126.

Gómez Fernández, J. C., and N. Meléndez, 1991, Rhythmically laminated lacustrine deposits in the Lower Cretaceous of la Serranía de Cuenca basin (Iberian Ranges, Spain), *in* P. Anadón, Ll. Cabrera, and K. Kelts, eds., Lacustrine Facies Analysis, International Association of Sedimentologists Special Publication No. 13, p. 247-258.

Martín Closas, C., and M. C. Diéguez, 1998, Charophytes from the Lower Cretaceous of the Iberian Ranges (Spain), Paleontology, v. 41, part 6, p. 1133–1152.

Martínez Delclòs, X., and M. J. Ruiz de Loizaga, 1994, Les insectes des calcaires lithographiques du Crétacé inférieur d'Espagne. Faune et taphonomie, Geobios Memoire Special, v. 16, p. 195-201.

Martínez Delclòs, X., A. Nel, and Y. A. Popov, 1995, Systematic and functional morphology of *Iberonepa romerali* n.gen., n.sp. Belostomatidae Stygeonepinae from the Spanish Lower Cretaceous (Insecta, Heteroptera, Nepomorpha), Journal of Paleontology, v. 69, p. 496-508.

Meléndez, N., ed., 1995, Las Hoyas. A lacustrine Konservat-Lagerstätte, Cuenca, Spain, Editorial Universidad Complutense, Madrid, 89 p.

Pérez Moreno, B. P., J. L. Sanz, A.D. Buscalioni, J. J. Moratalla, F. Ortega, and D. Rasskin Gutman, 1994, A unique multitoothed ornithomimosaur from the Lower Cretaceous of Spain, Nature, v. 370, p. 363-367.

Poyato Ariza, F. J., 1994, A new Early Cretaceous gonorynchiform fish (Teleostei: Ostariophysi) from Las Hoyas (Cuenca, Spain), Occasional Papers of the Museum of Natural History, The University of Kansas, no. 164, p. 1-37.

Poyato Ariza, F. J., 1996, A revision of the Ostariophysan fish family Chanidae, with special reference to the Mesozoic forms, Palaeo Ichthyologica, v. 6, p. 52.

Poyato Ariza, F. J., 1997, A new assemblage of Spanish Early Cretaceous teleostean fishes, formerly considered "leptolepids": phylogenetic relevance, Comptes Rendus de l'Académie des Sciences de Paris, v. 325, p. 373-379.

Poyato Ariza, F. J., M. R. Talbot, M. A. Fregenal Martínez, N. Meléndez, and S. Wenz, 1998, First isotopic and multidisciplinary evidence for nonmarine coelacanths and pycnodontiform fishes: paleoenvironmental implications, Palaeogeography, Palaeoclimatology, Palaeoecology, v. 144, p. 65-84.

Sanz, J. L., J. F. Bonaparte, and D. Lacasa, 1988, Unusual Early Cretaceous bird from Spain, Nature, v. 331, p. 433-435.

Sanz, J. L., S. Wenz, A. Yébenes, R. Estes, X. Martínez Delclòs, E. Jiménez-Fuentes, C. Diéguez, A. D. Buscalioni, L. J. Barbadillo, and L. Vía, 1988, An Early Cretaceous faunal and floral continental assemblage: Las Hoyas fossil site (Cuenca, Spain), Geobios, v. 21 (5), p. 611-635.

Sanz, J. L., L. M. Chiappe, J. F. Bonaparte, and A. D. Buscalioni, 1995, The osteology of *Concornis lacustris* (Aves: Enantiornithes) from the Lower Cretaceous of Spain and a reexamination of its phylogenetic relationships, American Museum Novitates, no. 3133, p. 1-23.

Sanz, J. L., L. M. Chiappe, B. P. Pérez Moreno, A. D. Buscalioni, J. J. Moratalla, F. Ortega, and F. J. Poyato Ariza, 1996, An Early Cretaceous bird from Spain and its implications for the evolution of avian flight, Nature, v. 382, p. 442-445.

Sanz, J. L., M. A. Fregenal Martínez, N. Meléndez, and F. Ortega, in press, Las Hoyas lake, *in* D. E. K. Briggs and P. R. Crowther, eds., Paleobiology II. Blackwell Science, London.

Talbot, M. R., N. Meléndez Hevia, M. A. Fregenal Martínez, F. J. Poyato Ariza, and S. Wenz, 1995, Hydrology of the Las Hoyas paleolake (Lower Cretaceous, Cuenca, Spain): An integrated strontium and stable isotope study, Abstracts, First International Limnogeological Congress, Copenhagen, p. 133.

Section V

Late Cretaceous to Paleocene

Cojan, I., 2000, Late Cretaceous and early Paleocene lacustrine episodes in the Provence Basin, southern France, *in* E. H. Gierlowski-Kordesch and K. R. Kelts, eds., Lake basins through space and time: AAPG Studies in Geology 46, p. 317–322.

Chapter 26

Late Cretaceous and Early Paleocene Lacustrine Episodes in the Provence Basin, Southern France

Isabelle Cojan
Ecole Nationale Supérieure des Mines de Paris, CGES Sédimentologie
Fontainebleau, France

GENERAL SETTING

During Late Cretaceous and Early Tertiary times, southern France was characterized by continental sedimentation. In Provence, the continental episode extended from the end of the Santonian to the Lutetian. The formations now crop out in east-west oriented synclines such as the Aix-en-Provence, Rians, and Salernes synclines (Figure 1).

The depositional environment is interpreted as floodplain sediments that pass into lacustrine carbonates or interfinger with proximal alluvial fan deposits. This pattern is repeated several times through the continental sedimentation period (Durand, 1980; Lapparent, 1938; Collot, 1890). The two lacustrine episodes considered here are the Calcaire de Rognac Formation and the Calcaire de Vitrolles Formation, which were respectively deposited during the middle Maastrichtian and early Paleocene (Figure 2).

Regional paleogeography corresponded to a drainage network of parallel valleys, flowing into a major valley (Aix-en-Provence). Fluvial deposits are characteristic of a distal braided system in a low floodplain, and the lacustrine formations testify to relatively shallow lakes or palustrine environments (Cojan, 1993).

STRATIGRAPHIC FRAMEWORK

Lithostratigraphy correlates the Calcaire de Rognac Formation to the middle Rognacian, and the Calcaire de Vitrolles Formation to the lower Vitrollian (Babinot and Durand, 1984; Babinot and Freytet, 1983; Babinot, 1980; Babinot and Durand, 1980; Feist-Castel, 1975; Corroy, 1957; Dughi et Sirugúe, 1957, 1958; Vasseur, 1898). On the basis of the continental fauna, the Rognacian is generally related to upper Maastrichtian and the Vitrollian to Dano-Montian (Durand et al., 1984); however, recent paleomagnetic results bring out a new framework for the stratigraphic scale. The Rognacian extends over the entire Maastrichtian and, therefore, the Calcaire de Rognac Formation would correspond to the middle Maastrichtian (Jaeger and Westphal, 1989; Westphal and Durand, 1989; Krumsiek and Hahn, 1989) and the Calcaire de Vitrolles Formation would be attributed to the Danian (Cojan et al., 1998). The age result concerning the Calcaire de Rognac Formation is largely in agreement with the paleontologic data (giving a late, but not terminal, Maastrichtian age). Unfortunately, due to the lack of fossils, there is no paleontologic control over any determination of the Vitrollian age.

DETAILED SECTION OF THE CALCAIRE DE ROGNAC FORMATION

Inferred from the study outcrops, three ancient lakes are thought to have existed (Figure 3a–c). In the Aix and Rians regions, each lake can be estimated to be over 1500 km^2. In the Salernes region, the lake was probably more restricted, but present outcrop conditions do not allow any tentative evaluation.

In the Aix-en-Provence area, the Calcaire de Rognac Formation shows four major asymmetric depositional sequences whose individual thicknesses do not exceed 15 m (Figure 4). In each sequence, interpreted paleoenvironments range from shallow water-depth to subaerial conditions. In these relatively shallow lakes, lake level variations greatly influenced the lake size and shoreline migration.

Long desiccation and subaerial weathering of the upper part of the carbonate deposits are evidenced by dissolution, pseudo-karstification, erosion, and iron encrusting (Cojan, 1989a; Freytet and Plaziat, 1982). When the lake level rose, sedimentation was characterized by fine-grained sediments such as silt, shale, and organic matter accumulation (total organic carbon or TOC maximum 8.5%). The spatial distribution of these facies was controlled by the balance between the rate of the lake rise and the sediment influx. When the lake level was stable, the clastics prograded into the lake and coarsening-upward deposits are observed.

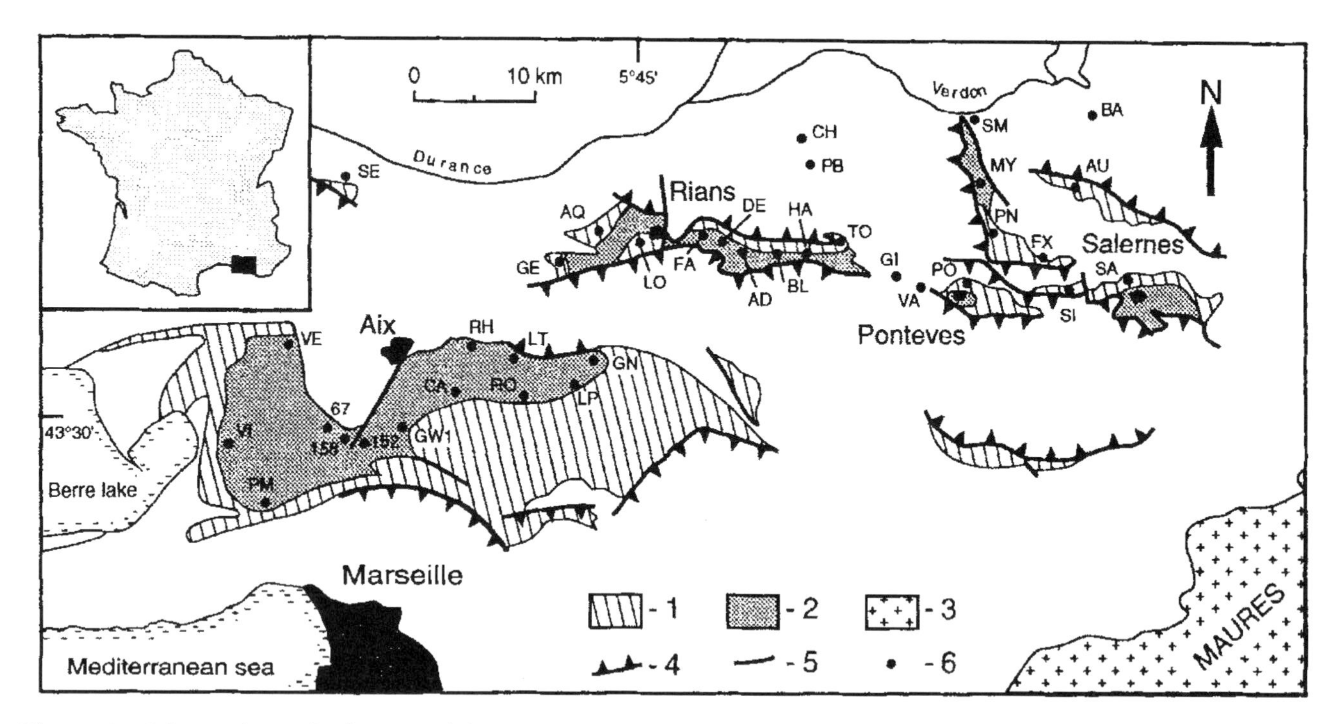

Figure 1—Schematic geologic map of the Provence continental deposits. 1 = Upper Cretaceous; 2 = Paleocene–Eocene; 3 = Permian and metamorphic rocks; 4 = thrust fault; 5 = normal fault; 6 = measured section. Aix-en-Provence area: CA = Canet; GN = Genty; LP = Les Prés; LT = Le Trou; PM = Pennes Mirabeau; RH = Roques Hautes; RO = Rousset; VE = Ventabren; VI = Vitrolles; 67, 158, 152, GW1, sondages. Rians-Salernes area: AD = Adret; AQ = Aqueduc; AU = Aups; BA = Bauduen; BL = Blacasse; CH = Chaberte; DE = Désidère; FA = Fabresse; FX = Fox-Amphoux; GE = Gerle; GI = Ginaservis; HA = Haut Adret; LO = Louvière; MY = Montmeyan; PB = Petite Bastide; PN = Petit Nans; PO = Ponteves; SA = Salernes; SI = Sillans; SM = St. Meme; TO = Touars; VA = Varage.

Isolated sandstone channel fills indicate the maximum energy deposits; however, because the landscape was relatively flat and the pedogenesis quite extensive, the availability of siliciclastic material was limited. The sedimentation was then dominated by carbonates from an algal or micriobial origin, as is often observed in modern sediment, and probably also from an eolian input from the surrounding reliefs of Jurassic carbonates. The lacustrine carbonates contain abundant freshwater ostracods, gastropods, and charophytes (Chatelet, 1972) and some organic matter (TOC from 0.01 to 0.2%).

In the Rians area, although they display the same type of organization, the lacustrine deposits show greater lateral variations (Figure 3b). In this eastward

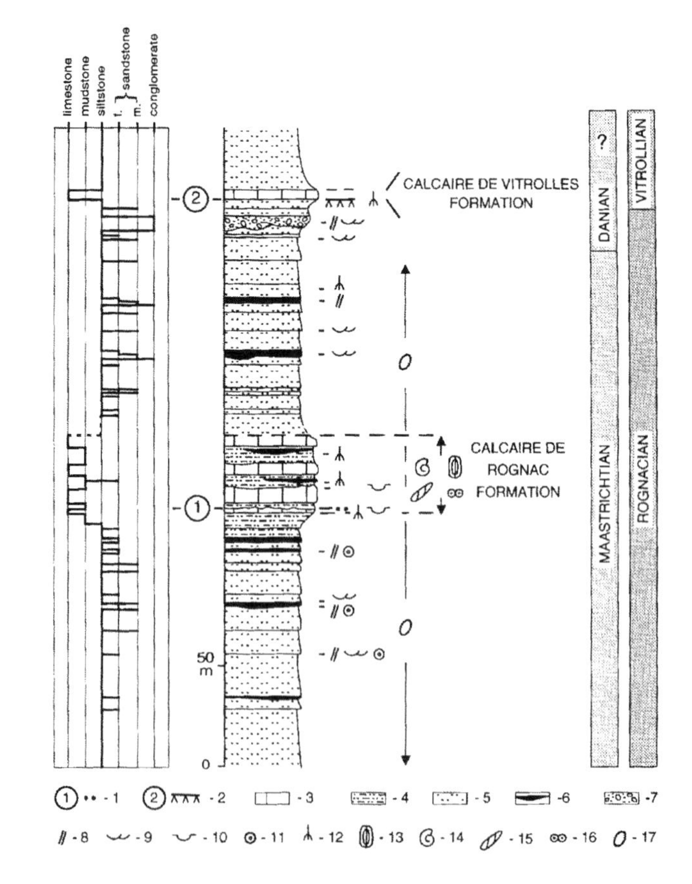

Figure 2—Section of the Calcaire de Rognac Formation, Calcaire de Vitrolles Formation Interval (eastern part of Aix-en-Provence area). Stratigraphy is based on Westphal and Durand, 1989. 1 and 2 = surfaces with a well-developed meteoric diagenesis that have been used as correlation horizons; 3 = limestone; 4 = dark mudstone; 5 = red siltstones; 6 = sandstones; 7 = well-rounded conglomerates; 8 = planar cross-stratification; 9 = trough cross-bedding; 10 = channel ; 11 = oncolites; 12 = major paleosols; 13 = ostracodes; 14 = gastropods; 15 = charophytes; 16 = pollens; 17 = dinosaur eggs.

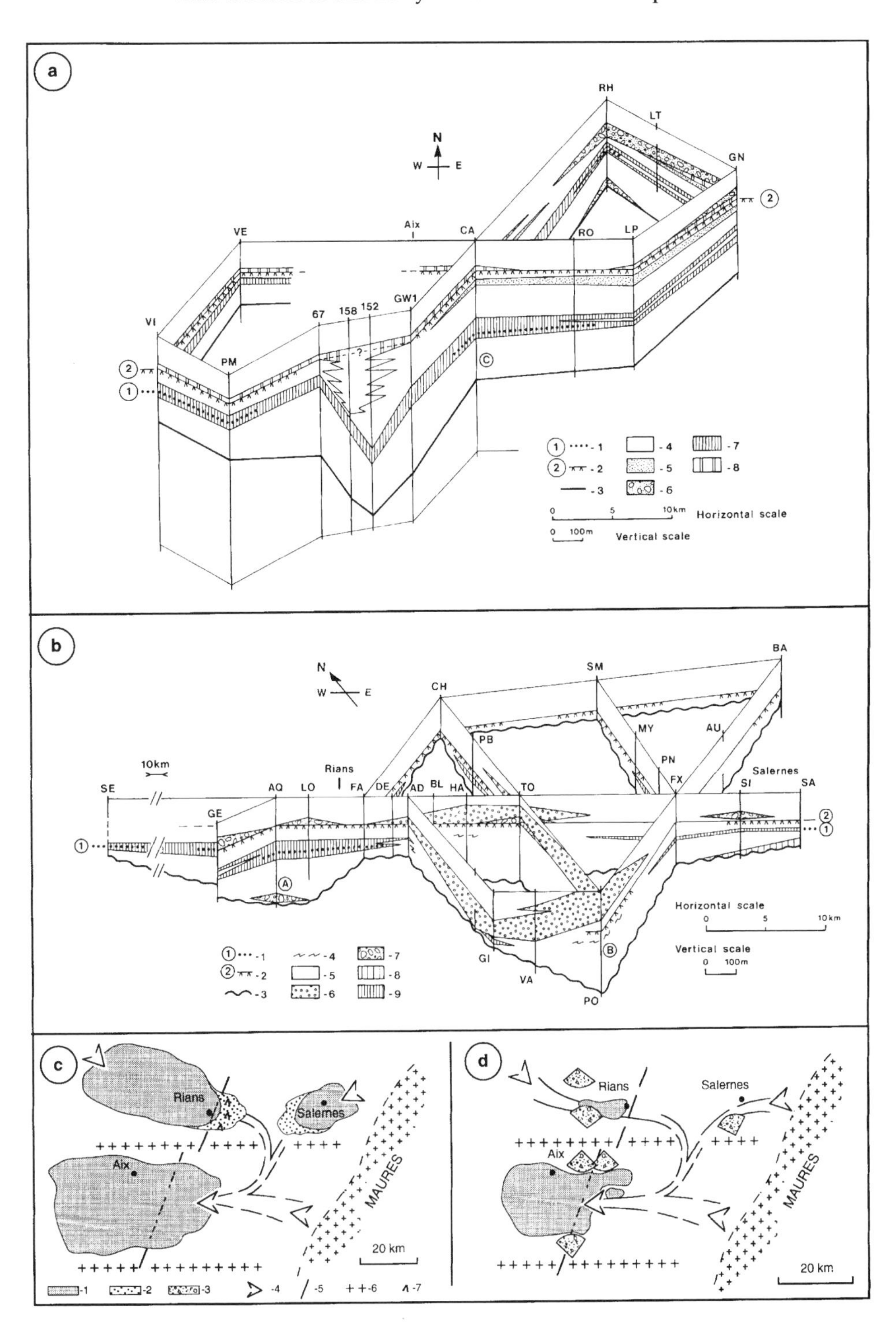

Figure 3—(a, b): Lithofacies fence diagrams of the Rognacian Vitrollian series in the Provence region. Correlation horizon 2 is considered as horizontal and horizontal scale has been palinspastically restored. (a) Aix-en-Provence area: 1 and 2 = surfaces with a well-developed meteoric diagenesis that have been used as correlation horizons; 3 = approximate Begudian/ Rognacian boundary; 4 = silt, sandstone; 5 = conglomeratic episode of Poudingue de la Galante Formation; 6 = breccia, proximal fan; 7 = Calcaire de Rognac Formation; 8 = Calcaire de Vitrolles Formation. (b) Rians area: 1 and 2 = surfaces with a well-developed meteoric diagenesis that have been used as correlation horizons; 3 = unconformity; 4 = reworked lacustrine material; 5 = siltstone, sandstone; 6 = conglomeratic episode of Poudingue des Touars Formation; 7 = breccia, proximal fan; 8 = Calcaire de Rognac Formation; 9 = other lacustrine deposits. (c, d) Approximate lake extension during the two considered lacustrine episodes. (c) Calcaire de Rognac Formation; (d) Calcaire de Vitrolles Formation. 1 = lake; 2 = reworked lacustrine facies; 3 = proximal fans; 4 = fluvial network; 5 = fault; 6 = topographic high; 7 = diapir.

oriented valley, lake development was permitted through the existence of a natural dam generated by fault activity [local structural high, probably a diapiric structure (Cojan, 1989b)]. Relative to this morphological divide, upstream deposits (around 70 m thick fossil-rich lacustrine carbonates) contrast with downstream sedimentation (slumped lacustrine carbonate beds filling channels and conglomeratic sheets of reworked lacustrine material).

In the Salernes area, the same type of setting as in the Rians zone is suggested from the abundant reworked material.

DETAILED SECTION OF THE CALCAIRE DE VITROLLES FORMATION

Following the Calcaire de Rognac episode, the paleolake was restricted to the area of greatest subsidence controlled by the transverse fault system (Figure 3a, cores 152 and 158; Figure 3c). The next extension of this perennial lake was during the Calcaire de Vitrolles episode. The lacustrine lithofacies correspond to a shallow lake environment with more restrictive conditions than during the Calcaire de Rognac episode (Figure 3d).

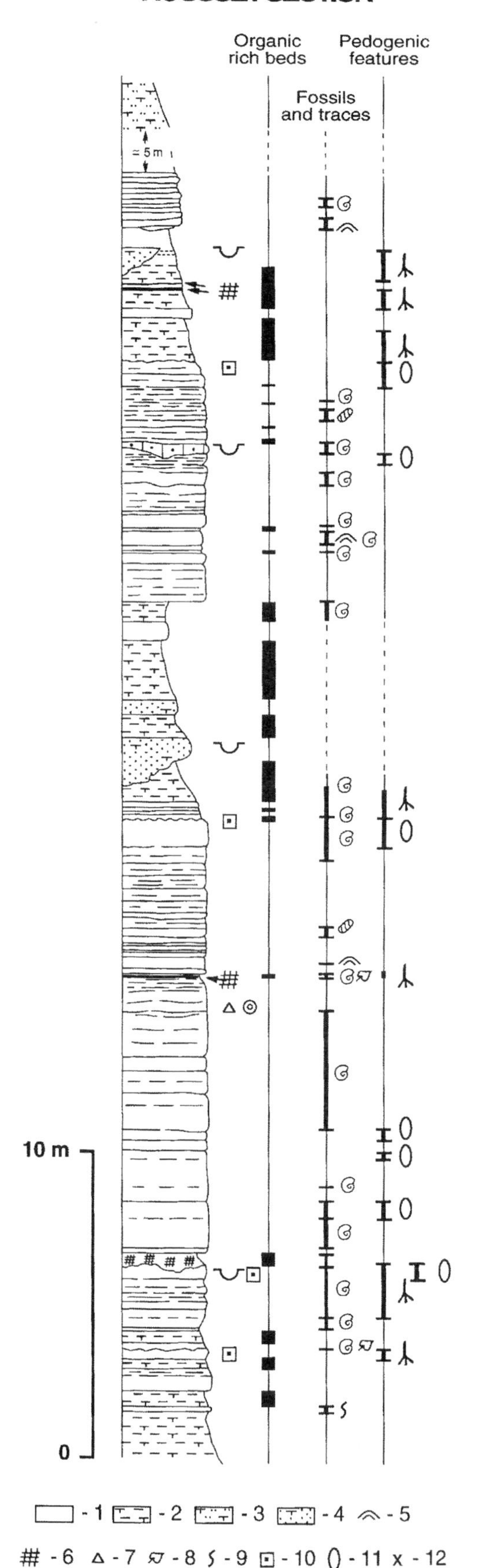

Figure 4—Detailed section of the Calcaire de Rognac Formation (section RO on Figure 1, Aix-en-Provence area). 1 = limestone; 2 = dark mudstone; 3 = red siltstone; 4 = sandstone; 5 = stromatolite; 6 = lignite; 7 = graded bed; 8 = leaves; 9 = bioturbation; 10 = ferruginous nodules; 11 = pseudo-karstification; 12 = calcrete.

The Calcaire de Vitrolles episode was preceded over the entire area by a well-developed horizon of complex calcimorphic pedogenesis (Figure 3a, b) (Colson and Cojan, 1996). Identification of palygorskite in this horizon and at the base of the lacustrine limestone testifies to a period of a chemically confined conditions that developed under a hot and semiarid climate (Cornet, 1977; Sittler, 1965; Colson et al., 1998).

In the Aix-en-Provence region, the lacustrine carbonate accumulation does not exceed 20 m (Figure 5) and hardly any fossils have ever been found in this formation (Fabre-Taxy, 1959). The mottled and marmorized carbonate mudstones show many pedologic features, such as recrystallization in voids, root traces, and desiccation cracks. Three main horizons of well-developed pseudo-karstification are identified.

In the Rians region, the lacustrine facies are nearly restricted to algal mats and sometimes interfinger with the breccia from proximal alluvial fan deposits.

INTERPRETATION

Basin Architecture

The general pattern of drainage was controlled by an east-west oriented fault system (Durand and Guieu, 1980; Angelier, 1974). A transverse fault system played only a local role by inducing rapid facies and thickness variations over limited areas (Figure 3). In the Aix-en-Provence region, a perennial lake existed between the Calcaire de Rognac and the Calcaire de Vitrolles episodes in the vicinity of the transverse fault system where the subsidence was maximum. In the Rians and Salernes areas, topographic highs, generated by the transverse fault system, allowed the development of lakes. Such a setting, existing in Provence since the Jurassic, probably involved basement structures. A strike-slip or thrust-fault system might be a good model for explaining basin characteristics.

Sedimentation Context

The regional paleogeography was relatively flat as attested by the fluvial sedimentary structures. Influence of long-term variations, such as tectonic activity and climatic evolution, have been investigated. The period that extended from the deposition of the Calcaire de Rognac Formation to the deposition of the Calcaire de Vitrolles Formation was characterized by a relatively low rate of subsidence (20 m/m.y., estimated from decompacted data). Reconstruction of climate from pollen analysis indicated clear tropical affinities (Asharf and Erben, 1986). Fluctuations of the

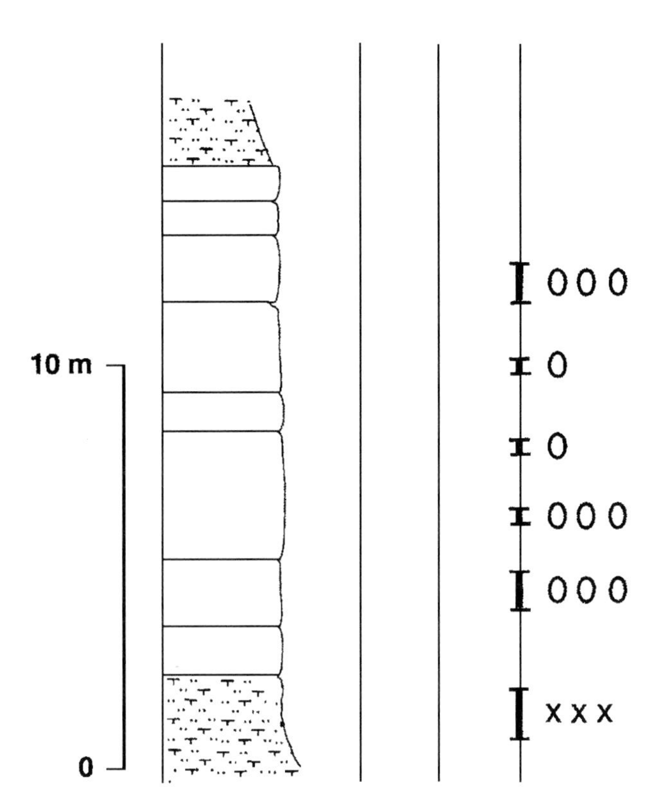

Figure 5—Detailed section of the Calcaire de Vitrolles Formation (section CA on Figure 1, Aix-en-Provence area). See Figure 4 caption for legend.

temperature-evaporation in the run-off waters showed that the highest values were obtained around the Calcaire de Rognac Formation (Iatzoura et al., 1991) and around the Calcaire de Vitrolles Formation (Colson and Cojan, 1996); therefore, it is proposed that the biochemical sedimentation occurred during relatively arid conditions.

Interpretation

Such an example shows that climatic variations played an important role in the development of facies sequences. Indeed, because the rate of evaporation is high in tropical climates, any fluctuation in the evaporation/precipitation ratio may induce large changes in the hydrologic balance. Then, any variation in the lakestand of these shallow lakes led to wide migrations of the shoreline in such a flat topography.

Some surfaces deeply affected by meteoric diagenesis were used as correlation horizons, and their equivalents could be traced in other southwestern European basins (Cojan, 1989a). These surfaces that extend over such a large area may suggest that most of the climatic fluctuations recorded in these formations find their origin in events that are at least at the scale of the southwestern European province.

REFERENCES CITED

Asharf, A. D., and H. K. Erben, 1986, Palynogische Untersuchungen an der Kreide/Tertiar Grenze West-Mediterraner Regionen: Palaeontographica Abteilung B 200, v. 1–6, p. 111–163.

Babinot, J. F., and J. P. Durand, 1980, Les étages français et leurs stratotypes : Valdonnien, Fuvélien, Bégudien, Rognacien, Vitrollien: Mémoire Bureau Recherches Géologiques Minières, France, v. 109, p. 171–192.

Babinot, J. F., and P. Freytet, (Coordonateurs), 1983, Colloque sur les étages Coniacien à Maastrichtien. Marseille: Géologie Méditerranéenne, v. 10, (3–4), p. 245–268.

Chatelet, H., 1972, Etude stratigraphique et paléoécologique du Rognacien en Provence: Thèse, Marseille, Travaux Laboratoire Géologie Historique et de Paléontologie, v. 3, 95 p.

Cojan, I., 1989b, Structure diapirique contrôlant la sédimentation. Séries continentales de Provence (Rians-K/T): Publication Association Sédimentologistes Français, Paris, v. 10, p. 79–80.

Cojan, I., 1993, Alternating fluvial and lacustrine sedimentation: tectonic and climatic controls (Provence Basin, S. France, Upper Cretaceous/Palaeocene): Special Publication International Association Sedimentologists, v. 17, p. 425–438.

Cojan, I., M. G. Moreau, and L. Stott, 1998, An integrated $\delta^{13}C$ and magneto-stratigraphy in the early Cenozoic continental series of the Provence Basin (France): Strata, Toulouse, France, p. 34–38.

Colson, J., and I. Cojan, 1996, Groundwater dolocretes in a lake marginal environment : an alternative model for the dolocretes formation in the continental setting (Danian of the Provence basin, France): Sedimentology, v. 43, (1), p. 175–188.

Colson, J., I. Cojan, and M. Thiry, 1998, A hydrological model for palygorskite formation in the Danian continental facies of the Provence basin (France): Clay Minerals, v. 33, p. 333–347.

Cornet C., 1977, Etude préliminaire des minéraux argileux des séries continentales du Crétacé supérieur et du Tertiaire des fossés nord-varois (Provence): Géologie Méditerranéenne, v. 4, p. 379–382.

Durand, J. P., P. Gaviglio, J. F. Gonzales, and R. Monteau, 1984, Upper Cretaceous, Paleocene and Eocene fluvio-lacustrine sediments in the Arc syncline (Aix-en-Provence District): 5th International Association Sedimentologists European Meeting, Marseille, Fieldtrip Guidebook, 44 p.

Fabre-Taxy, S., 1959, Faunes lagunaires et continentales du Crétacé supérieur de Provence. III le Maastrichtien et le Danien: Annales Paléontologie, v. 45, p. 55–124.

Iatzoura, A., I. Cojan, and M. Renard, 1991, Géochimie

des coquilles d'oeufs de dinosaures: essai de reconstitution paléoenvironnementale (Maastrichtien, Bassin d'Aix-en-Provence, France): Comptes Rendus Académie Sciences, Paris, v. 312, II, p. 1343–1349.

Krumsiek, K., and G. G. Hahn, 1989, Magnetostratigraphy near the Cretaceous/Tertiary boundary at Aix-en-Provence (southern France): Cahier de la Réserve Géologique de Haute-Provence, Digne, France, v. 1, p. 38–41.

Sittler, C., 1965, Le Paléogène des fossés rhénan et rhodanien. Etudes sédimentologiques et paléoclimatiques: Mémoire Service Carte Géologique Alsace Lorraine, Strasbourg, v. 24, 392 p.

Westphal, M., and J. P. Durand, 1989, Magnétostratigraphie des séries continentales fluvio-lacustres du Crétacé supérieur dans le synclinal de l'Arc (région d'Aix-en-Provence, France): Bulletin Société Géologique France, v. 8, (4), p. 609–621.

ADDITIONAL REFERENCES

Angelier, J., 1974, L'évolution continentale de la Provence septentrionale au Crétacé terminal et à l'Eocène inférieur : la gouttière de Rians-Salernes: Bulletin Bureau Recherches Géologiques Minières, France, v. 2, p. 65–83.

Babinot, J. F., 1980, Les ostracodes du Crétacé supérieur de Provence: Thèse, Marseille, 287 p.

Babinot, J. F., and J. P. Durand, 1984, Crétacé supérieur fluvio-lacustre. *In*: Synthèse Géologique du Sud-Est de la France (Ed.; Debrant-Passart, S.): Mémoire Bureau Recherches Géologiques Minières, France, v. 125, p. 363–371.

Cojan, I. ,1989a, Discontinuités majeures en milieu continental. Proposition de corrélation avec des évènements globaux (Bassin de Provence, S. France, Passage Crétacé/Tertiaire): Comptes Rendus Académie Sciences Paris, v. 309, II, p. 1013–1018.

Collot, L., 1890, Description du terrain crétacé dans une partie de la Basse Provence. 2ème partie: couches d'eau douce et généralités: Bulletin Société Géologique France, v. 19, (3), p. 39–92.

Corroy, G., 1957, La limite entre le Crétacé et le Tertiaire en Provence occidentale: Comptes Rendus Sommaires Société Géologique France, p. 286.

Dughi, R., and F. Sirugue, 1957, La limite supérieure des gisements d'oeufs de dinosaures du Bassin d'Aix en Provence: Comptes Rendus Académie Sciences, Paris, v. 245, p. 907–909.

Dughi, R., and F. Sirugue, 1959, La limite du Crétacé continental dans le bassin d'Aix en Provence: Comptes Rendus Académie Sciences, Paris, v. 249, p. 2370–2372.

Durand, J. P., 1980, Les sédiments fluvéliens du synclinal de l'Arc (Provence): Revue Industrie Minérale, Supplément juin 1980, p. 13–25.

Durand, J. P., and G. Guieu, 1980, Cadre structural du bassin de l'Arc): Revue Industrie Minérale, Supplément juin 1980, p. 3–12.

Feist-Castel, M., 1975, Répartition des charophytes dans le Paléocène du Bassin d'Aix en Provence: Bulletin Société Géologique France, v. 7, (17), p. 88–97.

Freytet, P., and J. C. Plaziat, 1982, Continental carbonate sedimentation and pedogenesis. Late Cretaceous and Early Tertiary of Southern France: Contributions to Sedimentology, v. 12, 213 p.

Jaeger, J. J., and M. Westphal (editeurs), 1989, La limite Crétacé-Tertiaire dans le Bassin de l'Arc (Sud-Est, France): Cahier de la Réserve Géologique de Haute-Provence, Digne, France, v. 1, 60 p.

Lapparent (De), A. F., 1938, Etudes géologiques dans les régions provençales et alpines entre le Var et la Durance: Bulletin Service Carte Géologique France, v. 40, (198), 302 p.

Medus, J., 1972, Palynological zonation of the Upper Cretaceous southern France and northeastern Spain: Review of Paleobotany and Palynology, Elsevier, Amsterdam, p. 287–295.

Penner, M., 1985, The problem of dinosaurs extinction. Contribution of the study of terminal Cretaceous eggshells from southeast France: Geobios, v. 18, (5), p. 665–669.

Plaziat, J. C., 1981, Late Cretaceous to late Eocene palaeogeographic evolution of southwest Europe: Palaeogeography, Palaeoclimatology, Palaeoecology, v. 36, p. 263–320.

Vasseur, G., 1898, Sur la découverte de fossiles dans les assises qui constituent en Provence la formation dite: "étage de Vitrolles" et sur la limite des terrains crétacés et tertiaires dans le bassin d'Aix en Provence (Bouches du Rhône): Comptes Rendus Académie Sciences, Paris, v. 127, p. 890.

Palma, R. M., 2000, Lacustrine facies in the Upper Cretaceous Balbuena subgroup (Salta Group): Andina Basin, Argentina, *in* E. H. Gierlowski-Kordesch and K. R. Kelts, eds., Lake basins through space and time: AAPG Studies in Geology 46, p. 323–328.

Chapter 27

Lacustrine Facies in the Upper Cretaceous Balbuena Subgroup (Salta Group): Andina Basin, Argentina

R. M. Palma
Facultad de Ciencias Exactas y Naturales-Departamento de Geología, Universidad de Buenos Aires
Buenos Aires, Argentina

REGIONAL SETTING AND STRATIGRAPHY

The Cretaceous–Early Tertiary Andina Basin in northwestern Argentina was formed during an initial phase of rifting (syn-rift), followed by a long period of thermal subsidence (post-rift). The basin contains a thick continental succession (7500 m) that comprises siliciclastic and carbonate rocks (Table 1). The Cretaceous–lower Tertiary sequence belongs to the Salta Group (Figure 1), which is exposed in vast areas of the Salta and Jujuy provinces, with different recognizable depocenters (Lencinas and Salfity, 1973). The Salta Group is divided into three subgroups (Pirgua, Balbuena, and Santa Bárbara; Table 1), which represent different stages in the evolution of the basin (Moreno, 1970; Gómez Omil et al., 1989).

The Pirgua subgroup consists of reddish gray conglomerates, sandstones, and claystones, interpreted as alluvial fan and fluvial deposits, which are associated in various outcropping zones with alkaline basalt (Galliski and Viramonte, 1988). The Balbuena subgroup lies conformably above these rocks and contains the Lecho, Yacoraite, and Olmedo Formations. Lithologically, these formations consist of extensive beds of sandstones, interpreted as fluvial and eolian deposits with a mixed carbonate/siliciclastic sequence of a probable playa-lake complex. Units of the Santa Bárbara subgroup lie conformably above the Balbuena subgroup and contain a monotonous succession of sandstones, claystones, and limestones interpreted as fluvial and lacustrine in origin. The Santa Bárbara subgroup is assigned to the Paleocene–lower Eocene based on the presence of mammalian fossils (Pascual, 1984).

YACORAITE FORMATION

This contribution describes the lacustrine facies of the Yacoraite Formation (Figure 2) of the Balbuena subgroup (Lencinas and Salfity, 1973). Late Cretaceous (Maastrichtian) in age, the Yacoraite Formation is lithostratigraphically and biostratigraphically correlated with another Upper Cretaceous carbonate sequence in Bolivia (El Molino Formation) (Lencinas and Salfity, 1973; Camoin et al., 1997). The Yacoraite Formation lies conformably over the Lecho Formation, except in the northeastern sector of the Andina Basin where it lies unconformably over lower Paleozoic rocks.

The Yacoraite Formation is a mixed carbonate/siliciclastic sequence (Figure 3). The siliciclastic lithofacies include massive medium-grained sandstones, laminated to symmetric to asymmetric ripple cross-laminated fine-grained sandstones, and laminated to massive siltstones and claystones. Carbonate lithofacies of the Yacoraite Formation include grainstones to packstones, wackestones, and mudstones containing freshwater fossils, stromatolites, coated grains, and polygenetic intraclasts (Palma, 1984). Fossils include gastropods (Parodiz, 1969), bivalves (Lencinas and Salfity, 1973), reptile remains and tracks (Powell, 1979; Alonso, 1980), fish fragments (Cione et al., 1985), charophytes (Musacchio, 1981; Palma, 1984), and cyanobacterial remains in oncolites and stromatolites (Palma, 1984; Palma and Andreis, 1988; Marquillas and del Papa, 1993). An assemblage of typical freshwater ostracods associated with extremely rare foraminifera was reported by Mendez and Viviers (1973); these foraminifera are represented by scattered and anomalous forms of miliolids common in lake deposits (Anadón, 1989, 1992).

Although the stratification in both the carbonates and siliciclastics facies is markedly tabular with planar contacts (sharp or transitional), there are some lenticular sandstones and intraformational conglomerate bodies. Normally, the thickness of the beds are thin to medium (10–90 cm). Yellow, brown, and olive-gray colors predominate over red. The reddish color is restricted to some localities in the southeast of the basin. The Yacoraite Formation has its greatest thickness (250 m) in the

Table 1. Stratigraphic Framework of the Cretaceous–Lower Tertiary Deposits of the Andina Basin (Salta Group) in Northwestern Argentina.

AGE		SUBGROUP	FORMATION
EARLY PALEOCENE EOCENE		SANTA BÁRBARA (2000 m)	LUMBRERA (fluvial and lacustrine facies) MAIZ GORDO (fluvial and lacustrine facies) MEALLA (fluvial facies)
PALEOCENE - MAASTRICHTIAN	SALTA GROUP	BALBUENA (1450 m)	OLMEDO (lacustrine, playa-lake facies) YACORAITE (LACUSTRINE FACIES) LECHO (fluvial and eolian facies)
CAMPANIAN - VALANGINIAN		PIRGUA (4000 m)	LOS BLANQUITOS (fluvial facies) LAS CURTIEMBRES (fluvial facies) LA YESERA (alluvial facies)

Adapted from Salfity and Marquillas, 1981.

Tres Cruces depocenter (Figure 2), where the stratotype has been defined (Lencinas and Salfity, 1973). Also, similar thicknesses were measured in the Alemanía depocenter (Marquillas, 1985). In the northeast of Tucumán Province, in the southernmost part of the basin (Figure 2), the thickness of the Yacoraite Formation reaches 35 m.

SEDIMENTARY FACIES

Different sedimentary lithofacies, both siliciclastic and carbonate, were identified in the Yacoraite Formation. These lithofacies within an environmental interpretation are listed as follows.

Lacustrine Shoreline Association

This association contains the following lithofacies: (1) laminated to massive fine- to medium-grained sandstones, mostly reddish brown, with mudcracked surfaces with purple to red mottling, root traces, and calcareous nodules, (2) symmetric to asymmetric ripple cross-laminated fine-grained sandstones, (3) planar cross-bedded, medium-grained sandstones with mudstone intraclasts, (4) laminated and ripple cross-

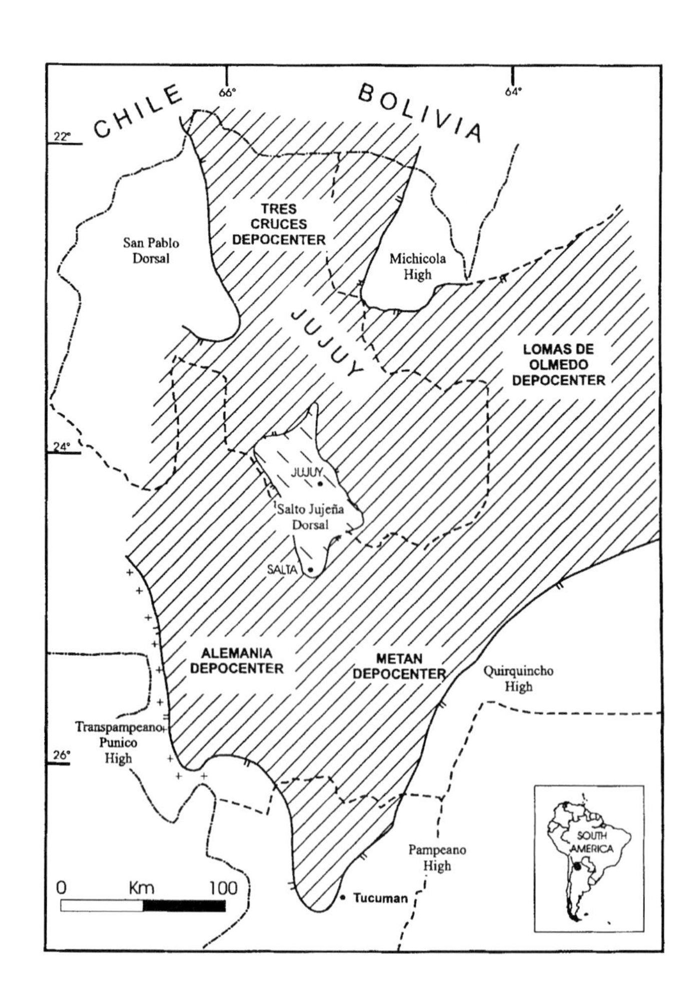

Figure 1—Map of northwestern Argentina showing the distribution of the Salta Group within the Cretaceous–lower Tertiary Andina Basin (adapted from Salfity and Marquillas, 1981).

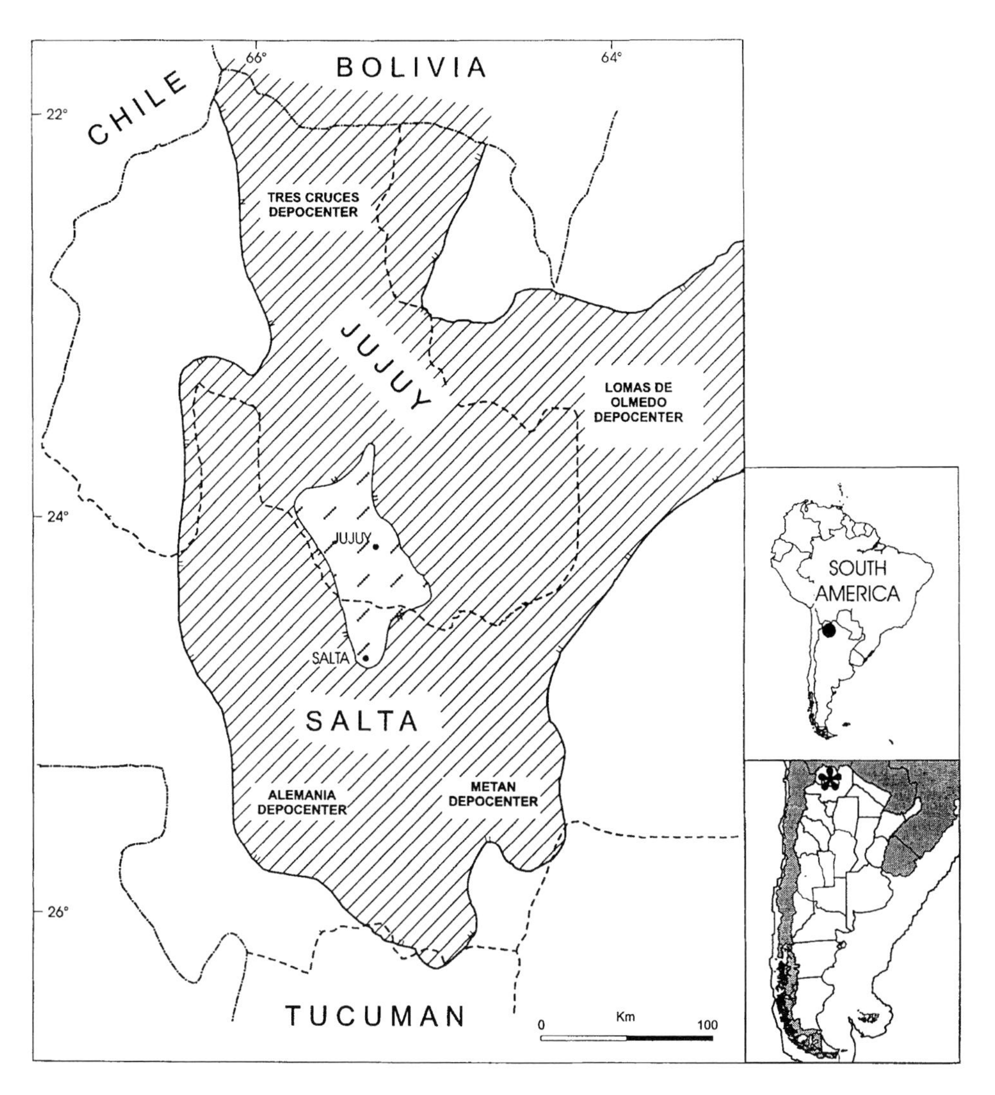

Figure 2—Locality map showing the distribution and main depocenters of the Yacoraite Formation of the Salta Group and Balbuena subgroup in the Andina Basin of Argentina (adapted from Salfity and Marquillas, 1981).

laminated siltstones deformed by convolute structure, and (5) packstones/wackestones. These lithofacies are characterized by tabular or lenticular beds interbedded with calcareous facies. The average thickness of siliciclastic beds is about 40 cm. The existence of mudcracked surfaces and related pedogenic structures, associated with asymmetric ripples, suggest a shoreline environment that was periodically exposed and vegetated; moreover, locally small coarsening-upward intervals (Sp) with associated massive medium sandstones are interpreted as delta progradation into a very shallow lake margin. The presence of syndepositional deformational structures suggests a high rate of sedimentation. Packstones/wackestones represent seasonal or tectonic changes affecting detrital influx into areas of low relief.

Nearshore Lacustrine Assemblage

This association exhibits the greatest lithologic diversity, containing the following lithofacies: (1) very fine-grained sandstones with wave ripple-marks, (2) fine-grained sandstones and siltstones with current ripple cross-lamination, (3) mud-supported flat pebble conglomerates, (4) oolitic to bioclastic grainstones/packstones, and (5) stromatolitic limestones. The siliciclastic lithofacies with wave ripples and current ripple bedding, 10–30 cm in thickness, suggest wave reworking processes during deposition. Oolitic to bioclastic grainstones/packstones contain oncolites associated with well-preserved fossil assemblages (molluscs, ostracods, charophytes) and intraclasts. Oolites are spheroidal, ovoid to ellipsoidal in shape (Palma and Pazos, 1993). Oncolites show spheroidal and ellipsoidal shapes with smooth surfaces. Their size ranges from less than 0.20 to 1.66 mm in diameter. Two types of oncolites are recognized: (1) concentrically stacked (type C) and (2) randomly stacked (type R). Continuous layer configuration suggests agitated shallow water conditions with rare discontinuous layers reflecting short periods of tranquility (Palma and Andreis, 1988). Stromatolitic limestone beds range in thickness from 0.2 to 2.0 m. Stromatolites vary in size and morphology (types LLH-C, SH/LLH); they commonly coat mollusc fragments, ostracods, and oolites (Palma, 1984). Microstructure is well preserved with recognizable branching tube colonies of cyanobacteria (Palma, 1994). Siliciclastic material occurs within stromatolitic

Figure 3—Schematic columnar section of the Yacoraite Formation of the Andina Basin of Argentina, showing lithology, sedimentary features, fossils, and interpretations. Sedimentary structures key: M = mud grain size, W = wackestone, SI = siltstone, FS = fine-grained sandstone, MS = medium-sized sandstone, GP = grainstone/packstone, C = conglomerate. See GGLAB legend for lithologic and biogenic key.

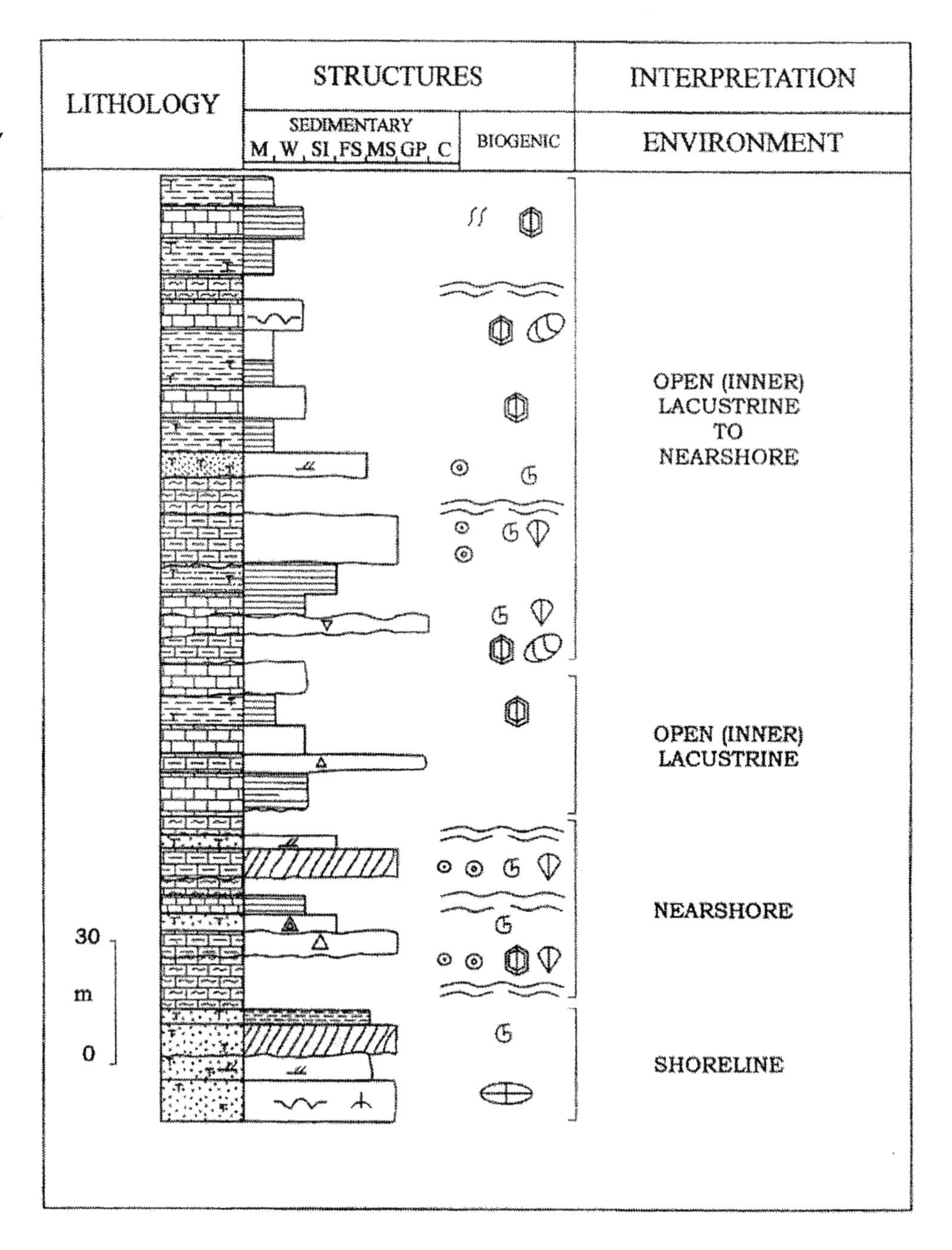

structures, and polygonal cracking is present on the top of stromatolite packages. Grainstone/packstone units contain symmetric ripple marks, thin conglomerate lenses, intraclasts (clastic carbonates), mudcracks, and coated grains (Palma and Pazos, 1993) and are interbedded with claystones and mudstones/wackestones of the open lacustrine association on a 20-cm-scale interval.

This association is interpreted as nearshore lacustrine with occasional exposure based on the coated grain morphology, symmetric ripple-marks, conglomerate lenses, mudcracks interpreted as desiccation cracks, and intraclasts (Palma and Pazos, 1993) in the carbonates and the ripple cross-lamination in the siliciclastics. Interbedding of grainstones and packstones with open lacustrine facies suggests a progradation of oolitic bars into the deeper lake sediments.

Open Lacustrine Assemblage

This association contains (1) massive to laminated carbonate mudstones and wackestones, (2) massive to laminated siltstones and claystones, and (3) rare oolitic grainstones/packstones with beds 4–20 cm thick. Carbonates (facies 1) and siliciclastics (facies 2) are interbedded as single beds varying from a few to 40 cm thick. Upper contacts of the beds are sharp to undulate, whereas bases are sharp. The fossil assemblage contains ostracods (*Candona* sp., *Illyocypris* sp., and *Cytheura* sp.), gastropods (*Cynerida* sp., and *Corbicula* sp.), as well as charophytes (*Porochara* sp., *Platychara* sp., and *Ambyochara* sp.). Horizontal, slightly sinuous burrows are present, as well. This assemblage is interpreted as deposits below wave base. Fossils suggest a typical aerobic lacustrine environment, probably with a water depth of less than 10 m. In addition, thin oolitic grainstone/packstone are interpreted as deposits of storm-generated transport.

SEDIMENTATION PROCESSES AND BASIN EVOLUTION

Characteristics of sedimentation in northwestern Argentina during the Cretaceous–Early Tertiary was a response to two stages of evolution of a rift system (Bianucci and Homocv, 1982; Gómez Omil et al., 1989; Uliana et al., 1989). Basin infilling began with alluvial and fluvial sediments associated with alkaline basaltic volcanism in the Alemanía depocenter (Galliski and Viramonte, 1988). This phase of evolution corresponds to deposition of the Pirgua subgroup interval (synrift). The second stage of sedimentation began in Late Cretaceous and continued into the early Tertiary, and was controlled by thermal subsidence (Gómez Omil et al., 1989). During this final sag-basin phase of deposition, fluvial, lacustrine, and playa-lake environments of the Balbuena and Santa Bárbara subgroups were formed. The Yacoraite Formation represents the climax of lacustrine deposition within the basin. The mixture of carbonates and siliciclastics in this paleolake system may be due to climatic influences (Platt, 1989), as well as changes in source area drainage (Gierlowski-Kordesch, 1998).

REFERENCES CITED

Alonso, R. N., 1980, Icnitas de dinosuarios (Ornithopoda, Hadrosauridae) en el Cretácico superior del norte de Argentina. Acta Geológica Lilloana, v. 15 (2), p. 55–63.

Anadón, P., 1989, Los lagos salinos interiores (atalásicos) con faunas de afinidad marina del Cenozoico de la Península Ibérica. Acta Geológica Hispánica, v. 24 (2) p. 83–102.

Anadón, P., 1992, Composition of inland waters with marine-like fauna and interferences for a Miocene lake in Spain. Palaeogeography, Palaeoclimatology, Palaeoecology, v. 19, p. 1–8.

Bianucci, H. A., and J. F. Homovc, 1982, Tectogénesis de un sector de la Cuenca del Subgrupo Pirgua, noroeste argentino. V Congreso Geológico Argentino, v. 1, p. 539–546.

Camoin, G., J. Casanova, J. M. Rouchy, M. M. Blanc-Valleron, and J. F. Deconinck, 1997, Environmental controls on perennial and ephemeral carbonate lakes: the central palaeo-Andean Basin of Bolivia during Late Cretaceous to early Tertiary times. Sedimentary Geology, v. 113, p. 1–26.

Cione, A., S. M. Pereira, R. N. Alonso, and J. Arias, 1985, Los bagres (Osteichthyes y Siluriformes de la Formación Yacoraite-Cretácico tardío del norte argentino). Consideraciones biogeográficas y bioestratigráficas. Ameghiniana, v. 21 (2–4), p. 294–304.

Galliski, M. A., and J. G. Viramonte, 1988, The Cretaceous paleorift in northwestern Argentina: a petrologic approach. Journal of South American Earth Sciences, v. 1 (4), p. 329–342.

Gierlowski-Kordesch, E. H., 1998, Carbonate deposition in an ephemeral siliciclastic alluvial system: Jurassic Shuttle Meadow Formation, Newark Supergroup, Hartford Basin, USA. Palaeogeography, Palaeoclimatology, Palaeoecology, v. 140, p. 161–184.

Gómez Omil, R., A. Boll, and R. M. Hernández, 1989, Cuenca Cretácico–Terciario del Noroeste Argentino (Grupo Salta). Cuencas Sedimentarias Argentinas. Serie de Correlación Geológica, v. 6, p. 43–64.

Lencinas, A. N., and J. A. Salfity, 1973, Algunas características de la Formación Yacoraite en el oeste de la Cuenca Andina, provincias de Salta y Jujuy. V Congreso Geológico Argentino, v. 3, p. 253–267.

Marquillas, R. A., 1985, Estratigrafía, sedimentología y paleoambientes de la Formación Yacoraite (Cretácico superior) en el tramo austral de la cuenca, norte argentino. Ph.D. Dissertation, University of Salta, 139 p.

Marquillas, R. A., and C. del Papa, 1993, Las calizas estromatolíticas del Cretácico-Terciario del noroeste argentino. XII Congreso Geológico Argentino y II Congreso Exploración de Hidrocarburos, v. 1, p. 263–273.

Mendez, I., and M. C. Viviers, 1973, Estudio micropaleontológico de sedimentitas de la Formación Yacoraite-Provincias de Salta y Jujuy. V Congreso Geológico Argentino, v. 3, p. 467–470.

Moreno, J. A., 1970, Estratigrafía y paleogeografía del Cretácico superior en la cuenca del noroeste argentino, con especial mención de los Subgrupos Balbuena y Santa Bárbara. Revista Asociación Geológica Argentina, v. 24 (1), p. 9–44.

Musacchio, E., 1981, South American Jurassic and Cretaceous Foraminifera, Ostracoda and Charophyta of Andean and subandean regions. Cuencas Sedimentarias del Jurásico y Cretácico de América del Sur, v. 2, p. 461–498.

Palma, R. M., 1984, Características sedimentológicas y estratigráficas de las formaciones en el límite Cretácico superior-Terciario inferior de la Cuenca Salteña. Ph.D. Dissertation, University of Tucumán, 256 p.

Palma, R. M., and R. R. Andreis, 1988, Oncolitos en la Formación Yacoraite y su significado ambiental-Valle del Tonco, Salta-Argentina. II Reunión Argentina de Sedimentología, v. 1, p. 207–211.

Palma, R. M., and P. J. Pazos, 1993, Interpretatión y diagénesis de un depósito de tormentas en la Formación Yacoraite, Yavi, provincia de Jujuy, Argentina. XII Congreso Geológico Argentino y II Congreso Exploración de Hidrocarburos, v. 1, p. 258–262.

Palma, R. M., 1994, Colonias de Cianofitas (*Rivularia haematites*) en la Formación Yacoraite, Río Azul, provincia de Jujuy. V Reunión Argentina de Sedimentología, v. 1, p. 59–64.

Parodiz, J. J., 1969, The Tertiary non-marine Mollusca of South America. Annals Carnegie Museum of Pittsburgh, v. 40, p. 1–242.

Pascual, R. M., 1984, La sucesión de las edades mamífero, de los climas y del diastrofismo sudamericano

durante el sudamericano durante el Cenozoico: fenómenos concurrentes Academia Nacional de Ciencias Exactas, Físicas y Naturales, Anales, v. 36, p. 15–36.
Platt, N. H., 1989, Climatic and tectonic controls on sedimentation of a Mesozoic lacustrine sequence: The Purbeck of the western Cameros Basin-Northern Spain. Palaeogeography, Palaeoclimatology, Palaeoecology, v. 70, p. 187–197.
Powell, J. E., 1979, Sobre una asociación de dinosaurios y otras evidencias de vertebrados del Cretácico superior de la región de La Candelaria, provincia de Salta, Argentina. Ameghiniana, v. 16 (1–2), p. 191–204.
Salfity, J. A., and R. A. Marquillas, 1981, Las unidades estratigráficas Cretácicas del norte de la Argentina. Cuencas Sedimentarias del Jurásico y Cretácico de América del Sur, *in* W. Volkheimer and E. A. Musacchio, eds., II Congreso Latinoamericano de Paleontología, v. 1, p. 303–318.
Uliana, M. A., K. T. Biddle, and J. Cerdan, 1989, Mesozoic extension and the formation of Argentine sedimentary basins. AAPG Memoir 46, p. 599–614.

Owen, R. B., 2000, Late Cretaceous–early Tertiary continental lacustrine basins of Hong Kong and southeast China, *in* E. H. Gierlowski-Kordesch and K. R. Kelts, eds., Lake basins through spaceand time: AAPG Studies in Geology 46, p. 329–334.

Chapter 28

Late Cretaceous–Early Tertiary Continental and Lacustrine Basins of Hong Kong and Southeast China

R. Bernhart Owen
Department of Geography, Hong Kong Baptist University
Kowloon Tong, Hong Kong

GENERAL GEOLOGY

The coastal region of southern China and Hong Kong contains several Late Cretaceous–early Tertiary continental basins (Figure 1) with red beds and laminated, lacustrine mudstones. Probable early Tertiary lake sediments crop out on the island of Ping Chau (Figure 2) in northeastern Hong Kong (Lai, 1991; Lai et al., 1996; Lee et al., 1991b), although their precise age has been the subject of a long debate (Heanley, 1923; Williams, 1943; Davis, 1953; Ruxton, 1960; Allen and Stephens, 1971; Peng, 1971; Lee, 1985, 1987). The Ping Chau Basin covers 45 km^2 and is an asymmetric half graben bounded by normal faults to the northwest and northeast (Lai, 1985). Two major sedimentary units are present. The Ping Chau Formation (about 210 m exposed on the island of Ping Chau) lies at the basin center (Figure 2) and consists of thinly laminated, calcareous and dolomitic mudstone and siltstone, which dip 5–20° to the northeast. These overlie Port Island Formation red beds composed of siltstone, sandstone, and conglomerate.

A third sedimentary unit, the Kat O Formation, occurs in a smaller basin to the west of Ping Chau (Figure 1). This unit consists of red, massively bedded, sedimentary breccia and conglomerate laid down in fault-controlled alluvial fans. The precise stratigraphic relationships between the Kat O, Port Island, and Ping Chau Formations are uncertain due to poor exposure of boundaries and a lack of dating control.

Several similar sedimentary basins of Late Cretaceous–early Tertiary age, with red beds or lacustrine mudstone and shale, also occur in southeast China (Gu and Renaut, 1994; Lai 1991). These include the Sanshui Dongguan and Xinhui Basins of Guangdong (Figure 1). Wang et al. (1985) suggested that the Guangdong Basins may have formed a single paleolake during the early Tertiary, centered on the modern city of Guangzhou (Figure 1).

The record in Guangdong is well preserved by the rocks of the Sanshui Basin. Lai (1991) notes that the lower Tertiary Buxin Group rests on red beds of the Upper Cretaceous Dalang Shan Formation (Figure 3) and can be divided into three parts. The Paleocene First formation (the Xinzhang Formation of Lee et al., 1991b) consists of 80 m of conglomerate overlain by 100–300 m of mudstone, marlstone and siltstone red beds with gray intercalations. The Second formation (Eocene) consists of 100–300 m of gray to black lacustrine mudstone, siltstone, and marlstone. The Third formation consists of 200–350 m of Oligocene sandstone, siltstone, and mudstone.

Various attempts have been made to correlate the Sanshui (Guangdong) succession with the sequence recognised in Hong Kong at Ping Chau (Figure 3). Lai (1991) suggests that reddened mudstone, sandstone, and conglomerate of the Port Island Formation (Hong Kong) correlate with the Upper Cretaceous Dalang Shan Formation and that the Paleocene First formation of the Buxin Group (Tang et al., 1980) (Figure 3) correlates with sediments (visible on seismic sections) that are hidden below the sea floor near Ping Chau. According to Lai (1991), laminated mudstone and siltstone exposed on Ping Chau Island correlate with the Second formation, whereas the upper Ping Chau Formation sediments (also hidden below sea level) may equate to the Oligocene Third formation. The Kat O Formation's red conglomerate remains problematic, with Lai (1985) suggesting an Late Cretaceous age and Lee et al. (1991b) preferring a correlation with the Oligocene red beds of the Huayong or Sanshui Formations of the Sanshui Basin.

Similar continental successions are reported from other parts of Guangdong Province by Sun et al.(1981) and Song et al. (1986). Lee et al. (199lb) note that this suggests the widespread development of oil shales, coals, and lake basins during the late Paleocene–Eocene in the southern coastal regions of south China.

THE PING CHAU FORMATION

The Ping Chau Formation (Figure 4) is characterized by parallel finely laminated (<0.5 mm) sediments, with

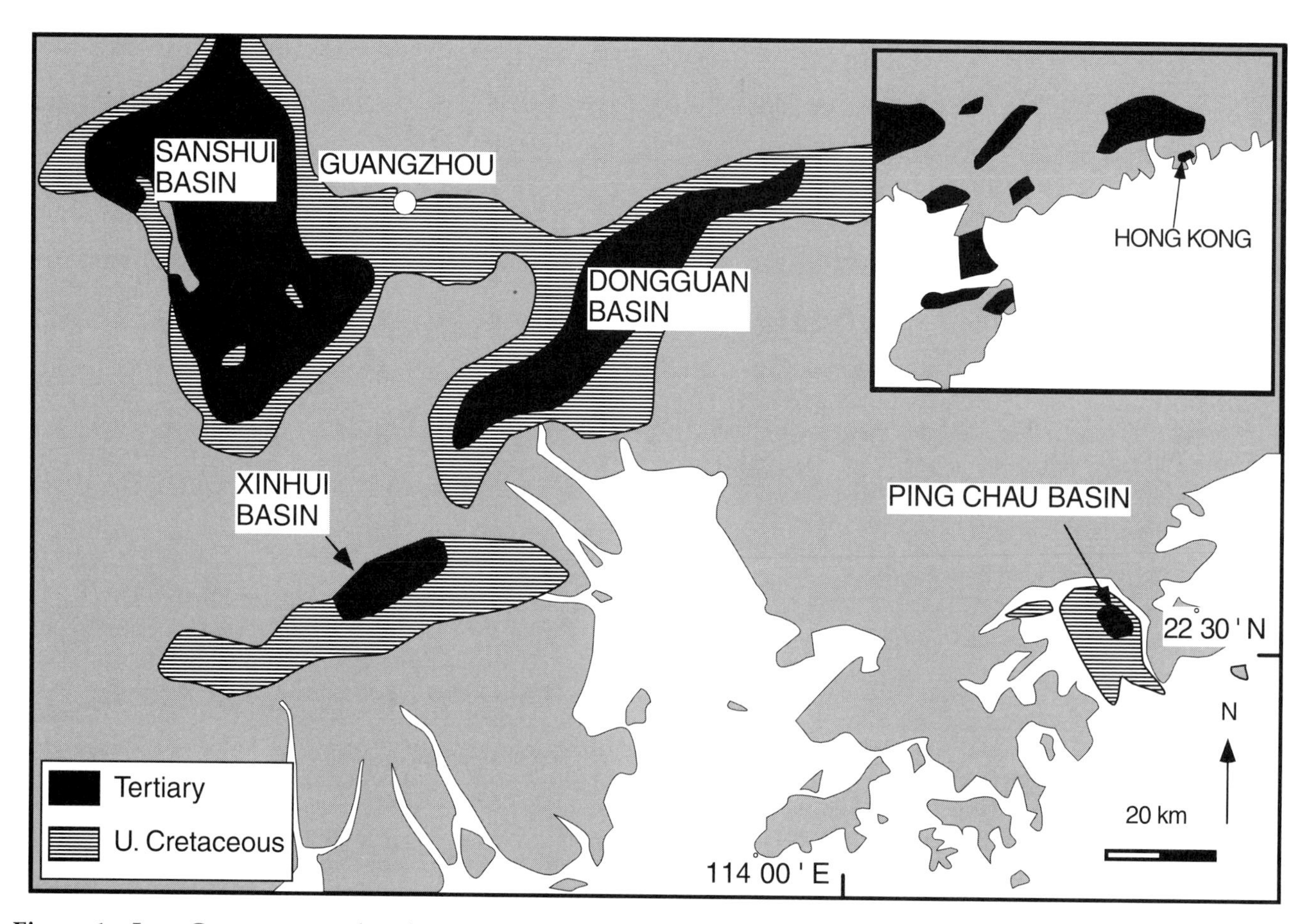

Figure 1—Late Cretaceous and early Tertiary sedimentary basins in the Hong Kong–South China region (modified after Lai, 1991). Inset shows a simplified palaeogeographic reconstruction of possible early Tertiary (Eocene) lake basins (after Wang et al., 1985).

local cross-lamination (ripples), tabular cross-bedding, slumps, and mudcracks. There is little or no bioturbation and laminae can be traced for many tens of meters. The rocks consist of mudstone, siltstone, and marlstone, although Taylor et al. (1990) note that calcite and dolomite predominate at least locally. They comment that there are very few terrigenous particles and that the rocks should be referred to as limestone and dolostone.

The lower and upper Ping Chau Formation are largely concealed below sea level (with lithologies in Fig 3 being partially inferred). The middle Ping Chau Formation (about 200 m) is well exposed and was further subdivided by Lai (1991) into three parts based on rock type and authigenic mineralization, which was first mapped (Figure 2) and recognized by Nau (1979).

The lower part (62 m) recognised by Lai is dominated by laminated, aegirine-bearing (<1 mm radial clusters) and dolomitic siltstone (10–45% dolomite), with mudstone and marlstone. A distinctive cherty tuff (Wu and Nau, 1989) forms a prominent marker horizon (Figure 3). These rocks give way to a middle part (46 m) dominated by laminated and thinly bedded dolomitic siltstone, siltstone and mudstone, all of which contain abundant pyrite. The upper part (92 m) includes thinly bedded, laminated, zeolitic (natrolite and analcime) siltstone with local pyrite and dolomitic and calcareous siltstone. Locally, laminae drape over zeolite nodules (up to 5 cm in diameter). The zeolites and aegirine occur as syngenetic crystals parallel to lamination (Taylor et al., 1990), as replacements of gypsum (Wu and Nau, 1989) and as epigenetic veins and lenses that cut across laminae (Taylor et al., 1990). Taylor et al. (1990) also note the presence of felsic volcanics and their possible relationship to the growth of zeolites and aegirine.

Figure 4 shows a detailed composite section for a 114 m portion of the middle Ping Chau Formation (part of the base and several tens of metres of the uppermost zeolitic sequence are not recorded). Distinctive features include aegirine and pyrite nodules toward the base and the presence of wispy calcite laminae and carbonate infilled pseudomorphs (after evaporites?) at two distinct levels (14–22 m and 70–74 m). Zeolites are present at about 80–82 m. Massive mudstone layers, generally <5 cm thick, are common below the tuff horizon at 40 m.

The occurrence of calcite (and dolomite) suggest a saline-alkaline setting rich in Ca and Mg, whereas the zeolites and aegirine (as well as albite) indicate a Na-rich environment (Taylor et al., 1990). The occurrence of pyrite implies reducing conditions in the sediments (Lee et al., 1991a). The excellent lamination suggests

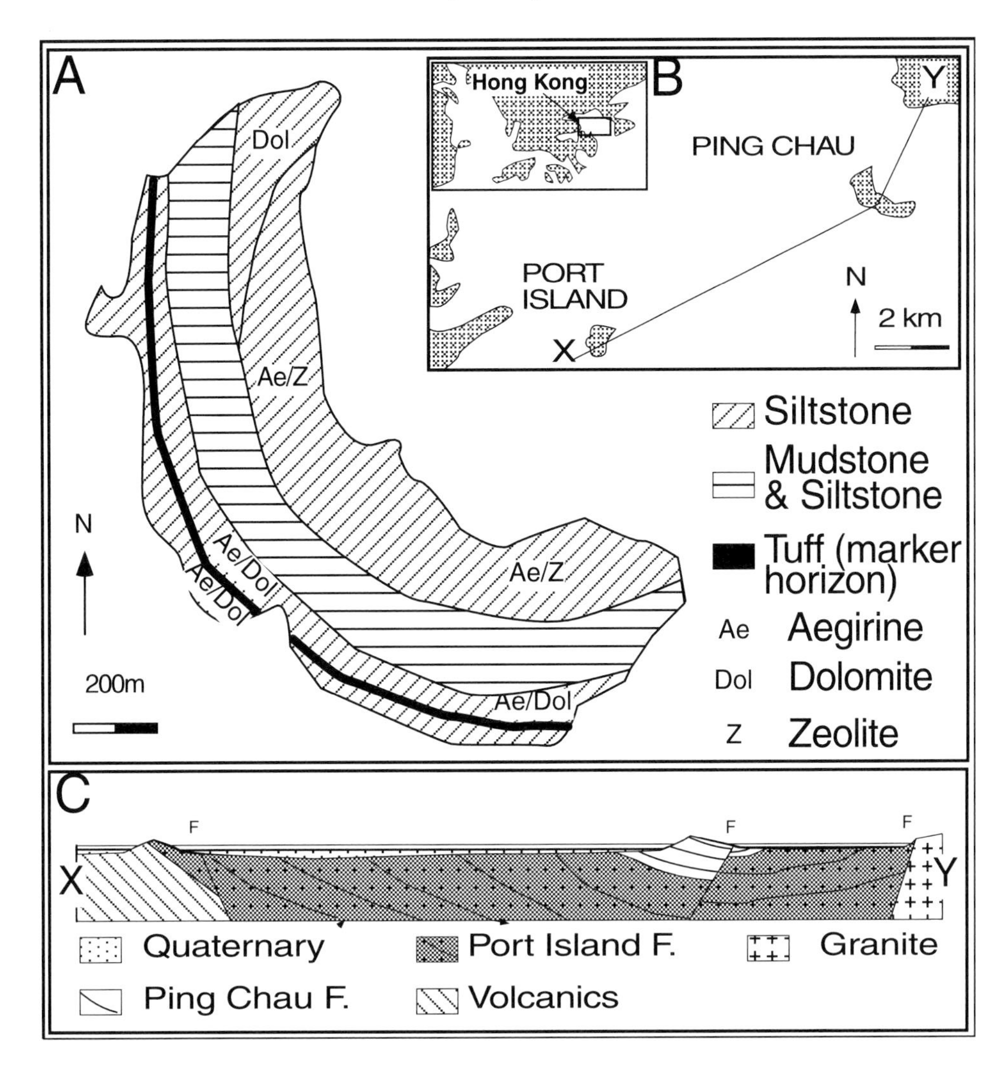

Figure 2—Location and geology of Ping Chau Island. (A) Distribution of major lithofacies and authigenic minerals on Ping Chau Island (modified from Nau, 1979; Lee et al., 1991b; Lai, 1991). (B) Location of Ping Chau and cross section shown in (C). (C) Simplified cross section (after Lai, 1991), inferred from seismic data and outcrops on islands.

low energy conditions with little sediment disruption due to organisms and perhaps deposition in the open waters of a stratified saline lake with bottom anoxia.

Fragmentary carbonized plants, insects, and sporo-pollen have been recovered from the Ping Chau Formation in a series of studies (Williams, 1943; Wu and Nau, 1989; Nau et al., 1990; Zhou et al., 1990; Lee et al., 1991a). Wu and Nau (1989) and Lee et al. (1991a) collected fossil plants that included several hundred specimens belonging to the angiosperms (17 genera), gymnosperms (15 genera), and pteridophytes (3 genera). Dates inferred from the floras are contradictory, with some species that are restricted to the Mesozoic occurring together with forms suggesting the Tertiary.

According to Nau et al. (1990), the palynological flora is dominated by pteridophytes and to a lesser extent by gymnosperms. Lee et al. (1991a) report 17 genera of sporo-pollen of which five species are restricted to the Tertiary or early Tertiary.

PALEOGEOGRAPHY

The Late Cretaceous–early Tertiary rocks of southern China and Hong Kong suggest a period of continental deposition. Wang et al. (1985) suggest that the Upper Cretaceous red beds of Guangdong and Hong Kong (Dalangshan and Port Island formations) formed in an arid zone that extended across much of southern and central China. Wang et al. (1985) also note that climatic conditions changed and produced a humid subtropical to tropical belt across southern China during the early Tertiary, which Gu and Renaut (1994) suggest occurred by the early late Eocene and signified the onset of the Asian monsoon. This humid belt gave way northward to arid environments in central and northern China.

Lee et al. (1991b) suggest that the Eocene Ping Chau Formation developed close to the boundary between these arid and humid climatic zones in a saline lake environment. Wang et al. (1985) suggest that there were, in fact, several lake basins present during this period in southern China, with the three Guangdong Basins of Sanshui, Dongguan, and Xinhui forming a single palaeolake (inset, Figure 1).

REFERENCES CITED

Allen, P., and E. A. Stephens, 1971, Report of the Geological Survey of Hong Kong 1967–1969, Hong Kong Government Press, 116 p.

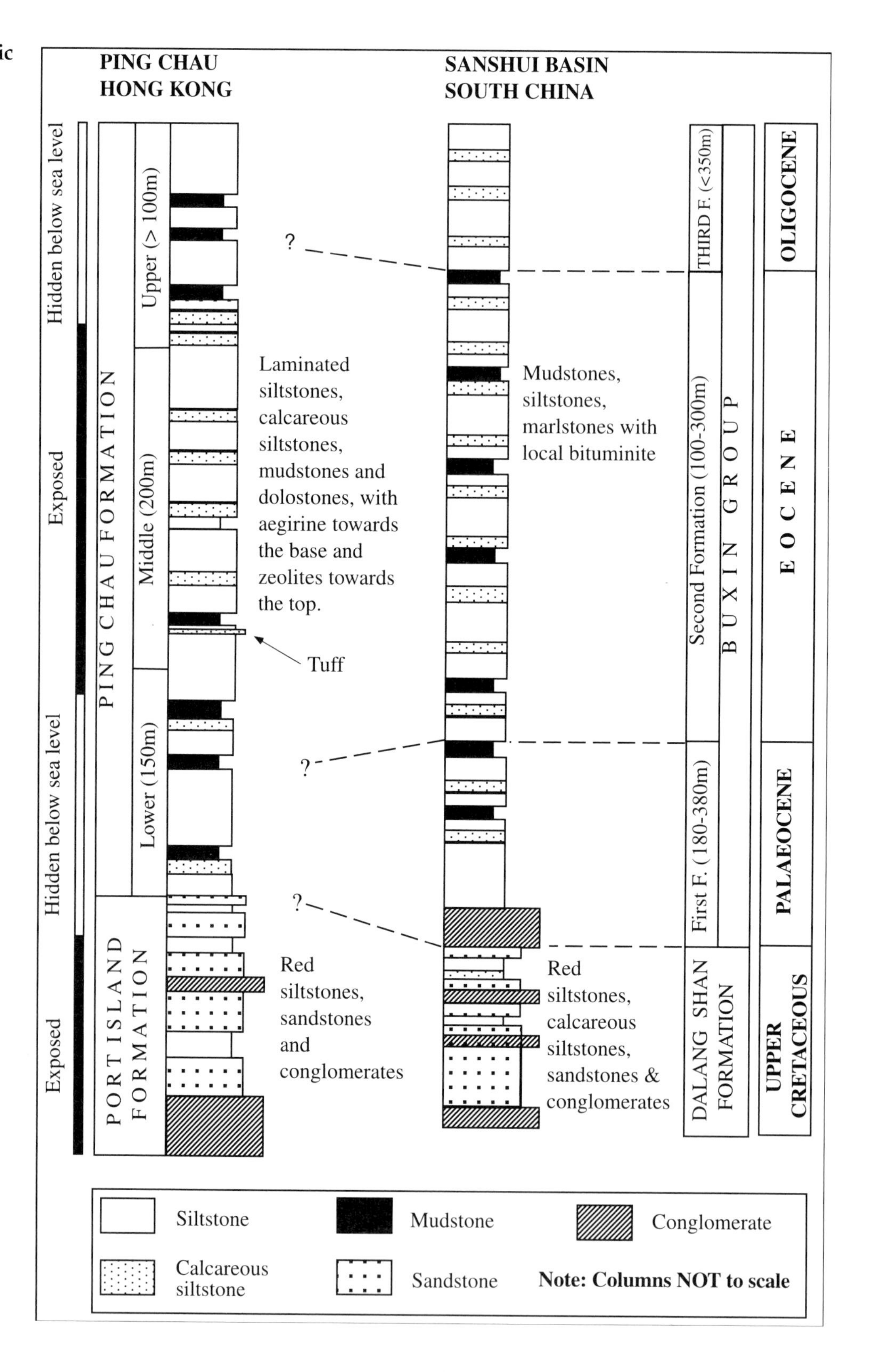

Figure 3—Lithostratigraphic comparison between the Ping Chau Formation and the Buxin Group (modified after Lai, 1991). Correlation is based on lithology and pollen data and is tentative. Note that the lithology for sections "hidden below sea level" is based on seismic data (Lai et al., 1996) and is inferred only.

Davis, S. G., 1953, The geology of Hong Kong, Hong Kong Government Press, 210 p.

Gu, C., and R. W. Renaut, 1994, The effect of Tibetan uplift on the formation and preservation of Tertiary lacustrine source-rocks in eastern China, Journal of Paleolimnology, v. 11, p. 31–40.

Heanley, C.M., 1923, Some geological excursions around Hong Kong, The Caduceous, v. 2, p. 85–94.

Lai, K. W., 1985, A review of the late Cretaceous–Palaeogene fault basins around Hong Kong, *in* P. G. D. Whiteside and R. S. Arthurton, eds., Marine geology of Hong Kong and the Pearl River mouth, Geological Society of Hong Kong, Hong Kong, p. 39–46.

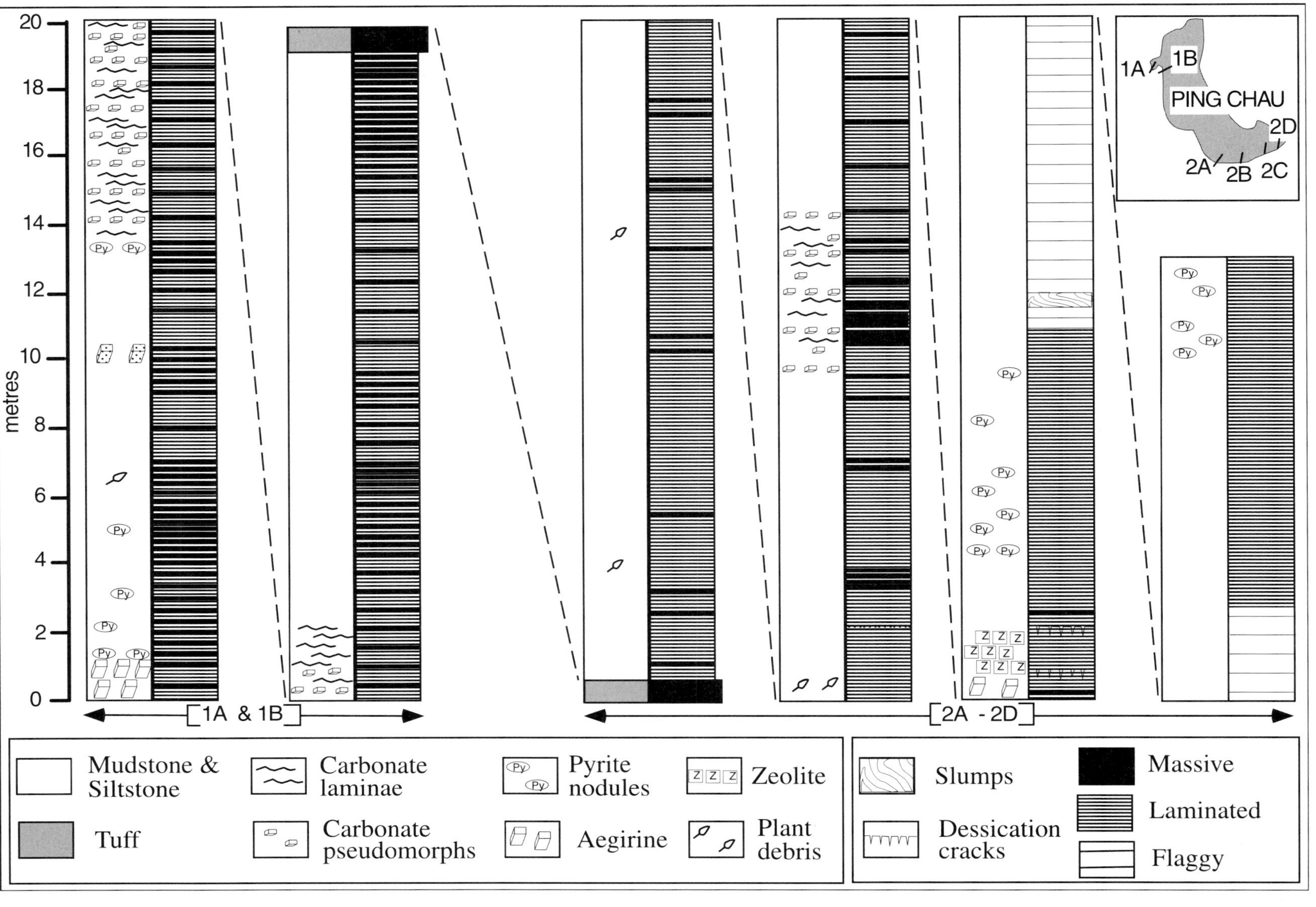

Figure 4—Detailed composite section for the exposed part of the middle Ping Chau Formation. The massive beds are mostly mudstones.

Lai, K. W., 1991, Stratigraphy of the Ping Chau Formation, Geological Society of Hong Kong Newsletter, v. 9 (2), p. 3–23.

Lai, K. W., S. D. G. Campbell, and R. Shaw, 1996, Geology of the northeastern New Territories, Hong Kong Geological Survey Memoir No. 6, Geotechnical Engineering Office, Hong Kong, 144 p.

Lee, C. M., 1985, Recent developments in Hong Kong stratigraphy, Geological Society of Hong Kong Newsletter, v. 3 (4), p. 7–9.

Lee, C. M., 1987, A summarised introduction on Hong Kong geology, Geology of Guangdong, v. 2 (1), p. 29–48.

Lee, C. M., M. J. Atherton, S. Q. Wu, G. X. He, J. H. Chen, and P. S. Nau, 1991a, Discovery of angiosperm fossils from Hong Kong—with discussion on the age of the Ping Chau Formation, Geological Society of Hong Kong Newsletter, v. 9 (1), p. 50–60.

Lee, C. M., J. H. Chen, G. X. He, M. J. Atherton, and K. W. Lai, 1991b, On the age of the Ping Chau Formation, Geological Society of Hong Kong Newsletter, v. 9 (1), p. 34–49.

Nau, P. S., 1979, Geological notes on the sedimentary rocks of Ping Chau Island, Mirs Bay, Hong Kong, Annals of the Geography, Geology and Archaeology Society, University of Hong Kong, v. 7, p. 33–41.

Nau, P. S., Q. Q. Jiang, and Q. J. Wu, 1990, Spore-pollen flora from Ping Chau Island, Mirs Bay, Hong Kong, Geological Society of Hong Kong Newsletter, v. 8 (2), p. 25–29.

Peng, C. J., 1971, Acmite and zeolite-bearing beds of Hong Kong, a new and unusual occurrence, Proceedings of the Eighth International Sedimentological Congress, Heidelberg, p. 76–77.

Ruxton, B. P., 1960, The geology of Hong Kong, Quarterly Journal of the Geological Society of London, v. 115, p. 233–260.

Song, Z. C., M. Y. Li, and L. Zhong, 1986, Cretaceous and early Tertiary sporo-pollen assemblages from the Sanshui Basin, Guangdong Province, Science Press, Beijing, 204 p.

Sun, X. J., M. X. Li, Y. Y. Zhang, Z. Q. Lei, Z. C. Kong, P. Li, Q. Ou, and Q. N. Liu, 1981, Sporo-pollens, Tertiary palaeontology of the northern continental shelf the South China Sea, Guangdong Science and Technology Press, 58 p.

Tang, T. F., Y. S. Xue, Y. K. Zhou, W. R. Yang, and Q. L. Zhang, 1980, The characteristics of carbonate rocks and their sedimentary environments of the lower Tertiary Buxin Group in the Sanshui Basin, Guangdong, Acta Geologica Sinica, v. 54, p. 249–259.

Taylor, G., D. R. Workman, and M. R. Peart, 1990, The rocks of Ping Chau, Mirs Bay, Hong Kong, Geological Society of Hong Kong Newsletter, v. 8 (4), p. 46–50.

Wang, H. Z., X. C. Chu, B. P. Liu, H. F. Hou, and L. F. Ma, 1985, Atlas of the palaeogeography of China, Cartographic Publishing House, Beijing, 45 p.

Williams, M. Y., 1943, The stratigraphy and palaeontology of Hong Kong and the New Territories, Transactions of the Royal Society of Canada. Third Series, v. 37 (4), p. 93–117.

Wu, Q. and P. Nau, 1989, The characteristics of fossil plants and age of the strata of Ping Chau Island, Mirs Bay, Hong Kong, Geological Society of Hong Kong Newsletter, v. 7 (4), p. 2–11.

Zhou, Z. Y., H. M. Li, Z. Y. Cao, and P. S. Nau, 1990, Some Cretaceous plants from Pingzhou (Ping Chau) Island, Hong Kong, Acta Palaeontologica Sinica, v. 29 (4), p. 415–430.

Lucas, S. G., R. J. Emry, V. Chkhikvadze, B. Bayshashov, L. A. Tyutkova, P. A. Tleuberdina, and A. Ahamangara, 2000, Upper Cretaceaous–Cenozoic lacustrine deposits of the Zaysan basin, eastern Kazakhstan, *in* E. H. Gierlowski-Kordesch and K. R. Kelts, eds., Lake basins through space and time: AAPG Studies in Geology 46, p. 335–340.

Chapter 29

Upper Cretaceous–Cenozoic Lacustrine Deposits of the Zaysan Basin, Eastern Kazakhstan

Spencer G. Lucas
New Mexico Museum of Natural History
Albuquerque, New Mexico, U.S.A.

Robert J. Emry
Department of Paleobiology, National Museum of Natural History, Smithsonian Institution
Washington, D.C., U.S.A.

Viacheslav Chkhikvadze
Institute of Paleobiology, National Academy of Sciences
Tblisi, Georgian Republic

Bolat Bayshashov
Lyubov A. Tyutkova
Pyruza A. Tleuberdina
Institute of Zoology, Akademgorodok
Almaty, Kazakhstan

Ayzhan Zhamangara
Institute of Botany
Almaty, Kazakhstan

INTRODUCTION

The Zaysan Basin of northeastern Kazakhstan–northwestern China (Figure 1) is a collisional successor basin that formed during the Late Cretaceous (Graham et al., 1993; Allen et al., 1995). Beginning during the late Campanian–Maastrichtian, a lake basin developed in the Zaysan depression and has been there until the present. Lake Zaysan is now one of the large freshwater lakes of central Asia. Located between the Altai and Tarbagatay ranges at an altitude of 386 m, it is fed by the Cherny Irtysh River to the east and historically was about 100 km long, 32 km wide, and 8 m deep. Its original surface area of 3200 km^2, however, was enlarged, and the lake was deepened (to 14 m) in the 1980s when Lake Zaysan became part of Bukhtarma Reservoir to the west. The new waterbody is 440 km long and has a total area of 5504 km^2. Here, we present a brief overview of the lithostratigraphy, paleontology, and depositional history of the Zaysan lacustrine basin, the majority of which is located in eastern Kazakhstan (Figure 1).

LITHOSTRATIGRAPHY

Two lithostratigraphic (actually lithochronologic) schemes are generally applied to the Upper Cretaceous–Pleistocene strata of the Zaysan Basin (Figure 2). These schemes, of Lavrov and Yerofeyev (1958) and Borisov (1963, 1983, 1984), employ the Soviet stratigraphic term "svita" as the basic unit of lithostratigraphic subdivision. These svitas are recognized not only by their lithologic characteristics, but they also represent (at least in theory) isochronously bounded time stratigraphic units defined by their fossil content; furthermore, the base of each svita is supposed to correspond to the initiation of a separate cycle of deposition. Each svita thus has lithologic, biochronologic, and genetic

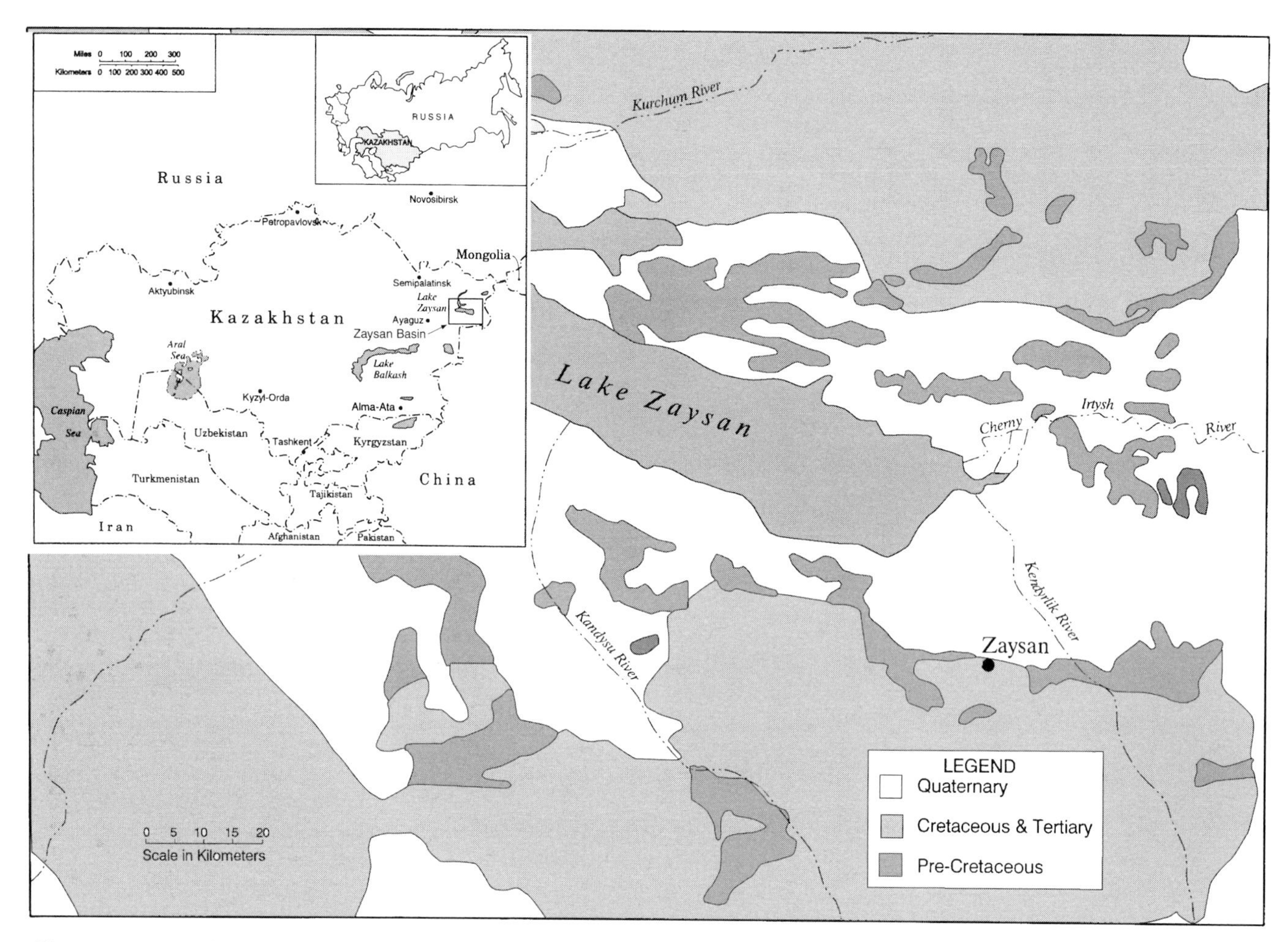

Figure 1—Location map and generalized geologic map of Kazakhstan portion of the Zaysan Basin.

(sedimentologic) significance so that the svita has no precise equivalent in Western stratigraphic theory and terminology.

We will publish a strict lithostratigraphic scheme of the Upper Cretaceous–Cenozoic strata in the Zaysan basin, but for now we use informal designations (formations A–J) to summarize the section (Figure 2). Formation A is mostly red-bed mudstones and lesser amounts of deeply weathered kaolinitic paleosols, trough-crossbedded quartzose sandstones, silica-pebble conglomerates, and travertine limestones (Lucas et al., 1995). The entire formation is as much as 135 m thick. The six svitas recognized by Borisov (1963) can be viewed as member-level units. Lacustrine mudstones and shales are intermittently present throughout formation A.

The lower part of formation B (Chapaktas and Obayla svitas) is as much as 52 m thick and dominated by brown and dark yellowish-orange smectitic mudstones, siltstones, and sandstones. The upper part of formation B (Sargamys svita) is yellowish-gray siltstones and mudstones with interbedded selenite, as much as 36 m thick. The Konurkura and Chaybulak svitas are shale-dominated, lateral equivalents of the Sargamys svita.

One of the most distinctive and persistent lithostratigraphic units in the Zaysan basin is formation C. Its lower part (Kyzylkain svita) is as much as 19 m thick and consists of quartzose sandstones and conglomerates, many of which show extensive evidence of deep paleoweathering. Strata of the upper part of formation C (Aksyir svita) are dominantly grayish-yellow to yellowish-orange siltstones and fine-grained sandstones that are typically laminated and ripple laminated. A few hummocky bedded or trough-crossbedded sandstones are present, and extensive bioturbation is common. Thin, laterally persistent beds of red, orange, or green shale in the Aksyir svita yield numerous charophytes and ostracods. Maximum thickness is 123 m, and the unit grades upward into formation D.

Formation D is a shale/mudstone-dominated package as much as 140 m thick. Its lower part (Kusto svita) is mostly red and green variegated smectitic shale, whereas its upper part (Buran svita) is mostly yellowish-brown and gray variegated mudstones and siltstones. Locally, the upper part (Buran svita) of formation D is absent, suggesting the presence of an unconformity at the base of formation E.

The lower part of formation E (Oshagande svita) is as much as 23 m thick and consists of yellowish-gray to yellowish orange micaceous sandstone, siltstone, and quartzite-cobble conglomerate. Ripple lamination and trough crossbedding are the dominant bedforms.

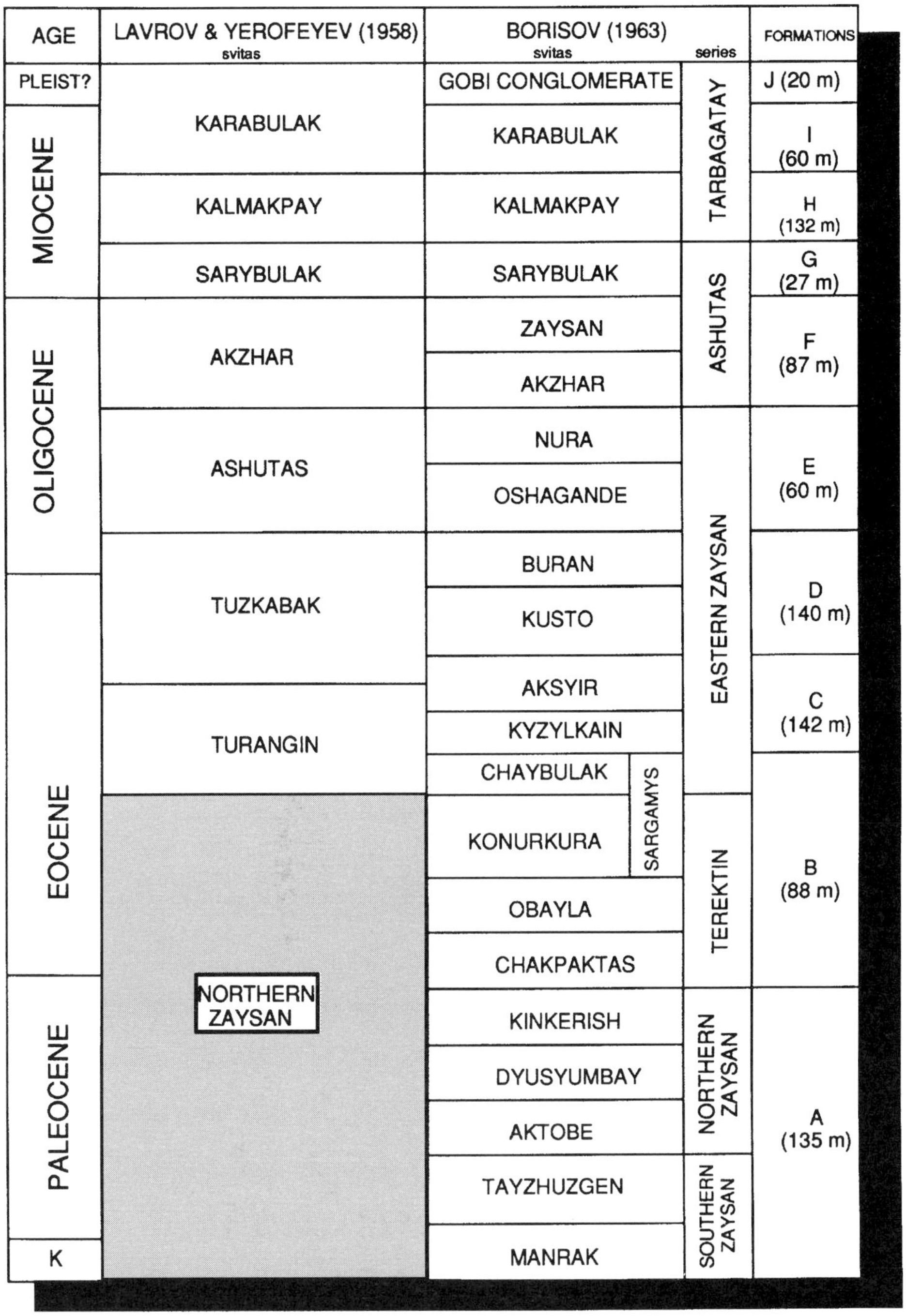

AGE	LAVROV & YEROFEYEV (1958) svitas	BORISOV (1963) svitas		BORISOV (1963) series	FORMATIONS
PLEIST?	KARABULAK	GOBI CONGLOMERATE		TARBAGATAY	J (20 m)
MIOCENE		KARABULAK			I (60 m)
	KALMAKPAY	KALMAKPAY			H (132 m)
	SARYBULAK	SARYBULAK		ASHUTAS	G (27 m)
OLIGOCENE	AKZHAR	ZAYSAN			F (87 m)
		AKZHAR			
	ASHUTAS	NURA		EASTERN ZAYSAN	E (60 m)
		OSHAGANDE			
	TUZKABAK	BURAN			D (140 m)
EOCENE		KUSTO			
	TURANGIN	AKSYIR			C (142 m)
		KYZYLKAIN			
		CHAYBULAK	SARGAMYS		B (88 m)
	NORTHERN ZAYSAN	KONURKURA		TEREKTIN	
		OBAYLA			
		CHAKPAKTAS			
PALEOCENE		KINKERISH		NORTHERN ZAYSAN	A (135 m)
		DYUSYUMBAY			
		AKTOBE			
		TAYZHUZGEN		SOUTHERN ZAYSAN	
K		MANRAK			

Figure 2—Comparison of stratigraphic schemes for Upper Cretaceous–Cenozoic rocks of the Zaysan Basin, Kazakhstan.

These strata become more shaly upward and grade into a succession of black, gray, and brown smectitic shales and mudstones as much as 37 m thick (Nura svita).

The base of formation F is pedogenically modified sandstones that grade upward into pink, gray, and green siltstones and shales up to 54 m thick (Akzhar svita). The upper part of formation F (Zaysan svita) is laterally extensive beds of green calcareous shale with minor limestone interbeds as much as 33 m thick.

Formation G is as much as 27 m of yellow-to-gray, trough crossbedded micaceous sandstones and quartzite conglomerates that locally scour deeply into formation F. Formation H is a thick (up to 132 m) succession of red-bed mudstones. Formation I is about 60 m of trough-crossbedded sandstone, conglomerate, and sedimentary breccia, and formation J is up to 20 m of similar conglomerate and sedimentary breccia that rests with angular unconformity on underlying strata.

PALEONTOLOGY

The Upper Cretaceous–Cenozoic deposits of the Zaysan Basin have received intensive study, especially because of their rich fossil record. Borisov (1963), Martinson and Kyansep-Romashkina (1980), Gabuniya et

Figure 3—Distribution of major fossil groups in the Zaysan Basin Upper Cretaceous–Cenozoic section.

STRATIGRAPHY		megafossil plants	charophytes	ostracods	molluscs	fishes	turtles	crocodilians	birds	mammals
GOBI CONGLOMERATE	J									
KARABULAK	I								■	■
KALMAKPAY	H					■	■			■
SARYBULAK	G	■			■	■	■			■
ZAYSAN	F		■	■	■	■	■		■	■
AKZHAR	F	■	■	■	■	■	■			■
NURA	E	■	■	■	■	■	■			■
OSHAGANDE	E	■								
BURAN	D	■	■	■	■	■	■		■	■
KUSTO	D		■	■	■	■	■	■	■	■
AKSYIR	C	■	■	■	■	■	■	■	■	■
KYZYLKAIN	C	■			■					■
CHAYBULAK (SARGAMYS)	B	■			■	■	■	■		■
KONURKURA (SARGAMYS)	B	■			■	■	■	■		■
OBAYLA	B		■	■	■	■	■	■		■
CHAKPAKTAS	B				■	■	■	■		■
KIINKERISH	A	■								
DYUSYUMBAY	A									
AKTOBE	A	■			■					
TAYZHUZGEN	A	■								
MANRAK	A				■					

al. (1983), Ilinskaya et al. (1983), Gabuniya (1984), and Russell and Zhai (1987) reviewed much of this record. Lacustrine or lacustrine margin deposits in the Zaysan Basin produce extensive fossil assemblages of megaflora, charophytes, ostracods, unionid bivalves, nonmarine gastropods, freshwater fishes, turtles, and mammals (Figure 3). Crocodilian fossils are restricted to the middle–late Eocene part of the section, and there is a sparse fossil bird record.

The fossil mammals are particularly important because they provide the most precise age control of the section (Russell and Zhai, 1987; Emry et al., 1998). They also indicate the presence of major temporal gaps at the bases of formations C, E, and G.

DEPOSITIONAL SYSTEMS

Stratigraphic, sedimentologic, and structural data from the Zaysan Basin allow a preliminary interpretation of cycles of sedimentation during the Late Cretaceous–Cenozoic (Figure 4). We model deposition in the Zaysan Basin as base-level driven, in part due to differential rates of tectonic subsidence that produced five major lake highstands (Figure 4). We also recognize that climate cycles affected sedimentation in the basin.

Rocks that underlie the Upper Cretaceous red beds are folded and to varying degrees metamorphosed strata of late Paleozoic (Carboniferous–Permian) and Late Triassic–Early Jurassic age. The Upper Cretaceous red beds rest on a deeply eroded and complex paleotopography that developed on these rocks between the Early Jurassic and Late Cretaceous. Extensive pedogenesis of the older rocks and several meters of local stratigraphic relief at the base of the Upper Cretaceous strata support this conclusion. The Early Jurassic–Late Cretaceous depositional hiatus was followed by a rise in base level that led to the deposition of lacustrine red beds, stable landscape paleosols, and gravelly channel sandstones that initiated evolution of the Zaysan depositional basin during the Late Cretaceous (late Campanian–Maastrichtian). This deposition may have continued into the Paleocene (Lucas et al., 1995).

Pedogenesis, a marked change in lithology, and the absence of late Paleocene–early Eocene fossils indicate the presence of a second basinwide unconformity at the base of formation B. This unconformity suggests sustained low base level during the late Paleocene and early Eocene. Base level then rose and the first large Zaysan lake basin formed during the middle Eocene. High base level persisted until Sargamys svita deposition, when another drop in base level produced an erosional unconformity associated with the base of the overlying Kyzylkain svita.

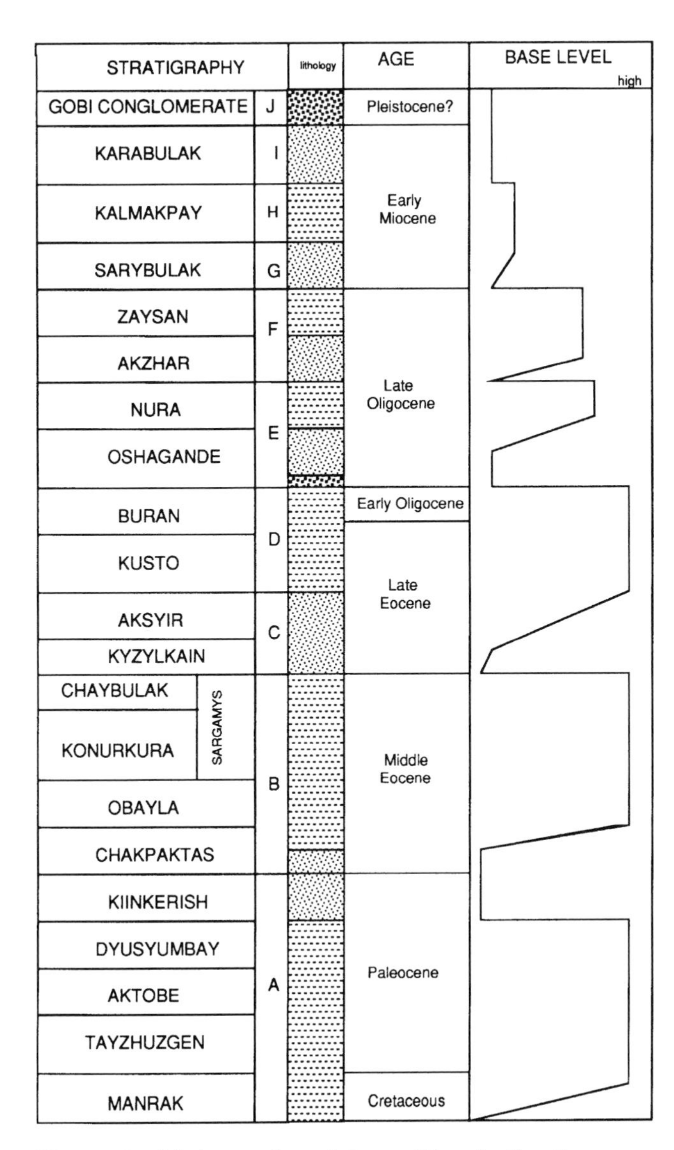

Figure 4—Major cycles of deposition in the Zaysan Basin during the Late Cretaceous–Pleistocene.

Kyzylkain and overlying Aksyir svita deposits represent the beginning of the next cycle. These two svitas are river channel, river delta, and shallow, silty pond deposits that filled the incised topography that developed during pre-Kyzylkain erosion. The cycle culminated in extensive lacustrine shales and mudstones of formation D (Kusto-Buran svitas).

The succeeding two cycles of sedimentation are similar to that just outlined. A drop in base level produced erosion and incision prior to the deposition of formations E and F. The lower portions of these formations are sandstones, siltstones, and conglomerates of fluvial origin that filled in and buried the incised topography during base-level rise. Lake highstands followed during deposition of the upper parts of formations E and F.

By the time formation G was deposited, Lake Zaysan must have assumed its current size and configuration because all early Miocene and younger strata exposed around the current lake are fluvial (formation G), river floodplain (formation H), or alluvial fan (formations I and J) deposits, not lacustrine strata.

ACKNOWLEDGMENTS

The National Geographic Society and the Smithsonian Institution's Charles D. Walcott Fund supported research in the Zaysan Basin.

REFERENCES CITED

Allen, M. B., A. M. C. Sengör, and B. A. Natal'in, 1995, Junggar, Turfan and Alakol basins as Late Permian to Early Triassic extensional structures in a sinistral shear zone in the Altaid orogenic collage, Central Asia. Journal of the Geological Society of London, v. 152, p. 327–338.

Borisov, B. A., 1963, Straigrafya verkhnevo mela i paleogena-neogena Zaysanskoy vpadiny [Stratigraphy of the Upper Cretaceous and Paleogene–Neogene of the Zaissan depression]. Trudy VSEGEI i Gosudarstveno Geologicheskiy Komineta, new series, v. 94, p. 11–75.

Borisov, B. A., 1983, Biostratigrafiya kontinentalnykh paleogenovykh otlozheniy Zaytysanskoy vpadiny [Biostratigraphy of the continental Paleogene sediments of the Zaysan depression], *in* K. O. Rostovtsev and A. I. Zhamoyda, eds., Stratigrafiya Fanerozoya SSSR: Leningrad, Ministrerstvo Geologii SSSR, 110 pp.

Borisov, B. A., 1984, Stratigrafiya nizhneye-srednyeeotsenovykh otlozheniy Zaysanskoy vpadiny [Stratigraphy of the lower and middle Eocene sediments of the Zaissan depression, *in* L. K. Gabuniya, ed., Flora y Fauna Zaisanskoiy Vpadniny: Tbilisi, Akademiya Nauk Gruzinskoiy SSR: "Metsniereba," 199 pp.

Emry, R. J., S. G. Lucas, L. Tjutkova, and B. Wang, 1998, The Ergilian–Shandgolian (Eocene–Oligocene) transition in the Zaysan basin, Kazakhstan. Bulletin of the Carnegie Museum, v. 34, p. 298–312.

Gabuniya, L. K., ed., 1984. Flora y Fauna Zaisanskoiy Vpadniny [Fauna and Flora of the Zaysan Basin]: Tbilisi, Akademiya Nauk Gruzinskoiy SSR: "Metsniereba," 199 pp.

Gabuniya, L. K., A. A. Gureyev, M. B., Efimov, E. K. Sycheskaya, N. V. Tolstikova, V. M. Chkhikvadze, N. S. Shevyreva, and B. A. Borisov, 1983, Fauna paleogena Zaysanskoy vpadiny [Paleogene fauna of the Zaissan basin], *in* K. O. Rostovtsev and A. I. Zhamoyda, eds., Stratigrafiya Fanerozoya SSSR: Leningrad, Ministrerstvo Geologii SSSR, 119 pp.

Graham, S. A., M. S. Hendrix, L. B. Wang, and A. R. Carroll, 1993, Collisional successor basins of western China: Impact of tectonic inheritance on sand composition. Geological Society of America Bulletin, v. 105, p. 323–344.

Ilinskaya, I. A., N. P. Kyansep-Romaskina, L. A. Panova, and B. A. Borsov, 1983, Paleogenova flora Zaysanskoy vpadiny [Paleogene flora of the Zaiysan basin], *in* K. O. Rostovtsev and A. I. Zhamoyda, eds., Stratigrafiya Fanerozoya SSSR: Leningrad, Ministrerstvo Geologii SSSR, 119 pp.
Lavrov, V. C., and V. S. Yerofeyev, 1958, Stratigrafiya tretichnykh tolshch Zaisanskoi vpadiny [Stratigraphy of Tertiary beds of the Zaysan basin]. Vestnik Akademiya Nauk Kazakhskoi SSR, v. 11, p. 68–82.
Lucas, S. G., E. G. Kordikova, and R. J. Emry, 1995, Upper Cretaceous–Paleocene stratigraphy and vertebrate paleontology in the Zaysan basin, Kazakhstan. Journal of Vertebrate Paleontology, v. 15 (3), p. 41A.
Martinson, G. G. and N. P. Kyansep-Romashkina, eds., 1980, Paleolimnologiya Zaysana [Zaysan Paleolimnology]: Leningrad, Akademiya Nauk SSR, Institut Ozerovedeniya, 184 pp.
Russell, D. E., and R. Zhai, 1987, The Paleogene of Asia: mammals and stratigraphy. Mémoires du Muséum National d'Histoire Naturelle Série C Sciences de la Terre, v. 52, p. 1–487.

Section VI

Eocene

Rasplus, L., and F. Ménillet, 2000, Upper Bartonian (Eocene) lacustrine limestone of the Touraine Basin, France, in E. H. Gierlowski-Kordesch and K. R. Kelts, eds., Lake basins through time and space: AAPG Studies in Geology 46, p. 343–348.

Chapter 30

Late Bartonian (Eocene) Lacustrine Limestone of the Touraine Basin, France

Léopold Rasplus
Laboratoire de Géologie des Systèmes sédimentaires, Faculté des Sciences et Techniques, Université de Tours
Tours, France

François Ménillet
BRGM, SGN/ALS, Lingolsheim
Strasbourg, France

INTRODUCTION

The Touraine Basin is a syncline basin formed in remote association with the Alpine orogeny. Basin fill for the lower Eocene is represented by alluvial sandstones or silcretes. This sequence is topped by the Calcaire lacustre de Touraine (lacustrine limestones of Touraine), which is given a late Bartonian (late Ludian) age (Cavelier et al., 1979), although some thin overlying limestones yield a late Stampian age. Overlying these limestones are either the sandstones and clays of the uppermost Eocene Brenne Formation in the southern part of the basin, Neogene marine deposits (faluns), or Neogene alluvial clayey sands (Rasplus, 1987). This contribution focuses on the sedimentary details of the Calcaire lacustre de Touraine (CLT).

STRATIGRAPHY AND SEDIMENTOLOGY

The lacustrine limestones of the Touraine Basin (CLT) are exposed mainly on the Mettray and Champeigne plateaus near Tours (Figure 1). To the south, near the southern boundary of the Touraine Basin, near Descartes and Preuilly-sur-Claise, smaller outcrops occur on both sides of the Brenne Formation, which is best exposed south of Preuilly-sur-Claise. A maximum thickness of 33 m for the CLT occurs in the middle of the central Champeigne subbasin (at Cormery, well log) (Lecointre, 1943).

Age of the CLT is based on the *Nystia plicata*, *Nystia duchasteli*, and *Psilochara repanda* association. Near Fondettes (western Mettray subbasin), the uppermost thin limestone of the CLT contains *Potamides Lamarcki*, which is assigned a late Stampian age.

Seven lithofacies have been identified in the CLT (Table 1) after Dollfus (1901, 1904, 1920), Denizot (1927), Georges (1942), Lecointre (1947), Vatan (1947), Rasplus (1968, 1978, 1982), Rasplus and Alcayde (1974), Rasplus et al. (1978, 1982, 1989a, b), Saugrin (1982), Ménillet (1988b), and Alcayde (1990). Early diagenetic features of the CLT include micritization, partial cementation and recrystallization, karstification, calcitization of clays, *Microcodium*, and partial reworking. Later diagenetic processes include groundwater silicification (replacement of calcite by quartzine, lussatite, or microquartz) in hydromorphic paleosols or meteoritic silicification (chert development) that took place in the early Quaternary (Ménillet, 1987, 1988a, c).

Two representative detailed lithologic sections of the CLT are shown in Figure 2. The paleoenvironment of the CLT is interpreted as marshlands, ponds, and shallow lakes (see Figure 3). The Truyes Terrages Quarry section in the center of the Champeigne subbasin (Figure 2) represents marshy and algal facies and the Saint-Quentin-sur-Indrois section (Figure 2) from the margin of the subbasin represents lacustrine shore facies. In both cases, water depth is hypothesized to be less than 10 m because of common subaerial features. With a highly variable water table, bioturbation traces signal wet periods, whereas desiccation and pedogenic structures (Saugrin, 1982; Valleron and Dulau, 1983) indicate drier times. Climate may have been seasonal to subarid.

Although rare, invertebrate fossils are associated with the shoreface of the ancient paleolake (Calcaire vermiculé and nodular facies). The fauna (Dollfus, 1906, 1920, 1926; Jodd, 1942) is mainly represented by gastropods and charophytes (Table 2).

GEOCHEMISTRY

From bulk sampling, mineralogy of the CLT is calcite, chert in tiny amounts of detritic quartz (<1g/1000g), quartzine, lussatite, and clay minerals

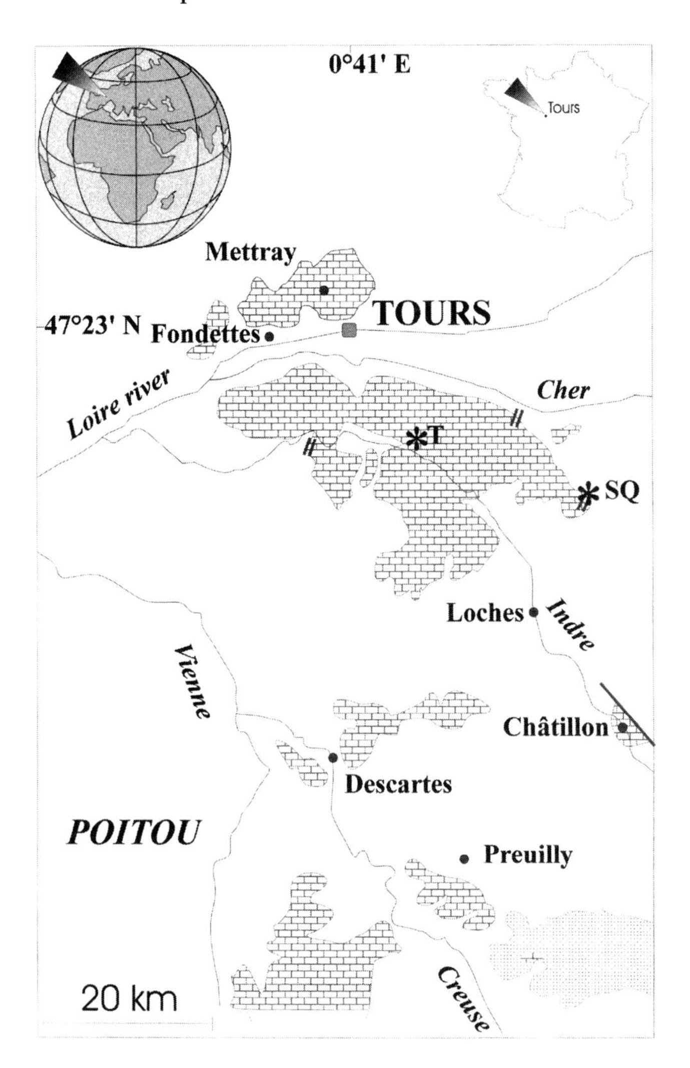

Figure 1—Location map of the exposures of the Eocene (late Bartonian) Calcaire lacustre de Touraine (CLT) in the Touraine basin of France. The basin is divided into the Champeigne, Mettray, Descartes, and Brenne subbasins. The CLT exposure is indicated by a block pattern and the Brenne Formation is illustrated by a dot pattern in the southernmost area of the basin. Location of the measured lithologic sections of Figure 3 are shown as asterisks in the Champeigne subbasin: *T = the Truyes section of the Terrages Quarry and *SQ = the Saint-Quentin-sur-Indrois section.

(palygorskite, sepiolite, attapulgite, kaolinite, smectites, and illites). In the Champeigne subbasin, pink sparitic cement in gravelly limestones or porous silica in karstic cavities rarely contain a sepiolite called quincyte, which is colored by organic pigments (carotenoids) (Louis et al., 1968; Saugrin, 1982). Basic chemical analysis of bulk samples by Saugrin (1982) produced the following results (% by weight of 20 analyses) : CaO 50.61; SiO_2 3.24; Al_2O_3 1.18; Fe_2O_3 0.59; MgO 0.56; Na_2O 0.14; K_2O 0.14, and traces of Sr at 323 ppm.

REFERENCES CITED

Alcayde, G., 1990, Carte géologique de la France à 1/50000. Feuille Châtillon-sur-Indre et notice explicative. Bureau de Recherches Géologiques et Minières. Orléans, France.

Cavelier, C., C. Guillemin, G. Lablanche, L. Rasplus and J. Riveline, 1979, Précisions sur l'âge des

Table 1. Limestones Types of the Touraine Basin.

Lithofacies	Color	Bedding - Thickness	Subenvironment
"Calcaire vermiculé" (micritic limestone with rhizoliths) interbedded with	White or light brown	Well-stratified 10-50 cm-thick beds	Shoreface of a shallow lake
marls or clays	White Green	1-10 cm-thick layers	Shallow lake
Powdery clayey limestone	White	irregular meter-thick lenses	Paleosol
Marl	White	meter-thick lenses	Marsh
Gravelly, pisolitic, granular, and nodular limestone	White or light brown	10-100 cm-thick lenses	Shoreface of a shallow lake
Intraclastic breccias	White	cm to dm-thick lenses (cm-size clasts)	Marsh
Algal limestones	White	irregularly thinly laminated	Lake
Flint and chert	Gray	irregular lenses	diagenetic

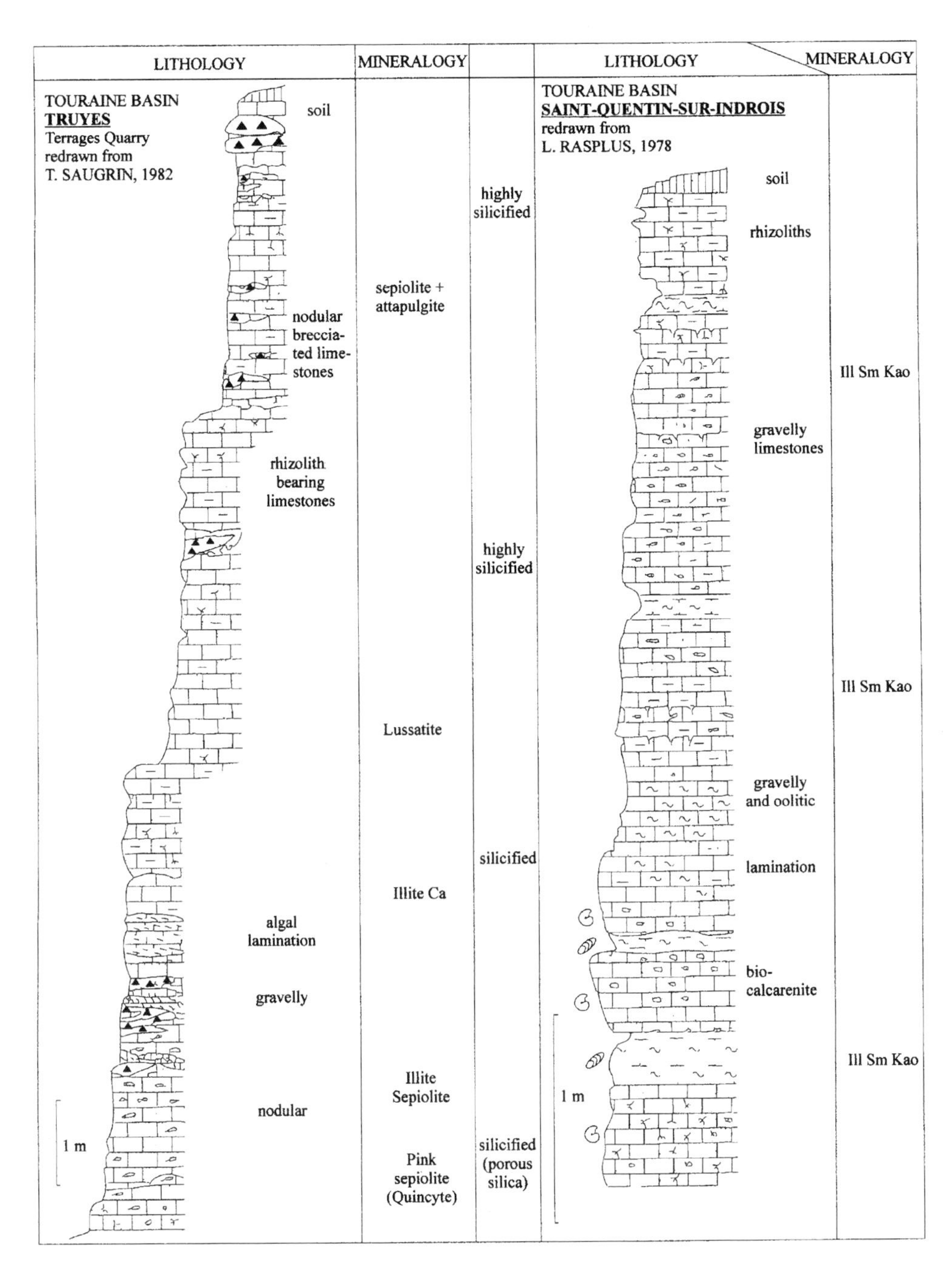

Figure 2—Detailed lithologic sections of the Eocene Calcaire lacustre de Touraine at two localities, in the Champeigne s ubbasin (see Figure 1 for exact location): the Truyes section of the Terrages Quarry and the Saint-Quentin-sur-Indrois section.

calcaires lacustres du Sud du Bassin de Paris d'après les characées et les mollusques. Bull. Bureau de Recherches Géologiques et Minières (2nd series, sec. 1), no. 1, p. 27–30.

Denizot, G., 1927, Les formations continentales de la région orléanaise. Thèse. Ann. Fac. Sci. Marseille (2nd series), v. 3, 582 p.

Dollfus, G. F., 1901, Feuille de Bourges au 1/320000. Terrains tertiaires. Bull. Serv. Carte Géol. France, v. 11 (80), p. 22–29.

Dollfus, G. F., 1904, Feuille de Bourges au 1/320000. Calcaires lacustres de Touraine. Bull. Serv. Carte Géol. France, v. 15 (98), p. 149–156.

Dollfus, G. F., 1906, Révision des faunes de mollusques terrestres et fluviatiles du Tertiaire des bassins de Paris et de la Loire. Bull. Soc. Géol. France, (4th series), v. 6, p. 249–252.

Dollfus, G. F., 1920, Calcaires lacustres du département de l'Indre-et-Loire. Bull. Serv. Carte Géol. France, v. 24 (140), p. 6–12.

Dollfus, G. F., 1926, Observations sur le mode de formation des dépôts lacustres. Comptes Rendues Somm. Soc. Géol. France, v. 15, p. 143–145.

Georges, P., 1942, Les sédiments lacustres de la Champeigne tourangelle. Contribution à l'étude des meulières et de leur altération. Bull. Soc. Géol. France (5th series), v. 12, p. 251–260.

Jodot, P., 1942, Sur le bassin lacustre ludien de Châtillon-sur-Indre. Comptes Rendues Somm. Soc. Géol. France, v. 14, p. 174–176.

Lecointre, G., 1943, Révision de la feuille de Loches à 1/80000. Bull. Serv. Carte Géol. France, v. 43 (2), p. 83–88.

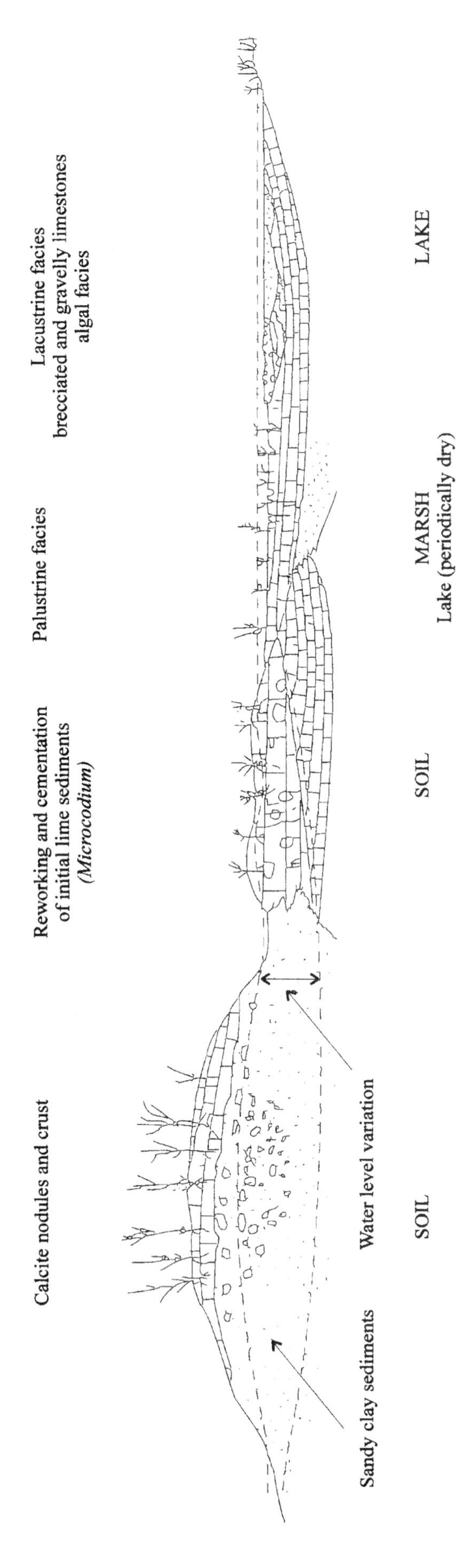

Figure 3—Schematic diagram illustrating the interpreted paleoenvironment of the Eocene lacustrine limestone of the Touraine Basin (central Champeigne subbasin). Vertical scale = 10 m maximum; horizontal scale = 100 m to 1 km. Adapted from Saugrin (1982).

Lecointre, G., 1947, La Touraine. Hermann, Paris, 250 p.

Louis, M., C. Guillemin, S. C. Goni, and J. P. Ragot, 1968, Coloration rose carmin d'une sépiolite éocène, la quincyte, par les pigments organiques. 4th International Meeting on Organic Geochemistry, Univ. Amsterdam, Proceedings, p. 553–566.

Ménillet, F., 1987, Les meulières du Bassin de Paris (France) et les faciès associés. Rôle des altérations supergènes néogènes à Quaternaire ancien dans leur genèse. Thèse Sciences. Univ. Strasbourg I, 536 p.

Ménillet, F., 1988a, Meulière, argile à meulière et meulièrisation. Historique, évolution des termes et hypothèses génétiques. Bull. Information des Géologues du Bassin de Paris, France, v. 25 (4), p. 71–79.

Ménillet, F., 1988b, Les accidents siliceux des calcaires continentaux à lacustres du Tertiaire parisien. Bull. Information des Géologues du Bassin de Paris, France, v. 25 (4), p. 57–70.

Ménillet, F., 1988c, Désilicifications et silicifications au Plio–Quaternaire dans les karsts de calcaires tertiaires du Bassin de Paris. Bull. Information des Géologues du Bassin de Paris, France, v. 25 (4), p. 81–91.

Rasplus, L., 1968, Carte géologique de la France à 1/50000. Feuille Loches et notice explicative. Bureau de Recherches Géologiques et Minières. Orléans, France.

Rasplus, L., and G. Alcayde, 1974, Carte géologique de la France à 1/50000. Feuille Tours et notice explicative. Bureau de Recherches Géologiques et Minières. Orléans, France.

Rasplus, L., 1978, Contribution à l'étude géologique des formations continentales détritiques tertiaires de la Touraine, de la Brenne et de la Sologne. Thèse d'Etat Sciences, Univ. d'Orléans, texte: Vol. 1 and 2, 454 p., Vol. 3: 132 p.

Rasplus, L., G. Alcayde, and J. J. Macaire, 1978, Carte géologique de la France à 1/50000. Feuille Preuilly-sur-Claise et notice explicative. Bureau de Recherches Géologiques et Minières. Orléans, France.

Rasplus, L., 1982, Contribution à l'étude géologique des formations continentales détritiques tertiaires de la Touraine, de la Brenne et de la Sologne. Thèse condensée. Sciences géologiques, Univ. Strasbourg, Memoir 66, 227 p.

Rasplus, L., J. J. Macaire, and G. Alcayde, 1982, Carte géologique de la France à 1/50000. Feuille Bléré et notice explicative. Bureau de Recherches Géologiques et Minières. Orléans, France.

Rasplus, L., 1987, Anjou, Maine, Touraine et Brenne : la marge sud-ouest du Bassin Parisien. Bull. Information des Géologues du Bassin de Paris, France, Memoir v. 6, p 181–202.

Rasplus, L., G. Alcayde, G. Lablanche, and J. J. Macaire, 1989a, Carte géologique de la France à 1/50000. Feuille Buzançais et notice explicative. Bureau de Recherches Géologiques et Minières. Orléans, France.

Rasplus, L., J. Lorenz, C. Lorenz, and J. J. Macaire, 1989b, Carte géologique de la France à 1/50000. Feuille Saint-Gaultier et notice explicative. Bureau de Recherches Géologiques et Minières. Orléans, France.

Saugrin, T., 1982, Les calcaires lacustres de Touraine :

Table 2. Fossils of the Touraine Basin Limestones.

Gastropods (Dollfus, G.F., 1904; Lecointre, G., 1947; Rey, R. in Rasplus, L., 1978)	**Charophytes** (Riveline, J., in Rasplus L., 1978)
Bithinella terebra (may be Bithinia monthiersi)	*Chara oehlerti*
Bithinella rhodanica	*Gravesichara distorta*
Bithinia epiedensis	*Gyriogona medicaginula*
Bithinia subrotunda (ex Bithinia monthiersi)	*Gyrogona cf wrighti*
Bithinia ugernansis	*Harrisichara tuberculata*
Galba longiscata ostrogallica	*Rhabdochara cf stockmansi*
Gyraulus acuticarinatus	*Stephanochara sp.*
Hydrobia elongta	*Sphaerochars sp.*
Hydrobia infalta	*Tolypella sp.*
Lymnea fusiformis (= L. Pseudopyramidalis)	
Lymnea morini	
Nystia duchasteli	
Nystia plicata	
Palaeoxestina serpentinites	
Planobarius cornu	
Planorbis planulatus (= P. lens)	
Planorbis lendonensis	
Planorbis depresus	
Pseudoamnicola oxyspiriformis	
Radix ioutonensis	
Radix fabulum	
Radix briarensis	
Sphaerium berthereaue	
Stenothyrella nicolasi	
Ludian	**Late Ludian** (upper part of Lower Bembridge charophyte zone)

sédimentation et diagénèse. Thèse 3rd cycle, Univ. d'Orléans, 248 p.

Valleron, M., and N. Dulau, 1983, Calcitisations et opalisations dans l'Eocène du Sud-Est de la France. Comparaison avec des faciès analogues d'Alsace et de Touraine. Bull. Soc. Géol. France (7th series), v. 25 (1), p. 11–18.

Vatan, A., 1947, La sédimentation continentale tertiaire dans le bassin de Paris méridional. Editions toulousaines de l'Ingénieur, Toulouse, France, 215 p.

Evans, J. E., and L. C. Welzenbach, 2000, Lacustrine limestones and tufas in the Chadron Formation (late Eocene), Badlands of South Dakota, U.S.A., *in* E. H. Gierlowski-Kordesch and K. R. Kelts, eds., Lake basins through space and time: AAPG Studies in Geology 46, p. 349–358.

Chapter 31

Lacustrine Limestones and Tufas in the Chadron Formation (Late Eocene), Badlands of South Dakota, U.S.A.

James E. Evans
Department of Geology, Bowling Green State University
Bowling Green, Ohio, U.S.A.

Linda C. Welzenbach
National Museum of Natural History, Public Programs Sector, Smithsonian Institution
Washington DC, U.S.A.

GEOLOGIC SETTING

Regional Geology

The geology of western South Dakota and the surrounding areas consists of Precambrian basement overlain by a thick succession of Paleozoic and Mesozoic sedimentary rocks, terminating in the Late Cretaceous Pierre Shale. Early Cenozoic weathering led to the formation of the Yellow Mounds paleosol series on the exposed marine sedimentary rocks (Retallack, 1983). Subsequent early Cenozoic uplift and unroofing of the Black Hills, as well as incision and backfill in adjacent areas, led to the accumulation of up to 250 m of nonmarine sediments of the Eocene–Oligocene White River Group in the study area (Clark et al., 1967; Stanley, 1976; Evans and Terry, 1994; Terry et al., 1995; Evans, 1996; LaGarry, 1998; Terry, 1998; Terry and LaGarry, 1998; Evans and Welzenbach, 1998).

White River Group

The study area is located in southwestern South Dakota, east of the Black Hills (Figure 1). In this region, the White River Group consists of the Chamberlain Pass, Chadron, and Brule Formations (Figure 2). The Chamberlain Pass Formation is a newly recognized unit more than 16 m thick that is comprised of multistory channel deposits and overbank mudstones (Terry, 1991, 1998; Evans and Terry, 1994). The overbank mudstones have been modified by pedogenesis to form the interior paleosol series (Retallack, 1983), whereas the channel sandstones have been pedogenically altered to form the Weta paleosol series (Terry and Evans, 1994).

The overlying Chadron Formation is about 100 m thick and fills and overtops a more than 70 m paleovalley incised through the Chamberlain Pass Formation and into the Pierre Shale (Harksen and Macdonald, 1969; Evans and Terry, 1994; Terry, 1998). Inside the paleovalley the Chadron Formation is subdivided into the Ahearn, Crazy Johnson, and Peanut Peak Members (Clark, 1954). Outside the paleovalley the Chadron Formation is finer grained and was undifferentiated by Clark et al. (1967), but has recently been assigned to the Peanut Peak Member (Terry et al., 1995; Terry, 1998). In Nebraska, the Peanut Peak Member is overlain by the newly defined Big Cottonwood Creek Member (Terry and LaGarry, 1998).

The Brule Formation is the upper unit of the White River Group. The Brule Formation is about 140 m thick and consists of fluvial channel and overbank deposits with interstratified air-fall tuffs and paleosols (Wanless, 1923; Bump, 1956; Clark et al., 1967; Harksen and Macdonald, 1969; Retallack, 1983; LaGarry, 1998). Prominent "nodular zones" in the Brule Formation have been recently reinterpreted as fluvial sheet sandstones that were cemented after burial and later jointed, rather than being pedogenic concretions (Wells et al., 1995).

CARBONATES IN THE CHADRON FORMATION

Lithology

The Chadron Formation is almost entirely fluvial, but several varieties of carbonates can be found in the unit, including lacustrine limestone, tufa, travertine, pedogenic calcrete, and nonpedogenic calcrete. The presence of limestone was long recognized (e.g., Wanless, 1923; Clark et al., 1967; Harksen and Macdonald, 1969), but sedimentological studies were not initiated until fairly recently (Welzenbach, 1992; Welzenbach and Evans, 1992; Evans and Welzenbach, 1998; Evans, 1998).

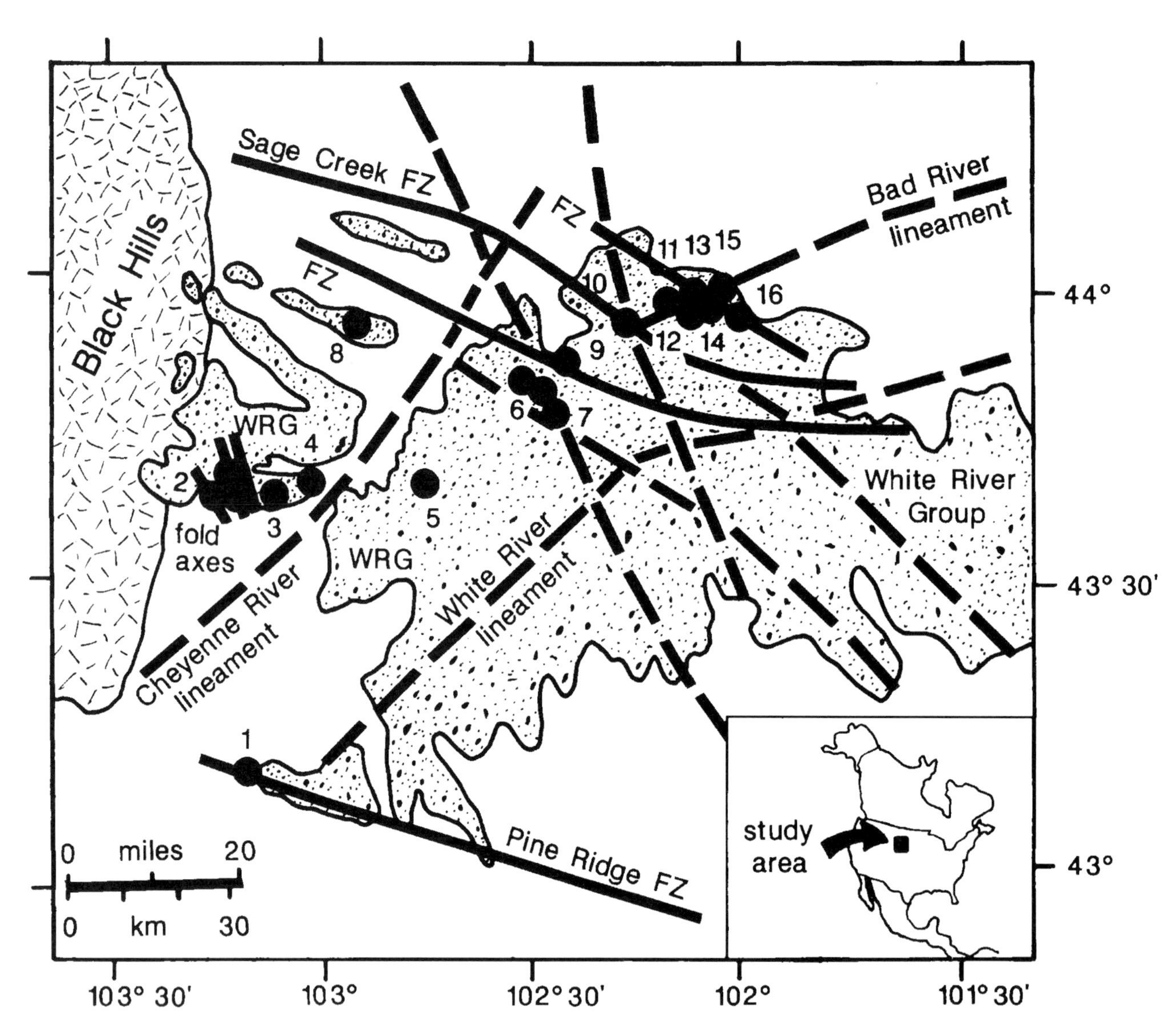

Figure 1—Map showing the location of field sites east of the Black Hills (in the Badlands of South Dakota). Numbers refer to (1) Limestone Butte, (2) Fairburn area, (3) French Creek, (4) Red Shirt table, (5) Cottonwood Pass, (6) Bear Creek, (7) Scenic, (8) Railroad Buttes, (9) Beaver Creek, (10) Sage Creek, (11) Mile Post 114, (12) Bloom Basin, (13) Mile Post 118, (14) Mile Post 121, (15) Street Hill, and (16) Walker Hill (see Evans and Welzenbach, 1998, for location data). Structural features shown include faults (solid line), fold axis fracture systems (solid line as marked), and LANDSAT lineaments (dashed lines). Note correspondence of paleogroundwater deposits and structural features (from Evans, 1999).

The calcretes in the Chadron Formation have been a source of controversy with regard to their pedogenic versus groundwater origin (Retallack, 1983; Lander, 1991). Although some of the calcretes show characteristic pedogenic morphologies (Retallack, 1983; Evans and Welzenbach, 1998; Evans, 1999), other calcretes are structureless. Lander (1991) showed that stable isotope compositions of some calcretes indicate mixing of meteoric and groundwater sources.

Tufas, travertines, and other paleogroundwater features have only recently been recognized in the Chadron Formation (Evans, 1996, 1998, 1999). Such features are locally abundant and diverse. Using the classification system of Ford and Pedley (1996), the Chadron Formation contains fluvial tufas (braided stream and barrage types), lacustrine tufas, perched springline tufas, paludal tufas, and isolated travertine resurgences with rimstone (Evans, 1999). Characteristic features in the tufas include pisolites, oncolites, coated grains, bacterioherms (stromatolites), phytoclast accumulations, and travertine banding. It is argued elsewhere that all of the carbonate in the Chadron Formation should be evaluated using a tufa (or paleogroundwater) facies model (Evans, 1999).

Stratigraphy and Age

The age of the Chadron Formation is late Eocene (about 37–33.7 Ma), based on $^{40}Ar/^{39}Ar$ dating of tuffs (Swisher and Prothero, 1990) and magnetostratigraphy (Prothero and Swisher, 1992; Prothero and Whittlesey,

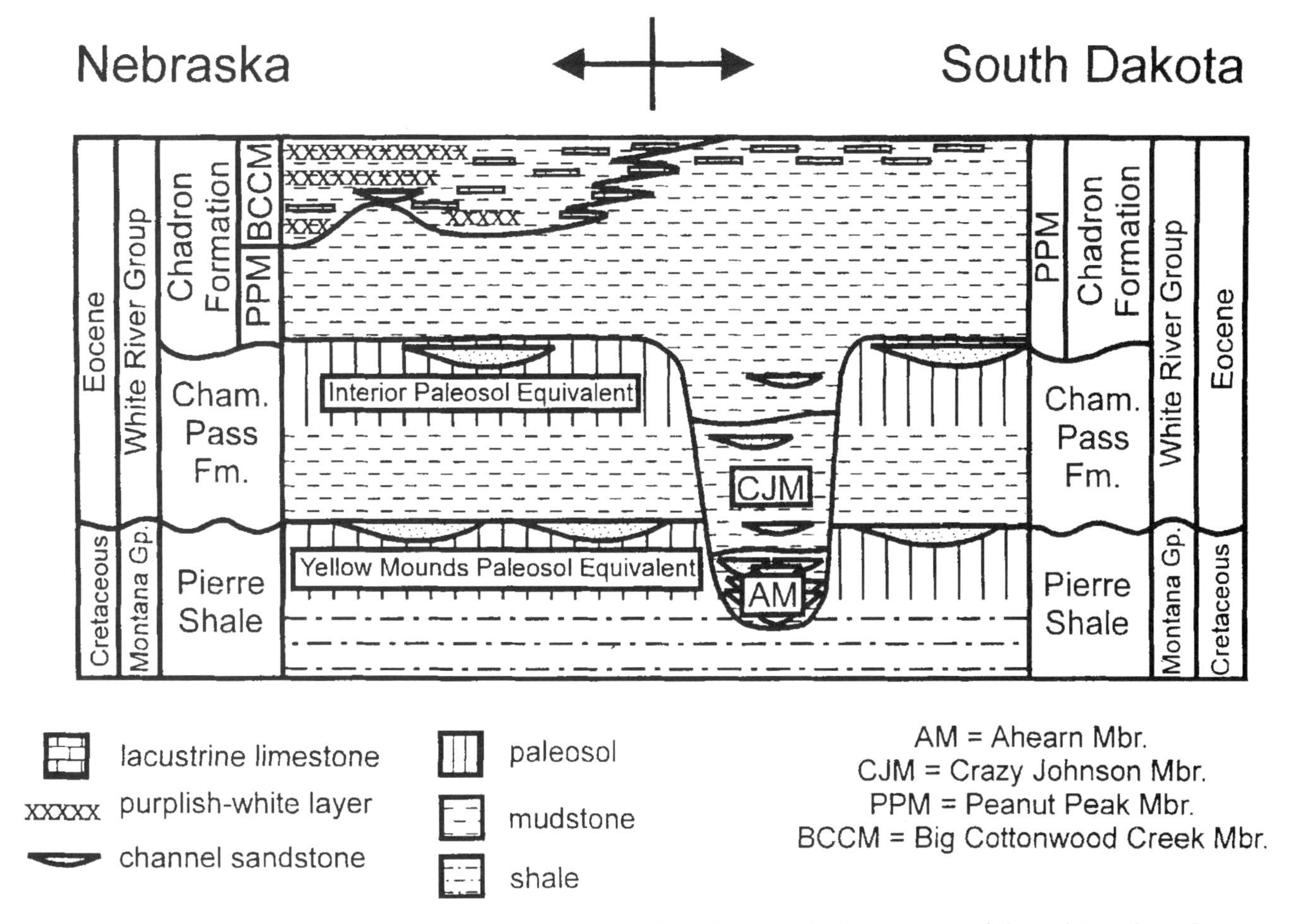

Figure 2—Schematic diagram showing stratigraphic relationships in the lower part of the White River Group in Nebraska and South Dakota (from Terry, 1998). Cham. Pass Fm. means Chamberlain Pass Formation.

1998). None of the carbonates have been directly dated, but southwest of Badlands National Park, they are interstratified with rocks that are in the upper part of Chron 13R, close to the Chron 13R–13N (Eocene–Oligocene) boundary (Prothero and Swisher, 1992; Prothero and Whittlesey, 1998).

Carbonates in the Chadron Formation are locally abundant and laterally discontinuous (Figure 3). Localized occurrences of abundant, vertically stacked carbonate deposits are mostly found in fault zones, fracture systems, or along the trend of paleovalleys (Figure 1). Evidence is presented elsewhere that these structural features were syndepositional or older (Evans, 1998, 1999). It is hypothesized that the tufas, travertines, and limestones are found on structure because their origin is linked to paleogroundwater flow and discharge at springs and seeps (Evans, 1999).

Limestones and tufas often form the caprock of small buttes and mesa in the Badlands area, appearing to be one or more continuous stratigraphic horizons. In earlier work (Welzenbach, 1992; Welzenbach and Evans, 1992), we informally proposed to call some of these lacustrine limestones near Badlands National Park the "Bloom Basin Limestone Bed." However, as we have expanded the study area (and remapped the original study area), it has become apparent that the individual limestones and other carbonates are laterally discontinuous deposits typically less than 1 km^2 in area. Accordingly, we no longer use this name and recommend its abandonment.

Lithofacies

Welzenbach (1992) recognized two common field units—massive and laminated limestone. Later work expanded the field area (Evans and Welzenbach, 1998; Evans, 1998) and led to the following elaboration. The lacustrine limestones typically consist of a shallowing-upward succession of five facies (from bottom to top): (1) thinly bedded marl and sapropel, (2) stromatolitic limestone (bacterioherms), (3) laminated limestone ("algal mat"), (4) undulose-bedded (wave-reworked) limestone, and (5) charophyte-rich, massive limestone. These shallowing-upward sequences range from about 40 cm to 2.5 m thick (Figure 4). At many locations, only the upper two or three lithofacies can be observed. Also at many locations, repetitive shallowing-upward sequences can be vertically stacked (Evans, 1998). Other carbonate lithofacies that can be recognized in the field include tufa, travertine, and nonpedogenic calcrete (Evans, 1998).

Microfacies analysis has been an important part of interpreting units observed in the field. Microfacies in

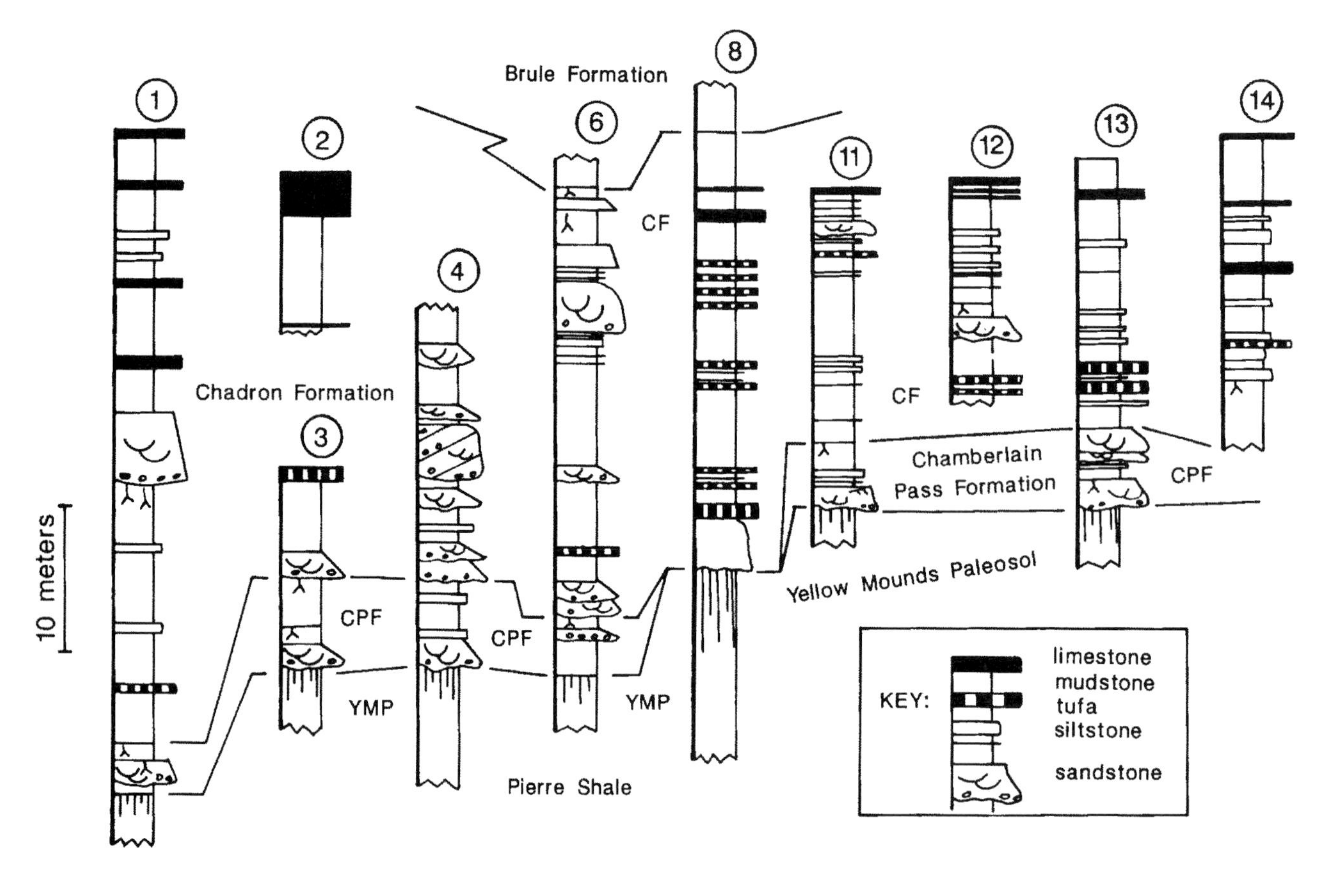

Figure 3—Stratigraphic correlation of sections discussed in this study (from Evans, 1999). Numbers refer to locations in Figure 1.

these deposits include porous and clotted micrite, dense micrite, ostracode-charophyte wackestone, oncolite-peloidal packstone, gastropod-charophyte-vertebrate floatstone, algal bindstone, and pisolite-intraclast wackestone (Evans, 1998).

Paleontology

Specimens identified include Amiidae and Centrachidae fish (*Plioplarchus* sp.), turtle plastron plates, gastropods (*Galba* sp. and *Planorbis* sp.), bivalve shell fragments, ostracodes (*Cypris* sp.), calcified roots of aquatic macrophyte, charophytes (stems and gyrogonites), and algal filaments (Welzenbach, 1992). The assemblage represents a high-abundance, low-diversity fauna with evidence of environmental stress, including the "fish-kill" horizons and death horizons of *Cypris* ostracodes (Alison J. Smith, written communication, 1991). Common trace fossils include *Taenidium, Skolithos, Palaophycus, Planolites,* and probable bird tracks (Evans and Welzenbach, 1998).

DISCUSSION

Depositional Environment

The lacustrine limestones are interpreted as shallowing-upward, carbonate pond deposits (Figure 5). Facies relationships suggest that these ponds were similar to low-energy, bench-margin carbonate lakes of Platt and Wright (1991). The Chadron lakes formed in a humid temperate paleoclimate and share many apparent similarities to other groundwater-fed hardwater lakes described from temperate regions (e.g., Dean, 1981; Brunskill, 1969; Eggleston and Dean, 1976; Murphy and Wilkinson, 1980; Treese and Wilkinson, 1982; McConnaughey et al., 1994; Smith et al., 1997). In addition, some of these modern groundwater-fed lakes, such as Elk Lake, Minnesota (Smith et al., 1997), are roughly equivalent to the inferred size and depth of the Chadron lakes.

The limestones are found in fault zones and are spatially associated with paleogroundwater deposits (various types of tufas, travertines, and nonpedogenic calcretes). Evidence is presented elsewhere that some of these carbonate ponds formed behind fluvial tufa barrages (Evans, 1999). The close spatial relationship between limestones and fluvial tufas suggests it is appropriate to evaluate the Chadron Formation limestones as lacustrine tufas (e.g., Ford and Pedley, 1996; Pedley et al. 1996). Other features of interest include weak development of shoreline features, lack of siliciclastic detritus in the limestones, presence of encrusted aquatic macrophytes, presence of spring resurgences, and "tufa breccias" interpreted as erosion of tufa barrages (Evans, 1999). These groundwater-fed depositional systems were isolated from adjacent, siliciclastic fluvial systems (Figure 6).

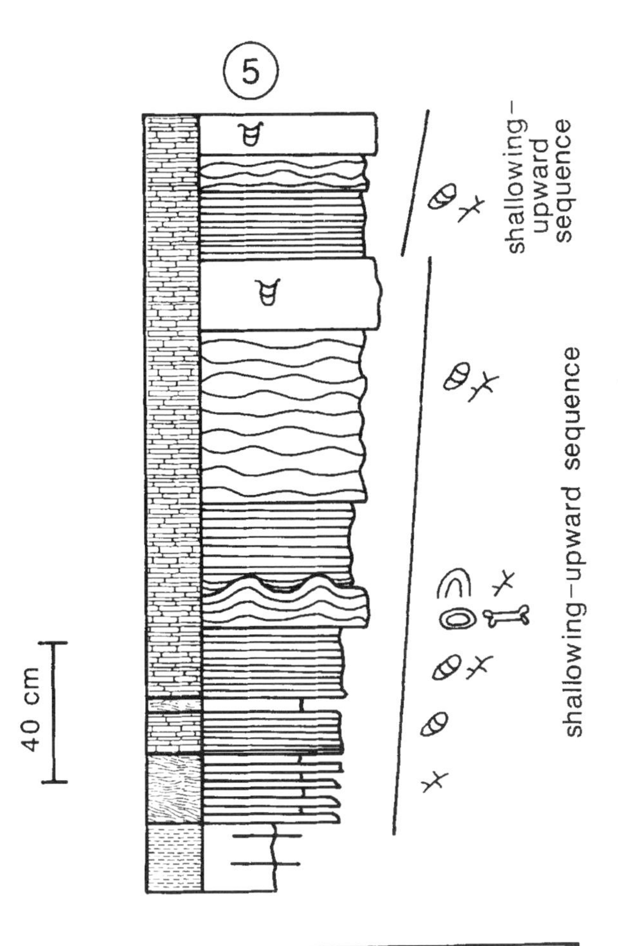

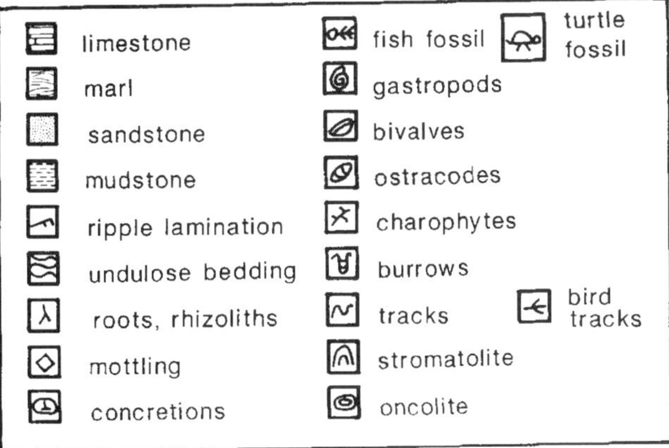

Figure 4—Stratigraphic section through two shallowing-upward carbonate lake sequences near Fairburn (location 5), South Dakota (from Evans, 1998).

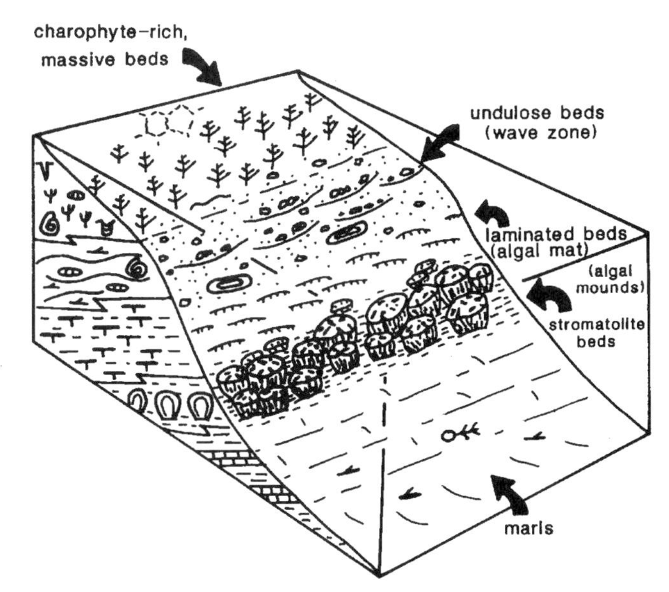

Figure 5—Block diagram showing facies relationships in the Chadron Formation carbonate lakes, an example of a low-energy, bench-margin type of freshwater carbonate system (from Evans, 1998).

Implications

The formation of carbonates in the Chadron Formation tells an important tectonic and paleoclimatic story. Uplift of the Black Hills dome commenced during the Laramide orogeny (Figure 7A). Fission-track data indicates that unroofing of the Black Hills dome was greatest during the early Eocene (Strecker, 1996). By the late Eocene, erosion of approximately 2.1–2.5 km of sedimentary section exposed Paleozoic carbonate bedrock units and the underlying Precambrian crystalline rocks (Evans, 1996). During the late Eocene, karstification of the Paleozoic carbonates in the Black Hills, under a humid paleoclimate, generated carbonate-rich groundwaters that flowed along the structural features described and discharged into the Chadron depositional environment, precipitating tufas, travertines, and lacustrine limestones near springs (Figure 7B) (Evans and Welzenbach, 1998; Evans, 1998, 1999). In other words, the tectonic evolution of the Black Hills uplift provided the source for carbonate, a topographically elevated recharge area, and established the hydraulic gradients and the groundwater centripetal flow system that persists to the present day.

The carbonate pond deposits formed low-energy, bench-margin type lakes in regions near springs, or upstream of fluvial tufa barrages. Each carbonate pond was relatively small and short lived. Pond faunas were high-abundance, low-diversity assemblages, with evidence of environmental stress. In some cases, evidence suggests that ponds dried up or were dewatered suddenly (perhaps due to erosion of a fluvial tufa barrage downstream). The presence of these ponds is in accord with terrestrial faunal and paleosol morphologies indicative of a humid, temperate paleoclimate in this region.

Carbonate production is abruptly reduced in the overlying lower Oligocene Brule Formation. These depositional changes coincide with well-documented paleoclimatic trends toward increased aridity in this region across the Eocene–Oligocene boundary, including changes in paleosols (Retallack, 1992), land snail faunas (Evanoff et al., 1992), and reptile and amphibian faunas (Hutchison, 1992). It is probable that increasing aridity affected regional recharge-discharge relationships, flow rates, water table positions, and

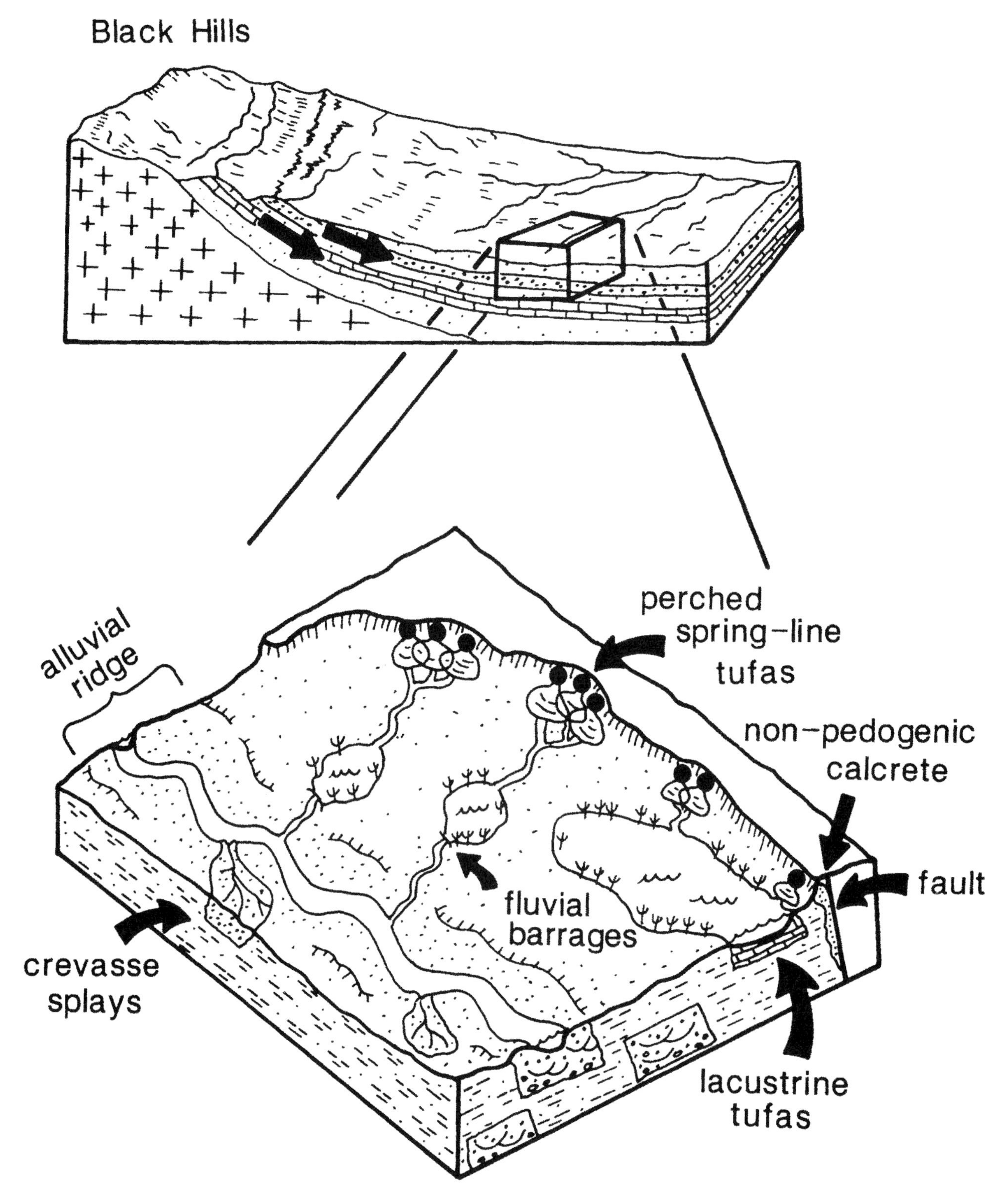

Figure 6—Facies model for tufa and travertine deposition in the late Eocene Chadron Formation showing paleogroundwater flow system and recharge (top), and discharge in the Chadron depositional basin along structural features (bottom) (from Evans, 1999).

rates of carbonate precipitation (Evans and Welzenbach, 1998; Evans, 1998, 1999). Similar mechanisms have been called upon to explain changes in Holocene tufa production from a variety of climatic settings (e.g., McGannon, 1975; Heimann and Sass, 1989; Goudie et al., 1993; Pedley et al., 1996). The modern regional groundwater flow system (Figure 7C), under an arid climate, includes deep aquifer recharge (rather than discharge) along the structural

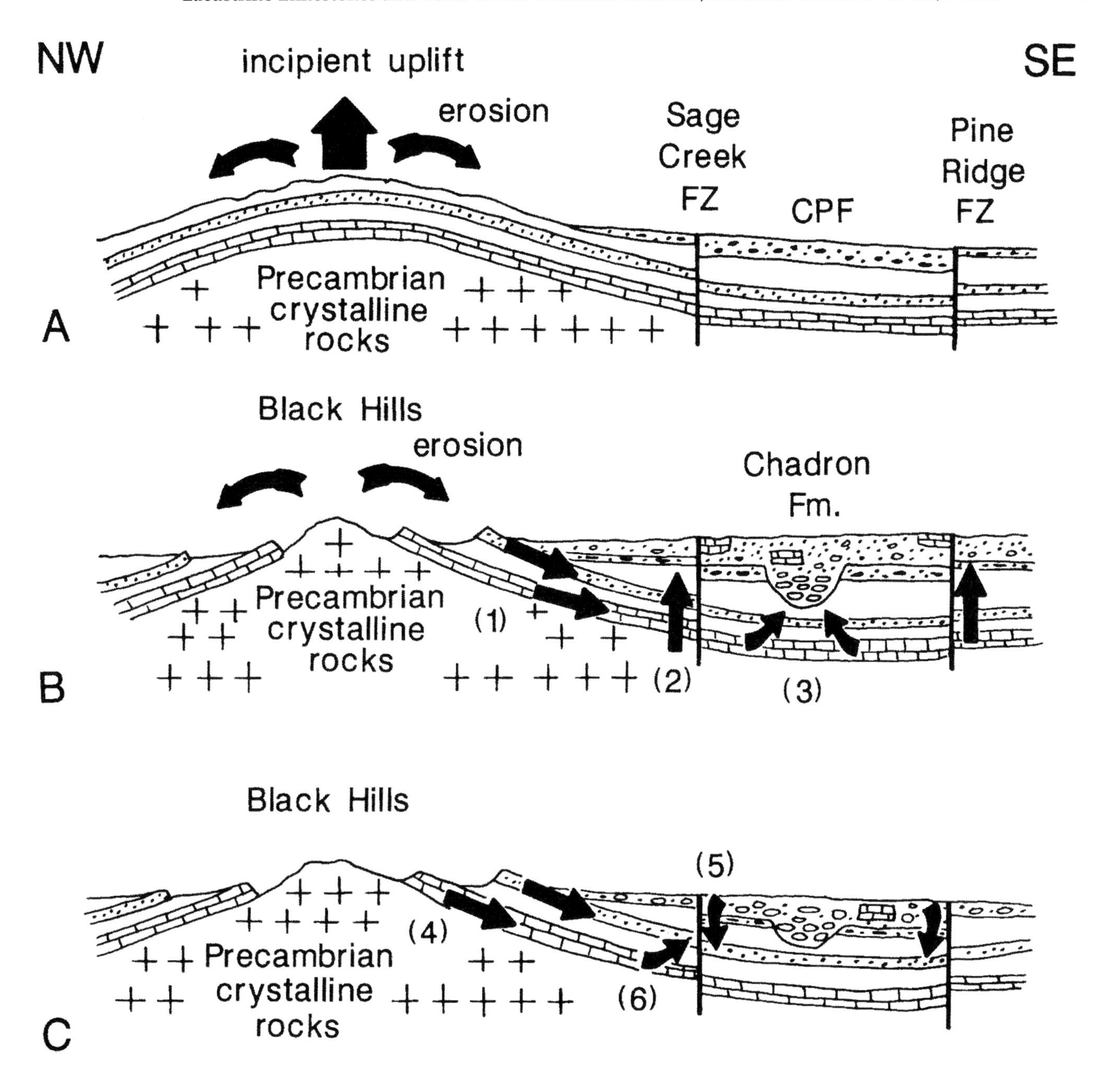

Figure 7—Diagram showing evolution of the paleohydrologic system (from Evans, 1999). (A) Late Eocene unroofing of the Black Hills uplift; age is indicated by changes in sandstone petrology in the White River Group. (B) Late Eocene flow systems showing source area in the karsted Paleozoic limestones exposed near the Black Hills crystalline core. (C) Modern flow systems (modified from Rahn and Hayes, 1996). Numbers refer to (1) recharge in the Black Hills and flow through the Dakota Sandstone (upper aquifer) and Pahasapa Limestone-Minnelusa Formation (lower aquifer); (2) discharge along faults and fractures; (3) discharge into paleovalleys; (4) modern recharge and flow systems; (5) recharge of the Dakota Sandstone aquifer from the Pahasapa Limestone-Minnelusa Formation aquifer along faults and fractures; and (6) surface seepage recharge of the Dakota Sandstone aquifer through the Cretaceous shale aquicludes along faults and fractures.

features that had served to localize carbonate deposition in the Chadron Formation (Evans, 1998, 1999).

ACKNOWLEDGMENTS

We wish to thank Alison J. Smith, Richard M. Forester, Richard Hoare, Gerald R. Smith, Scott Foss, and John Howe for their assistance with paleontological identifications. We thank Charles F. Kahle for assistance with carbonate microfacies interpretations and Dennis O. Terry, Jr. for assistance in the field. An earlier manuscript was improved by reviews from Blas L. Valero Garces and Elizabeth Gierlowski-Kordesch. Finally, we greatly appreciate the efforts by the staffs of Badlands National Park and Buffalo Gap National

Grasslands for collecting permits, logistical support, and landowner contacts.

REFERENCES CITED

Brunskill, G. J., 1969, Fayetteville Green Lake, New York. II. Precipitation and sedimentation of calcite in a meromictic lake with laminated sediments: Limnology & Oceanography, v. 14, p. 830–847.

Bump, J. D., 1956, Geographic names for the members of the Brule Formation in the Big Badlands of South Dakota: American Journal of Science, v. 254, p. 429–432.

Clark, J., 1954, Geographic designation of the members of the Chadron Formation in South Dakota: Annual Reports of the Carnegie Museum, v. 33, p. 197–198.

Clark, J., J. P. Beerbower, and K. K. Kietzke, 1967, Oligocene sedimentation, stratigraphy, paleoecology, and paleoclimatology in the Big Badlands of South Dakota: Fieldiana Geology Memoir 5, 158 pp.

Dean, W. E., 1981, Carbonate minerals and organic matter in sediments of modern temperate hardwater lakes, *in* F. G. Ethridge and R. M. Flores, editors, Recent and Ancient Nonmarine Depositional Environments: Models for Exploration: SEPM Special Publication 31, p. 213–231.

Egglestone, J. R., and W. E. Dean, 1976, Freshwater stromatolitic bioherms in Green Lake, New York, *in* M. R. Walter, editor, Stromatolites, Developments in Sedimentology no. 20: New York, Elsevier, p. 479–488.

Evanoff, E., D. R. Prothero, and R. H. Lander, 1992, Eocene–Oligocene climatic changes in North America: The White River Formation near Douglas, east-central Wyoming, *in* D. R. Prothero and W. A. Berggren, editors, Eocene–Oligocene Climatic and Biotic Evolution: Princeton University Press, Princeton, N.J., p. 116–130.

Evans, J. E., 1996, Evidence for late Eocene unroofing of the Black Hills uplift [abstract]: Geological Society of America, Abstracts with Programs, v. 28, p. A373.

Evans, J. E., 1998, Facies associations of freshwater carbonates in the Eocene–Oligocene Chadron Formation, White River Group, Nebraska and South Dakota, U.S.A., *in* J. K. Pittman and A. Carroll, editors, Modern and Ancient Lakes: Utah Geological Association Special Publication, Guidebook 26, p. 209–231.

Evans, J. E., 1999, Recognition and implications of Eocene tufas and travertines in the Chadron Formation, White River Group, Badlands of South Dakota: Sedimentology, v. 46, p. 771–789.

Evans, J. E., and D. O. Terry, 1994, The significance of incision and fluvial sedimentation in the basal White River Group (Eocene–Oligocene), Badlands of South Dakota: Sedimentary Geology, v. 90, p. 137–152.

Evans, J. E. and L. C. Welzenbach, 1998, Episodes of carbonate deposition in a siliciclastic-dominated fluvial sequence, Eocene–Oligocene White River Group, South Dakota and Nebraska, U.S.A., *in* D. O. Terry, Jr., H. E. LaGarry, and R. M. Hunt, Jr., editors, Depositional Environments, Lithostratigraphy, and Biostratigraphy of the White River and Arikaree Groups (Late Eocene to Early Miocene, North America): Geological Society of America, Special Paper 325, p. 93–116.

Ford, T. D., and H. M. Pedley, 1996, A review of tufa and travertine deposits of the world: Earth Science Reviews, v. 41, p. 117–175.

Goudie, A. A., H. A. Viles, and A. Pentecost, 1993, The late Holocene tufa decline in Europe: Holocene, v. 3, p. 181–186.

Harksen, J. C., and J. R. Macdonald, 1969, Type section for the Chadron and Brule formations of the White River Oligocene in the Big Badlands, South Dakota: Report of Investigations, South Dakota Geological Survey, no. 99, 23 pp.

Heimann, A., and E. Sass, 1989, Travertines in the northern Hula Valley, Israel: Sedimentology, v. 36, p. 95–108.

Hutchison, J. H., 1992, Western North America reptile and amphibian record across the Eocene/Oligocene boundary and its climatic implications, *in* D. R. Prothero and W. A. Berggren, editors, Eocene–Oligocene Climatic and Biotic Evolution: Princeton University Press, Princeton, N.J., p. 149–164.

LaGarry, H. E., 1998, Lithostratigraphic revision and redescription of the Brule Formation (White River Group) of northwestern Nebraska, *in* D.O. Terry, Jr., H. E. LaGarry, and R. M. Hunt, Jr., editors, Depositional Environments, Lithostratigraphy, and Biostratigraphy of the White River and Arikaree Groups (Late Eocene to Early Miocene, North America): Geological Society of America, Special Paper 325, p. 63–91.

Lander, R. H., 1991, White River Group diagenesis [Ph.D. dissertation]: University of Illinois, Urbana-Champaign, IL, 143 pp.

McConnaughey, T., J. W. LaBaugh, D. O. Rosenberry, R. G. Striegl, M. M. Reddy, P. F. Schuster, and V. Carter, 1994, Carbon budget for a groundwater-fed lake calcification supports summer photosynthesis: Limnology & Oceanography, v. 39, p. 1319–1332.

McGannon, D. E., 1975, Primary fluvial oolites: Journal of Sedimentary Petrology, v. 45, p. 719–727.

Murphy, D. H., and B. H. Wilkinson, 1980, Carbonate deposition and facies distribution in a central Michigan marl lake: Sedimentology, v. 27, p. 123–135.

Pedley, H. M., J. Andrews, A. Ordonez, M. A. Garcia del Cura, J.-A. Gonzales Martin, and D. Taylor, 1996, Does climate control the morphological fabric of freshwater carbonates? A comparative study of Holocene barrage tufas from Spain and Britain: Palaeogeography, Palaeoclimatology, Palaeoecology, v. 121, p. 239–257.

Platt, N. H., and V. P. Wright, 1991, Lacustrine carbonates: facies models, facies distributions, and hydrocarbon aspects, *in* P. Anadon, Ll. Cabrera, and K. Kelts, editors, Lacustrine Facies Analysis: IAS Special Publication 13, p. 57–74.

Prothero, D. R., and C. C. Swisher III, 1992, Magnetostratigraphy and geochronology of the terrestrial Eocene–Oligocene transition in North America, *in* D. R. Prothero and W. A. Berggren, editors, Eocene–Oligocene Climatic and Biotic Evolution: Princeton University Press, Princeton, N.J., p. 46–73.
Prothero, D. R., and K. E. Whittlesey, 1998, Magnetic stratigraphy and biostratigraphy of the Orellan and Whitneyan land mammal "ages" in the White River Group, *in* D. O. Terry, Jr., H. E. LaGarry, and R. M. Hunt, Jr., editors, Depositional Environments, Lithostratigraphy, and Biostratigraphy of the White River and Arikaree Groups (Late Eocene to Early Miocene, North America): Geological Society of America, Special Paper 325, p. 39–61.
Rahn, P. H., and T. S. Hayes, 1996, Hydrogeology of the central Black Hills: Road Log, Field Trip 3, *in* C. J. Paterson and J. G. Kirchner, editors, Guidebook to the Geology of the Black Hills, South Dakota: South Dakota School of Mines and Technology Bulletin, v. 19, p. 19–29.
Retallack, G. J., 1983, Late Eocene and Oligocene fossil soils from Badlands National Park, South Dakota: Geological Society of America, Special Paper 193, 82 pp.
Retallack, G. J., 1992, Paleosols and changes in climate and vegetation across the Eocene/Oligocene boundary, *in* D. R. Prothero and W. A. Berggren, editors, Eocene–Oligocene Climatic and Biotic Evolution: Princeton University Press, Princeton, N.J., p. 382–398.
Smith, A. J., J. J. Donovan, E. Ito, and D. R. Engstrom, 1997, Groundwater processes controlling a prairie lake's response to middle Holocene drought: Geology, v. 25, p. 391–394.
Stanley, K. O., 1976, Sandstone petrofacies in the Cenozoic high plains sequence: eastern Wyoming and Nebraska: Geological Society of America Bulletin, v. 87, p. 279–309.
Strecker, U., 1996, Studies on Cenozoic continental tectonics: Part I. Apatite fission-track thermochronology of the Black Hills uplift, South Dakota [Ph.D. Dissertation]: University of Wyoming, Laramie, Wyoming, 69 pp.
Swisher, C. C., III, and D. R. Prothero, 1990, Laser-fusion $^{40}Ar/^{39}Ar$ dating of the Eocene–Oligocene transition in North America: Science, v. 249, p. 760–762.
Terry, D. O., 1991, The study and implications of comparative pedogenesis of sediments from the White River Group, South Dakota [M.S. Thesis]: Bowling Green State University, Bowling Green, Ohio, 183 pp.
Terry, D. O., 1998, Lithostratigraphic revision and correlation of the lower part of the White River Group: South Dakota to Nebraska, *in* D. O. Terry, Jr., H. E. LaGarry, and R. M. Hunt, Jr., editors, Depositional Environments, Lithostratigraphy, and Biostratigraphy of the White River and Arikaree Groups (Late Eocene to Early Miocene, North America): Geological Society of America, Special Paper 325, p. 15–37.
Terry, D. O., and J. E. Evans, 1994, Pedogenesis and paleoclimatic implications of the Chamberlain Pass Formation, basal White River Group, Badlands of South Dakota: Palaeogeography, Palaeoclimatology, Palaeoecology, v. 110, p. 197–215.
Terry, D. O., H. E. LaGarry, and W. B. Wells, 1995, The White River Group revisited: vertebrate trackways, ecosystems, and lithostratigraphic revision, redefinition, and redescription, *in* R. F. Diffendal and C. A. Flowerday, editors, Geologic Field Trips in Nebraska and Adjacent Parts of Kansas and South Dakota: Nebraska Conservation and Survey Division, Guidebook, v. 10, p. 43–57.
Terry, D. O., and H. E. LaGarry, 1998, The Big Cottonwood Creek Member: A new member of the Chadron Formation in northwestern Nebraska, *in* D. O. Terry Jr., H. E. LaGarry, and R. M. Hunt, Jr., editors, Depositional Environments, Lithostratigraphy, and Biostratigraphy of the White River and Arikaree Groups (Late Eocene to Early Miocene, North America): Geological Society of America, Special Paper 325, p. 117–141.
Treese, K. L., and B. H. Wilkinson, 1982, Peat-marl deposition in a Holocene paludal lacustrine basin—Sucker Lake, Michigan: Sedimentology, v. 29, p. 375–390.
Wanless, H. R., 1923, The lithology and stratigraphy of the White River beds of South Dakota: American Philosophical Society Proceedings, v. 62, p. 190–269.
Wells, W. B., D. O. Terry, Jr., and H. E. LaGarry, 1995, Stratigraphic implications of a fluvial origin for the "nodular zones," Brule Formation (Orella Member), northwestern Nebraska [abstract]: Geological Society of America, Abstracts with Programs, v. 27 (3), p. 95.
Welzenbach, L. C., 1992, Limestones in the Lower White River Group (Eocene–Oligocene), Badlands of South Dakota: Depositional environment and paleoclimatic implications [M.S. Thesis]: Bowling Green State University, Bowling Green, Ohio, 131 pp.
Welzenbach, L. C., and J. E. Evans, 1992, Sedimentology and paleolimnology of an extensive lacustrine unit, White River Group (Eocene–Oligocene), South Dakota [abstract]: Geological Society of America, Abstracts with Programs, v. 24 (6), p. A54.

Evans, J. E., 2000, Lacustrine facies in an Eocene wrench-fault step-over basin, Cascade Range, Washington, U.S.A. *in* E. H. Gierlowski-Kordesch and K. R. Kelts, eds., Lake basins through space and time: AAPG Studies in Geology 46, p. 359–368.

Chapter 32

Lacustrine Facies in an Eocene Wrench-Fault Step-Over Basin, Cascade Range, Washington, U.S.A.

James E. Evans
Department of Geology, Bowling Green State University
Bowling Green, Ohio, U.S.A.

GEOLOGIC SETTING

Regional Geology

The Nahahum Canyon Member (late middle Eocene) is a lacustrine and fluvial unit 1.9 km thick that was deposited in a releasing bend (step-over basin) between two parallel, partly overlapping, dextral oblique-slip faults in central Washington state (Figure 1). The Nahahum Canyon Member is part of the middle–late Eocene Chumstick Formation, which is more than 12 km thick and is additionally subdivided into the Clark Canyon Member, Tumwater Mountain Member, and Deadhorse Canyon Member (Evans, 1994).

The pre-Tertiary bedrock in this region consists of several major lithostratigraphic terranes that became joined to North America as part of late Mesozoic accretion (Tabor et al., 1982, 1987). The Chumstick Basin formed above the suture zone between two of these major terranes. The Chumstick Basin initially formed a half-graben open to the west, but the onset of oblique-slip faulting in this region at about 44–42 Ma resulted in partitioning of the basin and development of both open-basin and closed-basin drainages, with sediment derived from sources both east and west. By the late Eocene, tectonic relief in the fault zones had been reduced, and fluvial systems flowed across earlier structures. The basin was deformed in a zone of dextral transpression during 37–34 Ma (Evans, 1994).

The Chumstick Formation disappears to the southeast beneath the extensive flows of the Miocene Columbia River Basalt. Similar deposits have been found at distances up to 150 km away in exploration wells through the basalt (Lingley and Walsh, 1986). Lacustrine facies of the Chumstick Formation may be the source rocks for significant natural gas resources beneath the Columbia Plateau, whereas the adjacent fluvial facies may serve as reservoirs.

Chumstick Formation

The basin is divided into an eastern subbasin (between the Entiat fault zone and Eagle Creek fault zone) and a western subbasin (between the Eagle Creek fault zone and the Leavenworth fault zone). Each basin had a separate depositional history, as indicated by radiometric and paleobotanical age relationships, facies patterns (Figure 2), paleocurrent (Figure 3) and provenance relationships, and thermal maturity indicators (Evans, 1994; Evans and Johnson, 1989).

The western subbasin opened at prior to 51 Ma and was initially filled by the Clark Canyon Member (Evans, 1994). This unit consists of gravel- and sand-bedload stream deposits that are interpreted as part of a humid-region alluvial-fan system (Evans, 1991a). Paleobotanical data indicate that the paleoclimate supported montane tropical rainforest in the proximal parts of the fan and riparian vegetation along channels and on the rapidly shifting alluvial braidplain in the distal parts of the fan (Evans, 1991b). The fluvial depositional system flowed generally southwest (Figure 3), at times forming continuous depositional systems with deposits in the Swauk Formation and at other times flowing parallel to the Leavenworth fault zone (i.e., flowing locally southeast), depending on episodes of tectonic uplift in the fault zone (Evans, 1994; Evans and Johnson, 1989).

A major change of facies patterns, paleocurrent directions, and provenance relationships occurred starting at about 44–42 Ma (Figure 3). Sustained uplift in the Leavenworth fault zone disrupted drainage, formed new source areas, and resulted in coarse-grained fluvial systems that flowed eastward into the western subbasin (the Tumwater Mountain Member). At the same approximate interval, the older, southwestward-flowing fluvial systems of the Clark Canyon Member were beheaded (i.e., detached from their source areas) by the formation of the eastern subbasin. The formation of the eastern subbasin resulted

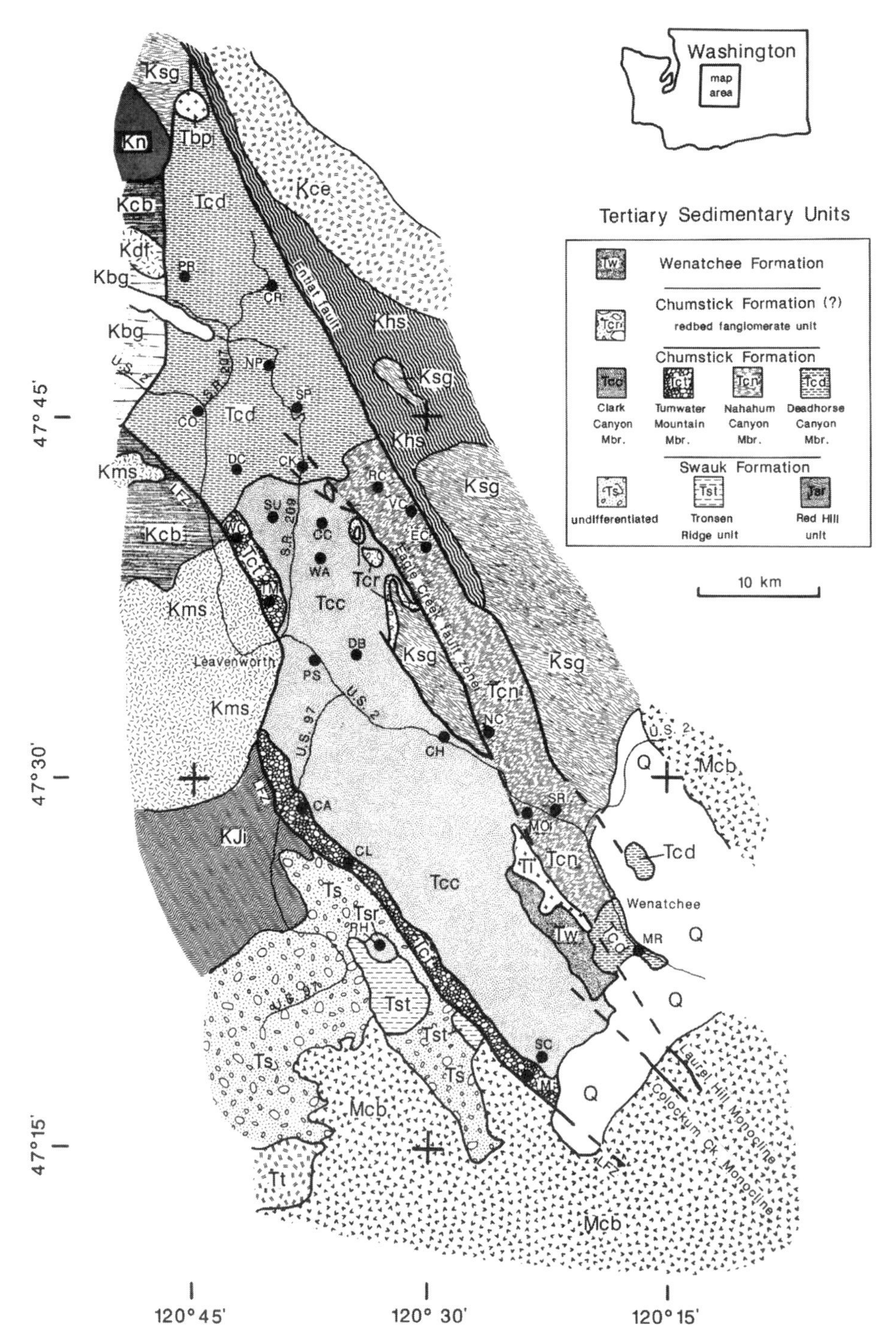

Figure 1—Geologic map of the Chumstick Basin and related rocks, showing locations of measured sections (from Evans, 1994). The lacustrine facies discussed in this paper are in the southeastern part of the Nahahum Canyon Member (Tcn) outcrop area (locations MO and SR). Geologic units: Kbg = banded gneiss, Kcb = Chiwaukum schist, Kce = Entiat pluton, Kch = banded hornblende and biotite gneiss, Kdf = Dirty Face pluton, Khs = heterogeneous schist, Kms = Mt. Stuart batholith, Kn = rocks of the Napeequa River area, Ksg = Swakane biotite gneiss, KJi = Ingalls tectonic complex , Mcb = Columbia River basalt , Q = alluvium, Tbp = Basalt Peak pluton, Tc = Chumstick Formation, Tcc = Clark Canyon Member, Tcd = Deadhorse Canyon Member, Tcn = Nahahum Canyon Member, Tcr = redbed fanglomerate unit, Tct = Tumwater Mountain Member, Ti = Wenatchee dome, Ts = Swauk Formation, Tsr = Red Hill unit, Tst = Tronsen Ridge unit, Tt = Teanaway Formation, Tw = Wenatchee Formation. Locations: CA = Camas Creek, CC = Clark Canyon, CH = Cashmere, CK = Chumstick Creek, CL = Camasland, CO = Cole's Corner, CR = Chiwawa River, DB = Derby Canyon, DC = Deadhorse Canyon, EC = Eagle Creek, MA = Malaga Road, MO = Monitor, MR = Mission Ridge, NC = Nahahum Canyon, NP = North Plain, PR = Pole Ridge, PS = Peshastin, RC = Railroad Canyon, RH = Red Hill, SC = Squilchuck Canyon, SP = South Plain, SR = Sunnyslope Road, SU = Sunitsch Canyon, TM = Tumwater Mountain, VC = Van Canyon, WA = Walker Canyon, WC = Wright Canyon.

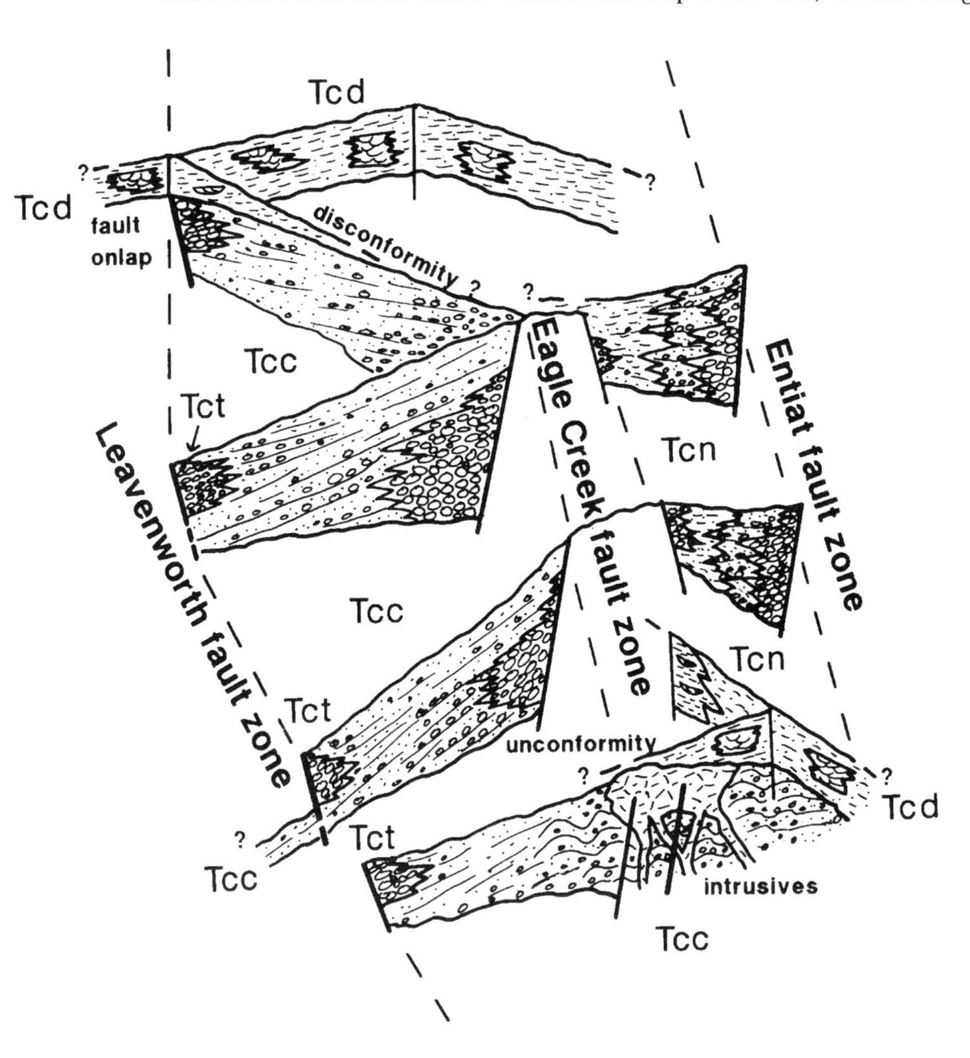

Figure 2—Schematic fence diagram showing stratigraphic relationships of the four members of the Chumstick Formation (from Evans, 1994). The lacustrine facies are in the southeastern part of the eastern subbasin, between the Eagle Creek fault zone and Entiat fault zone. Abbreviations used: Tcc = Clark Canyon Member, Tcd = Deadhorse Canyon Member, Tcn = Nahahum Canyon Member, Tct = Tumwater Mountain Member.

in streams flowing into the basin from both east and west and the development of an axial-drainage system that flowed southeast into a lacustrine-deltaic depositional system (the Nahahum Canyon Member). Dextral movement on all three fault zones would explain these events: (1) dextral movement on the left-stepping portions of the Leavenworth fault zone would result in transpression and uplift, while (2) dextral movement on the Entiat and Eagle Creek fault zones (two linear, parallel, and partly overlapping faults), would create a step-over basin in the position of the eastern subbasin.

The youngest part of the Chumstick Formation is fine-grained fluvial deposits (the Deadhorse Canyon Member) that disconformably overlie the three older units and show no evidence for tectonic control of facies or paleocurrent patterns (Figure 3). Basin deformation and erosion preceded prior to deposition of the lower Oligocene Wenatchee Formation.

NAHAHUM CANYON MEMBER

Nomenclature

The Nahahum Canyon Member was defined by Gresens et al. (1981) as a finer grained, fluvial and lacustrine unit found near the top of the Chumstick Formation. Evans (1988, 1994) has both geographically and stratigraphically restricted the Nahahum Canyon Member to refer to the fill of the eastern subbasin of the Chumstick Formation (Figure 2).

Stratigraphy and Age

The Nahahum Canyon Member comprises conglomerate, sandstone, and mudrocks with a net thickness of approximately 1900 m (Figure 4). The Nahahum Canyon Member is late middle Eocene in age because (1) the deposits are interstratified with flows and eruption breccias from a rhyodacite dome complex, which has numerous radiometric ages between 44 and 40 Ma, and (2) the Nahahum Canyon Member is partially overlain by the Deadhorse Canyon Member, which is late Eocene in age based on palynology (Evans, 1988, 1994; Gresens, 1983; Gresens et al., 1981; Margolis, 1987; Newman, 1981).

Lithofacies

There are five facies associations in the Nahahum Canyon Member: gravel-bedload stream deposits, sand-bedload stream deposits, mixed-load stream

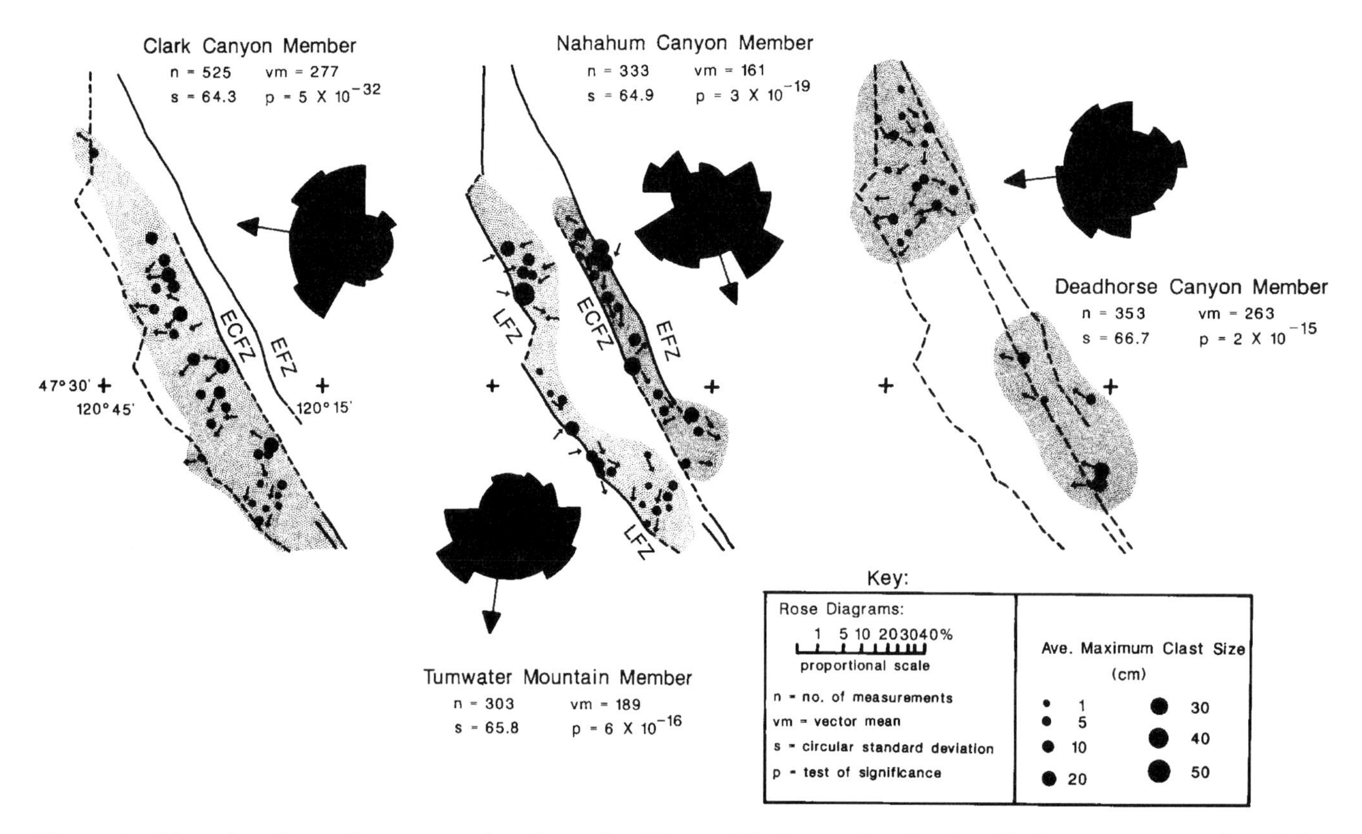

Figure 3—Map showing paleocurrent data from the Chumstick Formation for the Clark Canyon Member (left), Tumwater Mountain and Nahahum Canyon Members (center), and Deadhorse Canyon Member (right). The lacustrine facies are the southeastern portion of the Nahahum Canyon Member (center). Plots are on areally proportional rose diagrams (Nemec, 1988) showing vector mean (Curray, 1956), standard circular deviation (Krause and Geijer, 1987), and Rayleigh goodness-of-fit test (Curray, 1956).

deposits, delta-front deposits, and lake-basin deposits (Evans, 1991a, 1994). Gravel-bedload stream deposits consist of massive to stratified pebble-cobble conglomerates and planar-bedded sandstone, with minor crossbedded sandstone and rare mudrocks (Figure 4). These deposits are found as tabular bodies that are vertically stacked along the basin margins. Sand-bedload stream deposits consist of multistory channel sequences of trough-crossbedded sandstone, and related flood-basin deposits (Figure 4). Mixed-load stream deposits consist of single-story and multistory channel sequences of trough-crossbedded sandstone with minor laminated and rippled sandstone, interstratified with thick sequences of overbank deposits (Figure 5). Paleosols and bioturbation structures *(Skolithos, Sinusites)* are common in overbank deposits and upper parts of channel-fills. All three types of fluvial deposits contain an unusual diversity of soft-sediment deformation structures (convoluted bedding, overturned crossbedding, clastic dikes, and frondescent marks), which have been interpreted as paleo-seismicity indicators (Evans, 1988, 1994).

The delta-front deposits are organized into isolated, lenticular channels that are separated by intervals of thin-bedded turbidites and syndepositional slumps. Most channels are lenticular in shape with an erosional base. Hansen slip-line analysis of syndepositional slump-fold axes indicates that paleoslopes faced southeast, which was also the direction of paleoflow (Evans, 1988). Channel-fill sequences include turbidites, massive sandstones, and channel-margin slumps, with rare examples of dune deposits (Figure 6). All channel-fills contain abraded plant fragments, rip-up clasts, and soft-sediment deformation structures, such as flame structures, load casts, convoluted bedding, mud diapirs, and clastic dikes. Typical turbidite channel-fills are Bouma sequences Tade or Tabde, with abundant sole marks (flute casts and groove casts). Thin-bedded turbidites (adjacent to or overlying channel-fills) are composed of Bouma sequences Tde, Tcde, or Tbde. Bedding is tabular with sharp, planar contacts. Slumps into channels consist of blocks of thin-bedded turbidites that range from coherent to recumbently folded.

The lake-basin deposits are poorly exposed on the hillsides surrounding the city of Wenatchee. They consist of repetitive thin-bedded turbidite sequences. The individual depositional units are typically less than 1 cm thick, tabular, with planar bases, fine upward, and consist of Bouma sequences Tde and Tcde. Trace fossils are rare and consist of *Skolithos*.

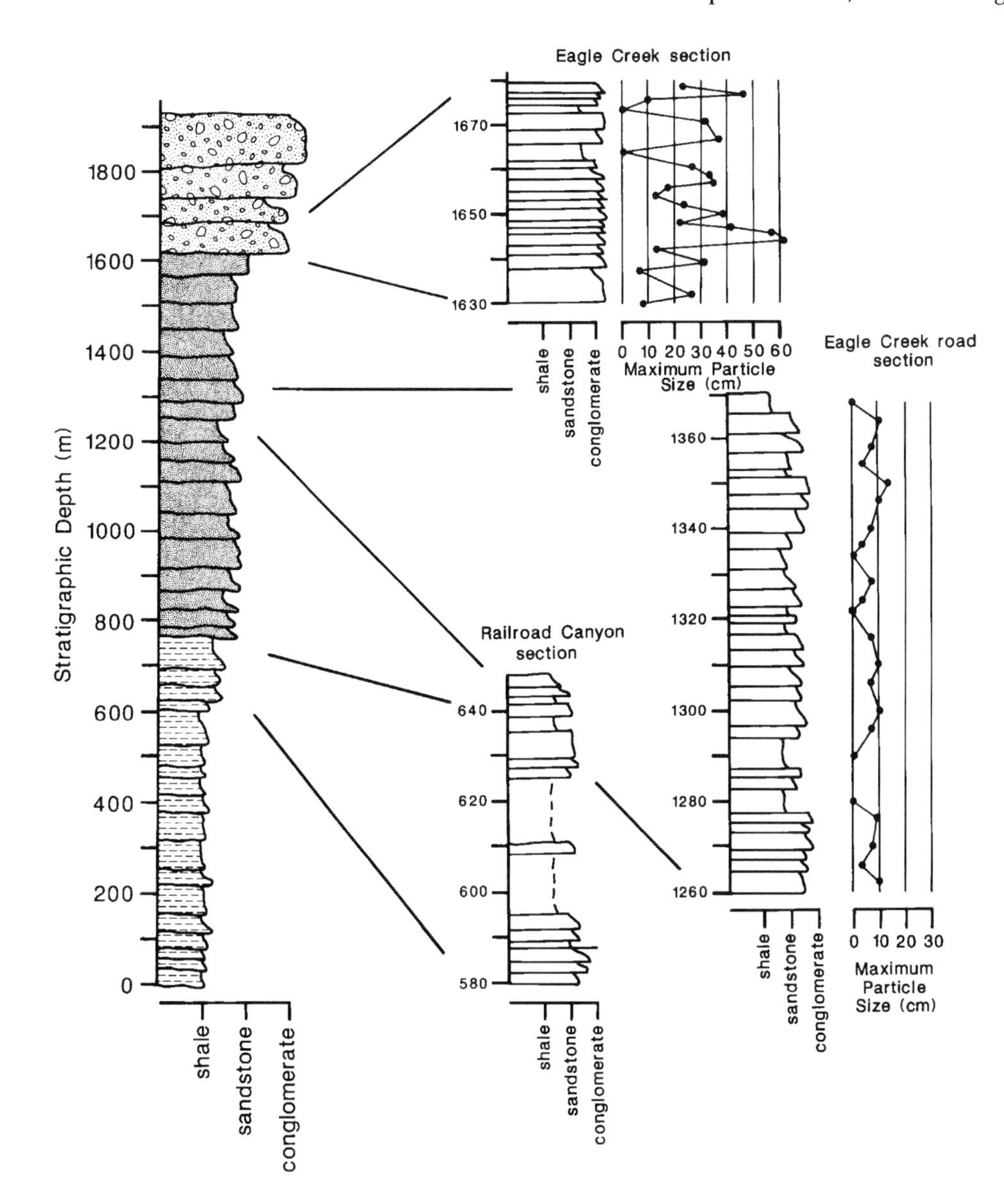

Figure 4—Stratigraphy of the Nahahum Canyon Member showing the composite column (left) and some individual sections (right). The coarsest grained deposits are in the center of a syncline that is a parallel forced fold adjacent to the Entiat fault zone. The locations of sections are given in Figure 1 (from Evans, 1994).

DISCUSSION OF LACUSTRINE FACIES

Paleolimnology

The ancient lake was a long, linear body, fault-bounded to the northwest and the southeast, with a maximum width of about 5 km and length of at least 20 km. There are few recognizable shoreline features, including rare examples of mudcracks, crenulated surfaces, and reworked (planed-off) tops of ripplemarks. The lack of well-developed shoreline features, and the vertical stacking of delta-front deposits at the northwest corner of the lake (without evidence for progradation), suggest tectonic control and rapid subsidence of the lake margin (Evans, 1988, 1994).

Evidence for high turbidity and rapid sedimentation in the ancient lake includes abraded organic debris, lack of faunal remains, and rarity of trace fossils. Compared to the excellent preservation of leaf fossils in fluvial overbank deposits, organic materials in the lake deposits consist only of indistinct fragments of leaves and rounded pieces of stems. Pollen grains are similarly abraded and corroded (Evans, 1988, 1991b).

The lake-water chemistry of this system is less well understood. Many humid-tropical lakes are characterized by annual or seasonal stratification, and some develop anoxic bottom water conditions. The lacustrine deposits in the Nahahum Canyon Member have between 3 and 5% total organic carbon (TOC), suggesting either rapid burial or stratification conditions (Evans, 1988). Water chemistry may have been affected by syndepositional hydrothermal activity. Evidence for such includes subaqueous hot-spring deposits, vents, and eruption breccias observed in mines adjacent to the Eagle Creek fault zone (Margolis, 1987).

Thermal Maturity

Vitrinite reflectance data (Figure 7) show that maximum thermal maturity (about 0.6% R_O) is achieved in the delta-front deposits. The isoreflectance contours

Figure 5—Stratigraphic column from the Monitor section (Figure 1), showing mixed-load fluvial deposits in the region adjacent to the lacustrine-deltaic facies in the Nahahum Canyon Member.

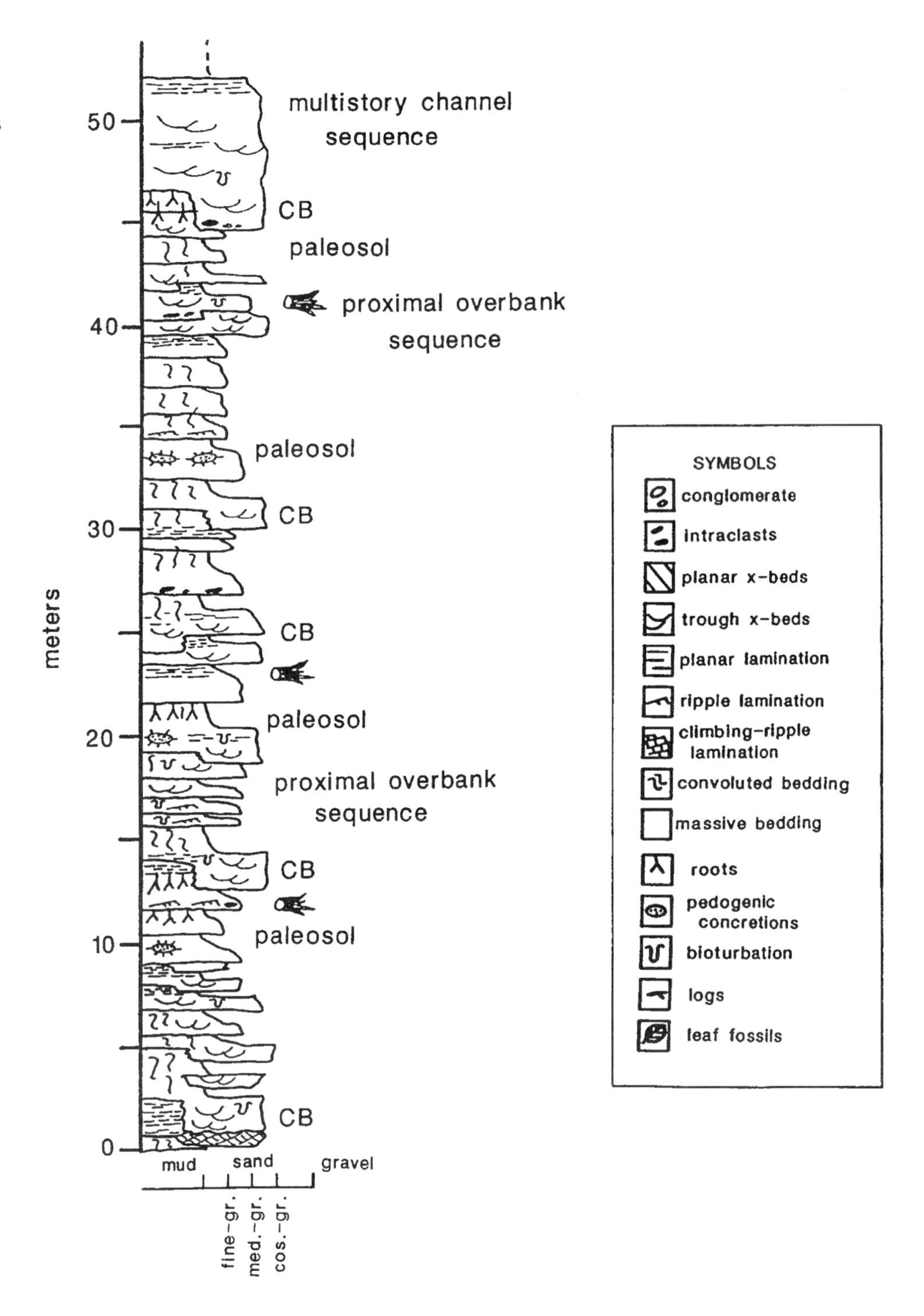

are perpendicular to the two basin-bounding faults, which is in accord with maximum subsidence in a step-over basin and suggests tectonic control of basin subsidence.

Evans (1994) calculated basin burial temperatures from the Chumstick Formation, as well as paleogeothermal gradients from the NORCO-1 well (located in the Eagle Creek fault zone). The average maximum burial temperature was between 180° and 210°C, which is consistent with independent paleotemperature calculations from zeolite minerals and from discordant apatite and zircon fission-track ages (Evans, 1994); however, Eocene–Oligocene hydrothermal activity associated with gold mineralization in the Eagle Creek fault zone was clearly responsible for locally much higher paleotemperatures. For example, fluid inclusion studies in the mineralized zone suggest that mean homogenization temperatures were about 285°C (Klisch, 1994). Thus, given the 3–5% TOC content of the shales and the indicated thermal maturity, the Nahahum Canyon Member is a potential source rock for natural gas resources.

It has been speculated that the Chumstick Formation extends beneath the Columbia River Basalt (e.g.,

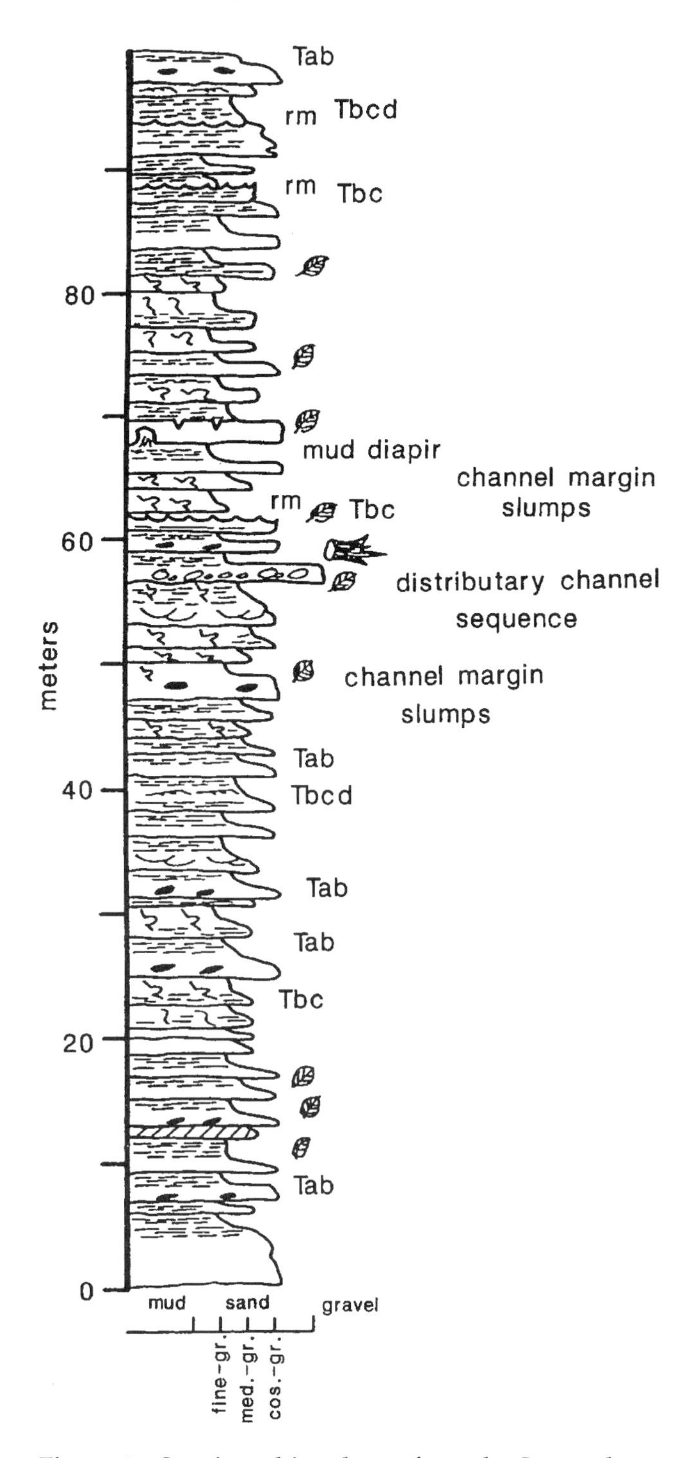

Figure 6—Stratigraphic column from the Sunnyslope Road section (Figure 1) showing delta-front deposits and lake-basin deposits of the Nahahum Canyon Member. Sandstone bodies are broadly lenticular and represent channel filled by channel-margin slumps, turbidites, or distributary channel deposits. Symbols are the same as in Figure 5. Abbreviations: rm = ripple marks; Tab, Tbcd, Tbc, etc. = turbidite with respective Bouma divisions listed.

Gresens and Stewart, 1981) and may represent a potential natural gas reservoir that has been observed in exploration wells (Lingley and Walsh, 1986; Campbell, 1987). The few wells drilled in the Columbia Plateau have mostly been on structure, which may be the wrong strategy. The Eagle Creek fault zone is continuous with the Colockum Creek and Laurel Hill monoclines in the Columbia River Basalt. These monoclines have been interpreted as fault propagation folds generated by subbasalt structures (Plescia and Golombek, 1986; Reidel, 1984). Thus the exploration targets may be the narrow, elongate basins located between structures under the Columbia River Basalt (Evans, 1988).

Implications

The Nahahum Canyon Member represents a fluvial and lacustrine-deltaic depositional system that filled a step-over basin that opened between two parallel, oblique-slip faults. Tectonic control of the basin margin resulted in vertical stacking of delta-front deposits, minimal development of shoreline features, and rapid sedimentation. The high organic carbon content, thermal maturity, basin structure, and basin orientation suggest that the Nahahum Canyon Member is a potential source rock for natural gas reservoirs that have been identified beneath the Columbia River Basalt.

ACKNOWLEDGMENTS

I wish to thank Joanne Bourgeois, Samuel Johnson, William Lingley, Timothy Walsh, Jacob Margolis, Robyn Burnham, Estella Leopold, and Karl Newman for assistance and advice with this research. This study received financial support from the Geological Society of America, Shell Oil Company, the Atlantic Richfield Company, Sigma Xi, and the University of Washington.

REFERENCES CITED

Campbell, N. P., 1987, Structural and stratigraphic relationships between the northwestern Columbia River Basalt margin and recent sub-basalt gas wells [abstract]: Geological Society of America, Abstracts with Programs, v. 19, p. 364.

Curray, J. R., 1956, The analysis of two-dimensional orientation data: Journal of Geology, v. 64, p. 117–131.

Evans, J. E., 1988, Depositional environments, basin evolution, and tectonic significance of the Eocene Chumstick Formation, Cascade Range, Washington [Ph.D. dissertation]: University of Washington, Seattle, Washington, 325 p.

Evans, J. E., 1991a, Facies relationships, alluvial architecture, and paleohydrology of a Paleogene, humid-tropical alluvial-fan system: Chumstick Formation, Washington State, U.S.A.: Journal of Sedimentary Petrology, v. 61, p. 732–755.

Evans, J. E., 1991b, Paleoclimatology and paleobotany of the Eocene Chumstick Formation, Cascade Range, Washington (U.S.A.): a rapidly subsiding

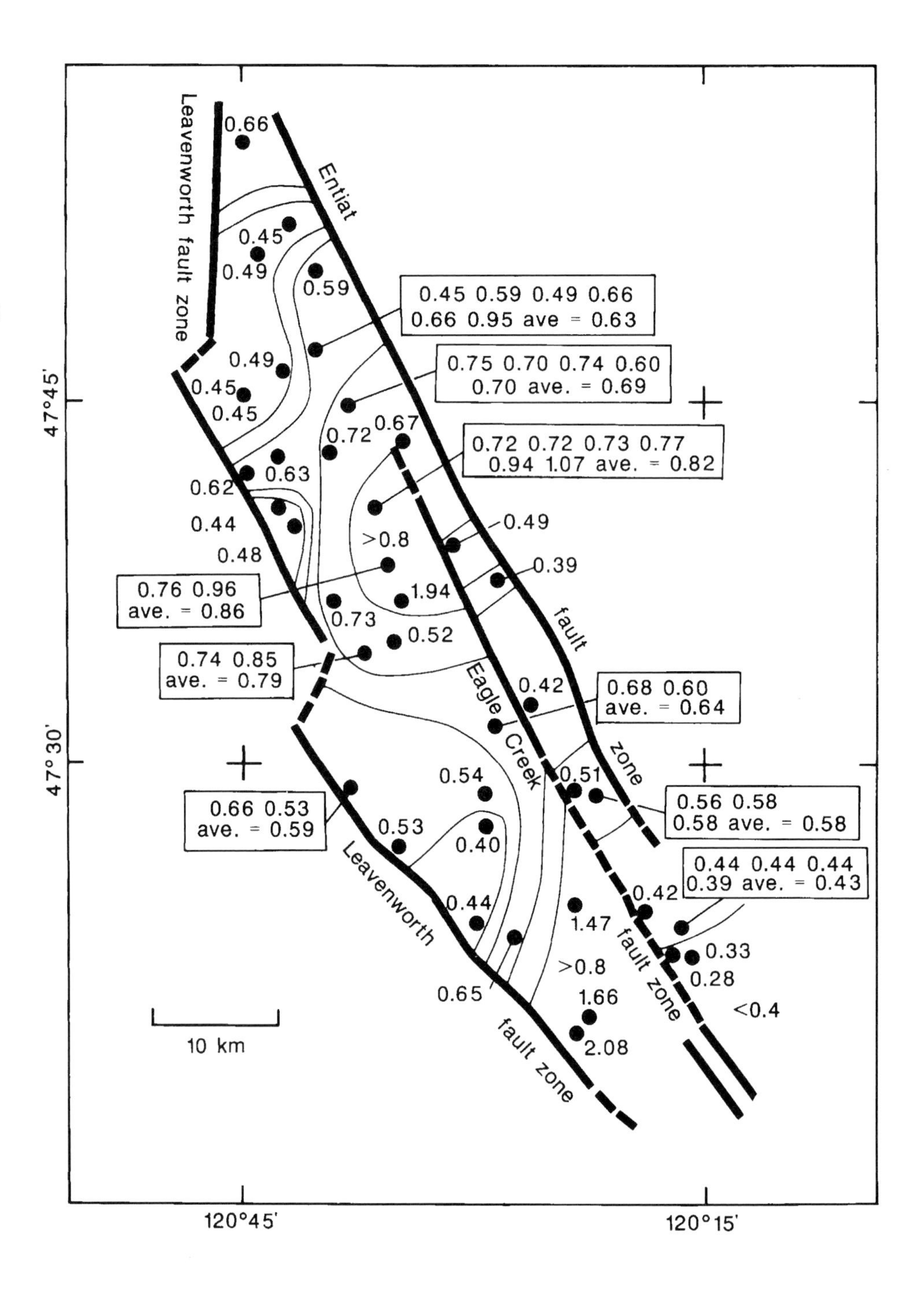

Figure 7—Vitrinite reflectance values from surface outcrops in the Chumstick Formation showing the different distribution patterns in the eastern and western subbasins. The western subbasin has a half-bull's-eye pattern showing the highest thermal maturity in the oldest deposits adjacent to the Eagle Creek fault zone. The eastern subbasin (Nahahum Canyon Member) has a linear pattern with the highest thermal maturity in the vertically stacked delta-front deposits near the center of the subbasin (from Evans, 1994).

alluvial basin: Palaeogeography, Palaeoclimatology, Palaeoecology, v. 88, p. 239–264.

Evans, J. E., 1994, Depositional history of the Eocene Chumstick Formation: Implications of tectonic partitioning for the history of the Leavenworth and Entiat-Eagle Creek fault systems, Washington: Tectonics, v. 13, p. 1425–1444.

Evans, J. E., and S. Y. Johnson, 1989, Paleogene strike-slip basins of central Washington: Swauk Formation and Chumstick Formation, *in* N. L. Joseph and 13 others, editors, Geologic Guidebook for Washington and Adjacent Areas: Washington Division of Geology and Earth Resources, Information Circular 86, p. 215–237.

Gresens, R. L., 1983, Geology of the Wenatchee and Monitor quadrangles, Chelan and Douglas counties, Washington: Washington Division of Geology and Earth Resources Bulletin 75, 75 p.

Gresens, R. L., and R. J. Stewart, 1981, What lies beneath the Columbia Plateau?: Oil and Gas Journal, v. 79, p. 157–164.

Gresens, R. L., C. W. Naeser, and J. T. Whetten, 1981, Stratigraphy and age of the Chumstick and Wenatchee Formations—Tertiary fluvial and lacustrine rocks, Chiwaukum graben, Washington: Geological Society of America Bulletin, v. 92, p. 841–876.

Klisch, M. P., 1994, A study of fluid inclusions in mineralized veins at the Cannon Mine, Wenatchee, Washington, *in* J. Margolis, editor, Epithermal Gold Mineralization, Wenatchee and Liberty Districts,

Central Washington: Society of Economic Geologists, Guidebook Series, v. 20, p. 55–57.

Krause, R. G. F., and T. A. M. Geijer, 1987, An improved method for calculating the standard deviation and variance of paleocurrent data: Journal of Sedimentary Petrology, v. 57, p. 779–780.

Lingley, W. S., Jr., and T. J. Walsh, 1986, Issues relating to petroleum drilling near the proposed high-level nuclear waste repository at Hanford: Washington Geologic Newsletter, Washington Division of Geology and Earth Resources, v. 14 (3), p. 10–19.

Margolis, J., 1987, Hydrothermal alteration and structure associated with epithermal gold mineralization, Wenatchee Heights, Washington [M.S. Thesis]: University of Washington, Seattle, Washington, 120 p.

Nemec, W., 1988, The shape of the rose: Sedimentary Geology, v. 59, p. 149–152.

Newman, K. R., 1981, Palynologic biostratigraphy of some early Tertiary nonmarine formations in central and western Washington, *in* J. M. Armentrout, editor, Pacific Northwest Cenozoic Biostratigraphy: Geological Society of America Special Paper 184, p. 49–65.

Plescia, J. B., and M. P. Golombek, 1986, Origin of planetary wrinkle ridges based upon a study of terrestrial analogs: Geological Society of America Bulletin, v. 97, p. 1289–1299.

Reidel, S. P., 1984, The Saddle Mountains: the evolution of an anticline in the Yakima fold belt: American Journal of Science, v. 284, p. 942–978.

Tabor, R. W., R. B. Waitt, Jr., V. A. Frizzell, Jr., D. A. Swanson, G. R. Byerly, and R. D. Bentley, 1982, Geologic map of the Wenatchee 1:100,000 Quadrangle, Washington: U.S. Geological Survey, Miscellaneous Geologic Investigations, I–1311, 26 p.

Tabor, R. W., V. A. Frizzell, Jr., J. T. Whetten, R. B. Waitt, Jr., D. A. Swanson, G. R. Byerly, D. B. Booth, M.J. Hetherington, and R. E. Zartman, 1987, Geologic map of the Chelan 30′ × 60′ quadrangle, Washington: U.S. Geological Survey, Miscellaneous Geologic Investigations, I–1661, 29 p.

Daley, B., N. Edwards, and I. Armenteros, 2000, The upper Eocene Bembridge Limestone Formation, Hampshire Basin, England, *in* E. H. Gierlowski-Kordesch and K. R. Kelts, eds., Lake basins through space and time: AAPG Studies in Geology 46, p 369–378.

Chapter 33

The Upper Eocene Bembridge Limestone Formation, Hampshire Basin, England

Brian Daley
Nicholas Edwards
*School of Earth, Environmental and Physical Sciences, University of Portsmouth
Portsmouth, UK*

Ildefonso Armenteros
*Departamento de Geologia, Facultad de Ciencias, Universidad de Salamanca
Salamanca, Spain*

LOCATION AND GEOLOGIC CONTEXT

The northwest-southeast oriented Hampshire-Dieppe Basin, approximately 280 km long (Hamblin et al., 1992), is a structural basin formed by Miocene inversion of the underlying Mesozoic Hampshire-Dieppe High and the adjacent Jurassic–Lower Cretaceous extensional sedimentary basins (Chadwick, 1993; Hibsch et al., 1993). The onshore portion in southern England, the Hampshire Basin (Figure 1), covers much of the county of Hampshire and extends into the adjacent counties of the Isle of Wight, Dorset, and West Sussex. Its southern boundary is formed by east-west, north-younging monoclinal folds involving Upper Cretaceous and Paleogene strata, draped over reversed normal faults in Jurassic–Lower Cretaceous strata (Chadwick, 1993). The other boundaries correspond to the Paleogene–Upper Cretaceous (Chalk) junction and were exposed as a result of post-Miocene erosion. The Paleogene succession (Daley, 1999) was deposited unconformably on weakly folded and eroded Upper Cretaceous (Campanian) Chalk (Melville and Freshney, 1982) and subsequently uplifted and deformed during mid-Paleocene (Laramide) crustal compression (Hamblin et al., 1992). The Paleogene strata dip steeply in the vicinity of the monocline, whereas to the north, dips are shallow in low-amplitude folds.

At the beginning of Paleogene time, the British area was some 12° south of its present position (Irving, 1967) and considerably warmer than at present (Reid and Chandler, 1933; Daley, 1972). Britain lay on the western margins of what has been called the Northwest European Tertiary Basin (Vinken et al., 1988). At the western end of this basin, the Tertiary North Sea Basin extended northward between the British and Scandinavian land masses. The present Hampshire (tectonic) Basin was at the southeastern end of this sedimentary basin. The Paleogene succession here is up to 652 m thick (Edwards and Freshney, 1987a) (Figure 2). It begins with upper Paleocene coastal plain sediments and continues with lower to mid-Eocene (Ypresian–Bartonian, NP10-NP17), cyclically alternating marine, nearshore, and coastal plain sediments.

Extreme shallowing and possibly subaerial exposure (Edwards and Freshney, 1987b) at the mid–upper Eocene (NP17-NP18) boundary preceded the onset of predominantly argillaceous, strongly river-influenced, shallow-marine, laguno-lacustrine and lacustrine sedimentation (represented by the Solent Group), which continued at least into the early Oligocene (Rupelian, NP23). This major change in sedimentary style signifies the establishment of a better-defined sedimentary basin of unknown extent. The change coincided with gradual uplift of the Weald-Artois axis (coincident with the major Midi basement fault), linking extreme southeastern England to northeastern France (Hamblin et al., 1992). As a result of post-Miocene erosion, the middle and upper formations of the three comprising the Solent Group are now confined to the Isle of Wight.

THE BEMBRIDGE LIMESTONE

Although the Solent Group comprises predominantly fine-grained siliciclastic sediments, sections in the oldest of its three formations (the Headon Hill Formation) at western localities contain poorly lithified freshwater-to-well-limestones, a few centimeters to several meters thick, interbedded with the siliciclastic sediments.

The overlying Bembridge Limestone Formation is much thicker and more laterally persistent, although confined to the Isle of Wight, where it outcrops intermittently at the coast in the northern part of the island and inland just north of the steep limb of the central

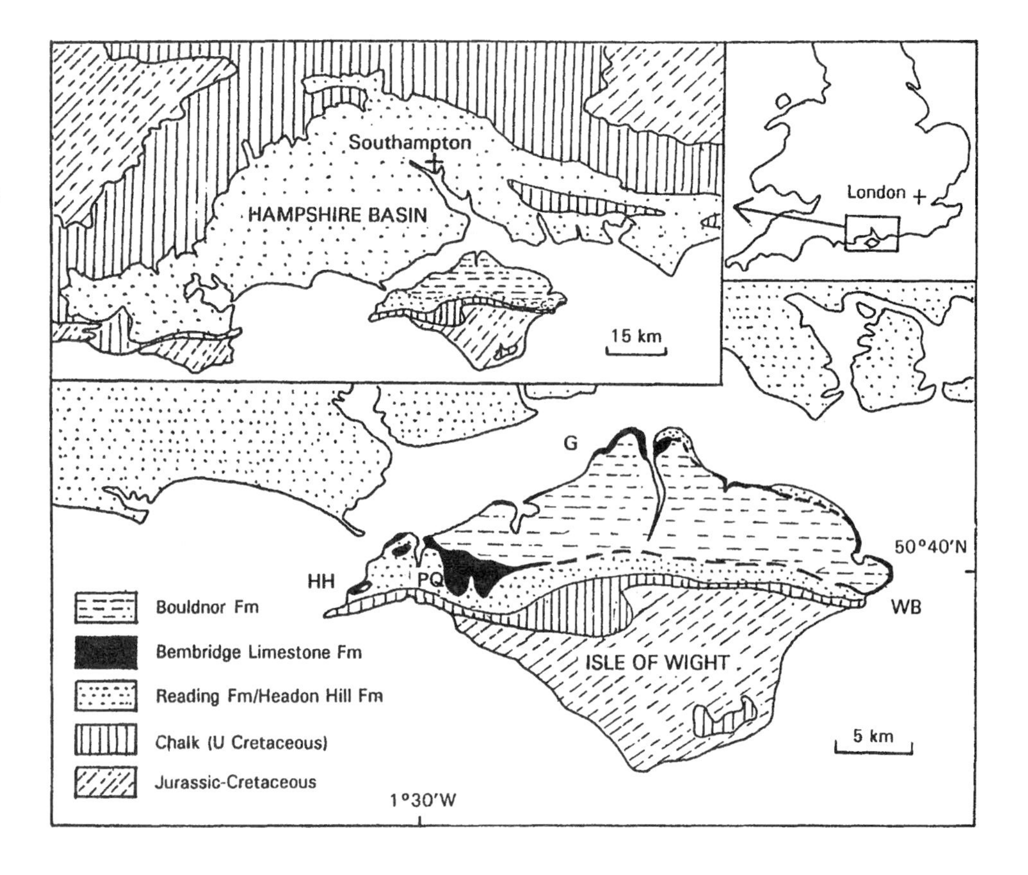

Figure 1—Map of the Isle of Wight, southern England, showing the outcrop of the Bembridge Limestone and important localities in this formation: Whitecliff Bay (WB), Gurnard Ledge (G), Prospect Quarry (PQ), and Headon Hill (HH).

monocline. It reaches its maximum thickness of 9 m in Whitecliff Bay but elsewhere thins considerably (Insole and Daley, 1985, their figure 23). The formation comprises an "upper limestone" and "lower limestone" sequence separated by a central sequence of green and black siliciclastic muds. The greater importance of siliciclastic muds at localities on the northern coast suggests northward gradation from limestone-dominated to mud-dominated successions. This suggests that a carbonate source area (Upper Cretaceous Chalk?) lay to the south and a clastic source area (older Paleogene or Mesozoic sediments?) lay to the north. Although the geographic limits of carbonate deposition are unknown, isotopic and paleontologic evidence for episodes of slightly raised salinities suggest it occurred in a laguno-lacustrine, rather than land-locked, basin (Armenteros et al., 1992, 1997).

FACIES AND CYCLICITY IN THE BEMBRIDGE LIMESTONE

That the Bembridge Limestone is predominantly of freshwater origin was recognized many years ago from the common occurrence of freshwater pulmonate gastropods, particularly *Lymnaea* (*Galba*) and to a lesser extent *Planorbina*. Early studies showed, however, that where complete, the mainly fine-grained siliciclastics separating the upper limestones and lower limestones contain fossils with brackish water affinities (White, 1921). Later, Daley and Edwards (1990) recognized that the presence of hydrobiid gastropods at some horizons within the limestone lithologies implied periodic brackishness, whereas the identification by Murray and Wright (1974) of foraminiferids within the formation indicated the proximity of the sea. That terrestrial conditions were nearby was implied by the presence of fossil land snails (Pain and Preece, 1968; Preece, 1976) and, more recently, terrestrial mammalian remains (Hooker et al., 1995).

Although the literature on the formation indicates that earlier workers recognized lithologic variation both laterally and vertically, it is only relatively recently that explanations for such variations have been published. Daley and Edwards (1990) were the first to record the cyclic nature of the succession. They recognized four cycles at the type locality in Whitecliff Bay, which, where complete, comprise an ascending sequence with the following components: (1) intraformational limestone conglomerate, resting on an erosion surface, (2) mud, marls, or marly limestones, (3) purer, essentially biomicritic, well-indurated limestones, and (4) a pedogenic alteration zone.

Armenteros et al. (1997) (Figure 3) recognized seven cycles in the Whitecliff Bay succession, of which four were attenuated cycles equivalent to cycle 3 of Daley and Edwards (1990). Armenteros et al. (1997) considered them to be transgressive/regressive cycles, representing, when completely developed, inundation (associated with erosion and the development of the intraformational, intraclastic limestone conglomerates), initial deepening, and then progressive shallowing followed by subaerial exposure and limited pedogenic alteration. Armenteros et al. (1997) were

Period	Epoch	Age	Group	Formation	Member	Thickness (m)	Lithology	Interpretation
TERTIARY PALEOGENE	OLIGOCENE	Rupelian	Solent	Bouldnor	Cranmore	9		Marine-lagoonal
					Hamstead	78		Fluvio-lacustrine
	EOCENE	Priabonian			Bembridge Marls	34		Laguno- & fluvio-lacustrine
								Hiatus (intra-Priabonian folding & erosion)
				Bembridge Limestone		9		Lacustrine/palustrine
				Headon Hill	Seagrove Bay	0-15		Fluvio-lacustrine
					Osborne Marls	15-34		Fluvio-lacustrine
					Fishbourne	3-11		Fluvio-lacustrine
					Lacy's Farm Limestone	7		Lacustrine/palustrine
					Cliff End	9-35		Laguno-lacustrine
					Hatherwood Limestone	9		Lacustrine/palustrine
					Linstone	0-5		Intertidal/subtidal bay
					Colwell Bay	3-32		Shallow-marine/lagoonal
					Totland Bay	4-32		Fluvio-lacustrine
								Hiatus (emergence)
		Bartonian	Barton	Becton Sand		28-60		Upper shoreface
				Chama Sand		7-15		Lower shoreface
				Barton Clay		45-83		Marine shelf
								Hiatus (emergence)
		Lutetian	Bracklesham	Selsey		30-65		Marine shelf
				Marsh Farm		35-43		Lagoonal/intertidal
				Earnley Sand		25		Marine shelf
				Wittering		53-100		Shallow-marine/paludal
								Hiatus (erosion)
		Ypresian	Thames	London Clay	Division D	14-38		
					Division C	10-30		4 cyclical sequences:
					Division B	25-35		Marine shelf-inner
					Division A	10-45		sublittoral-littoral
					Basement Bed	3-4		Shallow-marine
								Hiatus (erosion)
	PALEOCENE	Thanetian	Lambeth	Reading		26-46		Laguno-lacustrine
					Basement Bed	0-1		Shallow-marine
								Major hiatus (mid-Paleocene uplift & erosion)
CRETACEOUS		Campanian	Chalk					

Dating method: Paleontology (calcareous nannoplankton, charophytes, dinoflagellate cysts, Foraminifera, mammals, molluscs, ostracods, palynology), magnetostratigraphy
Reliability index: B
Sources: Ali et al. (1992), Aubry (1985, 1986), Aubry et al. (1986), Bujak et al. (1980), Castel (1968), Costa & Downie (1976), Costa et al. (1976, 1978), Eaton (1971a & b, 1976), Fiest-Castel (1977), Hooker (1987, 1992), Liengjaren et al. (1980), Martini (1970), Townsend & Hailwood (1985)

Figure 2—The Paleogene succession in the Isle of Wight, southern England, after various authors.

uncertain about the causes of the cyclicity, suggesting that it may have resulted from rises and falls of sea level or an alternation of wetter and drier climatic phases.

In the course of their stratigraphic revision of the Bembridge Limestone, Daley and Edwards (1990) briefly commented on lateral variations. There is an apparent northward gradation from limestone-dominated to mud-dominated successions, and in the northwest, there is faunal (hydrobiid) evidence of brackish-water incursions during the deposition of the limestones. Pedogenic alteration is particularly marked at the westerly localities where there is also a stronger terrestrial component in the fossil molluscan assemblages.

More recently, Armenteros et al. (1997) have assigned the succession to three facies. The two main facies are a lacustrine facies and a palustrine facies. The former is represented mainly by marls, marly limestones and harder limestones (mainly biomicrites), and thin intraclastic limestones (intraformational limestone conglomerates). Pedogenic alteration is limited, with thin developments of brecciated-nodular limestone and vuggy porosity indicative of periods of subaerial exposure. The palustrine facies, dominant in the west of the island, comprises a complex association of lithologies from pedogenically unaltered biomicrites to clotted-peloidal-ooidal limestones and thin laminar (crustose) limestones, which together reflect repeated exposure and pedogenic modification. The third facies, the gypsiferous lake margin facies, has a restricted development and appears to represent the postdepositional development of microlenticular gypsum following the evaporation of pore fluids from sediments having some degree of lateral proximity to more saline waters.

LITHOLOGIES OF THE BEMBRIDGE LIMESTONE FORMATION

The lithologies associated with the cyclic development and facies are described in more detail. These are summarized as follows.

(1) Intraclastic limestones
(2) Marls and marly limestones
(3) Biomicrites
(4) Brecciated-nodular limestones
(5) Clotted-peloidal-ooidal limestones
(6) Laminar (crustose) limestones
(7) Limestones and silicified limestones with microlenticular gypsum pseudomorphs
(8) Muds and other siliciclastic lithologies

Intraclastic Limestones

Gray-green marly limestones composed of silty micrite and fine sparite, enclosing gray subangular limestone intraclasts of lacustrine and pedogenically modified limestones, 0.05–5 cm diameter. This lithology overlies erosion surfaces at the base of lacustrine transgression-regression units in the Whitecliff Bay section. The poor sorting and subangularity of intraclasts is evidence for minimal transport.

These intraclastic limestones represent erosion when earlier lithified carbonates that had been exposed during regressive phases were subject to erosion during resubmergence as lacustrine water levels rose once more.

Marls and Marly Limestones

These graduate from slightly green or greenish-white marls to gray to dark gray gastropod-rich nonconsolidated marly limestones. Beds are massive to diffusely laminated, 0.10–0.50 m thick and laterally extensive. They usually succeed intraclastic limestones and pass upward into biomicrites. This lithologic association is best developed in the lower part of the succession in Whitecliff Bay (Figure 3).

The absence of any features indicating exposure shows that these deposits accumulated under a permanent water cover at a time when the lakes were at their deepest. The marls may reflect times of maximum influx of fine siliciclastics, and the marly limestones increasing organic activity.

Biomicrites

Cream-colored and gray to dark gray, compact, well-lithified limestones, displaying a mudstone to wackestone texture and comprising micrites with "floating" shells and shell fragments of chiefly pulmonate water snails, charophyte stem fragments, and gyrogonites. Fossil land snails are present in the succession in Whitecliff Bay, but are a more important part of the molluscan assemblages at the western localities (Pain and Preece, 1968). Beds are 0.5–1.0 m thick, are massive, or display alternating diffuse light and dark bands on a 10-cm scale. Vuggy porosity is usually present and root-moldic porosity can also occur. Porosity is at least partly filled by pellety clasts 0.04–1.00 mm in diameter, derived from pore walls, in fine to medium sparitic cement. Drusy calcite mosaic or calcite silt cemented by fine sparite fills the upper parts of many pores. This lithology occurs in all exposed sections and at Whitecliff Bay characterizes the upper part of the cycles where best developed.

The lithologic and paleontologic characteristics of the biomicrites suggest low-energy sedimentation in shallow holomictic lakes with oxygenated bottom waters and mainly biogenic carbonate production. Although the sediments became reduced, they were subject to subaerial exposure and minor to major solutional effects after consolidation (Figure 3), indicating palustrine conditions, when wide areas of the lake bed became emergent (cf. Freytet and Plaziat, 1982).

Brecciated-Nodular Limestones

These lithologies represent a bed-top modification. They typically underlie erosion surfaces and are succeeded by intraclastic limestones. They characteristically grade up from biomicrites. Commonly, they display more or less interconnected, sediment-filled fissures centimeters to millimeters wide, which define the brecciated (or more strictly pseudobrecciated) character of the lithology. Networks of cracks 50–200 μm across are also present (Armenteros et al., 1997). In some hand specimens from Prospect Quarry, multiple episodes of brecciation and reconsolidation are apparent. Vugs, channels, and moldic porosity are also rarely present. Freytet and Plaziat (1982) termed analogous lithologies from French Paleogene lacustrine limestones "desiccation breccias." This lithology reflects the earliest stage in the subaerial modification of the underlying carbonate sediment.

Clotted-Peloidal-Ooidal Limestones

These lithologies, represented in the more western outcrops of the formation, such as that of Prospect Quarry, characterize the palustrine facies of Armenteros et al. (1997) and reflect increased pedogenic modification. The textures developed represent different grades of pedality (sensu Bullock et al., 1985) and may be related to two end members: clotted-peloidal texture (where development of peds is incipient) and peloidal-ooidal texture (with well-differentiated elements). Although the clotted-peloidal lithologies are essentially wackestones in texture, the peloidal-ooidal lithologies are packstone/grainstones. There is, nevertheless, a continuum from clotted biomicrite through a clotted-peloidal microstructure, in which the peloids are bounded with circumgranular cracks, to various ooidal packstone/grainstone textures.

Peloids range in size from 10 to 100 μm, whereas the ooids range from 0.05 mm to 1 cm in diameter. Sorting is often poor and packing varies considerably. Fabrics with dense packing, where small particles (peloids) fill the spaces between ooids, contrast with open fabrics

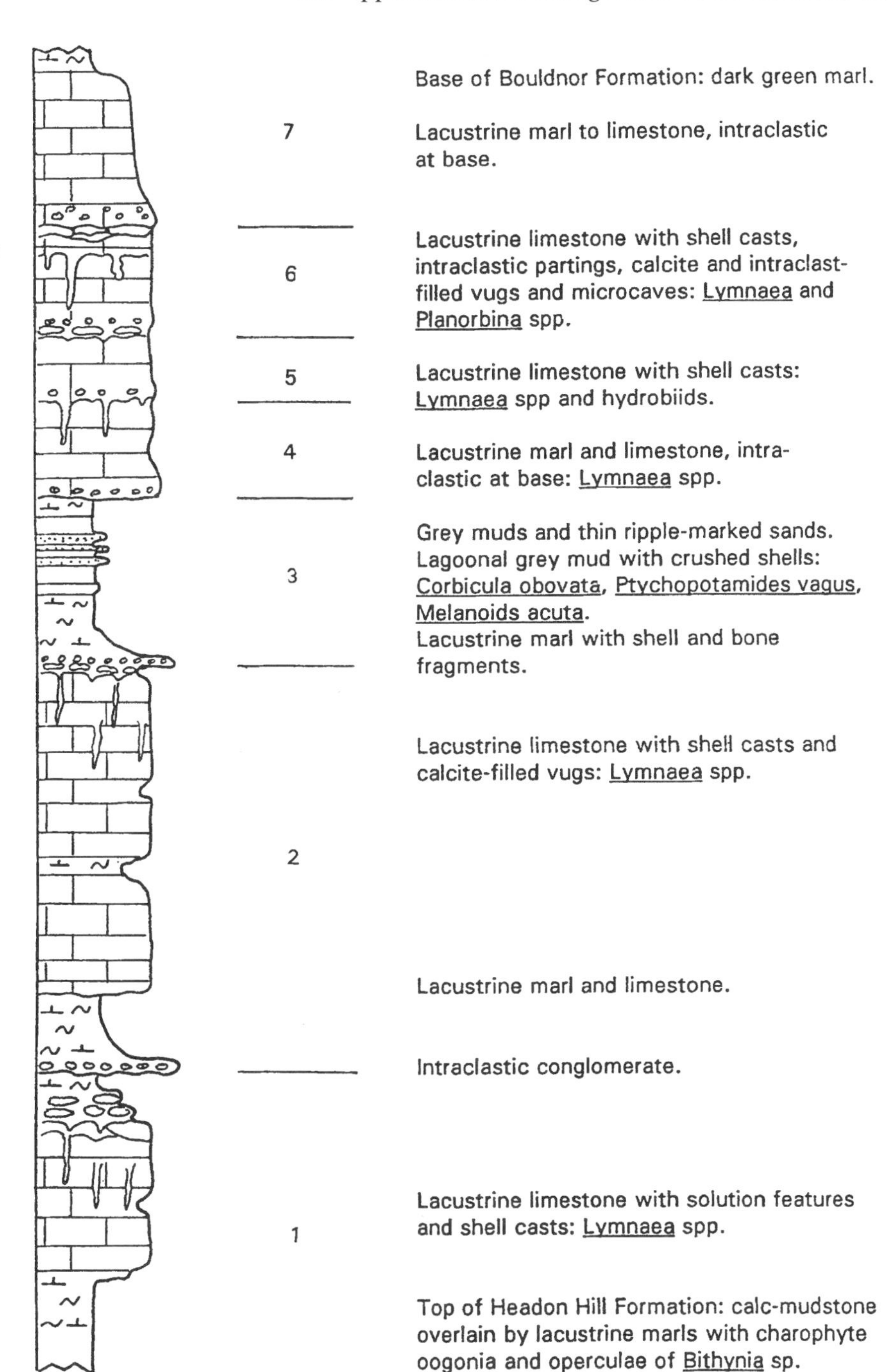

Figure 3—The Bembridge Limestone succession in Whitecliff Bay, Isle of Wight, southern England; after Armenteros et al. (1997).

with considerably greater interparticle porosity. As well as the latter, other porosities are present: vug, moldic, and intraparticle, and larger, downward-narrowing solution cavities. The smaller cavities may be infilled both mechanically by silt and/or by sparry cement and frequently show geopetal structure. The larger cavities may be filled mechanically by larger debris that commonly shows reverse grading (upward coarsening).

Within these markedly altered lithologies, fossils such as gastropods and charophyte oögonia are preserved mainly as molds. Other primary remnants include scattered "pseudointraclasts" of mainly soft, white, only slightly altered original lacustrine sediment.

This spectrum of lithologies reflects a further stage in the pedogenic modification of the lacustrine sediments. They resemble examples from French Paleogene lacustrine limestones, which developed by glaebular differentiation of carbonate muds, as a result of alternate wetting and drying, on low-energy lake margins subject to prolonged subaerial exposure (Freytet and Plaziat, 1982, their pl. 34A–D; and Armenteros and Daley, 1998). Armenteros et al. (1997) interpreted these lithologies from the Bembridge Limestone similarly. They suggested a comparison with calcretes formed in semi-arid warm to hot temperatures while conceding that other evidence (the solution structures, the presence of land snails, and evidence from mammalian remains for wooded hinterland presented by Hooker et al., 1995) suggests that such dry conditions alternated with wetter climatic phases over a longer term.

Laminar (Crustose) Limestones

These limestones represent only a small (<1%) part of the succession but are a significant diagnostic component. They form thin, laterally impersistent bands from 0.5 to 10 cm in thickness, have an essentially localized development, and occur unpredictably. They display alternate lighter and darker submillimetric to millimetric-scale subhorizontal wavy lamination, in discordant sets 1–3 cm thick. Some of the laminar limestones have been disrupted by brecciation. Microscopic examination indicates a range of fine detail from an intensely spongy/porous fabric to one comprising finely laminated dense micrites.

These laminar limestones vary in their lithologic relationships at different exposures of the Bembridge Limestone in the Isle of Wight. At Whitecliff Bay, the lithology occurs as poorly lithified, thin brown crusts coating hummocky crack surfaces within and the upper surfaces of two biomicrite beds. At Prospect Quarry, a thick laterally impersistent crust near the base of the exposed section is interbedded with and appears to be replacing the clotted-peloidal-ooidal and nodular-brecciated limestones. At Headon Hill, the lithology, dominated by the intensely porous fabric, forms lenses up to 8 cm thick and 60 cm long at the top of beds of chalky and clotted-peloidal/ooidal limestones, with slabby clasts of the lithology also occurring in overlying conglomerates.

As interpreted by Armenteros et al. (1997), this lithology has features indicating that it represents calcified root mats, as reported from ancient and Holocene calcretes (cf. Wright et al., 1988, their figure 5) and palustrine environments (Menillet, 1980; Platt, 1989; Alonso Zarza et al., 1992). That the laminar limestone sometimes, as at Prospect Quarry, appears to truncate adjacent lithologies is consistent with the findings of Klappa (1979) that laminated limestones appear to represent a replacement of other pedogenic lithologies in recent calcretes.

Limestones and Silicified Limestones with Microlenticular Gypsum Pseudomorphs

At Gurnard, both the upper limestones and lower limestones of the formation contain pseudomorphs after microlenticular gypsum (Armenteros et al., 1997). Here, these limestones are thinner than at Whitecliff Bay and contain both freshwater and more saline faunal elements. The upper limestones have muds with brackish-water faunas both stratigraphically above and below them. In these upper limestones, calcite pseudomorphs after microlenticular gypsum occur in biomicrites containing hydrobiid gastropods and foraminifers. Toward the top of the lower limestones, similar pseudomorphs occur but, in addition, silicified pseudomorphs are present in chert nodules developed at this level within ferruginized, ochrous limestones (Daley, 1989).

These lithologies are thought to have developed where primary lacustrine carbonates (albeit deposited in slightly variable salinities) were modified by the growth of microlenticular gypsum following their exposure and the evaporation of brackish pore waters. Such an occurrence suggests the proximity of the Bembridge Limestone lake to more saline waters. Armenteros et al. (1997) suggested that the cherts may be related to a paleoexposure surface immediately above, or may reflect the evaporite pumping of siliceous ground water moving slowly under high watertable conditions within the carbonate sediment flats toward the lake (Summerfield, 1983; Armenteros et al., 1997).

Muds and Other Siliciclastic Lithologies

The lower limestones and the upper limestones are separated by siliciclastics. These thicken northward and reach a maximum thickness of some 3.3 m in the Thorness Bay to Gurnard section where they contain green to black muds. Here, brackish water and freshwater molluscs occur at different horizons, and land plant fruits and seeds are present. At Whitecliff Bay, the siliciclastics include thin sands but are less than 1 m in thickness. Brackish water fossils (*Corbicula, Potamides*) occur in a black mud here. This siliciclastic sequence is absent at the western localities, except at Headon Hill, where a sequence of siliciclastic (sometimes carbonaceous) muds and marls accumulated in conditions considered lacustrine (Hooker et al., 1995).

This siliciclastic sequence, representing variable salinities from those of fresh to brackish waters, supports the existence of a marine connection; consequently, the Bembridge Limestone lake was part of a laguno-lacustrine complex rather than a land-locked body of water.

CONCLUSIONS

The Bembridge Limestone comprises two main carbonate facies. Its lacustrine facies, best seen at Whitecliff Bay, is mainly a freshwater deposit, although higher (brackish) salinities are represented at some horizons. Its palustrine facies, found at the westerly localities in the Isle of Wight, represents a pedogenic modification of the lacustrine sediments and is characterized by features present in Holocene and ancient calcretes described elsewhere.

Armenteros et al. (1997) were uncertain as to the cause of the Bembridge Limestone cycles. One possibility was a climatic interpretation. During periods of relatively higher rainfall, when lake levels were high, the influx of land-derived, fine siliciclastics led to the development of marls and marly limestones. During drier periods, lake levels were low, the basin was isolated, and sedimentation was characterized by biogenic/evaporative and, at times of exposure, pedogenic carbonates. Long-term fluctuations in rainfall are supported by evidence for alternately forested and more open conditions in the hinterland, based on mammal assemblages from the Bembridge Limestone at Headon Hill (Hooker et al., 1995), and are consistent with paleoclimatic instability resulting from the location of northwest Europe at this time on the fluctuating boundary between the northeast trade winds to the south and the prevailing humid westerlies to the north (Krutsch, 1992).

Their alternative suggestion was that it may have resulted from rises and falls of sea level (cf. Vail et al., 1977), which could have influenced lake levels because it certainly appears that the lake waters were, from paleontologic, mineralogic, and stable isotope evidence, part of a laguno-lacustrine complex rather than a land-locked basin. Under such circumstances, there would be the possibility that the epeirogenic fluctuations that had effected the cyclic nature of the pre–Solent Group Paleogene succession could continue to be significant at the time when the Bembridge Limestone was accumulating.

Maybe there was yet a third possible factor involved. The influence that earth movements might have played should not be disregarded, because it is known that tectonic warping has affected this part of the local Paleogene succession (Daley and Edwards, 1971); furthermore, evidence is now emerging from the older part of the local Paleogene succession that locally, intra-Paleogene pulses of uplift and erosion had alternated with deposition on several previous occasions (Huggett and Gale, 1997).

REFERENCES

Ali, J. R., C. King, and E. A. Hailwood, 1993, Magnetostratigraphic calibration of early Eocene depositional sequences in the southern North Sea Basin, *in* E. A. Hailwood and R. B. Kidd, eds., High Resolution Stratigraphy, Geological Society Special Publication no. 70, p. 99–125.

Alonso Zarza, A. M., J. P. Calvo, and M. A. Garcia del Cura, 1992, Palustrine sedimentation and associated features—grainification and pseudo-microkarst—in the middle Miocene (Intermediate Unit) of the Madrid Basin, Spain, Sedimentary Geology, v. 76, p. 43–61.

Armenteros, I., B. Daley, and E. Garcia, 1997, Lacustrine and palustrine facies in the Bembridge Limestone (late Eocene, Hampshire Basin) of the Isle of Wight, southern England, Palaeogeography, Palaeoclimatology, Palaeoecology, v. 128, p. 111–132.

Armenteros, I. and B. Daley, 1998, Pedogenic modification and structure evolution in palustrine facies as exemplified by the Bembridge Limestone (Late Eocene) of the Isle of Wight, southern England. Sedimentary Geology, 119, p. 275–295.

Armenteros, I., C. Recio, and B. Daley, 1992, Sedimentology and stable isotope study of the lacustrine Bembridge Limestone (Upper Eocene), Isle of Wight, southern England, III Congresso Geologico de España, VIII Congresso Latinamericano de Geologia, Salamanca, 1992, Simposios tomo 1, p. 27–38.

Aubry, M-P., 1985, Northwestern European Paleogene magneto-stratigraphy, biostratigraphy and paleogeography: calcareous nannofossil evidence. Geology, 13, p. 198–202.

Aubry, M-P., 1986, Palaeogene calcareous nanno-plankton biostratigraphy of northwestern Europe. Palaeogeography, Palaeoclimatology, Palaeoecology, 55, p. 267–334.

Aubry, M-P, E. A. Hailwood, and H. A. Townsend, 1986, Magnetic and calcareous-nannofossil stratigraphy of the lower Palaeogene formations of the Hampshire and London Basins. Journal of the Geological Society of London, 143, p. 729–735.

Bujak, J. P., C. Downie, G. L. Eaton, and G. L. Williams, 1980, Dinoflagellate cysts and acritarchs from the Eocene of Southern England. Palaeontology Special Paper, 24, p. 1–100.

Bullock, P., N. Federoff, A. Jongerius, G. Stoops, and T. Tursina, 1985, Handbook for soil thin section description, Albrighton, Waine Research, 152 p.

Castel, M., 1968, Zones de Charophytes pour l'Oligocène d'Europe Occidentale, Compte rendu sommaires des séances de la société géologiques de France, 4, p. 121–122.

Chadwick, R. A., 1993, Aspects of basin inversion in southern Britain, Journal of the Geological Society of London, v. 150, p. 311–322.

Costa, L. I. and C. Downie, C., 1976, The distribution of the dinoflagellate Wetzeliella in the Palaeogene of north-western Europe, Palaeontology, 19, p. 591–614.

Costa, L.I., C. Denison, and C. Downie, 1978, The Palaeocene/Eocene boundary in the Anglo-Paris Basin, Journal of the Geological Society of London, 135, p. 261–264.

Costa, L.I., C. Downie, and G. L. Eaton, 1976, Palinostratigraphy of some Middle Eocene sections from the Hampshire Basin (England), Proceedings of the Geologists' Association, 87, p. 273–284.

Daley, B., 1972, Some problems concerning the early Tertiary climate of southern Britain, Palaeogeography, Palaeoclimatology, Palaeoecology, v. 11, p. 177–190.

Daley, B., 1989, Silica pseudomorphs from the Bembridge Limestone (upper Eocene) of the Isle of Wight, southern England, and their palaeoclimatic significance, Palaeogeography, Palaeoclimatology, Palaeoecology, v. 69, p. 233–40.

Daley, B. , 1999, Hampshire Basin: Isle of Wight localities, in British Tertiary Stratigraphy, B. Daley and P. Balson, Joint Nature Conservation Committee, Peterborough, England, p. 85–158.

Daley, B., and N. Edwards, 1971, Palaeogene warping in the Isle of Wight, Geological Magazine, v. 108, p. 399–405.

Daley, B., and N. Edwards, 1990, The Bembridge Limestone (Late Eocene), Isle of Wight, southern England: a stratigraphical revision, Tertiary Research, v. 12, p. 51–64.

Eaton, G.L., 1971a, The use of microplankton in resolving stratigraphical problems in the Eocene of the Isle of Wight, Journal of the Geological Society of London, 127, p. 281–283.

Eaton, G.L., 1971b, A morphogenetic series of dinoflagellate cysts from the Bracklesham Beds of the Isle of Wight, Hampshire, England, Proceedings II Planktonic Conference, Roma, p. 355–379.

Eaton, G.L., 1976, Dinoflagellate cysts from the Bracklesham Beds (Eocene) of the Isle of Wight, southern England, Bulletin of the British Museum (Natural History) Geology, 26, p. 225–332.

Edwards, R. A., and E. C. Freshney, 1987a, Lithostratigraphical classification of the Hampshire Basin Palaeogene deposits (Reading Formation to Headon Formation), Tertiary Research, v. 8, p. 43–73.

Edwards, R. A., and E. C. Freshney, 1987b, Geology of the country around Southampton, Memoirs of the Geological Survey of the United Kingdom, London, HMSO, 111 p.

Feist-Castel, M., 1977, Evolution of the Charophyte floras in the Upper Eocene and Lower Oligocene of the Isle of Wight. Palaeontology, 20, p. 143–157.

Freytet, P., and J-P. Plaziat, 1982, Continental carbonate sedimentation and pedogenesis—late Cretaceous and early Tertiary of southern France, Contributions to Sedimentology, v. 12, 213 p.

Garcia Gomez, E., 1990, Sedimentología de la Formacíon Calizas de Bembridge, Eoceno superior continental de la Isla de Wight (Reino Unido), Grado de Salamanca, Universidad de Salamanca (unpublished degree course dissertation), 82 p.

Grambast, L., 1958, Etude sur les Charophytes tertiaires d´Europe occidentale et leurs rapports avec les formes actuelles (2 volumes), Thése d´Sciences, Université de Paris, 150 p.

Hamblin, R. J. O., A. Crosby, P. S. Balson, S. M. Jones, R. A. Chadwick, I. E. Penn, and M. J. Arthur, 1992, The geology of the English Channel, British Geological Survey Offshore Regional Report, No. 10, London, HMSO, 106 p.

Hibsch, C., E. M. Cushing, J. Cabrera, J. Mercier, P. Prasil, and J. J. Jarrige, 1993, Paleostress evolution in Great Britain from Permian to Cenozoic: a microtectonic approach to the geodynamic evolution of the southeastern UK basins, Bulletin des Centres de Recherches, Exploration-Production Elf-Aquitaine, v. 17, p. 303–30.

Hooker, J.J., 1987, Mammalian faunal events in the English Hampshire Basin (late Eocene-early Oligocene) and their application to European biostratigraphy, Münchener Geowissentschaftliche Abhandlungen, 10, p. 109–116.

Hooker, J.J., 1992, British mammalian paleocommunities across the Eocene-Oligocene transition and their environmental implications, in Eocene-Oligocene Climatic and Biotic Evolution, D.R. Prothero and W.A. Berggren, eds., Princeton University Press, p 494–515.

Hooker, J. J., M. E. Collinson, P. F. Van Bergen, R. L. Singer, J. W. de Leeuw, and T. P. Jones, 1995, Reconstruction of land and freshwater palaeoenvironments near the Eocene–Oligocene boundary, southern England, Journal of the Geological Society of London, v. 152, p. 449–68.

Insole, A. N., 1972, Upper Eocene and lower Oligocene mammal faunas from southern England, Ph.D. thesis, University of Bristol (unpublished), 384 p.

Insole, A. N., and B. Daley, 1985, A revision of the lithostratigraphical nomenclature of the late Eocene and early Oligocene strata of the Hampshire Basin, southern England, Tertiary Research, v. 7, p. 67–100.

Irving, E., 1967, Palaeomagnetic evidence for shear along the Tethys, *in* C. G. Adams and D. V. Ager, eds., Aspects of Tethyan biogeography, London, Systematics Association, p. 59–76.

Klappa, C. F., 1979, Cacified filaments in Quaternary calcretes; organo-mineral interaction in the subaerial vadose environment, Journal of Sedimentary Petrology, v. 49, p. 955–968.

Lefévre, J., 1880, Note sur le Bulimus ellipticus Sow., fossile des Calcaires de Bembridge, Ile de Wight, Anales de la Société Malacologique de la Belge, v. 14, p. 82–87.

Liengjaren, M.. L. Costa, and C. Downie, 1980, Dinoflagellate cysts from the Upper Eocene-Lower Oligocene of the Isle of Wight, Palaeontology, 23, p. 475–499.

Martini, E., 1970, Standard Palaeogene calcareous nannoplankton zonation, Nature 226, no 5245, May 9, 1980, p. 560–561.

Melville, R. V., and E. C. Freshney, 1982, British regional geology: the Hampshire Basin and adjoining areas (4th ed.), London, HMSO, 146 p.

Menillet, F., 1980, Le lithofaciès des Calcaire de Beauce (Stampien supérieur et Aquitanien) du Bassin de Paris (France), Mémoires du Bureau de Recherches géologiques et minières, Series 2, v. 4 (1), p. 15–55.

Murray, J. W., and C. A. Wright, 1974, Palaeogene foraminifera and palaeoecology, Hampshire and Paris basins and the English Channel, Palaeontology Special Paper, No. 14, 129 p.

Owen, R., 1846, Description of an upper molar tooth of Dichobune cervinum, from the Eocene marl at Binstead, Isle of Wight, Quarterly Journal of the Geological Society of London, v. 2, p. 420–421.

Pain, T., and R. C. Preece, 1968, The land Mollusca of the Bembridge Limestone, Proceedings of the Isle of Wight Natural History and Archaeological Society (for 1967), v. 6, p. 101–111.

Platt, N. H., 1989, Lacustrine carbonates and pedogenesis: sedimentology and origin of palustrine deposits from the Early Cretaceous Rupelo Formation, W. Cameros Basin N. Spain, Sedimentology, v. 36, p. 665–684.

Plaziat, J. -C., 1981, Late Cretaceous to late Eocene palaeogeographic evolution of southwest Europe, Palaeogeography, Palaeoclimatology, Palaeoecology, v. 36, p. 263–320.

Pratt, S. P., 1831, Remarks on the existence of Anoplotherium and Palaeotherium in the Lower Freshwater Formation at Binstead, near Ryde, in the Isle of Wight, Proceedings of the Geological Society of London, v. 1, p. 239.

Preece, R. C., 1976, A note on the terrestrial Gastropoda of the Bembridge Limestone (upper Eocene/lower Oligocene) of the Isle of Wight, England, Tertiary Research, v. 1, p. 17–19.

Preece, R. C., 1980, The occurrence of Proserpina in the British Tertiary, with the description of a new species, Archiv fur Molluskenkunde, v. 111, p. 49–54.

Reid, E. M., and M. E. J. Chandler, 1926, A note on certain plants from a clay-bed in the Bembridge Limestone near Gurnard, Isle of Wight, Proceedings of the Isle of Wight Natural History and Archaelogical Society (for 1925), v. 1, p. 378.

Reid, E. M., and M. E. J. Chandler, 1933, The Flora of the London Clay, London, British Museum (Natural History), 561 p.
Singer, R. L., 1993, Palynological organic matter from the late Eocene of the Isle of Wight, England, and the Holocene of the Mobile delta, Alabama, U.S.A., Ph.D. thesis, University of London (unpublished), 136 p.
Summerfield, M.A. (1983) Silcrete. *in* A.S.Goudie and K. Pye, eds., Geochemical Sediments and geomorphology, Academic Press, London, p. 59-91.
Townsend, H.A. and E. A. Hailwood, 1985, Magnetostratigraphic correlation of Palaeogene sediments in the Hampshire and London Basins, southern UK, Journal of the Geological Society of London, 142, p. 957–982.
Vail, P. R., R. M. Mitchum, and S. Thompson, 1977, Seismic stratigraphy and global changes of sea level, part 4: Global cycles of relative changes of sea level, Memoir of the American Association of Petroleum Geologists, Number 26, Seismic stratigraphy-application to hydrocarbon exploration, section 2, p. 83–97.
Vinken, R., C. H. Von Daniels, F. Gramman, A. Köthe, R. W. O'B. Knox, F. Kockel, K. J. Meyer, and W. Weiss, (eds.), 1988, The Northwest European Tertiary Basin. Results of the IGCP Project No. 124, Geologisches Jahrbuch, A 100.
White, H. J. O., 1921, A short account of the geology of the Isle of Wight, Memoirs of the Geological Survey of Great Britain, 201 p.
Wright, V. P., N. H. Platt, and W. A. Wimbledon, 1988, Biogenic laminar calcretes: evidence of calcified root-mat horizons in paleosols, Sedimentology, v. 35, p. 603–620.

Section VII

Oligocene-Miocene

Nury, D., 2000, Lacustrine Oligocene basins in southern Provence, France, *in* E. H. Gierlowski-Kordesch and K. R. Kelts, eds., Lake basins through space and time: AAPG Studies in Geology 46, p. 381–388.

Chapter 34

Lacustrine Oligocene Basins in Southern Provence, France

Denise Nury
Laboratoire de Géologie structurale et appliqueé, Provence Université
Marseille, France

GEOGRAPHIC LOCATION AND TECTONIC SITUATION

The Oligocene basins of southern Provence (Nury, 1988 ; Nury and Schreiber, 1997) extend from Fos-sur-mer in the west, to the Luberon massif in the north, and as far east as Toulon (Figure 1). Major basins include the Marseilles Basin, Aix-en-Provence Basin, and the Nerthe Chain area, which contains the Saint-Pierre-les-Martiques Basin and the Le Rouet Basin. The small Bandol and Ollioules Basins lie near Toulon in the southernmost study area. These irregular basins were formed during the extension of the European platform in the Oligocene resulting from the formation of the West-European rift system (Illies, 1975; Arthaud et al., 1980; Bergerat, 1985, 1987) and the Liguro-Provenal rift, which evolved into an oceanic basin (Orsini et al., 1980; Rehault, 1981; Guieu and Roussel, 1990). The extensional tectonics directly affected the patterns of sedimentation within the continental basins.

STRATIGRAPHY AND SEDIMENTOLOGY

Stratigraphy of these Oligocene basins was established through fossils of mammals (Vianey-Liaud, 1979; Hugueney and Ringeade, 1990), charophytes (Riveline, 1986), gastropods (Nury, 1988; Nury and Schreiber, 1997), spores, pollen, and dinocysts (Châteauneuf, 1977; Châteauneuf and Nury, 1995), nannoplankton (Martini, 1988), and other marine data (Bureau de Recherches Géologiques et Minières, 1972; Magne et al., 1987) (Table 1). The Provençal series of the Oligocene has been correlated to the Stampian Parisian series (Cavelier, 1979) and the upper Chattian Dutch series (Sittler, 1965).

With a basement of essentially Mesozoic carbonates, the sediment fill of the Oligocene basins consists mainly of siliciclastics and limestones associated with some lignites and gypsum (Nury, 1988; Nury and Schreiber, 1997). Siliciclastics were derived from a distant source area. The following facies were recognized (Figure 2), whose thickness varies, depending on their age and paleogeographic location.

(1) Massive to bioturbated limestones and claystones (with fine siltstone intercalations) in repetitive sequences (meters to 10 m thick), containing fossils of charophytes, gastropods, and pelecypods).
(2) Laminated to thin-bedded limestones to mudstones (no fossils or bioturbation) (decimeters to 1 m thick).
(3) Porous carbonates (decimeters to 1 m thick).
(4) Microbial biohermal accumulations (decimeters to 10 m thick) composed of meter-scale, branching, cylindrical masses to decimeter-scale, nodular to concentric masses in carbonates described above, or in siliciclastic rocks.
(5) Lignites (decimeters to meters thick) interbedded with claystones, containing rare tree stumps.
(6) Gypsum (decimeters to meters thick) interbedded with carbonates (gypsum is primary as well as diagenetic in origin).
(7) Carbonate breccias and megabreccias (decimeters to 100 m thick) containing olistolithic clasts the size of which range from centimeters to decimeters in diameter and matrix material of heterogeneous carbonates, marls, marlstones, and sandy marls.
(8) Fine to coarse size sands and sandstones containing trough cross-stratification, decimeters to 100 m thick, interbedded with claystones and conglomerates or breccias.
(9) Carbonate or siliciclastic conglomerates containing clasts reaching 50 cm in size, with matrix material composed of carbonates, marls, or sandy marls. Meter-scale channels are common. The mixed series of conglomerates, sands, sandstones, and marlstones reaches 700 m thickness (Poudingues de Marseille, late Chattian).

The top of most sedimentary sequences exhibits diagenetic alteration attributed to paleosol development (and nondeposition episodes), including desiccation surfaces, *Microcodium*, calcrete, and Terra Rossa-like paleosols.

Figure 1—Geographic location of Oligocene exposures/basins in southern Provence. Key: 1 = Oligocene outcrops, 2 = overthrust mountains' limits, 3 = dip-slip fault, 4 = normal fault. Adapted from Nury and Schreiber, 1997.

The depositional environment for the Oligocene is interpreted as fluviolacustrine in origin (Nury, 1988; Nury and Schreiber, 1997) with episodic marine incursions (Figure 2). Freshwater evidence includes the presence of charophytes, lacustrine gastropods *(Limnea, Planorbis),* and sessile freshwater gastropods *(Viviparus, Pseudamnicola, Bithynia, Melanoides)* who normally inhabit shallow pools or slow-moving rivers. Rare pelecypods *(Sphaerium, Pisidium)* could have lived in quiet water (5–10 m depth). The fossil assemblages imply a seasonal tropical climate. Estuarine to marine fossils at certain horizons include euryhaline dinoflagellates (Châteauneuf and Nury, 1995), marine to estuarine Peneides (Nury, 1988), euryhaline fishes (Gaudant, 1978), pelagic and planktonic coelenterates *(Siphonophora, Chondrophorides, Discallioides* nov. gen; Nel et al., 1987). To the south (Le Rouet basin), continental to brackish sediments laterally change into open-marine deposits with coral reefs to brackish deposits (Nury and Thomassin, 1994).

The first six facies listed are interpreted as lacustrine to palustrine deposits, whereas the other facies are interpreted as fluvial to alluvial in origin. Typical facies pattern distribution within the Oligocene basins is shown in Figure 3.

Table 1. Oligocene Stratigraphy of Southern Provence Based on Vertebrates, Charophytes, Gastropods, Pollen, Spores, and Dinocysts. Adapted from Nury and Schreiber, 1997.

EPOCH	STAGE	AGE	MAMMALIAN REFERENCE UNITS	CHAROPHYTE ZONES	GASTROPOD FAUNAS	POLLEN, SPORES and DINOCYSTS	OTHER CHARACTERISTIC DATA
OLIGOCENE	CHATTIAN	Late Chattian	MP 30	*Chara notata*	*Wenzia ramondi* *Tympanotonos margaritaceus* *Tympanotonos labyrinthum* *Potamides lamarckii*	*Armeria* *Cedrus* *Dicolpollis kockelli*	NP 25 (Martini) N 4 (Blow) -23,MY
		Middle Chattian		*Stephanochara ungeri*	*Wenzia ramondi minor*	Cold coniferous *(Abies, Picea, Tsuga)* *Slovakipollis hypophaeoides*	
		Early Chattian	MP 26	*Chara microcera*			
	RUPELIAN (*sensu strictu* STAMPIAN)	Late Rupelian latest			*Potamides lamarckii* *Tympanotonos labyrinthum*	Graminaceae, advent of plenty of Chenopodiaceae *Boehlensipollis hohli* association, *S. hypophaeoides* *Wetzeliella gochti* *Abies, Picea*	
		Late Rupelian early	MP 23	*Rhabdochara major*			
		Middle Rupelian late			Last *Melanoides tourainei* Impoverished striatelles grouping *Brotia laurae*	*Boehlensipollis hohli*	
		Middle Rupelian early					
		Early Rupelian late	MP 22				
		Early Rupelian earliest	MP 21	*Stephanochara pinguis*	Striatelles grouping *(Melanoides tourainei, M. nysti* *M. fasciatus, M. acutus, Nystia chastelli, Viviparus soricinensis* *Pseudamnicola angulifera)*		
EOCENIAN	PRIABONIAN	late middle Ludian	MP 19-20	*Harrisichara tuberculata, H. lineata*	*Galba longiscata* *Nystia hedonensis*		

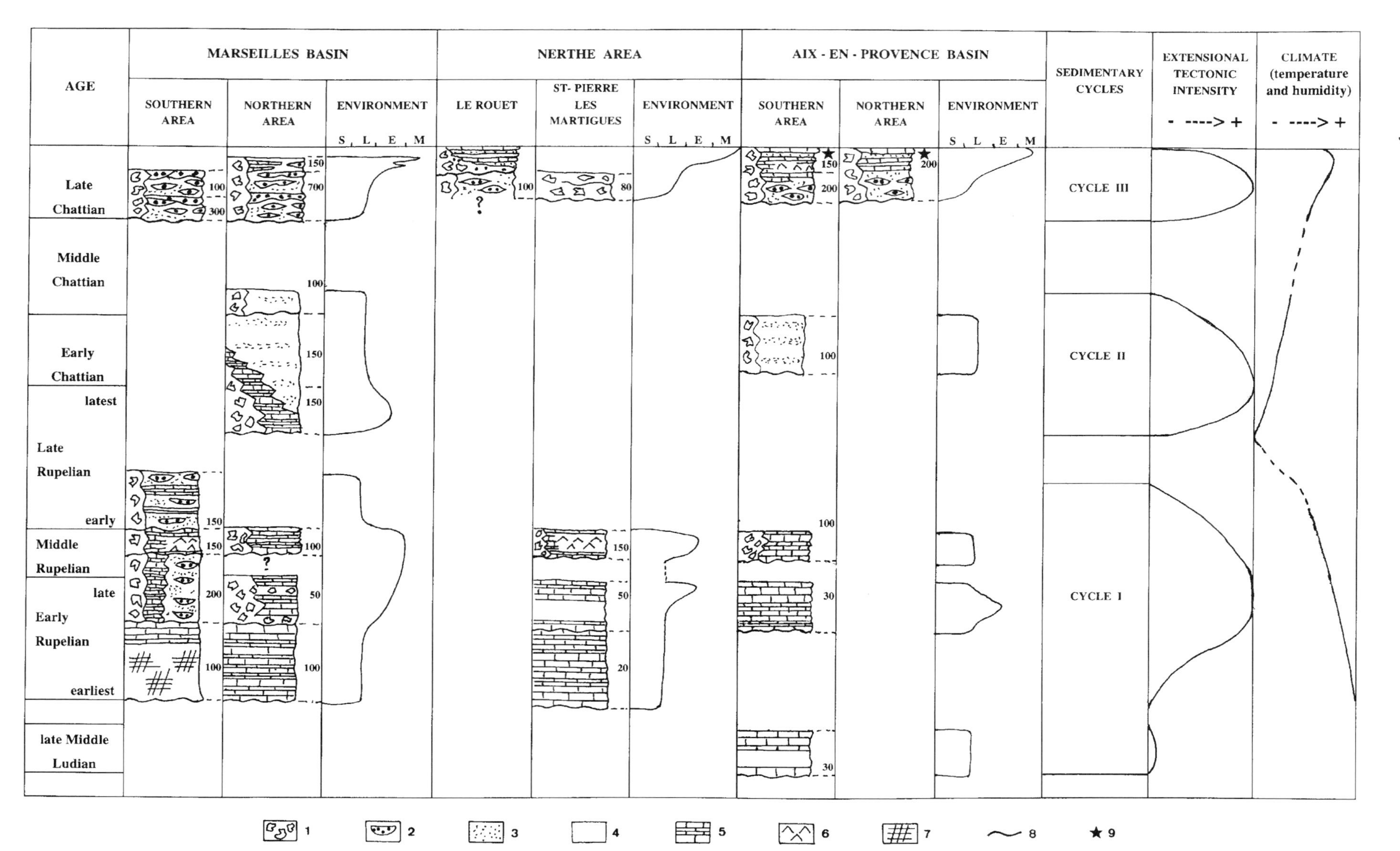

Figure 2—Schematic lithologic columns within the three principal sedimentary areas of the Oligocene showing their relative intensities of tectonic extension, evolution of their depositional paleoenvironments, and inferred changes in temperature and humidity. Key: 1 = breccias, 2 = conglomerates, 3 = sands and sandstones, 4 = claystones, 5 = limestones, 6 = gypsum, 7 = lignite, 8 = discordance, 9 = location of alpine heavy minerals. Depositional paleoenvironment key: S = paleosols, L = lacustrine and fluvial setting, E = estuaries and lagoons, M = marine shorelines. The numbers to the right of sections indicate general thickness in meters of each sequence. Adapted from Nury and Schreiber, 1997.

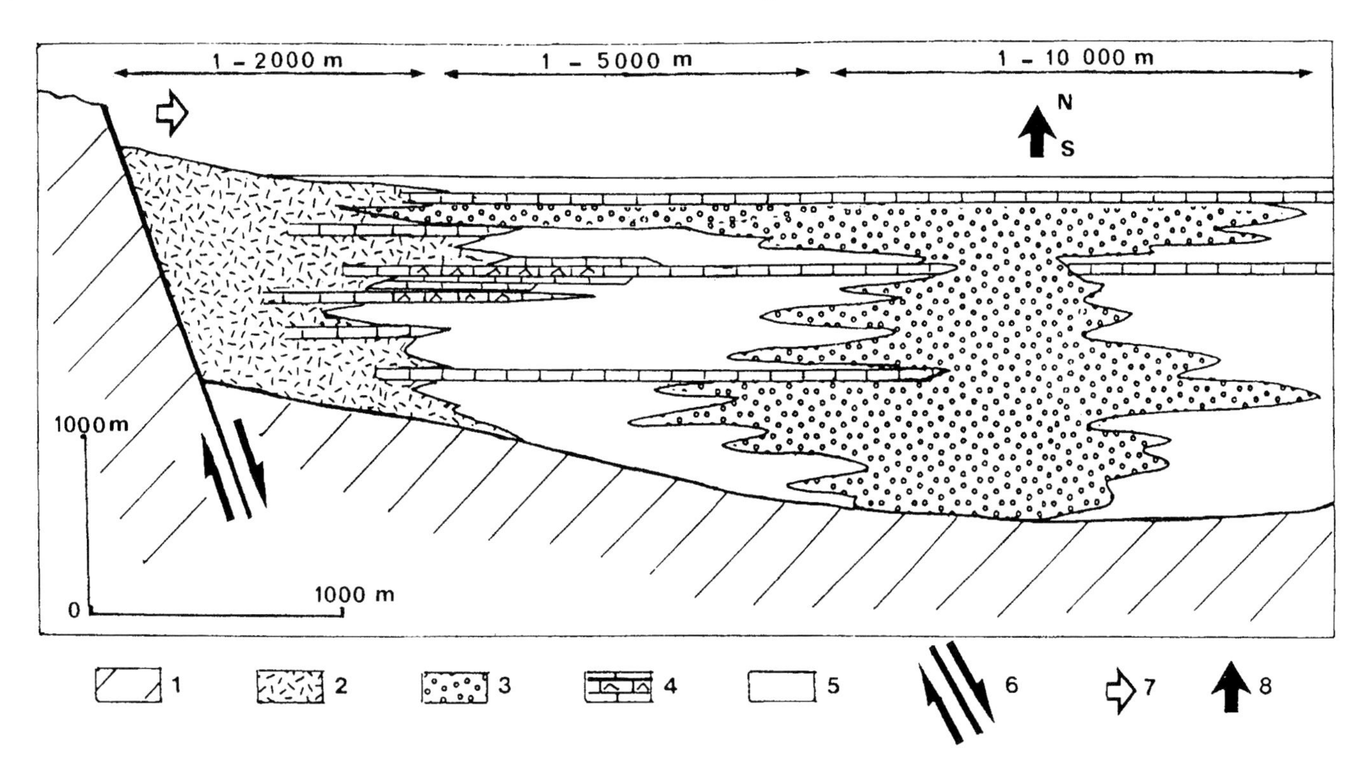

Figure 3—A schematic cross-section of the sedimentary pattern through time in the Oligocene basins of southern Provence. Key: 1 = substrate (Mesozoic limestones), 2 = breccias and megabreccias, 3 = fluvial conglomerates, 4 = limestones (and locally gypsum), 5 = claystones and sandstones, 6 = normal faulting, 7 = transport direction from local source area (carbonate basement), 8 = transport direction of detritus from southern source area. The directions of the faulted boundary range from N20° to 160°.

SEDIMENTATION PATTERNS AND TECTONIC CONTROLS

The active extensional regime controlled cyclic sedimentation within the fault-bounded basins of the Oligocene (Nury, 1988; Hippolyte et al., 1991, 1993; Nury and Schreiber, 1997). Successive sedimentary sequences unconformably rest on each other, and areas of sedimentation appear to have migrated through time. Three main sedimentary cycles are identified in the Oligocene basins (Figure 2), defining times of sedimentation separated by nondeposition and paleosol developments. These sedimentary breaks are postulated to be related to active extensional episodes.

Breccias and megabreccias along the boundary of the basins are often interpreted as alluvial fan sediments deposited along the active normal fault system near carbonate highs. In the siliciclastic fluviatile deposits, paleocurrent markers, grain-size analysis, and mineralogy of the basin fills indicate that source areas lay to the south from crystalline Paleozoic basement approximately 250–350 km distant. Similar mineralogies (andalusite and heavy minerals) in the fluvial deposits of the adjacent Pyrenees and Sardinia (predrift positions) indicate that the common southern source area delivered sediments all through the Oligocene until the middle late Chattian. During the remaining Oligocene, sediment sources (related to the Alpine orogeny) appear to have shifted from the northeast (glaucophane and heavy minerals), a prelude to the opening of the Mediterranean Sea (Le Douaran et al., 1984; Guieu and Arlhac, 1986; Nury, 1988; Nury and Schreiber, 1997).

Marine incursions into the Oligocene basins may have occurred in association with the most active periods of extension (Figure 2) through connections with the southern crystalline source area and the adjacent Ligurian Tethys. The final marine transgression (late Chattian) is timed with the opening of the western Mediterranean Sea (Nury, 1988, 1994; Nury and Thomassin, 1994; Châteauneuf and Nury, 1995).

There is a concordance between the first influx of marine waters (fossil evidence) and the first appearance of gypsum deposits during each gypsum deposition period. Another source for gypsum may also be recycled Triassic sulfates, as concluded by Fontes et al. (1980) in their stable isotope study. With possible marine and recycled Triassic gypsum sources, lake bodies turned into lagoons or estuaries, which readily precipitated and recycled gypsum. Perhaps a more arid climate at this time also promoted gypsum formation.

SUMMARY

The Oligocene basins of southern Provence formed in association with the extensional tectonics resulting from the opening of the western Mediterranean Sea and the West-European rift system. Sedimentary cycles and patterns within the various basins were related to episodic tectonic movements that altered source areas and hydrology of the basins through time and their geographic nearness to the Tethys, which determined frequency and amount of marine water influx through time. A change in the aridity of the climate may have promoted the precipitation of gypsum in a system saturated with probable recycled Triassic gypsum in addition to the periodic influx of marine waters. The carbonate-dominated lacustrine systems in the basins are the result of Mesozoic limestone basement rocks.

REFERENCES CITED

Arthaud, F., M. Ogier, and M. Séguret, 1980, Géologie et géophysique du golfe du Lion et de sa bordure Nord. Bull. Rech. Géol. Min. v. 2 (3), p. 175–193.

Bergerat, F., 1985, Déformations cassantes et champs de contraintes tertiaires dans la plateforme européenne. Thèse d' Etat des Sciences, Université Paris VI, Paris, 317 p.

Bergerat, F., 1987, Stress fields in the European platform at the time of the Africa-Eurasia collision. Tectonics, v. 6 (2), p. 99–132.

Bureau de Recherche Géologoques et Minières, 1972, Contribution à l'étude de l'Aquitanien. La coupe de Carry-Le-Rouet (Bouches-du-Rhône)—V°congrès du Néogène méditerranéen—Vol. III. Bull. Bur. Rech. Géol. Min., Orléans (2), section I, 135 p.

Cavelier, C., 1979, La limite Eocène–Oligocène en Europe occidentale. Mém. Sciences Geologiques (Université Louis Pasteur, Strasbourg), v. 54, 280 p.

Châteauneuf, J. J., 1977, Etude palynologique de l'Oligocène du bassin de Marseille. Géologie Méditerranéenne v. 4 (1), p. 37–46.

Châteauneuf, J. J., and D. Nury, 1995, La flore de l'Oligocène de Provence méridionale: implications stratigraphiques, environementales et climatiques. Géologie de la France v. 2, p. 43–55.

Fontes, J. C., J. Gaudant, and G. Truc, 1980, Données paléoécologiques, teneur en isotopes lourds et paleohydrologie du bassin gypsifère oligocène d'Aix-en-Provence. Bulletin de la Sociéte de Géologique de France (Ser. 7), v. 22 (3), p. 491–500.

Gaudant, J., 1978, Sur les conditions de gisement de l'ichthyofaune oligocène d'Aix-en-Provence (bouches-du-Rhône)—Essai de définition d'un modèle paléoécologique et paléogéographique. Geobios, v. 11, p. 393–397.

Guieu, G., and P. Arlhac, 1986, Hypothèse d'un bombement crustal Golfe du Lion-Sud Provence entre le Crétacé supérieur et l'Oligocène. Mécanismes, conséquences. Comptes rendus de' la Académie des Sciences, Paris (Ser. II), v. 303D, p. 1691–1696.

Guieu, G., and J. Roussel, 1990, Arguments for the pre-rift uplift and rift propagation in the Ligurian-Provenal Basin (northwestern Medditerranean) in the light of Pyrenean Provenal orogeny. Tectonics, v. 9 (5), p. 1113–1142.

Hippolyte, J. C., D. Nury, J. Angelier, and F. Bergerat, 1991, Relation entre tectonique extensive et sédimentation continentale: exemple des bassins oligocènes de Marseille et de Basse Provence. Bulletin de la Sociéte Géologique de France, v. 162 (6), p. 1083–1094.

Hippolyte, J. C., J. Angelier, F. Bergerat, D. Nury, and G. Guieu, 1993, Tectonic-stratigraphic record of paleostress time changes in the Oligocene basins of the Provence, Southern France, Tectonophysics, v. 226, p. 15–35.

Hugueney, M., and M. Ringeade, 1990, Synthesis on the "Aquitanian" lagomorph and rodent faunas of the Aquitaine basin (France), *in* E. H. Lindsay, V. Falbusch, and P. Mein eds., European Neogene Mamal Chronology, NATO ASI Séries A: Life sciences, v. 180, Plenum Press, New York, p. 139–156.

Illies, J. H., 1975, Recent and paleo-intraplate tectonics in stable Europe and Rhinegraben rift system. Tectonophysics, v. 29, p. 251–264.

Le Douaran, S., J. Burrus, and F. Avedik, 1984, Deep structures of the North Western Mediterranean basin—Results of the two ship seismic survey. Marine Geology, v. 55, p. 325–345.

Magne, J., Y. Gourinard, and M. -J., Wallez, 1987, Comparaison des étages du Miocène inférieur définis par stratotypes ou par zones paléontologiques. Strata (Ser. 1), v. 1, (3), p. 95–107.

Martini, E., 1988, Late Oligocene and Early Miocene calcareous nannoplankton (Remarks on French and Moroccan sections). Newletter of Stratigraphy, v. 18 (2), p. 75–80.

Nel, A., G. A. Gill, and D. Nury, 1987, Découvertes d'empreintes attribuables à des coelentérés siphonophores dans l'Oligocène de Provence. Comptes rendus de' la Académie des Sciences, Paris (Ser. II), v. 35, p. 635–641.

Nury, D., 1988, L'Oligocène de Provence meridionale. Stratigraphie, dynamique sédimentaire, reconstitutions paléogéographiques. Documents du Bulletin du Bureau de Recherches Géologiques et Miniéres (Orleans), v. 163, 411 p. (Thèse d'État ès Sciences, 1987, Université Provence, Marseille, France).

Nury D., 1994, Relations géométriques entre carbonates et évaporites. Exemple de l'Oligocène terminal de la région Marseillaise. Géologie méditerranéenne, v. 21 (1–2), p. 85–94.

Nury D., and B. A. Thomassin, 1994, Paléoenvironnement tropicaux, marins et lagunaires d'un littoral abrité (fonds meubles à bancs coralliens, lagune évaporitique) à l'Oligocène Terminal (région d'Aix-en-Provence/Marseille, France). Géologie méditerranéenne, v. 21 (1–2), p. 95–108.

Nury D., and B. C. Schreiber, 1997, The Paleogene basins of southern Provence, *in* G. Busson and B. C. Schreiber, eds., Sedimentary Deposition in Rift and Foreland Basins in France and Spain (Paleogene and Lower Neogene), Columbia University Press, New York, p. 240–300.

Orsini, J. B., C. Coulon, and T. Cocozza, 1980, La dérive cénozoïque de la Corse et de la Sardaigne. Géologie Alpine, v. 56, p. 169–202.

Sittler, C, 1965, Le Paléogène des fossés Rhénan et

Rhodanien. Etudes sédimentologiques et paléoclimatiques. Mémoires du Services de la Carte géologique d' Alsace et d'Lorraine, v. 24, p. 1–392.
Rehault, J. P., 1981, Evolution tectonique et sédimentaire du bassin ligure (Mediterranée occidentale). Thèse de Doctorat d'Etat, Université Paris VI, 132 p.
Riveline, J., 1986, Les Charophytes du Paléogène et du Miocène inferieur d'Europe occidentale—Biostratigraphie des formations continentales. Cahiers de Paléontologie, Centre National de la Recherche Scientifique, Paris, 227 p.
Vianey-Liaud, M., 1979, Evolution des rongeurs à l'Oligocène en Europe occidentale. Paleontographica, v. 166A, p. 136–236.

Ménillet, F., and L. Rasplus, 2000, The Calcaire du Berry, Paris Basin, France, *in* E. H. Gierlowski-Kordesch and K. R. Kelts, eds. Lake basins through space and time: AAPG Studies in Geology 46, p. 389–394.

Chapter 35

The Calcaire du Berry, Paris Basin, France

François Ménillet
BRGM, SGR/ALS, Lingelsheim
Strasbourg, France

Léopold Rasplus
Laboratoire de Géologie des Systèmes sédimentaires, Faculté des Sciences et Techniques, Université de Tours
Tours, France

INTRODUCTION

The Calcaire du Berry (CDB) is a formation exposed near the southern and eastern border of the Paris Basin of France (Figure 1). The CDB fills a series of small half-grabens in the area of Mehun-sur-Yèvre and Châteauneuf-sur-Cher. These half-grabens formed within the Jurassic bedrock of the Paris Basin during the Priabonian–Rupelian (Lablanche, 1982). The CDB is dated as middle Priabonian to late Rupelian (late Eocene to middle Oligocene) through mollusks and charophytes (Jodot, 1947; Guillemin, 1976; Cavelier et al., 1979; Riveline, 1986) (see Table 1).

STRATIGRAPHY

The CDB in the eastern portion of the study area (Figure 2) rests on Jurassic limestones and Eocene kaolinitic clays (containing iron pisolites) found as ferricrusts or in karstic structures. To the southwest, the CDB overlies the sandstones and clays of the Brenne Formation (late Ypresian to early Rupelian). Late Pleistocene–Early Quaternary sands cover the CDB in some parts of the study area. Thickness of the CDB ranges from 20 to 40 m, with a maximum of 42 m at Saint-Ambroix (see Figure 1).

SEDIMENTOLOGY

The CDB is mainly composed of limestones and cherts and served as the building stone for Bourges Cathedral. Eight lithofacies (Table 2) have been described from the CDB (Lablanche, 1982). Early diagenetic features include rhizolith and glaebule formation (pedogenesis), cementation, in situ brecciation, recrystallization, and karstification.

The paleoenvironments of the CDB are interpreted to be lacustrine, palustrine, and pedogenic in origin (Table 2). Figure 3 illustrates a detailed lithologic section of the upper part of a lacustrine to palustrine sequence at Mehun-sur-Yèvre (location in Figure 1). Figure 4 presents limestone facies relationships in the interpreted depositional paleoenvironment of the CDB in a cross-section across the Mehun-sur-Yèvre area (see Figure 2 for exact location). The paleoclimate during the deposition of the CDB is inferred to have been dry, with seasonal contrasts and cool periods (Chateauneuf, 1980).

GEOCHEMISTRY

Bulk analysis of the CDB yielded calcite, quartz (chert and minor amounts of detrital quartz), clays (mainly smectite, with palygorskite in nodules) in the "calcaire vermiculé" facies, and some amounts of kaolinite and illite in breccia and vesicular facies. For a 3–4 km radius around Mehun-sur-Yèvre and near the village of Quincy (Figure 4), limestone and breccia cements contain quincyte, a pink sepiolite colored by organic pigments (Louis et al., 1968; Youde, 1971).

REFERENCES CITED

Cavelier, C., C. Guillemin, G. Lablanche, L. Rasplus, and J. Riveline, 1979, Précisions sur l'âge des calcaires lacustres du sud du Bassin de Paris d'après les characées et les mollusques. Bull. Bureau de Recherches Géologiques et Minières (2ème série, sect. 1), no. 1, p. 27–30.

Chateauneuf, J. J., 1980, Palynostratigraphie et paléoclimatologie de l'Eocène supérieur et de l'Oligocène du Bassin de Paris (France). Mém. BRGM, Orléans, no. 110, 310 p.

Guillemin, C. B., 1976, Les formations carbonatées dulçaquicoles tertiaires de la région Centre: Briare, Château-Landon, Berry, Beauce. Master's thesis (Thèse 3ème Cycle), Univ. Orléans, 258 p.

Jodot, P., 1947, L'âge des formations continentales

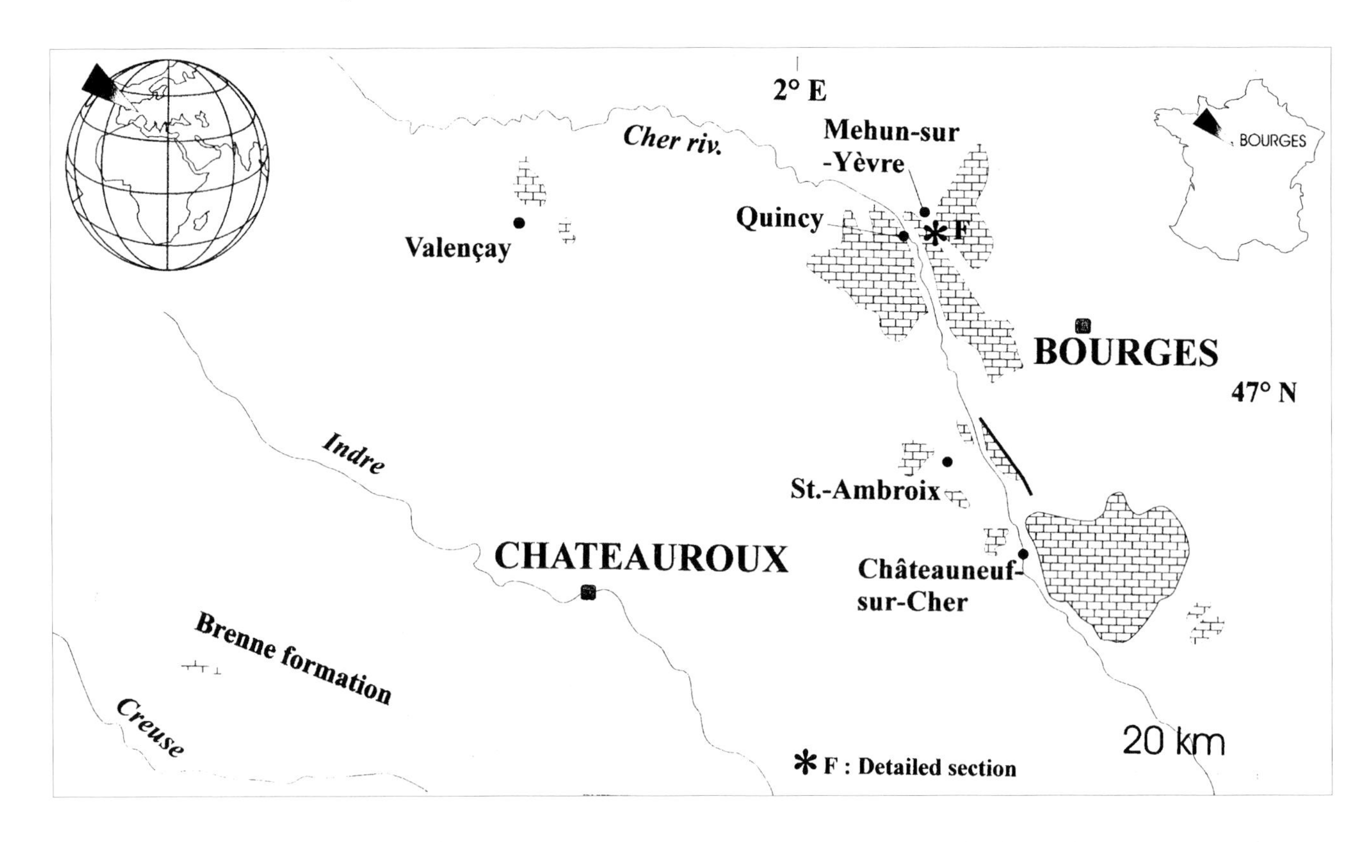

Figure 1—Outcrop exposures of the Calcaire du Berry (Eocene–Oligocene) in the Paris Basin (adapted from Lablanche, 1982).

Table 1. Fossils of the Calcaire du Berry (Eocene–Oligocene), Paris Basin, France.

Stratigraphy	Mollusks	Charophytes
Chara microcera zone upper Rupelian (Cavelier et al., 1979)		(Riveline, 1986) *Nitellopsis (Tectochara) meriani* *Rhabdochara major* *Psilochara acuta* *Chara microcera* *Sphaerochara humeri* var. *longuscula* *Sphaerochara* cf. *haedonensis* *Chara* cf. *cylindrica* *Chara* cf. *tornata*
Middle to upper Ludian	(Guillemin, 1976) *Bithinia monthiersi* *B. epiedensis* *Radix fabulum*	
	(Jodot, 1947) *Lymnea (Stagnicola) ostrogallica* *L. (St.) hordlensis* *L. (St.) duranti* *Planorbis (Planorbina) goniobasis* *P. (Pl.) vasseuri* *P. (Coretus) lendonensis* *P. (Hippeutis) headonensis*	

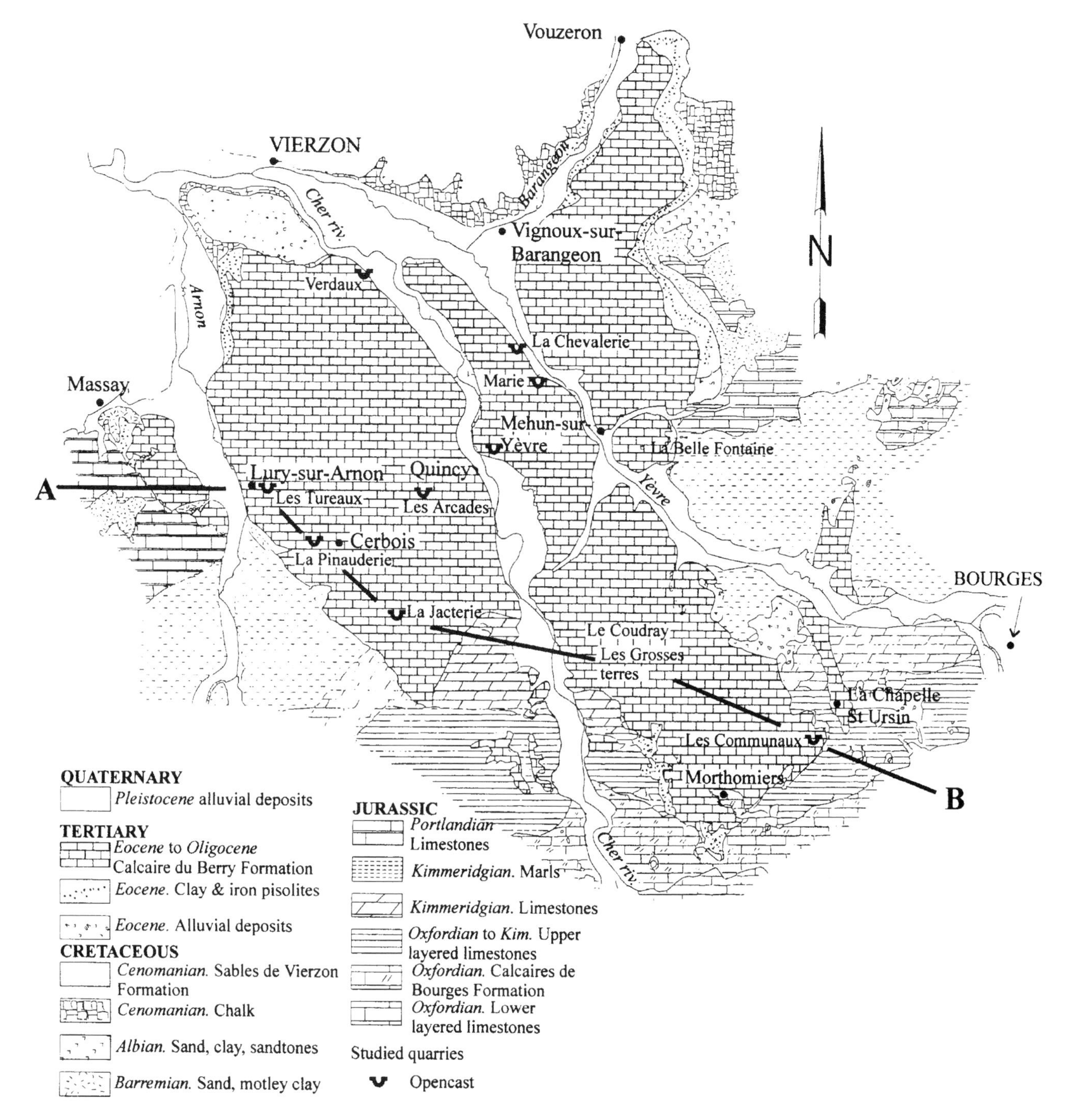

Figure 2—Detailed geologic map and stratigraphy of the Berry area (Mehun-sur-Yèvre Basin of Figure 1) within the Paris Basin of north-central France (adapted from Lablanche, 1982).

nummulithiques de la Brenne. C.R. Somm. Soc. Géol. France (5th series), v. 17, p. 327–329.

Lablanche, G., 1982, Les calcaires lacustres paléogènes de la Champagne berrichonne (étude cartographique et pétrographique, reconstitution du milieu de sédimentation). Diplôme d'Etudes Supérieures, Univ. Paris VI et Doc. BRGM, no. 49, 127 p.

Louis, M., C. Guillemin, S. C. Goni, and J. P. Ragot, 1968, Coloration rose carmin d'une sépiolite éocène, la quincyte, par les pigments organiques. 4th International Meeting on Organic Geochemistry, Univ. Amsterdam, Proceedings, p. 553–566.

Riveline, J., 1986, Les Charophytes du Paléogène et du Miocène inférieur d'Europe occidentale. Cahiers de paléontologie, Centre National de la Recherche Scientifique, 221 p.

Youde, B., 1971, Investigations of a geologic pigment, quincyite. Master's thesis, Organic Geochemistry Unit, Bristol, Univ. Bristol.

Table 2. Limestone Types of the Calcaire du Berry (Eocene–Oligocene), Paris Basin, France.

Lithofacies	Distribution	Allochems and Fossils	Interpretation	Average Thickness (m)
marls and clays	mainly at the base of the CDB*	calcareous nodules	lacustrine	2–3 range: cm to m
chalky limestone	localized	calcareous nodules	lacustrine	0.5–1 range: cm to m
massive limestones	localized		lacustrine	2–15 range: cm to m
laminated limestones	rare	algal remains glaebules in lenses	stromatolites calcrete	0.50–1 laminae: mm to 5 cm thick
"calcaire vermiculé"	most abundant	rootlet molds, freshwater molluscs	palustrine	2–4 range: cm to m
vesicular limestones	in small lenses or patches in other facies	solution channels, birds' eyes, freshwater mollusks, root molds	palustrine	0.50–1 range: cm to m
nodular limestones	abundant in Châteauneuf Basin	glaebules	calcrete	0.2–1 range: cm to m
brecciated limestones	in lenses, mainly in massive limestones	puzzle fabrics, rare black pebbles, clasts of Jurassic limestones	calcrete karst features	0.1–0.2 range: cm to dm

*CDB = Calcaire du Berry

ADDITIONAL REFERENCES OF INTEREST

Abrard, R., 1950, Géologie régionale du Bassin de Paris. Payot, Paris, 397 p.

Aufrère, L., 1930, Les formations continentales éocènes du Berry oriental et méridional (calcaire lacustre et sidérolithique des auteurs) et leur signification morphologique. Assoc. Française Avancement des Sciences, Algiers, p. 152–155.

Cavelier, C., and J. P. Ragot, 1973, Excursion à Quincy organisée à l'occasion du 6ème congrès de géochimie organique. Rueil-Malmaison, Livret Guide, 13 p.

Chateauneuf, J. J., 1977, Nouvelle contribution de la palynologie à la datation du Tertiaire continental de Brenne. Bull. BRGM, sect. 1, no. 4, p. 353–355.

Closier, L., 1982, Carte géologique de la France à 1/50000, feuille no. 458, Sancoins et notice explicative, BRGM, Orléans.

Cuvier, G., 1825, Recherche sur les ossements fossiles, vol. 2 (1), 232 p.

Dagincourt, E., L. de Launay, and M. Busquet, 1941, Carte géologique de la France à 1/80000, feuille Saint-Pierre-le-Moutiers et notice explicative, 2nd édition. Service de la Carte géologique de la France, Paris.

Debrand-Passard, S., C. Cavelier, G. Lablanche, D. Flamand, and J.P. Soulas, 1977, Carte géologique de la France à 1/50000, feuille no. 519, Bourges et notice explicative, BRGM Orléans.

Debrand-Passard, S., G. Lablanche, and F. Bavouzet, 1975, Carte géologique de la France à 1/50000, feuille no. 545, Issoudun et notice explicative, BRGM Orléans.

Debrand-Passard, S., G. Lablanche, B. Martin, and B. Petitfils, 1978, Carte géologique de la France à 1/50000, feuille no. 518, Vatan et notice explicative, BRGM Orléans.

Dealance, J. H., S. Debrand-Passard, L. Clozier, J. F. Ingargiola, G. Lablanche, J. C. Menot, and B. Roy, 1977, Carte géologique de la France à 1/50000, feuille no. 521, Nevers et notice explicative, BRGM Orléans.

Dollfus, G. F., 1901, Feuille de Bourges au 1/320000. Terrains tertiaires. Bull. Serv. Carte Géol. France, 11 (80), p. 22–29.

Dollfus, G. F., 1906, Révision des faunes de mollusques terrestres et fluviatiles du Tertiaire des bassins de Paris et de la Loire. Bull. Soc. Géol. France (4th series), v. 6, p. 249–252.

Dollfus, G. F., 1926, Observations sur le mode de formation des dépôts lacustres. C. R. Somm. Soc. Géol. France, v. 15, p. 143–145.

Donnadieu, J. P., 1976, Données nouvelles sur les formations de l'Eocène continental (Bartonien sens large) du Sud-Ouest du Bassin de paris, les dépôts de Brenne et des confins du Poitou. Bull. Soc. Géol. France (7th series), v. 18, p. 1647–1658.

Douvillé, H., 1875, Note sur le système du Sancerrois et le terrain sidérolithique du Berry. Bull. Soc. Géol.

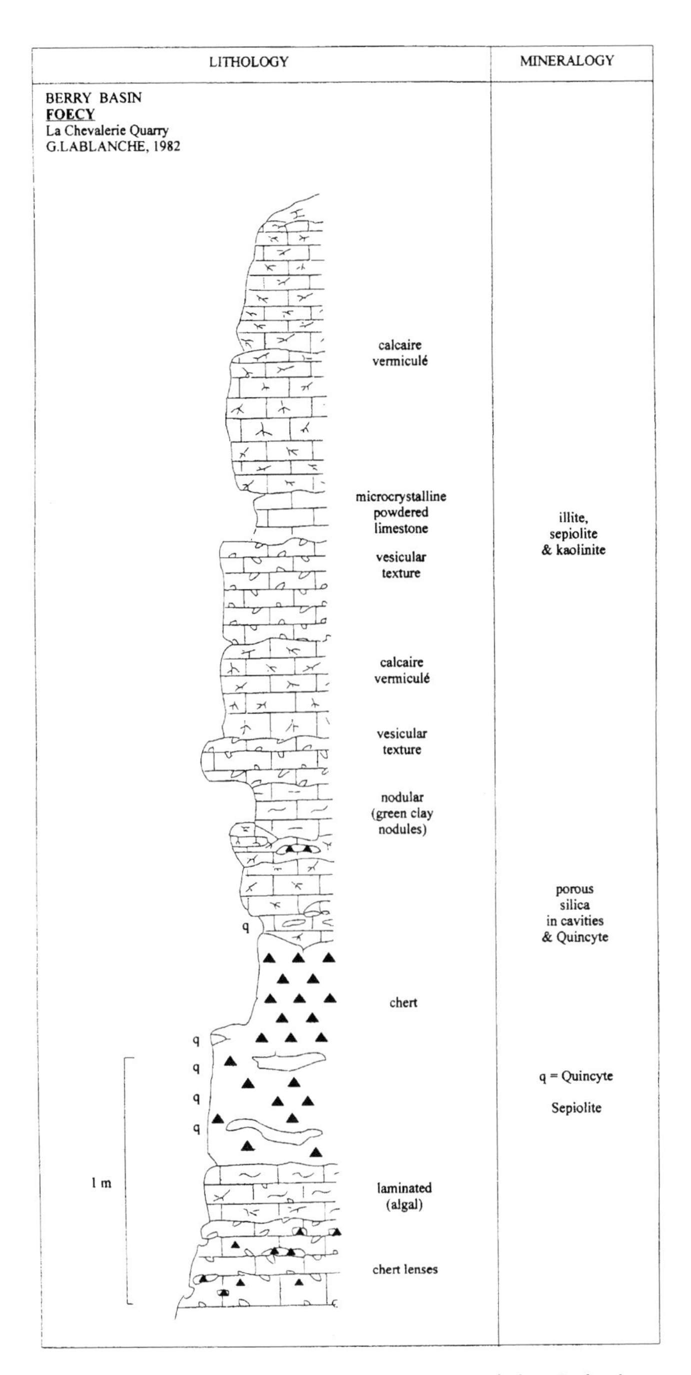

Figure 3—Detailed lithologic section of the Calcaire du Berry at Foecy, La Chevalerie Quarry (adapted from Lablanche, 1982). Exact location of quarry shown in Figures 1 and 2.

France (3rd series), v. 4, p. 104–111.

Douvillé, H., 1926, Sur quelques sédiments de l'époque tertiaire: calcaires lacustres et argiles tertiaires. C.R. Somm. Soc. Géol. France, p. 142–145.

Fabre, J. M., 1838, Description physique du département du Cher et considération géologique sur le mode de formation des terrains métazoïques. Mémoire pour servir à l'explication à la statistique du département du Cher. Jollet-Souchois, Bourges, 192 p.

Gras, J., 1963, Le Bassin de Paris méridional. Etude morphologique. Thèse lettres, Université de Rennes, France, p. 344–373.

Grossouvre, A. de, 1906, Feuille de Bourges au 1/320000. Bull. Serv. Carte Géol. France, v. 17 (115), p. 114–120.

Grossouvre, A. de, 1907, Sur l'âge des calcaires lacustres du Berry. Assoc. Française Avanc. Sciences, v. 2, p. 400–402.

Grossouvre, A. de, 1909, Feuille de Bourges au 1/320000. Bull. Serv. Carte Géol. France, v. 19 (122), p. 25–43.

Grossouvre, A. de, 1912, Sur l'âge des calcaires lacustres et d'autres dépôts superficiels de la Nièvre et du Cher. C.R. Somm. Soc. Géol. France, p. 179–180.

Grossouvre, A. de, M. Busquet, L. de Launay, 1894, Carte Géologique de la France à 1/80000. Feuille Nevers, 1st edition, Service Carte Géol. France, Paris.

Klein, C., 1975, Massif armoricain et Bassin parisien. Contribution à l'étude géologique et géomorphologique d'un massif ancien et de ses enveloppes sédimentaires. Fondation Baulig, Strasbourg, 12, 882 p.

Lablanche, G., 1984, Carte géologique de la France à 1/50000, feuille no. 546, Châteauneuf-sur-Cher et notice explicative. BRGM, Orléans.

Manivit, J., 1977, Carte géologique de la France à 1/50000, feuille no. 490, Selles-sur-Cher et notice explicative. BRGM, Orléans.

Mégnien, F., (Editor), 1980, Lexique des noms de formations (Synthèse géologique du Bassin de paris, vol. III), Mém. BRGM, no. 103, 467 p.

Omalius d'Halloy, J. J., 1812, Sur le gisement du calcaire d'eau douce dans les départements du Cher, de l'Allier et de la Nièvre. Annales des Mines France, v. 32, p. 43–64.

Rabate, P., 1926, Le Berry géologique, climatologique et économique. Langlois, Châteauroux, 164 p.

Rasplus, L., 1978, Contribution à l'étude géologique des formations continentales détritiques tertiaires de la Touraine, de la Brenne et de la Sologne. Ph.D. dissertation (Thèse d'Etat Sciences), Univ. D'Orléans, Vol. 1 and 2, 454 p., Vol. 3, 132 figures, 25 photos, 10 maps. (Thèse condensée. Sciences géologiques, Strasbourg, mém. 66, 227 p.)

Vacher, A., 1908, Le Berry, contribution à l'étude géographique d'une région française. A. Colin, Paris, 548 p.

Valensi, L., 1953, Sur une meulière sphérolithique du calcaire lacustre du Berry. Bull. Soc. Géol. France (6th series), v. 9, p. 841–845.

Vatan, A., 1935, Résumé d'observations sur les conditions de dépôts et la lithologie des terrains sidérolithiques du Berry. C. R. Somm. Soc. Géol. France, (5th series), v. 5, p. 64–65.

Vatan, A., 1936, Remaniements sous-lacustres dans les Calcaires éocènes du Berry. C. R. Somm. Soc. Géol. France (5th series), v. 6, p. 43–45.

Vatan, A., 1939a, Les brêches dans les terrains tertiaires du centre de la France. C. R. Somm. Soc. Géol. France, (5th series), v. 9, p. 54–56.

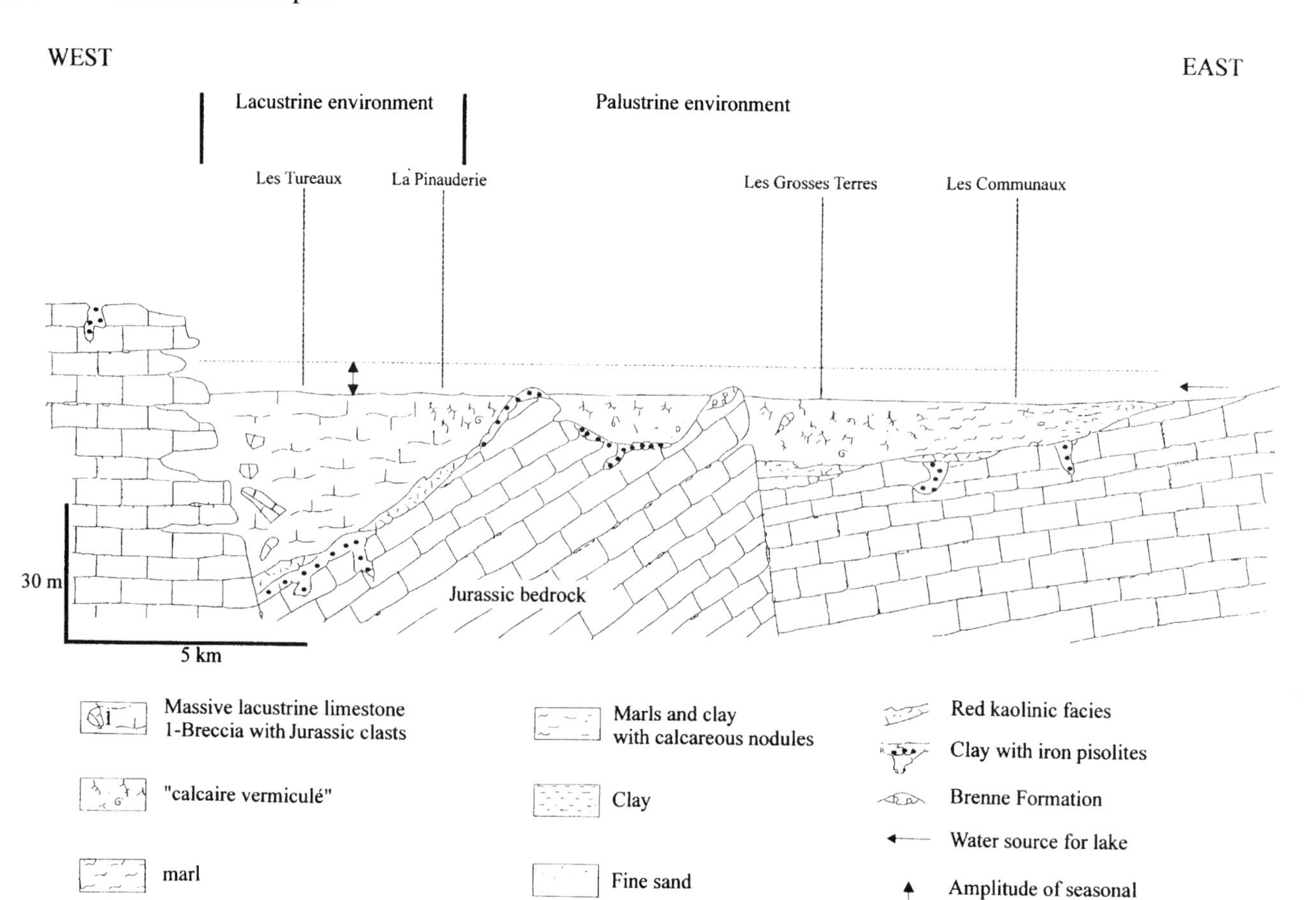

Figure 4—Paleoenvironmental interpretation of the depositional regime of the Calcaire du Berry showing the relationships among the various facies and interpreted water levels in the lacustrine and palustrine paleoenvironment (adapted from Lablanche, 1982). Location of cross-section illustrated in Figure 2 as A-B.

Vatan, A., 1939b, Les structures oolithiques et pisolithiques dans les terrains tertiaires continentaux du centre de la France. C. R. Somm. Soc. Géol. France, (5th series), v. 9, p. 89–91.

Vatan, A., 1947a, La sédimentation continentale tertiaire dans le bassin de Paris méridional. Editions toulousaines de l'Ingénieur, Toulouse, France, 215 p.

Vatan, A., 1947b, Remarques sur la silicification. C. R. Somm. Soc. Géol. France (5th series), v. 17, p. 99–101.

Arenas, C., and G. Pardo, 2000, Neogene lacustrine deposits of the north-central Ebro Basin, northeastern Spain, *in* E. H. Gierlowski-Kordesch and K. R. Kelts, eds., Lake basins through space and time: AAPG Studies in Geology 46, p. 395–406.

Chapter 36

Neogene Lacustrine Deposits of the North-Central Ebro Basin, Northeastern Spain

C. Arenas
G. Pardo
Departamento de Ciencias de la Tierra, Universidad de Zaragoza
Zaragoza, Spain

GEOLOGIC SETTING

The Ebro Basin is located in the northeastern sector of the Iberian peninsula and is bounded by the Pyrenean Ranges to the north, Iberian Ranges to the south, and Catalan Coastal Ranges to the east (Figure 1). This basin originated during the Paleogene, and its evolution was mainly linked to the development of the Pyrenean orogeny (Riba et al., 1983), thus the basin represents the southern foreland basin of the Pyrenean Ranges. The Iberian Ranges and Catalan Coastal Ranges were also active margins but of less importance. From the latest Eocene to the late Miocene, the Ebro Basin was a closed intermontane basin. The depositional systems comprised alluvial fans sourced from the margins and connected to wide areas in the center of the basin occupied by lakes.

During the late Oligocene and Miocene, Pyrenean source areas north of the Ebro Basin depositional systems mainly included Paleozoic–early Oligocene siliciclastic formations, Cretaceous and Paleogene calcareous formations, and Paleozoic igneous and metamorphic rocks. Iberian source areas in the south consisted of Paleozoic and Mesozoic siliciclastic and calcareous rocks; Triassic and Liassic formations contain large volumes of evaporites.

The sedimentary record of the related lacustrine systems mainly consists of sulfate and carbonate deposits. In the Aragonese central part of the basin, between the Pyrenean Ranges and the Iberian Ranges, Neogene carbonate deposits form prominent uplands in the center of the basin (Figure 1): La Muela, La Plana, and La Muela de Borja south of the Ebro River, and the Sierra de Alcubierre, Montes de Castejón, and Sancho Abarca-Plana Negra north of the Ebro River. Sulfate facies mostly crop out at the foothills of these uplands and at the lowest parts of the basin (Figure 2). These lacustrine deposits may also intercalate with distal alluvial sediments.

The aim of this contribution is to show the paleogeographic evolution of the latest Oligocene–late Miocene lacustrine systems of the north-central part of the Ebro basin (Sierra de Alcubierre, Montes de Castejón, and their foothills) and the related northern alluvial systems, directly linked to the Pyrenean margin. This study also contributes to the correlation between the northern and southern areas of the Ebro Basin, the Pyrenean and Iberian domains, respectively. The main stratigraphic and sedimentologic data of the Iberian area are shown in Pérez et al. (1994), as well as in Pérez et al. (1988), Pérez (1989), and Villena et al. (1992; 1996a, b).

STRATIGRAPHY

In the north-central study area (Figure 1) within the Ebro basin, three tecto-sedimentary units (TSU[1]), named N1, N2, and N3, have been characterized (Arenas et al., 1989; Arenas, 1993) (Figures 2–4). Of these three units, only N1 has continuous outcrops throughout the study area and from the northern to the southern margins of the basin. For this reason, it is possible to determine complete relationships between the alluvial and lacustrine systems for unit N1. Boundaries among units are represented by shifts or trend changes in the sequential evolution. At the basin margins, these boundaries are correlated with progressive unconformities, although at the Pyrenean margin only the boundary between N1 and older units can be reported (Figure 4). Units N2 and N3 crop out in the center of the Ebro basin and do not have equivalent alluvial deposits at the studied part of the Pyrenean margin.

The main stratigraphic and evolutionary characteristics of these units are illustrated in Figures 2–4. Ages of units are based on the correlation with the Iberian margin Neogene sequences and on the presence of

[1]A tecto-sedimentary unit is a type of allostratigraphic unit made up of a succession of strata deposited within a concrete interval of geologic time and under a tectonic and sedimentary dynamic of definite polarity (modified from Garrido-Megías, 1982). Its boundaries have basinal extent and are generated by inflections or sharp changes in the rate of allocyclic factors that controlled basin fill dynamics (modified from Pardo et al., 1989, and Villena et al., 1996a).

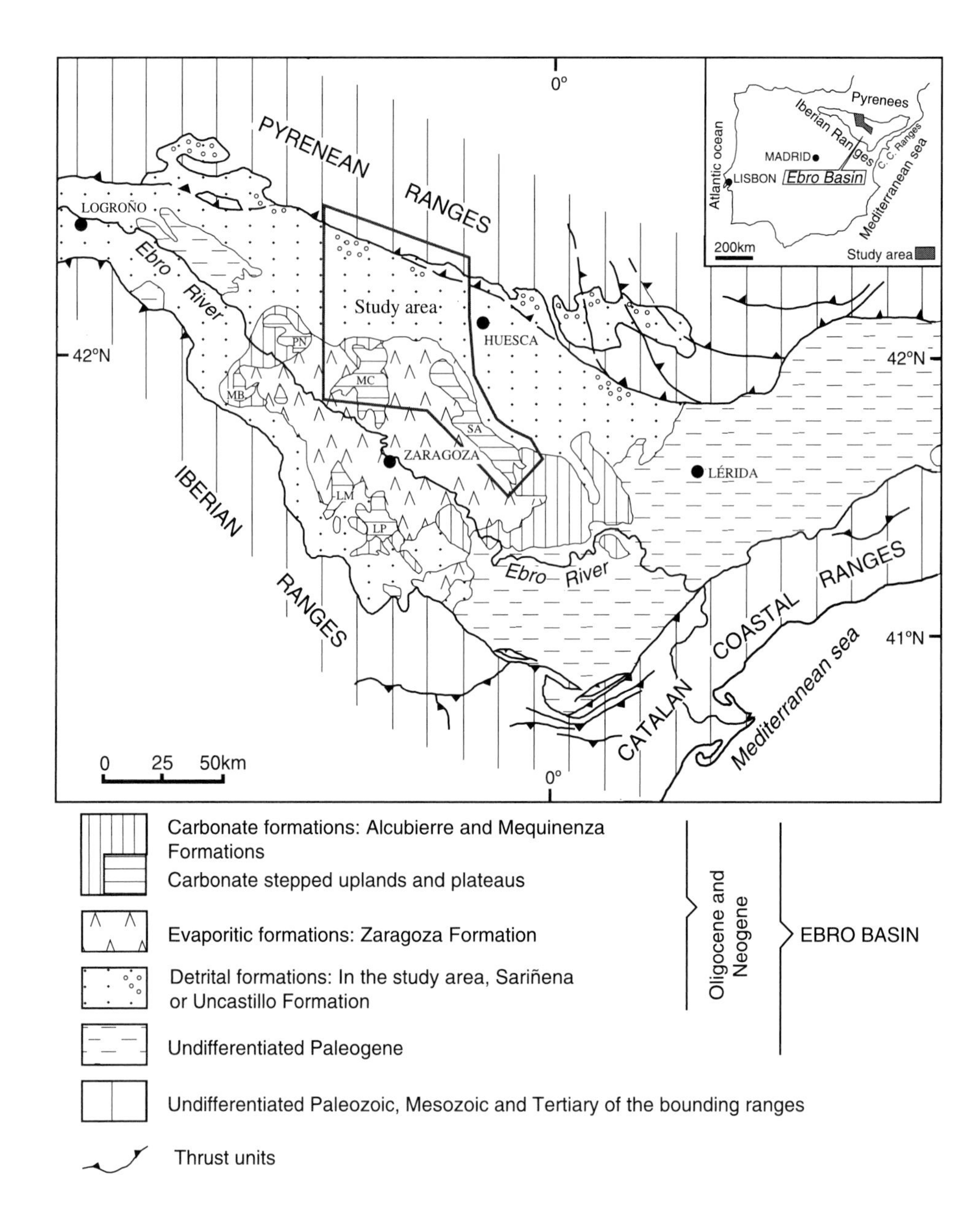

Figure 1—General and geologic map of the Ebro Basin showing the study area on the Iberian peninsula. Uplands: PN = Plana Negra, MC = Montes de Castejón, SA = Sierra de Alcubierre, LP = La Plana, LM = La Muela, MB = Muela de Borja. (Modified from Instituto Tecnológico y Geominero de España, 1995.)

fossil mammal sites within the research region (Agustí et al., 1994).

SEDIMENTOLOGY AND PALEOGEOGRAPHY: LACUSTRINE EVOLUTION

Sedimentologic analysis of the study area in the north-central sector of the Ebro Basin has identified several alluvial and lacustrine facies (Table 1) (Arenas, 1993). In the lacustrine areas, carbonate and sulfate lacustrine facies, as well as distal alluvial facies, are arranged in simple vertical associations or facies sequences, decimeters to 4 m thick. The facies assemblages characterized allow two main sedimentary lacustrine environments to be differentiated; these correspond to two distinct lacustrine situations or lake-systems that alternated through time (Arenas and Pardo, 1999): carbonate and sulfate systems.

Carbonate depositional environments or freshwater lacustrine systems correspond to high lake levels associated with surface waters as permanent supply. The lacustrine area corresponds to a single body of fresh water yielding rich fossiliferous massive limestones surrounded by palustrine margins with bioturbated limestones. Sulfate depositional environments or saline lacustrine systems correspond to low lake levels associated with intermittent, sheet flow supply. It is possible that more than one ponded area existed (Arenas et al., 1999). Gypsum is the main sulfate precipitate. The surrounding saline mud flats underwent evaporative processes, as suggested by sedimentary features, consequent to the lowering of the water table. Transitions from carbonate to sulfate depositional environments and vice versa led to fluctuations in salinity marked by moderate to low salinity phases that are documented by the presence of laminated

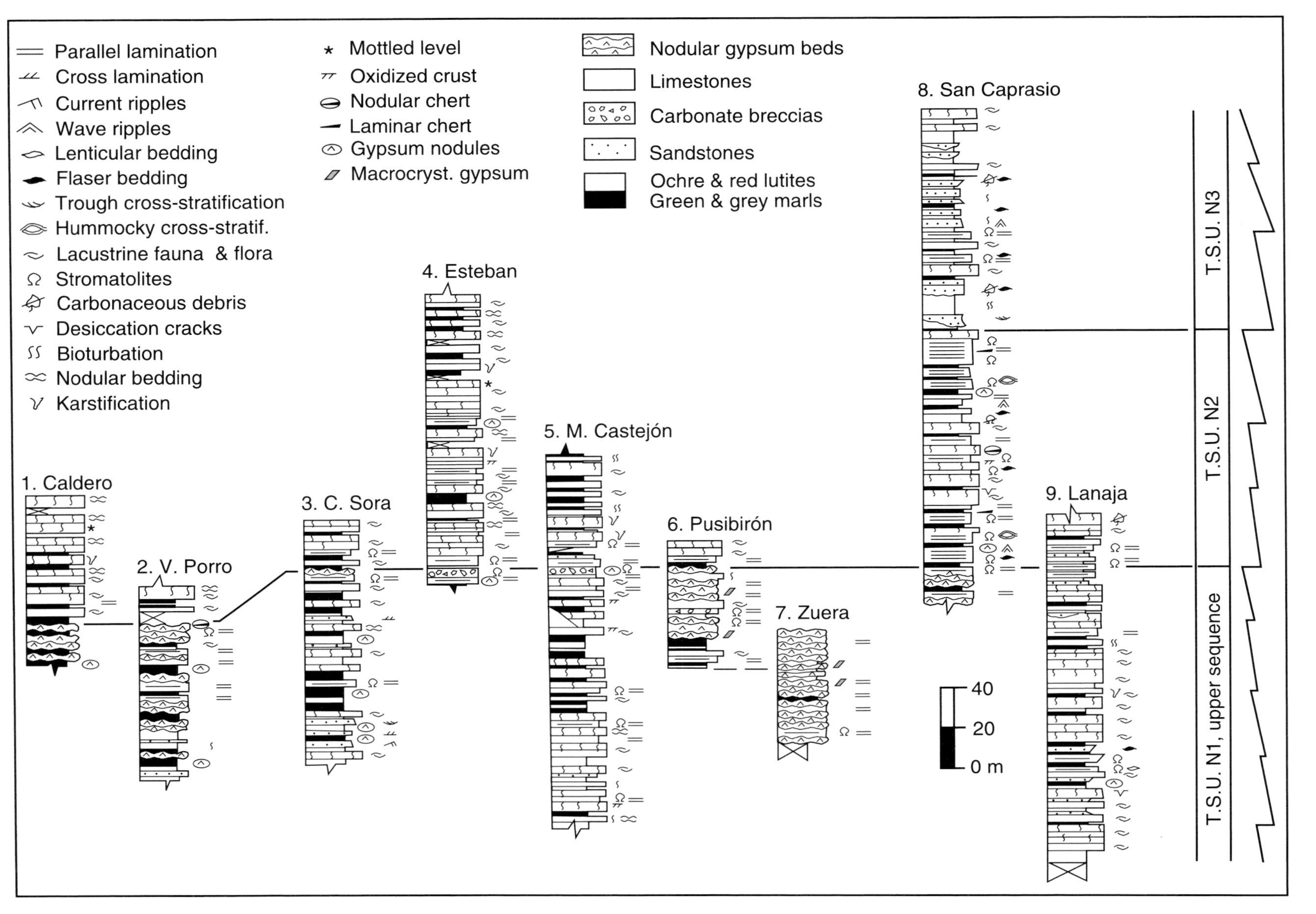

Figure 2—Correlation sketch of the most representative sections of upper Oligocene–upper Miocene rocks in the lacustrine study area in the north-central sector of the Ebro Basin. Location of sections in Figure 3. TSU = tecto-sedimentary unit (see Figure 4 for stratigraphic positions).

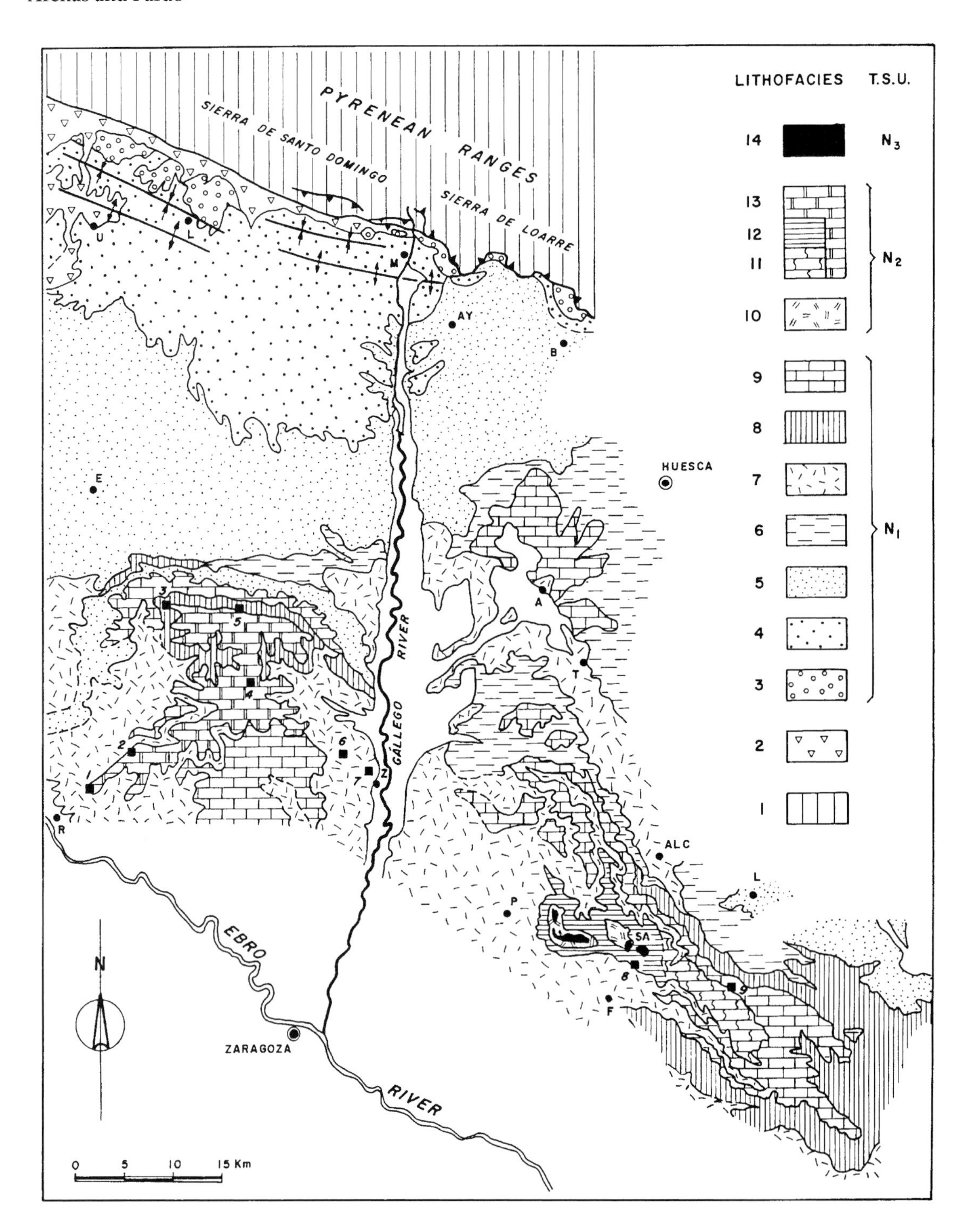

Figure 3—Simplified geologic map of the study area (from Arenas, 1993). Lithofacies: 1 = undifferentiated Mesozoic and Tertiary deposits of the Pyrenees, 2 = undifferentiated upper Oligocene tecto–sedimentary unit, 3 = conglomerates and sandstones (proximal alluvial facies), 4 = sandstones and mudstones (middle alluvial facies), 5 = mudstones and sandstones (distal alluvial facies), 6 and 14 = mudstones, sandstones, and limestones (distal alluvial and lacustrine facies), 7 and 10 = gypsum, marls or mudstones, and limestones ± sandstones (saline mud flats and sulphate lacustrine facies), 8 = marls or mudstones and limestones, 9 and 13 = limestones and marls (carbonate lacustrine facies), 11 = mostly massive and bioturbated limestones and marls, 12 = mostly laminated limestones and marls. 2–5: Uncastillo or Sariñena Formation, 7: Zaragoza Formation, 8–14: Alcubierre Formation, 10: Perdiguera Member of the Alcubierre Formation. Quaternary formations within the study area are white areas. Localities: Ay = Ayerbe, B = Bolea, U = Uncastillo, L = Luesia, E = Ejea de los Caballeros, Z = Zuera, A = Almudévar, Alc. = Alcubierre, L = Lanaja, P = Perdiguera, R = Remolinos, MC = Montes de Castejón, SA = Sierra de Alcubierre. Black squares represent location of sections shown in Figure 2.

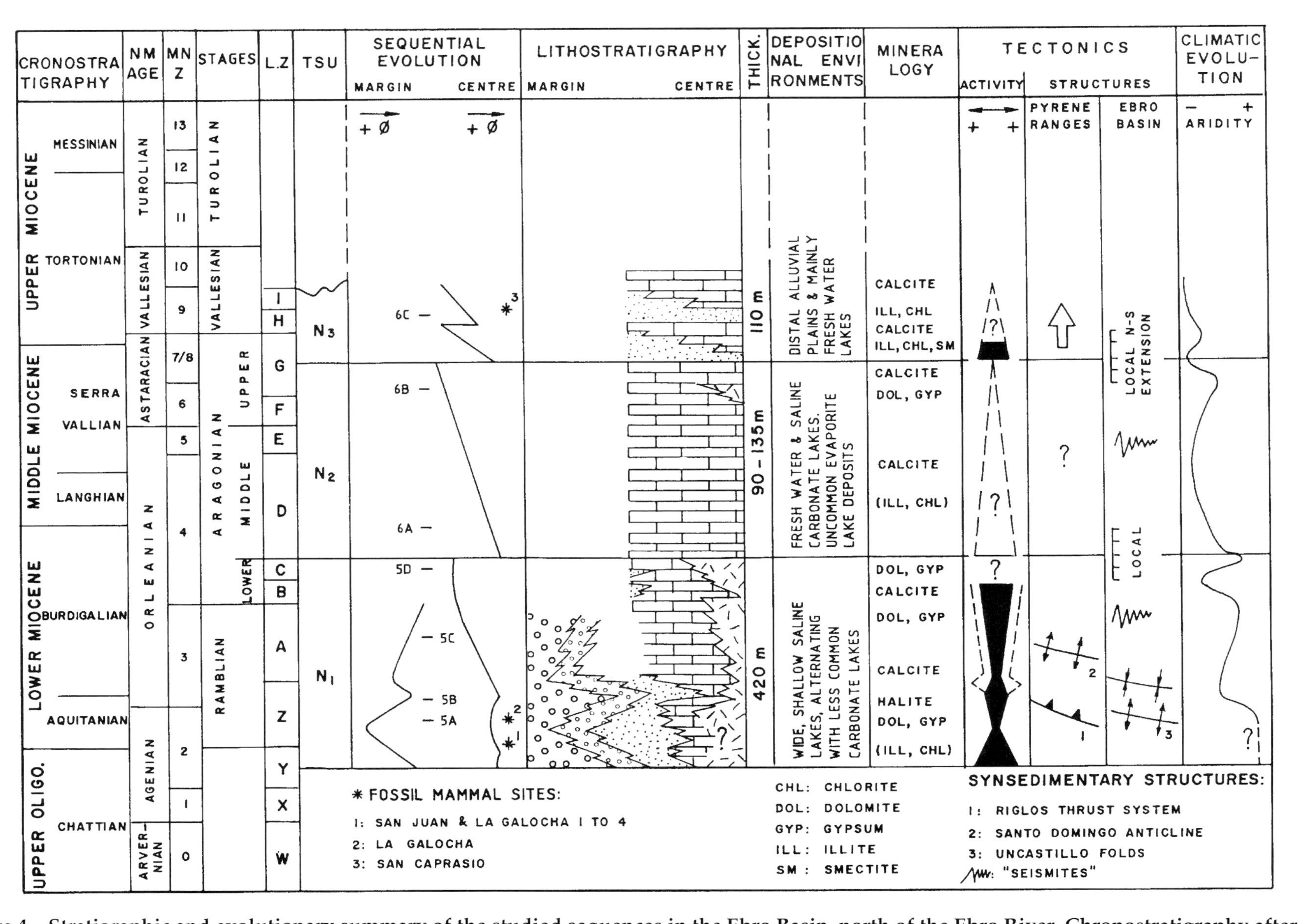

Figure 4—Stratigraphic and evolutionary summary of the studied sequences in the Ebro Basin, north of the Ebro River. Chronostratigraphy after Calvo et al. (1993). N.M. Age = Neogene Mammal Age, M.N.Z. = Mammal Zones, L.Z. = Local Zones. For legend of lithostratigraphic column, see Figure 3. Curves of sequential trends of units show time position of sketches in Figures 5 and 6. Tectonic activity evolution is expressed by black areas, with maximum (wide) and minimum (narrow) relative values. Dashed lines indicate probable or alternative tectonic activity trends. The arrow indicates uplift of Pyrenean source areas. Associated active structures are unknown.

Table 1. Main Characteristics and Depositional Interpretations of the Alluvial and Lacustrine Facies of the Oligocene–Miocene Deposits of the North-Central Part of the Ebro Basin of Spain.*

Facies	Subfacies	Sedimentary Structures	Biologic and Diagenetic Features	Interpretation
Conglomerates	(A) Organized tabular stratification. (B) Poorly organized massive or crude stratification.	Clast-supported. Massive, horizontal, low-angle, trough or planar cross-stratification. Fining or coarsening upward.	None	(A) Proximal fluvial fans. Braided conglomeratic systems. (B) Proximal alluvial fans. Mostly sheet flood deposits.
Sandstones	Depending on internal and external geometries.	Massive, parallel, rippled, or trough cross-stratification.	Bioturbation Mottling #	Meandering and low-sinuosity sandy channels. Sheet flow deposits are more common near or within lacustrine areas.
Green, Gray, and Ochre Mudstones		Massive or parallel lamination.	Bioturbation Mottling G Ø Ô #	Floodplains. Basinward, alluvial plains surrounding lacustrine areas or nearshore lake areas.
Marls	Laminated marls	Parallel lamination or lenticular strata.	#	Mostly in quiet offshore lake areas.
	Massive marls	None.	G Ø Ô #	
Laminated Limestones	Depending on the sedimentary structures.	Lenticular or wavy stratification.	Ω Associated	Moderate salinity waters.
		Hummocky cross-stratification.	Ω Associated	Wave influence above storm surge level.
		Graded parallel lamination.	Ω Associated	Shore sheet flows or turbiditelike currents.
Stromatolitic Limestones	Planar crusts Bioherms Biostromes Coated grains	Microscopic alternations of light and dark micrite laminae.	Rod and coccoid calcite bodies. Rare filamental algae and fungi.	Moderate salinity waters. Shallow and/or marginal lacustrine areas.
Massive Limestones		Uncommon and poor, irregular lamination.	If present, weak bioturbation G Ø Ô #	Freshwater, shallow lacustrine areas. Permanent water supply.
Bioturbated Limestones	Depending on type of pedogenesis.	Root traces Desiccation cracks	Nodules, breccias, etc. G Ø Ô #	Palustrine conditions. Shallowing of previous freshwater ponded areas.
Calcareous Crusts	Depending on diagenetic processes.	Massive or laminated character, according to the remaining precursor facies.	Salt lenticular or cubic growths. Nodular gypsum Dolomite Breccias	Saline diagenesis of preexisting carbonate facies.
Nodular Gypsum		Commonly, massive	Alabastrine mm–dm nodules, as beds or isolated	Evaporative processes in saline mudflats.
Lenticular Gypsum		Massive, rarely forming laminae	Alabastrine texture	Gypsum precipitation in water and interstitially within the sediment.
Rippled and Laminated Gypsum		Parallel, lenticular, and rippled lamination	Alabastrine texture	Gypsum precipitation in hypersaline lake water. Wave reworking.
Halite		Cubic and chevron textures		Precipitation of NaCl in very shallow salinas.

G = gastropods, Ø = ostracods, Ô = charophytes, Ω = stromatolites, # = carbonaceous debris.
* Simplified from Arenas, 1993.

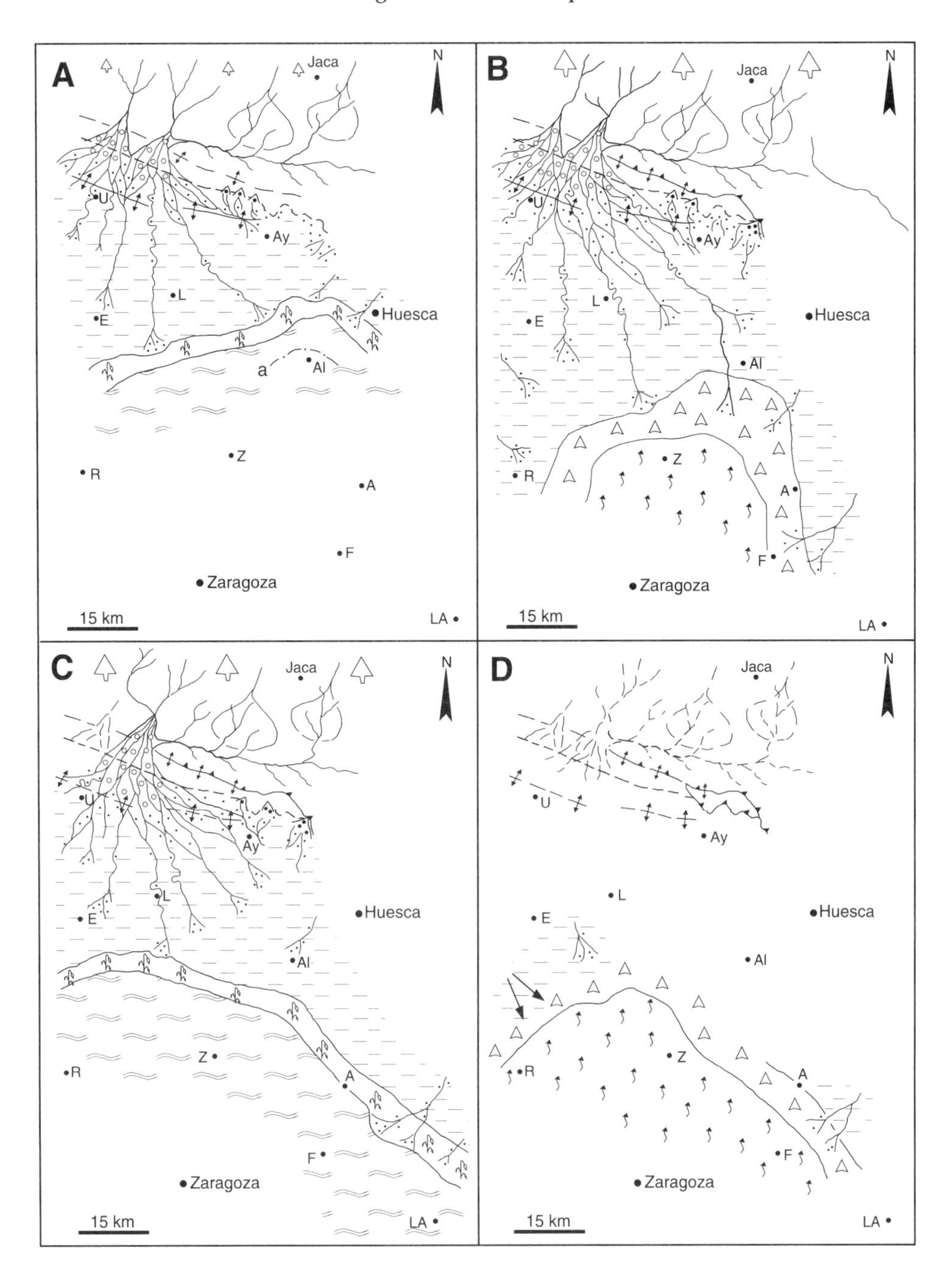

Figure 5—Paleogeographic sketches for unit N1 through time (see time position of the different sketches in trend curves of Figure 4). (A) and (C) correspond to high lake levels (freshwater carbonate lakes) and (B) and (D) to low lake levels (saline lakes). Line a indicates the approximate northern boundary of sulfate deposition areas. See legend in Figure 6 and Figure 6 caption for names of villages.

limestones, in part corresponding to storm events, and stromatolitic limestones (Arenas et al., 1993a, b; Arenas and Pardo, 1999).

Apart from these two main lacustrine situations, during the lowermost lake level phases halite precipitation took place mainly in areas north of the Ebro River, in particular in the Remolinos-Zuera-Zaragoza area. Glauberite has not been found in outcrop in the study area; nevertheless, glauberite deposits associated with halite have been observed in western and central areas of the Ebro Basin (Salvany and Ortí, 1994), as well as in some cores north of the city of Zaragoza (J. Mandado, personal communication, 1997). Thus it is possible that glauberite occurred in central areas of the basin, forming at intermediate situations between calcium sulfate and halite depositional environments. Diagenetic transformations into other sulfates, mainly gypsum, could have obliterated its presence.

Sedimentologic tools necessary to reconstruct paleogeography are the mappable facies associations or lithofacies differentiated in each TSU (Figure 3). Each lithofacies is defined by its dominant facies and facies sequences; thus each lithofacies represents sedimentation in a particular lacustrine subenvironment that extended over a part of the study area during discrete time intervals. Their distribution through space and time will allow interpretation of the evolution of the basin during the late Agenian (latest Oligocene)–Vallesian (late Miocene). Figures 5 and 6 show the main paleogeographic features of the study area of the Ebro Basin. These sketches have been made for specific

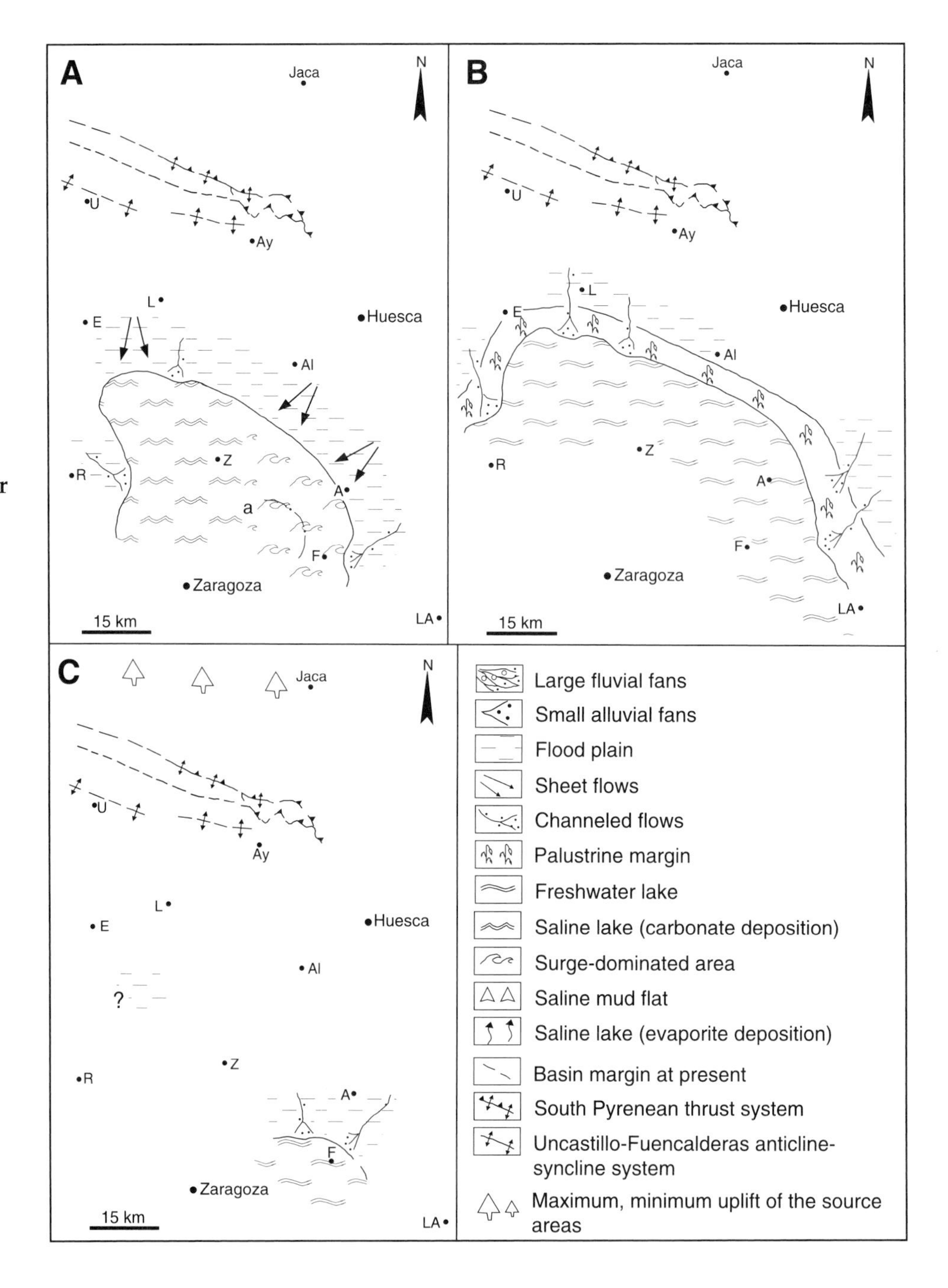

Figure 6—Paleogeographic sketches for the lower (A) and upper (B) parts of unit N2 and for unit N3 (C, simplified from Arenas, 1993) (see time position of the different sketches in trend curves of Figure 4). (A) represents low-salinity carbonate lake conditions. Line a in A indicates the approximate northern boundary of sulfate depositional areas during the upper part of unit N2. (B) and (C) correspond to high lake levels (freshwater carbonate lakes). U = Uncastillo, Ay = Ayerbe, E = Ejea, Al = Almudévar, R = Remolinos, Z = Zuera, A = Alcubierre, F = Farlete, LA = La Almolda.

times and different lake level conditions. Tectonic structures in the north have been simplified to keep the sketches clear. According to these sketches, lacustrine systems of the north-central part of the Ebro Basin evolved from predominantly saline conditions during N1 to freshwater carbonate depositional environments through N2 and during N3.

Unit N1 (Upper Agenian–Lower Aragonian)

This period was characterized by the extensive development of wide playa lake complexes. Saline lakes were important in areal extent and duration at the end of the unit (Figure 5D), giving rise to thick sulfate deposits. Lower lake levels led to halite deposition in salinas in the lowest areas. Common, moderate to low salinity lake phases are recorded by important laminated carbonate deposits and stromatolite build-ups. Freshwater carbonates were also present during times of high lake levels (Figure 5A, C). Carbonate facies produced during low salinity and freshwater lake phases are rare or absent in areas surrounding saline lake deposits (evaporite facies) in the Iberian domain of the basin (see Pérez et al., 1994; Villena et al., 1996b).

In the study area, two fluvial systems from the Pyrenees (Luna and Huesca fluvial systems) (after nomenclature of Hirst and Nichols, 1986) discharged into these lacustrine areas. The Luna system is entirely represented in the study area, but the northern part of the Huesca system (northeast of the city of Huesca) is outside the study area. Two main, large fluvial fans (Luesia and Uncastillo fans) composed the Luna system during deposition of unit N1 and older systems (Figure 5A, B) (Arenas, 1993; Arenas and Pardo, 1996), but during sedimentation of the upper part of unit N1, only the Luesia fan was present (Figure 5C). There were also small, marginal alluvial fans originating along the Pyrenean thrust front that coalesced into the large Luesia fluvial fan (Figure 5A–C).

Unit N2 (Middle and Upper Aragonian)

Carbonate deposition in saline lake environments prevailed at the beginning of the deposition of this unit, giving rise to stromatolitic and laminated limestones (Figure 6A). Saline lakes producing evaporites were uncommon and smaller than those of unit N1 and were restricted to the southern part of the study area (Figure 6A). Freshwater conditions became increasingly more extensive and persistent through time, as proven by the predominantly massive and bioturbated limestones. Thus freshwater lakes dominated and, consequently, the lake area became larger through time (Figure 6B); however, in the Iberian domain of the basin, limestones that resulted from freshwater conditions were scarce, and carbonate facies formed mostly during the predominantly moderate to low salinity lake phases (see Pérez, 1989; Pérez et al., 1994; Villena et al., 1996b).

Unit N3 (Upper Aragonian-Vallesian)

Extensive progradation of distal alluvial deposits is recorded during this unit (Arenas and Pardo, 1991). These facies occupied the previous carbonate deposition areas of N2, and from the beginning of unit N3 deposition, the lacustrine system was located farther south. Dilute conditions prevailed and carbonate deposition predominated throughout this time. Some rare saline carbonate periods occurred. In the Iberian domain of the basin, lacustrine deposits also recorded inputs from the north (see Pérez et al., 1994; Villena et al., 1996b).

Analysis of the information available on units N1 and N2 of the entire central part of the Ebro basin (Pérez, 1989; Pérez et al., 1994; González, 1989; Arenas, 1993; Arenas et al., 1993b, 1997, among others) shows that for a particular time with high lake levels, the lacustrine area consisted of one large single body of fresh water (about 90 km north-south × 130 km east-west, based on present outcrops). It is possible that more than one ponded area existed during the saline-lake system situations, and particularly during halite precipitation episodes when the lacustrine areas would have been restricted to the topographically lowest zones of the basin. It is suggested that the paleolakes of units N1 and N2 possessed an increasing salinity gradient toward the south, giving rise to the predominance of saline lake deposits in the Iberian domain of the basin (Arenas and Pardo, 1999). This asymmetric distribution of salinity could have been favored by, or in part caused by, a marked hydrologic contrast between the Pyrenean and Iberian domains, such as the existing contrast today. Although the northern margin of the lacustrine system received perennial surface inputs from the Pyrenees, scarce surface supply was provided by the Iberian Ranges into the southern margin of the lacustrine system. Most inputs to the southern margin were very probably provided as saline groundwaters (see modern hydrologic model proposed by Martínez et al., 1988, for that region).

From sedimentologic data, mineralogic analyses (Arenas et al., 1993c), O^{18} isotopic data (Arenas et al., 1997), and evolution of the study sequences (Arenas, 1993) north of the Ebro River, a general climatic evolution toward wetter conditions can be inferred for the lacustrine systems of the Ebro basin from the latest Oligocene to the late Miocene (Figure 4). This general trend was modulated by shorter, alternating wet and arid cycles.

As stated, in the Pyrenean domain of the Ebro basin, the sequential trend of unit N1 is cyclic (fining-coarsening upward) and complex (Figure 4) and was controlled by Pyrenean tectonics, as shown by progressive unconformities affecting the alluvial system deposits along the basin margin; however, in the Iberian domain and basin center, unit N1 has a fining-upward trend (Figure 4). Such a difference may reflect the contrasting tectonic patterns of the Iberian and Pyrenean ranges during the studied time range: weak, decreasing compressive tectonic activity in the Iberian Range (after Pérez, 1989) and intense, continuous compressive tectonic activity in the Pyrenean Range (Figure 5). In the basin center, during unit N1 deposition, climatic tendency toward greater precipitation inferred made the lacustrine system larger. As a result, the Pyrenean tectonic signature on the sequential trend of the deposits was masked or partially obliterated in the central part of the basin.

ACKNOWLEDGMENTS

This work was supported by the projects PB89-0342 and PB93-0580 of the Dirección General de Investigación Científica y Técnica and by a scholarship of the Consejo Asesor de Investigación of the Diputación General de Aragón. The editors and reviewers are greatly thanked for the meticulous and helpful revision of the paper.

REFERENCES CITED

Agustí, J., C. Arenas, Ll. Cabrera, and G. Pardo, 1994, Characterisation of the latest Aragonian–early Vallesian (late Miocene) in the central Ebro Basin (NE Spain), Scripta Geologica, v. 106, p. 1-10.

Arenas, C., 1993, Sedimentología y paleogeografía del Terciario del margen pirenaico y sector central de la Cuenca del Ebro (zona aragonesa occidental), Unpublished Ph.D. dissertation, University of Zaragoza, 858 p.

Arenas, C., and G. Pardo, 1991, Significado de la ruptura entre las Unidades Tectosedimentarias N2 y N3 en el centro de la Cuenca del Ebro, Geogaceta, v. 9, p. 67-70.

Arenas, C., and G. Pardo, 1996, Late Oligocene–early Miocene syntectonic fluvial sedimentation in the Aragonese Pyrenean domain of the Ebro Basin: facies models and structural controls, Cuadernos de Geología Ibérica, v. 21, p. 277-296.

Arenas, C., and G. Pardo, 1999, Latest Oligocene–late Miocene lacustrine systems of the north-central part of the Ebro Basin (Spain): sedimentary facies model and paleogeographic synthesis, Palaeogeography, Palaeoclimatology, Palaeoecology, v. 151 (1-3), p. 127-148.

Arenas, C., G. Pardo, and J. Casanova, 1993a, Bacterial stromatolites in lacustrine Miocene deposits of the Ebro Basin (Aragón, Spain), Bolletino della Società Paleontologica Italiana, special vol. 1, p. 9-22.

Arenas, C., G. Pardo, and J. Villena, 1993b, Sistemas lacustres neógenos de la Sierra de Alcubierre (sector central de la Cuenca del Ebro). Guía de campo de la II Reunión del Grupo Español del Proyeto IGCP 324 (GLOPALS), Secretariado Publicaciones Universidad Zaragoza, 139 p.

Arenas, C., J. Casanova, and G. Pardo, 1997, Isotopic characterization ($\delta^{13}C$ and $\delta^{18}O$) of the Miocene lacustrine systems of Los Monegros (Ebro Basin, Spain): paleogeographic and paleoclimatic implications, Palaeogeography, Palaeoclimatology, Palaeoecology, v. 128, p. 133-155.

Arenas, C., A. M. Alonso-Zarza, and G. Pardo, 1999, Dedolomitization and other early diagenetic processes in Miocene lacustrine deposits, Ebro Basin (Spain), Sedimentary Geology, v. 125, p. 23-45.

Arenas, C., G. Pardo, J. Villena, and A. Pérez, 1989, Facies lacustres carbonatadas de la Sierra de Alcubierre (sector central de la Cuenca del Ebro). XII Congreso Español Sedimentología, Bilbao, comunicaciones, p. 71-74.

Arenas, C., C. Fernández-Nieto, M. González López, and G. Pardo, 1993c, Evolución mineralógica de los materiales miocenos de la Sierra de Alcubierre (sector central de la Cuenca del Ebro): Implicaciones en la evolución paleogeográfica, Boletín de la Sociedad Española Mineralogía, v. 16, p. 51-64.

Calvo, J. P. and 24 co-authors, 1993, Up-to-date Spanish continental Neogene synthesis and paleoclimatic interpretation, Revista de la Sociedad Geológica de España, v. 6 (3-4), p. 29-40.

Garrido-Megías, A., 1982, Introducción al análisis tectosedimentario: Aplicación al estudio dinámico de cuencas, V Congreso Latinoamericano de Geología, Argentina, Actas, v. 1, p. 385-402.

González, A., 1989, Análisis tectosedimentario del Terciario del borde SE de la Depresión del Ebro (sector bajoaragonés) y cubetas ibéricas marginales, Unpublished Ph.D. dissertation, University of Zaragoza, 507 p.

Hirst, J. P. P., and G. J. Nichols, 1986, Thrust tectonic controls on the Miocene alluvial distribution patterns, southern Pyrenees, *in* P. A. Allen and P. Homewood, eds., Foreland Basins, International Association of Sedimentologists Special Publication No. 8, p. 247-258.

Instituto Tecnológico y Geominero De España, 1995, Mapa Geológico de la Peninsula Ibérica, Baleares y Canarias, escala 1:1,000,000, Madrid.

Martínez, F. J., J. A. Sánchez, J. L. De Miguel, and J. San Román, 1988, El drenaje subterráneo de la Cordillera Ibérica en la Cuenca del Ebro como proceso de movilización y transporte de substancias en disolución: Sus implicaciones en el aporte de sulfatos, *in* A. Pérez, A. Muñoz, and J. A. Sánchez, eds., Sistemas lacustres neógenos del margen ibérico de la Cuenca del Ebro, Guía Campo III Reunión Grupo Español IGCP 219, Secretariado Publicaciones Universidad Zaragoza, p. 57-80.

Pardo, G., J. Villena, and A. González, 1989, Contribución a los conceptos y a la aplicación del análisis tectosedimentario. Rupturas y unidades tectosedimentarias como fundamento de correlaciones estratigráficas, *in* J.A. Vera, ed., División de unidades estratigráficas en el análisis de cuencas, Revista de la Sociedad Geológica de España, v. 2 (3-4), p. 199-221.

Pérez, A., 1989, Estratigrafía y sedimentología del Terciario del borde meridional de la Depresión del Ebro (sector riojano-aragonés) y cubetas de Muniesa y Montalbán. Unpublished Ph.D. dissertation, University of Zaragoza, 525 p.

Pérez, A., A. Muñoz, G. Pardo, and J. Villena, 1994, Lacustrine Neogene deposits of the Ebro Basin (southern margin), northeastern Spain, *in* E. Gierlowski-Kordesch and K. Kelts, eds., Global Geological Record of Lake Basins, Cambridge, Cambridge University Press, vol. 1, p. 325-330.

Pérez, A., A. Muñoz, G. Pardo, C. Arenas, and J. Villena, 1988, Las unidades tectosedimentarias del Neógeno del borde ibérico de la Depresión del Ebro (sector central), *in* A. Pérez, A. Muñoz, and J. A. Sánchez, eds., Sistemas lacustres neógenos del margen ibérico de la Cuenca del Ebro, Guía Campo III Reunión Grupo Español IGCP 219, Secretariado Publicaciones Universidad de Zaragoza, p. 7-20.

Riba, O., S. Reguant, and J. Villena, 1983. Ensayo de síntesis estratigráfica y evolutiva de la cuenca terciaria del Ebro, *in* Libro Jubilar J. M. Ríos, ed., Geología de España, vol. II, Instituto Geológico y Minero de España, p.131-159.

Salvany, J. M. and F. Ortí, F., 1994, Miocene glauberite deposits of Alcanadre, Ebro Basin, Spain: Sedimentary and diagenetic processes, *in* R. W. Renaut and W. M. Last, eds., Sedimentology and Geochemistry of Modern and Ancient Saline Lakes, SEPM Special Publication No. 50, p. 203-215.

Villena, J., A. González, A. Muñoz, G. Pardo, and A. Pérez, 1992, Síntesis estratigráfica y sedimentológica del Terciario del borde Sur de la Cuenca del Ebro: Unidades genéticas, Acta Geológica Hispánica, v. 27 (1-2), p. 225-245.

Villena, J., G. Pardo, A. Pérez, A. Muñoz, and A. González, 1996a, The Tertiary of the Iberian margin of the Ebro basin: sequence stratigraphy, *in* P. Friend and C.J. Dabrio, eds., Tertiary Basins of Spain: the Stratigraphic Record of Crustal Kinematics, World and Regional Geology 6, Cambridge,

Cambridge University Press, p. 77-82.

Villena, J., G. Pardo, A. Pérez, A. Muñoz, and A. González, 1996b, The Tertiary of the Iberian margin of the Ebro basin: paleogeography and tectonic control, *in* P. Friend and C.J. Dabrio, eds.,Tertiary Basins of Spain: the Stratigraphic Record of Crustal Kinematics, World and Regional Geology 6, Cambridge, Cambridge University Press, p. 83-88.

Luzón, A., and A. Gonzalez, 2000, Sedimentology and evolution of a Paleogene–Neogene shallow carbonate lacustrine system, Ebro Basin, northeastern Spain, *in* E. H. Gierlowski-Kordesch and K. R. Kelts, eds., Lake basins through space and time: AAPG Studies in Geology 46, p. 407–416.

Chapter 37

Sedimentology and Evolution of a Paleogene–Neogene Shallow Carbonate Lacustrine System, Ebro Basin, Northeastern Spain

Arantxa Luzón
Angel González
Departamento de Ciencias de la Tierra, Facultad de Ciencias, Universidad de Zaragoza
Zaragoza, Spain

GEOLOGIC SETTING

This work focuses on the central eastern sector of the Ebro Basin, which is the largest Cenozoic basin on the northeastern side of the Iberian plate. The Ebro Basin is triangular in shape and is bounded by the Iberian Range to the south–southwest, the Catalonian Coastal Range to the east–southeast, and the Pyrenean Range to the north (Figure 1). During the Alpine cycle, as a consequence of the convergent movement of both the Iberian and European plates and the end of subduction of the Iberian under the European plate, the aforementioned three ranges uplifted. The Ebro Basin is the youngest foreland basin of the Pyrenean Range and in cross-section is an asymmetric sedimentary trough, the thickest sections of which are adjacent to the bounding range.

From the Late Cretaceous to the Miocene, the Pyrenean Range was tectonically active and its evolution was critical for the Ebro Basin. At the beginning of the Tertiary (Paleocene and Eocene), the Ebro Basin was linked to the sea, whereas from the early Oligocene to the late Miocene, it became an endoreic basin. During this second stage, alluvial systems developed and expanded from the three basin margins toward the central area, where their distal deposits interfingered with wide freshwater carbonate or saline lacustrine systems. A series of alluvial and lacustrine units deposited during the late Oligocene to the early Miocene crop out in the eastern central sector of the Ebro Basin (Figures 1, 2). The description and interpretation of one of them, the Torrente de Cinca unit, follows.

THE TORRENTE DE CINCA UNIT

Outcrops of the Torrente de Cinca unit are continuous, covering about 60 × 35 km in area (Figure 2), with thicknesses ranging between 20 m and 120 m (Figure 3). Fossil mammals demonstrate that the unit is late Oligocene–early Miocene in age (Figure 4). The Torrente de Cinca unit is mainly calcareous with interbedded lutites and marls, and rare sandstones and coal layers (Figure 3). Interpreted as a shallow freshwater carbonate lake, the unit is laterally equivalent to several siliciclastic units mainly composed of reddish lutites and yellow sandstones deposited in both alluvial plains and mudflats with sediments derived from northern and southern provenance areas. Carbonate input and precipitation were the main sedimentary processes in the lake system because the sediment source area was the Iberian Range and the Pyrenees, where thick sequences of carbonate rocks (more than 65% of provenance) are present. Two main lithofacies are differentiated in the Torrente de Cinca Unit: limestones, and lutites and marls (Figure 5).

Limestones

Limestones are grey to brown in color, and only rarely black. According to the Dunham (1962) classification, they are composed of mudstones, wackestones, and packstones. The most common fossils are gastropods, charophytes, and ostracods. Bedding is either tabular or lenticular in shape and thickness ranges from centimeter to diameter in scale. Commonly, limestone beds, with interbedded layers of lutites and marls a few millimeters thick, are grouped in tabular packages (decimeters to meters in thickness), which, in turn, are separated by layers of lutites or marls (centimeters to decimeters in thickness). Microscopic and macroscopic features of these limestones allow four groups of facies to be distinguished: laminated, massive, bioturbated, and nodular limestones.

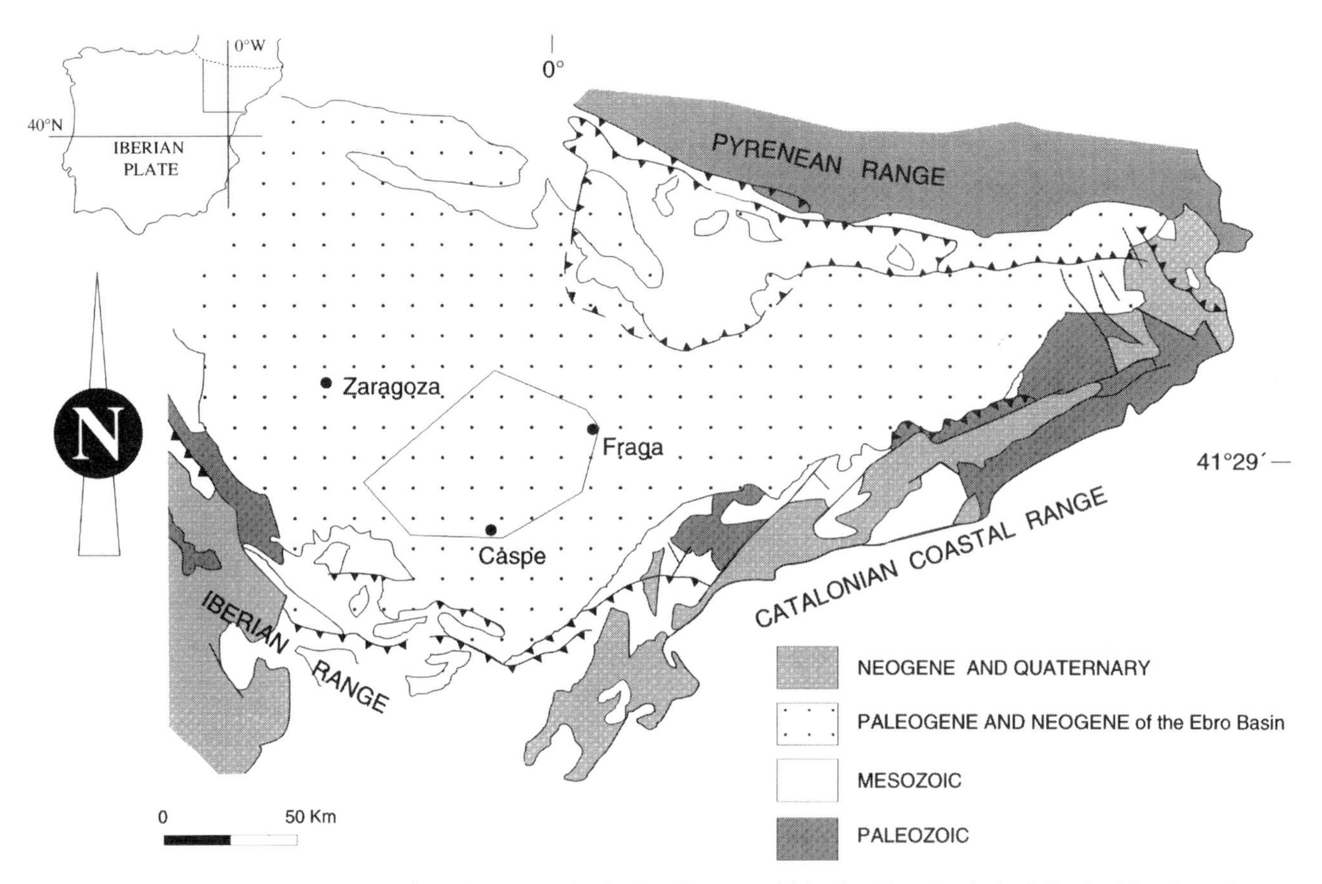

Figure 1—Simplified geologic map of northeastern Iberia. Study area within the Ebro Basin is delimited by the polygon.

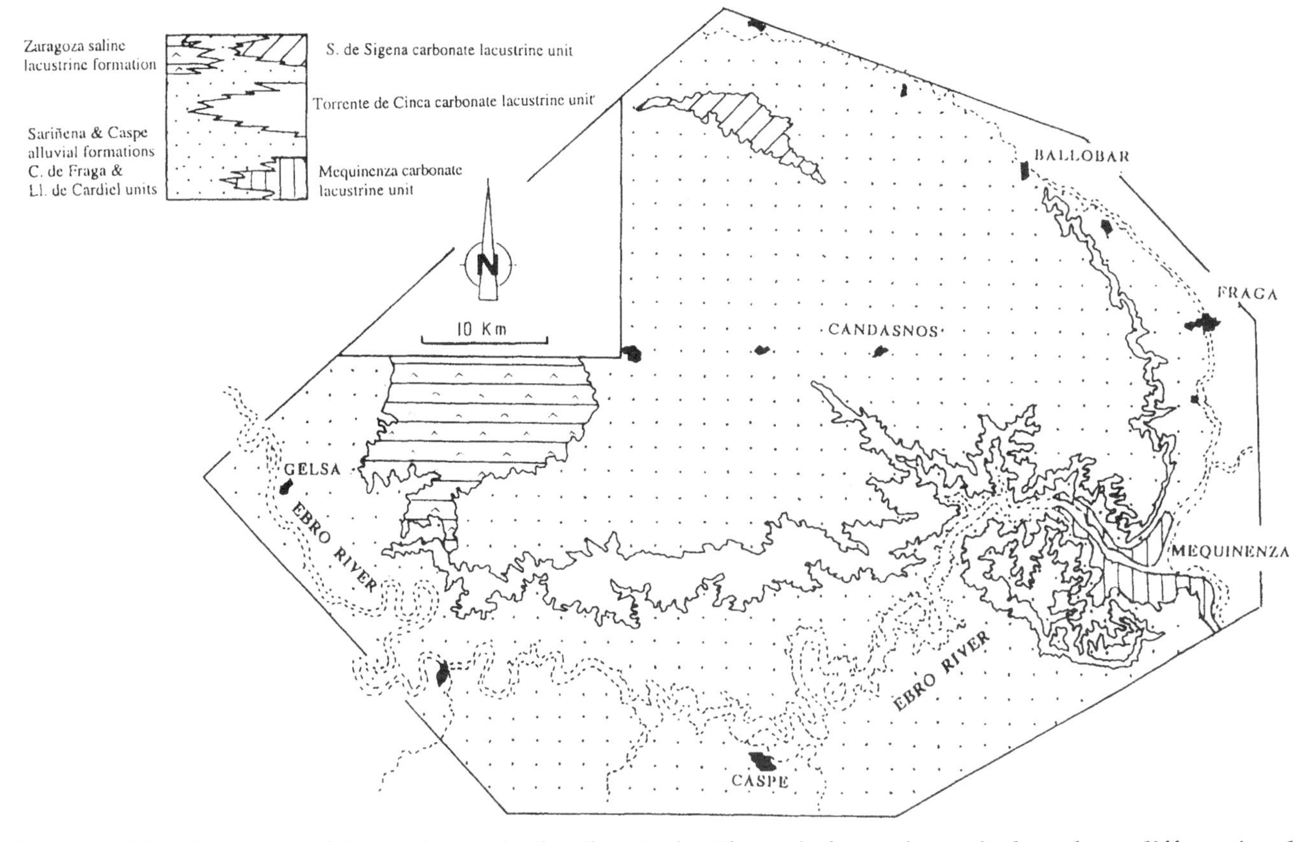

Figure 2—Lithologic map of the study area in the Ebro Basin. The main lacustrine units have been differentiated from the alluvial units. See exact location of study area in Figure 1.

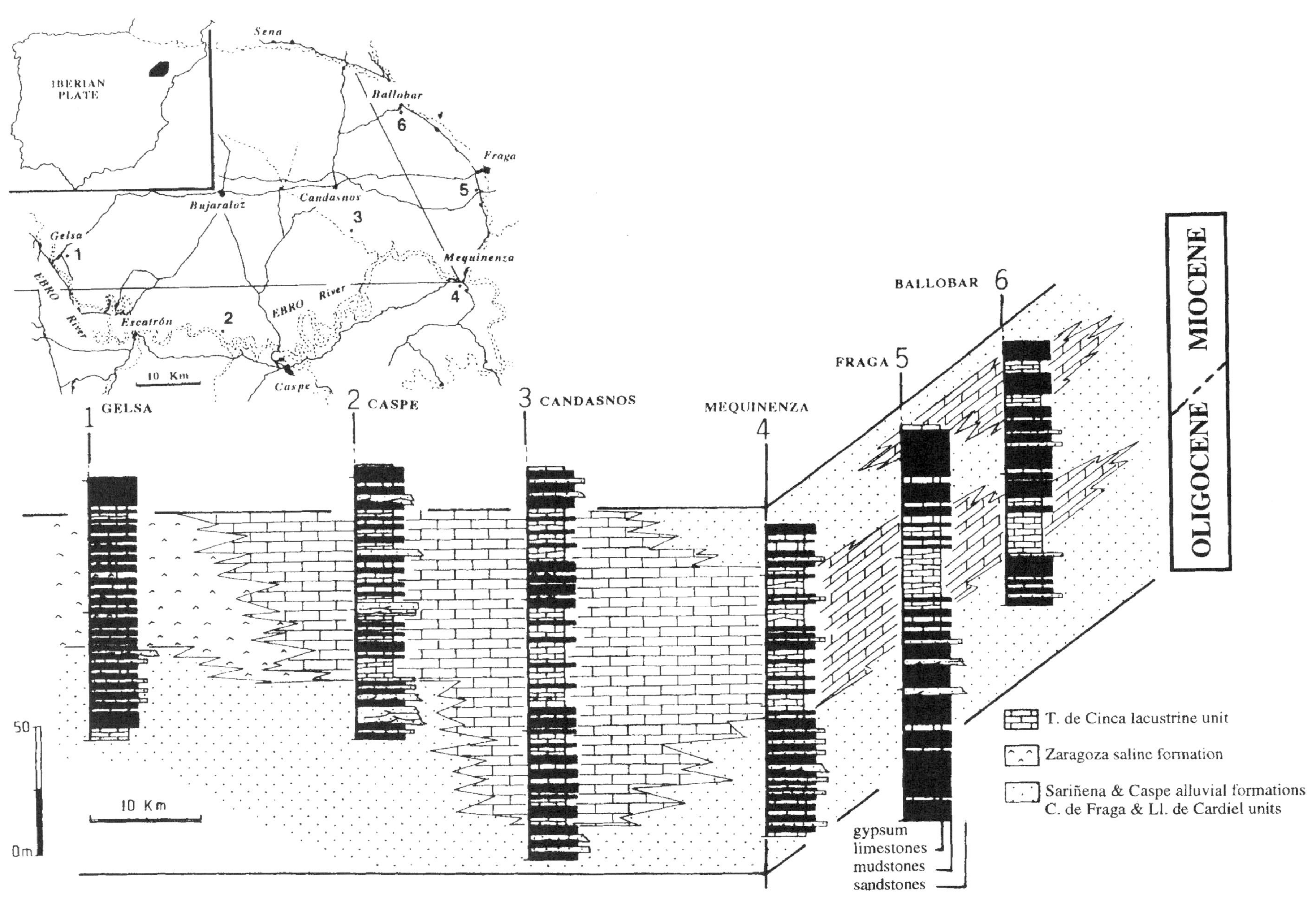

Figure 3—Stratigraphic profiles of Oligocene–Miocene rocks from several localities in the study area in the central eastern sector of the Ebro Basin.

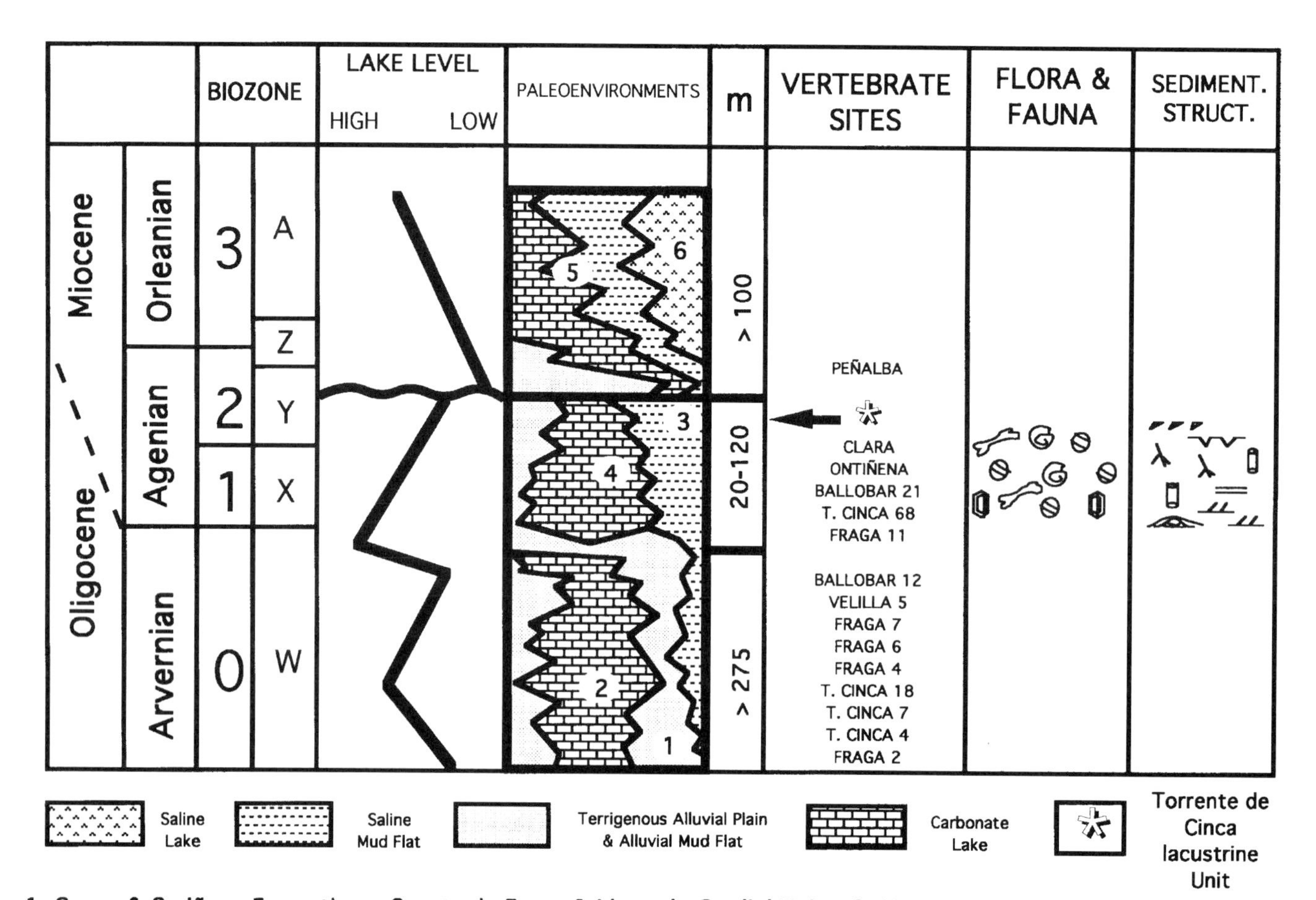

Figure 4—Chronostratigraphy of the Oligocene and Miocene rocks of the central eastern sector of the Ebro Basin. For fossils and sedimentary structures, see GGLAB legend.

Laminated limestones contain alternating sparitic and micritic, mostly discontinuous, horizontal laminae. Mudstones, wackestones, and packstones are present. Tabular bedding is the most common geometry. Common sedimentary structures include cross-lamination and hummocky cross-stratification. Hummocky cross-sets are decimeters thick. Ostracods and charophytes are the dominant fossil remains. These facies are interpreted as a result of alternating stormy and calm periods in littoral lacustrine areas.

Massive limestones are composed of both tabular and lenticular beds. Both wackestones and packstones are present with small vertical or horizontal tubes (millimeters to a few centimeters wide and a few centimeters long), but mudstone does occur. Charophytes and gastropods are by far the most common fossils, although ostracods are present, as well. These facies are interpreted as carbonate deposition in littoral or eulittoral lacustrine areas, where bioturbation was common.

Bioturbated limestones are characterized by vertical traces, but horizontal traces are rare. Bedding shows tabular or lenticular morphologies, and, in some places, mud mounds (a few centimeters to 20 cm thick and up to 1 m long) are present. Mudstones, wackestones, and packstones are the most common textures. Charophytes and gastropods are very common; chert nodules and gypsum-filled rhizoliths are also abundant. Mudcracks and ferruginous surfaces at the top of the beds, as well as marmorized layers, are rare. The vertical tubes, up to 2 cm in diameter and 30 cm long can divide downward into several smaller branches; they have been interpreted as rhizoliths. The horizontal traces are attributed to invertebrate activity. This facies is interpreted as carbonate deposition either in eulittoral or palustrine areas where plants were very common. This zone could have been subaerially exposed on occasion.

Nodular limestones are the rarest facies, disturbing the texture of every stratigraphic level. Mudstones to packstones are common textures. Carbonate crusts and ferruginous surfaces are rare. These limestones were formed initially either in the shallowest zone of the lake (eulittoral) or in the palustrine area. They have been interpreted as early diagenetic alteration of palustrine-lacustrine carbonate sediments (Alonso Zarza et al., 1992).

FACIES	COLORS	MAIN TYPES	MICROSCOPIC FEATURES	OTHERS	THICKNESS	SEDIMENTARY SETTING
Mudstone to Packstone	Grey, Brown, Black, Dun, Beige	Laminated	Mudstone-wackestone-packstone, rare siliciclastics	Tabular bedding, grey and dun color, cross-lamination, hummocky cross-stratification, ostracods, charophytes	cm-dm	Alternation of both storm and quiet episodes in the lacustrine area. LITTORAL
		Massive	Wackestone-packstone, some siliciclastics	Tabular and lenticular bedding, grey and dun color, mudcracks, vertical and horizontal tubes,gastropods, charophytes, some ostracods	cm-dm	Representative of both shallow and deep lacustrine areas with bioturbation. EULITTORAL-LITTORAL
		Bioturbated	Mudstone-packstone, some siliciclastics	Tabular and lenticular bedding, grey and black color, ramified tubes chert nodules, mud mounds, rhizoliths filled with gypsum or mudstone, mudcracks, ferruginous surfaces, marmorized levels gastropods, charophytes	cm-dm	Occasionally exposed shallow lacustrine and palustrine areas with vegetation and an oscillating water table. EULITTORAL-PALUSTRINE
		Nodular	Mudstone-packstone common siliciclastics	Irregular bedding, beige and dun color, gastropods, charophytes carbonate crust, ferruginous surfaces brecciation, nodulization	cm-dm	Palustrine and shallow lacustrine areas with an oscillating water table. EULITTORAL-PALUSTRINE
Lutites-marls	Grey, Reddish, Black, Orange, Brown	Grey	Laminated or massive	Tabular bedding, mottled yellow high fossil content	cm-m	Resulted from the sedimentation of the finest detrital fraction in the inner areas of the lake. LITTORAL-EULITTORAL
		Reddish	Massive or bioturbated	Tabular bedding, gypsum nodules, bioturbation, marmorization	cm-m	Fine detrital fraction accumulation in either palustrine or alluvial mud flat areas. PALUSTRINE-MUDFLAT

Figure 5—Main sedimentologic features of the differentiated lithofacies of the Torrente de Cinca Unit of Oligocene–Miocene age in the central eastern sector of the Ebro Basin.

Lutites and Marls

Lutite and marl layers are tabular in shape, and their thicknesses vary from centimeters to meters in scale. Two facies are differentiated: grey lutites or marls, and reddish lutites or marls.

Grey lutites and marls can be either laminated or massive and rarely mottled yellow. They contain ostracods, charophytes, gastropods and, rarely, vertebrate remains. These facies developed under subaqueous conditions, with rare exposure, and are interpreted as proximal to distal lacustrine (littoral-eulittoral area). Laminated levels formed in more distal areas (without bioturbation).

Reddish lutites and marls are either massive or bioturbated. Marmorized layers are very common, and gypsum nodules centimeters to decimeters in diameter occur. Gastropods and vertebrate remains are present. Reddish lutites or marls were deposited either in palustrine areas or in the alluvial mudflats surrounding the lake. These areas were frequently affected by oscillations of the water table, as evidenced by marmorization.

SEDIMENTARY SEQUENCES OF THE TORRENTE DE CINCA UNIT

The Torrente de Cinca Unit is organized into sedimentary sequences ranging from centimeters to meters in thickness. They are grouped into two main types: type A and type B (Figure 6). There are other types of sequences, but they are the result of either the combination of the two main types or by the absence of some of their members.

Type A sequences have thicknesses from centimeters to decimeters and contain two main members. The lower member is composed of laminated or massive grey lutites or marls, which have a high fossil content (ostracods and charophytes) and are rarely mottled yellow. The upper member is calcareous with laminated limestones overlain by massive limestones and then by bioturbated limestones. This upper member contains a high density of fossils, mainly gastropods and charophytes.

The thicknesses of Type B sequences range from decimeters to meters and are composed of three

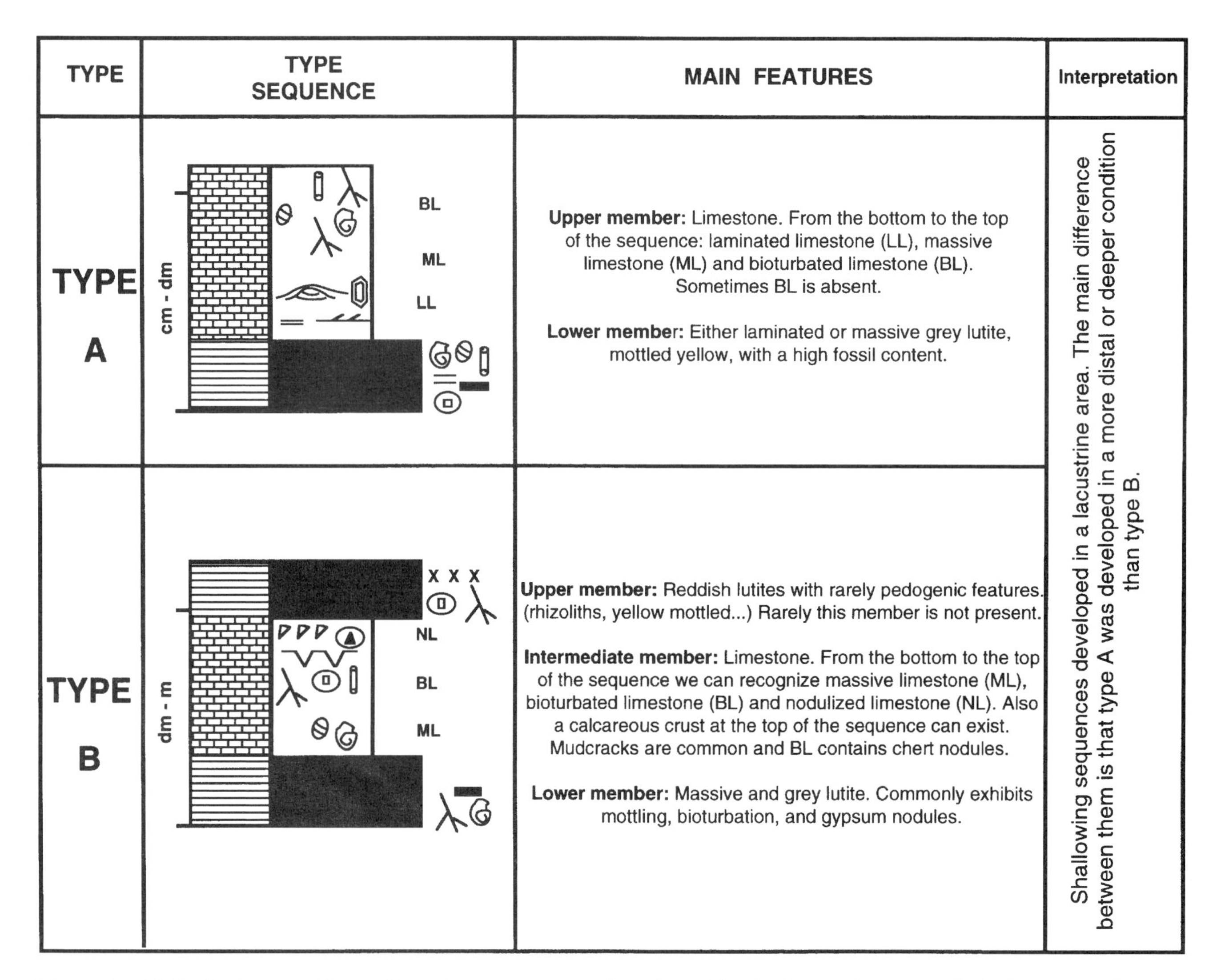

Figure 6—Main features of the sedimentary sequences developed in the lacustrine area of the Torrente de Cinca unit of Oligocene–Miocene age in the central eastern sector of the Ebro Basin. See GGLAB legend for fossils and sedimentary structures.

members. The lower member is composed of massive grey lutites or marls that are mottled yellow with bioturbation and gypsum nodules. The intermediate member is calcareous with massive limestones overlain by bioturbated limestones and, finally, nodular limestones. The upper member consists of reddish lutites or marls with evidence of pedogenic alteration, such as root traces and a mottled appearance. In some cases, this member is absent.

Type A and type B sequences have been interpreted as two different kinds of shallowing-upward sequences. The main difference is that type B sequences are representative of more marginal areas of the lake (eulittoral-palustrine) than type A (littoral). The distal or central area remained under subaqueous conditions, whereas the marginal area, covered with vegetation, underwent frequent episodes of subaerial exposure as a consequence of water table oscillations.

MEGASEQUENTIAL ARRANGEMENT OF THE TORRENTE DE CINCA UNIT

Stratigraphic analysis of the Torrente de Cinca Unit distinguishes a lower member and an upper member. The lower member of the unit, 5–100 m thick, is characterized by the predominance of both laminated and massive limestones associated with grey lutites or marls. Type A and type B sequences are present in this member, but type A predominates. The upper member of the unit, 10–60 m thick, is characterized by the general absence of laminated limestones, with common massive and bioturbated limestones. Grey and reddish lutites or marls are also present. Both bioturbated limestones and especially reddish lutites or marls are much more common at the top of this member. Type B sequences predominate in this member.

The megasequential arrangement of the Torrente de Cinca Unit (Figure 7) indicates that the lacustrine system had two main stages, evolving toward shallower conditions through time. In the first stage, the lake was at a high relative level (deeper conditions), and in the second stage, the lake evolved toward shallower conditions. Repetitive oscillations of the water level favored pedogenic processes in marginal areas of the lake. Finally, the lake was completely filled and development of mudflats took place in the area. Small oscillations in the lake water level could be a direct consequence of varying climatic conditions, but the general evolution of the lacustrine system was also closely related to that of the surrounding alluvial areas and existed as a consequence of tectonism at the basin margins (Figure 8). After a period of progradation (middle–late Arvernian) of the alluvial systems toward the center of the basin, development of the Torrente de Cinca lacustrine system took place in the central eastern sector of the Ebro Basin (Agenian). There is no clear evidence of how the alluvial systems from the north evolved during this period, but those from the south disappeared suddenly (González, 1989), simultaneously with the development of the lake (Luzón, 1994). This alluvial retrogradation occurred during a period of increasing tectonic activity in the Iberian Range while various thresholds developed in the southeastern sector of the Ebro Basin. One of these thresholds (La Ginebrosa thrust) partially limited the passage of the alluvial system coming from the southern source area, provoking a decrease in the detrital supply toward the center of the basin and favoring the expansion of the lacustrine area southward (Figure 8, Agenian). Finally, La Ginebrosa threshold was overcome and a new detrital supply from the south was shed toward the basin center, giving rise to filling of the lacustrine area (lowermost Orleanian).

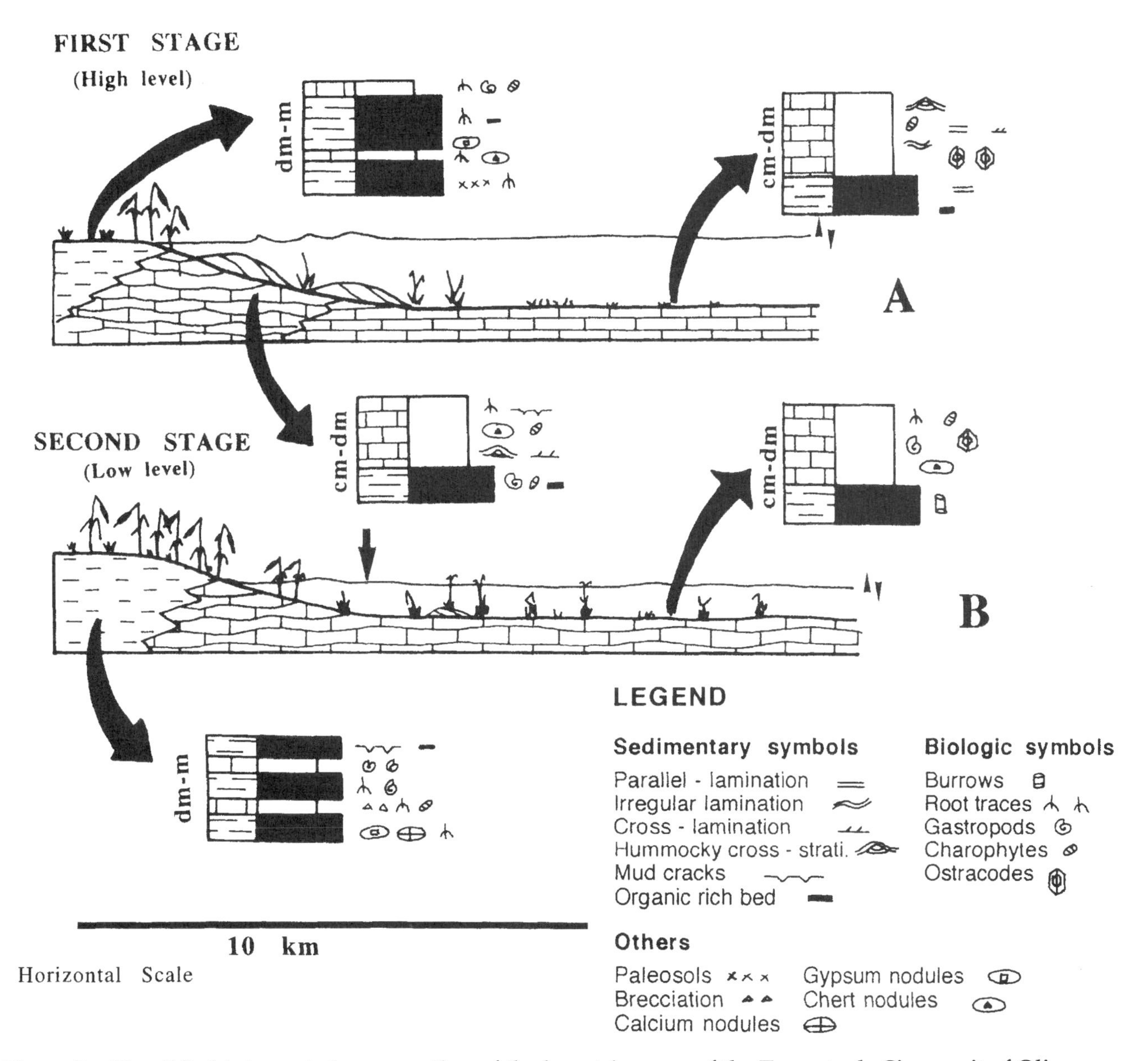

Figure 7—Simplified interpreted cross-section of the lacustrine area of the Torrente de Cinca unit of Oligocene–Miocene age. There were two main stages in the lake evolution: (A) a deeper stage, and (B) a shallower stage.

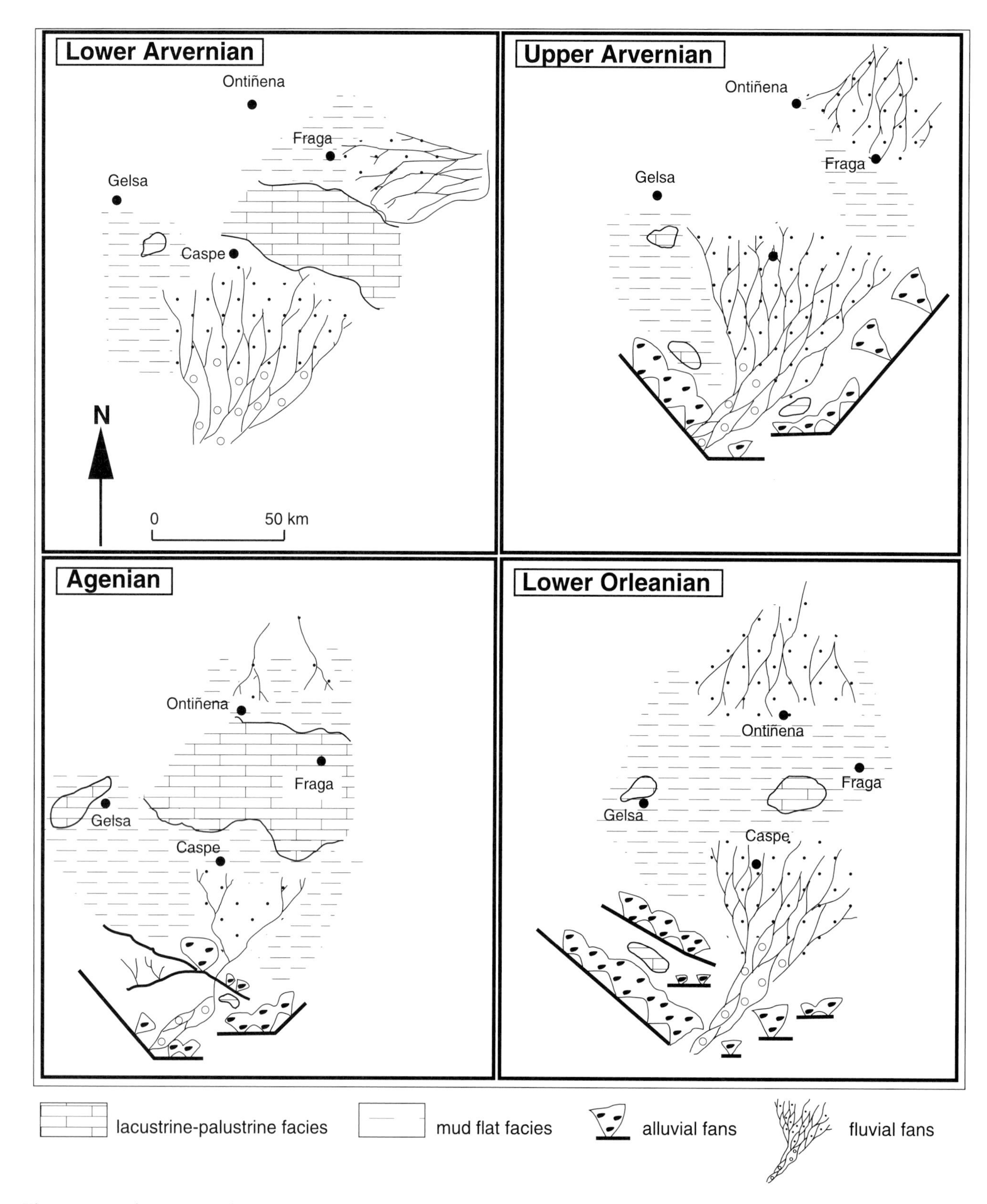

Figure 8—Paleogeographic sketches of the southeastern Ebro Basin between the upper Arvernian and the lowermost Orleanian (upper Oligocene–lower Miocene). Torrente de Cinca Unit is represented in the most northern area (lacustrine-palustrine facies) of the Agenian sketch. From the source area to the distal parts of the fluvial systems there is a gradual transition from coarse to sandier facies. See Figure 4 for chronostratigraphy.

ACKNOWLEDGMENTS

We are grateful to the reviewers and editors for their valuable and informative reviews of English. Thanks are due to V. Luzón, C. Arenas, and J. Stamp. This work has been supported by project PB93/0580 of the DGICYT (Dirección General de Investigación Científica y Técnica).

REFERENCES CITED

Alonso Zarza, A. M., J. P. Calvo, and M. A. García del Cura, 1992, Palustrine sedimentation and associated features—grainification and pseudo-microkarst—in the Middle Miocene (Intermediate Unit) of the Madrid Basin, Spain. Sedimentary Geology, v. 76, p. 43–61.

Dunham, R. J., 1962, Classification of carbonate rocks according to depositional texture, *in* W.E. Hern, ed., Classification of Carbonate Rocks, AAPG Memoir 1, p. 108–121.

González, A., 1989, Análisis tectosedimentario del Terciario del borde SE de la Depresión del Ebro (sector bajoaragonés) y Cubetas Ibéricas Marginales. Ph.D. dissertation, Universidad de Zaragoza, 507 p. (unpublished).

Luzón, A., 1994, Los materiales del tránsito Oligoceno–Mioceno del sector centro-oriental de la Depresión del Ebro: Análisis estratigráfico e interpretación evolutiva. Master's thesis, Universidad de Zaragoza, 259 p. (unpublished).

ADDITIONAL REFERENCES OF INTEREST

Agustí, J., P. Anadón, S. Arbiol, Ll. Cabrera, F. Colombo, and A. Sáez, 1987, Biostratigraphical characteristics of the Oligocene sequences of Northeastern Spain (Ebro and Campins Basins), Müncher Geowissenschaften Abhandlung, v. 10A, p. 34–42.

Agustí, J., X. Barberà, Ll. Cabrera, J. M. Parés, and M. Llenas, 1994, Magnetobiostratigraphy of the Oligocene–Miocene transition in the Ebro Basin (Eastern Spain): state of the art. Münchner Geowissenschaften Abhandlung, v. 26A, p. 161–172.

Agustí, J., Ll. Cabrera, P. Anadón, and S. Arbio, 1988, A late Oligocene–early Miocene rodent biozonation from the SE Ebro Basin (NE Spain): A potential mammal stage stratotype. Newsletter Stratigraphy, v. 18, p. 81–97.

Anadón, P., Ll. Cabrera, B. Colldeforns, F. Colombo, J. L. Cuevas, and M. Marzo, 1989, Alluvial fan evolution in the SE Ebro Basin: Response to tectonics and lacustrine level changes, 4th International Conference on Fluvial Sedimentology, Excursion Guidebook no. 9, 91 p.

Anadón, P., Ll. Cabrera, B. Colldeforns, and A. Sáez, 1989, Los sistemas lacustres del Eoceno superior y Oligoceno del sector oriental de la Cuenca del Ebro. Acta Geológica Hispánica, v. 24, p. 205–230.

Anadón, P., Ll.Cabrera, B. Colldeforns, and A. Sáez, 1989, Sistemas deposicionales lacustres paleógenos en el sector oriental de la Cuenca del Ebro, Geogaceta, v. 6, p. 102–103.

Anadón, P., Ll. Cabrera, S. J. Choi, F. Colombo, M. Feist, and A. Sáez, 1992, Biozonación del Paleógeno continental de la zona oriental de la Cuenca del Ebro mediante carófitas: implicaciones en la biozonación general de carófitas de Europa occidental. Acta Geológica Hispánica, v. 27, p. 69–94.

Azanza, B., J. I. Canudo, and G. Cuenca, 1988, Nuevos datos bioestratigráficos del Terciario continental de la Cuenca del Ebro (sector centro-occidental). II Congreso Geológico de España (Granada, 1988), Comunicaciones, v. I, p. 261–264.

Cabrera, Ll., 1983, Estratigrafía y sedimentología de las formaciones lacustres del tránsito Oligoceno–Mioceno del SE de la Cuenca del Ebro. Ph.D. dissertation, Universidad de Barcelona, 443 p. (unpublished).

Cabrera, Ll., 1983, Procesos de sedimentación y diagénesis temprana en un sistema lacustre endorreico somero: características de los depósitos lacustres del Oligoceno superior del SE de la Cuenca del Ebro, X Congreso Nacional de Sedimentología (Menorca, 1983), Comunicaciones, p. 1.57–1.61.

Cabrera, Ll., and F. Colombo, 1986, Las secuencias de abanicos aluviales Paleógenos del Montsant y su tránsito a sucesiones lacustres someras (sistemas de Scala Dei y Los Monegros, sector SE de la Cuenca del Ebro). XI Congreso Español de Sedimentología (Barcelona, 1986), Guía de excursiones, Excursión no. 7, 53 p.

Cabrera, Ll., F. Colombo, and S. Robles, 1985, Sedimentation and tectonics interrelationships in the Paleogene marginal alluvial system of the SE Ebro Basin. Transition from alluvial to shallow lacustrine environnents. 6th European International Association of Sedimentologists Meeting (Lleida, Spain), Excursion Guidebook, Excursion no. 10, 98 p.

Cabrera, Ll., and J. Gaudant, 1985, Los Ciprínidos (Pisces) del sistema lacustre Oligocénico–Miocénico de Los Monegros (Sector SE de la Cuenca del Ebro, povincias de Lleida, Tarragona, Huesca y Zaragoza). Acta Geológica Hispánica, v. 20, p. 219–226.

Canudo, J. I., G. Cuenca, G. S. Odin, M. Lago, E. Arranz, and M. Cosca, 1994, Primeros datos radiométricos de la base del Rambliense (Mioceno inferior) en la Cuenca del Ebro. II Congreso Grupo Español del Terciario (Jaca, 1994), Comunicaciones, p. 73–76.

Colombo, F., 1980, Estratigrafía y sedimentología del Terciario inferior continental de los catalánides. Ph.D. dissertation, Universidad de Barcelona, 690 p. (unpublished).

Colombo, F., 1986, Estratigrafía y sedimentología del Paleoceno continental del borde meridional occidental de Los Catalánides (Provincia de Tarragona, España). Cuadernos de Geología Ibérica, v. 10, p. 295–334.

Cuenca, G., 1991, Nuevos datos bioestratigráficos del Mioceno del sector central de la Cuenca del Ebro. I Congreso del Grupo Español del Terciario (Vic,

1991), Comunicaciones, p. 101–104.
Cuenca, G., B. Azanza, J. I. Canudo, and V. Fuertes, 1989, Los micromamíferos del Mioceno Inferior de Peñalba (Huesca), Implicaciones bioestratigráficas. Geogaceta, v. 6, p. 75–77.
Glenn, C., and K. Kelts, 1991, Sedimentary rhythms in lake deposits, *in* G. W. Einsele, S. Ricken, and A. Seilacher, eds., Cycles and Events in Stratigraphy: Berlin, Springer Verlag, p. 128–221.
González, A., J. Guimerà, and A. Luzón, 1994, Relaciones tectónica-sedimentación en la Zona de Enlace y borde sur-oriental de la Depresión del Ebro. II Congreso del Grupo Español del Terciario (Jaca, 1994), Guía de excursiones, p. 1–134.
Kelts, K., 1988, Environments of deposition of lacustrine petroleoum source rocks: an introduction, *in* A. Fleet, K. Kelts, and M. Talbot, eds., Lacustrine Petroleum Source Rocks, Geological Society (London) Special Publication No. 40, p. 3–26.
Luzón, A., and A. González, 1994, Evolución megascuencial de los materiales del Oligoceno superior y Mioceno inferior del sector centro-oriental de la Cuenca del Ebro. II Congreso Grupo Español del Terciario (Jaca, 1994), Comunicaciones, p. 141–144.
Luzón, A., and A. González, 1995, Sedimentary supply as control of the sedimentary evolution of a lacustrine area, example from the NE of Iberia (Ebro Basin). 16th Regional European Meeting of the International Association of Sedimentologists and 5ème Congrès Francais de Sedimentologie (Aix Les Bains, 1995), Abstracts, p. 96.
Luzón, A., and A. González, 1995, Sedimentología de una unidad lacustre carbonatada del tránsito Oligoceno–Mioceno (Sector central de la Cuenca del Ebro, España). XIII Congreso Español de Sedimentología (Teruel), Comunicaciones, p. 151–152.
Parès, J. M., X. Barberà, P. Anadón, and Ll. Cabrera, 1993, High resolution magnetic stratigraphy in a late Oligocene succession, 7th meeting of the European Union of Geosciences (Strasbourg, 1993), Abstract Supplement to Terra Nova, v. 5, p. 79.
Quirantes, J., 1971, Las calizas en el Terciario continental de los Monegros. Estudios Geológicos, v. 27, p. 355–362.
Quirantes, J., 1978, Estudio sedimentológico y estratigráfico del Terciario Continental de los Monegros, *in* Instituto Fernando el Católico (C.S.I.C.), ed., Diputación Provincial de Zaragoza, 207 p.
Riba, O., S. Reguant, and J. Villena, 1983, Ensayo de síntesis estratigráfica y evolutiva de la Cuenca terciaria del Ebro, *in* I.G.M.E. ed., Geología de España, Libro Jubilar J.M. Ríos-Tomo II, p. 131–159.

Ménillet, F., and N. Edwards, 2000, The Oligocene–Miocene Calcaires de Beauce (Beauce Limestones), Paris Basin, France, *in* E. H. Gierlowski-Kordesch and K. R. Kelts, eds., Lake basins through space and time: AAPG Studies in Geology 46, p. 417–424.

Chapter 38

The Oligocene–Miocene Calcaires de Beauce (Beauce Limestones), Paris Basin, France

François Ménillet
BRGM Alsace
Strasbourg, France

Nicholas Edwards
School of Earth, Environmental & Physical Sciences, University of Portsmouth
Portsmouth, UK

GEOLOGY

The Permian–Tertiary sedimentary fill of the intracratonic Paris Basin, northern France (Figure 1), is 3180 m thick in the zone of maximum subsidence (Pomerol, 1980). The Tertiary succession (Figure 2), unconformable on the Late Cretaceous Chalk, comprises many named units, but can be resolved into 21 transgression-regression sequences separated by hiatuses recording lowered sea levels (Gély and Lorenz, 1991). During the Tertiary the depocenter migrated southward and by the late Oligocene–early Miocene lay between Étampes and Orléans (Cavelier and Pomerol, 1979). As the basin became increasingly isolated from the open sea to northwest, laguno-lacustrine to palustrine limestones accumulated on the southern and eastern margins (Pomerol and Feugueur, 1968; Gély and Lorenz, 1991). The most extensive, the Oligocene–Miocene Calcaires de Beauce Formation (Beauce Limestones Formation), onlaps older limestones southward onto the Late Cretaceous (Campanian) Chalk. The formation is overlain by Miocene fluviatile sediments in the southern part of the basin. Pliocene–Quaternary tectonic uplift of northern France caused extensive erosion of the Tertiary succession (Pomerol, 1980).

THE CALCAIRES DE BEAUCE FORMATION

This formation underlies an intensely cultivated, dissected plateau of moderate relief, centered on the city of Orléans. There are no natural outcrops, but good exposures occur in working and abandoned roadstone and dimension-stone quarries and in roadcuts on the eastern margin of the plateau. The original thickness of the formation is uncertain, perhaps 100 m. In borehole logs its thickness varies laterally between 30 to 65 m over a few tens of kilometers (Manivit, 1979). This lenticularity reflects the paleodunal paleotopography on the underlying Fontainebleau Sands in the northern part of the outcrop (Pomerol and Feuguer, 1968), synsedimentary tectonic warping (Ménillet, 1980), and synsedimentary and Tertiary–Quaternary erosion. The depositional limits are unknown.

The three major limestone units (Figure 3) are separated by siliciclastic sediments and marls/marly limestones. One siliciclastic/marly unit, the Molasse du Gâtinais, apparently overlies a paleotopographic erosion surface with up to 20 m relief (Manivit, 1979). Except near the base, where bands of hydrobiid and cerithid gastropods indicate occasionally raised salinities, the fossil assemblage is dominated by pulmonate pond-snails, plus or minus land-snails. On this evidence, Roys (1869) and more recent authors regarded it as lacustrine.

LITHOLOGIES

The limestones are predominantly low-Mg calcitic (mean 97% $CaCO_3$), with traces of attapulgite, sepiolite and smectite, minor detrital quartz, and diagenetic silica. The Late Cretaceous Chalk was the probable carbonate source (Ménillet, 1974a, 1980). The carbonate lithologies and their diagenetic modifications, described in detail by Ménillet (1974a, 1980), are summarized in the following section, with emphasis on evidence for pedogensis.

Major Primary Lithologies

Marly to Cemented Micrites and Microsparites

These rocks range from friable, very porous marls to well-lithified limestones with a flat to subconchoidal fracture. They form planar lenticular beds, 0.1–2.5 m thick, interbedded with and interpenetrated by rhizolithic micrites. Cemented micrites display traces of sedimentary laminations, indicating early cementation. The

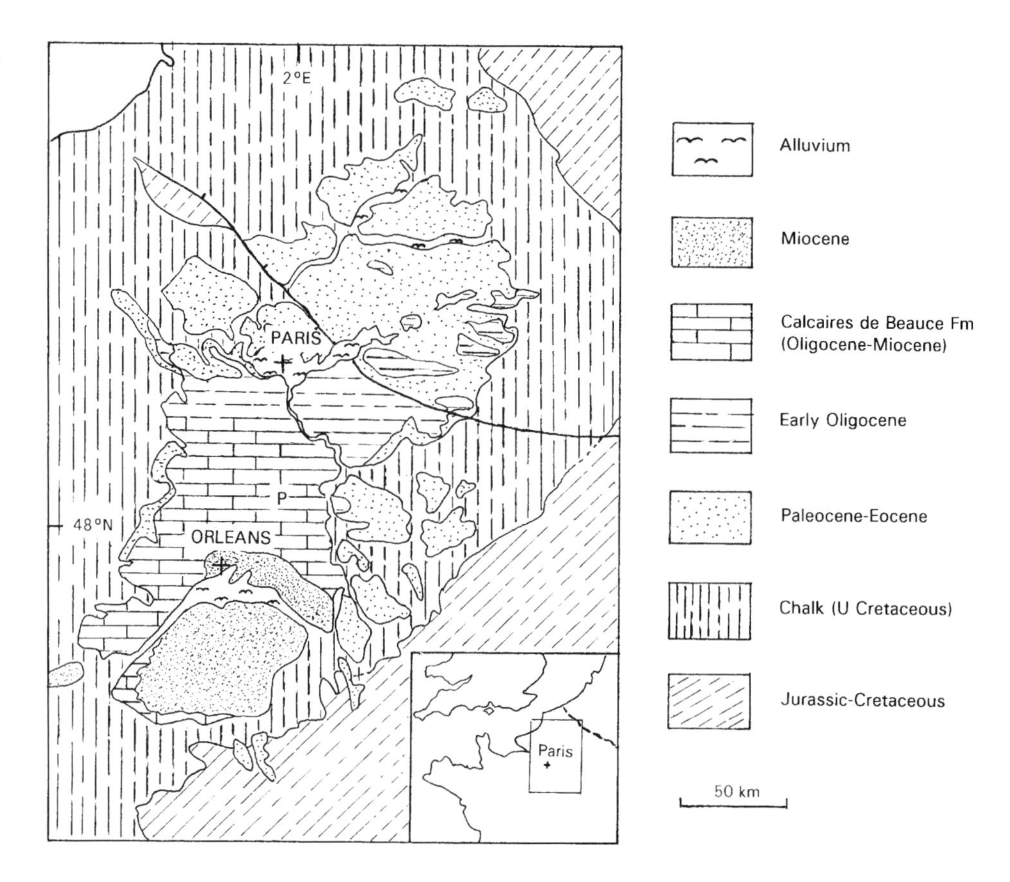

Figure 1—Geological sketch map of the Paris Basin, showing major geological boundaries. P = Pithiviers.

presence in marly micrites of tufa-lined rootlet molds and, in the basal member, the tidal mudflat snails *Hydrobia* and *Potamides*, suggests extremely shallow-water deposition. In cemented micrites/microsparites, the presence of circumgranular cracks delimiting isolated peloids and ooids (cf. Platt, 1989; Alonso-Zarza et al., 1992) and synsedimentary breccias indicates subaerial exposure and desiccation.

Biomicrites

The biomicrites are gray to dark gray, compact, well-lithified micrite to clotted micrite, pelmicrite, and microsparite. Beds are planar lenticular, varying in thickness over short distances, but mostly 0.3–2 m thick. Shallow-water, possibly paludal deposition is indicated by the abundant pond-snails and the presence of root molds, with and without carbonaceous linings. Desiccation cracks, circumgranular cracks, and breccia intraclasts are evidence for subaerial exposure. Spar-lined or filled molds of pond-snail shells, charophyte stems and gyrogonites, reticular and rhizolithic micrites, pseudomicrokarst (see Freytet, 1984), and geodic calcite-filled solution hollows indicate longer periods in the vadose zone, perhaps concurrent with cementation.

Disaggregational Lithologies

By analogy with published interpretations of similar palustrine carbonate formations, surficial disaggregation of these lithologies resulted either from the effects of subaerial exposure and mechanical weathering (Freytet and Plaziat, 1982), or incipient pedogenesis (Armenteros and Daley, 1998). Reticular cracking in irregular patches and veins, often adjacent to paleokarst cavities, progressed to the development of distinct peloidal and brecciated lithologies. These total 25% of the aggregate thickness of the formation, indicating the importance of the process in eliminating primary lithologies and depositional sequences.

Thicknesses of individual lithologies vary considerably, even in hand specimens, and are usually in the millimetric to centimetric range, but may be up to 1 m thick in some sections.

Reticular Cracking

Networks of straight to curved cracks define incipient intraclasts approximately 1.0–20.0 mm in diameter, paler or darker than the dark brown or pale micrite, microsparite, or sparite crack-fill micrite (Ménillet, 1980, photo 1). The initial cracking resulted either from desiccation (Freytet and Plaziat, 1982) or sediment shrinkage during repeated diurnal draining and flooding (cf. Mazzullo and Birdwell, 1989, their figure 2B). The presence of rootlet molds, crystalline concentric rhizoliths, micrite-coated peloidal host rock intraclasts, and alveolar-septal structure indicates pedogenic modification of the crack-fill micrite.

Matrix-Supported Autochthonous Breccias

Larger intraclasts are angular and little disturbed, but smaller intraclasts, usually more abundant, are angular or subrounded to rounded (Ménillet, 1980, photo 4). Some intraclasts are defined by spar-filled circumgranular cracks, or coated by brownish, vaguely laminar micrite. Cements,

Period	Epoch	Age	Formation	Thick-ness (m)	Lithology	Interpretation
TERTIARY NEOGENE	MIO-CENE	Aquitanian	Calcaires de Beauce	50-100		Lacustrine/ palustrine
TERTIARY PALEOGENE	OLIGOCENE	Chattian				
		Rupelian				Hiatus (emergence)
			Sables de Fontainebleau s.l.	50		Shallow-marine
						Hiatus
			Marnes à Huitres	2		Laguno-marine
			Calcaire de de Sannois / Calcaire de Brie	3-10		Laguno-marine/lacustrine
			Argiles vertes de Romainville	8		Laguno-lacustrine
						Major hiatus
	EOCENE	Priabonian	Marnes blanches de Pantin	7		Laguno-lacustrine
			Marnes bleues d'Argenteuil	10		Laguno-lacustrine
			Gypse	15		
			Marnes entre deux masses / Calcaire de Champigny/ Château Landon	25		Laguno-lacustrine
			Gypse	15		Lacustrine
			Marnes à Lucines	4		Marine-lagoonal
			Gypse	4		Laguno-lacustrine
			Marnes à P. ludensis / Calcaire de Ludes	1-3		Marine/lacustrine
						Hiatus
		Bartonian	Sables de Marines / Calcaire de St Ouen	3-12		Shallow-marine/ lacustrine
			Sables de Cresnes	20		
			Sables de Beauchamp / Calcaire de Jaignes	2-10		Shallow marine/ lacustrine
			Sables d'Auvers	10		
		Lutetian	Marnes et Caillasses Banc vert / Calcaire de Morancez	10-15		Shallow-marine/ lacustrine
			Calcaire grossier s.l.	43		Shallow-marine
		Ypresian	Fausses glaises	5		
			Marnes à Huitres et Cyrènes	4		Laguno-lacustrine
			Sables de Sinceny	6		Shallow-marine
	PALEOCENE	Thanetian	Argile à Lignites			
			Argiles plastiques	20		Laguno-lacustrine
			Marnes de Cheney / Sables de Bracheux	13		Shallow-marine
						Hiatus
		Danian	Argiles et Lignites	10		Laguno-lacustrine
			Marnes de Sinceny	7		
			Conglomerat de Meudon	2		Shallow-marine
						Major hiatus (uplift and erosion)
CRETA-CEOUS		Campanian	Craie (Chalk)			

Dating method: Paleontology (calcareous nannoplankton, charophytes, dinoflagellate cysts, Foraminifera, mammals, molluscs, ostracods, palynology)
Reliability index: B
Sources: Bignot & Le Calvez (1969), Blondeau & Curry (1963), Bronnimann et al. (1968), Cavelier et al. (1979), Châteauneuf (1980), Châteauneuf & Gruas-Cavagnetto (1978), Ginsburgh & Hugueney (1980, 1987), Le Calvez (1970)

Figure 2—General Tertiary stratigraphy in the Paris Basin south of Paris showing important named units and the lateral equivalence of marine/laguno-lacustrine sediments and basin-margin lacustrine limestones.

often a mosaic of crystallites of more than one generation, include or are dominated by pedogenic textures (rhizoliths or alveolar-septal structure and microlaminar micrite, also complex combinations of these) (see Ménillet, 1980, photos 15, 16). *Microcodium* prisms, of a type illustrated by Freytet and Plaziat (1982, their figure 18A2), can also be present (see also Guillemin, 1976). Although in laguno-lacustrine micrites generally, a gradation from reticular cracking to matrix-supported breccias may result from concurrent desiccation and sedimentation (Ainardi and Champetier, 1976), in Quaternary–Holocene calcretes, pedogenic breccias can develop by rounding and separation of intraclasts, either by micritization of the host rock (Millot et al., 1977), combined micritization and development of rootlet-rich micritic cements (Klappa, 1978a), or active solution-weathering and substrate degradation (Harrison and Steinen, 1978, their figures 16H, I).

Peloidal Micrites

These overlie erosion surfaces and fill solution pockets, fissures, and pseudomicrokarst. Four major varieties are noted here.

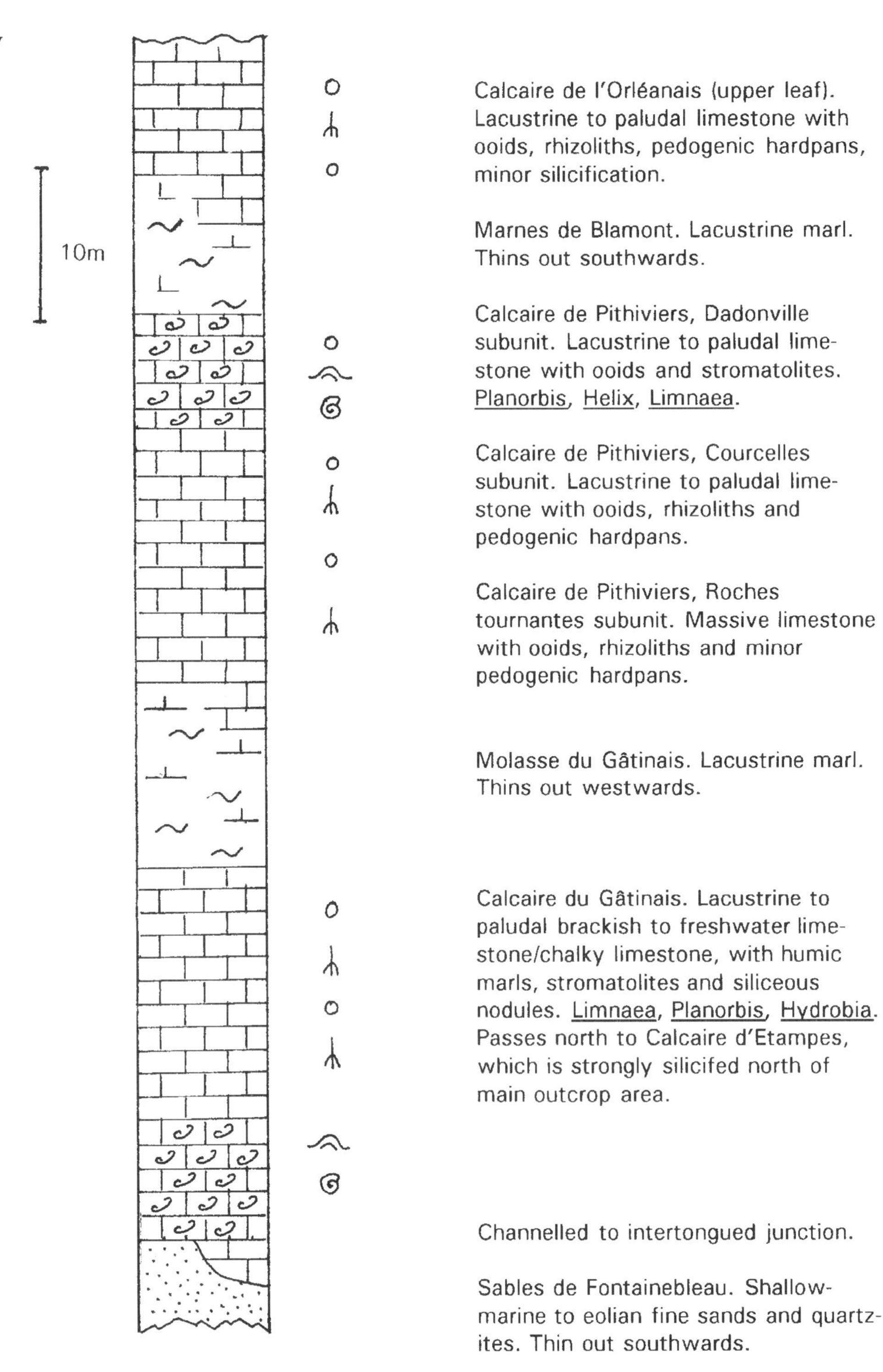

Figure 3—Detailed stratigraphy of the Calcaires de Beauce Formation in the Pithiviers area showing named units and gross lithologies.

Peloid-pisoid wackestones are subrounded to rounded intraclasts of host and first-generation micrite cement in a crumby-granular to minutely pelleted or microbrecciated micrite, cemented by microspar. More highly modified matrices display alveolar-septal structure of varying complexity. Rhizoliths are present or dominant. Outer micritic laminae of pisoids can form meniscus bridges, indicating vadose-zone diagenesis.

Peloid-crumb packstones are unsorted, subrounded, dense micrite peloids 10–50 μm in diameter and micrite crumbs 3–5 μm in diameter, the latter usually in groups, in a microspar or microspar lath matrix. Peloids show marginal alteration to darker brown micrite, which forms meniscus bridges and has totally replaced most crumbs. Reworked *Microcodium* prisms occur in irregular groups and are scattered through the matrix. This lithology seems to have developed by microbrecciation of the micrite matrix of a peloidal wackestone. Reverse-graded (upward-coarsening) peloid-crumb grainstones secondarily derived from this lithology occur in cracks and solution cavities and above crusts (cf. Freytet and Plaziat, 1982, pl. 37-E).

Peloidal grainstones are micrite peloids freed of their original matrix, possibly by vadose solution or erosion, which can be linked by micrite meniscus bridges and can be coated by sparite.

Peloid packstones develop when sparite fills voids and partly replaces peloids and earlier matrices or cements. Peloids with adherent earlier matrix resemble pisoids; some peloids are composed entirely of an earlier matrix (cf. Freytet and Plaziat, 1982m ok, 35A-C).

Peloids can develop in carbonate sediments during diurnal draining and flooding, producing fenestral pores (Mazzullo and Birdwell, 1989), and by microbrecciation during repeated desiccation fracturing, physicochemical weathering, and the activity of plant roots and associated fungi (Harrison, 1977; Alonso Zarza et al., 1992). The occurrence of rhizoliths and alveolar-septal structure in a peloidal matrix is analogous to a Holocene peloidal calcrete (Harrison, 1977, pl. 9-1, 9-2).

Breccio-Conglomerates

These transgressive lag deposits, composed of carbonate clasts in a carbonate matrix, form relatively flat or irregular lenses, often interbedded with crusts. They thicken into shallow topographic depressions but are not associated with any marked paleorelief. The clasts can represent single or composite carbonate lithologies. Breccio-conglomerates are of two broad types.

Clast-supported (Ménillet, 1980, photo 7) breccio-conglomerates, are unsorted to moderately well sorted, but can be imbricated. They are angular to subrounded, partly angular and subrounded, embayed, and tabular.

In matrix-supported (Ménillet, 1980, photo 17) breccio-conglomerates, the matrix can resemble marly micrite/microsparite, usually has a peloidal wackestone texture, can contain root casts or crystalline rhizoliths, or rarely is rhizolithic. In situ spalling and fragmentation of clasts, which can have partial compact or laminar micritic coats, are evidence either for dessication or, as in calcrete hardpans, in situ micritization (Millot et al., 1977, their figure 3), or volume increase of the matrix (e.g., Rutte, 1953, figure 21).

Pedogenic Lithologies

These occur at and are interdigitated with limestone bed tops and are commonly associated with disaggregational lithologies.

Rhizolithic Micrites

Intensely rhizolithic micrites to pelmicrites with 30–50% porosity occur as bunlike mounds, lenticular beds up to 0.75 m thick, as the fill of horizontal to oblique cracks, solution cavities, and root molds in other lithologies, and as the matrix of some reticular breccias and breccio-conglomerates. Simple root-moldic micrites are probably paludal in origin, but many display pedogenic textures and resemble simple Quaternary–Holocene root-moldic calcrete hardpans described by Harrison (1977), Perkins (1977, figures 7, 9, 10), Klappa (1980, figure 9a), and Rossinsky et al. (1992). Pedogenic textures include crystalline concentric rhizoliths and concentric tubules (probable inward-younging root mold linings). By analogy with Holocene soils (Bal, 1975; Davies, 1991), dark brown micritic aureoles surrounding many rhizoliths (cf. Freytet, 1965) resulted from cementation in the vadose zone.

Crusts and Stromatolites

Micritic crusts of 2 cm average thickness (30 cm maximum) occur at or close to bed tops, sometimes discordantly. They are commonly interbedded with or enclose lenses of peloidal-pisoidal lithologies, rhizolithic micrites, and breccio-conglomerates, and line solution cavities and fissures. The four major types of micritic crust noted here are part of a broad continuum.

Distinctly laminated to massive micritic crusts coat and fill surface irregularities (Ménillet, 1980, photos 4, 9; Freytet, 1965, pl. 1, 2) and resemble Quaternary–Holocene North African calcrete crusts described by Ruellan (1967). The alternating, plicated, paler and darker brown laminae, 0.1–2.0 mm thick, can be emphasized by differential iron-staining. Pedogenic textures include microbrecciation, crystalline concentric rhizoliths, and micritic aureoles.

Indistinctly laminated micritic crusts display undulose dark brown laminae 0.5–2 mm thick with indistinct boundaries and include lines and bands of sparry calcite-filled equant to elongate fenestrae, resembling concentric tubules. They resemble the rather broadly defined, moderately root moldic, Quaternary densely laminated "soilstone" crusts (Shinn and Lidz, 1988) on the Florida Keys (e.g., Robbin and Stipp, 1979; Shinn and Lidz, 1988), especially in thin section.

Porous laminated micritic crusts are approximately 0.5 mm thick, form sets approximately 2 mm thick, and are more-or-less root-moldic. They can include widely distributed, rounded micritic clasts. The most root-moldic sets are least obviously laminated, probably due to bioturbation. These crusts may resemble Quaternary–Holocene subsurface root-moldic laminar hardpans and sheet calcretes (Klappa, 1978b).

Cryptalgal stromatolites occur as columns (Ménillet, 1974b, 1980), 0.5 cm wide and 1–2 cm high and often are capped by linked laminae and domes displaying upward-facing unconformities and laminations at acute angles to the substrate (cf. Shinn and Lidz, 1988), and stacked hemispheroids, coating upward-facing topographic surfaces. In thin section they are micritic, resembling distinctly laminated crusts, and lack algal filament traces or root molds.

All crust types, especially when they line solution cavities and vertical fractures (cf. Ward, 1975, his figure 10b), accord with different aspects of biogenic laminated calcrete crusts, including the development of small stromatolites (Wright, 1989). Similar crusts in the upper Eocene Bembridge Limestone formed by modification of peloidal lithologies (Armenteros and Daley, 1998). Many, therefore, probably formed within the weathering or pedogenic zone, where they were affected by episodic subsurface desiccation and rhizobrecciation. Most also suffered surface erosion and disaggregation during lake transgressions.

LITHOLOGIC RELATIONSHIPS

Lenticular bedding on a scale of tens to hundreds of meters, and vertical and lateral lithologic changes involving all combinations of lithologies, including interpenetration of one or more lithologies in another, preclude detailed correlation among sections. Composite lithologies from this formation, displaying multiple episodes of sedimentation, brecciation, pelleting, pedogenesis, solution, and erosion (Ménillet, 1980, photo 18), are much more complex than composite palustrine lithologies illustrated by Freytet and Plaziat (1982) and Platt and Wright (1992, their figure 2). They are condensed sequences from sediment-starved, extremely shallow-water environments.

DISCUSSION

The formation was cited by Freytet and Plaziat (1982) as displaying many characteristic lithologies and textures of palustrine limestones. These include pseudomicrokarst, small-scale reticular breccias, and simple to polycyclical peloidal wackestones, grainstones, and packstones; however, many textures and fabrics described from calcretized marine limestones (e.g., Davies, 1991) and from calcretes generally (Esteban and Klappa, 1983) are also present, although volumetrically not as important. Crystalline concentric rhizoliths resemble rootlet pseudomorphs from Holocene calcareous soils and calcretes (Bal, 1975; Klappa, 1980, figures 5f, 6f; Calvet and Julia, 1983, their figure 2a).

Concentric rhizoliths occur in Holocene calcretes (Harrison, 1977, pl. 6-I; Klappa, 1980, figure 5d; Semeniuk and Meagher, 1981; Arakel, 1982, figure 6b). The occurrence of vaguely, and especially definitely, concentric micritic coats on peloids is attributable to episodic precipitation of coating micrite in the vadose zone, possibly promoted by fungae (cf. Calvet and Julia, 1983; Alonso Zarza et al., 1992). Alveolar-septal structure has a well-documented association with Quaternary–Holocene calcretes (Harrison and Steinen, 1978, figure 8f; Klappa, 1980, figures 9b, c; Calvet and Julia, 1983, figures 3f, g; Esteban and Klappa, 1983). *Microcodium*, common in calcretes, is either a crystalline ?soil-fungal structure (Klappa, 1978b; Esteban and Klappa, 1983, figure 70; Monger et al., 1991; Wright, 1994), or a form of rhizolith (Alonso Zarza et al., 1998). This combination of palustrine and pedogenic modifications contributed to the characteristic lateral and vertical lithostratigraphic variability of the formation.

The Calcaires de Beauce shares lithologic characteristics of all three climatic types of palustrine limestone as defined by Platt and Wright (1992) but corresponds chiefly to the intermediate climatic type. This agrees with paleobotanic and palynologic evidence (Koeniguer et al., 1985) for a Mediterranean-type paleoclimate in the Paris Basin during the Oligocene.

ACKNOWLEDGMENTS

BRGM and the University of Portsmouth generously provided laboratory and technical facilities during this study.

REFERENCES CITED

Ainardi, R., and Y. Champetier, 1976, Processus de formation d'intraclastes par dessication en milieu margino-littoral; exemple dans le "Purbeckien" du Jura: Bulletin de la Société Géologique de France, Série 7, v. 18, p. 159–164.

Alonso Zarza, A. M., J. P Calvo, and M. A. Garcia del Cura, 1992, Palustrine sedimentation and associated features—grainification and pseudo-microkarst—in the middle Miocene (Intermediate Unit) of the Madrid Basin, Spain: Sedimentary Geology, v. 76, p. 43–61.

Alonso Zarza, A. M., M. E. Sanz, J. P. Calvo, and P. Estévez, 1998, Calcified root cells in Miocene pedogenic carbonates of the Madrid Basin: evidence for the origin of *Microcodium* b: Sedimentary Geology, v. 116, p. 81–97.

Arakel, A. V., 1982, Genesis of Quaternary soil profiles, Hutt and Leman Lagoons, Western Australia: Journal of Sedimentary Petrology, v. 52, p. 109–125.

Armenteros, I., and B. Daley, 1998, Pedogenic modification and structure evolution in palustrine facies as exemplified by the Bembridge Limestone (late Eocene) of the Isle of Wight, southern England: Sedimentary Geology, v. 119, p. 275–295.

Bal, L., 1975, Carbonate in soil: a theoretical consideration on, and proposal for its fabric analysis. 2. Crystal tubes, intercalary crystals, K fabric: Netherlands Journal of Agricultural Science, v. 23, p. 163–176.

Bignot, G., and Y. Le Calvez, 1969, Contribution à l'étude des foraminifères planctonique de l'Eocène du Bassin de Paris: Proceedings of the First International Conference on Planktonic Microfossils, 1967, p. 161–166.

Blondeau, A., and D. Curry, 1963, Sur la présence de *Nummulites variolarius* (Lmk.) dans les diverse zones du Lutétien des Bassins de Paris, de Bruxelles et du Hampshire: Bulletin de la Société Géologique de France, Série 7, v. 5, p. 275–277.

Bronnimann, P., D. Curry, C. Pomerol, and E. Szöts, 1968, Contribution à la connaissance des foraminiféres planctoniques de l'Elocéne (incluant le Palépvène) du bassin anglo-franco-belge: Mémoire du Bureau de Recherches Géologiques et Minières, v. 58, pl 101–108.

Calvet, F., and R. Julia, 1983, Pisoids in the caliche profiles of Tarragona (NE Spain), *in* T. M. Peryt, ed., Coated grains: Berlin, Springer-Verlag, p. 456–473.

Cavelier, C., C. B. Guillemin, G. Lablanche, L. Rasplus, and J. Riveline, 1979, Précisions sur l'âge des calcaires lacustres du Sud du bassin de Paris d'aprés les Characées et les Mollusques: Bulletin du Bureau de Recherches Géologiques et Minières, Série 2, v. 1, section 1, p. 27–30.

Cavelier, C., and C. Pomerol, 1979, Chronologie et interprétation des évènements tectoniques cénozoïques dans le bassin de Paris: Bulletin de la Société Géologique de France, Série 7, v. 21, p. 33–48.

Châteauneuf, J. J., 1980, Palynostratigraphie et paléoclimatologie de l'Eocène supérieur et de l'Oligocène du Bassin de Paris: Mémoire du Bureau de Recherches Géologiques et Minières, v. 116, 360 p.

Châteauneuf, J. J., and C. Gruas-Cavagnetto, 1978, Les zones de *Wetziella* (Dinophyceae) du Bassin de Paris: Bulletin du Bureau de Recherches Géologiques et Minières, Série IV, v. 2, p. 59–93.

Davies, J. R., 1991, Karstification and pedogenesis on a late Dinantian carbonate platform, Anglesey, North Wales: Proceedings of the Yorkshire Geological Society, v. 48, p. 297–321.

Esteban, M., and C. F. Klappa, 1983, Subaerial exposure environment, *in* P. P. Scholle, D. G. Bebout, and C. H. Moore, eds., Carbonate depositional environments: AAPG Memoir, v. 33, p. 25–54.

Freytet, P., 1965, Sédimentation microcyclothémique avec croûtes zonaires à algues dans le Calcaire de Beauce de Chauffour-Etréchy (Seine-et-Oise): Bulletin de la Société Géologique de France, Série 7, v. 2, p. 309–313.

Freytet, P., 1984, Les sédiments lacustres carbonatés et leurs transformations par émersion et pédogenèse. Importance de leur identification pour les reconstitutions paléogéographiques: Bulletin des Centres de Recherches Exploration-Production Elf-Aquitaine, v. 8, p. 223–47.

Freytet, P., and J.-C. Plaziat, 1982, Continental carbonate sedimentation and pedogenesis–Late Cretaceous and Early Tertiary of southern France, Contributions to Sedimentology, 12: Stuttgart, Schweizerbart'sche, 213 p.

Gély, J.-P., and C. Lorenz, 1991, Analyse séquentielle de l'Eocène et de l'Oligocène du bassin de Paris (France): Revue de l'Institut Français du Pétrole, v. 46, p. 713–747.

Ginsburg, L., and M. Hugueney, 1980, La faune de mammifères terrestres dans le Stampien marin d'Etampes (Essonne): Comptes Rendus de l'Académie des Sciences, Paris, Série D, v. 268, p. 1266–1268.

Ginsburg, L., and M. Hugueney, 1987, Aperçu sur les faunules de mammifères du Stampien du Bassin de Paris: Bulletin d'Information des Géolologues du Bassin de Paris, v. 24, p. 19–22.

Guillemin, C., 1976, Les formations carbonatées dulçaquicoles tertiaires de la région centre (Briare, Château-Landon, Berry, Beauce): Doctoral Thesis, Université d'Orléans, 258 p.

Harrison, R. S., 1977, Caliche profiles: indicators of near-surface subaerial diagenesis, Barbados, West Indies: Bulletin of Canadian Petroleum Geology, v. 25, p. 123–173.

Harrison, R. S., and R. P. Steinen, 1978, Subaerial crusts, caliche profiles and breccia horizons: comparison of some Holocene and Mississippian exposure surfaces, Barbados and Kentucky: Bulletin of the Geological Society of America, v. 89, p. 385–389.

Klappa, C. F., 1978a, Morphology, composition and genesis of Quaternary calcretes from the western Mediterranean: a petrographic approach (2 vols): Unpublished Ph.D. Thesis, University of Liverpool, 446 p.

Klappa, C. F., 1978b, Biolithogenesis of *Microcodium*: elucidation: Sedimentology, v. 25, p. 489–522.

Klappa, C. F., 1980, Rhizoliths in terrestrial carbonates: classification, recognition, genesis and significance: Sedimentology, v. 27, p. 613–629.

Koeniguer, J.-C., M. Laurain, J. Mouton, J.-C. Plaziat, R. Wyns, and E. Boureau, 1985, Sur les nouveaux gisements Cénozoiques à végétaux fossiles dans le Bassin de Paris: végétations et paléoclimats: Comptes Rendus de l'Académie des Sciences, Paris, Série II, v. 301, no. 7, p. 509–514.

Le Calvez, Y., 1970, Contribution à l'étude des foraminifères paléogènes du Bassin de Paris: Cahiers de Paléontologie du CNRS, 326 p.

Manivit, J., 1979, Carte géologique de France (1/50,000), Feuille Malesherbes (293), Notice explicative, Bureau de Recherches Géologique et Miniéres, Orléans, 34 p.

Mazzullo, S. J., and B. A. Birdwell, 1989, Syngenetic formation of grainstones and pisolites from fenestral carbonates in peritidal settings: Journal of Sedimentary Petrology, v. 59, p. 605–611.

Ménillet, F., 1974a, Etude pétrographique et sédimentologique des Calcaires d'Etampes et de Beauce. Formations dulçaquicoles du Stampien supérieur à l'Aquitanien dans le bassin de Paris: Doctoral Thesis, Université de Paris Sud, 136 p.

Ménillet, F., 1974b, Sur la présence de structures stromatolitiques dans les Calcaires de Beauce, formations dulçaquicoles du Stampien supérieur à l'Aquitanien dans le bassin de Paris: Comptes Rendus de l'Académie des Sciences, Paris, v. 278, p. 3173–3176.

Ménillet, F., 1980, Les lithofaciès des Calcaires de Beauce (Stampien supérieur et Aquitanien) du bassin de Paris (France): Bulletin du Bureau de Recherches Géologiques et Minières, Série 2, v. 4, no. 1, p. 15–45.

Millot, G., D. Nahon, H. Paquet, A. Ruellan, and Y. Tardy, 1977, L'épigénie calcaire des roches silicatées dans les encroûtements carbonatés en pays subaride Antiatlas, Maroc: Sciences Géologiques, Bulletin (Strasbourg), v. 30, p. 129–152.

Monger, H. C., L. A. Daugherty, W. C. Lindemann, and C. M. Lidell, 1991, Microbial precipitation of pedogenic calcite: Geology, v. 19, p. 997–1000.

Perkins, R. D., 1977, Depositional framework of Pleistocene rocks in south Florida, *in* P. Enos and R. D. Perkins, eds., Quaternary sedimentation in south Florida: Geological Society of America Memoir, no. 147, p. 131–198.

Platt, N. H., 1989, Lacustrine carbonates and pedogensis: sedimentology and origin of palustrine deposits from the Early Cretaceous Rupelo Formation, W. Cameros Basin, N. Spain: Sedimentology, v. 36, p. 665–684.

Platt, N. H., and V. P. Wright, 1992, Palustrine carbonates and the Florida Everglades: towards an exposure index for the fresh-water environment?: Journal of Sedimentary Petrology, v. 62, p. 1058–1071.

Pomerol, C., 1980, Geology of France with twelve itineraries and a geological map at 1: 1,250,000: Paris, Masson, 256 p.

Pomerol, C., and L. Feugueur, 1968, Bassis de Paris, Ile-de-France: Paris, Masson, 174 p.

Robbin, D. M., and J. J. Stipp, 1979, Depositional rate of laminated soilstone crusts, Florida Keys: Journal of Sedimentary Petrology, v. 49, p. 175–180.

Rossinsky, V. Jr., H. R. Wanless, and P. K. Swart, 1992, Penetrative calcretes and their stratigraphic implications: Geology, v. 20, p. 331–334.

Roys, Marquis de, 1869, Note sur les formations d'eau douce supérieures aux sables de Fontainebleau: Bulletin de la Société Géologique de France, Série 2, v. 26, p. 376–380.

Ruellan, A., 1967, Individualisation et accumulation du calcaire dans les sols et les dépôts quaternaires du Maroc: Cahiers de l'ORSTOM, série Pédologie, v. 5, p. 421–462.

Rutte, E., 1953, Kalkkrusten in Spanien: Neues Jahrbuch für Geologie und Paläontologie, Abhandlungen, v. 106, p. 52–138 + pls 3–4.

Semeniuk, V., and T. D. Meagher, 1981, Calcrete in Quaternary coastal dunes in southwestern Australia: a capillary-rise phenomenon associated with plants: Journal of Sedimentary Petrology, v. 51, p. 47–68.

Shinn, E. A., and B. H. Lidz, 1988, Blackened limestone pebbles: fire at subaerial unconformities, *in* N. P. James and P. W. Choquette, eds, Paleokarst: New York, Springer-Verlag, p. 117–131.

Ward, W. C., 1975, Carbonate eolianites of northeastern Yucutan peninsula, Mexico, *in* K. F. Wantland and W. C. Pusey, III, eds., Belize shelf-carbonate sediments, clastic sediments, and ecology, AAPG Studies in Geology, no. 2, p. 500–571.

Wright, V. P., 1989, Terrestrial stromatolites and laminar calcretes: a review: Sedimentary Geology, v. 65, p. 1–13.

Wright, V. P., 1994, Paleosols in shallow marine carbonates: Earth Science Reviews, v. 35, p. 367–395.

Larsen, D., 2000, Upper Eocene and Oligocene lacustrine deposits of the southwestern United States with emphasis on the Creede and Florissant Formations, in E. H. Gierlowski-Kordesch and K. R. Kelts, eds., Lake basins through space and time: AAPG Studies in Geology 46, p. 425–438.

Chapter 39

Upper Eocene and Oligocene Lacustrine Deposits of the Southwestern United States, with Emphasis on the Creede and Florissant Formations

Daniel Larsen
Department of Geological Sciences, University of Memphis
Memphis, Tennessee, U.S.A.

INTRODUCTION

The purpose of this paper is to briefly review the upper Eocene and Oligocene lacustrine deposits in the southwestern part of the United States and to discuss how they have been used to understand the paleoclimate, tectonic history, and volcanic environment in this region. Two well-studied lacustrine deposits (upper Eocene Florissant and Oligocene Creede Formations) will be a primary focus because they represent two major types of lake basins that formed in the region during this time. The Florissant Formation was deposited in a lake basin formed by a blocked stream drainage (McElroy and Anderson, 1966). The Creede Formation was deposited in a lake basin formed by caldera collapse (Steven and Ratté, 1965; Larsen and Crossey, 1996). Other examples of these two types of lake basins and other lake deposits in the region are also discussed (see Figure 1 and Table 1).

Much of the southwestern part of the United States (Arizona, Utah, Colorado, and New Mexico) was elevated relative to surrounding regions during the latest Eocene through Oligocene (Figure 1) (Christiansen and Yeats, 1992; Elston and Young, 1991; Gregory and Chase, 1992; Chapin and Cather, 1994; Wolfe et al., 1998). This was largely the result of Laramide-style high-angle faulting and filling of adjacent Laramide-style basins (Miller et al., 1992; Dickinson et al., 1988). Prominent erosional surfaces existed adjacent to tectonic highlands in the Rocky Mountains (Evanoff, 1990; Chapin and Cather, 1994) and northern Arizona–southern Utah area (Elston and Young, 1991). Voluminous volcanism initially produced intermediate-composition composite volcanoes in numerous areas. This was followed in the Sawatch Range, San Juan, Marysvale, Latir, and Mogollon-Datil volcanic fields by silicic calderas (see locations on Figure 1) (Steven, 1975; Elston, 1984; Christiansen and Yeats, 1992). Latest Eocene and early Oligocene sedimentation occurred largely in a few broad, low-relief Laramide basins (Claron: southwestern Utah, Goldstrand, 1994; Baca, central New Mexico, Cather and Johnson, 1986; Uinta: northeastern Utah, Andersen and Picard, 1972) and in volcaniclastic aprons and proto-rift basins adjacent to the volcanic regions (Anderson et al., 1975; Ingersoll et al., 1990). Few other late Eocene–Oligocene deposits are present in the region, and most of those are associated with initial extension in the southern Rio Grande rift (Mack et al., 1994) and southern Basin and Range (Grover, 1984; Eberly and Stanley, 1978; Christiansen and Yeats, 1992). Small, localized accumulations of fluvial and lacustrine deposits in volcanic depressions and blocked stream drainages (see Figure 1 and Table 1) preserve the only sedimentary records for understanding the tectonic, climatic, and paleoenvironmental history of this region during this time.

FLORISSANT FORMATION

The upper Eocene Florissant Formation was deposited in a blocked stream drainage on the Rocky Mountain erosion surface of the central Rocky Mountains (Figure 2) (Epis and Chapin, 1975; Evanoff and Chapin, 1994). A debris flow (lahar?) that flowed from the contemporaneous Thirtynine Mile volcanic field apparently blocked the drainage (Figure 2) and initiated lacustrine sedimentation (McElroy and Anderson, 1966; Evanoff, 1994). The age of the Florissant Formation is constrained by the underlying 36.7 Ma Wall Mountain Tuff (McIntosh and Chapin, 1994), a regional ash-flow tuff, and an overlying volcaniclastic debris flow with 34.9 Ma biotite grains (Epis and Chapin, 1974). Based on the thicknesses of seasonal laminae in the lake beds, the lake probably existed for only between 2500 and 5000 yr (McElroy and Anderson, 1966).

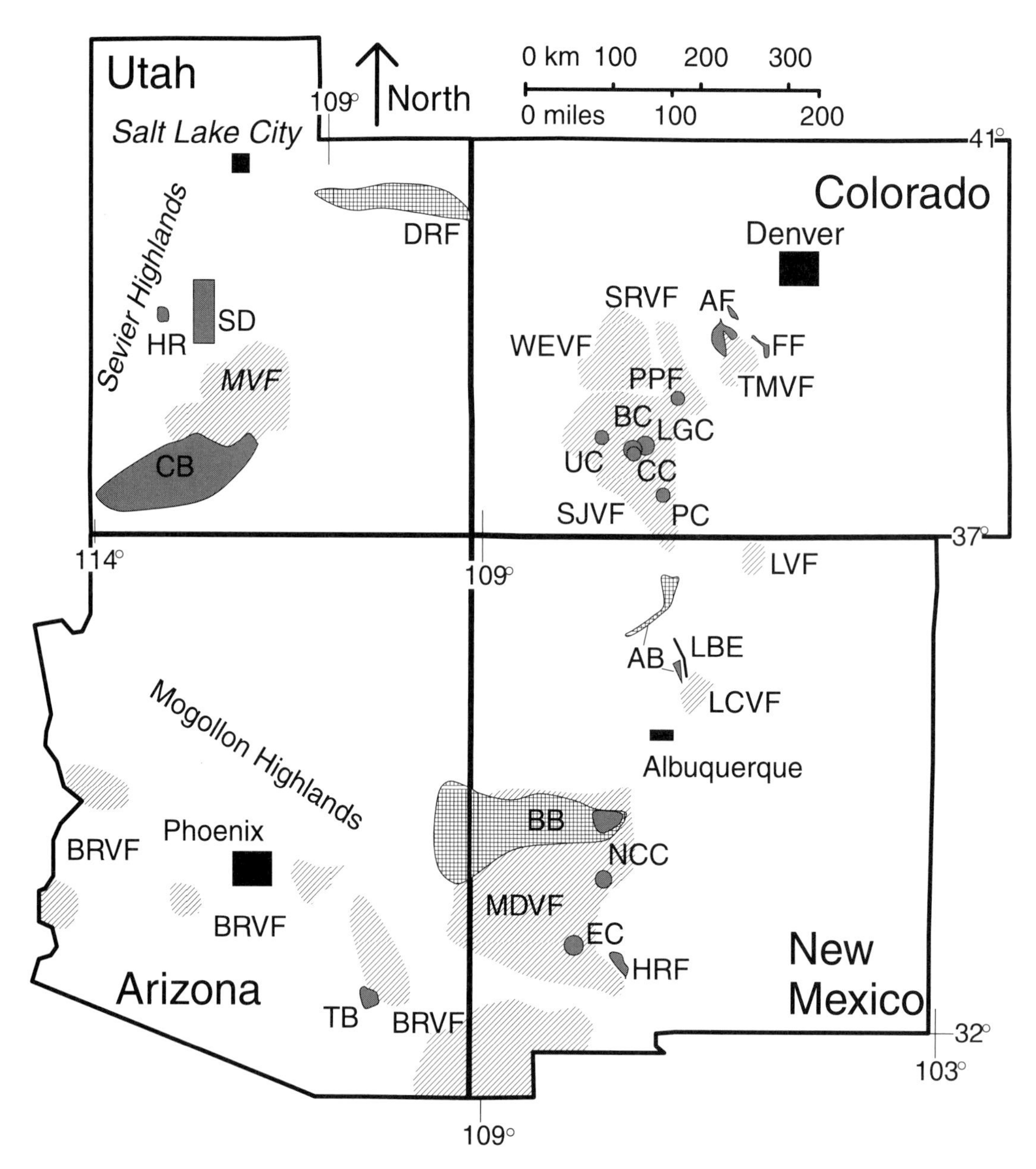

Figure 1—Map showing the locations of deposits, basins, and other features discussed in the text. Gray shaded pattern represents lacustrine deposits. Checkered pattern represents basins and formations discussed in text. Diagonal hatch pattern represents volcanic fields. Abbreviations are as follows: AF = Antero Formation, AB = Abiquiu Formation, BB = Baca Basin, BC = Bachelor caldera, BRVF = Basin-and-Range volcanic field, CB = Claron Basin, CC = Creede caldera, DRF = Duchesne River Formation, EC = Emory cauldron, FF = Florissant Formation, HR = House Range, HRF = Hayner Ranch Formation, LBE = La Bajada escarpment, LGC = La Garita caldera, LCVF = Los Cerrillos volcanic field, LVF = Latir volcanic field, MDVF = Mogollon-Datil volcanic field, MVF = Marysvale volcanic field, NCC = Nogal Canyon caldera, PC = Platoro caldera, PPF = Pitch=Pinnacle Formation, SD = Sevier Desert, SJVF = San Juan volcanic field, SRVF = Sawatch Range volcanic field, TB = Teran basin, TMVF = Thirtynine Mile volcanic field, UC = Uncompahgre caldera, WEVF = West Elk volcanic field.

The Florissant Formation comprises approximately 53 m of fluvial and lake deposits, of which the middle 40 m contain evidence for lacustrine sedimentation (Figure 3). Lake sedimentation appears to have followed emplacement of a 5- to 10-m-thick ash-rich debris-flow that buried *Sequoia* and angiosperm stumps. These stumps are now mineralized and preserved in the lake beds. The subsequent 20 m of lake beds are mostly thinly-bedded to laminated, tuffaceous, fossiliferous shales and mudstones, although a thick (as much as 7.9 m) debris-flow bed intervenes in the sequence. The lake shales also include diatomite-sapropel laminae that McElroy and Anderson (1966) interpreted as seasonal laminae. The upper 10 m of the lake beds include abundant sandstone and conglomerate that contain sphaeriid clam fossils, attesting to a lacustrine origin of the beds (Evanoff, 1994).

The fossil content of the Florissant Formation has been studied extensively for more than 100 yr (see references in McElroy and Anderson, 1966). Although the number of fossil insect species identified dwarfs that of the fossil plants (Dexter, 1994), the fossil flora has received more attention. MacGinitie's (1953) monograph is the most complete study of the fossil plant organs and paved the way for modern paleoclimatic and paleoenvironmental analysis using fossil plants (Wolfe, 1978; Gregory and Chase, 1992). The most recent estimates of the paleoenvironment at Florissant have been obtained using the Climate-Leaf Analysis Multivariate Program (CLAMP) (Wolfe, 1995). These estimates indicate a mean annual temperature (MAT) of 11.8°C and paleoaltitude of 3.8 km, based on the moist-air enthalpy method (Gregory, 1994; Wolfe et al., 1998).

The lake environment was most completely studied by McElroy and Anderson (1966). They concluded that the lake was meromictic, but not saline, based on the mineralogy and lithology of the various laminae. Meromixis was probably maintained by some combination of chemical stratification and thermal density differences maintained by the physiographic setting of the lake in a mountain valley. The epilimnion was likely neutral to slightly alkaline based on the abundance of diatoms (McElroy and Anderson, 1966). In contrast, the monimolimnion was probably mildly acidic and oxygen-deficient as judged by the presence of pyrite in sapropel (McElroy and Anderson, 1966) and preservation of fossils. Although carbonate is largely absent in the lake beds, algal stromatolites are found above one of the marker tuff beds (Evanoff, 1994). These suggest that ash falls into the lake may have been sufficient to alter the chemistry of the lake for a short period and to permit stromatolites to algal mats to grow (Evanoff, 1994).

OTHER LAKE BASINS SIMILAR TO FLORISSANT IN THE REGION

The only other well-documented example of a late Eocene–Oligocene lake basin that clearly formed by blockage of a stream valley is the Antero Formation. The Antero Formation was also deposited on the Rocky Mountain erosion surface, similar to the Florissant Formation, but is slightly younger (Figure 1) (Stark et al., 1949; Evanoff, 1994). Stark et al. (1949) interpreted that the Antero lake basin was either formed or modified by eruption of trachyte and andesite lava flows from the neighboring Thirtynine Mile volcanic field. The age of Antero Formation is known to be Oligocene based on fossil evidence (Stark et al., 1949) and interbedded 33.7 Ma nonwelded tuffs near the southwest extent of Antero Formation exposures (Evanoff, 1994; McIntosh and Chapin, 1994). The lacustrine deposits include limestone, shale, tuffaceous mudstone and siltstone, and fall-out tuffs. In the main exposures adjacent to the Thirtynine Mile volcanic field, stromatolitic limestone is common, and ostracode and gastropod fossils are present in the shale. In the northern exposures, fossil plant organs are present in the shales and tuffs and include pine, oak, and shrub species similar to those in the Florissant and Creede floras. Wolfe (1992) considers this flora to represent a subalpine community, similar to that at Creede (see discussion).

Stark et al. (1949) interpreted the Antero Formation to represent deposits of a laterally continuous freshwater lake with an area of approximately 166 km^2. The lake shorelines and deltas were composed of sands and gravels. The time-transgressive character of the shore-zone deposits suggests that the lake expanded and contracted numerous times during its existence.

OLIGOCENE CREEDE FORMATION

The Creede Formation is located in the San Juan volcanic field and is an excellent example of lacustrine deposition in a caldera basin (Figure 4). The Creede caldera formed at 26.9 Ma (Lanphere, 1994) by an eruption-related collapse of a nearly circular, 12-km-diameter region and was concurrently partially filled with the dacitic Snowshoe Mountain Tuff (Steven and Ratté, 1965). The caldera was set into the high-altitude, moderate-relief volcanic plateau that was established during the later stages in the development of the Tertiary San Juan volcanic field (Steven, 1968). The structural moat of the caldera was subsequently filled with intracaldera lava flows of the Fisher Quartz Latite and intercalated sediments of the Creede Formation (Figure 4).

More than 1 km of sediment was deposited in the north-central part of the basin (Larsen and Crossey, 1996) during the <660 k.y. duration of sedimentation (Lanphere, 1994). Following an initial period of subaerial sedimentation, lacustrine sedimentation by sublittoral sediment-gravity-flow and suspension processes dominated the basin-fill history (Figure 5). Thick sublacustrine slump deposits enclosed by intervals of turbidites and suspension laminae indicate water depths in excess of 50 m. Carbonate travertine mounds were also precipitated around the margin of the basin. Although most of these deposits were precipitated in the lake, the term "travertine" is applied because of their dominant microbial and physico-chemical origins (see Ford and Pedley, 1996) and historical precedence (Steven and Ratté, 1965). Regional faulting following Creede sedimentation displaced beds by as much as 100 m.

Creede Formation sedimentation occurred in a topographically closed basin of limited lateral extent that was influenced by concurrent volcanic activity within and nearby the Creede caldera (Larsen and Crossey, 1996). Because of this setting, sediment sources were largely limited to (1) sediment weathered and eroded from the caldera walls and valleys, (2) pyroclastic materials from local eruptions, (3) biotic and abiotic carbonate sediment, and (4) intrabasinal material. Detrital sediment entered the lake via steep (as much as 30% gradient) caldera-wall drainages and

Table 1. General Information About Lacustrine Deposits Discussed in the Text.

Name of Deposit or Formation	Location of Lacustrine Strata (lat., long.)	Age of Lacustrine Strata	Thickness of Lake Beds (m)	Major Facies (in Order of Decreasing Abundance) and Documented Fossil Remains
Abiquiu Fm.	35°30´ N, 106°15´ W	Late Oligocene	~ 75 m	Limestone, interbedded tuffaceous, calcareous horizontal-bedded and rippled cross-laminated sandstone and calcareous mudstone, tuff.
Antero Fm.	38°30´ N to 39°30´ N, 105°30´ W to 106°0´ W	Late Eocene	~ 500 m	Tuffaceous shale and normal-graded sandstone, interbedded with limestone marl in benches along bedrock shorelines, travertine; gastropod, ostracode, fish, and carbonized plant and insect fossils, and stromatolites.
Baca Fm.	34°0´ N to 34°25´ N, 107°30´ W to 107°45´ W	Late Eocene	~ 200 m	Calcareous mudstone and claystone, thin-bedded sandstone, inclined sets of beds common; ostracodes and fossil pollen.
Bachelor caldera	37°47´ N to 107°10´ W	27.35 Ma, late Oligocene	< 50 m	Laminated tuffaceous sandstone and siltstone.
Claron Basin (Brian Head Fm.)	37°40´ N to 38°0´ N, 112°35´ W to 113°0´ W	Late Eocene to (middle) Oligocene	178 m	Limestone partially replaced by chalcedony, tuffaceous sandstone, siltstone, claystone, and tuff; gastropod, pelecypod, and charophyte fossils.
Creede Fm.	37°40´ N to 37°50´ N, 106°50´ W to 107°0´ W	26.9 to 26.2 Ma, late Oligocene	691 m measured, ~ 1000 m inferred	Tuffaceous siltstone laminae, tuffaceous normal-graded sandstone and pebbly sandstone, travertine, limestone laminae, cross-bedded pebbly sandstone, and matrix-supported conglomerate; diatom fossils, carbonized plant and insect fossils, and stromatolites.
Emory Cauldron (lake beds interbedded with Bear Canyon basalt)	32°55´ N, 107°35´W	Late Oligocene	< 50 m	Tuffaceous siltstone laminae, normal-graded pebbly sandstone, cross-bedded sandstone, and tuff; carbonized plant fossils.
Florissant Fm.	38°50´ N to 39°0´ N, 105°15´ W	Late Eocene	40 m	Shale, mudstone, tuffaceous mudstone and sandstone, pumiceous sandstone, and volcaniclastic "debris-flow" conglomerate; carbonized plant, insect, and arachnid fossils, and pelecypod, gastropod, ostracode and fish fossils, and stromatolites.
Tertiary limestone in the House Range, Utah	39°17´ N, 113°15´W	Late Eocene	>10 m	Limestone; gastropod and carbonized plant fossils.
Hayner Ranch Fm.	32°41´ N, 107°06´W	Late Oligocene to middle Miocene	~ 4 m	Limestone; ostracode fossils and stromatolites.

Table 1. (continued)

Name of Deposit or Formation	Location of Lacustrine Strata (lat., long.)	Age of Lacustrine Strata	Thickness of Lake Beds (m)	Major Facies (in Order of Decreasing Abundance) and Documented Fossil Remains
La Garita caldera	37°46′ N, 106°47′ W	27.6 Ma, late Oligocene	<10 m	Tuffaceous siltstone laminae, tuffaceous normal-graded sandstone and pebbly sandstone, and limestone laminae; carbonized plant fossils.
Pitch-Pinnacle Fm.	38°22′ N, 106°19′ W	Late Eocene to early Oligocene	178 m	Tuffaceous siltstone laminae, tuffaceous normal-graded sandstone and pebbly sandstone, and matrix-supported conglomerate; carbonized plant and insect fossils.
Platoro caldera (Summitville Andesite)	37°17′ N to 37°25′ N, 106°22′ W to 106°35′ W	29.5 to 28.2 Ma, late Oligocene	>25 m	Tuffaceous normal-graded sandstone and pebbly sandstone, cross-bedded pebbly sandstone, tuffaceous siltstone laminae, matrix-supported conglomerate, and limestone laminae; carbonized plant fossils.
Sevier Desert (Tertiary basin fill)	39°30′ N, 112°40′ W	Late Oligocene	>215 m in core	Siltstone, claystone, nodular anhydrite, and tuffaceous sandstone; fossil pollen.
Teran basin (Mineta Fm.)	32°17′ N, 110°17′ W	Late Oligocene	?	Mudstone, siltstone, normal-graded sandstone, laminated and cross-bedded sandstone, limestone and gypsum; stromatolites.
Uncompahgre caldera (Burns Fm.)	38°00′ N to 38°02′ N, 107°18′ W	28.1 Ma, late Oligocene	< 100 (?) m	Tuffaceous siltstone laminae, tuffaceous normal-graded sandstone and pebbly sandstone, cross-bedded pebbly sandstone, and matrix-supported conglomerate.

less-steep (as much as 20% gradient) valleys draining from the resurgent dome and surrounding volcanic plateau (Figure 4). Considerable quantities of ash entered the lake via direct pyroclastic fallout; however, most of the tuffaceous sediment in the Creede Formation is reworked. Most of the detrital sediment was deposited in the lake via sediment-gravity-flow processes or suspension settling.

Carbonate sediments were precipitated largely by algal and bacterial processes in the photic zone (Larsen, 1994a; Finkelstein, 1997), but bacterial and abiotic precipitation dominated at depth. The most prominent carbonate deposits in both the open-lake and spring environments are suspension laminae, although stromatolites are common in the travertine and shallow-water facies (Larsen and Crossey, 1996). Calcite pseudomorphs with pseudotetragonal habit and porous replacement textures are common in detrital facies and travertine. Based on the mode of occurrence in laminae and travertine, pseudomorph morphology, and texture and chemistry of the replacement calcite, Larsen (1994b) interpreted the pseudomorphs to be after ikaite ($CaCO_3 \bullet 6H_2O$). Ikaite is a rare cold-water, carbonate mineral known to form in cold spring and organic-rich mud environments (Shearman and Smith, 1985; Bischoff et al., 1993).

Because the Creede caldera basin is regarded to have been a topographically closed basin, the major controls on sedimentary facies distributions were the morphology and geomorphic evolution of the basin and climatically driven changes in lake level (Larsen and Crossey, 1996). Volcano-tectonic subsidence may have contributed to basinal-thickening trends in the lacustrine deposits, especially in the north-central and northwestern parts of the basin. Patterns of sedimentation in the Creede Formation show many similarities to those in modern caldera lakes (Crater Lake, Oregon, U.S.A.: Nelson et al., 1994; Lake Atitlan, Guatemala: Newhall et al., 1987) and ancient crater and caldera basins (Cretaceous crater deposits, Bushmanland, South Africa: Smith, 1986; Miocene Whitehorse caldera, Oregon, U.S.A.: Rytuba et al., 1981). The abundance of travertine is anomalous and

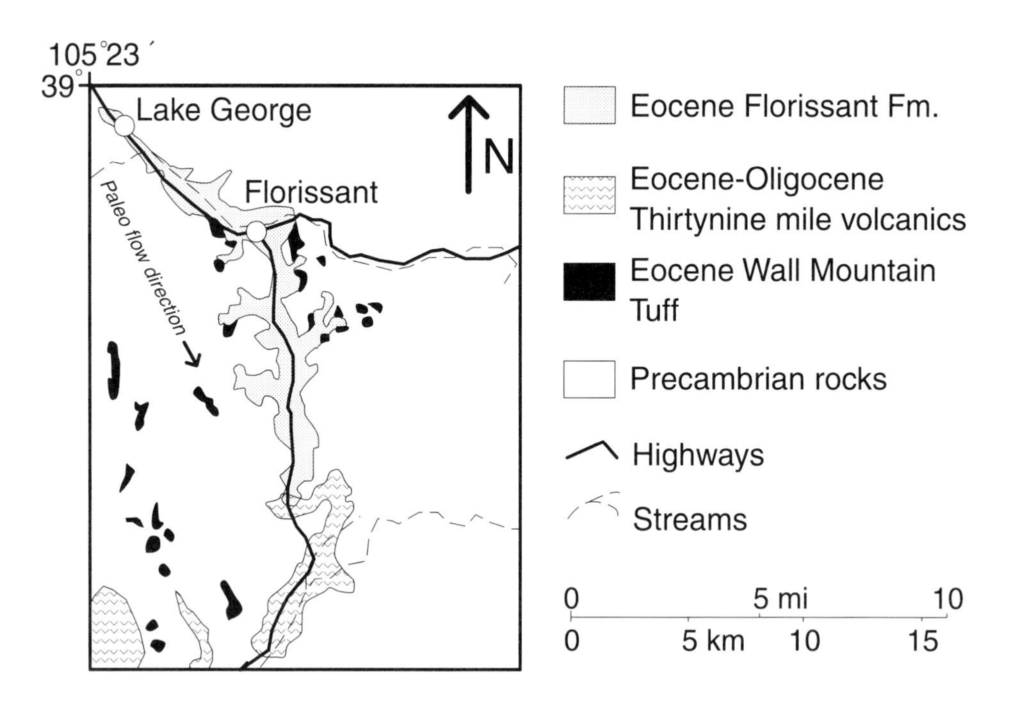

Figure 2—Generalized geologic map of the Florissant Formation and vicinity (modified from Evanoff, 1994).

probably reflects the importance of shallow groundwater discharge into the caldera basin (Larsen, 1994b).

Characteristics of the laminated facies and carbonate minerals in the Creede Formation indicate that Lake Creede was cold and stratified (Larsen, 1994b; Larsen and Crossey, 1996). Calcite pseudomorphs after ikaite in the lacustrine deposits (Larsen, 1994b) and paleofloral analysis based on leaf physiognomy (Wolfe and Schorn 1989; Wolfe et al., 1998) indicate a cold climate (MAT = 4.2°C), similar to that present in the Creede area today. In contrast, Axelrod (1987) indicated a much warmer and drier climate for Creede based on comparison of the Creede fossil flora to nearest modern relatives. Stratification is indicated by the absence of bioturbation, preservation of organic carbon, abundance of pyrite, and pseudomorphs after ikaite across much of the basin. These characteristics suggest the presence of an anoxic, concentrated hypolimnion (monimolimnion) (Larsen and Crossey, 1996; Finkelstein, 1997). The presence of algal remains and evidence for thriving lacustrine shrimp populations require an oxygenated epilimnion, as would be expected in most lake environments. Considering the abundance of calcareous deposits and the thermodynamic stabilities of interpreted precursors (aragonite, ikaite, etc.), Larsen and Crossey (1996) interpreted the hypolimnion waters to be brackish to saline, but not highly alkaline. Recent observations of trace quantities of sulfate minerals, interpretation of rice-grain-shape calcite pseudomorphs after gypsum, and other textural observations by Finkelstein et al. (1996) and Finkelstein (1997) suggest that the lake contained saline, Na-SO_4-Cl waters of near-neutral pH. These authors and Larsen (1995) suggest that volcanic gases were important in controlling water chemistry and mineral precipitation in the lake.

OTHER LAKE DEPOSITS ASSOCIATED WITH VOLCANOES IN THE REGION

Lake deposits are present within numerous other mid-Tertiary calderas and volcanic depressions in the region. In most cases these deposits have not been examined in great detail; thus, only limited information is available regarding their extent and character. The deposits are discussed in order from north to south in the region.

Gregory and McIntosh (1996) investigated the Oligocene Pitch-Pinnacle Formation in the Sawatch Range, central Colorado (Figure 1), to establish the paleoclimate and paleoaltitude represented by the fossil flora. The Pitch-Pinnacle Formation lies in a depression that may represent either (1) the collapse structure from the eruption of the 33.7 Ma Thorn Ranch Tuff or (2) a paleovalley that was blocked by thick andesite flows. The age of the formation is constrained to be between 32.9 and 29 Ma based on $^{40}Ar/^{39}Ar$ dating. The deposits include light and dark (seasonal?) laminae, graded siltstone and sandstone, matrix-supported conglomerate, mudstone, and ash. Fossil plant and insect remains are common along bedding contacts. No aquatic fossils are observed. These deposits appear to represent deposition in sublittoral depositional environments similar to those represented by the Creede Formation. Using CLAMP analysis (Wolfe, 1995), Gregory and McIntosh (1996) determined a MAT of 12.7°C. Using moist static enthalpy of the atmosphere (Wolfe et al., 1998), they determined a paleoaltitude of 2–3 km, assuming that the flora were coeval with pre-Oligocene climate deterioration conditions, or 1 km, assuming the flora were coeval with post-Oligocene climate deterioration. Supporting geologic evidence favors the high paleoaltitude estimate (Gregory and McIntosh, 1996).

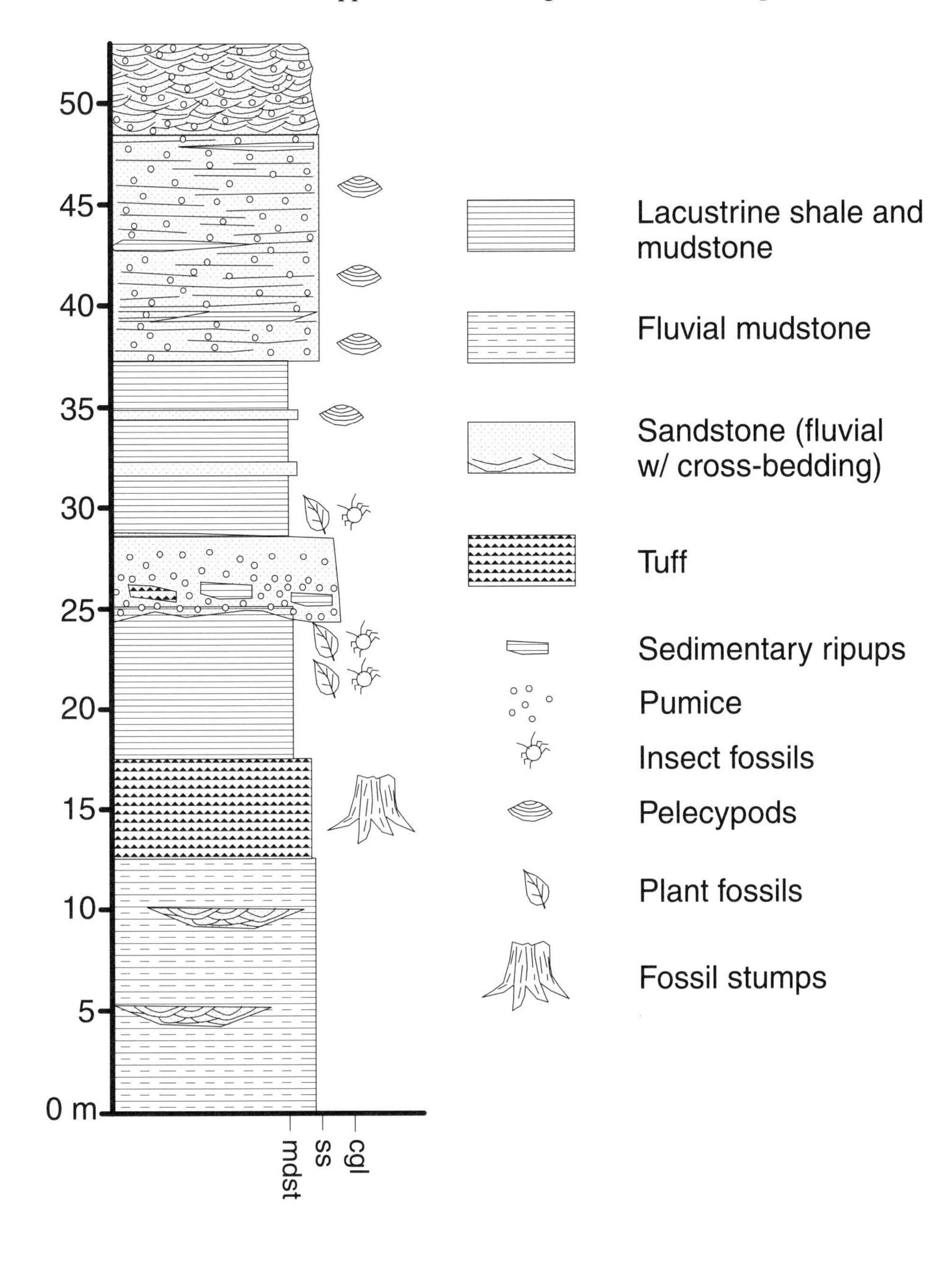

Figure 3—Generalized stratigraphy of the Florissant Formation (modified from Evanoff, 1994).

Aside from Creede, four other calderas in the San Juan volcanic field of southern Colorado contain lake deposits. Lacustrine and fluvio-deltaic sediments are interbedded with post-collapse andesite lavas in the Platoro caldera (Figure 1) (Lipman, 1975). The Platoro caldera is a complex caldera that experienced multiple collapse events and associated ash-flow-tuff eruptions (Lipman et al., 1996). Beds comprising graded tuffaceous sandstone and matrix-supported volcanic conglomerate and minor low-angle-crossbedded and planar, tabular-crossbedded sandstone are present below intracaldera lavas in the collapse structure associated with the 29.3 Ma La Jara Canyon Tuff. Beds comprising graded tuffaceous sandstone, volcanic conglomerate, tuffaceous siltstone and sandstone laminae, tuff, and low-angle and trough-crossbedded sandstone are present in the collapse structure associated with the 28.4 Ma Chiquito Peak Tuff. In both cases, a spectrum of shallow-lacustrine, deltaic, and deep-lacustrine environments is represented. Evidence for transitions from basin-margin and deltaic facies to sublittoral facies are observed over distances of 0.5 km, suggesting rather steep lake slopes. Pollen are preserved in the tuffaceous siltstone laminae in the Chiquito-Peak collapse structure and record a flora intermediate between that at Creede and Florissant (Lipman, 1975).

Lacustrine and fluvio-deltaic deposits are also interbedded with andesitic lavas in the Burns Formation, the intracaldera fill of the Uncompahgre caldera in the San Juan volcanic field (Figure 1) (Lipman et al., 1973). The Uncompahgre caldera was probably

Figure 4—Generalized geologic map of the Creede Formation (modified from Larsen, 1994b). The locations of identified (I) and hypothesized (H) paleovalleys and paleocanyons are indicated.

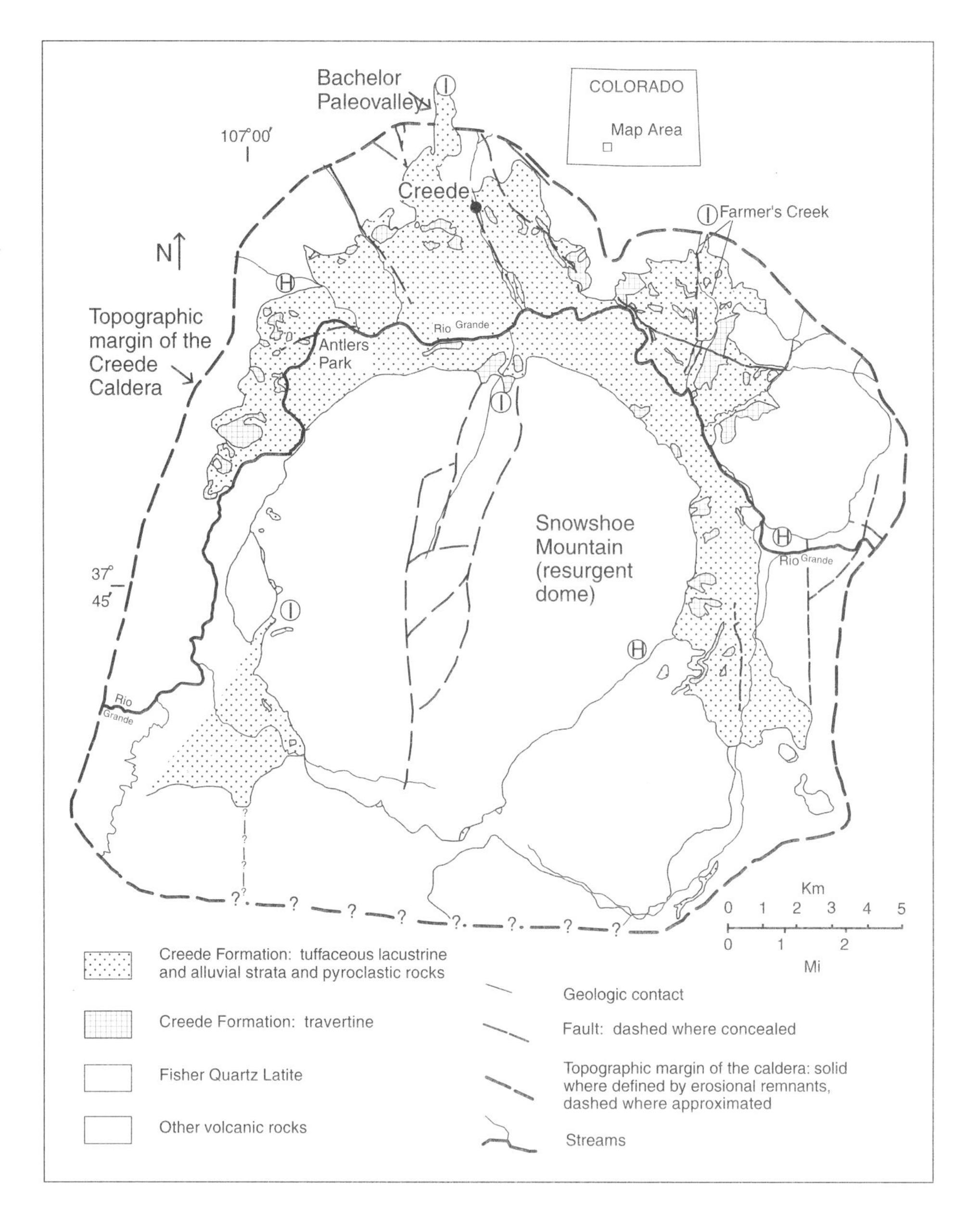

formed in association with eruption of the 28.1 Ma Dillon Mesa Tuff (Lipman et al., 1996). Exposures along the eastern margin of the caldera include beds of tuffaceous sandstone, volcanic conglomerate, and tuffaceous siltstone. The deposits represent sublittoral environments and resemble sublacustrine-slope and -fan facies at Creede (Larsen and Crossey, 1996). Exposures in the northeastern part of the caldera include beds of tuffaceous sandstone, tuffaceous siltstone and sandstone laminae, tuff, and oscillatory-rippled tuffaceous sandstone. These deposits represent nearshore environments, possibly influenced by deltaic processes. No information is available about the fossil content of the Burns Formation deposits.

A few scattered exposures of lake deposits are present within the La Garita (27.6 Ma) and Bachelor (27.35 Ma) calderas in the San Juan volcanic field (Lipman, in press), but little work has been completed on these deposits. One point of note, however, is that calcite pseudomorphs after ikaite are present in laminated sediments of the La Garita caldera (Larsen, 1994b). This suggests that the conditions needed for formation of this unusual mineral also developed in an older caldera neighboring the Creede caldera.

Although lacustrine deposits are present within calderas of the Mogollon-Datil volcanic field of southwestern New Mexico, they are the least studied of the lake beds discussed here. The best documented are those within the eastern moat basin of the 34 Ma Emory cauldron (Figure 1) (Elston et al., 1987). Near Hillsboro, New Mexico, the lake beds are interbedded with and overlain by 28.1 Ma Bear Springs Basaltic Andesite. The beds contain tuffaceous sandstone, siltstone, and shale and planar, tabular, and scour-and-fill crossbedded sandstone. The beds are organized into coarsening-upward cycles suggesting either shallowing-deepening

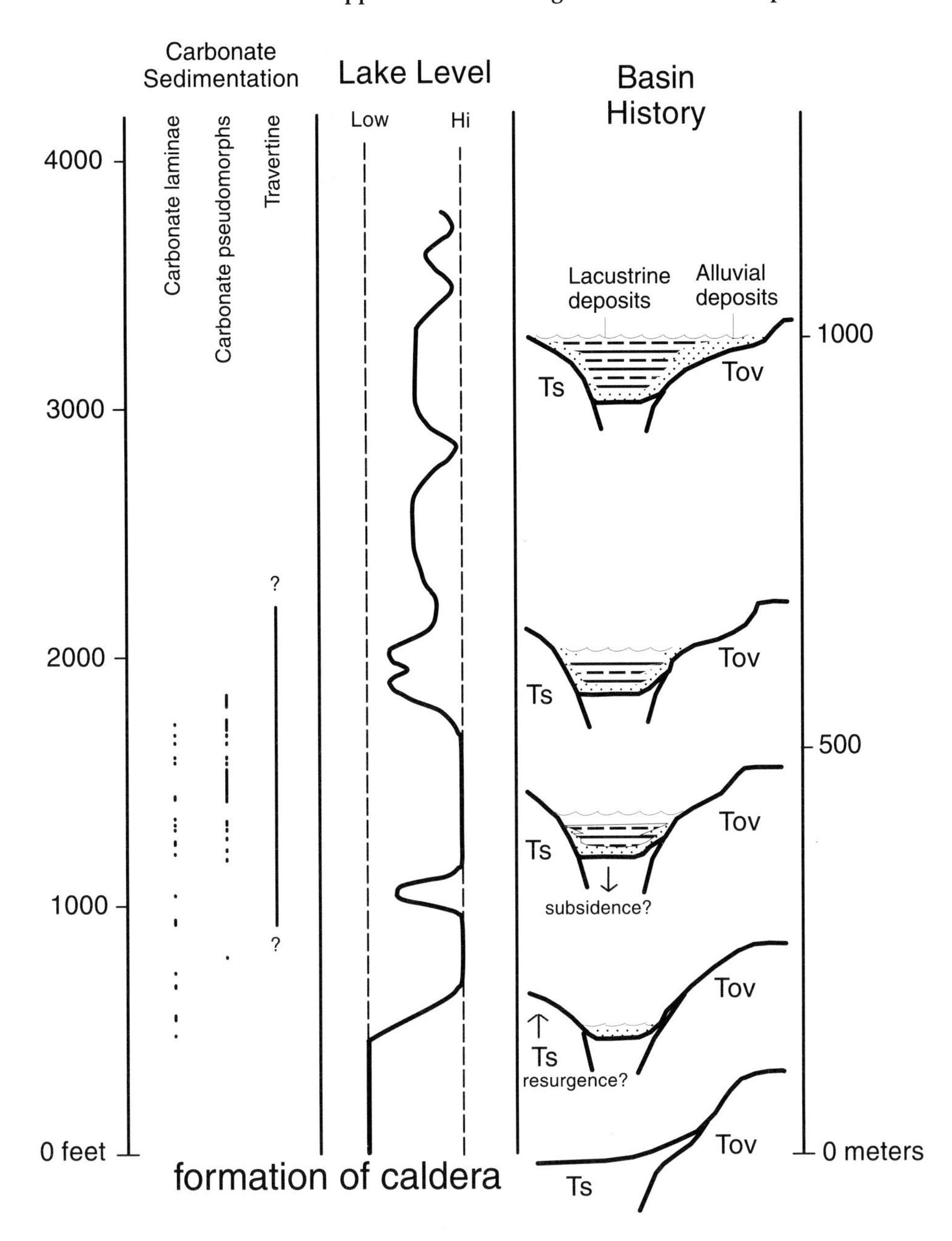

Figure 5—Depositional history of the Creede Formation (modified from Larsen and Crossey, 1996). The stratigraphic distributions of carbonate laminae and pseudomorphs are based on occurrence in Continental Scientific Drilling Program core CCM=2 (see Figure 4). The stratigraphic distribution of travertine is based on field studies and correlations in Larsen and Crossey (1996). The upper part of the lake-level curve is based solely on mining industry cores from and exposures in the Bachelor paleovalley (see Figure 4) and is more speculative.

cycles or delta-lobe switching in a lake. Varved lake beds are also present in the moat of the Emory cauldron 24 km northwest of Hillsboro, near the town of Hermosa (Axelrod and Bailey, 1976). Axelrod and Bailey (1976) examined fossil floras from both locations and, based on living conditions of nearest modern relatives of the fossils, interpreted the floras as representing a subalpine zone, under a cold temperate climate (MAT = 7°C, altitude of 2500 m). Lacustrine deposits are also reported in the Nogal Canyon caldera (Figure 1) (S. Cather, personal communication, 1998). Given the presence of 20–25 calderas and cauldrons in southwestern New Mexico (Elston, 1984), the lack of caldera-lake deposits is somewhat anomalous. It is not known whether this is because of basin-and-range faulting and erosion, physical volcanology, or climatic factors.

OTHER UPPER EOCENE AND OLIGOCENE LAKE DEPOSITS IN THE REGION

Other upper Eocene and Oligocene lake deposits are present in the region. Lacustrine sediments were deposited during late Eocene and, in some cases, early Oligocene in the Uinta (Utah), Claron (Utah), and Baca (New Mexico) Laramide-style basins (Figure 1) (Andersen and Picard, 1972; Goldstrand, 1994; Cather and Johnson, 1986). A final phase of lacustrine sedimentation in the Uinta Basin is represented by the middle–upper Eocene Uinta Formation (Bruhn et al., 1986). Lacustrine deposits were reported by Clark (1957) in the upper Eocene–Oligocene Duchesne River Formation, although no recent information is available on these and Andersen and Picard (1972) describe only fluvial deposits in the unit. Lacustrine sedimentation in the Claron Basin occurred from as early as the Late Cretaceous–early Oligocene (Anderson et al., 1975). The upper Eocene and lower Oligocene strata are now distinguished from the Claron Formation and a revised Brian Head Formation is applied (Sable and Maldonado, 1997). Studies of these beds indicate deposition in fluvial and shallow lacustrine environments that were progressively affected by volcanic sedimentation from the Marysvale volcanic field. Lacustrine deposits reflecting internal drainage are also present in the Baca Basin, but these are mostly of middle Eocene age (Cather, 1983; Cather and Johnson, 1986). In the cases of the Claron and Baca Basins, and to a lesser extent, the Uinta Basin, volcanic and volcaniclastic rocks are interbedded with and overlie the lacustrine deposits.

Limited exposures of upper Eocene–lower Oligocene lacustrine limestone are present in the House Range, west-central Utah (Figure 1) (Gregory-Wodzicki, 1997). CLAMP analysis of the flossil flora suggests a MAT of 13.2°C and a paleoaltitude of 3.6 ±0.7 km, using the moist-air-enthalpy method (Gregory-Wodzicki, 1997). Upper Oligocene lacustrine evaporites were deposited in a closed-basin lake in the Sevier Desert region, west-central Utah (Lindsey et al., 1981). Lindsey et al. (1981) note that fossil pollen suggest an arid environment consistent with the presence of evaporites.

Lacustrine deposits were described in the Oligocene Abiquiu Formation of north-central New Mexico (Figure 1) by Stearns (1953) and are indicated in Oligocene paleogeographic reconstructions of north-central New Mexico (Ingersoll et al., 1990). Stearns (1953) described a poorly exposed series of limestone overlain by cliff-forming, thin-bedded tuff at the base of the La Bajada escarpment (Figure 1) and attributed these to the Abiquiu Formation. Preliminary field observations of the cliff-forming tuff beds indicate approximately 25 m of tuffaceous, dolomitized shallow-lacustrine and lacustrine margin deposits. The tuffaceous deposits grade upward into buff-colored siliciclastic alluvial and eolian deposits typical of the Miocene Santa Fe Group.

Upper Oligocene lacustrine deposits are present in extensional basins associated with the Rio Grande rift (Albuquerque Basin: Lozinsky, 1994; Hayner Ranch Formation: Mack et al., 1994) and Basin and Range in southern Arizona (southwestern Arizona: Eberly and Stanley, 1978; Nations et al., 1982; southeastern Arizona-Teran Basin: Grover, 1984). These lacustrine deposits are mostly shallow-lake and playa deposits, although nonevaporitic lacustrine limestone is present in the Hayner Ranch Formation in south-central New Mexico (Mack et al., 1994).

PALEOCLIMATIC AND TECTONIC IMPLICATIONS

Lower Eocene and Oligocene lacustrine rocks deposited in the southwestern United States provide a record of the paleoenvironment during a time of climatic, tectonic, and volcanic change. A rapid global climate change toward cooler oceanic and terrestrial temperatures near the Eocene–Oligocene boundary is well documented (e.g., Savin, 1977; Miller, 1992; Wolfe, 1995). Also during this time, the tectonic regime in western North America was changing from crustal-thickening associated with Laramide-style tectonics (Bird, 1984; Dickinson et al., 1988; Cather, 1990) to incipient extension associated with the Rio Grande rift (Ingersoll, et al., 1990; Chapin and Cather, 1994) and development of the Basin and Range (Christiansen and Yeats, 1992). Distributed volcanism across the southwestern part of North America was also initiated during this time (Steven, 1975; Elston, 1984; Christiansen and Yeats, 1992).

CLAMP analyses of fossil flora in upper Eocene–lower Oligocene lacustrine rocks in Colorado and Utah indicated mean annual temperatures of between 11.8 and 13.2°C (Gregory-Wodzicki, 1997; Wolfe et al., 1998). The upper Oligocene Creede Formation yielded a substantially lower value (4.2°C) (Wolfe et al., 1998). Mean annual precipitation estimates vary from site to site, but the minimum value cited from statistical and physiognomic studies is 65 cm/yr (Wolfe and Schorn, 1989; Gregory and Chase, 1994; Gregory and McIntosh, 1996; Gregory-Wodzicki, 1997). Based on the dominance of subalpine fossil flora, Axelrod and Bailey (1976) estimated a low MAT (7°C) for the lake beds in the Emory Cauldron. The close association of the Emory lake beds with the 28.1 Ma Bear Springs Basaltic Andesite suggests that they are probably of late Oligocene age. It is not clear at present whether the substantial decrease in temperatures in the late Oligocene lake beds is due to higher altitudes created by volcanism or an effect of global climate deterioration in the Oligocene or both. If lower temperatures were associated solely with volcanic construction and uplift, one would expect the Pitch-Pinnacle Formation, which is clearly in a volcanic setting, to show a similar low MAT. Gregory and McIntosh (1996) have demonstrated that this is not the case. Further analysis of fossil flora in lake deposits in the San Juan and Mogollon-Datil calderas and the Antero Formation may help to resolve this issue and clarify possible paleoclimatic changes in the region.

Aspects of the physical sedimentology of the lake beds are consistent with interpretations of the fossil flora. Little evidence for evaporative conditions is present in most of the upper Eocene–Oligocene deposits throughout the region. The Florissant, Antero, Pitch-Pinnacle, and caldera-lake deposits (excluding Creede) all contain deposits typical of relatively freshwater lakes, some of which probably had water depths of tens of meters or more. These characteristics suggest moist, cool conditions that would limit evaporative concentration. Playa depositional settings and evaporites are present in upper Oligocene deposits in west-central Utah (Lindsey et al., 1981), upper Oligocene deposits in southeastern Arizona (Grover, 1984), and upper Oligocene deposits in the Rio Grande rift (Lozinsky, 1994). These deposits reflect sedimentation in extensional basins that were presumably both drier and warmer (e.g., Lindsey et al., 1981; Meyer, 1983).

Currently, no consensus exists regarding the presence of evaporitic conditions in the Creede Formation. Larsen's (1994b) interpretation that the calcite pseudomorphs are after ikaite (a nonevaporitic carbonate salt) is consistent with the paleotemperatures obtained from the paleofloral analysis of Wolfe and Schorn (1989) and Wolfe et al. (1998). Finkelstein et al. (1996) interpreted some of the calcite pseudomorphs to have formed after gypsum, which they believe is more consistent with the warmer, more arid paleoclimate estimated from Axelrod's (1987) paleofloral analysis. In either case, the chemistry of Lake Creede appears to have been affected by volcanic emissions and responded to local climate in a complex manner. A problem with interpreting the Creede fossil flora, and possibly other floras from caldera lakes, is that the sides of the lake basin were likely very steep and rose to as much as 1 km above the lake surface at any given time. Thus, mixing of plant remains from multiple botanical zones is probable. More detailed analysis of the Creede fossil flora and integration with recent stratigraphic work (Larsen and Crossey, 1996) is needed.

Prior to 10 years ago, much of the southwestern region of the United States was considered to have been a relatively low-relief, moderate-elevation (approximately 1 km) surface of erosion with several large accumulations of volcanic deposits during much of the late Eocene and Oligocene (Epis and Chapin, 1975; Dickinson et al., 1988). Reevaluation of the basis for this hypothesis, considering both geologic and paleofloral data, suggests that much of the region was at a considerably higher elevation (2–3 km) (Molnar and England, 1990; Gregory and Chase, 1992, 1994; Elston and Young, 1991; Chapin and Cather, 1994; Gregory, 1994; Gregory-Wodzicki, 1997; Wolfe et al., 1998) and comprised a composite of multiple surfaces (Elston and Young, 1991; Evanoff and Chapin, 1994). The upper Eocene and Oligocene lacustrine deposits discussed in this contribution provide much of the paleofloral data for the revised tectonic interpretation. The depositional character of most of the upper Eocene–lower Oligocene lake deposits and associations with blocked drainages are consistent with high-elevation plateaus and associated incised drainage systems. The upper Oligocene evaporite and playa deposits of western Utah, southern Arizona, and Rio Grande rift are interpreted to reflect incipient topographic collapse of these regions due to Neogene extension. The regional implication of such high mid-Tertiary elevations is that late Cenozoic regional uplift of the Rocky Mountains and part of the intermontane west is no longer required to produce the present-day topography. This has required reconsideration of continental denudation rates (Gregory and Chase, 1994), modeling paleoclimates (Sloan and Barron, 1992), and the timing of development of regional drainage systems (Elston and Young, 1991). Recent modeling suggests that much of the relief observed in mountain ranges associated with the Laramide-age uplift is a result of preferential valley erosion driven by Pliocene–Pleistocene climate change (Small and Anderson, 1998). Other studies have shown that local uplift during the late Tertiary and early Quaternary has influenced the current elevations of some Laramide-age ranges (Formento-Trigilio and Pazzaglia, 1998).

Middle Tertiary uplift has important implications for burial histories of potential hydrocarbon source rocks throughout the region. Uplift and erosion of overburden is typically modeled to have occurred in the late Cenozoic (e.g., Law, 1992; Crossey and Larsen, 1992). Initiation of uplift and erosion in the late Eocene would require higher heat flow values than are typically invoked or convective heat flow (e.g., Law, 1992) to obtain observed degrees of thermal maturation. This suggests either more complex thermal histories or thermal transfer mechanisms or both in the region during the middle–late Tertiary.

Many of the upper Eocene and Oligocene lakes in the region provide geologic snapshots of paleoenvironmental conditions in ancient volcanic regions. Few, if any, modern volcanic regions can compare to the extent or magnitude of volcanic activity indicated by the combined Marysvale, San Juan, Thirtynine Mile, and Mogollon-Datil volcanic fields. Although ash fallout was commonplace in these volcanic areas, fossil evidence suggests that vibrant plant and insect communities existed around the lakes. The vertebrate record, both mammal and fish, from the lake deposits and associated fluvial deposits is minimal or nonexistent in most cases. Could this be a result of the continual volcanic activity, especially in the cases of caldera lakes or lakes affected by nearby volcanism, or simply poor fossil preservation? At Florissant, ash fallout appears in one case to have changed water chemistry (Evanoff, 1994), but did not have much of an effect on the vegetation patterns in the watershed (Cross et al., 1994). At Creede, ash fallout is known to have affected water chemistry in the lake (Larsen and Crossey, 1996; Finkelstein et al., 1996); however, a study of the effects on the plant communities has not been done. In numerous cases, volcanism and volcaniclastic sedimentation either displaced lacustrine deposition (Platoro caldera, Colorado) or changed local and regional drainage patterns enough to end lacustrine deposition (Brian Head Formation, Utah; Emory cauldron, New Mexico). Thus, it is clear that volcanism was an important influence on the lacustrine environments in the region during late Eocene and Oligocene.

In summary, although lower Eocene–Oligocene lacustrine deposits are not abundant in the southwestern part of the United States, they provide critical information needed to reconstruct the paleoclimatic and tectonic history of the region. They also provide an important record of paleoenvironmental response to changes in climate, tectonic regime, and volcanic activity during the late Eocene and Oligocene. Further work may help to resolve the response of high-altitude regions to the well-known climate deterioration at the Oligocene–Eocene boundary.

REFERENCES CITED

Anderson, D. W., and M. D. Picard, 1972, Stratigraphy of the Duchesne River Formation (Eocene–Oligocene?), northern Uinta Basin, northeastern Utah: Utah Geological and Mineralogical Survey Bulletin, v. 97, p. 1–29.

Anderson, J. J., P. D. Rowley, R. J. Fleck, and A. E. M. Nairn, 1975, Cenozoic Geology of the Southwestern High Plateaus of Utah: GSA Special Paper 160, 88 p.

Axelrod, D. I., 1987, The Late Oligocene Creede Flora, Colorado: Berkeley, University of California Press, 234 p.

Axelrod, D. I., and H. P. Bailey, 1976, Tertiary vegetation, climate, and altitude of the Rio Grande depression, New Mexico-Colorado: Paleobiology, v. 2, p. 235–254.

Bird, P., 1984, Laramide crustal thickening in the Rocky Mountain foreland and Great Plains: Tectonics, v. 3, p. 741–758.

Bischoff, J. L., S. Stine, R. J. Rosenbauer, J. A. Fitzpatrick, and T. W. Stafford, 1993, Ikaite precipitation by mixing of shoreline springs and lake water, Mono Lake, California: Geochimica et Cosmochimica Acta, v. 57, p. 3855–3866.

Bruhn, R. L., M. D. Picard, and J. S. Isby, 1986, Tectonics and sedimentology of the Uinta Arch, western Uinta Mountains, and Uinta Basin, *in* J. A. Peterson, ed., Paleotectonics and Sedimentation in the Rocky Mountain Region, United States: AAPG Memoir 41, p. 333–352.

Cather, S. M., 1983, Lacustrine deposits of the Eocene Baca Formation, western Socorro County, New Mexico, *in* C. E. Chapin, ed., Socorro Region II, New Mexico Geological Society 34th Field Conference: Socorro, New Mexico, New Mexico Geological Society, p. 179–186.

Cather, S. M., 1990, Stress and volcanism in the northern Mogollon-Datil volcanic field, New Mexico: Effects of the post-Laramide tectonic transition: GSA Bulletin, v. 102, p. 1447–1458.

Cather, S. M., and B. D. Johnson, 1986, Eocene tectonics and depositional setting of west-central New Mexico and eastern Arizona, *in* J. A. Peterson, ed., Paleotectonics and Sedimentation in the Rocky Mountain Region, United States: AAPG Memoir 41, p. 623–652.

Chapin, C. E., and S. M. Cather, 1994, Tectonic setting of the axial basins of the northern and central Rio Grande rift, *in* G. R. Keller and S. M. Cather, eds., Basins of the Rio Grande Rift: Structure, Stratigraphy, and Tectonic Setting: GSA Special Paper 291, p. 5–26.

Christiansen, R. L., and R. S. Yeats, 1992, Post-Laramide geology of the U.S. Cordilleran region, *in* B. C. Burchfiel, P. W. Lipman, and M. L. Zoback, eds., The Cordilleran Orogen: Conterminous U.S., The Geology of North America, Volume G–3: Boulder, Geological Society of America, p. 261–406.

Clark, J., 1957, Geomorphology of the Uinta Basin: Intermountain Association of Petroleum Geologists Guidebook, 8th Annual Field Conference, p. 17–20.

Cross, A. T., R. E. Taggart, and A. P. Hascall, 1994, Paleobotanical and palynological reconstruction of the vegetation of Oligocene Lake Florissant (abs.): GSA Abstracts with Programs, v. 26, p. 9.

Crossey, L. J., and D. Larsen, 1992, Authigenic mineralogy of sandstones intercalated with organic–rich mudstones: Integrating diagenesis and burial history of the Mesaverde Group, Piceance basin, NW Colorado, *in* D. W. Houseknecht and E. D. Pittman, eds., Origin, Diagenesis, and Petrophysics of Clay Minerals in Sandstones: SEPM Special Publication no. 47, p. 125–144.

Dexter, W. A., 1994, Life styles of the rich and famous Florissant fossils (abs.): GSA Abstracts with Programs, v. 26, p. 9.

Dickinson, W. R., M. A. Klute, M. J. Hayes, S. U. Janecke, E. R. Lundin, M. A. McKittrick, and M. D. Olivares, 1988, Paleogeographic and paleotectonic setting of Laramide sedimentary basins in the central Rocky Mountain region: GSA Bulletin, v. 100, p. 1023–1039.

Eberly, L. D., and T. B. Stanley, Jr., 1978, Cenozoic stratigraphy and geologic history of southwestern Arizona: GSA Bulletin, v. 89, p. 921–940.

Elston, D. P., and R. A. Young, 1991, Cretaceous–Eocene (Laramide) landscape development and Oligocene–Pliocene drainage reorganization of the Transition Zone and Colorado Plateau, Arizona: Journal of Geophysical Research, v. 96, p. 12,389–12,406.

Elston, W. E., 1984, Mid-Tertiary ash flow tuff cauldrons, southwestern New Mexico: Journal of Geophysical Research, v. 89, p. 8733–8750.

Elston, W. E., W. R. Seager, and R. J. Abitz, 1987, The Emory resurgent ash-flow tuff (ignimbrite) cauldron of Oligocene age, Black Range, southwestern New Mexico: GSA Centennial Field Guide, Rocky Mountain Section, p. 441–445.

Epis, R. C., and C. E. Chapin, 1974, Stratigraphic nomenclature of the Thirtynine Mile volcanic field, central Colorado: USGS Bulletin 1395–C, 23 p.

Epis, R. C., and C. E. Chapin, 1975, Geomorphic and tectonic implications of the post-Laramide, late Eocene erosional surface in the southern Rocky Mountains, *in* B. F. Curtis, ed., Cenozoic History of the Southern Rocky Mountains: GSA Memoir 144, p. 45–74.

Evanoff, E., 1990, Early Oligocene paleovalleys in southern and central Wyoming: evidence of high local relief on the late Eocene unconformity: Geology, v. 18, p. 443–446.

Evanoff, E., 1994, Late Paleogene geology and paleoenvironments of central Colorado: Rocky Mountain Section GSA Field Trip Guide, 99 p.

Evanoff, E., and C. E. Chapin, 1994, Composite nature of the "late Eocene surface" of the Front Range and adjacent regions, Colorado and Wyoming (abs.): GSA Abstracts with Programs, v. 26, p. 12.

Finkelstein, D. B., 1997, Origin and diagenesis of lacustrine sediments of the late Oligocene Creede Formation, southwestern Colorado: Ph.D. dissertation, University of Illinois, Urbana, Illinois, 144 p.

Finkelstein, D. B., R. L. Hay, and S. P. Altaner, 1996, Unmasking the effects of diagenesis in ancient lacustrine rocks of the late Oligocene Creede Formation, southwestern Colorado (abs.): GSA Abstracts with Programs, v. 29, p. A–214.

Ford, T. D., and H. M. Pedley, 1996, A review of tufa and travertine deposits of the world: Earth-Science Reviews, v. 41, p. 117–175.

Formento-Trigilio, M. L., and F. J. Pazzaglia, 1998, Tectonic geomorphology of the Sierra Naciemento: traditional and new techniques in assessing long-term landscape evolution in the southern Rocky Mountains: Journal of Geology, v. 106, p. 433–453.

Goldstrand, P. M., 1994, Tectonic development of Upper Cretaceous to Eocene strata of southeastern Utah: GSA Bulletin, v. 106, p. 145–154.

Gregory, K. M., 1994, Paleoclimate and paleoelevation of the 35 Ma Florissant flora, Front Range, Colorado: Paleoclimates, v. 1, p. 23–57.

Gregory-Wodzicki, K. M., 1997, The late Eocene house Range flora, Sevier Desert, Utah: paleoclimate and paleoelevation: Palaios, v. 12, p. 552–567.

Gregory, K. M., and C. G. Chase, 1992, Tectonic significance of paleobotanically estimated climate and altitude of the late Eocene erosion surface, Colorado: Geology, v. 20, p. 581–585.

Gregory, K. M., and C. G. Chase, 1994, Tectonic and climatic significance of a late Eocene low-relief, high-level geomorphic surface, Colorado: Journal of Geophysical Research, v. 99, p. 20,141–20,160.

Gregory, K. M., and W. C. McIntosh, 1996, Paleoclimate and paleoelevation of the Oligocene Pitch-Pinnacle flora, Sawatch Range, Colorado: GSA Bulletin, v. 108, p. 545–561.

Grover, J. A., 1984, Petrology, depositional environments and structural development of the Mineta Formation, Teran basin, Cochise County, Arizona: Sedimentary Geology, v. 38, p 87–105.

Ingersoll, R. V., W. Cavazza, W. S. Baldridge, and M. Shafiqullah, 1990, Cenozoic sedimentation and paleotectonics of north-central New Mexico: implications for initiation and evolution of the Rio Grande Rift: GSA Bulletin, v. 102, p. 1280–1296.

Lanphere, M. A., 1994, Duration of sedimentation of the Creede Formation (abs.): GSA Abstracts with Programs, v. 26, p. A–400.

Larsen, D., 1994a, Depositional Environments and Diagenesis in the Creede Formation, San Juan Mountains, Colorado [unpublished Ph.D. Dissertation]: University of New Mexico, Albuquerque, 315 p.

Larsen, D., 1994b, Origin and paleoenvironmental significance of calcite pseudomorphs after ikaite in the Oligocene Creede Formation, Colorado: Journal of Sedimentary Research, v. A64, p. 593–603.

Larsen, D., 1995, Hydrochemical modeling of an ancient caldera lake: The Oligocene Creede Formation, Colorado (abs.): First SEPM Congress on Sedimentary Geology, Congress Program and Abstracts, v. 1, p. 81–82.

Larsen, D., and L. J. Crossey, 1996, Depositional environments and paleolimnology of an ancient caldera lake: Oligocene Creede Formation, Colorado: GSA Bulletin, v. 108, p. 526–544.

Law, B. E., 1992, Thermal maturity patterns of Cretaceous and Tertiary rocks, San Juan Basin, Colorado and New Mexico: GSA Bulletin, v. 104, p. 192–207.

Lindsey, D. A., R. K. Glanzman, C. W. Naeser, and D. J. Nichols, 1981, Upper Oligocene evaporites in basin fill of Sevier Desert region, western Utah: AAPG Bulletin, v. 65, p. 251–260.

Lipman, P. W., 1975, Evolution of the Platoro caldera complex and related volcanic rocks, southeastern San Juan Mountains, Colorado: USGS Professional Paper 852, 128 p.

Lipman, P. W., (in press), The central San Juan caldera cluster: Regional volcanic framework, *in* P. M. Bethke and R. L. Hay, eds., Ancient Lake Creede: its volcano-tectonic setting, history of sedimentation and relation to mineralization in the Creede Mining District: GSA Special Paper.

Lipman, P. W., T. A. Steven, R. G. Luedke, and W. S. Burbank, 1973, Revised history of the San Juan, Uncompahgre, Silverton, and Lake City calderas in the western San Juan Mountains: USGS Journal of Research, v. 1, p. 627–642.

Lipman, P. W., M. A. Dungan, L. L. Brown, and A. Deino, 1996, Recurrent eruption and subsidence at the Platoro caldera complex, southeastern San Juan volcanic field, Colorado: New tales from old tuffs: GSA Bulletin, v. 108, p. 1039–1055.

Lozinsky, R. P., 1994, Cenozoic stratigraphy, sandstone petrology, and depositional history of the Albuquerque basin, central New Mexico, *in* G. R. Keller and S. M. Cather, eds., Basins of the Rio Grande Rift: Structure, Stratigraphy, and Tectonic Setting: GSA Special Paper 291, p. 5–26.

MacGinitie, H. D., 1953, Fossil Plants of the Florissant Beds, Colorado: Carnegie Institute Washington Publication 599, 198 p.

Mack, G. H., W. R. Seager, and J. Kieling, 1994, Late Oligocene and Miocene faulting and sedimentation, and evolution of the southern Rio Grande rift, New Mexico, USA: Sedimentary Geology, v. 92, p. 79–96.

McElroy, C. A., and R. Y. Anderson, 1966, Laminations of the Oligocene Florissant lake deposits, Colorado: GSA Bulletin, v. 77, p. 605–618.

McIntosh, W. C., and C. E. Chapin, 1994, $^{40}Ar/^{39}Ar$ geochronology of ignimbrites in the Thirtynine mile volcanic field, Colorado, *in* E. Evanoff, ed., Late Paleogene Geology and Paleoenvironments of Central Colorado: Rocky Mountain GSA Field Trip Guide, p. 23–26.

Meyer, H. W., 1983, Fossil plants from the early Neogene Socorro flora, central New Mexico, *in* C. E. Chapin, ed., Socorro Region II, New Mexico Geological Society 34th Field Conference: Socorro, New Mexico, New Mexico Geological Society, p. 193–196.

Miller, D. M., T. H. Nilsen, and Bilodeau, 1992, Late Cretaceous to early Eocene geologic evolution of the U.S. Cordillera, *in* B. C. Burchfiel, P. W. Lipman, and M. L. Zoback, eds., The Cordilleran Orogen: Conterminous U.S., The Geology of North America, Volume G–3: Boulder, Geological Society of America, p. 205–260.

Miller, K. G., 1992, Middle Eocene to Oligocene stable isotopes, climate, and deep-water history: the terminal Eocene event?, *in* D. R. Prothero and W. A. Berggen, eds., Eocene–Oligocene Climatic and Biotic Evolution: Princeton, New Jersey, Princeton University Press, p. 160–177.

Molnar, P., and P. England, 1990, Late Cenozoic uplift of mountain ranges and global climate change: chicken or egg?: Nature, v. 346, p. 29–34.

Nations, J. D., J. J. Landye, and R. H. Hevly, 1982, Location and chronology of Tertiary sedimentary deposits in Arizona: a review, *in* R. V. Ingersoll and M. O. Woodburne, eds., Cenozoic Nonmarine Deposits of California and Arizona: SEPM, p. 107–122.

Nelson, C. H., C. R. Bacon, S. W. Robinson, D. P. Adam, J. P. Bradbury, J. H. Barber, D. Schwartz, and G. Vagenas, 1994, The volcanic, sedimentologic, and paleolimnologic history of the Crater Lake caldera floor, Oregon: evidence for small caldera evolution: GSA Bulletin, v. 106, p. 684–704.

Newhall, C. G., C. K. Paull, J. P. Bradbury, A. Higuera-Gundy, L. J. Poppe, S. Self, N. Bonar Sharpless, and J. Ziagos, 1987, Recent geologic history of Lake Atitlan, a caldera lake in western Guatemala: Journal of Volcanology and Geothermal Research, v. 33, p. 81–107.

Rowley, P. D., H. H. Mehnert, C. W. Naeser, L. W. Snee, C. G. Cunningham, T. A. Steven, J. J. Anderson, E. G. Sable, and R. E. Anderson, 1994, Isotopic ages and stratigraphy of Cenozoic rocks of the Marysvale volcanic field and adjacent areas, west-central Utah: USGS Bulletin 2071, 35 p.

Rytuba, J. J., S. A. Minor, and E. H. McKee, 1981, Geology of the Whitehorse caldera and caldera-fill deposits, Malheur County, Oregon: USGS Open-File Report 81–1092, 19 p.

Sable, E. G., and F. Maldonado, 1997, The Brian Head Formation (revised) and selected Tertiary sedimentary rock units, Markagunt Plateau and adjacent areas, southwestern Utah: USGS Bulletin 2153, p. 7–26.

Savin, S. M., 1977, The history of the Earth's surface temperature during the past 100 million years: Annual Reviews in Earth and Planetary Science, v. 5, p. 319–355.

Shearman, D. J., and A. J. Smith, 1985, Ikaite, the parent mineral of jarrowite-type pseudomorphs: Proceedings of the Geologists Association, v. 96, p. 305–314.

Sloan, L. C., and E. J. Barron, 1992, Paleogene climatic evolution: a climate model investigation of the influence of continental elevation and sea-surface temperature upon continental climate, *in* D. R. Prothero and W. A. Berggen, eds., Eocene–Oligocene Climatic and Biotic Evolution: Princeton, New Jersey, Princeton University Press, p. 202–217.

Small, E. E., and R. S. Anderson, 1998, Pliestocene relief production in Laramide mountain ranges, western United States: Geology, v. 26, p. 123–126.

Smith, R. M. H., 1986, Sedimentation and palaeoenvironments of Late Cretaceous crater-lake deposits in Bushmanland, South Africa: Sedimentology, v. 33, p. 369–386.

Stark, J. T., J. H. Johnson, C. H. Behre, Jr., W. E. Powers, A. L. Howland, D. B. Gould, and others, 1949, Geology and Origin of South Park, Colorado: GSA Memoir 33, 188 p.

Stearns, C. E., 1953, Tertiary geology of the Galisteo-Tonque area, New Mexico: Geological Society of America Bulletin, v. 64, p. 459–508.

Steven, T. A., 1968, Critical review of the San Juan peneplain, southwestern Colorado: USGS Professional Paper 594–I, 19 p.

Steven, T. A., 1975, Middle Tertiary volcanic field in the southern Rocky Mountains, *in* B. F. Curtis, ed., Cenozoic History of the Southern Rocky Mountains: GSA Memoir 144, p. 75–94.

Steven, T. A., and J. C. Ratté, 1965, Geology and structural control of ore deposition in the Creede district, San Juan Mountains, CO: USGS Professional Paper 487, 90 p.

Wolfe, J. A., 1978, A paleobotanical interpretation of Tertiary climates in the Northern Hemisphere: American Scientist, v. 66, p. 694–703.

Wolfe, J. A., 1992, Climatic, floristic, and vegetational changes near the Eocene/Oligocene boundary in North America, *in* D. R. Prothero and W. A. Berggen, eds., Eocene–Oligocene Climatic and Biotic Evolution: Princeton, New Jersey, Princeton University Press, p. 421–436.

Wolfe, J. A., 1995, Paleoclimatic estimates from Tertiary leaf assemblages: Annual Reviews in Earth and Planetary Sciences, v. 23, p. 119–142.

Wolfe, J. A., C. E. Forest, and P. Molnar, 1998, Paleobotanical evidence of Eocene and Oligocene paleoaltitudes in midlatitude western North America: GSA Bulletin, v. 110, p. 664–678.

Wolfe, J. A., and H. E. Schorn, 1989, Paleoecologic, paleoclimatic, and evolutionary significance of the Oligocene Creede flora, Colorado: Paleobiology, v. 15, p. 180–198.

Section VIII

Miocene–Pliocene

Ramos-Guerrero, E., I. Berrio, J. J. Fornós, and L. Moragues, 2000, The middle Miocene Son Verdera lacustrine-palustrine system (Santa Margalida Basin, Mallorca), *in* E. H. Gierlowski-Kordesch and K. R. Kelts, eds., Lake basins through space and time: AAPG Studies in Geology 46, p. 441–448.

Chapter 40

The Middle Miocene Son Verdera Lacustrine-Palustrine System (Santa Margalida Basin, Mallorca)

Emilio Ramos-Guerrero
Departament d'Estratigrafia i Paleontologia, Facultat de Geología, Universitat de Barcelona Barcelona, Spain

Igone Berrio
Departamento de Paleontología, Facultad de Ciencias, Universidad del País Vasco Bilbao, Spain

Joan Josep Fornós
Departament de Ciencies de la Terra, Facultat de Ciencies, Universitat Illes Balears Palma de Mallorca, Spain

Lluis Moragues
Lignitos, S.A., c/ Joan Maragall Palma de Mallorca, Spain

GEOLOGIC SETTING

The island of Mallorca forms part of the Balearic promontory, which, in turn, forms part of the alpine Betic Range extending from the southern Iberian Peninsula as a northeastward submarine prolongation into the western Mediterranean. The Betic orogen underwent its phase of maximum deformation in Mallorca during the early–middle Miocene (Burdigalian–Langhian), when the area developed a series of imbricate thrust sheets transported in a northwest direction (Fallot, 1922; Alvaro, 1987; Sábat et al., 1988). After this compressive stage, post-Langhian extension resulted in a series of horsts and grabens bounded by northeast-southwest and northwest-southeast oriented faults. The Santa Margalida Basin is one of these small grabens filled by marine, as well as continental, postorogenic sediments, ranging from Serravallian to Quaternary in age (Figure 1).

The postorogenic sedimentary record in Mallorca is split into 11 sequences (Figure 2). The thickness of these successions is highly variable due to the strong fault activity, which generated several grabens and half-grabens. From borehole and geophysical data, Benedicto et al. (1993) estimated a postorogenic fill thickness of 1500 m for the Inca Basin, which lies a few kilometers to the west, whereas the Santa Margalida Basin, one of the smallest grabens, is filled by up to 300 m of postorogenic sediments. Approximately 150 m of this infill belongs to Sequence V (Figure 2).

The lowermost unit of this postorogenic filling (Sequence V according to Fornós et al., 1991) consists of 150 m of clastic nonmarine carbonates (Figure 2). Three informal lithostratigraphic units can be identified in this nonmarine sequence. Two of these units (the Pina Marls unit and the Son Verdera Limestone unit (Pomar et al., 1983); units 1 and 2 respectively in Figure 2) consist mainly of lacustrine and palustrine carbonate rocks that interfinger laterally into the third terrigenous unit (marginal terrigenous complex), which extends along the basin edges (Manacor Silt unit; Pomar et al., 1983).

The lowermost Pina Marls unit lies unconformably over the folded and thrusted pre- and syn-orogenic substratum (Figure 2). The unit is composed of grey marls with interbedded gypsum and sandy layers, and contains abundant resedimented marine microforaminifera attributed to the Burdigalian (Colom, 1967). The Son Verdera Limestone unit conformably overlies the Pina Marls and is composed of limestones and fine-grained siliciclastic rocks with variable amounts of organic matter.

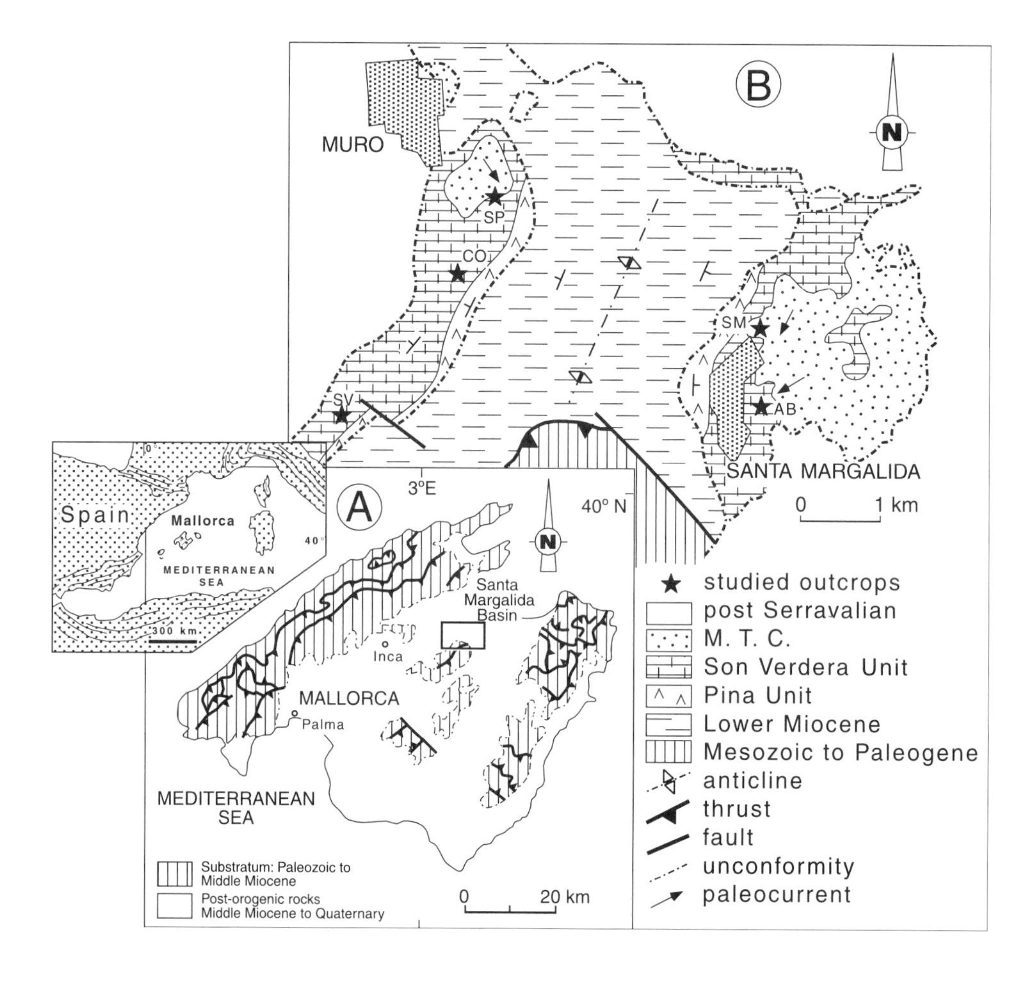

Figure 1—(A) Location and structural sketch of Mallorca showing the position of the Santa Margalida Basin. (B) Location map and geologic sketch of outcrops studied in the Santa Margalida Basin. AB = S'Abellà section; CO = Coscoia section; SM = Santa Margalida section; SV = Son Verdera section; and SP = Son Poquet section.

The Sequence V successions in Mallorca are widespread, and they are recognized both in outcrops and the subsurface. This summary refers only to the lacustrine-palustrine Miocene sedimentation in the Santa Margalida Basin, in the central sector of Mallorca (Figure 1).

THE SON VERDERA LIMESTONE UNIT

In the Santa Margalida Basin, the thickness of the Son Verdera Limestone unit ranges from 11 m in the Santa Margalida (SM) profile to 70 m in the Son Verdera (SV) profile (Figure 3). On the basis of regional criteria, Fornós et al. (1991) attributed this unit to the upper Serravallian. The unit consists of clastic carbonates interbedded with 1- to 3-cm-thick gypsarenite beds. Organic matter content is variable, and locally the unit contains layers of coal and coaly facies up to a few centimeters thick. Ramos-Guerrero et al. (1992) distinguished two facies assemblages in the Son Verdera Limestone unit: lacustrine and palustrine, plus a series of laterally related siliciclastic facies that constitutes the marginal terrigenous complex (MTC) (see Figure 3).

Lacustrine Facies Assemblage

The facies in this assemblage are well stratified and are either internally laminated or massive, but neither pedoturbation nor any other features indicative of edaphic (subaerial) processes have ever been observed. The most characteristic lithologies are (1) massive marls, (2) organic matter–bearing siltstones, and (3) well-stratified limestones.

Massive Marls

White to greyish massive marls are interbedded with limestone layers, which together can constitute beds up to 2.5 m thick, although their thickness range is on average from 0.15 to 0.5 m. These facies contain benthic and nektonic fossil remains, mainly gastropods, ostracodes, and rare charophytes and fish otoliths. Scattered quartz grains, coarse sand in size (ø ≤ 1 mm) are present; however, tractive sedimentary structures have not been observed. Low organic matter content has been reported from these facies with total organic carbon (TOC) ranging from 0.03 to 0.17%.

The massive greyish marls record deposition of suspended fine-grained mixed sediment under a lacustrine water column. Their massive framework is presumably attributable to bioturbation processes involving benthic fauna. The low organic carbon content of these facies, as well as the occurrence of nektonic fauna remains, suggest the existence of a stable oxygenated lacustrine water column.

Organic-Matter–Bearing Siltstones

These are silts with variable amounts of carbonate and scattered, sand-size quartz grains. When carbonate content decreases, the content of quartz grains increases.

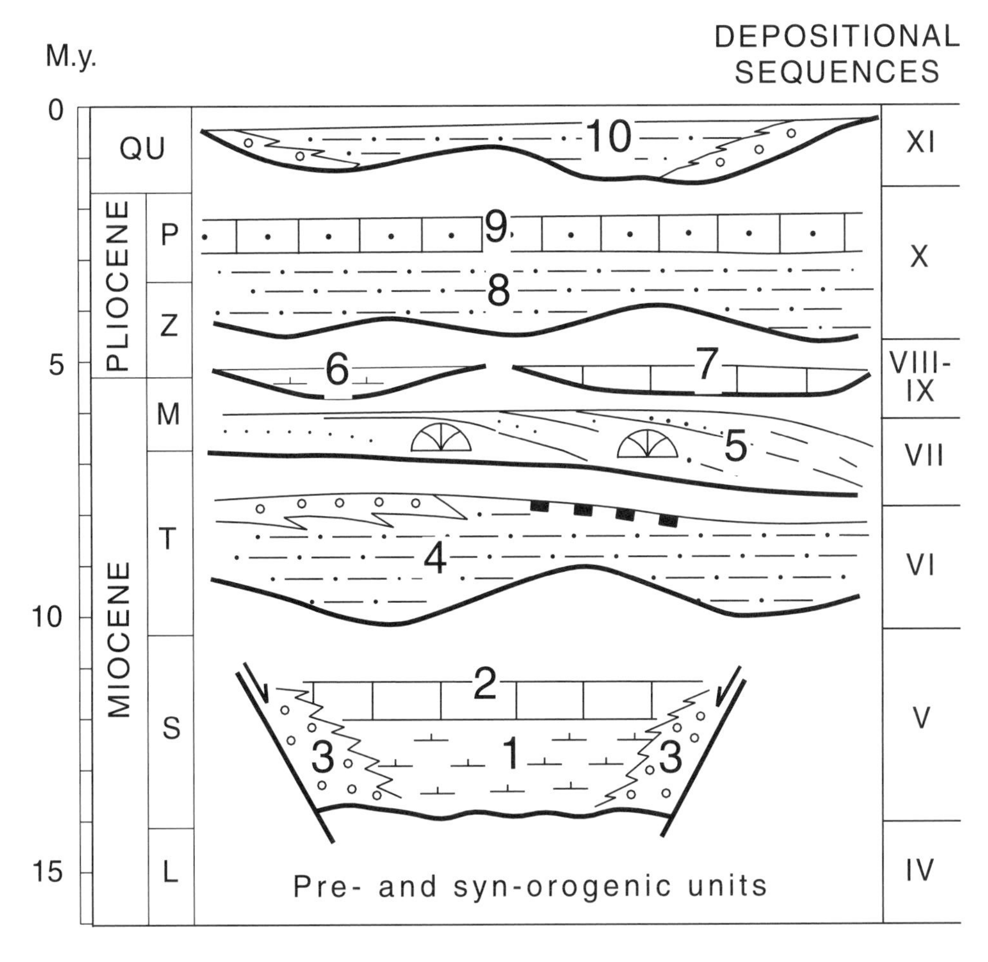

Figure 2—Chronostratigraphic sketch of the postorogenic Neogene of Mallorca. 1 = Pina Marls unit, 2 = Son Verdera Limestone unit, 3 = Marginal Terrigenous Complex (MTC), 4 = *Heterostegina* Calcisiltites unit, 5 = Reef Complex unit, 6 = Bonanova Marls unit, 7 = Santanyí Limestone unit, 8 = Son Mir Calcisiltites unit, 9 = Sant Jordi Calcarenites unit, 10 = Palma Lutites unit. Chronostratigraphic scale: QU = Quaternary, P = Piacenzian, Z = Zanclian. M = Messinian, T = Tortonian, S = Serravalian, and L = Langhian, Depositional sequences from Fornós et al. (1991). The thickness of the Son Verdera Limestone Unit in the Santa Margalida Basin varies from 11 to 70 m (see text).

These facies are generally massive and greyish in color. Their organic matter content is the highest of the lacustrine facies, although TOC content constitutes no more than 0.65%. This facies contains an abundance of well-preserved fossil remains, mainly gastropods and ostracodes, and minor bivalves, charophytes, fish teeth and otoliths, and exotic rodent remains.

These facies were formed by the deposition of mixed fine-grained sediments under anoxic conditions; however, the abundance and state of preservation of benthic and nektonic fauna suggest that these anoxic conditions only occurred below the sediment/water interface with a well-oxygenated water column. Also, the massive texture suggests periods of oxic conditions to allow for bioturbation of sediment. The inverse relationship between the content of carbonate and quartz grains suggests that the relative increase in carbonate sedimentation was controlled by decreasing siliciclastic input into the lacustrine area.

Well-Stratified Limestones

Two subfacies are present in this facies: massive and finely laminated limestones. The massive, well-stratified limestones are mudstones, wackestones, and locally bioclastic packstones that form beds ranging from 10 to 20 cm in thickness, and which in places constitute thickening-upward sequences up to 1.5 m thick. Elongated in shape, parallel to the bedding chert nodules are frequent.

The finely laminated, well-stratified limestones are white to brownish bioclastic wackestones-packstones in which reworked gastropoda, ostracoda, and blue-green algal filaments are the major components. They make up beds ranging in thickness from 10 to 25 cm and internally show coarse plano-parallel lamination due to the selective accumulation of fossil remains. Laminae are commonly bounded by erosive surfaces, and the bioclastic components show positive grain-size classification. Loading structures also occur.

Despite the fact that the bulk of the Son Verdera limestone sequences were deposited under shallow lacustrine-palustrine conditions, both laminated and massive, well-stratified limestones record the development of relatively deeper lacustrine environments. The massive, well-stratified limestones were the result of carbonate input (and minimal siliciclastic input) into the inner lacustrine basin zones, whereas the laminated limestones were the result of subaqueous transport of carbonate sediment from the marginal lacustrine areas to deeper areas.

Palustrine Facies Assemblage

Unlike the lacustrine assemblage, palustrine facies are generally massive and show evidence of pedoturbation by plants as evidenced by rhizocretions and pedogenic textures. The most characteristic palustrine lithofacies are (1) greenish marls, (2) coaly facies, (3) brecciated limestones, and (4) laminated limestones.

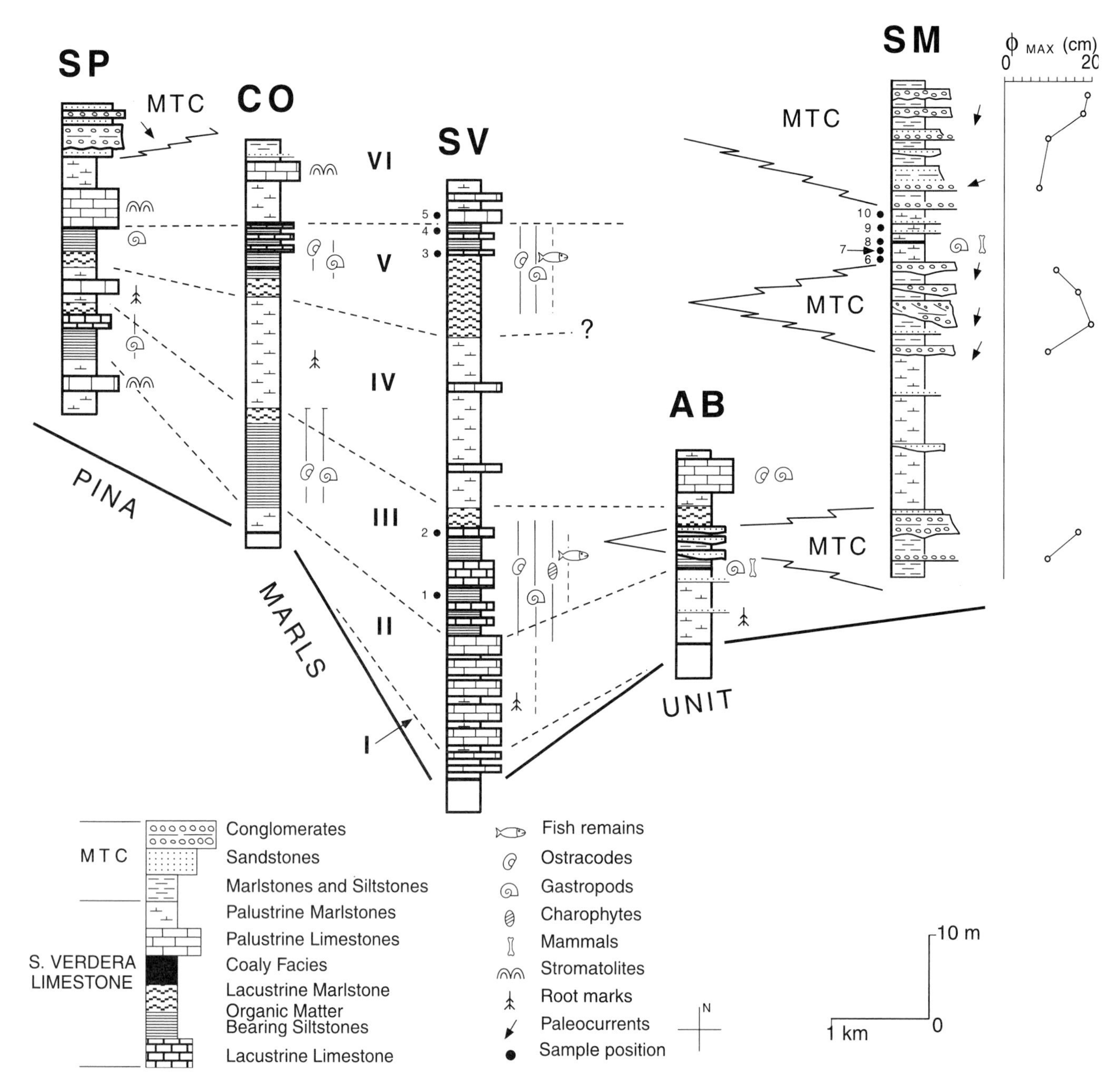

Figure 3—Lithostratigraphic correlation among the Santa Margalida Basin logs. See locations on Figure 1B. MTC = marginal terrigenous complex. The log to the left of the Santa Margalida section (SM) indicates the variation of the maximum conglomerate clast size for the MTC. I to VI = cycles shown in Figure 5 (see text). Numbers 1 to 10 indicate the location of the organic matter samples studied (see values in Table 1).

Greenish Marls

These marls are massive, locally nodular, greyish to greenish in color, and commonly mottled. They contain numerous, small-size (few mm up to 2 cm) carbonate and ferruginous nodules and glaebules, which are either dispersed or constitute thin layers, as well as small lenticular gypsum crystals or their pseudomorphs. Paleontologic content is scarce, and limited to sparse coalified macrophytic debris, although root-traces are common.

These facies are interpreted as the result of the deposition of carbonates in palustrine areas and ponds adjacent to the lake. These sediments subsequently underwent pedogenic alteration and subaerial exposure. They may have been covered by vegetation.

Coaly Facies

This set of facies contains a relatively high organic matter content, varying from carbonaceous lutites to coal. This facies is interbedded with the greenish marls facies and form beds up to 15 cm thick. The coal seams do not show any signs of reworking and their lower boundary is usually a gradual transition from greenish marls. Some root traces start in the coal beds and penetrate into the lower marls. The upper limit is abrupt and commonly is topped by a discontinuous layer 0.5–2 cm thick of yellowish clays. Coaly beds often include remains of hydrophilous vegetation (*Phragmites* sp.) and, to a lesser extent, gastropod and rodent remains.

The coal seams are interpreted as in situ peat accumulation in swamps/marshes of the peripheral lacustrine belt. The volumetric significance of the coal and coaly facies in the sections studied is negligible.

Brecciated Limestones

These limestones are massive, pale-colored mudstones-wackestones with beds ranging in thickness from 10 to 25 cm. They display a brecciated framework due to pervasive root penetration; vertical root and subhorizontal rhizome traces are widespread. Mudcracks have been observed at the base of a few limestone beds. Fossil remains are scarce and limited to some gastropod or ostracode shells.

The sedimentologic interpretation of this facies is similar to that of the massive greenish marls. The main difference is that here the carbonate content is higher than in the marly facies. This might be because carbonate input overwhelmed any siliciclastic influx. Brecciation occurred as an early diagenetic process as the result of root bioturbation that developed over the lacustrine carbonate mud.

Laminated Limestones

These white limestones form 15- to 50-cm-thick beds composed of a bindstone of cyanobacteria (blue-green algae). The limestone beds commonly contain mudcracks at their bottom and top surfaces. These facies show a finely laminated (submillimetric) stromatolitic texture, composed of alternating couplets of algal filaments and micrite. one to three-cm-thick gypsarenite and other evaporite casts intercalations are frequently interbedded in both brecciated and laminated limestones.

Laminated limestones are believed to be stromatolites resulting from the organic building action of cyanobacteria colonies. These stromatolites reached maximum development in the lacustrine littoral zone, where subaerial exposure was frequent.

Paleobiologic Content

The Son Verdera limestones are characterized by an abundance of nonmarine gastropods and ostracodes with minor charophyte, fish otolith, and rodentia remains.

Fossil mammal assemblages from the unit have been studied by Mein and Adrover (1982) and Adrover et al. (1983–1984). They recognize the presence of *Gymnesicolagus gelaberti* (a giant Ochotonid) and three forms of Glirids: *Carbomys sacaresi, Margaritamys llulli,* and *Peridyromys ordinasi.* According to these authors, these micromammal associations are endemic and developed from an ancestral upper Paleogene fauna.

A study of the abundant gastropod fauna has been carried out by Colom (1975), who reported the presence of *Hydrobia dubuissoni, H. sandbergeri, H. inflata, Planorbis* sp., and *Limnaea* sp. in the Son Verdera section (SV) (Figure 3). This author also cited the presence of *Chara maioricensis* among other charophyte remains.

The most significant paleoenvironmental data come from the study of the ostracode fauna, which comprise largely *Heterocypris salina* and *Leptocythere* sp., accompanied by forms of *Candona angusta, C. compressa, Pseudocandona* sp., *Darwinula* sp., and *Hemicyprinotus* sp.

Heterocypris salina, the most abundant and widely distributed form, is a recent species that normally lives in both permanent and ephemeral chloride or chloride-sulfate saline waters and tolerates a salinity content of up to 20%. *Candona compressa* is another present-day species that commonly inhabits restricted littoral environments with slightly (up to 5.7%) saline waters, whereas *Darwinula* is a present-day genus living in lacustrine bottoms related to the littoral zone and brackish waters. Thus, the association found in the Son Verdera unit characterized an environment with slightly chloride or chloride-sulfate saline waters, often related to lacustrine littoral environments.

A high percentage of the ostracode fauna studied (89%) have articulated carapaces. This taphonomic condition is characteristic of low-energy sedimentary environments with a high sedimentation rate.

Organic Matter

Pyrolysis (Rock-Eval) analysis of scattered organic matter in the limestones was carried out in 10 samples, taken from the Son Verdera (SV) and Santa Margalida (SM) sections. Their exact location is shown in Figure 3. Samples 1–4 are lacustrine facies, whereas samples 5–10 belong to palustrine facies. The TOC was low in general, ranging between 0.03 and 2.37% (Table 1).

The highest TOC values belong to the palustrine deposits. Greenish palustrine marls (samples 6, 7, 9, and 10) show the lowest values, ranging between 0.03 and 0.17% TOC. Similar values were obtained for palustrine limestones (sample 5) with a 0.11% TOC. The carbonaceous lutites that make up the transition between the greyish marl beds and the coal seams (sample 8) have the highest TOC value (2.37%). Pyrolysis analysis was not carried out on coals.

The lacustrine facies show intermediate values in their TOC content. Lacustrine, well-stratified limestones (samples 2 and 3) have values of 0.36 and 0.27%, respectively, whereas the organic matter–bearing siltstones (samples 1 and 4) show TOC values of 0.61 and 0.65%, respectively.

Due to the low TOC content of the bulk of the samples studied, there is not sufficient data available to evaluate the maturity (T_{max}) of these samples.

In contrast to the Oligocene and Eocene lacustrine systems, which contain gas-prone kerogens type II to III and have good source potential (Ramos-Guerrero et

Table 1—Organic Matter Data from Pyrolysis Analysis of Studied Samples 1 to 10.*

Sample	Facies	T.O.C.% w	H.I.	O.I.
Lacustrine				
1	org. mat. bear. silts	0.61	0	98
2	well-strat. limestones	0.36	25	300
3	well-strat. limestones	0.27	15	307
4	org. mat. bear. silts	0.65	3	194
Palustrine				
5	brecciated limestones	0.11	0	218
6	greenish marls	0.03	—	—
7	greenish marls	0.17	0	212
8	carbonaceous lutites	2.37	5	63
9	greenish marls	0.09	0	411
10	greenish marls	0.14	0	236

*Note that the TOC content in palustrine samples is lower than that in lacustrine samples, except for the coaly facies (sample no. 8), which shows the highest TOC content values. See sample location in Figure 3.

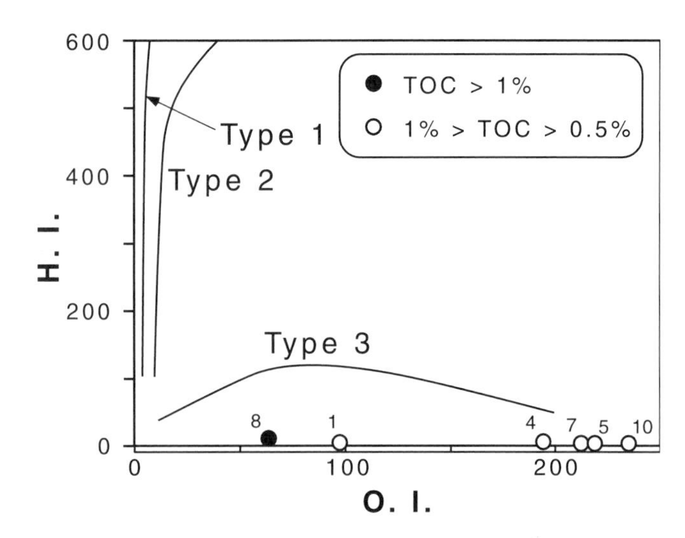

Figure 4—Hydrogen index (H.I.) versus oxygen index (O.I.) diagram of the organic matter of the samples studied (1 to 10, Figure 3 and Table 1). Samples with O.I. >250 (samples 2, 3, 9) have not been plotted.

al., 1994 a, b), the source potential for hydrocarbon generation in the lacustrine-palustrine Miocene sediments is very poor to barren. The low hydrogen index values (Table 1, Figure 4) suggest a type IV inertinitic kerogen, which has no generative potential and may be the result of severely oxidized organic matter.

Marginal Terrigenous Complex (MTC)

The marginal terrigenous complex (MTC) displays a marginal and peripheral basinal position and grades laterally into the lower Pina Marls unit and the upper Son Verdera Limestone unit. This unit is only exposed in the marginal sections of Son Poquet (SP) and Santa Margalida (SM) (Figure 3). The MTC is composed by conglomerates and sandstones, lenticular and tabular in shape, interbedded with massive siliciclastic mudstones. These sequences interfinger with the palustrine facies of the Son Verdera unit.

The clasts in the conglomerates are subangular to angular, indicating a local provenance. Composition in clast spectra shows a dominance of Mesozoic limestones, although Paleogene limestones have also been recognized. The vertical evolution of the maximum clast size shows a poorly defined coarsening-upward trend.

Paleocurrent analysis shows a relatively persistent trend for each area studied, with the paleocurrent directions converging toward the inner part of the basin (Figure 1B).

Mudstones are massive and display evidence of pedogenetic features (mottling and carbonate nodules). The Santa Margalida section (SM) includes a 10-m-thick packet of palustrine facies.

The MTC redbeds record sedimentation on small alluvial fans, where channel and sheetlike gravel and sand deposits interbedded between the flood plain lutites.

SEQUENTIAL ORGANIZATION AND PALEOGEOGRAPHY

The Son Verdera stratotype section (Figure 5) shows the existence of three lacustrine deepening-shallowing cycles. These three deepening-shallowing lacustrine cycles can also be recognized in the bulk of the sections studied (Figure 3). The lowest lying of these represents a sudden lacustrine spread with a brief, but deep episode (I, Figure 5) followed by a longer shallow filling episode (II, Figure 5). The transition between deep and shallow facies is sharp and the shallow facies are the thickest (episode II).

Geologic mapping (Figure 1B) and stratigraphic correlation (Figure 3) show the shallow lacustrine-palustrine rocks to be extensive, overlying both the deeper lacustrine and the underlying lower Pina Marls unit. Thus, episode II of the lower cycle represents a lacustrine basin expansion, which developed under a situation of equilibrium between subsidence and sedimentary accumulation.

The later successive phases of lacustrine deepening were also sharp (III and V, Figure 5), and each was followed by its respective lacustrine fill episodes under shallow conditions (IV and VI, Figure 5).

The outcropping conditions do not allow a very precise correlation to be drawn between the logs studied, but a certain relation between the sudden lacustrine deepening and basinward progradating wedges of the marginal terrigenous complex is suggested here (Figure 3).

Lacustrine deepening must have been controlled by climatic change or tectonism. Thus, Benedicto et al. (1993) calculated a high tectonic subsidence rate for this unit in the neighboring Inca Basin. A tectonic control for lacustrine deepening would also agree with the marginal alluvial reactivation.

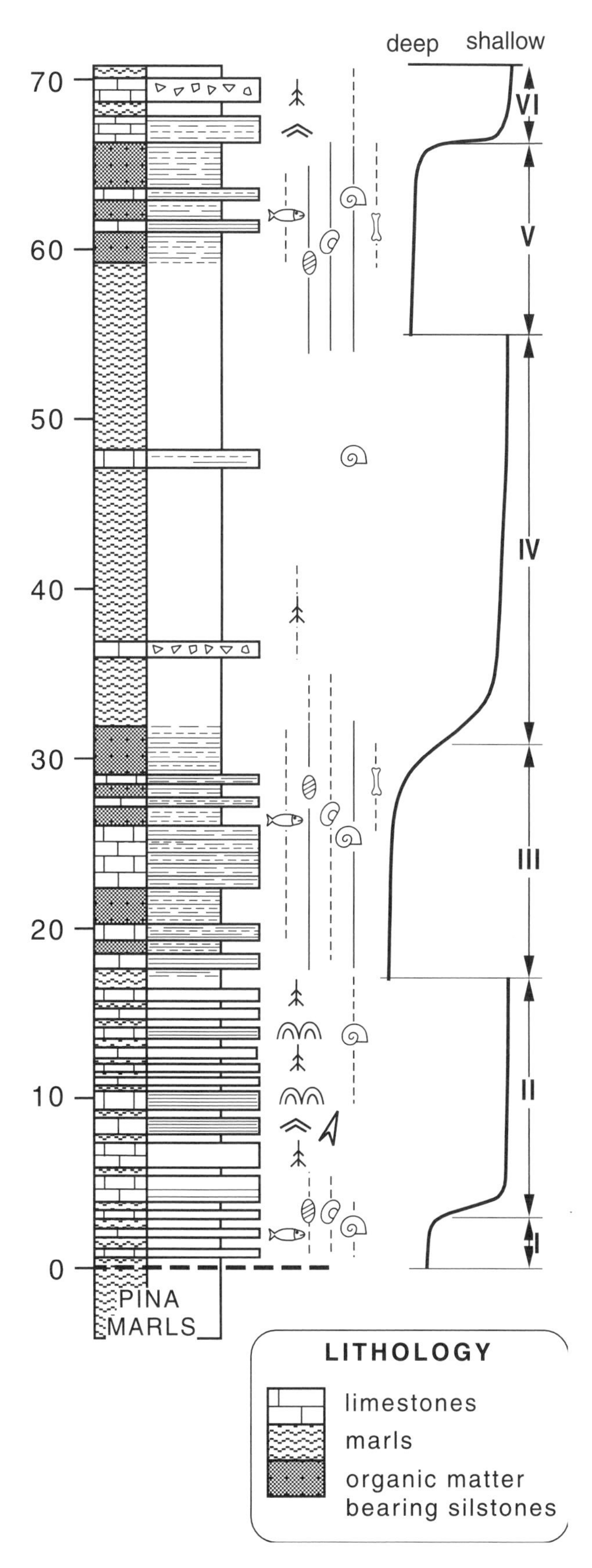

Figure 5—Detailed log of the Son Verdera section (SV). Legend for paleobiologic content in Figure 3. I to VI illustrate the three deepening and shallowing cycles found in the Son Verdera Limestone.

All the evidence suggests that this high tectonic subsidence rate was accompanied by a high lacustrine sedimentation rate, giving rise to the accumulation of thick, shallow facies successions (episodes II, IV, and VI, Figure 5). This and the taphonomic conditions deduced from ostracode preservation suggest a considerable sediment supply into the lacustrine basin.

The three deepening-shallowing lacustrine cycles recorded in the Son Verdera succession (Figure 5) can also be recognized in the bulk of the sections studied (Figure 3).

The paleocurrents recorded show convergent trends toward the inner lacustrine basin depocenter (Figure 1B). This, together with the paleogeographic control exercised on the lacustrine basin by the bounding faults and the absence of any signs of marine connection, suggests that the Son Verdera Limestone unit is a record of the filling of a small internal drainage lacustrine basin.

The saline facies identified in the lacustrine basin fill, as well as the brackish hydrochemistry recorded by the ostracode assemblages, appear to be a consequence of the recycling of Triassic evaporites (Ramos-Guerrero et al., 1992) that cropped out during the middle Miocene as part of the contractive relief generated during the early Miocene.

ACKNOWLEDGMENTS

Financial support has been provided by the Spanish DSICYT Project PB 94-0914 and Comisionat per Universitats i la Recerca, Grup de Qualitat GRQ 94-1048. The manuscript has been improved by critical comments and suggestions from L. Cabrera and E. Gierlowski-Kordesch.

REFERENCES CITED

Abell, P. I., S. M. Awramik, R. H. Osborne, and S. Tomellini, 1982, Plio–Pleistocene lacustrine stromatolites from Lake Turkana, Kenya: morphology, stratigraphy and stable isotopes: Sedimentary Geology, v. 32, p. 1–26.

Adrover, R., J. Agustí, S. Moyà, and J. Pons, 1983–84, Nueva localidad de micromamíferos insulares del mioceno medio en las proximidades de San Lorenzo en la isla de Mallorca: Paleontologia i Evoluciò, v. 18, p. 121–129.

Alvaro, M., 1987, La tectónica de cabalgamientos de la Sierra Norte de Mallorca (Islas Baleares): Boletín Geológico y Minero, v. 95, p. 3–25.

Barrón, E., M. R. Rivas Carballo, and M. F. Valle, 1996, Síntesis bibliográfica de la vegetación y clima de la Península Ibérica durante el Neógeno: Revista Española de Paleontología, v. extraordinario, p. 225–236.

Benedicto, A., E. Ramos-Guerrero, A. Casas, F. Sàbat, and A. Baron, 1993, Evolución tectonosedimentaria de la cubeta neógena de Inca (Mallorca): Revista de la Sociedad Geológica de España, v. 6, p. 167-176.

Colom, G., 1967, Los depósitos lacustres del Burdigaliense superior de Mallorca: Memorias de la Real Academia de Ciencias y Artes de Barcelona, v. 38, p. 327–395.

Colom, G., 1975, Geología de Mallorca, 2 vols.; Palma de Mallorca, Institut d'Estudis Baleàrics, 519 p.

Fallot, P., 1922, Etude Géologique de la Sierra de Majorque; París, Libr. Polytech. Béranger ed., 481 p.

Fornós, J. J., M. Marzo, L. Pomar, E. Ramos-Guerrero, and A. Rodríguez-Perea, 1991, Evolución tectonosedimentaria y análisis estratigráfico del Terciario de la isla de Mallorca; Libro-Guia excursión no. 2 del I Congreso del Grupo Español del Terciario. Vic, 145 p.

Freytet, P., and J. C. Plaziat, 1982, Continental carbonate sedimentation and pedogenesis- Late Cretaceous and Early Tertiary of Southern France. Contributions to Sedimentology 12; Stuttgart, E. Schweizerbart'sche Verlagsbuchhanlung, 213 p.

Mein, P., and R. Adrover, 1982, Une faunule de Mammifères insulaires dans le Miocène moyen de Majorque (Iles Baléares): Geobios, Mém. Spéc. 6, p. 451–463.

Pedley, H. M., 1990, Classification and environmental models of cool freshwater tufas: Sedimentary Geology, v. 68, p. 143–154.

Pomar, L., M. Marzo, and A. Baron, 1983, El Terciario de Mallorca, *in* L. Pomar, A. Obrador, J. J. Fornós, and A. Rodriguez-Perea, eds., El Terciario de las Baleares (Mallorca-Menorca): Palma de Mallorca, Institut d'Estudis Baleàrics, p. 21–45.

Ramos-Guerrero, E., I. Berrio, J. J. Fornós, and L. Moragues, 1992, Depósitos lacustres-palustres del Mioceno medio de Mallorca: La Unidad de Calizas de Son Verdera; Simposios del III Congreso Geológico de España y VIII Congreso Latinoamericano de Geología, v. 1, p. 128–136.

Ramos-Guerrero, E., Ll. Cabrera, and M. Marzo, 1994a, The Eocene lacustrine-palustrine system of Peguera (Mallorca Island, western Mediterranean), *in* E. Gierlowski-Kordesch and K. Kelts, eds., Global Geological Record of Lake Basins, v. I, Cambridge University Press, p. 241–244.

Ramos-Guerrero, E., Ll. Cabrera, and M. Marzo, 1994b, Oligocene lacustrine deposits in the Cala Blanca Formation (Mallorca Island, western Mediterranean), *in* E. Gierlowski-Kordesch and K. Kelts, eds., Global Geological Record of Lake Basins, v. I, Cambridge University Press, p. 285–288.

Sábat, F., J. A. Muñoz, and P. Santanach, 1988, Transversal and oblique structures at the Serres de Llevant thrust belt (Mallorca Island): Geologische Rundschau, v. 77, p. 529–538.

Suc, J. P., G. Clauson, M. Bessedik, S. Leroy, Z. Zheng, A. Drivaliari, P. Roiron, P. Ambert, J. Martinell, R. Domenech, I. Matías, R. Julià, and R. Anglada, 1992, Neogene and Lower Pleistocene in Southern France and Northeastern Spain, Mediterranean environments and climate: Cahiers de Micropaléontologie, v. 7, p. 165–186.

Türkmen, I., and I. E. Kerey, 2000, Alluvial and lacustrine facies of the Yeniçubuk Formation (lower–middle Miocene), Upper Kizilirmak Basin, Türkiye, *in* E. H. Gierlowski-Kordesch and K. R. Kelts, eds., Lake basins through space and time: AAPG Studies in Geology 46, p. 449–464.

Chapter 41

Alluvial and Lacustrine Facies of the Yeniçubuk Formation (Lower-Middle Miocene), Upper Kizilirmak Basin, Türkiye (Turkey)

Ibrahim Türkmen
I. Erdal Kerey
Jeoloji Müh. Bölümü, Mühendislik Fakültesi, Firat üniversitesi
Elazig, Türkiye (Turkey)

INTRODUCTION

The Upper Kizilirmak Basin (UKB) of Türkiye contains middle Eocene–Quaternary basin fill. This strike-slip basin in the central Anatolian Province is associated with the Paleocene collision of the Arabian plate (Sengör et al., 1985). The UKB is surrounded (Figure 1) by the Akdag Mountains to the northwest (composed of Paleozoic metamorphics including marble) and the Çaldag and Hinzir mountains to the south–southwest (composed of Paleozoic and Paleocene–Oligocene sedimentary rocks). The basin fill (Figure 2) contains a thick lower–middle Miocene sedimentary sequence called the Yeniçubuk Formation, which is interpreted as alluvial to lacustrine in origin. This contribution focuses on the sedimentology of this formation in the area around Gemerek (Figure 2) within the UKB, where a 50-km-long outcrop belt (Figure 1) trends northeast-southwest.

SEDIMENTOLOGY

Past work on the geologic mapping, sedimentology, and coal geology of the Yeniçubuk Formation includes Lebküchner (1956), Soytürk and Birgül (1971, 1992), Gökten (1983), Sümengen et al. (1987), and Türkmen (1993). The formation has been mapped as containing five members (Figure 2): Lalelik Dagi, Kavga Beli, Küçükkamisili dere, Tatili, and Kizilsardag Members. Through detailed sedimentologic analysis of nine measured sections (Figures 3–7) in the Gemerek area (see Figure 2 for locations), 19 facies (Table 1) were identified. These facies were interpreted as alluvial to lacustrine in nature; facies associations and their interpretation are discussed in following sections.

FACIES ASSOCIATIONS

Association A: Alluvial Systems

Ephemeral Stream Facies Association (A1)

This facies association comprises the whole of the Lalelik Dagi Member and the Tatili Member. This association contains facies Gm, Gp, Sm, St, Sl, Fl_1, and Fm (Figures 3, 4b, 7a). It ranges in thickness from 100 to 200 m and forms fining-upward interbedded megasequences. Coarse-grained rocks (massive conglomerate, stratified conglomerate, massive sandstone, trough cross-stratified sandstones, epsilon cross-stratified sandstone) occur in the lower part of the sequences and fine-grained rocks (disrupted silty mudrock, red mudstone) are in the upper parts. Coarse-grained deposits have a lenticular geometry and vary from 4 to 12 m in thickness and 20 to 30 m in lateral extent. Fine-grained facies are up to 100 m thick. The ratio of coarse-grained sediments to fine-grained sediment is between 1:5–3:1.

The base of the upward-fining interbedded megasequences consists of channel fill conglomerates (Gm), longitudinal conglomerates (Gp), and trough and planar cross-bedded sandstones (St, Sp). Interbedded sandstone, mudrock, and silty mudrock, which are laterally and vertically intercalated with the coarse-grained rocks, are interpreted as extrachannel sheetflood sediments. Epsilon cross-stratified sandstones are interpreted as lateral accretion deposits. This facies association is interpreted as ephemeral braidplain stream deposits, perhaps associated with distal alluvial fan deposits (after Graf, 1988; Miall, 1996; Gibling et al., 1998).

Alluvial Plain Facies Association (A2)

This facies association comprises the Küçükkamisili dere Member and contains facies Sp, St, Sr, Fl_1, Fl_2, C,

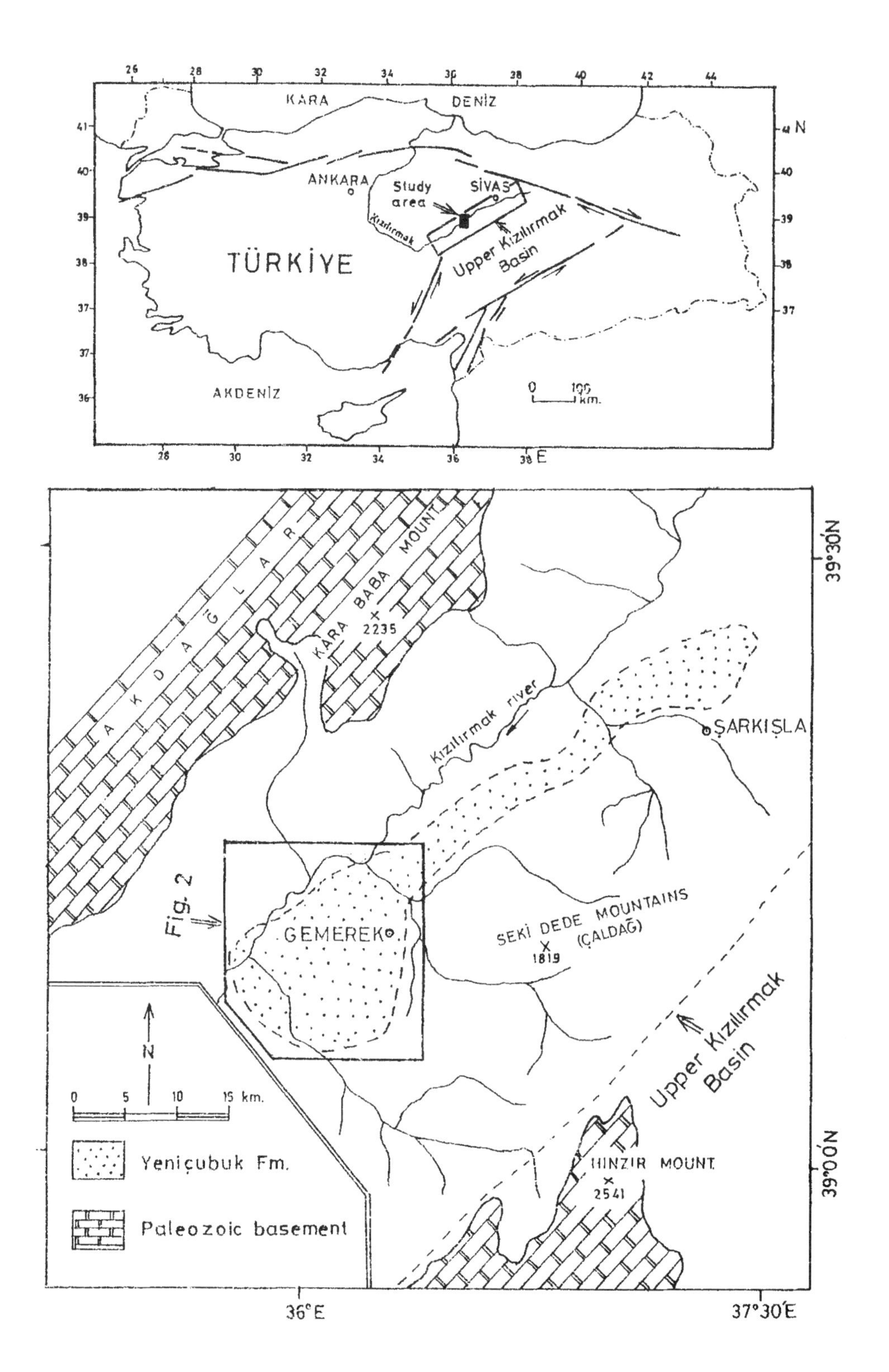

Figure 1—Location of the Upper Kizilirmak Basin in Türkiye and the Yeniçubuk Formation exposures in the Upper Kizilirmak Basin (lower-middle Miocene).

and Pf (Figure 3b). This member can be up to 260 m and is characterized by disrupted silty mudrock (35%), disrupted carbonaceous shale (30%), interbedded sandstone and mudrock (15%), trough cross-stratified sandstone (10%), and biomicrite (10%). The upper and lower parts of the association contain disrupted carbonaceous shale and biomicrite. Wave-rippled fine-grained sandstone beds occur within mudstones.

Alluvial plain facies in this association can be grouped into the following subassociations.

Channel/Levee Subassociation. Facies in this subassocation consist of trough cross-bedded sandstones (Sp, St) and cross-laminated sandstone (Sr). The sandstones are usually graded, have eroded bases and lenticular geometry, and are 2–4 m thick. In some places, siltstone intercalations occur. The base of sandstones contains scattered gravel, mudstone, and organic particles. These facies intercalate with SF facies, mostly siltstone-mudstone. Lenticular sandstones found within organic-bearing mudstones are regarded as channel deposits with siltstone interpreted

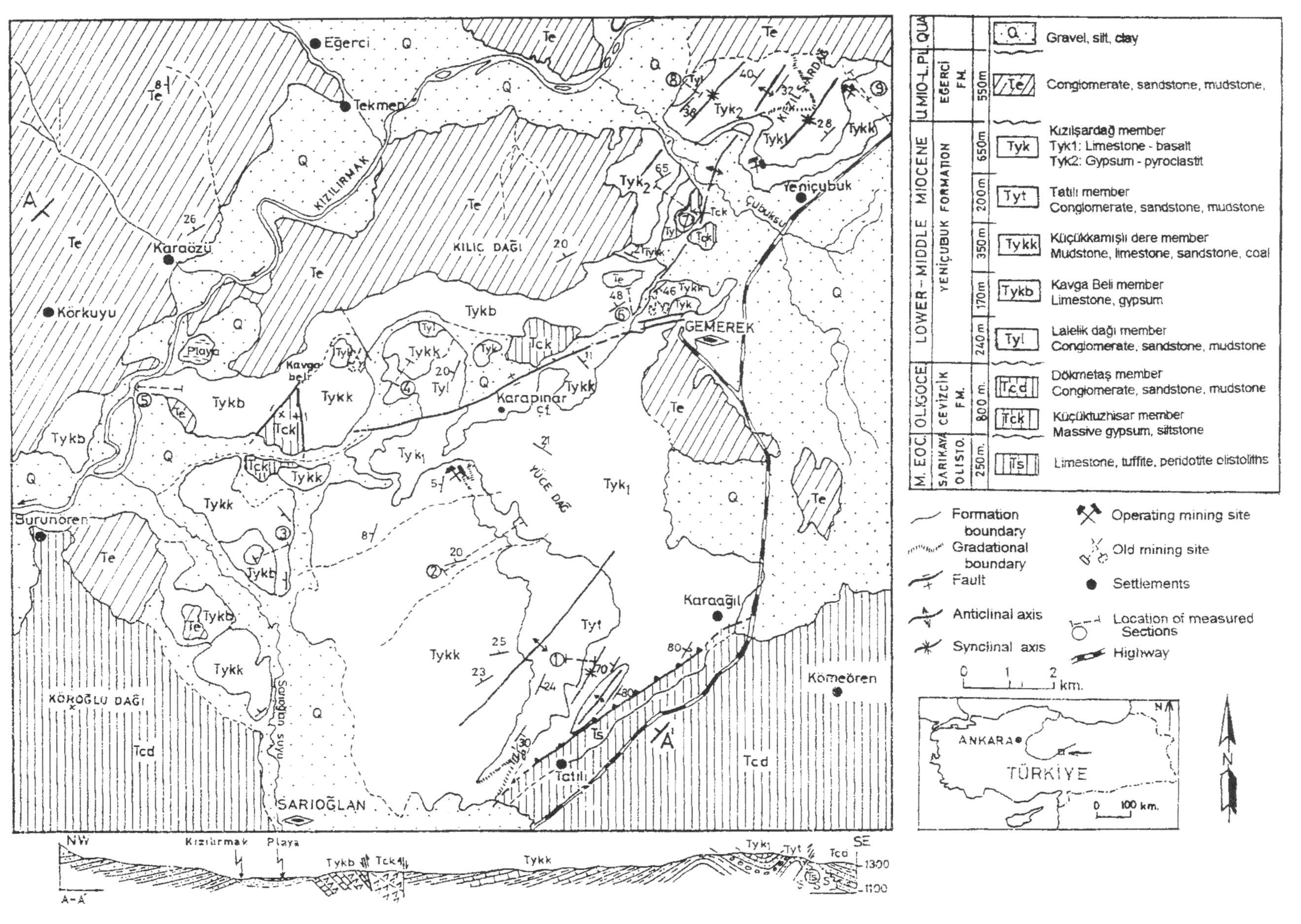

Figure 2—Geologic map and stratigraphic relations of study area in the Upper Kizilirmak Basin. Measured sections are featured in Figures 3–7.

Figure 3—Measured lithologic columns of section 1 (a) illustrating a sequence containing the Tatili and lower Kizilsardag Members and section 2 (b) showing a succession containing the Küçükkamisili dere, Tatili, and Kizilsardag Members of the Miocene Yeniçubuk Formation. For location of sections see Figure 2. For legend see Table 2.

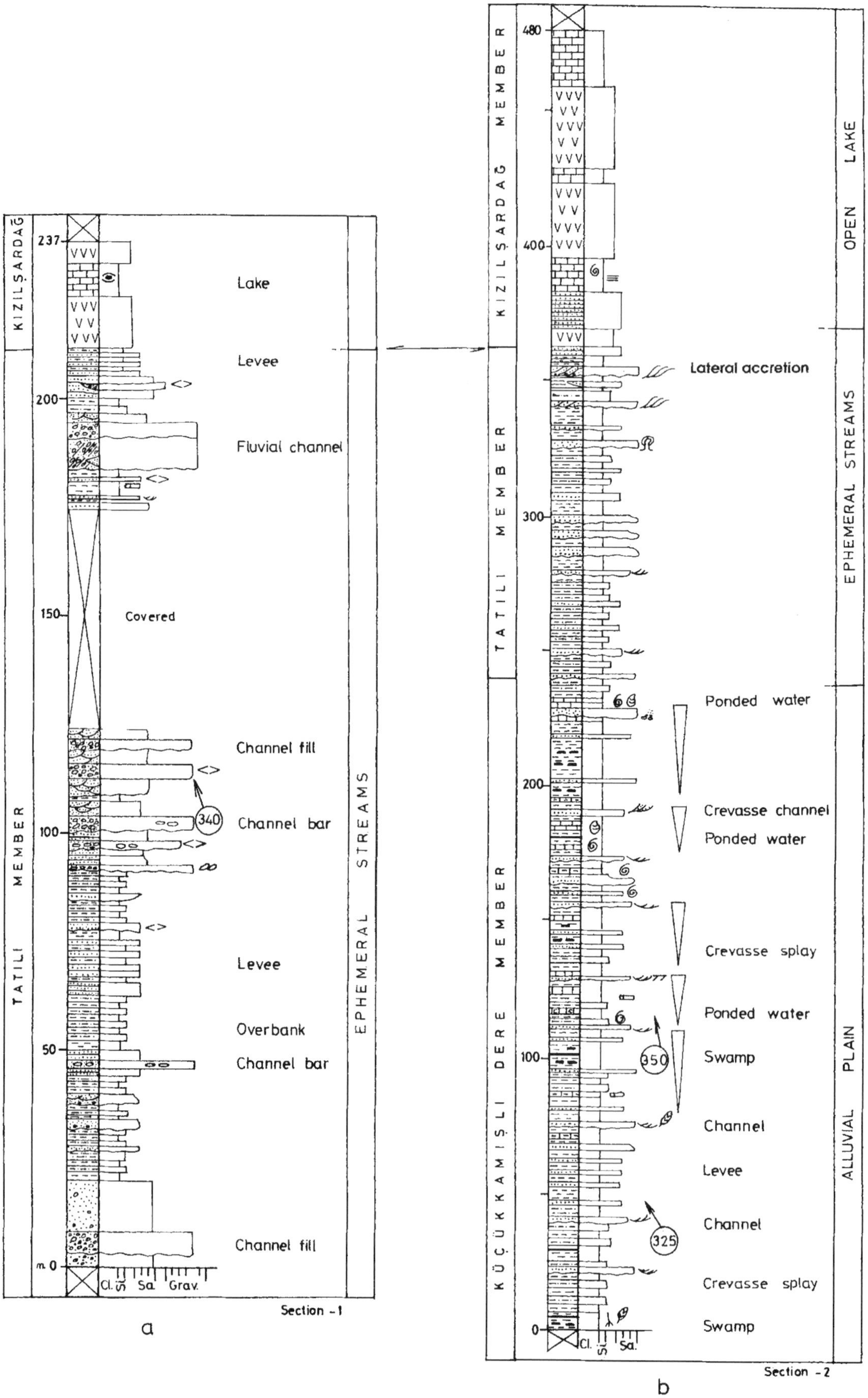

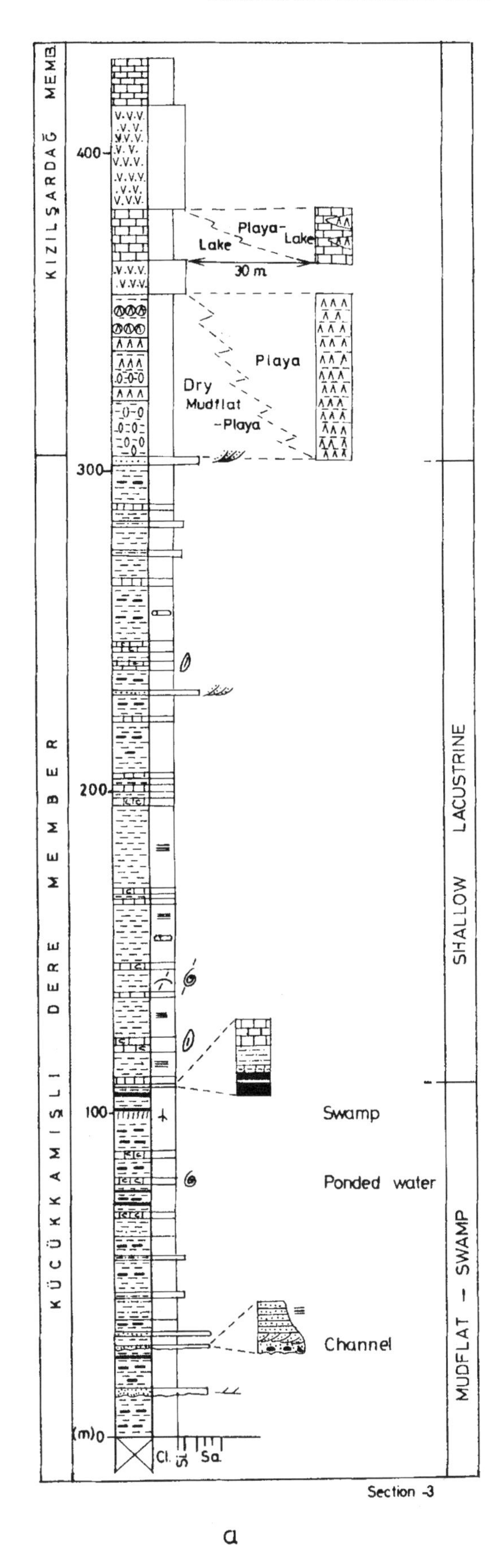

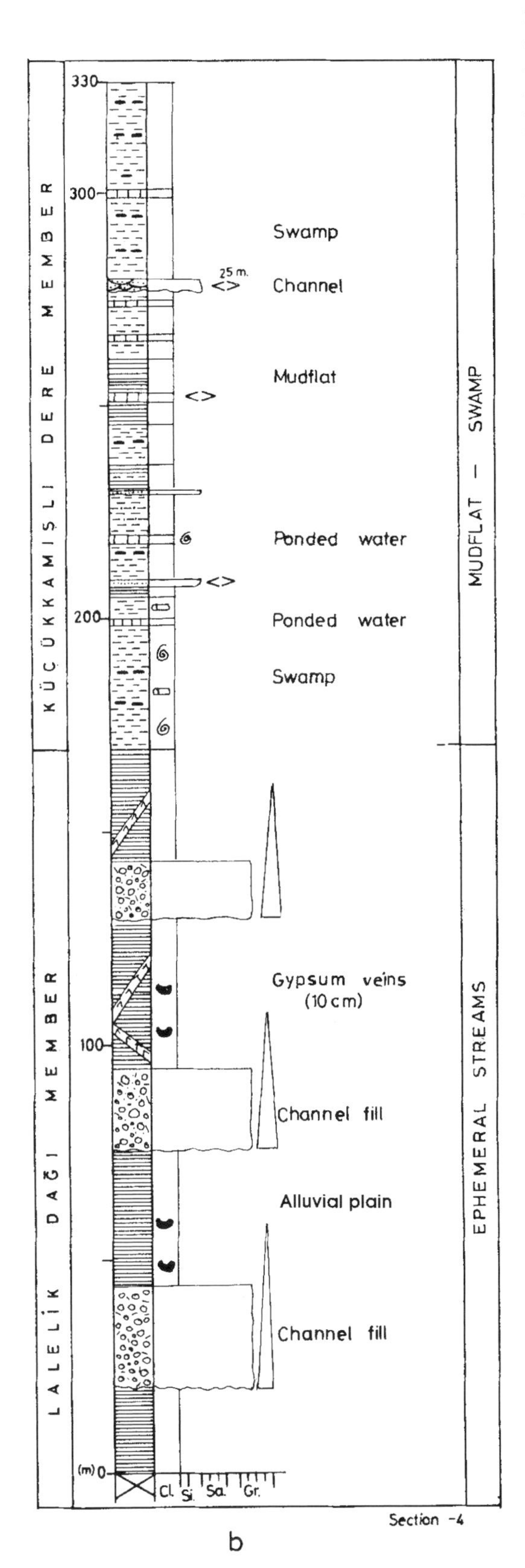

Figure 4—Measured lithologic columns of section 3 (a) illustrating a sequence containing the Küçükkamisili dere and Kizilsardag Members and section 4 (b) showing a succession containing the Laleklik Dagi and Küçükkamisili dere Members of the Miocene Yeniçubuk Formation. For location of sections see Figure 2. For legend see Table 2.

as levee deposits. These lenticular sandstones commonly erosively cross-cut other alluvial plain facies.

Wetland Subassociation. This subassociation contains disrupted silty mudrock (Fl1) and biomicrite (Pf) (Figure 3b) and grades laterally into swamp/mire facies. Its thickness ranges from 10 to 30 m. This mudstone-dominated facies contains bioclasts of gastropods, bivalves, and plants in carbonate as well as wave-rippled sandstone intercalations. This subassociation has been interpreted as lakes and ponds associated with crevasse splays developed on an anastomosing floodplain (Smith, 1983; Eberth and Miall, 1991; Nadon, 1994).

Swamp/Mire Subassociation. This subassociation is mostly composed of disrupted carbonaceous shale (Fl2) and coals (C) and is up to 5 m thick. Coals overlie massive mudstones (Fm); rhizoliths in these mudstones exist near the upper contact with the coals. Thickness of coal beds ranges from 5 to 10 cm. This subassociation is commonly cross-cut by channel/levee subassociation. The carbonaceous shale contains abundant leaves and other plant fossils and has upward and downward transitions into biomicrite (Pf).

This subassociation resembles facies associated with coal development within wetlands of anastomosed river floodplains (Rust et al., 1984; Flores and Hanley, 1984; Kirschbaum and McCabe, 1992; Valero Garcés et al., 1997). Facies in the study area have been interpreted as channel, ponded water, and swamp deposits, specifically as a delta plain (anastomosing) system (Kerey, 1982) because of the relationship between sandstone and interchannel facies. Deposition of nonmarine carbonate within this siliciclastic-dominated river system points toward a source area containing an abundance of Ca^+ (see Valero Garcés et al., 1997). Only in interchannel areas of anastomosing systems can carbonate precipitate in association with siliciclastic and coal-forming environments.

Association B: Siliciclastic-Carbonate Lacustrine Complex

Mudflat-Swamp Complex (B1)

This facies association occurs widely within the Küçükkamisili dere Member, containing facies Fm, Fl_1, C, and Pf (Figure 4) with thicknesses between 50 and 150 m. About 70% of this facies association is composed of disrupted silty mudrock (Fl1) with gastropod-bivalve-bearing biomicrite (Pf) intercalations. The biomicrite zones have a lenticular shape, are 1–2 m in thickness, and 8–30 m in lateral extent. This facies often has lateral and vertical transitions into disrupted carbonaceous shale (Fl2). This association differs from association A by the rarity of channel sandstone facies and the more distal nature of the deposits.

This association is interpreted as mudflat-swamp deposits (Belt et al., 1984; McCabe, 1984) on a distal alluvial plain. Coal layers overlying the rootlet-bearing mudstone are autochthonous and were deposited in a swamp/mire environment. Nonmarine limestones with abundant gastropod shells indicate protected areas where calcite precipitation occurred, as in wetlands receiving a high Ca^+ input from overland transport (Gierlowski-Kordesch, 1998). With a lateral transition to the shallow lake association (B2), this association is interpreted as marshes, swamps, or mires in wetlands draining into shallow lakes. Sagri et al. (1989) have interpreted similar deposits, green siltstone-mudstone, coal, limestone, and gravelly sandstone, of the Tertiary in the Daban basin (Somali) as being deposits of mudflat and swamps around lakes. The source area for these deposits also contained calcic rocks.

Shallow Lake (B2)

This facies association occurring within the Küçükkamisili dere Member contains facies Fl_1, Fl_2, Fl_3, C, and Pf (Figures 4a, 7b). Thickness of the association ranges from 50 to 125 m. Distribution of facies in this association changes locally. This facies association consists mainly of shale, siltstone, gastropod and mollusc? fragments, carbonate, and organic material. Main facies components include disrupted carbonaceous shale (60%), disrupted silty mudrock (30%), biomicrite (10%), and coal (C) and, in some sections, 70% grey shale and 30% biomicrite, but no coal. Discontinuous carbonate and clay laminae commonly alternate within fine-grained facies. Biomicrites include lamellibranch, gastropod, and charophyte remains. Grey shales are interbedded with biomicrite (Figure 4a). Thickness of coal seams ranges from 0.2 to 1.2 m. Rootlets occur at the base of some coal seams; other coal seams overlie siltstone (Fl_1) that contains no rootlets. Vertical and lateral transitions of this association to alluvial plain deposits indicate the distal nature of these deposits.

Presence of nonmarine limestone and interbedded shale, as well as the fine-grained, distal nature of the deposits in this facies association, indicates a shallow lacustrine environment associated with marshes and swamps. Anadón et al. (1989, 1991) interpreted similar facies succession from the Tertiary deposits of Campins, Mora, and Libros Basins in northeastern Spain as shallow lake or paludal deposits.

Association C: Evaporite Lake Complex

Dry Mudflat-Playa Complex (C1)

This association occurring within the Kavga Beli Member has a maximum thickness of 150 m. About 70% is gypsiferous mudstone (Fg) and 30% bedded gypsum (Eg) composed of interbedded nodular and bedded gypsum lenses and a gypsum-bearing red-green mudstone with abundant mudcracks (Fm) (Figures 5a, 6, 7a). Lenticular gypsum is 0.5–3 m in thickness and 20–50 m in lateral extent. There are also many gypsum lenses within the red-green mudstone. The peripheral portions of these lenses contain nodular gypsum with the central part as bedded gypsum. Red mudstones pass into green mudstone and bedded gypsum at upper levels. Discoidal gypsum crystals are common within green mudstones. This association shows lateral and vertical transitions into interbedded gypsum

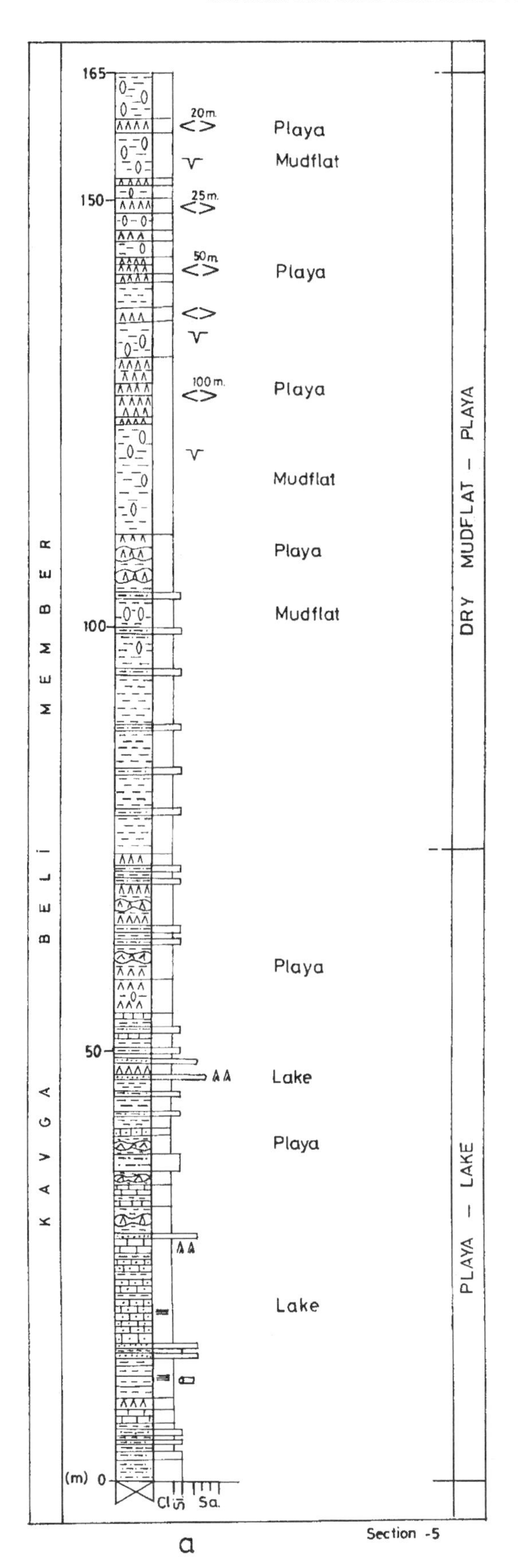

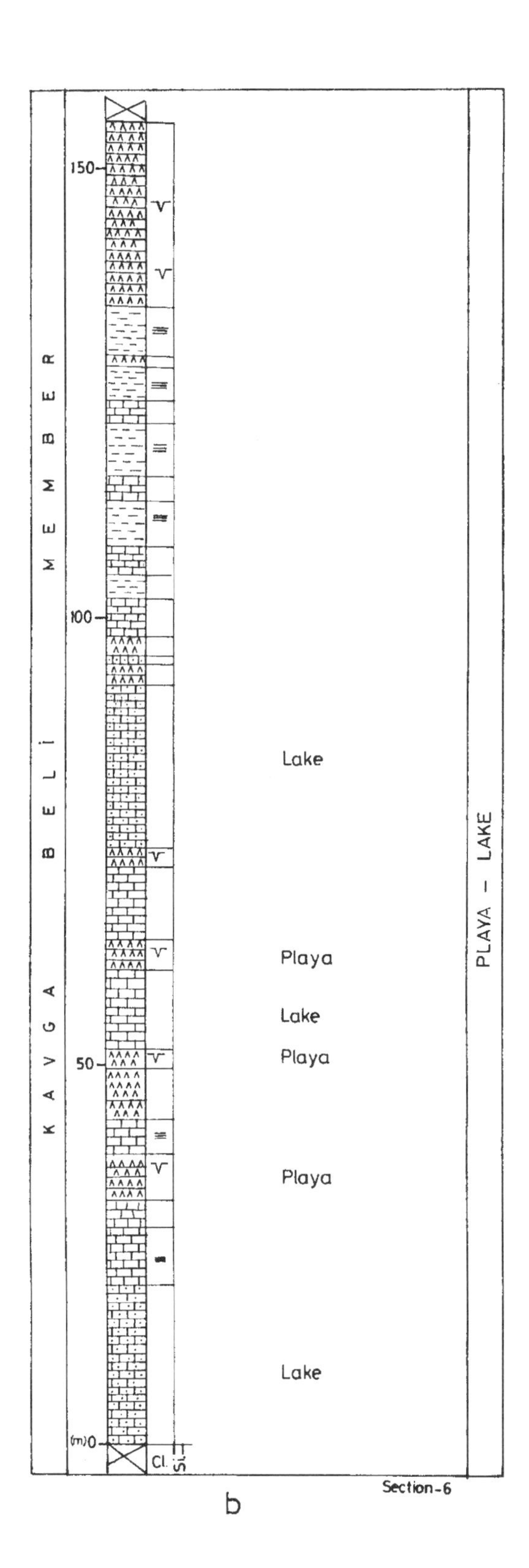

Figure 5—Measured lithologic columns of section 5 (a) and section 6 (b) illustrating a sequence containing the Kavga Beli Member of the Miocene Yeniçubuk Formation. For location of sections see Figure 2. For legend see Table 2.

Figure 6—Measured lithologic column of section 7 illustrating a sequence containing the Lalelik Dagi, transitional Küçükkamisili dere, and Kizilsardag Members of the Miocene Yeniçubuk Formation. For location of sections see Figure 2. For legend see Table 2.

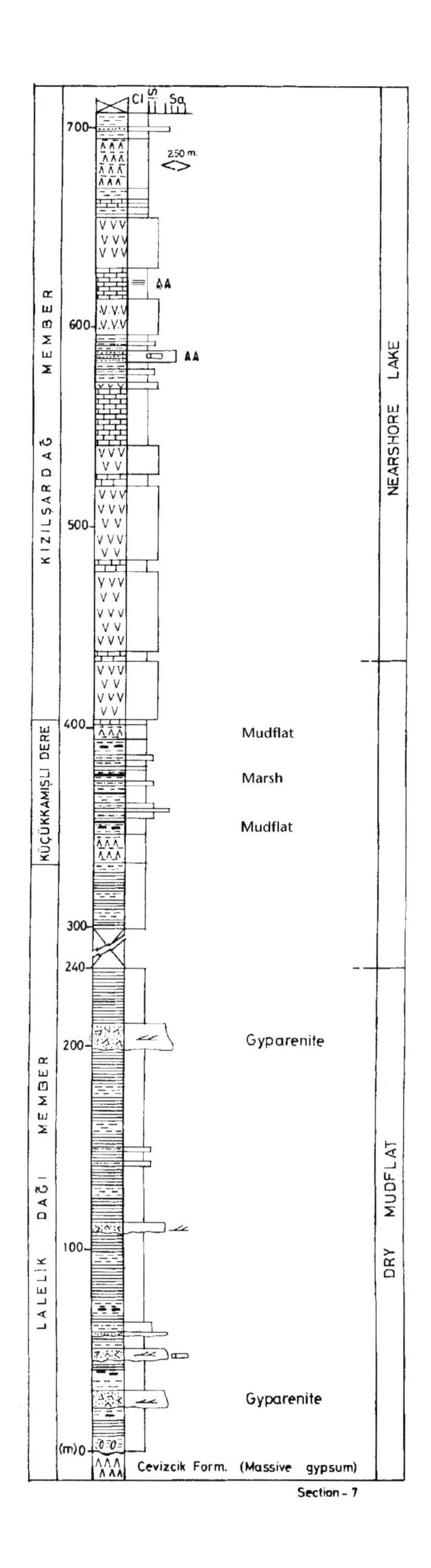

and limestone. Facies have lateral transitions into lake deposits in some places (association C3). Additional facies that are represented consist of well-sorted, cross-laminated sandstones and gyparenites (Sr) (Figure 6).

This association is interpreted as mudflat-playa complex deposits (Wolfbauer, 1973; Hardie et al., 1978; Sagri et al., 1989). Lenticular and bedded gypsum are thought to be deposited on playa mudflats. Lateral and vertical transition of this association to the playa-lake complex (association C2) indicates deposition on mudflats surrounding the lake.

Playa-Lake Complex (C2)

This association has extensive outcrops in the Kavga Beli Member and is composed of bedded limestone (Pm), oolitic limestone (Po), and bedded gypsum (Eg). Maximum measurable thickness of the association is 160 m and the ratio of limestone to gypsum ranges from 2:1 to 1:5 (Figure 5). Concave-faced subhedral calcite crystals and meniscus calcite overgrowth occur within the oolitic limestone, indicating vadose diagenesis. Claystone layers (Fm) between gypsum beds exhibit mudcracks. Gypsum commonly contains porphyroblastic texture. This association laterally and vertically interbeds with dry mudflat-playa deposits (association C1).

This facies is interpreted as deposits of ephemeral lakes (Wolfbauer, 1973; Surdam and Wolfbauer, 1975; Lowenstein and Hardie, 1985; Warren, 1989). Alternation of limestone with gypsum indicates possible fluctuations in salinity or provenance changes. Mudcracks near the gypsum levels indicates desiccation periods associated with gypsum precipitation. This association is similar to the playa-lake complex of the Green River Formation (Surdam and Wolfbauer, 1975; Smoot, 1983).

Open Lake (C3)

This facies association constitutes the whole Kizilsardag Member and contains dismicrite (bedded limestone) (Pm), oolitic limestone (Po), and bedded gypsum (Eg). At the eastern margin of the basin (around Kizilsardag and Yücedag), the association is composed of interbedded dismicrite (60%) and volcanics [basalt and pyroclastics (tuff?)] (40%) (Figures 3, 7b). At the western margin, the association grades into thin-bedded gypsum (80%) and tuff? interbeds (Figure 7a). Basalt-limestone interbeds crop out extensively in the basin. Bedded limestone also intercalates with oolitic limestone and contains gypsum seams, bird's-eye textures, and intraclasts. Gypsum is thin-bedded and shows microfolding. Very thin carbonate and clay laminae (Fl3) occur between gypsum beds; alabastrin texture is common in the gypsum.

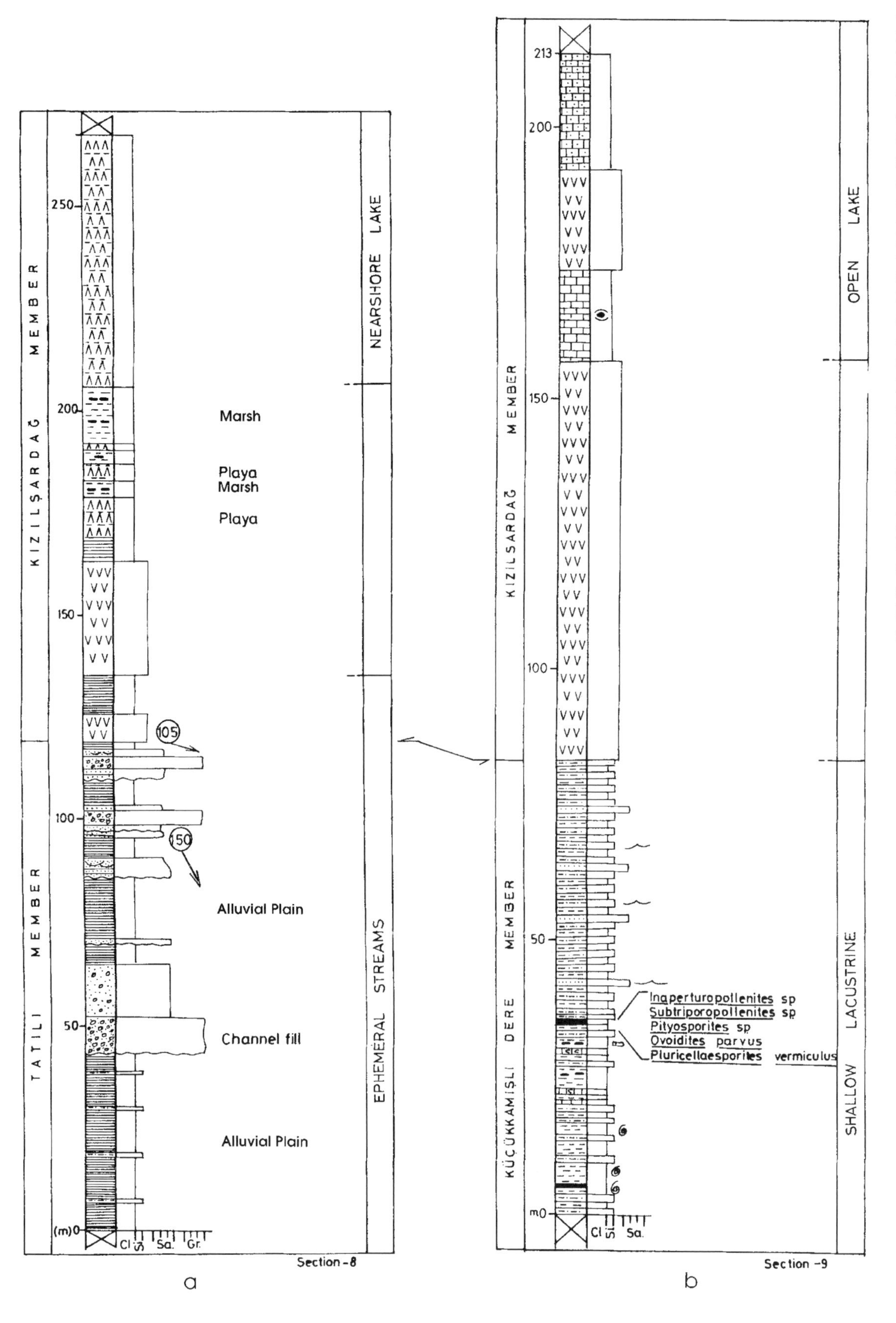

Figure 7—Measured lithologic columns of section 8 (a) showing a succession containing the Tatili and Kizilsardag Members and section 9 (b) illustrating a sequence containing the Küçükkamisilidere and Kizilsardag Members of the Miocene Yeniçubuk Formation. For location of sections see Figure 2. For legend see Table 2.

Table 1. Summary of Facies, Their Descriptions, and Their Interpretation for the Yeniçubuk Formation (Miocene) of the Upper Kizilirmak Basin in Türkiye.*

Facies	Code	Description	Interpretation
Massive conglomerate	Gm	Matrix- to clast-supported conglomerate, massive, well-sorted clasts of gravel up to 15 cm long, normal grading, in lenses up to 2 m thick, erosive base, unit average thickness: 10 m	Winnowed debris flow to channel fills of ephemeral streams
Stratified conglomerate	Gp	Planar cross-stratified conglomerate, clast-supported, clasts of well-sorted pebbles to gravel, imbrication, units up to 1 m thick	Longitudinal and side bars of ephemeral streams
Massive sandstone	Sm	Very coarse to granular sandstone containing dewatering structures, units up to 15 dm thick, grey color	Scour fills and sheet flood deposits
Epsilon cross-stratified sandstone	Sl	Medium- to coarse-grained sandstone, very low-angle planar cross-stratification, lenses of 4 m thickness, erosive base	Lateral accretion deposits
Horizontally stratified sandstone	Sh	Fine- to medium-grained sandstone, horizontal lamination, lenses 10–30 cm thick and 20 m width, unit average thickness: 2 m	Sheetflood to channel deposits
Cross-laminated sandstone	Sr	Fine- to medium-grained sandstone, moderately well sorted, lenticular geometry of beds with cosets 15–40 cm thick, non-erosive base with incorporated intraformational mud chips, unit average thickness: 60 cm. Clasts: quartz, feldspar, and gypsum (gyparenites)	Sheetflood to channel deposits
Planar cross-stratified sandstone	Sp	Medium- to coarse-grained sandstone, moderately well sorted, planar cross-stratification: cosets: 2–2.5 m, sets: 10–12 dm	Sheet flood to channel deposits
Trough cross-stratified sandstone	St	Medium-grained sandstone, grey color, well sorted, trough cross-stratification, lenses 3-4 m thick, sets 10–75 cm thick, cosets 35–150 cm thick—commonly upwards fining, erosive bases	Sheetflood to channel deposits
Interbedded sandstone, siltstone and mucrock	SF	Uneven alternations of fine sandstone, siltstone, and mudrock, laminated siltstone, and mudrock to rippled, grey color, unit average thickness: 2 m	Sheetflood to channel deposits
Disrupted silty mudrock	Fl_1	Laminated to disrupted to massive mudrock/siltstone with rippled fine sandstone lenses, thin units 5–10 cm thick, rhizoliths, mudcracks, unit average thickness: 7 m	Shallow lacustrine or peripheral marsh
Disrupted carbonaceous shale	Fl_2	Finely laminated shale to massive mudstone, laminations to layers, carbon-rich with macrophytic remains and freshwater mollusc fossils, bioturbation, unit average thickness: 5 m	Shallow lacustrine
Grey shale	Fl_3	Fine continuous horizontal lamination, alternations of clay and carbonate laminae, grey color, unit average thickness: 30 m	Open lacustrine
Massive mudstone	Fm	Mudcracks, rarely carbonate nodules (avg. 5 cm in diameter), rhizoliths, unit average thickness: 30 m. Colors: grey, red, and green	Paleosol
Coal	C	Lignite, 0.2–1 m thick layers	Marsh/swamp/mire
Biomicrite	Pf	Thin-bedded limestone with interbeds of mudstone and coal, units 2–10 m thick, fossils floating in micritic matrix: lamellibranchs, gastropods, and charophytes	Lacustrine carbonate
Dismicrite	Pm	Bedded limestone, layers 5–40 m thick (avg. 9 m), bird's eye structures, rhizoliths, and gypsum veins, rare gastropod fossils	Palustrine carbonate
Oolitic limestone	Po	Carbonates containing oolites, micritic intraclasts, and gypsum cements, interbedded with gypsum layers, lenses average 16 m thick	Carbonate shoals and bars
Bedded gypsum	Eg	Gypsum layers (up to 5 cm thick) intercalated with mudstone and dismicrite, commonly alabastrin gypsum texture, average unit thickness: 8 m	Saline playa
Gypsiferous mudstone	Fg	Siliciclastic mudstone containing 20% gypsum as discoidal crystals or in nodular form, abundant mudcracks, grey to red in color, average unit thickness: 5 m	Saline mudflat/paleosol

*Subfacies codes are adapted from Miall (1977).

Table 2. Legend for Measured Sections in Figures 3–8.*

KEY FOR MEASURED SECTIONS

SYMBOL	FACİES CODE
	Gm
	Gp
	Sm
	Sp
	St
	Sl
	Sh
	SF
	Fl_1
	Fm
	Fl_2
	Fl_3
	C
	Pf
	Pm
	Po
	Eg
	Fg
	Pyroclastics
	Basalt

SEDIMENTARY STRUCTURES
Trough cross-bedding
Epsilon cross-bedding
Planar cross-bedding
Low angle cross-bedding
Ripple cross-lamination
Horizontal lamination
Flaser bedding
Wave ripples
Bird's eye structure
Soft sediment deformation
Grading

BIOTA
Gastropoda
Lamellibranch
Macrophyte leaves
Charophytes
Rhizoliths
Gastropod fragments
Macrophyte

Lenticular shape
Paleocurrent direction (random)
Caliche
Mudcrack
Gravel orientation
İmbrication
Fining and coarsening upward sequence

*Grain size scale: Cl. = Clay, Si. = Silt, Sa. = Sand, and Gr. = Gravel.

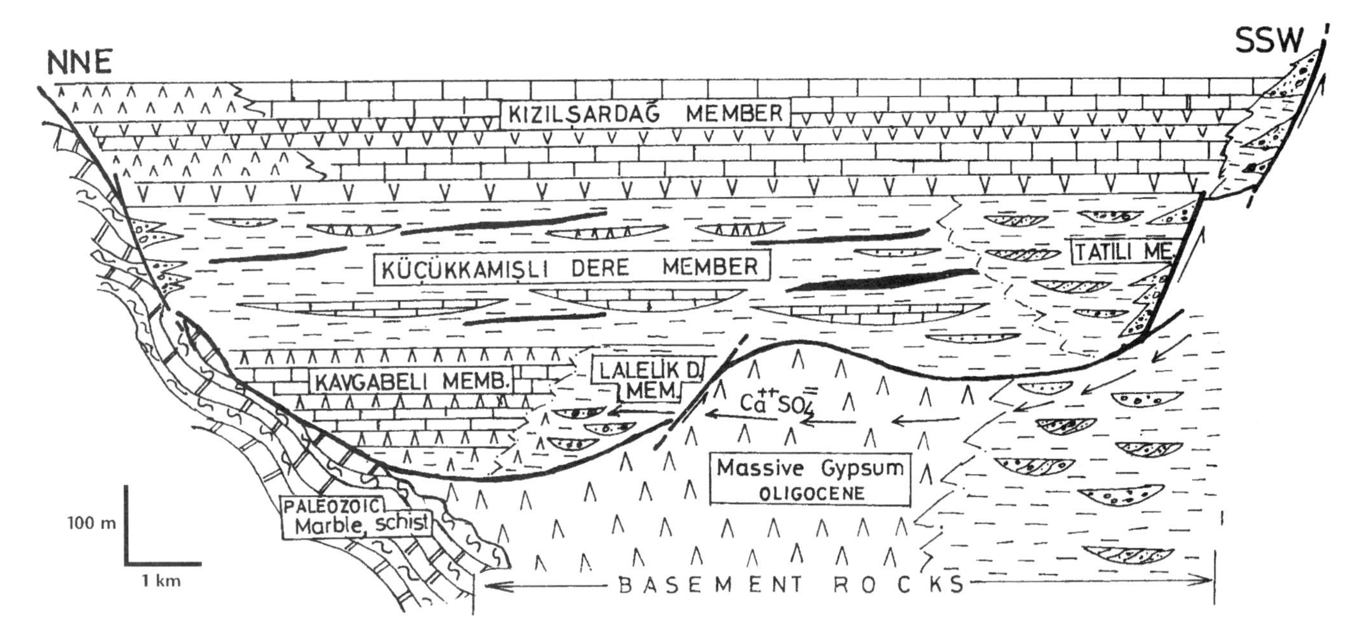

Figure 8—Schematic cross-section of the Yeniçubuk Formation across the Upper Kizilirmak Basin of Türkiye illustrating the stratigraphic relationships of its members. Three phases of deposition linked to possible tectonic influences include (1) Lalelik Dagi/Kavga Beli deposition, (2) Küçükkamisili dere/Tatili deposition, and (3) Kizilsardag deposition.

This association, with its volcanics and limestone interbeds at the upper level of the Yeniçubuk Formation, is interpreted as perennial lake deposits. Kinsman (1969) named the paleoenvironment a continental sabkha. Volcanic eruptions took place episodically while carbonate was deposited in the basin. Gypsum appears to have been the main marginal evaporite in nearshore environments while carbonate was deposited in distal open lake areas. Similar facies transitions have been described from the upper Miocene deposits in the Libros basin of Spain (Anadón et al., 1989) with carbonates laterally and vertically linked to thin-bedded gypsum.

PALEOGEOGRAPHIC EVOLUTION

Tertiary-age sediments in the study area and nearby crop out in an approximately northeast-southwest direction with the Akdag Massive to the northwest and Çaldag-Hinzirdag Group to the southeast. This intramontane corridor was filled by a very thick sequence of Tertiary sediments interbedded with volcanics. The lower and middle Miocene Yeniçubuk Formation, forming the lower part of the Neogene sediments near Gemerek, uncomformably covers the Oligocene Küçüktuzhisar member of the Cevizcik Formation.

Climate during the middle Miocene is documented by floral data. In western Anatolia, the dominance of pollen from *Pinus*, *Quercus*, *Castanea*, and *Alnus*, as well as the presence of subtropical plant taxa such as *Taxodium*, *Cupressacease*, *Cyrillaceae*, *Carya*, and *Engelhardia,* suggest a subtropical (moist and hot) paleoclimate (Akgün and Akyol, 1992). In central Anatolia, middle Miocene paleoclimate is interpreted as still moist but more cool in temperature based on the floral diversity of the presence of *Pinus*, *Quercus*, *Alnus*, *Taxodium*, *Carya*, and *Engelhardia* in continental deposits similar to those of the Yeniçubuk Formation (Akgün et al., 1995). A global cooling trend in the middle Miocene has been documented in the marine and, more recently, in continental sediments in the western Mediterranean (Spain) (Krijgsman et al., 1994).

The Yeniçubuk Formation is interpreted as accumulating during three phases of deposition (Figure 8), controlled mainly by tectonic subsidence. Source rocks include the gypsum of the Oligocene Küçüktuzhisar Member in the basement and gravels of andesite, basalt, ophiolite, and limestone of the Paleozoic Çaldag Group. In the first phase, ephemeral streams of the Lalelik Dagi Member flowed from the east and southeast into the shallow lakes and playas of the Kavga Beli Member. In the second phase of deposition, the Küçükkamisili dere Member represents the wetlands (lakes, marshes, ponds, mudflats, swamps) of an anastomosed river system with the proximal Tatili Member representing ephemeral stream/alluvial fan deposits. The coal seams in the study area are associated with the interpreted mudflat-swamp and shallow lake deposits. During the third phase of deposition, an extensive lake covered the basin, as represented by the gypsiferous and carbonate Kizilsardag Member.

Evidence for tectonic influence in altering basin drainage and deposition involves volcanic signatures. These include (1) basalt flows associated with the Küçükkamisili dere Member on the western slope of Kizilsardag Mountain and (2) the basalt and tuff deposits within the open-lake facies of the uppermost

Kizilsardag Member of the Yeniçubuk Formation. After his study in the east and southeast part of the study area (around Çaldag), Gökten (1983) concluded that there was no deposition during the Miocene, only erosion. This is also supporting evidence for tectonic uplift in the surrounding region.

CONCLUSIONS

The Upper Kizilirmak Basin is interpreted as an intermontane, foreland basin with a northeast-southwest axial direction surrounded by the Akdag Mountains to the northwest and Çaldag and Hinzir Mountains to the south-southwest. The Yeniçubuk Formation, a part of its Neogene fill, has been mapped as being composed of five members: Lalelik Dagi, Kavga Beli, Küçükkamisili dere, Tatili, and Kizilsardag, constituting three main phases of deposition within the Upper Kisilirmak Basin. Eighteen facies and three facies associations were recognized.

Facies associations have been interpreted as ephemeral stream/lower alluvial fan, alluvial plain, mudflat-swamp, shallow lake, playa-lake, and offshore lake. Ions necessary for gypsum in the playa-lake deposits are thought to come from the Oligocene sediments of the basement and are not necessarily due to concentration as a result of an arid climate. This is supported by floral data indicating a more humid climate. Carbonates associated with the coal swamp deposits are also a result of the high carbonate content of the source area.

ACKNOWLEDGMENTS

We thank G. Nadon and E. Gierlowski-Kordesch for reviews that significantly improved this manuscript, and Hükmü Orhan and Baki Varol for their helpful suggestions.

REFERENCES CITED

Akgün, F., and E. Akyol, 1992, Paleoecology and correlative palyno-stratigraphy of Yukarikasikara and Yarikkaya (Isparta-Türkiye). Bulletin of the Turkish Association of Petroleum Geologists, v. 4 (1), p. 129–139.

Akgün, F., E. Olgun, I. Kuscu, V. Toprak, and C. Göncüoglu, 1995, New data on the stratigraphy, depositional environment, and real age of the Oligo–Miocene cober of the central Anatolian crystalline complex. Bulletin of the Turkish Association of Petroleum Geologists, v. 6 (1), p. 51–68 (in Turkish).

Anadón, P., Ll. Cabrera, R. Julía, E. Roca, and L. Rosell, 1989, Lacustrine oil-shale basins in Tertiary grabens from NE Spain (Western European Rift Systems). Palaeogeography, Palaeoclimatology, Palaeoecology, v. 70, p. 7–28.

Anadón, P., Ll. Cabrera, R. Julia, and M. Marzo, 1991, Sequential arrangement and asymmetrical fill in the Miocene Rubielos de Mora Basin (Northeast Spain), *in* P. Anadón, Ll. Cabrera, and K. Kelts, eds., Lacustrine Facies Analysis. International Association of Sedimentologists Special Publication No. 13, p. 257–275.

Belt, E. S., R. M. Flores, P. D. Warwick, K. M. Conway, K. R. Johnson, and R.S. Waskowitz, 1984, Relationship of fluviodeltaic facies to coal deposition in the lower Fort Union Formation (Palaeocene), southwestern North Dakota, *in* R. A. Rahmani and R. M. Flores, eds., Sedimentology of Coal and Coal Bearing Sequences. International Association of Sedimentologists Special Publication No. 7, p. 177–195.

Eberth, D. A., and A. D. Miall, 1991, Stratigraphy, sedimentology, and evolution of a vertebrate-bearing, braided to anastomosing fluvial system, Cutler Formation (Permian-Pennsylvanian), north-central New Mexico. Sedimentary Geology, v. 72, p. 225–252.

Flores, R. M., 1979, Coal depositional models in some Tertiary and Cretaceous coal in the U.S. Western Interior. Organic Geochemistry, v. 1, p. 225–235.

Flores, R. M., and J. H. Hanley, 1984, Anastomosed and associated coal-bearing fluvial deposits: Upper Tongue Member, Palaeocene Fort Union Formation, northern Powder River basin, Wyoming, U.S.A., *in* R. A. Rahmani and R. M. Flores, eds., Sedimentology of Coal and Coal-bearing Sequences International Association of Sedimentologists Special Publication No. 7, p. 85–103.

Gibling, M., G. C. Nanson, and J. C. Maroulis, 1998, Anastomosing river sedimentation in the Channel Country of central Australia. Sedimentology, v. 45, p. 595–619.

Gierlowski-Kordesch, E. H., 1998, Carbonate deposition in an ephemeral siliciclastic alluvial system: Jurassic Shuttle Meadow Formation, Newark Supergroup, Hartford Basin, U.S.A. Palaeogeography, Palaeoclimatology, Palaeoecology, v. 140, p. 161–184.

Gökten, E., 1983 , Stratigraphy and geological evolution of the south-southeast of Sarkisla (Sivas). Bulletin of the Geological Survey of Türkiye, v. 26 (2), p. 167–176. (in Turkish).

Graf, W. L., 1988. Fluvial Processes in Dryland Rivers. Springer-Verlag, Berlin, 346 p.

Hardie, L. A., J. P. Smoot, and H. P. Eugster, 1978, Saline lakes and their deposits: a sedimentological approach, *in* A. Matter and M. E. Tucker, eds., Modern and Ancient Lake Sediments. International Association of Sedimentologists Special Publication No. 2, p. 7–41.

Kerey, I. E., 1982, Stratigraphical and sedimentological studies of upper carboniferous rock in Northwestern Turkey. Ph.D. thesis, Keele University, 238 p.

Kinsman, D. J. J., 1969, Modes of formation of sedimentary associations and diagenetic features of shallow water and supratidal evaporites. AAPG Bulletin, v. 53, p. 830–840.

Kirschbaum, M. A., and P. J. McCabe, 1992, Controls on the accumulation of coal and on the development of anastomosed fluvial systems in the Cretaceous Dakota Formation of southern Utah. Sedimentology, v. 39, p. 581–598.

Krijgsman, W., C. G. Langereis, R. Daams, and A. J.

van der Meulen, 1994, Magnetostratigraphic dating of the middle Miocene climate change in the continental deposits of the Aragonian type area in the Catalayud-Teruel basin (cental Spain). Earth and Planetary Science Letters, v. 128, p. 513–526.

Lebküchner, R. F., 1956, Report on the results of geological studies around Gemerek where lignite occurs. Mineral Research and Exploration Institute (Ankara) Report No. 2850, 38 p. (in Turkish).

Lowenstein, T., and Hardie, L. A., 1985. Criteria for the recognition of salt-pan evaporites. Sedimentology, v. 32, p. 627–644.

McCabe, P. J., 1984, Depositional environments of coal and coal-bearing strata, *in* R. A. Rahmani and R. M. Flores, eds., Sedimentology of Coal and Coal-bearing Sequences. International Association of Sedimentologists Special Publication No. 7, p. 13–42.

Miall, A. D., 1977, A review of the braided river depositional environment. Earth Science Reviews, v. 13, p. 1–62.

Miall, A. D., 1996, The Geology of Fluvial Deposits: Sedimentary Facies, Basin Analysis, and Petroleum Geology. Springer-Verlag, New York, 582 p.

Nadon, G. C., 1994, The genesis and recognition of anastomosing fluvial deposits: data from the St. Mary River Formation, southwestern Alberta, Canada. Journal of Sedimentary Reserach, v. B64, p. 451–463.

Rust, B. R., M. R. Gibling, and A. S. Legun, 1984, Coal deposition in an anastomosing-fluvial system: the Pennsylvanian Cumberland Group south of Joggins, Nova Scotia, Canada, *in* R. A. Rahmani and R. M. Flores, eds., Sedimentology of Coal and Coal-bearing Sequences. International Association of Sedimentologists Special Publication No. 7, p. 105–120.

Sagri, M., T. Abbate, and P. Bruni, 1989, Deposits of ephemeral and perennial lakes in the Tertiary Daban Basin (northern Somalia). Palaeogeography, Palaeoclimatology, Palaeoecology, v. 70, p. 225–233.

Smith, D. G., 1983, Anastomosing river deposits: modern and ancient examples in Alberta, Canada, *in* J. D. Collinson and J. Lewin, eds., Modern and Ancient Fluvial Systems. International Association of Sedimentologists Special Publication No. 6, p. 155–168.

Smoot, J. P., 1983, Depositional subenvironments in an arid closed basin; the Wilkins Peak Member of the Green River Formation (Eocene) Wyoming, USA, *in* A. Matter and M. E. Tucker , eds., Modern and Ancient Lake Sediments. International Association of Sedimentologists Special Publication No. 2, p. 109–127.

Soytürk, N., and A. Birgül, 1971, Geology of the Bünyan–Gemerek region. Turkish Petroleum Corporation (Ankara) Report No. 528, 83 p. (in Turkish).

Surdam, R. C., and C. A. Wolfbauer, 1975, Green River Formation, Wyoming: A playa-lake complex. Geological Society of America Bulletin, v. 83, p. 148–165.

Sümengen, M., I. Terlemez, T. Bilgiç, M. Gürbüz, E. Unay, S. Ozaner, and K. ve Tüfekçi, 1987, Stratigraphy, sedimentology, and geomorphology of the Sarkisla–Gemerek basin in the Tertiary. Mineral Research and Exploration Institute (Ankara) Report No. 8118, 240 p.

Sengör, A. M. C., N. Görür, and G. Saroglu, 1985, Strike-slip faulting and related basin formation in zones of tectonic escape: Turkey as a case study, *in* K. T. Biddle and N. Christie-Blick, eds., Strike-Slip Deformation, Basin Formation, and Sedimentation. SEPM (Society of Economic Paleontologists and Mineralogists) Special Publication No. 37, p. 227–64.

Soytürk, N., and A. Birgül, 1972, Geology of the Sarkisla–Kaynar–Kaleköy region. Turkish Petroleum Corporation (Ankara) Report No. 703, 70 p. (in Turkish).

Türkmen, I., 1993, Gemerek Dolaylarinin Neojen Çökelleri üzerinde Sedimentolojik Incelemeler. Doctoral Dissertation, Firat University, Graduate School of Science and Technology, Elazig (Türkiye), 158 p. (in Turkish).

Valero Garcés, B. L., E. Gierlowski-Kordesch, and W. A. Bragonier, 1997, Pennsylvanian continental cyclothem development: no evidence of direct climatic control in the upper Freeport Formation (Allegheny Group) of Pennsylvania (northern Appalachian Basin). Sedimentary Geology, v. 109, p. 305–320.

Warren, J. K., 1989, Evaporite sedimentology: importance in hydrocarbon accumulation. Prentice Hall, Englewood Cliffs, 285 p.Wolfbauer, C. A., 1973, Criteria for recognizing paleoenvironments in a playa lake complex, the Green River Formation of Wyoming. 25th Field Conference, Wyoming Geological Assocation Guidebook, p. 87–91.

ADDITIONAL REFERENCES OF INTEREST

Andrews, J. E., S. T. Michael, N. Ghulam, and B. Spiro, 1991, The anatomy of an early Dinantian terraced floodplain: Palaeoenvironment and early diagenesis. Sedimentology, v. 38, p. 271–287.

Besley, B. M., and J. D. Collinson, 1991, Volcanic and tectonic controls of lacustrine and alluvial sedimentation in the Stephanian coal-bearing sequences of the Malpas-Short Basin, Catalonian Pyrenees. Sedimentology, v. 38, p. 3–26.

Fielding, C. R., 1984, Upper delta plain lacustrine and fluviolacustrine facies from the Wesphalian of the Durham coalfield, NE England. Sedimentology, v. 31, p. 547–567.

Flores, R. M., 1979, Coal depositional models in some Tertiary and Cretaceous coal in the U.S. Western Interior. Organic Geochemistry, v. 1, p. 225–235.

Flores, R. M., 1983, Basin facies analysis of coal-rich Tertiary fluvial deposits, Northern Powder River Basin, Montana and Wyoming, *in* J. D. Collinson and J. Lewin, eds., Modern and Ancient Fluvial Systems. International Association of Sedimentologists Special Publication No. 6, p. 501–515.

Gierlowski-Kordesch, E., and B. R. Rust, 1994, The Jurassic East Berlin Formation, Hartford Basin, Newark Supergroup (Connecticut and Massachusetts): a saline lake-playa-alluvial plain, *in* R. W. Renaut and W. M. Last, eds., Sedimentology and

Geochemistry of Modern and Ancient Saline Lakes. SEPM (Society of Economic Paleontologists and Mineralogists) Special Publication No. 50, p. 249–265.

Heward, A. P., 1978, Alluvial fan and lacustrine sediments from the Stephanian A and B (La Magdalena, Cinera-Matallana and Sabero) coalfields, northern Spain. Sedimentology, v. 25, p. 451–488.

MacCarthy, I. A. J., 1990, Alluvial sedimentation patterns in the Munster Basin, Ireland. Sedimentology, v. 37, p. 685–712.

Platt, N. H., and V. P. Wright, 1991, Lacustrine carbonates: facies models, facies distribution and hydrocarbon aspects, *in* P. Anadón, Ll. Cabrera, and K. Kelts, eds., Lacustrine Facies Analysis. International Association of Sedimentologists Special Publication No. 13, p. 57–74.

Owen, R. B., and R. W. Renaut, 2000, Miocene and Pliocene diatomaceous lacustrine sediments of the Tugen Hills, Baringo District, Central Kenya rift, *in* E. H. Gierlowski-Kordesch and K. R. Kelts, eds., Lake basins through space and time: AAPG Studies in Geology 46, p. 465–472.

Chapter 42

Miocene and Pliocene Diatomaceous Lacustrine Sediments of the Tugen Hills, Baringo District, Central Kenya Rift

R. Bernhart Owen
Department of Geography, Hong Kong Baptist University
Kowloon Tong, Hong Kong, China

Robin W. Renaut
Department of Geological Sciences, University of Saskatchewan
Saskatoon, Canada

GENERAL GEOLOGY

The central Kenya Rift extends from about 1°N to 0°10′N and is bounded by the Elgeyo fault scarp in the west and the Laikipia fault scarp 70 km to the east (Figure 1). Within this semi-arid region, there are a series of geographically and temporally distinct fault-bounded basins with fluvial and lacustrine sediments that have developed and been destroyed because of combined tectonic, volcanic, and climatic controls. These Neogene basins have tended to shift eastward with time as rift extension has progressed (Chapman and Brooke, 1978; Chapman et al., 1978). Their sediments include the Miocene Kimwarer and Tambach Formations along the Elgeyo border-fault (Murray-Hughes, 1933; Shackleton, 1951; Lippard, 1972; Ego, 1994; Renaut et al., 1999), several Miocene–Pliocene successions exposed in the Tugen Hills (King and Chapman, 1972; Chapman et al., 1978; Hill et al., 1985; Hill, 1995), and the Pleistocene Kapthurin Formation and modern Lake Baringo in the eastern axial part of the rift (Figure 1) (Tallon, 1978; Renaut et al., this volume).

The Tugen Hills, also known as Kamasia, is a large north-south trending tilt block that dips westward and rises to more than 1000 m above the rift floor. Pleistocene faulting and erosion have exposed a sequence of more than 3000 m (Hill et al., 1986) of interbedded fluvial and lacustrine sedimentary rocks and lavas, principally phonolites, trachytes, and basalts (Figures 1, 2). Lacustrine sedimentary rocks, including diatomaceous shales, silts, and weakly lithified siltstones, are found in the Miocene Ngorora and Lukeino formations and the Pliocene–Pleistocene Chemeron Formation (Figure 1). To date, few detailed mineralogical and paleolimnological studies of these sediments have been undertaken. In this paper, we provide results of a reconnaissance study of the diatom floras found in these formations. This evidence shows that the Neogene paleolakes of the central Kenya rift ranged from fresh to saline and alkaline, much like the modern Kenya rift lakes.

THE NGORORA FORMATION

The Ngorora Formation was defined formally by Bishop and Chapman (1970), with further details provided by Chapman (1971), Bishop and Pickford (1975), and Pickford (1975a, 1975b, 1978a). The Ngorora Formation, with a maximum thickness of approximately 450 m, comprises a series of sediments that were deposited over about 4 m.y. in a group of separate fault-controlled basins (Figures 1–3). The formation lies on the Tiim Phonolite Formation (Figure 2), dated at 13.15 Ma (Deino et al., 1990), and is overlain by the Ewalel Phonolite Formation (Pickford, 1978a; Hill et al., 1985, 1986). K-Ar ages of tuffs and magnetostratigraphy of the type-section give ages ranging from 12.7 Ma to 10.52 Ma (Tauxe et al., 1985). Elsewhere, the Ngorora sediments may be as young as 8 Ma (Hill, 1995).

Pickford (1978a) suggested that the basin was limited to the north by lavas of the Tiati volcanic center, which lies 60 km north of the type-section at Kabarsero (Figure 1A), to the west by the Elgeyo fault scarp (Figure 1B), and to the south and east by a general rise in the rift floor. The basin was also split by north-south trending faults that controlled local sedimentation patterns. Bishop and Chapman (1970) recognized five units, which were raised to member status by Pickford (1975a). The oldest member, Member A (104 m thick), consists of subaerial volcaniclastic sediments, lahars,

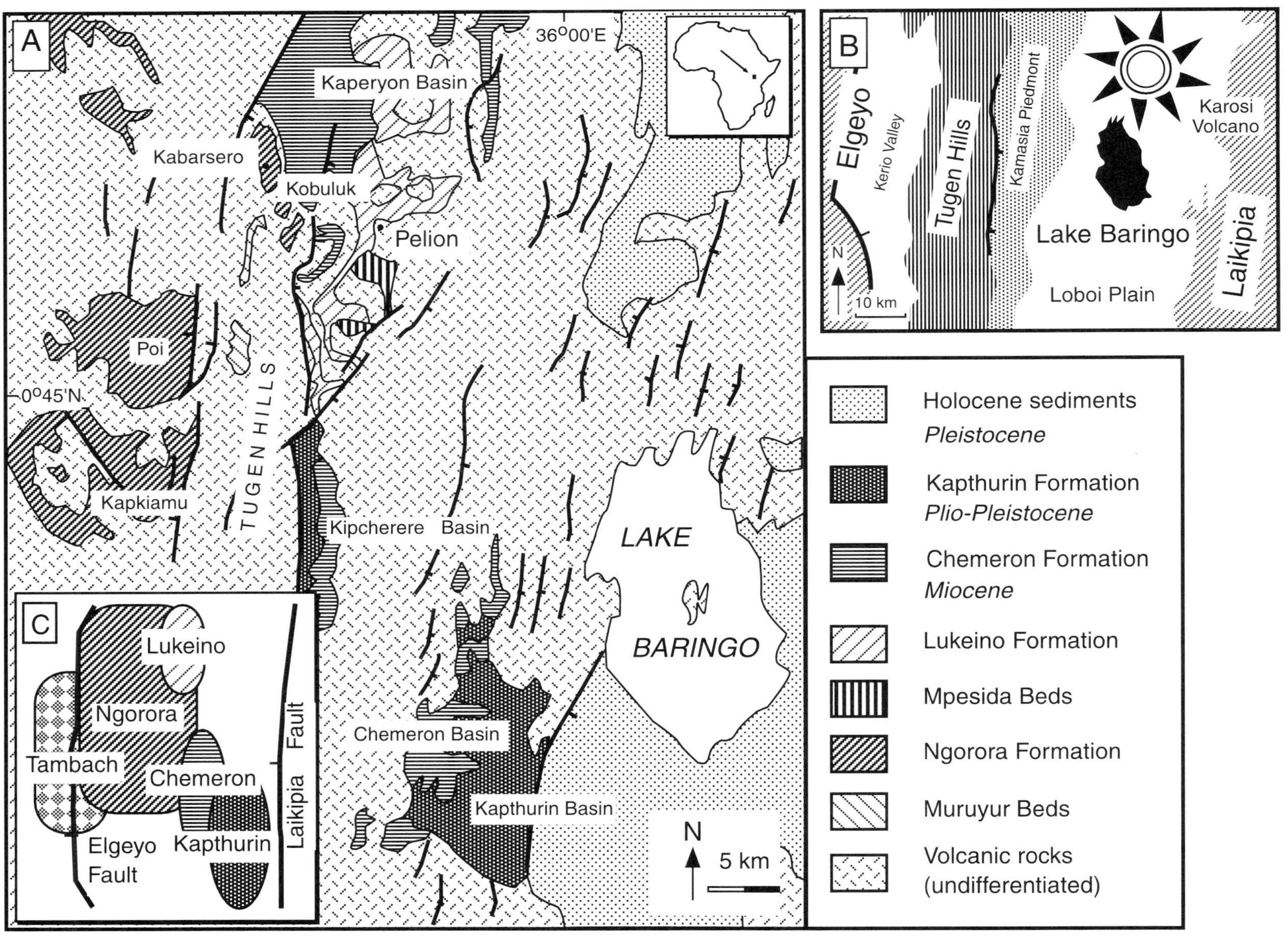

Figure 1—(A) Simplified geological map of the major sedimentary formations of the Tugen Hills, Kenya Rift Valley. (B) The main morphostructural elements of the Tugen Hills–Baringo region. The unshaded area containing Lake Baringo is the modern axial rift depression. The Tugen Hills fault-block separates the latter from the Kerio Valley, a sedimentary basin adjacent to the Elgeyo escarpment border fault. The Laikipia highlands represent the eastern margin of the rift. (C) Schematic diagram showing the eastward migration of the main depositional with lacustrine sediments referred to in the text. The Tambach Basin is not shown on (A).

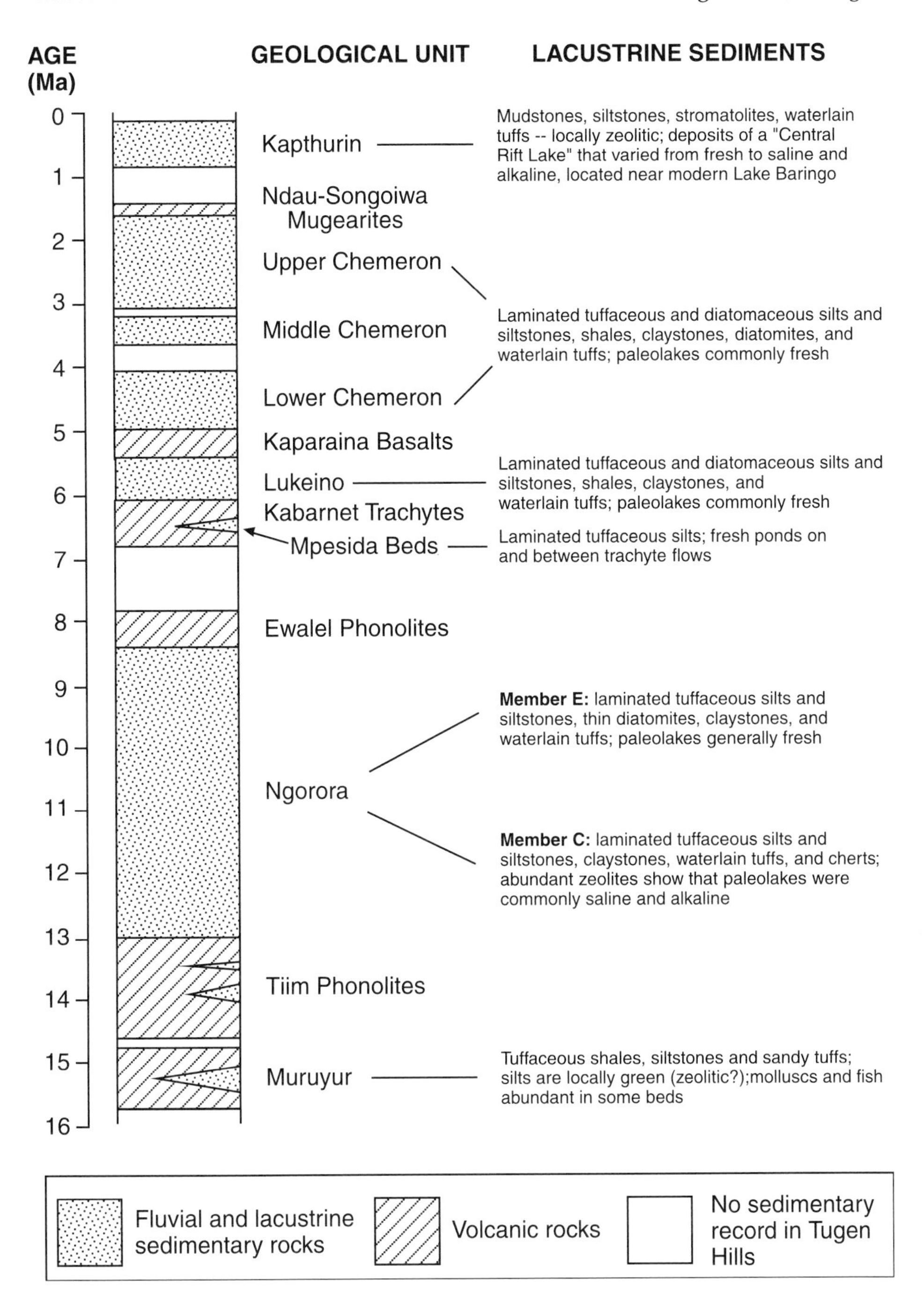

Figure 2—Summary stratigraphic column for the Tugen Hills, Kenya (stratigraphic and chronology modified from Kingston et al., 1994; Hill, 1995). The ages of most lacustrine units within each formation are not tightly constrained. The ages of the paleolakes may also vary from subbasin to subbasin.

and fluvial deposits (Figure 3). Members B (42 m) and D (66 m) are dominated by fluvial sandstones and paleosols. Members C (62 m) and E (125 m) are predominantly lacustrine, consisting of white laminated shales with thin paleosols. Desiccation cracks, bird footprints, clastic dikes, and other structures indicating subaerial exposure are common in Member C; rare diatomites are found in Member E. Minor waterlain tuffs, with local fluviatile sandstones and conglomerates, are intercalated with the lacustrine deposits.

Diatoms are absent from Members A, B, and D, and from Member C at the Kapkiamu subbasin (Figure 1) in the western Tugen Hills (Owen, 1981). Member C at Kapkiamu and Poi contains a zonation of authigenic minerals from K-feldspar at the center of the subbasin, through analcime and clinoptilolite to smectite at the margins (Ego, 1994). This finding is consistent with the absence of diatoms and indicates a saline, alkaline lake (Renaut et al., 1999). In contrast, scarce diatoms are found at several levels in both members C and E at Kabarsero (Figures 1, 3). Diatomites are present locally in Member E, with one prominent unit toward the top of this member.

Two diatom assemblages are present in the Ngorora lake sediments. *Aulacoseira granulata* forms 100% of the flora in rare diatomite units of Member E. The second assemblage is dominated by *A. granulata-agassizi* with less common *A. agassizi* v. *malayensis*(?)

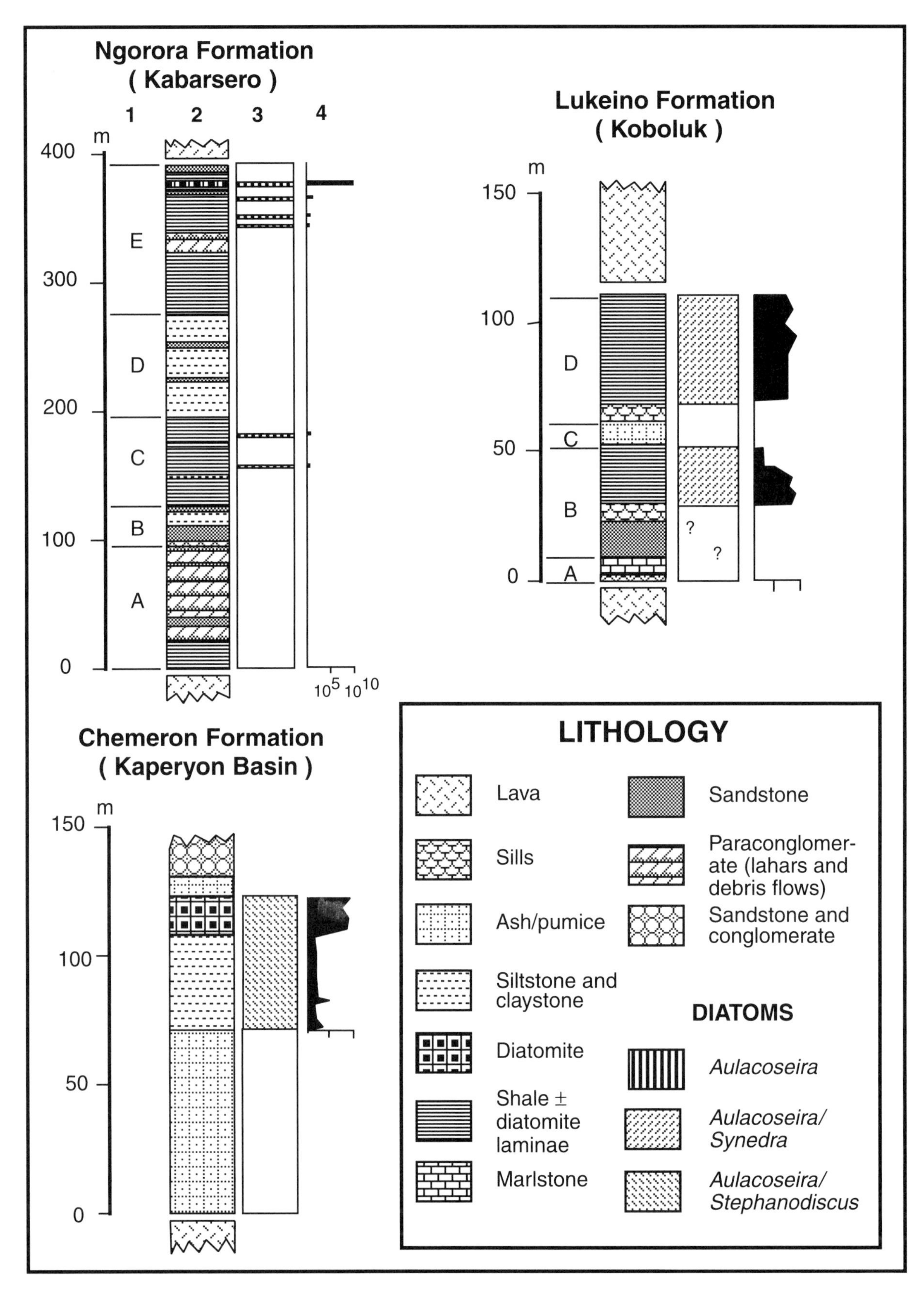

Figure 3—Diatom assemblages and distributions in the Ngorora, Lukeino, and Chemeron Formations, Tugen Hills, Kenya. 1 = members (for Ngorora and Lukeino Formations); 2 = lithology; 3 = diatom assemblage; 4 = diatom abundance.

and *A. granulata*. These taxa are archaic in appearance with coarser ornamentation than modern forms. *A. granulata* resembles *A. praegranulata* of Servant (1973). Other rare taxa include *A. granulata* v. *angustissima f. curvata, Synedra ulna, Fragilaria construens, F. brevistriata,* and *Nitzschia* spp.

Modern Aulacoseira generally indicate fresh waters with fairly high silica levels. The presence of zeolites in several parts of the formation implies saline, alkaline waters (Renaut et al., 1999). The paleolakes may have alternated between being mainly saline, alkaline, and fresh, or perhaps became stratified occasionally with a dilute mixolimnion.

THE LUKEINO FORMATION

The sedimentary rocks that constitute the Lukeino Formation (Bishop et al., 1971; Pickford, 1975a, 1978b) were originally described by Martyn (1969), Chapman (1971), and McClenaghan (1971) as the Lukeino Member of the Kaparaina Basalts Formation. They lie between the Kabarnet Trachyte and Kaparaina Basalts formations in the eastern Tugen Hills (Figures 1, 2). Hill et al. (1986) obtained bracketing ages of 6.20 Ma for the upper part of the Kabarnet Trachyte Formation and 5.65 Ma for lower flows in the Kaparaina Basalts Formation. Hill et al. (1985) reported a date of 6.06 Ma for the Lukeino sediments.

The sediments were deposited in an asymmetric fault-controlled trough with the main Tugen Hills fault (Figure 1B) forming a western barrier (Pickford, 1978b). Impoundment of the paleolake was completed by gently sloping rift floors to the east, south, and north. The upper and lower bounding surfaces are unconformable. Pickford (1975a, 1978b) recognized four members (Figure 3). Member A (12–30 m), the oldest unit, consists of subaerial red marls, gray silty tuffs, channel sandstones, and microbial limestones. Member B (50 m) is dominated by lacustrine deposits and includes diatomaceous and tuffaceous siltstones and shales, red siltstones, and calcretes. An 11.5-m-thick sill is present within this member. Member C (11.2 m) consists of lacustrine and subaerial bedded pumice-tuffs, red siltstones, and gray tuffs. Member D (12 m) consists of lacustrine diatomaceous shales and thin tuffs.

Diatom samples from members B and D at the type section, located at Koboluk (Figure 3), contain abundant diatoms. Two assemblages are present: Assemblage 1 is dominated by *A. granulata* (71–99%), with lesser *A. granulata* v. *angustissima*, *A. granulata* v. *muzzanensis*, *Nitzschia frustulum* (?), *Navicula cryptocephala*, and *Synedra ulna*. Assemblage 2 is dominated by *Synedra acus* v. *angustissima* (71–80%) with subdominant *A. granulata* (1–22%). Less common taxa include *Synedra ulna* (and var. *aequalis*), *Fragilaria construens* (and var. *venter*), and *Nitzschia frustulum* (?). The *Aulacoseira* appear archaic with coarse ornamentation (Owen, 1981).

Large parts of members B and D consist of varve-like laminae of alternating white diatomite (0.8–1 mm) and dark gray diatomaceous smectitic clay (0.1–0.2 mm). The white laminae contain 1.8×10^{10} diatom valves per gram of sediment, and a flora dominated by *A. granulata*. Dark laminae contain 4.2×10^5 valves per gram and are dominated by *A. granulata* and *Synedra* spp. This regular variation may be evidence of periodic (seasonal?) blooms of diatoms alternating with times of increased siliciclastic sediment influx.

Kilham et al. (1986) suggested that *Synedra ulna* and *S. acus* indicate high Si/P ratios, and that *Aulacoseira* reflect high Si concentrations, with *A. granulata* being favored by relatively low light/P ratios, and turbulent and dilute waters. These data and the well-preserved laminae suggest that the Lukeino paleolake was fresh, moderately deep, and perhaps stratified. The differing requirements of *Synedra* and *Aulacoseira* may reflect seasonal succession patterns that were stimulated by a seasonal paleoclimate. This suggestion is also supported by the presence of varve-like sediments.

THE CHEMERON FORMATION

Gregory (1921) referred to the Chemeron deposits as "Kamasia sediments," which later became the basis for the concept of pluvial stages in east Africa. McCall et al. (1967) referred to these deposits as the "Chemeron Beds." Martyn (1967, 1969) recognized two distinct depositional basins that were termed the Chemeron and Kipcherere Basins (Figure 1). Parts of what had previously been termed the Kaperyon Formation, located near Pelion in the northern Tugen Hills (Chapman, 1971), were later assigned by Pickford (1975a) to the Chemeron Formation, although problems remain with respect to their stratigraphic position (Hill et al., 1986; Williams and Chapman, 1986).

The Chemeron Formation crops out between the Tugen Hills and Lake Baringo (Figure 1). The Chemeron sequence rests on the Kaparaina Basalts (Figure 2) dated at 5.9 and 5.1 Ma (Hill et al., 1985, 1986). Recent tephrostratigraphic analyses indicate an age range of approximately 4 Ma to less than 1.6 Ma for the Chemeron Formation (Nawamba, 1993).

The formation is 230 m thick and includes both lacustrine and fluvial sediments. Martyn (1967) divided the formation into five members. Member A (0–50 m), the lowest unit of the formation, consists of fluvial sandstones and rare impure diatomites. Member B (100 m) is dominated by lacustrine silts, diatomites, and local fluvial sandstones. Fossil fish are locally present in the finer lacustrine deposits. Member C (3–5 m) is a lacustrine waterlain tuff. Member D (80 m) includes fluvial sandstones and conglomerates and lacustrine silts, diatomites, and fish-bearing siltstones. Member E (6–30 m) is composed of lacustrine tuffs.

Figure 3 shows the succession in the Kaperyon Basin (Figure 1). There, the sequence consists of fluvial and lacustrine tuffs that are overlain by lacustrine diatomaceous siltstones and diatomites. The diatomaceous sediments are dominated by two floras. Assemblage 1 consists of *A. granulata* (60–70%), with less common *Stephanodiscus rotula*, *S. minutula*, and rare *S. hantzschii* v. *pusilla*. Assemblage 2 consists of *A. granulata* (45–80%), with less common *Synedra ulna*, *S. ulna* v. *aequalis*,

S. rumpens, Cocconeis placentula, Epithemia argus, A. ambigua, and *S. rotula.* Floras collected from members B and D in the Chemeron Basin (Figure 1) were similar.

The coexistence of *Aulacoseira* and *Stephanodiscus* indicates fresh waters with the former taxon occurring during periods of turbulence and high silica concentrations. *Stephanodiscus* may have replaced *Aulacoseira* during periods of silica depletion and low Si/P ratios. Assemblage 2, with its higher proportion of *Synedra*, implies higher Si/P ratios.

OTHER LACUSTRINE ROCKS

Other lacustrine units in the Tugen Hills region include the Miocene (~15.5 Ma) Muruyur Beds (Pickford, 1988; Hill et al., 1991) and the Mpesida Beds (~7–6.3 Ma). M. Pickford (1977, personal communication) reported *Aulacoseira* from the latter unit. Diatomaceous silts and poorly lithified siltstones are also present between lava flows of the Kaparaina Basalts Formation (5.4–6.5 Ma) near Pelion (Figure 1). These sediments were deposited in local ponds or small temporary lakes. Their fossil diatom floras are dominated by *Fragilaria (=Staurosira) construens* and *F. construens* v. *venter*. Less common are *Fragilaria (=Pseudostaurosira) brevistriata, Synedra ulna, Cocconeis placentula,* and *Aulacoseira granulata.*

PALEOGEOGRAPHIC EVOLUTION OF THE TUGEN HILLS

The Miocene and Pliocene paleolakes of the Tugen Hills have tended to form, develop, and be destroyed as a consequence of volcanic and tectonic interaction associated with rifting (Bishop et al., 1971; Pickford, 1975a; Tiercelin and Vincent, 1987). Throughout the Miocene and Pliocene, the loci of these basins shifted in an easterly direction as rift extension took place (Chapman et al., 1978) (Figure 1B), moving progressively away from the Elgeyo border fault. A Miocene saline, alkaline Tambach lake (Ego, 1994; Renaut et al., 1999) formed adjacent to the Elgeyo escarpment in the western part of the rift. Sedimentation ceased when flood phonolites infilled the contemporary rift depression, but was renewed to the east by the formation of several fault-controlled lake basins in which the Muruyur Beds and the Ngorora Formation were deposited. The Ngorora paleolakes were ponded by lava flows and fault scarps that periodically led to hydrologic closure; consequently, the Ngorora paleolakes commonly became saline and alkaline, but with periods of fresh waters, as confirmed by their diatom floras. Phonolites again flooded the rift floor and destroyed the Ngorora lake basins.

During the latter part of the Miocene, the Kabarnet trachytes were erupted, with deposition of the Mpesida lake beds in the eastern Tugen Hills. The mainly freshwater Lukeino lake (Figure 1C) developed in the eastern Tugen Hills at about 6 Ma. The Kaparaina Basalts were erupted both contemporaneously with and after deposition of the Lukeino Formation sediments (Williams and Chapman, 1986), with temporary shallow lakes forming locally on lava flows, probably in fault-bounded basins. The Pliocene Chemeron Formation again demonstrates a further shift in position of lacustrine basins to the east (Figure 1). The Chemeron lake waters were mainly fresh. Major faulting again destroyed the lake basin and lacustrine sedimentation was renewed to the east, giving rise to the Kapthurin Formation, which also varied between saline, alkaline, and fresh (Renaut et al., 1999). The same pattern of lacustrine sedimentation continues in the modern axial rift depression, represented by freshwater Lake Baringo (Tiercelin and Vincens, 1987; Renaut et al., this volume) and saline, alkaline Lake Bogoria (Renaut and Tiercelin, 1994), which lies 22 km south of Lake Baringo.

This pattern of successive formation and destruction of small lakes that have lifespans of thousands to a few million years is typical for the volcanic regions of the eastern branch of the East African rift (mainly Kenya and Ethiopia). The short duration of the paleolakes is in marked contrast to the large, generally deep lake basins of the western branch, where volcanism is far less important (Yuretich, 1982). There, many of the lakes have long histories that span millions of years (Cohen et al., 1993).

The location of the small paleolakes of the Tugen Hills–Elgeyo region appears to have been controlled or strongly influenced by two dominant tectonic lineaments—the north-south trend of the modern (Tertiary–Holocene) rift and the northwest-southeast lineaments that are inherited from the Precambrian basement. The interplay of these two lineaments led to compartmentalization of the depositional basins, as well as controlling the location of central volcanoes and hydrothermal activity (cf. Le Turdu et al., 1998, in press). Similar segmentation under basement influence is present in the large deep rift lakes (e.g., Coussement et al., 1994; Wescott et al., 1996). Unlike the deep lakes of the western branch, volcanic rocks and tephra commonly filled many of the small basins in the Tugen Hills, periodically terminating lacustrine sedimentation. Faulting can lead to drainage diversion, which can have a major impact on hydrology and recharge of small basins. The compartmentalization also creates possibilities for hydrologic closure, given suitable (i.e., semi-arid) climatic conditions; consequently, the late Neogene depositional history of the Tugen Hills, recorded by diatoms and authigenic minerals, was one of frequent and commonly rapid changes in lake water chemistry, punctuated by periods of comparative stability when tectonic and volcanic activity were subdued and the climate remained relatively stable.

REFERENCES CITED

Bishop, W. W., and G. R. Chapman, 1970, Early Pliocene sediments and fossils from the northern Kenya Rift Valley: Nature, v. 226, p. 914–918.
Bishop, W. W., G. R. Chapman, A. Hill, and J. A.

Miller, 1971, Succession of Cainozoic vertebrate assemblages from the northern Kenya Rift Valley: Nature, v. 233, p. 389–394.

Bishop, W. W., and M. H. L. Pickford, 1975, Geology, fauna and palaeoenvironments of the Ngorora Formation, Kenya Rift Valley: Nature, v. 254, p. 185–192.

Chapman, G. R., 1971, The geological evolution of the northern Kamasia Hills, Baringo District, Kenya: Ph.D. dissertation, University of London, 360 p.

Chapman, G. R., and M. Brook, , 1978, Chronostratigraphy of the Baringo Basin, Kenya Rift Valley, *in* W. W. Bishop, ed., Geological background to fossil man; recent research in the Gregory Rift Valley, East Africa: Edinburgh, Scottish Academic Press, p. 207–223.

Chapman, G. R., S. J. Lippard, and J. E. Martyn, 1978, The structure of the Kamasia Range, Kenya Rift Valley: Journal of the Geological Society, London, v. 135, p. 265–281.

Cohen, A. S ., M. J. Soreghan, and C. A. Scholz, 1993, Estimating the age of ancient lakes: an example from Lake Tanganyika, East African Rift system: Geology, v. 21, p. 511–514.

Coussement, C., P. Gente, J. Rolet, J.-J. Tiercelin, M. Wafula, and S. Buku, 1994, The North Tanganyika hydrothermal fields: their tectonic control and relationship to volcanism and rift segmentation: Tectonophysics, v. 237, p. 155–173.

Deino, A., L. Tauxe, M. Monaghan, and R. Drake, 1990, $^{40}Ar/^{39}Ar$ age calibration of the litho- and paleomagnetic stratigraphies of the Ngorora Formation, Kenya: Journal of Geology, v. 98, p. 567–587.

Ego, J. K., 1994, Sedimentology and diagenesis of Neogene sediments in the Central Kenya Rift Valley: M.Sc. dissertation, University of Saskatchewan, 149 p.

Gregory, J. W., 1921, The rift valleys and geology of East Africa: Seeley Service, London, 479 p.

Hill, A., 1995, Faunal and environmental change in the Neogene of East Africa: evidence from the Tugen Hills sequence, Baringo District, Kenya, *in* E. S. Vrba, G. H. Denton, T. C. Partridge, and L. H. Burkle, eds., Paleoclimate and evolution with emphasis on human origins: New Haven, Yale University Press, p. 178–193.

Hill, A., A. K. Behrensmeyer, B. Brown, A. Deino, M. Rose, J. Saunders, S. Ward, and A. Winkler, 1991, Kipsaramon: a lower Miocene hominoid site in the Tugen Hills, Baringo District, Kenya: Journal of Human Evolution, v. 20, p. 67–75.

Hill, A., G. Curtis, and R. Drake, 1986, Sedimentary stratigraphy of the Tugen Hills, Baringo, Kenya, *in* L. E. Frostick, R. W. Renaut, I. Reid, and J. J. Tiercelin, eds., Sedimentation in the African Rifts: Geological Society Special Publication 25, p. 285–295.

Hill, A., R. Drake, L. Tauxe, M. Monaghan, J. C. Barry, A. K. Behrensmeyer, G. Curtis, B. Fine Jacobs, L. Jacobs, N. Johnson, and D. Pilbeam, 1985, Neogene palaeontology and geochronology of the Baringo Basin, Kenya: Journal of Human Evolution, v. 14, p. 759–773.

Kilham, P., S. S. Kilham, and R. E. Hecky, 1986, Hypothesized resource relationships among African planktonic diatoms: Limnology and Oceanography, v. 31, p. 1169–1181.

King, B. C., and Chapman, G. R., 1972, Volcanism of the Kenya Rift Valley: Philosophical Transactions of the Royal Society, London, Series A, v. 271, p. 185–208.

Kingston, J. D., B. D. Marino, and A. Hill, 1994, Isotopic evidence for Neogene hominid paleoenvironments in the Kenya Rift Valley: Science, v. 264, p. 955–959.

Le Turdu, C., J. J. Tiercelin, C. Coussement, J. Rolet, R. W. Renaut, J. P. Richert, J. P. Xavier, and D. Coquelet, 1995, Basin structure and depositional patterns interpreted using a 3D remote sensing approach: the Baringo-Bogoria basins, central Kenya Rift, East Africa: Bulletin des Centre de Recherche Exploration-Production Elf-Aquitaine, v. 19, p. 1–37.

Le Turdu, C., J.-J. Tiercelin, J. P. Richert, J. Rolet, J.-P. Xavier, R. W. Renaut, K. E. Lezzar, and C. Coussement, in press, Influence of pre-existing oblique discontinuities on the geometry and evolution of extensional fault patterns: evidence from the Kenya Rift using SPOT imagery: American Association of Petroleum Geologists Studies in Geology, v. 45.

Lippard, S. J., 1972, The stratigraphy and structure of the Elgeyo Escarpment, southern Kamasia Hills and adjoining areas, Rift Valley Province, Kenya: Ph.D. dissertation, University of London.

Martyn, J. E., 1967, Pleistocene deposits and new localities in Kenya: Nature, v. 215, p. 476–477.

Martyn, J. E., 1969, The geological history of the country between Lake Baringo and the Kerio River, Baringo District, Kenya: Ph.D. dissertation, University of London, 195 p.

McCall, G. J. H., B. H. Baker, and J. Walsh, 1967, Late Tertiary and Quaternary sediments of the Kenya Rift Valley, *in* W. W. Bishop and J. D. Clark, eds., Background to evolution in Africa: Chicago, University of Chicago Press, p. 191–220.

McClenaghan, M. P., 1971, Geology of the Ribkwo area, Baringo District, Kenya: Ph.D dissertation, University of London, 212 p.

Murray-Hughes, R., 1933, Notes on the geological succession, tectonics and economic geology of the western half of Kenya Colony: Report of the Geological Survey of Kenya, 3.

Nawamba, F. L., 1993, Tephrostratigraphy of the Chemeron Formation, Baringo Basin, Kenya: M.Sc. dissertation, University of Utah, 78 p.

Owen, R. B., 1981, Quaternary diatomaceous sediments and the geological evolution of Lakes Turkana, Baringo and Bogoria, Kenya Rift Valley: Ph.D. dissertation, University of London, 465 p.

Pickford, M. H. L., 1975a, Stratigraphy and palaeoecology of five late Cainozoic formations from the northern Kenya Rift Valley: Ph.D. dissertation, University of London, 456 p.

Pickford, M. H. L., 1975b, Late Miocene sediments and fossils from the northern Kenya Rift Valley: Nature, v. 256, p. 279–284.

Pickford, M. H. L., 1978a, Geology, palaeoenvironments and vertebrate faunas of the mid-Miocene Ngorora Formation, Kenya, *in* W. W. Bishop, ed., Geological background to fossil man; recent research in the Gregory Rift Valley, East Africa: Edinburgh, Scottish Academic Press, p. 237–262.

Pickford, M. H. L., 1978b, Stratigraphy and mammalian palaeontology of the late Miocene Lukeino Formation, Kenya, *in* W. W. Bishop, ed., Geological background to fossil man; recent research in the Gregory Rift Valley, East Africa: Edinburgh, Scottish Academic Press, p. 263–278.

Pickford, M. H. L., 1988, Geology and fauna of the middle Miocene hominoid site at Muruyur, Baringo District, Kenya: Journal of Human Evolution, v. 3, p. 381–390.

Renaut, R. W., J. Ego, J. J. Tiercelin, C. Le Turdu, and R. B. Owen, 1999, Saline, alkaline palaeolakes of the Tugen Hills-Kerio Valley region, Kenya Rift Valley, *in* P. Andrews and P. Banham, eds., Late Cenozoic environments and hominid evolution: a tribute to Bill Bishop: London, Geological Society Publications, p. 41–58.

Renaut, R. W., and Tiercelin, J.-J., 1994, Lake Bogoria, Kenya Rift Valley: a sedimentological overview, *in* R. W. Renaut and W. M. Last, eds., Sedimentology and geochemistry of modern and ancient saline lakes: Society for Sedimentary Geology (SEPM) Special Publication 50, p. 101–123.

Servant, M., 1973, Séquences continentales et variations climatiques: évolution du bassin du Tchad au Cenozoic Superieur: Thèse, Université de Paris, 348 p.

Shackleton, R. M., 1951, A contribution to the geology of the Kavirondo Rift Valley: Quarterly Journal of the Geological Society, v. 106, p. 345–392.

Tallon, P. W. J., 1978, Geological setting of hominid fossils and Acheulian artefacts from the Kapthurin Formation, Baringo District, Kenya, *in* W. W. Bishop, ed., Geological background to fossil man: Edinburgh, Scottish Academic Press, p. 361–378.

Tauxe, L., M. Monaghan, R. Drake, G. Curtis, and H. Staudigel, 1985, Paleomagnetism of Miocene East African rift sediments and the calibration of the geomagnetic reversal time scale: Journal of Geophysical Research, v. 90, p. 4639–4646.

Tiercelin, J. J., and A. Vincens, eds., 1987, Le demi-graben de Baringo-Bogoria, Rift Gregory, Kenya: Bulletin des Centres de Recherche Exploration-Production Elf-Aquitaine, v. 11, p. 249–540.

Wescott, W. A., C. K. Morley, and F. M. Karanja, 1996, Tectonic controls on the development of rift-basin lakes and their sedimentary character: examples from the East African Rift System, *in* T. C. Johnson and E. O. Odada, eds., The limnology, climatology and paleoclimatology of the East African lakes: Amsterdam, Gordon and Breach, p. 3–21.

Williams, L. A. J., and G. R. Chapman, 1986, Relationship between major structures, salic volcanism and sedimentation in the Kenya Rift from the equator northwards to Lake Turkana, *in* L. E. Frostick, R. W. Renaut, I. Reid, and J. J. Tiercelin, eds., Sedimentation in the African Rifts: Geological Society Special Publication 25, p. 59–74.

Yuretich, R. F., 1982, Possible influences upon lake development in the East African rift valleys: Journal of Geology, v. 90, p. 329–327.

Yang, H., 2000, The Shanwang Basin (Niocene) in Shandong Province, eastern China, *in* E. H. Gierlowski-Kordesch and K. R. Kelts, eds., Lake basins through space and time: AAPG Studies in Geology 46, p. 473–480.

Chapter 43

The Shanwang Basin (Miocene) in Shandong Province, Eastern China

Hong Yang
Department of Science and Technology, Bryant College
Smithfield, Rhode Island, U.S.A.

INTRODUCTION

The Shanwang Basin (36°30′N, 118°43′E) is located 22 km east of Linqu Shandong Province, eastern China (Figure 1). The Shandong peninsula is on the Sino-Korean platform, which stabilized during the Proterozoic (Yang et al., 1986; Ren et al., 1987). The Tancheng-Lujiang fault zone, a shearing deep fracture system extending more than 1000 km in a north–northeast-south–southwest direction in eastern China, is east of the Shanwang area (Zhao et al., 1983; Jin, 1985; Ren et al., 1987) (see Figure 1). The Miocene Shanwang lacustrine basin was one of the major Tertiary diatomite mining districts in China until 1987; however, the excellent fossil material associated with the diatomite is now the major interest of this lacustrine basin. The lake deposits contain a highly diversified and well-preserved Miocene fossil biota that has provided important paleontologic and paleoenvironmental information since the site was discovered in the early 1930s. The Chinese government has declared a 1.5 km^2 area in the Shanwang Basin as a National Major Natural Protection Area, and the Shanwang fossil site is also on the UNESCO World Heritage List.

GEOLOGIC SETTING

The Shanwang Basin is one of the small volcanic-related lake basins within the Tertiary lake basin system in the Shandong Province. The Shanwang Miocene lake rested on a round-shaped lava depression created by the eruptions of the Niushan Formation basalts (Yang, 1993); however, Li and Zhao (1988) have proposed that the ancient lake originated from the infilling of a small volcanic crater. Figure 2 summarizes the generalized stratigraphy of the basin (Young, 1936a; Yan et al., 1983; Yang, 1988, 1993; Li, 1991; Yang and Yang, 1994). The Tertiary olivine tholeiite (the Niushan Formation) is composed of the lake basin floor and overlies Cretaceous pyroclastic rocks and Precambrian metamorphic rocks (the Taishan Group) (Wang et al., 1981; Zhao et al., 1983; Wang, 1986). The Shanwang Formation, deposited as the main body of the Shanwang Miocene lacustrine system, consists of breccia, sandstone, diatomite, mudstone, volcanic ash, phosphatic nodules, and interbedded basalt. The diatomaceous shale hosts a diverse soft-tissued fossil biota (Yang and Yang, 1994). The Yaoshan Formation, consisting mainly of Tertiary alkalic olivine basalt, tops the Miocene sequence in the study area (Wang, 1986; Zhi, 1990).

The radiometric dates ($^{40}K/^{40}Ar$ method) of basalt samples from the Niushan Formation and the Yaoshan Formation bracket the Shanwang Formation at between 10 and 16 Ma (Wang and Jin, 1985; Chen and Peng, 1985; Jin, 1985; Zhu et al., 1985). Mammalian fossils indicate that the Shanwang fossil deposit was likely formed between 15.5 and 17 m.y. ago (Qiu, 1990; Yang, 1993).

SHANWANG FORMATION: THE LACUSTRINE DEPOSITS

The Shanwang Formation represents the main deposition of the Shanwang Miocene lake. Informally, the Shanwang Formation can be subdivided into three parts (lower, middle, and upper) (Figure 2). The lower part (25 m) consists of yellowish sandstone and breccia, and unconformably overlies the Niushan Formation basalt. The poorly sorted clasts in the breccia are supported by coarse sandstone matrix, and most clasts (up to 90 cm in diameter) have the same composition as the Niushan Formation basalt below. The coarse sandstone contains lamination, cross-bedding, and isolated Miocene vertebrate fossils. The lower part of the formation has been interpreated as lake shore or braided channel deposits (Yan et al., 1983; Yang et al., 1991). The middle part of the Shanwang Formation (30 m) consists of pale to yellowish gray diatomaceous earth with interbedded thin volcanic ash, phosphatic nodules, and yellowish claystone. Both the diatomaceous earth and the phosphatic nodules are commercially minable ore deposits. The ore deposits (about 30 m in thickness) contain up to 70% diatoms that are mixed with clay minerals (Juan, 1937; Zhang, 1982). Rich and well-preserved Miocene fossils with soft tissues have

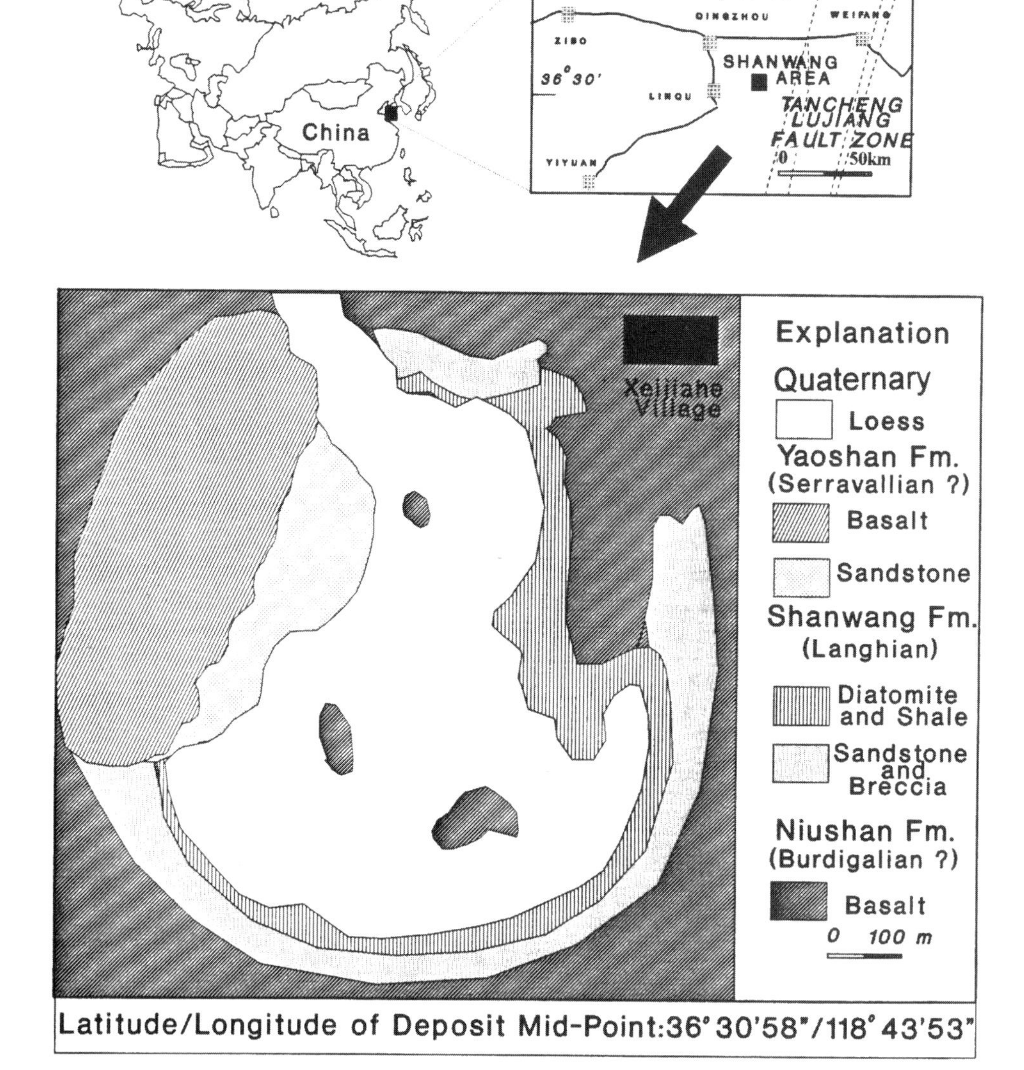

Figure 1—Map showing the location of the Shanwang area (upper left map), relative position to the Tancheng–Lujiang fault system (dash lines in upper right map), and outline and lithofacies of the Shanwang Miocene basin (map below).

been reported from the diatomaceous earth. The upper part of this formation (47 m) consists of an oil shale, a layer of brownish peat, and two layers of basalt flows (olivine basalt) (Liu and Shi, 1989).

The Shanwang Formation contains five distinctive lithofacies (Yang, 1993). The sandstone at the bottom of the formation contains isolated vertebrate bones and teeth and was interpreted as near-shore lacustrine deposits with possible fluvial influence (Yang et al., 1991). The laminated diatomaceous earth, phosphatic nodules, and shales are continuous across the basin and represent deposits in a low-energy, shallow lacustrine setting. They contain rich aquatic plant, pollen, fish, insect, and vertebrate fossils. The interbedded thin basalt layers and volcanic ash indicate local volcanic eruptions during the lake deposition.

The diatom blooms, abundant aquatic plants, and phosphatic deposits indicate a eutrophic water column for the ancient Miocene lake. The water column of the Shanwang ancient lake was probably too shallow (less than 10 m based on the study of freshwater aqueous plants) to be stratified (Yang, 1988, 1993). A substantial portion of the ancient lake sediments were endogenic particles produced by biological activities. Preservation of delicate biological organs indicates a low-energy environment in the lake, and the diatom assemblage suggests a nearly neutral pH for the lake water (Li, J., 1982; Yang, 1993). A warm-temperate climate with mild winters, but distinct seasonality, is indicated for the Miocene Shanwang area by both fossil assemblages and sediment types; the weather was similar to that of the present-day lower Yangtze River (Changjiang) valley in southeastern China or in the southeastern United States (Li, H., 1982; Yang, 1988, 1993, 1996).

RECORDED PALEOBIOTA IN THE LACUSTRINE DEPOSITS

The diverse Shanwang paleobiota in diatomite deposits has yielded more than 500 fossil species from 11 major fossil groups (including trace fossils) (Figure 3A) (Yang and Yang, 1994). Taxonomy of the following fossil groups has been documented and refined since

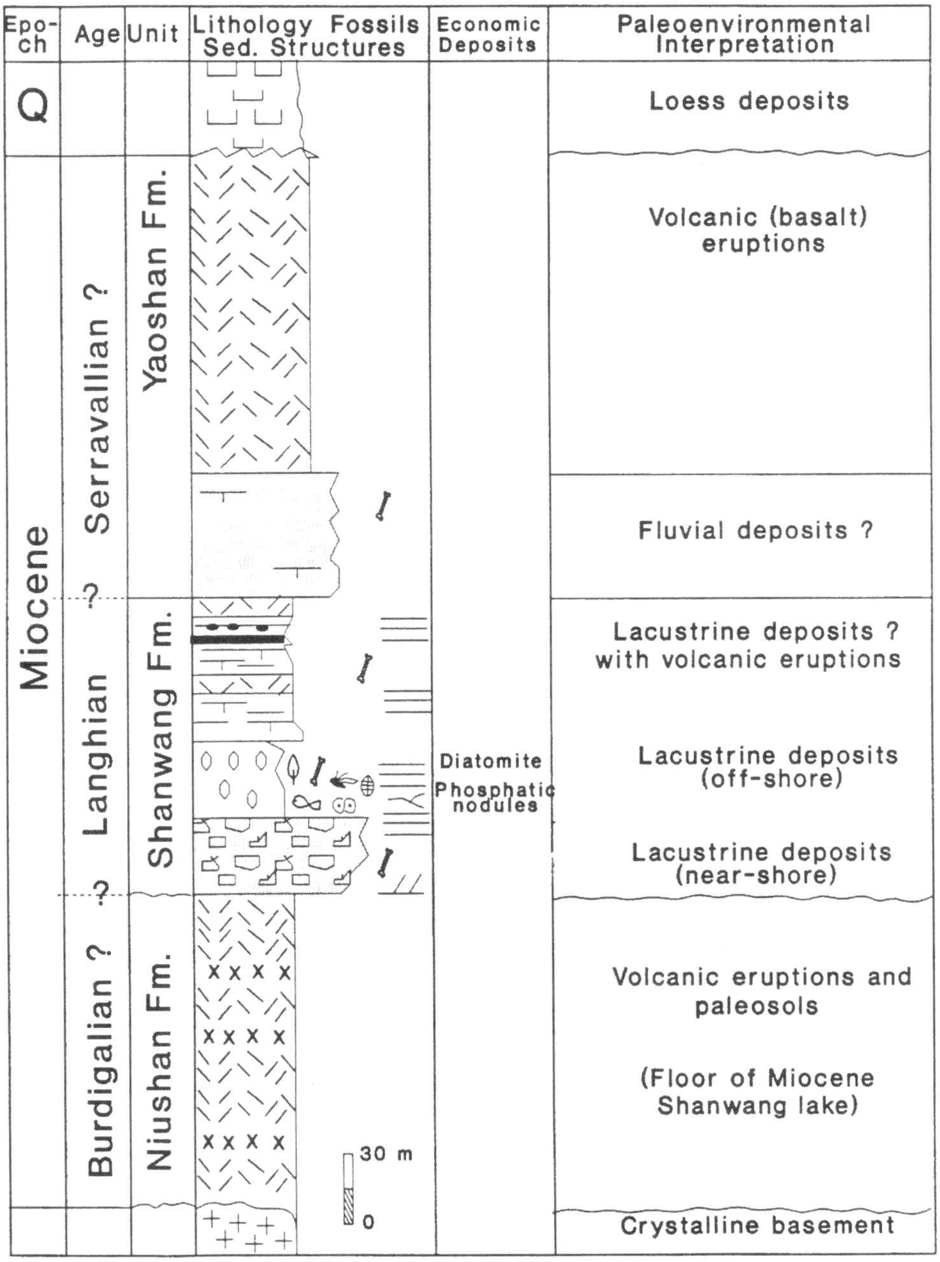

Figure 2—Generalized stratigraphic sequence of the Shanwang Basin showing chronostratigraphy, lithostratigraphic units, and facies; note the sandstone and fossiliferous diatomaceous units in the lower part of Shanwang Formation.

the 1930s: fungal fossils (Wang, 1991), diatoms (Skvortzov, 1937; Li, J., 1982; Shi, 1990), spores and pollen (Sung, 1959; Song et al., 1964; Wang, 1981; Liu, 1986; Liu and Leopold, 1992), plant megafossils (Hu and Chaney, 1940; Sze, 1951; Academia Sinica, 1978; Li, H., 1982; Li and Zheng, 1986; Yang, 1988), insects (Hong, 1979, 1983, 1985; Zhang, 1986, 1989, 1990), ostracodes (Zheng, 1986), fish (Young and Tchang, 1936; Zhou, 1990), amphibians (Young, 1936b; Yang, 1977; Gao, 1986), reptiles (Sun, 1961; Young, 1965; Li and Wang, 1987), birds (Yeh, 1977, 1980, 1981; Yeh and Sun, 1984), mammals (Young, 1937; Teilhard de Chardin, 1939; Hu, 1957; Wang, 1965; Li, 1974; Zhou and Shih, 1978; Qiu, 1981; Xie, 1982; Yan, 1983; Yan et al., 1983; Qiu et al., 1985a, 1986; Qiu and Sun, 1988; Qiu, 1990), and trace fossils (Guo, 1991).

Both plant megafossils and microfossils found in the Shanwang Formation are comparable to middle Miocene floras in other parts of China, Japan, and North America (Hu and Chaney, 1940; Academia Sinica, 1978; Li, H., 1982; Yang, 1988). Abundant mammalian fossils from the site provide detailed biochronological correlations among mammalian stages of Shanwang, Europe, and North America. The Shanwang fauna is believed to be comparable to the European early middle Miocene mammalian stage (MN5, 15.5–17 Ma, Langhian) (Wu and Chen, 1978; Yan et al., 1983; Qiu, 1990).

The Shanwang fossil biota is an excellent example of lacustrine *Konservat-Lagerstatten*. Fine preservation of fossil material has yielded nonmineralized organic tissues of animals and plants, detailed morphology of delicate organs, original coloration of plants and

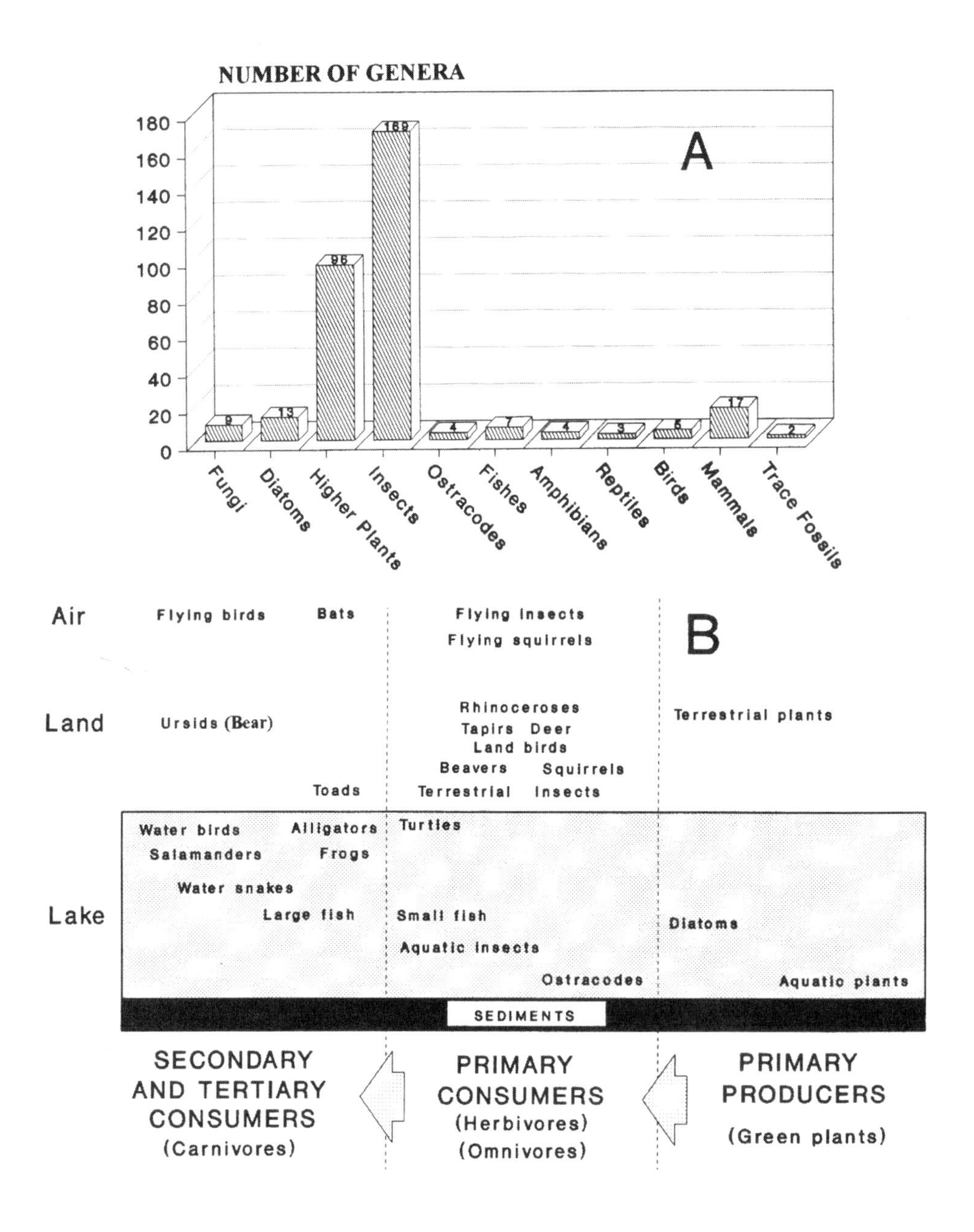

Figure 3—(A) Taxonomic diversity (at the genus level) of the Shanwang fossil biota. (B) Schematic trophic structure for the Shanwang fossil biota represented by recorded fossils found in different habitats.

insects, and completely articulated animal bodies with in situ teeth, skin, and stomach contents. A detailed description of the taphofacies can be found in Yang (1993) and Yang and Yang (1994). The lacustrine deposit recorded a broad ecologic spectrum of Miocene organisms that lived within and around the ancient lake. Figure 3B is a conceptual model intending to show the complex relationship among Miocene organisms co-occuring at the site. The trophic implications of the ancient lake are based on analysis of taphonomic conditions and functional morphology of fossil material, and comparison of modern organisms in similar lacustrine settings (Yang, 1993). Diatoms and aquatic plants, joined by fragments of terrestrial plants, served as primary producers in the lake. The aquatic insects, small fish, and turtles were primary consumers and fed on plant detritus in the nutrient-rich shallow water. The Shanwang fauna apparently lacked a large benthic scavenger community. Benthic invertebrate are represented by only a few ostracodes. Small carnivorous animals in the water column, such as large fish, frogs, and salamanders, made up the secondary consumers. The highest trophic level in the lake was represented by water snakes, water birds, and alligators.

The terrestrial trophic structure of the Miocene Shanwang area around the lake was also well stratified. The diverse Shanwang flora served as primary producers on land near the lake, providing suitable food and habitats for various herbivores. Besides a diverse insect fauna, different deer populations acted as major herbivores living immediately around the lake basin despite other primary consumers such as land birds, squirrels, beavers, tapirs, and rhinoceroses that also were heavily dependent on the higher plants. At the secondary consumer level, toads, bats, and birds, either on the forest floor or at the forest canopy level, fed on the diverse insect fauna. The highest trophic level in the terrestrial paleoenvironment, the tertiary consumers (top carnivores), consisted of large ursids preying on other mammals. Various amphibians, reptiles, and birds also preyed on both land and water.

REFERENCES CITED

(Please note the words "Shantung" and "Shandong" refer to the same place due to change of spelling system in early 1970s.)

Academia Sinica, 1978, Cenozoic Plants of China. Scientific Publishing House, Beijing, 232 p. (in Chinese).

Chen, D-g., and Z-c. Peng, 1985, K-Ar ages and Pb, Sr isotopic characteristics of Cenozoic volcanic rocks in Shandong, China. Geochemica, v. 4, p. 303–311 (in Chinese with English abstract).

Gao, K-q., 1986, Middle Miocene Pelobatidae fossils from Linqu, Shandong with a revision of toad fossils. Vertebrata PalAsiatica, v. 24, p. 63–71 (in Chinese with English abstract).

Guo, S-x., 1991, A Miocene trace fossil of insect from Shanwang Formation in Linqu, Shandong. Acta Palaeontologica Sinica, v. 30, p. 739–742 (in Chinese with English abstract).

Hong, Y-c., 1979, *Oxycephala* gen. nov. a Miocene Homoptera (Insecta) from Linqu of Shandong. Acta Palaeontologica Sinica, v. 18, p. 301–307 (in Chinese with English abstract).

Hong, Y-c., 1983, Fossil insects in the diatoms of Shanwang. Bulletin of Tianjin Geologic Institute, v. 8, p. 1–15 (in Chinese with English abstract).

Hong, Y-c., 1985, Discovery of Miocene scorpius from the diatoms of Shanwang in Shandong Province. Bulletin of Tianjin Geologic Institute, v. 8, p. 17–21 (in Chinese with English abstract).

Hu, C-k., 1957, An antler fragment of *Stephanocemas* and some teeth of *Aceratherium* from Linchu, Shantung. Vertebrata PalAsiatica, v. 1, p. 163–166 (in Chinese with English abstract).

Hu, H. H., and R. W. Chaney, 1940, A Miocene Flora from Shantung Province, China. Palaeontologia Sinica, New Series A, p. 1–112.

Jin, L-y., 1985, K-Ar ages of Cenozoic volcanic rocks in the middle segment of the Tancheng-Lujiang fault zone and stages of related volcanic activity. Geological Review, v. 31, p. 309–315 (in Chinese with English abstract).

Juan, V. C., 1937, Diatomaceous earth in Shanwang, Linchu, Shantung. Bulletin of Geological Society of China, v. 17, p. 183–187.

Li, C-k., 1974, A probable geomyoid rodent from Middle Miocene of Linchu, Shantung. Vertebrata PalAsiatica, v. 12, p. 43–53 (in Chinese with English abstract).

Li, F-l., 1991, Reconsiderations of the Shanwang Formation in Linqu, Shandong. Journal of Stratigraphy, v. 15, p. 123–129 (in Chinese with English abstract).

Li, H-m., 1982, The age of the Shanwang flora, Shandong, *in* Selected Papers from the 12th Annual Convention of Palaeontological Society of China. Scientific Publishing House, Beijing, p. 159–162 (in Chinese with English abstract).

Li, H-m., and Y-h. Zheng, 1986, Some fossil catkins and a male cone with their pollen in situ from Shanwang flora in Shandong. Shandong Palaeontology, v. 1, p. 42–46 (in Chinese with English abstract).

Li, J-l., and B-z. Wang, 1987, A new species of *Alligator* from Shanwang, Shandong. Vertebrata PalAsiatica, v. 25, p. 199–207 (in Chinese with English abstract).

Li, J-y., 1982, Miocene diatom assemblage of Shanwang, Shandong Sheng. Acta Botanica Sinica, v. 24, p. 456–467 (in Chinese with English abstract).

Li, T-l., and T. Zhao, 1988, Discovery of Shanwang palaeovolcanic orifice and its significance for the preservation of fossils. Geoscience, v. 2, p. 489–494 (in Chinese with English abstract).

Liu, G-w., 1986, A Late Tertiary palynological assemblage from the Yaoshan Formation of Shanwang, Linqu County, Shandong. Acta Palaeobotanica and Palynologica Sinica, v. 1, p. 65–84 (in Chinese with English abstract).

Liu, G-w., and E. B. Leopold, 1992, Paleoecology of a Miocene flora from the Shanwang Formation, Shandong Province, northern China. Palynology, v. 16, p. 187–212.

Liu, H-f., and N. Shi, 1989, Paleomagnetic study of Shanwang Formation, Shandong Province. Acta Sinentiarum Naturalium Universitatis Pekinensis, v. 25, p. 576–593 (in Chinese with English abstract).

Qiu, Z-d., 1981, A new sciuroptere from Middle Miocene of Linqu, Shandong. Vertebrata PalAsiatica, v. 19, p. 228–238 (in Chinese with English abstract).

Qiu, Z-d., and B. Sun, 1988, New fossil micromammals from Shanwang, Shandong. Vertebrata PalAsiatica, v. 26, p. 50–58 (in Chinese with English abstract).

Qiu, Z-x., 1990, The Chinese Neogene mammalian biochronology—Its correlation with European Neogene mammalian zonation, *in* E. H. Lindsay, V. Fahlbusch, and P. Mein, eds., European Neogene Mammal Chronology. Plenum Press, New York, p. 627–556.

Qiu, Z-x., Yan, D-f., Jia, H., and B. Sun, 1985a, Preliminary observations on the newly found skeletons of *Palaeomeryx* from Shanwang, Shandong. Vertebrata PalAsiatica, v. 23, p. 173–195 (in Chinese with English abstract).

Qiu, Z-x., Yan, D-f., Jia, H., and B-z. Wang, 1985b, Dentition of the Ursavus skeleton from Shanwang, Shandong Province. Vertebrata PalAsiatica, v. 23, p. 264–275 (in Chinese with English abstract).

Qiu, Z-x., Yan, D-f., Jia, H., and B. Sun, 1986, The large-sized ursid fossils from Shanwang, Shandong. Vertebrata PalAsiatica, v. 24, p. 182–194 (in Chinese with English abstract).

Ren, J., Jian, C., Zhang, Z., and D. Qin, 1987, Geotectonic Evolution of China. Science Press Beijing, Springer-Verlag, Berlin, Heidelberg, New York, London, Tokyo, 203 p.

Shi, L., 1990, Miocene diatoms from the Shanwang basin of Shandong Province and analysis by fuzzy mathematics of the paleoenvironment. Acta Botanica Sinica, v. 32, p. 888–895 (in Chinese with English abstract).

Skvortzov, B. V., 1937, Neogene diatoms from eastern Shantung. Bulletin of Geological Society of China,

v. 17, p. 193–208.
Song, Z-c., Tsao, L., and M-y. Li, 1964, Tertiary spore-pollen complex of Shantung. Memoir of Institute of Geology and Paleontology (Academia Sinica), v. 3, p. 177–190 (in Chinese with English abstract).
Sun, A-l., 1961, Notes on fossil snakes from Shanwang, Shantung. Vertebrata PalAsiatica, v. 4, p. 307–312 (in Chinese with English abstract).
Sung, T-s., 1959, Miocene spore-pollen samples of Shanwang, Shantung. Acta Palaeontologica Sinica, v. 7, p. 99–109 (in Chinese with English abstract).
Sze, X-j, 1951, Review of H. H. Hu and R. W. Chaney's "A Miocene Flora from Shantung Province, China." Science Sinica, v. 2, p. 245–252 (in Chinese).
Teilhard de Chardin, P., 1939, The Miocene cervids from Shantung. Bulletin of Geological Society of China, v. 19, p. 269–278.
Wang, B-y., 1965, New vertebrate fossils from Shanwang, Linqu, Shandong. Vertebrata PalAsiatica, v. 9, p. 109–118 (in Chinese with English abstract).
Wang, F-z., 1986, The characteristics of mineral chemistry and origin of rock-forming minerals in volcanic rocks on Shanwang, Linju of Shandong Province. Acta Petrologica Sinica, v. 2, p. 41–54 (in Chinese with English abstract).
Wang, F-z., and L-y. Jin, 1985, Petrological and geochemical characteristics of Cainozoic volcanic rocks of Shanwang, Shandong. Bulletin of Minerology and Petrology of China, v. 2, p. 43–65 (in Chinese with English abstract).
Wang, H-f., Zhu, B-q., Zhang, Q-f., Fan, C-y., and L-m. Dong, 1981, A study on K-Ar isotopic ages of Cenozoic basalts from Linqu area, Shandong Province. Geochimica, v. 4, p. 321–328 (in Chinese with English abstract).
Wang, X-z., 1981, An introductory survey of the paleosurroundings of Miocene Shanwang lake in Linqu, Shantung Province. Beijing University Bulletin, v. 4, p. 100–110 (in Chinese with English abstract).
Wang, X-z., 1991, The discovery and significance of the fungal spores from Shanwang Formation in Linqu, Shandong Province, *in* Beijing University, ed., Lithospheric Geoscience. Beijing University Press, Beijing, p. 19–25 (in Chinese with English abstract).
Wu, W-y., and G-f. Chen, 1978, Vertebrate fossil from Shanwang of Shandong Province and a review of the Miocene mammalian fossil of China. Proceedings of Palaeontological Society of China, v. 14, p. 25–41 (in Chinese).
Xie, W-m., 1982, New discovery on aceratherine rhinoceros from Shanwang in Linchu, Shandong. Vertebrata PalAsiatica, v. 20, p. 133–137 (in Chinese with English abstract).
Yan, D-f., 1983, The morphology and classification of *Plesiaceratherium*. Vertebrata PalAsiatica, v. 21, p. 134–143 (in Chinese with German abstract).
Yan, D-f., Qiu, Z-d., and Z-y. Meng, 1983, Miocene stratigraphy and mammals from Shanwang, Shandong. Vertebrata PalAsiatica, v. 21, p. 211–221 (in Chinese with English abstract).
Yang, C-j., 1977, On some Salientia and Chiroptera from Shanwang, Linqu, Shandong. Vertebrata PalAsiatica, v. 15, p. 76–79 (in Chinese with English abstract).
Yang, H., 1988, The Miocene Shanwang flora and its paleoclimatology from Shandong, China. M.S. Thesis, China University of Geosciences, Beijing, 65 p. (in Chinese with English summary).
Yang, H., 1996, Comparison of Miocene fossil floras in lacustrine deposits: implications for paleoeclimatic interpretations at the middle latitudes of the Pacific rim. Palaeobotanist, v. 45, p. 416–429.
Yang, H., 1993, Miocene Lake Basin Analysis and Comparative Taphonomy: Clarkia (Idaho, U.S.A.) and Shanwang (Shandong, P.R. China). Ph.D. Dissertation, University of Idaho, Moscow, Idaho, 272 p.
Yang, H., and S-p. Yang, 1994, The Shanwang fossil biota in eastern China: A Miocene *Konservat-Lagerstatte* in lacustrine deposits: Lethaia, v. 27, p. 345–354.
Yang, M-h., Li, F-l., and Z-h. Luo, 1991, The discovery of volcanic agglomerate and braided channel deposits of Shanwang Formation, Shandong and its significance. Geoscience, v. 5, p. 273–278 (in Chinese with English abstract).
Yang, Z-y., Chen, Y-c., and H-z. Wang, 1986, The Geology of China. Clarendon Press, Oxford, 303 p.
Yeh, H-k., 1977, First discovery of Miocene bird in China. Vertebrata PalAsiatica, v. 15, p. 244–248 (in Chinese with English abstract).
Yeh, H-k., 1980, Fossil birds from Linqu, Shandong. Vertebrata PalAsiatica, v. 18, p. 116–125 (in Chinese with English abstract).
Yeh, H-k., 1981, Third note on fossil bird from Miocene of Linqu, Shandong. Vertebrata PalAsiatica, v. 19, p. 149–155 (in Chinese with English abstract).
Yeh, H-k., and B. Sun, 1984. New materials of fossil Phasianid bird from Linqu, Shandong. Vertebrata PalAsiatica, v. 22, p. 208–212 (in Chinese with English abstract).
Young, C. C. (Yang, C-j.), 1936a, On the Cenozoic geology of Itu, Changlo and Linchu districts (Shantung). Bulletin of Geological Society of China, v. 15, p. 171–188.
Young, C. C. (Yang, C-j.), 1936b, A Miocene fossil frog from Shantung. Bulletin of Geological Society of China, v. 15, p. 189–196.
Young, C. C. (Yang, C-j.), 1937, On a Miocene mammalian fauna from Shantung. Bulletin of Geological Society of China, v. 17, p. 209–238.
Young, C. C. (Yang, C-j.), 1965, On the first occurrence of the fossil salamanders from the upper Miocene of Shantung, China. Acta Palaeontologica Sinica, v. 13, p. 455–459 (in Chinese with English abstract).
Young, C. C. (Yang, C-j.), and T.L. Tchang, 1936, Fossil fishes from the Shanwang Series of Shantung. Bulletin of Geological Society of China, v. 15, p. 197–205.
Zhang, F-f., 1982, Geologic characteristics of the diatomite ore deposits in Linqu, Shandong. Geological Reconstruction Materials, v. 4, p. 24–29 (in Chinese).
Zhang, J-f., 1986, Ecological and biogeographical analysis on Miocene Shanwang insect fauna, *in* Palaeontological Society of China, ed., Selected Papers from the 13th and 14th Annual Conventions of Palaeontological Society of China. Anhui Science and Technology

Publishing House, Hefei, p. 237–246 (in Chinese with English abstract).
Zhang, J-f., 1989, Fossil Insects from Shanwang, Shandong, China. Shandong Science Press, Jinan, 394 p. (in Chinese with English abstract).
Zhang, J-f., 1990, New fossil species of Apoidea (Insecta: Hymenoptera). Acta Zootaxonomica Sinica, v. 15, p. 83–90 (in Chinese with English abstract).
Zhao, D-s., Xiao, Z-y., and Y-t. Wang, 1983, Petrological characteristics and genesis of the Cenozoic volcanic rocks in Tancheng-Lujiang fault belt and the adjacent areas. Acta Geologica Sinica, v. 2, p. 129–141 (in Chinese with English abstract).
Zheng, S-y., 1986, Miocene ostracodes from Shanwang, Shandong. Shandong Palaeontology, v. 1, p. 47–53 (in Chinese with English abstract).
Zhi, X-c., 1990, Trace element geochemistry and petrogenesis of Cenozoic alkalic basalts from the Penglai and Lingju areas. Shandong Province Geological Review, v. 36, p. 385–393 (in Chinese with English abstract).
Zhou, B-x., and M-z. Shih, 1978, A skull of Lagomeryx from Middle Miocene of Linchu, Shantung. Vertebrata PalAsiatica, v. 16, p. 111–122 (in Chinese with English abstract).
Zhou, J-j., 1990, The Cyprinidae fossils from Middle Miocene of Shanwang basin. Vertebrata PalAsiatica, v. 28, p. 95–127 (in Chinese with English abstract).
Zhu, M., Hu, H-g., Zhao, D-z., Liu, S-l., Hu, X-z., Ma, Z-q., and W-y. Jian, 1985, Potassium-Argon dating of Neogene basalt in Shanwang area, Shandong Province. Petrological Research, v. 5, p. 47–59 (in Chinese with English abstract).

Tanner, L. H., 2000, Miocene–Pliocene lacustrine and marginal lacustrine sequences of the Furnace Creek Formation, Furnace Creek and Death Valley Basins, Death Valley region, U.S.A., *in* E. H. Gierlowski-Kordesch and K. R. Kelts, Lake basins through space and time: AAPG Studies in Geology 46, p. 481–490.

Chapter 44

Miocene–Pliocene Lacustrine and Marginal Lacustrine Sequences of the Furnace Creek Formation, Furnace Creek and Central Death Valley Basins, Death Valley Region, U.S.A.

Lawrence H. Tanner
Department of Geography and Earth Science, Bloomsburg University
Bloomsburg, Pennsylvania, U.S.A.

INTRODUCTION

The Furnace Creek Basin is a northwest-southeast oriented structural trough, bordered to the northeast by the Furnace Creek fault zone (FCFZ), which separates the basin from the Funeral Mountains, and to the southwest by the Greenwater Range and the Grand View fault, which borders the Black Mountains (Figure 1). Subsidence of the Furnace Creek Basin was controlled by movement on the pre-existing dextral strike-slip FCFZ beginning in the Miocene, coincident with regional extension (Cemen et al., 1985). The Central Death Valley basin is a pull-apart basin with the form of an east-tilting half-graben (Ellis and Trexler, 1991) formed largely by dip-slip movement along the Death Valley fault zone concurrent with northwest extension starting in the Miocene (Christie-Blick and Biddle, 1985; Cemen and Wright, 1994; Wright, 1994).

Tertiary strata of these two basins comprise three formations: the Miocene Artist Drive Formation, Miocene–Pliocene Furnace Creek Formation (FCF), and the Pliocene Funeral Formation (Figure 2). The Artist Drive Formation is an assemblage of variegated conglomerates, sandstones, and minor siltstones and claystones with a total thickness of 1300 m (McAllister, 1970; Cemen et al., 1985; Cemen and Wright, 1988). Up to 2100 m of FCF conglomerates, basalt flows and breccias, sandstones, mudstones, claystones, marls, and borates unconformably overlie the Artist Drive Formation. Dates of 7.5 Ma for tuff in the upper Artist Drive Formation (Cemen and Wright, 1988) and 4.03 $\pm$ 0.12 Ma for a basalt flow in the overlying Funeral Formation (Cemen et al., 1985) partially constrain the age of the FCF, which spans the Miocene–Pliocene boundary established by the Decade of North American Geology (DNAG). Diatoms of early Pliocene age have been collected from the lower part of the FCF (McAllister, 1970). The FCF extends beyond the Furnace Creek fault zone, onlapping the southern end of the Funeral Mountains (Figure 1). The FCF underlies a total area of approximately 1600 km^2 in the combined Furnace Creek and Central Death Valley Basins (Barker and Barker, 1985). In the Furnace Creek Basin, the FCF is well exposed along much of the length of Furnace Creek Wash and in canyons incised into the formation from the floor of Death Valley. Quaternary sediments now cover much of the floor of Death Valley, but outcrops of the FCF occur in the Central Death Valley basin along the western flank of the central Funeral Mountains and near the north end of the basin in the Salt Creek Hills (Figure 1) (Wright and Troxel, 1973). The FCF is overlain by the Funeral Formation, comprising well-organized to disorganized conglomerates, sandstones, and interfingering basalt flows totaling up to 700 m thickness. Clast composition indicates reworking of the underlying Furnace Creek Formation and older units (Cemen et al., 1985). In the southeastern Furnace Creek Basin, the Funeral Formation consists almost entirely of basalt flows up to 160 m thick.

LITHOSTRATIGRAPHY OF THE FURNACE CREEK FORMATION

The basal unit of the FCF in the Furnace Creek Basin is defined by McAllister (1970) as the lower conglomeratic unit (Tfc), subdivided by Savoca and Cemen (1989) into three subunits on the basis of clast composition and matrix. The lower conglomeratic unit interfingers and grades vertically into the laminated marlstones and mudstones that comprise the main body of the formation (Figure 3) (unit Tf of McAllister, 1970). Basalt flows and breccias (units Tfb and Tfa of McAllister, 1970) interfinger with the mudstones and marlstones of the main member. These volcanics reach considerable thickness in the northern Black Mountains, and Greene (1997) elevates this unit to formation

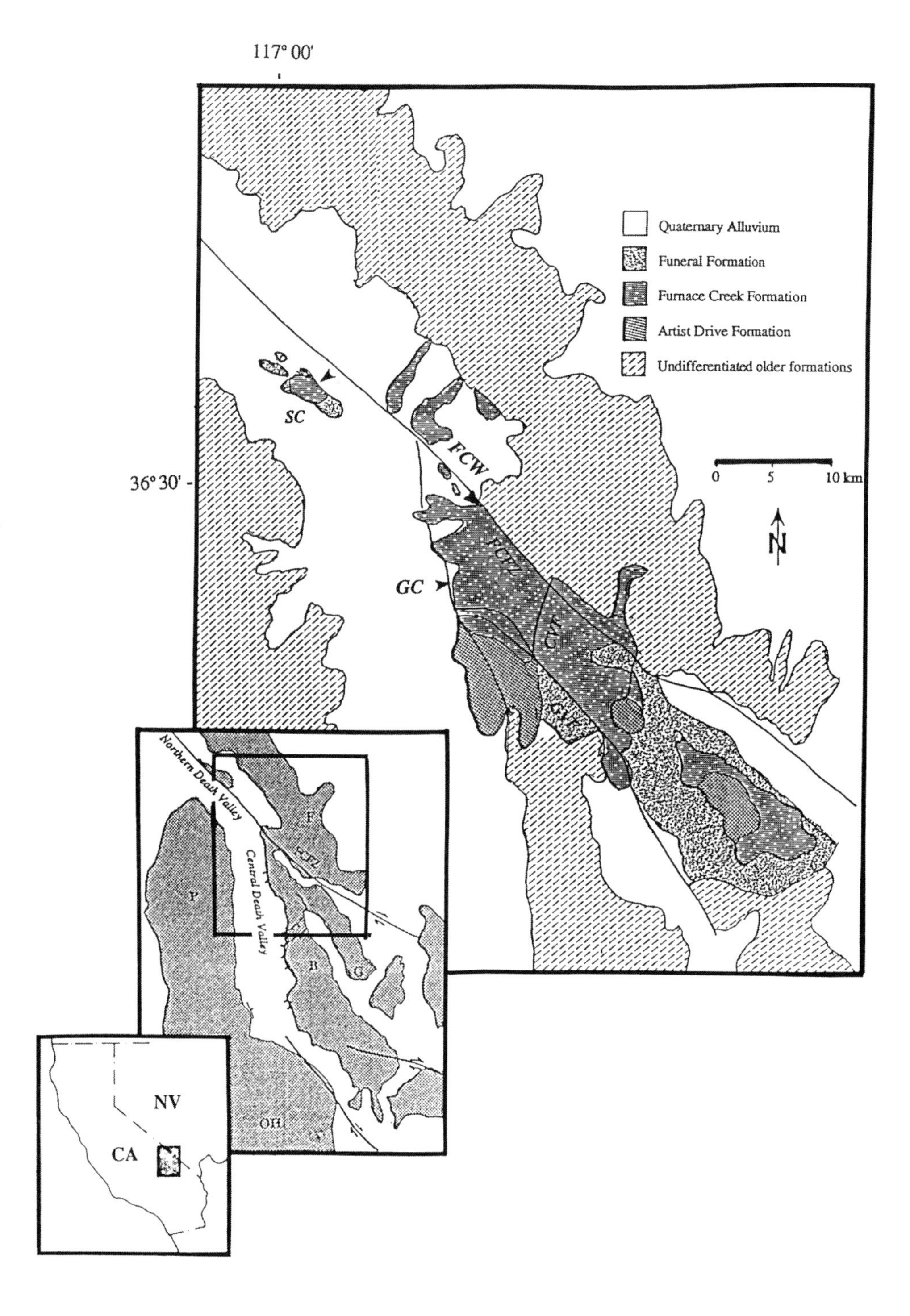

Figure 1—Area of outcrop of the Furnace Creek Formation (FCF) and main structural features of the Furnace Creek Basin. On the main map, FCFZ = Furnace Creek fault zone, GVF = Grand View fault, CVF = Cross Valley fault. Sections were measured at Furnace Creek Wash (FCW), Golden Canyon (GC), and Salt Creek (SC). On the inset map, F = Funeral Mountains, B = Black Mountains, G = Greenwater Range, P = Panamint Range, and OH = Owlshead Mountains.

status, the Greenwater Formation. The main unit coarsens vertically and interfingers with the upper conglomerate member (Tfcu of McAllister, 1970). Interbedded laminated mudstones and pebbly sandstones overlying the upper conglomerate member onlap the south end of the Funeral Mountains (Tfu of McAllister, 1970). Outcrops of laminated mudstones and cross-bedded sandstones in the Salt Creek Hills at the northern end of the Central Death Valley basin are correlated to the FCF by stratigraphic position beneath conglomerates of the Funeral Formation.

FURNACE CREEK FORMATION FACIES

Conglomerates

Poorly sorted polymict orthoconglomerates with cobble- to boulder-size clasts comprise beds 0.5–4 m thick. These beds commonly display basal inverse grading and clast fabric within beds ranging from random to largely flat-lying. These beds resulted from the deposition of clast-rich noncohesive debris flows (Gloppen and Steel, 1981; Nemec and Steel, 1984; Cole and Stanley, 1995). Meter-scale paraconglomerate

Epoch	Formation	thickness and lower boundary
Pleistocene	*Alluvium*	
	Funeral Fm. *(4.03+0.12 Ma)*	*700 m* *conformable to unconform.*
Pliocene	*Furnace Creek Fm.*	*~2100 m* *angular unconformity*
Miocene	*(7.5 Ma)* *Artist Drive Fm.*	*~1300 m* *angular unconformity*
Paleozoic to Proterozoic Formations		

Figure 2—Tertiary stratigraphy of the Furnace Creek Basin. Radiometric dates are from volcanics in the upper Artist Drive and lower Funeral Formations (Cemen and Wright, 1988; Cemen et al., 1985).

beds with pebble- to cobble-size clasts in a sandy, clay-rich matrix also display strong inverse grading, and commonly contain outsize clasts at the bed tops. In contrast, some beds display normal grading in the top few centimeters. These beds display features associated with cohesive debris-flow deposits. In particular, normal grading at the flow top is associated with debris flow in a subaqueous environment (Gloppen and Steel, 1981; Nemec and Steel, 1984). Massive to horizontally stratified pebble conglomerate beds up to 1 m thick may have resulted from deposition of hyperconcentrated flows or density-modified grain flows (Smith and Lowe, 1991; Cole and Stanley, 1995). A lack of soil features (e.g., caliche) supports the interpretation of largely, if not entrirely, subaqueous deposition. Clast compositions and paleocurrent measurements indicate derivation of material from erosion of older formations in the Funeral Mountains to the northeast of the basin and the Black Mountains to the southwest, as well as the Panamint and Cottonwood mountains to the west (McAllister, 1970, 1973; Cemen and Wright, 1991, 1994; Cemen, 1996; Prave, 1996).

Sandstones

Sandstone beds 0.2–1.1 m thick are interbedded with conglomeratic facies and mostly display a massive fabric but lack evidence of bioturbation. Dish structures occur locally, indicating rapid deposition. These beds are regarded as the deposits of high-density turbidites or subaqueous grain flows (Cole and Stanley, 1995). Horizontally stratified to wave-ripple laminated sandstones contain clay-filled root traces up to 10 cm long and are locally capped by desiccation-cracked mud drapes. These features indicate sediment aggradation up to a shoreline position with intermittent subaerial exposure. Collectively, the interbedded conglomerate and sandstone facies are arranged in fining-upward sequences 2–12 m thick (Figure 4) and record deposition primarily on the subaqueous portions of fan deltas prograding into the Furnace Creek lacustrine basin from the basin margins. A similar facies distribution is observed in the modern Walker Lake basin, Nevada (Link et al., 1985). Centimeter- to decimeter-scale sandstone beds interbedded with the laminated facies are pebbly to fine-grained. Pebbly sandstone beds are graded to massive, whereas fine-grained beds are typically massive. These beds are presumed to result from turbidity currents originating near the basin margins, possibly as the fluidized distal extensions of subaqueous debris flows (Hampton, 1972; Stow, 1994).

Outcrops in the Salt Creek Hills in the Central Death Valley basin display coarsening-upward sequences, up to 8 m thick, comprising centimeter-scale beds of ripple-laminated sandstone interbedded with decimeter- to meter-scale beds of trough to planar cross-bedded or horizontally laminated sandstone, rarely pebbly, containing fluid-escape structures, such as convolute bedding and pillow structures (Figure 5). This facies occurs interbedded with meter-scale sequences of laminated marlstone and claystone and records rapid progradation of a succession of thin deltas from the western margin of the basin into a lake occupying the Central Death Valley basin (Dean and Fouch, 1983).

Carbonate/Siliciclastic Laminites

The "main member" of McAllister (1970) is dominated by fine-grained laminites that comprise predominantly carbonates, siliciclastics, or laminae of alternating lithologies. Most commonly, this facies consists of alternating mudstone/marlstone to marlstone/limestone laminae, in places dolomitic, with

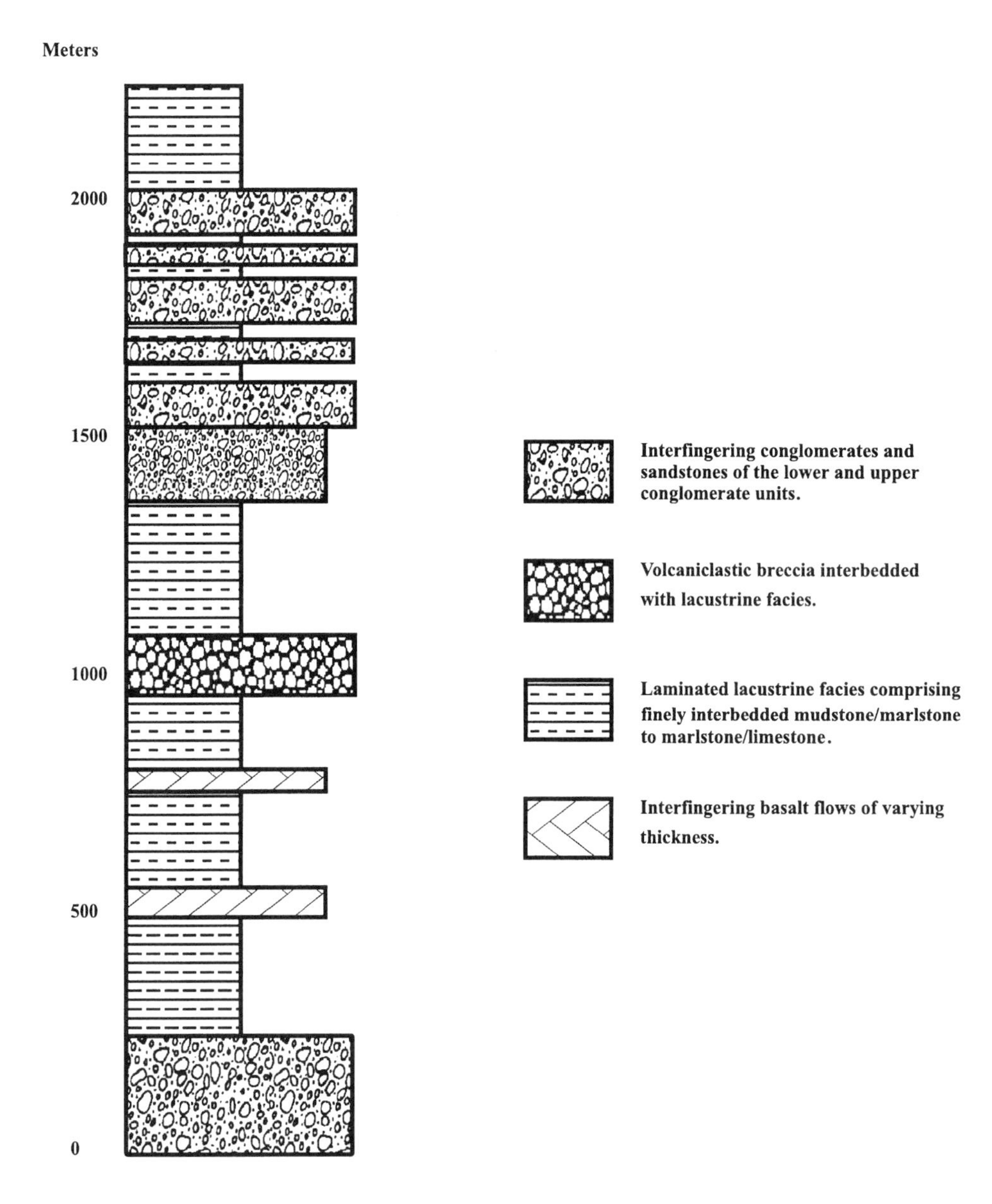

Figure 3—Representative lithologic column for the Furnace Creek Formation, although thickness and lithology vary widely within the Furnace Creek Basin.

interbedded centimeter-scale tuffs (Figure 6). The laminae range in thickness from less than 1 mm to 1 cm or greater, varying by an order of magnitude or more within small (<10 cm) stratigraphic intervals. The composition of the laminae alternates between carbonate-rich and siliciclastic-rich. The carbonate laminae comprise dense fine-grained micrite to coarser micrite containing varying amounts of silt- and clay-size siliciclastic material and exhibit abrupt to gradational upper and lower contacts. The carbonate is generally low-Mg calcite but is locally dolomitic. The laminae generally lack body fossils or charophyte debris but locally contain clotted fabrics and lenticular peloidal layers. Laminae are typically laterally continuous, although dense micrite laminae commonly exhibit abrupt lateral thickness variations. Micro-faulting, small-scale slumps, and scalloping commonly disrupt the lateral continuity of the dense micritic layers. Clayey or marly laminae contain varying proportions of carbonate mixed with silt and terrigenous clay, usually smectitic, and are predominantly ungraded, but graded laminae also occur. Small amounts of organic matter are disseminated in these layers. Gypsum and borate minerals occur locally as interbedded layers, typically less than 1 cm thick. The color of the mudstone generally varies from light greenish-gray to yellowish-gray. Dark gray mudstone containing plant remains is exposed in the southeastern part of the basin. This facies weathers to badlands topography typical of the FCF outcrop area, interfingering with fan-delta and deltaic facies in the central areas of both the Furnace Creek and Central Death Valley Basins.

Similarly rhythmic carbonate laminae have been described by other authors (Dean and Fouch, 1983; Glenn and Kelts, 1991; Gómez Fernández and Meléndez, 1991). These authors draw an analogy to modern temperate lakes in which seasonal turnover provides nutrients for increased photosynthetic activity, inducing rapid (fine-grained) carbonate precipitation (Kelts and Hsü, 1978). Decreasing productivity as the nutrients are

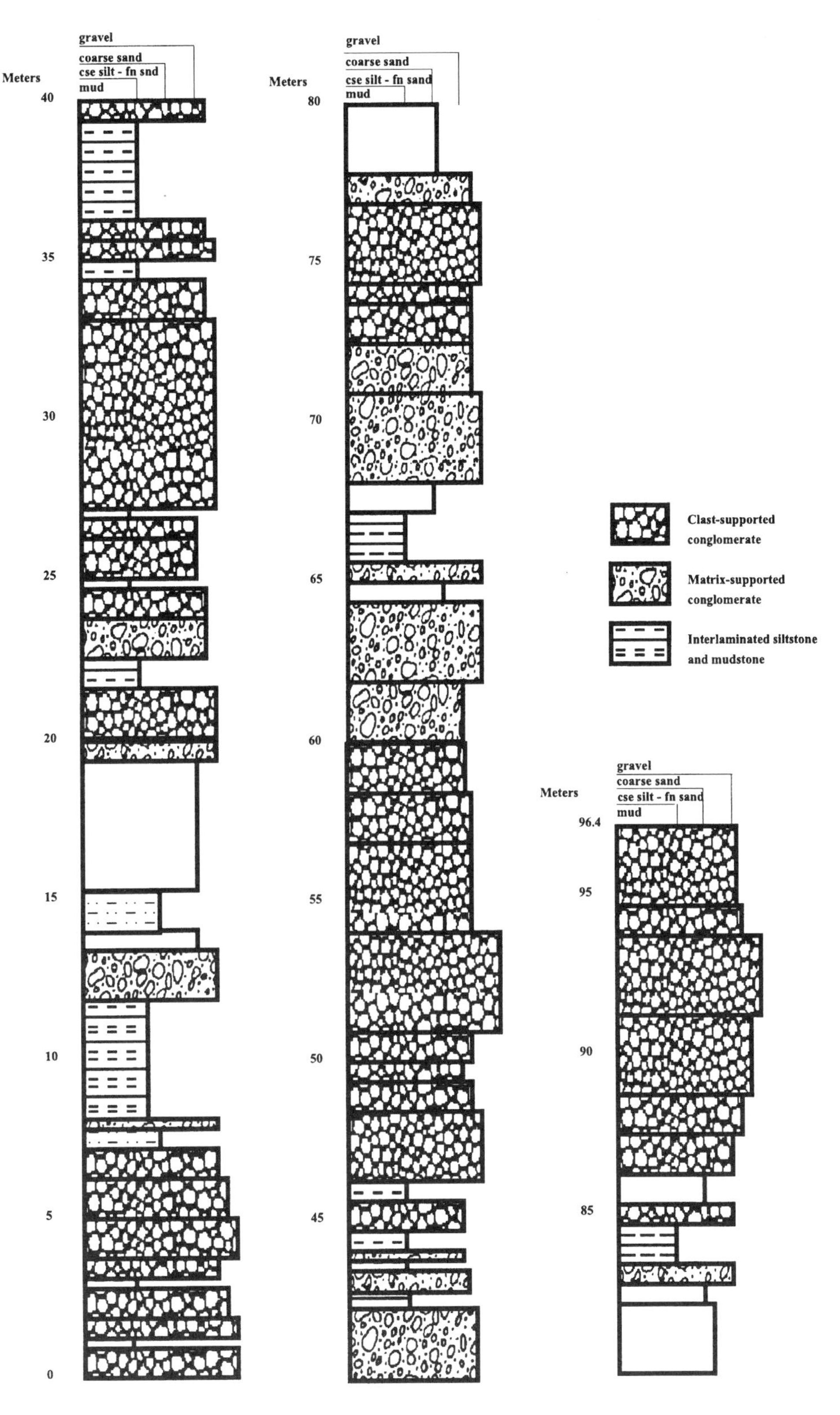

Figure 4—Section of the lower conglomerate facies measured at Furnace Creek Wash (see Figure 1) in the Furnace Creek Basin, comprising clast- and matrix-supported conglomerate interbedded with sandstone, siltstone, and mudstone.

depleted leads to slower (coarser) carbonate precipitation. Dean and Fouch (1983) also suggest temperature control of bio-induced carbonate precipitation, with peak productivity during the warmer (summer) months and clay fallout during the colder (winter) months. The claystone/marlstone or marlstone/limestone couplets in the FCF are presumed to reflect alternating depositional conditions, perhaps seasonal, in a carbonate lake. The dense micrite member of each couplet formed during a period of rapid, perhaps bio-induced, carbonate precipitation within the closed basin, while deposition of the siliciclastic-rich component, commonly containing a substantial silt fraction, may have coincided with seasonal runoff during the wet season. Dilution of

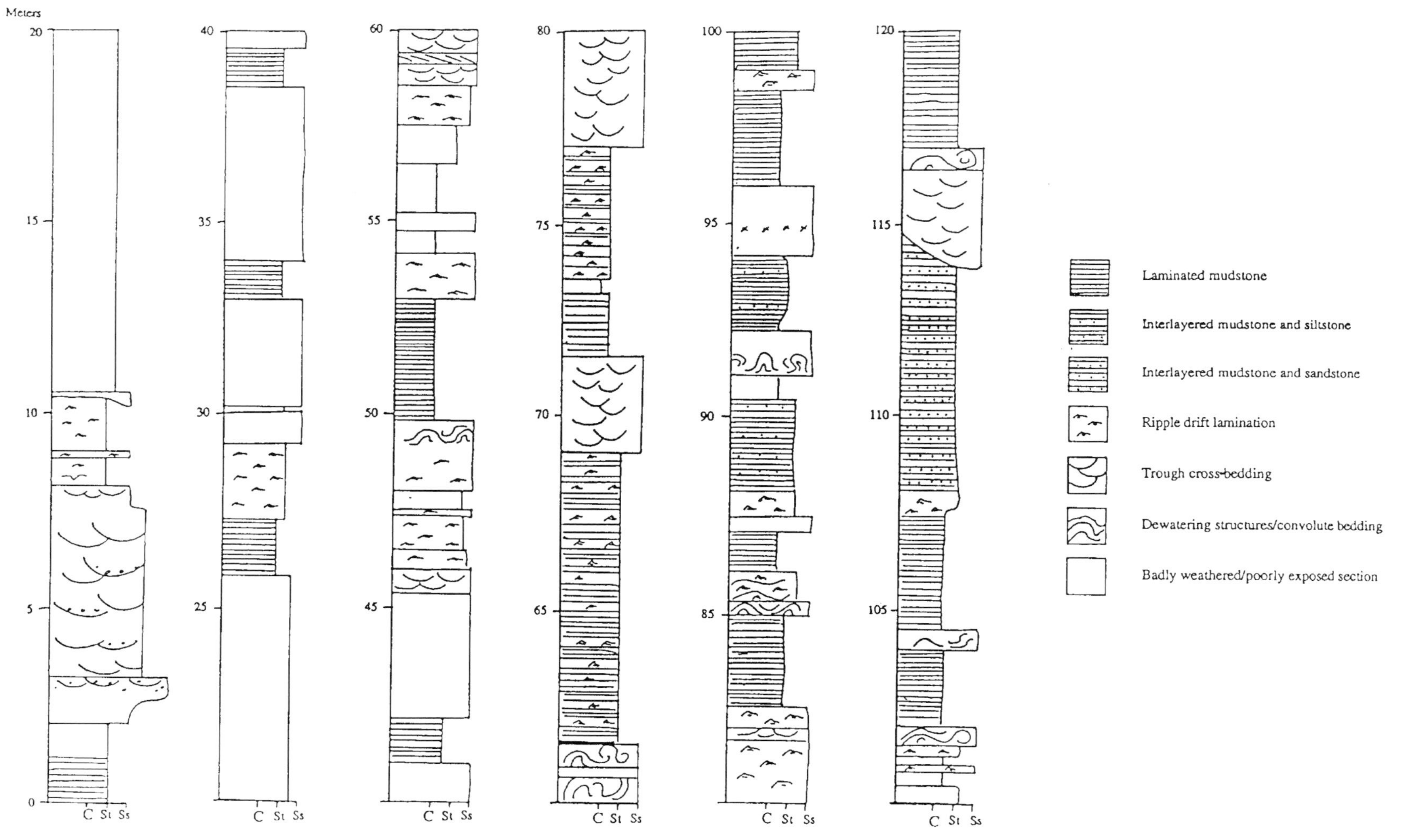

Figure 5—Section of lacustrine facies measured at Salt Creek (see Figure 1 for location) in Central Death Valley Basin comprising laminated fine-grained siliciclastics interbedded with meter-scale sandstone beds.

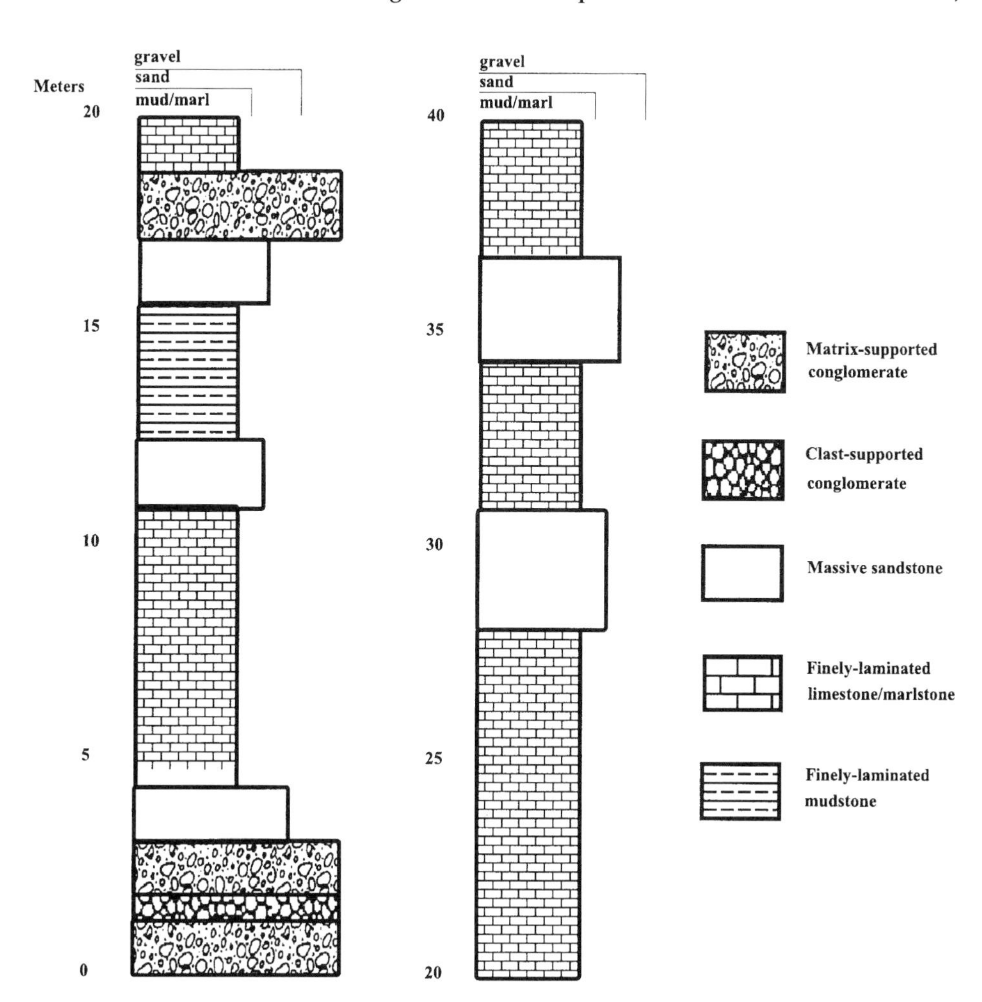

Figure 6—Representative section of lacustrine facies measured in Golden Canyon in the Furnace Creek Basin (see Figure 1) above lower conglomerate facies.

the lake water by runoff would also predictably result in slower crystal growth and the greater calcite crystal size associated with the siliciclastic-rich laminae. Grading within some centimeter-scale marlstone beds suggests deposition by density flows in the profundal zone, which occasionally eroded and reworked the dense micrite laminae (Glenn and Kelts, 1991). Stratification of the lake and anoxic bottom conditions prevented growth of a community of burrowing organisms, preserving the laminae. Features indicating a littoral position, such as wave reworking and desiccation cracks, are largely lacking in this facies, suggesting perennial deep water (below wave influence) conditions. In the Salt Creek Hills, the laminated facies is coarser and thicker, comprising centimeter-scale interbedded mudstone and calcareous mudstone beds and ripple-laminated fine-grained sandstone, reflecting a more proximal sediment source and probable deposition by density currents (Ashley, 1975). Two probable sources existed for carbonate enrichment of the lake waters: (1) hydrothermal springs that built tufa mounds on the lake floor and (2) weathering of Paleozoic-age carbonate rocks that crop out within the catchment basin.

Borates

Centimeter-scale bedding-parallel layers of colemanite, probertite, and ulexite are commonly interbedded in the laminated lacustrine facies in the lower one-third of the formation. Economically significant deposits of borates, in beds that are meters to tens of meters thick, are localized in several areas in the Furnace Creek Basin, particularly to the southeast, where they occur as stratiform deposits within the laminated facies (Barker and Wilson, 1976; Evans et al., 1976; Barker, 1980; Barker and Barker, 1985). Barker and Barker (1985) propose a model ascribing the origin of borates to precipitation in a thermally stratified (heliothermal) perennial lake with hydrothermal leaching of the volcanic country rock as a source of the boron. This conclusion is supported by isotopic studies of Swihart et al. (1993).

Massive borate deposits in the FCF contain clasts of calcareous mudstone and limestone and commonly occur adjacent to irregularly textured porous limestone bodies that are several meters high and up to 10 m wide. The limestone in these bodies consists of low-Mg calcite with a clotted micritic texture. These bodies are interpreted as spring-deposited limestones or tufas. Chandler (1996) interprets deposition of the borates in the FCF as resulting from vigorous outflow of the hot springs, brecciating and redepositing spring-apron carbonates.

Volcanics

Basalt flows, massive basalt breccias, and tuffs are interbedded with various facies of the FCF at several stratigraphic intervals (Figure 3). Basalt breccias comprise basalt boulders up to 1 m wide and smaller metasedimentary clasts in a nonbasalt sandy matrix, forming massive beds up to 10 m thick. The beds have irregular bases and tops, defined by interbedded centimeter-scale mudstone/marlstone laminites, and uncommonly display inversely graded bases, but are predominantly nongraded. These breccias resulted from subaqueous deposition of mass flow deposits, probably subaqueous debris avalanches and debris flows, originating from the collapse of synsedimentary basalt flows updip.

DEPOSITIONAL MODEL OF THE FURNACE CREEK FORMATION

Facies of the FCF comprise coarse siliciclastic basin margin facies (subaerial/subaqueous fan-delta, shoreline, and deltaic sandstones and conglomerates) grading and interfingering basinward with finer basinal lacustrine facies. Miocene–Pliocene lacustrine sedimentation took place in the rapidly subsiding Furnace Creek Basin during which episodic high-energy deposition, primarily of mass flow deposits, was controlled by tectonic reactivation of the FCFZ and associated faults and resulted in fan-delta progradation into the lake that occupied the Furnace Creek trough. Between episodes of tectonism, near quiescent conditions allowed episodic carbonate-siliciclastic sedimentation, perhaps seasonally controlled, on the floor of a hydrologically closed carbonate lake, punctuated periodically by volcanic activity. Hydrothermal springs discharged borate-charged waters, brecciating previously deposited tufa and lacustrine sediment layers, and causing precipitation of borate minerals within the resulting breccia.

REFERENCES CITED

Ashley, G. M., 1975, Rhythmic sedimentation in glacial Lake Hitchcock, Massachusetts-Connecticut, *in* A. V. Jopling and B. C. McDonald, eds., Glaciofluvial and glaciolacustrine sedimentation: SEPM Special Publication 23, p. 304–320.

Barker, C. E., 1980, The Terry borate deposit, Amargosa Valley, Inyo County, California: California Geology, v. 33, p. 181–187.

Barker, C. E., and J. M. Barker, 1985, A re-evaluation of the origin and diagenesis of borate deposits, Death Valley region, California, *in* J. M. Barker and S. J. Lefond, eds., Borates: economic geology and production: Society of Mining Engineers, American Institute of Mining, Metallurgy, and Petroleum Engineering, p. 101–135.

Barker, J. M. and J. L. Wilson, 1976, Borate deposits in the Death Valley region, *in* K. G. Papke, J. H. Schilling, J. M. Barker, J. L. Wilson, and R. A. Walters, eds., Guidebook: Las Vegas to Death Valley and return: Nevada Bureau of Mines and Geology, Report 26, p. 22–23.

Cemen, I., 1996, Evolution of the Furnace Creek Basin, Death Valley, California: Geological Society of America Abstracts with Programs, v. 28, p. A–437.

Cemen, I., and L. A. Wright, 1988, Stratigraphy and chronology of the Artist Drive Formation, Furnace Creek Basin, Death Valley, California, *in* J. L. Gregory and E. J. Baldwin, eds., Geology of the Death Valley Region: South Coast Geological Society Field Book, Santa Ana, California, p. 63–77.

Cemen, I., and L. A. Wright, 1991, Neogene conglomerates of the Furnace Creek Basin, Death Valley, California: origin and tectonic significance: Geological Society of America Abstracts with Programs, v. 23, p. 82.

Cemen, I., and L. A. Wright, 1994, Tertiary basinal sedimentary deposits in the area of the southern Funeral Mountains, Death Valley, California: their stratigraphy, structure, and tectonic significance: Geological Society of America Abstracts with Programs, v. 26, p. 44.

Cemen, I., L. A. Wright, R. E. Drake, and F. C. Johnson, 1985, Cenozoic sedimentation and sequence of deformational events at the southeastern end of Furnace Creek strike-slip fault zone, Death Valley region, California, *in* K. T. Biddle and N. Christie-Blick, eds., Strike-slip deformation, basin deformation, and sedimentation: SEPM Special Publication 37, p. 127–141.

Chandler, C. D., 1996, The Billie mine, Death Valley, California: Mineralogical Record, v. 27, p. 35–46.

Christie-Blick, N., and K. T. Biddle, 1985, Deformation and basin formation along strike-slip faults, *in* K. T. Biddle and N. Christie-Blick, eds., Strike-slip deformation, basin deformation, and sedimentation: SEPM Special Publication 37, p. 1–34.

Cole, R. B., and R. G. Stanley, 1995, Middle Tertiary extension recorded by lacustrine fan-delta deposits, Plush Ranch basin, western Transverse Ranges, California: Journal of Sedimentary Research, v. B65, p. 455–468.

Dean , W. E., and T. D. Fouch, 1983, Lacustrine environment, *in* P. A. Scholle, D. G. Bebout, and C. H. Moore, eds., Carbonate depositional environments: AAPG Memoir 33, p. 97–130.

Ellis, M. A., and J. H. Trexler, Jr., 1991, Basin-margin development in pull-apart settings: an example from Death Valley, California: Geological Society of America Abstracts with Programs, v. 23, p. 82.

Evans, J. R., G. C. Taylor, and J. S. Rapp, 1976, Mines and mineral deposits in Death Valley National Monument, California: California Division of Mines and Geological Special Report 125, 61 p.

Glenn, C. R., and K. Kelts, 1991, Sedimentary rhythms in lake deposits, *in* G. Einsele, W. Ricken, and A. Seilacher, eds., Cycles and events in stratigraphy: Berlin, Springer-Verlag, p. 188–221.

Gloppen, T. G., and R. J. Steel, 1981, The deposits, internal structure and geometry of six alluvial fan-fan delta bodies (Devonian, Norway)—a study of the significance of bedding sequences in conglomerates, *in* F. G. Ethridge and R. M. Flores, eds., Recent and ancient nonmarine depositional environments: models for exploration: SEPM Special Publication 31, p. 49–69.

Gómez Fernández, J. C, and N. Meléndez, 1991, Rhythmically laminated lacustrine carbonates in the Lower Cretaceous of La Serranía de Cuenca basin (Iberian Ranges, Spain), *in* P. Anadon, L. Cabrera, and K. Kelts, eds., Lacustrine facies analysis: International Association of Sedimentologists Special Publication 13, Blackwell Scientific, p. 245–256.

Greene, R. C., 1997, Geology of the Northern Black Mountains, Death Valley, California: United States Geological Survey Open-File report OF-97-79, 110 p.

Hampton, M. A., 1972, The role of subaqueous debris flow in generating turbidity currents: Journal of Sedimentary Petrology, v. 42, p. 775–793.

Kelts, K., and K. J. Hsü, 1978, Freshwater carbonate sedimentation, *in* A. Lerman, ed., Lakes: Chemistry, Geology, Physics: Berlin, Springer-Verlag, p. 295–323.

Link, M. H., M. T. Roberts, and M. S. Newton, 1985, Walker Lake basin, Nevada: an example of Late Tertiary(?) to Recent sedimentation in a basin adjacent to an active strike-slip fault, *in* K. T. Biddle and N. Christie-Blick, eds., Strike-slip deformation, basin deformation, and sedimentation: SEPM Special Publication 37, p. 105–125.

McAllister, J. F., 1970, Geology of the Furnace Creek borate area, Death Valley, Inyo County, California: California Division of Mines and Geology, Map Sheet 14.

McAllister, J. F., 1971, Preliminary geologic map of the Funeral Mountains in the Ryan quadrangle, Death Valley region, Inyo County, California: United States Geological Survey Open-File Report, Scale 1:62,500.

McAllister, J. F., 1973, Geologic map and sections of the Amargosa Valley borate area—southeast continuation of the Furnace Creek area—Inyo County, California: United States Geological Survey Miscellaneous Geologic Investigations Map I-782, 1:24,000.

McAllister, J. F., 1976a, Geologic maps and sections of a strip from Pyramid Peak to the southeast end of the Funeral Mountains, Ryan Quadrangle, California, *in* B. W. Troxel and L. A. Wright, eds., Geologic features of Death Valley, California: California Division of Mines and Geology Special Report 106, p. 63–65.

McAllister, J. F., 1976b, Columnar sections of the main part of the Furnace Creek Formation of Pliocene (Clarendonian and Hemphillian) age across Twenty Mule Canyon, Furnace Creek borate area, Death Valley, California: United States Geological Survey Open-File Report 76–261.

McAllister, J. F., and R. J. Ross, Jr., 1978, Steeply inclined algal columns in late Tertiary Furnace Creek Formation, Death Valley, California: AAPG Bulletin, v. 62, p. 541.

Nemec, W., and R. J. Steel, 1984, Alluvial and coastal conglomerates: their significant features and some comments on gravelly mass-flow deposits, *in* E. H. Koster and R. J. Steel, eds., Sedimentology of gravels and conglomerates: Canadian Society of Petroleum Geology Memoir 10, p. 1–31.

Prave, A. R., 1996, Pebbles and cross-beds: data to constrain the Miocene tectonic evolution of the Furnace Creek Basin, Death Valley, California: Geological Society of America Abstracts with Programs, v. 28, p. 309–310.

Savoca, M., and I. Cemen, 1989, Uplift of the central Funeral Mountains, Death Valley, California: evidence from a conglomerate exposed east of Texas Spring: Geological Society of America Abstracts with Programs, v. 21, p. 286.

Smith, G. A., and D. R. Lowe, 1991, Lahars: volcano-hydrologic events and deposition in the debris flow-hyperconcentrated flow continuum, *in* R. V. Fisher and G. A. Smith, eds., Sedimentation in volcanic settings: SEPM Special Publication 45, p. 59–70.

Stow, D. A. V., 1994, Deep sea processes of sediment transport and deposition, *in* K. Pye, ed., Sediment transport and depositional processes: Boston, Blackwell Scientific Publications, p. 257–291.

Swihart, G. H., E. H. McBay, D. H. Smith, and S. B. Carpenter, 1993, Boron isotopic study of the Tertiary bedded borate deposits of Furnace Creek, California: Geological Society of America Abstracts with Programs, v. 25, p. 319.

Wright, L. A., 1994, Overview of the Tertiary basinal deposits of the Death Valley region, California: Geological Society of America Abstracts with Programs, v. 26, p. 105.

Wright, L. A., and B. W. Troxel, 1973, Shallow fault interpretation of Basin and Range structure, southwestern Great Basin, *in* K.A. De Jong and R. Scholten, eds., Gravity and tectonics: New York, John Wiley, p. 397–407.

Alonso-Zarza, A. M., J. P. Calvo, J. van Dam, and L. Alcalá, 2000, Northern Teruel graben (Neogene), northeastern Spain, *in* E. H. Gierlowski-Kordesch and K. R. Kelts, eds., Lake basins through space and time: AAPG Studies in Geology 46, p. 491–496.

Chapter 45

Northern Teruel Graben (Neogene), Northeastern Spain

Ana M. Alonso-Zarza
José P. Calvo
Departamento de Petrología y Geoquímica, Universidad Complutense de Madrid
Madrid, Spain

Jan van Dam
Institute of Earth Sciences, Utrecht University
Utrecht, The Netherlands

Luis Alcalá
Museo Nacional de Ciencias Naturales, CSIC
Madrid, Spain

INTRODUCTION

The Teruel graben, also named Teruel fosse (Mein et al., 1990) and Teruel basin, is located on the northeastern side of the Iberian Peninsula (Figure 1). The basin is oriented north–northeast-south–southwest and occupies an area of approximately 100 km in length and 15 km in width. It is filled by a rather complete Neogene succession reaching 500 m in thickness (Moissenet, 1983, 1989). The Teruel basin is regarded as a half-graben bounded by several en echelon north–northeast-south–southwest normal faults that are mainly located in the eastern part of the basin (Anadón and Moissenet, 1996). Hanging-wall subsidence is westward of the faults, whereas footwall uplift is eastward of the faults. This structure resulted in tilting of the Neogene deposits that, in general, dip toward the east-southeast. The main tectonic lineation that controlled the formation of the basin is in contrast with the typical northwest-southeast Paleogene compressional structures of the Iberian Chain (Figure 1), thus providing evidence of a gradual change in stress regime throughout the upper Oligocene in this area of the northeast Iberian Peninsula (Simón, 1984). Formation of the basin is envisaged as related to the extensional movements linked to the rifting of the western Mediterranean during the Miocene (Anadón et al., 1989; Roca, 1996).

In the northern part of the Teruel graben (Teruel-Alfambra region) (Figure 1), the Neogene succession is bounded by siliciclastic and evaporite formations of Triassic age, as well as by Jurassic carbonate deposits. The oldest Neogene deposits, which unconformably overlie Upper Jurassic formations in the eastern margin of the basin, have been dated as lower Aragonian (Montalvos mammal locality) (MNT, Figure 2); however, most of the basin fill sequence cropping out in the region comprises sediments ranging from Lower Vallesian to Pliocene in age. These deposits show rapid lateral and vertical facies changes, which is typical of a sedimentary context characterized by internal drainage. The dating method is mainly based on mammal faunas found across the basin (Figure 2) (Alcalá, 1994; van Dam, 1997). The composite Vallesian–Turolian mammal succession of the area consists of 90 assemblages, all of them except one containing micromammals and 27 containing both micro- and macromammals (van Dam, 1997). The majority of the mammal localities are in stratigraphic superposition, and various key sections have provided the basis for magnetostratigraphic analysis (Krijgsman et al., 1996; Garcés et al., in press; van Dam and Weltje, 1999), which allows correlation of the mammal succession to the geomagnetic polarity time scale (GPTS) (van Dam, 1997). The Pliocene sedimentary record of the northern part of the Teruel graben is also well dated on the basis of more than 25 mammal localities (Alcalá, 1994).

LITHOSTRATIGRAPHY OF THE NORTHERN TERUEL GRABEN

Thickness of Neogene sediments in the northern part of the Teruel graben reaches 500 m. Due to the rapid lateral stratigraphic changes, it is difficult to show a general composite lithostratigraphic log for the whole area. The lowermost deposits of the Miocene

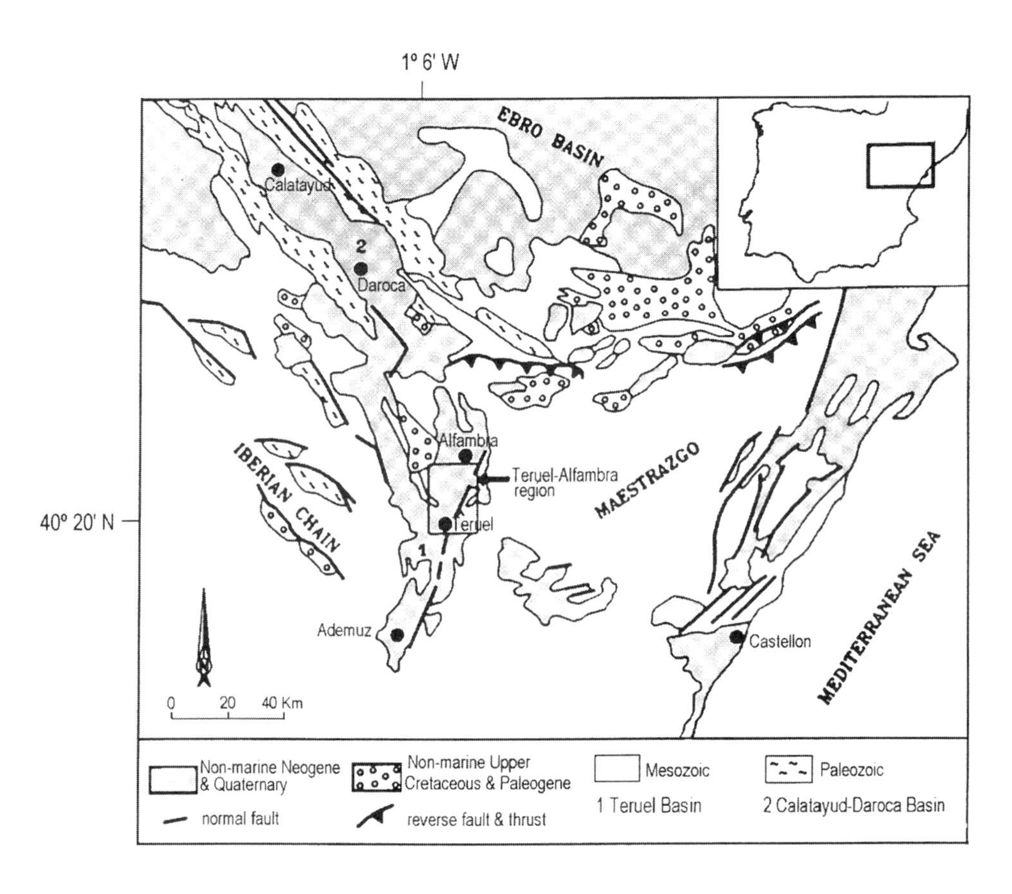

Figure 1—Schematic map of northeastern Spain showing the location of the Teruel graben (boxed area corresponds to the northern part of the basin, the Teruel-Alfambra region). Note that the basin stretches north–northeast south–southwest, which is in contrast with northwest-southeast and east-west tectonic lineations generally observed in the Iberian Chain, but coincident with the main tectonic lineations close to the Mediterranean Sea.

sedimentary fill and the marginal facies surrounding the basin consist of reddish alluvial facies that have been referred to by several names (Peral Formation, Tejares Formation, Los Monotos Series) (Alcalá, 1994). These facies will be named the Red Clastic facies in this paper; they are found at several levels throughout the Neogene sedimentary record of the basin. The clastic facies grade laterally into carbonate successions, including abundant organic-rich marlstone beds that have furnished most of the Miocene mammal localities found in the basin. These carbonates belong to the Alfambra Formation (van de Weerd, 1976) (Figure 3). Evaporite, mainly gypsum, sediments occur at two distinct levels of the general Neogene succession (Figure 3). Some minor gypsum beds occur interbedded with clastic deposits of Vallesian age, yet the bulk of gypsiferous sediments developed during the Turolian and the lower Pliocene, reaching up to 150 m in thickness (Tortajada Formation). The youngest sediments are especially well exposed in the northernmost part of the basin and comprise a variety of siliciclastic, carbonate, and marlstone lithofacies grouped into the Escorihuela Formation, which reaches up to 40 m in thickness.

LITHOFACIES ASSOCIATIONS AND FACIES INTERPRETATION

Neogene deposits of the northern area of the Teruel basin were deposited in three different depositional systems. The Red Clastic facies were deposited in alluvial systems. Shallow lake carbonate systems accounted for the deposition of the Alfambra and Escorihuela Formations, whereas the Tortajada Formation reflects the development of evaporite lake systems.

Alluvial Deposits

Alluvial systems are well developed in the eastern margin of the basin where they are entrenched within Mesozoic rocks, expanding toward the west. The alluvial deposits include alluvial fans, floodplains, and paleosols.

Well-developed alluvial fan systems occur in the Valdecebro area (Figure 2), as well as to the east of the village of Tortajada. Smaller fan-colluvial deposits occur in several localities adjacent to the basin margins. Proximal alluvial facies consist of horizontally bedded gravel beds, up to 3 m thick. Gravel lithofacies are massive and vary from massive, matrix-supported to crudely bedded, clast-supported, imbricated gravels. Medial fan facies consist of red mudstones with interbedded gravel and sandstone channels. Channels are erosively incised in the mudstones and show typical fining-upward sequences. Proximal and medial fan facies were deposited as debris and channelized flows, the former being predominant in the proximal alluvial fan areas, whereas braided channels developed in the medial and distal fan areas.

Distal fan and floodplain deposits are mudstones that intercalate deeply entrenched gravel and sandstone channels as well as carbonate-rich paleosols ranging from stage 1 to stage 3 of Machette (1985). Light brown to red mudstones occur in beds up to 7 m thick. They show drab green haloes due to hydromorphism, scattered carbonate nodules, and root traces.

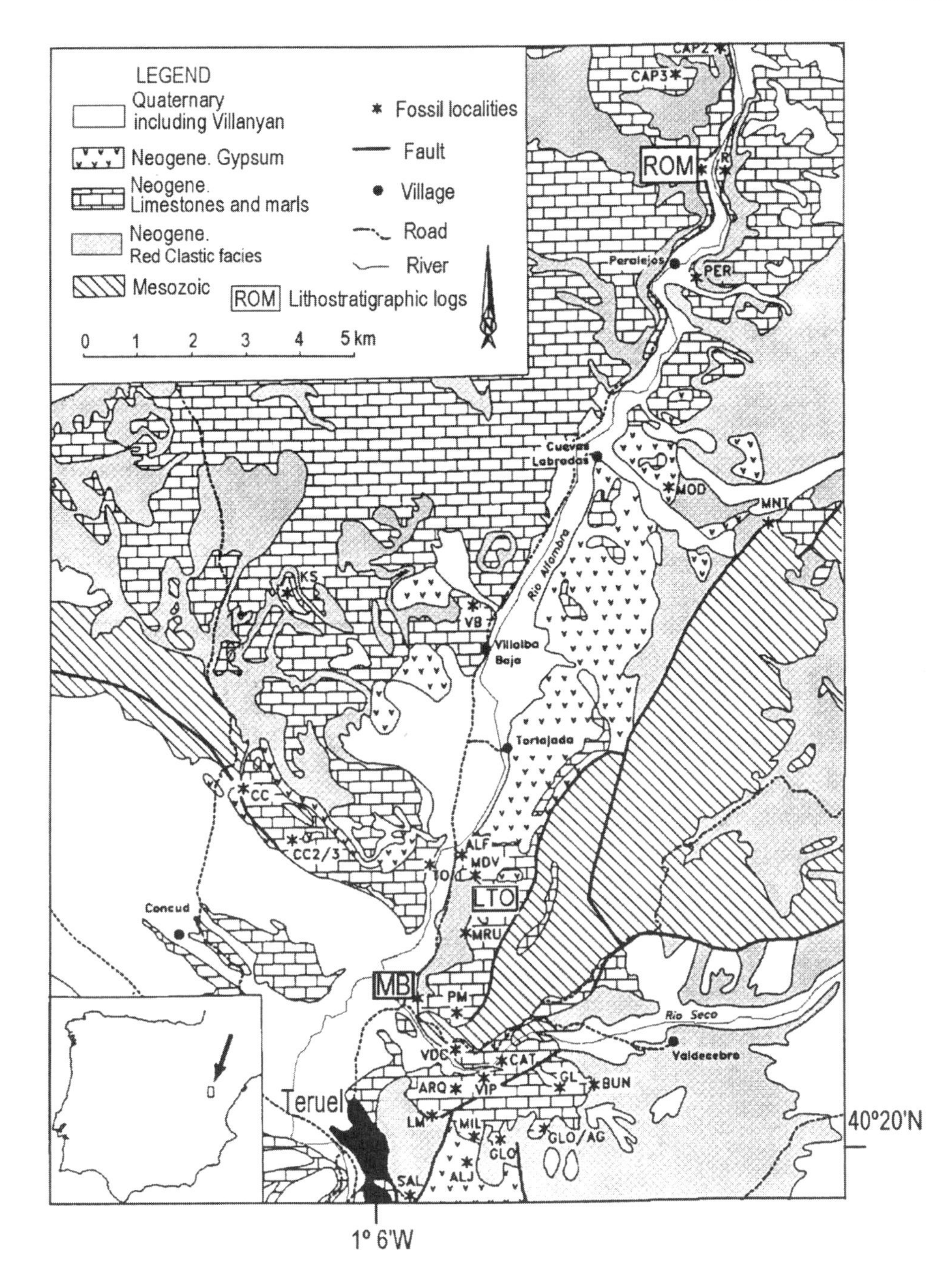

Figure 2—Schematic geologic map of the northern part of the Teruel graben (Teruel-Alfambra region) with location of the lithostratigraphic logs shown in Figure 4 (boxed sections codes: MB, LTO, ROM) and main mammal localities found throughout the basin (stars).

Channels occur very rarely within the mudstones; they are deeply erosive (up to 1 m) and narrow (20 m width). They are filled by gravels and sands arranged in fining-upward sequences that may grade progressively into carbonate deposits. Gravels commonly exhibit clast-supported fabrics and imbrication. Sandstones show both planar and trough cross-bedding.

Paleosols are clearly recognized in the floodplain and distal fan mudstones. The paleosols occur in two different situations (Figure 4): (1) at the foot of the fan systems and lateral to fluvial channels and (2) in vertical transition from floodplain/distal fan facies to lacustrine deposits.

At the foot of the fan systems and lateral to fluvial channels, paleosol maturity increases with distance to the areas of higher sedimentation rates. Carbonate paleosols reach up to stage 3 of Machette (1985). Paleosol features include drab green haloes, carbonate nodules, root traces, and desiccation cracks. Thickness of these paleosols may reach up to 80 cm, with gradual basal contacts and sharp top contacts.

In the vertical transition from floodplain/distal fan facies to lacustrine deposits, a thick paleosol grading to shallow lake carbonate deposits develops in the gradual transition between the Red Clastic facies and the Alfambra Formation, whose age is MN-10. Paleosol thickness may reach up to 3 m below the palustrine carbonates. Soils show pedogenic and phreatic features developed with rising groundwater. Pedogenic features include root traces, alveolar-septal structures, peloids, and micritic coatings around carbonate nodules. Phreatic features

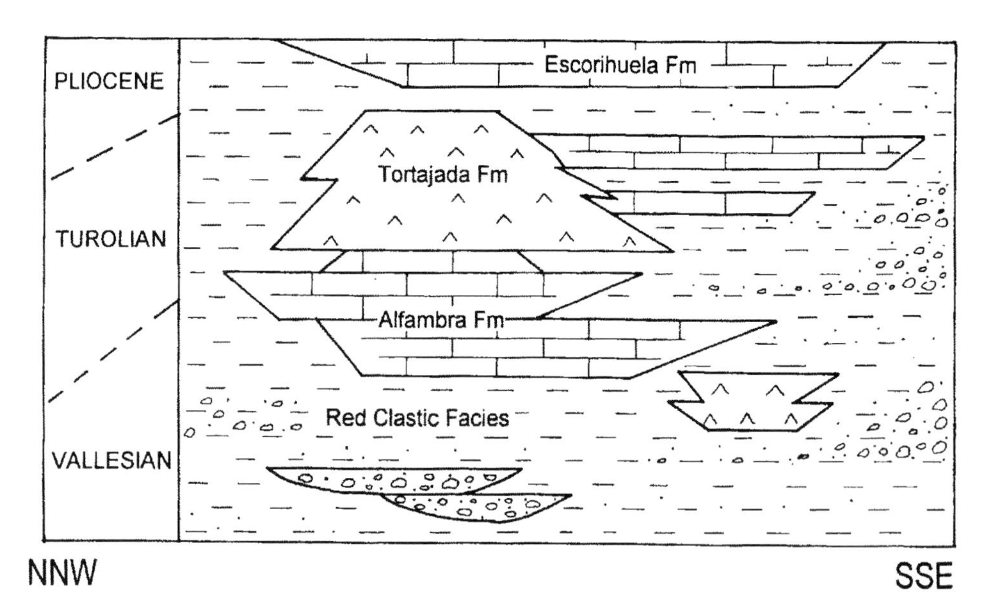

Figure 3—Stratigraphic sketch for the Neogene sedimentary record of the northern Teruel graben with indication of the names used for the stratigraphic units defined in the area.

include change in color from red to yellow and green and carbonate cementation of either micrite or spar.

Shallow Lake Systems

Shallow lake deposits of the Alfambra and Escorihuela Formations include palustrine carbonates, channels infilled by phytoclasts and gastropod remains, biomicrites, organic-rich marlstones, and evaporite lake deposits.

Palustrine carbonates occur in three different situations: (1) at the top of some paleosols, which may or may not grade vertically into lacustrine deposits, (2) at the top of lacustrine carbonates, and (3) interbedded with siliciclastic deposits. Palustrine carbonates show a variety of facies that include very indurated limestones with prismatic structure due to root bioturbation, brecciated micrite with coated grains, and nodular limestones. These facies are commonly mottled and indicate short subaerial exposure periods and subsequent organic activity, desiccation, and hydromorphism. Average thickness of the palustrine beds is 1 m.

Channels infilled by phytoclasts and gastropod fragments of about 0.5 m thick occur as amalgamated beds to form units several meters in thickness. Width of the channels reaches up to 50 m. Lower surfaces of the channels are slightly erosive. Internally, the channels show crude oblique lamination and much of their infill is made of fragments of tufaceous mounds, fragments of bioclasts, and carbonate intraclasts. A particular case is observed in Masia del Barbo where the channels include large amounts of gastropod opercula (Figure 4). The occurrence of these channels reveals the extensive reworking of the carbonate deposits formed in the lake margin flats by ephemeral floods. This type of lithofacies characterizes the base of the Masia de la Roma section (Figure 4).

Biomicrites occur as massive tabular beds, ranging from a few decimeters to 1 m in thickness. Fenestral structures and micritic intraclasts are very common within the biomicrite beds. Desiccation cracks are also common. Main bioclasts are gastropods, bivalves, and charophytes. Biomicrites reflect deposition in shallow hardwater lakes scarcely modified by subaerial exposure and representing the more stable lacustrine facies.

Green to black organic-rich marlstones occur either interbedded among other lacustrine deposits or within floodplain mudstones. They are about 0.5 m thick on average and display massive to roughly laminated internal structure. They commonly contain gastropod debris and plant fragments. These facies contain much of the mammal sites recognized in the area (Figure 4). The marlstones accumulated in ponded areas (marshes) in which development of extensive vegetation accounted for stability of reduced conditions that favored preservation of mammal remains.

Evaporite lake environments were present in the region mainly during the late Turolian and accounted for deposition of the Tortajada Formation. Some minor gypsum occurrences are also recognized within the Alfambra Formation and the Red Clastic facies. Two main lithofacies types indicate evaporite lake environments: (1) bioturbated gypsum beds and (2) massive to laminated dolostone beds.

Bioturbated gypsum beds intercalated within marly dolostones are the main lithofacies of the Tortajada Formation and consist of about 1.5-m-thick, white to beige beds formed of a dense mosaic of lenticular gypsum crystals. The gypsum displays a peculiar fabric characterized by strong burrowing of the gypsum [tangle-patterned small burrows of Rodríguez-Aranda and Calvo (1998)]. Locally, the matrix between lenticular crystals is dolomicrite mixed with variable amounts of clays. Strong burrowing of the lenticular gypsum is thought to be due to chironomid larvae, reflecting evaporite lake systems of moderate salinity (Rodríguez-Aranda and Calvo, 1998).

Massive to laminated dolostone beds of about 2 m thick occur at the top of the Alfambra Formation and in the carbonates laterally equivalent to the Alfambra Formation. They consist of relatively porous and hardened dolomicrite beds with scattered gypsum molds.

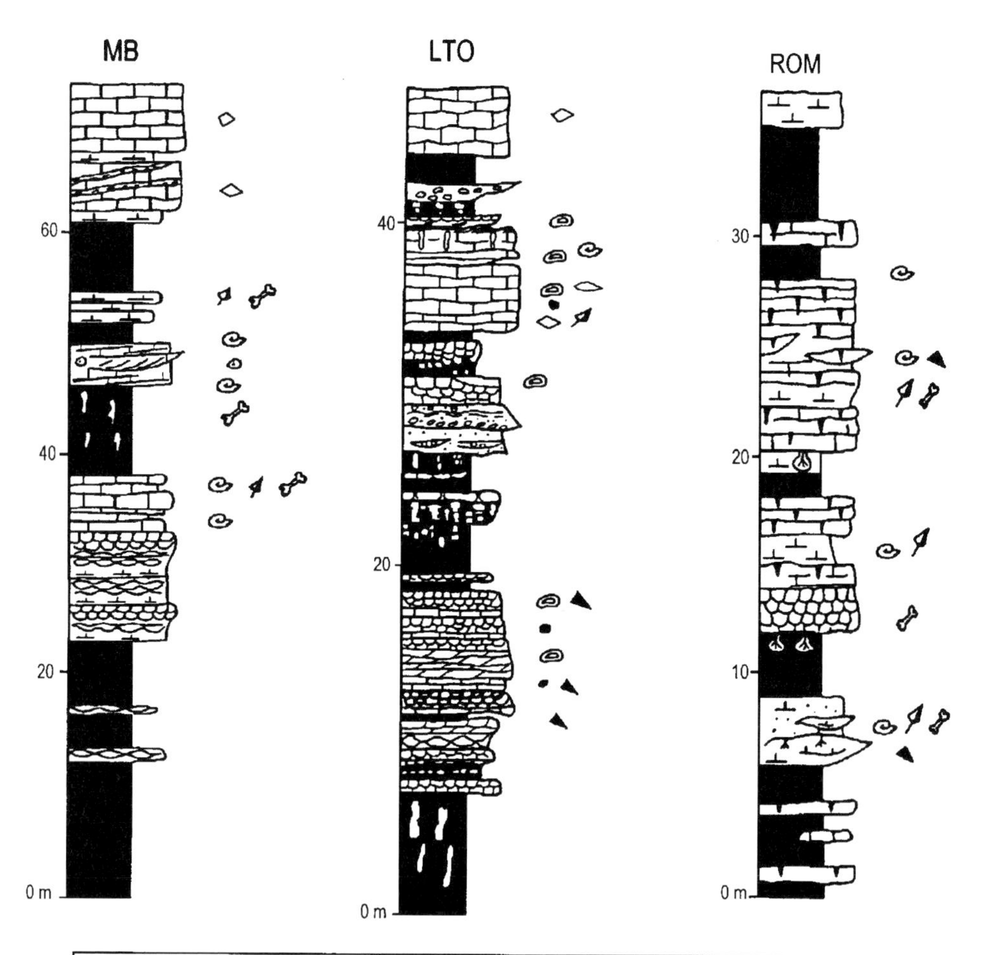

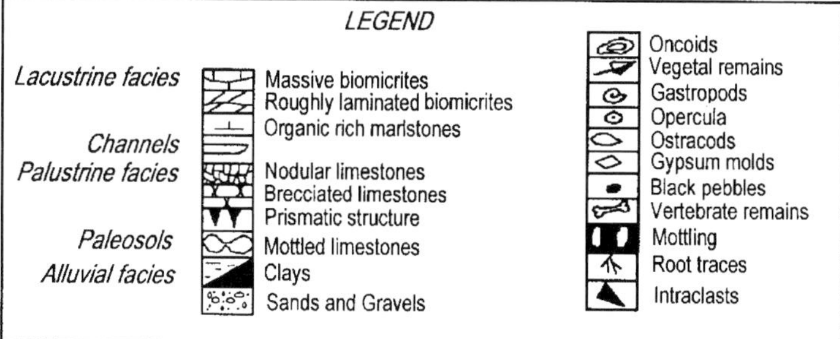

Figure 4—Partial measured logs of the Neogene in the northern Teruel graben. The Masía del Barbo (MB) section shows the vertical evolution from distal fan facies containing paleosols to palustrine-lacustrine environments. A thick channel rich in gastropod opercula can be observed at about 46 m from the base. Several sequences distal fan-paleosol-palustrine facies are shown throughout the Laguna de Tortajada (LTO) section. Biomicrites characterize the uppermost part of the section, indicating more permanent lake systems. Marlstones and clays rich in organic matter and containing microvertebrate remains are recognized at several levels of the Masía de La Roma (ROM) section. Channels filled with carbonate fragments are also common. See Figure 2 for locations.

MAIN CONSTRAINTS ON SHALLOW LAKE SEDIMENTATION IN THE TERUEL GRABEN

The lacustrine lithofacies described are typical of deposition under shallow to very shallow water conditions. Chemistry of the lake water, clastic input, vegetation, and morphology of the lake depression controlled occurrence and distribution of the different types of deposits. Surficial and/or groundwaters draining the adjacent Mesozoic rocks, enriched in Ca^{+2} and HCO_3^-, favored carbonate deposition in the lakes. Palustrine carbonates and biomicrites were deposited in freshwater, low-gradient lakes. Differences between these two lithofacies reveal the permanence of the water body and duration of subaerial exposure processes. Occurrence of channels filled with carbonates indicates local reworking of the lake margin carbonate flats by ephemeral floods entering the lake.

Highly vegetated ponded areas (marshes) were totally or partially isolated from the major water body. Thus, with reduced circulation or renewal of their water column, poorly oxygenated shallow lakes resulted, in which organic matter was preserved due to the reduced conditions. These marshes were favorable sites for the concentration of vertebrate remains in organic-rich marlstones and clays.

The influence of source rock as a control for the development of evaporite lake systems is clear in areas adjacent to the Triassic evaporites, especially after uplift of the basin margins caused exposure of these rocks. Both ground and surficial waters carried larger amounts of sulfate to the lakes during some periods of the evolution of the basin.

EVOLUTION OF THE LAKE SYSTEMS

Development of the Neogene lake systems was controlled by the location, lithology, and tectonic regime of the basin margins. Red Clastic facies were deposited all along the basin in the initial stages of its infilling; location of the major alluvial fan systems was controlled by morphology of the basin margins. Coarse-grained alluvial deposits evolved vertically and distally into red mudstones on which carbonate paleosols developed. Extensive paleosol development during the late Vallesian (MN-10 mammal zone) marks the base of the Alfambra Formation. During the late Vallesian and early Turolian, carbonate shallow lake systems and floodplains were dominant in the region. Locally, in areas adjacent to the basin margins formed by Triassic evaporites, some evaporite systems developed. Movements of the faults bounding the basin controlled the development of the carbonate sequences (Alonso-Zarza and Calvo, in press). Increasing uplift of the basins margins, especially in the eastern side of the basin, resulted in the exposure of Triassic evaporites, thus allowing the subsequent deposition of evaporite deposits of the Tortajada Formation during the late Turolian. Carbonate lake systems and floodplains formed in other areas of the basin during the same period. Extensive deposition of lacustrine carbonates took place at the end of the Turolian with the establishment of more humid and cooler conditions as inferred from vertebrate faunas of the middle MN13 mammal zone (van Dam, 1997). These conditions also prevailed throughout the early Pliocene.

ACKNOWLEDGMENTS

The work in the Teruel basin has been supported by Project PB95-0114, financed by the Spanish DGICYT (Dirección General de Investigación Científica y Técnica). Thanks are also given to the Netherlands Research School of Sedimentary Geology. The authors wish to thank E. Gierlowski and K. Kelts for their editorial effort on our paper and in the whole volume.

REFERENCES CITED

Alcalá, L., 1994, Macromamíferos neógenos de la fosa de Alfambra-Teruel. Inst. Estudios Turolenses-Museo Nacional de Ciencias Naturales, CSIC, Teruel-Madrid, 554 p.

Alonso-Zarza, A. M., and J. P. Calvo, (in press), Palustrine sedimentation in an episodically subsiding basin: the Miocene of the Northern Teruel Graben. Palaeogeography, Palaeoclimatology, Palaeoecology.

Anadón, P., L. Cabrera, R. Juliá, E. Roca, and L. Rosell, 1989, Lacustrine oil-shale basins in Tertiary grabens from NE Spain (Western European rift system). Palaeogeography, Palaeoclimatology, Palaeoecology, v. 70, p. 7–28.

Anadón, P., and E. Moissenet, 1996, Neogene basins in the Eastern Iberian Range, *in* P. F. Friend and C. J. Dabrio, eds., Tertiary Basins of Spain. The stratigraphic record of crustal kinematics. Cambridge University Press, World and Regional Geology Series 6, Cambridge, p. 68–76.

Dam, J. A., van, 1997, The small mammals from the upper Miocene of the Teruel-Alfambra region (Spain): paleobiology and paleoclimatic reconstructions. Geologica Ultraiectina, Utrecht, 156, 204 p.

Dam, J. A., van, and G. J. Weltje, 1999, Reconstruction of the Late Miocene climate of Spain using rodent palaeocommunity successions: an application of end-member modelling. Paleogeography, Palaeoclimatology, Palaeoecology, v. 151, p. 267–305.

Garcés, M., W. Krijgsman, J. A. van Dam, J. P. Calvo, L. Alcalá, and A. M. Alonso-Zarza, in press, Late Miocene alluvial sediments from the Teruel area: Magnetostratigraphy, magnetic susceptibility, and facies organisation. Acta Geológica Hispánica.

Krijsman, W., M. Garcés, C. G. Langereis, R. Daams, J. van Damm, A. J. van der Meulen, J. Agustín, and L. Cabrera, 1996, A new chronology for the middle to late Miocene continental record in Spain. Earth and Planetary Sciences Letters, v. 142, p. 367–380.

Machette, M. N., 1985, Calcic soils of southwestern United States, *in* Weide, C. L., ed., Soil and Quaternary Geology of the Southwestern United States. Spec. Paper. Geological Society of America, v. 203, p. 1–21.

Mein, P., E. Moissenet, and R. Adrover, 1990, Biostratigraphie du Néogène supérieur de Teruel. Paleontologia i Evolució, v. 23, p. 121–139.

Moissenet, E., 1983, Aspectos de la neotectónica en la Fosa de Teruel, *in* Geología de España, t. II, Instituto Geológico y Minero de España, Madrid, p. 427–446.

Moissenet, E., 1989, Les fossés néogènes de la Chaîne Ibérique: leur évolution dans le temps. Bulletin Societé Géologique de France, sér. 8, 5, p. 919–926.

Roca, E., 1996, La evolución geodinámica de la Cuenca Catalano-Balear y áreas adyacentes desde el Mesozoico hasta la Actualidad. Acta Geológica Hispánica, v. 29, p. 3–25.

Rodríguez-Aranda, J. P., and J. P. Calvo, 1998, Trace fossils and rhizoliths as a tool for sedimentological and palaeoenvironmental analysis of ancient continental evaporite successions. Palaeogeography, Palaeoclimatology, Palaeoecology, v. 140, p. 383–399.

Simón, J. L., 1984, Compresión y distensión alpinas en la Cadena Ibérica oriental. Instituto de Estudios Turolenses, Teruel, 269 p.

Weerd, A., van de, 1976, Rodent faunas of the Mio–Pliocene continental sediments of the Teruel-Alfambra region, Spain. Utrecht Micropaleontological Bulletin, Special Publication 2, p. 1–217.

Anadón, P., F. Ortí, and L. Rosell, 2000, Neogene lacustrine systems of the southern Teruel graben (Spain), *in* E. H. Gierlowski-Kordesch and K. R. Kelts, eds., Lake basins through space and time: AAPG Studies in Geology 46, p. 497–504.

Chapter 46

Neogene Lacustrine Systems of the Southern Teruel Graben (Spain)

P. Anadón
Institut de Ciències de la Terra "J. Almera" (CSIC)
Barcelona, Spain

F. Ortí
L. Rosell
Departament de Geoquímica, Petrologia i Prospecció Geològica, Facultat de Geologia, Universitat de Barcelona
Barcelona, Spain

INTRODUCTION

The Teruel graben, up to 15 km wide, extends more than 100 km and cuts across the central part of the northwest-southeast compressional Alpine Iberian Chain (Figure 1). A detachment level within this chain is formed by thick Upper Triassic evaporitic and lutitic formations that separate two different structural levels: the Hercynian basement (with a Permian–Lower Triassic thin covering) and a thick Jurassic–Paleogene carbonate-dominated cover. The Teruel graben displays en echelon north–northeast-south–southwest oriented faults and a half-graben structure with hanging-wall subsidence westward of each fault and footwall uplift eastward of each fault (Moissenet, 1980; Anadón and Moissenet, 1990). This graben is the result of the gravitational collapse of a part of the Iberian Chain that experienced the largest crustal thickening during the Paleocene and early Miocene compression (Guimerà, 1997). Rifting that created the Teruel graben must be considered within the framework of the Western European Rift System and the late Oligocene–Neogene extensional deformation that created the Catalan-Balearic Basin (Anadón et al., 1989; Anadón and Roca, 1996; Guimerà, 1996; Roca, 1996).

The Neogene basin-fill sequence in the Teruel graben is up to 500 m thick and consists of a complex arrangement of alluvial siliciclastic facies and lacustrine carbonates and evaporites (Hernández et al., 1985). Age of the basin-fill sequence ranges from early Aragonian (early Miocene) to Pliocene (Adrover et al., 1978). Dating of the diverse units has been based exclusively on abundant mammal fossil sites (Gautier et al., 1972; Adrover et al., 1978; Besems and van der Weerd, 1983; Mein et al., 1983; Adrover, 1986; Alcalá, 1994, and references within).

BASIN-FILL SEQUENCE IN THE SOUTHERN TERUEL GRABEN (CASCANTE-ADEMUZ ZONE)

Several stratigraphic subdivisions have been proposed for the Neogene sequences in the southern Teruel graben (Bakx, 1935; Gautier et al., 1972; Broekman, 1983; Broekman et al., 1983; Arce et al., 1983); however, most of them are not supported by detailed mapping or lack biostratigraphic control for correlation, and because of common lateral facies changes, they have only local significance. For the Libros-Ademuz area, Anadón and Moissenet (1990) proposed a stratigraphic subdivision supported by detailed mapping that has been refined by Anadón et al. (1997) for the Libros-Cascante area. In this paper, we use a stratigraphic subdivision based on Anadón et al. (1997), modified by adding lateral relationships to the north–northeast and to the south–southwest of the Libros-Riodeva zone (Figures 2, 3).

The lower alluvial unit mainly consists of red alluvial deposits (conglomerates, sandstones, and mudstones) with minor limestone and gypsum intercalations (i.e., El Campo Gypsum, up to 3 m thick). Total thickness is up to 150 m. The upper part of the sequence consists of red mudstones and a gypsum unit up to 25 m thick (El Morrón Gypsum).

The El Bolage Limestone, up to 80 m thick, consists of thin-bedded limestones with minor mudstones and lignites (section 1 in Figure 4). To the north it passes into a thinner succession of massive and bioturbated limestones and marls. Locally, some dark mudstone beds and lignite-bearing sequences have yielded lower Aragonian (middle Miocene) mammalian fossils (Dupuy de Lôme and Fernández de Caleya, 1918; Adrover et al., 1978).

The Libros Gypsum is up to 120 m thick in Las Minas de Libros section (6 km to the east of Libros village)

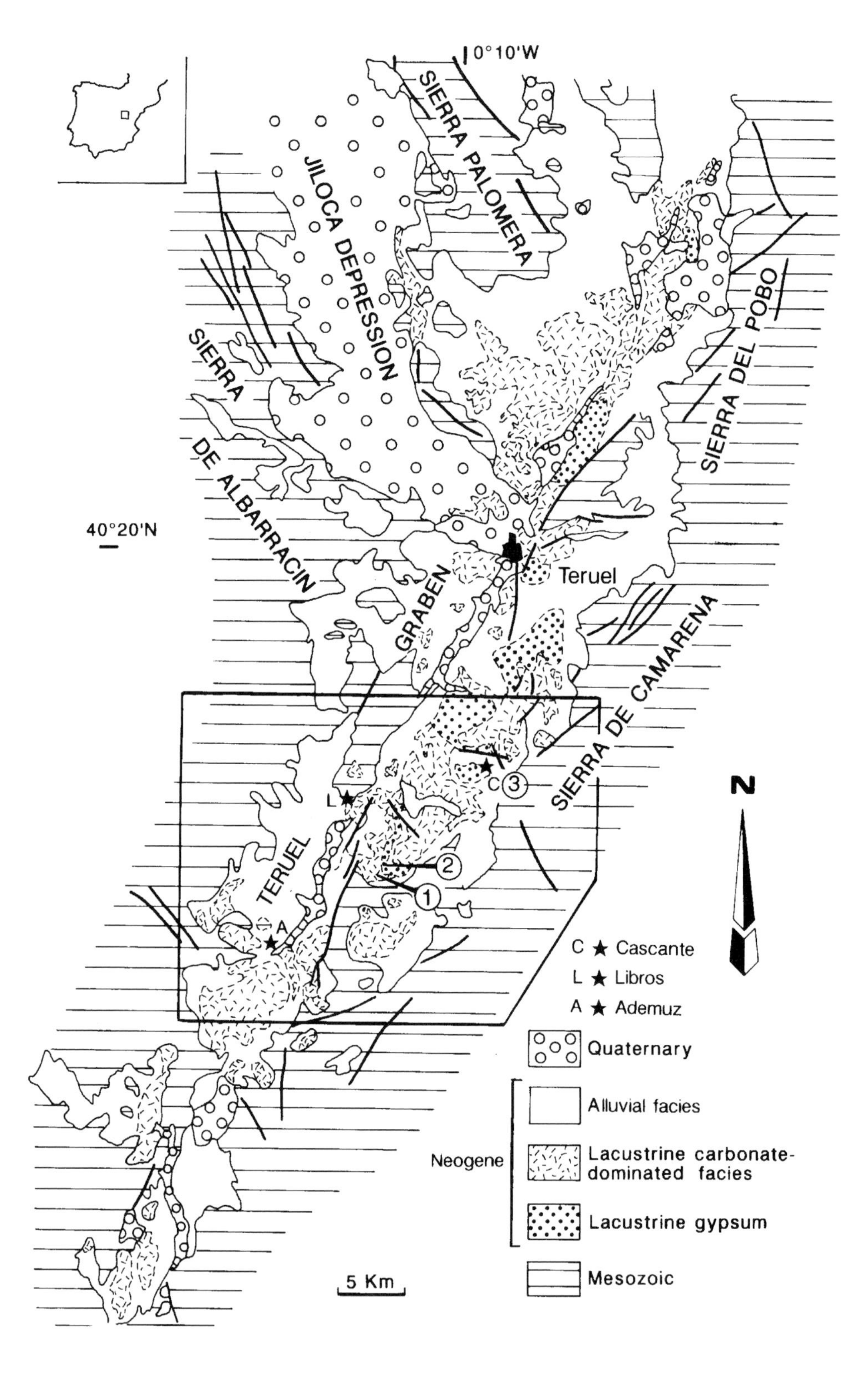

Figure 1—Facies distribution in the Teruel graben. Encircled numbers refer to the location of detailed lithologic sections in Figure 4. Stars indicate the main localities (villages). Inset shows the studied area in the southern Teruel graben.

where it consists, in its lower part, of an alternation of laminated mudstones (oil shales), limestones, and laminated gypsum (sections 2 and 2a in Figure 4). The upper part is mainly formed by laminated gypsum with minor intercalations of mudstones and limestones. Native sulfur occurs associated with some gypsum and limestone beds. Late Miocene mammal fossils have been reported from this unit (Roman, 1927; Gautier et al., 1972).

The upper alluvial unit consists of red mudstones with interbedded sandstones and minor conglomerates and limestones up to 40 m thick.

The La Nava (or Loma de La Nava) Limestone is mainly composed of massive and bioturbated limestones and interbedded carbonate mudstones (section 3 in Figure 4). Tufa limestones are common. This unit is up to 60 m thick and contains a number of lower and middle Turolian (uppermost Miocene) mammal sites (Anadón and Moissenet, 1990).

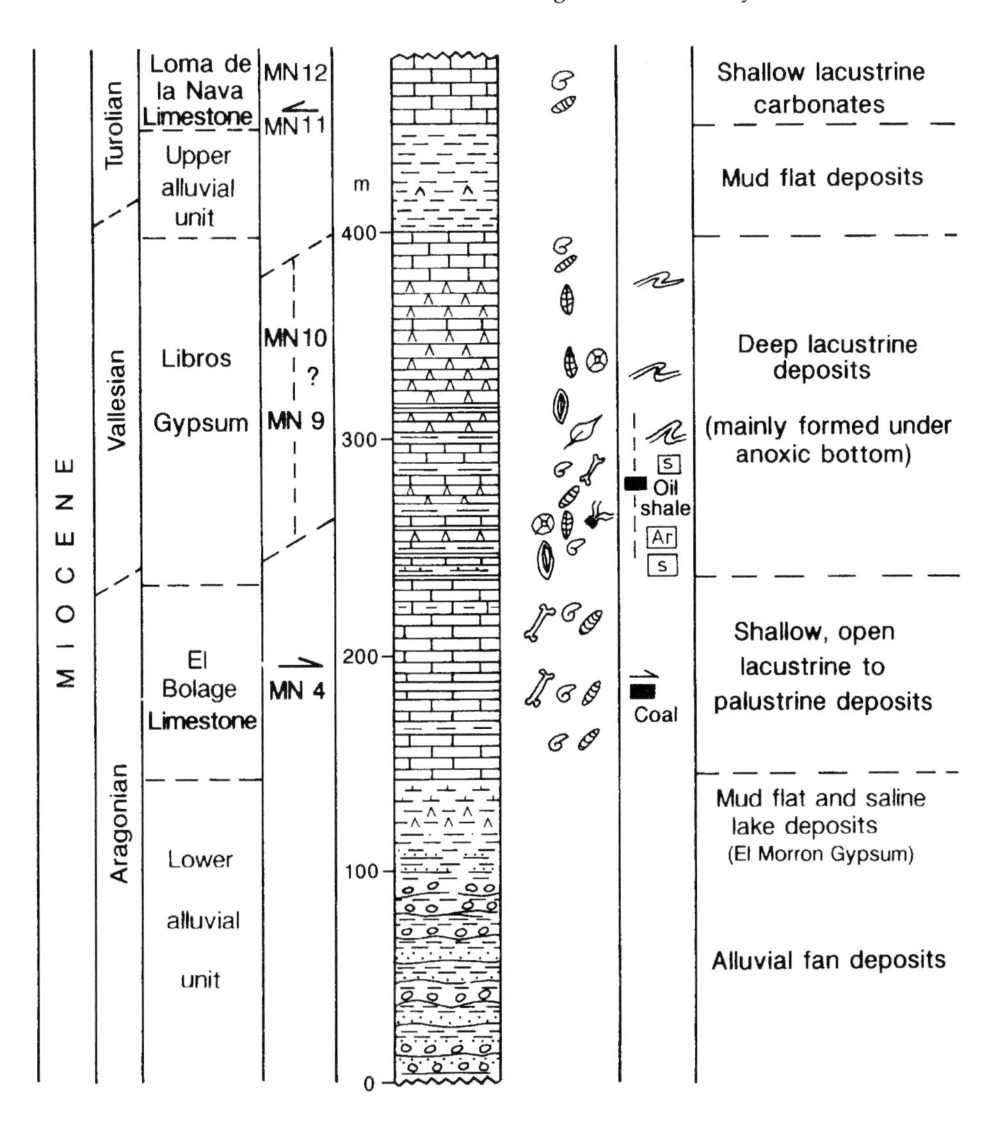

Figure 2—General stratigraphic log for the Neogene sequence in the Libros area (southern Teruel graben) showing age, formations, and mammal biozones (MN zones). See legend in Figure 4.

To the south (Ademuz zone), overlying the Turolian limestones, a succession of red mudstones and sandstones, carbonate mudstones, and tufa limestones has yielded lower Pliocene mammalian fossils (Adrover et al., 1978). To the north, near Cascante, an upper Turolian succession, up to 100 m thick, consisting of an alternation of red siliciclastic deposits, evaporites (Cubla Gypsum), and lacustrine limestones (El Pañuelo Limestone) overlies La Nava Limestone (Figure 3).

Lacustrine Facies

The main primary lithofacies of the lacustrine deposits consist of carbonates, evaporites (gypsum), and mudstones.

Carbonates

Massive limestones usually consist of massive beds, up to 1 m thick, of micritic mudstones-wackestones or packstones with charophytes (oogonia and stems), gastropods, and ostracods. Moldic porosity is common and bioturbation traces are abundant, especially at the tops of beds. Textures display a nodular aspect and pedogenic features. Massive limestones alternate with carbonate siltstones or constitute thick massive limestone sequences.

Laminated limestones are composed of dark gray to brown, finely laminated wackestones, and micritic mudstones and packstones. Skeletal grains are charophyte calcifications and fragments of gastropod and ostracod shells. Laminated limestones occur at the base of some massive limestone or gypsum beds or, more commonly, form thick sequences.

Tufas formed by carbonate encrustations on reeds or stems of aquatic plants are common. In some places the reeds are vertically oriented, indicating that the encrustations formed in situ, although commonly tufa beds consist of accumulations of transported, broken encrustations.

Evaporites

Gypsum is the unique primary evaporite deposit found in the Teruel graben. Massive gypsum consists of thin to thick beds of subhedral to euhedral microlenticular gypsum, mixed with micritic carbonate, clays, or silts in variable proportions. Bioturbation structures, probably produced by insects or polychaete worms (cf. Rodriguez Aranda, 1992), are

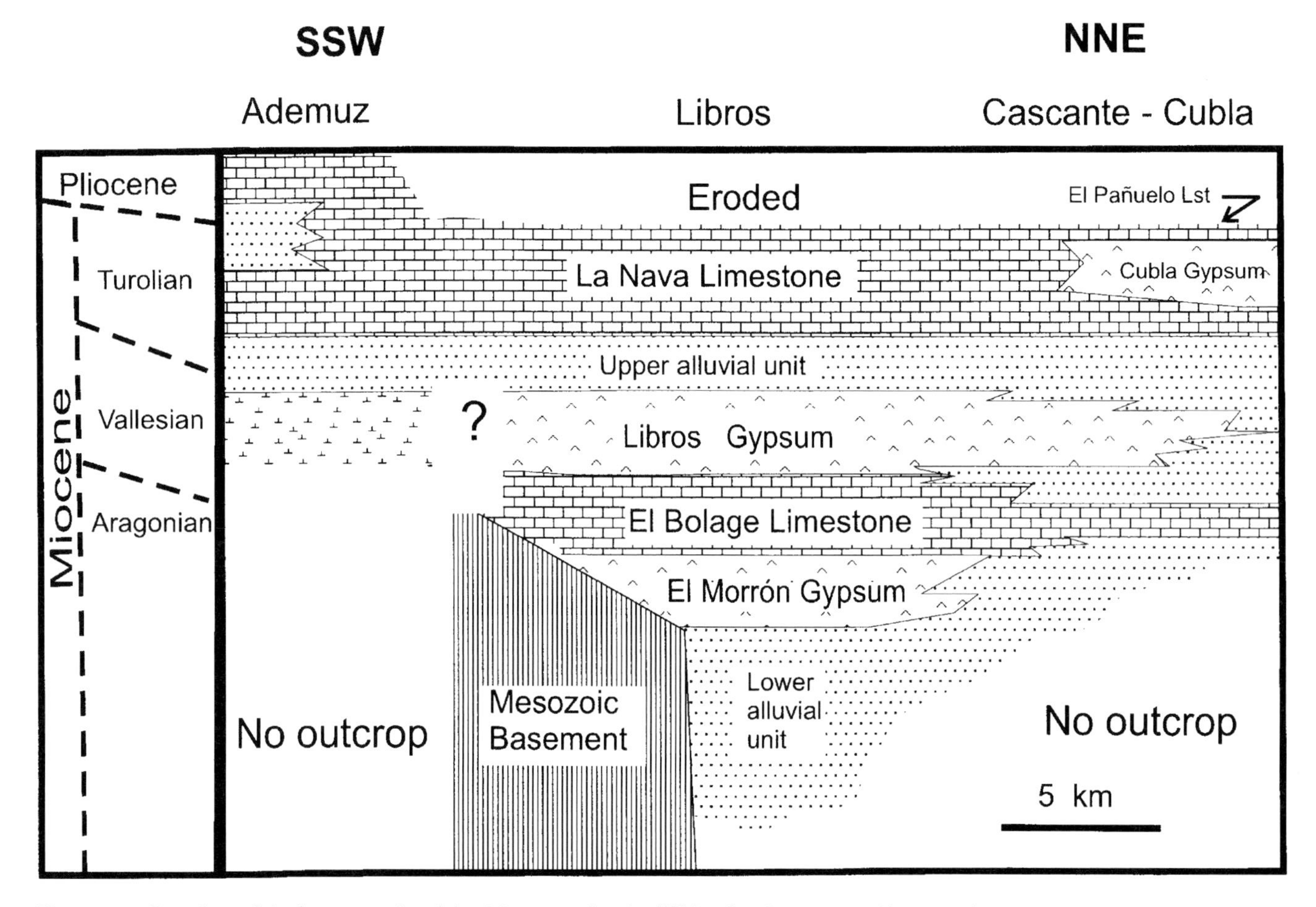

Figure 3—Stratigraphic framework of the Neogene basin fill in the Cascante-Libros-Ademuz area. No vertical scale.

abundant. Massive gypsum beds usually alternate with massive, gray, green, or red mudstones or are grouped in sequences several meters thick.

Laminated gypsum consists of thin-bedded to laminated, fine-grained microlenticular or anhedral gypsum, which, in some cases, contains diatoms and sponge spicules.

Some diagenetic gypsum deposits, formed by irregular bodies that partially replace limestone beds, probably were derived as by-products from the oxidation processes linked to sulfate reduction that produced native sulfur in the Minas de Libros zone. This gypsum exhibits anhedral textures.

Mudstones

Lacustrine mudstones (siliciclastic mudrocks to lutites) are associated with carbonate or gypsum beds. Massive mudstones are green, brown or pale red, typically display a high carbonate content (carbonate mudstones), and contain charophyte oogonia and stems, as well as gastropod and ostracod shells. Bioturbation features are abundant.

Dark, laminated mudstones (including oil-shales) are usually associated with laminated limestones or laminated gypsum beds. Laminated mudstones, up to 10 m thick, are common in the Minas de Libros zone where they are organic-rich (oil shales) and contain abundant diatoms and sponge spicules. Well-preserved remains of exotic fossils occur, such as amphibians, birds, and snakes (Navás, 1922a, b; Luque et al., 1996). Kiefer (1988) reports that the organic matter (OM) in this facies consists mainly of algal debris plus amorphous kerogen, with vitrinite reflectance R_o = 0.42 and TOC (total organic carbon) values up to 35%; pyrolysis data indicate type I kerogen; nevertheless, other analyses of Libros oil shales showed lower OM content. TOC from oil-shale outcrops ranged from 1 to 2.6% and mean oil yield from well samples attained 180 l/MT (liter/metric ton) (IGME, 1981). Anadón et al. (1988) indicated that laminated diatomitic muds yield 10 kg/MT, a hydrogen index of 340 mg HC/g TOC, and that immature organic matter derived from waxy terrestrial plants with relatively minor algal/bacterial contribution. A similar origin for this oil-prone organic matter has been deduced from biomarker studies (De las Heras, 1991).

Coal (Lignite)

The lignite beds are up to 3 m thick (Dupuy de Lome and Fernández de Caleya, 1918; Arce et al., 1983) and are associated with carbonate mudstones and limestones that contain abundant gastropods, ostracods, and charophytes (Gautier et al., 1972; Ludwig, 1987). They are located in the lower part of El Bolage Limestone (cf. Gautier et al., 1972), in which the thickest coal bed was mined at the beginning of the twentieth century.

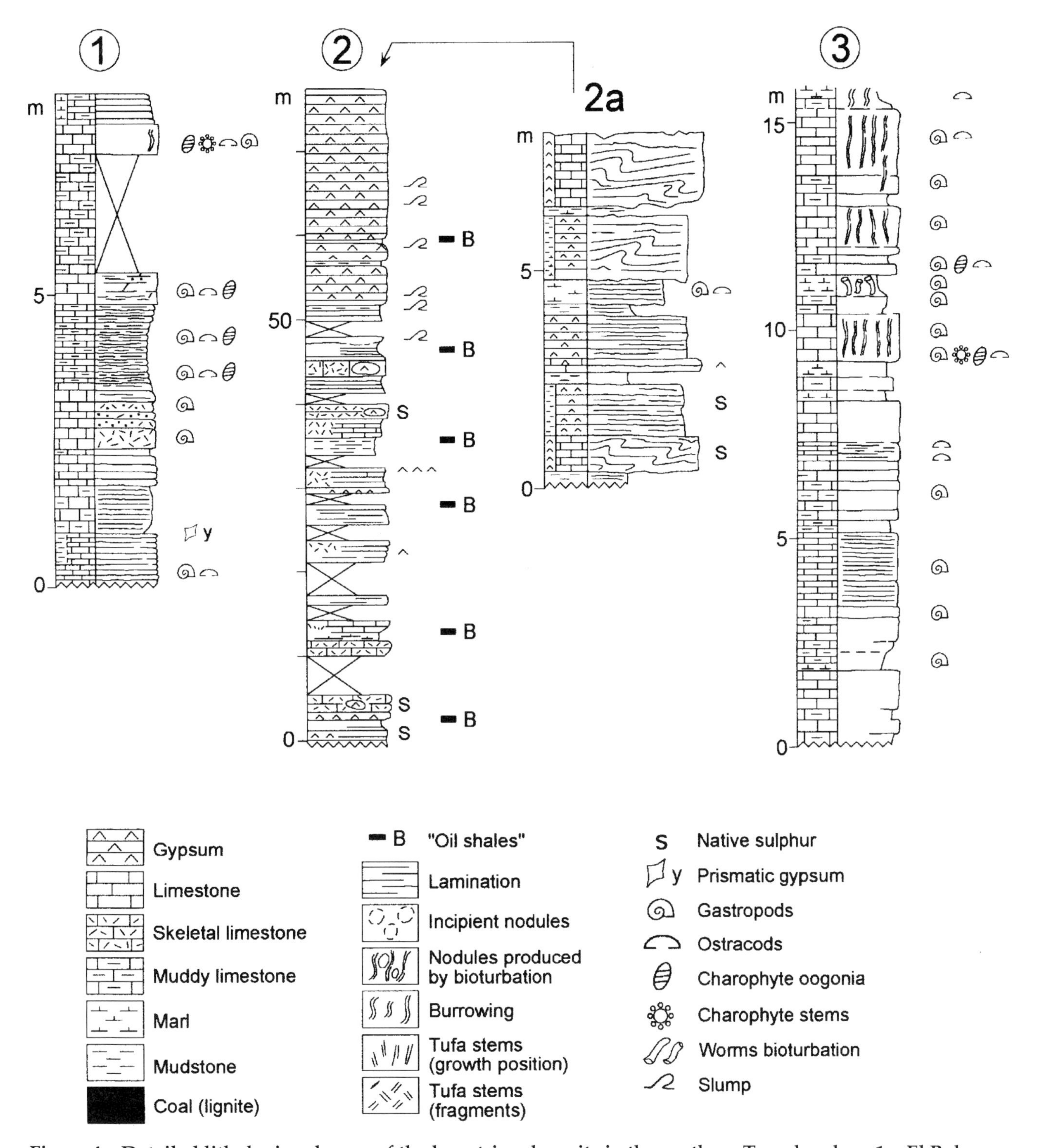

Figure 4—Detailed lithologic columns of the lacustrine deposits in the southern Teruel graben. 1 = El Bolage Limestone. 2 = Libros Gypsum (Minas de Libros section). 2a = Detailed log for Libros Gypsum deposits 14 m stratigraphically over top beds of section 2 (covered interval, Minas de Libros section). 3 = La Nava Limestone. See Figure 1 for locations.

Lacustrine Systems

Carbonate Lacustrine Systems

The main facies that characterizes the Aragonian lacustrine system (El Bolage Limestone) in the Minas de Libros area are laminated limestones, 50–80 m thick, with minor mudstone intercalations. Preservation of lamination, rare bioturbation, and fossil content (abundant charophyte remains, ostracods, and freshwater mollusks, e.g., Planorbidae) indicate deposition in open areas of freshwater lakes. To the north, these deposits pass laterally to a thinner succession (15 m near Cascante) of bioturbated limestones and carbonate mudstones with pedogenic features suggesting shallow water and palustrine environments. To the south, several thin coal beds and mudstones located in the lower part of the succession indicate shallow, swampy areas (Ludwig, 1987).

The Turolian lacustrine system (La Nava Limestone) is mainly characterized by bioturbated limestones with minor intercalations of mudstones and laminated limestones. Shallowing-upward cycles are common, with increasing bioturbation and pedogenic features to the top of beds. These features and the fossil content (pulmonate gastropods, ostracods, and charophytes) indicate a freshwater, shallow lacustrine environment.

Other carbonate lacustrine systems developed during the Vallesian and early Pliocene in the southern part of the basin (Ademuz area); nevertheless, they are not well understood because of scarce stratigraphic and sedimentologic data.

Evaporite Lacustrine Systems

Gypsum deposits formed in shallow evaporite lacustrine systems (El Campo Gypsum, Cubla Gypsum, El Morrón Gypsum, and other minor gypsum units) are up to several meters thick and are exclusively or predominantly formed by massive, bioturbated microlenticular gypsum. Minor mudstone and thin carbonate intercalations also occur. The gypsum deposits precipitated in shallow evaporite ponds and are related to red mudstones and sandstones that formed in mudflat environments. The inner lacustrine facies of El Morrón Gypsum are formed by massive gypsum with minor laminated gypsum intercalations. Locally, at the base of a sequence, some cycles formed by red or green mudstone and laminated gypsum also occur. These features suggest that evaporite deposition took place in a shallow lacustrine environment with an oscillating water level.

A deep evaporite lacustrine system (Libros Gypsum) is interpreted for the Libros Gypsum (Vallesian evaporite lacustrine system of Libros-Cascante zone). The inner areas of this system are represented by the Minas de Libros succession that is up to 120 m thick. It is characterized by laminated gypsum locally associated to massive gypsum, massive mudstones, laminated carbonates, or oil shales. The presence of laminated gypsum and oil shales indicates deeper depositional conditions, although depth is difficult to assess. An abundance of laminated facies (gypsum, limestones, oil shales) and slumps and preservation of organic matter and exotic fossils indicate a relatively deep lacustrine environment with an anoxic bottom (Anadón et al., 1989). Salinity fluctuations were frequent, and diatoms and sponge spicules, typical of fresh to slightly saline waters, are present in the oil shales (Margalef, 1947; Gregor and Günther, 1985; Gil, 1986). The high sulfate concentration in the lake can be attributed to the recycling of the Upper Triassic marine evaporites (mainly gypsum) that crop out extensively near the basin margin. This main solute acquisition process has been documented by isotopic analyses of the Libros Gypsum (Utrilla et al., 1992). Diagenetic processes linked to sulfate reduction formed the native sulfur deposits and diagenetic aragonite and gypsum that replace primary gypsum or carbonates (Anadón et al., 1992). To the north, this succession changes laterally to cyclic sequences of massive mudstones, thin-bedded limestones and massive gypsum (Cascante zone) (Ortí, 1987), and to red alluvial deposits.

EVOLUTION OF THE LAKE SYSTEMS

During the early Miocene, the southern Teruel graben was dominated by alluvial environments. Lacustrine environments were poorly developed and consisted of small saline (sulfate-dominated) ponds, developed in mudflats, where bioturbated, massive gypsum was deposited (e.g., El Campo Gypsum). Near the lower–middle Miocene boundary, a larger saline lacustrine environment developed. The inner, deeper areas were located east of Libros (El Morrón) where massive and laminated gypsum alternations were deposited.

Lacustrine sedimentation, during the middle Aragonian (middle Miocene) occurred in a freshwater carbonate lake system. The open lacustrine areas, where laminated facies originated, were located east of Libros. In marginal, palustrine areas (Libros village and Cascante zones) bioturbated limestones and carbonate mudstones were formed. To the south, swampy areas led to the accumulation of terrestrial and lacustrine organic-rich deposits (lignites).

During the Vallesian (late Miocene), a lacustrine expansion took place in the southern Teruel graben and an extensive saline (sulfate-rich) lacustrine system developed in the Libros-Cascante zone. The inner, deeper areas of this saline lacustrine system correspond to the Minas de Libros, where anoxic bottom conditions developed leading to formation of laminated facies (gypsum, carbonates, and oil shales). Instability in the lake margin produced several slumps. Both salinity and lake levels experienced significant variations, as suggested by the cyclic sequences involving oil shales with oligosaline diatoms, laminated gypsum, and biogenic carbonates. To the north, in marginal zones of the evaporite system (i.e., west of Cascante), lake level variations produced cycles characterized by bioturbated massive mudstones overlain by laminated and massive gypsum with minor laminated carbonate. To the south of Ademuz, a shallow carbonate lacustrine system developed. This was characterized by marl and tufa formation, and its links with the Vallesian evaporite system are unknown (Adrover et al., 1978).

A notable lacustrine regression took place during the late Vallesian and early Turolian (late Miocene): alluvial sedimentation was widespread and mudflat environments dominated the inner parts of the southern Teruel graben. This alluvial phase was followed during the early–middle Turolian (late Miocene) by the onset of a large, shallow freshwater lacustrine system where bioturbated massive limestones formed.

During the late Turolian and early Pliocene, shallow freshwater lacustrine environments persisted in the southern areas (Ademuz zone). Small saline ponds developed during the late Turolian in the northern areas (Cubla zone).

ACKNOWLEDGMENTS

This work has been partially supported by projects PB 90-0080 and PB 94-0085. We would like to thank Elizabeth Gierlowski-Kordesch and two anonymous reviewers who provided constructive comments and suggestions.

REFERENCES CITED

Adrover, R., 1986, Nuevas faunas de roedores en el Mio-plioceno continental de la región de Teruel (España). Interés biostratigráfico y paleoecológico: Publicaciones del Instituto de Estudios Turolenses, 433 p.

Adrover, R., P. Mein, and E. Moissenet, 1978, Nuevos datos sobre la edad de las formaciones continentales neógenas de los alrededores de Teruel: Estudios Geológicos, v. 34, p. 205–214.

Alcalá, L., 1994, Macromamíferos neógenos de la fosa de Alfambra-Teruel: Instituto de Estudios Turolienses–Museo Nacional de Ciencias Naturales, CSIC, 554 p.

Anadón P., and E. Moissenet, 1990, Part II Excursion guidebook to the Neogene grabens of the Eastern Iberian Chain, *in* J. Agustí and J. Martinell, eds., Iberian Neogene basins: field guidebook: Paleontología i Evolució, Memoria Especial 2, p. 103–130.

Anadón, P., and E. Moissenet, 1996, Neogene basins in the Eastern Iberian Range, *in* P. F. Friend and C. Dabrio, eds., Tertiary basins of Spain: Cambridge, Cambridge University Press, p. 68–76.

Anadón, P., and E. Roca, 1996, Geological setting of the Tertiary basins of Northeast Spain, *in* P. F. Friend and C. Dabrio, eds., Tertiary basins of Spain: Cambridge, Cambridge University Press, p. 43–48.

Anadón, P., J. Cawley, and R. Julià, 1988, Oil source rocks in lacustrine sequences from tertiary grabens, western Mediterranean rift system, northeast Spain: AAPG Bulletin, v. 72, p. 983.

Anadón, P., F. Ortí, and L. Rosell, 1997, Unidades evaporíticas de la zona de Libros-Cascante (Mioceno, Cuenca de Teruel): Características estratigráficas y sedimentológicas: Cuadernos de Geología Iberica, v. 22, p. 283–304.

Anadón, P., L. Rosell, and M. R. Talbot, 1992, Carbonate replacement of lacustrine gypsum deposits in two Neogene continental basins: Sedimentary Geology, v. 78, p. 201–216.

Anadón, P., L. Cabrera, R. Julià, E. Roca, and L. Rosell, 1989, Lacustrine oil-shale basins in Tertiary grabens from NE Spain (Western Eurpean rift system): Palaeogeogaphy, Palaeoclimatology, Palaeoecology, v. 70, p. 7–28.

Arce, M., J. Boquera, V. Calderón, C. Dabrio, and M. A. Zapatero, 1983, Posibilidades lignitíferas de la cuenca Neógena de Ademuz (Fosa de Teruel): Boletín Geológico y Minero, v. 94, p. 415–425.

Bakx, L. A. J., 1935, La Géologie de Cascante del Río et de Valacloche (Espagne): Leidse Geologie Medelingen, v. 7, p. 157–220.

Besems, R. E., and A. van de Weerd, 1983, The Neogene rodent biostratigraphy of the Teruel-Ademuz Basin (Spain): Koninklije Nederlandse Akademie van Wetenschappen, Proceedings, ser. B, v. 86, p. 17–23.

Broekman, J. A., 1983, Environments of deposition, sequences and history of the Tertiary continental sedimentation in the Basin of Teruel-Ademuz (Spain): Koninklije Nederlandse Akademie van Wetenschappen, Proceedings, ser. B, v. 86, p. 25–37.

Broekman, J. A., R. E. Besems, P. van Daalen, and K. Steensma, 1983, Lithostratigraphy of Tertiary continental deposits in the Basin of Teruel-Ademuz (Spain): Koninklije Nederlandse Akademie van Wetenschappen, Proceedings, ser. B, v. 86, p. 1–16.

Dupuy de Lome, E., and C. Fernández de Caleya, 1918, Nota acerca de un yacimiento de mamíferos fósiles en el Rincón de Ademúz (Valencia): Boletín del Instituto Geológico y Minero de España, v. 19, p. 299–348.

De las Heras, X., 1991, Geoquímica orgànica de conques lacustres fòssils: Institut d'Estudis Catalans, Arxius de la Secció de Ciències, v. 97, 324 p.

Gautier, F., E. Moissenet, and P. Viallard, 1972, Contribution à l'étude stratigraphique et tectonique du fossé néogène de Teruel (Chaînes Ibériques, Espagne): Bulletin du Muséum National d'Histoire Naturelle, v. 77, p. 179–207.

Gil, M. J., 1986, Organic microscopic remains in Miocene lacustrine sediments near Libros (Teruel, Spain): Hydrobiologia, v. 143, p. 209–212.

Gregor, H-J., and T. Günther, 1985, Neue Pflanzenfunde aus dem Vallesium (jüngeres Neogen) von Libros (Becken von Teruel, Spanien): Mitteilungen des Badischen Landesvereins für Naturkunde und Naturschutz, v. 13, p. 297–309.

Guimerá, J., 1996, Cenozoic evolution of eastern Iberia: structural data and dynamic model: Acta Geologica Hispanica, v. 29 (1994), p. 57–66.

Guimerá, J., 1997, Las fosas neógenas de Teruel y el Jiloca: su relación con la estructura cortical, *in* J. P. Calvo and J. Morales eds., Avances en el conocimiento del Terciario Ibérico: Madrid, Museo Nacional de Ciencias Naturales, p. 105–108.

Hernández, A., A. Godoy, M. Alvaro, et al., 1985, Memoria del Mapa Geológico de España, Escala 1:200.000, (2nd serie): Hoja de Teruel, v. 47, 192 p.

IGME, 1981, Ampliación de la investigación de pizarras bituminosas en el sector de Libros (Teruel): Unpublished Report, Instituto Geológico Minero de España, Madrid, 48 p.

Kiefer, E., 1988, Facies development of a lacustrine tecto-sedimentary cycle in the Neogene Teruel-Ademuz Graben (NE Spain): Neues Jahrbuch für Geologie und Paläontologie Monatshefte, v. 1988 (6), p. 327–360.

Luque, L., L. Merino, B. Sanchiz, and L. Alcalá, 1996, Procesos de fosilización de las ranas miocenas de Libros (Teruel): II Reunión de Tafonomía y fosilización, Comunicaciones, p. 169–174.

Ludwig, K., 1987, Mikropaläontologische Hinweise für

die autochthone Entstehung der miozänen Braunkohle mi Becken von Teruel-Ademuz (NE-Spanien): Paläontologische Zeitschrift, v. 61, p. 17–27.

Margalef, R., 1947, Observaciones micropaleontológicas sobre los yacimientos lacustres miocénicos de Libros (Teruel): Estudios Geológicos, v. 5, p. 171–177.

Mein, P., E. Moissenet, and R. Adrover, 1983, L'extension et l'âge des formations continentales pliocènes du fossé de Teruel (Espagne): Compte Rendues de l'Academie des Sciences de Paris, II série, v. 296, p. 1603–1610.

Moissenet, E., 1980, Relief et déformations récentes: trois transversales dans les fossés internesdes chaînes ibériques orientales: Revue géographique des Pyrénées et du Sud-Ouest, v. 51, p. 315–344.

Navás. L., 1922a, Algunos fósiles de Libros (Teruel): Boletín de la Sociedad Ibérica de Ciencias Naturales, v. 21, p. 52–61.

Navás. L., 1922b, Algunos fósiles de Libros (Teruel), Adiciones y correcciones: Boletín de la Sociedad Ibérica de Ciencias Naturales, v. 21, p. 172–175.

Ortí, F., 1987, La zona de Vilel-Cascante- Javalambre. Introducción a las formaciones evaporíticas y al volcanismo jurásico, *in* M. Gutierrez and A. Meléndez, eds., XXI Curso de Geología Práctica de Teruel, Teruel: Colegio Universitario de Teruel, p. 56–95.

Roca, E.,1996, La evolución geodinámica de la Cuenca Catalano-Balear y áreas adyacentes desde el Mesozoico hasta la actualidad: Acta Geologica Hispanica, v. 29 (1994), p. 3–25.

Rodríguez Aranda, J. P., 1992, Significado de bioturbaciones en un medio evaporítico continental (Mioceno de la Cuenca de Madrid): Geogaceta, v. 12, p. 113–115.

Roman, F., 1927, Sur quelques restes de mammifères découverts par le R. P. Longinos Navas dans les argiles pontiques de Libros (Teruel): Bulletin de la Societé Géologique de France, v. 27, p. 379–385.

Utrilla, R., C. Pierre, F. Ortí, and J. J. Pueyo, 1992, Oxygen and sulphur isotope composition as indicators of the origin of Mesozoic and Cenozoic evaporites from Spain: Chemical Geology (Isotope Geoscience Section), v. 102, p. 229–244.

Basilici, G., 2000, Pliocene lacustrine deposits of the Tiberino Basin (Umbria, central Italy), *in* E. H. Gierlowski-Kordesch and K. R. Kelts, eds., Lake basins through space and time: AAPG Studies in Geology 46, p. 505–514.

Chapter 47

Pliocene Lacustrine Deposits of the Tiberino Basin (Umbria, Central Italy)

Giorgio Basilici
Departmento de Administração e Política de Recursos Minerais, Instituto de Geociências, Universidade Estadual de Campinas Campinas, Brazil

INTRODUCTION

Pliocene lacustrine deposits are found in the Tiberino Basin, a depositional basin located in Umbria (central Italy) about 100 km north of Rome. The basin extends for 125 km in the north-northwest direction, with a width of up to 20 km (Figure 1). The Tiberino Basin is an extensional basin situated at the western margins of the Apennines Chain. Its origin is older than the middle Pliocene (sensu Rio et al., 1994) and it is connected with the tectonic extensive regime that brought about the opening of the Tyrrhenian Sea (Lavecchia, 1988). It is an asymmetric basin delimited on the eastern margin by a master normal fault (Martana fault) and by anthitetic normal faults on the western margins (Figure 2A). About 450 m of sediments are exposed in the study area, but a gravimetric survey (Ambrosetti et al., 1993) identified the depocenter to the east of the basin axis with maximum sediment thickness of 2300 m near the village of Collevalenza (Figure 2B).

The Tiberino Basin fill consists principally of siliciclastic deposits. There are four lithostratigraphic formations recognized in the area between Marsciano and Terni (Figure 3): the Fosso Bianco Formation (FBF), the Ponte Naja Formation (PNF), the Santa Maria di Ciciliano Formation (SMCF), and the Acquasparta Formation (AF) (Basilici, 1992).

The Fosso Bianco and the Ponte Naja Formations were deposited in an extensive lacustrine system; their exposed thicknesses are 250 m and 140 m, respectively. The description of these formations is the object of this contribution. Pollen grains, as well as mammal and malacofauna remains, date these sediments as middle–late Pliocene (Follieri, 1977; Esu and Girotti, 1991; Basilici et al., 1995; Basilici, 1997).

The Santa Maria di Ciciliano Formation, 180 m thick, consists of sandy bodies, formed in meandering channels, alternated with silty clays in an alluvial plain system. This formation unconformably overlies the Fosso Bianco Formation, with normal faults separating it from the Ponte Naja Formation near Todi (Figure 2A). Mammal remains and pollen grains (Follieri, 1977; Ambrosetti et al., 1995) date the Santa Maria di Ciciliano Formation as early Pleistocene in age. The Acquasparta Formation, 30 m thick, is composed of carbonate deposits and travertine. This formation represents sedimentation in small and shallow closed lakes on the eastern margin of the basin. Malacofauna remains place the Acquasparta in the early Pleistocene (Ambrosetti et al., 1987).

LACUSTRINE DEPOSITIONAL SYSTEM

Three major facies associations (A, B, C) have been identified in the Fosso Bianco Formation and one facies association (D) in the Ponte Naja Formation from a total of 12 measured sections and 10 drilling cores.

Facies association A is the most extensive; it constitutes more than 80% of the measured sections and can be 150 m or more in thickness (Figure 4). Bluish gray, marly clay is the most common facies; planar parallel, continuous, sandy laminae and dark bluish gray, marly clay laminae are interlayered with the massive marly clay. To a lesser extent, thin rhythmic alternations of marly clay with silty, calcareous laminae are present. This association contains rare fossil remains. Most common are leaves with preserved organic cuticles; freshwater ostracods and gastropods are also rare and found in coarser laminae. Rare thin cylindrical burrows, millimeters in size, also are present.

Facies association A has been interpreted as deep offshore lacustrine deposits. This interpretation is supported by the fine-grained sediments, sedimentary structures, fossils, and vertical and lateral continuity of the facies. The rhythmic facies evidence variable depositional processes (settling of mud, organic material, or carbonates, as well as low-density turbidity currents) alternating in time on an anoxic lacustrine bottom that preserved sedimentary structures (Glenn and Kelts, 1991).

Facies association B consists of coarse-grained deposits (sands, gravelly muds, and sandy gravels) alternating with bluish gray, marly clays (Figure 5). Thickness of this association ranges from 10–50 m in

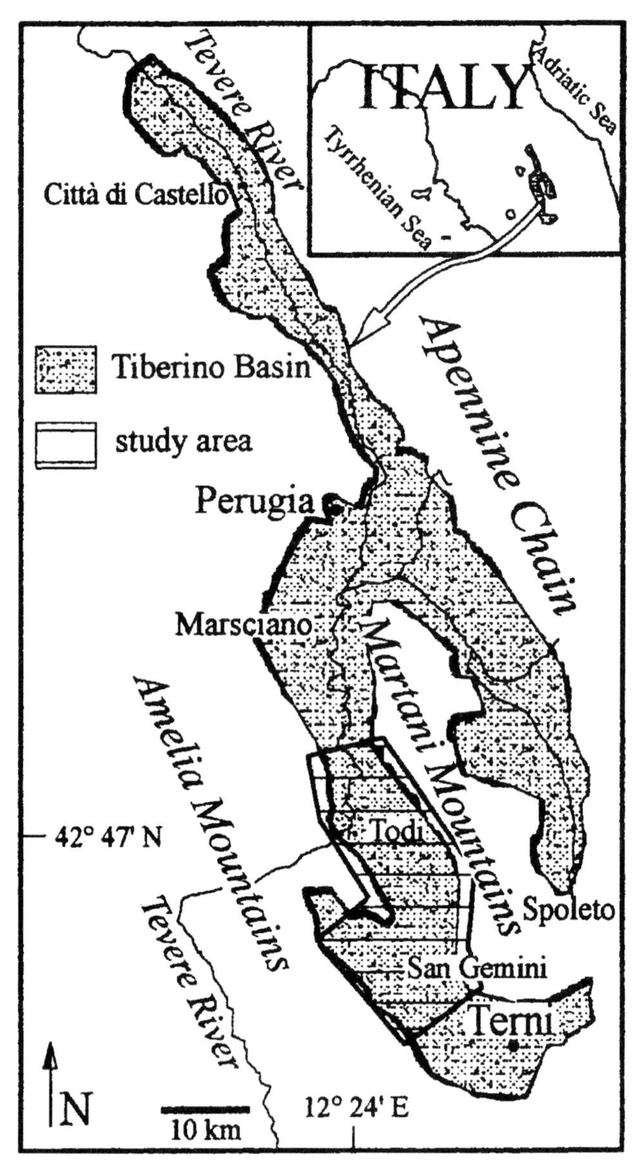

Figure 1—Location of the Tiberino Basin, including the principal extensional basins and the principal thrust fronts in central Italy. The study area is defined by horizontal lines.

the measured sections. Very fine to medium sands form layers 2–25 cm thick. Perpendicular to paleocurrent direction, these sands are several tens of meters wide and show a faintly lenticular shape, with planar bottom and convex top. Each single layer is graded and organized into planar-parallel laminations, overlain by current ripple cross-lamination (commonly climbing) and by mud/silt laminae; in some cases, only lenticular layers of asymmetric ripples can be observed. Matrix-supported gravelly muds (Figure 5) constitute rectangular beds 50 cm to 3 m thick. Disorganized fabric is the principal characteristic of this facies; contorted fragments of resedimented strata may be found within these beds. Sandy, gravelly bodies crop out in two sites at the margins of the basin; they overlie marly clays alternated with sands and gravelly muds. These sandy gravels contain grain-supported disorganized pebbles with sandy matrix; they rarely show blade-like clasts with a (p) a (i) imbrication. Strata are organized into foresets that are inclined up to 20°, tangential toward the bottom, and up to 50 m high. Paleocurrent data show current directions perpendicular to the pre-Pliocene margins.

These sandy strata are very similar to those described as facies D by Mutti and Ricci Lucchi (1972), or as facies C.2.2 by Pickering et al. (1986). These strata may be interpreted as turbidity current sediments deposited from intermediate concentration flows (Pickering et al., 1986), and the gravelly muds represent cohesive debris flows (sensu Lowe, 1979). Density-modified grain flows (Lowe, 1976) are invoked as depositional mechanisms for the sandy gravelly bodies, representing the frontal portion of Gilbert-type deltas.

In the measured sections, sands and gravelly muds, alternated with marly clays, lie above facies association A and are mostly distributed near the pre-Pliocene substratum. Sedimentary features and paleocurrent data support the interpretation that facies association B consists of prodelta deposits.

Facies association C is characterized by two facies subassociations and is found only along the western margins of the Tiberino Basin.

Facies subassociation C1 (Figure 6A) is represented by alternating sandy and marly clayey layers with thicknesses reaching about 15 m. The sands, 1–110 cm thick, form planar-convex lenses with abrupt, rarely erosive, lower contacts. These sand layers are graded and characterized by low-angle undulated or parallel laminations, which are organized into sets that cut into analogous underlying sets. The upper part of each sandy sedimentation unit shows wave ripple lamination, sometimes with spill-over ripple crests (Seilacher, 1982). The marly clays, 1–15 cm thick, have sharp contacts with the sands and show interbedded planar, parallel, and continuous laminae of silts or very fine sands.

Each couplet of sand and marly clay constitutes a depositional sequence similar to the "HFXM hummocky stratification type" of Dott and Bourgeois (1982). Sandy strata represent lacustrine storm wave deposits and marly clays the normal background sedimentation. Facies subassociation C1 is interpreted as a lacustrine wave-dominated coastline.

Facies subassociation C2 (Figure 6B) is approximately 40 m thick; it is composed of interfingered terrigenous fine deposits (massive silty clays and laminated clays) and to a lesser extent by lignites and sands. Massive silty clays, occurring in beds from 30 cm to 4 m thick, contain root traces, fossil remains of marsh gastropods, bioturbation features, calcareous or siderite nodules, and in situ trunks of the swamp tree *Glyptostrobus* (Martinetto, 1994). Laminated clays, 60–80 cm thick and not more than 20 m in lateral extent, exhibit freshwater fauna and flora, including remains of fishes, gastropods, and aquatic plants. Lignite layers, from a few centimeters

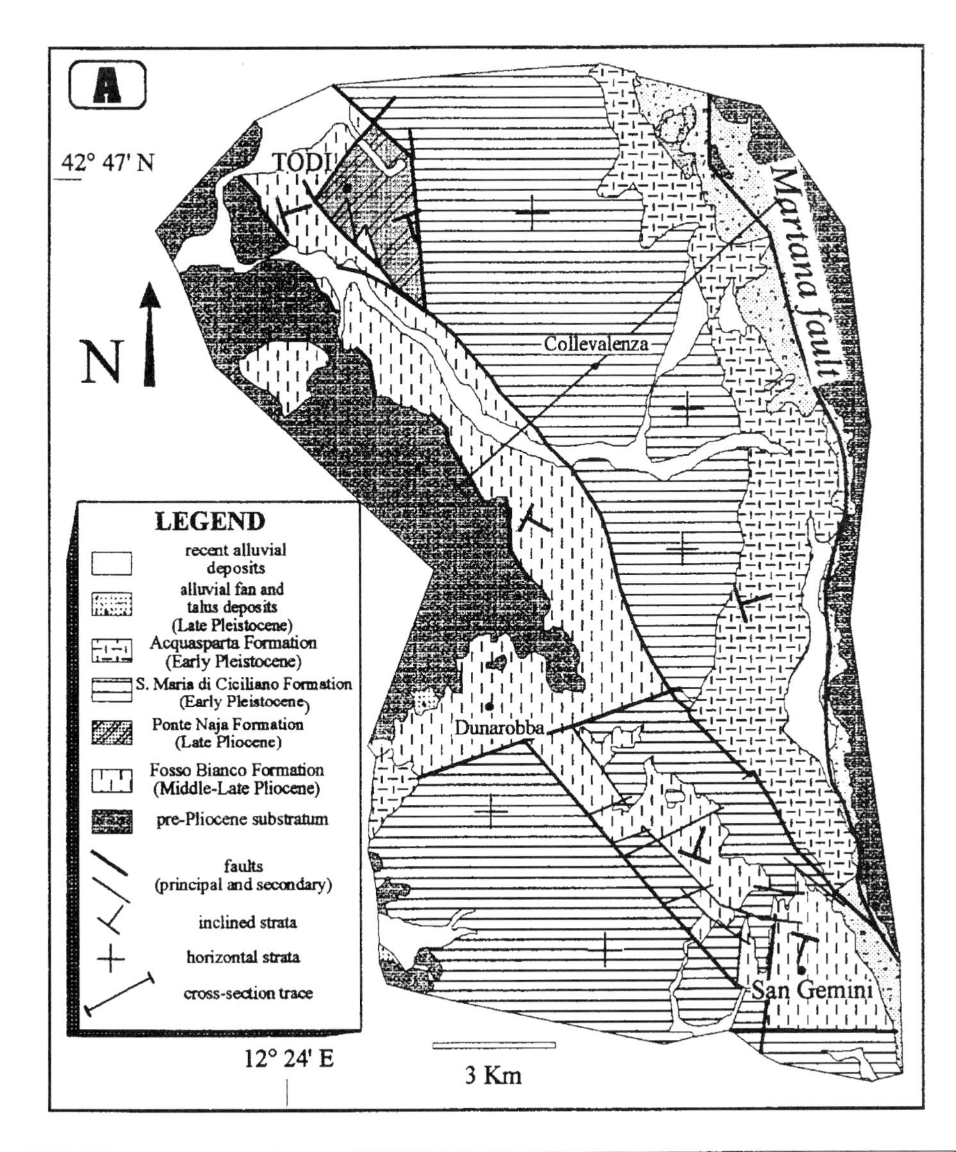

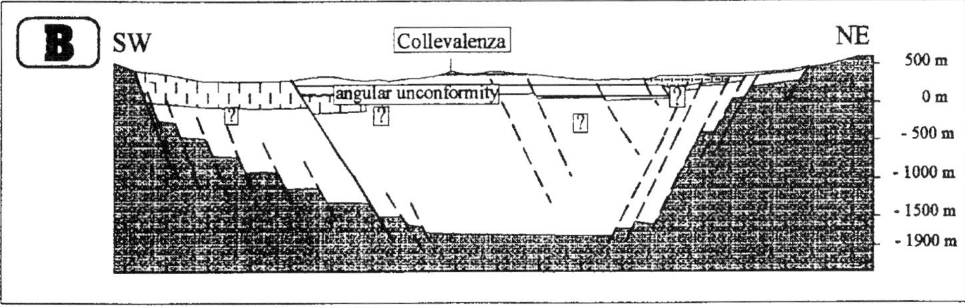

Figure 2—(A) Geologic map of the southwestern branch of the Tiberino Basin. (B) Schematic cross-section of the Tiberino Basin. Surface distribution data of the lithostratigraphic formations was done by the geologic survey, whereas the position of the pre-Pliocene substratum was derived through gravimetric data (Ambrosetti et al., 1993). See Figure 1 for study area location.

to 70 cm thick, consist of arboreal remains. Sparse sandy lenses, up to 40 cm thick, are massive or show cross-stratification.

The features and fossils of the silty clays are consistent with waterlogged anoxic conditions and represent hydromorphic paleosols (Duchaufour, 1977; Retallack, 1990). The laminated clays can be interpreted as coastal pond deposits with lignite as the remains of marginal swamp vegetation. The sandy lenses could represent storm deposits. Facies subassociation C2 evidence indicates an emerged or a shallow-water wetland coastline, in contrast to the wave-dominated open coastline of subassociation C1.

Facies association D has a limited areal distribution near Todi as Ponte Naja Formation with a thickness of more than 140 m (Figures 2A, 3). The exposed succession (Figure 7) consists of cyclic alternations of clayey sandy silts, on average 1 m thick, and dark bluish gray, silty clays, on average 15 cm thick, with rare sandy gravel lenses (1% of the measured sections). The clayey sandy silts are exposed as rectangular beds more than 50 m long in lateral extent, are commonly

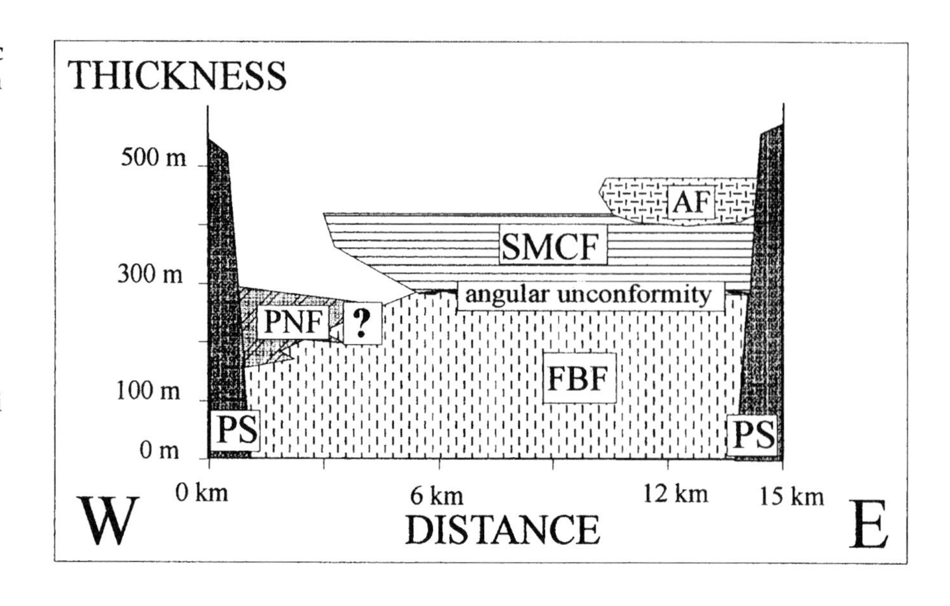

Figure 3—Synthetic stratigraphic framework of the Tiberino Basin infilling between Todi and San Gemini (see Figure 2). PS = pre-Pliocene substratum; FBF = Fosso Bianco Formation (Middle–Late Pliocene); PNF = Ponte Naja Formation (Late Pliocene); SMCF = Santa Maria di Ciciliano Formation (Early Pleistocene); AF = Acquasparta Formation (Early Pleistocene). From Basilici (1997).

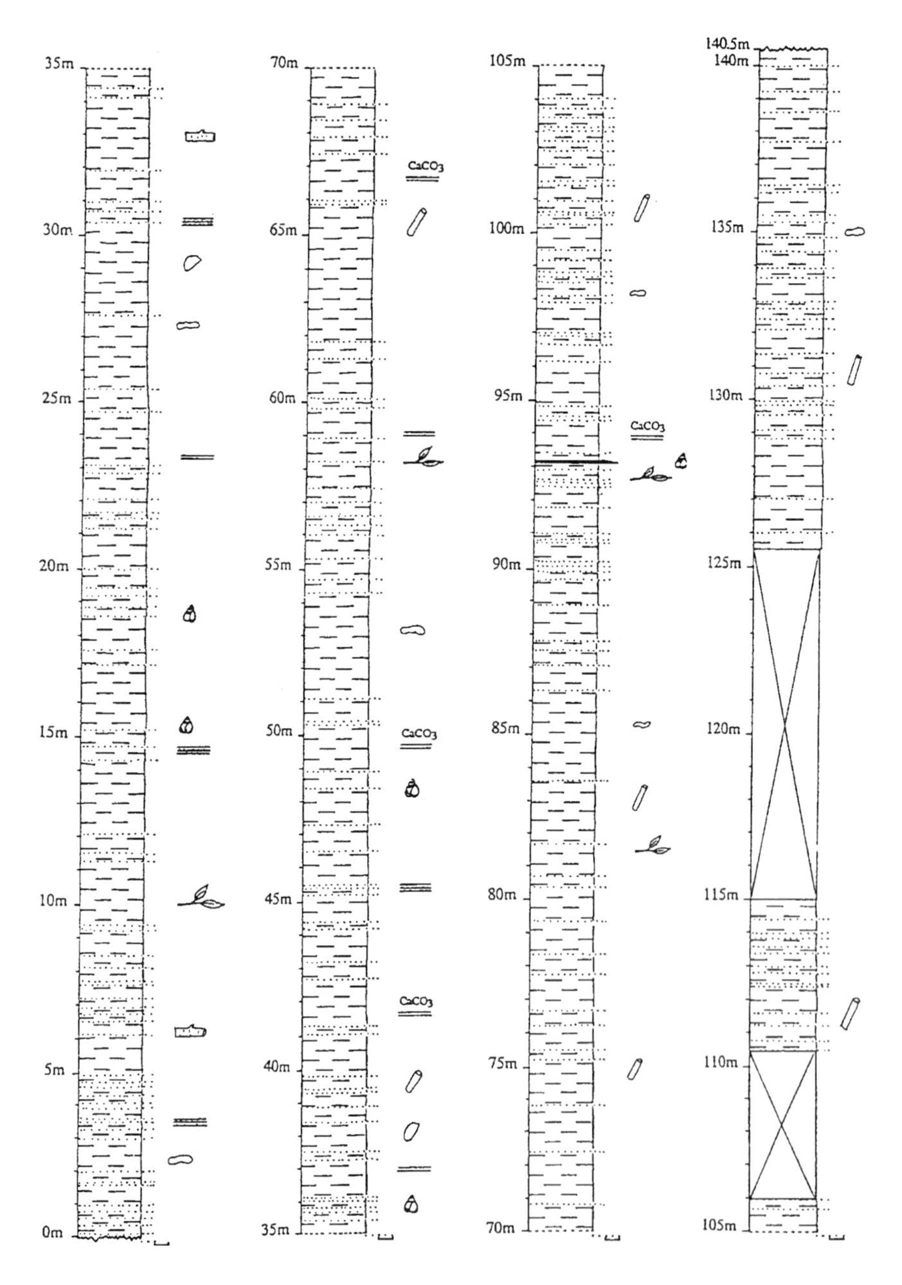

Figure 4—Detailed lithologic column of facies association A cropping out near Todi. The succession contains marly clays interpreted as offshore lake bottom. For legend, see Figure 5.

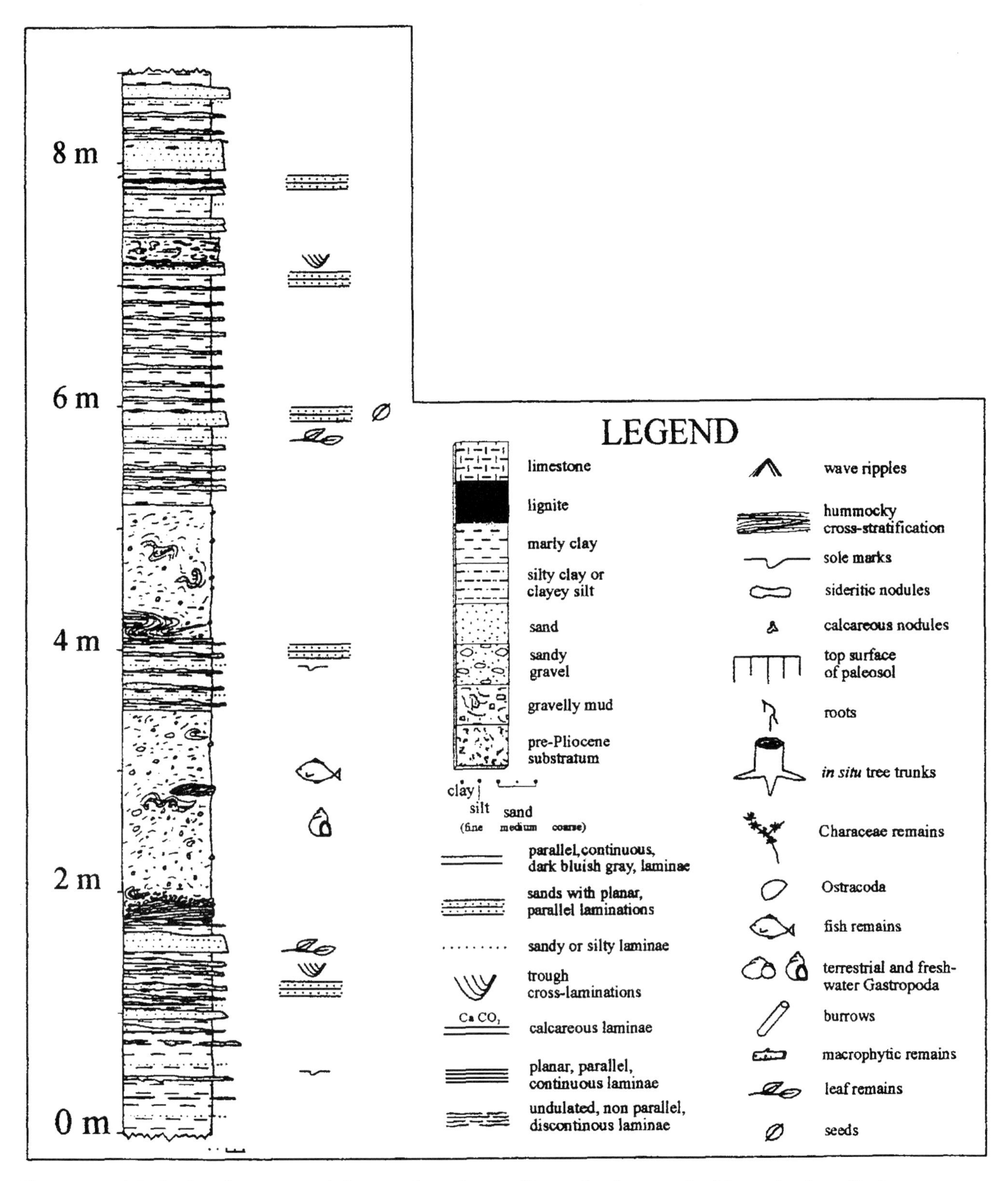

Figure 5—Lithologic column containing sands and gravelly muds alternated with marly clays (facies association B). These facies are interpreted as deposits of a prodelta lacustrine systems. From Basilici (1997).

graded, and contain irregular and noncontinuous laminae of silts or very fine sands alternating with silty clays; moreover, climbing current ripples, convolute laminations, and fugichnia (escape structures) characterize this facies. The dark bluish gray, silty clays contain root traces, terrestrial gastropods, mammal remains, bioturbation, massive fabric, and calcareous nodules. Rare planar-concave lithosomes (height × width: 2 m × 6 m) of sandy gravels (Figure 7) are unsorted, massive, clast-supported, and poorly organized with rare clast

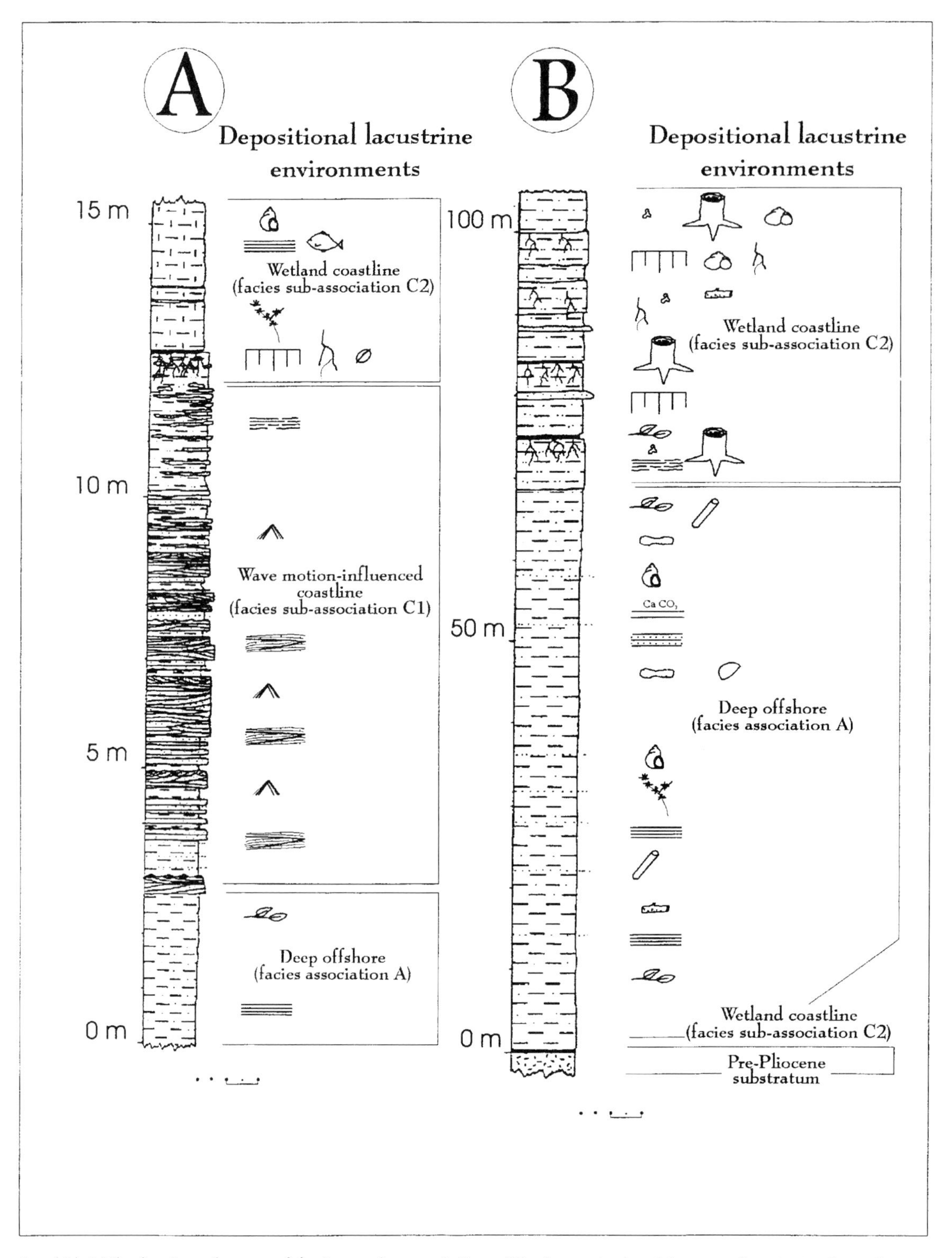

Figure 6—(A) Lithologic column of facies subassociation C1 characterized by sandy strata that alternate with marly clays. This subassociation represents wave-dominated coastal sediments that overlie offshore lacustrine deposits (facies association A). (B) Lithologic column of the facies subassociation C2 characterized by massive silty clays with pedogenetic features and lignites; it overlies offshore lacustrine deposits (facies association A). For legend, see Figure 5. From Basilici (1997).

imbrications. Clasts, with an average pebble size, are poorly rounded and composed of rocks from the nearby mountain margin. Paleocurrents show east–northeast paleoflows, opposite to the pre-Pliocene margin of the basin (Figure 2A).

Characteristics of the clayey sandy silts suggest deposition from nonchannelized flows with high concentration of suspended load and rapid deposition (Hubert and Hyde, 1982; Gierlowski-Kordesch and Rust, 1994). The dark bluish gray, silty clays are interpreted as

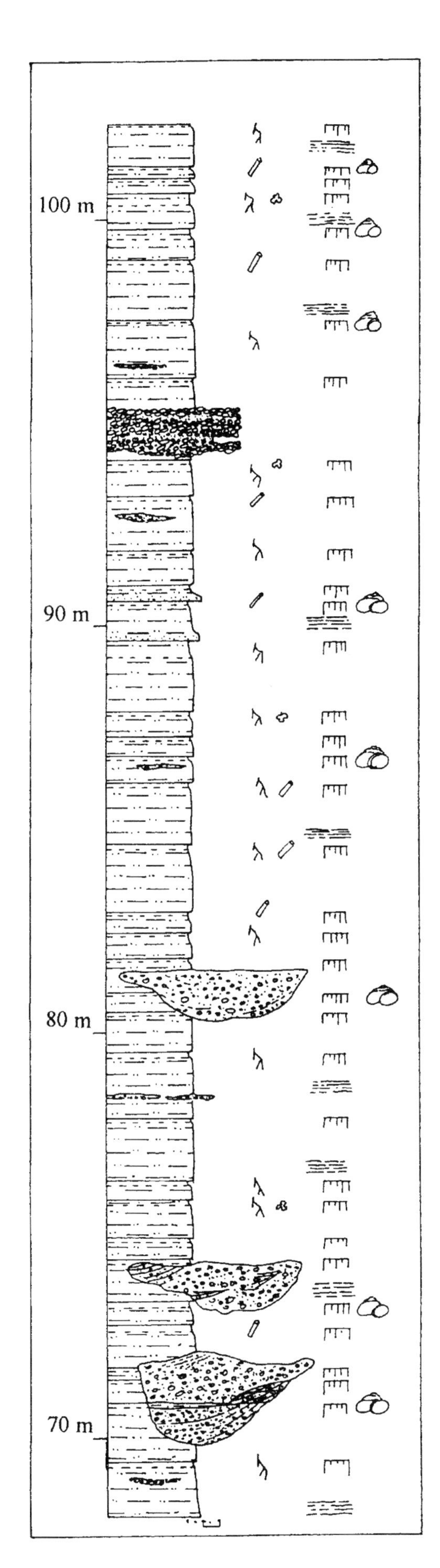

Figure 7—Part of the stratigraphic section of facies association D. Most of the sediments are interpreted as sheetflood deposits alternating with incipient paleosols; rare ribbon channel deposits can be found. This facies association is interpreted as the distal part of an alluvial fan. For legend, see Figure 5.

incipient paleosols. Each couplet clayey sandy silts/dark bluish gray, silty clays corresponds, respectively, to rapid sedimentation processes (sheet floods) followed by short periods of no sedimentation and pedogenesis (Basilici, 1995). The planar-concave sandy gravels can be identified as ribbon channel deposits characterized by "flashy" and highly concentrated depositional processes (Friend et al., 1979). Facies association D is interpreted as the distal part of an alluvial fan, probably at the transition of the mudflat onto the lacustrine margin.

PALEOENVIRONMENTAL RECONSTRUCTION AND TECTONIC-SEDIMENTATION RELATIONSHIP

The Tiberino Basin contained a lacustrine system during the Pliocene (Figure 8); facies analysis allows an interpretation of the following characteristics: size, coastline morphology, depth, and hydrodynamics.

Fine offshore sediments (facies association A) show a large areal and temporal extent (80% of the measured sections) and are interpreted as a single lacustrine paleoenvironment. The marginal deposits (facies associations B, C, and D) crop out near the pre-Pliocene margin of the Tiberino Basin, defining the shape of the Pliocene lake as elongated toward the north–northwest, with a length of at least 50 km and a width of not more than 10 km.

The narrow range of marginal deposits and presence of Gilbert-type deltas and sedimenty gravity flows (mud flows, grain flows, and turbidites) suggest steep lacustrine margins. Gilbert-type delta foresets 50 m high testify to water depths of at least 50 m. In the fine sediments, macrophytic remains (leaves, wood), siderite nodules, and laminated to rarely bioturbated layers suggest anoxic conditions, implying a more-or-less permanent stratification (meromictic lake). In addition to the depth and morphology of the lake, climate may have been an important factor in lake water stratification. In fact, pollen analyses (Follieri, 1977; Basilici et al., 1995) indicate typical subtropical vegetal association, suggesting medium-high external temperatures allowing long periods of thermal water stratification with colder and denser waters in the lower part of the lake (e.g., Lake Tanganyika) (Cohen, 1989; Cohen et al., 1997).

Normal faults, having main directions of north–northwest-south–southeast and secondary directions of east–northeast-west–southwest, delimit the Tiberino Basin (Barchi et al., 1991). Distribution of marginal lacustrine deposits follows these tectonic patterns and paleocurrent data (channel flows, prograding delta

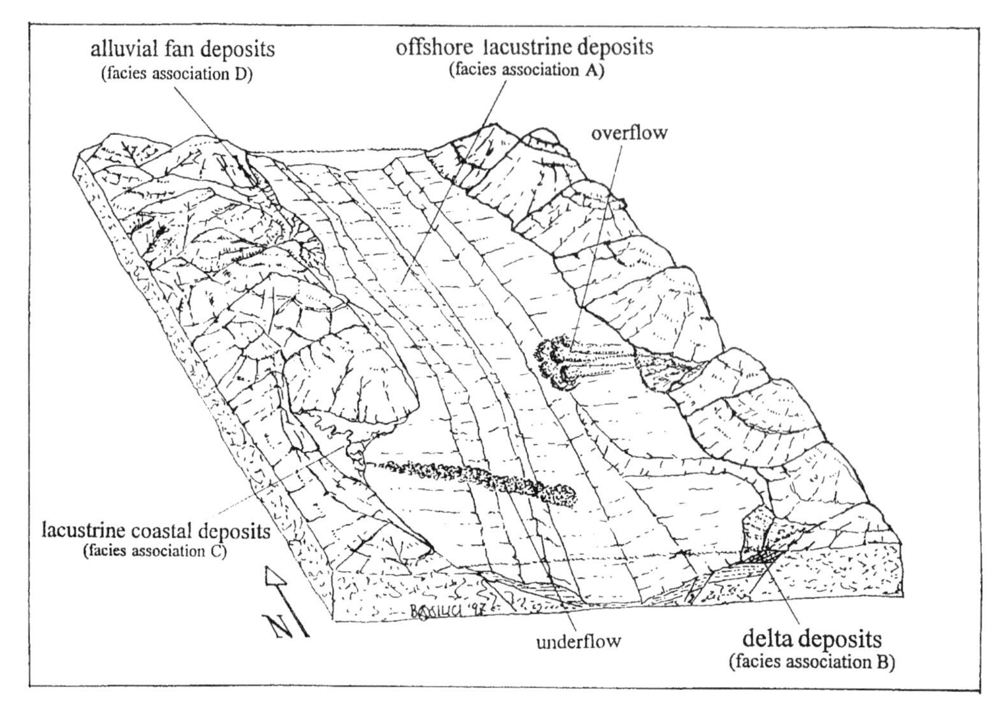

Figure 8—Paleogeographic reconstruction of the lacustrine depositional system of the Tiberino Basin during the late Pliocene showing locations of facies associations. The area corresponds to that of Figure 2A.

foresets, sediment gravity flows) are perpendicular to these faults. It can be postulated that the shape and size of the lake, as well as parts of the depositional processes, were principally controlled by tectonics.

REFERENCES CITED

Ambrosetti, P., M. G. Carboni, M. A. Conti, D. Esu, O. Girotti, G. B. La Monica, B. Landini, and G. Parisi, 1987, Il Pliocene ed il Pleistocene inferiore del Bacino del fiume Tevere nell'Umbria meridionale: Geografia Fisica e Dinamica del Quaternario, v. 10, p. 10–33.

Ambrosetti, P., M. Barbieri, G. Basilici, F. Bozzano, P. De Pari, M. Di Filippo, R. Di Maio, G. Duddridge, G. Etiope, P. Gambino, P. Grainger, S. Lombardi, A. Mottana, D. Patella, E. Pennacchioni, T. Ruspandini, G. Scarascia Mugnozza, G. Sordoni, S. Tazioli, B. Toro, G. Valentini, and G. Zuppi, 1993, Analysis of geoenvironmental conditions as morphological evolution factors of the sand-clay series of the Tiberino valley and Dunarobba Forest preservation: migration of radionuclides in the geosphere (Mirage project—third phase): Proceedings of the progress meeting (work period 1992), Report EUR 15914 EN, Bruxelles 7–8 October 1993, p. 171–180.

Ambrosetti, P., G. Basilici, L. Capasso Barbato, M. G. Carboni, G. Di Stefano, D. Esu, E. Gliozzi, C. Petronio, R. Sardella, and E. Squazzini, 1995, Il Pleistocene inferiore del ramo sud-occidentale del Bacino Tiberino (Umbria): aspetti litostratigrafici e biostratigrafici: Il Quaternario, v. 8, no. 1, p. 19–36.

Barchi, M., F. Brozzetti, and G. Lavecchia, 1991, Analisi strutturale e geometrica dei bacini della media Valle del Tevere e della Valle Umbra: Bollettino della Società Geologica Italiana, v. 110, p. 65–76.

Basilici, G., 1992, Il Bacino continentale Tiberino (Pliocene–Pleistocene, Umbria): analisi sedimentologica e stratigrafica: Ph.D. Dissertation, University of Bologna, 323 p.

Basilici, G., 1995, Sedimentologia della parte distale di una conoide alluvionale del Pliocene superiore (Bacino Tiberino, Umbria): Il Quaternario, v. 8, no. 1, p. 37–52.

Basilici, G., 1997, Sedimentary facies in an extensional and deep-lacustrine depositional system: the Pliocene Tiberino Basin, central Italy: Sedimentary Geology, v. 109, p. 73–94.

Basilici, G., A. Bertini, and M. R. Pontini, 1995, Oscillations climatiques dans le Pliocene de l'Italie Centrale: l'example de la section de Fosso Bianco (Bassin Tiberino, Ombrie): 14th Symposium of Association des Palynologues de langue francaise,. Volume des Résumés, p. 18.

Cohen, A. S., 1989, Facies relationship and sedimentation in large rift lakes and implication for hydrocarbon exploration: example from lakes Turkana and Tanganyika: Palaeogeography, Palaeoclimatology,Palaeoecology, v. 70, p. 65–80.

Cohen, A. S., M. R. Talbot, S. M. Awramik, D. L. Dettman, and P. Abell, 1997, Lake level and paleoenvironmental history of the Lake Tanganyika, Africa, as inferred from late Holocene and modern stromatolites: Geological Society of America Bulletin, v. 109, p. 444–460.

Dott, R. H., Jr., and J. Bourgeois, 1982, Hummocky stratification: significance of its variable bedding sequences: Geological Society of America Bulletin, v. 93, p. 663–680.

Duchaufour, P., 1977, Precis de pedologie: Masson, Paris, 481 p.

Esu, D., and O. Girotti, 1991, Late Pliocene and Pleistocene assemblages of continental molluscs of Italy.

A survey: Il Quaternario, v. 4, no. 1a, p. 137–150.

Follieri, M., 1977, Evidence on the Pliocene–Pleistocene palaeofloristic evolution in central Italy: Rivista Italiana di Paleontologia e Stratigrafia., v. 83, p. 925–930.

Friend, P. F., M. J. Slater, and R. C. Williams, 1979, Vertical and lateral building of river sandstone bodies, Ebro Basin, Spain: Journal of Geological Society (London), v. 136, p. 39–46.

Gierlowski-Kordesch, E., and B. R. Rust, 1994, The Jurassic East Berlin Formation, Hartford Basin, Newark Supergroup (Connecticut and Massachusetts): a saline lake-playa-alluvial plain system, *in* R. W. Renault, and W. M. Last, eds., Sedimentology and Geochemistry of Modern and Ancient Saline Lakes: SEPM Special Publication, v. 50, p. 249–265.

Glenn, C. R., and K. R. Kelts, 1991, Sedimentary rhythms in lake deposits, *in* G. Einsele, W. Ricken, and A. Seilacher, eds., Cycles and Events in Stratigraphy, Springer Verlag, Berlin, p. 188–221.

Hubert, J. F., and M. G. Hyde, 1982, Sheet-flow deposits of graded beds and mudstones on an alluvial sandflat-playa system: Upper Triassic Blomidon redbeds, St Mary's Bay, Nova Scotia: Sedimentology, v. 29, p. 457–474.

Lavecchia, G., 1988, The Tyrrhenian-Apennines system: structural setting and seismotectogenesis: Tectonophysics, v. 147, p. 263–296.

Lowe, D. R., 1976, Grain flow and grain flow deposits: Journal of Sedimentary Petrology, v. 46, p. 188–199.

Lowe, D. R., 1979, Sediment gravity flows: their classification and some problems of application to natural flows and deposits, *in* L. J. Doyle and O. H. Pilkey, Jr., eds., Geology of the Continental Slope: SEPM Special Publication No. 27, p. 75–82.

Martinetto, E., 1994, Paleocarpology and the "in situ" ancient plant communities of a few Italian Pliocene fossil forests, *in* R. Matteucci, M. G. Carboni, and J. S. Pignatti, eds., Studies on Ecology and Palaeoecology of Benthic Communites: Bollettino della Società Paleontologica Italiana Volume Speciale, v. 2., p. 189–196.

Mutti, E., and F. Ricci Lucchi, 1972, Le torbiditi dell'Appennino settentrionale: introduzione all'analisi di facies: Memorie della Società Geologica Italiana, v. 11, p. 161–199.

Pickering, K., D. Stow, M. Watson, and R. Hiscott, 1986, Deep-water facies, processes and models: a review and classification scheme for modern and ancient sediments: Earth-Science Reviews, v. 23, p. 75–174.

Retallack, G. J., 1990, Soils of the past. An introduction to paleopedology: Unwin Hyman, Boston, 520 p.

Rio, D., R. Sprovieri, and E. Di Stefano, 1994, The Gelasian stage: a proposal of new chronostratigraphic unit of the Pliocene series: Rivista Italiana di Paleontologia e Stratigrafia, v. 100, p. 103–124.

Seilacher, A., 1982, Distinctive features of sandy tempestites, *in* G. Einsele and A. Seilacher, eds., Cyclic and Event Stratification: Springer-Verlag, New York, p. 333–349.

Section IX

Quaternary

Cavinato, G. P., and E. Miccadei, 2000, Pleistocene carbonate lacustrine deposits: Sulmona Basin (Central Opennines, Italy), *in* E. H. Gierlowski-Kordesch and K. R. Kelts, eds., Lake basins through space and time: AAPG Studies in Geology 46, p. 517–526.

Chapter 48

Pleistocene Carbonate Lacustrine Deposits: Sulmona Basin (Central Apennines, Italy)

Gian Paolo Cavinato
C.N.R., Centro di Studio per il Quaternario e l'Evoluzione Ambientale, c/o Dip. Scienze della Terra Università di Roma "La Sapienza", Rome, Italy

E. Miccadei
Dipartimento di Scienze Geologiche, Università G. D'Annunzio Chieti, Italy

GEOGRAPHIC AND GEOLOGIC SETTING OF THE SULMONA BASIN

Pleistocene carbonate lacustrine deposits are located in the Sulmona Basin, which is 150 km to the north–northeast of Rome (eastern part of the Apennines chain, Abruzzi area, central Italy, see Figure 1). The Apennines is a collisional fold and thrust belt developed along the Burdigalian trough since early Pliocene along the Adriatic ensialic microplate in response to west dipping subduction (Royden et al., 1987; Cipollari and Cosentino, 1996). The central Apennines chain consists of a Mesozoic–Tertiary stack of northeastly verging folds and southwesterly-dipping thrust systems, juxtaposed with ramp and basinal sequences of the same age as well as piggy-back flysch basins of Messinian to early Pliocene age (Figure 1) (Bigi et al., 1992; Ghisetti et al., 1992; D'Andrea et al., 1992; Patacca et al., 1992; Cipollari and Cosentino, 1996).

From the late Pliocene to Quaternary, the compressional tectonic setting of the Apennine chain was progressively modified and dissected, from the Tyrrhenian margin to the Adriatic Sea, by extensional and transtensional tectonic events. These tectonic events produced northwest-southeast and east-west high-angle normal, oblique, and transfer fault systems, southwest to south dipping, mainly developed along thrust ramps. Along the intersections of these main fault systems, several northwest-southeast intramontane graben and half-graben basins of different shapes and widths were formed (i.e., Fucino, L'Aquila, Sulmona Basins, Figure 1) (Patacca et al., 1992; Cavinato et al., 1994). The sedimentary evolution of these basins has been controlled since the late Pliocene by tectonic activity along their margins, as testified by present-day crustal seismicity (Vittori et al., 1995). A general overview of the sedimentary infill of the basins shows the presence of basal coarse-grained deposits (alluvial and alluvial fan deposits) overlain by lacustrine, palustrine, and alluvial deposits.

The Sulmona Basin has a subrectangular shape and is 15–20 km long and 3–4 km wide (Figure 2). The basin is surrounded by an imbricate thrust composed of Mesozoic–Tertiary carbonate sequences (Marsica and Morrone thrusts), juxtaposed on Messinian and lower Pliocene flysch deposits (Figure 2). The eastern part of the Sulmona Basin is characterized by a very steep margin (Morrone Mt., 2000 m a.s.l.) in contrast to the flat interior of the basin (360 m a.s.l.). The eastern margin is bounded by two main northwest–southeast normal fault systems dipping southwest displaced by east–west normal or oblique fault systems dipping southward (Figures 2, 3) (Miccadei et al., 1993; Vittori et al. 1995; Cavinato and Miccadei, 1995; Miccadei et al., 1999). The Popoli–Pacentro fault system (Figure 2) represents the most recent fault scarp that divides the Mesozoic Morrone flank from the Sulmona Basin (Figure 2). The Schiena d'Asino fault system (Figure 2) divides the northen flank of Morrone Mountains. These fault systems were superimposed onto the Neogene Morrone thrust systems and since the late Pliocene have produced progressive down-faulting of the Morrone flank toward the basin. Vertical offset may exceed 1000–1500 m.

In addition, Pleistocene deposits are deformed by recent east-west, northwest-southeast, and northeast-southwest fault and joint systems. Those deformations are mainly distributed along the Morrone fault boundary zone (Miccadei et al., 1993; Vittori et al., 1995; Cavinato and Miccadei, 1995; Miccadei et al., 1999) and in the Pratola Peligna area (Figure 2). Geophysical subsurface data obtained for hydrocarbon exploration (seismic and gravimetric data) (Figure 2) (Mostardini and Merlini, 1988; Di Filippo and Miccadei, 1997) indicate a thickness of continental deposits in the Sulmona Basin of

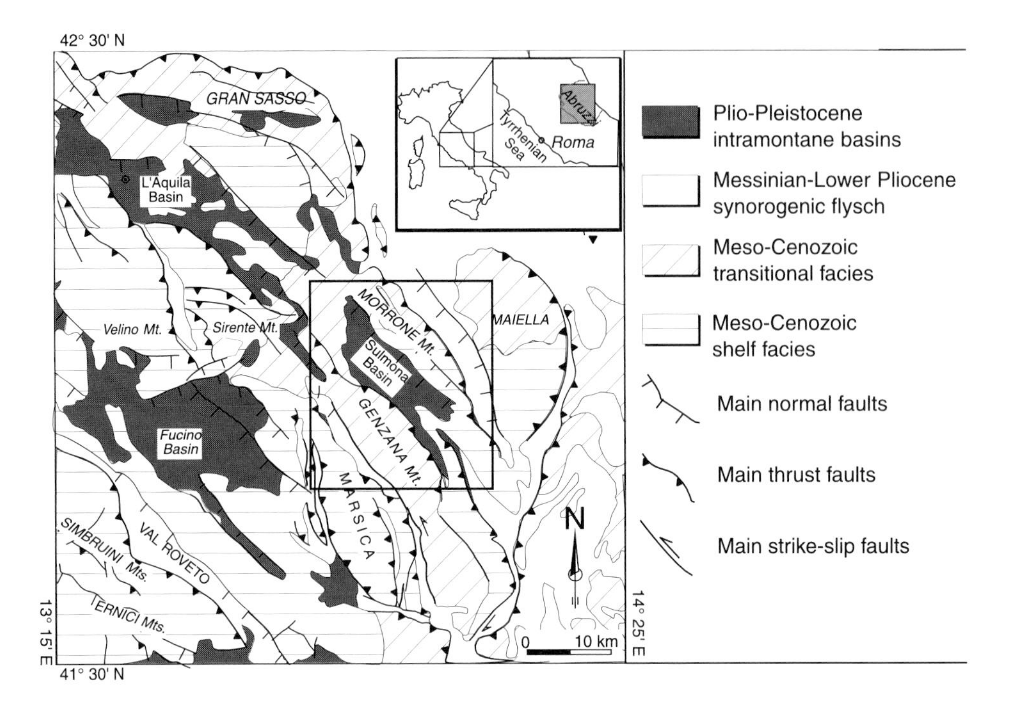

Figure 1—Tectonic framework of the eastern central Apennines, Italy, showing the location of the Sulmona Basin.

at least 300–500 m, with a maximum at the depocenter in the northen part of the basin. The whole Quaternary sequence gently dips toward the northeast (Morrone fault zone); an angular unconformity separates alluvial strata from lower lying lacustrine Pleistocene deposits in the western part of the basin (Figure 2).

PLEISTOCENE LACUSTRINE DEPOSITS

In the early Pleistocene, the Sulmona Basin first received sediments that were dominated by coarse-grained deposits derived from alluvial fan systems along the western and eastern mountain flanks (S. Venanzio, Orsa Mt. Formations, Figures 2, 3). During the middle Pleistocene, subsidence decreased along with the sedimentation rate. The southeastern part of the basin was tranformed into an alluvial plain with axial drainage (Sagittario, Gizio, Vella Rivers, Figure 2) and several scattered ephemeral springs and pools along the basin boundaries. The northern and central part of the basin is interpreted as a closed lake system with shallow water lacustrine (open and marginal lacustrine deposits) and palustrine deposits (Figures 2, 3). Fluctuations in lake level appeared to have been controlled by a travertine barrier located along the Popoli gorge (Figures 2, 3), which acted as a threshold during cold periods triggering fluvial erosion (Figure 3) (Carrara, 1998, personal communication).

From the uppermost part of the middle Pleistocene to the upper Pleistocene, the geologic evolution of the Sulmona Basin was controlled by the effects of northwest-southeast and east-west extensional tectonic events and glacial and interglacial climatic changes. Sedimentary evolution was characterized by filling and erosion that eroded previous lacustrine and palustrine deposits with deposition on irregular surfaces ["Terrazza alta di Sulmona" (TAS); "Terrazza bassa di Sulmona" (TBS) of Beneo, 1940, 1943] (Figure 3) by fluvial deposits with axial drainage toward the northwest (Figure 3).

The lowermost portion of the Pleistocene lacustrine deposits is not exposed and sequences reach 50–75 m maximum thickness with 5–10 m exposed in the western part of the basin to 50–70 m in the northeastern. Only in the northwestern central part of the basin is detailed stratigraphy of solely lacustrine sequences possible. In the west and east, the lacustrine deposits interfinger with Pleistocene alluvial fan deposits of the Aterno River (S. Venanzio Gorge) and alluvial fan systems developed along the flank of Mt. Morrone (Figures 2, 3). In the central and southern part of the basin, lacustrine deposits with lateral discontinuities interfinger with the Pratola Peligna palustrine deposits (Figures 2, 3). On top of the lacustrine and palustrine deposits, an unconformity marks the contact with the middle–upper Pleistocene progradational fluvial deposits that form the flat surface of the present-day Sulmona plain (Figures 2, 3).

Lacustrine sequences of the Sulmona Basin contain an abundance of fossils, including freshwater gastropods and ostracods, charophytes, invertebrate trace fossils, plant fragments, and some mammalian fauna. Tephra layers (5–10 cm thick) interbedded with lacustrine deposits are common. They consist of sandstone and siltstone pyroclastic deposits (tuff), related to the Pleistocene alkaline eruptive activities of the Tyrrhenian volcanic complexes (Cavinato et al., 1994). Dating of the Sulmona lacustrine deposits as middle Pleistocene has been obtained by bio-, chrono-, and lithostratigraphic correlation with some mammal and human sites as well as with K/Ar and Ar/Ar dating of volcaniclastic deposits. In particular, the alluvial deposits of the Terrazza alta di Sulmona (TAS) containing bones of *Mammuthus chosaricus* indicate the uppermost part of the middle Pleistocene (isotopic stage 6; Figure 3) (Leuci and Scorziello, 1972, 1974; Esu et al., 1992; Gliozzi et al. 1997). At Le Svolte (near

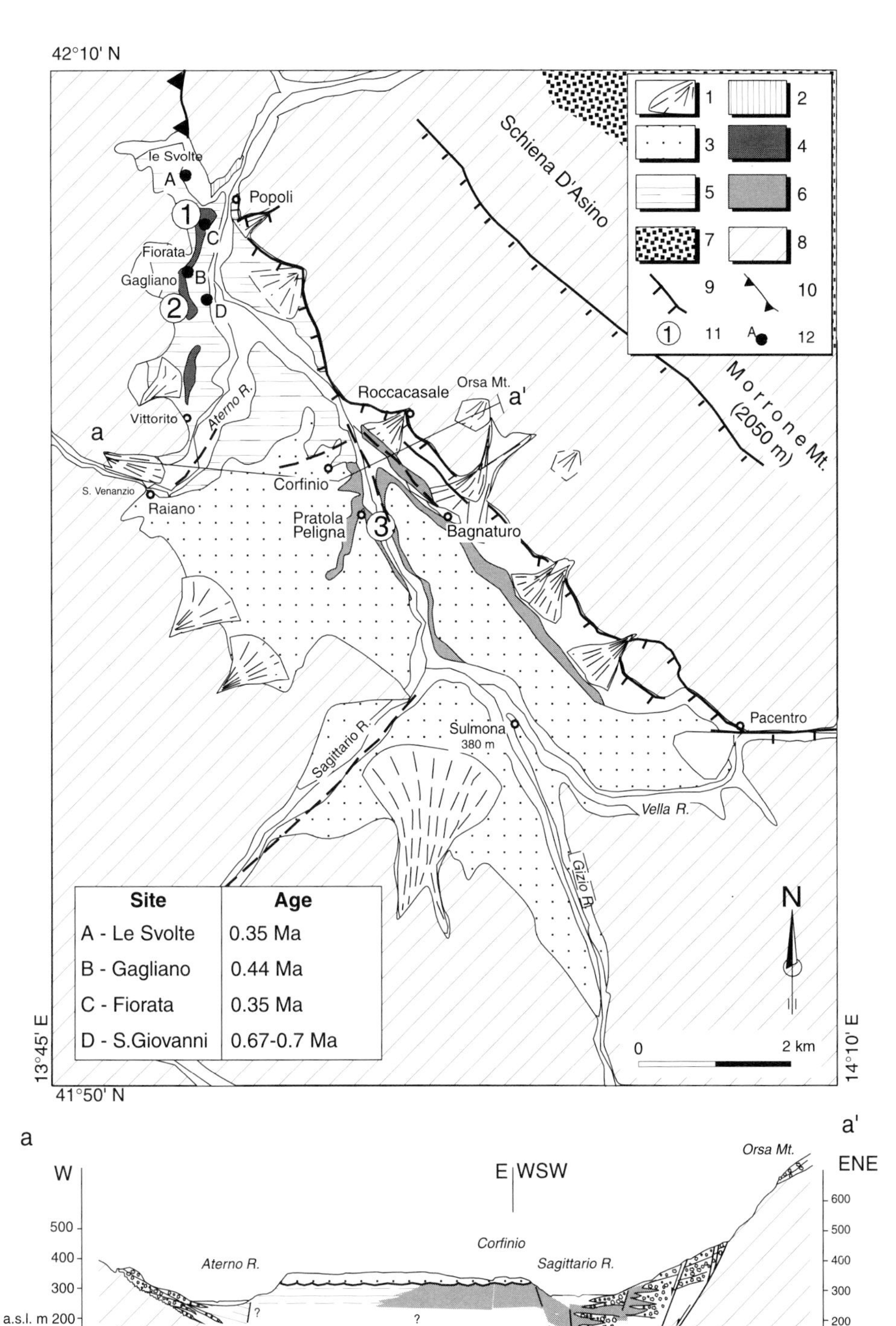

Site	Age
A - Le Svolte	0.35 Ma
B - Gagliano	0.44 Ma
C - Fiorata	0.35 Ma
D - S.Giovanni	0.67-0.7 Ma

Figure 2—Geologic map of the Sulmona Basin. Legend: 1 = Alluvial deposits, recent and ancient alluvial fan deposits (Holocene–lower Pleistocene); 2 = travertine deposits (upper Pleistocene); 3 = fluvial deposits (middle–upper Pleistocene); 4 = marginal lacustrine deposits (Fiorata unit) (middle Pleistocene); 5 = open lacustrine deposits (Gagliano units) (middle Pleistocene); 6 = Pratola Peligna palustrine deposits (Lower Pleistocene); 7 = flysch sequences (upper Miocene–lower Pliocene); 8 = marine carbonate bedrock (Lias–upper Miocene); 9 = normal fault; 10 = thrust fault; 11 = location of stratigraphic logs of Figure 5; 12 = radiometric data sites; see table for ages. At the bottom of the figure, a geologic cross-section of the Sulmona Basin illustrates the half-graben structure. The dotted line represents the top of the Mesozoic carbonate bedrock (from Di Filippo and Miccadei, 1997).

Popoli town; Figures 2, 3) Mousterian human sites were found (Radmilli, 1964, 1984). K/Ar and Ar/Ar dating of volcaniclastic layers interbedded in the lacustrine deposits assign a middle Pleistocene age to the lacustrine sequences (from 0.7 to 0.4 Ma, Figures 2, 3).

Stable isotope analyses were carried out on the carbonate of the lacustrine facies associations interpreted as produced by primary precipitates due to the activities of blue-green algae and associated organisms that form encrustations of crystal aggregates (Catalano, 1964) (Figure 4). The data show typical nonmarine values of $\delta^{13}C$ from –1 to 4‰ and $\delta^{18}O$ from –6 to 10‰ (Figure 4). The carbon and oxygen isotopic values show a scattered distribution; in particular, data

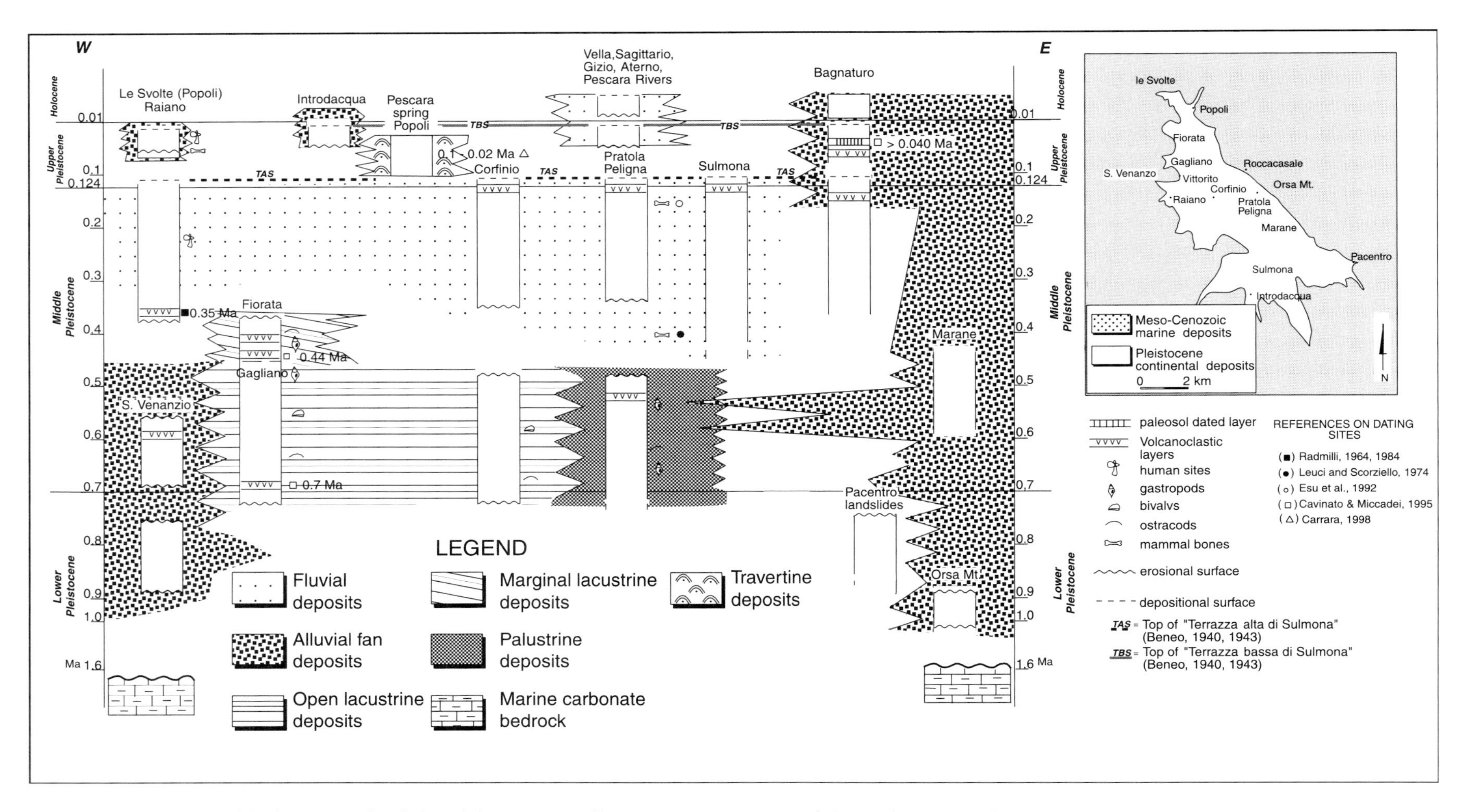

Figure 3—Stratigraphic framework of the Pleistocene sedimentary sequences of the Sulmona Basin.

distribution shows a relative large variation in the $\delta^{18}O$ value, with a covariant distribution (Figure 4). This feature is a common feature of hydrologically closed lakes but also reflects a lake's particular hydrologic and climatic setting (Kelts and Talbot, 1990; Talbot, 1990).

On the basis of lithostratigraphic analysis, three main facies associations have been identified in the lacustrine deposits represented in the stratigraphic sections of Figure 3: the Gagliano, Fiorata, and Pratola Peligna facies associations.

GAGLIANO FACIES ASSOCIATION

The Gagliano facies association mainly crops out in the northwestern and central part of the basin with a thickness that ranges from a few meters in the west to 35–50 m in the east. In the southern and eastern part of the basin (Pratola Peligna and Morrone Mts. flank areas, Figure 2), on the basis of 20–60-m-deep drillholes along the highway Rome-Pescara from Corfinio to Popoli (Figure 2), a vertical rapid change of Gagliano facies interbedded with the palustrine deposits of Pratola Peligna facies and alluvial fan deposits is evident. This facies association mainly consists of alternation of carbonate mudstone with white siltstone and clays (Figure 5) (Catalano, 1964; Cavinato and Miccadei, 1995). The carbonate mudstone is well-bedded with layers 5–20 cm thick containing up to 60–70% calcium carbonate interbedded with 10–40-cm-thick layers of white siltstone. The mudstone contains poorly preserved planar and parallel lamination. Fauna assemblages are represented by rare freshwater gastropods and bivalves (*Bithynia tentaculata, Dreissena polymorpha*), diatom assemblages (*Melosira arenaria Moore, Diploneis elliptica genuina*), and Cypridae ostracods. The siltstone bodies consist of 5–10-cm-thick irregular silty layers alternating with 2–10-cm-thick laminated clayey silts; the laminae are planar, graded, and commonly disrupted by rhizoliths and bioturbation (Figure 5). The lateral extent of these beds is unknown. Siliciclastic material is minimal. At the top of the sequence, the mudstone and siltstone layers pass gradually (over 5–10 m) into siltstone and sandstone beds containing vertebrate bones.

Several dark-brown sandy tuff layers (5–10 cm thick) are interbedded with the Gagliano facies association. The layers consist of mineral associations rich in altered micas, pyroxene, and sanidine. Ar/Ar and K/Ar radiometric dating of the volcaniclastic deposits indicates that the age of the Gagliano facies association is between 0.7 and 0.44 Ma (early to middle Pleistocene) (Figure 3) (Catalano, 1964).

The Gagliano facies association is interpreted as a basinal lacustrine environment, as shown by its fine grain size, fine lamination, and presence of charophytes, freshwater molluscs, and ostracods. The plentiful supply of calcium carbonate into the lake system was derived from surface transport (overland transport or dissolved load) (Platt and Wright, 1991; Gierlowski-Kordesch, 1998) from the surrounding marine

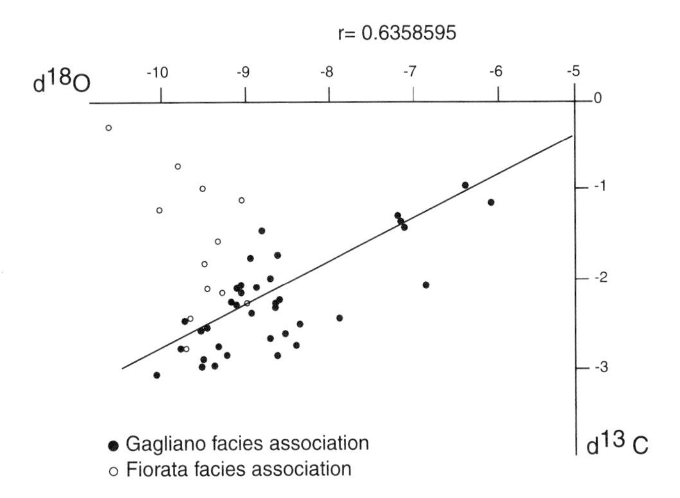

Figure 4—Isotopic analyses of carbonate from middle Pleistocene of the Gagliano and Fiorata lacustrine facies associations of the Sulmona Basin using bulk samples of micrite and associated fossils. The carbon and oxygen isotopic data show a covariant distribution typical of hydrologically closed lakes (r = regression coefficient).

Mesozoic carbonate sequences; moreover, the apparent absence of coarse siliciclastic input during the lower–middle Pleistocene indicates that the northern and central part of the Sulmona Basin was a very distal lacustrine environment (Figure 6).

This restriction in the depositional area within the basin could be related to a modification of the structure of this portion of the basin during the lower–middle Pleistocene. Evidence includes tilting of 15–20° toward the east–northeast of older alluvial fan sequences and its interfingered lacustrine deposits along the Aterno gorge; these deposits are overlain by younger horizontal alluvial and lacustrine deposits (Figure 2). Gravimetric and seismic data place the basin depocenter in the central and northern part, bounded by east–northeast-west–southwest and northwest-southeast fault systems (Figure 2). These modifications produced a progressive steepening of the lake floor and a basinal environment. The rapid facies change at the top of the lake sequence from mudstones to rudstone layers suggests changes of lake level during the middle Pleistocene or tectonic perturbations altering facies distributions.

FIORATA FACIES ASSOCIATION

The Fiorata facies association crops out only along the northwestern margin of the basin (Figure 2). Its thickness ranges from 2 to 2 m and the facies is characterized by massive whitish bioclastic carbonates interbedded with carbonate mudstones and rudstones, siliciclastic deposits, peat, and paleosols; on the top of the sequence, laminated carbonate is present (Figure 5) (Catalano, 1964; Miccadei et al., 1999). Rudstone bodies are 30–100 cm thick and 20–30 m wide with a lenticular

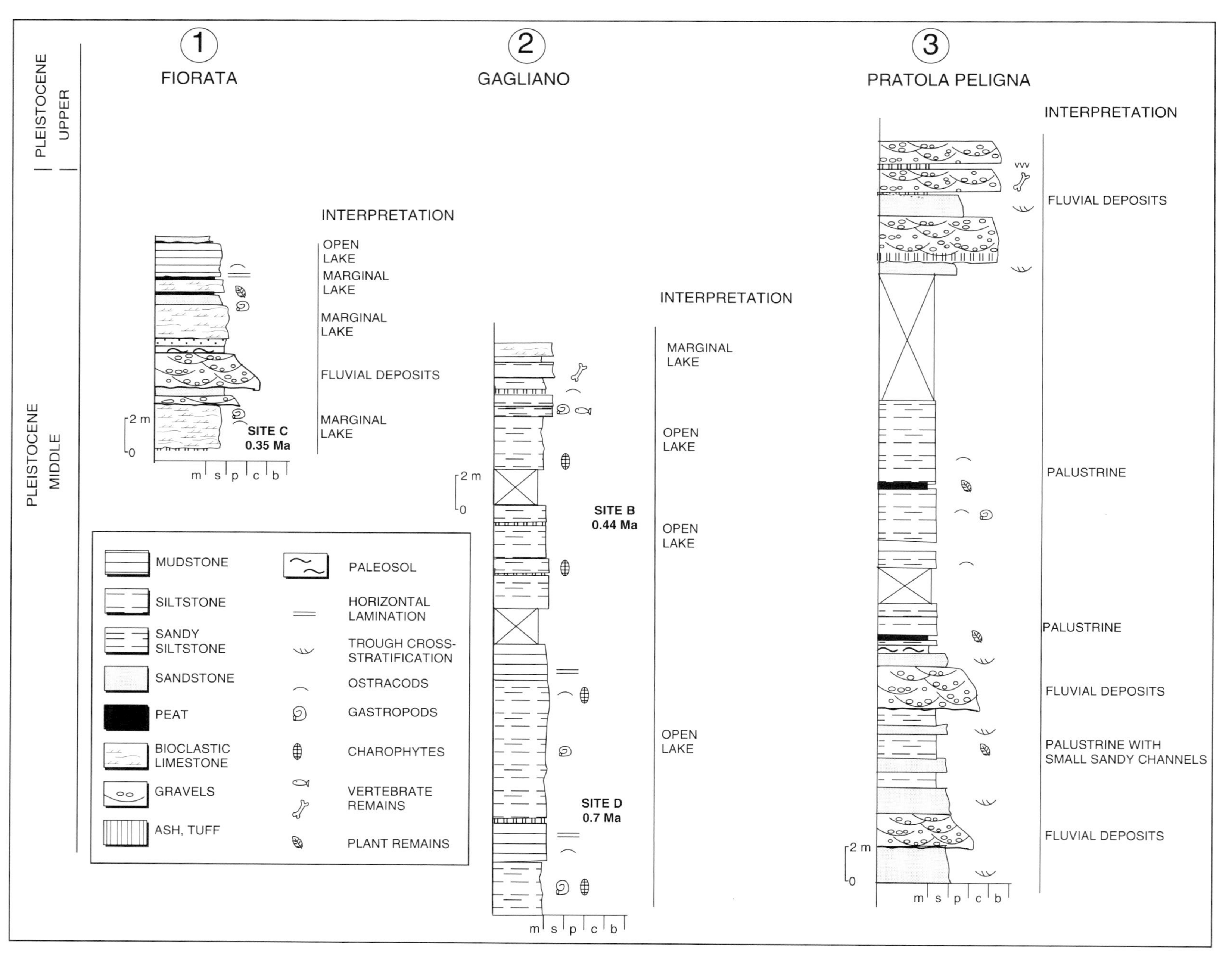

Figure 5—Detailed measured logs of the lacustrine and palustrine deposits in the Sulmona Basin. Localities shown in Figure 2.

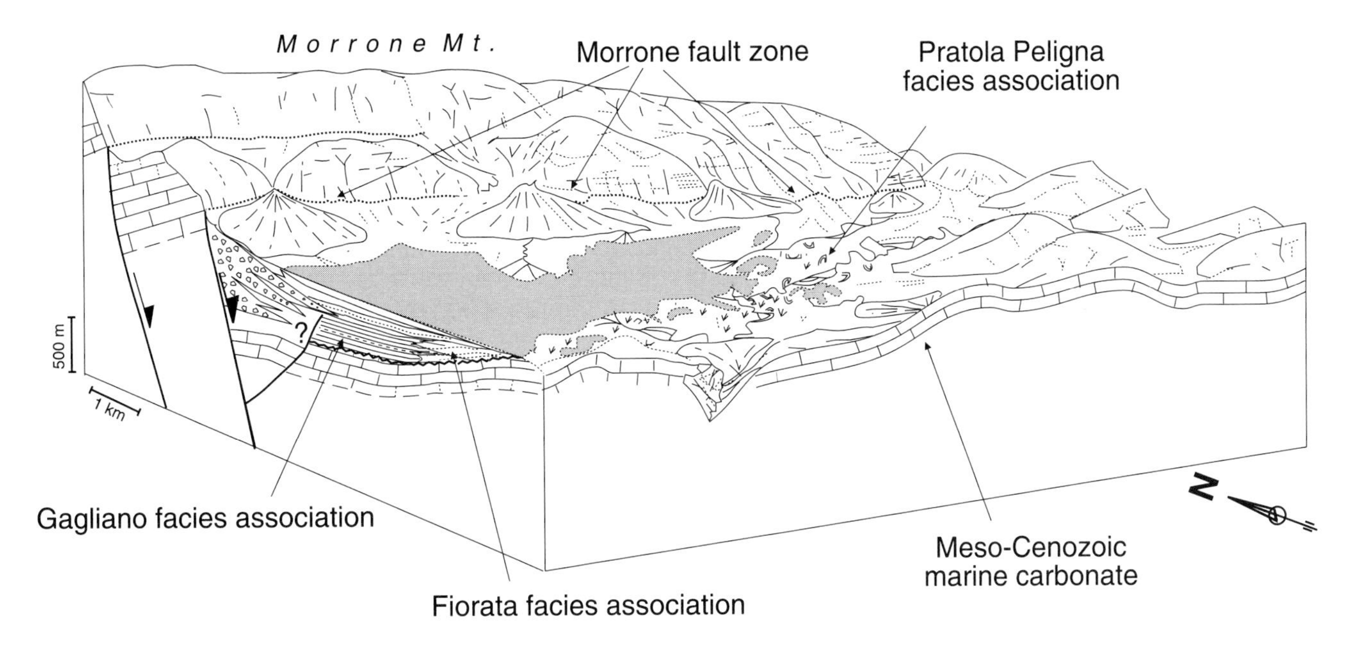

Figure 6—Paleogeographic schematic diagram of the Sulmona Basin during the middle Pleistocene illustrating the distribution of the three main facies associations.

shape, passing laterally into mudstone layers. The rudstone contains clasts that are angular to subrounded in shape, composed of micrite containing fossil remains of charophytes, including gyrogonites and stems (Figure 5); rare sedimentary structures are represented by gently dipping foresets. The fine-grained deposits (20–40 cm thick) are represented by well to poorly laminated carbonate mudstones with charophytes, diatoms, and Cypridae ostracods, as well as common mollusks and bivalves represented by *Bithynia tentaculata* and *Dreissena polymorpha*.

Locally, fine-grained carbonate deposits (10–20 cm thick) commonly contain lamination; these layers are interbedded with the rudstone deposits. The laminae are characterized by organic-rich, brown-gray, clay laminae (1–2 mm thick) alternating with whitish silty carbonates (2–3 mm thick) (Figure 5).

Planar-concave gravel and sandy gravel channel fill deposits are common; they display erosive bases and crude planar beds up to 3 m thick and 2–10 m wide (Figure 5). Interbedded with the lacustrine rudstone deposits and gravel bodies are peat layers, dark-brown or gray-brown silty clays, and clayey sand beds containing small accumulations of organic matter and plant remains, as well as siderite and calcareous nodules (Figure 5). Textures and the lack of horizonation suggest that these three layers represent the development of immature paleosols. Volcaniclastic layers are interbedded in this sequence (5–10 cm thick), and radiometric ages of these layers indicate a middle Pleistocene age (0.44–0.35 Ma) (Figures 3, 5).

The Fiorata facies association is interpreted as sediments in a marginal lake setting. Rudstone deposits essentially made of charophyte fragments indicate the development of extensive colonization by charophytes on a shallow lake with a flat or gentle bench slope margin. The rudstone layers are largely composed of angular to subrounded chara stem fragments; no evidence of features diagnostic of wave motion or exposure are present (Figure 6).

The conglomeratic bodies represent episodes of siliciclastic input related to the progradation of drainage systems from the western source area (Aterno, Sagittario rivers, Figure 2). These fluvial channels developed along the lake margins during low lake stands (Figure 6). The lens of immature paleosols and peat layers indicates that protected ponds developed with an extensive wetland associated with the fluvial floodplain.

PRATOLA PELIGNA FACIES ASSOCIATION

The Pratola Peligna facies association crops out in the central and southern part of the Sulmona Basin and consists of interbedded sandstones, siltstones, marlstones, peats, and gravels. The transition between the Pratola Peligna and the Gagliano facies associations is not well exposed. Bore-holes drilled along the highway from Pratola Peligna to Popoli (Figure 2) show the two facies associations interfingering. Along the eastern margin of the basin, the Pratola Peligna facies association covers the toe of the alluvial fan deposits.

Sandstone beds are 30–100 cm thick and 10–20 m wide with planar to trough cross-bedding and interbedded gravel channels (0.5–1 m thick, Figure 3). The siltstone contains only rare gastropod fragments. The light-grayish marl bodies are 20–100 cm thick with planar parallel laminations. Abundant freshwater gastropods (*Valvata piscinalis*, *Pisidium*, *Lymnea palustris*, *Bithynia tentaculata*, *Planorbis*) (Cavinato and Miccadei, 1995) and *Candona* and *Cyprideis* ostracods are present. Peat layers (10–20 cm thick) are interbedded more commmonly at the top of the facies association. On the top of the facies association sequence, rare blue-dark grayish massive mudstone layers (10–20 cm thick) contain features indicative of pedogenic alteration (immature paleosols). The gravel bodies are characterized by planar to concave lithosomes (1–2 m thick and 5–20 m wide) diagnostic of tractive currents; they are clast-supported, well sorted, and with common imbrication, and contain imbricated clasts and trough cross-stratification. The clasts are composed of Mesozoic-Cenozoic carbonate and sandstone derived from the surrounding area.

The Pratola Peligna facies association is interpreted as deposits of a palustrine/wetland environment. The dominance of shallow lacustrine marlstones associated with peat, anastomosing fluvial channels, and the absence of significant exposure and brecciation suggest humid climatic conditions in a wetland environment. The presence of common peat layers toward the top of the Pratola Peligna association suggests progressive infilling of wetlands due to decreasing subsidence or the lack of significant sediment input (Figure 6).

CONCLUDING REMARKS

Pleistocene carbonate lacustrine deposits are present in the Sulmona Basin, a north–northwest-south–southeast narrow extensional intramontane basin located in the easternmost part of the Neogene fold and thrust belt of the central Apennines. The sedimentary evolution of the Sulmona Basin shows the typical evolution of a lake in an extensional basin (half-graben) (Sladen, 1994). The early stage began in the late Pliocene? with the formation of a tectonic depression along northwest-southeast and east-west high-angle normal fault systems and filling of the basin by alluvial deposits (Figure 7). In the central and northern part of the basin during the middle Pleistocene, thick distal lacustrine mudstone sequences and marginal palustrine carbonate accumulated (Figure 7). During the early–middle Pleistocene, subsidence decreased and the basin gradually filled with alluvial deposits. Finally, new northwest-southeast extensional tectonic events and the glacial and interglacial climatic periods eroded portions of the Sulmona facies associations.

ACKNOWLEDGMENTS

We wish to thank P. G. Catalano (Società Romane e Abruzzesi SARA) for making data available. This work is part of a research program that has been financially supported by CNR (Centro di Studio per il Quaternario e l'Evoluzione Ambientale, Roma) and CROP-11 Project (Marina di Tarquinia-Vasto; Director M. Parotto; Dipartimento di Scienze Geologiche, Terza Università di Roma).

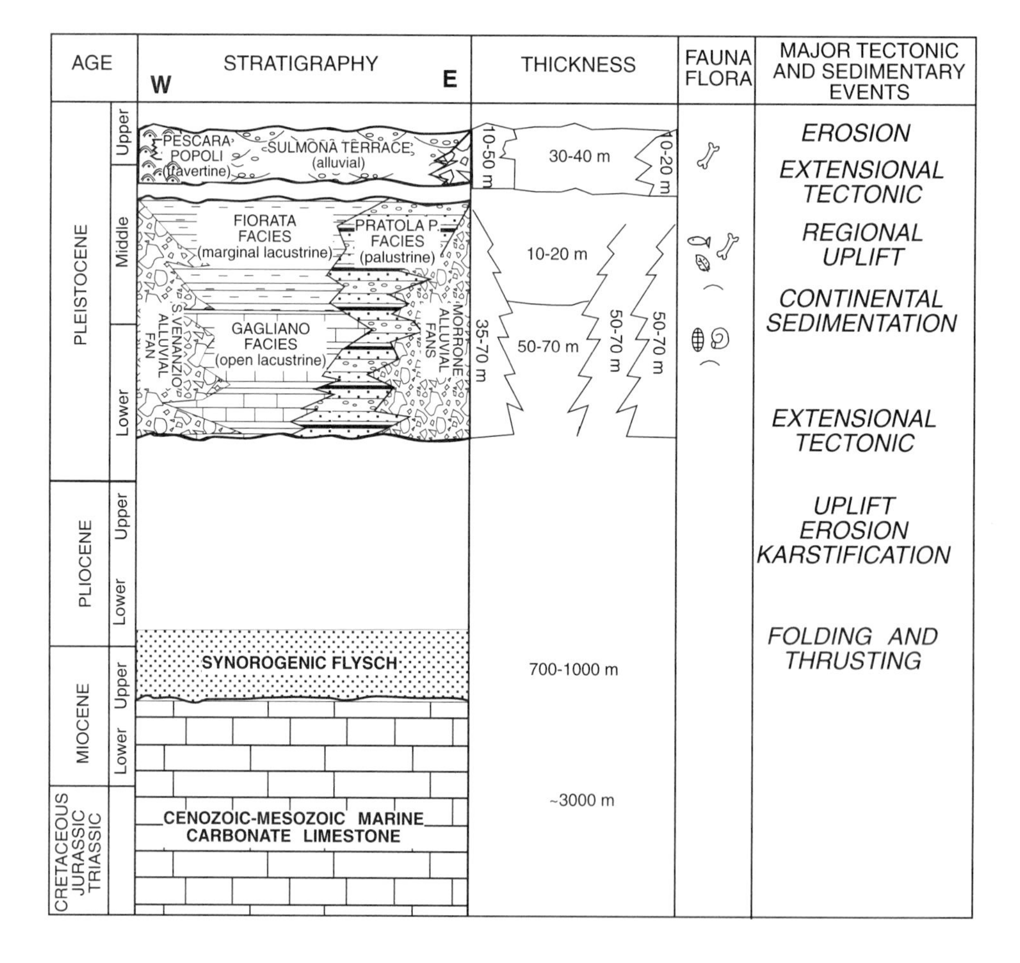

Figure 7—Overview of tectonic-sedimentary evolution of the Sulmona Basin. Legend is shown in Figure 5.

REFERENCES CITED

Beneo, E., 1940, Tettonica della Valle dell'Aterno (Abruzzo). Bollettino Regio Ufficio Geologico Italiano, v. 55, p. 1–7.

Beneo, E., 1943, Note illustrative del Foglio "Sulmona" (no.146) della Carta Geologica d'Italia al 100,000. Regio Ufficio Geologico Italiano, 41 pp.

Bigi, G., D. Cosentino, M. Parotto, R. Sartori, and P. Scandone, eds., 1992, Structural Model of Italy, scale 1:500,000. C.N.R., Quaderni de "La Ricerca Scientifica," v. 114 (3), sheet map No. 4.

Cavinato, G. P., D. Cosentino, R. Funiciello, M. Parotto , F. Salvini, and M. Tozzi, 1995, Constraints and new problems for geodynamical modelling of the Central Italy (CROP 11 Civitavecchia—Vasto deep seismic lines): Bollettino Geofisica Teorica Applicata (PAG) "Crop Project offshore crustal seismic profiling in the Central Mediterranean," v. 36 (141–144), p. 159–174.

Catalano, P. G.,1964, Il quaternario lacustre del Bacino di Sulmona. Consiglio Nazionale delle Ricerche, internal report.

Cavinato, G. P., D. Cosentino, D. De Rita, R. Funiciello, and M. Parotto, 1994, Tectonic-sedimentary evolution of intra-apenninic basins and correlation with the volcano-tectonic activity in central Italy: Memorie Descritive Carta Geologica d'Italia, v. 49, p. 49–62.

Cavinato, G. P., and E. Miccadei, 1995, Sintesi preliminare delle caratteristiche tettoniche e sedimentarie dei depositi quaternari della conca di Sulmona (L'Aquila). Il Quaternario, v. 8, p. 129–141.

Cipollari, P., and D. Cosentino, 1997, Il sistema Tirreno-Appennino: segmentazione litosferica e propagazione del fronte compressivo, *in* Cipollari, P., D. Cosentino, and M. Parotto, 1996, Modello cinematico-strutturale dell'Italla centrale, *in* G. Cello, G. Deiana, and P. P. Pierantoni, eds., Geodinamica e tettonica attiva del sistema Tirreno-Appennino, Studi Geologici Camerti, Special Vol. 1995/2, p. 135–143.

D'Andrea, M., E. Miccadei, and A. Praturlon, 1992, Rapporti tra il margine orientale della piattaforma laziale-abruzzese ed il margine occidentale della piattaforma Morrone-Pizzalto-Rotella, *in* M. Tozzi, G. P. Cavinato, and M. Parotto, eds., Studi preliminari all'acquisizione dati del profilo CROP 11 Civitavecchia-Vasto. Studi Geologici Camerti, Spec. Vol. 1991/2, p. 389–395.

Di Filippo, M., and E. Miccadei, 1997, Analisi gravimetriche della Conca di Sulmona. Il Quaternario, v. 10, p. 483–486.

Esu, D., O. Girotti, and T. Kotsakis, 1992, Molluschi e vertebrati di alcuni bacini continentali dell'Appennino centrale: indicazioni biostratigrafiche e paleoecologiche *in*, M. Tozzi, G. P. Cavinato, and M. Parotto, eds, Studi preliminari all'acquisizione dati del profilo CROP 11 Civitavecchia-Vasto. Studi Geologici Camerti, Spec. Vol. 1991/2, p. 295–299.

Ghisetti, F., U. Follador, R. Lanza, and L. Vezzani, 1992, La zona di taglio Rigopiano-Bussi-Rivisondoli: svincolo transpressivo al margine nord-orientale della piattaforma laziale-abruzzese, *in* M. Tozzi, G.P. Cavinato and M. Parotto, eds., Studi preliminari all'acquisizione dati del profilo CROP 11 Civitavecchia-Vasto. Studi Geologici Camerti, Special Vol. 1991/2, p. 215–220.

Gierlowski-Kordesch, E. H., 1998. Carbonate deposition in an ephemeral siliciclastic alluvial system: Jurassic Shuttle Meadow Formation, Newark Supergruop, Hartford Basin, USA. Palaeogeography, Palaeoclimatology, Palaeoecology, v. 140, p. 161–184.

Gliozzi, E., L. Abbazzi, P. Argenti, A. Azzaroli, L. Caloi, L. Capasso Barbato, G. di Stefano, D. Esu, Ficcarelli., O. Girotti, T. Kotsakis, F. Masini, P. Mazza, C. Mezzabotta, M. R. Palombo, C. Petronio, L. Rook, B. Sala, R. Sardella, E. Zanalda, and D. Torre, 1997, Biochronology of selected mammals, molluscs and ostracods from middle Pliocene to the late Pleistocene in Italy. The state of art. Rivista Italiana Paleontologia e Stratigrafia, v. 103, p. 369–388.

Kelts, K., and M. R. Talbot, 1990, Lacustrine carbonates as geochemical archives of environmental change and biotic/abiotic interactions, *in* M. M. Tilzer, and C. Serruya, eds., Large Lakes. Ecological Structure and Function. Springer, Berlin, p. 288–315.

Leuci, G., and R. Scorziello, 1972, Su un molare di *Elephas antiquus* sp. rivenuto nelle alluvioni terrazzate della Conca di Sulmona (Pratola Peligna, L'Aquila, Abruzzo). Bollettino Società Naturalisti Italiani, v. 81, p. 303–312.

Leuci, G., and R. Scorziello, 1974, Su un molare di *Elephas trogontherii* sp. Bollettino Società Naturalisti Italiani, v. 83, p. 1–12.

Miccadei, E., G. P. Cavinato, and E. Vittori, 1993, Elementi neotettonici della Conca di Sulmona, *in* P. Farabollini, C. Invernizzi, A. Pizzi, G. P. Cavinato, and E. Miccadei, eds., Evoluzione geomorfologica e tettonica quaternaria dell'Appennino centro-meridionale. Studi Geologici Camerti, Spec. Vol. 1992/1, p. 165–174 .

Miccadei, E., R. Barberi, G. P. Cavinato, 1999, La geologia quaternaria della Conca di Sulmona (Abruzzo, Italia Centrale). Geologia Romana, v. 34, p. 59–86.

Mostardini, F., and S. Merlini, 1988, Appennino centromeridionale: sezioni geologiche e proposta di modello strutturale. Memorie Società Geologica Italiana, v. 35, p. 177–202.

Patacca, E., P. Scandone, M. Bellatalla, N. Perilli,and U. Santini, 1992, La zona di giunzione tra l'arco appenninico settentrionale e l'arco appenninico meridionale nell'Abruzzo e nel Molise, *in* M. Tozzi, G. P. Cavinato, and M. Parotto, eds., Studi preliminari all'acquisizione dati del profilo CROP 11 Civitavecchia-Vasto. Studi Geologici Camerti, Spec. Vol. 1991/2, p. 417–441.

Platt, N. H., and V. P. Wright, 1991, Lacustrine carbonates: facies models, facies distributions and hydrocarbon aspects, *in* P. Anadón, Ll. Cabrera, and K. Kelts, eds., Lacustrine Facies Analysis. Special Publication International Association of Sedimentologists, v. 13, p. 57–74.

Radmilli, A. M., 1964, Abruzzo preistorico, il Paleolitico inferiore-medio abruzzese. Sansoni, Firenze, 117 p.

Radmilli, A. M., 1984, Le Svolte di Popoli (Abruzzo), *in*

Ministero Beni Culturali e Ambientali-Sovrintendenza Speciale Musco Preistorico Etnografico " L. Pigorini," I primi abitanti d'Europa, De Luca, Roma, p. 141–143.

Royden, L., E. Patacca, and P. Scandone, 1987, Segmentation and configuration of subducted lithosphere in Italy: an important control on thrust belt and foredeep basin evolution. Geology, v. 15, p. 714–717.

Sladen, C. P., 1994, Key elements during the search for hydrocarbons in the lake systems, *in* E. Gierlowski-Kordesch and K. Kelts, eds., Global Geological Record of Lake Basins, Vol. 1. Cambridge Univesity Press, Cambridge, p. 3–17.

Talbot, M. R., 1990, A review of the paleohydrological interpretation of carbon and oxygen isotopic ratios in primary lacustrine carbonates. Chemical Geology, v. 80, p. 261–279.

Vittori, E., G. P. Cavinato, and E. Miccadei, 1995, Active faulting along the Northeastern Edge of the Sulmona Basin (Central Apennines). Bulletin American Association Engineering Geologists Special Issue, v. 6, p. 115–126.

Cavinato, G. P., E. Gliozzi, and I. Mazzini, 2000, Two lacustrine episodes during the late Pliocene–Holocene evolution of the Rieti Basin (Central Apennines, Italy), *in* E. H. Gierlowski-Kordesch and K. R. Kelts, eds., Lake basins through space and time: AAPG Studies in Geology 46, p. 527–534.

Chapter 49

Two Lacustrine Episodes During the Late Pliocene–Holocene Evolution of the Rieti Basin (Central Apennines, Italy)

Gian Paolo Cavinato
Elsa Gliozzi
C.N.R., Centro di Studio per il Quaternario e l'Evoluzione Ambientale, Dip. Scienze della Terra, Univ. "La Sapienza" Rome, Italy

Ilaria Mazzini
Via M. Menghini
Rome, Italy

INTRODUCTION

Two main lacustrine episodes punctuated the continental evolution of the Pliocene–Pleistocene Rieti intraapennine basin (Central Apennines, Italy): the most ancient during the early Pleistocene (Case Strinati-Madonna della Torricella lake) and the more recent during the Holocene. These lacustrine episodes show different characteristics both sedimentologically and paleontologically. The Case Strinati-Madonna della Torricella lake, located in the northern sector of the basin, was a shallow saline-to-freshwater lake initially influenced by the nearby lower Pleistocene sea. The lake was formed during the second filling phase of the basin, when the subsidence rate decreased allowing deposition of fine-grained sediments. The Holocene lacustrine system contains at least two freshwater large lakes (Southern lake and Lago Lungo-Ripa Sottile lake), located in the south and north of the basin, respectively. They are divided by a travertine barrier (Rieti-Colarieti travertine). The setting of the Holocene lakes is strictly linked to the changes in the recent global climatic pattern.

GEOLOGIC SETTING AND EVOLUTION OF THE RIETI BASIN

The Rieti Basin is a half-graben located in the Central Apennine chain between the Sabini and Reatini Mountains (Figures 1, 2). Its origin is related to the late Pliocene northwest-southeast and east-west extensional tectonic phases that reactivated the main compressional faults (Cavinato, 1993; Cavinato et al., 1994) (Figure 3). During the late Pliocene–early Pleistocene, the depression was a catchment area filled by alluvial fan systems linked to the northeastern active basin margin (first filling phase). In the southern sector of the basin, mostly coarse-grained deposits crop out, while in the north they are fine-grained. Two lithostratigraphic units have been recognized in the first filling phase: the lower depositional unit (LDU) and the "Calcariola-Fosso Canalicchio" unit (CFCU) (Cavinato, 1993; Barberi and Cavinato, 1993; Barberi et al., 1995). The LDU crops out both in the southern and northern sector, and the CFCU is found only in the south. Both the LDU and CFCU have been interpreted as alluvial fan systems characterized by conglomerates with both clast- and matrix-supported fabrics. The clasts were derived from transitional and carbonate sedimentary successions of the Mesozoic–Cenozoic Umbria-Sabine and Latium Abruzzi sequences that surround the basin (Figure 2). Based on their depositional character, these conglomerates have been interpreted as semiarid alluvial fans related both to activity of the fault boundary zone and to a climatic cooling possibly at the end of the late Pliocene, when a marked sea-level lowering is recorded on the western side of the Narnesi-Amerini Mountains (Girotti and Piccardi, 1994), correlatable with the "Aullan erosional phase" as redefined by Azzaroli et al. (1988).

During the early Pleistocene, the subsidence rate decreased, as well as the sedimentation rate. The southern sector was transformed into a rather flat braidplain in which several scattered pools were more or less ephemeral; these sediments are recognized as part of the second filling phase (Figure 3). Near the southeastern border of the basin, two fine-grained successions were sampled for ostracod analyses showing the presence of an oligothermophilous and crenophil association (*Ilyocypris bradyi* Sars, *Potamocypris zschokkei*

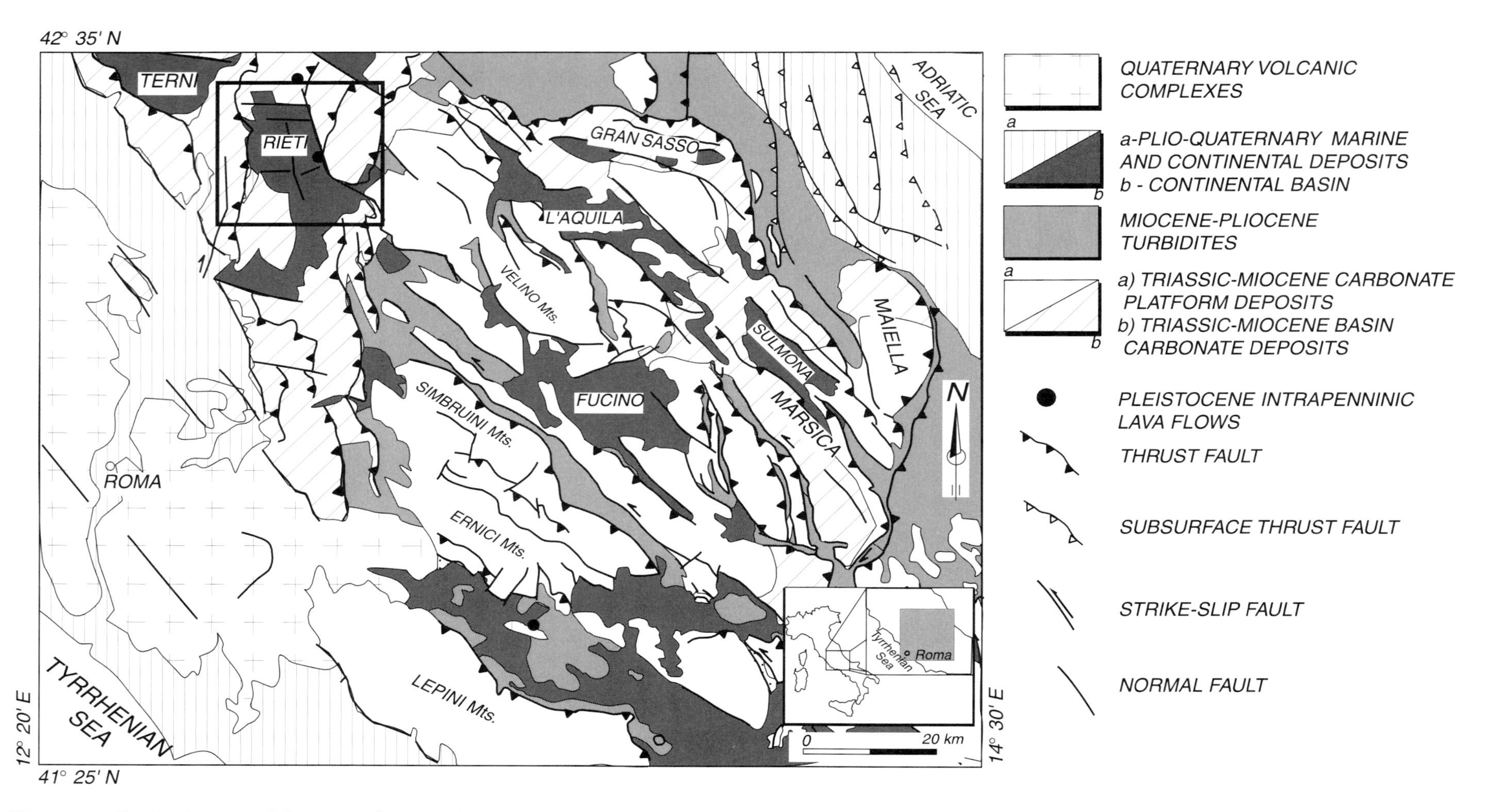

Figure 1—Geologic map of the central Apennines in Italy showing the location of the Rieti Basin.

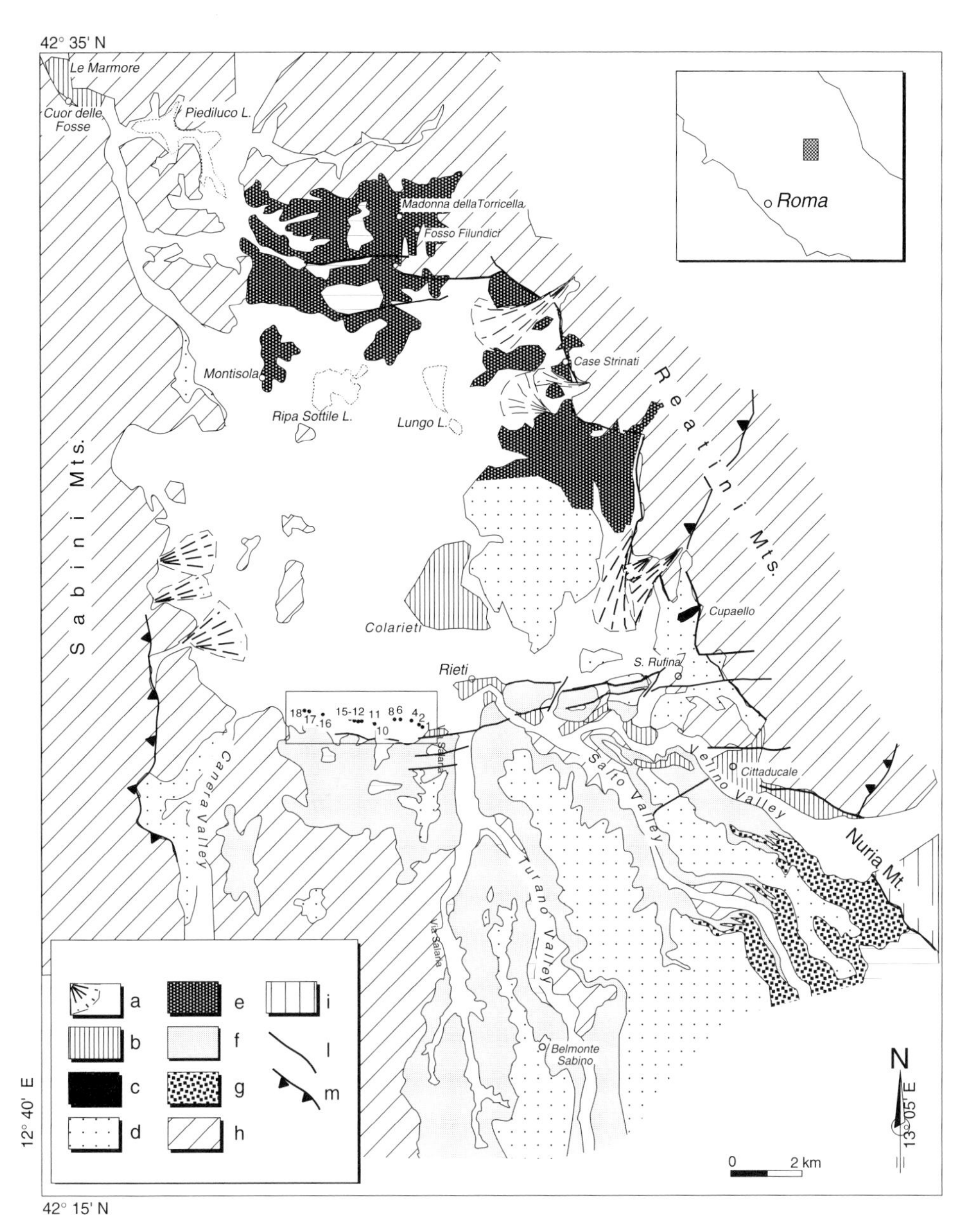

Figure 2—Geologic scheme of the Rieti Basin. Legend: a = Holocene lacustrine deposits and recent talus fans; b = middle and late Pleistocene travertines; c = Cupaello lava flow; d = upper depositional unit (UDU) (early Pleistocene); e = early Pleistocene deposits of the northern sector; f = lower depositional unit (LDU) (?late Pliocene–early Pleistocene); g = "Calcariola-Fosso Canalicchio" unit (CFCU) (?late Pliocene–early Pleistocene); h = Mesozoic–Cenozoic units of the Umbro-Sabina domain; i = Mesozoic–Cenozoic units of the Latium-Abruzzi domain; l = main normal faults; m-thrust faults. The numbered dots located west of the town of Rieti are the locations of sediment cores taken of the Holocene age Southern lake.

Kaufmann), *Candona neglecta* Sars), accompanied by euryplastic species such as *Cyclocypris laevis* (Müller), *Fabaeformiscandona* cf. *F. fabaeformis* (Fischer) and *Pseudocandona marchica* (Hartwig). The sediments recording this phase are included in the upper depositional unit (UDU), which is characterized essentially by intercalations of conglomerates, sands, and marly clays. The conglomerates commonly contain cross-bedding and are lens-shape, whereas the finer grained deposits exhibit cross- and horizontal lamination. In the northern sector, a lake developed (Case Strinati-Madonna della Torricella lake) due to a preexisting paleomorphologic depression (Ciccolella et al., 1995) (Figure 3). This lake was initially oligohaline, influenced by the nearby early Pleistocene sea. After a regional tectonic uplift around 1.5 Ma causing the regression of the sea, Case Strinati-Madonna della Torricella lake changed into a shallow freshwater lake (Cavinato, 1993; Barberi and Cavinato, 1993; Barberi et al., 1995).

At the end of the early Pleistocene, the Rieti Basin was affected by a northwest-southeast and east-west extensional tectonic phase that caused the collapse of its central area, forming the "Conca di Rieti" (Cavinato, 1993) (Figure 3). This event was almost coeval with the activity of the intrapenninic volcanic center of Cupaello (0.64–0.54 Ma, Laurenzi et al., 1994). This collapse produced an abrupt change in subsidence rate and depocenter position so that the basin became roughly asymmetric toward the north–northwest.

During the middle and late Pleistocene, the geologic evolution of the Conca di Rieti was dominated by several filling/erosional events due to stream activity (mainly that of the paleo-Velino River) (Figure 3). Sedimentation on the alluvial braidplain of the Conca was mainly controlled by climatic factors that were responsible for the

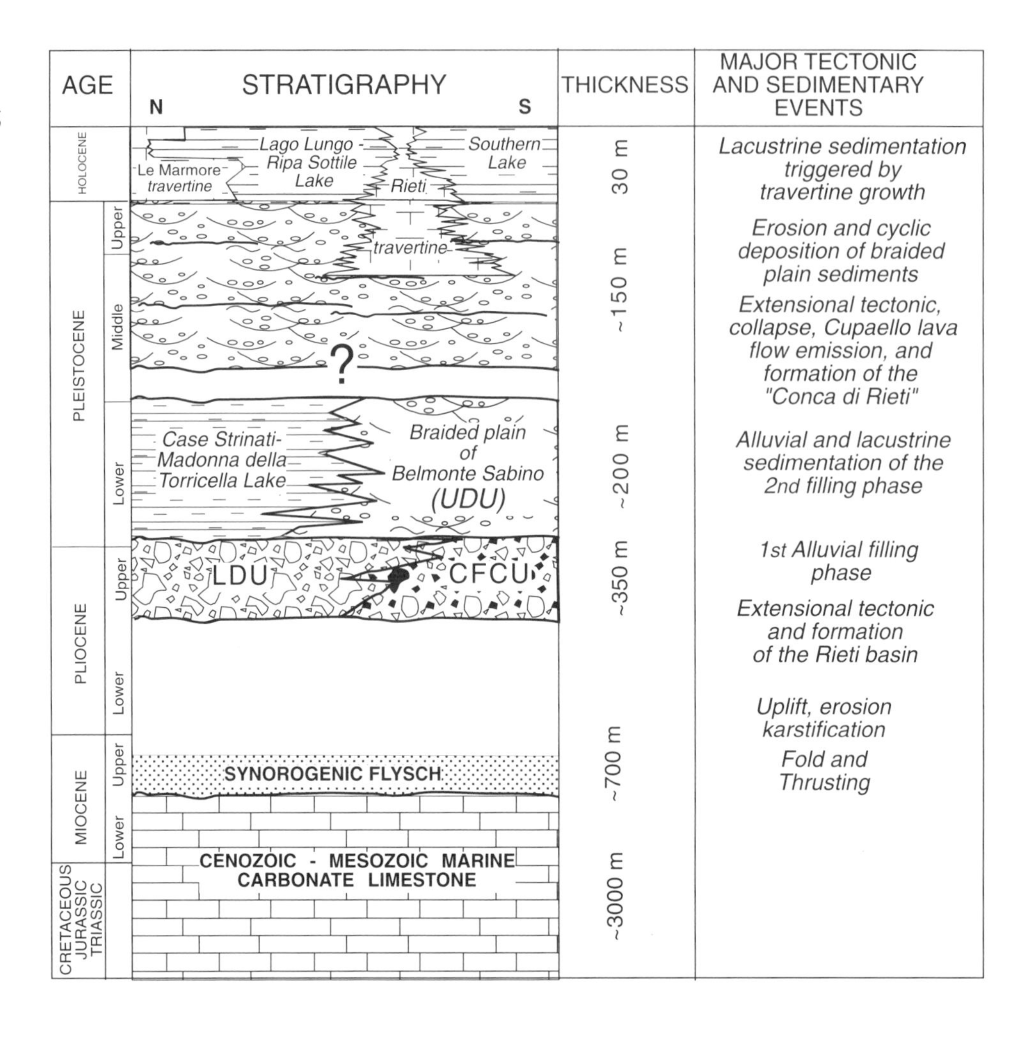

Figure 3—Stratigraphic overview of the deposits of the Rieti Basin showing the Pliocene–Pleistocene tectono-sedimentary evolution.

formation of upstream fluvial terraces. In particular, two travertine dams, located respectively in the southern and northern sectors of the Conca (Colarieti-Rieti and Tre Strade/Le Marmore travertines) (Figure 2), acted as thresholds during warm periods causing river flooding and collapsed during cold periods triggering fluvial erosion (Brunamonte et al., 1993; Carrara et al., 1992; Ferreli et al., 1992; Barberi et al., 1995; Calderini et al., 1998). Relicts of middle and upper Pleistocene fluvial sediments, cropping out along the Velino valley and inside the Conca di Rieti, are assigned to six orders of fluvial terraces spanning the early Middle Pleistocene to ancient Holocene. Carrara et al. (1992), Ferreli et al. (1992), Brunamonte et al. (1993), Carrara et al. (1995), and Calderoni et al. (1994) provided a chronologic framework, based on radiometric dating, for most of the terrace orders. In particular, $^{39}Ar/^{40}Ar$ dating of a pyroclastic level yielded 0.4 Ma for the fourth order, and the travertine on top of the fifth order was dated by $^{234}U/^{230}Th$ as between 0.18 and 0.03 Ma.

CASE STRINATI LAKE

The brackish-freshwater lake of Case Strinati-Madonna della Torricella developed in the northern sector of the Rieti Basin during the second filling phase. Here, a psammitic/pelitic sequence representing a lacustrine/marshy depositional environment overlies LDU or gravel and sandy bodies that represent a fluvial channel facies. Three localities show good exposures of this succession: Case Strinati, Fosso Filundici, and Madonna della Torricella (Barberi et al., 1995) (Figures 2–4). At Case Strinati, a sequence of diatomitic silts crops out at 470 m a.s.l. with a thickness of about 3 m. They are characterized by thin parallel lamination and are rich in plant frustules, well-preserved leaves, crushed mollusc shells, and abundant ostracods. At Fosso Filundici (650 m a.s.l.), a dark clayey paleosol marks the change from siliciclastic to carbonate mud sedimentation. This paleosol is rich in lignite fragments and small carbonate clasts with angular shapes. Pelitic sedimentation consists of calcareous silts at the base and top, containing parallel lamination with laminae some centimeters thick. In the middle part of the sequence, the silts are massive and bear plant frustules, gastropod opercules, and abundant ostracods. At Madonna della Torricella, the upper part of the lacustrine/marshy sequence contains laminated gray calcareous silts interbedded with lignite-rich levels. Molluscs and ostracods are abundant.

Paleontologic analyses of ostracods and mollusks in the lower part of the lacustrine succession evidence the existence of a brackish marsh that underwent changes in water body chemistry, probably related to local water dilution from precipitation. In particular,

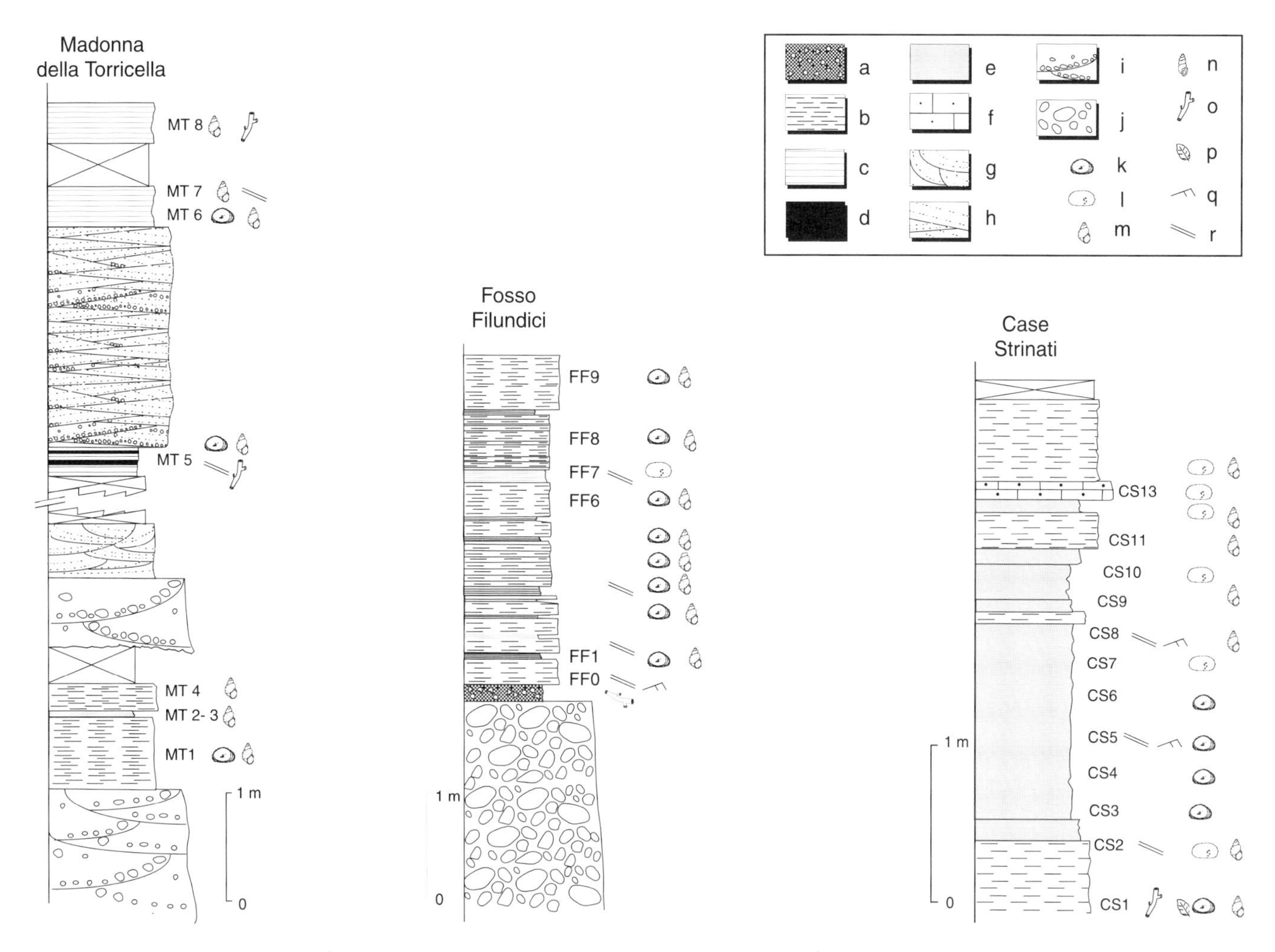

Figure 4—Lithologic logs of the sections sampled in the northern sector of the Rieti Basin (Case Strinati-Madonna della Torricella lower Pleistocene lake). Geographic locations shown in Figure 2. Legend: a = dark clayey paleosol; b = calcareous silts; c = clays; d = lignites; e = diatomitic silts; f = cemented calcarenites and calcareous muds; g = sands with trough cross-stratification; h = sands with planar stratification; i = cross-bedded conglomerates; j = massive conglomerates; k = freshwater ostracod associations; l = oligo-mesohaline ostracod associations; m = freshwater mollusk associations; n = terrestrial mollusks; o = plant frustules; p = leaf prints ; q =cross-lamination; r = horizontal lamination.

two ostracod faunas have been recognized: an oligohaline fauna dominated by *Cyprideis torosa* (Jones), with *Ilyocypris gibba* (Ramdohr), *Candona angulata* Müller, and *Cypria ophtalmica* (Jurine), and a second fauna characterized by an assemblage dominated by Candoninae [*Candona candida* Müller, *Fabaeformiscandona* cf. *F. levanderi* (Hirshmann), *Pseudocandona marchica* (Hartwig), and *Candonopsis kingsleii* (Brady and Robertson)] (Barberi et al., 1995). A statistical analysis of the sieve-pores of the *Cyprideis torosa* shells showed possible salinity variations from freshwater up to 1.45‰ (Gliozzi and Mazzini, 1998). In the upper part of the Madonna della Torricella section, ostracod assemblages recorded an evolution toward a shallow freshwater lake, with abundant vegetation [*Ilyocypris bradyi* Sars, *Candona* cf. *C. candida* Müller, *Pseudocandona marchica* (Hartwig), *Fabaeformiscandona* cf. *F. fragilis* (Hartwig), and *Cyclocypris laevis* (Müller)]. Molluscs, absent in the lower part of the lacustrine succession, except for the rare *Bithynia* opercules, are more abundant in the upper part [*Bithynia* sp., *Valvata cristata* (Müller), *Acroloxus lacustris* (Linnaeus), and *Belgrandia* sp.], particularly in association with the lignite-rich levels [*Teodoxus (Neritea) groyanus* (Férussac), *Emmericia umbra* De Stefani, *Prososthenia* sp., *Melanopsis affinis* Férussac, *Hydrobia* cf. *H. stagnorum* (Gmelin), *Hydrobia* sp., *Valvata piscinalis* (Müller), *Valvata cristata* (Müller), and *Viviparus* sp.].

The oligohaline interval of the Case Strinati-Madonna della Torricella shallow lake (Case Strinati and Fosso Filundici outcrops) has been stratigraphically correlated to the maximum transgression event of the Tyrrhenian Sea during the earliest part of the early Pleistocene (Santernian substage of the Mediterranean marine chronostratigraphy) when the coastline was located against the western side of the Narnesi-Amerini and Sabini Mountains. The overlying freshwater interval of this lake (Madonna della Torricella outcrop) has been connected to the subsequent regression (Emilian substage) linked to a regional tectonic uplift (Barberi et al., 1995).

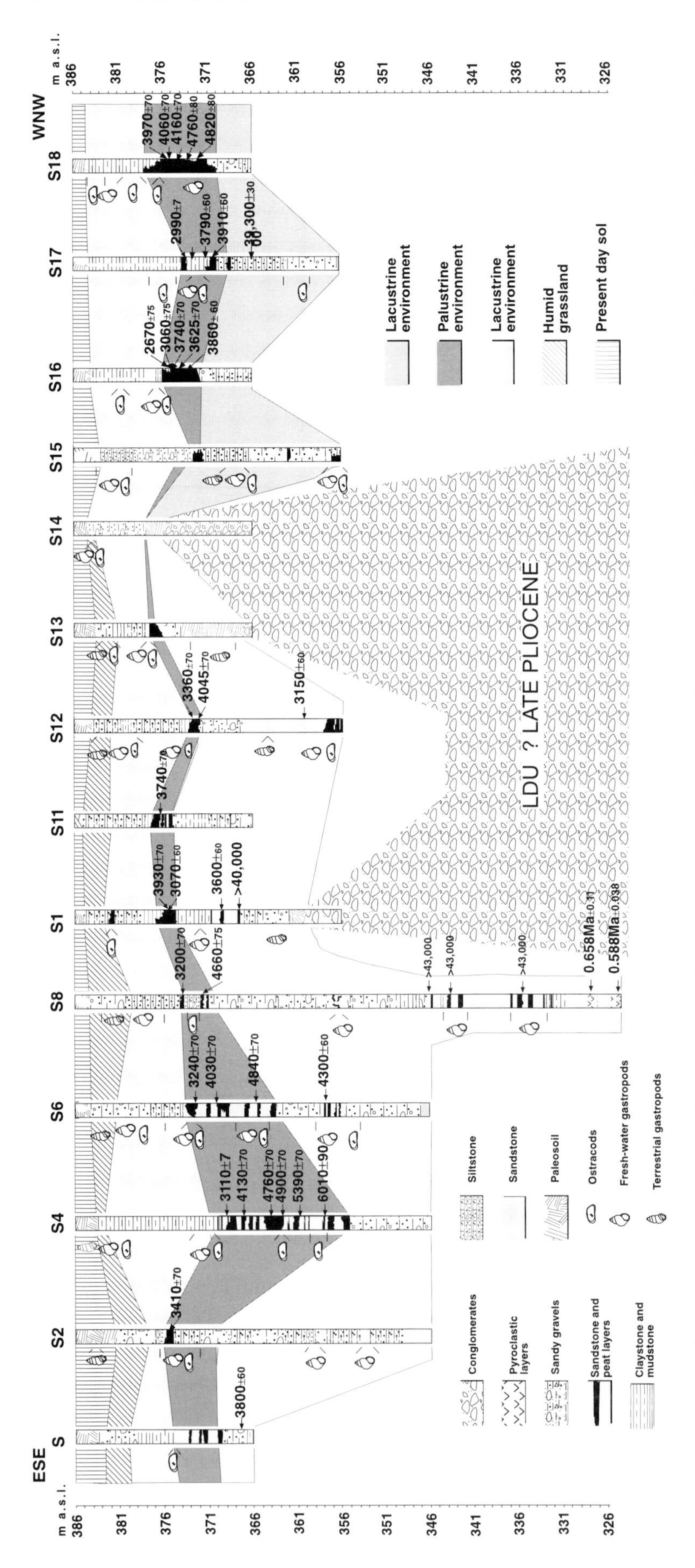

Figure 5—Lithologic logs across Southern lake (Holocene) with paleoenvironmental interpretation (modified from Calderini et al., 1998). Geographic locations of the logs shown in Figure 2.

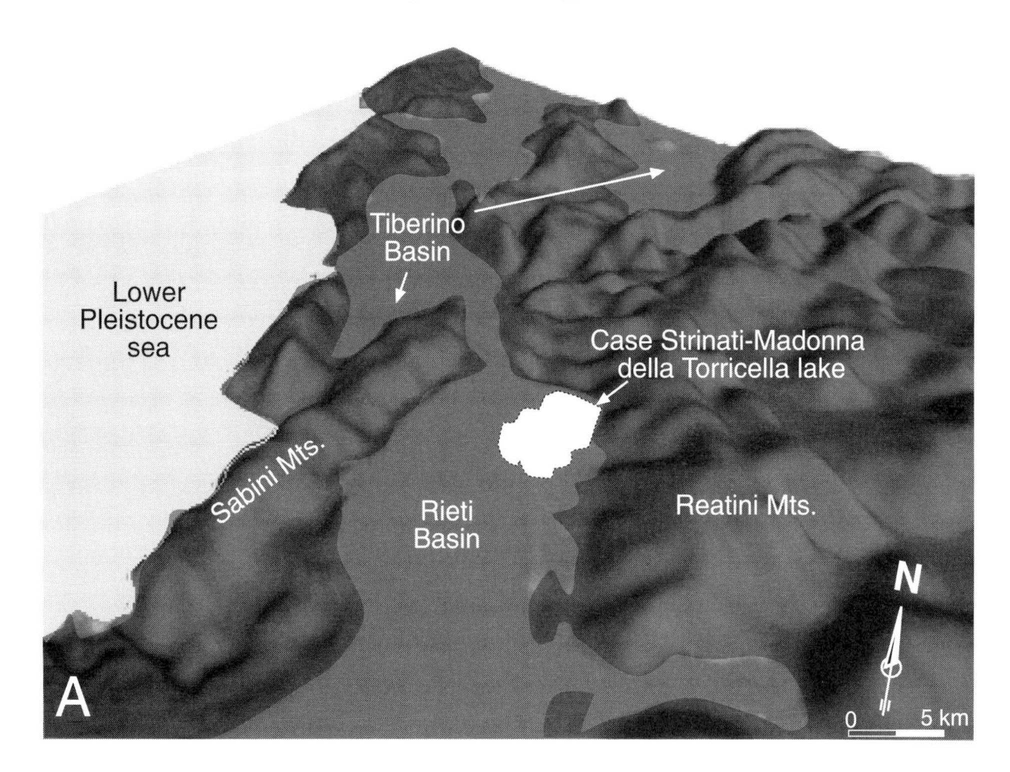

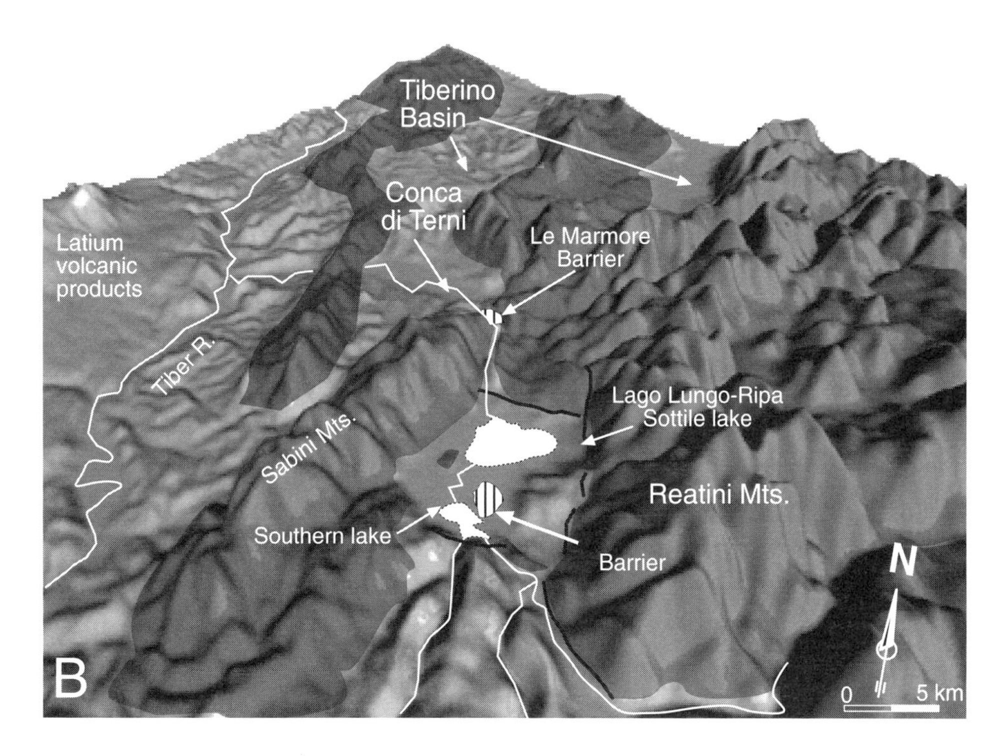

Figure 6—Three-dimensional reconstruction of the Rieti Basin. (A) Early Pleistocene: the coastline was located against the Apennines and a narrow connection allowed the formation of transitional brackish conditions in the continental environment in the Tiberino Basin. Similar conditions were recorded in the Rieti Basin showing that the marine influence reached even the innermost areas. (B) Late Pleistocene–Holocene: during the early Pleistocene, a regional tectonic uplift caused the westward regression of the sea to a coastline similar to the present coastline. The Rieti Basin was restricted to the present day Conca di Rieti after the tectonic collapse of the middle Pleistocene. Travertine formation during warmer times dammed the rivers producing two lakes: the Lago Lungo-ripa Sottile lake and the Southern lake. The northern lake is represented by the Lago Lungo and Ripa Sottile lakes today.

HOLOCENE LAKES

The alternating erosional and depositional phases of Holocene streams appear to have been controlled by the travertine natural dams located in the Conca. It is inferred (Carrara et al., 1995) that noticeable Holocene travertine deposition at Le Marmore caused a partial obstruction of the valley resulting in the aggradation of a sixth-order terrace that represents the present plain surface (Figure 3).

In the southern part of the Conca di Rieti (Calderini et al., 1998), a study of 16 sediment cores has identified a Holocene shallow lacustrine/palustrine environment (Southern lake) (Figure 5) through the presence of clays and silty clays with abundant plant debris, rare charophytes, and abundant, fragmentary, aquatic molluscs, such as *Acroloxus lacustris* (Linnaeus), *Valvata piscinalis* (Müller), *Valvata cristata* (Müller), *Bithynia tentaculata* (Linnaeus), and *Bithynia leachi* (Sheppard). Among the ostracods, Candoninae with *Candona neglecta* Sars (dominant), *Candona angulata* Müller, *Candona candida* Müller, *Pseudocandona marchica* (Hartwig), and *Candonopsis kingsleii* (Brady and Robertson) predominate, whereas *Ilyocypris* species are subordinate. Both ostracod faunas and molluscs suggest a shallow freshwater, carbonate-rich environment developed over a soft muddy bottom rich in aquatic plants. Several ^{14}C age determinations have shown that this lacustrine basin underwent a dramatic and sudden change around 4000–2600 yr BP toward a partially emerged environment under humid conditions in which peat developed (Figure 5) (Calderini et al., 1998).

A comparable Holocene history has also been recognized for the northern sector of the "Conca," in both sediment cores [the Lago Lungo drillhole (Calderoni et al., 1994) and the Ripa Sottile drillhole (E. Gliozzi, unpublished data)], and at outcrop sections at Cuor delle Fosse (Carrara et al., 1995) and Montisola sites (Ferreli et al., 1992) (see Figure 2 for locations). In particular, during the Holocene, the northern sector seems to have been characterized by a wide, shallow to deep freshwater lake (Lago Lungo-Ripa Sottile lake). Ostracod assemblages from the Ripa Sottile core are dominated by *Cytherissa lacustris* Sars indicating a freshwater and rather deep and cold water body. Around 4000–3000 yr BP, a significant lowering of the lake level and perhaps its division into separate smaller lakes is suggested by all four studied sites. Ferreli et al. (1992), Calderoni et al. (1994), and Carrara et al. (1995) provided evidence that the Holocene evolution of the Le Marmore travertine dam strictly affected the drainage of the whole Conca di Rieti. The ancient lakes that occupied the Conca di Rieti in the Holocene were likely formed in response to the fast growth of the travertine dam at Le Marmore during the climatic amelioration of Boreal and Atlantic chrons (Figures 3, 6). Subsequent progressive cooling at approximately 4000–3500 yr BP caused lowering of the dam and the consequent decrease of lake levels. Timing of this climatic event matches that of a global regressive event that occurred during mid-Subboreal chron (Calderini et al., 1998).

REFERENCES CITED

Azzaroli A., C. De Giuli, G. Ficcarelli, and D. Torre, 1988, Late Pliocene to early mid-Pleistocene mammals in Eurasia: faunal succession and dispersal events, Palaeogeography, Palaeoclimatology, Palaeoecology, v. 66, p. 77–100.

Barberi, R., and G. P. Cavinato, 1993, Analisi sedimentologiche ed evoluzione paleogeografica del settore meridionale del bacino di Rieti, Studi Geologici Camerti, spec. vol. 1992/1, p. 45–54.

Barberi, R., G. P. Cavinato, E. Gliozzi, and I. Mazzini, 1995, Late Pliocene–early Pleistocene palaeoenvironmental evolution of the Rieti Basin (Central Apennines), Il Quaternario, v. 8 (2), p. 515–534.

Brunamonte, F., C. Carrara, G. P. Cavinato, L. Ferreli, L. Serva, A. M. Michetti, and M. Raglione, 1993, La conca di Rieti, Il Quaternario, v. 6 (2), p. 396–401.

Calderini, G., G. Calderoni, G. P. Cavinato, E. Gliozzi, and P. Paccara, 1998, The upper Quaternary sedimentary sequence at the Rieti Basin (Central Italy): a record of sedimentation response to environmental changes, Palaeogeography, Palaeoecology, Palaeoclimatology, v. 140, p. 97–111.

Calderoni G., C. Carrara, L. Ferreli, M. Follieri, E. Gliozzi, D. Magri, B. Narcisi, M. Parotto, L. Sadori, and L. Serva, 1994, Palaeoenvironmental, palaeoclimatic and chronological interpretations of a late-Quaternary sediment core from Piana di Rieti (Central Apennine, Italy), Giornale di Geologia, v. 56 (2), p. 43–72.

Carrara, C., F. Brunamonte, L. Ferreli, P. Lorenzoni, L. Margheriti, A. M. Michetti, M. Raglione, M. Rosati, and L. Serva, 1992, I terrazzi della medio-bassa valle del F. Velino, Studi Geologici Camerti, Special vol. 1992/1, p. 97–102.

Carrara, C., D. Esu, and L. Ferreli, 1995, Lo sbarramento di travertino delle Marmore (Bacino di Rieti, Italia centrale): aspetti geomorfologici, faunistici ed ambientali, Il Quaternario, v. 8 (1), p. 111–118.

Cavinato, G. P., 1993, Recent tectonic evolution of the Quaternary deposits of the Rieti Basin (Central Apennines, Italy): southern area, Geologica Romana, v. 29, p. 411–434.

Cavinato, G. P., D. Cosentino, D. De Rita, R. Funiciello, and M. Parotto, 1994, Tectonic-sedimentary evolution of intra-apenninic basins and correlation with the volcano-tectonic activity in central Italy, Memorie Descrittive Carta Geologica d' Italia, v. 49, p. 49–62.

Ciccolella, A., M. Di Filippo, S. Jacovella, and B. Toro, 1995, Prospezioni ed analisi gravimentriche della Piana di Rieti, Il Quaternario, v. 8 (1), p. 141–148.

Ferreli, L., F. Brunamonte, G. Filippi, L. Margheriti, and A. M. Michetti, 1992, Riconoscimento di un livello lacustre della prima età del Ferro nel Bacino di Rieti e possibili implicazioni neotettoniche, Studi Geologici Camerti, vol. spec. 1992/1, p. 127–135.

Girotti, O., and E. Piccardi, 1994, Linee di riva del Pleistocene, Il Quaternario, v. 7 (2), p. 525–536.

Gliozzi, E., and I. Mazzini, 1998, Palaeoenvironmental analysis of early Pleistocene brackish marshes in the Rieti and Tiberino intrapenninic basins (Latium and Umbria, Italy) using ostracods, Palaeogeography, Palaeoclimatology, Palaeoecology, v. 140, p. 325–333.

Laurenzi, M., F. Stoppa, and I. Villa, 1994, Eventi ignei monogenici e depositi piroclastici nel distretto ultraalcalino umbro-laziale (ULUD): revisione, aggiornamento e comparazione dei dati cronologici, Plinius, v. 12, p. 61–66.

Basilici, G., 2000, Floodplain lake deposits on an early Pleistocene alluvial plain (Tiberino Basin, central Italy), *in* E. H. Gierlowski-Kordesch and K. R. Kelts, eds., Lake basins through space and time: AAPG Studies in Geology 46, p. 535–542.

Chapter 50

Floodplain Lake Deposits on an Early Pleistocene Alluvial Plain (Tiberino Basin, Central Italy)

Giorgio Basilici
Departmento de Administração e Politica de Recursos Minerais, Instituto de Geociências, Universidade Estadual de Campinas Campinas, Brazil

INTRODUCTION

Floodplain lakes are small depressions on river floodplains. Among the many factors that can contribute to the formation of these features, the most important include: (1) differential lowering of the floodplain surface in comparison with the fluvial belt due to major compaction of floodplain deposits (e.g., muddy or organic deposits), sedimentation rate within the fluvial belt, and local fault control; (2) cut of the meander belt (oxbow lakes), abandonment of fluvial branches, or splay deposition; (3) water table variations due to changes in climate and topography; and (4) permeability of the floodplain deposits (Leopold et al., 1964; Nanson and Croke, 1992; Willis and Behrensmeyer, 1994). This paper deals with early Pleistocene floodplain lake deposits within an alluvial plain lithostratigraphic unit designated as the S. Maria di Ciciliano Formation (SMCF) (Basilici, 1992) (Figure 1A).

During the early(?)–middle Pliocene, extensional tectonic activity linked to the opening of the Tyrrhenian Sea (Lavecchia, 1988; Martini and Sagri, 1993) formed the large Tiberino Basin, located in central Italy (Umbria) about 130 km south of Firenze. This rift basin reached a length of 125 km and a maximum width of 20 km; in plan view it has an "overturned Y" shape and is oriented north–northwest-south–southeast (Figure 1A). In the southeastern branch of the Tiberino Basin, the infill deposits consist of four lithostratigraphic units (Figure 1B) (Basilici, 1992). The Fosso Bianco Formation (FBF) (middle–late Pliocene) is made up principally of marly clays and interpreted as a deep lacustrine system (Basilici, 1997). The Ponte Naja Formation (PNF) (late Pliocene) is composed of sandy clayey silts with rare lenses of planar-concave gravelly-sandy ribbon bodies interpreted as deposits of the distal part of an alluvial fan (Basilici, 1995). The SMCF (early Pleistocene) is interpreted as alluvial plain deposits. The Acquasparta Formation (AF) (early Pleistocene) is composed of carbonate and travertine sediments and interpreted as a shallow lake and marsh environment.

DEPOSITIONAL AND PALEOGEOGRAPHIC RECONSTRUCTION OF THE S. MARIA DI CICILIANO FORMATION

The SMCF is exposed in the southeastern branch of the Tiberino Basin, covering an area 45 km long and 10 km wide. The SMCF is, on average, 150 m thick and overlies the FBF and PNF on an angular unconformity. Its general lithology is composed of sandy strata alternating with silty clays. The architectural organization of the SMCF consists of these sandy bodies encompassed by muddy sediments; in vertical succession, the sandy lithosomes are spaced 15 m apart on average (Figure 2) (Ambrosetti et al., 1995).

The sandy facies association is composed of sandy bodies that have a lenticular shape (concave bottom and flat top), are 6–7 m thick and 400–600 m wide, and average 1.5 km long, parallel to paleocurrent direction. These sandy bodies have erosive bottoms, are graded, and contain trough cross-stratification with sets up to 1 m thick, which decrease in thickness upward; epsilon cross- bedding (Allen, 1963), underlain by thin beds of clayey silts, appears in the upper 2 m of the sandy bodies. These sedimentary characteristics and fossil remains (freshwater molluscs, mammal bones, and tree trunk casts) allow the sandy bodies to be interpreted as infill of high-sinuosity fluvial channels.

The muddy facies association consists mainly of silty clays and subordinately of sands and lignites. These deposits record a floodplain, where four principal lithofacies can be recognized. (1) Silty clay lithofacies contain well-defined horizons, pedogenic structures (commonly granular and subangular blocky textures), calcareous nodules, root traces, terrestrial gastropods, mammal bones, and a general lack of sedimentary structures (Catt, 1990). These silty clays are interpreted as paleosol profiles. Poorly evolved paleosol profiles can be distinguished from more evolved paleosols by the scarcity of pedogenic

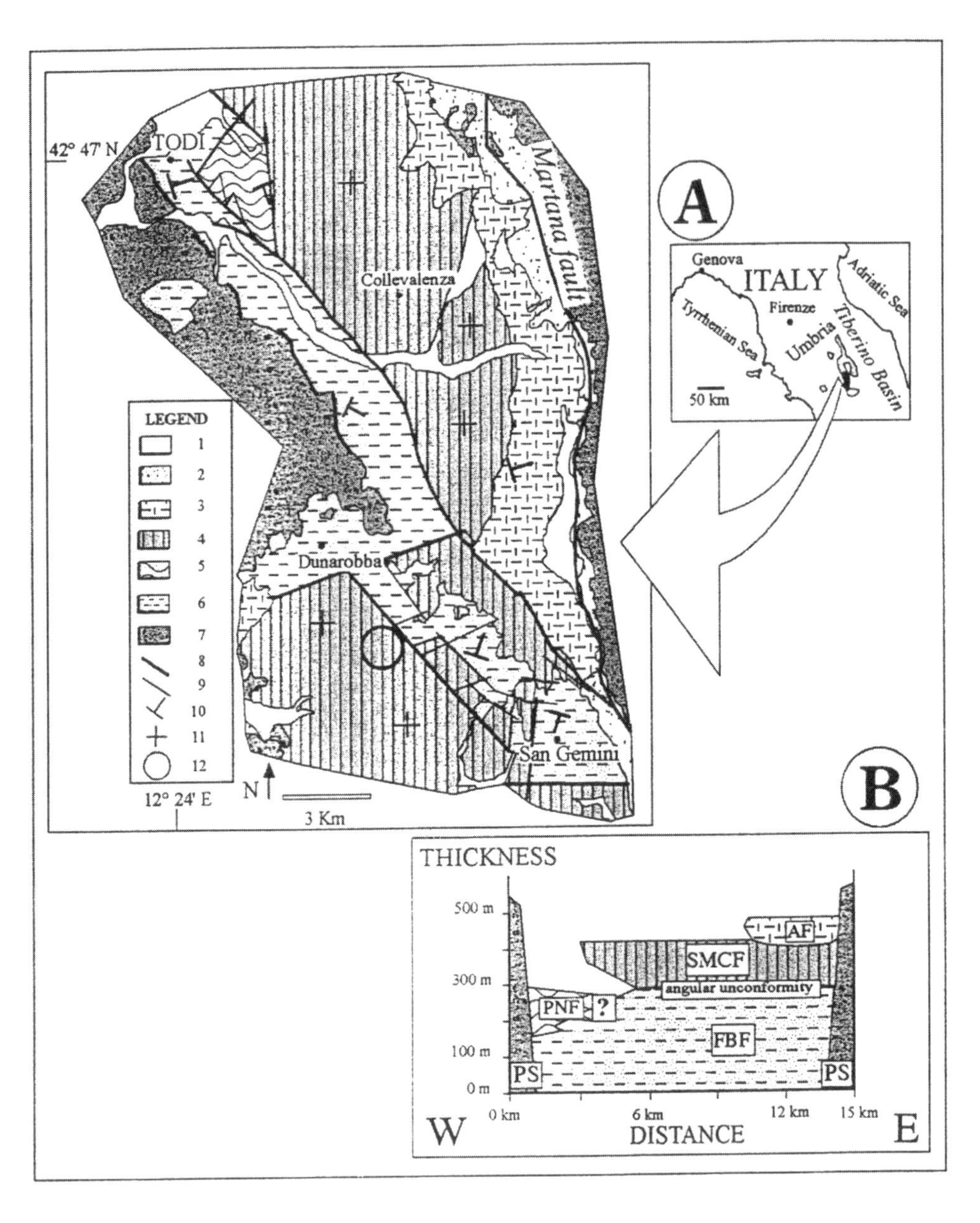

Figure 1—(A) Location and geologic map of the study area in the Tiberino Basin, central Italy. Legend: 1 = recent alluvial deposits; 2 = alluvial fan and talus deposits (Late Pleistocene); 3 = Acquasparta Formation (early Pleistocene); 4 = S. Maria di Ciciliano Formation (early Pleistocene); 5 = Ponte Naja Formation (late Pliocene); 6 = Fosso Bianco Formation (middle–late Pliocene); 7 = pre-Pliocene substratum; 8 = principal faults; 9 = secondary faults; 10 = inclined strata; 11 = horizontal strata; 12 = stratotype area of S. Maria di Ciciliano Formation. The study area is in the southwestern portion of the Tiberino Basin. (B) Generalized stratigraphic sketch of the Pliocene–Pleistocene deposits exposed in the southwestern part of the Tiberino Basin. PS = pre-Pliocene substratum; FBF = Fosso Bianco Formation (middle–late Pliocene); PNF = Ponte Naja Formation (late Pliocene); SMCF = S. Maria di Ciciliano Formation (early Pleistocene); AF = Acquasparta Formation (early Pleistocene).

features and by reduced thickness. Most proximal paleosols contain features interpreted as formed under waterlogged conditions (hydromorphic paleosols), whereas distal paleosols exhibit drier conditions of development. These last are indicated by a high concentration of calcium carbonate nodules within subsurface (Bk) horizons. (2) Laminated silty clays are a lithofacies interpreted as small and shallow lake deposits, which are described in the next section. (3) Lenticular sandy bodies (up to 1.3 m high) show some trough cross-stratification and cross-lamination and alternate with thin strata (0.5–10 cm thick) of silty clays. This lithofacies is interpreted as crevasse splay deposits. (4) Autochthonous lignites overlying paleosols with herbaceous remains are interpreted as ancient wetland deposits.

In the study area, the SMCF has been interpreted as a rectangular-shape alluvial plain 7–10 km wide and 3 km long, oriented north–northwest-south–southeast, and bounded at the eastern and western margins by the Martani Mountains and the Amelia Mountains, respectively (Figure 3) (Basilici, 1992; Ambrosetti et al., 1995). Paleocurrents (reconstructed from trough cross-stratification) indicate that the rivers flowed southward (Figure 3). Common occurrence of hydromorphic paleosols, lignites, and lacustrine deposits, as well as the recovery of brackish molluscs (e.g., *Cerastoderma edule*, *Potamides* sp.) on the southern part of the alluvial plain, suggest that this area was the distal part of an alluvial system occurring close to the marine coastline whose deposits crop out toward the southeast.

THE LACUSTRINE DEPOSITS OF THE S. MARIA DI CICILIANO FORMATION

Description

The laminated silty clay lithofacies of the SMCF are interpreted as deposits of small and shallow lakes on

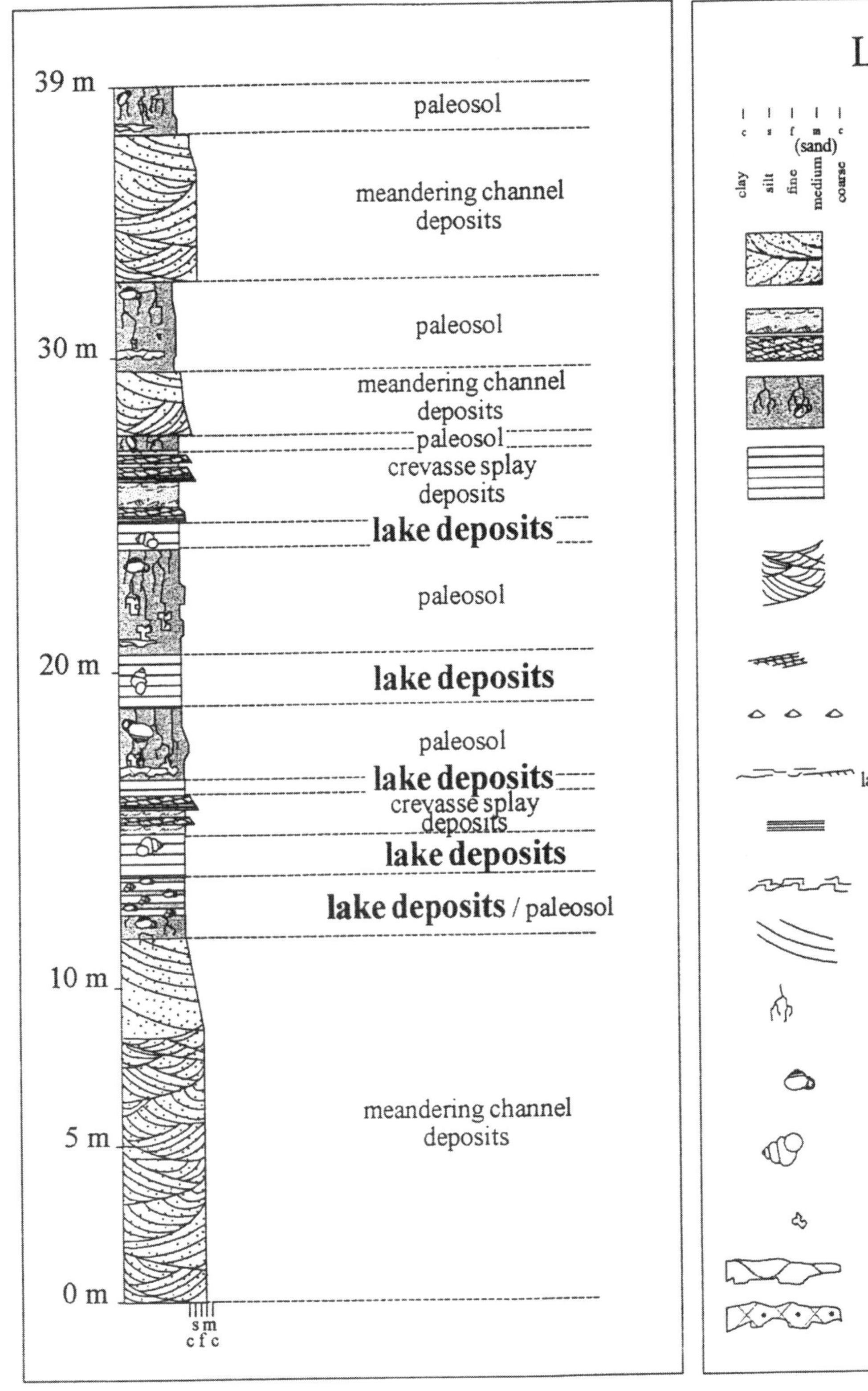

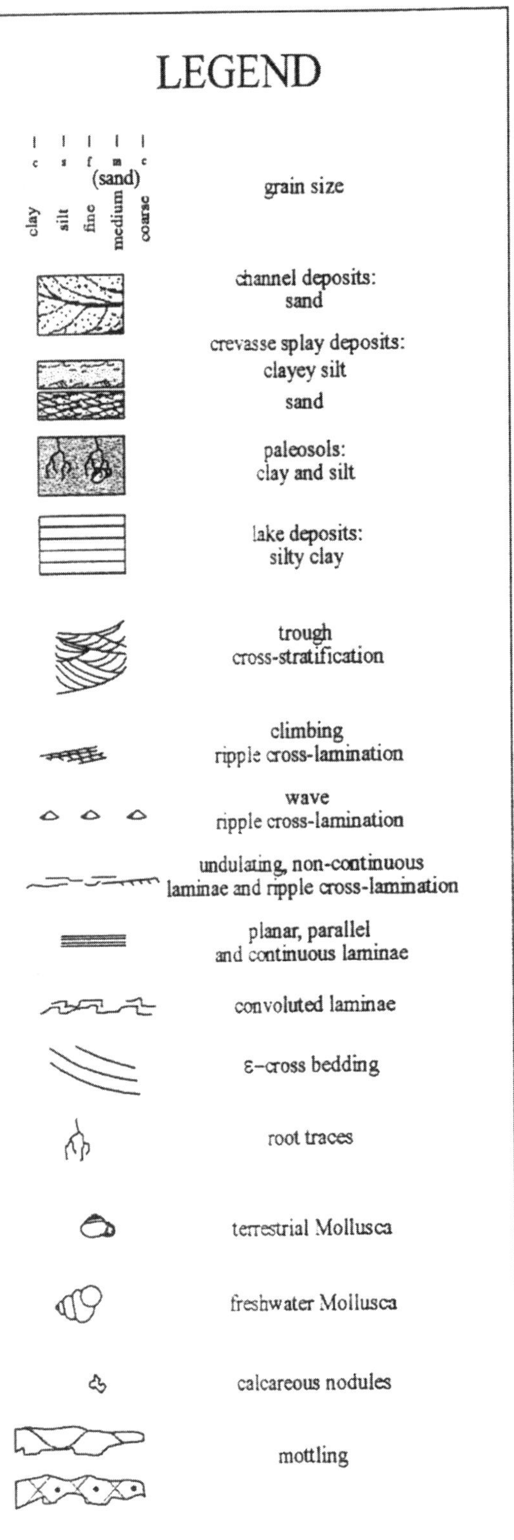

Figure 2—Stratigraphic section of the S. Maria di Ciciliano Formation (SMCF) from the stratotype area in the Tiberino Basin, southeast of Dunarobba (see Figure 1). The principal sandy bodies consist of river channel deposits alternating with muddy, subordinately sandy sediments deposited on a floodplain (lake deposits, crevasse splay deposits, and paleosols).

an alluvial plain. The lacustrine lithofacies is made up of bluish-gray silty clays with planar, parallel, and continuous laminae or thin beds, as defined by grain size or color variations (Figure 4). The most common intercalations consist of less than 1 mm to 3 cm thick clays and clayey silts (or very fine sands). Very fine sand or silt layers have nonerosive, abrupt lower contacts and grade upward into clays, testifying to suspension settling from currents. Dark bluish-gray and light bluish-gray laminae are probably related to different organic contents; each lamina or thin bed can be observed to extend laterally for more than 25 m. Only rarely do these lithofacies show small-scale cross-lamination up to 1 cm thick with symmetric profiles, which on bedding surfaces are exposed as straight-lined and parallel crests interpreted as wave ripples. Thickness of the lacustrine deposits ranges from 15 cm to 2.3 m, with an average of 1 m. In a few well-exposed outcrops, it has been possible to verify that this lithofacies does not change its characteristics laterally for at least 200 m; therefore, these deposits have a rectangular shape up to several hundred meters long. Lake deposits principally overlie paleosols (60% of the cases) and subordinately crevasse splay deposits or lignites; paleosols (54% of the cases) and crevasse splay deposits (30%) lie above lake deposits.

Although the laminated silty clay lithofacies makes up 20% of the stratigraphic successions, a variable distribution exists: 10% in the proximal part of the alluvial plain (northern part) and 25–40% in the distal areas (southern part). Lateral continuity of this lithofacies shows that the areal dimensions were equal to or greater than 10^4 m^2. Macrofossil remains are rare, of which the better represented classes are freshwater ostracods and gastropods (Esu and Girotti, 1991; Ambrosetti et al., 1995). The most common, sometimes only, freshwater gastropods species is *Viviparus belluccii*; according to D. Esu (personal communication, 1995), this species indicates stagnant waters. Other gastropod species (*Teodoxus groyanus*, *Melanopsis affinis*, *Planorbarius corneus*, *Valvata interposita*, some *Limnea* genera), pelecypods (*Unio*), and fish remains are rare. In addition, fossil leaves of trees and seeds or stems of vascular plants (*Trapa*, *Pseudoeuryale*, *Potamogeton*) and oogonia of freshwater algae (Characeae) occur rarely. Macroscale bioturbation is absent in all of the lacustrine deposits outcrops.

Interpretation

As indicated by the fine-grained sediments and its structures, the laminated silty clay lithofacies were deposited from settle out from turbulent suspension in an aqueous environment. Slow and continuous settling of clay contributed to the bulk of the strata, whereas interflows or low-density turbidity currents deposited widespread laminae (or thin beds) of clayey silts or very fine sands. Sometimes very weak wave action reworked very fine sands on the bottom, forming small-scale wave ripples. The dark bluish-gray laminae were probably linked to a high organic productivity (seasonal?) in paleolake waters. The undisturbed

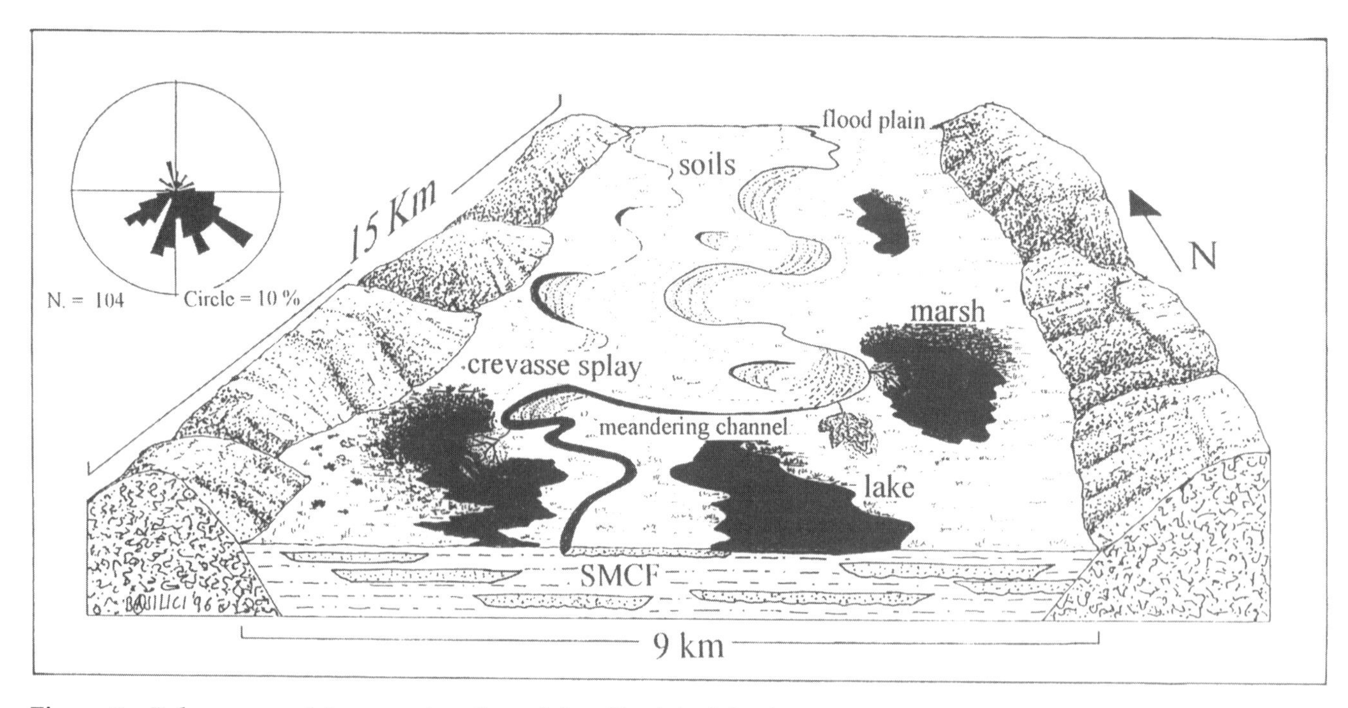

Figure 3—Paleogeographic reconstruction of the alluvial plain depositional system, as represented by the S. Maria di Ciciliano Formation (SMCF) of the Tiberino Basin during the early Pleistocene, showing facies relationships. The area corresponds to that of Figure 1A. Rose diagram to the left shows paleocurrent data (N = 104) based on trough cross-sets from fluvial channel sands. Circle = 10% means that the radius length is equivalent to 10% of the data.

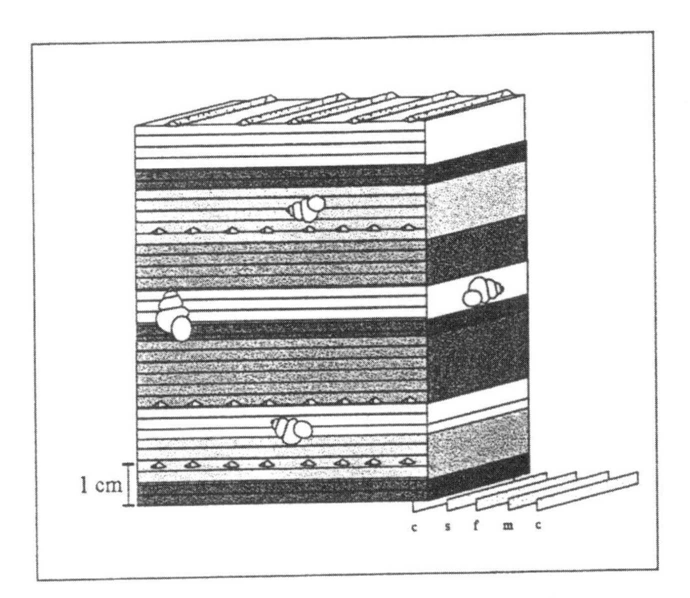

Figure 4—Floodplain lacustrine deposits of the S. Maria di Ciciliano Formation in the Tiberino Basin. This lithofacies is characterized by planar, parallel, and continuous laminae; their thickness ranges from 15 cm to 2.3 m with lateral continuity exceeding 200 m. Different grain sizes (c = clay, s = silt, f = fine sand, m = medium sand, c = coarse sand) and different shades of gray (probably linked to variable organic content) make up the laminae; rarely very small-scale wave ripples are present. See legend in Figure 2.

fine lamination of the bluish-gray silty clays, leaf remains with preserved cuticles, and absence of endobiotic fauna are all evidence for anoxic conditions below the sediment/water interface. It is likely that the anoxic conditions extended above the lake bottom, as the absence of any autochthonous macrofauna or monotypic *Viviparus* assemblages (an adaptable gastropod) seems to indicate. Rare, rich fossil assemblages indicate occasional oxygenated waters, but as a rule, the dearth of fossils suggests anoxic bottom waters.

Relationships with Other Lithofacies and Lake Evolution

According to the model of Bridge and Leeder (1979), the architectural organization of the SMCF indicates a high rate of sedimentation; moreover, biostratigraphic data (Ambrosetti et al., 1995) outline a short time interval (100,000–200,000 yrs) within the early Pleistocene for sedimentation of the SMCF. For these reasons, the thin (1 m on average) lacustrine sediments suggest that the lakes had a short life on the floodplain. This observation is also confirmed by the common alternations of thin immature paleosols and lake deposits.

Vertical and lateral relationships with the other lithofacies, crevasse splay deposits, and paleosols give clues on sedimentary evolution. Above lacustrine sediments, two types of crevasse splay deposits are recognized. The first type (Figure 5A) consists of triangular sandy bodies, up to 1.3 m high and more than 100 m long with thin (2–10 cm thick) muddy beds dividing different sandy sedimentation events. Trough cross-stratification units up to 20 cm thick appear in the lower part of the triangular sand strata, whereas overlying thinner beds show cross-lamination (commonly climbing). These sandy bodies are interpreted as the proximal parts of crevasse splay systems (Plint and Browne, 1994). The second type of crevasse splay deposits is composed of laminae or thin beds of silts and sands alternating with laminated silty clays, which are organized packages of coarsening- and fining-upward sequences, approximatively 60–100 cm thick (Figure 5B). The silty laminae are planar or undulating, nonparallel, and noncontinuous; the sandy beds, averaging 20 cm thick, show cross-lamination (commonly climbing) or parallel lamination overlain by cross-lamination, which suggests low-density currents as the principal depositional mechanism. Elliot (1974) and Fielding (1984a, b) have called analogous lithofacies "minor mouth bars" and interpreted them as the distal deposition of crevasse splay systems. Proximal and distal crevasse splay deposits can be overlain by paleosols or lacustrine deposits (Figure 5); however, in the studied sequences (Figures 2, 5), crevasse splay deposits commonly are overlain by lacustrine deposits, suggesting that the bulk of sediments transported by the crevasse systems was not enough to fill the small lakes, testifying to the stability of the aqueous paleoenvironments. Commonly in the SMCF, the top of paleosols in floodplain sediments is abruptly covered by laminated lacustrine deposits, recording a sudden drowning of an emergent surface (Figure 6A). Alternations of thin, poorly evolved paleosols and lacustrine deposits are observed (Figure 6B); these sequences testify to frequent fluctuations of the lake level that caused emersion or flooding of neighboring areas.

The relative position of the water table principally controlled the evolution of the floodplain lakes, although the lack of detailed stratigraphic control within the SMCF makes detailed analysis difficult. The rise (drowning of soils) and fall (emersion and pedogenesis of lake deposits) of the water table did not involve widespread areas; it was likely connected to local phenomena related to floodplain dynamics. Paleobiologic information from mammal and pollen fossils unfortunately do not record meaningful climate variations. More detailed isotopic work on the laminated lacustrine deposits may perhaps shed more light on a possible climatic cause of lake level fluctuations recorded in the SMCF. Tectonic influences (such as changes in sediment input) cannot be ruled out either (see Plint and Browne, 1994).

REFERENCES CITED

Allen, J. R. L., 1963, The classification of cross-stratified units, with notes on their origin: Sedimentology, v. 2, p. 93–114.

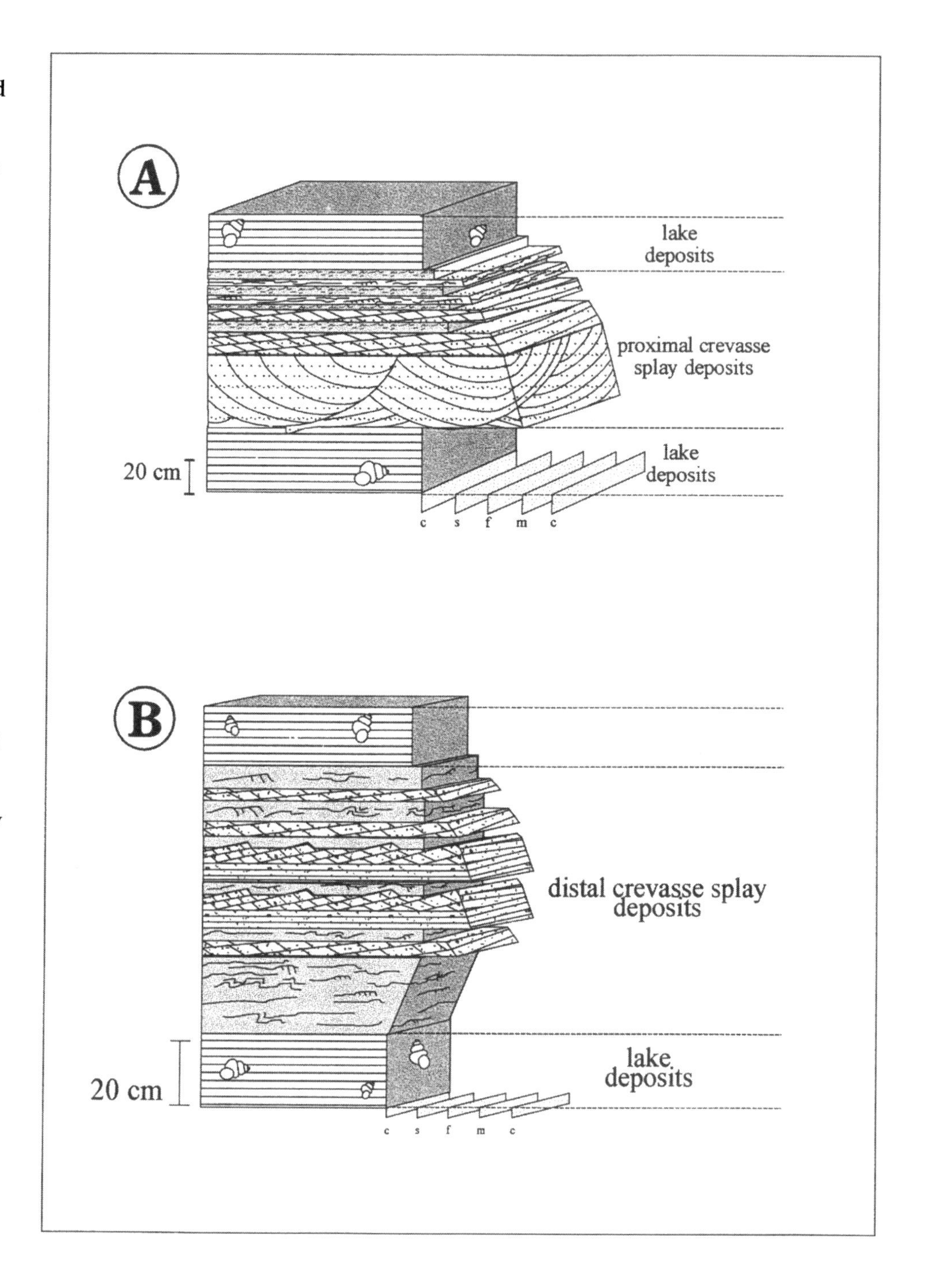

Figure 5—Alternations between lake deposits and crevasse splay deposits of the S. Maria de Ciciliano Formation in the Tiberino Basin. (A) Fining-upward sandy sequences interlayered with very thin muddy beds make up the proximal part of the crevasse splay complex. At the base of the sequence, the sandy beds were formed by 3D migrating dunes with current ripples in the upper sandy beds. Thin (2–10 cm thick) mud beds testify to the episodic nature of the flood deposits. (B) On the distal part of the crevasse complex thin sandy beds alternate with muddy beds. Depositional processes, similar to turbidity currents, formed the sandy beds. See legend in Figure 2 for lithology and sedimentary structures, and figure caption in Figure 4 for grain size parameters.

Ambrosetti, P., G. Basilici, L. Capasso Barbato, M. G. Carboni, G. Di Stefano, D. Esu, E. Gliozzi, C. Petronio, R. Sardella, and E. Squazzini, 1995, Il Pleistocene inferiore del ramo sud-occidentale del Bacino Tiberino (Umbria): aspetti litostratigrafici e biostratigrafici: Il Quaternario, v. 8 (1), p. 19–36.

Basilici, G., 1992, Il Bacino continentale Tiberino (Pliocene–Pleistocene, Umbria): analisi sedimentologica e stratigrafica: Ph.D. Dissertation, University of Bologna, 323 p.

Basilici, G., 1995, Sedimentologia della parte distale di una conoide alluvionale del Pliocene superiore (Bacino Tiberino, Umbria): Il Quaternario, v. 8 (1), p. 37–52.

Basilici, G., 1997, Sedimentary facies in an extensional and deep-lacustrine depositional system: the Pliocene Tiberino Basin, Central Italy: Sedimentary Geology, v. 109, p. 73–94.

Catt, J. A., 1990, Paleopedology manual: Quaternary International: Journal of the International Union for Quaternary Research, v. 6, p. 1–31.

Bridge, J. S., and M. R. Leeder, 1979, A simulation model of alluvial stratigraphy: Sedimentology, v. 26, p. 617–644.

Elliot, T., 1974, Interdistributary bay sequences and their genesis: Sedimentology, v. 21, p. 611–622.

Esu, D., and O. Girotti, 1991, Late Pliocene and Pleistocene assemblages of continental molluscs of Italy. A survey: Il Quaternario, v. 4 (1a), p. 137–150.

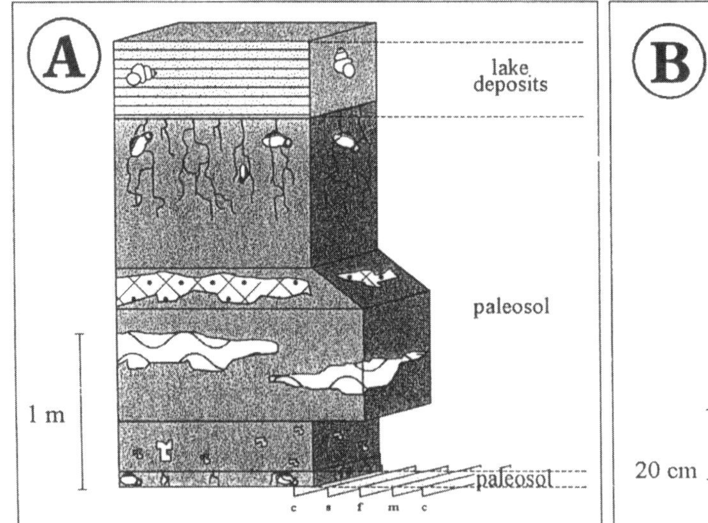

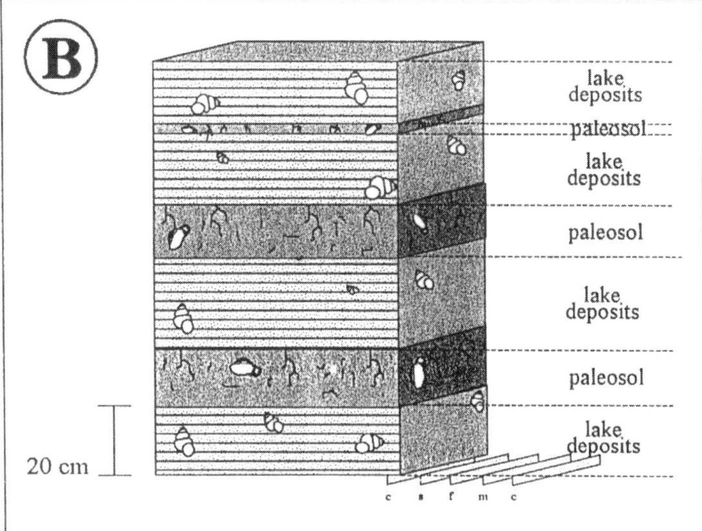

Figure 6—Alternations between lake deposits and paleosols of the S. Maria di Ciciliano Formation in the Tiberino Basin. (A) In most cases, the tops of paleosols are overlain by lake deposits, testifying to the sudden drowning of emergent areas on the floodplain. (B) Thin alternations of lake deposits and poorly evolved paleosols suggest rapid and frequent oscillations of lake level. See legend in Figure 2 for lithology and sedimentary structures, and figure caption in Figure 4 for grain size parameters.

Fielding, C. R., 1984a, Upper delta plain lacustrine and fluviolacustrine facies from the Westphalian of the Durham coalfield, NE England: Sedimentology, v. 31, p. 547–567.

Fielding, C. R., 1984b, A coal depositional model for the Durham coal measures of NE England: Journal of Geological Society of London, v. 141, p. 919–931.

Lavecchia, G., 1988, The Tyrrhenian-Apennines system: structural setting and seismotectogenesis: Tectonophysics, v. 147, p. 263–296.

Leopold, L. B., M. G. Wolman, and J. P. Miller, 1964, Fluvial Processes in Geomorphology: W. H. Freeman and Company, New York, 522 p.

Martini, I. P., and M. Sagri, 1993, Tectono-sedimentary characteristics of the Late Miocene–Quaternary extensional basins of the Northern Apennines, Italy: Earth Science Reviews, v. 34, p. 197–223.

Nanson, G. C., and J. C. Croke, 1992, A genetic classification of floodplains: Geomorphology, v. 4, p. 459–486.

Plint, A. G., and G. H. Brown, 1994, Tectonic event stratigraphy in a fluvio-lacustrine strike-slip setting: the Boss Point Formation (Westphalian A), Cumberland Basin, maritime Canada: Journal of Sedimentary Research, v. B64 (3), p. 341–364.

Willis, B. J., and A. K. Behrensmeyer, 1994, Architecture of Miocene overbank deposits in Northern Pakistan: Journal of Sedimentary Research, v. B64 (1), p. 60–67.

Sabato, L., 2000, A lower–middle Pleistocene lacustrine system in late evolutionary stages of the Sant'Arcangelo Basin (southern Italy), *in* E. H. Gierlowski-Kordesch and K. R. Kelts, eds., Lake basins through space and time: AAPG Geology Studies 46, p 543–552.

Chapter 51

A Lower–Middle Pleistocene Lacustrine System in Late Evolutionary Stages of the Sant'Arcangelo Basin (Southern Italy)

Luisa Sabato
Dipartimento di Geologia e Geofisica, Università di Bari
Bari, Italy

GEOLOGIC SETTING

The Sant'Arcangelo Basin is a Pliocene–Pleistocene intramontane basin; it is located near the outer margin of the Apennines thrust belt (Figure 1A) west of the Bradanic trough (southern Italy) (Figure 1B, C). The latter is a northwest-southeast elongated basin representing the southern part of the Pliocene–Pleistocene Apennines foredeep. The Apennines thrust belt is an east-verging accretionary wedge made up of deformed Mesozoic and Tertiary units (Ippolito et al., 1975). The outer front of the belt is buried beneath the sediments of the Bradanic trough (Figure 1D). The eastern margin of the Bradanic trough is represented by the Apulian foreland, part of the Adriatic foreland.

Turco and Malito (1988) and Turco et al. (1990) believed the Sant'Arcangelo Basin developed during the Pliocene in pull-apart conditions. On the contrary, Caldara et al. (1988), Roure et al. (1988), and Hippolyte et al. (1991) defined the basin as a piggyback basin. Development occurred on the inner front of the last allochthonous Apennines thrust sheet since the late Pliocene. According to Pieri et al. (1996), the basin originated as an open piggyback basin bounded to the west and south by Apennines relief and partially separated to the east–northeast from the most extensive foredeep area by a submerged ridge (2 in Figure 1C).

Overall basin-fill is siliciclastic and some thousands of meters thick (Mostardini and Merlini, 1986); the exposed sequences (Vezzani, 1967; Lentini and Vezzani, 1974) are essentially interpreted as fluvio-deltaic systems strongly controlled by synsedimentary tectonic activity (Caldara et al., 1988; Pieri et al., 1994; Mutti et al., 1996; Vitale, 1996). In the northern part of the basin, Pieri et al. (1996) recognized four depositional sequences (A–D), late Pliocene–middle Pleistocene in age (Figure 2), separated from each other by unconformities with some being syntectonic unconformities (sensu Riba, 1976). These sequences were mainly affected by northwest-southeast oriented folds and faults that were active during sedimentation. Some of these structures determined paleogeographic change through time.

The first sequence (A) was deposited during the late Pliocene. This sequence lies unconformably on the deformed pre-Pliocene units and is made up of conglomerates, sandstones, and marly claystones representing a complete sedimentary marine cycle with transgressive and regressive deposits with a thickness of up to 1000 m. Its sedimentologic features suggest that the basin can be recognized in this stage as an open piggyback (Pieri et al., 1996).

Sequence B lies with a syntectonic unconformity on sequence A; it developed in the late Pliocene–early Pleistocene and is composed of silty claystones, sandstones, and conglomerates up to 1000 m thick, forming a marine fan delta system. Sequence C lies unconformably on sequence A and through a syntectonic unconformity on sequence B.

Sequence C, outcropping only in the northern part of the basin, is early–middle Pleistocene in age (Caggianelli et al., 1992; Pieri et al., 1993, 1994; Marino, 1994; Sabato et al., 1998). Sedimentologic features of sequence C are different, its development having been controlled by the synsedimentary growth of a ramp fold (Alianello anticline, Figure 2). This structure split the basin into a western and an eastern part. In the eastern part, which maintained a connection with the Bradanic trough conserving the characteristics of an open piggyback basin, marine conditions lasted until the middle Pleistocene and a fan delta system was deposited (Sauro cycle, composed of silty claystones, sandstones and conglomerates, up to 1300 m thick). On the contrary, the western part of the basin was separated from the sea and assumed characteristics of a closed piggyback basin, where exclusively continental sediments were deposited (San Lorenzo cycle in Pieri et al., 1994, composed of silty claystones and conglomerates, 500 m in thickness).

The fourth sequence (D), continental in origin and middle Pleistocene in age, lies unconformably on the previous sequences; it is made up of silty sandstones

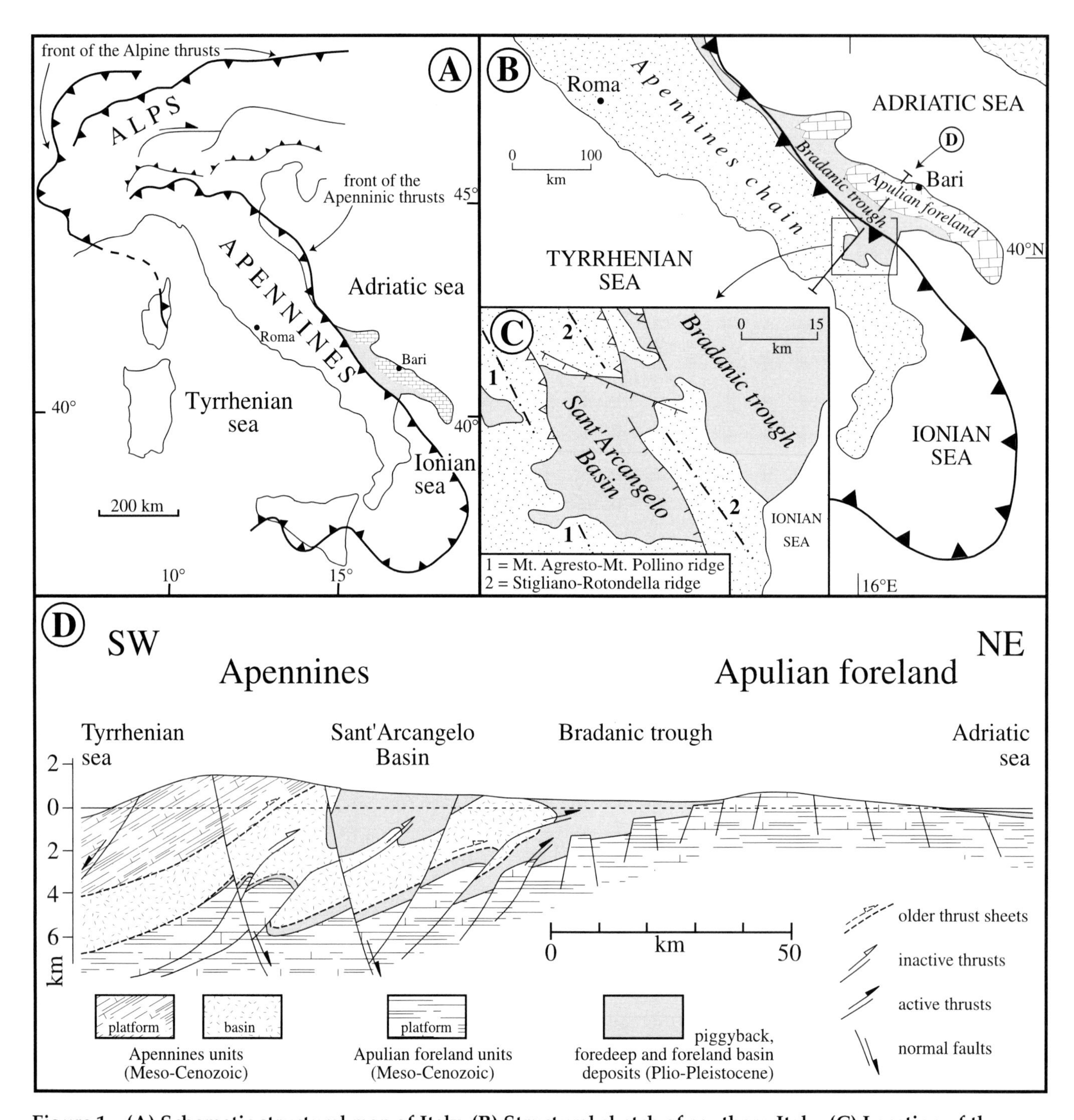

Figure 1—(A) Schematic structural map of Italy. (B) Structural sketch of southern Italy. (C) Location of the Sant'Arcangelo Basin and its main structural elements; in particular, the Stigliano-Rotondella ridge was partially submerged until the middle Pleistocene. (D) Schematic cross-section through the area from southwest to northeast in (C).

and conglomerates about 150 m thick and is interpreted as an alluvial fan deposit. This sequence represents the last evolutionary stage of the basin and was brought about by the emergence of an outer thrust (2 in Figure 1C), which caused physical isolation of the basin from the rest of the foredeep in which sedimentation continued for the entire Pleistocene. According to Pieri et al. (1996), only in this last stage can the entire Sant'Arcangelo Basin be defined as a closed piggyback basin.

THE SAN LORENZO CYCLE (SEQUENCE C)

The San Lorenzo cycle deposits of sequence C outcrop in various areas (Figure 2). These deposits are up to

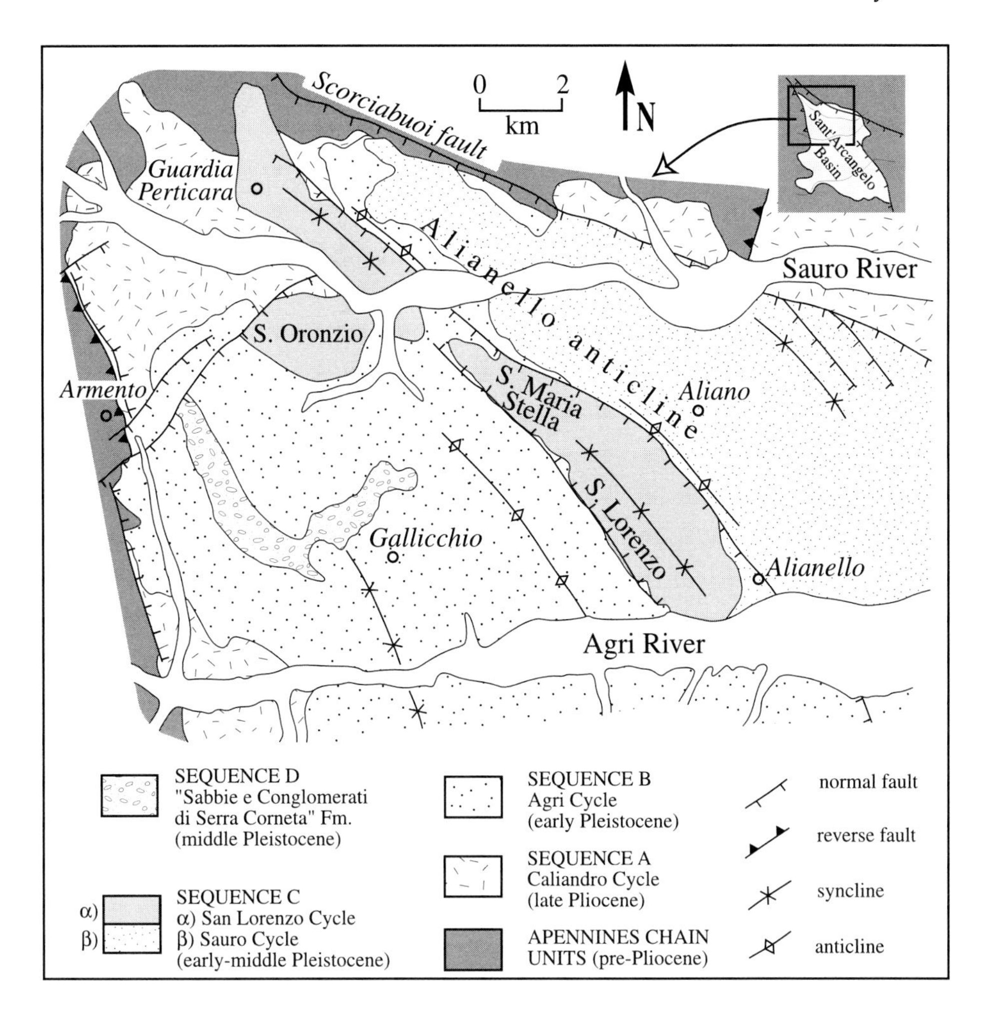

Figure 2—Geologic map and stratigraphic details of the northern part of the Sant'Arcangelo Basin (simplified from Pieri et al., 1993).

500 m thick, have a synclinal arrangement, and can be divided into three stratigraphic units (a, b, c in Figure 3):

(1) a lower sandy conglomerate (alluvial-dominated) unit
(2) a middle silty claystone (lacustrine or palustrine) unit
(3) an upper sandy conglomerate (alluvial-dominated) unit

The lower unit has a thickness of 150 m and is generally composed of poorly sorted, clast-supported conglomerates stratified into thick beds (over 10 m) and locally containing some decimeter-thick clayey beds and massive or laminated fine sandstone layers (up to 2 m thick) with rare pebbles. Locally this facies is overlain by meter-scale sandstones bodies containing lenticular planar cross-bedded conglomerates, some decimeters thick. Measurements from imbrication indicate a general paleoflow toward the southeast. This unit is interpreted as an alluvial fan (Figure 3B).

The middle unit has a variable thickness up to 200 m; it contains stratified units of claystones and silty claystones tens of centimeters thick, commonly thinly laminated, that alternate with ripple-laminated sandy packages, all interbedded with calcareous and volcaniclastic layers, several decimeters thick. In the lower and upper parts of the middle unit, pebbles and lenticular conglomerates, some decimeters thick, are present. Horizontal lamination and soft-sediment deformation structures are common in the finer facies; ostracodes, gastropods, plants, and vertebrate remains occur locally (Sabato, 1992). This unit is lacustrine or palustrine in origin (Figure 3B); preliminary facies analysis shows that the middle unit close to the Guardia Perticara and Sant'Oronzio locality (Figure 2) can be interpreted as prevalently palustrine sediments, while the southernmost outcrop, located between the Sauro and Agri rivers (Figures 2, 3A), as lacustrine deposits.

The upper unit is over 150 m thick and contains generally poorly sorted and massive reddish conglomerates up to over 10 m thick, prevalently clast-supported with a sandy matrix and meter-scale, gravelly, trough cross-bedded sandstones, with localized ripple cross-lamination. This unit is also interpreted as an alluvial fan system (Figure 3B) with paleocurrent measurements indicating a south–southeast trend.

LACUSTRINE MIDDLE UNIT OF THE SAN LORENZO CYCLE

The middle unit of the San Lorenzo cycle (b in Figure 3) is well exposed close to the Agri river valley,

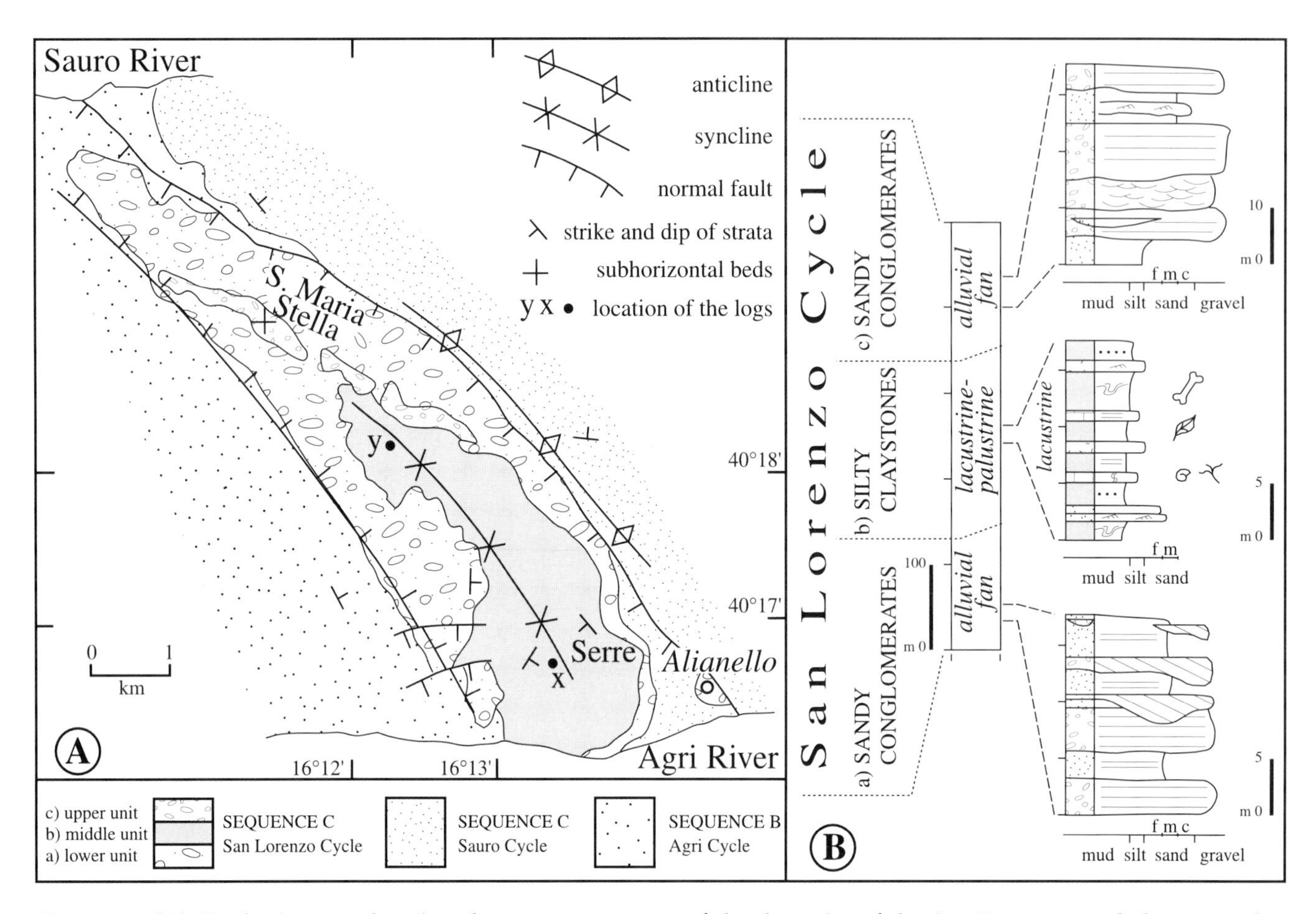

Figure 3—(A) Geologic map showing the outcrop pattern of the deposits of the San Lorenzo cycle between the Agri and Sauro Rivers (modified from Pieri et al., 1993). (B) Lithostratigraphic logs and main facies characteristics of the entire San Lorenzo cycle. Legend in Figure 4.

where it crops out over a 5-km-long and 2-km-wide area along a northwest-southeast trending synclinal basin (Figure 3A). This lacustrine unit contains a sequence containing mainly siliciclastics with minor carbonates. In the depocenter sectors and along the eastern side of the exposure limit of the lacustrine unit, deposits are generally composed of up to several decimeter-thick stratified claystones and silty claystones with interbedded sandstones and calcareous and volcaniclastic interlayers, which can be several centimeters to decimeters thick. Along the western side of the exposed lacustrine unit, sandstones and conglomerates are interbedded with claystones. Sandstone beds are up to 1–2 m thick, are commonly normally graded, and show horizontal and ripple lamination; conglomerates have erosive bases, are trough cross-bedded or horizontally stratified, and locally form thin lenticular bodies, up to over 1 m thick, that pass upward into horizontal or ripple-laminated sandstones. Paleocurrents indicate a source area from the north; transverse contributions along the western side toward the eastern sectors have also been measured. Thickness of this lacustrine system progressively decreases from the 200 m in the southeast to only tens of meters in the northwest.

In the southern area, the dip of the beds is about 10° toward the axis of the syncline; the dip angle decreases toward the north, until it becomes subhorizontal. To define the different stratigraphic and sedimentologic features of the lacustrine unit, two successions located in the southern and northern sectors have been studied (logs x and y in Figures 3A, 4). Both successions lie on the lower sandy conglomerate unit (a in Figures 3, 4); only the northern sequence (log y) passes upward into the upper sandy conglomerate unit (c in Figures 3, 4).

In the depocenter area, the southern succession of the lacustrine middle unit (log x in Figure 4, northwest of Alianello, close to the Serre locality, Figure 3A) is 200 m thick and is represented by thinly bedded claystones and silty claystones that alternate with sandstone packages and interbeds of calcareous and volcaniclastic layers. Claystones and silty claystone beds have a variable thickness, from several centimeters to 1–2 m. Lacustrine laminites are very common in these facies and are predominantly made up of couplets of lighter and darker clayey laminae; commonly laminae are volcaniclastic or carbonate in composition. Thin (up to 10 cm thick) organic-rich beds and carbonate layers also occur in association with the laminite-dominated facies. Sandstone packages are up to 1 m

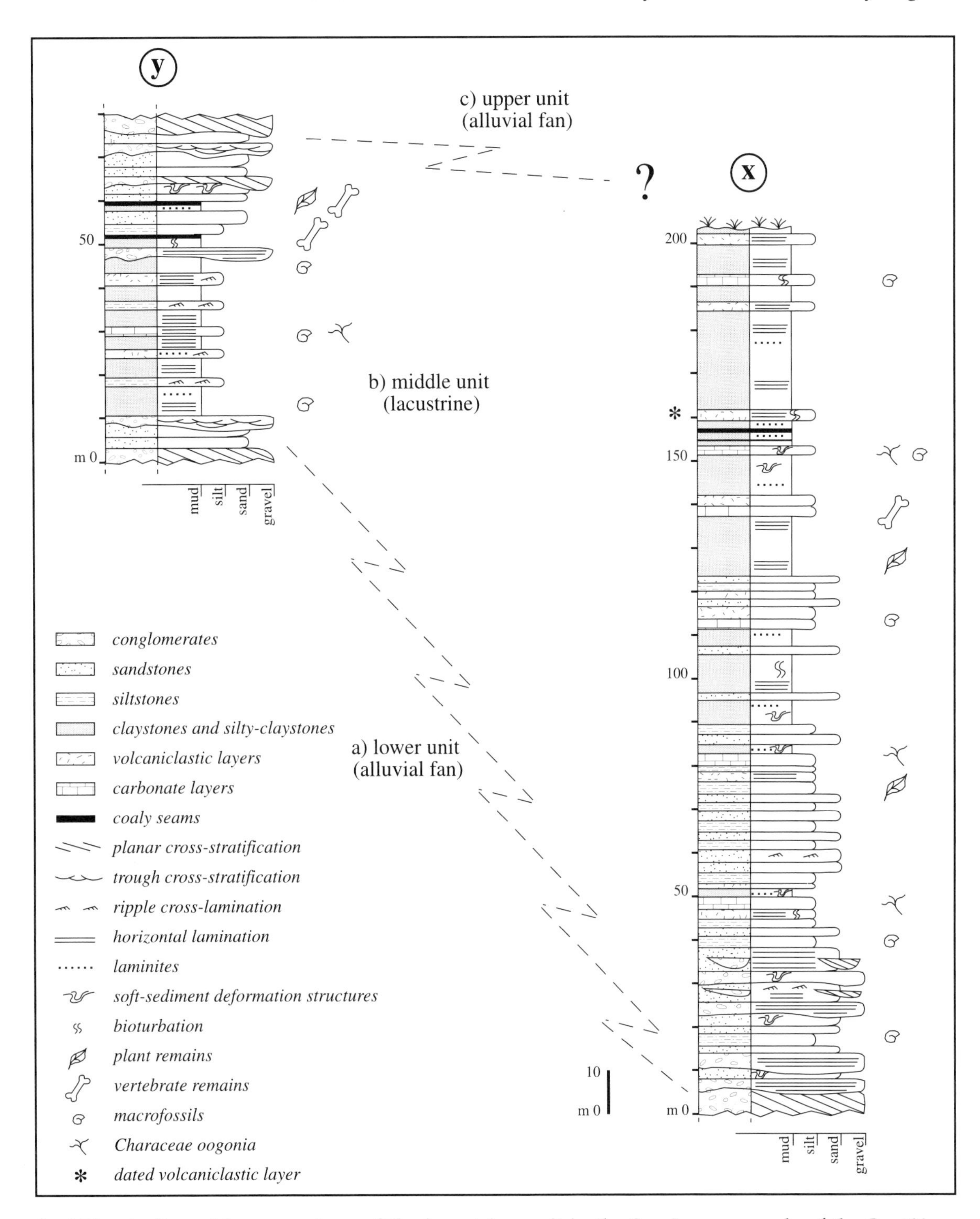

Figure 4—Lithostratigraphic successions of the lacustrine unit in the San Lorenzo cycle of the Sant'Arcangelo Basin measured in the depocenter in the SE (x), and at a proximal location to the northwest (y). Location of logs in Figure 3A.

thick, commonly normally graded, horizontal or ripple cross-laminated, and rarely massive. Carbonate layers are several centimeters thick, laterally continuous, and some dolomitic in composition; thin sections of these carbonates exhibit horizontal and ripple lamination, fecal pellets, and microbioturbation. Volcaniclastic intercalations, up to several tens of centimeters thick, are present throughout the southern succession, but they become more common in its upper part. They are horizontal to ripple-laminated and, in the upper part of the beds, microbioturbation is found. Graded bedding is common. The volcaniclastic layers are phonolitic in composition; a K-Ar age of 1.1 ± 0.3 Ma was obtained for the volcanic glasses (Caggianelli et al., 1992), in agreement with the age of the marine Sauro cycle (Marino, 1994; Pieri et al., 1994).

Fossils of the middle lacustrine unit are represented by ostracodes and gastropods (*Planorbis* sp.), common in the laminite facies, as well as by plant and vertebrate remains. In the interbedded carbonate layers, Characeae and gastropods commonly occur.

Different types of soft-sediment deformation, including contorted bedding, slumps, and load casts (millimeter to decimeter in scale) are observed throughout the southern succession and in the laminite-dominated sequence. Slumps indicate a paleoslope dipping toward the axis of the syncline.

The laminite-dominated sequence changes laterally and vertically into a sand-mudstone facies assemblage, which is better developed in the northwestern side of the basin. The lacustrine middle unit in this northern area (log y in Figure 4, near Santa Maria Stella locality, Figure 3A) contains only 40 m of thinly laminated claystones and silty claystones with interbedded sandstones, in its lower part. The upper 20 m of this northern succession contains some decimeter-thick sandstone-claystone beds, lenticular conglomerates (decimeter-scale), and thin coal seams bearing fossil plant remains (leaves, seams, and fragments of wood) and vertebrate bones (Cervidae indet., Agustí, 1993, personal communication). In these upper beds, up to 1-m-thick silty and volcaniclastic layers also occur. This succession is finally capped by a sandy conglomerate body (20–150 m thick), representing the upper unit of the San Lorenzo cycle (c in Figure 3). In this northern area, the lacustrine middle unit succession (log y) records the late evolutionary stage of encroachment of the fan delta-alluvial fan and proximal lacustrine environments over the distal lacustrine zones.

The lacustrine middle unit succession in the south area (log x) records sedimentation in a distal zone of a relatively deep, siliciclastic-dominated freshwater lake (Sabato, 1997). The laminated facies assemblage was dominant due to the tectonic setting of the paleolake and its close relationship with alluvial fan–fan delta systems that fed the basin. In fact, rhythmic laminae, very characteristic deposits of the lake bottoms, need a variable mechanism of sediment input for their formation (Glenn and Kelts, 1991). Also, climate can induce a variable sediment input signal and may reflect episodic events on widely varying scales (e.g., organic productivity is linked to seasonal wind-driven circulation or to annual solar radiation, which can be linked with glacial melting) (Kelts and Hsü, 1978; Glenn and Kelts, 1991; among others). Well-preserved volcaniclastic deposits suggest low depositional energy levels. Centimeter-thick graded sandstones and volcaniclastic beds may represent the product of low-density turbidity currents (Sturm and Matter, 1978).

Minor carbonate deposits may record either carbonate mud and bioclastic contributions from marginal lacustrine zones or the periodic development of a more concentrated lake water body. Climate controls the rate and nature of organic productivity and influences runoff, erosion, and chemical weathering rates in the catchment area. Also, tectonic activity can lead to the deposition of carbonate through changes in the drainage network and geometry of the basin and its facies distribution. All these factors are strictly related to the carbonate composition of the catchment area rocks that ultimately control the nature of detrital input (by rivers) and the lake chemistry (through groundwater) (Platt and Wright, 1991; Gierlowski-Kordesch, 1998).

Rare occurrence of organic-rich laminites may suggest water-stratified conditions (Demaison and Moore, 1980; Boyer, 1982). In fact, the dark color and good preservation of plant remains are indicative of anoxic conditions (Kelts, 1988) that can be reached below a thermocline with stagnant waters or near the sediment/water interface (Picard and High, 1981; Ramos-Guerrero et al., 1989; Hamblin, 1992; among others). This can be related to changes in climate (change in biological activity or water level) or in tectonic activity (change in sediment input or basin geometry affecting water level) (Anadón et al., 1991; Gierlowski-Kordesch et al., 1991; Demaison and Moore, 1980).

DISCUSSION

A lacustrine system developed in a northwest-southeast trending synclinal basin formed during the late tectonic-sedimentary evolutionary phases of the Sant'Arcangelo piggyback basin (southern Apennines) as a result of the growth of a ramp fold (Alianello anticline, Figures 2, 5). This lake system was influenced by thin to coarse alluvial fan–fan delta contributions that resulted in the development of small-scale, proximal, lacustrine fan delta gravels and sands related to distal muddy laminite-dominated lacustrine deposits. Minor carbonate, volcaniclastic, and organic-rich sediments were also deposited.

The most extensive part of the lake was composed of offshore, relatively deep waters containing bottom clay and silty clay sediments. Limited occurrence of organic-rich laminites suggests the formation of meromictic conditions in the deepest waters. Presence of minor carbonate layers may suggest changes in climate or in tectonic activity.

Stratigraphic features and facies relationships observed in the diverse basin sectors allowed the identification of a major alluvial feeding axis at the northwestern end of the basin, whereas the inner lacustrine depocenter was located to the southeast. Transverse alluvial contributions have been measured along the western side of the basin.

Stratigraphic data also suggest that tectonics had been active during sedimentation. This is confirmed also by sedimentologic features (e.g., slumps), and by the variability of the thickness of the lacustrine sequences; therefore, it is suggested that lacustrine sedimentation was synchronous to syncline generation. The major normal fault bounding the lacustrine deposits to the west (Figure 5) controlled geometries and facies development, forming an asymmetric arrangement of the entire lacustrine system with reduced sediment thickness eastward. Similar situations have been recognized in various lacustrine basins in different tectonic settings (e.g., Anadón et al., 1991; Changsong et al., 1991).

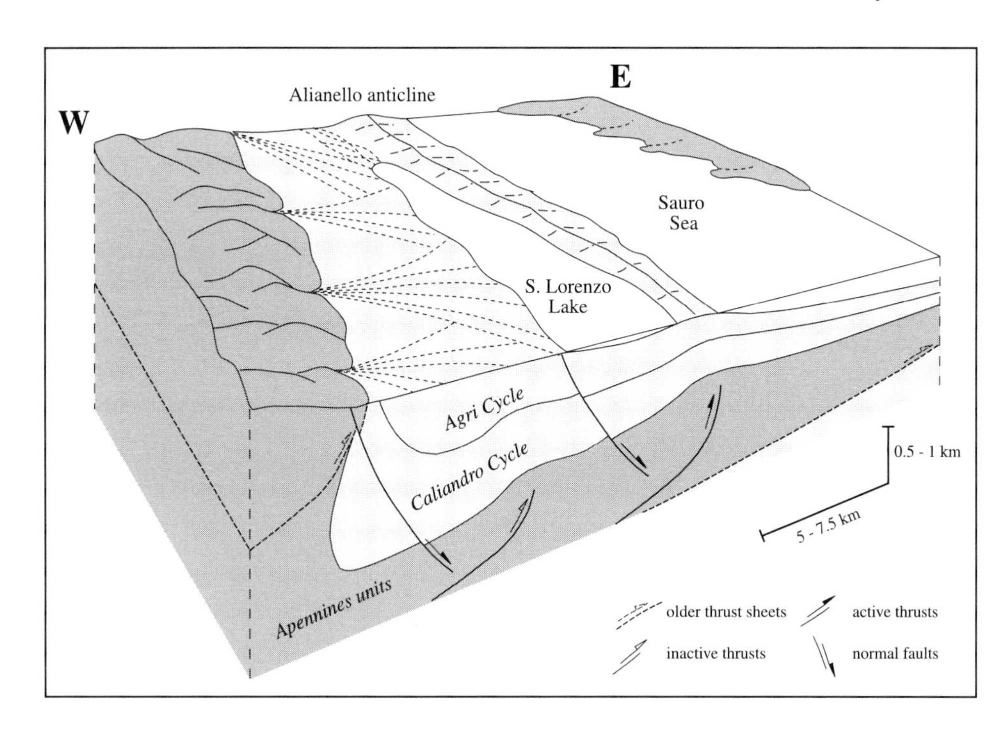

Figure 5—Block diagram showing the dissection of the early Pleistocene in the Sant'Arcangelo Basin into two sectors as a result of the formation of a ramp fold (Alianello anticline, Figures 2, 3A). In the western sector, lacustrine sedimentation began and continued until the middle Pleistocene, when uplift of the Stigliano-Rotondella Ridge (2 in Figure 1C) caused the complete emergence and deactivation of the entire Sant'Arcangelo Basin.

In addition, the deactivation of the whole Sant' Arcangelo piggyback basin was contemporaneous with the end of lacustrine system evolution. Evidence includes the complete emergence of the Stigliano-Rotondella ridge (2 in Figure 1C) coincident with the activity of the Scorciabuoi fault (Figure 2), and occurrence of an important northwest-southeast normal fault, middle Pleistocene in age (the Alianello fault, close to the Alianello anticline, Figures 2, 3). The Alianello fault may have affected lacustrine sedimentation, as well. Thus, the lacustrine basin records the relationships between tectonics and sedimentation during the lower and middle Pleistocene in the Sant'Arcangelo piggyback basin.

Finally, the occurrence of some lithologic features (carbonates, organic-rich deposits) and facies characters (laminites) could be related to a climatic influence, also linked to worldwide climatic events, such as changes in solar radiation (Kutzbach and Street-Perrott, 1985) and ice-volume fluctuation. In fact, the lacustrine deposits span a time between the early and the middle Pleistocene that record four glacial/interglacial phases.

ACKNOWLEDGMENTS

Research was supported by National Council of Research (CNR) grants through G. Piccarreta, ministerial founds (MURST 40%) and ASI (Italian Space Agency) grants through P. Pieri. I wish to thank Lluís Cabrera for his comments on the manuscript, for useful discussion in the field, and for his advice on the future work on the lacustrine Sant'Arcangelo deposits. I am very grateful to Piero Pieri and Marcello Tropeano for their comments and critical revision of the manuscript, and to Carlo Doglioni for discussion about the geodynamic context. Pere Anadón made helpful comments on the first version of the paper; final comments by Elizabeth Gierlowski-Kordesch were essential in improving the manuscript.

REFERENCES CITED

Anadón, P., Ll. Cabrera, R. Julià, and M. Marzo, 1991, Sequential arrangement and asymmetrical fill in the Miocene Rubielos de Mora Basin (northeast Spain), *in* P. Anadón, Ll. Cabrera, and K. Kelts, eds., Lacustrine Facies Analysis: International Association of Sedimentologists Special Publication 13, p. 257–275.

Boyer, B. W., 1982, Green River laminites: does the playa-lake model really invalidate the stratified lake model?: Geology, v. 10, p. 321–324.

Caggianelli, A., P. Dellino, and L. Sabato, 1992, Depositi lacustri infrapleistocenici con intercalazioni vulcanoclastiche (Bacino di S. Arcangelo, Basilicata): Il Quaternario, v. 5, p. 123–132.

Caldara, M., F. Loiacono, E. Morlotti, P. Pieri, and L. Sabato, 1988, I depositi pliopleistocenici della parte Nord del Bacino di S. Arcangelo (Appennino lucano): caratteri geologici e paleoambientali: Memorie della Società Geologica Italiana, v. 41, p. 391–410.

Changsong, L., Y. Qi, and L. Sitian, 1991, Structural and depositional patterns of the Tertiary Baise Basin, Guang Xi autonomous Region (southeastern China): a predictive model for fossil fuel exploration, *in* P. Anadón, Ll. Cabrera, and K. Kelts, eds., Lacustrine Facies Analysis: International Association of Sedimentologists Special Publication 13, p. 75–92.

Demaison, G. J., and G. T Moore, 1980, Anoxic environments and oil source bed genesis: AAPG Bullettin, v. 64, p. 1179–1209.

Gierlowski-Kordesch, E., 1998, Carbonate deposition

in an ephemeral siliciclastic alluvial system: Jurassic Shuttle Meadow Formation, Newark Supergroup, Hartford Basin, USA: Palaeogeography, Palaeoclimatology, Palaeoecology, v. 140, p. 161–184.

Gierlowski-Kordesch, E., J. C. Gomez Fernandez, and N. Melendez, 1991, Carbonate and coal deposition in an alluvial-lacustrine setting: Lower Cretaceous (Weald) in the Iberian Range (east-central Spain), *in* P. Anadón, Ll. Cabrera, and K. Kelts, eds., Lacustrine Facies Analysis: International Association of Sedimentologists Special Publication 13, p. 109–125.

Glenn, C. R., and K. Kelts, 1991, Sedimentary rhythms in lake deposits, *in* G. Einsele, W. Richen, and A. Seilacher, eds., Cycles and Events in Stratigraphy: Springer-Verlag, Berlin, p. 188–221.

Hamblin, A. P., 1992, Half-graben lacustrine sedimentary rocks of the lower Carboniferous Strathlorne Formation, Hort Group, Cape Breton Island, Nova Scotia, Canada: Sedimentology, v. 39, p. 263–284.

Hippolyte, J. C., J. Angelier, F. Roure, and C. Müller, 1991, Géométrie et mécanisme de formation d'un bassin "piggyback": le bassin de Sant'Arcangelo (Italie méridionale): C. R. Acad. Sci. Paris, v. 312 (II), p. 1373–1378.

Ippolito, F., B. D'Argenio, T. Pescatore, and P. Scandone, 1975, Structural-stratigraphic units and tectonic framework of southern Apennines, *in* C. Squyres, ed., Geology of Italy: The Earth Sciences Society of the Libyan Arab Republic, Tripoli, p. 317–328.

Kelts, K., 1988, Environments of deposition of lacustrine petroleum source rocks: an introduction, *in* A. J. Fleet, K. Kelts, and M. R. Talbot, eds., Lacustrine Petroleum Source Rocks: Special Publication of Geological Society of London, v. 40, p. 3–26.

Kelts, K., and K. J. Hsü, 1978, Freshwater carbonate sedimentation, *in* A. Lerman, ed., Lakes: chemistry, geology and physics, p. 295–323.

Kutzbach, J. E., and F. A. Street-Perrott, 1985, Milankovitch forcing of fluctuations in the level of tropical lakes from 18 to 0 ka BP: Nature, v. 317, p. 130–134.

Lentini, F., and L. Vezzani, 1974, Note illustrative del Foglio 506 S. Arcangelo: Consiglio Nazionale delle Ricerche, Istituto di Ricerca per la Protezione Idrogeologica, Cosenza, 46 p.

Marino, M., 1994, Biostratigrafia integrata a nannofossili calcarei e foraminiferi di alcune successioni terrigene pliocenico-superiori del Bacino di S. Arcangelo: Bollettino della Società Geologica Italiana, v. 113 (2), p. 329–354.

Mostardini, F., and S. Merlini, 1986, Appennino centro meridionale. Sezioni geologiche e proposta di modello strutturale: Agip, 59 p.

Mutti, E., G. Davoli, R. Tinterri, and C. Zavala, 1996, The importance of ancient fluvio-deltaic systems dominated by catastrophic flooding in tectonically active basins: Memorie di Scienze Geologiche, v. 48, p. 233–291.

Picard, M. D., and L. R. High, Jr., 1981, Phisical stratigraphy of ancient lacustrine deposits, *in* F. G. Ethridge and L. R. High, eds., Recent and ancient nonmarine depositional environments: models for exploration: SEPM Special Publication 31, p. 233–259.

Pieri, P., L. Sabato, and F. Loiacono, 1993, Carta geologica del Bacino di Sant'Arcangelo (tra il Torrente Sauro e il Fiume Agri): V. Paternoster printing office, Matera, Italy.

Pieri, P., L. Sabato, F. Loiacono, and M. Marino, 1994, Il bacino di piggyback di Sant'Arcangelo: evoluzione tettonico-sedimentaria: Bollettino della Società Geologica Italiana, v. 113 (2), p. 465–481.

Pieri, P., L. Sabato, and M. Marino, 1996, The Pliocene–Pleistocene piggyback Sant'Arcangelo Basin: tectonic and sedimentary evolution: Notes et Mémoires du Service géologique du Maroc, v. 387, p. 195–208.

Platt, N. H., and V. P. Wright, 1991, Lacustrine carbonates: facies models, facies distributions and hydrocarbon aspects, *in* P. Anadon, Ll. Cabrera, and K. Kelts, eds., Lacustrine facies analysis: International Association of Sedimentologists Special Publication 13, p. 57–74.

Ramos-Guerrero, E., Ll. Cabrera, and M. Marzo, 1989, Sistemas lacustres paleógenos de Mallorca (Mediterraneo Occidental): Acta Geologica Hispanica, v. 24, p. 185–203.

Riba, O., 1976, Syntectonic unconformities of the Alto Cardener, Spanish Pyrenees: a genetic interpretation: Sedimentary Geology, v. 15, p. 213–233.

Roure, F., P. Casero, and R. Vially, 1988, Evolutive geometry of ramps and piggy-back basins in the Bradanic trough: Atti del 74° Congresso Nazionale della Società Geologica Italiana, September 13–17, Sorrento, Italy, v. B, p. 360–363.

Sabato, L., 1992, Pleistocene lacustrine deposits in the S. Arcangelo Basin (southern Italy): Abstracts, 29th International Geological Congress, 24 August–3 September, Kyoto, Japan, v. 2, p. 308.

Sabato, L., 1997, Sedimentary and tectonic evolution of a lower-middle Pleistocene lacustrine system in the Sant'Arcangelo piggyback basin (southern Italy): Geologica Romana, v. 33, p. 137–145.

Sabato, L., A. Bertini, F. Masini, A. Albianelli, and G. Napoleone, 1998, A lower-middle Pleistocene lacustrine sequence in the Sant'Arcangelo piggyback basin (Southern Italy): Abstracts, International Association of Sedimentologists, 15th International Sedimentological Congress, April 12–17, Alicante, Spain, p. 687–688.

Sturm, M., and A. Matter, 1978, Turbidites and varves in Lake Brienz (Switzerland): deposition of clastic detritus by density currents, *in* A. Matter and M. E. Tucker, eds., Modern and Ancient Lake Sediments: International Association of Sedimentologists Special Publication 2, p. 147–168.

Turco, E., and M. Malito, 1988, Formazione di bacini e rotazione di blocchi lungo faglie trascorrenti nell'Appennino Meridionale: Atti del 74° Congresso Nazionale della Società Geologica Italiana, September 13–17, Sorrento, Italy, v. B, p. 424–426.

Turco, E., R. Maresca, and P. Cappadona, 1990, La tettonica plio–pleistocenica del confine calabro-lucano: modello cinematico: Memorie della Società

Geologica Italiana, v. 45, p. 519–529.

Vezzani, L., 1967, Il bacino pliopleistocenico di S. Arcangelo (Lucania): Atti della Accademia Gioenia di Scienze Naturali in Catania, v. 18, p. 207–227.

Vitale, G., 1996, Evoluzione tettonica e stratigrafica dal passaggio Miocene–Pliocene all'attuale del sistema catena-avanfossa lungo il fronte bradanico-ionico: Ph.D. dissertation, Bari University, Italy, 159 p.

Owen, R. B., and R. W. Renaut, 2000, Spatial and temporal facies variations in the Pleistocene Olorgesailie Formation, southern Kenya Rift Valley, *in* E. H. Gierlowski-Kordesch and K. R. Kelts, eds., Lake basins through space and time: AAPG Studies in Geology, 46, p. 553–560.

Chapter 52

Spatial and Temporal Facies Variations in the Pleistocene Olorgesailie Formation, Southern Kenya Rift Valley

R. Bernhart Owen
Department of Geography, Hong Kong Baptist University, Kowloon Tong
Hong Kong, China

Robin W. Renaut
Department of Geological Sciences, University of Saskatchewan
Saskatoon, Canada

INTRODUCTION

Lake Olorgesailie was one of several large paleolakes known to have existed in the Kenya Rift Valley during the Pleistocene. The Olorgesailie Basin covers an area of about 100 km in the southern Kenya Rift, 25 km northeast of Lake Magadi (Figure 1). The Olorgesailie Formation includes a wide range of lacustrine, fluvial, and colluvial sedimentary rocks. Gregory (1921) first described the deposits from the area, which he attributed to a Miocene "Lake Kamasia." In 1943 Louis and Mary Leakey discovered abundant Acheulian hand axes in the deposits (Leakey, 1952), which led to further archeological studies by Isaac (1968), and more recently by Potts (1989, 1994). Shackleton (1955, 1978) undertook detailed mapping of the main archeological site. Some of his results were included in the Kenya Geological Survey report of Baker (1958), who assigned the sedimentary rocks to the "Olorgesailie Lake Beds." Baker and Mitchell (1976) later referred to them as the "Legemunge Beds." Isaac (1977) placed the deposits in the "Olorgesailie Formation," which he defined as "a series of well-stratified diatomites, pale yellowish volcanic siltstones, and claystones, and subordinate quantities of brown siltstones and volcanic sands." Marsden (1979) described the general setting and history of the Olorgesailie Formation. Owen (1981) and Owen and Renaut (1981) described the sedimentology, facies, and diatom stratigraphy of the Olorgesailie Formation, and attempted to reconstruct the sequence of depositional environments. Bye (1984) and Bye et al. (1987) examined aspects of the geochronology and diagenesis of the formation. In this paper, we summarize the main features of the Olorgesailie Formation, incorporating new data on the lateral variations in facies and the diatom distribution in the paleolake.

GEOLOGIC SETTING

The Olorgesailie sediments were deposited in several small north-south trending grabens. They are up to 60 m thick and rest unconformably on trachytes, basalts, and phonolites of Pleistocene and Pliocene age. Fault movements continued during sedimentation and, together with tilting of the rift floor, ultimately caused lacustrine sedimentation to cease. Much of the basin fill is concealed below later alluvial and colluvial deposits of the Legemunge and Oltepesi plains, but excellent exposures are present along an east-west trending scarp on the southern margins of the plains (Figure 1). Mount Olorgesailie, an eroded Pliocene central volcano, served as a southern barrier to drainage during sedimentation. To the east and west of this volcano lie the Kwenia and Koora grabens, respectively. In both grabens, the sedimentary rocks of the Olorgesailie Formation underlie a thin cover of younger, weakly lithified fluviatile and colluvial deposits.

A wide range of dates was reported for the formation by Evernden and Curtis (1965) and Miller (1967). Isaac (1977) considered that K-Ar dates of 0.42 and 0.48 Ma were compatible with the faunal and archeological evidence. Bye et al. (1987) later reported older ages of 0.93–0.70 Ma for tuffs from the lower part of the formation, and 0.7–0.6 Ma for the upper part. More recently, the age of the formation was bracketed by single-crystal ^{40}Ar-^{39}Ar dates of 0.99 Ma for the lowest member (1) and 0.49 Ma for the uppermost member (14) of the formation (Deino and Potts, 1990, 1992).

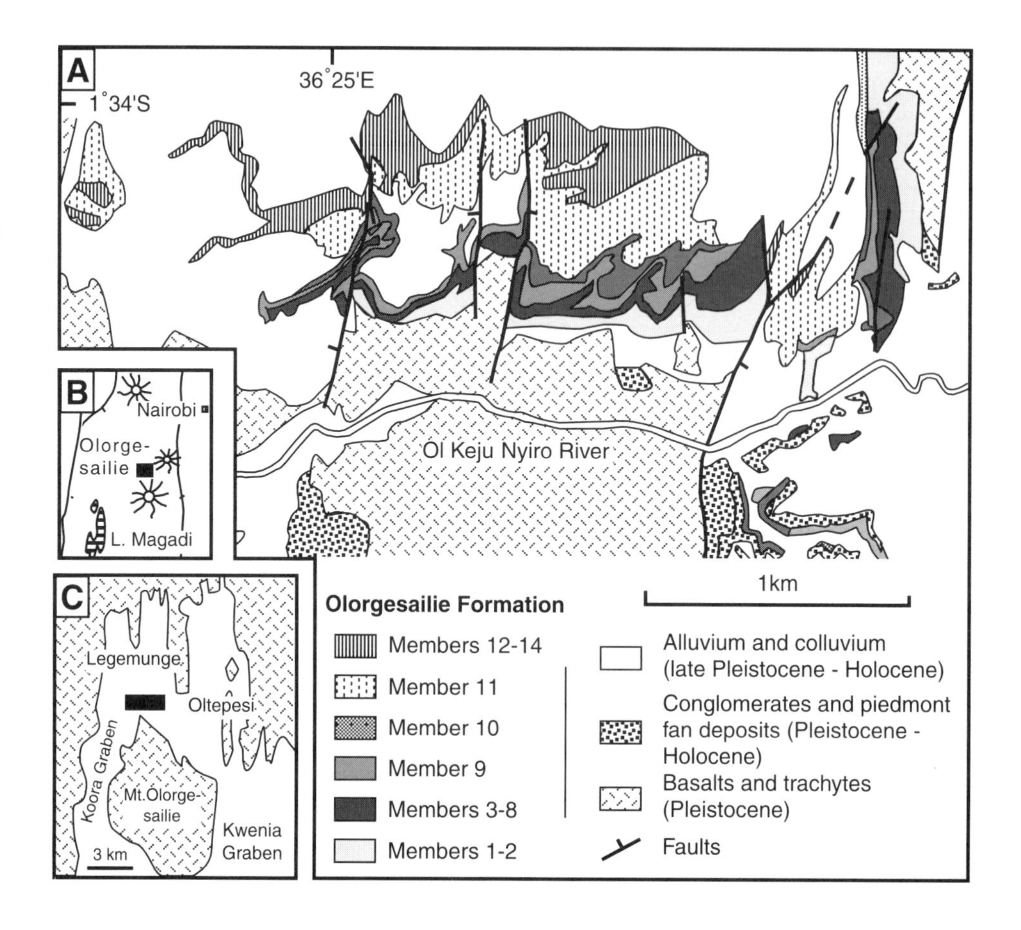

Figure 1—(A) Simplified geological map of the Pleistocene Olorgesailie Formation (based on detailed mapping by Shackleton, 1978). (B) Location of Olorgesailie in the southern Kenya Rift. (C) Location of the Koora and Kwenia grabens.

STRATIGRAPHY AND SEDIMENTARY FACIES

Shackleton (1944, 1978) divided the Olorgesailie lake deposits into 14 informal units that Isaac (1968, 1977) later raised to member status within the Olorgesailie Formation. These divisions cannot be recognized throughout the area of outcrop due to lateral facies variations (particularly Members 5–8); however, lithologic correlation among outcrops is possible using several marker beds, mainly pumice horizons, tuffs, and distinctive diatomites. The type section is shown in Figure 2, together with representative sections for other parts of the formation.

Owen and Renaut (1981) recognized and described the lithologic characteristics of ten main sedimentary facies in the Olorgesailie Formation. These include four lacustrine facies: (1) laminated, diatomaceous siltstones that formed in relatively deep water; (2) structureless, tuffaceous, diatomaceous claystones, siltstones, and fine sandstones that were deposited in shallower water; (3) reworked diatomaceous siltstones and claystones that were deposited in littoral flats; and (4) massive diatomites that were deposited in shallow waters that were locally vegetated by reed beds.

Three fluvial facies are also present: (5) interbedded volcanic sandstones, siltstones, and conglomerates that are interpreted to be braided, ephemeral stream deposits; (6) pumiceous sandstones and conglomerates that are mainly channel and overbank deposits; and (7) volcanic ash and tuff representing both primary airfall tuffs into the lake and volcaniclastic sediments reworked by ephemeral streams. Paleochannel orientations and paleocurrent indicators, such as imbrication, show that the fluvial sediments were deposited from both major streams and local channel systems.

Owen and Renaut (1981) also distinguished three subaerial facies: (8) brown siltstones and claystones encompassing both floodplain (overbank) sediments and colluvial soils; (9) poorly sorted, sandy conglomerates interpreted mainly as colluvial and unchannelled wash deposits; and (10) calcretes of several genetic types.

MINERALOGY

Optical microscopy and x-ray diffraction (XRD) analyses of the sand, silt, and clay fractions from all facies demonstrate clear textural variations in the mineralogical composition of detrital particles (Owen, 1981). Excluding pumiceous units, the sand and gravel fractions are dominated by lithic fragments (mostly trachyte) and anorthoclase, with subsidiary clinopyroxene, volcanic glass, and quartz. Micritic limestone (reworked calcrete lithoclasts?) and chert grains are also present.

The silt fraction has the greatest mineralogical variability. Alkali feldspars are common, together with fresh volcanic glass, sponge spicules, and diatoms. Quartz, aenigmatite, and augite are locally abundant. Rarer minerals include amphiboles, pyroxenes, sphene, nepheline, and plagioclase. The purest diatomites and tuffs contain little extraneous material.

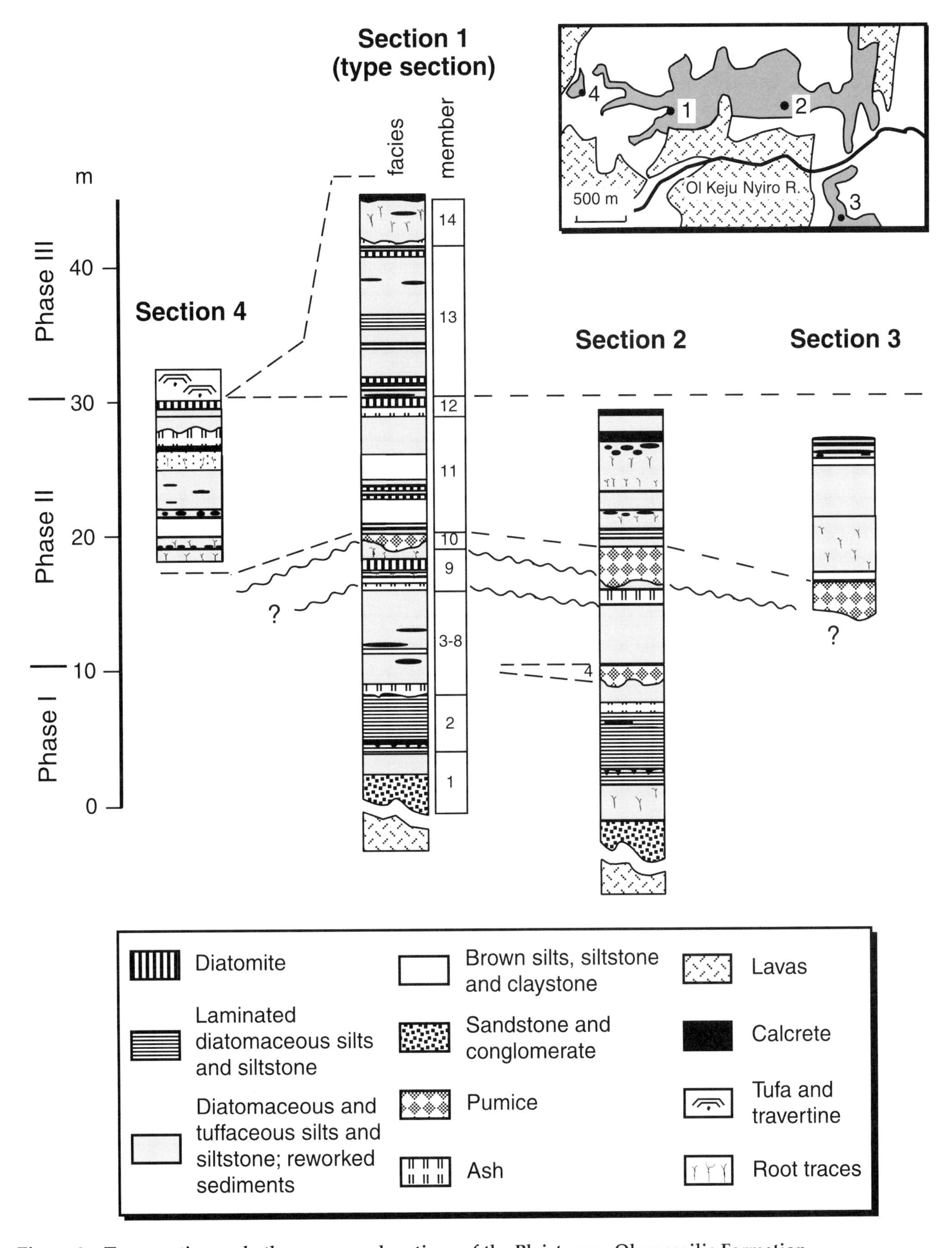

Figure 2—Type-section and other measured sections of the Pleistocene Olorgesailie Formation.

Clay minerals are generally dominated by smectites (70–100%), with kaolinite (0–20%) being most common in Facies 8. Illite and chlorite (0–10%) are locally present. Volcanic glass and silica (opal-A) are locally important components of some claystones and siltstones

Zeolites are generally absent, but minor authigenic analcime was confirmed by XRD in a tuffaceous, clayey siltstone near the base of Member 11 (Owen and Renaut, 1981). Smectite commonly replaces glass shards, especially in root-marked paleosols. Calcite is a common cement in sandstones and conglomerates, and commonly fills desiccation cracks and fractures. The presence of calcite-cemented intraclasts shows that some calcite precipitated soon after sedimentation. Various types of calcretes, some of which have been partially silicified (Isaac, 1977), are also present. Bye (1984) discussed aspects of the mineralogy and petrography, including the possible role of hydrothermal fluids in diagenesis.

FACIES DISTRIBUTION AND DIATOM VARIABILITY

The 14 members of the Olorgesailie Formation show variable degrees of lateral facies variation and in the distribution of diatoms (Figures 3, 4). Finer grained and more diatomaceous units are generally found in the western outcrops, reflecting the deeper, more permanent water that existed in that direction.

The oldest unit, Member 1 (0–9 m thick), consists of fluvial, colluvial, and shallow lacustrine siltstones and calcretes that rest on basal conglomerates. Scarce epiphytic diatoms, such as *Epithemia zebra* v. *saxonica* and *E. sorex*, are present in the siltstones, together with locally variable amounts of *Surirella ovalis* and *Rhopalodia* spp. (Figure 3). A few *Aulacoseira granulata* are present near the top of Member 1 in the western outcrops.

Member 2 (5 m thick) formed during the maximum expansion of the paleolake and is dominated in the east by subaerial brown siltstones and claystones (Facies 8 of Owen and Renaut, 1981) that pass through shallow water sediments of Facies 2 into highly diatomaceous sediments of Facies 1 in the west. Diatoms are scarce or absent in the east, but increase in abundance westward. The planktonic freshwater assemblage is dominated by *A. granulata* (and vars. *valida* and *angustissima*), *A. ambigua*, and *A. distans*. *Thalassiosira rudolfi* and *Stephanodiscus rotula* are also present, together with rare benthic species.

Members 3 and 4 become progressively more difficult to distinguish from each other and from Members 5–8 in a westerly direction (Figure 2). Member 3 (3.7 m thick) consists of shallow water sediments of Facies 2 that coarsen upward. Diatoms are rare, mostly fragmentary, and are dominated by benthic taxa, but planktonic forms are common in the eastern outcrops. Member 4 (3.5 m thick) has an erosional base cut into Member 3, and consists of fluviatile coarse pumice (Facies 6) in the east and volcanic sandstones (Facies 7) to the west. Diatoms are absent except in the extreme west, where rare planktonic and dilute water indicators, including *Aulacoseira agassizi* and *A. ambigua*, coexist with benthic species, such as *Epithemia zebra* v. *saxonica*, *Rhopalodia gibba*, and *R. vermicularis*.

Isaac (1977) described Members 5–8 (11 m thick) from the eastern outcrops, where they contain abundant Acheulian artifacts. These members, however, are difficult to recognize in the western outcrops, where they are represented by diatomaceous siltstones, tuffaceous siltstones, claystones, and tuff horizons. Diatoms are absent except at a few horizons where fragmentary *Epithemia*, *Rhopalodia*, *Cyclotella*, and *Thalassiosira* are present (Figure 3). *Aulacoseira* are scarce but slightly more common in the extreme western outcrops. This indicates that subaerial conditions prevailed to the west, but with shallow, alkaline lake waters periodically expanding and contracting across the western part of the basin.

Member 9 (7.5 m thick) consists of a basal ash unit lacking diatoms that rests disconformably on older sediments (Figure 2). The ash is overlain by widespread diatomaceous silts (Facies 2) with a uniform species composition dominated by *Thalassiosira rudolfi*, *Cyclotella meneghiniana*, *Rhopalodia gibberula*, and *Anomoeoneis sphaerophora*. These taxa indicate alkaline lake waters. The upper part of Member 9 is dominated by impure diatomite (Facies 4), with an ash layer in mid-sequence. These diatomites yield a diverse flora with many broken frustules and show considerable lateral variation. *Aulacoseira granulata* v. *valida* and *A. granulata* v. *angustissima* dominate in western outcrops. *Epithemia sorex* and *Rhopalodia gibba* dominate to the east. Outcrops of Member 9 in the southeast, near the footslopes of Mount Olorgesailie, lack diatomites; the rare diatoms present are dominated by *Surirella ovalis*, *Campylodiscus clypeus*, and *C. clypeus* v. *bicostata*.

The sedimentary rocks of Member 10 (3 m thick) rest disconformably on those of Member 9. The former consist of fluviatile pumice gravels and conglomerates (Facies 6) in the east, fining westward (Figure 3) into volcanic sands and sandstones (Facies 6) and reworked lacustrine siltstones (Facies 3). Diatoms are absent.

Member 11 (18 m thick) contains diverse sedimentary rocks, dominated by Facies 2, 3, 4, and 7, that formed mainly in shallow lake environments. Facies 8 is also present, representing subaerial conditions (Figure 2). Many horizons are calcareous, with local development of calcrete (Facies 10) and calcareous rhizoliths. Diatoms show diverse floras but commonly are absent. Two assemblages are present; one includes freshwater species such as *Aulacoseira ambigua*, *A. agassizi*, and *Epithemia* spp., and the other assemblage is dominated by alkaline water indicators, including *Thalassiosira rudolfi*, *Cyclotella meneghiniana*, and *Anomoeoneis sphaerophora*. Planktonic taxa dominate in western outcrops, whereas benthic forms are more common in the east (Figure 4). In the southeast, *C. clypeus* v. *bicostata* is common. These floras indicate generally shallow waters, but with alternating dilute and alkaline phases, perhaps due to the periodic loss and gain of a lake outlet or climatic variations.

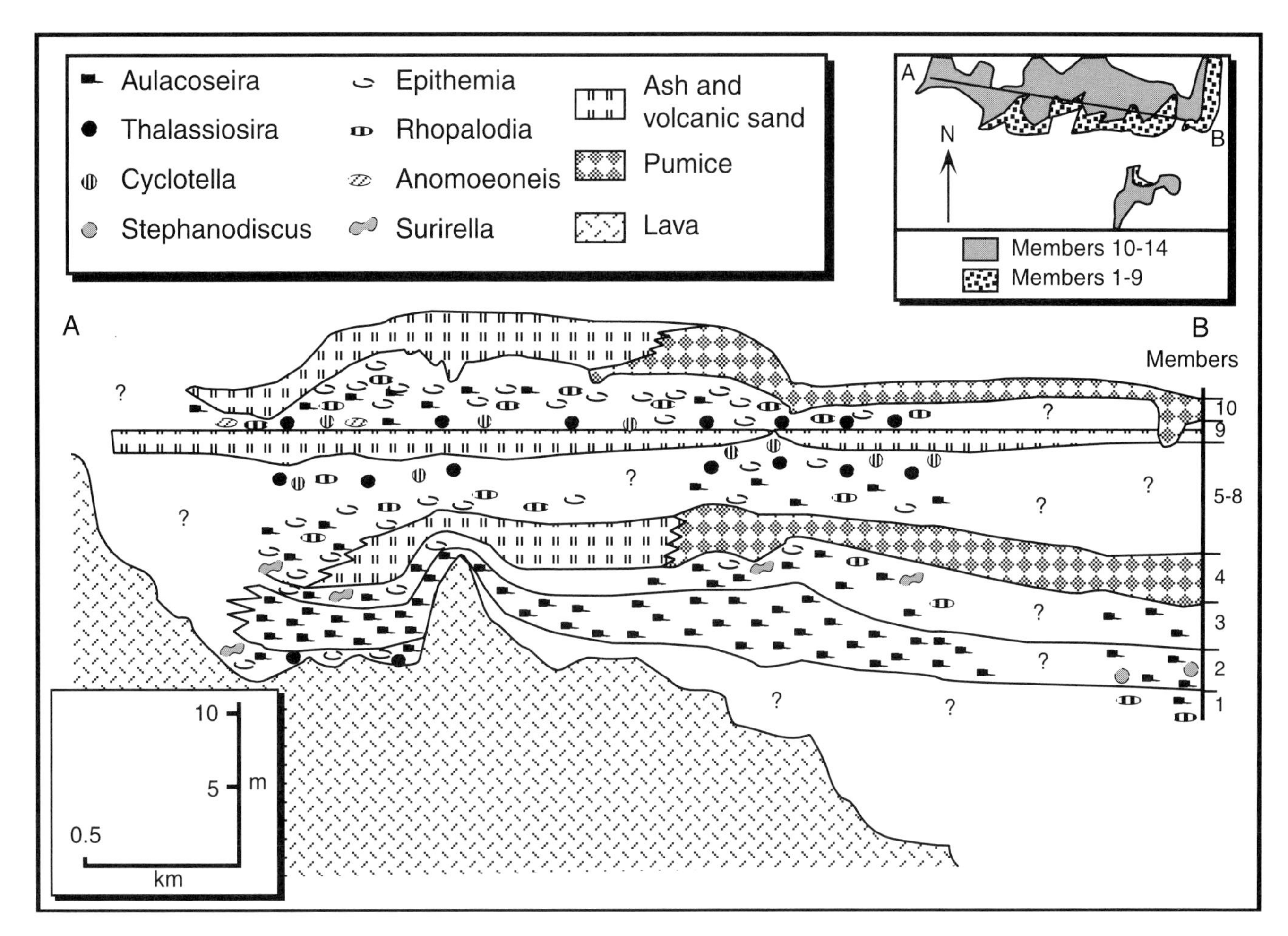

Figure 3—Schematic distribution of diatoms in Members 1–10 of the Olorgesailie Formation. The base of Member 9 has been redrawn to a horizontal datum plane from which thicknesses are measured (modified after Isaac, 1977).

Member 12 (2 m thick) comprises a basal ash and pumice unit overlain by a massive diatomite (Figure 2). In western outcrops, this diatomite consists of highly fragmented, benthic and epiphytic species dominated by *Cocconeis placentula* v. *euglypta* (50–70%), *Rhopalodia* spp., and *Epithemia* spp. Together with the common root traces, this flora indicates shallow lake water with abundant aquatic macrophytes. The eastern areas were not sampled.

Outcrops of Member 13 (12 m thick) are restricted to the western and central areas. Facies 3 and 8 predominate and diatoms are usually scarce. Where present, the diatoms are dominated by two assemblages—either *Aulacoseira granulata* (and varieties *valida* and *angustissima*), *A. agassizi*, and *A. ambigua*, or *Thalassiosira rudolfi*, *Cyclotella meneghiniana*, *Rhopalodia gibberula*, and *Epithemia* spp. These diatoms and the sediments indicate alternating freshwater and moderately alkaline lake waters with periods of emergence.

Member 14 (3 m thick) crops out only in the west, where it consists of Facies 3 and 10 (Figure 2). Diatoms are rare except at one level, where *Epithemia sorex* and *Rhopalodia gibberula* predominate.

BASIN PALEOGEOGRAPHY AND PALEOENVIRONMENTS

Sedimentation of the Olorgesailie Formation appears to have been at least partly contemporary with deposition of the Oloronga Beds (Baker, 1958; Eugster, 1980) in the greater Magadi-Natron basin (Bye et al., 1987). The latter sediments were deposited in former "Lake Oloronga," which was interpreted by Eugster (1980) to have been a large paleolake that covered much of the present Magadi basin, about 40 km to the southwest.

The Olorgesailie lake sediments derived from several sources. Most of the inferred paleocatchment lay to the north, thus much of the siliciclastic sediment influx probably originated from the north, but with local input from the adjacent fault-scarps and Mount Olorgesailie. Bye (1984) and Bye et al. (1987) showed from geochemical evidence that much of the volcaniclastic sediment derived from Ol Doinyo Nyoki, an extinct volcano that lies about 20 km south of the Olorgesailie basin.

Lake Olorgesailie probably received most of its dilute inflow from the north and east, periodically

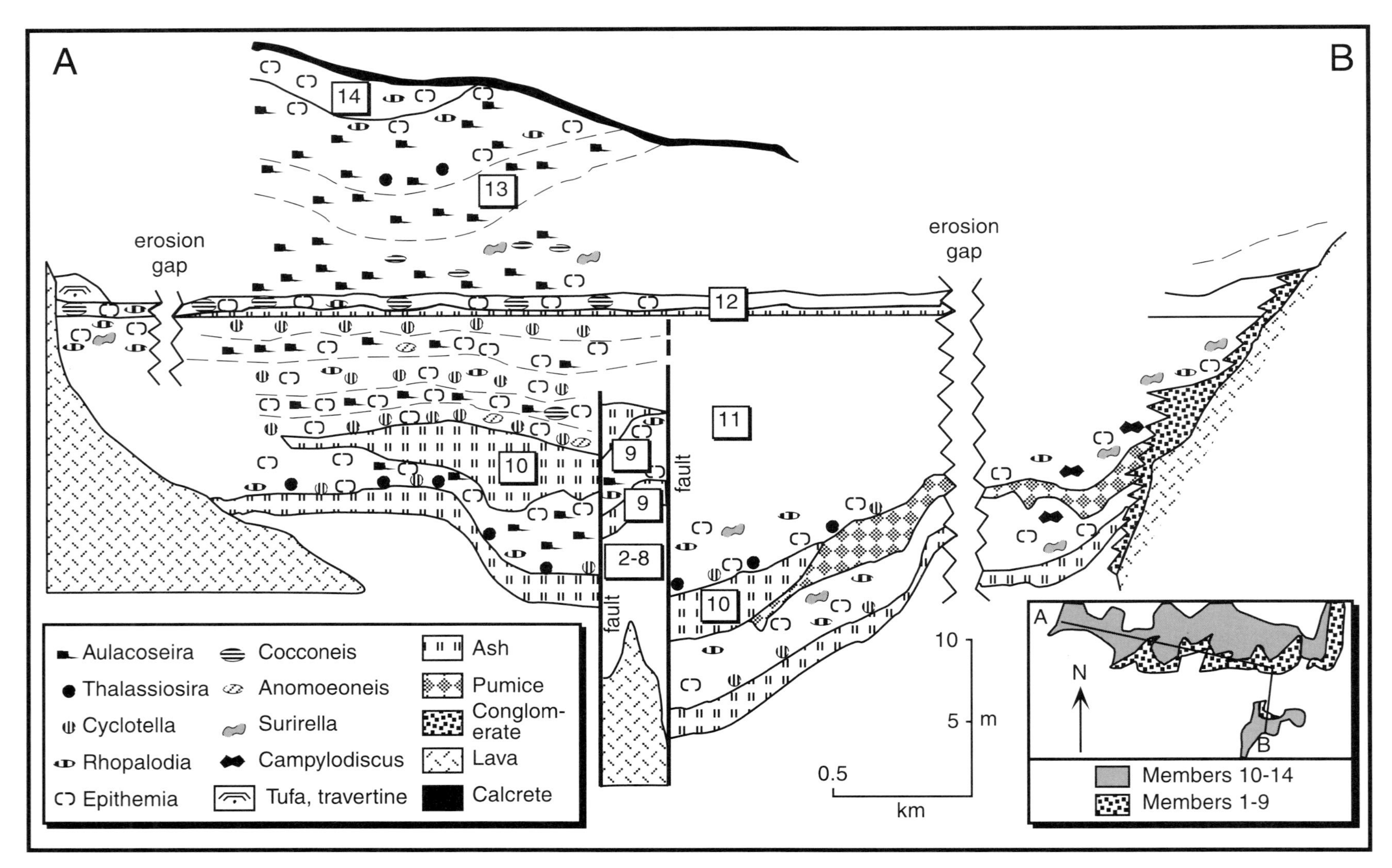

Figure 4—Schematic distribution of diatoms in Members 10–14 of the Olorgesailie Formation. The base of Member 12 has been redrawn to a horizontal datum plane from which thicknesses are measured (modified after Isaac, 1977).

overflowing to the southwest. Although not confirmed, hydrothermal springs probably also fed the paleolake. Younger tufa and travertine deposits cap the sediments on the faulted southwestern edge of the basin. Isaac (1977) suggested that the lowest point of the basin and the deeper waters were generally in the southwest, within the present Koora graben (Figure 1). From sedimentologic evidence, he also proposed that the lake waters briefly became deepest to the east, during deposition of Member 3. This was confirmed by Owen (1981) from diatom evidence.

Owen and Renaut (1981) recognized three main phases in the evolution of the Olorgesailie paleolake. Members 1–4 formed in and around a dilute lake with fluctuating, shallow to deep (Member 2) waters. The second stage occurred during deposition of Members 5–12, and was marked by variable alkalinities (5 to >80 meq l^{-1} $HCO_3 + CO_3$) and frequent periods of emergence and pedogenesis. The third stage (Members 13 and 14) was one of fresh to mildly alkaline waters with shallow to intermediate water depths and, locally, sedimentation under subaerial conditions. Absolute water depths during these stages are unknown.

The Olorgesailie paleolake remained fresh to moderately saline and alkaline throughout its history, which Isaac (1977) estimated to have been about 50,000 yr based on reasonable sedimentation rates. The diatom evidence indicates that the lake was marked by frequent changes in salinity and alkalinity, but except for the minor analcime in Member 11, evidence for sustained high salinity (e.g., evaporites or pseudomorphs of evaporite minerals) is not recorded in the sediments. The diatom flora in Member 11 also supports an alkaline lacustrine environment during its deposition. At several stages, the lake may have had an outlet to the southwest; at other times, the lake may have been a topographically closed basin, fluctuating in level and salinity in response to climatic changes (see Owen and Renaut, 1981). Given the evidence for syndepositional fault movements (Isaac, 1977), it is very difficult to determine the relative roles of climate and tectonics in controlling the geologic history of the lake. The disconformable contacts observed in the sequence (Figure 2) may partly result from changes in base-level due to tectonic movements (e.g., during deposition of Member 10) (see Isaac, 1977), and partly from climatic variations (Owen and Renaut, 1981).

REFERENCES CITED

Baker, B. H., 1958, Geology of the Magadi area: Report of the Geological Survey of Kenya 42, 81 p.

Baker, B. H., and J. G. Mitchell, 1976, Volcanic stratigraphy and geochronology of the Kedong-Olorgesailie area and the evolution of the south Kenya Rift Valley: Journal of the Geological Society, London, v. 132, p. 467–484.

Bye, B. A., 1984, Volcanic stratigraphy and red beds of the Olorgesailie Formation, southern Kenya: M. Sc. thesis, University of Utah.

Bye, B. A., F. H. Brown, T. E. Cerling, and I. McDougall, 1987, Increased age estimate for the lower Palaeolithic hominid site at Olorgesailie, Kenya: Nature, v. 329, p. 237–239.

Deino, A., and Potts, R., 1990, Single-crystal $^{40}Ar/^{39}Ar$ dating of the Olorgesailie Formation, southern Kenya Rift: Journal of Geophysical Research, Series B, v. 95, p. 8453–8470.

Deino, A., and Potts, R., 1992, Age-probability spectra for examination of single-crystal $^{40}Ar/^{39}Ar$ dating results: examples from Olorgesailie, southern Kenya Rift: Quaternary International, v. 13–14, p. 47–53.

Evernden, J. F., and G. H. Curtis, 1965, Potassium-argon dating of late Cenozoic rocks in East Africa and Italy: Current Anthropology, v. 6, p. 343–385.

Eugster, H. P., 1980, Lake Magadi, Kenya and its precursors, *in* A. Nissenbaum, ed., Hypersaline brines and evaporitic environments: Elsevier, Amsterdam, p. 195–232.

Gregory, J. W., 1921, The rift valleys and geology of East Africa: Seeley Service, London, 479 p.

Isaac, G. L., 1968, The Acheulian site complex at Olorgesailie, Kenya: contribution to the interpretation of Middle Pleistocene culture in East Africa: Ph.D. thesis, Cambridge University, U. K., 349 p.

Isaac, G. L., 1977, Olorgesailie, archeological studies of a mid-Pleistocene lake basin in Kenya: Chicago, University of Chicago Press, 272 p.

Leakey, L. S. B., 1952, The Olorgesailie Prehistoric Site, *in* L. S. B. Leakey and S. Cole, eds., Proceedings of the Pan-African Congress on Prehistory, 1947, p. 209.

Marsden, M., 1979, Origin and evolution of the Pleistocene Olorgesailie Lake series. Ph.D. thesis, McGill University, 360 p.

Miller, J. A., 1967, Problems of dating East African Tertiary and Quaternary volcanics by the potassium-argon method, *in* W. W. Bishop and J. D. Clark, eds., Background to evolution in Africa: Chicago, University of Chicago Press, p. 259–272.

Owen, R. B., 1981, Quaternary diatomaceous sediments and the geological evolution of Lakes Turkana, Baringo and Bogoria, Kenya Rift Valley: Ph.D. thesis, University of London, 465 p.

Owen, R. B., and R. W. Renaut, 1981, Paleoenvironments and sedimentology of the Middle Pleistocene Olorgesailie Formation, southern Kenya Rift Valley: Palaeoecology of Africa, v. 13, p. 147–174.

Potts, R., 1989, Olorgesailie: new excavations and findings in early and middle Pleistocene contexts, southern Kenya rift valley: Journal of Human Evolution, v. 18, p. 477–484.

Potts, R., 1994, Variables versus models of early Pleistocene hominid land use: Journal of Human Evolution, v. 27, p. 7–24.

Shackleton, R. M., 1944, Preliminary report on the Olorgesailie Prehistoric Site. Unpublished Report, Mines and Geological Department, Nairobi.

Shackleton, R. M., 1955, Pleistocene movements in the Gregory Rift Valley: Geologische Rundschau, v. 43, p. 257–263.

Shackleton, R. M., 1978, Geological map of the Olorgesailie Area, *in* W. W. Bishop, ed., Geological background to fossil man: Edinburgh, Scottish Academic Press, p. 171–172.

Renault, R. W., J.-J. Tiercelin, and R. B. Owen, 2000, Lake Baringo, Kenya Rift Valley, and its Pleistocene precursors, *in* E. H. Gierlowski-Kordesch and K. R. Kelts, eds., Lake basins through space and time: AAPG Studies in Geology 46, p. 561–568.

Chapter 53

Lake Baringo, Kenya Rift Valley, and its Pleistocene Precursors

Robin W. Renaut
Department of Geological Sciences, University of Saskatchewan
Saskatoon, Canada

Jean-Jacques Tiercelin
UMR 6538 "Domaines Océaniques," Institut Universitaire Européen de la Mer
Plouzané, France

R. Bernhart Owen
Department of Geography, Hong Kong Baptist University
Kowloon Tong, Hong Kong, China

INTRODUCTION

Lake Baringo is a shallow freshwater lake with predominantly siliciclastic sediments that is located in a semi-arid volcanic region of the central Kenya Rift Valley (Figure 1A, B) (Tiercelin and Vincens, 1987). The modern lake is the successor to a series of precursor lakes that have occupied the rift valley at this latitude (1°N) since the middle Miocene (Bishop et al., 1971; Chapman et al., 1978; Tiercelin, 1981; Hill, 1995, 1999; Renaut et al., 1999). Lake Baringo lies in the northern part of a rhomb-shaped half-graben basin, approximately 21 km long by 13 km wide. The shape of the lake and the basin morphology are controlled mainly by two regional tectonic trends—the north-south trend of the Tertiary to Holocene rift system, and a group of older northwest-southeast lineaments inherited from the Precambrian basement (Figure 1C) (Dunkley et al., 1993; Le Turdu et al., 1995, in press). The intersection of these two tectonic trends not only influences the patterns of drainage and sedimentation, but also has played a major role in controlling the recurrence of lake basins at this latitude. Rooney and Hutton (1977) showed from geophysical data that more than 1 km of sediments underlies the present lake basin. New geophysical evidence indicates that modern Lake Baringo and the Loboi Plain (Figure 1C) are the surface expression of a deep (more than 5 km) fault-controlled basin that was probably initiated during the Paleogene (Hautot et al., 1998, personal communication).

West of the lake, the land rises 1500 m above the rift floor to form the Tugen Hills, a large intrarift fault block composed of Neogene lavas and sedimentary rocks that lies between Lake Baringo and the Elgeyo escarpment border-fault (Figure 1B) (Chapman et al., 1978). Upper Neogene sedimentary rocks and lavas lie between the Tugen Hills and the lake, recording a gradual eastward migration of the depocenter. About 30 km east of Lake Baringo, the Laikipia escarpment forms the eastern edge of the rift (Hackman, 1988). To the north, Karosi, an extinct Pleistocene central volcano, forms the indented northern shoreline of Lake Baringo. To the south, upper Pleistocene trachyphonolites and basalts form the northward-dipping rift floor. Directly south of the lake, these lavas are covered by the Pleistocene and Holocene fluviatile and lacustrine sediments of the Loboi Plain. These sediments separate Lake Baringo from saline, alkaline Lake Bogoria, which fills a narrow asymmetric half-graben depression 16 km to the south (Figure 1).

LAKE BARINGO

Limnology and Recharge

Lake Baringo has a surface area of approximately 160 km^2 and drains a catchment of approximately 6,200 km^2, which is located mainly in wetter volcanic uplands to the southwest (Figure 2A). The annual rainfall is about 600–900 mm on the rift valley floor rising to more than 1000 mm/yr in the adjacent highlands. Mean annual evaporation on the rift floor is approximately 2600 mm. Most annual recharge comes from the Molo and Perkerra rivers (Figure 2A), which flow northward across the Loboi Plain, discharging into the lake at the Molo Delta (Figure 2B). These rivers are usually perennial, but in dry years recharge is strongly reduced because water is also abstracted

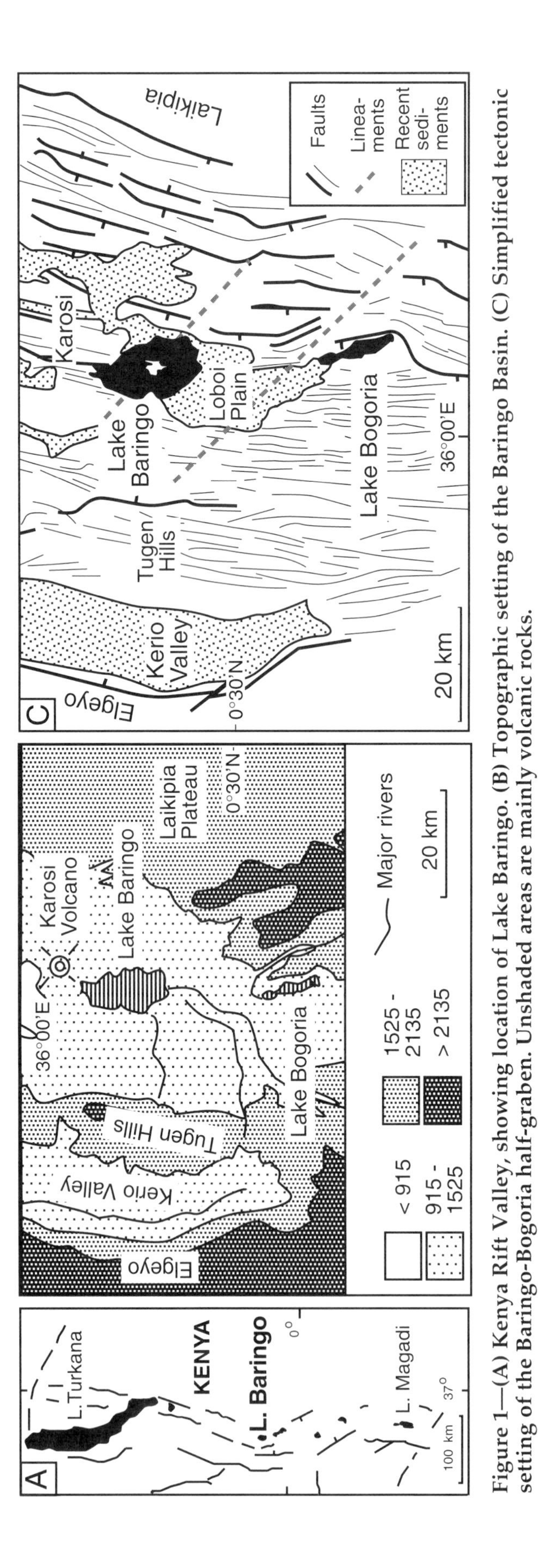

Figure 1—(A) Kenya Rift Valley, showing location of Lake Baringo. (B) Topographic setting of the Baringo Basin. (C) Simplified tectonic setting of the Baringo-Bogoria half-graben. Unshaded areas are mainly volcanic rocks.

for irrigation. Several large ephemeral streams (Ndau, Kapthurin) drain the Tugen Hills; other ephemeral streams drain the eastern shoreline. Small hot springs on Ol Kokwe Island contribute very little to the annual recharge (Dunkley et al., 1993). Some dilute groundwater inflow from the lake margins is possible given the elevated topography.

The lake changes in volume and chemical composition in response to seasonal and long-term changes in river inflow. The surface level lies at nearly 970 m elevation above sea level. During the past 20 yr its level has fluctuated over a vertical range of about 4 m in response to climatic changes. The average depth is about 4 m with the deepest waters lying in the northwestern quadrant (Figure 2B).

The lake waters have a Na-Ca-HCO_3 composition, similar to the dilute inflow (Beadle, 1932; Talling and Talling, 1965; Tiercelin, 1981; Tiercelin and Vincens, 1987). The lake waters range in salinity from 0.5 to 1.5 g l^{-1} TDS (total dissolved solids). The salinity of the surface waters varies both spatially and seasonally. In general, salinity increases northward with distance from the Molo Delta, with maximum salinities attained in the northwestern quadrant. The lake waters have evaporated by a factor of only 2–4 compared to inflow waters. Assuming chloride conservation, the lake waters become measurably depleted in Ca^{2+}, HCO_3^-, and SiO_2. This can be explained by calcite precipitation and buffering of silica levels by diatoms and sponges.

Lake Baringo remains fresh despite the lack of a surface outlet and its location in a region of high net evaporation. The explanation for its fresh waters has long been controversial (Gregory, 1921; Barton et al., 1987), but recent hydrologic and hydrogeologic evidence supports subsurface loss of water through the lake floor and its subsequent northward flow as groundwater (Allen and Darling, 1992; Dunkley et al., 1993). The lake is topographically, but not hydrologically, closed.

Recent Sedimentation

Fine-grained siliciclastics dominate the recent sediments of Lake Baringo. The recent lacustrine sediments are uniform in texture across most of the lake floor. Shallow cores reveal a gradual fining with distance from the Molo Delta and peripheral alluvial fans (Figure 3) (Tiercelin et al., 1980; Tiercelin, 1981). Very high rates of erosion in the catchment supply abundant silt and clay to the Molo and other deltas (Oostwoud Wijdenes and Gerits, 1994).

Most of the lake floor is covered by muds (feldspathic silts, clays), but the proportion of sand and coarser sediments is higher nearer the shorelines. Sands are uncommon but are present offshore from the Ol Arabel and other marginal alluvial fans and small deltas, where they were probably deposited from turbid flows during ephemeral stormfloods. Sandy shoreline deposits are rare; most modern shorelines have accumulations of plant debris, but few well-defined beaches. Lava gravel beaches are present along the rocky shorelines of the volcanic islands and peninsulas that project into the lake (e.g., Karosi, Ol Kokwe, and Kampi-Ya-Samaki), but coarse sediments are uncommon.

A broad zone of littoral marsh lies at the southern and southeastern margins of the lake near the mouths of the Molo, Ndau, Mukutan, and Ol Arabel rivers. This dense marsh extends outward into the lake for hundreds of meters during periods of relatively high lake level, locally forming floating islands. Sediments deposited in the marshy areas are mainly alternating muds and brown silty clays containing many ostracods, gastropods, diatoms, sponge spicules, *Chara* oogonia, and vegetal debris.

Sediments across most of the lake floor are muds and finely laminated clays, containing many gas escape vesicles, that alternate with more massive silty clays with millimeter-scale lamination and thin silts. These facies are common from prodelta settings, notably those distal from the mouths of the Molo and Ndau rivers, to the deepest parts of the lake. They are characterized by abundant bioclasts, including gastropod and ostracod shells, fish bones, sponge spicules and gemmules, diatoms, and vegetal debris. The bioclasts are commonly bound weakly by a diffuse organic matrix or by finely laminated microbial mats or biofilms. East of the Ol Kokwe, silty clays commonly contain small carbonate concretions of probable bacterial origin.

The clay fraction of the lake sediments is composed mainly of smectite (60–80%), with variable amounts of illite, chlorite, kaolinite, and interstratified kaolinite-smectite. Smectite is distributed across the lake floor, especially in the eastern part of the lake and north of the Molo Delta. The highest amounts of kaolinite and illite are present along the eastern and northeastern half of the lake. This distribution implies that the recent clay minerals are mostly detrital, reflecting their local provenance in different parts of the basin. Most organic matter in the recent sediments is allochthonous and derived from plants washed in by the peripheral streams. With a predominance of allochthonous organic debris, shallow fresh waters with high oxygenation, periodic desiccation, and a high siliciclastic sedimentation rate, the source rock potential of the modern lake is very low.

Thin (<2 cm) stromatolitic crusts and oncoids are present on littoral lava substrates mainly in the northern half of the lake, notably along the northern shores of Kokwob Murren peninsula north of Kampi-Ya-Samaki (Figures 3B, 4) and on the shorelines of the several volcanic islands. The stromatolites, which are

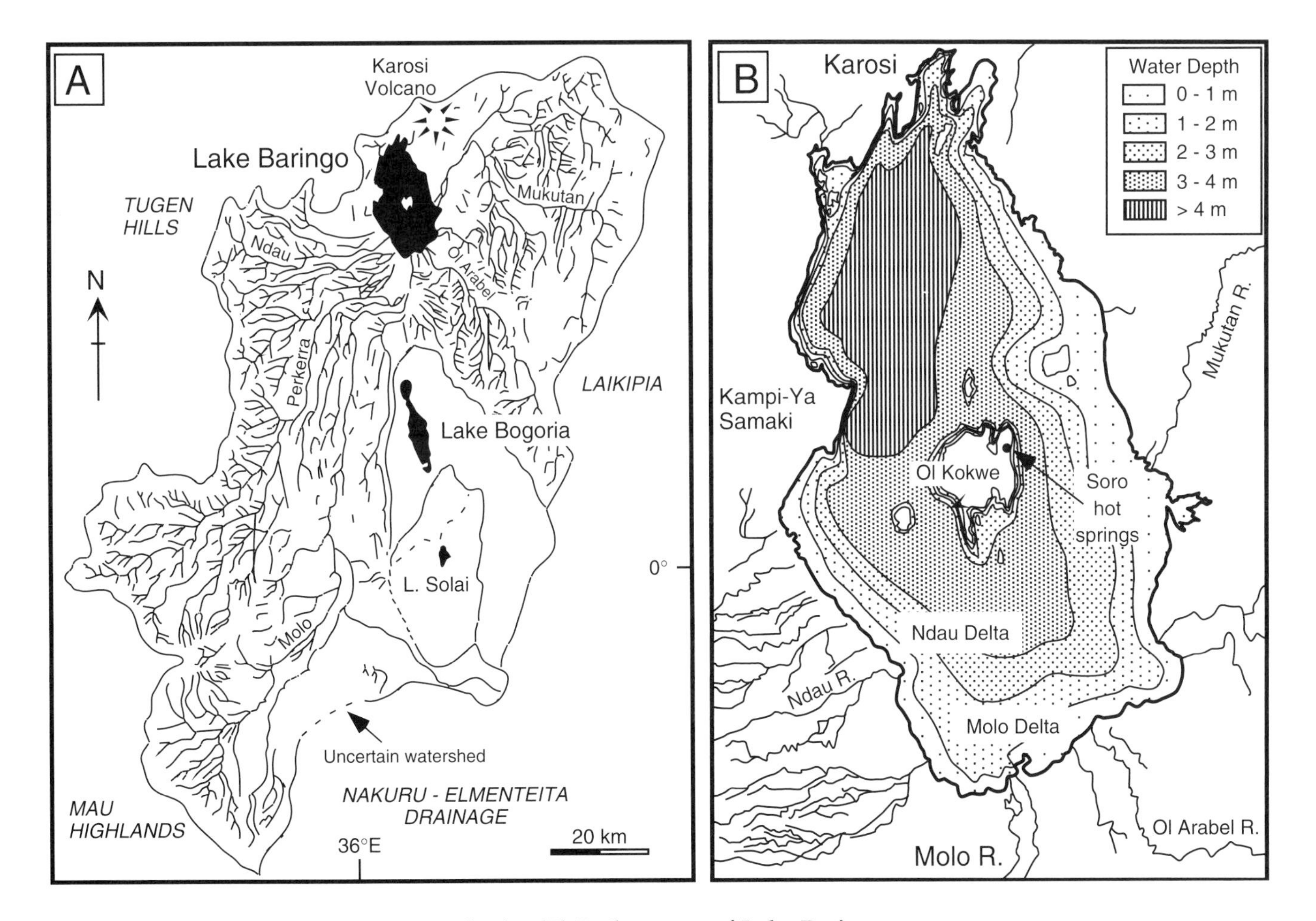

Figure 2—(A) Lake Baringo drainage basin. (B) Bathymetry of Lake Baringo.

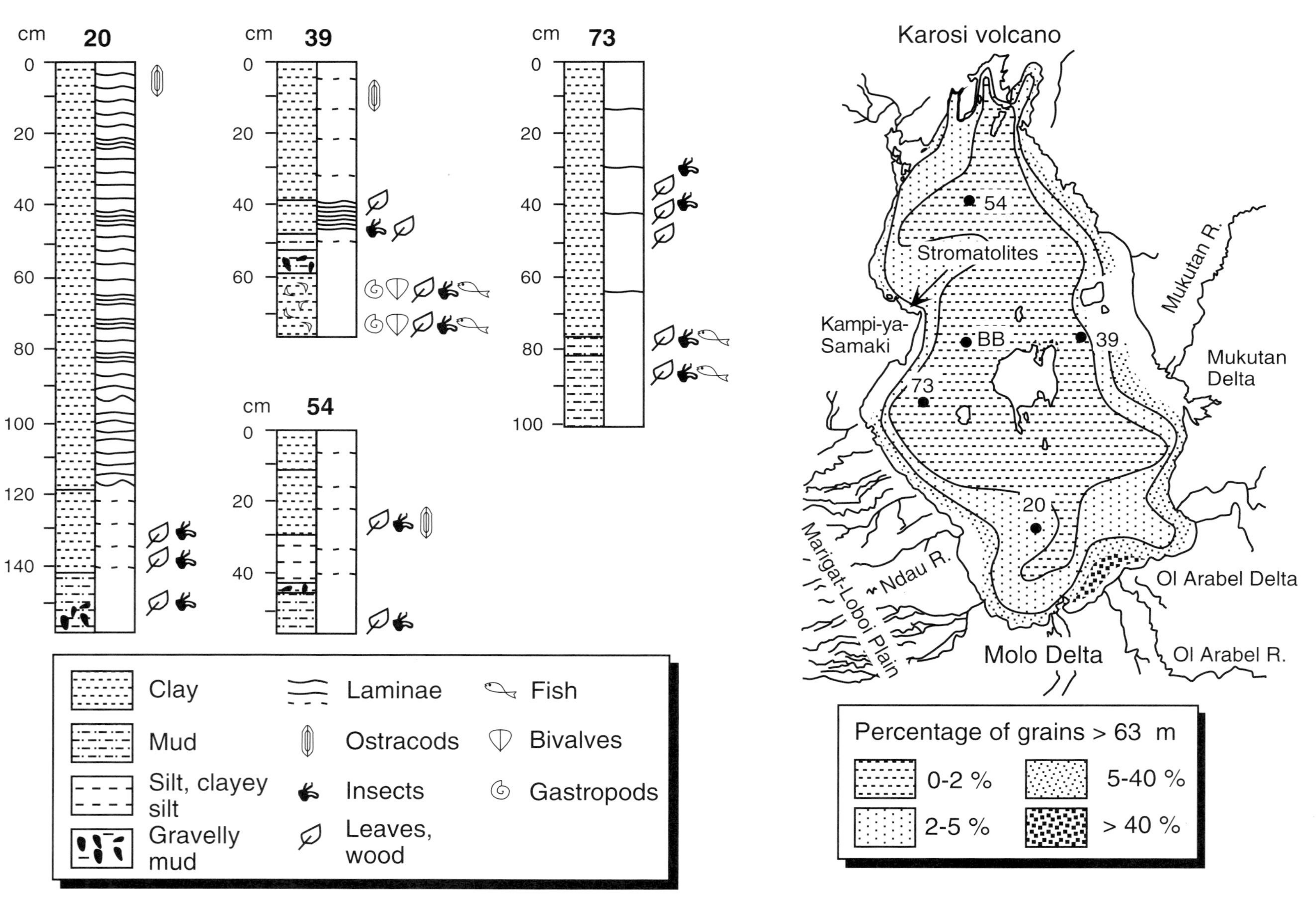

Figure 3—Logs of short cores from Lake Baringo. Map shows core locations and grain size distribution of modern lake-floor sediments.

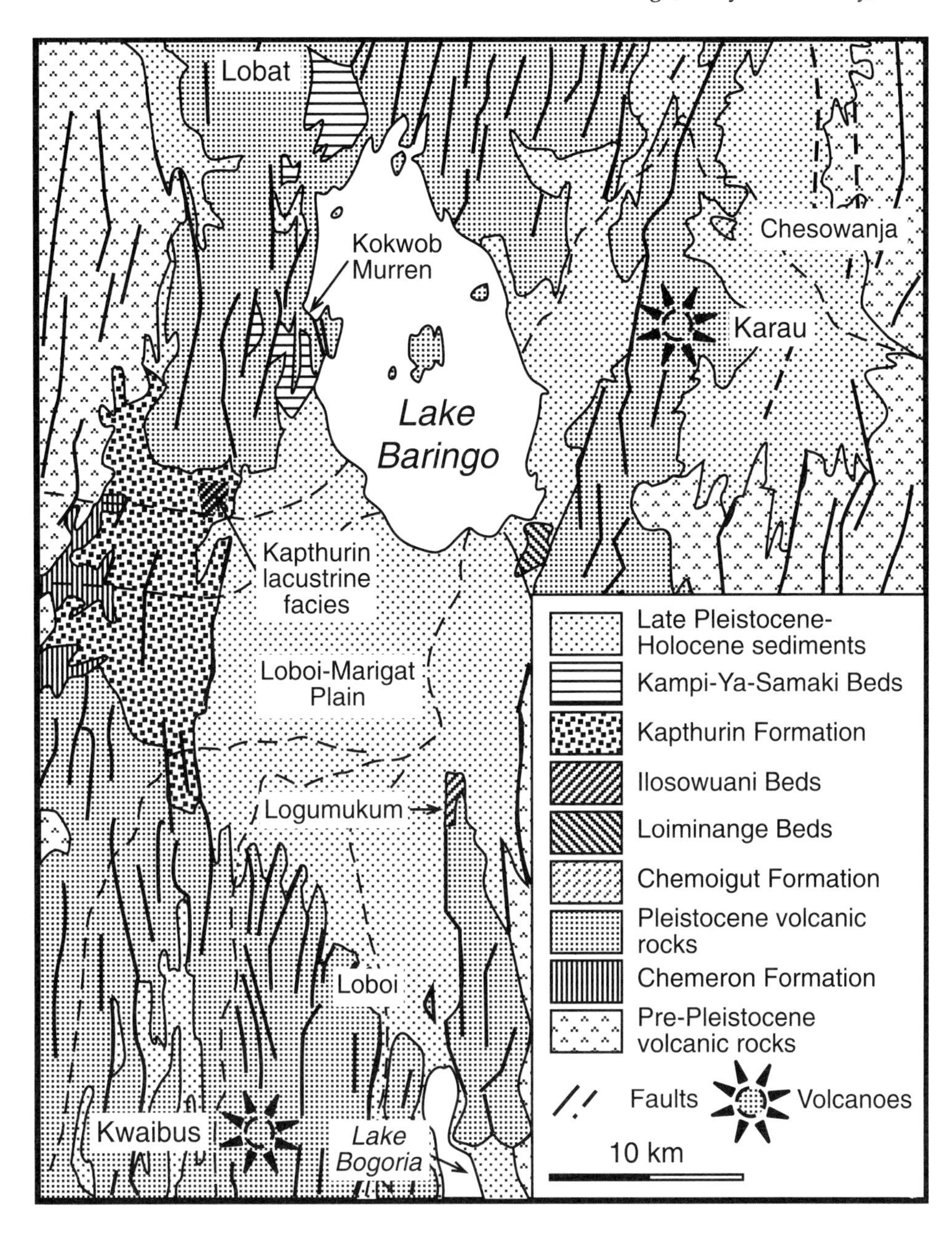

Figure 4—Simplified geologic map of Lake Baringo basin showing location of Pleistocene and Holocene sediments and volcanic rocks.

composed of calcite, contain well-preserved cyanobacterial filaments and diatoms. They form periodically when waters become supersaturated with $CaCO_3$ due to evaporation and microbial photosynthesis. The crusts are thickest in sheltered sites with relatively low turbidity. Some are actively forming. The stromatolites and charophytes probably account for most of the depletion of Ca^{2+} in the lake waters.

PLEISTOCENE PRECURSOR LAKES OF LAKE BARINGO

Late Pleistocene–Holocene Record

Lake Baringo has not yet been cored beyond a depth of 3 m, so the late Pleistocene–Holocene history of the lake is poorly known and dependent mainly on evidence from scattered outcrops around the lake margins (Renaut and Owen, 1980). A paleosol and diatoms that indicate elevated alkalinity (~80 meq l^{-1}) in a short core taken offshore from Kampi-Ya-Samaki (site BB in Figure 2B) show that the lake has nearly dried up and become moderately saline within the past few thousand years (Owen, 1981). Evidence for higher lake levels is seen in outcrops and paleoshoreline sediments of the Kokwob Formation, preserved mainly in a small graben at Kokwob Murren (Figures 4, 5). These comprise a series of lacustrine silts, sands, and mollusc coquinas at up to 985 m elevation that have yielded radiocarbon ages ranging from 13,850 to 8460 yr BP (Williams and Johnson, 1976). The highest paleoshorelines are about 15 m higher than the present level of Lake Baringo and 14 m lower than the maximum known highstand of Lake Bogoria (999 m). Unless there has been preferential subsidence in the Baringo basin since the late Pleistocene–early Holocene, it is unlikely that the two lakes combined as one lake at that time. Other evidence for former high lake levels includes incised alluvial terraces west of the lake. Although Lake Baringo has a Nilotic fauna that implies a recent connection with the Nile drainage system (Worthington and Ricardo, 1936; Vacelet et al., 1990), there is no obvious morphological evidence for a major former outflow to the north, thus it remains to be determined when this connection was severed.

Kapthurin Formation

During the Pleistocene, the Baringo region was tectonically and volcanically active. Major faulting lowered the north-south axial trough of the rift, creating the present topographic setting; several volcanoes were active, including Karosi and Ol Kokwe (Figures 1, 2B). Evidence for a paleolake that predated the last major phase of faulting, which occurred probably at <0.2 Ma, is preserved in the Pleistocene Kapthurin Formation, which crops out west of Lake Baringo (Fuchs, 1950; McCall et al., 1967; Martyn, 1969; Walsh, 1969; Tallon, 1976, 1978). The Kapthurin Formation, consists of a series of colluvial, fluvial, lacustrine, and volcaniclastic sediments, approximately 125 m thick, that were deposited following uplift of the Tugen Hills at about 1 Ma. The resultant rejuvenation of drainage resulted in a large alluvial fan–littoral plain complex east of the uplifted escarpment. Ephemeral, and possibly at times perennial, streams draining the uplifted fault-block deposited feldspathic silts, sands, and coarse volcanic gravels in the region directly west of Lake Baringo. The streams flowed into a contemporary Central Rift Lake (Tallon 1978), which probably had a depocenter near modern Lake Baringo. The Kapthurin sediments contain fossil hominids and artifacts (Leakey et al., 1969; McBrearty et al., 1996; McBrearty, 1999).

Most evidence for the Kapthurin paleolake or paleolakes is down-faulted and is buried below younger sediments of the Loboi Plain (Loboi Silts), but a small outcrop of lacustrine sediments is exposed along the northeastern edge of the outcrop (Figure 4). These consist of red, green, and black claystones, mudstones, and siltstones, some of which are tuffaceous (Figure 5). The smectitic black claystones contain fossil vertebrates (*Clarias*, *Hippopotamus*, *Crocodilus*) and were probably deposited in a shallow fresh lake. Much of the smectite may derive from alteration of volcanic

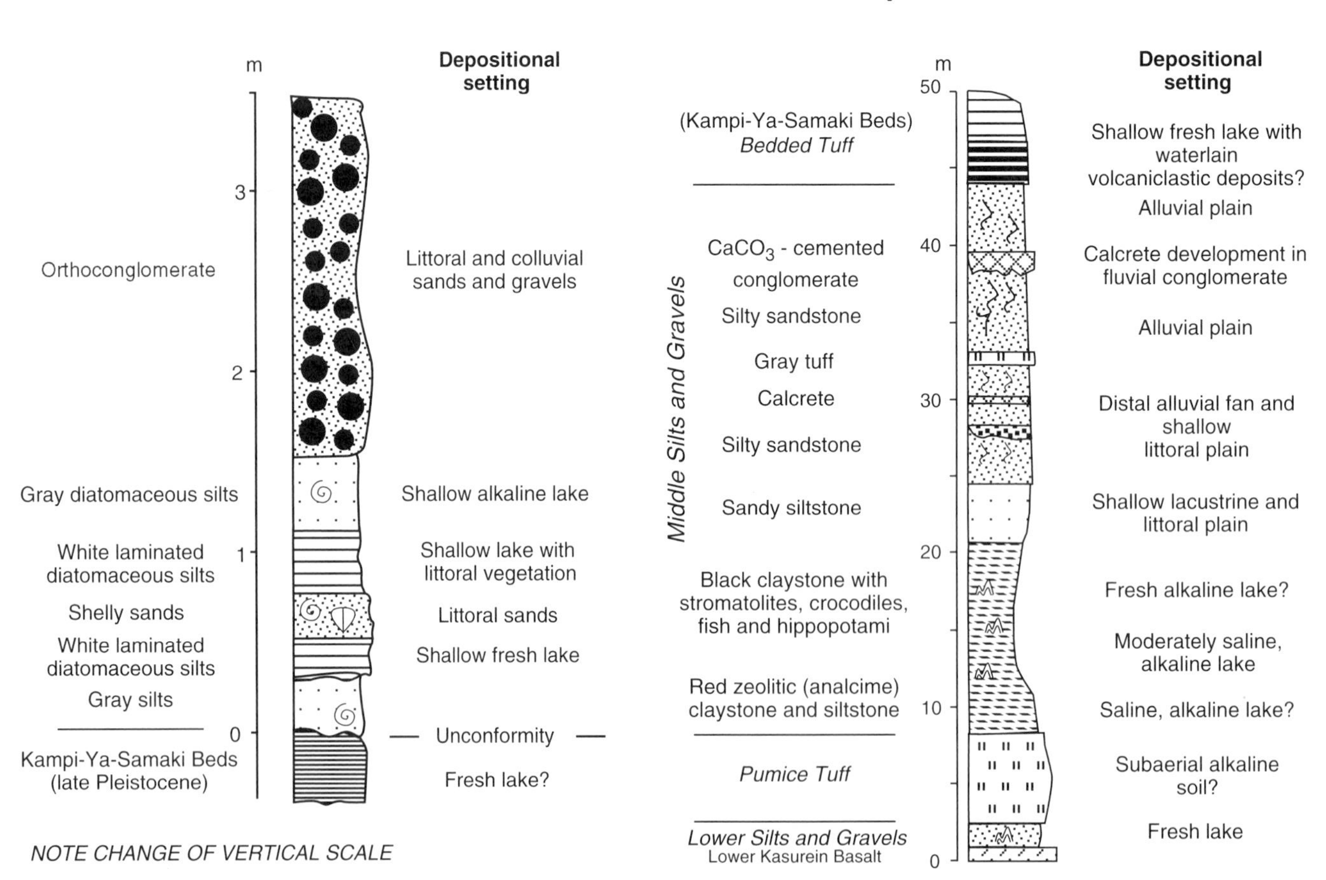

Figure 5—Stratigraphic sections of the lacustrine sequence of the Kokwob Formation at Kokwob Murren, and the Kapthurin Formation (locations of sections shown in Figure 4).

ash. In contrast, the red and green tuffaceous clays contain authigenic analcime and fluorite that indicate saline, alkaline waters in or on the margins of the paleolake (Renaut et al., 1999). This implies that for at least part of its history the Kapthurin Lake occupied a closed hydrologic basin, with outflow possibly limited to the north by the then-active Karosi Volcano. The Kampi-Ya-Samaki Beds (Figure 4) include laminated lacustrine volcaniclastic sediments, possibly derived from Karosi, that are probably equivalent to the upper part (Bedded Tuff Member) of the Kapthurin Formation (Tallon, 1978).

Other outlying pockets of lacustrine sediments are preserved on the eastern side of the basin. These include the Ilosowuani Beds, located at Logumukum (Figure 4), which are a series of lacustrine silts and clays capped by a zeolitic paleosol and calcrete that may be contemporary with part of the Kapthurin sequence (Tiercelin and Vincens, 1987). The lithologically similar Loiminange Beds (Figure 4) may be of equivalent age. Hominid-bearing Pleistocene lake deposits (Chemoigut Formation) are also present at Chesowanja, 12 km east of Lake Baringo (Bishop et al., 1978; Gowlett et al., 1981), but their paleogeographic relationship to time-equivalent sediments west of the lake remains poorly constrained due to late Pleistocene subsidence of the rift floor and the cover of younger sediments.

SUMMARY

Lake Baringo is a freshwater lake with predominantly siliciclastic fine-grained sediments that drains a large volcanic catchment in the central Kenya Rift Valley. The lake is polymictic and well oxygenated. Although in a closed topographic basin in a region with high net evaporation, the lake maintains dilute waters, probably by having subsurface discharge. The Holocene and late Pleistocene sedimentary record, although still poorly known, provides evidence of higher lake levels during the late Pleistocene and early Holocene, and lower lake levels than today probably during the past few thousand years. An earlier Pleistocene precursor lake is recorded in the sediments of the Kapthurin Formation. These lake beds provide mineralogical and faunal evidence of both fresh and saline, alkaline phases. Other Pleistocene lake beds are found in scattered outcrops around the basin, but their age and significance are still poorly known.

REFERENCES CITED

Allen, D. J., and W. G. Darling, 1992, Geothermics and hydrogeology of the Kenya Rift Valley between Lake Baringo and Lake Turkana: British Geological Survey Research Report SD/92/1, 39 p.

Barton, C. E., D. K. Solomon, J. R. Bowman, T. E. Cerling, and M. D. Sayer, 1987, Chloride budgets in transient lakes: Lakes Baringo, Naivasha, and Turkana: Limnology and Oceanography, v. 32, p. 745–751.

Beadle, L. C., 1932, Scientific results of the Cambridge expedition to the East African lakes, 1930–1, 4: The waters of some East African lakes in relation to their fauna and flora: Journal of the Linnean Society, v. 38, p. 157–211.

Bishop, W. W., A. Hill, and J. A. Miller, 1971, Succession of Cainozoic vertebrate assemblages from the northern Kenya Rift Valley: Nature, v. 233, p. 389–394.

Bishop, W. W., A. Hill, and M. H. L. Pickford, 1978, Chesowanja; a revised geological interpretation, *in* W. W. Bishop, ed., Geological background to fossil man; recent research in the Gregory Rift Valley, East Africa: Edinburgh, Scottish Academic Press, p. 309–327.

Chapman, G. R., S. J. Lippard, and J. E. Martyn, 1978, The stratigraphy and structure of the Kamasia Range, Kenya Rift Valley: Journal of the Geological Society, London, v. 153, p. 265–281.

Dunkley, P. N., M. Smith, D. J. Allen, and W. G. Darling, 1993, The geothermal activity and geology of the northern sector of the Kenya Rift Valley: British Geological Survey Research Report SC/93/1, 185 p.

Fuchs, V. E., 1950, Pleistocene events in the Baringo Basin: Geological Magazine, v. 87, p. 149–174.

Gowlett, J. A. J, J. W. K. Harris, D. Walton, and B. A. Wood, 1981, Early archaeological sites, hominid remains and traces of fire from Chesowanja, Kenya: Nature, v. 294, p. 125–129.

Gregory, J. W., 1921, The rift valleys and geology of East Africa: London, Seeley Service, 479 p.

Hackman, B. D., 1988, Geology of the Baringo-Laikipia area: Mines and Geology Department, Kenya, Report 104, 79 p.

Hill, A., 1995, Faunal and environmental change in the Neogene of East Africa: evidence from the Tugen Hills sequence, Baringo District, Kenya, *in* E. S. Vrba, G. H. Denton, T. C. Partridge, and L. H. Burkle, eds., Paleoclimate and evolution with emphasis on human origins: New Haven, Yale University Press, p. 178–193.

Hill, A., 1999, The Baringo Basin, Kenya: from Bill Bishop to BPRP, *in* P. Andrews and P. Banham, eds., Late Cenozoic environments and hominid evolution: a tribute to Bill Bishop: London, Geological Society Publications.

Le Turdu, C., J.-J. Tiercelin, C. Coussement, J. Rolet, R. W. Renaut, J. P. Richert, J. P. Xavier, and D. Coquelet, 1995, Basin structure and depositional patterns interpreted using a 3D remote sensing approach: the Baringo-Bogoria basins, central Kenya Rift, East Africa: Bulletin des Centre de Recherche Exploration-Production Elf-Aquitaine, v. 19, p. 1–37.

Le Turdu, C., J.-J. Tiercelin, J. P. Richert, J. Rolet, J. P. Xavier, R. W. Renaut, K. E. Lezzar, and C. Coussement, in press, Influence of pre-existing oblique discontinuities on the geometry and evolution of extensional fault patterns: evidence from the Kenya Rift using SPOT imagery: AAPG Studies in Geology, v. 44.

Leakey, M., P. V. Tobias, J. E. Martyn, and R. E. F. Leakey, 1969, An Acheulian industry with prepared core technique and the discovery of a contemporary hominid at Lake Baringo, Kenya: Proceedings of the Prehistoric Society, v. 35, p. 48–76.

Martyn, J. E., 1969, The geological history of the country between Lake Baringo and the Kerio River, Baringo District, Kenya: Ph. D. dissertation, University of London.

McBrearty, S., L. C., 1999, The archaeology of the Kapthurin Formation, *in* P. Andrews and P. Banham, eds., Late Cenozoic environments and hominid evolution: a tribute to Bill Bishop: London, Geological Society Publications.

McBrearty, S., L. C. Bishop, and J. D. Kingston, 1996, Variability in traces of Middle Pleistocene hominid behavior in the Kapthurin Formation, Baringo, Kenya: Journal of Human Evolution, v. 30, p. 563–580.

McCall, G. J. H., B. H. Baker, and J. Walsh, 1967, Late Tertiary and Quaternary sediments of the Kenya Rift Valley, *in* W. W. Bishop and J. D. Clark, eds., Background to evolution in Africa: Chicago, University of Chicago Press, p. 191–220.

Oostwoud Wijdenes, D. J., and J. Gerits, 1994, Runoff and sediment transport on intensively gullied, low-angle slopes in Baringo District, *in* R. B. Bryan, ed., Soil erosion, land degradation, social transition: Advances in GeoEcology, v. 27, p. 121–141.

Owen, R. B., 1981, Quaternary diatomaceous sediments and the geological evolution of lakes Turkana, Baringo and Bogoria, Kenya Rift Valley: Ph. D. dissertation, University of London, 465 p.

Renaut, R. W., and R. B. Owen, 1980, Late Quaternary fluvio-lacustrine sedimentation and lake levels in the Baringo Basin, northern Kenya Rift Valley: Recherches Géologiques en Afrique, v. 5, p. 130–133.

Renaut, R. W., J. Ego, J.-J. Tiercelin, C. Le Turdu, and R. B. Owen, 1999, Saline, alkaline palaeolakes of the Tugen Hills-Kerio Valley region, Kenya Rift Valley, *in* P. Andrews and P. Banham, eds., Late Cenozoic environments and hominid evolution: a tribute to Bill Bishop: London, Geological Society Publications, p. 41–58.

Rooney, D., and V. R. S. Hutton, 1977, A magnetotelluric and magnetovariational study of the Gregory Rift Valley, Kenya: Geophysical Journal of the Royal Astronomical Society, v. 51, p. 91–119.

Talling, J. F., and J. B. Talling, 1965, The chemical composition of African lake waters: Internationale Revue der Gesamten Hydrobiologie, v. 50, p. 421–463.

Tallon, P. W. J., 1976, The stratigraphy, palaeoenvironments and geomorphology of the Pleistocene Kapthurin Formation, Kenya: Ph. D. dissertation, University of London, 298 p.

Tallon, P. W. J., 1978, Geological setting of hominid fossils and Acheulian artefacts from the Kapthurin Formation, Baringo District, Kenya, *in* W. W. Bishop, ed., Geological background to fossil man: Scottish Academic Press, Edinburgh, p. 361–378.

Tiercelin, J.-J., 1981, Rifts continentaux, tectonique, climats, sédiments. Exemples: la sédimentation dans le Nord du Rift Gregory, Kenya, et dans le Rift de l'Afar, Ethiopie, depuis le Miocène: Ph.D. Dissertation, Université Aix-Marseille II, Marseille, 260 p.

Tiercelin, J.-J., and A. Vincens, eds., 1987, Le demi-graben de Baringo-Bogoria, Rift Gregory, Kenya: Bulletin des Centres de Recherches Exploration-Production Elf-Aquitaine, v. 11, p. 249–540.

Tiercelin, J.-J., J. Le Fournier, J. P. Herbin, and J. P. Richert, 1980, Continental rifts: modern sedimentation, tectonic and volcanic control. Example from the Bogoria-Baringo graben, Gregory Rift, Kenya, *in* Geodynamic Evolution of the Afro-Arabic Rift System: Rome, Accademia Nazionale dei Lincei, p. 143–164.

Vacelet, J., Tiercelin, J.-J., and Gasse, F., 1990, The sponge *Dosilia brouni* (Spongillidae) in Lake Baringo, Gregory Rift, Kenya: Hydrobiologia, v. 211, p. 11–18.

Walsh, J., 1969, Geology of the Eldama Ravine—Kabarnet area: Geological Survey of Kenya Report 84, 48 p.

Williams, R. E. G., and A. S. Johnson, 1976, Birmingham University radiocarbon dates X: Radiocarbon, v. 18, p. 249–267.

Worthington, E. B., and C. K. Ricardo, 1936, Scientific results of the Cambridge Expedition to the East African Lakes, 1930–1: The fish of Lake Rudolf and Lake Baringo: Journal of the Linnean Society (Zoology), v. 39, p. 353–389.

Ortega-Ramírez, J., J. Urrutia-Fucugauchi, and A. Valiente-Banuet, 2000, The Laguna de Babícora Basin: a late Quaternary paleolake in northwestern Mexico, *in* E. H. Gierlowski-Kordesch and K. R. Kelts, eds., Lake basins through space and time: AAPG Studies in Geology 46, p. 569–580.

Chapter 54

The Laguna de Babícora Basin: A Late Quaternary Paleolake in Northwestern Mexico

José Ortega-Ramírez
Jaime Urrutia-Fucugauchi
Departamento de Geomagnetismo y Exploración Geofísica
Instituto de Geofísica
Universidad Nacional Autonoma de México
Mexico City, México

Alfonso Valiente-Banuet
Departamento de Ecología Funcional y Aplicada
Instituto de Ecología
Universidad Nacional Autónoma de México
Mexico City, México

INTRODUCTION

The Laguna de Babícora lies in the subtropical high pressure climatic belt in the western part of the state of Chihuahua, northern Mexico (29°15´–29°30´N; 107°40´–108°00´W) (Figure 1). Atmospheric air mass circulation in the summer is dominated by air fluxes from the east, whereas during the winter, air fluxes move from the west and are associated with high atmosphere jet streams (Bryson and Lowry, 1955; Hales, 1974; Neilson, 1986). Annual mean precipitation is 450 mm and average annual temperature is 11.3°C. According to the Köppen classification C(E) (w1) (x´), climate is characterized by dry and cold winters (3.5°C mean temperature) and hot and humid summers (20.1°C mean temperature). Vegetation is characterized by arid tropical scrub below 2200 m and above this altitude, it is dominated by woodlands composed of *Juniperus, Quercus, Pinus cembroides,* and *Pinus oocarpa* with chaparral vegetation composed predominantly of *Arctostaphylos pungens.*

The Laguna de Babícora occupies a graben of approximately 1896 km^2 at 2100 m above sea level. This plain is surrounded by mountains ranging in altitude from 2500 m to approximately 3000 m above sea level (Figure 1B). These mountains contain Miocene–Pliocene rhyolitic rocks at their base interspersed with andesites and ignimbrites, which are capped by Pliocene–Pleistocene rhyolitic tuffs and basaltic lava flows. Previous studies (Ortega-Ramírez, 1990, 1995; Urrutia–Fucugauchi et al., 1997; Ortega-Ramírez et al., 1998) have shown that during the Quaternary, especially in the late Pleistocene and Holocene, various sedimentary environments prevailed in the region, such as lacustrine, bog, alluvial, fluvial, and eolian systems.

Origin of the Laguna de Babícora Basin is related to early Tertiary compressional and late Tertiary extensional tectonics, similar to basins in the Basin and Range province of the southeastern United States (Brand, 1937; Bryan, 1938; Urrutia-Fucugauchi, 1986; Morrison, 1991). The province containing the Laguna de Babícora in northern Mexico is characterized by a series of narrow, north-south oriented, steep mountain ranges. Within these, there are a large number of tectonic endorheic basins, which during the Pleistocene contained ephemeral lakes known in Mexico as "lagos-playas," "barriales," or "bolsones." The Laguna de Babícora occupies the southern part of this large province, within the subprovince named "Babícora-Bustillos" (Hawley, 1975).

Chronostratigraphic sequences of late Quaternary sediments within the Laguna de Babicora Basin, based on four stratigraphic profile descriptions selected according to geomorphologic, geologic, tectonic, and sedimentary criteria, are represented in Figure 2. These profiles are Las Varas in the northwestern, 318 cm depth; La Pinta in the southeast, 419 cm depth; El Diablo in the south, 243 cm depth; and El Cano in the east, 298 cm depth. Chronostratigraphic control is based on 11 conventional radiocarbon dates obtained from the profiles that provide a chronologic framework for the

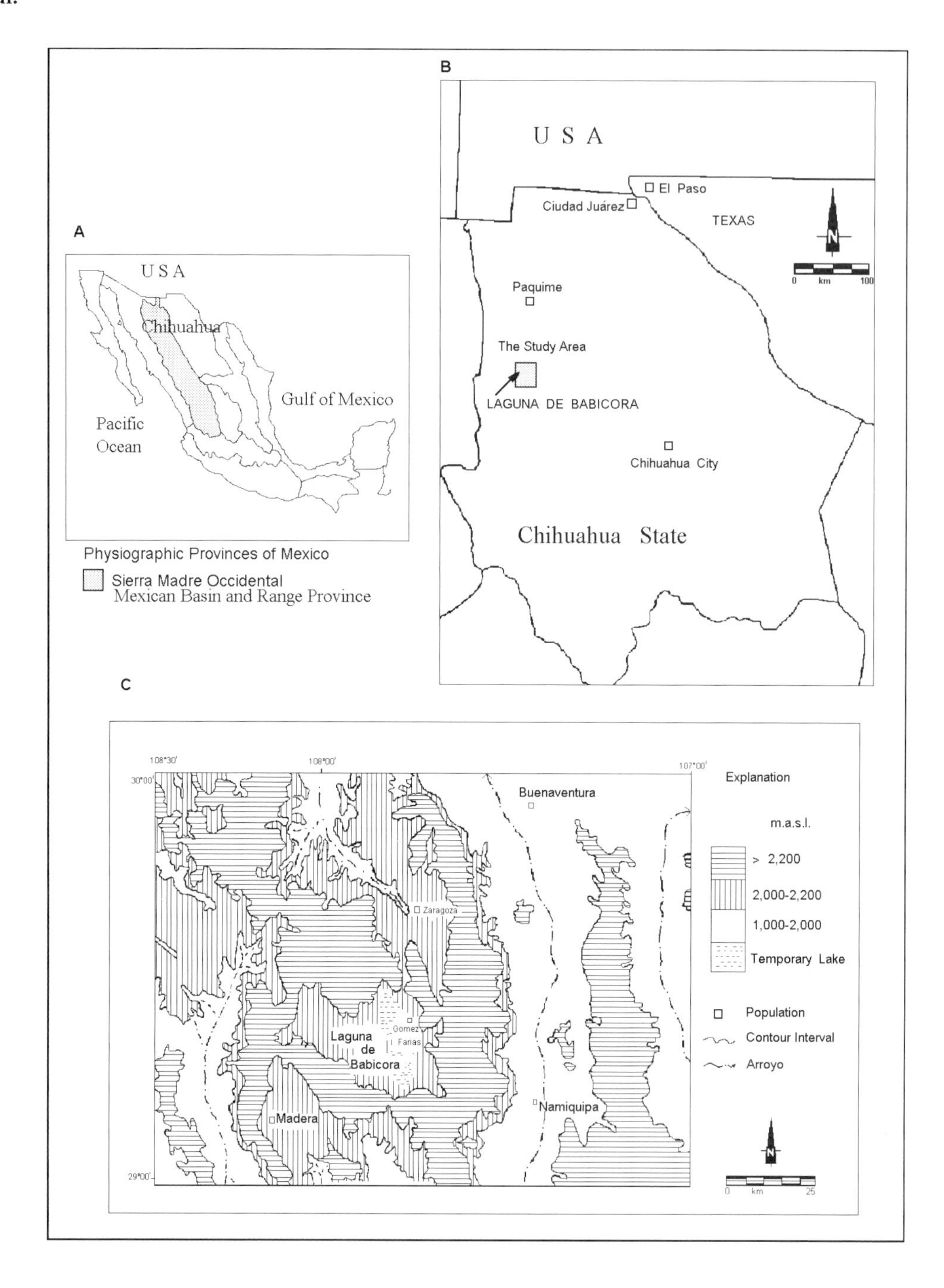

Figure 1—(A and B) Regional maps of Mexico and USA showing the location of the State of Chihuahua and the study area. (C) The Laguna de Babícora Basin and its hypsometric features.

available sedimentary data. All dates are given in ^{14}C-years BP (Table 1).

A total of 74 sediment samples from the profiles were collected and used for particle size analyses, which consisted of determining textural classes (six granulometric classes at intervals of 0.5 ϕ for sandy fractions between 1.25 to 0.0625 mm). An analytic method, including laser diffraction size, was employed to analyze the sediment fraction below 0.0625 mm. Loss-on-ignition (LOI) analysis was also conducted to determine the percentage of organic material. Sample color description using the Munsell Soil Color Chart (1975) and sedimentary analysis using the statistical four moment method (Griffiths, 1967) were applied to obtain a generalized picture of the nature of the deposits. Lastly, a principal components analysis was applied to sedimentary samples from Las Varas and La Pinta profiles, using as input the percentage of three variables: saltation, suspension, and traction load (Ortega-Ramírez, 1995). The discrimination index proposed by Greenwood (1969) was used to define the input values obtained by the four moment method of the upper part of the El Diablo profile (Ortega-Ramírez et al., 1998).

SEDIMENTARY FACIES DESCRIPTION AND INTERPRETATION

Sedimentary classification is based on primary structures, sediment grain sizes, and to some extent sorting.

Table 1. Radiocarbon dates of the profiles measured in the Laguna de Babícora basin of Mexico.

Profile N°	Depth in cm	Material dated	Date in yr B.P.	Laboratory
Las Varas	100	charcoal	3,305 ± 60	INAH-807
Las Varas	200	charcoal	3,840 ± 45	INAH-809
La Pinta	121	paleosol	2,890 ± 6	INAH-814
La Pinta	395	charcoal	8,070 ± 90	INAH-813
La Pinta	410	charcoal	8,120 ± 105	INAH-810
El Diablo	59	charcoal	4,346 ± 105	INAH-1215
El Diablo	100	humic material	10,976 ± 115	INAH-1214
El Diablo	166	humic material	16,342 ± 201	INAH-1212
Cano	43	humic material	3,503 ± 101	INAH-1211
Cano	78	humic material	7,965 ± 109	INAH-1210
Cano	168	humic material	9,614 ± 130	INAH-1209

Las Varas profile is 318 cm thick (Figure 3). Three major sedimentary units were recognized, from the oldest to the youngest, Unit III, Unit II, and Unit I. Unit III (318–199 cm depth) is composed of sand-silt interlayers. The base (63 cm thick) is characterized by an intercalated coarse and fine sand exhibiting planar lamination in which the laminae contacts are dominantly conformable and the sand is moderately sorted. These deposits are overlain by a 56-cm-thick layer of parallel-laminated siliclastic silts that are moderately sorted and contain rhizoconcretions and phytoliths. Sedimentary structures, cumulative curves of grain-size distribution, and the paleobiologic determination lead to an interpretation that the coarse- and fine-grained sands are load material, whereas the silty deposits are suspended load related to a turbulent flow (Ortega-Ramírez, 1990). High organic matter concentration (LOI), vertically oriented dissolution pipes (ranging between 5 mm and 3 cm in length) at the top of this unit, and the lack of unconformities suggest that after fluvial deposition, a marsh environment was established. The contact between Unit III and Unit II is depositional in nature.

Unit II (199–95 cm depth) consists of the following sequence from the base to the top: (1) 13 cm of massive, moderately sorted, fine sands; (2) 70 cm of laminated coarse silt containing rhizoconcretions, fragments of plants, mudcracks (in upper portion), and diatom frustules; and (3) 21 cm of massive, moderately sorted, fine sands containing fragments of organic matter. In applying the suite of sedimentary textures and structures to the C-M diagrams of Passega and Byramjee (1969), a single shallowing-upward cycle is interpreted. Evidence of subaerial exposure suggests that a shallow perennial lake or probably bog paleoenvironment continued through deposition of at least part of the upper part of the Unit II succession. The contact between Units II and I is unconformable characterized by a 0.67-cm-thick ferruginous layer .

Unit I (95–0 cm depth) contains poorly sorted coarse sands at the base, which grade upward into fine and very fine sands, with local iron oxide concentrations. Analysis of log-probability plots of grain-size distribution (Ortega-Ramírez, 1990) provide insight into depositional processes producing Unit I. From the base to the top, grain population is interpreted as transported by rolling and saltation to saltation and finally by suspension, typical of alluvial sheetflood deposits.

La Pinta profile is 419 cm thick (Figure 4). It is composed of three different sedimentary units, units III, II, and I. Unit III (419–344 cm depth) is composed, at the base, of moderately sorted, coarse silts characterized by planar stratification, locally containing abundant debris of plants, phytoliths, and diatom frustules. The plane-parallel beds range from 10 to 23 cm in thickness and show sharp contacts. These silty layers are overlain conformably by a 28-cm-thick, poorly sorted, very fine sand with a massive texture. Using flow regime theory and measured sedimentologic variables (Ortega-Ramírez, 1990), the lower silts are interpreted as graded suspension deposits and the upper sands as suspension settling from turbulent flows with very weak tractive transport. Lack of evidence for subaerial exposure of the siliciclastics suggests that marsh paleoconditions with fluvial influences prevailed during deposition. The transition between Unit III and Unit II is marked by a gradational contact.

Unit II (344–114 cm depth) contains sand-mud interlayers that comprise, from base to top (1) massive, poorly sorted, coarse to medium sand (20 cm thick); (2) laminated siliciclastic silts (50 cm thick) with moderate sorting, abundant syndepositional dissolution surfaces, and laminae thickness always less than 15 cm; (3) massive, moderately sorted, very fine sand (70 cm thick) containing vertical dissolution pipes (5 mm to 5 cm in length); and (4) massive, moderately sorted, siliciclastic silts and clays (90 cm thick), containing mudcracks and cutan concentrations. The lower sands are interpreted as fluvial deposits (Ortega-Ramírez, 1990), and evidence for subaerial exposures and pedogenesis suggests that shallow perennial lake or bog paleoconditions prevailed during deposition of the upper muds, followed by soil formation (Bt horizon). The transition with the overlying Unit I is marked by an erosional surface.

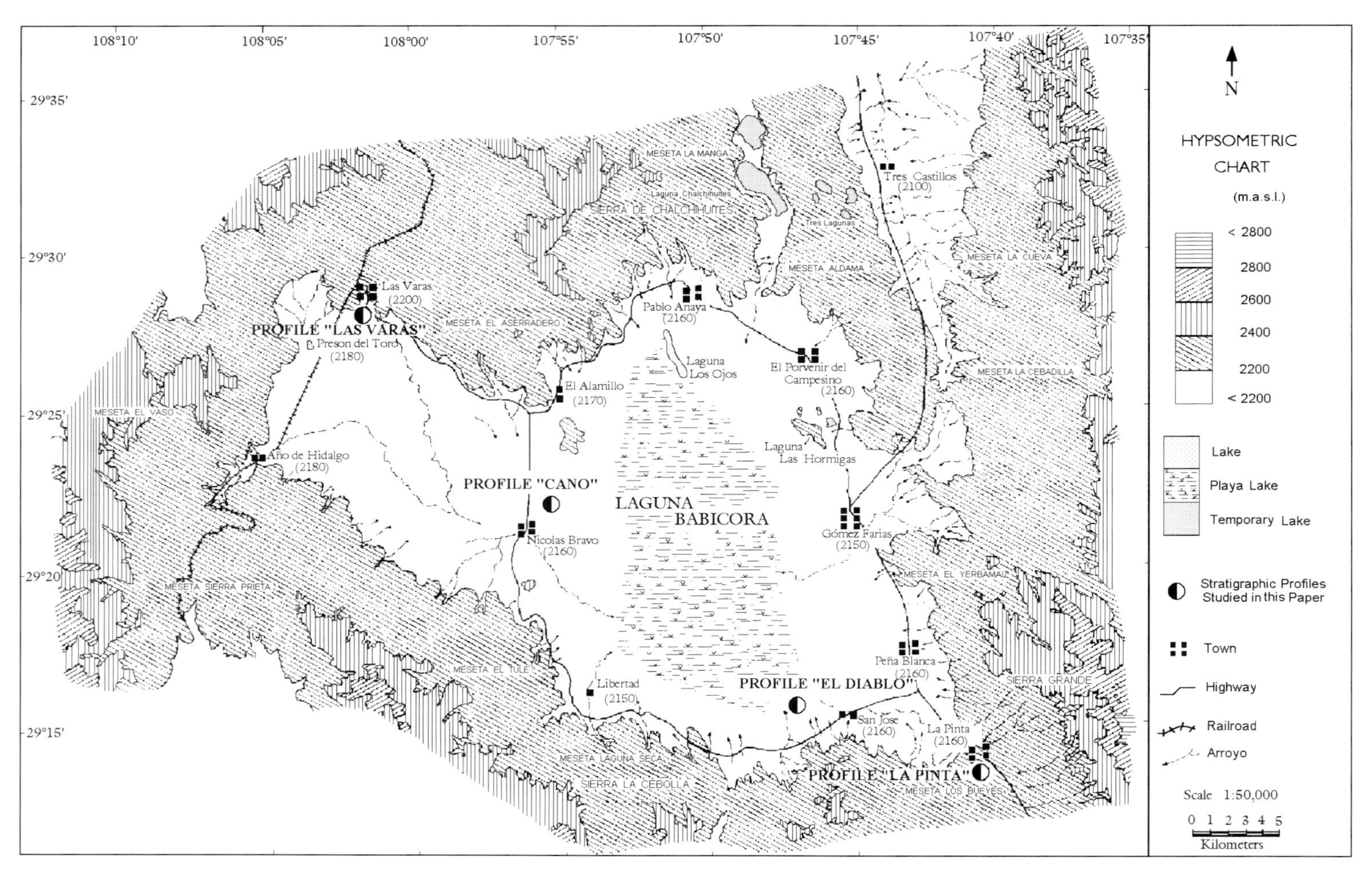

Figure 2—Detailed map of the the Laguna de Babícora basin, Mexico, showing elevations, present-day environments, and the location of the stratigraphic profiles.

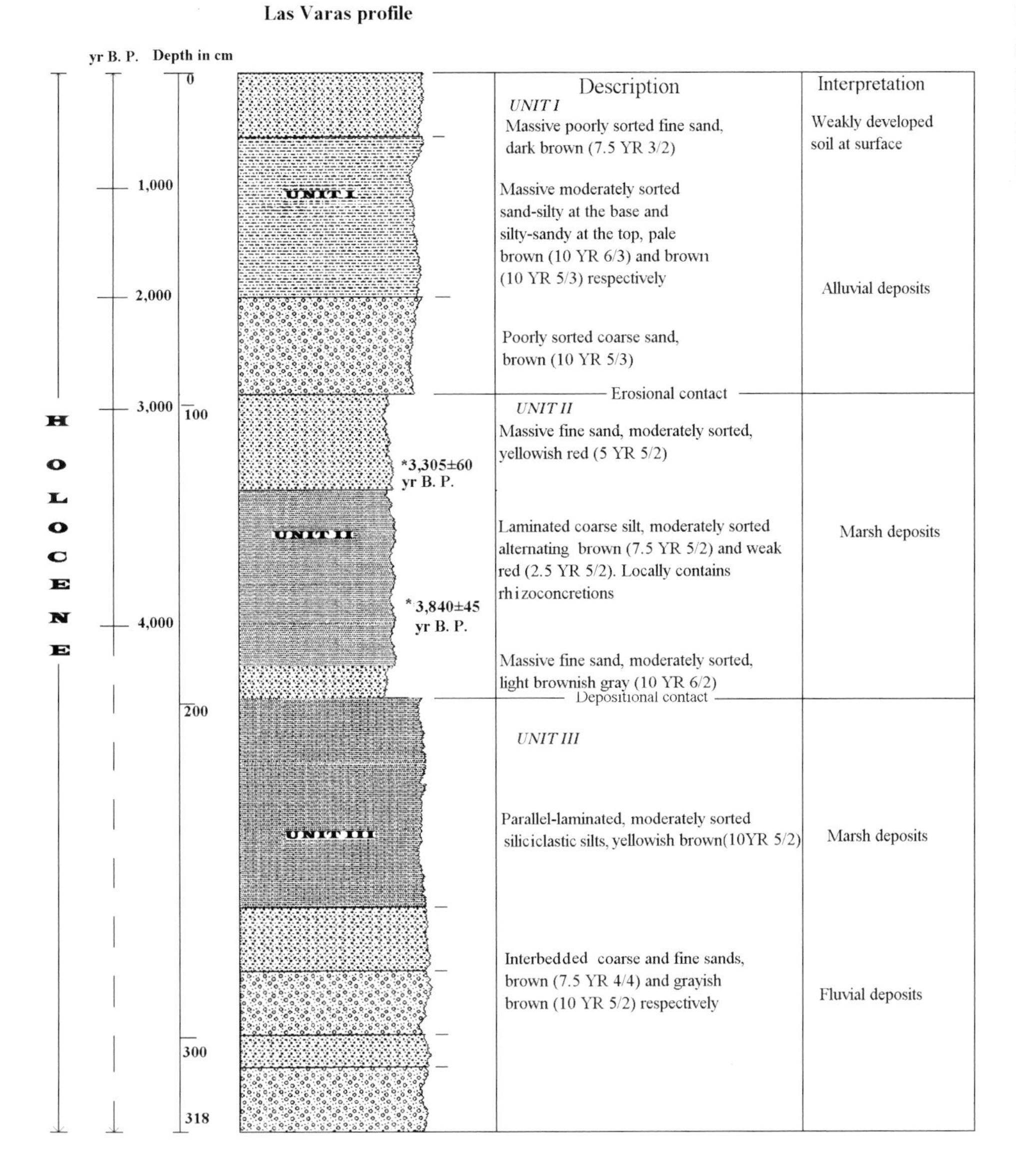

Figure 3—Stratigraphic section of the Las Varas profile in the Laguna de Babícora Basin of Mexico, including ^{14}C ages. Exact location shown in Figure 2.

Unit I (114–0 cm depth) is massive although roughly graded, ranging from very poorly sorted coarse sands to moderately sorted fine sands at the top. Although normal grading has been reported from deposits of muddy debris flows (Vallance and Scott, 1997), the exact mode of deposition may be related perhaps to grain flow (Lowe, 1976).

El Diablo profile is 243 thick (Figure 5), with four sedimentary units identified from Unit IV to Unit I. Unit IV (243–166 cm) is composed of moderately to well-sorted, sand-mud interlayers, characterized by plane-parallel lamination containing alternating fine sand and siliciclastic clayey silt layers ranging in thickness from 7 to 7.4 cm, respectively. High concentrations of Mg^{+} and LOI are associated with the clayey-silty beds, whereas Ca^{++} is mostly found in the fine sands granulometric fraction. The calcium curve shows four distinct peaks at 219, 218, 192, and 176 cm depth, whereas the Mg/Ca ratio curve is slightly higher in the basal unit and at 200 cm depth (Ortega-Ramírez et al., 1998). No evidence for subaerial exposure suggests that permanent lacustrine paleoconditions prevailed during deposition. The transition between this basal unit and overlying Unit III (166–154 cm depth) is gradational marked by a change in grain sizes.

Unit III (154–135 cm depth) consists of well-sorted, medium- to fine-grained, sandy sediments characterized by horizontal parallel stratification. Thin (3 cm) ostracode valve streaks are common at the top of Unit III. Mg and LOI decrease, while Ca^{++} increases significantly. The increase in calcium could be related to an influx of clastics from the catchment area (Burden et al., 1985). Alternately, the low abundance of organic matter in this unit might be the result of the formation of authigenic carbonates (Hayes et al., 1958). Lastly, the thin ostracode-rich layer suggests an alkali-enriched water, similar to categories 2 or 3 established by Forester (1986); consequently, the origin of Unit III can be interpreted as lacustrine with fluvial influence. The contact between Unit III and Unit II (135–122 cm depth) is conformable.

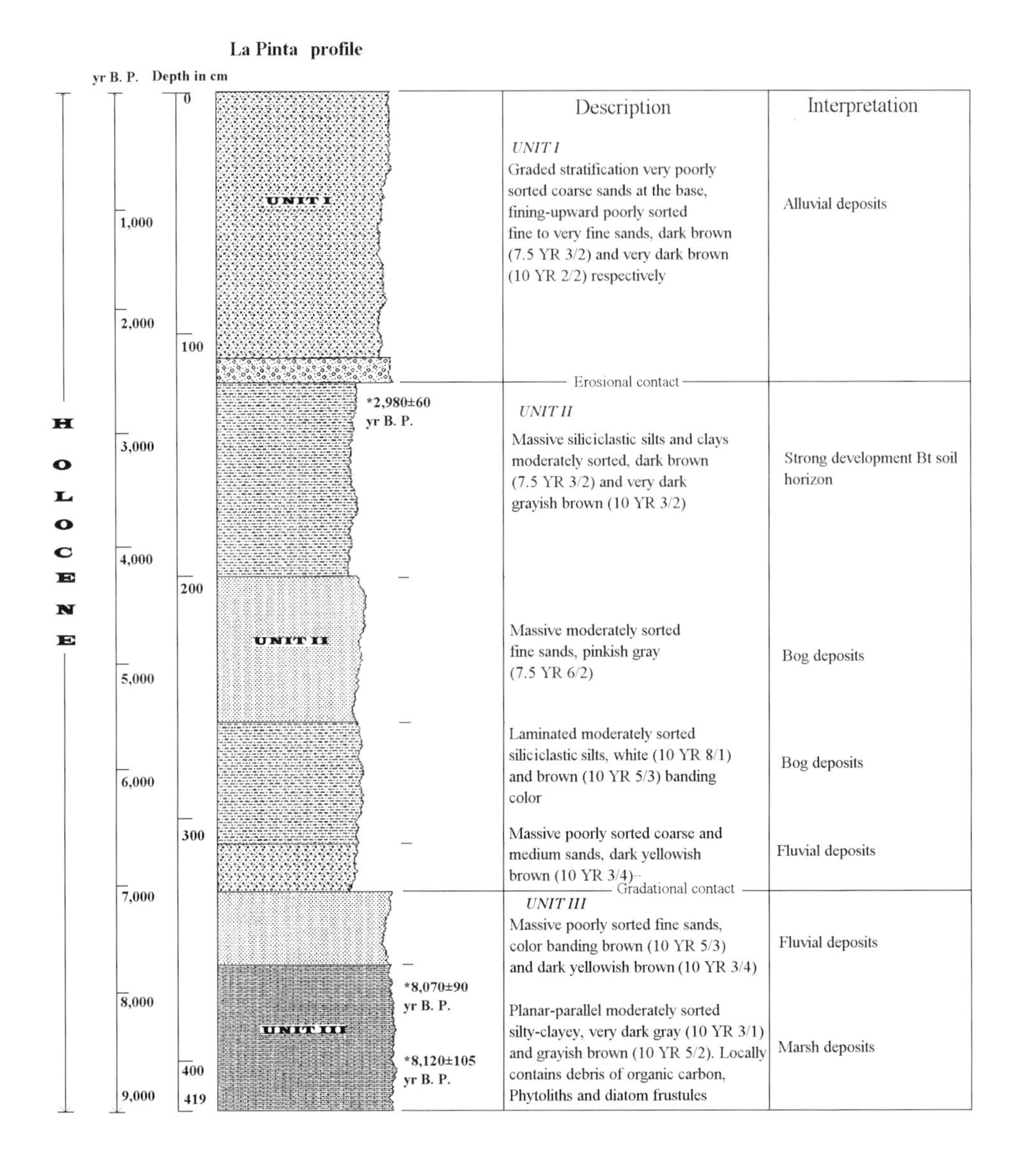

Figure 4—Stratigraphic section of the La Pinta profile in the Laguna de Babícora Basin of Mexico. Exact location shown in Figure 2.

Unit II (122–100 cm depth) is composed of thin parallel laminae up to 0.5 cm thick of fine sands and siliciclastic silts and clays. A significant increment of organic matter and Mg^{++} values, the dominance of Mg/Ca values over Ca values, and the undisrupted nature of the muds suggest that these are deposits of a permanent shallow perennial lake (Ortega-Ramírez et al., 1998). The transition between Unit II and Unit I (100–92 cm depth) is marked by a gradual increase of the siliciclastic fraction from silty-clayey-sandy to silty-sandy-clayey.

Unit I (92–0 cm depth) is divided into two subunits: Subunit I.1 and Subunit I.2. Subunit I.1 (92–59 cm depth) contains silty-sandy-clayey sediments with a massive texture. This deposit exhibits vertically oriented dissolution pipes ranging between 5 mm and 5 cm in length. Geochemical analysis documents high concentrations of Ca^{++} and Mg^{+} (Ortega-Ramírez et al., 1998), indicating an early brine evolution (Eugster and Hardie, 1978). Such high concentrations of Ca^{++} are attributed to the large number of ostracod and gastropod shell fragments and calcium carbonate cement in the sandy component; moreover, computing a linear discriminant function using as variables the moment measures in agreement with the discrimination index of Greenwood (1969), indicating that the sediments have a mean R value of –21.61 as estimated for all analyzed samples of this subunit, and an eolian origin. In summary, this unit is interpreted as bog to lake sediments that have dried out and been subjected to eolian processes in a dominantly arid environment (Ortega-Ramírez, 1995). These results point to an effective humidity reduction. Subunit I.2 (59–0 cm depth) gradually grades from a silty-sandy-clayey to sandy-silty-clayey sediment with a columnar structure. Calcium decreases, and magnesium and organic matter content increase. These characteristics suggest a bog paleoenvironment, similar to today's dry regime.

The Cano profile, 298 cm thick (Figure 6), consists of three sedimentary units, III–I. Unit III (298–293 cm depth) is composed of well-sorted, normally graded, gravel and sandy sediments, typical of traction fluvial deposits. The contact between Unit III and Unit II (293–290 cm depth) is gradational, marked by a sharp decrease of particle sizes from sandy to sandy-silty-clayey.

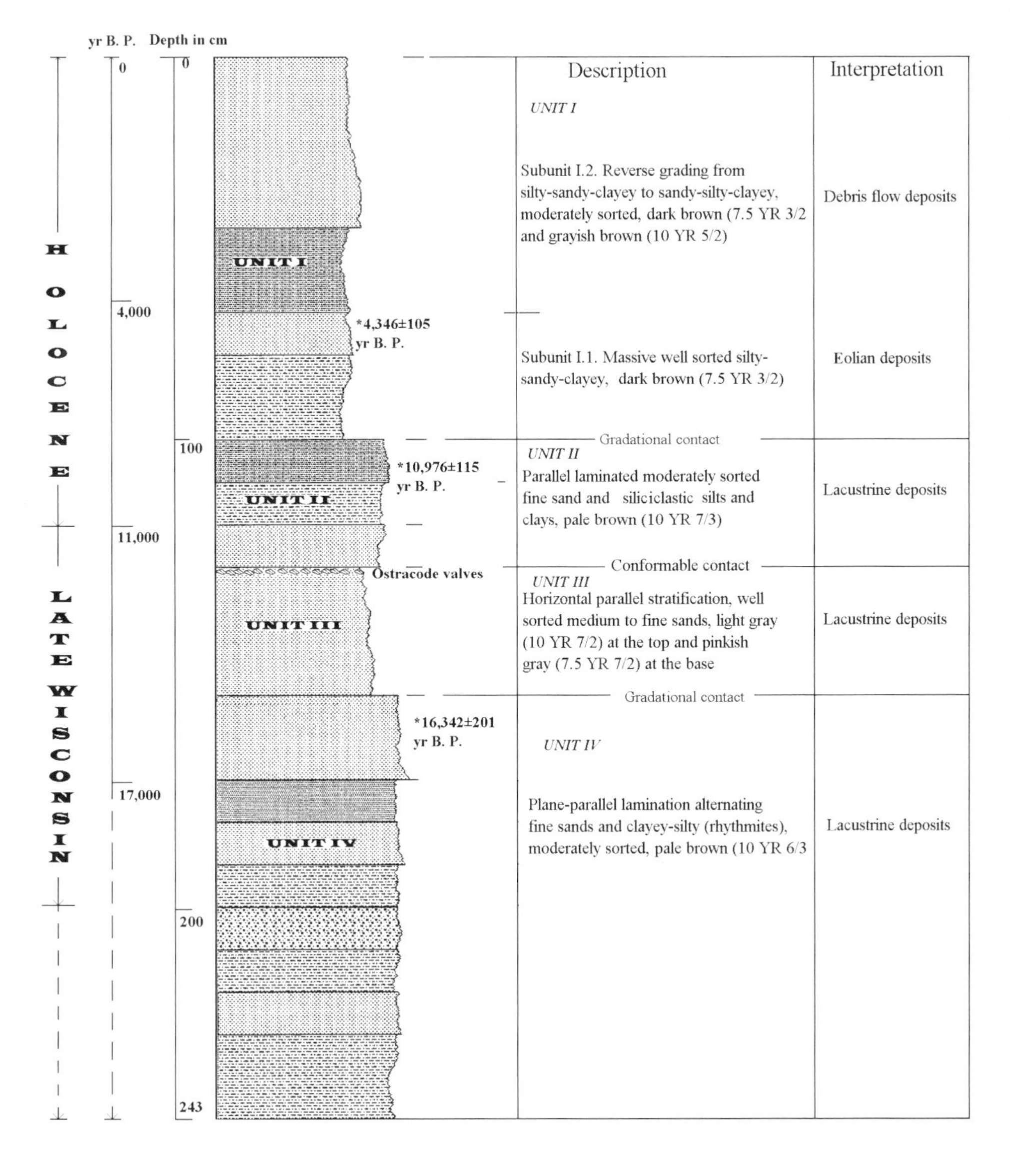

Figure 5—Stratigraphic section of the El Diablo profile in the Laguna de Babícora Basin of Mexico. Exact location shown in Figure 2.

Unit II (290–168 cm depth) is divided into three subunits (II.1, II.2, and II.3) according to facies relationships. Subunit II.1 (290–269 cm depth) contains at its base a 5-cm-thick lense of ostracod shells followed by a graded layer from sandy to silty and silty-clayey grains at the top. Normal grading here is interpreted as settling of turbulent suspension in a lacustrine paleoenvironment. The transition between subunits II.1 and II.2 (269–264 cm depth) is gradational with change in texture from silty-clayey-sandy to sandy-silty grains. Subunit II.2 (264–220 cm depth) is made of undisrupted laminated sandy and sandy-silty layers. No exposure features suggest that permanent lacustrine paleoconditions existed during deposition. The contact between Subunit II.2 and overlying Subunit II.3 (220–216 cm depth) is defined by a change from sandy-silty-clayey to silty-clayey-sandy grains. Subunit II.3 (216–168 cm depth) is composed of silty-clayey and sandy-silty sediments with a massive texture and dissolution cavities. These dissolution voids are vertically elongated pipes 5 mm to 5 cm in length that crosscut the deposits and are filled with muddy siliciclastic clays. These postdepositional features are interpreted as disruption by mudcracks through desiccation. The transition between Subunit II.3 and Unit I is depositional, marked by a grain size change from sandy-silty-clayey to clayey-silty.

Unit I (168–0 cm depth) is mainly composed of clayey-silty sediments with a massive texture and an abundance of organic matter, as exhibited by the predominance of black colors (10 YR 2/1).

PALEOENVIRONMENTAL INTERPRETATION

We compared our paleoenvironmental interpretation for the Laguna de Babícora with the chronostratigraphic divisions proposed for the American Southwest: late Wisconsinan 18–11 Ka, early Holocene 11–8.9 Ka, middle Holocene 8.9–4 Ka, and late Holocene

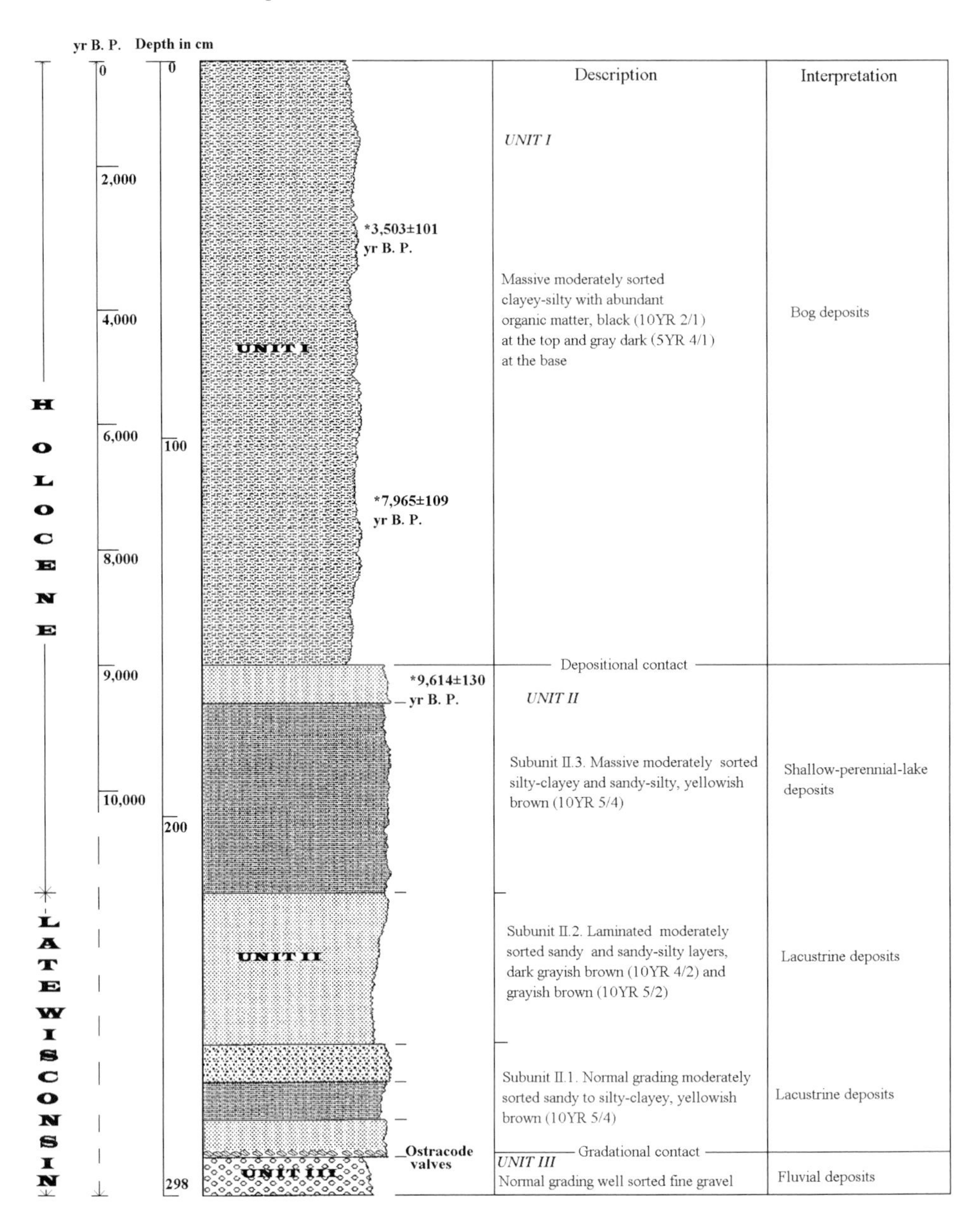

Figure 6—Stratigraphic section of the Cano profile in the Laguna de Babícora Basin of Mexico. Exact location shown in Figure 2.

4–0 Ka (Van Devender and Spaulding, 1979; Van Devender et al., 1987), based on Packrat middens studies (Van Devender and Spaulding 1979; Van Devender et al., 1987). These divisions are similar to those established by INQUA (Morrison, 1969) and used for climate simulation models (COHMAP, 1988; CLIMAP, 1976; Kutzbach and Street-Perrott, 1985). Chronologic and paleoenvironmental information reported for the neighboring areas in the Chihuahuan Desert are compared with our results and used to evaluate the late Quaternary paleoenvironments in the Laguna de Babícora. These results are schematically summarized in Figure 7.

Late Wisconsinan (~18–11 Ka)

Paleoenvironmental reconstructions based on stratigraphic, sedimentary, and radiocarbon data (one sample) from the El Diablo profile Unit IV (16 342 ±201 yr BP) indicate greater effective moisture than at present day; that is, the development of a large lake. This is supported by the sedimentary facies of units IV and III of the El Diablo sequence, units III and subunits II.1 and II.2 of the Cano profile, and the rock-magnetic properties from the lowest section of both profiles, as described by Urrutia-Fucugauchi et al. (1997). These results correlate with the paleoclimatic conditions reported for the northern sector of the Chihuahuan Desert (Van Devender, 1990a; Thompson et al., 1993), as well as in the more southerly extension of the Bolson de Mapimí (26°N), México (Van Devender and Burgess, 1985). Small differences may result from contrasts between both margins of the North Atlantic and North Pacific oceans and the Laurentian continental glacier (CLIMAP, 1981; COHMAP, 1988; Manabe and

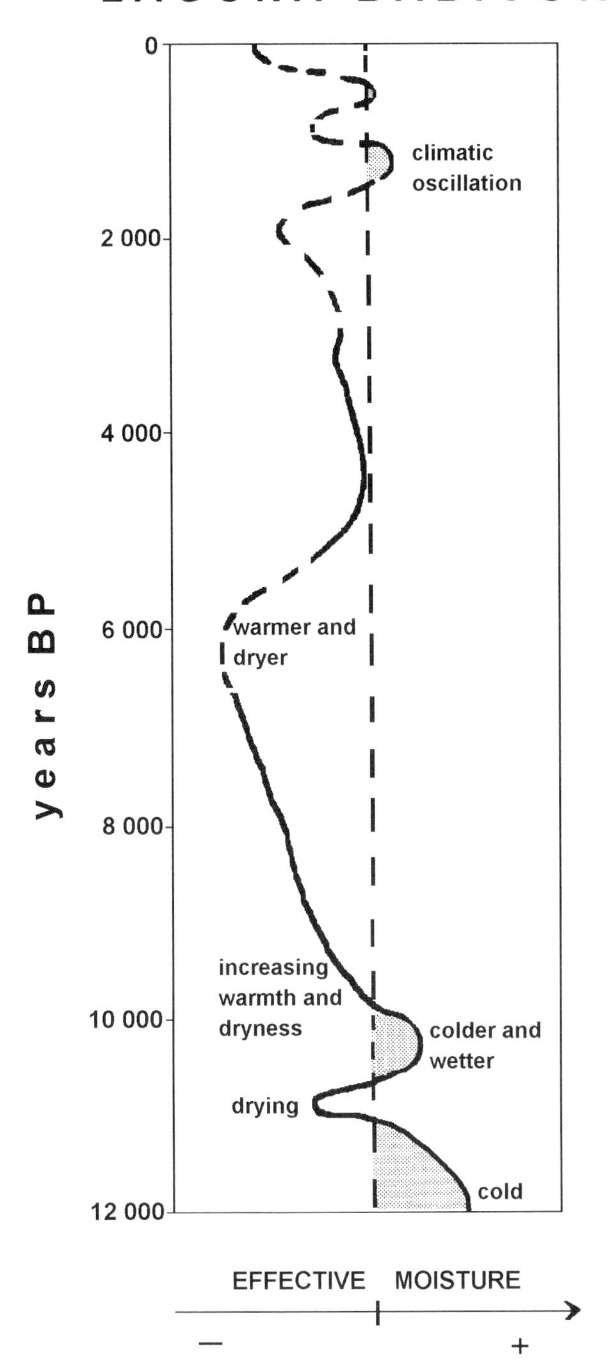

Figure 7—Schematic summary of the interpretation of the climatic changes from 12,000 yr BP until present as derived from the described cores of Figures 3–6. Summary interpretation: (1) a general decrease of effective moisture before 11 Ka, (2) a brief dry period centered on 11 Ka, (3) return of wetter and probably colder conditions between 11 and 10 Ka, and (4) a period of decreased effective moisture.

Broccoli, 1985; Kutzbach et al., 1993) and equatorial migration of the main climatic belts. Large changes in water balance observed in other pluvial lakes in western North America (Allen and Anderson, 1993) have been attributed to changes in precipitation and evaporation that were associated with changes in the position of the polar-front jet stream (Hostetler and Benson, 1990). This timing suggests that the major changes in climate recorded at Laguna de Babícora were part of a global as well as a regional response.

Early Holocene (11–8.9 Ka)

Sedimentary characteristics of Unit II at El Diablo and Subunit II.3 at Cano, as well as their geochemistry, depositional features, and organic matter contents, all indicate shallow perennial lakes. This period is represented by two dates 10976 ±115 yr BP and 9614 ±130 yr BP in the El Diablo and Cano profiles, respectively, indicating that generally moist conditions prevailed into the early Holocene. Similar climatic conditions are also inferred from the archaeological data for the Folsom Period in the American Southwest, which indicates a cold interval and increased effective moisture around 10.5–10 Ka (Irwin-Williams and Haynes, 1970). This period is marked by the southerly migration of prehistoric groups about 11 Ka. An occupation site has been found near Los Medanos de Samalayuca north of Laguna Babícora (Phelp, 1990). This probably correlates with conditions suggested for the American Southwest that are related to an increase in the summer solar radiation (Kutzbach and Wright, 1985; Spaulding and Graumlich, 1986; COHMAP, 1988; Spaulding, 1990, 1991), and an enhanced summer precipitation (monsoon type). Alternatively, taking into account the thermal inertia of the oceans and paleoceanographic conditions in the North Atlantic ocean (Broecker et al., 1990), this second pluvial event could be related to the climatic anomaly of the Younger Dryas from 11–10 Ka (Ruddiman and Duplessy, 1985; Berger et al., 1990).

Middle Holocene (8.9–4 Ka: Hypsithermal or Climatic Optimum)

This period can broadly be divided into two zones. The lower first zone, representing conditions during the early stages of the middle Holocene and based on sedimentary features of the lowest part of La Pinta Unit III (8070 ±90 yr BP and 8120 ±105 yr BP) and the middle part of Cano Unit I (7965 ±109 yr BP), corresponds to a period of decreased effective moisture conditions with the development of a shallow-perennial-lake (marsh) and bog environments, respectively. The sedimentary record from the two profiles can be correlated by stratigraphic position with Unit III of the Las Varas profile, in which a high organic matter concentration (LOI) confirms a marsh environment; furthermore, the Cano unit shows a decrease in magnetic minerals (Urrutia-Fucugauchi et al., 1997). The second zone is marked by eolian processes, particularly at the base of Unit I of the El Diablo profile (4346 ±105 yr BP), whereas in the Las Varas (top of Unit III and base of Unit II), La Pinta (basal and middle section of Unit II), and Cano (middle section of Unit I) profiles, the sediments indicate mainly alluvial and bog environments. These data point to an effective humidity reduction (Ortega-Ramírez, 1990, 1995) and are similar to those reported for Laguna Ascencion (30°N, 108°W), located northwest of the studied area, in which desert pavements have been reported (Ortega-Ramírez, unpublished data), indicating deflation processes and warmer temperatures than at present.

Increasing mid-Holocene aridity has also been reported for the northern sector of the Chihuahuan Desert (Elias and Van Devender, 1990; Van Devender et al., 1984; Van Devender, 1977) and its southern margin, in the Bolson de Mapimí (Van Devender and Burgess, 1985). Increased aridity is also temporally correlated with water-level decline in endorheic playa-lake basins in New Mexico, U.S.A. (Markgraf et al., 1984). This paleoclimatic change can be explained in terms of the Earth's tilt and orbital position, in which perihelion occurred in late July and the axial tilt was 24.2° compared with 23.4° today (Kutzbach and Guetter, 1986); therefore, maximum insolation values occurred in this period (Davis, 1984; Berger, 1978; Berger and Pestiaux, 1984), with climatic signals manifesting the effect of the different timing of seasonal maxima. They infer an increase of summer insolation for the northern hemisphere of 2–4°, strong thermal differences between the oceans, and continental shelves mainly in the littoral zones with continental interiors exhibiting deflation processes and eolian deposits. These observations are corroborated by paleoclimatic models (COHMAP, 1988) and paleolacustrine data of the American Southwest (Spaulding and Graumlich, 1986; Oviatt, 1988). In short, the mid-Holocene was characterized by reduced effective humidity and, therefore, by a widespread general aridity that probably caused the end of the paleo-Indian archaeological period in the American Southwest (Cordell, 1984).

Late Holocene (4 Ka–Present)

For the late Holocene, the seasonal insolation anomalies in both hemispheres declined considerably, weakening pressure and circulation patterns (Street-Perrott and Perrott, 1993). Van Devender (1990a) suggested that desertic climatic conditions in the Chihuahuan and Sonoran deserts prevailed from 4000 yr BP to the present; however, short-term climatic anomalies have been detected (Ortega-Ramírez et al., 1998). Thus, in Las Varas Unit II, abundant rhizoconcretions and laminations suggest marsh environments dated as 3870 ±45 yr, 3840 ±60 yr, and 3305 ±60 yr). Also, a paleosol dated as 2890 ±60 yr BP was found at the top of La Pinta Unit II, and in the upper part of Cano Unit I in a bog environment. Finally, in most of the top units of the sections, erosion surfaces and sediments interpreted as debris-flow and alluvial deposits typical of semi-arid environments (Smoot and Lowenstein, 1991) have been identified.

These documented climatic and environmental variations in the Laguna de Babícora basin perhaps can be correlated with global climatic changes during the late Holocene, as reported in paleobotanical, dendrochronologic, paleoceanographic, glaciological, and archaeological data in North America (e.g., Benedict, 1973; La Marche, 1973; Pisias, 1978; Street and Grove, 1979; Cordell, 1984; Grove, 1988; White et al., 1990; Davis, 1992; Hill, 1992).

ACKNOWLEDGMENTS

Economic support for field studies and data acquisition was provided through grants from the National Council of Sciences and Technology (CONACYT 3732-T9302) and the National University of Mexico (Internal project B-409). We greatly acknowledge useful critical comments by the two reviewers and the editors.

REFERENCE CITED

Allen, B. D., and R. Y. Anderson, 1993, Evidence from western North America for rapid shifts in climate during the Last Glacial Maximum. Science, v. 260, p. 1920–1923.

Allen, J. R. L., 1965, Fining-upwards cycles in alluvial successions. Geological Journal, v. 4, p. 229–246.

Berger, A., 1978, Long-term variations of daily insolation and Quaternary climatic changes. Journal of Atmospheric Sciences, v. 35, p. 2362–2367.

Berger, A., and P. Pestiaux, 1984, Accuracy and stability of the Quaternary terrestrial insolation, *in* A. Berger, J. Imbrie, J. Hays, G. Kukla, and B. Saltzman, eds., Milankovitch and Climate. D. Reidel Publications Company, Dordrecht, Holland, p. 83–112.

Berger, A., Th. Fichefet, H. Galle, I. Marsiat, Ch. Tricot, and P. Van Yperselej, 1990, Ice sheet and sea level change as a response to climatic change at the astronomical time scale, *in* R. Paepe, ed., Greenhouse Effect, Sea Level and Drought. Kluwer Academic Publishers, Dordrecht, Netherlands, p. 85–107.

Brand, D. C., 1937, The Natural Landscape of Northwestern Chihuahua. Univ. New Mexico Press, Albuquerque, New Mexico, 74 p.

Broecker, W. S., T. H. Peng, J. Jouzel, and G. Roussel, 1990, The magnitude of global fresh water transports of importance to ocean circulation. Climate Dynamics, v. 4, p. 73–79.

Bryan, K., 1938, Geology and groundwater conditions of the Rio Grande depression in Colorado, New Mexico, and Texas. National Resources Commission, Regional Planning, Washington, D. C., Part 6, p. 196–225.

Benedict, J. B., 1973, Chronology of cirque glaciation, Colorado Front Range. Quaternary Research, v. 3, p. 584–599.

Bryson, R. A., and W. P. Lowry, 1955, Synoptic climatology of the Arizona summer precipitation singularity. Bulletin of the American Meteorological Society, v. 36, p. 329–339.

Burden, E., G. Norris, and J. H. McAndrews, 1985, Geochemical indicators in lake sediments of upland erosion caused by Indian and European farming, Awenda Provincial Park, Ontario. Canadian Journal of Earth Sciences, v. 23, p. 55–65.

CLIMAP Project Members, 1976, The surface of the ice-age earth. Science, v. 191, p. 1131–1137.

CLIMAP Project Members, 1981, Seasonal reconstructions of the earth´s surface at the last glacial maximum. Geological Society of America Map and Chart Series, MC-36.

COHMAP Members, 1988, Major climatic changes of the last 18,000 years: observation and model simulations. Science, v. 241, p. 1043–1052.

Cordell, L. S., 1984, Prehistory of the Southwest. Academic Press, Orlando, Florida, 409 p.

Davis, O. K., 1984, Multiple thermal maxima during the Holocene. Science, v. 255, p. 617–619.

Davis, O. K., 1992, Rapid climatic change in coastal southern California inferred from pollen analysis of San Joaquín Marsh. Quaternary Research, v. 37, p. 89–100.

Davis, O. K., and D. S. Selters, 1987, Contrasting climatic histories for western North America during the Early Holocene. Current Research in the Pleistocene, v. 4, p. 87–89.

Davis, O. K., and D. S. Shafer, 1992, A Holocene climatic record for the Sonoran Desert from pollen analysis of Moctezuma well, Arizona, USA. Palaeogeography, Palaeoclimatology, Palaeoecology, v. 92, p. 107–119.

Elias, S. A., and T. R. Van Devender, 1990, Fossil insect evidence for late Quaternary climatic change in the Big Bend region, Chihuahuan Desert, Texas. Quaternary Research, v. 34, p. 249–261.

Eugster, H. P., and L. A. Hardie, 1978, Saline lakes, *in* A. L. Lerman., ed., Lakes: Chemistry, Geology, Physics. Springer-Verlag, Berlin, p. 237–294.

Forester, R. M., 1986, Determination of the dissolved anion composition of ancient lakes from fossil ostracodes. Geology, v. 14, p. 796–798.

Greenwood, B., 1969, Sediment parameters and environmental discrimination: an application of multivariate statistics. Canadian Journal of Earth Sciences, v. 6, p. 1347–1357.

Griffiths, J. C., 1967, Scientific Method on Analysis of Sediments. McGraw-Hill Company, New York, 508 p.

Grove, J. M., 1988, The Little Ice Age. Routledge, London and New York, 498 p.

Hales, J. R. E., Jr., 1954, Southwestern United States summer monsoon source—Gulf of Mexico or Pacific Ocean. Journal of Applied Meteorology, v. 13, p. 331–342.

Hawley, J. W., 1975, Notes on geomorphology and the late Cenozoic geology of northwestern Chihuahua. New Mexico Geological Society, 20th Annual Field Conference Guidebook, p. 131–142.

Hayes, F. R., B. L. Reid, and M. L. Cameron, 1958, Lake water and sediment: II oxidation-reduction relations at the mud-water interface. Limnology and Oceanography, v. 3, p. 308–317.

Hill, W. D., 1992, Chronology of the Zurdo site, Chihuahua. Masters Thesis, University of Calgary, Canada, 135 p.

Hostetler, S., and L. E. Benson, 1990, Paleoclimatic implications of the high stand of Lake Lahontan derived from models of evaporation and lake level. Climate Dynamics, v. 4, p. 207–217.

Irwin-Williams, C., and V. Haynes, 1970, Climatic change and early population dynamics in the southwestern United States. Quaternary Research, v. 1, p. 59–71.

Kutzbach, J. E., and F. A. Street-Perrott, 1985, Milankovitch forcing of the fluctuation in the level of tropical lakes from 18 to 0 k yr B.P. Nature, v. 317, p. 130–134.

Kutzbach, J. E., and H. I. Wright, Jr., 1985, Simulation of the climate of 18,000 yr B.P.: Results for the North American/North Atlantic/European sector and comparison with the geologic record of North America. Quaternary Science Reviews, v. 4, p. 147–187.

Kutzbach, J. E., and P. J. Guetter, 1986, The influence of changing orbital parameters and surface boundary condition on climate simulations for the past 18,000 years. Journal of the Atmospheric Sciences, v. 43, p. 1726–1759.

Kutzbach, J. E., P. J. Guetter, P. J. Behling, and R. Selen, 1993, Simulated climatic changes: results of the COHMAP climate-model experiments, *in* H. E. Wright, Jr., J. E. Kutzbach, T. Webb III, W. F. Ruddiman, F. A. Street-Perrott, and P. J. Bartlein, eds., Global climates since the Last Glacial Maximum. University of Minnesota Press, Minneapolis, p. 24–93.

La Marche, V. C., Jr., 1973, Holocene climatic variations inferred from treeline fluctuations in the White Mountains, California. Quaternary Research, v. 3, p. 632–660.

Lowe, D. R., 1976, Grain flow and grain flow deposits. Journal of Sedimentary Petrology, v. 22, p. 157–204.

Manabe, S., and A. J. Broccoli, 1985, The influence of continental ice sheets on the climate of an ice age. Journal of Geophysical Research, v. 90, p. 2167–2190.

Markgraf, V., J. P. Bradbury, R. M. Forester, G. Singh, and R. S. Sternberg, 1984, San Augustin Plains, New Mexico: age and paleoenvironmental potential reassessed. Quaternary Research, v. 22, p. 336–343.

Morrison, R. B., 1969, The Pleistocene–Holocene boundary: an evaluation of the various criteria used for determining it on a provincial basis, and suggestions for establishing it world-wide. Geologie en Mijnbouw, v. 48, p. 363–371.

Morrison, R. B., 1991, Quaternary geology of the southern Basin and Range, *in* R. B. Morrison, ed., Quaternary Nonglacial Geology Conterminous, U. S. The Geological Society of America, The Geology of North America, v. 2, p. 353–371.

Munsell Soil Color Chart, 1975. Kollmorgen Corporation, Baltimore.

Nielson, R. P., 1986, High-resolution climatic analysis and southwest biogeography. Science, v. 232, p. 27–34.

Ortega-Ramírez, J., 1990, Le sommet du remplissage Quaternaire de la Laguna Babícora (Etat de Chihuahua, Nord-Oest du Mexique): reconstitution des paléoenvironnements à partir de la sédimentologie et de la stratigraphie. Thèse de Doctorat, Université Louis Pasteur de Strasbourg, France, 345 p.

Ortega-Ramírez, J., 1995, Los paleoambientes holocénicos de la Laguna Babícora, Chihuahua, México. Geofísica Internacional, v. 34, p. 107–116.

Ortega-Ramírez, J., A. Valiente-Banuet, J. Urrutia-Fucugauchi, C. Mortera-Gutierrez, and G. Alvarado-Valdez, 1998, Paleoclimatic changes during the late Pleistocene-Holocene in Laguna Babícora, near the Chihuahuan Desert, México. Canadian Journal of Earth Sciences, v. 35, p. 1168–1179.

Oviatt, C. G., 1988, Late Pleistocene and Holocene lake fluctuations in the Sevier Lake Basin, Utah, USA. Journal of Palaeolimnology, v. 1, p. 9–21.

Passega, R., and R. Byramjee, 1969. Grain size image of clastic deposits. Sedimentology, v. 13, p. 233–252.

Phelp, A. L., 1990, A Clovis projectile point from the vicinity of Samalayuca, Chihuahua, Mexico. The Artifact, v. 28, p. 49–51.

Pisias, N. G., 1978, Paleoceanography of the Santa Barbara Basin during the last 8000 years. Quaternary Research, v. 10, p. 366–384.

Ruddiman, W. F., and J. C. Duplessy, 1985, Conference of the last deglaciation: timing and mechanism. Quaternary Research, v. 23, p. 1–17.

Smoot, J. P., and T. K. Lowenstein, 1991, Depositional evironments of non-marine evaporites, *in* J. L. Melvin, ed., Evaporites, petroleum, and mineral resources. Elsevier, Amsterdam, p. 189–347.

Spaulding, W. G., and L. J. Graumlich, 1986, The last pluvial climatic episodes in the deserts of southwestern North America. Nature, v. 230, p. 441–444.

Spaulding, W. G., 1990, Vegetational and climatic development of the Mojave Desert: The Last Glacial Maximum to the Present, *in* J. L. Betancourt, T. Van Devender, and P. Martin., eds., Packrat Middens: The Last 40,000 years of Biotic Change. University of Arizona Press, Tucson, p. 167–199.

Spaulding, W. G., 1991, Pluvial climatic episodes in North America and North Africa: types and correlation with global climate. Palaeogeography, Palaeoclimatology, Palaeoecology, v. 84, p. 217–227.

Street, F. A., and A. T. Grove, 1979, Global maps of lake-level fluctuations since 30,000 yrs BP. Quaternary Research, v. 12, p. 83–118.

Street-Perrott, F. A., and R. A. Perrott, 1993, Holocene vegetation, lake levels, and climate of Africa, *in* H. E. Wright, Jr., J. E. Kutzbach, T. Webb III, W. F. Ruddiman, F. A. Street-Perrott, and P. J. Bartlein, eds., Global climates since the Last Glacial Maximum. University of Minnesota Press, Minneapolis, p. 318–356.

Thompson, R. S., C. Whitlock, P. J. Bartlein, S. P. Harrison, and W. G. Spaulding, 1993, Climatic changes in the western United States since 18,000 yr B.P., *in* H. E. Wright, Jr., J. E. Kutzbach, T. Webb III, W. F. Ruddiman, F. A. Street-Perrott, and P. J. Bartlein, eds. Global Climates Since the Last Glacial Maximum. University of Minnesota Press, Minneapolis, p. 468–513.

Urrutia-Fucugauchi, J. 1986, Late Mesozoic-Cenozoic evolution of the northwestern Mexico magmatic arc zone. Geofísica Internacional, v. 1, p. 61–84.

Urrutia-Fucugauchi, J., J. Ortega-Ramírez, and R. Cruz-Gatica, 1997, Rock-magnetic study of late Pleistocene-Holocene sediments from the Babícora lacustrine basin, Chihuahua northern México. Geofísica Internacional, v. 10, p. 77–86.

Vallance, J. W., and K. M. Scott, 1997, The Oseola mudflow from Mount Rainier: sedimentology and hazard implications of a huge clay-rich debris flow. Geological Society of America Bulletin, v. 109, p. 143–163.

Van Devender, T. R., 1977, Holocene woodlands in the southwestern Desert. Science, v. 198, p. 189–192.

Van Devender, T. R., 1990a, Late Quaternary vegetation and climate of the Chihuahuan desert, United States and Mexico, *in* J. L. Betancourt, T. Van Devender, and P. Martin., eds., Packrat Middens: The Last 40,000 years of Biotic Change. University of Arizona Press, Tucson, p. 105–133.

Van Devender, T. R., 1990b, Late Quaternary vegetation and climate of the Sonoran Desert, United States and Mexico, *in* J. L. Betancourt, T. Van Devender, and P. Martin., eds., Packrat Middens: The Last 40,000 years of Biotic Change, University of Arizona Press, Tucson, p. 134–165.

Van Devender, T. R., and W. G. Spaulding, 1979, Development of vegetation and climate in the southwestern United States. Science, v. 204, p. 701–710.

Van Devender, T. R., J. B. Betancourt, and M. Wimberly, 1984, Biogeographic implications of a packrat midden sequence from the Sacramento Mountains, south-central New Mexico. Quaternary Reseach, v. 22, p. 344–360.

Van Devender, T. R., and T. L. Burgess, 1985, Late Pleistocene woodlands in the Bolson de Mapimi: A refugium for the Chihuahuan Desert biota? Quaternary Research, v. 24, p. 346–353.

Van Devender, T. R., R. S. Thompson, and J. L. Betancourt, 1987, Vegetation history of the deserts of the southwest of North America: the nature and timing of the late Wisconsin-Holocene transition, *in* W. F. Ruddiman, and H. E. Wright, Jr., eds., North America and adjacent oceans during the last glaciation. Geological Society of America, The Geology of North America, v. 3, p. 323–352.

White S. E., M. Reyes Cortés, J. Ortega-Ramírez, and S. Valastro Jr., 1990, El Ajusco: geomorfología volcánica y acontecimientos glaciales durante el Pleistoceno Superior y comparación con las series glaciales mexicanas y las de las Montañas Rocallosas. Colección Científica, v. 212, 77 p. (Instituto Nacional de Antropología e Historia.)

Whatley, R. C., and G. C. Cusminsky, 2000, Quaternary lacustrine ostracoda from northern Patagonia: a review, *in* E. H. Gierlowski-Kordesch and K. R. Kelts, eds., Lake basins through space and time: AAPG Studies in Geology 46, p. 581–590.

Chapter 55

Quaternary Lacustrine Ostracoda from Northern Patagonia: A Review

R. C. Whatley
Micropalaeontology Research Group, University of Wales Aberystwyth, UK

G. C. Cusminsky
PROGEBA & Centro Regional Universitario Bariloche San Carlos de Bariloche, Argentina

INTRODUCTION

Ostracoda are small entomostracan Crustacea that are abundant in marine and nonmarine aquatic environments and that are excellent microfossils in paleoenvironmental reconstruction (Whatley, 1983, 1988). They are extensively employed in this role in Quaternary nonmarine environments (Whatley, 1983; Aguirre and Whatley, 1995).

In this paper, we present a review of studies of late Quaternary lacustrine ostracods from northern Patagonia, based on the analysis of numerous samples from lake bed cores and from outcrops along an approximately west-east transect near Mount Tronador (lat. 41°00′S; long. 71°50′W) to Lake Cari-Laufquen (lat. 41°12′S; long. 69°25′W). This transect has been the subject of a multidisciplinary study called the Lagos Comahue Project, which has the objective of contributing to our knowledge of the evolution of the climate, the environment, and the biota from the late Pleistocene through the Holocene of the cordilleran and patagonian marginal areas as part of our understanding of global change.

There is a large and growing body of literature on the Ostracoda of Argentina, but only a small part of this work is devoted to the ostracods of Quaternary lake deposits. Among such papers are those by Daday (1902), Ramírez (1967), Whatley and Cholich (1974), Zabert (1981), Zabert and Herbst (1986), Bertels and Martínez (1990), Aguirre and Whatley (1995), Cusminsky (1994, 1995), Cusminsky and Whatley (1995,1996), and Whatley and Cusminsky (1995a, b).

THE AREA AND MATERIAL STUDIED

Physiographically, the area of the study comprises, in the west, Andean Patagonia with elevations below 1000 m covered with a permanent forest of *Nothofagus dombeyi* and *Austocedrus* with small trees of the Proteacea. Toward the east the Patagonian Steppe with *Austrocedrus* is dying out as the climate becomes progressively drier and is being replaced by grasses, Compositae, *Ephidia*, *Acaena*, and bushes of the families Rhannacea and Umbelliferae (Markgraf, 1983). The mean annual rainfall along the transect varies from west to east considerably as follows: Cerro Tronador, 3500 mm annual rainfall, 3400 m a.s.l. altitude; San Carlos de Bariloche, 823 mm annual rainfall, 750 m a.s.l. altitude; Estación Perito Moreno, 586 mm annual rainfall, 900 m a.s.l. altitude; and Ingeniero Jacobacci, 160 mm annual rainfall, 875 m a.s.l. altitude.

The study is based on the analysis of ostracods recovered from cores El Trébol II, Mallín Aguado, Lago Seco, Los Juncos, La Salina, and profile of Río Maquinchao during collecting excursions undertaken by the staff of PROGEBA (Programa en Gea Bariloche) and of the Instituto Antártico Argentino during 1991 and 1992. The samples from the beds of lakes and bogs were obtained using a Polish Wiecowsky corer. Table 1 gives the locality and date of the samples collected. Figure 1 shows the position of the lakes and Figure 2 shows the cores.

Lago Trébol

This lake is situated in the Nahuel Huapi National Park, approximately 20 km west–northwest of San Carlos de Bariloche. At the present day, the lake has a surface area of about 0.4 km^2 and its greatest depth is about 11 m. The lake is surrounded by steep slopes and high mountains and has a closed drainage basin (Valencio et al., 1982).

The 5.2-m-long El Trebol II core consists of gytja with tephra layers. The underlying one-third of the core comprises laminated clays with tephra horizons.

Table 1. Locality and Date of Samples.

Locality	Coordinates (lat.; long.)	Core Length (m)	Age and Depth
Laguna El Trébol	41°00′S; 71°50′W	11	10.480 ± 130[1] at 525 cm 13.120 ± 220[2] at 650 cm
Mallín Aguado	41°00′S; 71°29′W	14	10.160 ± 80[3] at 1125 cm 13.850 ± 100[4] at 1285 cm 14.635 ± 130[5] at 1300 cm
Lago Seco	41°01′S; 71°28′W	9	12.500 ± 220[6] at 870-875 cm
Los Juncos	41°03′S; 71°02′W	2	12050 ± 240[7] at 150-160 cm
La Salina	41°16′S; 69°32′W	-1	In process
Río Maquinchao Section	41°12′S; 69°25′W		13,200 BP base[8] of paleolake

Laboratory Numbers of the Datation:
[1] AMS ETH Nº 13227
[2] C14 Q-2951 Cambridge University
[3] AMS ETH Nº16.982
[4] AMS ETH Nº16.983
[5] AMS ETH Nº. 13226
[6] C14 Q-2953 Cambridge University
[7] C14 Q-2952 Cambridge University
[8] Thermoluminescence dating

Mallín Aguado

This bog is located 18 km north of San Carlos de Bariloche and is 10 ha in area. It is an area of grazing with some open, shallow ponds. Scattered bushes occur, and the area is surrounded by *Nothofagus* forest.

The core was taken in the center of the bog, 100 m from its eastern edge. This core consists of 13.5 m of peat becoming more compacted, laminated, and argillaceous downward. Several layers of tephra and sand occur throughout the core.

Lago Seco

This is a small, dry lakebed 10 km north of San Carlos de Bariloche whose surface is without vegetation. Virtually all 9 m of the core was of peat and peaty clay with a large number of tephra layers usually only 1 cm thick.

In the cores studied in this part of the transect, it was possible to identify the presence of two well-defined lithologies: a lower unit with rythmic laminated sediment clastics and an overlying unit composed of peat. The approximate contact between these two units (14,635 yr BP in Mallín Aguado and 13,340 yr BP in Mallín Boock) according to del Valle et al. (1996) indicates the beginning of postglacial time in the area. These workers claim that the sedimentary evidence from the cores taken from lakes and bogs suggests that during the deglaciation that followed the glacial maximum (oxygen isotope stage 2), a large lake called (in the language of the indigenous Mapuche Indians) "Elpalafquen" (lago del comienzo) was formed in the region some 16,000–18,000 yr ago. Later, during deglaciation and probably associated with tectono-volcanic activity, the natural dam at the eastern limit of this vast paleolake was breached and, in a cataclysmic event, an enormous volume of water escaped to the east and west, mostly probably via the paleo-River Limay and the Manso River. The level of the lake was drastically reduced, probably to that of the modern lake Nahuel Huapi. This event also brought about an increase in the trophic levels of some of the residual bodies of water, such as El Trébol, which at one time were arms of the paleolake but now are discrete and isolated lacustrine entities (del Valle et al., 1996).

Los Juncos

This lake is situated on the Patagonian Steppe adjacent to the small railway station at Perito Moreno, 28 km east–northeast of San Carlos de Bariloche. The shallow lake varies from being about 900 m in radius in winter to 200 m or less in summer. It is a nature reserve with willows and an abundant avifauna, including flamingos. A 2 m core was obtained, of which the top

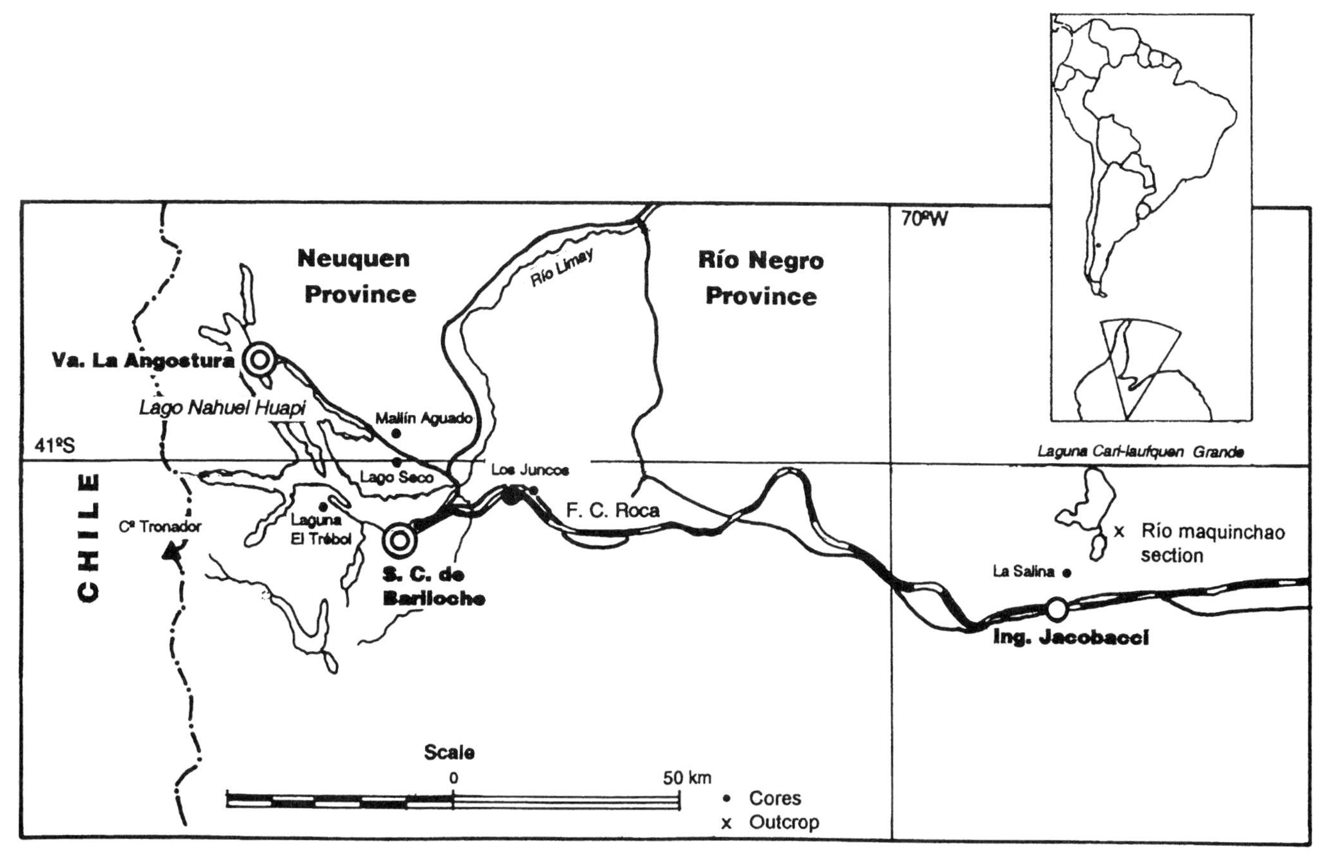

Figure 1—Localization (from Cusminsky, 1995).

1.55 m was dark and composed of gytja, peat, and clay and which overlies lighter colored clays. Salt minerals were seen as the core dried, and it contained thin levels of silt and tephra.

La Salina

This lake is one of three small lakes in the Maquinchao Basin, which is on the Patagonian Steppe 150 km east of San Carlos de Bariloche. The core was taken from the middle lake, which in a dry spell has a 2-cm-thick crust of salt: with rain this crust soon dissolves. Because of the muddy nature of the lake, it was not possible to core at its center. Instead, the core was taken some 30 m from the shore. The core comprised, from the top, 23 cm of bad-smelling black mud with a high organic content overlying some 20 cm of clay with salt minerals.

River Maquinchao Section

This section is situated in the Maquinchao Basin in the lower valley of the River Maquinchao near the two lakes Cari-Laufquen Grande and Cari-Laufquen Chica. The latter lake is situated 820 m above sea level and is a permanent water body that contains a considerable percentage of solutes (283 ppm) and has a pH of 7.8 (Galloway et al., 1988). When occasionally this lake overflows, the excess water is carried by the Río Maquinchao to lake Cari-Laufquen Grande. This lake is an ephemeral lake situated 800 m above sea level with a concentration of solutes of some 4000 ppm and a pH of 8.6 (Galloway et al., 1988).

Various authors (Volkheimer, 1973; Coria, 1979; Galloway et al., 1988; del Valle et al., 1993) have shown, based on the evidence of fossil strand lines, that in the past under more humid conditions, the two modern lakes were part of a larger, deeper entity referred to by del Valle et al. (1993) as Paleolago Maquinchao.

Lithologically, the section collected comprises, from base to top, deltaic sediments in at least four graded bedding cycles of gravel, sand, and mud (facies 1a), followed by lacustrine rythmites (facies 1b). The thickness of this lower unit, of which the base was not exposed, was at least 5.3 m. At the top of the succession, about 6 m of laminated clays with silty marls and sporadic gypsum occur (facies 2a). This upper part of the section represents the deeper water lacustrine sedimentation. An erosive unconformity with a volcanic ash with convoluted bedding planes and load structures divides the sediments of facies 2a and facies 3a. These sediments are coarser grained with gypsum and undulating irregular stratification.

According to del Valle et al. (1993), other facies that are absent in the studied section are present in the area. These are facies 2b, 2c, 3b, 3c, and 4. Facies 2b represents the shallow water facies. It is composed of

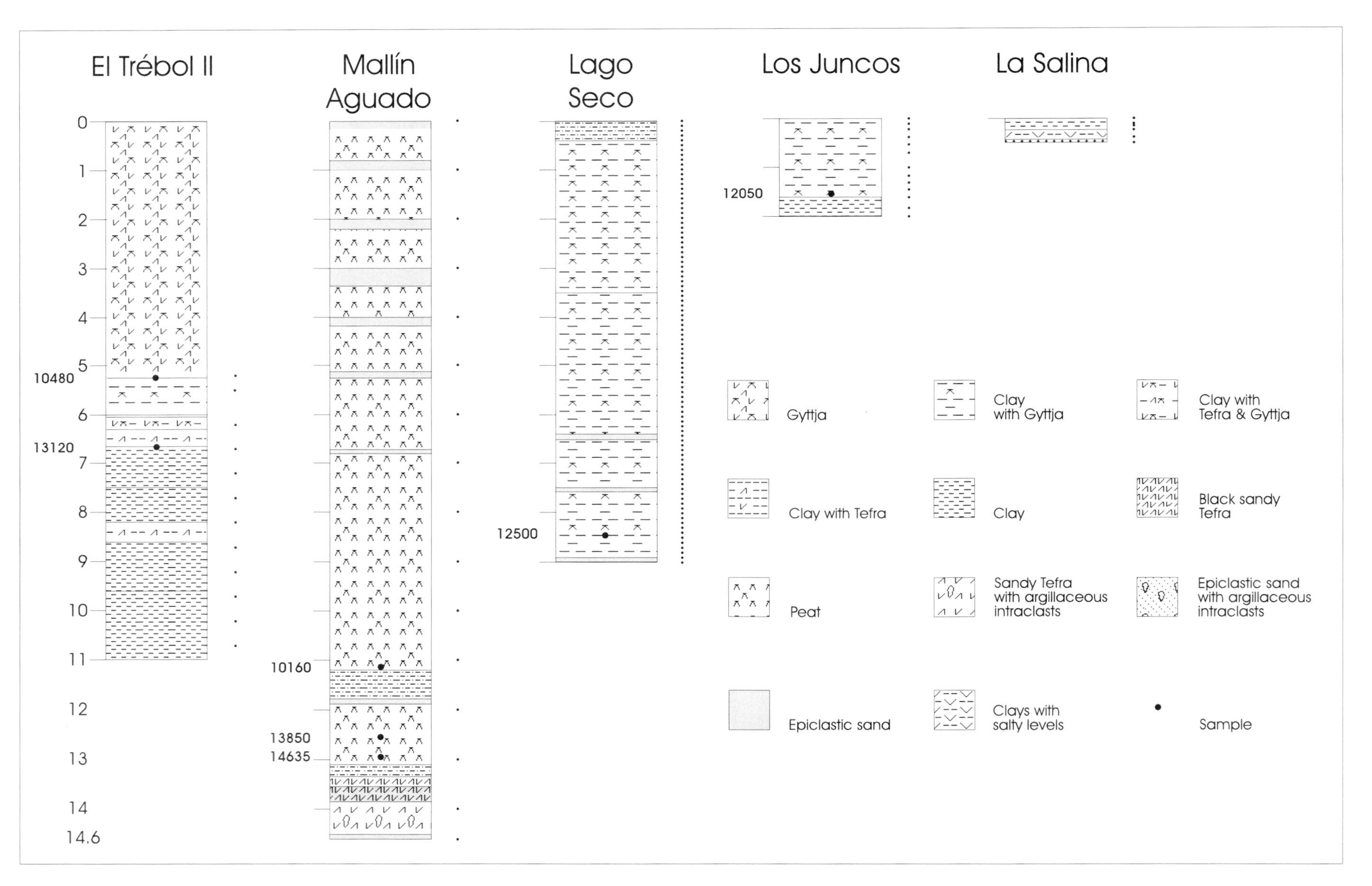

Figure 2—Analyzed cores (from Whatley and Cusminsky, 1995a).

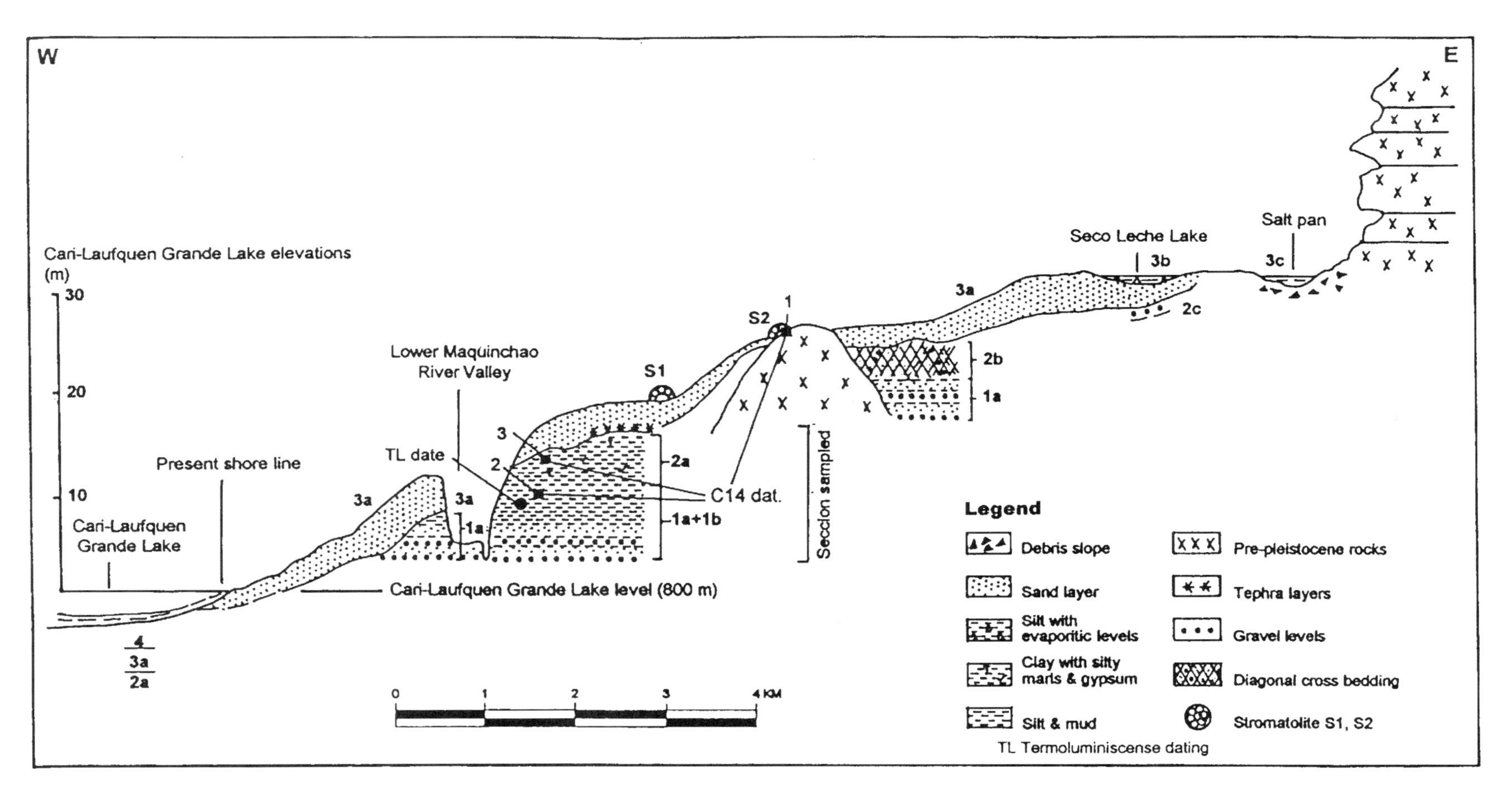

Figure 3—Maquinchao section facies distribution (from del Valle et al., 1993).

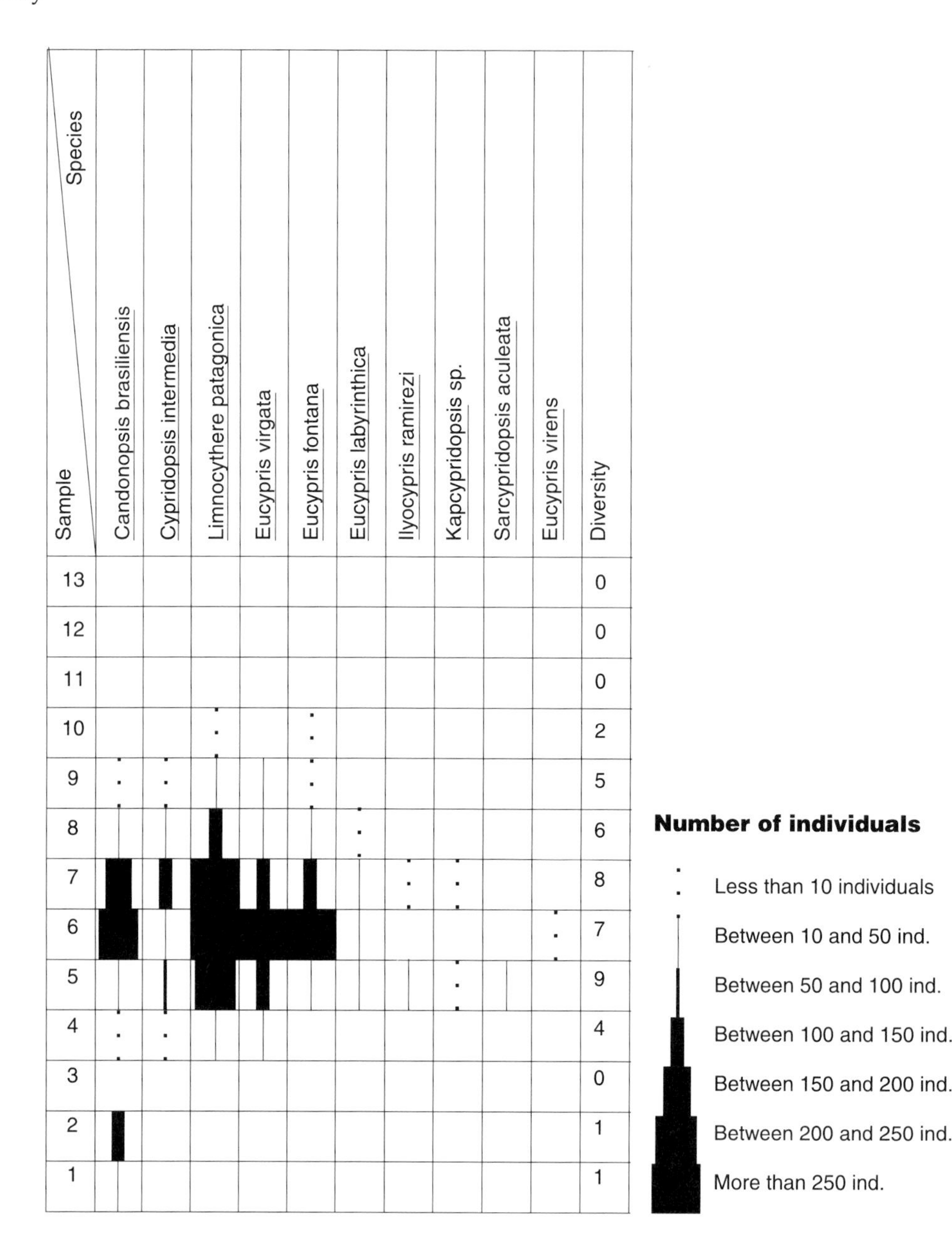

Figure 4—Los Juncos core (from Cusminsky and Whatley, 1996).

sandy mud with parallel and cross lamination, interbedded with sandy marls, and criss-crossed with fine gravel and coarse sand. The 2c facies represents the shoreline coastal facies and comprises mainly cross-laminated gravel, sand, and mud with occasional thin horizontal marl. Facies 3b is composed of clay sediments sometimes rich in soluble chlorites and sulfates accumulated in a retreated lake bay gradually isolated to ephemeral and saline lakes. Facies 3c represents recent sediments deposited in salt pan. Facies 4 represents the lacustrine recent bottom sediments composed mainly of alternating regular clay and silt deposited in the lake at 4 m depth.

Stromatolites were found in facies 3a, indicating stationary coastal waters of the paleolake (del Valle et al., 1993) (Figure 3).

This section yielded a number of samples that have been dated by radiocarbon or thermoluminescence techniques: sample LU 3677, 16,520 ± 120 yr BP (sample 1), ^{13}C (‰), –4.6; sample LU 3676, 15,220 ± 180 yr BP (sample 2), ^{13}C (‰), +3.9; sample LU 3792, 33,200 ± 260 yrs BP (sample 3), ^{13}C (‰), –0.3. (TL Proszynsky Laboratory Poland 13,200 yrs BP. Sample 1 stromatolites from the shallow place in paleolake (carbonates); sample 2 base of paleolake sediments (carbonatyes); sample 3 top of paleolake sediments (carbonates).)

The locations of the dated samples in the section is shown in Figure 3. These results lead us to believe that the ^{14}C ages do not reflect the true age of the sediments. The source of errors is probably the dissolution of old carbonates in the watershed. This process has affected the relic lake chemistry (^{14}C) and the top of the sediments section by infiltration of bicarbonates bearing solution throughout the soil profile (A. Tatur, 1995, personal communication). The thermoluminescence dating done in Poland seems to be more reliable.

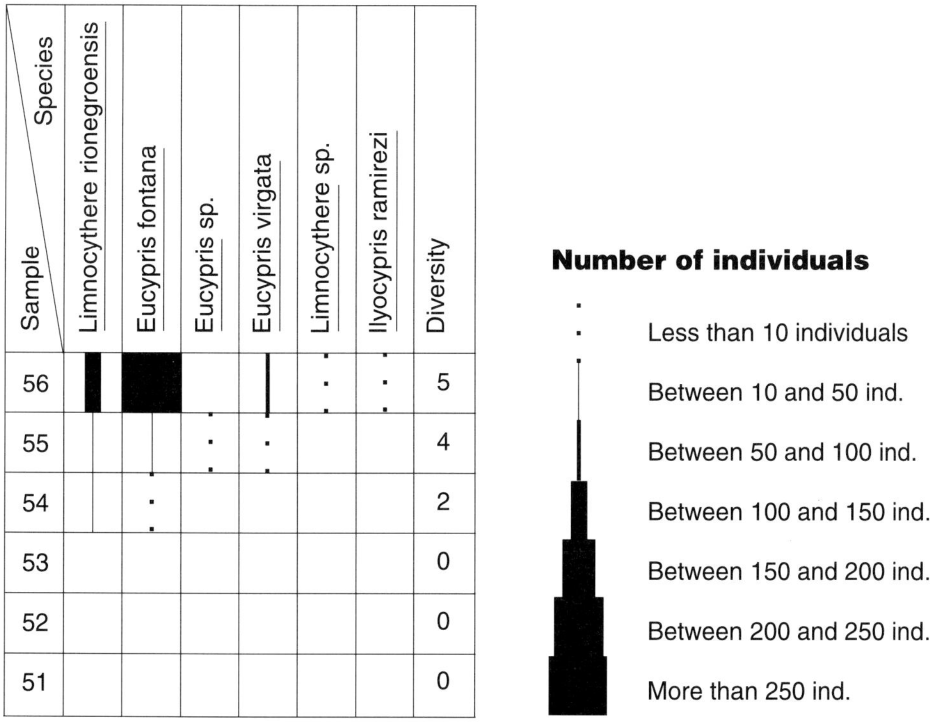

Sample \ Species	Limnocythere rionegroensis	Eucypris fontana	Eucypris sp.	Eucypris virgata	Limnocythere sp.	Ilyocypris ramirezi	Diversity
56							5
55							4
54							2
53							0
52							0
51							0

Number of individuals

- Less than 10 individuals
- Between 10 and 50 ind.
- Between 50 and 100 ind.
- Between 100 and 150 ind.
- Between 150 and 200 ind.
- Between 200 and 250 ind.
- More than 250 ind.

Figure 5—La Salina core (from Cusminsky and Whatley, 1996).

Projected further thermoluminescence dating and tephrochronological analysis should provide more precise ages for the sediments.

METHODS

The samples taken from the various cores were washed in water over a 63 µm (230 mesh/in) sieve. The identification of the ostracods is based on Cusminsky and Whatley (1996). Quantitative data, such as diversity and abundance, are based on calculations of the numbers of specimens and species present in 10 g of dry sediment.

RESULTS

Although we studied the laminated sediments between 5.20 and 11 m, the El Trébol core failed to yield any ostracods. This confirms Bertel's (in Valencio et al., 1982) opinion that ostracods were absent from this sequence. We have encountered only the ostracod genus *Darwinula* in modern lake.

Similarly, ostracods were absent throughout the Mallín Aguado core. The core from Lago Seco yielded a few specimens of *Eucypris fontana* (Graf) in its lower levels (8.80–9.00 m); the remainder of the core was entirely barren of Ostracoda.

The core taken at Los Juncos (Figure 4) yielded in its lower levels *Candonopsis brasiliensis* Sars; however, the overlying sediments were barren of ostracods, and the fact that they contain crystals of gypsum suggests that they were evaporites. Notwithstanding this, the overlying samples yield a progressively larger fauna predominantly of *Limnocythere patagonica* (Cusminsky and Whatley, 1996), associated with *Eucypris fontana*, *Eucypris virgata* (Cusminsky and Whatley, 1996), *Cypridopsis intermedia* Sars, and *Candonopsis brasiliensis*. The environmental conditions remained relatively stable until the 100 cm level in the core where the fauna begins to decline in abundance and diversity and to eventually disappear at the top of the core.

The La Salina core (Figure 5) was in its basal part entirely devoid of ostracods, probably because of high salinity as witnessed by the presence of evaporitic minerals; however, at about 13 cm up the core, *Limnocythere rionegroensis* (Cusminsky and Whatley, 1996) appears, associated with *E. fontana*. From here to the top of the core is a slow increase in ostracod diversity and a clear dominance of *E. fontana*.

Figure 6 gives the distribution of the Ostracoda through the Río Maquinchao section. At the base, the only ostracods are a few individuals of *L. rionegroensis* and *Eucypris virgata*. Although there is a gradual increase in diversity upward, this is mainly due to the appearance of a few specimens each of *E. virgata, E. fontana, C. brasiliensis,* and *L. patagonica*. The number of individuals, as well as the specific diversity, begins to increase gradually, which after an acme, seems a notable decline into the top of the core.

In all of the samples examined in the study, the ostracod fauna consists of adults (of both sexes in syngamic species) and a large suite of instars, conforming to a type A population age structure (Whatley, 1983, 1988) and suggesting deposition as a paleobiocoenosis in low-energy environments.

Many of the species in the study are of widespread occurrence across the transect, but it is notable that *L. rionegroensis* is absent from the Los Juncos core. This absence is possible evidence that, in the late Quaternary, as in the present day, the area around Los Juncos

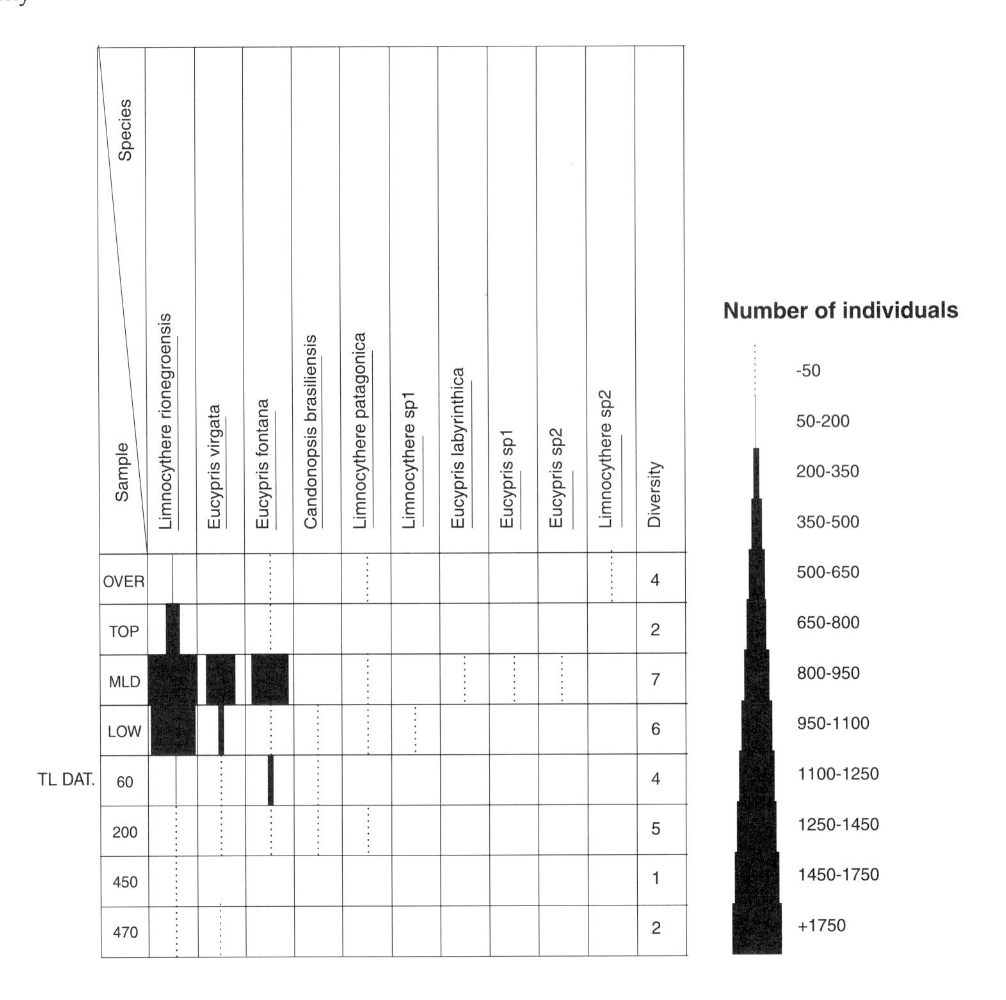

Figure 6—Maquinchao section.

enjoyed a climate of greater humidity than that of the Maquinchao Basin; therefore, *L. rionegroensis* may be used as an index species of more saline waters and drier environments in this sector of South America. Morphologically, *L. rionegroensis* resembles *L. bradburyi* Forester (Forester, 1985), from the southwestern United States and Mexico, which inhabits environments similar to those analyzed in this study (Cusminsky and Whatley, 1996).

On the basis of the results we have obtained so far, two of the cores (Laguna El Trébol II and Mallín Aguado) are totally devoid of ostracods, and they are rare in Lago Seco. In the case of Lago Seco and Mallín Aguado, most of the samples were of peat and, if there had been ostracods living in the lake, which is somewhat unlikely given the acid pH, the acidity of the forming peat would ensure that they would not be preserved as fossils. This general conclusion was outlined by Whatley (1983, 1988) and Neale (1964) and, more specifically, by Bianchi et al. (1993), Amos et al. (1993), and Cusminsky (1994, 1995). What is more difficult to understand is the absence of ostracods from the basal argillaceous sediments in the lower one-third of the El Trébol core. It has been suggested that this may represent an oligotrophic period in the evolution of the lake, with insufficient bioproductivity to support ostracod fauna (A. Tatur, 1994, personal communication; Bianchi et al., 1997).

In the case of Los Juncos, the absence of ostracods from the upper levels of the core is probably related to the increased trophic levels of the media and an associated decrease in oxygen levels, as well as increased competition for food and predation by other invertebrates.

In the Maquinchao Basin, the La Salinas core shows an increase in the species and diversity of fauna at the top, which may be related to an optimization of physicochemical and biological conditions entirely favorable to the ostracods. The same thing can be seen in the lacustrine samples of the Río Maquinchao section; the subsequent decline in the fortunes of the fauna may be attributed to volcanic events bringing about a deterioration in ambient parameters and a trend toward aridity. This is attested to by the abrupt disappearance of the remaining ostracods and the appearance of evaporitic minerals in the top of the sequence.

CONCLUSIONS

Along a transect of approximately 200 km in northern Patagonia, two probably nonsynchronous, lacustrine paleoenvironments are recognized. In the western sector of the transect, in the area of the large "Elpalafquen" palaeolake (del Valle et al., 1996), factors such as oligotrophy, high acidity, and taphonomy combined to ensure the complete or partial absence of

Ostracoda from the cores examined. In the eastern sector of the transect, however, conditions in the late Quaternary were more favorable for the existence of ostracods. This can be seen clearly in the Maquinchao basin on the Patagonian Steppe, where a large saline paleolake, Lake "Maquinchao" (del Valle et al., 1996), was developed during a pluvial period at around 13,200 yr BP. This confirms the model proposed by Garleff et al. (1994) on the basis of pollen data, which showed that at around 13,000 yr BP there was a more humid interval with heightened lake levels. These authors also indicate that, later in the Holocene, there was a change in conditions toward greater aridity that led to a shrinking of the paleolake, the precipitation of evaporitic minerals, and the formation of dunes around the lake shores.

All of our observations based on the study of lacustrine ostracods confirm the variability of the climatic changes that took place in northern Patagonia during the late Pleistocene and Holocene. Future studies are required to seek more detailed information on the nature of these climatic changes in this part of South America.

ACKNOWLEDGMENTS

Gabriella Cusminsky wishes to acknowledge her debt of gratitude to Alwine Bertels who introduced her to the study of microfossils and to Arturo Amos for critically reading the manuscript. We thank Gabriella Costa and Carmen Brogger for the preparation of samples and Victoria Amos for the artwork. Alicia Moguilevsky is thanked for all her help.

REFERENCES CITED

Aguirre, M. L., and R. C. Whatley, 1995, Late Quaternary marginal marine deposits and palaeoenvironments from northwestern Buenos Aires province, Argentina: A review. Quaternary Science Reviews, v. 14, p. 223–245.

Amos, A. J., M. M. Bianchi, G. C. Cusminsky, J. I. Masaferro, G. Roman Ross, and R. del Valle, 1993, Nuevas evidencia palaeoambientales post-pleistocénica en lagos y mallines del área de San Carlos de Bariloche (Lat. 41°00'S), República Argentina. Taller Internacional el Cuaternario de Chile. Santiago de Chile (Abstracts), p. 35.

Bertels, A., and D. E. Martínez, 1990, Quaternary ostracodes of continental and transitional littoral-shallow marine environments. Cour. Forsch., Inst. Senkenberg, v. 123, p. 141–159.

Bianchi, M. M., G. C. Cusminsky, C. Macchiavello, J. I. Masaferro, J. R. del Valle, M. S. Vigna, and G. Vobis, 1993, Análisis micropaleontológico preliminar de lagos y mallines de la región del Lago Nahuel Huapi. Proyecto Lagos-Comahue. Reunión Argentina de Ecología. Puerto Madryn 16 (abstracts).

Bianchi, M. M., J. I. Masaferro, G. Román Ross, R. del Valle, A. Tatur, and A. J. Amos, 1997, The Pleistocene–Holocene boundary from cores of lago El Trébol, Patagonia Argentina: Paleolimnological evidences. Verh. Internal Verein. Limnol, v. 26, p. 805–808.

Coria, B. L., 1979, Descripción geológica de la hoja 40 d, Ingeniero Jacobacci. Boletín No. 168 Servicio Geológico Nacional, 101 p.

Cusminsky, G. C., 1994, Variaciones climáticas cuaternarias en una transecta oeste-este a partir de ostrácodos lacustres, Patagonia Norte, Argentina. 2nd Annual Meeting Project 341, IGCP\IUGS\UNESCO, Rev. Mus. de Hist. Nat. de San Rafael, v. 12 (4), p. 251–262.

Cusminsky, G. C., 1995, Ostrácodos lacustres cuaternarios en la transecta Tronador-Laguna Cari-Laufquen, Patagonia Norte, Argentina. VI Cong. Arg. de Paleont. y Biostratigrafía, Trelew. Actas (1994), p. 99–105.

Cusminsky, G. C., and R. C. Whatley, 1995, Quaternary ostracods from Cari-Laufquen lake, Río Negro Province, Argentina. XXVI SIL International Congress. Sao Paulo. (Abstract), p. 302.

Cusminsky, G. C., and R. C. Whatley, 1996, Quaternary non-marine ostracods from lake beds in Northern Patagonia. Revista Española de Paleontología, v. 11 (2), p. 143–154.

Daday, E.V., 1902, Mikroskopische ausswasserthiere aus patagonian. Termeszetranzi Puzetek, v. 25, p. 201–313.

del Valle, R., A. Tatur, A. J. Amos, D. Ariztegui, M. M. Bianchi, G. C. Cusminsky, K. Hsu, J. M. Lirio, C. Martínez Macchiavello, J. I. Masaferro, H. J. Nuñez, C. A. Rinaldi, R. Valverdu, S. Vigna, G. Vobis, and R. C. Whatley, 1993, Laguna Cari-Laufquen Grande, registro de una fase climática húmeda del Pleistoceno tardío en la Patagonia septentrional. Pangea & Glopals Project, San Juan, p. 16–19, comunicaciones.

del Valle, R., J. M. Lirio, H. J. Nuñez, A. Tatur, C. A. Rinaldi, J. C. Lusky, and A. J. Amos, 1996, Reconstrucción paleoambiental Pleistoceno–Holoceno en latitudes medias al este de los Andes. XIII Cong. Geol. Argentino y III Cong. de Exp. de Hidrocarburos, Actas IV, p. 85–102.

Forester, R. M., 1985, Limnocythere bradburyi n sp. a modern ostracode from central Mexico and possible Quaternary paleoclimatic indicator. Journal of Paleontology, v. 59, p. 8–20.

Galloway, R. M., V. Markgraf, and J. P. Bradbury, 1988, Dating shorelines of lakes in Patagonia Argentina. Journal of South American Earth Sciences, v. 1 (2), p. 195–198.

Garleff, K., T. Reichert, M. Sage, F. Schabitz, and B. Stein, 1994, Períodos morfodinámicos y el paleoclima en el norte de la Patagonia durante los últimos 13000 años. 2nd Annual Meeting of Project 314 IGCP/IUGS/UNESCO. Rev. Mus. de Hist. Nat. de San Rafael 12 (4), p. 217–228.

Markgraf, V., 1983, Late and postglacial vegetational and paleoclimatic changes in subantarctic, temperate, and arid environments in Argentina. Palynology, v. 7, p. 43–70.

Neale, J. W., 1964, Some factors influencing the distribution of Recent British Ostracoda. Pubbl. Staz. Zool. Napoli, v. 33, suppl. p. 247–307.

Ramírez, F. C., 1967, Ostrácodos de lagunas de la

provincia de Buenos Aires. Rev. Mus. de La Plata (NS) Secc. Zoología, v. 10, p. 5–54.
Valencio, D. A., K. M. Creer, A. M. Sinito, E. J. Romero, and C. A. Fernández, 1982, Estudio paleomagnético y palinológico de ambientes lacustres. Parte 1, Lago Trébol. Rev. Asoc. Geol. Arg., v. 37 (2), p. 183–204.
Volkheimer, W., 1973, Observaciones geológicas en el área de Ingeniero Jacobacci y adyacencias (provincia de Río Negro). Rev. Asoc. Geol. Arg., v. 28 (1), p. 13–36.
Whatley, R. C.,1983, The application of Ostracoda to palaeoenvironmental analysis. in Maddocks, R. F., ed., Applications of Ostracoda. Houston, Geosciences, p. 51–77.
Whatley, R. C., 1988, Population structure of ostracods: some general principles for the recognition of palaeoenvironments, *in* De Deckker, P. et al., eds., Ostracoda in the Earth Sciences. Elsevier, p. 103–124.
Whatley, R. C., and T. Cholich, 1974, A new Quaternary ostracod genus from Argentina. Palaeontology, v. 17 (3), p. 669–684.
Whatley, R. C., and G. C. Cusminsky, 1995a, Quaternary lacustrine Ostracoda from Northern Patagonia, *in* Ostracoda and Biostratigraphy, Riha, J.,ed., Balkema, p. 303–310.
Whatley, R. C., and G. C. Cusminsky, 1995b, Nonmarine Ostracoda and Holocene palaeoenvironments from lake Cari-laufquen region, Río Negro province, Argentina. GLOPALS-IAS Meeting Chile 1995, (Abstracts), p. 16.
Zabert, L. L., 1981, Ostrácodos cuaternarios de Taco Pozo (provincia de Chaco, Argentina) con algunas consideraciones paleoecológicas. Facena, v. 4, p. 77–87
Zabert, L. L., and R. Herbst, 1986, Ostrácodos pleistocenos del arroyo Perucho Verna, Provincia de Entre Ríos, Argentina. Ameghiniana, v. 23 (3–4), p. 213–224.

Smith, G. I., 2000, Late Pliocene and Pleistocene Searles Lake, California, U.S.A., *in* E. H. Gierlowski-Kordesch and K. R. Kelts, eds., Lake basins through space and time: AAPG Studies in Geology 46, p. 591–596.

Chapter 56

Late Pliocene and Pleistocene Searles Lake, California, U.S.A.

George I. Smith
U.S. Geological Survey
Menlo Park, California, U.S.A.

Searles Valley, a closed basin in southeast California (U.S.A.), now contains a dry salt flat 100 km^2 in the area known as Searles Lake (Figure 1). Its subsurface saline layers and interstitial brines have been the source of approximately $4 billion of assorted industrial chemicals. During about 75% of the late Pliocene and Pleistocene periods, however, the valley contained a perennial body of water. This lake level was as much as 200 m above its present surface during the late Pleistocene; the depths of the earlier lakes are poorly constrained. Climatic fluctuations during the entire period caused major fluctuations in lake level, and during the late Pleistocene at least, they resulted in overflow from its southeast corner during wet periods, and in shrinkage or desiccation during dry periods.

Searles Valley is near the middle of a succession of five closed basins, four of which filled and overflowed at times during the past to make a chain of as many as five lakes. These were fed primarily by the south-flowing Owens River that drained the high-elevation, east-facing slopes of the Sierra Nevada. Upstream from Searles Lake were Owens Lake (terminus of the Owens River, 695 km^2 surface area when full), and China Lake (which coalesced with Searles Lake when that lake was nearly filled, forming a single body of water having an area of nearly 1000 km^2). At times, Owens Lake received overflow from Mono Lake (690 km^2, "Lake Russell") in an adjoining now-closed valley to the north. Downstream from Searles Lake were Panamint Lake (710 km^2, "Lake Gale") and Death Valley Lake (1600 km^2, "Lake Manly"). During the wettest periods, the lakes in all basins except Death Valley were overflowing, creating a continuous chain that had a combined surface area of as much as 4000 km^2.

Evidence from a long core in Owens Lake indicates that in this area, Owens River runoff during the late Holocene (0–5 ka) has been the lowest since at least 800 ka (Smith et al., 1997). When the runoff volume from the Owens River was produced by an "intermediate" hydrologic regime—about halfway between its estimated maximum and its late Holocene flow—Searles Lake's position in this chain of lakes made it susceptible to major fluctuations in lake levels and therefore combined surface areas. Lake-surface areas controlled evaporative losses from the chain of lakes, and calculations using modern evaporation rates show that when the Owens River flow was more than six times its late Holocene volume, Searles Lake would have been continuously overflowing; when that flow was less than 2.5 times late Holocene volume, Searles Lake would have been dry, although perennial water bodies could have existed in Owens and China Lakes (Smith and Street-Perrott, 1983). Maximum reconstructed Pleistocene flow volumes in the Owens River were possibly more than 10 times modern flow. If Pleistocene mean-annual temperatures were 5–10°C lower than at present, evaporation rates would have decreased and the river flow necessary to maintain lakes having any given distribution pattern would have been reduced by about 25–50%, respectively.

A core obtained from the center of Searles Lake penetrated bedrock at a depth of 915 m (Figure 2), and lacustrine sediments were recovered at all levels between the surface and 693 m (Smith et al., 1983). Those sediments range from marl and siltstone (deep-water deposits) to mono or multimineralic saline layers and playa sediments (very shallow or dry-lake deposits). Study of numerous cores representing deposits in the upper 50–100 m of the lake's fill (Smith, 1979; Bischoff et al., 1985) allows a more detailed reconstruction of the lake and climate history since about 150 ka (thousands of years before the present). An interpretation of the history of the lake (Figure 3), therefore, is readable as a history of climatic change in this mid-latitude part of the world; however, both similarities and differences are seen between this climatic record and the record of the globe's high-latitude glaciation as documented by ^{18}O variations in benthic foraminifera from deep-sea sediments (Smith, 1984).

Numerous outcrops of lacustrine and nonlacustrine sediments are preserved on the flanks of modern Indian Wells, Salt Wells, and Searles Valleys,

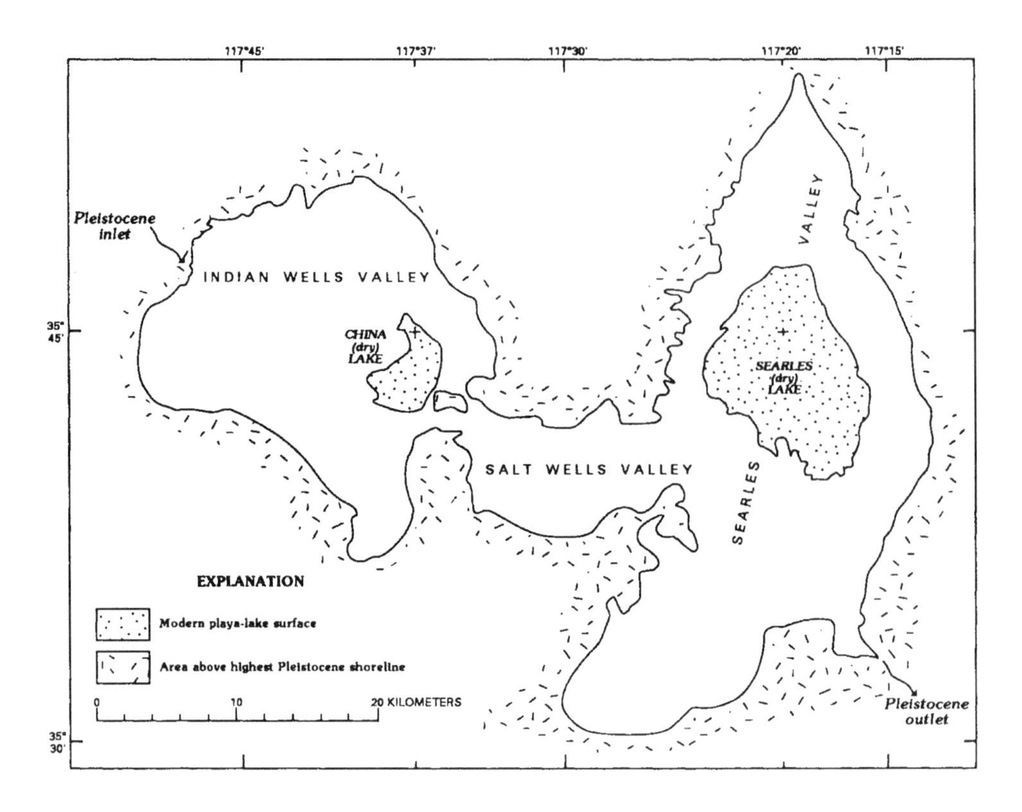

Figure 1—Map showing modern location of Searles Valley and Searles (dry) Lake, as well as Indian Wells Valley (and China Lake) and Salt Wells Valley, all of which were inundated during high-stands of Pleistocene Searles Lake, which covered the region within the line enclosed by random-dash pattern.

which contained the largest expansions of Pleistocene Searles Lake (Figure 1). Most of these deposits represent some part of the record of sedimentation since 150 ka, but no single exposure represents more than a fraction of that period of time. An example of one outcropping section is plotted in Figure 4. Many of these outcropping deposits, as in the majority of the sections exposed in these valleys, consist of massive to finely laminated aragonitic and dolomitic marl, but silt, sand, and gravel are common near the edges of the basin.

Chemical sedimentation was important in this basin. Thick beds of multimineralic salts occupy its center, and their interstitial brines have been the source of billions of dollars worth of industrial-mineral production. In surrounding areas of the valley (Smith, in press), massive deposits of calcite or aragonite "tufa" crop out as near-shore terraces and towers as much as 30 m high. The tufa terraces are believed to be products of near-shore losses of CO_2 caused by wave action and photosynthesis by algae; the tufa towers were formed where sublacustrine springs injected Ca-bearing waters into the alkaline, carbonate-rich lake waters. Algal remains incorporated into the tufa show that the springs were also favored sites for algal growth and carbonate precipitation due to CO_2 consumption. Deltaic deposits are mostly composed of coarse sand to gravel, but the largest of these, developed in the area that served as the inlet for the Owens River water, contains a high percentage of $CaCO_3$. This delta is therefore termed a "chemical delta." It formed where large amounts of Ca, introduced by the incoming fresh, relatively pH-neutral water, combined with existing CO_3 in the alkaline lake, to cause precipitation of most of the Ca as $CaCO_3$.

Whenever arid cycles caused the deep lakes to retreat and for saline horizons to be deposited in the center of the valley, the flanks of the valley became either sites of alluvial-gravel deposition or of erosion that created what are now disconformities in the peripheral sections of lake deposits. Distinctive soils developed on the surfaces of some of these alluvial layers. Geologic mapping of these outcropping deposits has allowed identification in some areas of more than two dozen late Pleistocene and Holocene lacustrine and nonlacustrine units, based on lithology, stratigraphic relations, and soil character (Smith, 1968, 1987, in press).

This valley has preserved an unusually long record of changes in regional precipitation and runoff. Deposition in the middle of the valley appears to have been continuous throughout the 3.2 m.y. period. Sequential changes in the saline mineralogy and sediment lithology provide a semi-quantitative record of hydrologic regime changes—and thus climatic changes—through this period. The elevations reached by the correlative lacustrine and alluvial sediments exposed on the flanks of the valley, most of which represent the time since 150 ka, provide additional detail on the elevations reached by the lacustrine advances and retreats during this period (Smith, in press).

REFERENCES CITED

Bischoff, J. L., R. J. Rosenbauer, and G. I. Smith, 1985, Uranium-series dating of sediments from Searles

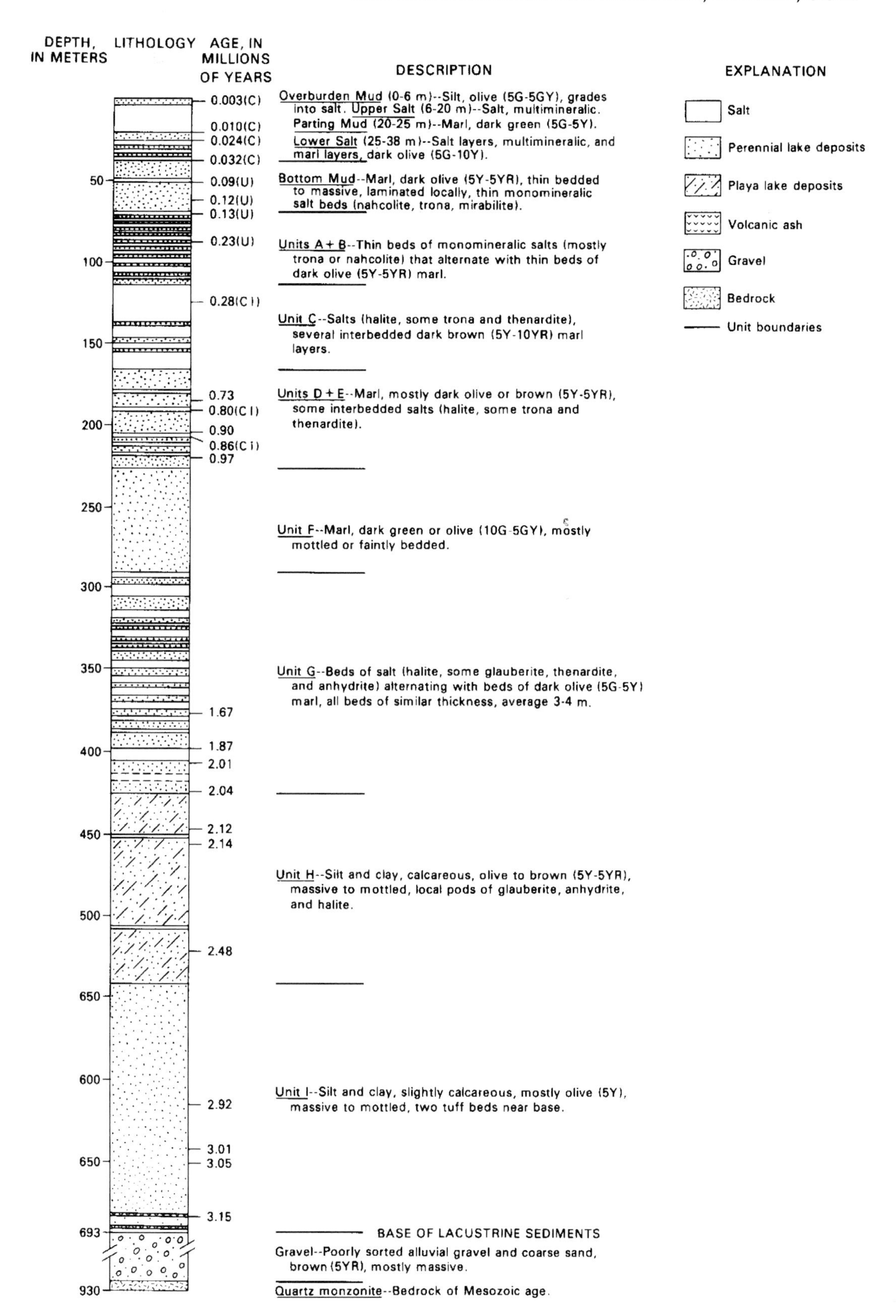

Figure 2—Graphic log of core KM–3 showing lithologies and ages of dated horizons (from Smith, 1991). Ages based on four methods: ^{14}C (indicated by C) (Stuiver and Smith, 1979), U/Th (U) (Bischoff et al., 1985), ^{36}Cl (Cl) (Phillips et al., 1983), and paleomagnetic orientation (ages only, no annotation) (Liddicoat et al., 1980).

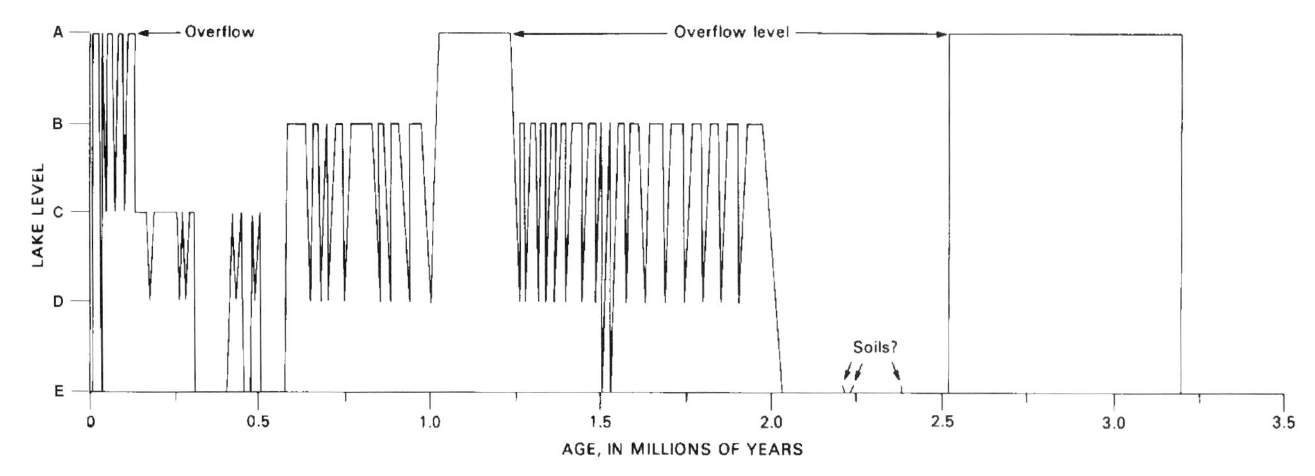

Figure 3—Reconstructed history of Searles Lake based primarily on core KM–3 (Smith et al., 1983). Inferred lake level or salinity indicated by letters. (A) Lake deep, fresh, overflowing most of the time. (B) Lake deep, slightly saline, overflowing only part of the time. (C) Lake shallow, moderately saline, overflowing rarely if at all; thin, mostly monomineralic salt beds deposited intermittently as result of winter cooling. (D) Lake shallow to dry, very saline; salts deposited during most of year. (E) Lake a playa or salt flat most of the year; flooded intermittently.

Lake: Differences between continental and marine climate records, Science, v. 227, p. 1222–1224.

Liddicoat, J. C., N. D. Opdyke, and G. I. Smith, 1980, Palaeomagnetic polarity in a 930-m core from Searles Valley, California, Nature, v. 286, p. 22–25.

Phillips, F. M., G. I. Smith, H. W. Bentley, D. Elmore, and H. E. Grove, 1983, Chlorine-36 dating of saline sediments: Preliminary results from Searles Lake, California, Science, v. 222, p. 925–927.

Smith, G. I., 1968, Late-Quaternary geologic and climatic history of Searles Lake, southeastern California, *in* R. B. Morrison and H. E. Wright, Jr., eds., Means of Correlation of Quaternary Successions, v. 8, Proceedings of VII Congress, International Association for Quaternary Research, Princeton University Press, p. 293–310.

Smith, G. I., 1979, Subsurface stratigraphy and geochemistry of late Quaternary evaporites, Searles Lake, California, U.S. Geological Survey Professional Paper 1043, 130 p.

Smith, G. I., 1984, Paleohydrologic regimes in the southwestern Great Basin, 0–3.2 m.y. ago, compared with other long records of "global" climate, Quaternary Research, v. 22, p. 1–17.

Smith, G. I., 1987, Searles Valley, California: Outcrop evidence of a Pleistocene lake and its fluctuations, limnology, and climatic significance, *in* M. L. Hill, ed., Centennial Field Guide Volume 1, Cordilleran Section of the Geological Society of America, Boulder, Colorado, p. 137–142.

Smith, G. I., 1991, Continental paleoclimatic records and their significance, part of Chapter 2, Quaternary Paleoclimates (T. L. Smiley, chap. leader), *in* R. B. Morrison, ed., The Geology of North America, v. K–2, Quaternary Non-glacial Geology: Conterminous U.S., The Geological Society of America, Boulder, Colorado, p. 35–41.

Smith, G. I., in press, Late Cenozoic geology of Searles Valley, Inyo and San Bernardino Counties, California, U.S. Geological Survey Professional Paper.

Smith, G. I., V. J. Barczak, G. F. Moulton, and J. C. Liddicoat, 1983, Core KM–3, a surface-to-bedrock record of late Cenozoic sedimentation in Searles Valley, California, U.S. Geological Survey Professional Paper 1256, 24 p.

Smith, G. I., and F. A. Street-Perrott, 1983, Pluvial lakes of the western United States, *in* H. E. Wright and S. C. Porter, eds., Late Quaternary Environments of the United States, University of Minnesota Press, Minneapolis, p. 190–212.

Smith, G. I., J. L. Bischoff, and J. P. Bradbury, 1997, Synthesis of the paleoclimatic record from Owens Lake core OL–92, *in* G. I. Smith and J. L. Bischoff, eds., An 800,000-year paleoclimatic record from core OL–92, Owens Lake, southeast California, Geological Society of America Special Paper 317, p. 143–160.

Stuiver, M.,and G. I. Smith, 1979, Radiocarbon ages of units, *in* G. I. Smith, ed., Subsurface stratigraphy and geochemistry of late Quaternary evaporites, Searles Lake, California, U.S. Geological Survey Professional Paper 1043, p. 68–78.

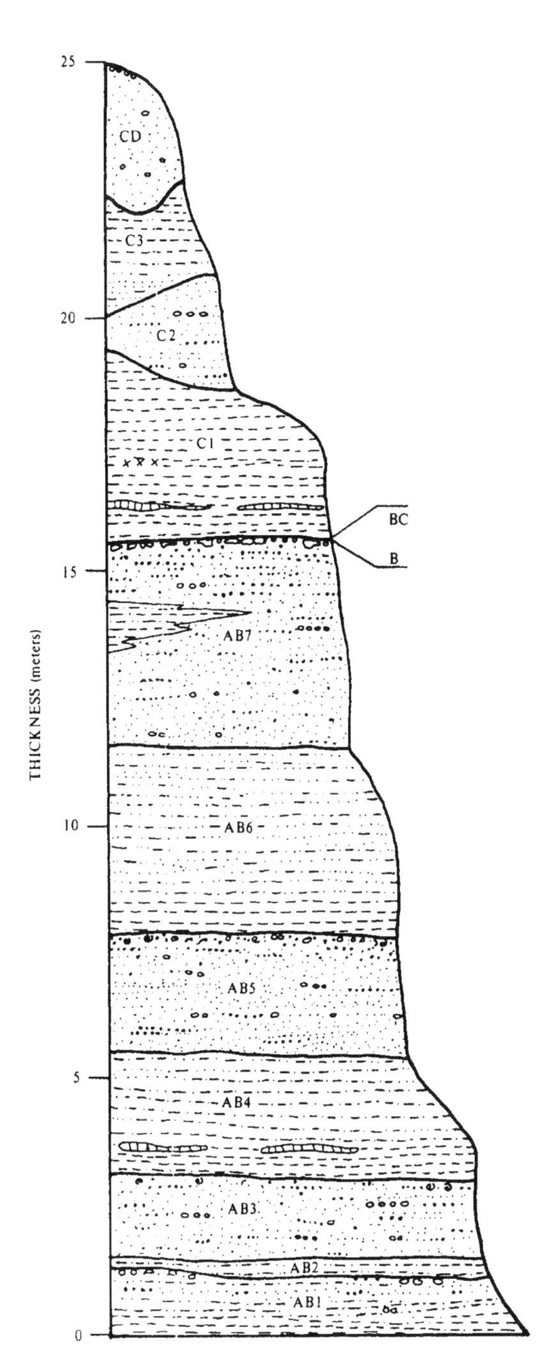

Unit CD: Pebbly sand, tan, top surface covered by scattered lag gravel and an incipient soil; fair to poor sorting, beds locally well developed; basal contact sharp

UNCONFORMITY--caused by subaerial erosion

Unit C3: Silt, mostly laminated or thinly bedded with up to 1 m of very fine sand at base; basal contact sharp

Unit C2: Sand, fine to coarse, tan or pinkish-tan; locally contains orange-stained pebbles reworked from unit AB7, fragments of nodose tufa and older lacustrine clay; a few molluscs; basal contact gradational

Unit C1: Clay and silt, mostly laminated, weathers to buff-colored surfaces; 1 m above base is a 3-cm-thick pinkish-tan layer of nodose tufa; about 2 m above the base of this unit is a discontinuous thin layer of orange-weathering tuff(?); gypsum crystals locally common; basal contact sharp

DISCONFORMITY--caused by sublacustrine erosion

Unit AB7: Sand, generally coarse to very coarse, mostly poorly sorted, especially in upper part, contains pebbles 2-3 cm in diameter, locally includes a 1-m bed of calcareous silt near middle of unit; in this area, sand is tan to slightly orange, but in area just west and south, unit is conspicuously orange and its uppermost zone is characterized by conspicuous lag gravel; basal contact gradational

Unit AB6: Silty sand and sand, lower 0.5 1 m is finer, more calcareous, and lighter colored; unit characterized by discontinuous beds 5-50 cm thick; tan to very pale orange, weathers with puffy surface; in this area has sink holes as much as 3 m deep; a small number of snails and clams noted; contact with underlying unit generally sharp

Unit AB5: Sand, fine in lower part and coarse in upper part; pebbles 2-3 cm across locally common, included are fragments of nodose tufa; bedding locally deltaic; molluscs scattered throughout, abundant at top; basal contact gradational in most places

Unit AB4: Silt and very fine sand, calcareous, basal 0.5 m is white laminated marl grading up into light-tan, less well-bedded silt and very fine sand; about 1 m up from base is a thin layer of white nodose tufa; gypsum crystals locally abundant; basal contact generally sharp

Unit AB3: Sand, medium to very coarse, local layers of pebbles 2-3 cm in diameter and clay fragments up to 10 cm in diameter; unit generally horizontally bedded but locally cross bedded, especially in lower half; tan orange to orange; notable concentrations of molluscs in upper 0.2 m; basal contact gradational

Unit AB2: Silt to very fine sand, some clay; faintly bedded; weathers to light tan-green puffy surface, contains a few molluscs; basal contact sharp, locally thins to a single white layer

Unit AB 1: Silt and sand; lower half is silt and very fine sand, calcareous, faintly bedded, weathers to light-tan puffy surfaces; upper half is medium to very coarse sand containing pebbles 2-3 cm in diameter, cross bedded, with a 2-cm orange-stained zone at the top; in places, top of unit represented by a conspicuous lag gravel composed of dark angular fragments of hypabyssal rocks; base not exposed

Figure 4—Columnar section representing a typical outcrop of exposed lacustrine sediments in Searles Valley (Smith, 1987). The age of the base of this section is about 32 ka, the top is about 6 ka, and two units ("BC" and "B") were removed by sublacustrine erosion, creating a discontinuity.

Rosen, M. R., 2000, Sedimentology, stratigraphy, and hydrochemistry of Bristol Dry Lake, California, U.S.A., *in* E. H. Gierlowski-Kordesch and K. R. Kelts, eds., Lake basins through space and time: AAPG Studies in Geology 46, p. 597–604.

Chapter 57

Sedimentology, Stratigraphy, and Hydrochemistry of Bristol Dry Lake, California, U.S.A.

Michael R. Rosen
Wairakei Research Centre, Institute of Geological and Nuclear Sciences
Taupo, New Zealand

INTRODUCTION

Bristol Dry Lake is situated in the Mojave Desert region of southeastern San Bernardino County near Amboy, California (Figure 1). It is the largest (155 km^2) in a system of three northwest-southeast trending dry lakes (playas) located in a structural trough between the Bristol and Sheephole Mountains to the north and the Bullion Mountains to the south. This location is one of the most arid places in the U.S.; consequently, Bristol Dry Lake is filled with over 500 m of sediment in the basin center, of which 260 m is almost pure halite.

Due to the position of the basins and the amount of sediment that these basins contain, the Bristol-Cadiz-Danby Dry Lake system is important in determining the structural, hydrological, and paleoclimatic development of the Mojave region since the Pliocene.

GEOLOGIC SETTING

Bristol Dry Lake is separated from Cadiz and Danby Dry Lakes to the south by projecting arms of Calument, Marble, and Ship Mountain Ranges. Cadiz Dry Lake is separated from Danby by the Kilbeck Hills. At present, each of the three subbasins has completely separate internal drainage; however, the elevation difference between Bristol Dry Lake and Cadiz Dry Lake is only 2–3 m. It has been suggested that water has flowed from one basin to the other during the pre-Holocene (Thompson, 1929).

A fourth basin, Alkali Dry Lake, is separated from Bristol Dry Lake by Amboy Crater, a relatively young cinder cone and basalt flow thought to be less than 6000 yr old (Parker, 1963). Recent work at the USGS suggests that it is much older than this and may be on the order of 50,000 yr old. Before the basalt flow blocked off the northern end of the drainage area, creating Alkali Dry Lake, this area drained into Bristol Dry Lake; therefore, for most of the history of the Bristol Dry Lake basin, the total drainage area into the playa was significantly greater than it is today (about 4000 km^2). The present total drainage area into Bristol Dry Lake, excluding drainage from Cadiz Dry Lake and Alkali Dry Lake, is just over 2000 km^2.

Thompson (1929) proposed that during the Pleistocene a large lake occupied all of the Bristol and Cadiz basins; however, no evidence for the existence of such a large lake has been documented (Basset et al., 1959; Brown and Rosen, 1995). Recent work from false-color thematic mapping (TM) satellite images by Jacobs Engineering (1991) has shown the presence of a small delta with wash-over sediments and barrier islands in the northeast section of Bristol Dry Lake; however, the elevation of this delta is below any possible overflow to Cadiz Lake. In fact, the position and alignment of the delta suggest that the flow of water was from the surrounding mountains into Bristol Dry Lake only.

The Bristol Dry Lake basin has been interpreted to have alternated between subaerially exposed periods and times when shallow ephemeral water bodies covered the playa surface (Handford, 1982a; Rosen, 1989, 1991). Overall, even during periods of regional high rainfall, evaporitic conditions seem to have been dominant (Rosen, 1991). Modern meteorological conditions for the central Mojave Desert and Bristol Dry Lake specifically indicate a mean annual rainfall of less than 100 mm. Thompson (1929) noted that there are periods of 2–3 yr when no precipitation has been recorded.

A regional geologic study by Darton (1915) and hydrologic studies conducted by the USGS on the Mojave Desert region (Thompson, 1929) were the first work done in the area. Gale (1951), Durrell (1953), and Bassett et al. (1959) first studied the geology of the area through trenching and core analyses. Recently, Handford (1982a, b) conducted a reconnaissance study of the gross sedimentology of the surface and shallow subsurface of Bristol Dry Lake and attempted the first integrated interpretation of the area. Rosen (1989, 1991), Rosen and Warren (1990), and Brown and Rosen (1992, 1995) have provided the most detailed work in

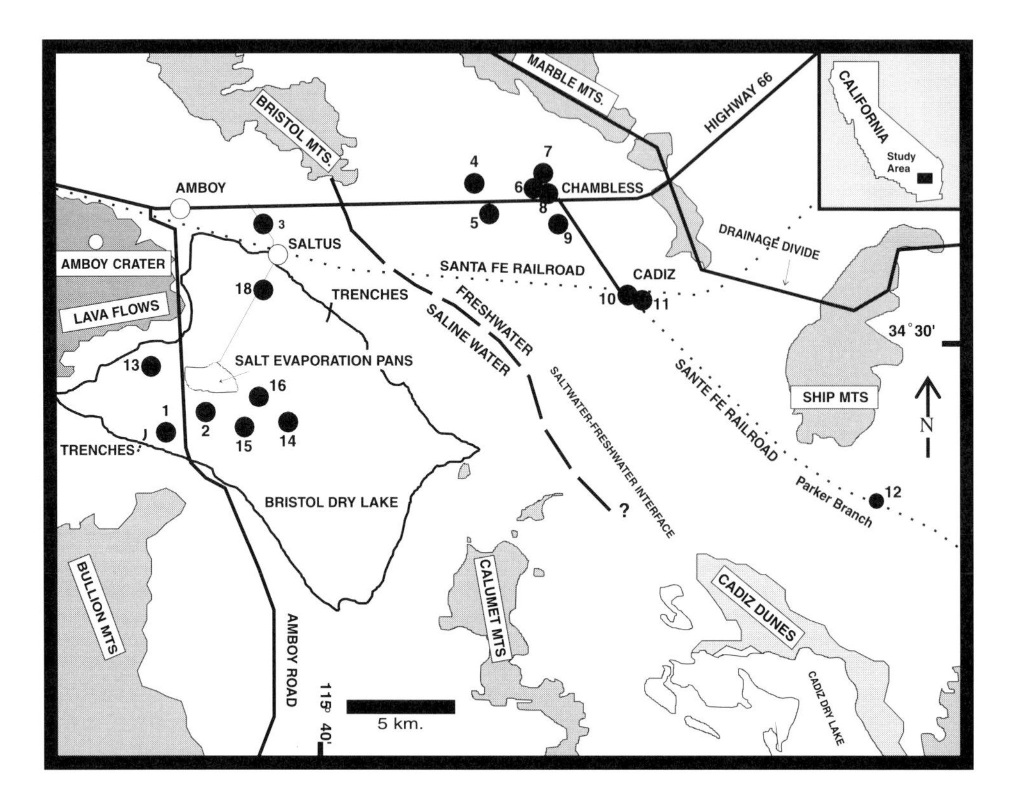

Figure 1—Location map of Bristol Dry Lake. Numbers correspond to water-wells and cores. CAES #1 = 17 and CAES #2 = 16 (see text and Figures 3 and 4). Alkali Dry Lake is off the map to the west and Danby Dry Lake is off the map to the southeast (modified from Rosen, 1991).

the basin on the sedimentology, geochemistry, and paleoclimate. An extremely detailed environmental impact assessment was made by Jacobs Engineering (1991) for the purposes of siting a landfill in the southern alluvial fan of the Bristol Mountains.

DEPOSITIONAL ENVIRONMENTS

Handford (1982a, b) recognized four broad subenvironments in Bristol Dry Lake. They are, from alluvial fan to basin center (1) the alluvial fan, (2) the playa-margin sand-flat sand wadi system, (3) the saline mud-flat, and (4) the salt pan at the basin center. These designations differ slightly from the Hardie et al. (1978) subenvironments but are useful for overall descriptions of the playa geometry. The four subenvironments recognized by Handford (1982a) are used in this paper to be consistent with his work; however, detailed field work indicates that these broad subenvironments are more diverse than previously described. The following descriptions detail further subdivision of the four broad depositional environments outlined by Handford (1982a) into more detailed sedimentologic units.

Alluvial Fan

The alluvial fan can be divided into three gradational subfacies based on the sediment grain-size. The terms used here are "proximal," "mid," and "distal fan." The proximal fan includes the part of the fan closest to the mountain source. At present, this area is dominated by channelled flow through arroyos incised through the older proximal and mid-fan deposits. The older fan deposits consist of coarse-grained gravel, cobble and boulder-size sediment interbedded with gravely-sands. Between the arroyos are cobble and boulder-covered surfaces darkly stained with desert varnish.

The mid-fan area is dominated by coarse-grained braided stream deposits. Broad, shallow channels are filled with low-angle trough cross-stratified cobbles, gravels, and sand. Extensive box-work calcrete is present in poorly sorted cobble strata of the upper part of the mid-fan. The calcrete zone (K-horizon) is best exposed on the north side of the basin in the Bristol Mountain fan arroyos (directly north of Amboy); however, the calcrete zone is widespread throughout the fans. The calcrete zone is approximately 1–3 m below the surface of the fan and is overlain by a dark red paleosol. The paleosol is disconformably overlain by a thin 1–3 m uncemented layer of low-angle trough cross-bedded braided stream deposits.

The distal fan is dominated by sheetflood deposits mostly composed of sand-size particles and fines. A thin veneer of the distal fan overlies, and grades into, the playa margin sediments. The well-developed paleosol and calcrete that extend up into the lower proximal fan are also exposed in the arroyo walls of the distal fan, approximately 1–3 m below the surface fan deposits.

In general, eolian deposits are ephemeral in Bristol Dry Lake because the playa surface lacks sufficient moisture to trap sediment. Although the wind regime in the basin is seasonal, the dominant wind direction is to the southeast. A large dune field, with some dunes up to 15 m high, is migrating to the southeast directly down-wind of the divide between the Bristol and Cadiz Basins. This suggests that a great deal of the sand is transported out of Bristol Dry Lake Basin and over the divide into Cadiz Basin. No unequivocally eolian deposits have been seen in core.

Playa Margin

Playa margin sediments are deposited in a transitional zone vegetated by sparsely populated halophyte shrubs, between the distal alluvial fan and the saline mud-flat. The sediments vary from silty sands to sandy muds that contain calcite-cemented nodules surrounding root holes of former halophyte shrubs.

Numerous wadi channels and distributary channels dissect the playa margin and bypass sediment and organic matter from the alluvial fans directly across the playa margin to the playa center. Some of these distributary channels are 1 m deep and 100 m across and extend well out onto the saline mud flat. Shallow trenches across these wadis and distributary channels reveal flaser-bedded sands and muds, as well as planar-bedded, sheetflow sands.

Where the playa margin passes into the saline mud-flat, a 0.15–0.5-km-wide zone of gypsum and celestite is present in a lens-shaped body in a ring around the entire basin. The details of this zone are presented in Rosen and Warren (1990). Field and chemical data indicate that discharging groundwater in this zone is saturated with respect to gypsum and celestite. Vertically aligned lens-shaped gypsum crystals grow displacively in the sediment near the top of the groundwater table where the sediment is still water-saturated and easy to move. Where the groundwater table has been constant for some time, the precipitation of gypsum has been greatest. In this area, gypsum may account for up to 90% of the volume of sediment in trenches that are 2 m deep. The matrix surrounding the gypsum consists of a mixture of detrital siliciclastics (quartz, feldspars, clays, and a large suite of heavy minerals), ranging from sand to clay-size particles, and authigenic rhombic calcite less than 1 μm in size (Rosen, 1989).

Saline Mud-Flat

The saline mud-flat is dominated by mostly homogeneous detrital mud. The dominantly grey brown mud is oxidized to a reddish brown color at the surface. Individual mud units are not thick, usually less than 0.5 m, and some are capped by mudcracked surfaces (Rosen, 1989). The modern saline mud-flat is up to 5 km across, and at times in the past extended across the entire basin. In addition to the muds, some of the larger wadis and distributary channels extend out into the mud-flat, creating flaser-bedded sand and mud out into the basin center. Eolian processes also bring some detrital sand and mud out to the basin-center, as well as some detrital gypsum and calcite. In addition, millimeter-laminated authigenic calcite is precipitated along with the detrital muds when there is a sufficient water body in the basin.

The surface of the saline mud-flat is hummocky and in many places water saturated. The hummocky nature is due to displacive halite growing just below the surface and pushing the sediment upward. Desiccation cracks and millimeter-thick efflorescent halite and calcium chloride crusts are also common on the mud-flat surface.

Where the groundwater table is below the surface of the playa, the water rises by capillary action to the surface and evaporates. Displacive halite crystals form in the sediment creating "self-rising ground." The water that rises to the surface keeps the surface cohesive enough so that the sediment is not deflated. Cementation by halite in the self-rising ground also helps retain sediment; however, where the surface lacks moisture because the groundwater table is too deep for water to rise to the surface by capillary action, the sediment dries out and is deflated. Deflation of the modern surface is particularly noticeable in the playa margin facies. Here, once buried displacive celestite and gypsum crystals are exposed on the surface of the playa.

Near the playa margin facies, the muds contain abundant molds of what were 5–10 mm displacive halite cubes. These molds are now empty but are rimmed with a manganese oxide stain. The molds form a network of cubic isolated pores in the mud and may be up to 35% of a given volume of mud. Toward the center of the playa, the molds are larger and are filled with halite. In the most basinward area of the saline mud-flat, giant hopper crystals of halite can be found, some up to 0.5 m across. Although the crystals are larger toward the basin center, they are also less abundant.

Small (5–30 mm) calcite concretions also occur in the saline mud-flat. Some of these concretions contain molds of halite crystals. These concretions are abundant at the surface in the saline mud-flat, in the basin center, and in the cores.

Salt Pan

The salt pans at the present surface (Figure 2) were about 0.4 m thick in 1988, and were composed of almost pure layers of 10–20 millimeter-size, vertically elongated chevron halite, separated by millimeter bands of detrital silt or mud. The halite is 99% pure, and both pans consist of approximately 30% chevrons, 40% clear diagenetic halite, and 30% porosity. Porosity decreases from 30% in the modern salt pan to 6% 3 m below the surface.

The most striking feature of the salt pan is the large halite tepee structures, caused by the force of crystallization of the halite (Rosen, 1991). Below the tepee crust is a water-saturated mud that is just below the point of halite precipitation. The mud is thixotropic and structureless and there are no displacive halite cubes in the mud.

Although the salt pan halite is only tens of millimeters thick at the surface, 3 m below the surface there is a 1-m-thick halite bed. In some cores, relatively pure (80–90%) halite may be tens of meters thick. The halite

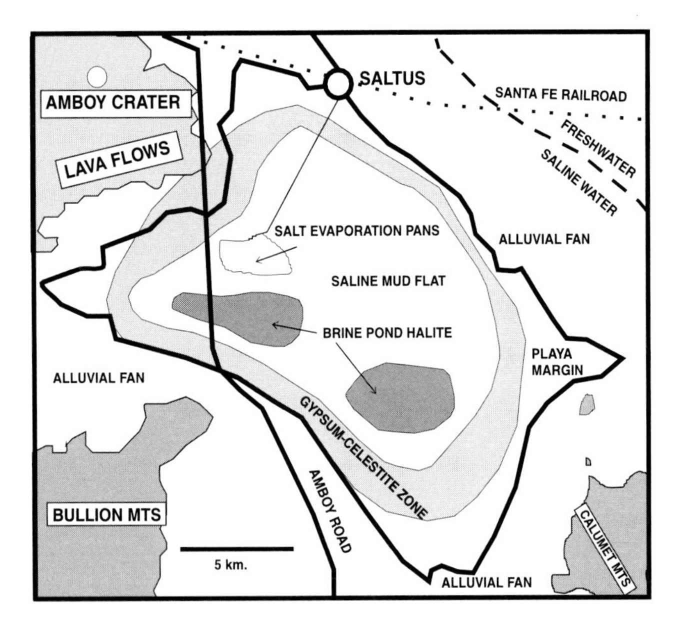

Figure 2—Facies diagram of the playa surface. The salt flat or brine pond halite is separated from the gypsum facies by the saline mud-flat. The saline mud-flat contains only detrital gypsum. Notice that in 1988 there were two salt pans on the surface of the playa (modified from Rosen, 1991).

beds in core have retained relatively little primary fabric. Although chevrons have been observed in the basin-center cores, they generally make up less than 2% of the fabric. Most of the subsurface halite is composed of interlocking centimeter-size crystals of clear equant halite containing less than 3% intergranular porosity. The amount of siliciclastic matrix mixed in with the halite varies from 3 to 50%.

FACIES DISTRIBUTION AND STRATIGRAPHIC FRAMEWORK

The distribution of evaporite minerals into concentric rings around the basin (Figure 2) implies a simple evaporation path of a relatively homogeneous groundwater as it moves toward the basin center (Rosen, 1989). The saline mud-flat, between the gypsum-celestite zone and the brine pan, is dominated by siliciclastic muds with only ephemeral displacive halite cubes and hopper crystals precipitating in the facies. The separation of the gypsum zone from the subaqueously deposited halite is a lateral characteristic of Bristol Dry Lake. The cause of this separation is most likely due to the limited availability of sulfate and the fact that the groundwater salinity must increase approximately three-fold before it reaches halite saturation. By the time this has occurred, the water has been ponded at the basin center. Only small amounts of sulfate minerals (gypsum or anhydrite) are found in the saline mud-flat and basin center, and much of this is wind transported from the deflating playa margin sediments.

In core, this spatial separation of the sulfate and halite zones is also evident in vertical succession. In the cores taken in the basin-center, brine pan and displacive halite alternates with muds from the saline mud-flat deposits for over 500 m and there is no appreciable accumulation of sulfate minerals (Figure 3a). In the basin margin, however, bedded, subaqueously deposited halite is not present in the sequence (Figure 3b); therefore, it appears that in any given vertical sequence it is unlikely that gypsum will be overlain by halite as one might expect in a normal prograding type of marine sequence.

Rosen (1991) argued that the bulk of the siliciclastic sediments in core were deposited in shallow, probably less than 10 m deep, perennial or ephemeral water bodies. The shallow lakes were either not particularly long-lived or hypersaline (or both) because no freshwater fauna was able to colonize these lakes. The red brown color of the muds indicates that the muds were subjected to repeated exposure during the development of the saline mud-flat stage after the lakes had dried up. Brown and Rosen (1995) provided additional evidence that Bristol Dry Lake has been a shallow lake or playa since the middle Pliocene (4 Ma).

BASIN HYDROCHEMISTRY

Hydrology of the Playa

The Bristol Dry Lake Basin has a remarkably diffuse groundwater flow into the basin. The consistent "bulls-eye" pattern distribution of the evaporite minerals is probably due to this diffuse groundwater input. The hydrology of playa basins also accounts for the lateral variability in evaporite beds.

The consistent bulls-eye pattern of evaporites in Bristol Dry Lake for over 500 m of sediment suggests that the hydrology of the basin has been largely the same for over 4 m.y. Variations in rainwater input produced by change in weather or climate has varied the amount of water in the basin. This variation was at times enough to produce standing water bodies, but essentially the unconfined nature of the aquifer system was the same. This created a predominantly aggrading facies pattern rather than a prograding pattern.

Chemistry of the Basin Waters

Groundwater data from Cadiz and Bristol dry lakes correlate well with brine data from the center of the Bristol Dry Lake (Figure 4). This suggests that Cadiz and Bristol dry lakes have similar sources for their ionic constituents, and possibly that the two basins are hydrologically connected. The Bristol Dry Lake brine is essentially Na-Ca-Cl, with equal but minor amounts of K and Mg. Chloride is by far the most abundant of all ions present in solution. In the groundwater from the alluvial fans, HCO_3 is actually 2 times more abundant than SO_4 and Cl, which occur in almost equal amounts, but by the time the water reaches the basin-center there is virtually none of either of these ions.

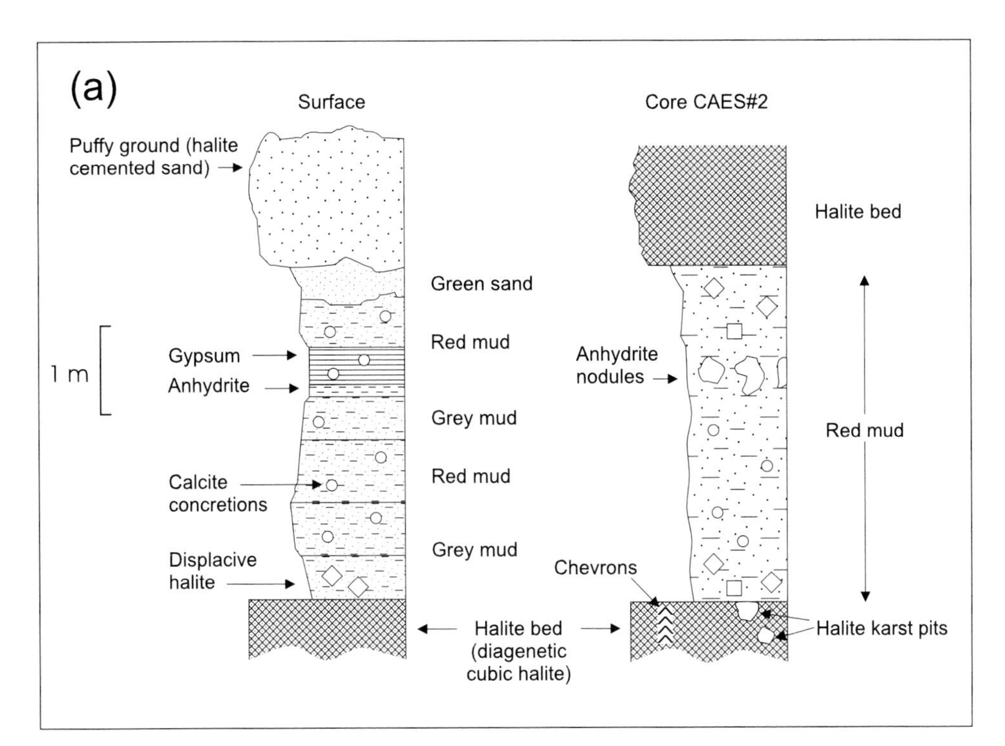

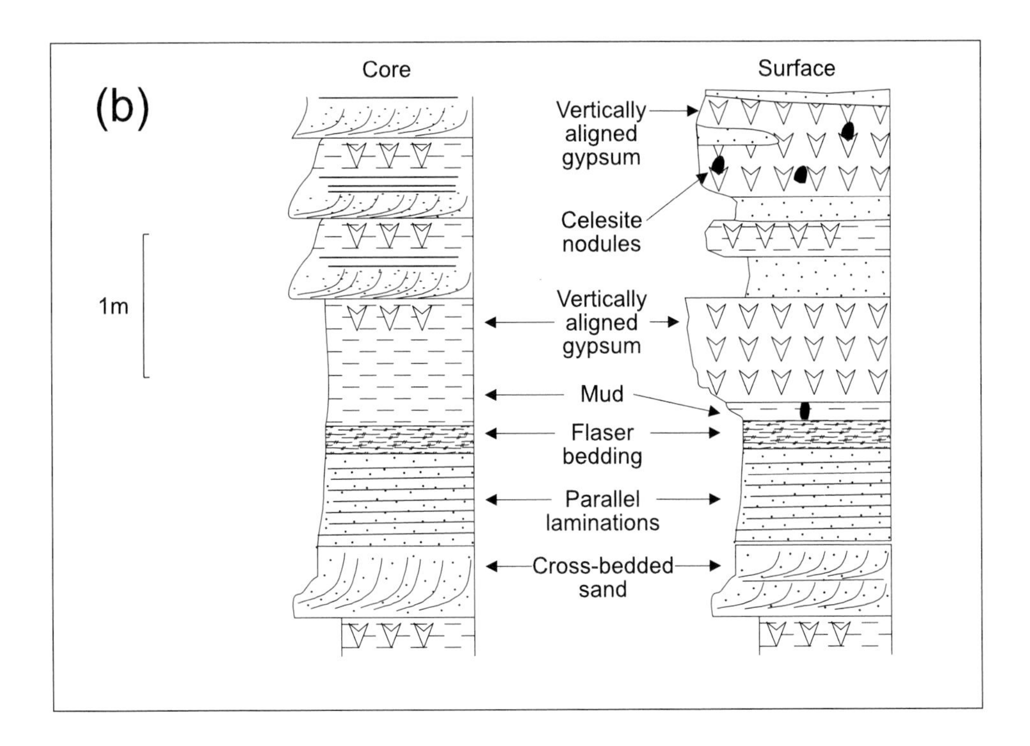

Figure 3—Comparison of surface and core stratigraphic sections (modified from Rosen, 1991). Sections are from (a) the saline mud-flat–salt pan transition and (b) the playa-margin–distal fan transition. Notice the lack of gypsum in the basin-center sequences.

The lack of SO_4 and HCO_3 in the basin center can be most easily attributed to the precipitation of calcite in the alluvial fans as calcrete and gypsum in the playa margin sediments. Figure 4 shows the increase in the groundwater sulfate concentration toward the distal part of the alluvial fan. Once gypsum saturation is reached, almost all the sulfate is used in the precipitation of gypsum.

Calcium concentrations plot closely with the sulfate concentrations, but bicarbonate is independent of both calcium and sulfate, and its concentration declines toward the basin center because bicarbonate is the limiting ion in calcite formation. Because calcium is relatively abundant in the groundwater and extremely abundant in the basin center, as soon as enough bicarbonate is in solution, saturation with respect to calcite

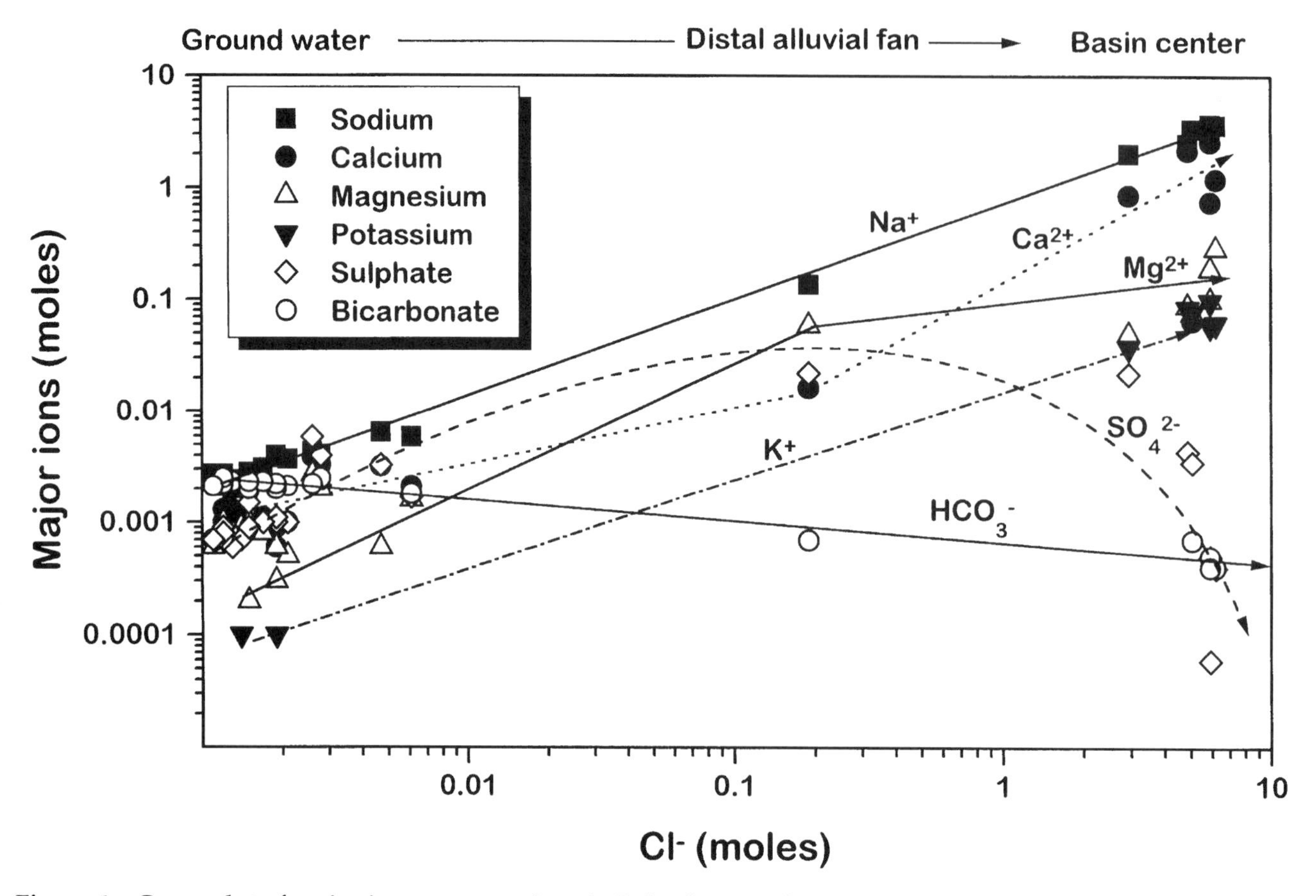

Figure 4—Cross-plot of major ion concentrations in Bristol Dry Lake groundwater as it flows from the basin margin to the center. Notice that sulfate is depleted by the time the water gets to the basin center, and the concentrations of bicarbonate and silica are always low.

is attained and bicarbonate is removed from solution, thus maintaining the constant low concentration profile. The presence of excess Ca in the brines accounts for the absence of dolomite in the basin.

AGE OF THE BASIN

The correlation of the tephra layers in Bristol Dry Lake with the tephra layers of known ages from other basins provides the basis for the chronology of the basin. In all, nine tephra layers from two cores have been correlated (Rosen, 1989). Two of the tephra layers can be correlated between the two cores (Figure 5).

There are no previous estimates or measurements of the age of the Bristol Dry Lake Basin. Although data from ^{36}Cl measurements in halite indicate that by 150 m depth, the chloride is approximately 2 m.y. old (N. Jannik, 1988, personal communication), this dating technique is still experimental and may have large errors associated with it (Phillips et al., 1983); nevertheless, there is an indication that the basin is relatively old. Correlations of the deepest tephra layers indicate that the basin is at least 3.7 ± 0.2 m.y. old at that those levels in the cores. In CAES #2 the lowest tephra is almost at the base of the core (500 m), but in CAES #1, there is still almost 270 m of sediment below the deepest ash. Using the lowest and highest calculated sedimentation rates for the core (Rosen, 1989), the bottom of CAES #1 would be somewhere between 6 and 10 m.y. old. Although no structural complications are apparent in CAES #1, faulting, such as seen in CAES #2, could greatly reduce the estimates.

Given that the bottom of CAES #1 is between 6 and 10 m.y. old, this is still not the maximum age limit for the basin. No cores taken in the center of Bristol Dry Lake have reached basement rocks; therefore, there is still an unknown depth of sediment below the cored interval that is older than the estimate. A crude estimate of the depth of the basin can be attempted by interpreting the Bouguer gravity map for the Needles Quadrangle. The total thickness of the basin-center is estimated to be approximately 680 m. If this calculation is correct, bedrock is approximately 100 m below the bottom of CAES #2; however, using the same methods, Biehler (1991) estimated the depth to basement to be over 1500 m.

ACKNOWLEDGMENTS

Thanks to Dianne Hewitt for typing the manuscript, and to Whitney Thordarson and Michelle Parks for redrafting the figures.

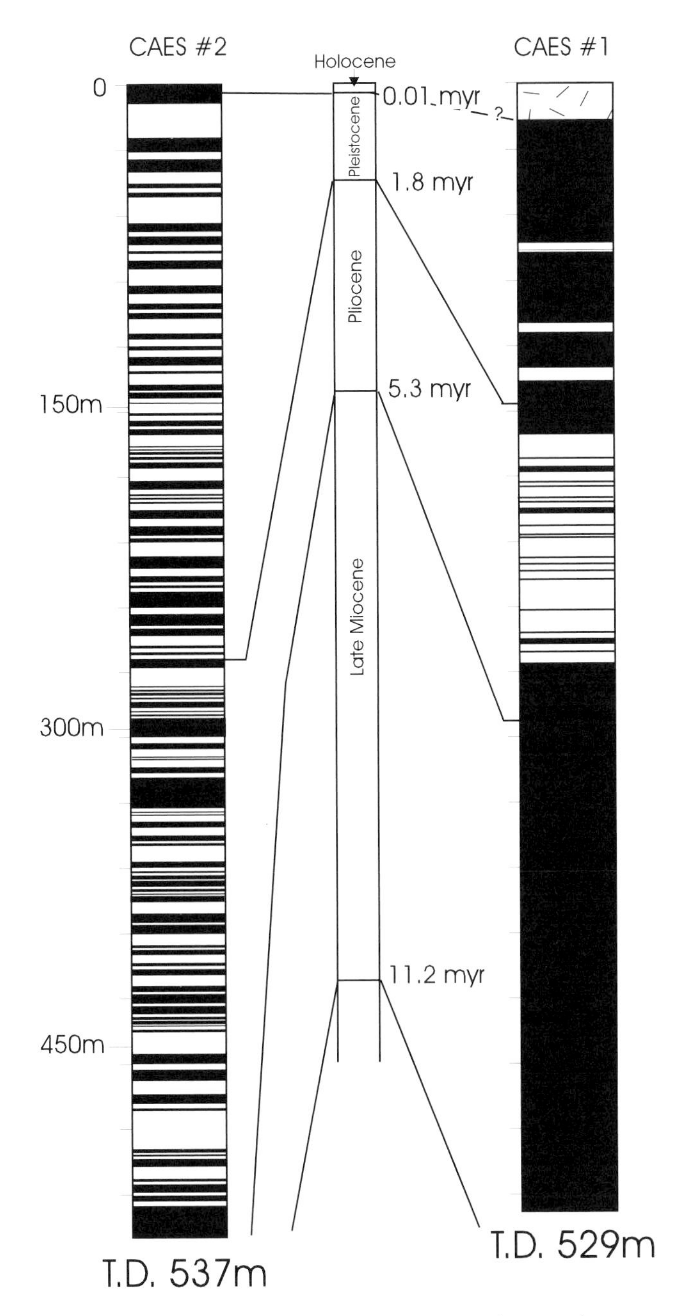

Figure 5—Preliminary age correlation of cores from the margin of the lake (CAES #1) and the center of the lake (CAES #2) based on tephra correlation (modified from Rosen, 1989). The black is mud and the white is halite. Crossed-line pattern at the top of CAES #1 is Amboy Crater lava flow. CAES #2 is 40% halite.

REFERENCES CITED

Bassett, A. M., D. H. Kupfer, and F. C. Barstow, 1959, Core logs from Bristol, Cadiz, and Danby Dry Lakes, San Bernardino County, California. U.S. Geological Survey Bulletin 1045-D, p. 97–138.

Biehler, S., 1991, Gravity and magnetic studies of the RAIL•CYCLE project site, San Bernardino County, California. Unpublished report prepared for Jacobs Engineering Group, Inc.

Brown, W. J., and M. R. Rosen, 1992, The depositional history of several desert basins in the Mojave Desert: Implications regarding a Death Valley–Colorado River hydrological connection, *in* R. E. Reynolds, ed., Old Routes to the Colorado, Special Pub. 92–2, San Bernardino County Museum Association, Redlands, p. 77–82.

Brown, W. J., and M. R. Rosen, 1995, Was there a Pliocene–Pleistocene fluvial-lacustrine connection between Death Valley and the Colorado River? Quaternary Research, v. 43, p. 286–296.

Darton, N. H., 1915, Guidebook of Western United States; part C, The Santa Fe Route. U.S. Geological Survey Bulletin 613, p. 152–155.

Durrell, C., 1953, Celestite deposits at Bristol Dry Lake, near Amboy San Bernardino County, California, *in* Geological investigation of strontium deposits in Southern California: Special Report California State Mining Bureau 32, p. 37–48.

Gale, G. K., 1951, Geology of the saline deposits, Bristol Dry Lake, San Bernardino County, California. Special Report California State Mining Bureau 13, 22 p.

Handford, C. R., 1982a, Sedimentology and evaporite genesis in a Holocene continental-sabkha playa basin—Bristol Dry Lake, California. Sedimentology, v. 29, p. 239–253.

Handford, C. R., 1982b, Terrigenous clastic and evaporite sedimentation in a Recent continental-sabkha playa basin, Bristol Dry lake, California. *in* C. R. Handford, R. G. Loucks, and G. R. Davies, eds., Depositional and Diagenetic Spectra of Evaporites—A core workshop. Society of Economic Paleontologists and Mineralogists Core Workshop No. 3, p. 65–74.

Hardie, L. A., J. P. Smoot, and H. P. Eugster, 1978, Saline lakes and their deposits: a sedimentological approach, *in* A. Matter and M. E. Tucker, eds., Modern and Ancient Lake Sediments, International Association of Sedimentologists Special Publication 2, p. 7–41.

Jacobs Engineering, 1991, Geological, hydrologic and geotechnical site characterisation study, Bolo Station Facilities, San Bernardino County, Phase II, Volume I.

Parker, R. B., 1963, Recent volcanism at Amboy Crater, San Bernardino County, California. Special Report 76, California Division Mines and Geology, 21 pp.

Phillips, F. M., G. I. Smith, H. W. Bentley, D. Elmore, and H. E. Grove, 1983, Chlorine-36 dating of saline sediments: preliminary results from Searles Lake, California. Science, v. 222, p. 925–927.

Rosen, M. R., 1989, Sedimentologic geochemical, and hydrologic evolution of an intracontinental, closed-basin playa (Bristol Dry Lake, Ca.): a model for playa development and its implications for paleoclimate. Ph.D. Dissertation, University of Texas–Austin, 226 p.

Rosen, M. R., 1991, Sedimentological and geochemical constraints on the evolution of Bristol Dry Lake Basin, California, USA. Palaeogeography,

Palaeoclimatology, Palaeoecology, v. 84, 229–257.
Rosen, M. R., 1994, The importance of groundwater in playas: a review of playa classifications and the sedimentology and hydrology of playas, *in* M. R. Rosen, ed., Paleoclimate and basin evolution of playa systems. Geological Society of America Special Paper 289, p. 1–18.
Rosen, M. R., and J. K. Warren, 1990, The origin and significance of groundwater-seepage gypsum from Bristol Dry Lake, California, USA. Sedimentology, v. 37, p. 983–996.
Thompson, D. G., 1929, The Mojave Desert region, California. US Geological Survey Water-Supply Paper 578, 759 p.

Cserny, T., and E. Nagy-Bodor, 2000, Limnogeology of Lake Balaton (Hungary), *in* E. H. Gierlowski-Kordesch and K. R. Kelts, eds., Lake basins through space and time: AAPG Studies in Geology 46, p. 605–618.

Chapter 58

Limnogeology of Lake Balaton (Hungary)

Tibor Cserny
Elvira Nagy-Bodor
Geological Institute of Hungary
Budapest, Hungary

A BRIEF HISTORY OF RESEARCH

Lake Balaton, an important and valuable national treasure, is the largest shallow lake in central Europe. Since the end of the last century, the lake and its environs have been investigated by a great number of specialists (Lóczy, 1884, 1894, 1913; Cholnoky, 1897, 1918; Entz and Sebestyén, 1942; Bulla, 1943, 1958; Zólyomi, 1953, 1987, 1995; Bendefy and V. Nagy, 1969; Rónai, 1969; Marosi and Szilárd, 1977, 1981; Müller and Wagner, 1978; Somlyódy and van Straten, 1983; Herodek and Máté, 1984; Vörös et al., 1984; Istvánovics et al., 1986, 1989; Máté, 1987). Despite the continuous work on Lake Balaton, modern aspects of mud development, eutrophication, and ecological disequilibrium are still not well understood. Specialists at the Geologic Institute of Hungary and all over Hungary have begun an extensive program to collect and analyze data from the lake and its catchment area to determine its age and ecological development. Aging of a lake is a natural process; however, its rate may be enhanced by environmental (particularly human) influences.

Since 1965, the Geologic Institute of Hungary has performed complex geologic surveying in order to assess the current state of the recreational area at Lake Balaton, to mitigate the existing hazardous situation caused by eutrophication and increase in fine-grained sedimentation, and to determine long-term goals. As a first step, from 1965 to 1979, an engineering geologic survey of some 780 km^2 of the shoreline zone was performed at the scale of 1:10,000. It was followed (1981–1990) by an environmental geologic mapping of the 5200 km^2 extended recreational area and a detailed investigation of the lake bed (Raincsák and Cserny, 1984; Cserny, 1987, 1990; Papp, 1992). Environmental geologic investigations of the lake are still in progress but are expected to be accomplished in the near future (Mihältz-Faragó, 1983; Bodor, 1987; Cserny, 1987; Cserny and Corrada, 1989; Cserny et al., 1991, 1995).

Research on Lake Balaton performed between 1981 and 1995 was carried out in three stages:

(1) From 1981 and 1986, a total of 17 boreholes were drilled into the lake bed and subsequently analyzed. Complex borehole logs show the results from sedimentologic, geophysical, and geochemical (organic and inorganic) tests as well as mineralogic, petrologic, and paleontologic studies of the sediments (see Cserny, 1994). This has highlighted the most important features of the more recent lacustrine deposits and carbonate mud.

(2) From 1987 to 1989, a continuous geophysical (seismo-acoustic and echograph-based) logging was performed, over a total length of 370 km. The study and evaluation of reflection logs covering Lake Balaton has led to compilation of a detailed map showing the thickness of the unconsolidated mud in the lake and a seismo-stratigraphic–tectonic map of the basement.

(3) From 1989 to 1995, a total of 16 new boreholes were drilled. By using isotopic geochemical analyses and by extending the range of paleontologic analyses (palynology, diatoms, ostracods, molluscs), the geohistory of Lake Balaton and its environment, including its paleoecologic and paleoclimatic features, have been outlined.

SUMMARY OF LIMNOGEOLOGIC RESULTS

A summary of the major characteristics of the sediments, their spatial distribution, as well as a reconstruction of the geologic and paleoecologic history of Lake Balaton and its immediate surroundings follows.

Geologic Background of Lake Balaton

Lake Balaton stretches in a southwest-northeast direction at the foothills of the Transdanubian Central Range in the western part of Hungary. At the northern shoreline of the lake, Paleozoic and Mesozoic basement rocks are partly covered by thin Tertiary and Quaternary sediments. On the southern side of the lake, upper Neogene sediments a few hundred meters thick define its border. Lake Balaton occupies a longitudinal piedmont depression along the southern margin of the Transdanubian Central Range, which is broken by mainly northwest-southeast directed cross-faults and subsidence basins (Rónai, 1969). The catchment area indicated in Figure 1 covers 5800 km^2.

Lake Balaton formed on loose Pannonian (upper Miocene) sediments formed in the one-time Pannonian

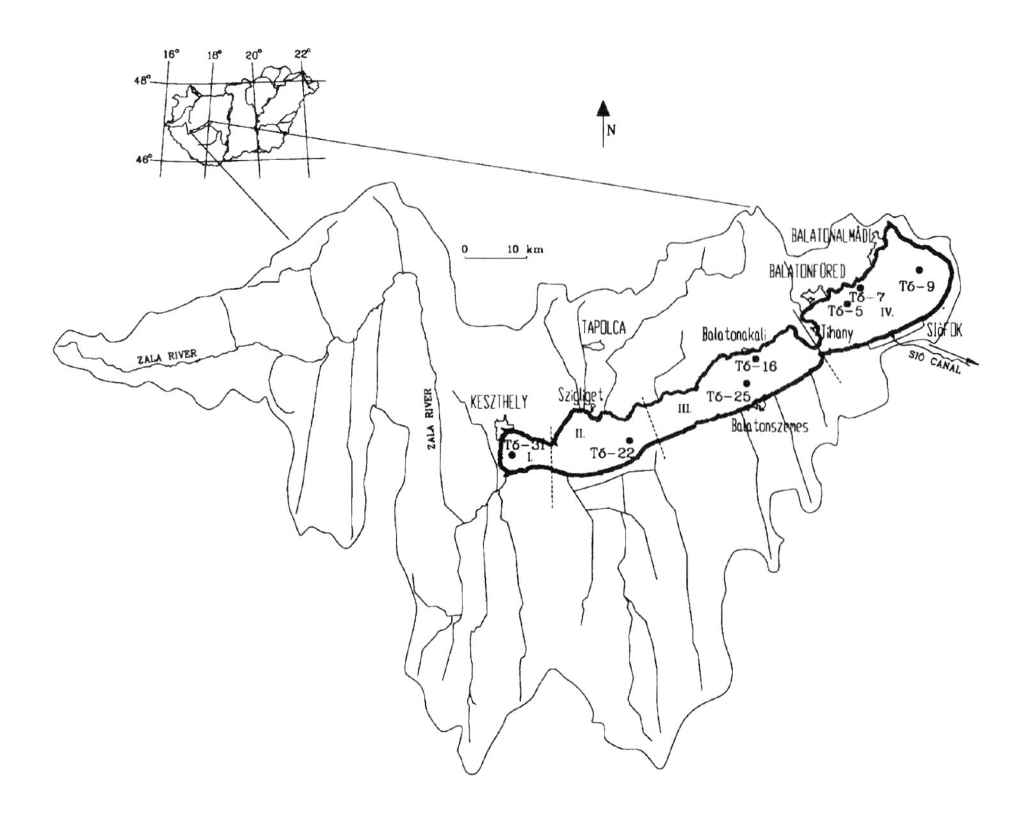

Figure 1—The catchment area, river system, and subbasins of Lake Balaton with a layout of reference boreholes. I = Keszthely subbasin, II = Szigliget subbasin, III = Szemes subbasin, IV = Siófok subbasin, Tó-9, shallow drillings.

(upper Miocene) lake representing a bay of the Tethys; consequently, the salinity of water gradually decreased becoming finally a freshwater environment. Two geomorphologic areas are distinguished: the comparatively high, so-called "Balaton basin unit" consisting of nonlacustrine deposits and the "lake basin unit" regarded as the area inundated at the lake's highest water level. The "Balaton basin unit" is substantially older than the "lake basin unit." Only sediments up to 15,000 yr BP in age are considered here as the lake basin unit. The "Balaton basin unit" was formed essentially through tectonic processes (uplift of separate blocks with different velocity), whereas the "lake basin unit" has a mixed tectonic-deflation origin.

Seismo-acoustic time sections identified seven zones, as well as their features and tectonic elements in the basement (Cserny, 1994). Geologic identification of the related zones was based on 33 underwater borehole cores.

Zone 1: Near the northern shore of the lake. Cemented debris derived from the Balaton Highlands (northern part of the lake) may be present; it is too hard to penetrate.

Zone 2: Located between the Tihany Peninsula and the western border of the Szigliget subbasin, a southward continuation of Zone 1. Basement consists of silt with very low sand and clay contents. Grain size distribution is similar to that of recent sediments of Lake Balaton but appears to be a bit more compact.

Zone 3: Limited to some spots in the central (Szemes) subbasin, with greater extension in the western subbasins. It has commonly a gradual transition southward from Zone 2. It includes a horizontally deposited layer consisting of clayey silt and clay.

Zone 4: Occurs in the central (Szemes) subbasin, and differs from the first three zones by being composed of several layers. Toward the north, it has a contact along a fault with Zone 3. The thickness of this zone may reach 30 m. The upper part is made up of compact clayey silt, whereas the lower hard "marker horizon" is identified as Miocene limestone or sandstone.

Zone 5: This zone can be found in the area west of Zone 4 and in the Siófok subbasin. In the north, the layers begin to dip toward south-southeast in a similar manner to Zone 4. Zone 5 differs from Zone 4 because it can clearly be classified into three layers above the "marker horizon": the deepest "O" layer is lithologically more homogeneous than the others; the second, the "A" layer, is heterogeneous and horizontally rather varied; and the uppermost layer, the "B," is very strongly microstratified and offers only little horizontal variability. The layers are silty clay and silt.

Zone 6: Located in the southern part of the Szigliget subbasin in a comparatively narrow strip, within Zone 5. It is separated from Zone 5 on the basis of its horizontal stratification pattern. The layers consist of the similar components as Zone 5, i.e., silty clay and silt.

Zone 7: Along the southern shore of the lake forming a 500–1500-m-wide strip. This zone is lithologically identified as a fine-grained sand redeposited in water and being rather compact.

In the basement beneath Lake Balaton, faults (with offsets greater than 10 m) and fracture zones (wider than 150 m) were identified. The main fault zone along the western shore of the Tihany Peninsula divides the basement into two parts. In the eastern part of the lake, longitudinal faults seem to be dilatational (secondary, normal) and become indistinct along their track. Rarely crossing geologic boundaries, transverse faults can be observed in two directions: north–northwest-south–southeast and approximately north-south. In the western part of the lake, several normal faults and some transcurrent faults were identified striking in north–northwest-south–southeast direction. The active nature of the faults is proven by earthquake activity taking place in the northeastern part of the lake in 1987 and 1989.

Features of the Lacustrine Sediments

A lithologic overview of the lake basement and lake sediments of the "lake basin unit" is represented in Figure 2. The late Pannonian (upper Miocene) basement of the lake was unconformably overlain by the sedimentary sequence of Lake Balaton that begins with a few-centimeter-thick, gravelly sand or sand that is generally overlain by a decimeter-scale peat bed, followed by massive calcareous mud with an average thickness of 4 m. The mud is mostly argillaceous silt with 50–70% carbonate. In the lower one-third of the mud sequence, which is typical for the entire lake, well-preserved molluscs *(Lythogliphus naticoides, Pisidium henslowanum, Valvata piscinalis, Bithynia tentaculata)* are present.

G. Müller's (1970, 1978) work on the carbonates of Lake Balaton illustrated that the lacustrine carbonates have a very high primary porosity and consist dominantly of magnesium-bearing calcite, and subordinately of dolomite and calcite and even of protodolomite at some sites. Our more recent work shows the constituent minerals to be very unstable chemically. For the magnesium-bearing calcite, the Ca/Mg ratio decreases with increasing depth. Borehole samples exhibiting two maxima in MgO content are essentially concentrated in the Szigliget, Szemes, and Siófok subbasins, whereas those containing no Mg-calcite are confined to the western Keszthely subbasin. This phenomenon is due to the diluting effect of inflow from the Zala River from the west (see Figure 1). Borehole samples with a low Mg-calcite content are observed at about the middle of the longitudinal axis of the lake, whereas those containing Mg-calcite and normal calcite have been found in its eastern basin. In the lower one-third of the lacustrine sequence, sulfur segregation and high enrichment in carbon have been observed. ^{18}O and ^{13}C isotope values of the carbonates record a gradual warming of the climate in the Holocene (Cserny et al., 1995).

Age of the peat bed encountered at the base of the Holocene sequence (see Figure 2) has been dated using radiocarbon, ranging from 12,500 to 10,500 yr BP. Peat development in the area of Lake Balaton began in the late Pleistocene, during the Bölling warming-up period following the Oldest Dryas (Figure 2). Peat formation persisted for a time (maximum 1500 yr) and was most widespread during the Alleröd.

Samples from a few boreholes have also been tested for radioactive isotopes. The ^{40}K nuclide and the three radioactive decomposition series (uranium, thorium, actinium) were of greatest importance. Activity values of radioactive isotopes show a similar tendency—a decrease with depth. Ratios of the three major radioactive isotopes (^{238}U, ^{232}Th, ^{40}K) show a good correlation all throughout the lake, which suggests that the samples from the boreholes were undisturbed.

Artificial ^{137}Cs isotope contamination has been detected in the uppermost mud layer (~0.5 m interval). This artificial isotope can only be traced in the atmosphere since 1951. Two maxima can generally be detected in the deposits, namely, the year prior to the nuclear test ban in 1964 and the Chernobyl disaster in 1984. This allowed for calculation of the rate of mud sedimentation in the lake. The lack of nuclear isotopes in lake sediments indicated subaquatic erosion, while their presence indicated mud accumulation. For example, during the past 40 yr the rate of mud deposition has been 14 mm per annum in the middle of the Szigliget subbasin under steady hydrological conditions, and 5 mm per annum at the eastern boundary of this subbasin. The rate of sedimentation throughout the lake exhibited local differences, as well as extreme differences in amounts, especially during contamination from the Chernobyl nuclear disaster. For example, over the past 5 yr, the average rate of deposition within the Szigliget subbasin was 2 cm per annum. In the middle of the Siófok subbasin, subaquatic erosion occurred over the same time period, whereas in the Szemes subbasin, an additional 20–30 cm on average was deposited. The fact that ^{137}Cs isotope peaks appear at a depth range of 2–3 cm indicates that the storms over the lake and bioturbation may have caused approximately 2–3-cm-thick layers to be reworked.

The unit density of lacustrine deposits gradually increases with increasing depth to 5 m from 1.0 to 1.4 g/cm^3 and from 1.7 to 2.0 g/cm^3 for argillaceous mud and fine sand, respectively. Density of calcareous deposits ranges from 2.2 to 2.3 g/cm^3, whereas their porosity, which is higher than 50% on the top of the mud sequence, decreases to 20–30%, as a function of compaction. Based on a consistency index, deposits changed from a "very soft" to a "soft" state in the lower part of the section. The plasticity index of these sediments varies as a function of the mud content, attaining in some cases 100%. This value is higher than expected (>10%), which is due to the intense weathering of the source rocks of the catchment area.

Trophic conditions of modern lakes are determined by the chlorophyll-a content per unit volume of water. Lake water quality can be thus characterized by low to high productivity, trophic conditions from oligotrophic, mesotrophic, eutrophic, to hypertrophic. In the samples of cores, percentage of algae, as well as the spores and pollen of water plants, were counted.

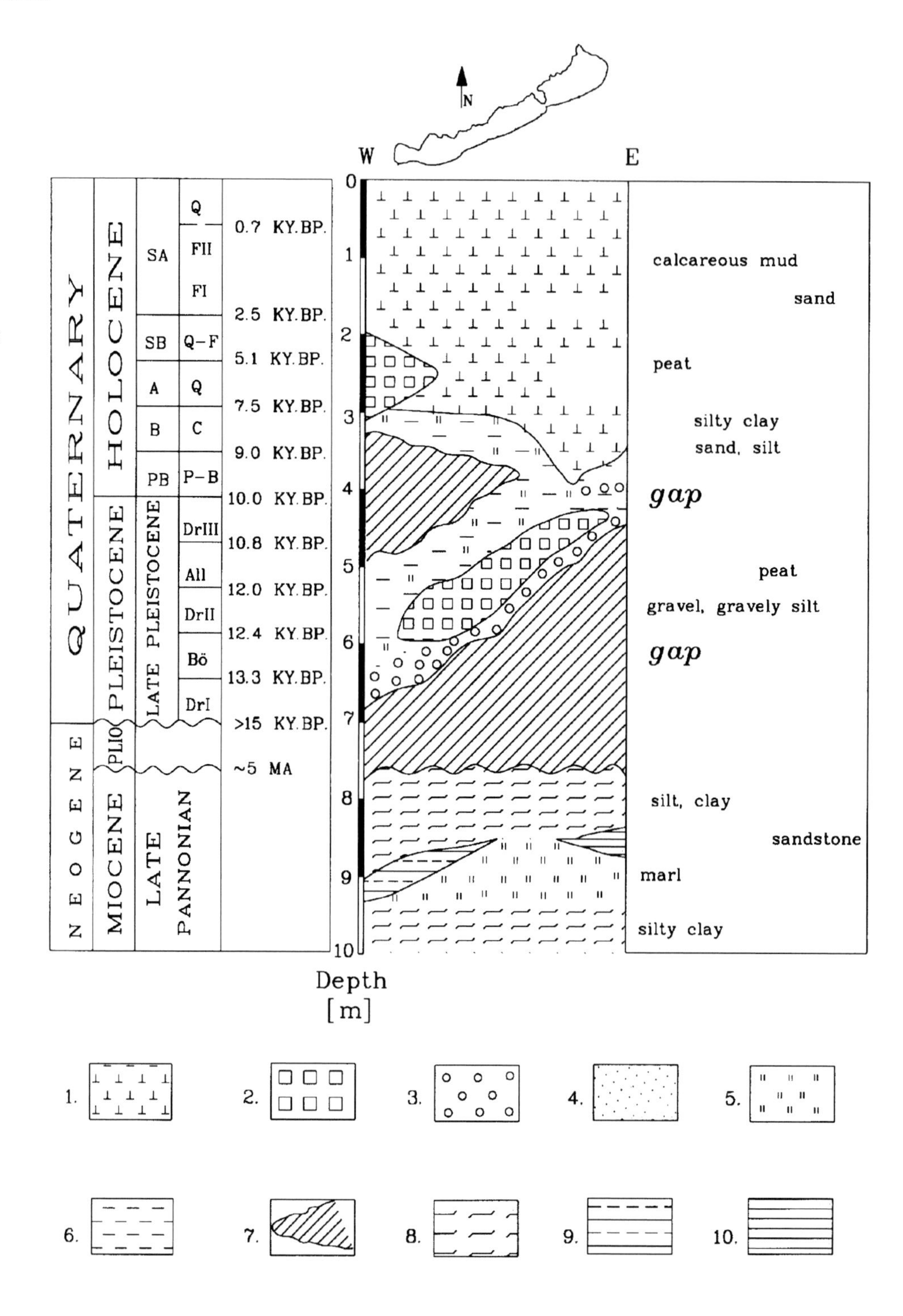

Figure 2—Geology of the Quaternary and pre-Quaternary sequence of Lake Balaton. Climatic zones: SA = subatlantic, SB = subboreal, A = Atlantic, B = boreal, PB = preboreal. Ages: DIII = younger Dryas, All = Alleröd, DII = old Dryas, Bö = Bölling, DI = oldest Dryas. Vegetation phases: Q = *Quercus*, F = *Fagus*, Q–F = *Quercus–Fagus*, C = *Corylus*, P–B = *Pinus–Betula*. Types of sediments: 1 = calcareous mud, 2 = peat, 3 = gravel, 4 = sand, 5 = silt, 6 - clay, 7 = gap, 8 = silty clay, 9 = marl, 10 = sandstone.

Ratios of algae to total number of plant taxa give a rough indication of trophic condition.

Spatial Distribution of the Lacustrine Sediments

Quaternary deposits in Lake Balaton have an average thickness of 5 m. The upper portion, ranging from 0.1 to 0.2 m in thickness, is very soft and colloidal. Mud thickness in the basin varies because basement morphology is also very variable. At higher elevations of the basement, mud thickness is reduced to 1.0 to 1.5 m,whereas in depressions, it increases to 8.0 m. A maximum thickness of 10 m has been found at the mouth of the Zala River (Cserny 1994). The average mud thickness is 6 m, 5 m, and about 4 m in the western, middle, and eastern subbasins of the lake, respectively. Radiocarbon dating (12,000 up to 13,000 yr BP) of the peat layers at the bottom of the Quaternary sequence and the variable mud thickness suggest a mud accumulation rate of 0.38–0.48 mm per annum for the entire Quaternary. The lowest value was observed in the Siófok subbasin, whereas the highest value was detected in the Keszthely subbasin.

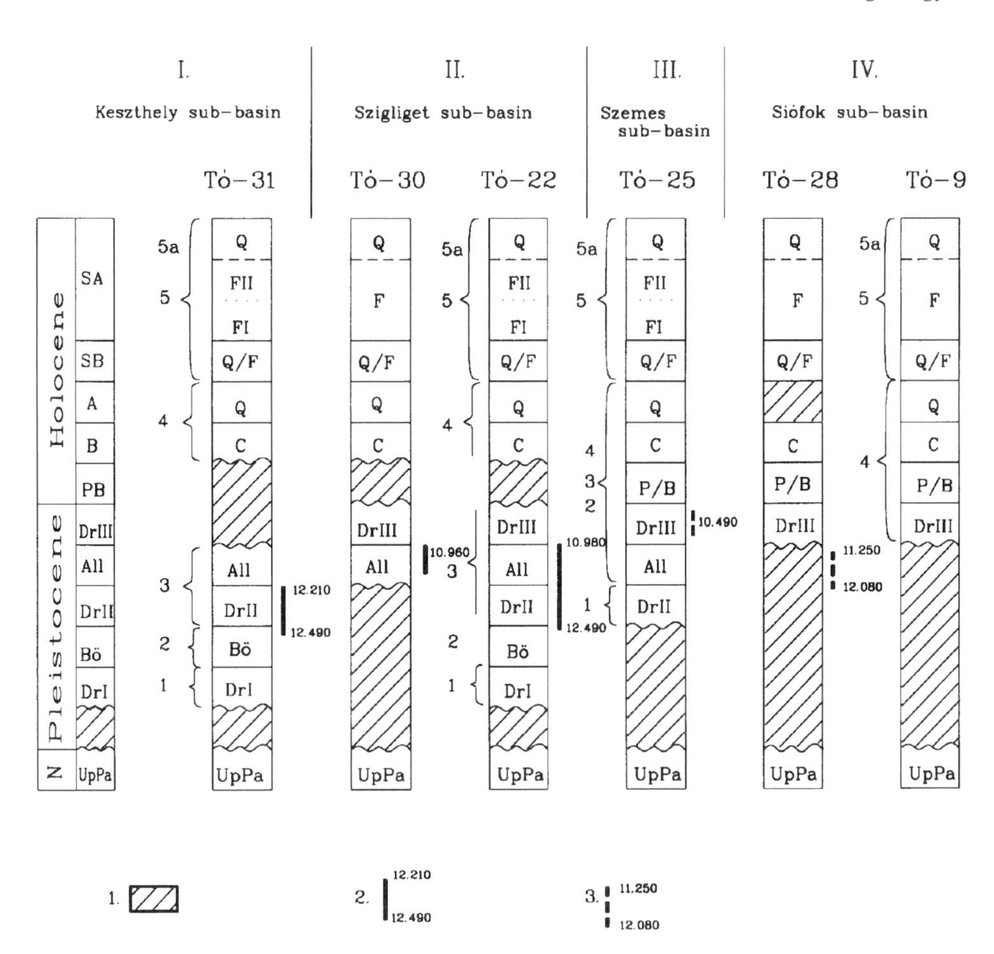

Figure 3—Correlation of palynologic data in the boreholes. Climatic zones: SA = subatlantic, SB = subboreal, A = Atlantic, B = boreal, PB = preboreal. Ages: DIII = younger Dryas, All = Alleröd, DII = old Dryas, Bö = Bölling, DI = oldest Dryas, N = Neogene, UpPa = upper Pannonian. Vegetation phases: Q = *Quercus*, F = *Fagus*, Q–F = *Quercus–Fagus*, C = *Corylus*, P–B = *Pinus–Betula*. Stages of lake development: 1 = preliminary (forming of ponds), 2 = shallow water, 3 = marsh, 4 = shallow water, 5 = open water, 5/a = recent vegetation. Tó-25, shallow drillings, 1. gap (hiatus), 2. peat and its radiocarbon age in the borehole, 3. peat and its radiocarbon age near the borehole.

Correlation of drilling profiles of lake sediments in different subbasins using detailed palynologic data is shown in Figure 3; in addition, interpreted climatic zones, ages of related sediments, vegetation phases, and stages of lake development are compared.

Formation of the Lake and its History of Evolution

Formation of the lake and its history is outlined in a number of studies conducted by Lóczy (1913), Cholnoky (1918), Kéz (1931), Bulla (1943), Zólyomi (1953, 1987, 1995), Erdélyi (1963), Rónai (1969), Mike (1980a, b), Marosi and Szilárd (1981), Bodor (1987), Cserny (1987, 1994), and Cserny et al. (1991). The evolution of Lake Balaton itself in the Late Pleistocene–Holocene is shown in Figures 4 and 5. Evaluation of samples taken from boreholes along the shore will take place in a future project to investigate the maximum areal extent of the lake.

Upper Miocene (Upper Pannonian)

Morphologically, Lake Balaton can be subdivided into four subbasins (Keszthely, Szigliget, Szemes, and Siófok subbasins), each having a different history of evolution during the late Pleistocene and early Holocene. Evidence for differences in the evolution of these four subbasins can be found as early as the upper Miocene. Drill 3–5 m deep into the Pannonian (upper Miocene) formations contain *Congeria balatonica,* and rarely *C. ungulacaprae* (Korpás-Hódi, personal communication, 1999). These formations can clearly be distinguished from Quaternary lacustrine deposits through analyses of sporomorphs and ostracods, allowing detailed stratigraphic classification and reconstruction of the paleoecologic history of each subbasin (Figures 6–9).

The unconformity between Pannonian (upper Miocene) and Quaternary lacustrine deposits (Figure 2) is attributed to a hiatus in sedimentation. Although the Pannonian (upper Miocene) lake covered the whole study area in the Upper Miocene, it cannot be regarded as the predecessor of Lake Balaton. The boundary between Pannonian (upper Miocene) and Quaternary lacustrine deposits can also be identified by the sharp change in the sporomorph faunas associated with a thin gravel band found in the majority of the boreholes. Sporomorphs of plants preferring warm climates and characterized by an abundance of species with few individuals vanished at this boundary and were replaced by taxa preferring cooler climate. The abundance of sporomorphs and ostracods through time show the fluctuation of salinity between the freshwater oligohaline and mio-mesohaline (Korpás-Hódi, personal communication, 1999). No diatoms were observed in boreholes penetrating the upper Miocene sediments (Cserny et al., 1991). For example, in the

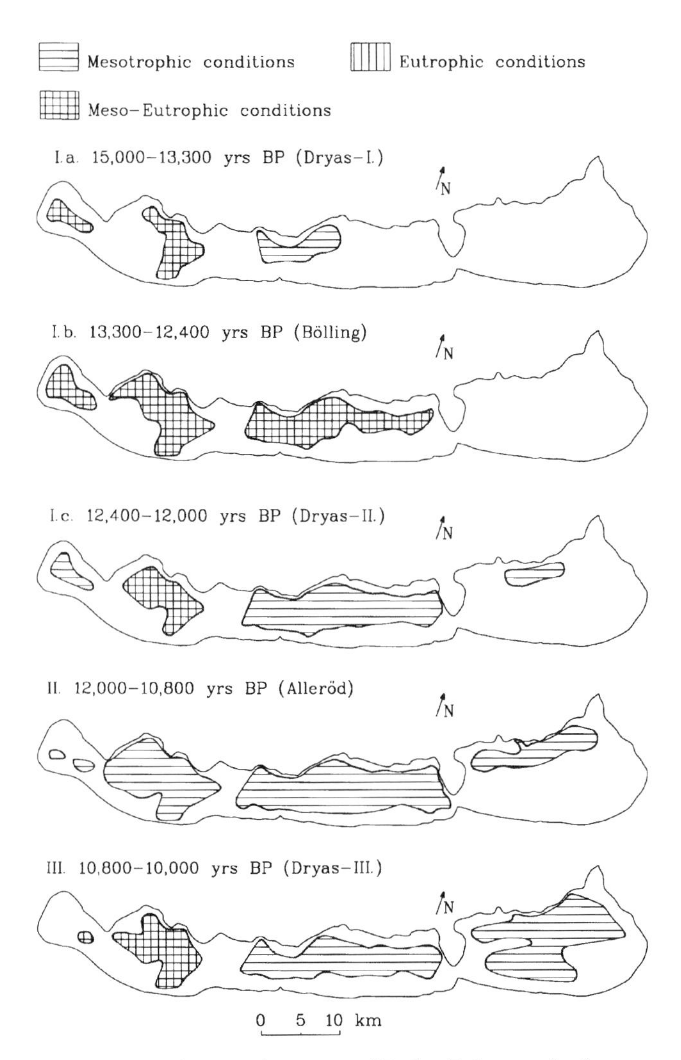

Figure 4—The areal extent of Lake Balaton during the late Pleistocene.

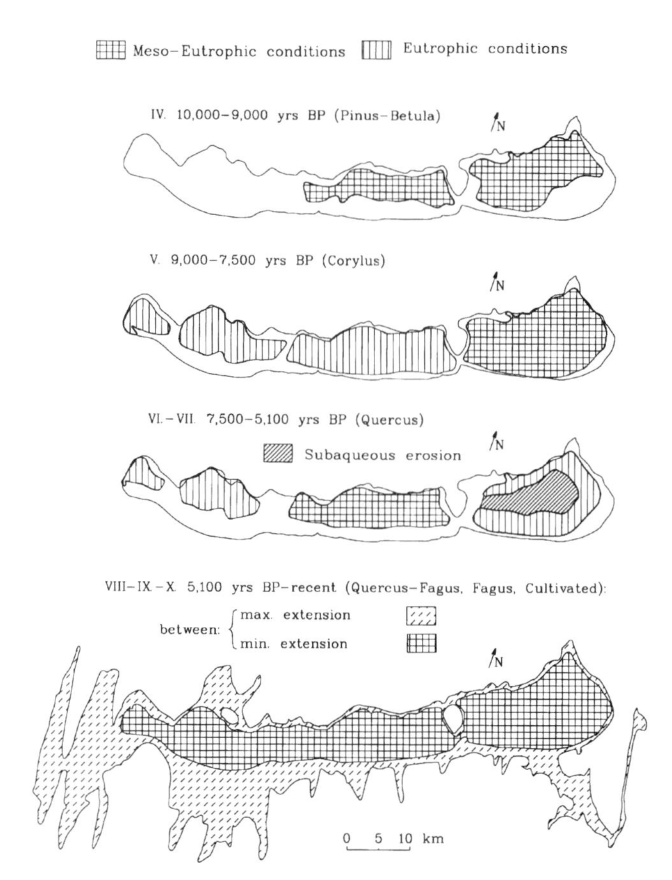

Figure 5—The areal extent of Lake Balaton during the Holocene.

Keszthely subbasin a sporomorph association indicating a paleoenvironment dominated progressively by freshwater was found associated with ostracods characteristic of extremely shallow water with frequent subaerial exposure. Ostracods found in the Szigliget and Szemes subbasins indicated somewhat deeper water, but the deepest water occurred in the easternmost Siófok subbasin.

Pleistocene

Erosion prevailed in the lake basin during the Pleistocene; however, alluvial sand and gravel, as well as eluvial red and variegated clay indicating comparatively abundant precipitation, were deposited during warmer periods in the environs. Differential vertical movements continued along ancient major reactivated faults resulting in the formation of some depressions. Toward the end of the Pleistocene, northern winds became stronger resulting in the formation of deflation depressions including the area covered by present-day Lake Balaton, at the base of the Transdanubian Central Range. Due to deflation and general erosion, as much as 100–150 m of sedimentary complex were removed from the Pannonian (upper Miocene) surface. History of the evolution of the present four subbasins was even more varied in the Late Pleistocene than in the Pannonian (Figures 6–9). Radiocarbon dating of peat pinpoints exactly when Lake Balaton first formed (13,500–15,000 yr BP).

Approximately 15,000–13,300 yr BP (Dryas I): Paleontologic investigations have not identified sediments older than Dryas I. Sporomorph associations and the ostracod fauna commonly contained redeposited elements from Pannonian (Upper Miocene) sediments. Inundation of the Keszthely and Szigliget subbasins started mainly in a meso-eutrophic, marshy environment. Fossils (pollens and ostracods) indicate a mesotrophic, marshy environment on the northern shore of the Szemes subbasin. Other parts of the Szemes and the Siófok subbasins are interpreted as terrestrial at this time.

13,300–12,400 yr BP (Bölling): Compared with the previous phase, territory covered by water increased, but shallow water environments were still predominant. Water quality deteriorated, but was still between meso- and eutrophic conditions; chemistry of the water was alkaline. Redeposited Pannonian (upper Miocene) sporomorphs and ostracods still occurred commonly in the Bölling. The area of the recent Siófok subbasin was still terrestrial.

12,400–12,000 yr BP (Dryas II): The extent and depth of water decreased substantially in the Keszthely and

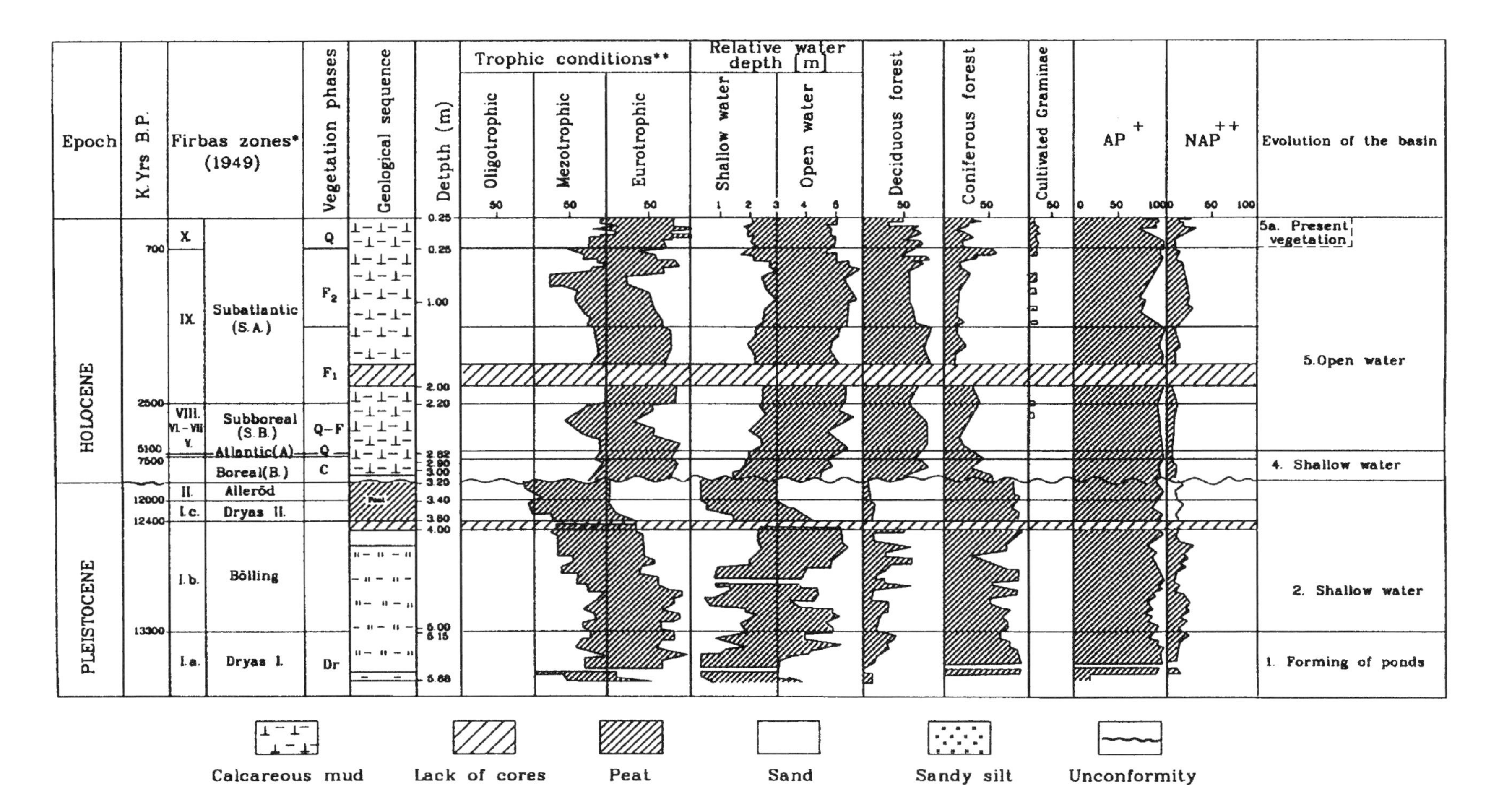

Figure 6—History of evolution of the Keszthely Basin based on data from the Tó-31 borehole. * Firbas zones: Pollen zones specified by Firbas (1949) from lacustrine sediments of the northern foreland of the Alps. ** Trophic conditions: percentage of algae as well as the spores and pollen of water plants are counted. Ratio of algae to total number of plant taxa gives a rough estimate of level of water productivity. +A.P.: Arbor pollen (trees), in percent. ++N.A.P.: Nonarbor pollen (herbs and graminids), in percent. Pollen sum: Total pollen sum counted in the slide.

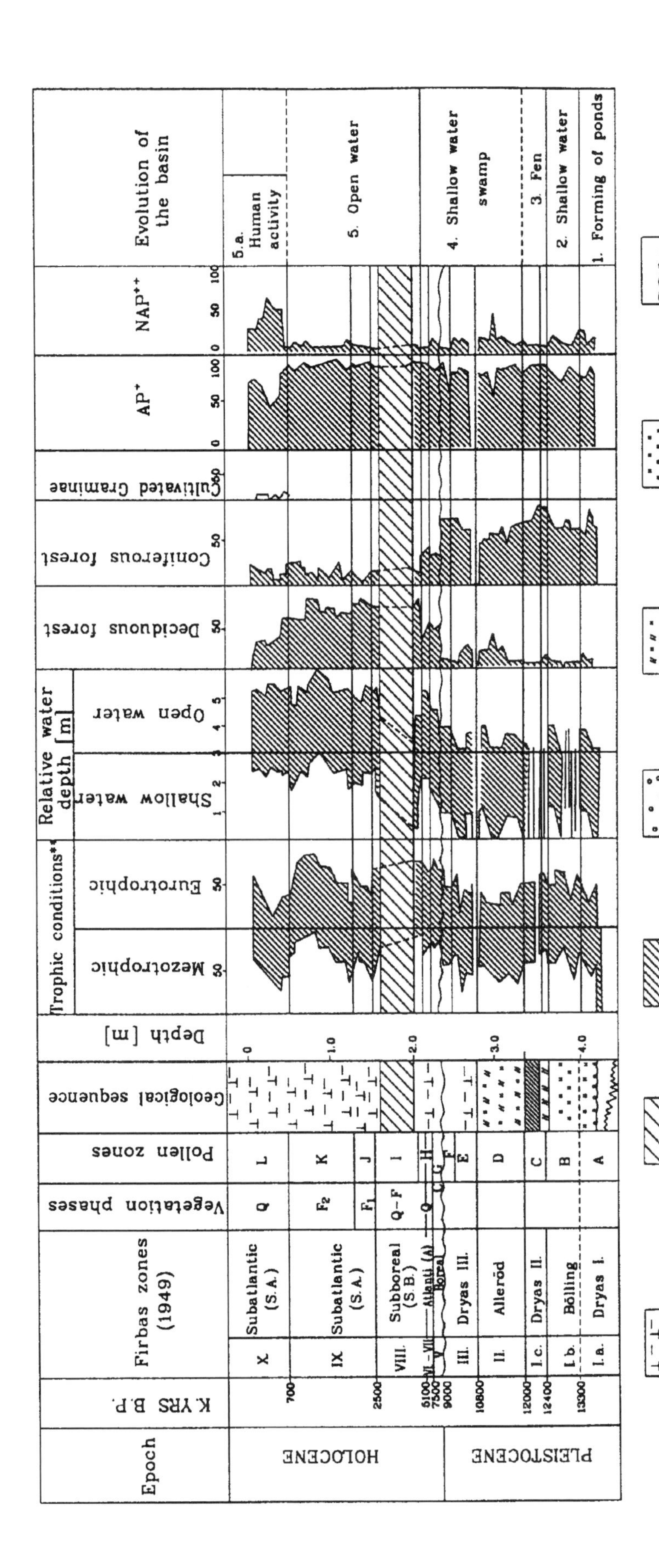

Figure 7—History of evolution of the Szigliget Basin based on data from the Tó-22 borehole. For explanation of headings, see Figure 6.

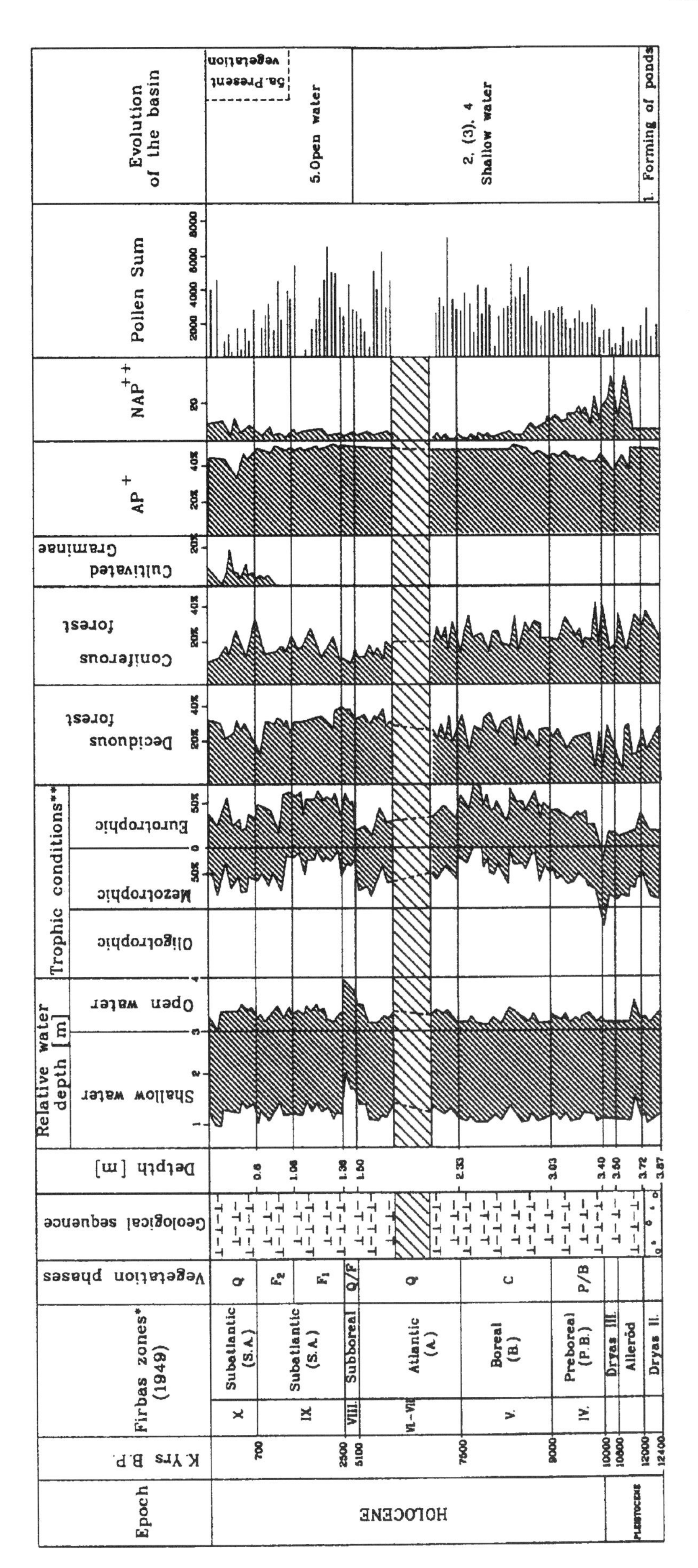

Figure 8—History of evolution of the Szemes Basin based on data from the Tó-25 borehole. For explanation of headings, see Figure 6.

Szigliget subbasins resulting in peat formation. Slightly acid marshland (pH = 5–6) formed in oligo-mesotrophic conditions in the Keszthely subbasin. At the same time, the Szigliget subbasin was occupied by a neutral (pH = 7–8) meso-eutrophic marsh. Difference in pH of the peat can be explained by the difference in bottom sediments (sand vs. clay); their morphology suggests that the two subbasins could have been separated. Peat formation started at the same time as the two former subbasins in the Szemes subbasin continuing up to the end of the Alleröd, and sporadically until Dryas III. Marshland in the Szemes subbasin was covered by slightly alkaline (pH = 8–9) and presumably deeper water than the area to the west. Based on the radiocarbon age of the peat strata in the Tó-5 and Tó-7 boreholes (Figure 1), the northern part of the Siófok subbasin was covered by water since Dryas II; on the basis of sporomorph analyses, water was interpreted as mesotrophic.

12,000–10,800 yr BP (Alleröd): Water depth decreased in the Keszthely subbasin; some parts even dried out, and an ephemeral marshland, covered by mesotrophic water, developed. The Szigliget and Szemes subbasins were also occupied by mesotrophic marshlands; the shoreline was skirted by dense reeds. In most of the Siófok subbasin, sedimentation first started at this time, which indicated mesotrophic conditions and a shallow water depth. Inundation progressed faster here engulfing larger areas as proven by the immediate and massive appearance of open-water weeds.

10,800–10,000 yr BP (Dryas III): By this time, the Keszthely subbasin was terrestrial; trophic conditions of the shallow water environment of the Szigliget subbasin increased to meso-eutrophy. Water depth also decreased in the Szemes subbasin, but it was still characterized by mesotrophic conditions. Inundation of the Siófok subbasin by mesotrophic water could have extended to a larger area as compared to the previous periods.

Summary of the late Pleistocene: Climatic changes (warming trends and increasing humidity) between 14,000 and 15,000 yr BP resulted in the formation of various small, shallow, clear, and cool ponds. Inundation of the basin took place progressively from west to east. The recharge area of the western Balaton basin was larger, and the near-surface karst and groundwater springs provided significant recharge to the lake. The western subbasins were marshy, commonly dry areas with calm, meso-eutrophic shallow water region surrounded by reeds. Marshy conditions moved progressively to the east and did not persist there as long as in the western subbasins. Except for a small bay on the northern shore, the area of the easternmost Siófok subbasin had been covered by water only since Dryas III. Water temperature had not risen significantly above 18°C. Water was generally mesotrophic to eutrophic.

Holocene

Two significant changes occurred in the evolution of aquatic and terrestrial environments of Lake Balaton: one at the late Pleistocene–Holocene boundary highlighting the beginning of expansion of deciduous forests and the other one at the beginning of the *Quercus–Fagus* vegetation phase with the culminating formation of modern Lake Balaton (see Figure 5).

10,000–9000 yr BP (Pinus–Betula vegetation phase): At this time, the western subbasins, Keszthely and Szigliget, were presumably terrestrial (see Figures 6, 7). In the Szemes and Siófok subbasins, a continuous Quaternary sequence was deposited (see Figures 8, 9). Based on the paleontologic data, shallow water of these areas was meso-eutrophic, while mixed deciduous forests developed on land.

9000–7500 yr BP (Corylus vegetation phase): Renewed sedimentation started in this period in the western (Keszthely and Szigliget) subbasins. Due to the continuous and long warming period, the water became extremely eutrophic (period of the first algal bloom). The Szemes subbasin was characterized by eutrophic shallow water, while water in the Siófok subbasin was meso-eutrophic. The different subbasins were still separated from each other.

7500–5100 yr BP (Quercus vegetation phase): Substantial warming and progressively drier conditions resulted in regression in the western (Keszthely and Szigliget) subbasins with water remaining eutrophic. Shallow water conditions prevailed; some areas became marshy. Rising evaporation rates were indicated by proliferation of siliceous algae favoring salty water. There were no more significant changes in the ostracod association: presently existing taxa have been the faunal assemblage in Lake Balaton since the *Quercus*. The central (Szemes) and eastern (Siófok) subbasins were spared the considerable shallowing and deterioration of water quality occurring in the western (Keszthely and Szigliget) subbasins. Intense subaqueous erosion interpreted from the Siófok subbasin suggests a comparative decrease in water depth (lack of sediments) (see Figure 9). The same shallowing trend of water is proven by ^{137}Cs isotope measurements (Cserny et al., 1995). Temperature achieved its maximum during this period: the average temperature in July was 24–25° C. The first archeological artifacts originated from this vegetation phase.

5100–2500 yr BP (Quercus–Fagus vegetation phase): The second important change in lake development during the Holocene occurred in this phase. As a result of the more humid climate, water extent and depth increased significantly bringing about abrasion of the subaqueous barriers among the subbasins. As a result, modern Lake Balaton formed. Open-water weeds propagated with increased water currents. As a result of the rising water volume, trophic conditions decreased significantly in the western subbasins, but later on it varied uniformly between mesotrophic and eutrophic for the whole lake. Rye appeared in the continental flora, indicating the beginning of agricultural activities.

2500–700 yr BP (Fagus vegetation phase): Water depth increased progressively in the western portion of Lake Balaton, but was mainly constant in the eastern subbasins. In the largest part of the lake (west of the Tihany peninsula, see Figure 1), the water quality ranged between meso-eutrophic and eutrophic with a

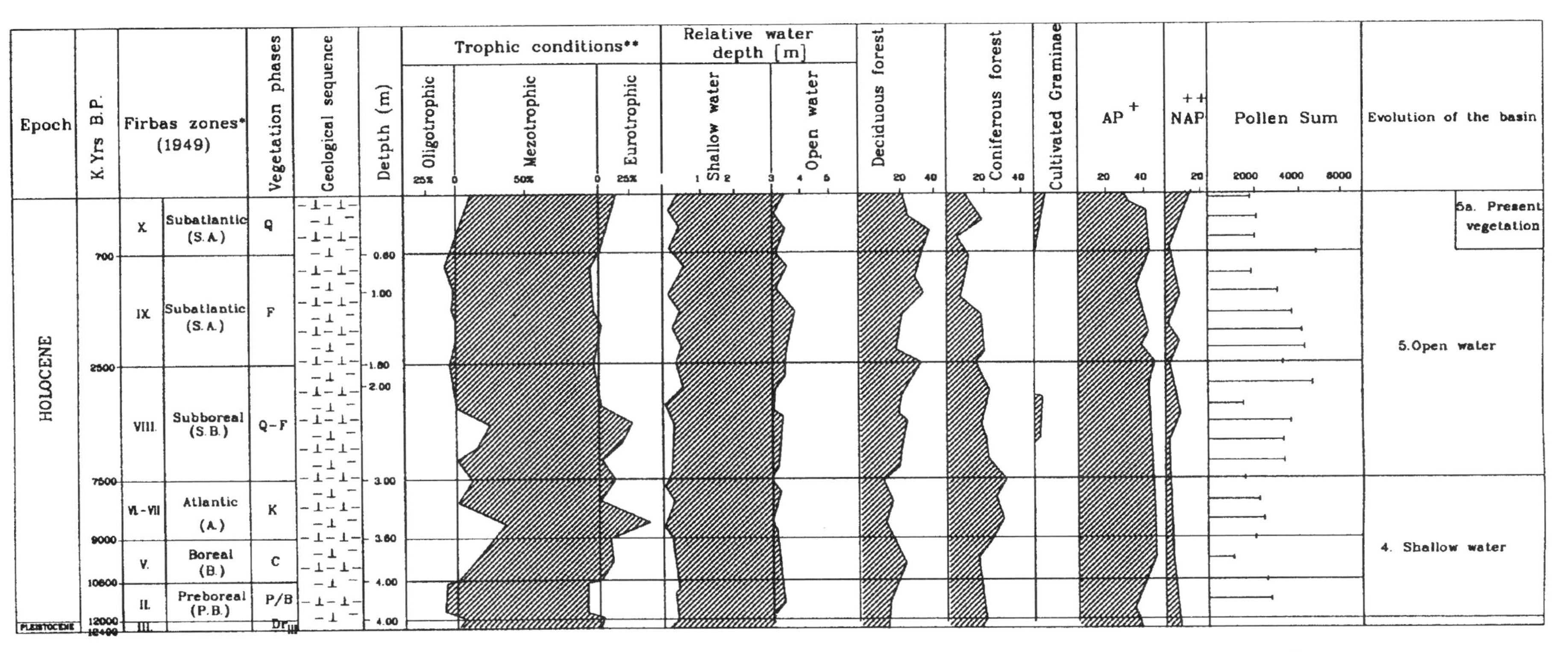

Figure 9—History of evolution of the Siófok Basin based on data from the Tó-9 borehole. For explanation of headings, see Figure 6.

decrease in pH. The Siófok subbasin was characterized by mesotrophic conditions.

700 yr BP to Holocene (cultivated forests): The floral assemblage of the western subbasins became similar to that of the *Corylus* vegetation phase; a new algae bloom indicated sudden deterioration of water quality. Increase in trophic conditions was less remarkable in the central and eastern subbasins. Water level and its extent changed as a function of climatic conditions and human impacts. Mesotrophic conditions, which characterized the lake until the 1960s, became eutrophic to hypertrophic due to the increasing human activity.

Summary of the Holocene: At the beginning of the Holocene (*Pinus–Betula* vegetation phase), the different subbasins were filled by mesotrophic and meso-eutrophic shallow waters. Water first became eutrophic during the *Corylus* vegetation phase. During the *Quercus–Fagus* vegetation phase, the climate became warmer and more humid resulting in an increase in water depth and the formation of present-day Lake Balaton. Subaqueous barriers between the subbasins disappeared due to water currents. Depth and trophic conditions of Lake Balaton changed several times. Comparatively regular fluctuation of the water level (rising for approx. 600 years and retreating for approx. 1100 years) between 104.6 m above sea level (a.s.l.) and 112.5 m a.s.l. can be traced along preserved lake terraces. As a function of climate, deciduous forests proliferated on land around the lake. The first appearance of humans can be traced back to the *Quercus* vegetation phase in this region. Agriculture and traces of other types of human activity can be dated back to the *Quercus–Fagus* vegetation phase.

ACKNOWLEDGMENTS

We would like to express our thanks to Corrada Ruben, Mária Földvári, Márta Hajós, Ede Hertelendi, Andrea Szuromi-Korecz, Sándor Tarján, Tibor Tullner, and Katalin Cserny-Mészáros for contributing to the success of the Lake Balaton study. We also would like to express our thanks to the Geologic Institute of Hungary for providing financial support as well as to OTKA (National Scientific Research Fund) Project No. 550 and T 022371 for funding some of the analyses and the final interpretation of the results.

REFERENCES CITED

Bendefy, L., and V. I. Nagy, 1969, Secular Changes in Lake Balaton's Shoreline. Müszaki Könyvkiadó, Budapest, 215 p. In Hungarian.

Bodor, E., 1987, Formation of the Lake Balaton palynological aspects, *in* M. Pécsi and L. Kordos, eds., Holocene Environment in Hungary. Geographical Research Institute Hungarian Academy of Sciences, Budapest, p. 77–80.

Bulla, B., 1943, Über die Ausbildung und das Alter des Balaton-Sees. Földrajzi Közlemények, 71 (1–2), p. 1–24.

Bulla, B., 1958, Geographic research of Lake Balaton and its surrounding, Földrajzi Közlemények, 6 (82), 4, p. 313–324. In Hungarian.

Cholnoky, J., 1897, Limnology of Lake Balaton, Magyar Földrajzi Társaság Balaton-bizottsága, p. 118. In Hungarian and German.

Cholnoky, J., 1918, Hydrography of Lake Balaton, Magyar Földrajzi Társaság Balaton-bizottsága, p. 316. In Hungarian and German.

Cserny, T., 1987, Results of recent investigations of the Lake Balaton deposits, *in* M. Pécsi and L. Kordos, eds., Holocene environment in Hungary. Geographical Research Institute Hungarian Academy of Sciences, Budapest, p. 67–76.

Cserny, T., 1990, Some results of engineering geologic mapping in the Lake Balaton region, *in* D. G. Price, ed., Proceedings Sixth International Congress International Association of Engineering Geology, 6–10 August 1990 Amsterdam, Netherlands, p. 79–84.

Cserny, T., 1994, Lake Balaton (Hungary), *in* E. Gierlowski-Kordesch and K. Kelts, eds., Global Geologic Record of Lake Basins. Vol. 1. Cambridge Univ. Press, p. 395–399.

Cserny, T., and R. Corrada, 1989, Complex geologic investigation of Lake Balaton (Hungary) and its results, Acta Geologica Hungarica 32 (1–2), p. 117–130.

Cserny, T., E. Nagy-Bodor, and M. Hajós, 1991, Contributions to the sedimentology and evolution history of Lake Balaton, *in* M. Pécsi and F. Schweitzer, eds, Quaternary environment in Hungary. Studies in Geography in Hungary, 26, Akadémiai Kiadó, p. 75–84.

Cserny, T., E. Hertelendi, and S. Tarján, 1995, Results of isotope-geochemical studies in the sedimentological and environmental geologic investigations of Lake Balaton. Acta Geologica Hungarica, 38 (4), p. 355–376.

Entz, G., and O. Sebestyén, 1942, The Life of Lake Balaton. A Királyi Magyar Természettudományi Társaság Kiadványa, p. 366. In Hungarian.

Erdélyi, M., 1963, Changes of Lake Balaton and its environment due to human activity, Hidrológiai Közlemények, 43 (3), p. 219–224. In Hungarian.

Firbas, F., 1949, Spat- und nachzeitliche Waldgeschichte Mitteleuropas nördlich der Alpen, Fischer Verl. Jena, p. 480.

Herodek, S., L. Laczkó, and Á. Virág, 1988, Lake Balaton: Research and Management, Nexus Nyomda Budapest, p. 110.

Herodek, S., and F. Máté, 1984, Eutrophication and its reversibility in Lake Balaton (Hungary), Proceedings of SHIGA Conference '84 on Conservation and Management of World Lake Environment, Lecs (Japan), p. 95–102.

Istvánovics, V., L. Vörös, S. Herodek, G. L. Tóth, and I. Tátrai, 1986, Changes of phosphorus and nitrogen concentration and of phytoplankton in enriched lake enclosures. Limnology and Oceanography, 31, p. 798–811.

Istvánovics, V., S. Herodek, and F. Szilágyi, 1989, Phosphorus adsorption by the sediments of shallow

Lake Balaton and its protecting reservoirs. Water Research, 23, p. 1357–1366.
Kéz, A., 1931, Sub-basins of Lake Balaton and the Zala River Valley, Természettudományi Közlemények pótfüzet, p. 49–61. In Hungarian.
Lóczy, L., 1884, Ancient terracces of Lake Balaton. Földrajzi Közlemények, 19 (9–10), p. 448–453. In Hungarian.
Lóczy, L., 1894, Geologic history and recent geologic importance of Lake Balaton. Földrajzi Közlemények, 22/3, p. 123–147. In Hungarian.
Lóczy, L., 1913, Results of the Scientific Investigations of the Lake Balaton Region. I.1.1. Geologic Formations of the Balaton Area and their Regional Distribution, Magyar Földrajzi Társaság Balaton-bizottsága, Budapest, p. 617. In Hungarian and German.
Marosi, S., and J. Szilárd, 1977, The Late Pleistocene Origin and Evolution of Lake Balaton, Földrajzi Közlemények, 25 (101) 1–3, p. 17–28.
Marosi, S., and J. Szilárd, 1981, Development of Lake Balaton, Földrajzi Közlemények, 29 (105) 1, p. 1–30. In Hungarian.
Máté, F., 1987, Mapping of the recent sediments of Lake Balaton. Földtani Intézet Évi Jelentése 1985-röl, p. 367–379. In Hungarian.
Máté, F., L. Vörös, S. Herodek, and B. Entz, 1981, Eutrophication and induced changes in lake Balaton, Man and Biosphere Programme, Survey of 10 Years Activity in Hungary, Hungarian National Committee for UNESCOMAB Program, Budapest, p. 167–196.
Miháltz-Faragó, M., 1983, Palynological Examination of Bottom Samples from Lake Balaton, Földtani Intézet Évi Jelentése 1981-röl, p. 439–448. In Hungarian.
Mike, K., 1980a, Neotectonics in the Lake Balaton Region, Földrajzi Közlemények, 28 (3), p. 185–204. In Hungarian.
Mike, K., 1980b, Fossil river beds around Lake Balaton, Földrajzi Értesítö, 29 (3), p. 313–334. In Hungarian.
Müller, G., 1970, High-magnesian Calcite and Protodolomite in Lake Balaton (Hungary) Sediments. Nature, 226 (5247), p. 749–750.
Müller, G., and F. Wagner, 1978, Holocene carbonate evolution in Lake Balaton (Hungary): a response to climate and impact of man, *in* A. Matter, and M. E. Tucker, eds., Modern and Ancient Lake Sediments. Blackwell Sci. Publ. 2., p. 57–81.
Papp, P., 1992, Two generations of the geologic mapping of the Lake Balaton Region, *in* P. Bíró, ed., Centenary of the Lake Balaton Research. The 33rd Hydrobiologic Days, Tihany, p. 130–139. In Hungarian.
Raincsák-Kosáry, Zs., and T. Cserny, 1984, Engineering geologic mapping of the Lake Balaton Region. Földtani Intézet Évi Jelentése 1982-röl, p. 49–54. In Hungarian.
Rónai, A., 1969, The geology of Lake Balaton and surroundings. Mitt. Internat. Verein Limnology, 17, p. 275–281.
Somlyódy, L., and G. van Straten, 1983, Modeling and Managing Shallow Lake Eutrophication: With Application to Lake Balaton. Springer-Verlag, Berlin, 386 p.
Vörös, L., V.K. Balogh, F. Máté, and L. Ligeti, 1984, Determination of the silting up by paleo-limnologic methods. Vízügyi Közlemények, 66 (1), p. 104–113. In Hungarian.
Zólyomi, B., 1953, Die Entwicklungsgeschichte der Vegetation Ungarns seit dem letzten interglatzial, Acta Biologica. Academia of Sciences, Hungary, 4, p. 367–430.
Zólyomi, B., 1987, Degree and rate of sedimentation in Lake Balaton, *in* M. Pécsi, ed., Pleistocene Environment in Hungary. Contribution of the INQUA Hungarian National Committee to the 12th INQUA Congress Ottawa, Canada 1987, p. 57–79.
Zólyomi, B., 1995, Opportunities for pollen stratigraphic analysis of shallow lake sediments: the example of Lake Balaton. GeoJournal, 36 (2–3), p. 237–241.

Brenner, M., T. J. Whitmore, Song Xueliang, Long Ruihua, D. A. Hodell, and J. H. Curtis, 2000, Sediment records from Qilu Hu and Xingyun Hu, Yunnan Province, China: late Pleistocene to present, *in* E. H. Gierlowski-Kordesch and K. R. Kelts, eds., Lake basins through space and time: AAPG Studies in Geology 46, p. 619–624.

Chapter 59

Sediment Records from Qilu Hu and Xingyun Hu, Yunnan Province, China: Late Pleistocene to Present*

Mark Brenner
Thomas J. Whitmore
Department of Fisheries and Aquatic Sciences, University of Florida
Gainesville, Florida, U.S.A.

Song Xueliang
Long Ruihua
Yunnan Institute of Geological Sciences
Yunnan, China

David A. Hodell
Jason H. Curtis
Department of Geology, University of Florida
Gainesville, Florida, U.S.A.

Qilu Hu (area = 37 km^2, z_{max} = 7 m) and Xingyun Hu (area = 39 km^2, z_{max} = 12 m) lie at 1797 m and 1723 m above mean sea level (Whitmore et al. 1997), respectively, on the Yunnan Plateau, southwest China (Figure 1). Analyses of water-column nutrients and Secchi disk measurements show that Qilu Hu is currently eutrophic and Xingyun Hu is meso-eutrophic (Moore et al., 1988; Whitmore et al., 1994a, 1997). Upper Tertiary deposits in Yunnan Province are rich in brown coal, but karsted limestone dominates the surface rock (Li and Walker, 1986). Valleys in the region often possess fluvial and lacustrine deposits of Quaternary age.

An 11.0-m sediment core was collected from Qilu Hu in 1987 and an 8.4-m section was obtained from Xingyun Hu in 1989. Chronologies for the sediment cores were established by standard ^{14}C dates on bulk organic matter and an AMS ^{14}C date on wood (Figure 2). Sediment profiles contain records of lacustrine deposition from the late Pleistocene to present (Brenner et al., 1991; Whitmore et al., 1994a, 1994b). The ^{14}C date on wood at 7.39 m in Qilu Hu is younger than the date on bulk organic matter from 5.99–5.81 m, suggesting that hard-water-lake error makes dates on lacustrine organic matter about 1800 yr older than true ^{14}C ages (Brenner et al., 1991).

Paleoecological studies in Yunnan Province have employed ostracod remains (Jiang, 1990) or magnetic measurements in lake sediments to infer paleoenvironmental conditions (Yu et al., 1990). Fang (1991) presented reconstructions of past lake levels. Paleoenvironmental inferences for the 1987 Qilu Hu core and the 1989 Xingyun Hu core were based on geochemical, palynological, and diatom analyses (Brenner et al., 1991; Deevey et al., 1988; Long et al., 1991; Song et al., 1994; Whitmore et al., 1994a, 1994b). Bottommost deposits from Qilu Hu (11.0–6.0 m) and Xingyun Hu (8.4–5.0 m) are dominated by gray silts and clays that possess low concentrations of both organic matter and carbonates (Figure 2). Coarse, sandy sediments below 9 m depth in the Qilu Hu core suggest depositional discontinuities. Possible late Pleistocene sedimentation hiatuses in the Qilu Hu profile are also inferred from the 17-k.y. age difference between basal deposits at 10.99–10.81 m and sediments less than 2 m above (9.19–9.01 m).

Previous palynological studies on Yunnan lake sediment sequences provided a picture of regional vegetation changes from the late Pleistocene to present (Li and Walker, 1986; Lin et al., 1986; Liu et al., 1986; Sun

*We retain the local names for these waterbodies and use the Chinese word for lake (hu).

et al., 1986; Walker, 1986). In the 1987 Qilu Hu core and 1989 Xingyun Hu core, *Pinus* grains dominate throughout the palynological records, comprising 50–90% of the total pollen (Long et al., 1991; Song et al., 1994). Deep sediments from Qilu Hu (~11.0–7.0 m) and Xingyun Hu (~8.4–6.0 m) possess *Abies* (fir) and *Picea* (spruce) pollen, indicating cool, moist conditions. Basal deposits from the 1987 Qilu Hu core are 31 k.y. old (Figures 2, 3) and thus predate the last glacial maximum (LGM), i.e., Dali glaciation. Oldest retrieved deposits from the 1989 Xingyun Hu core date to less than 12 Ka (Figures 2, 4) and postdate the LGM (last glacial maximum).

In the 1987 Qilu Hu core, diatoms were sparse or absent in sediments from 11.0 to 9.2 m depth, reflecting intermittent lacustrine conditions. Between 9.0 and 6.0 m, the Qilu Hu assemblage is dominated by planktonic, centric diatoms, suggesting relatively deeper water (Brenner et al., 1991). In the 1989 Xingyun Hu core, diatoms are present throughout the basal 3.5 m of the core, but shifts in the dominance of benthic versus planktonic assemblages may reflect fluctuating water levels in the late Pleistocene (Whitmore et al., 1994a).

Above 7.0 m in Qilu Hu and 6.0 m in Xingyun Hu, *Abies* and *Picea* pollen are nearly absent from the record, indicating climatic warming. Warmer conditions in the Holocene were associated with a shift in sediment type. Between about 6.0 m and 1.0 m in the Qilu Hu core, and 5.0 m and 1.8 m in the Xingyun Hu core, organic matter and carbonates comprise a greater proportion of the sediment dry weight, probably indicating increased lacustrine productivity. Throughout much of the Holocene, Qilu Hu was shallow, as inferred from the dominance of benthic diatom taxa and pollen of *Alisma*, an emergent macrophyte that inhabits swamps and shallow lakes. *Alisma* pollen was also found in Xingyun Hu deposits, particularly between 5.0 and 3.0 m, when benthic diatoms were prevalent. Overall, the Holocene records from Qilu Hu and Xingyun Hu indicate productive lakes with shallow, but fluctuating water levels.

Extrapolation of ^{210}Pb dates suggests that the topmost 85 cm of sediment in Qilu Hu accumulated during the past 240 yr. The uppermost 160 cm of sediment at Xingyun Hu accumulated over a period of about 530 yr. These clayey deposits are rich in aluminum, iron, and potassium, as well as lead and zinc, but are poor in organic carbon and carbonates that characterize the underlying sediments. Rapid accumulation of these clay-rich sediments is attributed to soil erosion from human activities in the watersheds, including deforestation, intensive agriculture, and lake bottom reclamation (Brenner et al., 1991; Whitmore et al., 1994b).

Environmental inferences based on pollen and diatoms in sediments from the Qilu Hu and Xingyun Hu cores suggest a transition from a cool climate and fluctuating lake levels during the latest Pleistocene, to warmer conditions and perhaps lower lake levels in the early Holocene (Figures 3, 4). Inferences for declining lake levels in the early Holocene, based largely on the lifeform preferences of dominant

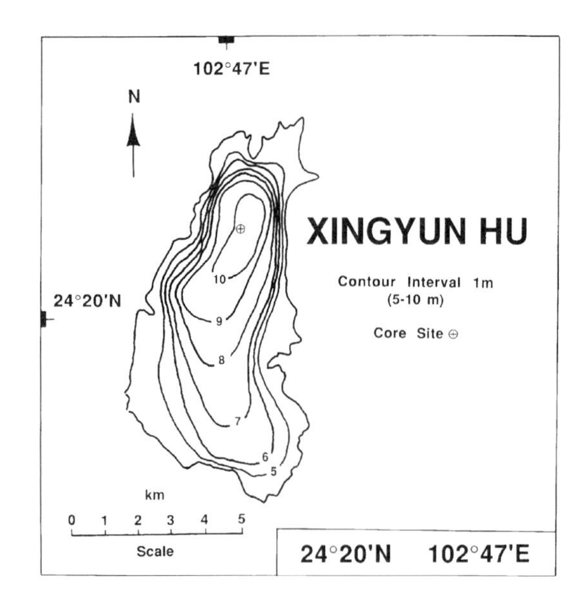

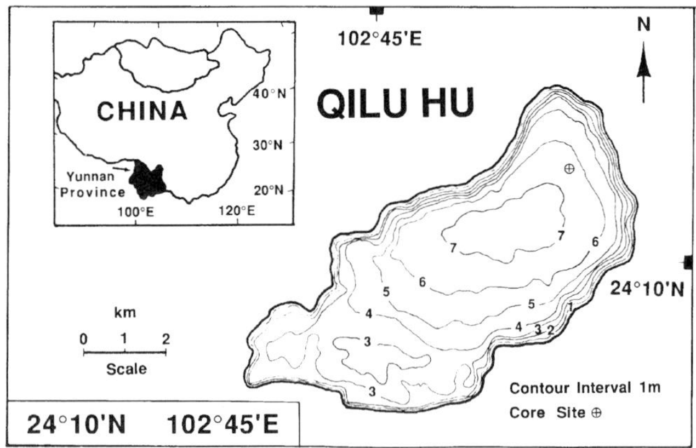

Figure 1—Bathymetric maps of Xingyun Hu and Qilu Hu, showing the 1987 (Qilu Hu) and 1989 (Xingyun Hu) sediment coring sites. Inset map shows the location of Yunnan Province in southwest China.

diatom taxa, are at odds with global climate models that predict greater moisture availability in the early Holocene for Yunnan basins under the influence of the developing Asian monsoon. The discrepancy may stem from the fact that sedimented diatom communities responded to shifts in nutrient availability as well as changes in lake stage.

Qilu Hu and Xingyun Hu were recored in 1994, and sections of 13.5 and 12.5 m, respectively, were retrieved. Ages in bottommost samples from the 1994 Qilu Hu and Xingyun Hu cores were based on AMS ^{14}C dates on bulk organic matter, and were 48,500 ^{14}C yr BP and 20,600 ^{14}C yr BP, respectively. The general lithostratigraphies of the 1994 sections were comparable to those reported for the 1987 and 1989 cores, but the 1994 profiles contain longer records of sediment accumulation. In the Qilu Hu core, all interglacial and interstadial stages (IS1-12) in the Greenland Summit ice core are marked by increases in the concentration of authigenic carbonate, indicating warming on the

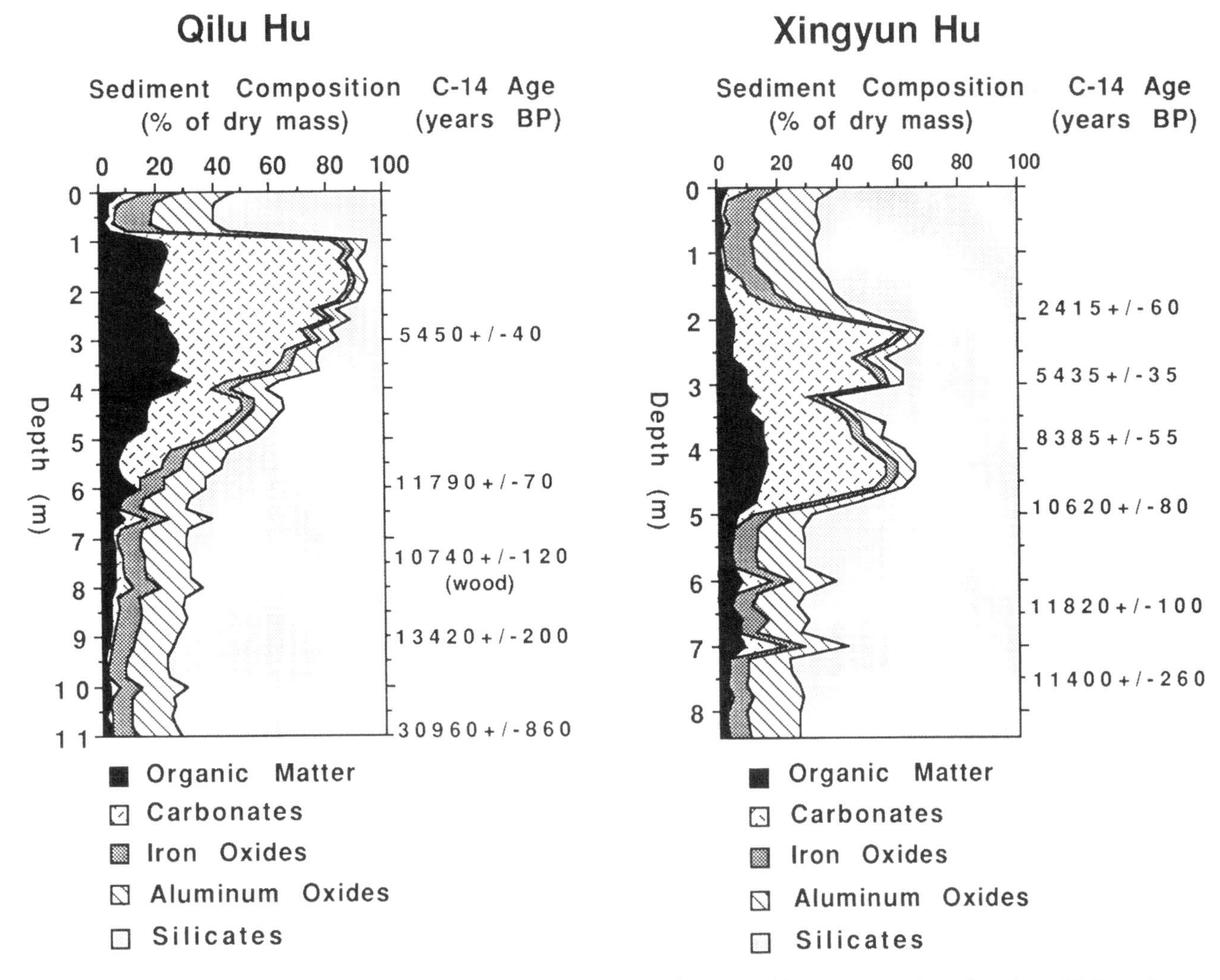

Figure 2—Composition of the 1989 Xingyun Hu and 1987 Qilu Hu sediment cores. Total carbon (TC) and carbonate carbon (IC) were measured coulometrically and organic carbon (OC) was figured as TC minus IC. Organic matter represents 2 × OC. Calcium concentrations greatly exceed magnesium concentrations, so carbonates were assumed to be bound primarily as $CaCO_3$. Carbonates were computed as 8.33 × IC. Iron and aluminum oxides were measured by x-ray fluorescence. Silicates represent the balance of the sediment matrix. Radiocarbon dates on bulk organic matter are from the University of Pittsburgh Applied Research Center Radiocarbon Laboratory. The AMS ^{14}C date on wood (AA-3070) was run at the University of Arizona-NSF Accelerator Facility.

Yunnan Plateau during Dansgaard-Oeschger events (Hodell et al., 1999).

High-resolution oxygen isotope measurements on carbonates from the 1994 Qilu Hu and Xingyun Hu cores demonstrate an abrubt decline in $\delta^{18}O$ values from around –6‰ in the latest Pleistocene, to much lower values in the early Holocene (Qilu Hu = ~10‰, Xingyun Hu = ~12‰), reflecting a shift in the isotopic signature of water entering the basins (Hodell et al., 1999). Seasonal measurements of $\delta^{18}O$ in modern rainwater indicate that values are most negative during summer months when the strong monsoon emanates from the Bay of Bengal to the southwest. Thus, the very negative $\delta^{18}O$ values for carbonates probably reflect intensification of the summer, southwest Asian monsoon during the early Holocene. Throughout the Holocene, $\delta^{18}O$ values for sedimented carbonates increase fairly steadily, reflecting a decrease in the intensity of the summer monsoon. $\delta^{18}O$ values for surface sediment carbonates are about –5‰, comparable to values recorded for the late Pleistocene.

ACKNOWLEDGMENTS

We thank Dr. Robert Stuckenrath for kindly providing the University of Pittsburgh radiocarbon dates. Drs. Michael W. Binford, Allen M. Moore, and Daniel R. Engstrom assisted with fieldwork. This project was supported by grants NSF INT-8802793 and NSF ATM-9305750. This paper is Journal Series No. R-04043 of the Florida Agricultural Experiment Station.

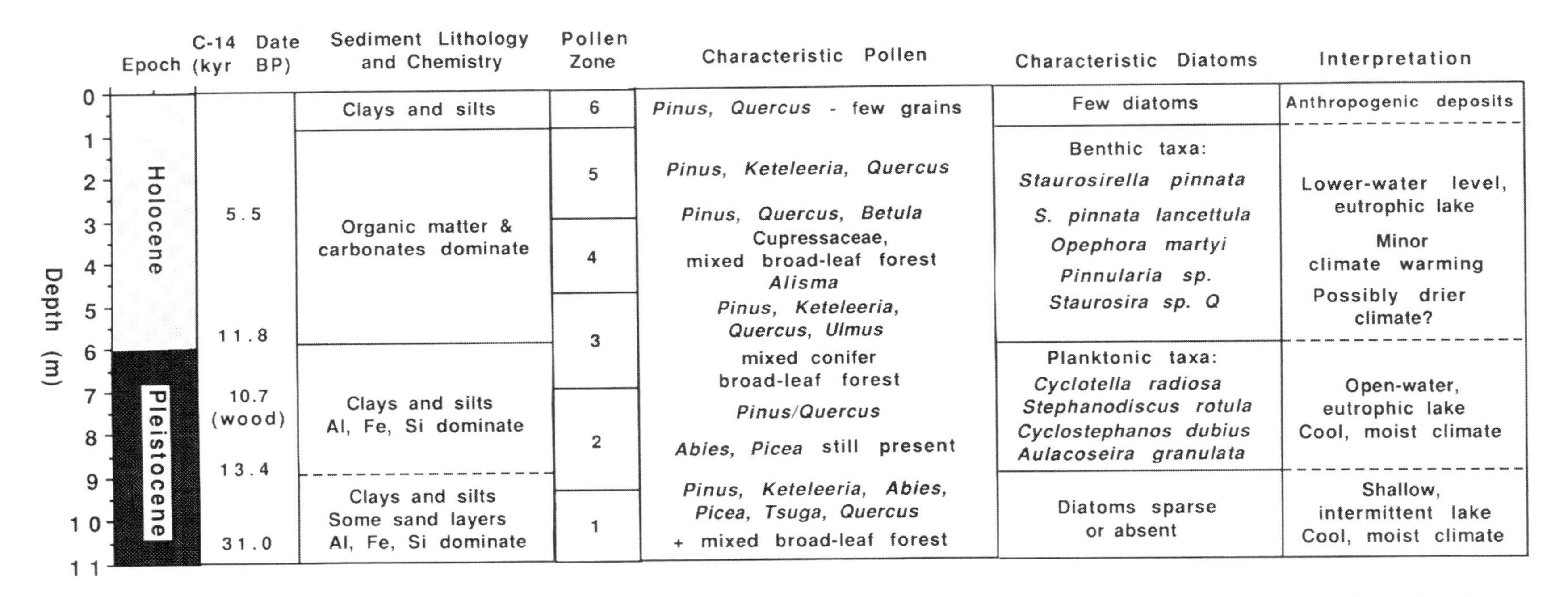

Figure 3—Stratigraphy of the sediment core collected from Qilu Hu in 1987. The Pleistocene–Holocene boundary is provisionally placed at 6.0 m depth, where there is a transition from clayey, silty deposits to organic- and carbonate-rich sediments. The lithologic shift generally coincides with the disappearance of cool, moist indicators *Picea* and *Abies* from the record. "Characteristic pollen" include taxa used to define pollen zones according to Long et al. (1991). "Characteristic diatoms" are principal taxa in the benthic and planktonic assemblages. Anthropogenic deposits are silts and clays from eroded riparian soils.

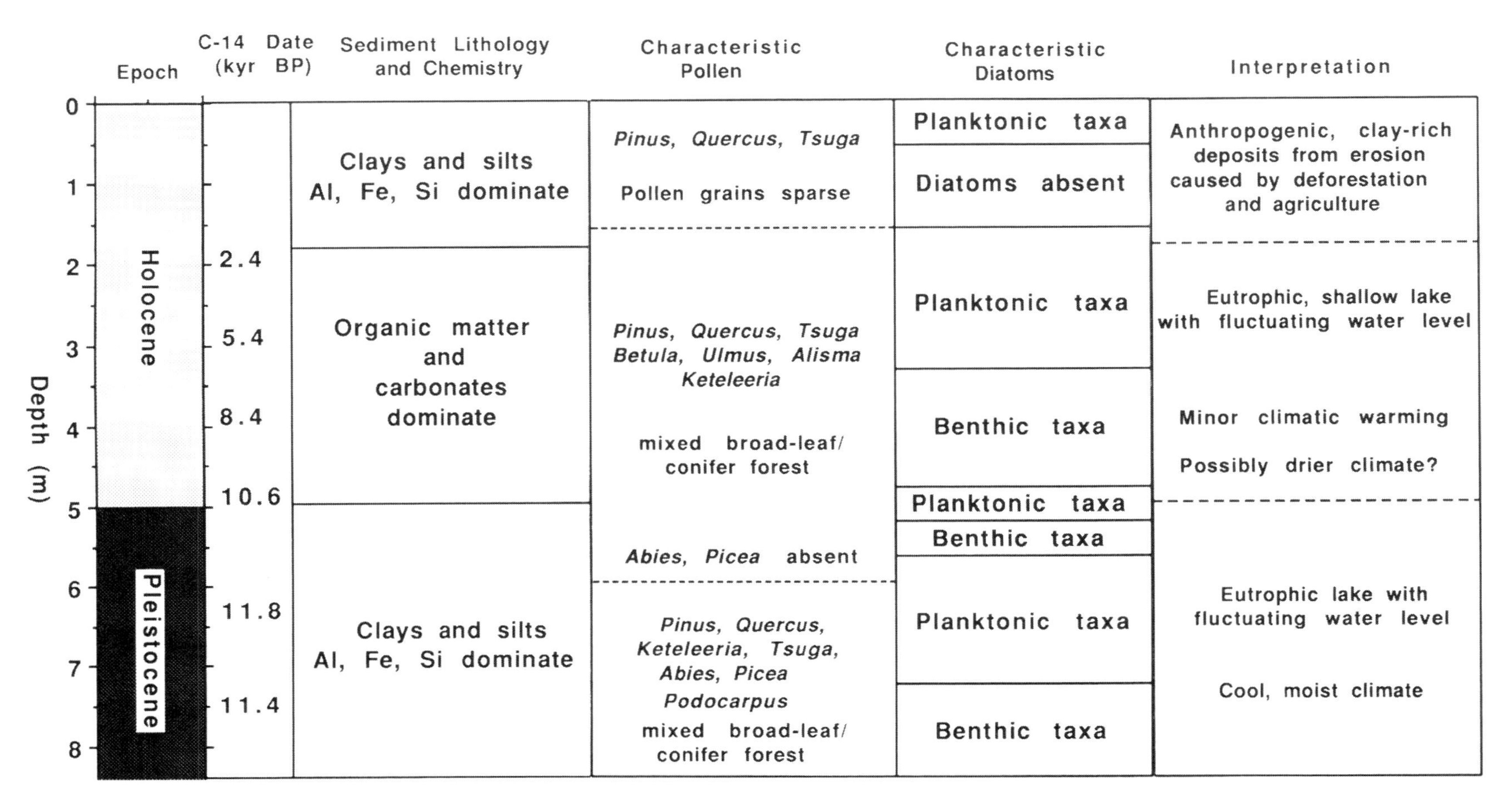

Figure 4—Stratigraphy of the sediment core collected from Xingyun Hu in 1989. The Pleistocene–Holocene boundary is provisionally placed at 5.0 m depth, where clayey, silty deposits are replaced by organic- and carbonate-rich sediments. "Characteristic pollen" include taxa used to define the prevailing vegetation. Pollen zones have not yet been assigned. "Characteristic diatoms" indicate prevalence of planktonic versus benthic diatom assemblages. Principal benthic taxa include *Opephora martyi*, *Staurosirella pinnata*, *Staurosira* sp. *Q*, and *Pseudostaurosira brevistriata*. Principal planktonic taxa include *Cyclotella radiosa*, *Cyclostephanos tholiformis*, *Aulacoseira ambigua*, and *Stephanodiscus rotula* var. *minutula*.

REFERENCES CITED

Brenner, M., K. Dorsey, Song X., Wang Z., R. Long, M. W. Binford, T. J. Whitmore, and A. M. Moore, 1991, Paleolimnology of Qilu Hu, Yunnan Province, China, Hydrobiologia, 214, p. 333–340.

Deevey, E. S., M. Brenner, A. M. Moore, M. W. Binford, K. T. Dorsey, Song X., Wang Z., and Hu Z., 1988, Paleolimnology and limnology of Qilu Hu, Tonghai County, Yunnan—preliminary results, Journal of Yunnan University, 10 (supplement), p. 57–62.

Fang, J., 1991, Lake evolution during the past 30,000 years in China, and its implications for environmental change, Quaternary Research, 36, p. 37–60.

Hodell, D. A., M. Brenner, S. L. Kanfoush, J. H. Curtis, J. S. Stoner, Song X., Wu Y., and T. J. Whitmore, 1999, Paleoclimate of southwestern China for the past 50,000 yr inferred from lake sediment records, Quaternary Research, 52, p. 369–380.

Jiang, Z., 1990, Microfaunas and their ecologic environment of Holocene sediments in plateau lakes, central Yunnan, Yunnan Geology, 9, p. 117–133.

Li, X., and D. Walker, 1986, The plant geography of Yunnan Province, southwest China, Journal of Biogeography, 13, p. 367–397.

Lin, S., Qiao Y., and D. Walker, 1986, Late Pleistocene and Holocene history at Hi Hu, Er Yuan, Yunnan Province, southwest China, Journal of Biogeography, 13, p. 419–440.

Liu, J., Tang L., Qiao Y., M. J. Head, and D. Walker, 1986, Late Quaternary vegetation history at Menghai, Yunnan Province, southwest China, Journal of Biogeography, 13, p. 399–418.

Long, R., Li B., M. Brenner, and Song X., 1991, A study of late Pleistocene to Holocene vegetation in the Jilu Lake area of central Yunnan, Yunnan Geology, 10, p. 105–118.

Moore, A. M., M. Brenner, M. W. Binford, E. S. Deevey, and Ou X., 1988, Fieldnotes on a paleoecological expedition to five lakes on the Yunnan Plateau, Journal of Yunnan University, 10 (supplement), p. 28–36.

Song, X., Wu Y., Jiang Z., Long R., Li B., M. Brenner, T. J. Whitmore, D. R. Engstrom, and A. Moore, 1994, Paleolimnological studies on the limestone district in central Yunnan, China, Yunnan Science and Technology Publishing, Kunming, PRC, 114 p.

Sun, X., Wu Y., Qiao Y., and D. Walker, 1986, Late Pleistocene and Holocene vegetation history at Kunming, Yunnan Province, southwest China, Journal of Biogeography, 13, p. 441–476.

Walker, D., 1986, Late Pleistocene–early Holocene vegetational and climatic changes in Yunnan Province, southwest China, Journal of Biogeography, 13, p. 477–486.

Whitmore, T. J., M. Brenner, and Song X., 1994a, Environmental implications of the late Quaternary diatom history from Xingyun Hu, Yunnan Province, China, Memoirs of the California Academy of Sciences, 17, p. 525–538.

Whitmore, T. J., M. Brenner, D. R. Engstrom, and Song X., 1994b, Accelerated soil erosion in watersheds of Yunnan Province, China, Journal of Soil and Water Conservation, 49, p. 67–72.

Whitmore, T. J., M. Brenner, Jiang Z., J. H. Curtis, A. M. Moore, D. R. Engstrom, and Wu Y., 1997, Water quality and sediment geochemistry in lakes of Yunnan Province, southern China, Environmental Geology, 32, p. 45–55.

Yu, L., F. Oldfield, Wu Y., Zhang S. and Xiao J., 1990, Paleoenvironmental implications of magnetic measurements on sediment core from Kunming Basin, Southwest China, Journal of Paleolimnology, 3, p. 95–111.

Valero-Garcés, B. L., M. Grosjean, B. Messerli, A. Schwalb, and K. Kelts, 2000, Later Quaternary lacustrine deposition in the Chilean Altiplano (18°–28°S), *in* E. H. Gierlowski-Kordesch and K. R. Kelts, eds., Lake basins through space and time: AAPG Studies in Geology 46, p. 625–636.

Chapter 60

Late Quaternary Lacustrine Deposition in the Chilean Altiplano (18°–28°S)

Blas L. Valero-Garcés
Instituto Pirenaico de Ecología-CSIC
Zaragoza, Spain

Martin Grosjean
Bruno Messerli
Department of Physical Geography, University of Bern
Bern, Switzerland

Antje Schwalb
Kerry Kelts
Limnological Research Center, University of Minnesota
Minneapolis, Minnesota, U.S.A.

GEOLOGIC AND GEOGRAPHIC SETTING

We summarize the Holocene sedimentary infilling of four high-altitude (>4000 m a.s.l.) lacustrine basins in the subtropical Chilean Altiplano (Central Andes, northern Chile): Laguna Chungará (18°S), Laguna Miscanti and Laguna Lejía (23°S), and Laguna del Negro Francisco (27°S) (Figure 1). These lakes are located within the active Andean convergent margin and formed by tectonic and volcanic activity during the Pliocene–Pleistocene. The Central Andes consists of several north-south mountain ranges and intermontane basins that resulted from the subduction of the Pacific Nazca plate from the Permian to the present. The Andes in northern Chile are characterized by a high fore-arc region (Coastal range, Longitudinal Valley, Chilean Precordillera, and Preandean Depression), an active magmatic arc (Western Cordillera and Altiplano), and a retro-arc belt (Eastern Cordillera and Chaco Foreland basin) (Börgel Olivares, 1983) (Figure 1). The Altiplano is a high ignimbrite plateau of some 100,000 km^2 from about 15°S to 28°S at an average altitude of 3800 m. The evolution of this complex unit is the result of the Miocene (Quechua tectonic phase) to recent volcanic and tectonic activity. Large, fault-bounded topographically closed basins developed in the Altiplano, including Titicaca and saline lake Poopó, the large Uyuni and Coipasa salares, and numerous small salares (Montti and Henriquez, 1970; Chong Díaz, 1988). The evolution of these alluvial and lacustrine depositional systems has been strongly controlled by the synsedimentary tectonic and volcanic activity.

Arid-semiarid climates in the region were already established in the middle Miocene as a result of the interplay of latitudinal location, rain shadow effect of the Andes, and cold oceanic currents (Parrish and Curtis, 1982); however, numerous proxy records have shown large climatic changes during the Quaternary in the Altiplano (Stoertz and Ericksen, 1974; Clapperton, 1993; Messerli et al., 1993; Grosjean, 1994; Grosjean et al., 1995 a, b; Valero-Garcés et al., 1996a). Within today's atmospheric circulation patterns, the Chilean Altiplano stretches from the tropical circulation zone with convective summer precipitation (Laguna Chungará) to the westerlies circulation zone with winter frontal cyclonic precipitation (Laguna del Negro Francisco) (Romero, 1985). The climate is dry and cold, characterized by larger daily than seasonal temperature variations. Average annual rainfall in the Chungará region is about 440 mm/yr (Cotacotani meteorological station), but the range is variable (100–500 mm/yr) (Baied, 1991). Potential evaporation in the nearby Salar de Surire is 4750 mm/yr (Alonso and Vargas, 1988). In the Miscanti-Lejía region (23°S), annual rainfall is about 200 mm and potential evaporation is 2040 mm (Grosjean, 1994; Vuille and Ammann, 1997). South of 27°S, fronts in the westerlies become more common in winter and precipitation in the highlands increases rapidly from barely 50 mm to about 300 mm at 31°S (Romero, 1985). Estimated precipitation in Laguna del Negro Francisco

Figure 1—Morphostructural units in the Chilean Central Andes (after Börgel Olivares, 1983) and location of the studied Chilean Altiplano lakes.

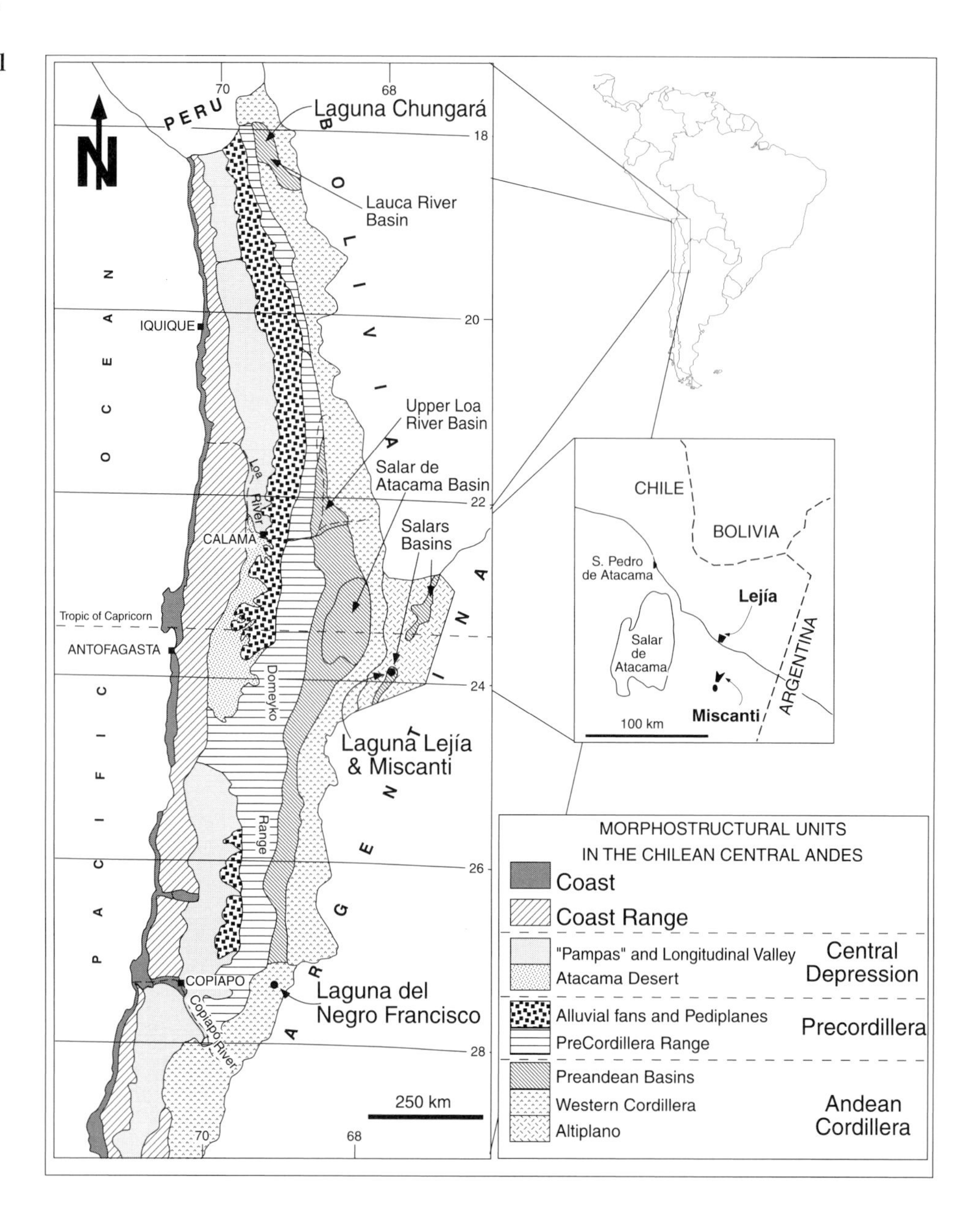

is about 250 mm/yr and evaporation is about 1500 mm/yr (Alonso and Vargas, 1988; Grosjean et al., 1997).

LAGUNA CHUNGARÁ

Geographic and Geologic Setting

Laguna Chungará lies at the northeastern edge of the Lauca basin, which is the westernmost and highest fluvio-lacustrine basin in the Altiplano (Figure 1). The Lauca basin is filled with a sequence of some 120 m of upper Miocene–Pliocene alluvial and lacustrine sediments (Kött et al., 1995). A Late Pleistocene collapse of the Nevado de Parinacota caldera (between 17,000 and 15,000 yr BP) (Baied, 1991; Baied and Wheeler, 1993; Kött et al., 1995) produced a volcanic avalanche responsible for the hummocky topography of the Cotacotani lakes area, west of Laguna Chungará. The presence of numerous moraine and glacio-fluvial deposits indicates that glaciers reached the high plain and probably as low as 4000 m during the Pleistocene glaciations, but never reached the Chungará-Cotacotani lake district after the Parinacota collapse (Baied, 1991).

Laguna Chungará (18°15′S, 69°10′W, 4520 m a.s.l.) stretches along the Chilean-Bolivian border (Figure 2A). Laguna Chungará lies in a tectonic basin, with a maximum water depth of 40 m, surface area of 21.5 km^2, and volume of about 385 million m^3. The main inlet is the Chungará River (300–500 l/s); the Ajata, Mal Paso, and Sopocalane creeks represent only about 20% of the inflow. There is no surface outlet, but groundwater outflow toward the Cotacotani lakes is about 6×10^6 m^3/yr. The lake loses about 1200 mm/yr to evaporation (Mladinic et al., 1987). The lake is polymictic, oligotrophic, contains 1.2 g/l TDS (total dissolved solids), and water chemistry is of Na-Mg-HCO_3-SO_4 type and alkaline (pH = 9). The phytoplankton community possesses few species, and diatoms are scarce. Macrophyte communities in the littoral zone form dense patches that contribute to primary productivity. The fauna includes endemic cyprinodontid fishes (e.g., 19 sp. of *Orestias*).

Stratigraphy

The late Holocene sedimentary history of Laguna Chungará was reconstructed based on high-resolution seismic profiling (Figure 3) and sedimentologic analyses of cores (Figure 2B) (Valero-Garcés et al., 1996b). Two seismic units, defined by two main reflectors (C1 and C2), were interpreted as lacustrine depositional units that overlie the massive volcanic substrate in the smaller, shallower eastern subbasin (sediment thickness up to 5 m; Figure 3, profile 1) and in the main, deeper, northwest-southeast trending basin (sediment thickness about 10 m; Figure 3, profile 2). An older seismic unit, with less well-defined reflectors, and interpreted as fluvial-lacustrine, occurs in the northwestern basin. A submerged platform in the western margin of the main basin provides evidence of former lower lake levels during the late Quaternary (Figure 3, profile 3).

In the lacustrine littoral platform of the eastern subbasin, two sedimentary facies associations were defined (Figure 2B): (1) macrophyte-dominated littoral deposits, composed of black muds with macrophyte remains and peaty muds and (2) Characeae-dominated lacustrine shelf deposits, composed of gray muds and sands with abundant Characeae remains. The two facies associations define three sedimentary cycles caused by oscillations in the lake level, from shallower (macrophyte) to deeper (Characeae) conditions. Five ^{14}C dates constrain this sedimentary sequence to the middle–late Holocene. Reservoir effects on Altiplano lake sediment chronology are discussed in detail elsewhere (Geyh et al., 1998).

Pollen stratigraphy in a nearby lake (Laguna Seca, Cotacotani area) (Baied, 1991; Baied and Wheeler, 1993) indicates a gradual transition toward drier and warmer climates since the late Pleistocene that culminates during the middle Holocene dry period, and is followed by a short, wet episode. Laguna Chungará lake level fluctuations reflect the hydrologic history of the basin and the interplay of active tectonics and volcanism, and changes in effective moisture (precipitation–evaporation) during the late Quaternary (Valero-Garcés et al., 1996b).

LAGUNA LEJÍA AND LAGUNA MISCANTI

Geographic and Geologic Setting

Laguna Lejía (23°30′S, 67°42′W, 4325 m a.s.l.) and Laguna Miscanti (23°44′S, 67°46′W, 4140 m a.s.l.) are located along a major tectonic fault striking N80°E (Quebrada Nacimiento fault) (Ramirez and Gardeweg, 1982) that separates the Precordillera and the Cordillera de los Andes and encompasses most of the Quaternary volcanic centers in the Atacama region (Figures 1, 4A, 5A). The lakes originated during the Pliocene–Pleistocene as a result of a reactivation of the Quebrada Nacimiento Fault after the Patao Ignimbrite emplacement and the eruptions of the Miñiques, Miscanti, and Chiliques volcanoes (Ramirez and Gardeweg, 1982).

The modern Laguna Lejía is a very shallow (Z_{max} < 1 m), small (2 km^2), very saline (62,800 µS/cm conductivity) system with Na-Mg-SO_4-Cl-(HCO_3) and alkaline (pH = 8.4) waters (Grosjean, 1994). Laguna Miscanti is a deeper (Z_{max} = 9 m), larger (15 km^2), and less saline (5 g/l; 6400 µS/cm conductivity) system with Na-Mg-SO_4-Cl and alkaline (pH = 8.8) waters (Valero-Garcés et al., 1996a). In both lakes, the hydrologic budget is controlled mainly by groundwater inflow from a large catchment area and surface evaporation. The Quebrada Nacimiento fault, however, permits some internal drainage in these topographically closed basins and prevents the lakes from becoming salars (Chong-Diaz, 1988; Grosjean, 1994, 1995).

Laguna Lejía Stratigraphy

Outcrops at the southern margin of Laguna Lejía show the following sequence (Grosjean, 1994; Grosjean, 1995; Grosjean et al., 1995b; Figure 4B): (1) pyroclastic and volcanic deposits, bentonites, and tuff layers (unit 4, >15,400 ^{14}C yr BP); (2) sublittoral laminated to banded carbonate muds with intercalated sandy layers and numerous carbonate tubes interpreted as calcite coating of macrophytes and Characeae (unit 3, 15,400–13,500 ^{14}C yr BP); (3) finely laminated carbonates composed of Mg-calcite, gypsum, and diatom remains (unit 2, 13,500–10,400 ^{14}C yr BP); and (4) banded to laminated carbonate sediments with oscillation ripples and higher amounts of allochthonous debris and sandy layers (unit 1, 10,400–8500 ^{14}C yr BP). Hydrated calcite needles, up to 6 cm long, occur in a 10-cm-thick layer in unit 1 and are interpreted as monocline ikaite ($CaCO_3 \times 6\ H_2O$) pseudormorphs. The presence of ikaite would suggest calcite precipitation at temperatures around 0°C. The Laguna Lejía sequence spans from late Pleistocene to early Holocene (Grosjean et al., 1995a, b; Geyh et al., 1998) and illustrates large water level changes: low during the Pleistocene (unit 4), high (up to 16–25 m higher than today) during the late Pleistocene–early Holocene humid phase (units 3 and 2), and declining lake levels around 8000 yr BP (unit 1). This humid phase is interpreted as a result of increased continental moisture brought to the Altiplano by strengthened tropical (monsoonal) circulation (Grosjean at al., 1995 a, b).

Laguna Miscanti Stratigraphy

Seismic data showed four main reflectors that defined five seismic units in the Laguna Miscanti basin (Valero-Garcés et al., 1996a) (Figure 3). The lowermost unit underneath the M4 reflector is acoustically massive and corresponds to Pliocene volcanic materials. The next reflector M3 is traceable throughout the

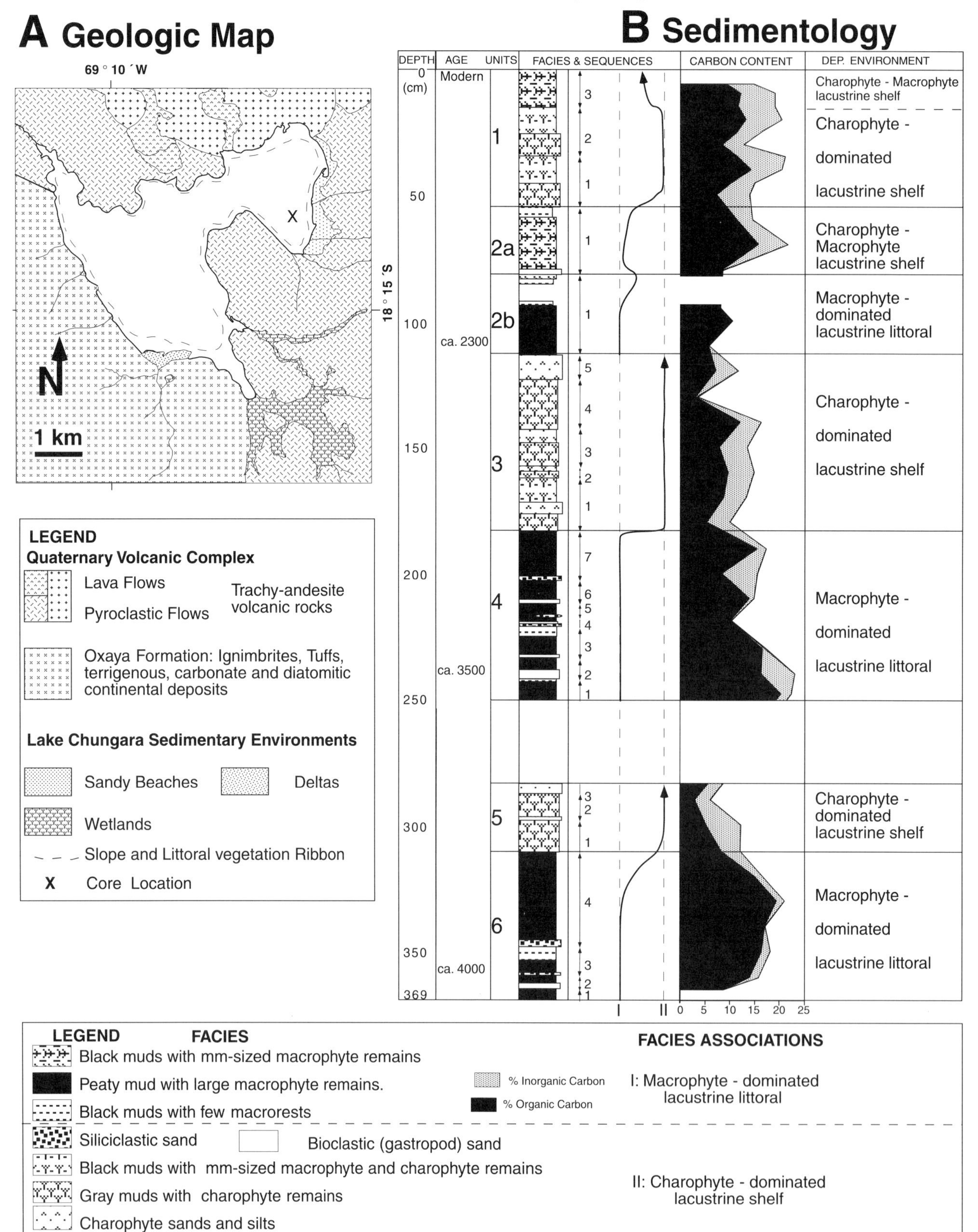

Figure 2—Laguna Chungará. (A) Geographic and geologic location. (B) Sedimentary facies, mineralogy, organic and inorganic carbon content, and interpretation of the sedimentologic units defined in the Laguna Chungará core (after Valero-Garcés et al., 1996b). The ages have been corrected for ^{14}C reservoir effects (about 2500 yr).

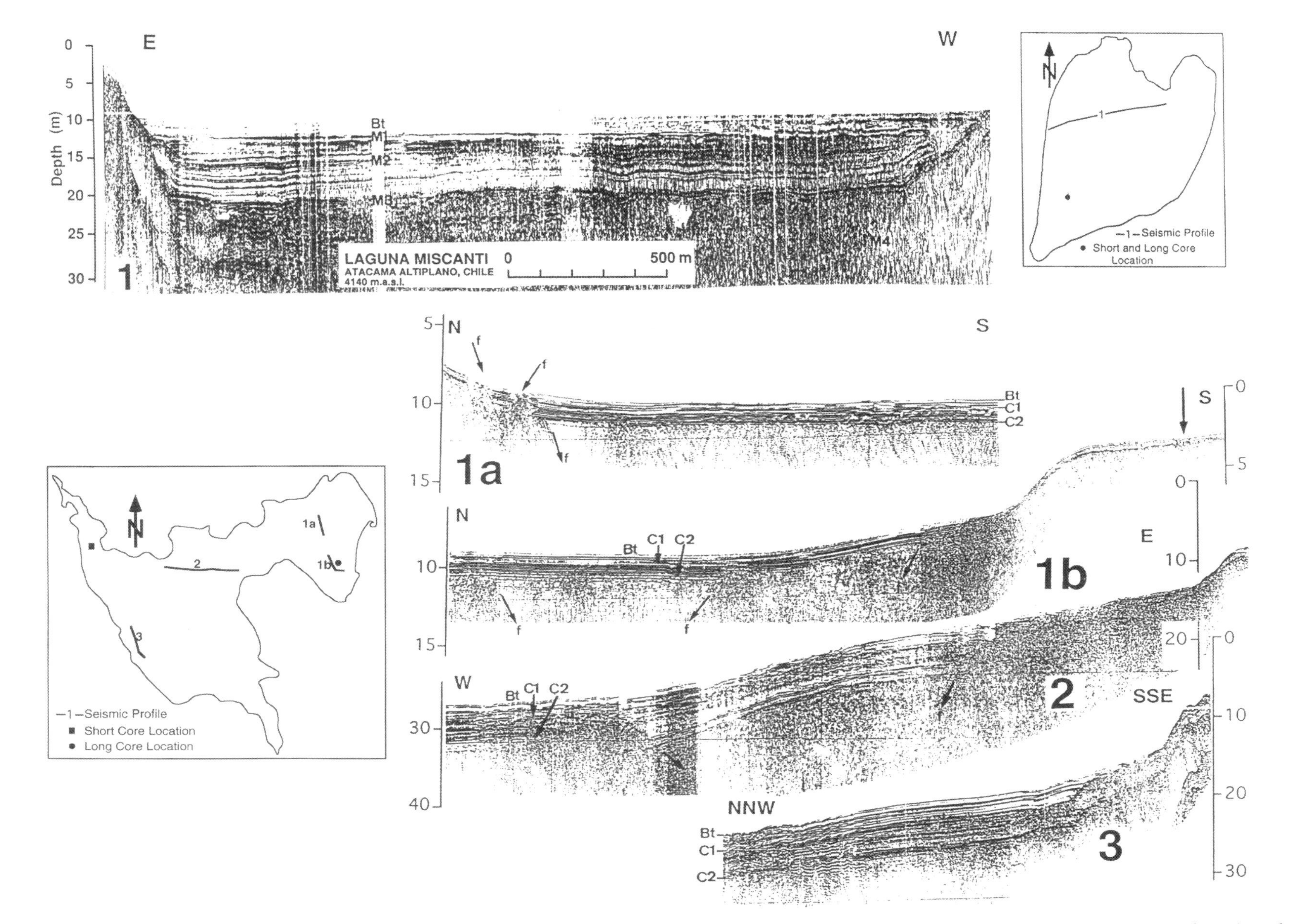

Figure 3—Seismic surveys of Miscanti (upper part) and Chungará (lower part) lakes. The profiles were recorded at 1/16 s sweep, 1–7 Khz, 1/2 s shot interval, and 100 lines/in. Location of the profiles and coring sites are shown in the inset maps. The Miscanti profile showed four main seismic reflectors (Bt, M1, M2, M3, and M4). Two main reflectors were identified in the Chungará Basin (C1 and C2); the core location is indicated by an arrow (seismic profile 1b) (after Valero-Garcés et al., 1996 a, b).

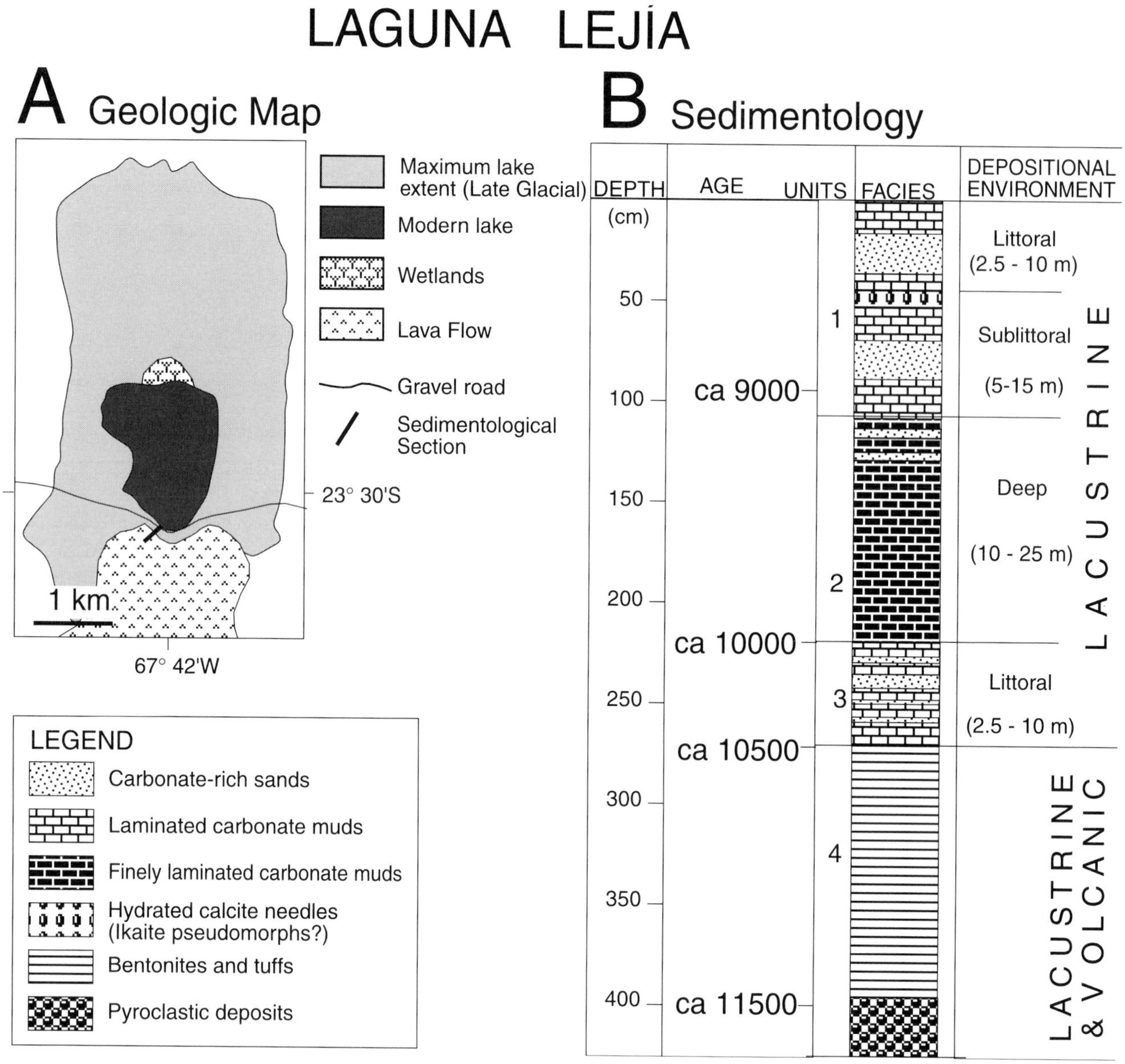

Figure 4—Laguna Lejía. (A) Geographic location and modern and late Glacial lake extent. (B) Sediment stratigraphy. The sediment sequence spans from 11,500 to about 9000 yr BP. The age framework is constrained by conventional ^{14}C ages (corrected for ^{14}C reservoir effects) and thermoluminescence (TL) (after Grosjean, 1994, 1995; Grosjean et al., 1995b).

whole basin and is interpreted as the boundary between lacustrine and alluvial sediments. The alluvial unit between seismic reflector M4 and M3 is more than 20 m thick and consists of coarse clastic sediments that correlate with the large alluvial fan deposits cropping out in the eastern part of the basin.

The upper three units correspond to the Pleistocene–Holocene lacustrine phases. The lower lacustrine seismic unit encompasses the sediments between the bottom unconformity with the alluvial unit (M3) and M2 (in the pelagial areas) or M1 (in the littoral areas) reflectors. This unit correlates with lacustrine sediments that outcrop around the lake and indicates

Figure 5—Laguna Miscanti. (A) Geographic and geologic location. (B) Sedimentary facies, mineralogy, organic and inorganic carbon, and interpretation of the sedimentologic units defined in the Miscanti core. The ages shown in the figure have been corrected for ^{14}C reservoir effects (after Valero-Garcés et al., 1996a).

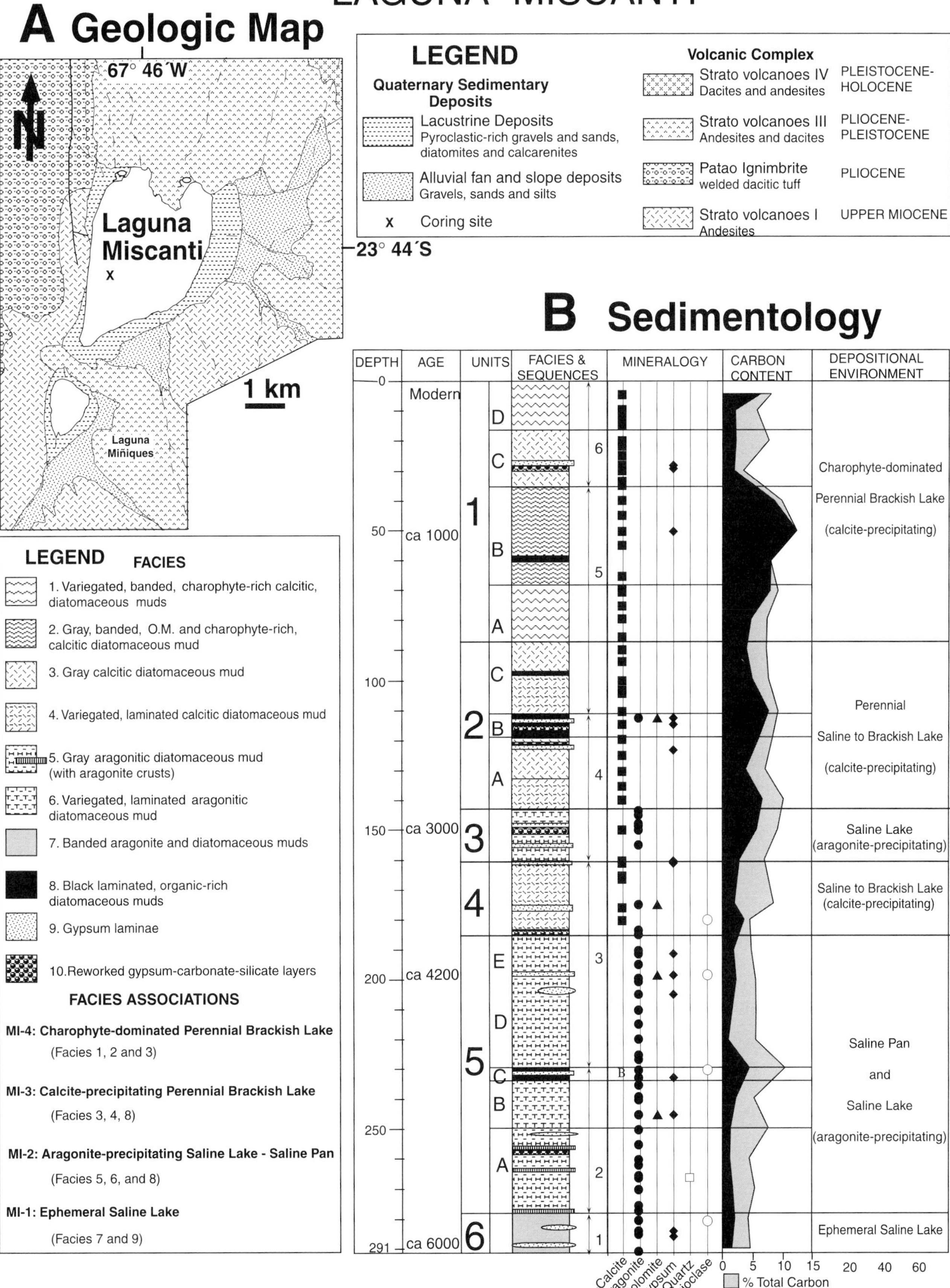
LAGUNA MISCANTI
A Geologic Map
67° 46´W
N
Laguna Miscanti
x
23° 44´S
1 km
Laguna Miñiques
LEGEND
Quaternary Sedimentary Deposits
Lacustrine Deposits
Pyroclastic-rich gravels and sands, diatomites and calcarenites
Alluvial fan and slope deposits
Gravels, sands and silts
x Coring site
Volcanic Complex
Strato volcanoes IV Dacites and andesites PLEISTOCENE-HOLOCENE
Strato volcanoes III Andesites and dacites PLIOCENE-PLEISTOCENE
Patao Ignimbrite welded dacitic tuff PLIOCENE
Strato volcanoes I Andesites UPPER MIOCENE
LEGEND FACIES
1. Variegated, banded, charophyte-rich calcitic, diatomaceous muds
2. Gray, banded, O.M. and charophyte-rich, calcitic diatomaceous mud
3. Gray calcitic diatomaceous mud
4. Variegated, laminated calcitic diatomaceous mud
5. Gray aragonitic diatomaceous mud (with aragonite crusts)
6. Variegated, laminated aragonitic diatomaceous mud
7. Banded aragonite and diatomaceous muds
8. Black laminated, organic-rich diatomaceous muds
9. Gypsum laminae
10. Reworked gypsum-carbonate-silicate layers
FACIES ASSOCIATIONS
MI-4: Charophyte-dominated Perennial Brackish Lake (Facies 1, 2 and 3)
MI-3: Calcite-precipitating Perennial Brackish Lake (Facies 3, 4, 8)
MI-2: Aragonite-precipitating Saline Lake - Saline Pan (Facies 5, 6, and 8)
MI-1: Ephemeral Saline Lake (Facies 7 and 9)
B Sedimentology
DEPTH AGE UNITS FACIES & SEQUENCES MINERALOGY CARBON CONTENT DEPOSITIONAL ENVIRONMENT
0 Modern
50 ca 1000
100
150 ca 3000
200 ca 4200
250
291 ca 6000
1 2 3 4 5 6
A B C D E
Charophyte-dominated Perennial Brackish Lake (calcite-precipitating)
Perennial Saline to Brackish Lake (calcite-precipitating)
Saline Lake (aragonite-precipitating)
Saline to Brackish Lake (calcite-precipitating)
Saline Pan and Saline Lake (aragonite-precipitating)
Ephemeral Saline Lake
Mg-Calcite Aragonite Dolomite Gypsum Quartz Plagioclase
0 5 10 15 20 40 60
% Total Carbon
% Organic Carbon

Figure 6—Isotope composition of ostracod valves from Laguna Miscanti and Laguna del Negro Francisco cores (after Schwalb et al., 1999).

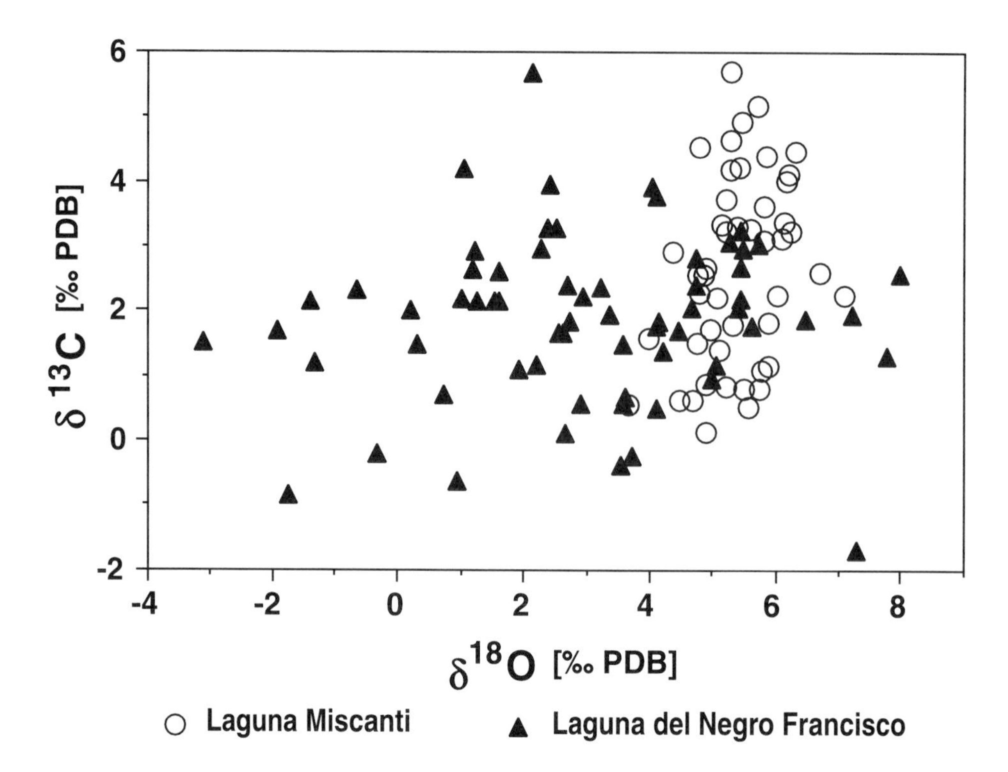

the highest lake stage during the late Glacial (15,545 ± 250 ^{14}C yr BP, with an unknown reservoir-effect) (Grosjean et al., 1995a; Valero-Garcés et al., 1996a). The middle lacustrine seismic unit (between reflectors M2 and M1) is characterized by irregular, poorly stratified reflectors with high amplitude. These irregularities indicate more heterogeneous deposition in fluctuating, but generally shallower water than before. Thickness of this unit ranges from 3 to 5 m, and the top is truncated by reflector M1. The upper lacustrine seismic unit (between M1 and Bt) is characterized by various well-stratified reflectors, a constant sediment thickness, and an onlap geometry over the lower seismic units in the littoral areas of the lake. The upper seismic unit represents deposition during higher lake levels than the previous unit, but lower than during the lower lacustrine seismic unit.

According to the age model constructed with four conventional radiocarbon dates (Valero-Garcés et al., 1996a; Geyh et al., 1998), the 291-cm-long Miscanti core (Figure 5B), comprising the upper lacustrine seismic unit, extends from the middle Holocene to present. The 10 sedimentary facies group into four facies associations (Figure 5B). Two facies associations are characterized by aragonite deposition: MI-1 groups facies 7 and 9 and is interpreted as an ephemeral saline lake; MI-2 consists of facies 5 and 6, with some intercalated aragonite crusts and organic-rich layers (facies 8), and represents deposition in a saline pan-saline lake complex. The other two associations contain facies with magnesium calcite as the only carbonate phase: MI-3 consists of an alternation of facies 3 and 4, with some intercalations of facies 8, and it is interpreted as a perennial brackish lake; MI-4 groups facies 1, 2, 3, and rarely 8 and represents a charophyte-dominated perennial brackish lake. Thin gypsum laminae (facies 9) and reworked layers (facies 10) occur in all facies associations. The presence of only one ostracod species (*Limnocythere* sp.) in varying abundances, as well as eusaline diatoms, indicate saline conditions (Schwalb et al., 1999). The heavy $\delta^{18}O$ values and the small range (3‰ PDB) suggests long residence time and ground water buffering (Figure 6).

The depositional history recorded in the core illustrates the evolution of a lacustrine system during several thousand years from an aragonite-producing, ephemeral saline lake (unit 6, facies association MI-1) to a charophyte-dominated perennial brackish lake (unit 1, facies association MI-4), with an aragonite saline pan-saline lake complex (units 5 and 3, facies association MI-3) and calcite-producing perennial brackish lake (units 4 and 2, facies associations MI-2) as intermediate stages. Succession of lacustrine environments was apparently abrupt, rather than gradual. The occurrence of extremely low lake levels during the middle Holocene is consistent with lower effective moisture recorded elsewhere in the Altiplano and the

Figure 7—Laguna del Negro Francisco. (A) Geographic and geologic location; geologic units after Mercado (1982). (B) Sedimentary facies, mineralogy, organic and inorganic carbon content, and interpretation of the sedimentologic units defined in the Negro Francisco cores (after Grosjean et al., 1997; Valero-Garcés et al., in press). The ages have been corrected for ^{14}C reservoir effects which is about 1000–3000 yr.

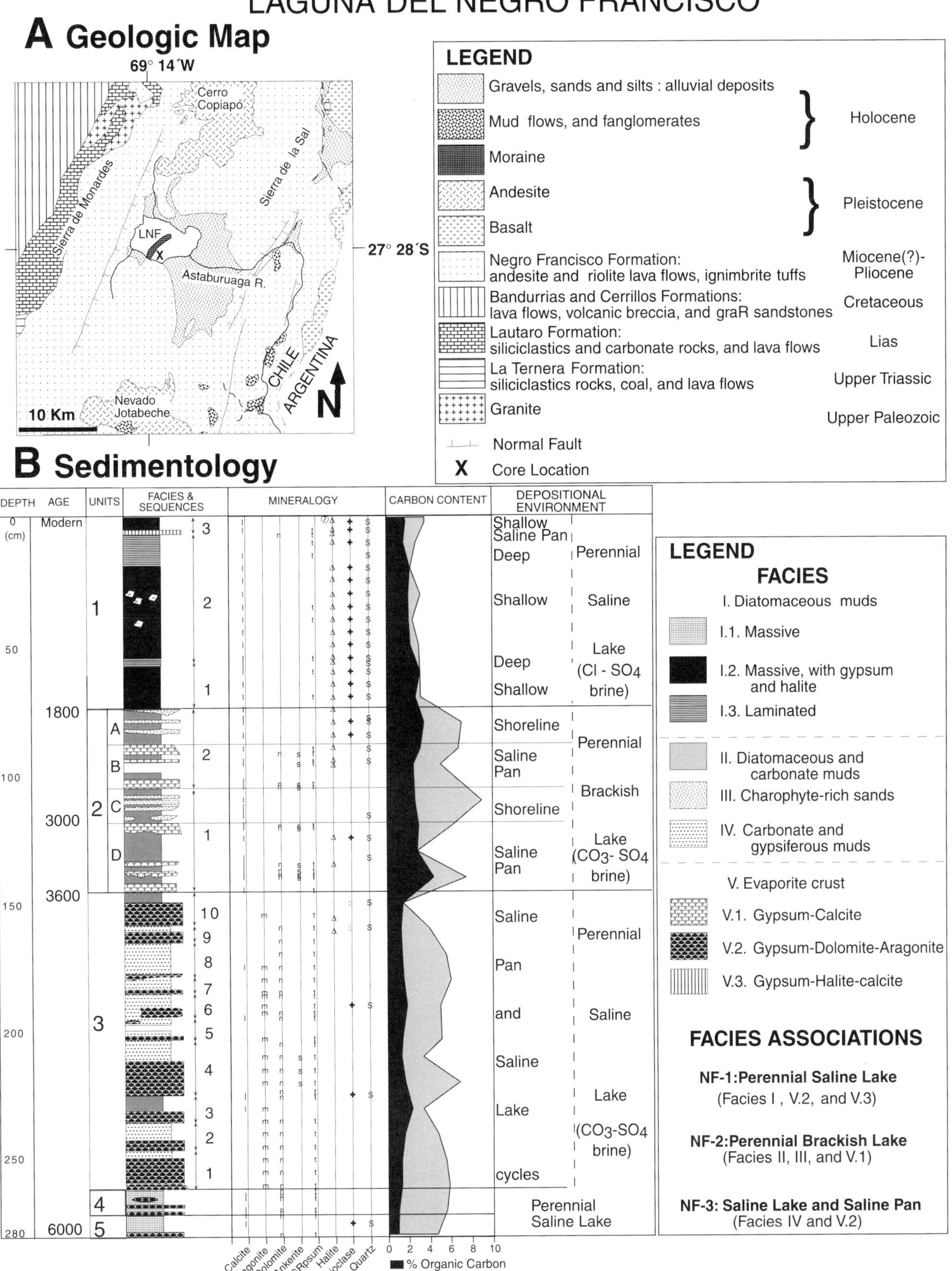
LAGUNA DEL NEGRO FRANCISCO
A Geologic Map
69° 14´W
Cerro Copiapó
Sierra de Monardes
Sierra de la Sal
LNF
X
Astaburuaga R.
27° 28´S
CHILE
ARGENTINA
N
Nevado Jotabeche
10 Km
LEGEND
Gravels, sands and silts : alluvial deposits
Mud flows, and fanglomerates
Moraine
Holocene
Andesite
Basalt
Pleistocene
Negro Francisco Formation: andesite and riolite lava flows, ignimbrite tuffs
Miocene(?)-Pliocene
Bandurrias and Cerrillos Formations: lava flows, volcanic breccia, and graR sandstones
Cretaceous
Lautaro Formation: siliciclastics and carbonate rocks, and lava flows
Lias
La Ternera Formation: siliciclastics rocks, coal, and lava flows
Upper Triassic
Granite
Upper Paleozoic
Normal Fault
Core Location
B Sedimentology
DEPTH (cm)
AGE
UNITS
FACIES & SEQUENCES
MINERALOGY
CARBON CONTENT
DEPOSITIONAL ENVIRONMENT
Modern
1800
3000
3600
6000
Shallow Saline Pan
Deep
Shallow
Deep
Shallow
Perennial Saline Lake (Cl - SO4 brine)
Shoreline
Saline Pan
Shoreline
Saline Pan
Perennial Brackish Lake (CO3- SO4 brine)
Saline Pan and Saline Lake cycles
Perennial Saline Lake (CO3-SO4 brine)
Perennial Saline Lake
Calcite
Aragonite
Dolomite
Ankerite
GRpsum
Halite
Plagioclase
Quartz
% Organic Carbon
% Inorganic Carbon
LEGEND
FACIES
I. Diatomaceous muds
I.1. Massive
I.2. Massive, with gypsum and halite
I.3. Laminated
II. Diatomaceous and carbonate muds
III. Charophyte-rich sands
IV. Carbonate and gypsiferous muds
V. Evaporite crust
V.1. Gypsum-Calcite
V.2. Gypsum-Dolomite-Aragonite
V.3. Gypsum-Halite-calcite
FACIES ASSOCIATIONS
NF-1:Perennial Saline Lake (Facies I , V.2, and V.3)
NF-2:Perennial Brackish Lake (Facies II, III, and V.1)
NF-3: Saline Lake and Saline Pan (Facies IV and V.2)

Amazon basin (Grosjean et al., 1995a; Valero-Garcés et al, 1996a). The late Holocene shows a progressive increase in effective moisture interrupted by numerous drier periods of various intensity and duration.

LAGUNA DEL NEGRO FRANCISCO

Geographic and Geological Setting

Laguna del Negro Francisco (27°28′S, 69°14′W, 4125 m a.s.l.) is located at the southern end of the Altiplano (Figure 1). The lake basin belongs to a north-south elongated tectonic depression, bounded to the west by the Cordillera de Domeyko and to the east by the Andean Cordillera that contains large salars (Atacama, Pedernales, Maricunga) and closed lakes. Oligocene and mostly Miocene volcanic materials filled the southern part of the depression, and the successive eruptions of the Copiapo volcano (6080 m) isolated the Negro Francisco basin from the Salar de Maricunga to the north (Figure 7A).

Quaternary filling of the Negro Francisco basin comprises alluvial, eolian, fluvial, glacial, and lacustrine deposits (Segerstrom, 1968; Mercado, 1982). Recent investigations have found glacial moraines in the Jotabeche area and in the Negro Francisco Plain and identified the east-west trending peninsula in the Laguna del Negro Francisco as a terminal moraine of the maximum Pleistocene glaciation (Jenny and Kammer, 1996). Most of the basin is covered by alluvial clastics deposited by the Astaburuaga river and other minor streams.

Laguna del Negro Francisco lies in an endorheic basin, fed mainly by the Astaburuaga River and secondarily by small streams and groundwater; however, in the past, the basin had an outlet that joined the Astaburuaga River and flowed northward toward the Salar de Maricunga (Segerstrom, 1968). Tectonic subsidence of the basin depressed the base level and, as a consequence, the flow was reversed and terraces were cut in the former alluvial deposits. Progressive erosion of the headwaters reached the Astaburuaga valley and the river was captured, and its water started to flow into the Negro Francisco basin instead of to the Salar de Maricunga. Former lake shorelines are mostly erosive terraces, devoid of sediments. The highest, at 25 m above modern lake level, is tentatively assigned to the late Pleistocene. Other paleoshorelines occur at 5 m above the modern lake surface, and three more are within 1 m above the current lake level.

The modern lake is very shallow (Z_{max} =1 m), with a total surface area of 20.7 km^2 and a catchment area of about 930 km^2. The east-west trending Pleistocene moraine divides the lake into two subbasins, each with a different brine composition. Waters of the southern subbasin, directly fed by the Astaburuaga River (830 µS/cm, conductivity), are alkaline (pH = 8.9), saline (15,300 µS/cm, conductivity), and of Na-(Mg)-Cl-SO_4 type. Water flows to the northern subbasin where it is further concentrated by evaporation (80,000 µS/cm, conductivity) (Grosjean et al., 1997).

Stratigraphy

Valero-Garcés et al. (1999) defined nine sedimentary facies in two Laguna del Negro Francisco cores retrieved from the southern subbasin. Sedimentary facies record deposition in a saline lake during the last six millennia. According to the age model developed using four conventional radiocarbon dates (Grosjean et al., 1997; Geyh et al., 1998), the 279-cm-long core extents to the middle Holocene (about 6000 yr BP). The core contains three facies associations (Figure 7B): NF-1 consists of diatomaceous muds (facies I) with interbedded gypsum-halite-calcite (facies V.3) or gypsum-dolomite-aragonite (facies V.2) crusts. NF-2 facies association groups diatomaceous and carbonate muds (facies II), charophyte-rich sands (facies III), and gypsum-calcite crusts (facies V.1). Finally, NF-3 consists on an alternation of carbonate and gyspsiferous muds (facies IV) and gypsum-dolomite-aragonite crusts (facies V.3). These facies associations represent deposition in a perennial saline lake (NF-1), perennial brackish lake (NF-2), and a saline lake–saline pan complex (NF-3). Laguna del Negro Francisco is also characterized by a single ostracod species (*Lymnocythere* sp.) and shallow-water saline diatom assemblages. The isotope record shows a $\delta^{13}C$ range similar to Laguna Miscanti, but a larger $\delta^{18}O$ range (Figure 6), indicating large effective moisture variability in shallow Laguna del Negro Francisco (Schwalb et al., 1999). Based on the presence of these facies associations, the two cores are divided into five sedimentary units, representing four depositional lake environments and brine compositions in the Laguna del Negro Francisco: units 5 and 4 illustrate a perennial saline carbonate-sulfate lake; unit 3 is interpreted as a saline pan–saline lake complex with a carbonate-sulfate brine; unit 2 represents a perennial brackish stage with high charophyte productivity and mostly calcite formation; finally, unit 1 represents the residual stage of the brine with precipitation of halite, after almost all carbonate has been consumed.

Mineralogical, chemical, and sedimentary facies analyses of the Negro Francisco cores confirm the regional significance of the middle Holocene arid period (6000–3600 yr BP). Similar to the Miscanti record, effective moisture increased after this arid period; however, the occurrence of humid phases between 3000 and 1800 yr BP that are not present in Miscanti, suggests different moisture sources in both areas (Grosjean et al., 1997).

ACKNOWLEDGMENTS

Research in the Altiplano lakes was part of the Swiss National Science Foundation Project "Climate Change in the Arid Andes" (SNF-20-36382.92) led by the University of Bern, Switzerland. Coring equipment, seismic surveys, and core lab facilities were provided by the Limnological Research Center, University of Minnesota. The "Universidad Católica del Norte" and "Universidad de Chile" helped organize the field expeditions. Numerous friends and colleagues from

Switzerland, Chile, and the United States helped during different stages of the project. We thank Mark Brenner (University of Florida) for his constructive review of the manuscript.

REFERENCES CITED

Alonso, H., and L. Vargas, 1988, Hidrogeoquímica de lagunas del Altiplano, Segunda Región, V Congreso Geológico Chileno, Abstract Book, V. II, p. 35–45.

Baied, C., 1991, Late-Quaternary environment, climate and human occupation of the South-Central Andes, Ph.D. Dissertation, University of Colorado, Boulder, 131 p.

Baied, C. A., and J. C. Wheeler, 1993, Evolution of High Andean Puna ecosystem environment, climate, and culture change over the last 12,000 years in the Central Andes, Mountain Research and Development, 13, p. 145–156.

Börgel Olivares, R., 1983, Geomorfología. Geografía de Chile, Volumen II, Instituto Geográfico Militar de Chile, 182 p.

Chong Díaz, G., 1988, The Cenozoic saline deposits of the Chilean Andes between 18°00′ and 27°00′ south latitude, *in* H. Bahlburg, Ch. Breitkreuz, and P. Giese, eds., The Southern Central Andes, Lecture Notes in Earth Sciences, 17, p. 1–17.

Clapperton, C., 1993, Quaternary geology and geomorphology of South America, Elsevier, Amsterdam, 779 p.

Geyh, M. A., U. Schotterer, and M. Grosjean, 1998, Temporal changes of the 14C reservoir effect in lakes, Radiocarbon, 40, p. 921–931.

Grosjean, M., 1994, Paleohydrology of the Laguna Lejía (north Chilean Altiplano) and climatic implications for late-glacial times, Palaeogeography, Palaeoclimatology, Palaeoecology, 109, p. 89–100.

Grosjean, M., 1995, Holocene lakes in the Altiplano. Field trip guide, *in* A. Saez, ed., Cenozoic and Quaternary lacustrine systems in northern Chile (Central Andes, Arc and Fore-Arc zones), Excursion guidebook, GLOPALS-IAS Meeting Antofagasta 1995, 77 p.

Grosjean, M., B. Messerli, C. Ammann, M. A. Geyh, K. Graf, B. Jenny, K. Kammer, L. Nuñez, U. Schotterer, H. Schreier, A. Schwalb, B. Valero, and M. Vuille, 1995a, Holocene environmental changes in the Atacama Altiplano and paleoclimatic implications, Bulletin de l'Institut Français des Etudes Andines, 24 (3), p. 585–594.

Grosjean, M., M. A. Geyh, B. Messerli, and U. Schotterer, 1995b, Late-glacial and early Holocene lake sediments, groundwater formation and climate in the Atacama Altiplano 22–24°S, Journal of Paleolimnology, AMQUA Special Issue, 14, p. 241–252.

Grosjean, M., B. Valero-Garcés, M. A. Geyh, B. Messerli, H. Schreier, and K. Kelts, 1997, Mid and Late Holocene Limnogeology of Laguna del Negro Francisco, northern Chile, and paleoclimatic implications, The Holocene, 7 (2), p. 151–159.

Jenny, B., and K. Kammer, 1996, Jungquartäre Vergletscherungen auf dem Chilenischen Altiplano, Geographica Bernensia G46, p. 1–80.

Kött, A., R. Gaupp, and G. Wörner, 1995, Miocene to Recent history of the western Altiplano in northern Chile revealed by lacustrine sediments of the Lauca Basin (18°15′–18°40′S/69°30′–69°05′W), Geol Rundschau, 84, p. 770–780.

Mercado, M., 1982, Geología de la Hoja Laguna del Negro Francisco, Carta geológica de Chile E. 1:100,000, n. 56, 73 p., 1 map.

Messerli, B., M. Grosjean, G. Bonani, A. Bürgi, M. Geyh, K. Graf, K. Ramseyer, H. Romero, U. Schotterer, H. Schreier, and M. Vuille, 1993, Climate change and natural resource dynamics of the Atacama Altiplano during the last 18,000 years: a preliminary synthesis, Mountain Research and Development, 13, p. 117–127.

Mladinic, P., N. Hrepic, and E. H. Quintana, 1987, Caracterización física y química de las aguas de los lagos Chungará y Cotacotani, Arch. Biol. Med. Exp., 20, p. 89–94.

Montti, S. C., and H. A. Henriquez, 1970, Interpretatión hidrogeológica de la génesis de salares y lagunas del Altiplano chileno. II Congreso Geológico Chileno, Arica. Abstract Book, p. 69–82.

Parrish, J. T., and R. L. Curtis, 1982, Atmospheric circulation, upwelling, organic-rich rocks in the Mesozoic and Cenozoic eras, Palaeogeography, Palaeoclimatology, Palaeoecology, 40, p. 31–66.

Ramirez, C., and M. Gardeweg, 1982, Hoja Toconao. Serv. Nac. Geol. Miner., Carta Geológica de Chile, No. 54, 119 p.

Romero, H., 1985, Geografía de los climas, IGM, Santiago, Chile, 243 p.

Schwalb, A., S. Burns, and K. Kelts, 1999, Holocene environments from stable isotope stratigraphy of ostracods and authigenic carbonate in Chilean Altiplano lakes, Palaeogeography, Palaeoclimatology, Palaeocology, 148, p. 153–168.

Segerstrom, K., 1968, Geología de las Hojas Copiapo y Ojos del Salado, Provincia de Atacama. Instituto de Investigaciones Geológicas Chile, Boletín 24, 58 pp.

Stoertz, G. E., and G. E. Ericksen, 1974, Geology of salars in Northern Chile, Geological Survey Professional paper 811, 65 p.

Valero-Garcés, B., M. Grosjean, A. Schwalb, M. A. Geyh, B. Messerli, and K. Kelts, 1996a, Limnogeology of Laguna Miscanti: evidence for mid to late Holocene moisture changes in the Atacama Altiplano (Northern Chile), Journal of Paleolimnology (special issue AMQUA, vol. II), 16, p. 1–21.

Valero-Garcés, B., M. Grosjean, A. Schwalb, K. Kelts, H. Schreier, and B. Messerli, 1996b, Limnogeología de Laguna Chungará y cambio climático durante el Holoceno tardío en el Altiplano Chileno septentrional, Cadernos Xeologico Laxe, 23, p. 271–280.

Valero-Garcés, B. L., M. Grosjean, H. Schreier, K. Kelts, and B. Messerli, 1999, Holocene lacustrine deposition in the Atacama Altiplano: facies models, climate and tectonic forcing, Palaeogeography, Palaeoclimatology, Palaeoecology, 151, p. 101–125.

Vuille, M., and C. Ammann, 1997, Regional snowfall patterns in the high, arid Andes (South America), Climatic Change, 36/3–4, p. 413–423.

Index